P9-DCP-846

B—Standard Normal Distribution

Numerical entries represent the probability that a standard normal random variable is between $-\infty$ and z where $z = \frac{x - \mu}{\sigma}$.

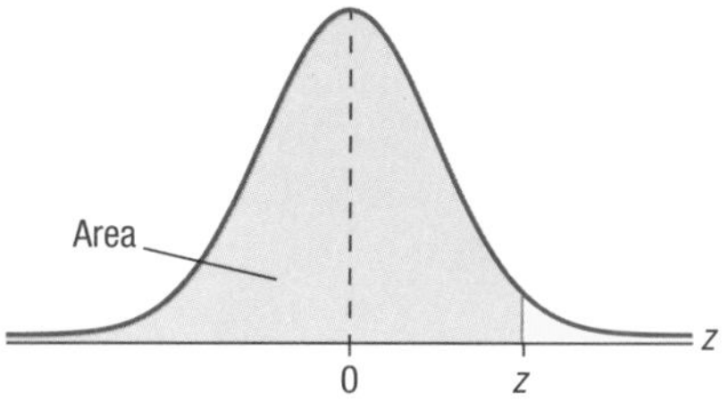

z	0.00	0.01	0.02	0.03	0.04	0.05	0.06	0.07	0.08	0.09
0.0	0.5000	0.5040	0.5080	0.5120	0.5160	0.5199	0.5239	0.5279	0.5319	0.5359
0.1	0.5398	0.5438	0.5478	0.5517	0.5557	0.5596	0.5636	0.5675	0.5714	0.5753
0.2	0.5793	0.5832	0.5871	0.5910	0.5948	0.5987	0.6026	0.6064	0.6103	0.6141
0.3	0.6179	0.6217	0.6255	0.6293	0.6331	0.6368	0.6406	0.6443	0.6480	0.6517
0.4	0.6554	0.6591	0.6628	0.6664	0.6700	0.6736	0.6772	0.6808	0.6844	0.6879
0.5	0.6915	0.6950	0.6985	0.7019	0.7054	0.7088	0.7123	0.7157	0.7190	0.7224
0.6	0.7257	0.7291	0.7324	0.7357	0.7389	0.7422	0.7454	0.7486	0.7517	0.7549
0.7	0.7580	0.7611	0.7642	0.7673	0.7704	0.7734	0.7764	0.7794	0.7823	0.7852
0.8	0.7881	0.7910	0.7939	0.7967	0.7995	0.8023	0.8051	0.8078	0.8106	0.8133
0.9	0.8159	0.8186	0.8212	0.8238	0.8264	0.8289	0.8315	0.8340	0.8365	0.8389
1.0	0.8413	0.8438	0.8461	0.8485	0.8508	0.8531	0.8554	0.8577	0.8599	0.8621
1.1	0.8643	0.8665	0.8686	0.8708	0.8729	0.8749	0.8770	0.8790	0.8810	0.8830
1.2	0.8849	0.8869	0.8888	0.8907	0.8925	0.8944	0.8962	0.8980	0.8997	0.9015
1.3	0.9032	0.9049	0.9066	0.9082	0.9099	0.9115	0.9131	0.9147	0.9162	0.9177
1.4	0.9192	0.9207	0.9222	0.9236	0.9251	0.9265	0.9279	0.9292	0.9306	0.9319
1.5	0.9332	0.9345	0.9357	0.9370	0.9382	0.9394	0.9406	0.9418	0.9429	0.9441
1.6	0.9452	0.9463	0.9474	0.9484	0.9495	0.9505	0.9515	0.9525	0.9535	0.9545
1.7	0.9554	0.9564	0.9573	0.9582	0.9591	0.9599	0.9608	0.9616	0.9625	0.9633
1.8	0.9641	0.9649	0.9656	0.9664	0.9671	0.9678	0.9686	0.9693	0.9699	0.9706
1.9	0.9713	0.9719	0.9726	0.9732	0.9738	0.9744	0.9750	0.9756	0.9761	0.9767
2.0	0.9772	0.9778	0.9783	0.9788	0.9793	0.9798	0.9803	0.9808	0.9812	0.9817
2.1	0.9821	0.9826	0.9830	0.9834	0.9838	0.9842	0.9846	0.9850	0.9854	0.9857
2.2	0.9861	0.9864	0.9868	0.9871	0.9875	0.9878	0.9881	0.9884	0.9887	0.9890
2.3	0.9893	0.9896	0.9898	0.9901	0.9904	0.9906	0.9909	0.9911	0.9913	0.9916
2.4	0.9918	0.9920	0.9922	0.9925	0.9927	0.9929	0.9931	0.9932	0.9934	0.9936
2.5	0.9938	0.9940	0.9941	0.9943	0.9945	0.9946	0.9948	0.9949	0.9951	0.9952
2.6	0.9953	0.9955	0.9956	0.9957	0.9959	0.9960	0.9961	0.9962	0.9963	0.9964
2.7	0.9965	0.9966	0.9967	0.9968	0.9969	0.9970	0.9971	0.9972	0.9973	0.9974
2.8	0.9974	0.9975	0.9976	0.9977	0.9977	0.9978	0.9979	0.9979	0.9980	0.9981
2.9	0.9981	0.9982	0.9982	0.9983	0.9984	0.9984	0.9985	0.9985	0.9986	0.9986
3.0	0.9987	0.9987	0.9987	0.9988	0.9988	0.9989	0.9989	0.9989	0.9990	0.9990
3.1	0.9990	0.9991	0.9991	0.9991	0.9992	0.9992	0.9992	0.9992	0.9993	0.9993
3.2	0.9993	0.9993	0.9994	0.9994	0.9994	0.9994	0.9994	0.9995	0.9995	0.9995
3.3	0.9995	0.9995	0.9995	0.9996	0.9996	0.9996	0.9996	0.9996	0.9996	0.9997
3.4	0.9997	0.9997	0.9997	0.9997	0.9997	0.9997	0.9997	0.9997	0.9997	0.9998

Critical Values of z

Level of Confidence, c	$\alpha = 1 - c$	$z_{\alpha/2}$
0.80	0.20	1.28
0.85	0.15	1.44
0.90	0.10	1.645
0.95	0.05	1.96
0.98	0.02	2.33
0.99	0.01	2.575

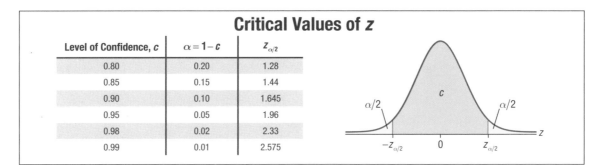

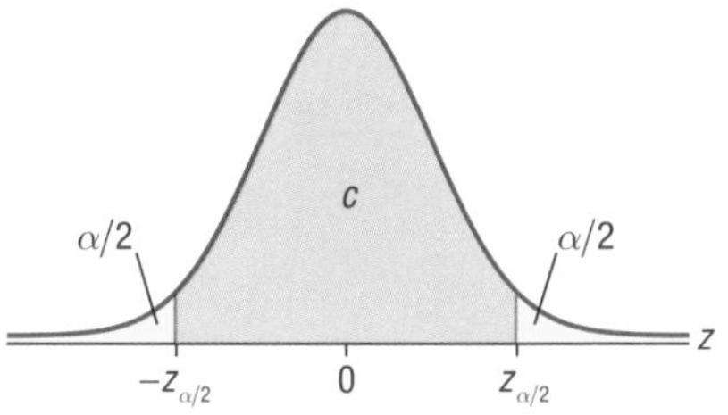

Beginning Statistics

Second Edition

Carolyn Warren · Kimberly Denley · Emily Atchley

Executive Editor: Danielle C. Bess
Executive Project Manager: Kimberly Cumbie
Vice President, Research and Development: Marcel Prevuznak
Editorial Assistants: Susan Fuller, Margaret Gibbs, Barbara Miller, Claudia Vance, Nina Waldron
Copy Editors: Jason Ling, Joseph Randich
Review Coordinator: Lisa Young

Senior Graphic Designers: Jennifer Moran, Tee Jay Zajac
Junior Designer: Kaylen Saxon
QSI (Pvt.) Ltd.: E. Jeevan Kumar, D. Kanthi, U. Nagesh, B. Syamprasad
Art: Jennifer Moran, Kaylen Saxon, Tee Jay Zajac
Cover Design: Jennifer Moran
Photography Credits: iStockphoto.com

HAWKES LEARNING
A division of Quant Systems, Inc.
546 Long Point Road, Mount Pleasant, SC 29464
Copyright © 2017, 2014 by Hawkes Learning/Quant Systems, Inc.
All rights reserved. No part of this publication may be reproduced, stored in a retrieval system, or transmitted in any form or by any means, electronic, mechanical, photocopying, recording, or otherwise, without the prior written consent of the publisher.
Library of Congress Control Number 2012951240
Printed in the United States of America

ISBN
Student Textbook 978-1-932628-67-8

1 2 3

Table of Contents

1 Introduction to Statistics

2 Graphical Descriptions of Data

3 Numerical Descriptions of Data

4 Probability, Randomness, and Uncertainty

5 Discrete Probability Distributions

6 Normal Probability Distributions

7 The Central Limit Theorem

8 Confidence Intervals

9 Confidence Intervals for Two Samples

10 Hypothesis Testing

11 Hypothesis Testing (Two or More Populations)

12 Regression, Inference, and Model Building

A Appendix A

B Appendix B

C Appendix C

AK Answer Key

I Index

Message from the Authors

Greetings:

We, the authors, are delighted to offer this second edition of *Beginning Statistics* to you. Following the warm reception of our first edition, we hope that you find that the newest edition contains the same features that made our text a favorite among students, as well as a host of additional features and improvements we feel sure you will love.

As before, our foremost goal has always been to make the study of statistics accessible to every student. To this end, we have kept the same conversational approach in presenting topics so that even the most difficult concepts can be easily understood. We pride ourselves on the readability of our text. We firmly believe that math textbooks do not have to be odious to read, and you will see that this one is not.

With a combined 34 years of teaching experience, we are also uniquely aware of common pitfalls and mistakes students make. Our examples and solutions are tailored with these common errors in mind, and are structured in a way we feel is most beneficial for our own students. In addition, we have expanded and improved our Side Notes and Memory Boosters in the margins, as well as the formula boxes within each chapter.

The most significant enhancement to *Beginning Statistics* has been the addition of technology solutions within the text and examples. Previously, technology was primarily limited to the technology section found at the end of each chapter. With the explosion of technology use that has occurred during the last few years, we feel strongly that calculator solutions deserve a greater level of attention within the text. We are aware that the tide is turning from the traditional focus on formulas and charts to a new focus on calculators and technology. With this in mind, our aim in the second edition is to present topics and methods in such a way that both the traditionally minded teacher and the technology minded teacher (and those in between) will have their needs met in this text. To do this, we have synthesized technology instructions alongside traditional instructions to allow teachers to bridge from one method to the other in a seamless fashion.

Another major change in this edition is the realignment of the presentation of confidence intervals and hypothesis testing for population means to focus on whether the population standard deviation is known when determining the choice of test statistic. This change will be reflected throughout Chapters 8–11, and in changes to the Hawkes Learning Systems software.

In addition to these changes, our second edition also contains improved and expanded exercise sets with additional real-world data, new graphics, and updated examples that are of interest to the traditional college student.

We sincerely hope that you will find the second edition of *Beginning Statistics* to be both an enjoyable read and a useful tool in your classrooms.

Best Regards,

Carolyn Warren, Kim Denley, and Emily Atchley

Acknowledgements

As a group, we would first like to thank Dr. Tristan Denley for his continued input and feedback on this text. We would also like to thank Emily's parents, Dr. and Mrs. Phillips, for the use of their home for our long weekend book summits. Thank you to Marcel Prevuznak, Emily Cook, Kim Cumbie and all of the people at Hawkes Learning Systems for their dedication to this project. In particular, we would like to thank our editor, Danielle Bess, and her team, who worked so diligently to make sure that our ramblings turned into a cohesive and consistent second edition.

Special thanks to our reviewers:

John Blackburn, East Georgia State College

Dr. Johnny Duke, Georgia Highlands College

Brent Griffin, Georgia Highlands College

Dr. David Hare, University of Louisiana at Monroe

Becky Hurley, Rockingham Community College

Dr. Victoria Ingalls, Tiffin University

Dr. Lana Ivanitskaya, Central Michigan University

Dr. Martin Jones, College of Charleston

Dr. Jeffrey Linek, Georgia Highlands College

Diana McCoy, Truckee Meadows Community College

Dr. Deanna Needell, Claremont McKenna College

Michael Papin, Yuba College

Laura Ralston, Georgia Highlands College

Suzanne Seeber, University of Louisiana at Monroe

Dr. Tom Short, John Carroll University

Michael Slade, Cornell University/CUNY - Baruch College

Dr. Karsten Stemmann, Yuba College

Scott Weir, Wake Technical Community College

Dr. Lynelle Weldon, Andrews University

The text was much improved by your remarks and suggestions, and we are most grateful for your input.

In addition, we want to thank all of our families and friends for your support and understanding throughout this long project. Carolyn would like to thank the Warrens and Shirleys, and to Amy, thanks for suggesting all those years ago that we write this book in the first place, I love you. Kim would like to thank her family for yet again understanding that my job is unconventional, unpredictable, and always a learning experience. Thanks for helping me succeed. I love you! Emily would like to thank her family for their extraordinary patience and encouragement throughout this journey. And especially for Peyton, who made it possible for her to keep perspective while juggling a new baby and book project—the second time around! Much love to you.

Features of

Beginning Statistics

Second Edition

Examples

Examples are presented in a step-by-step manner that is easy for students of any level to follow. Titles are given that alert the student to the concept they will be learning throughout the example. Examples utilizing technology such as a TI-83/84 Plus calculator are denoted with a calculator icon, and give detailed instructions on how to perform calculations using the calculator, including screenshots of the calculator screens. This will encourage students to perform calculations as they are working through the text for more thorough understanding. Examples feature a wide range of topics, including business, science, technology, and medicine. Many examples utilize real-world data sets, allowing students to make connections between the statistical concepts they are learning and the real world.

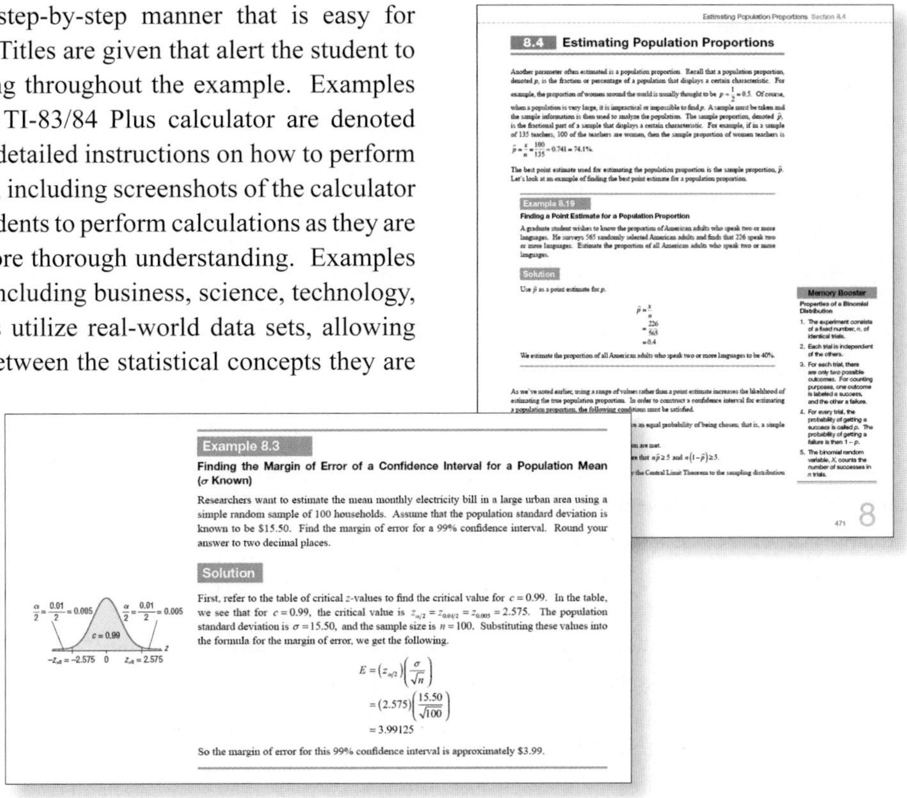

Margin Boxes

Throughout the text, informational aids are presented in the margins that reinforce student learning and understanding. Rounding Rules are presented to ensure that students have a clear understanding of when to employ rounding, and how to avoid sizable errors in final calculations. Memory Boosters serve to supply students with easy-to-remember shortcuts for statistical concepts. Math Symbols boxes highlight the importance of understanding the symbols used in many of the formulas being presented. Side Note boxes offer connections between statistical concepts and the real world to pique student interest and deepen their understanding of statistics in their everyday lives. Caution boxes serve to warn students of common mistakes that are made in statistical computations. These important tips draw students' attention to many of the most important fundamental aspects of statistics.

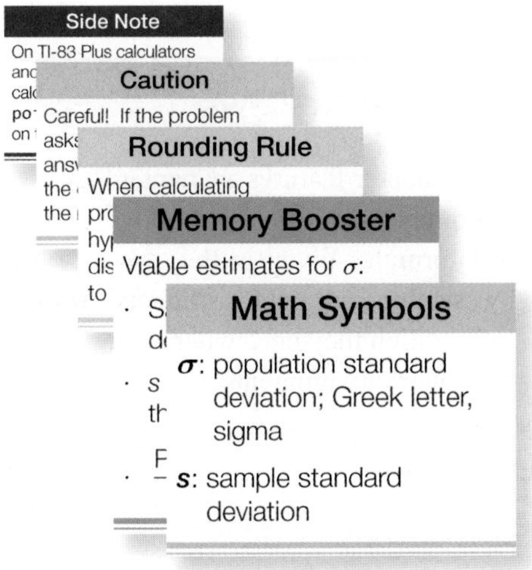

Definition, Formula, Properties, Procedure, and Theorem Boxes

Formulas, definitions, properties, procedures, and theorems are clearly presented for ease of reference. The boxes are color-coded to help students focus on the most important components of the chapters. Definition boxes are purple, formula and theorem boxes are orange, and properties and procedure boxes are green.

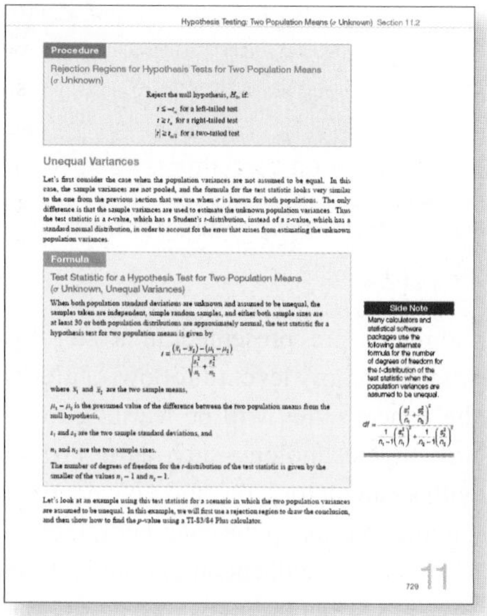

Definition

A **random variable** is a variable whose numeric value is determined by the outcome of a probability experiment.

A **probability distribution** is a table or formula that gives the probabilities for every value of the random variable X, where $0 \le P(X = x) \le 1$ and $\sum P(X = x_i) = 1$.

Procedure

Principles of Experimental Design

1. Randomize the control and treatment groups.
2. Control for outside effects on the response variable.
3. Replicate the experiment a significant number of times to see meaningful patterns.

Exercises

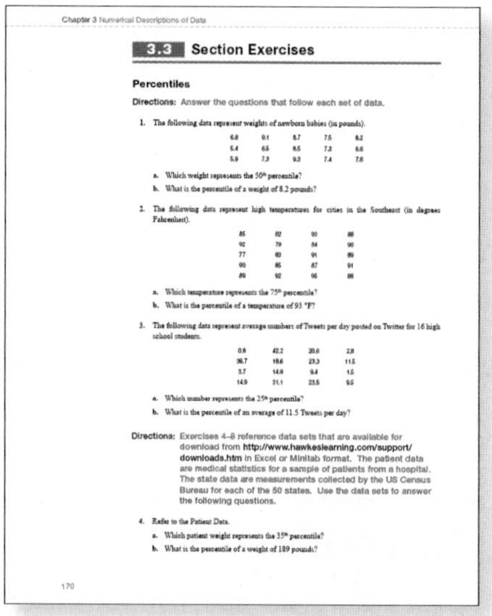

The text contains a wide variety of exercises at the end of each section, as well as at the end of each chapter. Section Exercises focus on material learned in a particular section, while Chapter Exercises challenge students to apply all of the statistical skills they have learned throughout the chapter. Exercises exhibit a wide range of difficulty levels and applications. As in the examples, real-world data sets are often employed to engage students with the statistical concepts they are learning. Additionally, the text features exercises and projects that utilize larger, more challenging data sets that are available for students to download and work with in the statistical analysis program of their choice.

Chapter Reviews

Each chapter features a Chapter Review that highlights important definitions, properties, theorems, processes, and formulas given in the chapter. Material is listed by section, allowing students to quickly find the information they need when studying for a test or doing homework assignments.

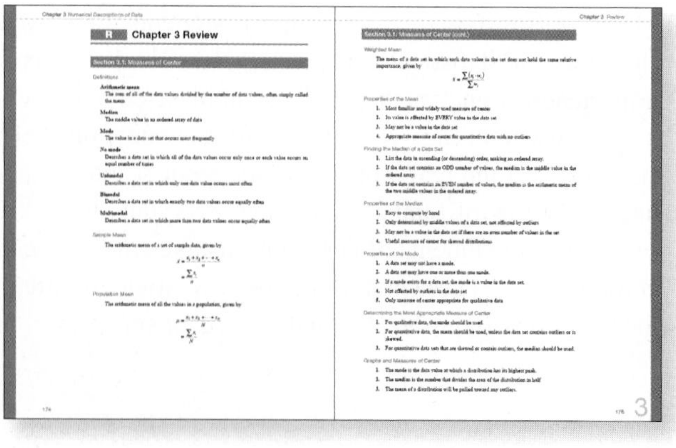

Chapter Projects

The text features Chapter Projects that give instructors the flexibility to have students work in groups or individually on a more challenging assignment. These projects allow students to collect and analyze their own data, while implementing statistical concepts that they have learned. The Chapter Projects extend beyond the Section and Chapter Exercises in that they employ a hands-on approach to learning statistics topics.

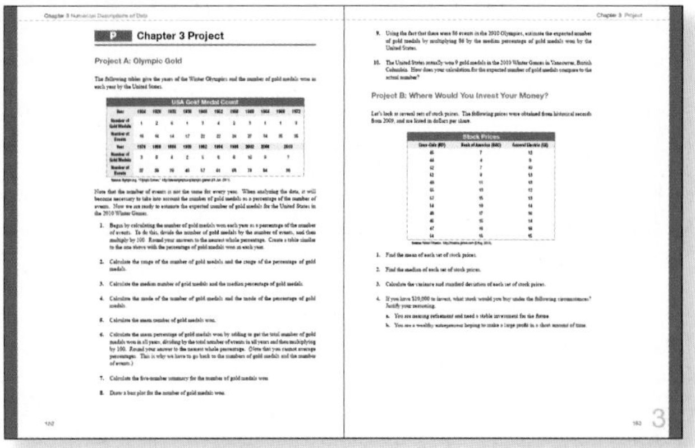

Chapter Technology

Technology sections are given throughout the text and feature easy-to-follow directions for using a TI-83/84 Plus calculator, Microsoft Excel, and MINITAB to apply the concepts taught throughout the chapters. Screenshots are presented throughout to aid in student learning. In addition, Getting Started guides are included as Appendices to familiarize students with the basic functions of Excel and MINITAB.

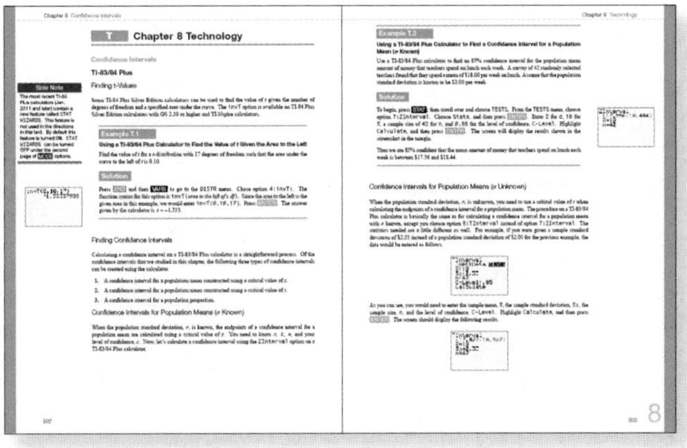

Answer Key and Formula Sheets

The Answer Key contains solutions for odd-numbered Section and Chapter Exercises for student reference. In addition, the text includes a tear-out formula sheet composed of all important formulas and the most commonly used probability distribution tables that students may use as a convenient reference and study tool throughout the course.

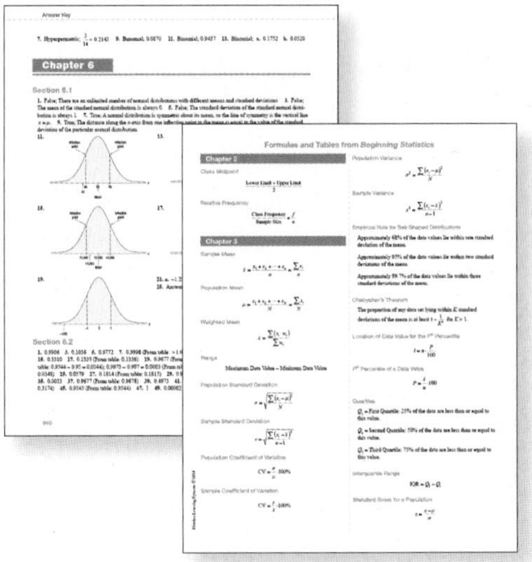

Major Changes for the Second Edition

New and more challenging exercises have been added.

Exercise sets throughout the text have been expanded to include more conceptual questions. In addition, the exercises feature a wider range of difficulty, and utilize real-world data sets when possible.

Larger, more challenging data sets are available for download.

New large data sets have been added, which are available for download so that students can use appropriate statistical technology to analyze the data. These data sets are referenced in the text, and can be used for more involved class projects.

Increased emphasis has been placed on technology.

More emphasis has been placed on using a TI-83/84 Plus calculator to solve problems within examples and exercises. Directions complete with screenshots for using a TI-83/84 Plus calculator have been added in many of the examples. Solutions to exercises are given for manual computations as well as calculator computations to give instructors the freedom to assign the method of their choice. Additionally, the Technology sections at the end of every chapter have been updated to include step-by-step instructions and screenshots for using a TI-83/84 Plus calculator, Microsoft Excel 2010, and MINITAB 15 Student Version to solve problems.

A discussion on dot plots was added.

A discussion on creating and interpreting dot plots was added in Section 2.2: Graphical Displays of Data.

The section on measures of relative position was updated to include alternate methods for calculating quartiles and a discussion of the interquartile range.

In Section 3.3: Measures of Relative Position, the 1st and 3rd quartiles are defined as the 25th and 75th percentiles, respectively. Additionally, information was added on finding quartiles using the method employed by the TI-83/84 Plus calculator. The definition for the interquartile range was also added to this section, along with information on interpreting box plots using the interquartile range.

Probability rules were split into two sections.

Probability rules have been separated into two sections for this edition to accommodate instructors who teach the material in more than one class period. Chapter 4 now contains Section 4.2: Addition Rules for Probability and Section 4.3: Multiplication Rules for Probability.

Chapter 6 focuses exclusively on the normal probability distributions.

Approximating a binomial distribution using a normal distribution is now covered in Chapter 6: Normal Probability Distributions. For this edition, all sections within Chapter 6 focus on normal distributions.

A z-statistic is used in constructing a confidence interval only when the population standard deviation is known.

The standard normal distribution (z) is only used to calculate the margin of error for a confidence interval for a population mean if the population standard deviation is known. Similarly, z is only used to calculate the margin of error for a confidence interval for the difference between two population means if the population standard deviations are known for both populations.

The Student's t-distribution is now introduced in Chapter 8.

The Student's t-distribution is now covered in Section 8.2: Student's t-Distribution, just before the discussion of constructing a confidence interval for a population mean when the population standard deviation is unknown.

A z-statistic is used in a hypothesis test only when the population standard deviation is known.

The standard normal distribution (z) is only used for the test statistic in a hypothesis test for a population mean if the population standard deviation is known. Similarly, z is only used for the test statistic in a hypothesis test for the difference between two population means if the population standard deviations are known for both populations.

Type I and Type II errors are now covered in Section 10.1.

Type I and Type II errors are now covered in Section 10.1: Fundamentals of Hypothesis Testing. Section 10.1 now covers the formulation of hypotheses, an introduction to test statistics, interpreting conclusions that result from hypothesis tests, and types of errors in hypothesis testing.

Content

This textbook assumes no previous knowledge of statistics. Furthermore, it includes enough topics ordered in such a way that it can be used in an introductory liberal arts course or a more rigorous business or psychology statistics course.

We would like to make a few comments on specific content. First, the overall layout of the book follows the order of our definition of statistics—the science of gathering, describing, and analyzing data. In Chapter 1 we discuss the details of how to gather data. In Chapters 2 and 3, we examine different ways of describing data. The remainder of the book explores, in detail, numerous ways to analyze data, such as estimating population parameters, testing hypotheses, and describing relationships between data.

CHAPTER 1: Introduction to Statistics

In this chapter, we introduce students to basic statistical vocabulary and the process of a statistical study. We want students to view newspaper articles and research reports critically, and not take them at face value; therefore we specifically point out common problems and weaknesses with different sampling methods and types of studies, as well as emphasizing under what circumstances to use which techniques. We devote an entire section of this chapter to how to critique a published study. Our goal here is to heighten students' awareness of possible problems in the research they read as well as to prepare them to think about these issues in their own research.

CHAPTER 2: Graphical Descriptions of Data

We begin our discussion of describing data with graphical representations of data, as this is probably more familiar to students than the numerical descriptions of data calculated in the next chapter. We include a section on analyzing graphs to encourage students to think critically about the information they receive.

CHAPTER 3: Numerical Descriptions of Data

In this chapter, we discuss calculating numerical descriptions of center, dispersion, and position. We demonstrate calculations by hand and through the use of technology. This approach will allow for teachers with different teaching styles to use this text for their purposes.

One interesting point we ran into while researching this chapter is how to calculate percentiles and quartiles. We have tried to be as accurate as possible while pointing out that there is not always agreement on how to calculate these values. We did reverse the order of the presentation of these topics, based on reviewers' comments. We hope this change will allow for a more natural flow to the topics in that section.

We have intentionally used the word "average" instead of the word "mean" in various places throughout the text. We want students to realize that the word "average" is the word used in everyday speech to indicate the mathematical concept of mean, but to also be aware of the fact that there are other mathematical calculations (median, mode) that can be expressed in laymen's terms as an average. We spend an entire subsection dealing with this topic. Our students are not statisticians, but nurses, teachers, journalists, social workers, etc. They will use and hear the word "average." We want them to equate that with the word mean, even when that is not what is explicitly said.

CHAPTER 4: Probability, Randomness, and Uncertainty

Our discussion of probability begins with terminology and simple probability calculations. We also include a discussion of the various types of probability. We introduce counting techniques as a means of counting the number of outcomes in the sample space or event for more complicated probability calculations. We end the chapter with more advanced counting techniques. This last section could easily be omitted for those courses not needing this more advanced discussion. The division of the sections has been adjusted from the first edition to better group similar topics.

CHAPTER 5: Discrete Probability Distributions

This chapter continues our discussion of probability, but focuses on discrete probability distributions. We discuss in detail the binomial, Poisson, and hypergeometric distributions. We chose to put these topics in a separate chapter, as we believe that the grouping together of these probability distributions into a single unit helps point out their similarities and differences, and thus helps clarify the concepts of this chapter.

CHAPTER 6: Normal Probability Distributions

In this chapter, we discuss the normal probability distribution. The discussion of the Student's *t*-distribution, which was in this chapter in the first edition, was moved to Chapter 8 to place it closer to the discussion of how it is used to estimate population parameters. The normal distribution forms the backbone for the rest of the text, so we spend an entire chapter describing this distribution and working with finding probabilities using the normal distribution tables and available technology. This chapter also lays the groundwork for discussions of other continuous probability distributions that appear later in the text.

CHAPTER 7: The Central Limit Theorem

In this chapter, we continue our discussion of the normal distribution with a discussion of the Central Limit Theorem. So much of statistics is based on this important theorem; we certainly feel that it deserves extensive treatment.

CHAPTER 8: Confidence Intervals

We introduce confidence intervals for population means, proportions, and variances in this chapter. We emphasize to students that a confidence interval is not guaranteed to contain the population parameter. We have moved the discussion of the Student's *t*-distribution to this chapter, so that instructors can introduce it right before students need to use it to estimate population parameters. We introduce the χ^2-distribution in this chapter as well.

One important note about this chapter is the way in which we decided to present the material in regards to the population standard deviation being known or unknown. Unlike in the first edition, we choose the method for calculating a confidence interval for a population mean based on whether or not σ is known. We continued this same practice throughout the rest of the text where applicable.

CHAPTER 9: Confidence Intervals for Two Samples

We continue our discussion of confidence intervals by extending our discussion to include comparing two population means, proportions, or variances. This chapter is separate from the chapter on hypothesis testing for two populations so that the discussion of confidence intervals is not interrupted by a discussion of hypothesis testing.

CHAPTER 10: Hypothesis Testing

We begin this chapter with an introduction to the fundamental concepts behind performing a hypothesis test. We discuss setting up the hypotheses, the burden of proof, the benefit of the doubt, and drawing conclusions. We use the illustration of the US legal system throughout the first section to tie this concept to something with which students are already familiar. In this edition, we moved the discussion of the types of errors and their relationship to the level of significance to the end of this first section. We have also included within the sections instructions for using appropriate technology to perform hypothesis tests.

CHAPTER 11: Hypothesis Testing (Two or More Populations)

Just as we discussed confidence intervals for one population, and then moved on to discuss two populations, we do the same for hypothesis testing. We now incorporate technology not only in the section on the ANOVA test, but in all appropriate places within this chapter.

CHAPTER 12: Regression, Inference, and Model Building

In this chapter, we discuss linear regression and regression models. We also emphasize when it is appropriate to use these models. Throughout the chapter, we tie linear regression to confidence intervals and hypothesis testing, using these techniques to discuss the appropriateness of the linear regression model for a given set of data. As was done in the first edition, we incorporate appropriate technology throughout this chapter.

Hawkes Learning:
Beginning Statistics Courseware

Overview

Hawkes Learning specializes in interactive mathematics courseware with a unique, mastery-based approach to student learning. Hawkes Learning's *Beginning Statistics* is designed to help students develop a solid foundation and understanding of statistics. Its mastery-based approach promotes and increases student success. The courseware consists of three learning modes: Learn, Practice, and Certify. Additional auxiliary instructional tools including videos, teaching slides, and eBook options are also available.

Exploring the Courseware

Our mastery-based courseware engages students in the learning process, so they successfully learn, understand, and demonstrate competencies for assigned topics. Just-in-time feedback and a student-centered learning environment offer a systematic approach to learning that includes the following learning modes, each with their own unique, differentiating features:

Learn

Learn is a multimedia presentation of each lesson that correlates with the content existing in the textbook.

Learn Features:

- Definitions, rules, and properties
- Example problems that are both similar to those contained in the text as well as supplemental problems
- Interactive questions and animations
- Audio narration
- Instructional video tutorials
- Clickable HotWords for additional information on key topics

Practice

Practice presents students with an unlimited number of algorithmically generated problems and intelligent feedback on incorrect answers. Within the Practice mode, the ***Interactive Tutor*** encourages active learning.

Practice Features:

- Explain Error: Gives specific feedback on students' mistakes that explains not only what is wrong, but why
- Step-By-Step: Breaks each problem down into smaller steps, offering feedback and solutions at each point
- Solution: Offers guided solutions for every problem
- Instructor Connect: Allows for monitoring of more challenging problems
- Performance Report: Tracks homework preparedness

Certify

Certify is an assignment that holds students accountable for learning the material at a defined proficiency level while removing learning aids.

Certify Features:

- Motivational, non-penalty approach
- Algorithmically generated, free-response assignments
- Comparable question sets to those offered in the Practice mode
- A mastery progress meter that allows students to visually track their progress toward topic mastery

Video

View instructional videos anytime, anywhere at **HawkesTV.com.**

Support

Please contact support for questions or technical help with Hawkes Learning's Beginning Statistics courseware.

Support Center: support.hawkeslearning.com
Email: support@hawkeslearning.com
Chat: chat.hawkeslearning.com
Phone: 843.571.2825

Chapter One

Introduction to Statistics

Sections

1.1 Getting Started

1.2 Data Classification

1.3 The Process of a Statistical Study

1.4 How to Critique a Published Study

Objectives

1. Learn the basic vocabulary of statistics.
2. Distinguish between population and sample; parameters and statistics.
3. Classify data as qualitative or quantitative; as discrete, continuous, or neither; and by the level of measurement.
4. Describe the process of a statistical study.
5. Identify various types of studies.
6. Learn the basic techniques for choosing a sample.
7. Understand the practical and ethical concerns that arise when conducting a study.

Introduction

Autism spectrum disorder, a neural development disorder, is characterized by impairments in communication, social interaction, or behavior. Children diagnosed with autism can suffer from severe developmental delays and may require significant therapy and special accommodations for life. For unknown reasons, the rates of autism diagnoses have been sharply rising for many years. In fact, in the spring of 2012, a disturbing report was issued by the Centers for Disease Control (CDC); based on data collected in 2008, the report estimates that 1 in 88 American children have been diagnosed with some form of autism. This rate is a staggering 76% increase between data collected in 2002 and data collected in 2008. Naturally, parents and researchers alike are searching for answers, trying to determine what could be causing this debilitating disorder.

Though many researchers have tirelessly searched for a link, the only possible cause for autism was given in a 1998 study published by British physician Dr. Andrew Wakefield. Dr. Wakefield's study of 12 autistic children showed a link between the childhood measles, mumps, rubella (MMR) vaccine and the autism disorder. His research even stated that autistic behavior in 8 of the 12 children studied began within two weeks of the administration of the vaccine. These results were powerful. Many fearful parents abandoned the longstanding and highly recommended vaccine schedules given by trusted pediatricians. MMR vaccination rates in Britain dropped to a dangerous low of 80% in 2004.

Only Dr. Wakefield was wrong. Not only that, his research was a *fraud.* When other researchers were unable to corroborate Dr. Wakefield's findings in subsequent studies, his methods were questioned. Upon further investigation it was determined that the doctor had either altered or misrepresented the medical histories of all 12 patients involved in the study. Furthermore, the motive for his fraud was found to be a sum of $674,000 paid to him by attorneys eager to sue the vaccine manufacturers. With the truth exposed, the autism study was retracted and Dr. Wakefield publicly disgraced. Unfortunately, the damage done by Dr. Wakefield and the vaccine scare he caused will take many years to undo.

This unsettling example shows not only the power of statistics, but the absolute necessity of a properly designed study. Let's now begin our study of the interesting and highly influential discipline called *statistics*.

1.1 Getting Started

Whether a tobacco study, the federal budget report, or even batting averages from local little league games, statistics are everywhere in today's world. Sources such as magazines, the Internet, and 24-hour cable news networks constantly give us information in the form of statistics. In order to be informed citizens and discerning consumers, we must understand exactly what these statistics mean. The goal of the branch of mathematics called statistics is to provide information so that informed decisions can be made. The goal of this text is to enable you to filter through the statistics you encounter so that you can be better prepared for the decisions you make in your daily life.

We begin our study of statistics with some basic vocabulary. The word **statistics** itself may refer to either the science of gathering, describing, and analyzing data or the actual numerical descriptions of sample data. To illustrate the difference in meanings, note the following examples:

- Section 1.1 is your first lesson in statistics (the science).
- Throughout this course, you will learn how to properly collect and analyze statistics (actual data).

Definition

Statistics is the science of gathering, describing, and analyzing data.

OR

Statistics are the actual numerical descriptions of sample data.

A statistical study centers upon a particular group of interest called a **population**. A study's population consists of all persons or things about which one is trying to make an inference or decision. For example, if, in order to market to targeted audiences, a software company wished to know the percentage of women who own a personal computer, the population would seem to be all women. However, we must consider if *all women* is the actual group in which we are interested because it implies all living women in the world. It is possible that our true intention is to know about all women, but more than likely, we want to know about women in a particular region or age group. Or consider an instructor who wants to know how many statistics students have red hair. What might be her population of interest? All statistics students on Earth? All statistics students in her country? All statistics students in her university? You can see that the choices might be endless for her. Hence, she must be clear when deciding on her population to study. As statisticians, it is imperative that we carefully define the population we are studying.

When studying a population, the values that change among members of the population are called **variables**. The information gathered about a specific variable is collectively called **data** (the singular form of data is datum), and it may be in a variety of forms including counts, measurements, or observations. For example, in a study looking at the average age of members at the local YMCA, the variable is the age of a member and the data are the actual ages gathered, such as 28, 32, 21, 45, and so on. If the population is small enough, data may be obtained from every member. This is called a **census**. You can easily remember this term by thinking of the US census that attempts to count each and every American.

The numerical description of a particular population characteristic is called a **parameter**. That is, the numbers that describe a population are called parameters. Suppose it is true that 76% of all college students own a computer. Then 76% is a population parameter. An important note is that although the parameter is a fixed number, often it is impossible or impractical to determine it precisely. For example, suppose that the population for a study includes all college students. Would it be possible to collect data from every individual college student? Probably not! Because of human limitations, usually the best we can do is to estimate a population's true parameter.

When the parameter cannot be determined, a subset of the population, called a **sample**, is chosen. Information is then collected only from the sample and then used to draw conclusions about the population. A number that describes a characteristic of a sample is called a **sample statistic**. For example, suppose that a research group wants to know the percentage of college students who do laundry on a weekly basis. If 100 college students are chosen for the study and 42% do laundry on a weekly basis, then the population is all college students and the 100 students chosen are the sample. The sample statistic is 42% of students. Therefore we might conclude that 42% of the population do laundry on a weekly basis.

Memory Booster

Population **P**arameter

Sample **S**tatistic

Notice that the first letters match!

Definition

A **population** is a particular group of interest.

A **variable** is a value or characteristic that changes among members of the population.

Data are the counts, measurements, or observations gathered about a specific variable in a population in order to study it.

A **census** is a study in which data are obtained from every member of the population.

A **parameter** is a numerical description of a population characteristic.

A **sample** is a subset of the population from which data are collected.

Sample statistics are numerical descriptions of sample characteristics.

Throughout this text, it is essential that you are mindful of the relationship between a population and a sample. Figure 1.1 is a picture to help you visualize this relationship. The large oval represents the entire population, and the smaller oval represents the sample chosen from the population. Note that the sample is a subset of the population. That is, the sample is a group from within the population.

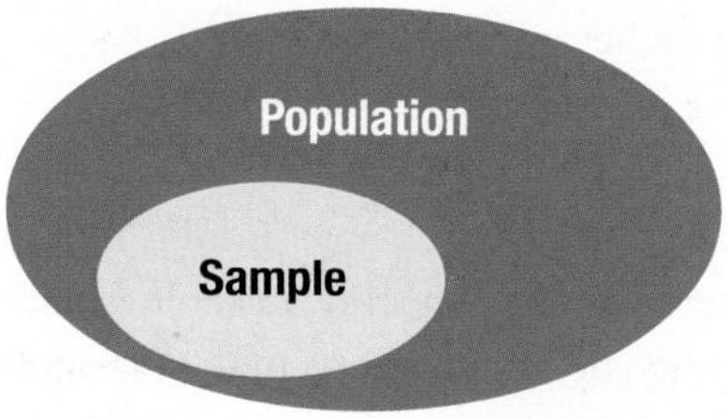

Figure 1.1: Population vs. Sample

However, when choosing a sample from the population, the old adage "anyone'll do" just isn't true. Care must be taken in choosing your sample so that the population is well-represented and the results of the study are meaningful. For instance, consider our earlier example of a study wishing to know the percentage of college students who do laundry on a weekly basis. If the 100 college students we chose for our study all lived off-campus because they were our friends and easy to ask, would this sample represent the population who you thought the study was about? What if they all were married? Or all over 40? Probably not the group you immediately thought of when we said "the percentage of college students who do laundry on a weekly basis." In Section 1.3, we will look more closely at how to choose a sample that is representative of the entire population being studied. Table 1.1 summarizes the differences between population parameters and sample statistics.

Table 1.1: Population vs. Sample

Population	Sample
Whole group	**Part** of the group
Group we **want to know** about	Group we **do know** about
Characteristics are called **parameters**	Characteristics are called **statistics**
Parameters are generally **unknown**	Statistics are always **known**
Parameters are **fixed**	Statistics **change** with the sample

Example 1.1

Identifying Population and Sample

Identify the population and the sample.

a. In a survey, 359 college students at the University of Jackson were asked if they had tried the October flavor of the month at the campus coffee shop. Eighty-three of the students surveyed said yes.

b. A survey of 1125 households in the United States found that 24% subscribe to satellite radio.

Solution

a. Population: All college students at the University of Jackson

Sample: The 359 college students who were surveyed

b. Population: All households in the United States

Sample: The 1125 households in the United States that were surveyed

Example 1.2

Identifying Population, Sample, Parameters, and Statistics

Read each of the shortened survey reports below. For each report:

a. Identify the population.

b. Identify the sample.

c. Determine whether the highlighted value is a parameter or statistic.

I. After an airplane security scare on Christmas day, 2009, the Gallup organization interviewed 542 American air travelers about increased security measures at airports. The report stated that 78% of American air travelers are in favor of United States airports using full-body-scan imaging on airline passengers.

Source: Jones, Jeffrey M. "In U.S., Air Travelers Take Body Scans in Stride." 11 Jan. 2010. http://www.gallup.com/poll/125018/Air-Travelers-Body-Scans-Stride.aspx (12 Dec. 2011).

II. Rasmussen Reports also conducted a survey in response to the airport security scare on Christmas day, 2009. The national telephone survey of 1000 adult Americans found that 59% of Americans surveyed favor racial profiling as a means of determining which passengers to search at airport security checkpoints.

Source: Rasmussen Reports. "59% Favor Racial, Ethnic Profiling For Airline Security." 7 Jan. 2010. http://www.rasmussenreports.com/public_content/lifestyle/general_lifestyle/january_2010/59_favor_racial_ethnic_profiling_for_airline_security (12 Dec. 2011).

Solution

I. **a.** Population: All American air travelers

b. Sample: The 542 American air travelers who were surveyed

c. The value 78% refers to *all* American air travelers; thus this is a population parameter.

II. a. Population: All adult Americans

b. Sample: The 1000 adult Americans who were surveyed

c. The value 59% refers to only those adult Americans who were surveyed, thus this is a sample statistic.

Branches of Statistics

Finally, there are two main branches of statistics that we need to distinguish. The first branch of statistics is called **descriptive statistics** as it gathers, sorts, summarizes, and generally describes the data that have been collected. Descriptive statistics involves the raw data, as well as the graphs, tables, and numerical summaries used to describe these data. You may think of this branch of statistics as "just the facts." Descriptive statistics refers to the sample data without making any assumptions about the population from which the sample was drawn. The Rasmussen Report in the previous example is an example of descriptive statistics.

The second branch of statistics is called **inferential statistics**. It deals with the interpretation of the information collected. These interpretations are typically inferences about the population based on sample data. The Gallup poll from the previous sample is an example of inferential statistics. The percentage is an inference about the entire population based on the sample of 542 air travelers who were actually interviewed. Of course, the accuracy of this inference is dependent on how well the 542 people who were interviewed represent the population of all air travelers.

Definition

The branch of **descriptive statistics**, as a science, gathers, sorts, summarizes, and displays the data.

The branch of **inferential statistics**, as a science, involves using descriptive statistics to estimate population parameters.

In this textbook, you will learn about both branches of statistics. Chapters 1 through 4 focus on descriptive statistics, showing you how to gather, sort, and organize data. The remaining chapters focus on inferential statistics, as you learn to analyze data and use descriptive statistics to make inferences about populations through confidence intervals and hypothesis tests.

Example 1.3

Identifying Descriptive and Inferential Statistics

In a news report on the state of the media by Tom Rosenstiel and Amy Mitchell, they write the following:

"AOL had 900 journalists, 500 of them at its local Patch news operation.... By the end of 2011, Bloomberg expects to have 150 journalists and analysts for its new Washington operation, Bloomberg Government."

Source: Rosenstiel, Tom and Amy Mitchell. "Overview." *The State of the News Media: An Annual Report on American Journalism.* Pew Research Center's Project for Excellence in Journalism. 2011. http://stateofthemedia.org/2011/overview-2/ (12 Dec. 2011).

Identify the descriptive and inferential statistics used in this excerpt from their article.

Solution

When the authors identify AOL as having "had 900 journalists, 500 of them at its local Patch news operation," they are describing the actual counts, not estimates; thus these numbers of journalists are descriptive statistics. On the other hand, when the authors state "By the end of 2011, Bloomberg expects to have 150 journalists and analysts for its new Washington operation," they are referring to an estimate based on past descriptive statistics. Therefore the estimate, 150 journalists and analysts, is an inferential statistic.

1.1 Section Exercises

Population and Sample

Directions: For each scenario, identify the population being studied and the sample chosen.

1. A national magazine wishes to determine America's favorite celebrities. A ballot is included in the November issue of the magazine. Readers are encouraged to mail in their ballots.

2. An education professor wants to gather information about parental involvement in early education for students attending a particular Ivy League university. She obtains a list of registered students from the registrar's office and randomly chooses 300 students to study.

3. A large discount store wants to determine the average income of its shoppers. A researcher chooses 100 shoppers at random between the hours of 1 and 5 p.m. on Thursday.

4. A hotel chain wants to build a new facility. The board of directors uses a list of the top 100 vacation spots in America and visits 20 of the cities on the list to determine their feasibility as the new hotel's location.

5. One Christmas season, you are interested in determining the average height of Christmas trees sold in your city. You visit two local Christmas tree farms and measure the height of 45 randomly selected trees.

Population, Sample, Parameters, and Statistics

Directions: For each scenario, determine the population and sample, if given. Also determine whether the highlighted value is a parameter or a statistic.

6. The manufacturer of Energy Bolt soda is testing for quality control. The company determines that 99.97% of all soda cans produced meet their quality standards.

7. Ms. Weigang wishes to determine the average ACT score for all students in her 11:00 algebra class. To do so, she anonymously collects information from each student. She calculates that the average ACT score for the class is 19.2.

8. The Nielsen Company wishes to determine which genre of TV shows is most popular for Americans aged 18–25. They survey a group of 1045 adults aged 18–25 about which TV shows they watch on a regular basis. It is determined that 53% of survey respondents say that their favorite shows are comedies.

9. A research group wishes to know the average salary of professors at public universities in the United States. It is determined from public records that professors at public universities earn an average salary of $78,300 per year.

10. A horticulturist is testing to see whether a new fertilizer produces plants that are significantly taller than those grown with traditional fertilizers. His greenhouse contains 520 plants treated with the new fertilizer. He measures 60 plants and determines that their average height is 22.9 inches.

11. A study is conducted in order to determine the percentage of high school seniors in the Atlanta area who plan to major in business upon entering college. For the study, 230 seniors from Atlanta area high schools are randomly chosen and surveyed. Of these, 42% say that they intend to major in business upon entering college.

12. For her dissertation, Maria needs to estimate the average number of hours per week that children aged 10–12 spend watching TV. She randomly surveys 250 schoolchildren in her area who are between 10 and 12 years old and calculates an average of 17.4 hours of TV per week.

13. For a news special, a reporter wishes to determine what percentage of adults in her viewing area are overweight. A random sample of 1067 adults from the area is chosen for the study. 40% of adults sampled are found to be overweight.

14. The governor of a Midwestern state wants to know his approval rating following a recent scandal. Using a telephone poll of registered voters in his state, 565 constituents are surveyed. The researchers determine that 37% of the governor's constituents that were surveyed currently approve of the job he is doing in office.

15. A large real estate firm wants to know the average price per square foot for condominiums in Okaloosa County, Florida. Based on real estate records for all condominiums in the county, the firm determines that condominiums in Okaloosa County sell for approximately $300 per square foot.

16. As part of a special report on marriage, Ebony is given the assignment of estimating the percentage of married men in Harbor City who would admit to having had an extramarital affair. Through an anonymous survey posted on a secure website, 338 married men are surveyed regarding their marriages. Of these men, 21% admit to having had an affair.

17. A consumer advocacy group wants to survey residents in the Northeast regarding hospital care. They mail out 10,500 surveys to randomly selected households in the Northeast. A total of 984 surveys are completed and returned. Out of this group, 64% say that their hospital care was above average. After analyzing all of the surveys, the consumer advocacy group determines that approximately 90% of people in the Northeast are satisfied with hospital care in their region.

18. The PTA of Brownsville is concerned about the number of hours that high school students spend each week using social networking websites. A group of 40 high school students is chosen at random and asked to take note of the amount of time spent using social networking websites during the following week. Out of these 40 students, 14 logged over 20 hours and another 9 logged between 15 and 20 hours on social networking sites that week. The average for the group was 18.9 hours.

Directions: Read the following fictitious studies. For each, determine the population, sample, parameter(s), and statistic(s).

19. Java Express is a company that sells high-end coffee appliances and gourmet coffee. Due to the current economic conditions, the company is looking into ways to change their strategy in order to appeal to more frugal consumers.

Before investing in costly setup expenses, Java Express needs to know the percentage of coffee consumers who would buy from a new line of less expensive products. To estimate this percentage, the company chooses 5 of their largest markets and surveys every registered customer from each of these chosen markets. A total of 6195 customers complete the survey.

The results show that 45% of those surveyed are very interested in the new line of products, 32% are somewhat interested, and the remaining 23% are not interested in the new products at all. Based on these results, the researchers estimate that approximately 77% of coffee consumers will be receptive to the new line of products. Java Express decides that it is worthwhile to introduce a new line of less expensive products to the market.

20. A study was conducted to investigate whether or not exercise is a factor in the overall outlook and positive spirit of Americans.

For the study, 400 Americans were selected at random from each of the following two categories: those who exercise on a regular basis and those who do not. For the purpose of the study, regular exercise was defined to be light to moderate activity at least three times per week. A total of 800 Americans completed the brief telephone survey. Of those who exercise regularly, 74% said that their overall outlook on life is positive. Of those who do not exercise regularly, 68% said that their overall outlook on life is positive.

Though not a definitive study, the researchers concluded that exercise does appear to influence the attitudes of Americans. In addition, the researchers estimated that approximately 71% of all Americans have a positive outlook. Further research would likely be conducted on the subject at a later time.

Descriptive vs. Inferential Statistics

Directions: Decide whether the following statements are examples of descriptive or inferential statistics.

21. Eighty-two percent of the employees from a small local company attended the annual company picnic.

22. Based on information from a recent survey, researchers estimate that the average American will spend $751 on gifts this Christmas.

23. The average age of entering freshmen at the University of Senatobia is 20.8 years old, based on information from the registrar's office.

24. Sixty-five percent of seniors at a local high school who are applying to college apply to at least one college out of state.

25. The average number of hours vacationers spend in national parks during the summer months is 4.5 hours, based on a survey of 1000 visitors in various national parks.

Directions: For each scenario, answer the questions that follow.

26. Since 1878, the Bureau of Labor Statistics has calculated the unemployment rate by surveying a group of randomly selected Americans (currently around 60,000) in regard to their employment status. A person is considered unemployed if he or she does not have a job and has looked for a job in the previous four weeks. Recently the method for compiling the unemployment rate has been criticized for not giving an accurate estimate of the true percentage of jobless Americans.

a. What is the population of the study?

b. What groups of jobless Americans might be left out based on the way "unemployed" is defined?

c. The unemployment rate is often used for comparison. For example, in recent times economists compared the unemployment rates computed each month during the recession of 2009 and 2010 to those of the Great Depression, when unemployment peaked at around 25%. Could comparisons like this be made if the method used to collect data for the unemployment rate was changed?

27. Suppose you are given the task of determining whether it would be profitable to open a new ladies' health club in the Atlanta suburb of Marietta, Georgia. Which of the following options would be the best choice for the population of your study? Explain your answer.

a. All women in metro Atlanta who exercise regularly

b. All women in Marietta, Georgia

c. All women in Marietta, Georgia who exercise regularly

d. All residents of the metro Atlanta area

28. A car manufacturer is in the design process for a new sedan that it plans to launch in the next year. Before finalizing the body type and other details about the car, the company wants to know the preferences of their target market. The manufacturer would like its new car to be appealing to all age groups, but its particular focus is on drivers in their 20s and 30s.

a. The company wants to know the preferences of drivers in their "20s and 30s." How might the company better define the population of the study?

b. Imagine that you are the researcher. Brainstorm different ways that you might collect data and obtain a sample that represents the population well.

1.2 Data Classification

Just as animals can be classified into phyla and then further into species, data collected in a statistical study can be classified into different categories. The different categories of group data are based on the type of statistical analysis that can be performed on the data. Therefore, knowing the classification of a set of data is the first step in any statistical process.

Qualitative vs. Quantitative Data

The first distinction we will make is between qualitative and quantitative data. **Qualitative data**, also known as categorical data, consist of labels or descriptions of traits of the sample. Examples of qualitative data include such information as favorite foods, hometowns, eye colors, or identification numbers.

Generally, qualitative data will be in terms of words; however, sometimes numbers act like words. For example, numbers on sports jerseys serve only to distinguish different players during games. In this case, the numbers are merely labels and must be classified as qualitative.

In contrast, **quantitative data** consist of counts or measurements, and therefore are numerical. Test scores, average amounts of rainfall, and weights are all examples of quantitative data. Notice that quantitative data can be manipulated in ways that qualitative data cannot. For example, you cannot add eye colors together, but you can add weights together.

Memory Booster

Qualitative data are descriptions (qualities).

Quantitative data are counts and measurements (quantities).

Example 1.4

Classifying Data as Qualitative or Quantitative

Classify the following data as either qualitative or quantitative.

a. Shades of red paint in a home improvement store

b. Rankings of the most popular paint colors for the season

c. Amount of red primary dye necessary to make one gallon of each shade of red paint

d. Numbers of paint choices available at several stores

Solution

a. Shades of paint are descriptions and cannot be measured, so these are qualitative data.

b. Rankings are numeric but not measurements or counts, so these are qualitative data.

c. The amounts of dye needed are measured and therefore are quantitative data.

d. The numbers of paint choices must be counted, so they are quantitative data as well.

It's important to take a moment and stress the difference between qualitative and quantitative data. Although the distinction may seem clear, often data may appear to be quantitative rather than qualitative simply because of the convenience of translating descriptive categories onto a numerical scale. For example, we often treat scales such as Never – Sometimes – Often – Very Often as numerical data represented by 1, 2, 3, 4. If we treat these numerical representations as quantitative, we infer that the distance between Never and Sometimes is the same as the distance between Sometimes and Often, which is subjective at best. In fact, we imply that the "average" of Never and Often is Sometimes. Is that always the case? Or just sometimes? Clearly this doesn't make sense—let this example caution you not to manipulate qualitative data inappropriately.

Memory Booster

Continuous data are usually measurements.

Discrete data are usually counts.

Continuous vs. Discrete Data

Quantitative data can be further classified as either continuous or discrete. **Continuous data** can take on any value in a given range of numbers. Continuous data are usually measurements, such as length and weight. If quantitative data are not continuous, then they are considered to be discrete. **Discrete data** can take on only particular values and cannot take on the values in between. Thus discrete data are usually counts. For example, the number of pets you have would be discrete because you can have either 2 pets or 3 pets, but not 2.75 pets.

Example 1.5

Classifying Data as Continuous or Discrete

Determine whether the following data are continuous or discrete.

a. Temperatures in Fahrenheit of cities in South Carolina

b. Numbers of houses in various neighborhoods in a city

c. Numbers of elliptical machines in every YMCA in your state

d. Heights of doors

Solution

a. Temperatures could be measured to any level of precision based on the thermometer used, so these are continuous data.

b. Numbers of houses are discrete data because houses are counted in whole numbers. A house under construction is still a house.

c. The numbers of elliptical machines are counts, so these are discrete data.

d. Heights are measurements and again, depending on the ruler, the heights could be measured to any level of precision, so they are continuous data.

Levels of Measurement

Another way that data may be classified is according to the level of measurement. There are four levels of measurement: nominal, ordinal, interval, and ratio. The higher the level of measurement is, the more mathematical calculations that can be performed on the data.

The first and lowest level of measurement is the nominal level. Data at the **nominal level** of measurement are qualitative, consisting of labels or names. Because we cannot add labels or names together, no calculations can be performed on data at the nominal level. You can think of the nominal level as labeling or categorizing a subject. For example, Maeve belongs in the category of females, and jersey number 15 labels Liam.

Example 1.6

Understanding the Nominal Level of Measurement

a. Suppose all students in a statistics class were asked what pizza topping is their favorite. Explain why these data are at the nominal level of measurement.

b. Suppose instead that you wish to know the number of students whose favorite pizza topping is sausage. Explain why this data value is not at the nominal level of measurement.

Solution

a. These data are nominal because the data simply describe or label the different toppings of the pizza.

b. In the second scenario, the data value is a *count* of students who prefer sausage. This data value is quantitative, not qualitative, so it is not a label and would not be considered to be at the nominal level of measurement.

The second level of measurement is the **ordinal level**, and data at this level are also qualitative. Data at the ordinal level can be arranged in a meaningful order, such as a ranking or ordered rating, but calculations such as addition or division do not make sense. In short, ordinal data have all the attributes of nominal data, but, in addition, they also have a natural order. For example, the 2011 rankings of SEC football teams would be at the ordinal level (1: LSU, 2: Alabama, and so on). However, note that there is no meaningful difference between any two data entries. For instance, in our SEC example, how much better is LSU than Alabama? Clearly the rankings give only an order, so these data are at the ordinal level. The survey responses of Excellent, Average, and Poor are another example of data at the ordinal level, qualitative data that definitely have a natural order.

Example 1.7

Classifying Data as Nominal or Ordinal

Determine whether the data are nominal or ordinal.

a. The seat numbers on your concert tickets, such as A23 and A24

b. The genres of the music performed at the 2013 Grammys

Solution

a. Seat numbers are ordinal because there is a meaningful order to the data, namely, the position in the theater.

b. Despite the fact that you may have your own personal preference for specific genres of music, there is no standard order. Therefore, music genres are nominal data.

The third level of measurement is the **interval level**, and data at this level are quantitative. Like data at the ordinal level, data at the interval level can also be ordered, but the interval level is distinguished because differences between data entries are meaningful. For example, if we compared the average temperatures of various cities, the data collected could be ordered. Furthermore, the differences between temperatures could be calculated and interpreted. Specifically, if Phoenix has an average temperature of 100 °F, and Stockholm's average temperature is 50 °F, we can say that Phoenix is hotter, and it is hotter by 50 °F.

One important note regarding the interval level is that the zero point associated with these data is a position on a scale, but does not actually mean the absence of something. In the given example regarding temperatures, zero on the Fahrenheit scale does not mean the absence of heat; it is simply a really cold temperature. Zeros in data at the interval level are merely placeholders. Some common examples of interval data include calendar dates, temperatures, and certain personality or intelligence test scores. One exception to temperatures being at the interval level of data is the Kelvin (K) scale of temperature. Absolute zero, 0 K, really does indicate the absence of heat, placing it in the last level of measurement, which we will cover next.

Example 1.8

Classifying Data by the Level of Measurement

The birth years of your classmates are collected. What level of measurement are these data?

Solution

Birth years can be ordered. It is also meaningful to subtract years to determine the difference in age. However, the year 0 A.D. does *not* mean the beginning of time. Therefore, birth years are at the interval level of measurement.

The fourth and final level of measurement is the **ratio level** of measurement. In this highest level of measurement, data are quantitative, can be ordered with meaningful differences, and the zero point indicates the absence of something. At this level, not only can we add or subtract data values, we can also multiply or divide. For example, suppose that we compare the prices of two cars. If one car costs $10,000 and the other $20,000, then the second car costs $10,000 more than the first and is twice as expensive. Note, though, that in the earlier example about interval data, it does not make sense to say that the average temperature of Phoenix is twice as hot as the average temperature of Stockholm.

Definition

Qualitative data consist of labels or descriptions of traits.

Quantitative data consist of counts or measurements.

Continuous data are quantitative data that can take on any value in a given interval and are usually measurements.

Discrete data are quantitative data that can take on only particular values and are usually counts.

Data at the **nominal level** of measurement are qualitative data consisting of labels or names.

Data at the **ordinal level** of measurement are qualitative data that can be arranged in a meaningful order, but calculations such as addition or division do not make sense.

Data at the **interval level** of measurement are quantitative data that can be arranged in a meaningful order, and differences between data entries are meaningful.

Data at the **ratio level** of measurement are quantitative data that can be ordered, differences between data entries are meaningful, and the zero point indicates the absence of something.

Example 1.9

Classifying Data by the Level of Measurement

Consider the ages in whole years of US presidents when they were inaugurated. What level of measurement are these data?

Solution

The ages of US presidents are measurable, can be ordered, and an age of zero indicates the absence of life. Therefore, ages are at the ratio level of measurement. In contrast to Example 1.8, involving birth years, you can be twice as old as someone else.

> **Memory Booster**
>
> Nominal ⇔ names
>
> Ordinal ⇔ order
>
> Interval ⇔ 0 is a placeholder
>
> Ratio ⇔ 0 means the absence of something

When categorizing data according to level of measurement, data should be associated with the highest level of measurement possible. The levels of measurement are like a staircase, and each level is the next stairstep. (See Figure 1.2.) You should try to "walk up" as many stairs as possible for each data set. For example, letter grades on a test categorize students' performances, thus satisfying the guidelines for nominal data, but we should not stop there. Letter grades can also be put in a meaningful order, so these data satisfy the guidelines for ordinal data as well. However, one cannot subtract a B from an A and have a meaningful calculation, so the data do not satisfy the guidelines for the interval level. Thus the highest level of measurement obtainable for letter grades is the ordinal level.

> **Memory Booster**
>
> **Qualitative Data**
>
> Natural order?

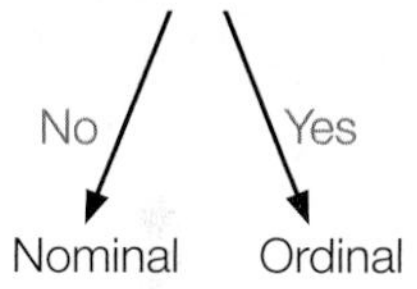

> **Quantitative Data**
>
> Meaningful zero?

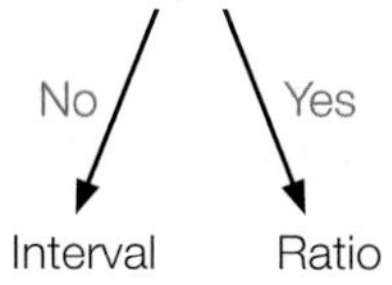

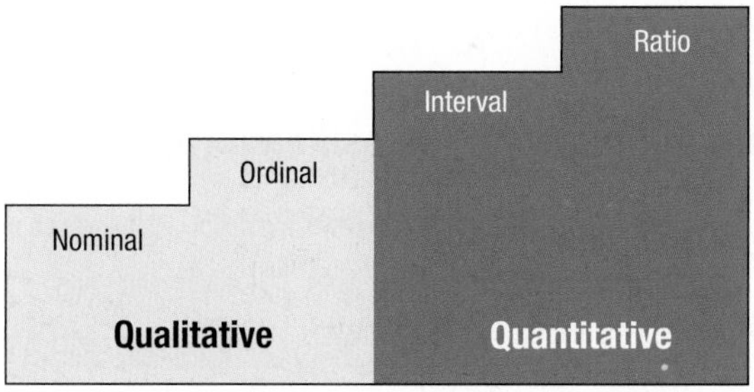

Figure 1.2: Levels of Measurement

Example 1.10

Classifying Data

Determine the following classifications for the given data sets: qualitative or quantitative; discrete, continuous, or neither; and level of measurement.

a. Finishing times for runners in the Labor Day 10K race

b. Colors contained in a box of crayons

c. Boiling points (on the Celsius scale) for various caramel candies

d. The top ten Spring Break destinations as ranked by MTV

Solution

a. The amount of time it takes for each runner to run the race is *quantitative* since calculations performed on these data are meaningful. A finishing time is a measurement, therefore the data are *continuous*. Differences between finishing times are meaningful, and a time of zero represents the absence of racing. We could also say that Andrew finished the race in half of Peyton's time; thus, the data are at the *ratio* level of measurement.

b. Colors are labels, so these data are *qualitative*. Qualitative data are *neither* discrete nor continuous. There are many ways to order colors, such as alphabetically or based on the color spectrum. However, when discussing colors of crayons, order is not the primary factor, as opposed to data such as rankings, in which order is important. Therefore, the data are at the *nominal* level of measurement.

c. Calculations can be performed on boiling points because they are measurements, making these data *quantitative*. Temperatures are measurements, so the data are *continuous*. For the Celsius scale, a temperature of zero degrees is simply a placeholder and does not indicate the absence of heat. Therefore, data from the Celsius scale are always at the *interval* level of measurement.

d. Since the rankings cannot be meaningfully added or subtracted, the data must be *qualitative*. Qualitative data are *neither* discrete nor continuous. The rankings are in a specific order, so the data are at the *ordinal* level of measurement.

1.2 Section Exercises

Data Classification

Directions: Determine the following classifications for the given data sets.

a. Qualitative or quantitative

b. Discrete, continuous, or neither

c. Level of measurement

1. Prices of a particular pair of jeans at various department stores

2. Widths of the doors in a home

3. Low temperatures in degrees Fahrenheit across the state for one evening in March

4. Total dollar value of all items placed in each safe deposit box at a local bank

5. Yearly amounts of snowfall in Cleveland over 10 years

6. Heights of orchids on a windowsill

7. Amount of weight gained by each person in a group of college freshmen

8. Number of antique cars collected by each member of a car club

9. Number of six-foot wooden boards it takes to build any given desk

10. Bank account PIN numbers

11. Heights of men entering the armed forces

12. Sizes of T-shirts on sale

13. Temperatures in Kelvin of various sites on the planet Mars

14. Numbers of siblings that students in Ms. Pitcock's third grade class have

15. Jersey numbers of players on a lacrosse team

16. Types of pets reported in a recent survey

17. Positions in line at the checkout counter of a grocery store

18. Letter grades on students' English essays

19. Birth order of children in a family

20. Titles that precede people's names (Dr., Mr., Ms., and so forth.)

21. Number of students enrolled in each section of College Algebra at The Ohio State University

22. Average temperatures in degrees Celsius of the water in the Bahamas for each month in 2011

23. Birth years of members of your immediate family

24. Numbers of people per household reported on census forms

25. Lengths of the yachts docked at the Beach Harbor yacht club

Directions: Respond thoughtfully to the following exercises.

26. Instead of classifying data into the groups qualitative and quantitative, some statisticians classify data as categorical or measurement data. **Categorical data** can be placed into categories, but no meaningful order can be assigned to the categories. **Measurement data** have numerical values assigned to them, and the data can be ordered according to those values.

a. Classify "genders of puppies adopted from an animal shelter" as categorical or measurement data.

b. Classify "high temperatures of Juneau, AK measured in degrees Fahrenheit" as categorical or measurement data.

c. What would be the problem with using this classification system for the data "T-shirt sizes listed as S, M, L, XL"?

27. Often, continuous data are measured in "discrete" units. In other words, the way that we express the units makes us think that we have discrete data because we are rounding the continuous values to whole number units. Give an example of a type of continuous data that might cause this confusion.

28. Explain why qualitative data are not further classified as discrete or continuous.

1.3 The Process of a Statistical Study

Now that we have learned how to classify data, let's turn to the process of a statistical study, which is a means for gathering, organizing, and analyzing data. A statistical study seeks to answer some question. When you read a statistical study, it is up to you to decide on the validity of the researcher's answer to the question being asked. Unfortunately, sometimes studies use faulty data or draw inappropriate conclusions, even those studies conducted by researchers who had good intentions. Being aware of how to design a good study helps in critiquing the results of studies you encounter in the news and professional journals.

The basic process of a statistical study is to state a question, collect data regarding the question, and then organize and analyze the data in order to answer the question. There are as many ways to perform each of these steps as there are questions that we might want to answer; however, there are some recognized methods of collecting, organizing, and analyzing data that you should be familiar with as you begin your study of statistics. We will go through each step in performing a statistical study and look at some of the most important points.

Procedure

Conducting a Statistical Study

1. Determine the design of the study.
   - **a.** State the question to be studied.
   - **b.** Determine the population and variables.
   - **c.** Determine the sampling method.
2. Collect the data.
3. Organize the data.
4. Analyze the data to answer the question.

A statistical study begins with a question that the researcher hopes to answer. Does taking 80 mg of aspirin each morning reduce the risk of heart attacks? What are students' opinions of the new layout of the student union? How do gorillas respond to captivity? These are just a few examples of questions that a researcher might be seeking to answer. The question being asked determines the population and the variables. Recall from Section 1.1 that variables are values that can change amongst members of the population. If the question is "Does taking 80 mg of aspirin each morning reduce the risk of heart attacks?" then the population of interest would be adults at risk for heart attacks. The variables would be the amount of aspirin taken and the occurrence of heart attacks. Notice how it is the question that tells us the population and the variables that are being studied.

Example 1.11

Identifying Population and Variables

Neurologists want to study the effect of vitamin C on nerve disorders. The goal of the study is to see if taking an intravenous dose of vitamin C will reduce the amount of nerve pain reported by patients. Identify the population of interest and the variables in this study.

Solution

Since the study seeks to determine if a new treatment will reduce nerve pain in patients with nerve disorders, the population of the study would be limited to patients with these specific types of disorders. Most likely, the study will further narrow the population to focus only on patients with a specific nerve disorder and not group all such patients in the same category. The variables of interest are the amount of vitamin C administered to a patient and the amount of nerve pain each patient reports.

Memory Booster

Use **experiments** to draw conclusions about **cause and effect**.

The second step of a statistical study is to collect data. The question that begins the study is often the best indicator of how to collect the data for the study. One way that we can gather data is through observing data that already exist. Studies that use this method to gather data are called **observational studies**. Alternately, we can generate data by performing an **experiment** or simulation. In situations where we wish to determine if one thing causes another to happen, it is best to use an experiment. Though observational studies can show that two variables are related in some meaningful way, they cannot be used to determine causality as an experiment can. For the research question "How do gorillas respond to captivity?" it would make sense to observe the behavior of gorillas that are already in captivity, thus we would perform an observational study. For our question about aspirin reducing the risk of heart attacks, it would be better to compare the heart attack risk of two groups, one that takes aspirin daily and one that does not. This scenario would be an experiment, which is necessary to draw the conclusion that one variable causes the change in the other variable.

Definition

An **observational study** observes data that already exist.

An **experiment** generates data to help identify cause-and-effect relationships.

Example 1.12

Identifying Observational Studies and Experiments

Which type of study would you conduct: an observational study or an experiment?

a. You want to determine the average age of college students across the nation.

b. Researchers wish to determine if flu shots actually help prevent severe cases of the flu.

Solution

a. An observational study would be used since you just need to consider existing records of college students to determine the average age of college students.

b. An experiment would need to be used in order to establish a cause-and-effect relationship between flu shots and flu prevention.

Observational Studies

An observational study observes data that already exist, such as students' SAT scores or favorite colors. No manipulation of the sample is performed by the researchers. One limitation of observational studies is that no cause-and-effect relationship can be determined between variables. For example, if we collect data on SAT scores and the GPAs of freshman college students, we might find a relationship between the variables, but it would be inappropriate to conclude that a cause-and-effect relationship exists.

In an observational study, a researcher collects data from the population. When data are collected from every member of the population, recall from Section 1.1 that the study is called a census. If it is not practical, or possible, to collect data from the entire population, then we would collect data from a sample of the population. As we said earlier, when choosing a sample, the researcher should be careful to choose one that represents the population well. A **representative sample** is one that has the same relevant characteristics as the population and does not favor one group from the population over another. Without a representative sample, it might not be possible to accurately generalize the results from the sample to the population as a whole.

There are several basic methods of choosing a sample from a population. The choice of sampling method depends on many factors, including the type of population, the study question, and practical considerations. Furthermore, some methods of choosing a sample generally produce more representative samples than others. We will discuss the advantages and disadvantages of the different sampling methods as we describe each of them.

Definition

A **representative sample** has the same relevant characteristics as the population and does not favor one group from the population over another.

Sampling Methods

Random Sampling: One of the most basic sampling techniques is a **random sample**. A random sample is one in which every member of the population has an equal chance of being selected. An example of this kind of sampling occurs when drawing names from a hat. Random sampling can also be performed by assigning identification numbers to each member of the population and using a random number generator to choose ID numbers. The technology section at the end of this chapter demonstrates how to use Microsoft Excel, Minitab, or a TI-83/84 Plus calculator to generate a series of random integers. This basic sampling technique is used as part of some of the more sophisticated methods described in this section.

Side Note

Smartphones and tablets now have many random number generator apps that can be downloaded for free.

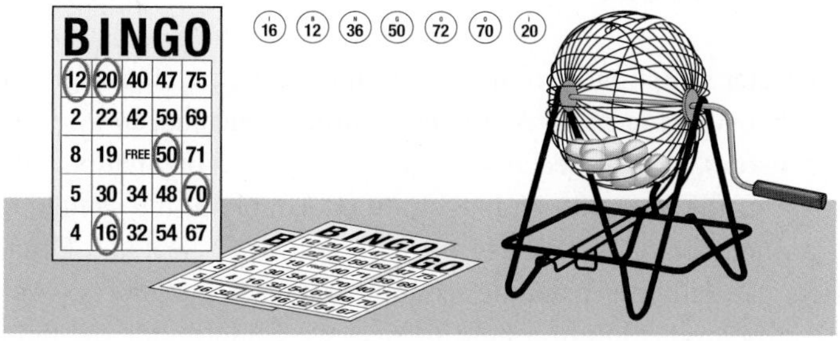

Figure 1.3: Random Sampling

Simple Random Sampling: Another sampling technique that uses a random sample is simple random sampling. There is a very important difference between these two types of sampling methods. In a random sample, every *member* of the population has an equal chance of being chosen. In a **simple random sample**, every *sample* from the population has an equal chance of being chosen. It can be confusing to distinguish between these two types of sampling techniques. Consider a researcher who needs to randomly select 10 retail stores in a metropolitan area. If the researcher randomly chooses a single shopping center, and surveys the 10 stores in that location, this *would* be a random sample. However, by using this method, not every sample of 10 stores has a possibility of being chosen. For instance, it would be impossible to choose a sample that contains stores located in different shopping centers. Thus, the described sampling technique would *not* produce a simple random sample. In order to choose a simple random sample of 10 stores, all retail stores in the metropolitan area would need to be listed, and then 10 stores from the list would be randomly chosen. In the future, it will be important to have a simple random sample to meet the assumptions of certain tests used to analyze data.

Stratified Sampling: A **stratified sample** is one in which members of the population are divided into two or more subgroups, called **strata**, that share similar characteristics like age, gender, or ethnicity. A random sample from *each* stratum is then drawn. For instance, if we divide the population of college students into strata based on the number of courses completed, then our strata would be freshmen, sophomores, juniors, and seniors. We would then select our sample by choosing a random sample of freshmen, a random sample of sophomores, and so on.

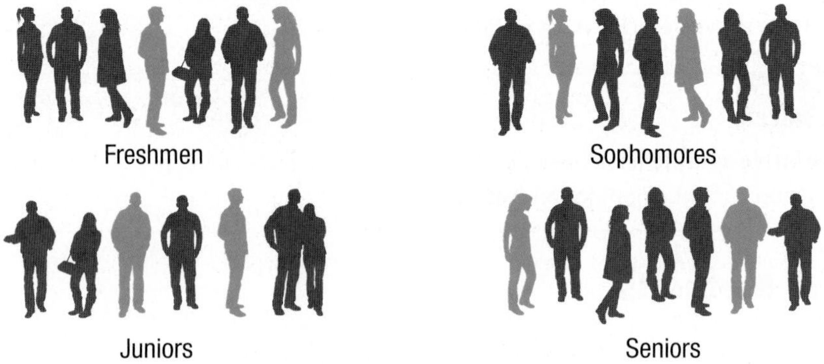

Figure 1.4: Stratified Sampling

This technique is used when it is necessary to ensure that particular subsets of the population are represented in the sample. Since a random sample of all students cannot guarantee that any sophomores would be chosen, we would use a stratified sample if it were important that sophomores be included in our sample. Furthermore, by using stratified sampling you can preserve certain characteristics of the population. For example, if freshmen make up 40% of our *population*, then we can choose 40% of our *sample* from the freshmen stratum. This type of stratification is also called **quota sampling**. Stratified sampling is one of the best ways to enforce representativeness on a sample.

Memory Booster

Stratified sample:
a *few* members of *each* group

Cluster sample:
each member of a *few* groups

Cluster Sampling: Cluster sampling is similar to stratified sampling in that groups of the population are considered when choosing a sample. A **cluster sample** is one chosen by dividing the population into groups, called **clusters**, that are each similar to the entire population. The researcher then randomly selects some of the clusters. Unlike stratified sampling, the sample consists of the data collected from *every* member of only those clusters that are selected. Often populations lend themselves to clusters naturally, for example, counties or voting precincts. As an example, if we need to estimate the crop production of a field of rice, we could divide the field of rice into small segments using a plot of the field. These segments would be our clusters. By randomly choosing segments of the field and measuring all of the rice produced by each of those segments, we could then estimate the total amount of rice the field would produce. However, you must use caution when choosing clusters to ensure that the clusters are representative of the entire population. For example, if all of the rice-field segments are on the edge of the field, or all in the middle of the field, there might be different water and nutrient levels making those samples different from the population as a whole.

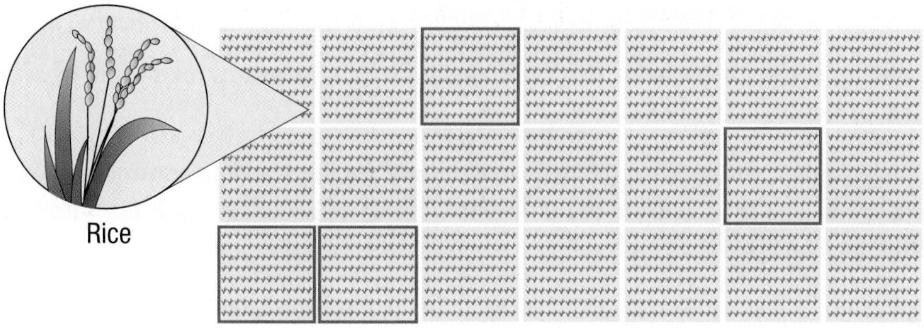

Figure 1.5: Cluster Sampling

Systematic Sampling: A **systematic sample** is one chosen by selecting every n^{th} member of the population. A classic example of this type of sampling is measuring every 100th item coming off of an assembly line. This sampling technique does provide a nice way to systematically choose the sample (as the name suggests) but there can be disadvantages. If the population has a pattern, systematic sampling can easily produce a sample that does not represent the population. For example, consider an assembly line that has a mechanism with a defect so that it causes an error in every 8th part made. If we start sampling every 8th part that is made by this assembly line, we will get either a sample of parts that have no errors or a sample of parts that all have errors. Neither of these samples are representative of the population of parts from that assembly line.

Figure 1.6: Systematic Sampling

Convenience Sampling: A **convenience sample** is one in which the sample is "convenient" to select. It is so named because it is convenient for the *researcher*. Because this method often leads to members of the sample all having a similar characteristic, it is prone to creating nonrepresentative samples. For example, a researcher standing in front of the campus cafeteria asking about how much students spend on eating out might get a different answer than if the researcher was standing at the door of a classroom on campus. However, in some instances convenience sampling is the preferred method. Take the case of a local grocer who wants to know what types of ice cream he should stock. Placing a sample-tasting table right on the ice cream aisle will yield a sample from the exact population he's looking for—ice cream consumers who shop in his store.

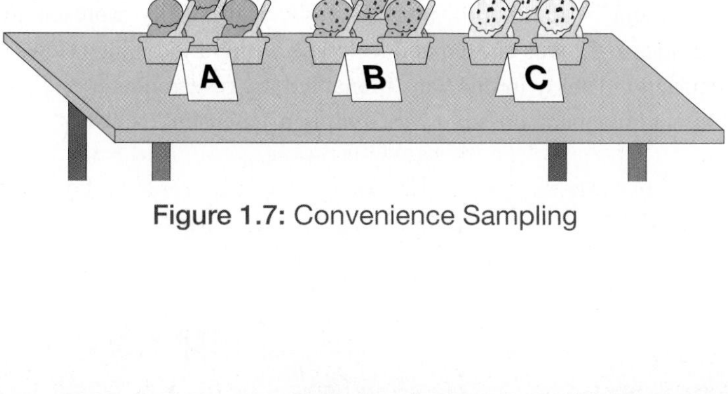

Figure 1.7: Convenience Sampling

Example 1.13

Identifying Sampling Methods

Identify the type of sampling used in each of the following scenarios.

a. A pollster surveys 50 people in each of a senator's 12 voting precincts.

b. The quality control department at a cereal manufacturer measures the weight of every 10th box off of the assembly line.

c. A female student walks down the halls in her dorm asking students how much money they would spend in a food court in the dorm lobby in an effort to persuade the administration to offer such an option.

d. An educator chooses 5 of the school districts in the Chicago area and asks each household in those districts how many school-age children are in the home.

e. To determine who will win a $100,000 shopping spree at the mall, the manager draws a name out of a box of entries.

Solution

a. Stratified sampling: The voting precincts are the strata.

b. Systematic sampling: The system of selecting the sample is to choose every 10th box.

c. Convenience sampling: This would be a very easy method of surveying for this particular scenario, and it would provide a representative sample of the dorm residents.

d. Cluster sampling: Every school district in the Chicago area is a cluster.

e. Random sampling: Every name in the box has an equal chance of being chosen.

Types of Observational Studies

Let's now consider several types of observational studies. The first, called a **cross-sectional study**, is characterized by data that are collected at a single point in time. This method of data collection gives the researcher a "snapshot" of the information. Most polls and surveys would fall into the category of cross-sectional studies. For example, a Washington Post-ABC News poll taken in September 2009, about a year after the start of a devastating decline in economic markets, showed that approximately 66% of Americans had been hurt financially by the recession. These results were current at the time the poll was taken.

In a **longitudinal study**, on the other hand, data are gathered by following a particular group over a period of time. For instance, the National Longitudinal Study of Adolescent Health was begun in 1994 in order to determine the effects that social environments and behaviors in adolescence have on health and achievement in adulthood. For the study, a nationally representative group of over 15,000 students in grades 7–12 was chosen and followed with periodic interviews over the course of 14 years. By structuring the study in this way, it enabled the researchers to see patterns and links in behavior that would not have been possible with simply a "snapshot" view.

Example 1.14

Classifying Studies as Cross-Sectional or Longitudinal

Categorize the following studies as either cross-sectional or longitudinal.

a. A group of 220 patients is followed for 15 years in order to determine the long-term health effects resulting from gastric bypass surgery.

b. A gastroenterologist surveys 130 of his patients six months after having gastric bypass surgery to determine the average amount of weight lost.

Solution

a. For this study, a group of gastric bypass patients is followed for a period of time. By definition, this is a longitudinal study.

b. In this study, a snapshot of the amount of weight lost at a specific point in time is gathered; thus, this is a cross-sectional study.

Instead of collecting data from a sample, some studies collect data from previous studies or analyze a single member of the population. When a study compiles information from previous studies, it is referred to as a **meta-analysis**. This type of study seeks to identify patterns across many studies on a similar topic that would not be discernible when looking at a single study. A benefit of meta-analysis is that many smaller samples can be combined into a single larger sample. A drawback is that the combined study is only as good as its weakest link. In other words, if you use poorly constructed studies, your study will not be able to draw strong conclusions.

In contrast to a meta-analysis that looks at one variable over several studies, a **case study** looks at multiple variables that affect a single event. When you desire to look at a single case in depth and all of the possible variables associated with that case, then it would be most appropriate to perform a case study. If you are interested in studying how administrators respond to hazing incidents on college campuses, you could perform a meta-analysis of all of the research done on hazing. In contrast, if you are interested in studying a particular hazing incident at a certain institution, you would perform a case study of that hazing incident.

Definition

In a **cross-sectional study**, data are collected at a single point in time.

In a **longitudinal study**, data are gathered by following a particular group over a period of time.

A **meta-analysis** is a study that compiles information from previous studies.

A **case study** looks at multiple variables that affect a single event.

Example 1.15

Classifying Studies as Meta-Analysis or Case Study

Categorize the following studies as either a meta-analysis or a case study.

a. Oceanographers study research on tsunamis dating from 1900 to 2000 to determine their effects on the ocean floor.

b. Meteorologists study the Indian Ocean tsunami of December 2004 to try to identify warning signs.

Solution

a. Because the oceanographers are looking at multiple studies relating to the single variable of tsunamis' effects on the ocean floor, this is a meta-analysis study.

b. In order to identify tsunami warning signs, meteorologists would most likely look at multiple variables relating to the 2004 tsunami. Because they are studying several aspects of a single tsunami, it is a case study.

Experiments

An experiment is another commonly used method of data collection. As opposed to observational studies that observe data, experiments seek to create data to help identify cause-and-effect relationships. In an experiment, researchers apply a **treatment** to a group of people or things, called **subjects**, and measure the response. (If they are people, they can also be referred to as **participants**.) A treatment is simply some condition that is applied to a group of subjects for experimental purposes, such as asking one group of people in an experiment to take a vitamin. The variable that responds to the treatment is called the **response variable**. The variable that causes the change in the response variable is called the **explanatory variable**. If researchers want to design an experiment to determine if using Chantix reduces a person's urge to smoke, then the use of the drug Chantix would be the explanatory variable and the urge to smoke would be the response variable.

> **Side Note**
>
> **Ethical Dilemma**
>
> Even though experiments are used to determine cause and effect, not all cause-and-effect relationships can be determined through an experiment for ethical reasons. For example, it is important to know if a new medication will cause birth defects if taken while a woman is pregnant. However, it would be unethical to perform an experiment to determine cause and effect because we would not want to administer medication to pregnant women if we do not know the possible effects on the unborn child. Therefore we would have to test the medication through other means.

Definition

A **treatment** is some condition that is applied to a group of subjects in an experiment.

Subjects are people or things being studied in an experiment.

Participants are people being studied in an experiment.

The **response variable** is the variable in an experiment that responds to the treatment.

The **explanatory variable** is the variable in an experiment that causes the change in the response variable.

The design of any experiment determines how reliably the results of the experiment can be used to make inferences about the population. Thus, when analyzing an experiment, you should analyze its design. The three main principles in experimental design are to **randomize**, **control**, and **replicate**. First, *randomization* is used in the initial stages of the design when separating the participants into different groups. Throughout the design, researchers must *control* for outside effects on the response variable. Lastly, the sample should be large enough that *replication* of the explanatory variable's effects on the response variable can be seen in a significant portion of the sample. In addition, the results of the experiment should be able to be *replicated* in future similar experiments.

Procedure

Principles of Experimental Design

1. Randomize the control and treatment groups.
2. Control for outside effects on the response variable.
3. Replicate the experiment a significant number of times to see meaningful patterns.

There are many ways to set up an experiment, but the goal of any experiment is to compare the results of the treatment to a control in order to establish a cause-and-effect relationship. The **treatment group** is the group of subjects to which the treatment is applied, while the **control group** does not receive the treatment. The control and treatment groups may be the same group of participants measured before and after the treatment is applied, such as in a pretest/posttest scenario, or they may be separate similar groups. In order to generalize the results, it is important that the two groups are as similar as possible so that any difference between the groups can be attributed to the treatment. This establishes that the treatment is the cause of any effects that are seen.

Memory Booster

Treatment group
does have a treatment applied.

Control group
does not have a treatment applied.

However, researchers cannot randomly select participants for an experiment, as they would for an observational study. Generally, participants volunteer to be part of an experiment. Therefore, techniques must be established to create two similar groups from the available participants. One method that researchers use to create similar groups is to randomly assign the volunteers to the two groups. Experience has shown that groups created in this way usually have similar characteristics; for example, it would be highly unlikely for all of the men to be randomly assigned to one group and all of the women randomly assigned to the other group. Another way that researchers create the treatment and control groups is to purposely assign volunteers to the groups based on important characteristics. Thus, the researcher would assign an equal number of women, smokers, diabetics, and so forth, to each group.

When researchers directly assign participants to the various groups, they are controlling for **confounding variables**, which are factors other than the treatment that cause an effect on the groups. This is just one of many ways that researchers can control for confounding variables. If we are studying the effect of a new weight-loss program, factors such as diet, heredity, exercise, and motivation might all affect the outcome of weight loss. These can all be confounding variables, and the design of the experiment should try to control for these factors. One way would be to specify the variables as much as possible (such as the amount of exercise) and then use randomization to control for the others.

Another effect that researchers need to control for is called the **placebo effect**. Because people respond to suggestion, giving someone a drug for the common cold and telling them that it will cure them will often produce effects caused by the suggestion alone and not the drug itself. This response is known as the placebo effect. To counteract the placebo effect, subjects in the control group are given a **placebo**, which appears identical to the actual treatment, but contains no intrinsic beneficial elements. For example, if the treatment were a small orange pill that contains vitamin C, then the placebo would be a small orange pill that does not contain vitamin C. With both groups taking what they believe to be the treatment, the placebo effect will be the same for both groups. Thus, differences in the groups can then be attributed to the actual treatment instead of the suggestion from the placebo.

Side Note

Placebo Effect

In a bizarre instance of the placebo effect, the *Archives of General Psychiatry* reported a study in which two groups of patients with Parkinson's disease underwent brain surgery. In the first group, human neurons were transplanted into the patients' brains. In the second group, the patients were *merely told by their doctor* that the transplant had taken place, when it had not. However, *both* groups showed significant postsurgical increases in body and brain functioning.

Thus, the power of suggestion was enough for the second group to show improvement!

Source: Newswise. "Placebo Effect Strong in New Parkinson's Study." 7 Apr. 2004. http://www.newswise.com/articles/placebo-effect-strong-in-new-parkinsons-study (12 Dec. 2011).

In experiments that use a placebo, the subjects do not know if they are in the control group or the treatment group. This is called a **single-blind** experiment. In a single-blind experiment, the people interacting with the subjects in the experiment know in which group each subject has been placed. The researchers knowing which subjects are taking the actual treatment can cause the researcher to subconsciously influence the results of the study through their interactions with the participants. In a **double-blind** experiment, however, neither the subjects nor the people interacting with the subjects, such as doctors or nurses, know to which group each subject belongs. A double-blind design would be preferable when the researchers have to make subjective decisions regarding the participants, such as in clinical trials for psychotropic drugs.

Lastly, when setting up an experiment it is important to make sure that the experiment is replicated a sufficient number of times. Performing an experiment on just two subjects is unlikely to yield meaningful results that can be generalized to any population. The design should also be described in such a way that future researchers can validate the findings of the study by replicating the original study.

Definition

A **control group** is a group of subjects to which no treatment is applied in an experiment.

A **treatment group** is a group of subjects to which researchers apply a treatment in an experiment.

Confounding variables are factors other than the treatment that cause an effect on the subjects of an experiment.

The **placebo effect** is a response to the power of suggestion, rather than the treatment itself, by participants of an experiment.

A **placebo** is a substance that appears identical to the actual treatment but contains no intrinsic beneficial elements.

In a **single-blind** experiment, subjects do not know if they are in the control group or the treatment group, but the people interacting with the subjects in the experiment know in which group each subject has been placed.

In a **double-blind** experiment, neither the subjects nor the people interacting with the subjects know to which group each subject belongs.

Example 1.16

Analyzing an Experiment

Consider the study from Example 1.11, in which neurologists want to determine if taking an intravenous dose of vitamin C will reduce the amount of nerve pain reported by patients. Suppose that the study was narrowed to focus only on patients with the nerve disorder, multiple sclerosis (MS). After study approval, the neurologists solicit volunteers who are patients with MS who are reporting nerve pain. The participants are then randomly assigned to two groups, each having 20 participants. Participants in Group A are administered intravenous doses of vitamin C, and their nerve pain is tracked. Participants in Group B are administered intravenous doses of saline (which has no active ingredients) and their pain levels are also tracked. The patients are not told which of the two groups they are in; however, the nurses administering the IVs are aware of the group assignments. After a predetermined length of time, the amounts of pain reported by the separate groups are compared to determine if an intravenous dose of vitamin C will reduce the amount of nerve pain.

a. Identify the explanatory and response variables.

b. What is the treatment?

c. Which group is the treatment group and which group is the control group?

d. What is the purpose of administering saline to Group B?

e. Is this a single-blind or double-blind study? Do you think this is the best choice for this study?

Solution

a. The explanatory variable is what "explains" the changes in the response variable. Since the neurologists are trying to determine if the dose of vitamin C can reduce nerve pain, the explanatory variable is the dose of vitamin C and the response variable is the amount of nerve pain reported by each patient.

b. The treatment is what is being applied to the group, so the treatment is the dose of vitamin C.

c. The group that received the treatment of vitamin C, namely Group A, is the treatment group. The group that did not receive the treatment, Group B, is the control group.

d. The saline that is administered to Group B is a placebo, and is administered to compensate for the placebo effect, so that all patients are responding to the same suggestion that they are receiving treatment.

e. Since the patients do not know the group assignments, but the nurses who are interacting with the patients do know to which group the patients were assigned, this is a single-blind study. As reported pain is a subjective measure, it would be easy for nurses to unintentionally influence the patients' responses; thus a double-blind study would probably be a better design for this experiment.

Institutional Review Boards

Once a researcher has formulated a question and designed a study to explore that question, there is one more step that must be taken before actually gathering data. The researcher must get approval to conduct the study from an **Institutional Review Board (IRB)**, particularly in the academic and medical communities. The IRB is usually made up of people from the institution with which the researcher is affiliated (such as a university or workplace), as well as people from the community where the institution is located. The job of the IRB is to review the design of the study to make sure that it is appropriate and that no unnecessary harm will come to the subjects involved. The IRB will require the researcher to fill out documents describing the proposed study in detail, including how issues of informed consent, human or animal subjects, and confidentiality will be handled. Once the IRB approves the design of a study, the researcher is ready to begin collecting data.

When collecting data, researchers must get the **informed consent** of participants. Informed consent involves completely disclosing to participants the goals and procedures involved in a study and obtaining their agreement to participate. However, getting a participant's informed consent is not as straightforward as it sounds. There are gray areas, such as studies involving child participants or people with intellectual disabilities, in which questions arise about the meaning of informed consent

Side Note

Unnecessary Harm?

As you read the paragraph on the IRB, you may have thought that the phrase "unnecessary harm" was strange. After all, isn't all harm unnecessary? Not really. When you have a sinus infection and go to the doctor, they inflict the pain of a shot in order to administer the medicine that will make you better. That "harm" is necessary.

and who should give it. Furthermore, not all studies involve human subjects. Since animal subjects cannot give their informed consent to participate, it is the job of the IRB to provide that consent on behalf of the animals. This means that the IRB will require more extensive documentation on study procedures involving animals and any possible harm that might happen to the animals in the course of the study.

Studies involve a high level of trust between the researcher and the participants. Part of that trust is that any information acquired will be kept confidential. An IRB will require evidence that any data gathered from participants will be kept in a secure location and that only people who need to see the raw data will be allowed to do so. However, this does not mandate that the data must be gathered anonymously. For example, an educator studying the numbers of class absences and their effect on students' final grades can keep the raw data confidential by keeping students' records in a locked file cabinet and only allowing authorized people to see the data; however, it is likely that the educator will know which student is associated with what data. Therefore, although the data are kept confidential, they are not collected anonymously.

Definition

An **Institutional Review Board (IRB)** is a group of people who review the design of a study to make sure that it is appropriate and that no unnecessary harm will come to the subjects involved.

Informed consent involves completely disclosing to participants the goals and procedures involved in a study and obtaining their agreement to participate.

In summary, a statistical study begins with a question to be answered. This question determines the population of the study and the variables of interest. Data are then collected by means of an observational study or an experiment, depending on whether the study hopes to determine a cause-and-effect relationship. Once data are collected, they are organized through the use of tables, graphs, or numerical summaries. Lastly, the results are analyzed to answer the original question. The remainder of the book, beginning with Chapter 2, will be devoted to these last two steps of a study: organizing and analyzing data.

1.3 Section Exercises

Vocabulary

Directions: Determine if each statement is true or false. Explain why.

1. The first step in any statistical study is to state the question to be answered.

2. Data collection must be complete before variables are chosen so that the researcher can be sure he has the data needed to answer the question.

3. If a researcher wishes to determine a cause-and-effect relationship, she should use an observational study.

4. A random sample is the same thing as a simple random sample.

5. An Institutional Review Board will require that a researcher disclose to participants the goals and procedures involved in her study and obtain their agreement to participate.

6. The question in a statistical study dictates the population and variables.

7. Participants in an experiment should always be allowed to choose which group they are placed in so that they feel as comfortable as possible for the duration of the experiment.

Directions: Complete each statement.

8. An experiment in which both the participant and the person administering the treatment are unaware of whether the participant is in the treatment group or the control group is referred to as a __________ experiment.

9. A member of the population that is being studied in an experiment is called a __________.

10. The group in an experiment that receives a placebo is called the __________ group.

11. An experiment is a type of statistical study in which a __________ is applied to a group of the population.

12. A pill that looks like the treatment pill but has no active ingredient is called a __________.

13. An experiment in which only the participant is unaware of whether they are in the treatment group or the control group is referred to as a __________ experiment.

14. When a subject believes that he has recovered from an illness because he is taking a treatment drug, when in reality he is in the control group of an experiment, it is referred to as the __________.

15. A human subject in an experiment is referred to as a __________.

16. A researcher gives an active drug to the __________ group in an experiment.

Observational Studies vs. Experiments

Directions: Determine which type of study you would conduct: an observational study or an experiment.

17. A football coach wants to know the average weight of his offensive linemen.

18. A doctor wants to study the effect of ginseng on patients' memories.

19. A city planner wants to know the average number of vehicles parked in downtown parking lots on any given business day.

20. A cell phone company wants to know the average total length of time teenage girls spend on the phone each day.

21. A dentist wants to look at the effects of a new dental material used for fillings.

Sampling Methods

Directions: Identify the sampling method used in each scenario.

22. The FDA chooses 15 hospitals around the country at random. Every doctor in the chosen hospitals is asked to participate in the study.

23. Every 4th dorm room is selected for a survey regarding study hours and campus security.

24. A state politician wants to gauge public opinion in his area before deciding to run for reelection. For the study, 200 registered voters are chosen at random from each county in his district.

25. A computer program is used to randomly generate a list of student ID numbers in order to gather a group to give feedback about the Greek system on campus.

26. In order to complete a psychology project, you pass out surveys to the first 25 people you find in the student union.

27. A student asks all the people living on the 1st, 5th, and 8th floors of his dorm to answer a survey about dorm life on your campus.

28. A local church wanted to know the average age of its morning congregation, so they asked every 10th person leaving the service to put their age in a box.

29. Ten students from each of the 15 sections of College Algebra were asked about the quality of the textbook used in the course.

30. One thousand phone numbers were selected by a computer to be called for a telephone survey.

31. A local politician asks 20 people in his neighborhood what they think about the new school board proposal.

Cross-Sectional vs. Longitudinal Studies

Directions: Classify each scenario as either a cross-sectional study or a longitudinal study.

32. A social worker wants to determine the number of current foster children in her district who were placed in foster care due to neglect.

33. A budget-conscious person wishes to find which gas station in his area has the cheapest gas on his way to work one morning.

34. A local teachers' group creates charts to demonstrate that pay raises over the last five years have not kept up with inflation.

35. The child welfare office keeps track of how many reports of child abuse are received each month over a two-year period to determine if there are certain times of the year that generally have a higher report rate.

36. An LGBT group gathers data from each state where same-sex marriage is legal regarding how many such marriages were performed in 2011.

37. A patient with HIV gets her blood tested every three months to check her viral load to make sure that it is not increasing.

Meta-Analysis vs. Case Studies

Directions: Classify each scenario as either a meta-analysis or a case study.

38. A child prodigy's home life is examined in order to determine environmental factors that may have shaped him intellectually.

39. Studies performed on four different airlines are compared in order to determine which provides the best customer care.

40. To prepare a story for the January issue, a magazine reporter analyzes eight different studies about dieting strategies in order to determine which is most effective over a long term.

41. For the purpose of studying sibling rivalry as affected by birth order, a typical American family is selected.

42. A medical researcher looks at multiple studies performed on a new drug in order to determine whether or not the drug is safe to put on the market.

Directions: Answer each question thoughtfully.

43. In the text we considered the research question: "Does taking 80 mg of aspirin each morning reduce the risk of heart attacks?" If we wanted to narrow our question to "Does taking 80 mg of aspirin each morning reduce the risk of heart attacks in African-American women over the age of 50?," how would this change the population of the study? If the results of the study showed that aspirin did indeed reduce the risk of heart attacks in this new population, would you be justified in recommending that your 52-year-old uncle begin taking aspirin daily? Does your answer change based on the ethnicity of your uncle? Explain.

44. Why not let a human choose the random sample? In reality, it is against our human nature to choose members of the population truly at random. To understand this phenomenon, take a moment to choose five random numbers between 1 and 100. Try to make sure they are truly random. Now, consider these questions. Are the numbers spaced out or grouped closely together? Are they all even, odd, or some of both? Did you have a reason for choosing any of the numbers? Did you alter any of your original responses and if so why? Did you repeat any of the numbers? In summary, would you say you were able to truly generate a set of random numbers?

45. In the text we considered an assembly line that has a mechanism with a defect so that it causes an error in every 8th part made. We said that if we start sampling every 8th part that is made by this assembly line, we will get either a sample of parts that have no errors or a sample of parts that all have errors. Explain how to get a sample with no errors. Is it possible to get an unbiased sample by choosing every 5th part from the assembly line? What about every 16th part? Why or why not?

46. One type of convenience sampling is a **self-selected sample**. A self-selected sample is one in which the survey participants volunteer to be a part of the study rather than having been chosen by the researcher. The problem with a self-selected sample is that usually only people with strong opinions will take the time to volunteer their time or information for the study. For instance, a popular American women's magazine wants to do research for a story regarding hospital care. The magazine lists a website in its June issue, inviting readers to log on and share stories about the care they received in the hospital. Describe the types of responses you could expect from people willing to log on and answer this survey. What would be the true population for the study described in this scenario? Can the results of this self-selected sample be generalized to describe hospital care for all patients in American hospitals?

1.4 How to Critique a Published Study

As we said in the previous section, when you read a statistical study, it is up to you to decide on the validity of the researcher's answer to the question being asked. In Section 1.3, we looked at design principles of a good study so that, in understanding the behind-the-scenes setup of a study, we may better identify valid conclusions or results from data. Let's take a look at some other things to keep in mind as we thoughtfully and critically filter through results that we come across.

Consider the *Source*

As you listen to sound bites on the radio or television spouting off statistics from studies, you should ask yourself pertinent questions: Who paid for the study? Where were the data collected? When was the information collected? Who published the study? Although that sound bite might not give you the who, what, when, and why about the study, a newspaper article or journal article often will. Even with the most well-constructed studies in the world, you should consider the source. Do the organizations funding or publishing the study have a vested interest in the results? The truth is that we need organizations and groups to promote different views, and it's perfectly reasonable for them to use statistics to back up their way of thinking. As a consumer, it's wise to consider the information outlet.

The source could also mean the population from which the data are collected. For instance, suppose you wanted to know whether it's dangerous to have radioactive material in airports. The truthful answer you would ideally gain from most of the general population should be "I have no idea." However, given an opportunity to express their own viewpoints, most people would not feel inhibited by answering the question. In this case, the population being surveyed is not truly qualified to answer the question. How much value should you place in their collective opinion? However, if you were to ask the same question at a physics conference, the "general population" there would have a better informed opinion to give.

Consider the *Variables*

Have you ever read a headline like "America's 100 Best Companies to Work For" or "Top 10 Things to Make You Happy"? You should read the fine print of the study to find out by what criteria *best* is defined or whose standard of *happy* they are talking about. Headlines like these can mean very different things to different people. You should not assume that your definition of *best* or *happy* is the one being measured. In fact, is it measured systematically at all, or were people left up to their own interpretations? These are much more elusive variables to measure than, say, average height of American teenagers.

Even a seemingly more concrete study has its complications. Consider the question, "Which toothpaste keeps your mouth most healthy?" On the face of it, *healthy* would seem like an easy thing to measure. However, which of the following characteristics would you say quantifies a healthy mouth?

- Having less bacteria overall
- Having less *harmful* bacteria
- Having a less acidic mouth
- Having a less acidic mouth for a longer period of time
- Having whiter teeth
- Contracting fewer cavities
- Reducing your dental bill
- Any combination of the above

The point is not that one of these is necessarily the correct way to measure health, but that you as the reader understand which definition is being used.

In addition to considering what the variable of the study is measuring, we also need to think about those variables not directly being measured—ones which may have an effect on the results. As we mentioned in the previous section, *confounding* variables play a role in any given study, even when you're not aware of them. Left unnoticed, they could run the risk of throwing a shadow of doubt over any conclusions that are drawn. A well-constructed study will make a point to account for as many confounding variables as possible.

Example 1.17

Considering the Variables

Consider the statement found on a popular bottle of shampoo: Makes hair 60% smoother.

Name some of the questions that would help you determine the validity and applicability of that claim.

Solution

- Smoother than what?
- Under what conditions?
- How did they measure "smoothness"?
- What types of hair were tested?
- In what type of weather was it tested?

Consider the *Setup*

As you process the information in a study, you should also critique its general setup. Does the study reflect any form of **bias**, or favoring of a certain outcome? One of the most important parts in setting up a study is choosing a sample. As we mentioned previously, the sample chosen for a study should accurately reflect the population. So an appropriate question to ask is, "Does it?" If it does not, we say the results are *biased* because they do not accurately represent the population being studied. This type of bias is called **sampling bias**. Often, having a representative sample depends on choosing a large enough sample; for example, it would be silly to predict the winner of the next presidential election based on the results in one small town. Another reason for choosing a large sample is the possibility of **dropouts**, or participants who begin the study but fail to finish. Dropouts can reduce the size of your sample, thus affecting how representative your sample is of the population.

Intentional or not, bias can occur in many different ways, not just from problems with how the sample was chosen. Probably the most accidental of these are **processing errors**—errors that occur simply from the data being processed, like typos when data are being entered or illegible handwriting on a survey. Although very unintentional, this could potentially sway the outcome and hence have a biasing effect. Other times, participants remain in the study until the end but stray from the directions they were given at the beginning. For example, consider a participant who was asked to exercise 30 minutes per day who really did not do that faithfully but claims they did to the researcher. These participants are called **nonadherents** because they did not adhere (think "stick") to the directions given.

Some other forms of bias may be caused, intentionally or unintentionally, by a survey itself. Consider filling out a survey. Does the wording of the survey influence your answers at all? For instance, what would you answer to the following: "Do you agree that severely punishing poor defenseless animals is cruel?" (This survey is attempting to gain support for eliminating *all* types of animal euthanasia—which you may or may not be for, but that's not what the question is asking, is it?) Questions that are deliberately leading in nature, and are intended to sway your opinion, are often referred to as "push polls." They usually are not intended for actual research but just as an excuse to expose you to an opinion someone wants you to know—political parties are often associated with these types of so-called "studies." Often survey questions can be tested for neutrality by having a psychologist read through them. In some cases, it might even be useful to administer the survey to a small trial sample to determine the types of responses that will be received. This will point out any questions that the sample finds confusing as well as any that might cause a certain response.

When it is the researcher who influences the results of the study to favor a certain outcome, the bias is called **researcher bias**. Researcher bias may also be intentional or unintentional. The researcher might intentionally choose a favorable sample or unintentionally influence the sample's responses by his or her actions. The researcher's facial expression, tone of voice, or physical proximity to the participant could all encourage the participant to respond with the answer they believe the researcher wants, instead of their true feelings. This type of answer is referred to as **response bias**. It is also worth noting that researchers should be mindful of how they handle very sensitive information in a setting that would make the subjects feel uneasy. Participants in an uncomfortable situation might be less likely to give truthful information. It is also possible for response bias to come from the participant, not the researcher. Study questions that ask participants to remember, for example, how many caffeinated beverages they have had in the last week, often lead participants to "make up" an answer in order to participate in the study.

Participation bias is created when there is a problem with the participation—or lack thereof—of those chosen for the study. Self-selected samples in which participants volunteer to be a part of the study rather than having been chosen by the researcher often fall into this category. Consider the websites, such as www.ratemyprofessors.com, that allow students to log on and rate their professors. Will students who have had professors that were "just OK" be likely to voluntarily rate their professors online? Probably not. Most likely only students who have had either an exceptionally good or an exceptionally bad experience with a particular professor will "self-select" themselves for the study. While these types of websites might be interesting to look at (especially as a professor), they rarely give a true picture of what is happening for the entire population. When participants must make an extra effort to be a part of the sample, usually only those with strong opinions will be included. In this case, the lack of participation among those with a more neutral opinion results in a sample that does not adequately represent the population. This lack of response is an example of **nonresponse bias**. Other examples of nonresponse bias include a person who refuses to participate in a survey or a respondent who omits questions when answering a survey. You should note that it is not the opposite of response bias (which we just discussed) where participants give a response that they perceive as being the one desired.

Side Note

Would You Be Fooled?

Consider the 14-year-old who reportedly conducted a science fair project in which he urged people to sign a petition demanding strict control or total elimination of the chemical "dihydrogen monoxide." After listing its detrimental properties, he asked 50 people if they supported a ban of the chemical.

- Forty-three said yes.
- Six were undecided.
- Only one knew that the chemical was...

Water! He titled his project, "How Gullible Are We?"

Source: Snopes.com. "Dihydrogen Monoxide." 1995-2011. http://www.snopes.com/science/dhmo.asp (12 Dec. 2011).

Definition

Bias is favoring of a certain outcome in a study.

Sampling bias occurs when the sample chosen does not accurately represent the population being studied.

Dropouts are participants who begin a study but fail to complete it.

Processing errors are errors that occur simply from the data being processed, such as typos when data are being entered.

Nonadherents are participants who remain in the study until the end but stray from the directions they were given.

Researcher bias occurs when a researcher influences the results of a study.

Response bias occurs when a researcher's behavior causes a participant to alter his or her response or when a participant gives an inaccurate response.

Participation bias occurs when there is a problem with the participation—or lack thereof—of those chosen for the study.

Nonresponse bias occurs when there is a lack of participation in a self-selected sample from certain segments of a population, when a person refuses to participate in a survey, or when a respondent omits questions when answering a survey.

Consider the *Conclusions*

In the end, it's the conclusions of the study that we should be most wise about. If we have a perfectly constructed, well-thought-out, unbiased study, we then want to make sure that what we are concluding from it is actually correct. Do the data support the conclusion? Most often, the answer is yes, but why not make sure yourself? What if you read the following headline: "Most Americans Don't Like to See Pajamas Worn in the Grocery Store"? The results given to support the headline were: Of those who responded, 60% do not like to see it and 30% don't mind. What isn't stated is that 95% of the people surveyed couldn't be bothered to answer the question! (Not hard to imagine, is it?)

Here are some other things to consider: Do the results present the whole picture or just a part? Could there be other conclusions drawn? Could there be other reasons for the same conclusion drawn? Does the study have any practical applications? It's not that you should distrust all studies and results that you run across; it's that you should be an informed consumer of information and process critically what you read or hear through the news, journals, or any other outlet of information.

1.4 Section Exercises

Vocabulary

Directions: Complete each statement.

1. A __________ occurs when results from a study are tabulated incorrectly.

2. __________ occurs when the person administering the study influences the participants' responses.

3. __________ occurs when the results of a study tend to favor one outcome over another.

4. A participant who begins a study but fails to complete it is referred to as a __________.

5. A participant who does not fully comply with the instructions for a study is called a __________.

6. If participants are asked to answer survey questions in an uncomfortable setting, there is potential for __________ to occur.

Considering the Variables

Directions: Respond thoughtfully to the following exercises.

7. Consider the following science fair question: Does the quality of air get better with more rain? Name the variables indicated by this question, some different ways one might measure these variables, and some of the terms that need more precise definitions.

8. How would you measure the "Ten Best Colleges of the Midwest"? Name at least five distinct measurements.

9. How would you measure "quality of life"? What are some difficulties you might face?

Considering the Setup

Directions: Describe as many potential sources of bias as you can for the following studies.

10. Star Crazy magazine wants to determine America's favorite celebrities. A survey is placed inside each subscription for readers to fill out and mail in to the company.

11. A struggling retailer wishes to improve sales. Store employees are given the task of polling local residents regarding their opinions about the store as well as asking them to make suggestions for improvement.

12. A major television network wants to know what TV shows people in one state are watching most often. For the study, televisions in 350 households in the four largest cities are monitored for one week.

13. American Star is a hit TV show in which Americans are asked to decide who has enough talent to be a star. Each week viewers call in and vote for their favorite performers. The contestant with the fewest votes is eliminated from the show.

14. A study regarding physical fitness is being conducted on campus. All exercise science majors are required to be a part of the study.

Directions: Respond thoughtfully to the following exercise.

15. The governor of Tennessee would like to know if Tennesseans want to move to a state income tax. Using Sections 1.3 and 1.4, construct a plan to gather the information you would need to answer the governor and explain how you would carry out the plan.

Mean Square for Treatments (MST)

$$\text{MST} = \frac{\text{SST}}{\text{DFT}} \text{ with DFT} = k-1$$

Mean Square for Error (MSE)

$$\text{MSE} = \frac{\text{SSE}}{\text{DFE}} \text{ with DFE} = n_T - k$$

Test Statistic for an ANOVA Test

$$F = \frac{\text{MST}}{\text{MSE}} \text{ with } df_1 = \text{DFT} = k-1 \text{ and } df_2 = \text{DFE} = n_T - k$$

Chapter 12

Pearson Correlation Coefficient

$$r = \frac{n\sum x_i y_i - \left(\sum x_i\right)\left(\sum y_i\right)}{\sqrt{n\sum x_i^2 - \left(\sum x_i\right)^2}\sqrt{n\sum y_i^2 - \left(\sum y_i\right)^2}}$$

$$\text{such that } -1 \le r \le 1$$

Test Statistic for a Hypothesis Test for a Correlation Coefficient

$$t = \frac{r}{\sqrt{\frac{1-r^2}{n-2}}} \text{ with } df = n-2$$

Slope of the Least-Squares Regression Line

$$b_1 = \frac{n\sum x_i y_i - \left(\sum x_i\right)\left(\sum y_i\right)}{n\sum x_i^2 - \left(\sum x_i\right)^2}$$

y-Intercept of the Least-Squares Regression Line

$$b_0 = \frac{\sum y_i}{n} - b_1 \frac{\sum x_i}{n}$$

Regression Line (Line of Best Fit)

$$y = \beta_0 + \beta_1 x \text{ (Population parameters)}$$

$$\hat{y} = b_0 + b_1 x \text{ (Sample statistics)}$$

Residual

$$y - \hat{y}$$

Sum of Squared Errors (SSE)

$$\text{SSE} = \sum \left(y_i - \hat{y}_i\right)^2$$

Standard Error of Estimate

$$S_e = \sqrt{\frac{\sum\left(y_i - \hat{y}_i\right)^2}{n-2}} = \sqrt{\frac{\text{SSE}}{n-2}}$$

Margin of Error for a Prediction Interval for an Individual y-Value

$$E = t_{\alpha/2} S_e \sqrt{1 + \frac{1}{n} + \frac{n\left(x_0 - \bar{x}\right)^2}{n\left(\sum x_i^2\right) - \left(\sum x_i\right)^2}}$$

Prediction Interval for an Individual y-Value

$$\hat{y} - E < y < \hat{y} + E$$

or

$$\left(\hat{y} - E,\ \hat{y} + E\right)$$

Multiple Regression Model

$$\hat{y} = b_0 + b_1 x_1 + b_2 x_2 + \cdots + b_k x_k$$

Critical Values of z

Level of Confidence, c	$\alpha = 1 - c$	$z_{\alpha/2}$
0.80	0.20	1.28
0.85	0.15	1.44
0.90	0.10	1.645
0.95	0.05	1.96
0.98	0.02	2.33
0.99	0.01	2.575

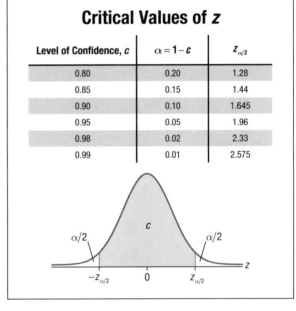

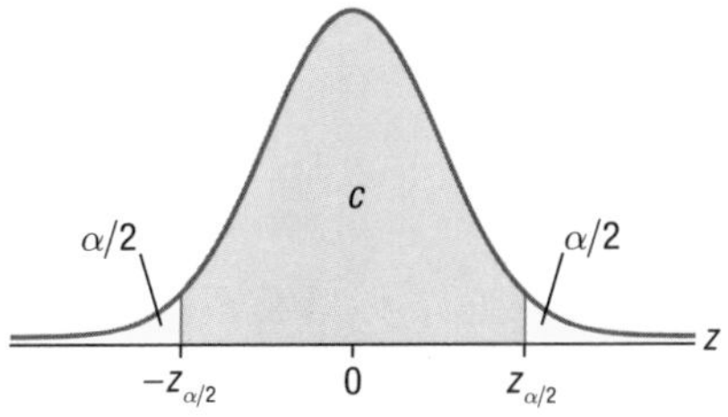

Hawkes Learning Systems © 2014

Test Statistic for a Hypothesis Test for a Population Mean (σ Unknown)

$$t = \frac{\bar{x} - \mu}{\left(\frac{s}{\sqrt{n}}\right)} \text{ with } df = n - 1$$

Test Statistic for a Hypothesis Test for a Population Proportion

$$z = \frac{\hat{p} - p}{\sqrt{\frac{p(1-p)}{n}}}$$

Test Statistic for a Hypothesis Test for a Population Variance or Population Standard Deviation

$$\chi^2 = \frac{(n-1)s^2}{\sigma^2} \text{ with } df = n - 1$$

Test Statistic for a Chi-Square Test for Goodness of Fit

$$\chi^2 = \sum \frac{(O_i - E_i)^2}{E_i} \text{ with } df = k - 1$$

Expected Value of a Frequency in a Contingency Table

$$E_i = \frac{(\text{row total})(\text{column total})}{n}$$

Test Statistic for a Chi-Square Test for Association

$$\chi^2 = \sum \frac{(O_i - E_i)^2}{E_i} \text{ with } df = (R-1)\cdot(C-1)$$

Chapter 11

Test Statistic for a Hypothesis Test For Two Population Means (σ Known)

$$z = \frac{(\bar{x}_1 - \bar{x}_2) - (\mu_1 - \mu_2)}{\sqrt{\frac{\sigma_1^2}{n_1} + \frac{\sigma_2^2}{n_2}}}$$

Test Statistic for a Hypothesis Test For Two Population Means (σ Unknown, Unequal Variances)

$$t = \frac{(\bar{x}_1 - \bar{x}_2) - (\mu_1 - \mu_2)}{\sqrt{\frac{s_1^2}{n_1} + \frac{s_2^2}{n_2}}}$$

$$\text{with } df = \text{smaller of the values } n_1 - 1 \text{ and } n_2 - 1$$

Test Statistic for a Hypothesis Test For Two Population Means (σ Unknown, Equal Variances)

$$t = \frac{(\bar{x}_1 - \bar{x}_2) - (\mu_1 - \mu_2)}{\sqrt{\frac{(n_1-1)s_1^2 + (n_2-1)s_2^2}{n_1 + n_2 - 2}}\sqrt{\frac{1}{n_1} + \frac{1}{n_2}}}$$

$$\text{with } df = n_1 + n_2 - 2$$

Test Statistic for a Hypothesis Test for the Mean of the Paired Differences for Two Populations (σ Unknown, Dependent Samples)

$$t = \frac{\bar{d} - \mu_d}{\left(\frac{s_d}{\sqrt{n}}\right)} \text{ with } df = n - 1$$

Test Statistic for a Hypothesis Test for Two Population Proportions

$$z = \frac{(\hat{p}_1 - \hat{p}_2) - (p_1 - p_2)}{\sqrt{\bar{p}(1-\bar{p})\left(\frac{1}{n_1} + \frac{1}{n_2}\right)}}$$

Weighted Estimate of the Common Population Proportion

$$\bar{p} = \frac{x_1 + x_2}{n_1 + n_2}$$

Test Statistic for a Hypothesis Test for Two Population Variances

$$F = \frac{s_1^2}{s_2^2} \text{ with } df_1 = n_1 - 1 \text{ and } df_2 = n_2 - 1$$

Grand Mean

$$\bar{\bar{x}} = \frac{\sum_{i=1}^{k}(n_i\bar{x}_i)}{\sum_{i=1}^{k} n_i}$$

Sum of Squares among Treatments (SST)

$$\text{SST} = \sum_{i=1}^{k} n_i\left(\bar{x}_i - \bar{\bar{x}}\right)^2$$

Sum of Squares for Error (SSE)

$$\text{SSE} = \sum_{j=1}^{n_1}\left(x_{1j} - \bar{x}_1\right)^2 + \sum_{j=1}^{n_2}\left(x_{2j} - \bar{x}_2\right)^2 + \cdots + \sum_{j=1}^{n_k}\left(x_{kj} - \bar{x}_k\right)^2$$

Total Variation

$$\sum_{j=1}^{n_1}\left(x_{1j} - \bar{\bar{x}}\right)^2 + \sum_{j=1}^{n_2}\left(x_{2j} - \bar{\bar{x}}\right)^2 + \cdots + \sum_{j=1}^{n_k}\left(x_{kj} - \bar{\bar{x}}\right)^2$$
$$= \text{SST} + \text{SSE}$$

Hawkes Learning Systems © 2014

Chapter 9

Margin of Error of a Confidence Interval for the Difference between Two Population Means (σ Known)

$$E = z_{\alpha/2}\sqrt{\frac{\sigma_1^2}{n_1}+\frac{\sigma_2^2}{n_2}}$$

Confidence Interval for the Difference between Two Population Means

$$(\bar{x}_1-\bar{x}_2)-E < \mu_1-\mu_2 < (\bar{x}_1-\bar{x}_2)+E$$

or

$$((\bar{x}_1-\bar{x}_2)-E,\ (\bar{x}_1-\bar{x}_2)+E)$$

Margin of Error of a Confidence Interval for the Difference between Two Population Means (σ Unknown, Unequal Variances)

$$E = t_{\alpha/2}\sqrt{\frac{s_1^2}{n_1}+\frac{s_2^2}{n_2}}$$

with df = smaller of the values n_1-1 and n_2-1

Margin of Error of a Confidence Interval for the Difference between Two Population Means (σ Unknown, Equal Variances)

$$E = t_{\alpha/2}\sqrt{\frac{(n_1-1)s_1^2+(n_2-1)s_2^2}{n_1+n_2-2}}\sqrt{\frac{1}{n_1}+\frac{1}{n_2}}$$

with $df = n_1+n_2-2$

Paired Difference

$$d = x_2 - x_1$$

Mean of Paired Differences

$$\bar{d} = \frac{\sum d_i}{n}$$

Sample Standard Deviation of Paired Differences

$$s_d = \sqrt{\frac{\sum(d_i-\bar{d})^2}{n-1}}$$

Margin of Error of a Confidence Interval for the Mean of the Paired Differences for Two Populations (σ Unknown, Dependent Samples)

$$E = (t_{\alpha/2})\left(\frac{s_d}{\sqrt{n}}\right) \text{ with } df = n-1$$

Confidence Interval for the Mean of the Paired Differences for Two Populations (Dependent Samples)

$$\bar{d}-E < \mu_d < \bar{d}+E$$

or

$$(\bar{d}-E,\ \bar{d}+E)$$

Margin of Error of a Confidence Interval for the Difference between Two Population Proportions

$$E = z_{\alpha/2}\sqrt{\frac{\hat{p}_1(1-\hat{p}_1)}{n_1}+\frac{\hat{p}_2(1-\hat{p}_2)}{n_2}}$$

Confidence Interval for the Difference between Two Population Proportions

$$(\hat{p}_1-\hat{p}_2)-E < p_1-p_2 < (\hat{p}_1-\hat{p}_2)+E$$

or

$$((\hat{p}_1-\hat{p}_2)-E,\ (\hat{p}_1-\hat{p}_2)+E)$$

Point Estimate for Comparing Two Population Variances

$$\frac{s_1^2}{s_2^2} \text{ with } s_1^2 \ge s_2^2$$

Confidence Interval for the Ratio of Two Population Variances

$$\left(\frac{s_1^2}{s_2^2}\cdot\frac{1}{F_{\alpha/2}}\right) < \frac{\sigma_1^2}{\sigma_2^2} < \left(\frac{s_1^2}{s_2^2}\cdot\frac{1}{F_{(1-\alpha/2)}}\right)$$

with $df_1 = n_1-1$ and $df_2 = n_2-1$

Confidence Interval for the Ratio of Two Population Standard Deviations

$$\left(\frac{s_1}{s_2}\cdot\frac{1}{\sqrt{F_{\alpha/2}}}\right) < \frac{\sigma_1}{\sigma_2} < \left(\frac{s_1}{s_2}\cdot\frac{1}{\sqrt{F_{(1-\alpha/2)}}}\right)$$

with $df_1 = n_1-1$ and $df_2 = n_2-1$

Chapter 10

Level of Significance

$$\alpha = 1-c$$

Test Statistic for a Hypothesis Test for a Population Mean (σ Known)

$$z = \frac{\bar{x}-\mu}{\left(\frac{\sigma}{\sqrt{n}}\right)}$$

Hawkes Learning Systems © 2014

A—Standard Normal Distribution

Numerical entries represent the probability that a standard normal random variable is between $-\infty$ and z where $z = \frac{x-\mu}{\sigma}$.

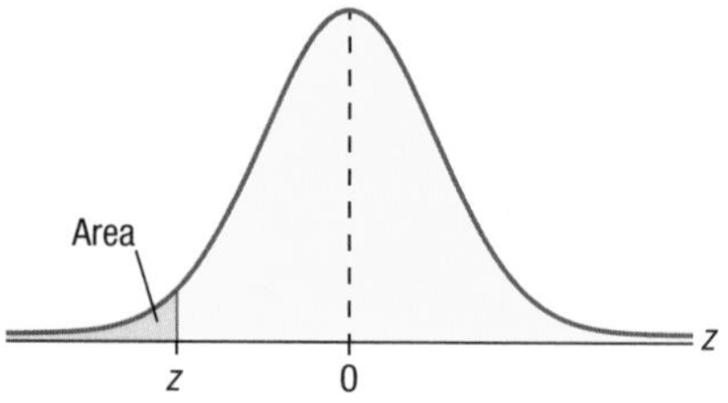

z	0.09	0.08	0.07	0.06	0.05	0.04	0.03	0.02	0.01	0.00
–3.4	0.0002	0.0003	0.0003	0.0003	0.0003	0.0003	0.0003	0.0003	0.0003	0.0003
–3.3	0.0003	0.0004	0.0004	0.0004	0.0004	0.0004	0.0004	0.0005	0.0005	0.0005
–3.2	0.0005	0.0005	0.0005	0.0006	0.0006	0.0006	0.0006	0.0006	0.0007	0.0007
–3.1	0.0007	0.0007	0.0008	0.0008	0.0008	0.0008	0.0009	0.0009	0.0009	0.0010
–3.0	0.0010	0.0010	0.0011	0.0011	0.0011	0.0012	0.0012	0.0013	0.0013	0.0013
–2.9	0.0014	0.0014	0.0015	0.0015	0.0016	0.0016	0.0017	0.0018	0.0018	0.0019
–2.8	0.0019	0.0020	0.0021	0.0021	0.0022	0.0023	0.0023	0.0024	0.0025	0.0026
–2.7	0.0026	0.0027	0.0028	0.0029	0.0030	0.0031	0.0032	0.0033	0.0034	0.0035
–2.6	0.0036	0.0037	0.0038	0.0039	0.0040	0.0041	0.0043	0.0044	0.0045	0.0047
–2.5	0.0048	0.0049	0.0051	0.0052	0.0054	0.0055	0.0057	0.0059	0.0060	0.0062
–2.4	0.0064	0.0066	0.0068	0.0069	0.0071	0.0073	0.0075	0.0078	0.0080	0.0082
–2.3	0.0084	0.0087	0.0089	0.0091	0.0094	0.0096	0.0099	0.0102	0.0104	0.0107
–2.2	0.0110	0.0113	0.0116	0.0119	0.0122	0.0125	0.0129	0.0132	0.0136	0.0139
–2.1	0.0143	0.0146	0.0150	0.0154	0.0158	0.0162	0.0166	0.0170	0.0174	0.0179
–2.0	0.0183	0.0188	0.0192	0.0197	0.0202	0.0207	0.0212	0.0217	0.0222	0.0228
–1.9	0.0233	0.0239	0.0244	0.0250	0.0256	0.0262	0.0268	0.0274	0.0281	0.0287
–1.8	0.0294	0.0301	0.0307	0.0314	0.0322	0.0329	0.0336	0.0344	0.0351	0.0359
–1.7	0.0367	0.0375	0.0384	0.0392	0.0401	0.0409	0.0418	0.0427	0.0436	0.0446
–1.6	0.0455	0.0465	0.0475	0.0485	0.0495	0.0505	0.0516	0.0526	0.0537	0.0548
–1.5	0.0559	0.0571	0.0582	0.0594	0.0606	0.0618	0.0630	0.0643	0.0655	0.0668
–1.4	0.0681	0.0694	0.0708	0.0721	0.0735	0.0749	0.0764	0.0778	0.0793	0.0808
–1.3	0.0823	0.0838	0.0853	0.0869	0.0885	0.0901	0.0918	0.0934	0.0951	0.0968
–1.2	0.0985	0.1003	0.1020	0.1038	0.1056	0.1075	0.1093	0.1112	0.1131	0.1151
–1.1	0.1170	0.1190	0.1210	0.1230	0.1251	0.1271	0.1292	0.1314	0.1335	0.1357
–1.0	0.1379	0.1401	0.1423	0.1446	0.1469	0.1492	0.1515	0.1539	0.1562	0.1587
–0.9	0.1611	0.1635	0.1660	0.1685	0.1711	0.1736	0.1762	0.1788	0.1814	0.1841
–0.8	0.1867	0.1894	0.1922	0.1949	0.1977	0.2005	0.2033	0.2061	0.2090	0.2119
–0.7	0.2148	0.2177	0.2206	0.2236	0.2266	0.2296	0.2327	0.2358	0.2389	0.2420
–0.6	0.2451	0.2483	0.2514	0.2546	0.2578	0.2611	0.2643	0.2676	0.2709	0.2743
–0.5	0.2776	0.2810	0.2843	0.2877	0.2912	0.2946	0.2981	0.3015	0.3050	0.3085
–0.4	0.3121	0.3156	0.3192	0.3228	0.3264	0.3300	0.3336	0.3372	0.3409	0.3446
–0.3	0.3483	0.3520	0.3557	0.3594	0.3632	0.3669	0.3707	0.3745	0.3783	0.3821
–0.2	0.3859	0.3897	0.3936	0.3974	0.4013	0.4052	0.4090	0.4129	0.4168	0.4207
–0.1	0.4247	0.4286	0.4325	0.4364	0.4404	0.4443	0.4483	0.4522	0.4562	0.4602
–0.0	0.4641	0.4681	0.4721	0.4761	0.4801	0.4840	0.4880	0.4920	0.4960	0.5000

Hawkes Learning Systems © 2014

B—Standard Normal Distribution

Numerical entries represent the probability that a standard normal random variable is between $-\infty$ and z where $z = \frac{x-\mu}{\sigma}$.

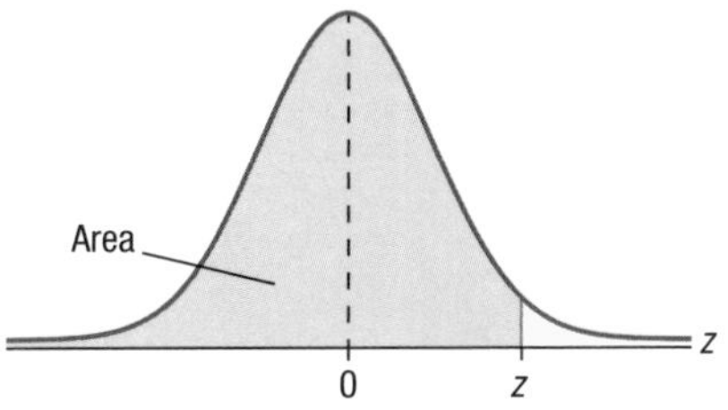

z	0.00	0.01	0.02	0.03	0.04	0.05	0.06	0.07	0.08	0.09
0.0	0.5000	0.5040	0.5080	0.5120	0.5160	0.5199	0.5239	0.5279	0.5319	0.5359
0.1	0.5398	0.5438	0.5478	0.5517	0.5557	0.5596	0.5636	0.5675	0.5714	0.5753
0.2	0.5793	0.5832	0.5871	0.5910	0.5948	0.5987	0.6026	0.6064	0.6103	0.6141
0.3	0.6179	0.6217	0.6255	0.6293	0.6331	0.6368	0.6406	0.6443	0.6480	0.6517
0.4	0.6554	0.6591	0.6628	0.6664	0.6700	0.6736	0.6772	0.6808	0.6844	0.6879
0.5	0.6915	0.6950	0.6985	0.7019	0.7054	0.7088	0.7123	0.7157	0.7190	0.7224
0.6	0.7257	0.7291	0.7324	0.7357	0.7389	0.7422	0.7454	0.7486	0.7517	0.7549
0.7	0.7580	0.7611	0.7642	0.7673	0.7704	0.7734	0.7764	0.7794	0.7823	0.7852
0.8	0.7881	0.7910	0.7939	0.7967	0.7995	0.8023	0.8051	0.8078	0.8106	0.8133
0.9	0.8159	0.8186	0.8212	0.8238	0.8264	0.8289	0.8315	0.8340	0.8365	0.8389
1.0	0.8413	0.8438	0.8461	0.8485	0.8508	0.8531	0.8554	0.8577	0.8599	0.8621
1.1	0.8643	0.8665	0.8686	0.8708	0.8729	0.8749	0.8770	0.8790	0.8810	0.8830
1.2	0.8849	0.8869	0.8888	0.8907	0.8925	0.8944	0.8962	0.8980	0.8997	0.9015
1.3	0.9032	0.9049	0.9066	0.9082	0.9099	0.9115	0.9131	0.9147	0.9162	0.9177
1.4	0.9192	0.9207	0.9222	0.9236	0.9251	0.9265	0.9279	0.9292	0.9306	0.9319
1.5	0.9332	0.9345	0.9357	0.9370	0.9382	0.9394	0.9406	0.9418	0.9429	0.9441
1.6	0.9452	0.9463	0.9474	0.9484	0.9495	0.9505	0.9515	0.9525	0.9535	0.9545
1.7	0.9554	0.9564	0.9573	0.9582	0.9591	0.9599	0.9608	0.9616	0.9625	0.9633
1.8	0.9641	0.9649	0.9656	0.9664	0.9671	0.9678	0.9686	0.9693	0.9699	0.9706
1.9	0.9713	0.9719	0.9726	0.9732	0.9738	0.9744	0.9750	0.9756	0.9761	0.9767
2.0	0.9772	0.9778	0.9783	0.9788	0.9793	0.9798	0.9803	0.9808	0.9812	0.9817
2.1	0.9821	0.9826	0.9830	0.9834	0.9838	0.9842	0.9846	0.9850	0.9854	0.9857
2.2	0.9861	0.9864	0.9868	0.9871	0.9875	0.9878	0.9881	0.9884	0.9887	0.9890
2.3	0.9893	0.9896	0.9898	0.9901	0.9904	0.9906	0.9909	0.9911	0.9913	0.9916
2.4	0.9918	0.9920	0.9922	0.9925	0.9927	0.9929	0.9931	0.9932	0.9934	0.9936
2.5	0.9938	0.9940	0.9941	0.9943	0.9945	0.9946	0.9948	0.9949	0.9951	0.9952
2.6	0.9953	0.9955	0.9956	0.9957	0.9959	0.9960	0.9961	0.9962	0.9963	0.9964
2.7	0.9965	0.9966	0.9967	0.9968	0.9969	0.9970	0.9971	0.9972	0.9973	0.9974
2.8	0.9974	0.9975	0.9976	0.9977	0.9977	0.9978	0.9979	0.9979	0.9980	0.9981
2.9	0.9981	0.9982	0.9982	0.9983	0.9984	0.9984	0.9985	0.9985	0.9986	0.9986
3.0	0.9987	0.9987	0.9987	0.9988	0.9988	0.9989	0.9989	0.9989	0.9990	0.9990
3.1	0.9990	0.9991	0.9991	0.9991	0.9992	0.9992	0.9992	0.9992	0.9993	0.9993
3.2	0.9993	0.9993	0.9994	0.9994	0.9994	0.9994	0.9994	0.9995	0.9995	0.9995
3.3	0.9995	0.9995	0.9995	0.9996	0.9996	0.9996	0.9996	0.9996	0.9996	0.9997
3.4	0.9997	0.9997	0.9997	0.9997	0.9997	0.9997	0.9997	0.9997	0.9997	0.9998

Hawkes Learning Systems © 2014

Chapter 7

The Central Limit Theorem (CLT)

1. Mean of a Sampling Distribution of Sample Means

$$\mu_{\bar{x}} = \mu$$

2. Standard Deviation of a Sampling Distribution of Sample Means

$$\sigma_{\bar{x}} = \frac{\sigma}{\sqrt{n}}$$

3. The shape of a sampling distribution of sample means will approach that of a normal distribution, regardless of the shape of the population distribution. The larger the sample size, the better the normal distribution approximation will be.

Standard Score for a Sample Mean

$$z = \frac{\bar{x} - \mu_{\bar{x}}}{\sigma_{\bar{x}}} = \frac{\bar{x} - \mu}{\left(\frac{\sigma}{\sqrt{n}}\right)}$$

Population Proportion

$$p = \frac{x}{N}$$

Sample Proportion

$$\hat{p} = \frac{x}{n}$$

Mean of a Sampling Distribution of Sample Proportions

$$\mu_{\hat{p}} = p$$

Standard Deviation of a Sampling Distribution of Sample Proportions

$$\sigma_{\hat{p}} = \sqrt{\frac{p(1-p)}{n}}$$

Standard Score for a Sample Proportion

$$z = \frac{\hat{p} - \mu_{\hat{p}}}{\sigma_{\hat{p}}} = \frac{\hat{p} - p}{\sqrt{\frac{p(1-p)}{n}}}$$

Chapter 8

Margin of Error of a Confidence Interval for a Population Mean (σ Known)

$$E = \left(z_{\alpha/2}\right)\left(\sigma_{\bar{x}}\right) = \left(z_{\alpha/2}\right)\left(\frac{\sigma}{\sqrt{n}}\right)$$

Confidence Interval for a Population Mean

$$\bar{x} - E < \mu < \bar{x} + E$$

or

$$\left(\bar{x} - E,\ \bar{x} + E\right)$$

Minimum Sample Size for Estimating a Population Mean

$$n = \left(\frac{z_{\alpha/2} \cdot \sigma}{E}\right)^2$$

Margin of Error of a Confidence Interval for a Population Mean (σ Unknown)

$$E = \left(t_{\alpha/2}\right)\left(\frac{s}{\sqrt{n}}\right) \text{ with } df = n-1$$

Margin of Error of a Confidence Interval for a Population Proportion

$$E = z_{\alpha/2}\sqrt{\frac{\hat{p}\left(1-\hat{p}\right)}{n}}$$

Confidence Interval for a Population Proportion

$$\hat{p} - E < p < \hat{p} + E$$

or

$$\left(\hat{p} - E,\ \hat{p} + E\right)$$

Minimum Sample Size for Estimating a Population Proportion

$$n = p(1-p)\left(\frac{z_{\alpha/2}}{E}\right)^2$$

Confidence Interval for a Population Variance

$$\frac{(n-1)s^2}{\chi^2_{\alpha/2}} < \sigma^2 < \frac{(n-1)s^2}{\chi^2_{(1-\alpha/2)}} \text{ with } df = n-1$$

Confidence Interval for a Population Standard Deviation

$$\sqrt{\frac{(n-1)s^2}{\chi^2_{\alpha/2}}} < \sigma < \sqrt{\frac{(n-1)s^2}{\chi^2_{(1-\alpha/2)}}} \text{ with } df = n-1$$

Hawkes Learning Systems © 2014

Chapter 4

Experimental Probability (or Empirical Probability)

$$P(E)=\frac{f}{n}$$

Classical Probability (or Theoretical Probability)

$$P(E)=\frac{n(E)}{n(S)}$$

Complement Rule for Probability

$$P(E)+P(E^c)=1$$

Addition Rule for Probability

$$P(E \textbf{ or } F)=P(E)+P(F)-P(E \text{ and } F)$$

Addition Rule for Probability of Mutually Exclusive Events

$$P(E \textbf{ or } F)=P(E)+P(F)$$

Multiplication Rule for Probability of Independent Events

$$P(E \textbf{ and } F)=P(E)\cdot P(F)$$

Multiplication Rule for Probability of Dependent Events

$$\begin{aligned} P(E \textbf{ and } F) &= P(E)\cdot P(F\mid E) \\ &= P(F)\cdot P(E\mid F) \end{aligned}$$

Conditional Probability

$$P(F\mid E)=\frac{P(E \textbf{ and } F)}{P(E)}$$

Fundamental Counting Principle

The total number of possible outcomes for the sequence of stages in a multistage experiment is $k_1\cdot k_2\cdot\cdots\cdot k_n$.

Factorial

$$n!=n(n-1)(n-2)\cdots(2)(1)$$

Combinations

$${}_nC_r=\frac{n!}{r!(n-r)!}$$

Permutations

$${}_nP_r=\frac{n!}{(n-r)!}$$

Special Permutations

$$\frac{n!}{k_1!k_2!\cdots k_p!}$$

Chapter 5

Expected Value

$$E(X)=\mu=\sum\left[x_i\cdot P(X=x_i)\right]$$

Variance for a Discrete Probability Distribution

$$\begin{aligned} \sigma^2 &= \sum\left[x_i^2\cdot P(X=x_i)\right]-\mu^2 \\ &= \sum\left[(x_i-\mu)^2\cdot P(X=x_i)\right] \end{aligned}$$

Standard Deviation for a Discrete Probability Distribution

$$\begin{aligned} \sigma &= \sqrt{\sigma^2} \\ &= \sqrt{\sum\left[x_i^2\cdot P(X=x_i)\right]-\mu^2} \\ &= \sqrt{\sum\left[(x_i-\mu)^2\cdot P(X=x_i)\right]} \end{aligned}$$

Probability for a Binomial Distribution

$$P(X=x)={}_nC_x\cdot p^x(1-p)^{(n-x)}$$

Probability for a Poisson Distribution

$$P(X=x)=\frac{e^{-\lambda}\lambda^x}{x!}$$

Probability for a Hypergeometric Distribution

$$P(X=x)=\frac{\left({}_kC_x\right)\left({}_{N-k}C_{n-x}\right)}{\left({}_NC_n\right)}$$

Chapter 6

Standard Score

$$z=\frac{x-\mu}{\sigma}$$

Finding the Value of a Normally Distributed Random Variable for a Given Probability

$$x=z\cdot\sigma+\mu$$

Normal Distribution Approximation of a Binomial Distribution

$$\mu=np$$

$$\sigma=\sqrt{np(1-p)}$$

Hawkes Learning Systems © 2014

C—Critical Values of t

	Area in One Tail				
	0.100	**0.050**	**0.025**	**0.010**	**0.005**
	Area in Two Tails				
df	**0.200**	**0.100**	**0.050**	**0.020**	**0.010**
1	3.078	6.314	12.706	31.821	63.657
2	1.886	2.920	4.303	6.965	9.925
3	1.638	2.353	3.182	4.541	5.841
4	1.533	2.132	2.776	3.747	4.604
5	1.476	2.015	2.571	3.365	4.032
6	1.440	1.943	2.447	3.143	3.707
7	1.415	1.895	2.365	2.998	3.499
8	1.397	1.860	2.306	2.896	3.355
9	1.383	1.833	2.262	2.821	3.250
10	1.372	1.812	2.228	2.764	3.169
11	1.363	1.796	2.201	2.718	3.106
12	1.356	1.782	2.179	2.681	3.055
13	1.350	1.771	2.160	2.650	3.012
14	1.345	1.761	2.145	2.624	2.977
15	1.341	1.753	2.131	2.602	2.947
16	1.337	1.746	2.120	2.583	2.921
17	1.333	1.740	2.110	2.567	2.898
18	1.330	1.734	2.101	2.552	2.878
19	1.328	1.729	2.093	2.539	2.861
20	1.325	1.725	2.086	2.528	2.845
21	1.323	1.721	2.080	2.518	2.831
22	1.321	1.717	2.074	2.508	2.819
23	1.319	1.714	2.069	2.500	2.807
24	1.318	1.711	2.064	2.492	2.797
25	1.316	1.708	2.060	2.485	2.787
26	1.315	1.706	2.056	2.479	2.779
27	1.314	1.703	2.052	2.473	2.771
28	1.313	1.701	2.048	2.467	2.763
29	1.311	1.699	2.045	2.462	2.756
30	1.310	1.697	2.042	2.457	2.750
31	1.309	1.696	2.040	2.453	2.744
32	1.309	1.694	2.037	2.449	2.738
34	1.307	1.691	2.032	2.441	2.728
36	1.306	1.688	2.028	2.434	2.719
38	1.304	1.686	2.024	2.429	2.712
40	1.303	1.684	2.021	2.423	2.704
45	1.301	1.679	2.014	2.412	2.690
50	1.299	1.676	2.009	2.403	2.678
55	1.297	1.673	2.004	2.396	2.668
60	1.296	1.671	2.000	2.390	2.660
70	1.294	1.667	1.994	2.381	2.648
80	1.292	1.664	1.990	2.374	2.639
90	1.291	1.662	1.987	2.368	2.632
100	1.290	1.660	1.984	2.364	2.626
120	1.289	1.658	1.980	2.358	2.617
200	1.286	1.653	1.972	2.345	2.601
300	1.284	1.650	1.968	2.339	2.592
400	1.284	1.649	1.966	2.336	2.588
500	1.283	1.648	1.965	2.334	2.586
750	1.283	1.647	1.963	2.331	2.582
1000	1.282	1.646	1.962	2.330	2.581
∞	1.282	1.645	1.960	2.326	2.576

Left Tail

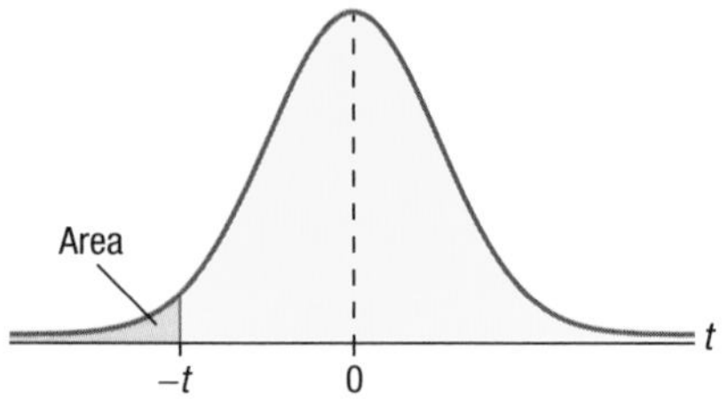

Right Tail

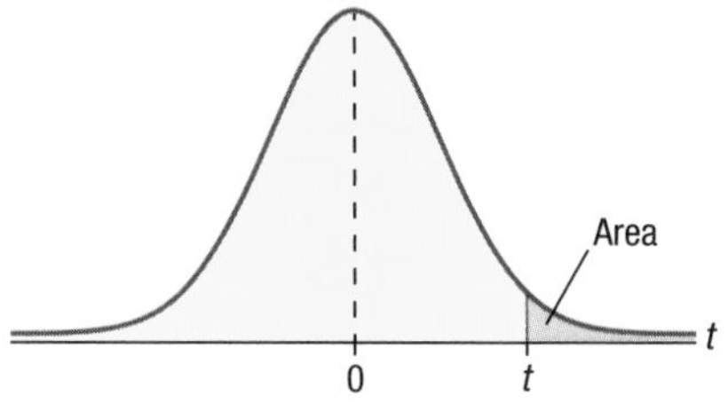

Two Tails

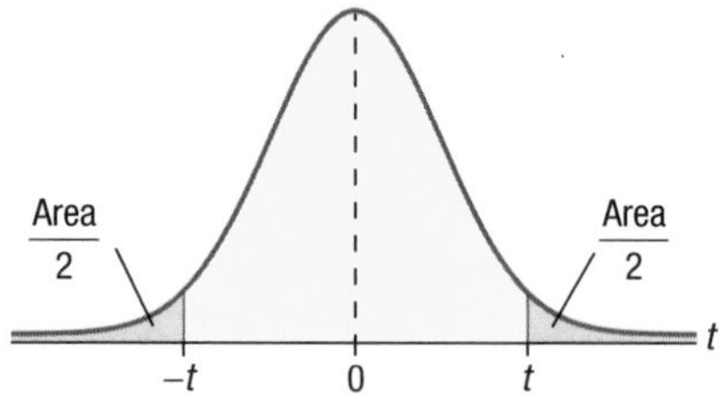

Hawkes Learning Systems © 2014

G—Critical Values of χ^2

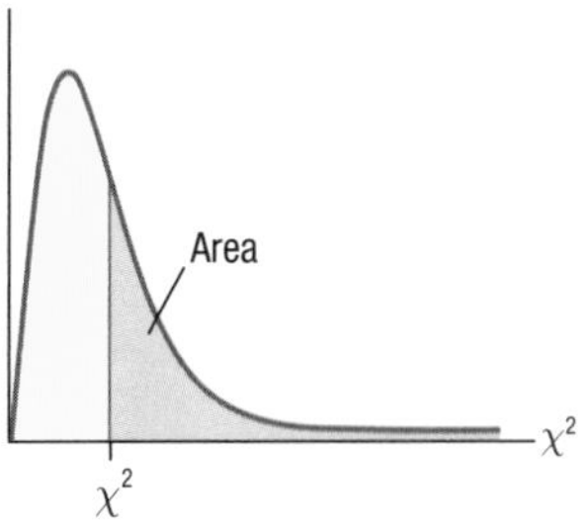

Area to the Right of the Critical Value of χ^2

df	0.995	0.990	0.975	0.950	0.900	0.100	0.050	0.025	0.010	0.005
1	0.000	0.000	0.001	0.004	0.016	2.706	3.841	5.024	6.635	7.879
2	0.010	0.020	0.051	0.103	0.211	4.605	5.991	7.378	9.210	10.597
3	0.072	0.115	0.216	0.352	0.584	6.251	7.815	9.348	11.345	12.838
4	0.207	0.297	0.484	0.711	1.064	7.779	9.488	11.143	13.277	14.860
5	0.412	0.554	0.831	1.145	1.610	9.236	11.070	12.833	15.086	16.750
6	0.676	0.872	1.237	1.635	2.204	10.645	12.592	14.449	16.812	18.548
7	0.989	1.239	1.690	2.167	2.833	12.017	14.067	16.013	18.475	20.278
8	1.344	1.646	2.180	2.733	3.490	13.362	15.507	17.535	20.090	21.955
9	1.735	2.088	2.700	3.325	4.168	14.684	16.919	19.023	21.666	23.589
10	2.156	2.558	3.247	3.940	4.865	15.987	18.307	20.483	23.209	25.188
11	2.603	3.053	3.816	4.575	5.578	17.275	19.675	21.920	24.725	26.757
12	3.074	3.571	4.404	5.226	6.304	18.549	21.026	23.337	26.217	28.300
13	3.565	4.107	5.009	5.892	7.042	19.812	22.362	24.736	27.688	29.819
14	4.075	4.660	5.629	6.571	7.790	21.064	23.685	26.119	29.141	31.319
15	4.601	5.229	6.262	7.261	8.547	22.307	24.996	27.488	30.578	32.801
16	5.142	5.812	6.908	7.962	9.312	23.542	26.296	28.845	32.000	34.267
17	5.697	6.408	7.564	8.672	10.085	24.769	27.587	30.191	33.409	35.718
18	6.265	7.015	8.231	9.390	10.865	25.989	28.869	31.526	34.805	37.156
19	6.844	7.633	8.907	10.117	11.651	27.204	30.144	32.852	36.191	38.582
20	7.434	8.260	9.591	10.851	12.443	28.412	31.410	34.170	37.566	39.997
21	8.034	8.897	10.283	11.591	13.240	29.615	32.671	35.479	38.932	41.401
22	8.643	9.542	10.982	12.338	14.041	30.813	33.924	36.781	40.289	42.796
23	9.260	10.196	11.689	13.091	14.848	32.007	35.172	38.076	41.638	44.181
24	9.886	10.856	12.401	13.848	15.659	33.196	36.415	39.364	42.980	45.559
25	10.520	11.524	13.120	14.611	16.473	34.382	37.652	40.646	44.314	46.928
26	11.160	12.198	13.844	15.379	17.292	35.563	38.885	41.923	45.642	48.290
27	11.808	12.879	14.573	16.151	18.114	36.741	40.113	43.195	46.963	49.645
28	12.461	13.565	15.308	16.928	18.939	37.916	41.337	44.461	48.278	50.993
29	13.121	14.256	16.047	17.708	19.768	39.087	42.557	45.722	49.588	52.336
30	13.787	14.953	16.791	18.493	20.599	40.256	43.773	46.979	50.892	53.672
40	20.707	22.164	24.433	26.509	29.051	51.805	55.758	59.342	63.691	66.766
50	27.991	29.707	32.357	34.764	37.689	63.167	67.505	71.420	76.154	79.490
60	35.534	37.485	40.482	43.188	46.459	74.397	79.082	83.298	88.379	91.952
70	43.275	45.442	48.758	51.739	55.329	85.527	90.531	95.023	100.425	104.215
80	51.172	53.540	57.153	60.391	64.278	96.578	101.879	106.629	112.329	116.321
90	59.196	61.754	65.647	69.126	73.291	107.565	113.145	118.136	124.116	128.299
100	67.328	70.065	74.222	77.929	82.358	118.498	124.342	129.561	135.807	140.169

Hawkes Learning Systems © 2014

Formulas and Tables from *Beginning Statistics*

Chapter 2

Class Midpoint

$$\frac{\text{Lower Limit} + \text{Upper Limit}}{2}$$

Relative Frequency

$$\frac{\text{Class Frequency}}{\text{Sample Size}} = \frac{f}{n}$$

Chapter 3

Sample Mean

$$\bar{x} = \frac{x_1 + x_2 + \cdots + x_n}{n} = \frac{\sum x_i}{n}$$

Population Mean

$$\mu = \frac{x_1 + x_2 + \cdots + x_N}{N} = \frac{\sum x_i}{N}$$

Weighted Mean

$$\bar{x} = \frac{\sum (x_i \cdot w_i)}{\sum w_i}$$

Range

Maximum Data Value – Minimum Data Value

Population Standard Deviation

$$\sigma = \sqrt{\frac{\sum (x_i - \mu)^2}{N}}$$

Sample Standard Deviation

$$s = \sqrt{\frac{\sum (x_i - \bar{x})^2}{n-1}}$$

Population Coefficient of Variation

$$\text{CV} = \frac{\sigma}{\mu} \cdot 100\%$$

Sample Coefficient of Variation

$$\text{CV} = \frac{s}{\bar{x}} \cdot 100\%$$

Population Variance

$$\sigma^2 = \frac{\sum (x_i - \mu)^2}{N}$$

Sample Variance

$$s^2 = \frac{\sum (x_i - \bar{x})^2}{n-1}$$

Empirical Rule for Bell-Shaped Distributions

Approximately 68% of the data values lie within one standard deviation of the mean.

Approximately 95% of the data values lie within two standard deviations of the mean.

Approximately 99.7% of the data values lie within three standard deviations of the mean.

Chebyshev's Theorem

The proportion of any data set lying within K standard deviations of the mean is at least $1 - \frac{1}{K^2}$ for $K > 1$.

Location of Data Value for the Pth Percentile

$$l = n \cdot \frac{P}{100}$$

P^{th} Percentile of a Data Value

$$P = \frac{l}{n} \cdot 100$$

Quartiles

Q_1 = First Quartile: 25% of the data are less than or equal to this value.

Q_2 = Second Quartile: 50% of the data are less than or equal to this value.

Q_3 = Third Quartile: 75% of the data are less than or equal to this value.

Interquartile Range

$$\text{IQR} = Q_3 - Q_1$$

Standard Score for a Population

$$z = \frac{x - \mu}{\sigma}$$

Standard Score for a Sample

$$z = \frac{x - \bar{x}}{s}$$

Hawkes Learning Systems © 2014

R Chapter 1 Review

Section 1.1: Getting Started

Definitions

Statistics
The science of gathering, describing, and analyzing data OR the actual numerical descriptions of sample data

Population
The particular group of interest

Variables
Values that can change amongst members of the population

Data
Counts, measurements, or observations gathered about a population in order to study it

Census
A study in which data are obtained from every member of the population

Parameter
A numerical description of a population characteristic

Sample
A subset of the population from which data are collected

Sample statistics
Numerical descriptions of sample characteristics

Branches of Statistics

1. **Descriptive statistics:** gathers, sorts, summarizes, and displays the data
2. **Inferential statistics:** uses descriptive statistics to estimate population parameters

Section 1.2: Data Classification

Definitions

Qualitative data
Consist of labels or descriptions of traits

Quantitative data
Consist of counts or measurements

Continuous data
Quantitative data that can take on any value in a given interval and are usually measurements

Discrete data
Quantitative data that can take on only particular values and are usually counts

Section 1.2: Data Classification (cont.)

Levels of Measurement

1. **Nominal:** qualitative data consisting of labels or names
2. **Ordinal:** qualitative data that can be arranged in a meaningful order, but calculations such as addition or division do not make sense
3. **Interval:** quantitative data that can be arranged in a meaningful order, and differences between data entries are meaningful
4. **Ratio:** quantitative data that can be ordered, differences between data entries are meaningful, and the zero point indicates the absence of something

Section 1.3: The Process of a Statistical Study

Definitions

Observational study
Observes data that already exist

Experiment
Generates data to help identify cause-and-effect relationships

Representative sample
Has the same relevant characteristics as the population and does not favor one group from the population over another

Treatment
Some condition applied to a group of people or things in an experiment

Subjects
People or things being studied in an experiment

Participants
People being studied in an experiment

Response variable
The variable that responds to the treatment

Explanatory variable
The variable that causes the change in the response variable

Treatment group
A group of subjects to which researchers apply a treatment in an experiment

Control group
A group of subjects to which no treatment is applied in an experiment

Confounding variables
Factors other than the treatment that cause an effect on the subjects of an experiment

Placebo effect
A response to the power of suggestion, rather than the treatment itself, by participants of an experiment

Section 1.3: The Process of a Statistical Study (cont.)

Placebo
A substance that appears identical to the actual treatment but contains no intrinsic beneficial elements

Single-blind
Subjects do not know if they are in the control group or the treatment group, but the people interacting with the subjects in the experiment know in which group each subject has been placed

Double-blind
Neither the subjects nor the people interacting with the subjects know to which group each subject belongs

Institutional Review Board (IRB)
A group of people who review the design of a study to make sure that it is appropriate and that no unnecessary harm will come to the subjects involved

Informed consent
Completely disclosing to participants the goals and procedures involved in a study and obtaining their agreement to participate

Conducting a Statistical Study

1. Determine the design of the study.
   a. State the question to be studied.
   b. Determine the population and variables.
   c. Determine the sampling method.
2. Collect the data.
3. Organize the data.
4. Analyze the data to answer the question.

Sampling Methods

1. **Random sampling:** Every member of the population has an equal chance of being selected
2. **Simple random sampling:** Every sample from the population has an equal chance of being chosen
3. **Stratified sampling:** Dividing the population into subgroups, called **strata**, that share similar characteristics and drawing a random sample from each stratum
4. **Quota sampling:** A type of stratification in which certain characteristics of the population are preserved in the sample
5. **Cluster sampling:** Dividing the population into groups, called **clusters**, that are each similar to the entire population and randomly selecting whole clusters to sample
6. **Systematic sampling:** Selecting every n^{th} member of the population
7. **Convenience sampling:** The sample is "convenient" for the researcher to select

Section 1.3: The Process of a Statistical Study (cont.)

Types of Observational Studies

1. **Cross-sectional study:** Data are collected at a single point in time
2. **Longitudinal study:** Data are gathered by following a particular group over a period of time
3. **Meta-analysis:** Study that compiles information from previous studies
4. **Case study:** Looks at multiple variables that affect a single event

Principles of Experimental Design

1. **Randomize** the control and treatment groups.
2. **Control** for outside effects on the response variable.
3. **Replicate** the experiment a significant number of times to see meaningful patterns.

Section 1.4: How to Critique a Published Study

Definitions

Bias
Favoring of a certain outcome in a study

Sampling bias
Occurs when the sample chosen does not accurately represent the population being studied

Dropouts
Participants who begin a study but fail to complete it

Processing errors
Errors that occur simply from the data being processed, such as typos when data are being entered

Nonadherents
Participants who remain in the study until the end but stray from the directions they were given

Researcher bias
Occurs when a researcher influences the results of a study

Response bias
Occurs when a researcher's behavior causes a participant to alter his or her response or when a participant gives an inaccurate response.

Participation bias
Occurs when there is a problem with the participation—or lack thereof—of those chosen for the study

Nonresponse bias
Occurs when there is a lack of participation in a self-selected sample from certain segments of a population, when a person refuses to participate in a survey, or when a respondent omits questions when answering a survey

E Chapter 1 Exercises

Directions: Respond thoughtfully to the following exercises.

1. Eighty-nine percent of Americans polled are in favor of arts education being taught in public schools.

   a. Does the above numerical value describe a parameter or a statistic?

   b. After reading the above statement, you decide to campaign for the arts in education and use the statement: "A majority of Americans are in favor of art education, which shows their love of arts. Americans support ballet teaching in public schools." What branch of statistics would you be applying?

2. The average lifespan of Americans in 1990 was 75.37 years.

   a. Does the above numerical value describe a parameter or a statistic?

   b. Giving a complete breakdown of the above statement into categories of gender, race, and geographical region would be an example of which branch of statistics?

3. Determine whether the following data sets are qualitative or quantitative and classify each set according to its level of measurement.

   a. The street addresses of houses along North Lamar Avenue

   b. The years in which Christmas falls on a Sunday

   c. The monthly rates for digital cable service for five competing companies

4. Jake wanted to know the most popular brand of cereal amongst teenagers. He stood in the center of the cereal aisle at the local grocery and surveyed every 10th teenager who approached him.

   a. How many biases can you name?

   b. In what ways could Jake better collect data to answer his question?

5. If all male residents are assigned odd-numbered rooms and female residents are assigned even-numbered rooms in a dorm, would choosing every 10th room give a representative sample of dorm residents? What about sampling every 5th room?

6. If someone is looking at the proportion of students who are left-handed, is your statistics class a good sample? Is this class a good indicator of the male-to-female ratio on a campus?

7. A supermarket wants to decide whether or not to carry a new brand of salsa and uses a free taste-test stand in the store.

a. What is the population being studied?

b. What sampling method is being used?

c. Is the sample obtained for the taste test representative of the population?

8. Explain how cluster sampling is different than stratified sampling. If you wanted to make sure to include certain characteristics of the population in the sample, would you use cluster sampling or stratified sampling?

9. How could you use cluster sampling if you want to know the average price of gasoline at gas stations located within a mile of rental car return locations at airports?

10. A researcher wants to determine the average age of people attending the state fair.

a. Give three different methods of sampling that could be used in this scenario. Which method is best and why?

b. Describe any potential for bias.

11. Suppose that a researcher wants to determine the optimal number of hours of sleep that the average adult requires each night.

a. Identify the following: the population and appropriate type of study to conduct.

b. Give a brief discussion on how you would set up and conduct this type of study.

12. Convenience sampling is the sampling technique most prone to bias. However, it is possible to obtain a representative sample using the convenience method. Give one example of how the convenience method could produce a representative sample and one example of it producing a biased sample.

13. Suppose that the governor of Nebraska wants to determine his approval rating among his constituents. A team from the governor's staff is assigned to the task. Four of the state's largest cities are chosen, and 100 registered voters from each city are chosen to be included in the sample.

a. Identify the sampling technique for this scenario.

b. Describe all potential sources of bias.

Directions: Answer the following questions for Exercises 14 and 15.

a. What is the population being studied?

b. What is the sample?

c. What are the sample statistics?

d. Describe any potential sources of bias.

14. In a 2009 poll of 1000 adults from Great Britain aged 18 or older, subjects were asked "How satisfied or dissatisfied do you feel about your standard of living at present?" The results were as follows.

Source: Ipsos MORI. "Standard of Living – Trends." Apr. 2009. http://www.ipsos-mori.com/researchpublications/researcharchive/poll.aspx?oItemId=2388 (12 Dec. 2011).

Very satisfied	19%
Fairly satisfied	52%
Neither satisfied nor dissatisfied	7%
Fairly dissatisfied	13%
Very dissatisfied	8%

15. According to a USA Today/Gallup poll conducted at the beginning of 2010, after the Haiti earthquake, 63% of Americans favored leaving American troops in Haiti until basic services were restored. Results were based on telephone interviews with 1067 American adults aged 18 or older.

Source: Morales, Lymari. "Americans Lean Against Letting More Haitians Into U.S." 25 Jan. 2010. http://www.gallup.com/poll/125372/Americans-Lean-Against-Letting-Haitians.aspx (12 Dec. 2011).

P Chapter 1 Project

Project A: Analyzing a Given Study

Poker on the Rise with Young Adults

For a national poll on casino visitation, a survey developed by TNS for Harrah's Entertainment was mailed to 100,000 American adults, of whom 57,205 participated. Of the respondents, 14,437 were identified as casino players. The survey revealed that 52.8 million US adults visited a casino in 2005, averaging 6.1 trips per person. It also showed that casino players were more affluent than the average American, with a median household income of \$56,663 for casino customers versus \$48,997 for the general population. They also had a slightly higher education level, with 56% of casino players having at least some college versus 53% of all American adults. This had been the norm over the past few years. What was surprising was the amount of young adults who played poker in 2005. To learn more about poker players, communications firm Luntz, Maslansky Strategic Research conducted a telephone poll of 800 American adults who were selected using random digit sampling. In this poll, 18% of all the adults surveyed said they played poker in 2005, up from 12% in 2003. Younger adults played poker more than any other age group, with 35% of those aged 21–39 reporting they played in 2005, an increase of 6% from the previous year. This was followed by 18% of adults aged 40–49, 15% of adults aged 50–64,and 11% of those aged 65 or older.

Source: American Gaming Association. *State of the States: The AGA Survey of Casino Entertainment.* 2006. http://www.americangaming.org/files/aga/uploads/docs/sos/aga-sos-2006.pdf (12 Dec. 2011).

Analyze the surveys described above using the following exercises as a guide.

1. Identify the populations being studied.

2. Identify the samples and the sample sizes. Are the samples representative of the populations? Explain your answer.

3. Describe how the samples were chosen. Is there any potential bias in the sampling methods? Explain your answer.

4. List the descriptive statistics given in the article.

5. What inferences does the article make from the descriptive statistics?

6. Who conducted the studies? Is there any potential researcher bias? Explain your answer.

Project B: Analyzing a Study You Find

Find a study of interest on the Internet, in a newspaper or magazine, or in an educational journal. If you are reading studies in another course, an excellent choice is to use an assignment from that class. This would also be a good opportunity to begin reading journal articles from your major field of study. The periodical section of the library or library Internet resources are wonderful starting places. Once you have found an article, make a copy to include with your analysis.

Note: Though it may be tempting to choose a very short article, if the article is too short, it might not contain all of the necessary elements.

Once you have found a study of interest, analyze it using the following exercises as a guide. Write a formal summary of your analysis.

1. Who conducted the study, when and where?
2. What question(s) does the study seek to answer?
3. Identify the population and variables being studied.
4. Identify the sample and the approximate sample size. Is the sample representative of the population? Explain your answer.
5. How do the researchers deal with the issues of confidentiality and informed consent?
6. Describe how the sample was chosen. Is there any potential bias in the sampling method? Explain your answer.
7. List the descriptive statistics given in the article.
8. Are there any confounding variables?
9. What inferences do the researchers draw from the descriptive statistics?
10. Is there any potential researcher bias?
11. Do you feel comfortable believing the results of the study based on your analysis? Explain your answer.

T Chapter 1 Technology

Generating Random Numbers

TI-83/84 Plus

To generate a random number using a TI-83/84 Plus calculator, use the `randInt(` function under the PRB (probability) menu. The function syntax is `randInt(`*lower*, *upper*, *number of trials*`)`. This will produce the specified number of random integers between the lower and upper values, inclusive. If you only want one random number, the formula can be modified to `randInt(`*lower*, *upper*`)`.

For example, if we want to get seven random numbers between 10 and 25, we would press MATH, then scroll over to PRB, and then scroll down to option `5:randInt(` and press ENTER. Then type `1Ø,25,7)` and press ENTER. The calculator will then produce a string of seven random whole numbers between 10 and 25, inclusive.

Example T.1

Using a TI-83/84 Plus Calculator to Generate Random Integers

Use a TI-83/84 Plus calculator to create a random sample of four integers between 2 and 9, inclusive.

Solution

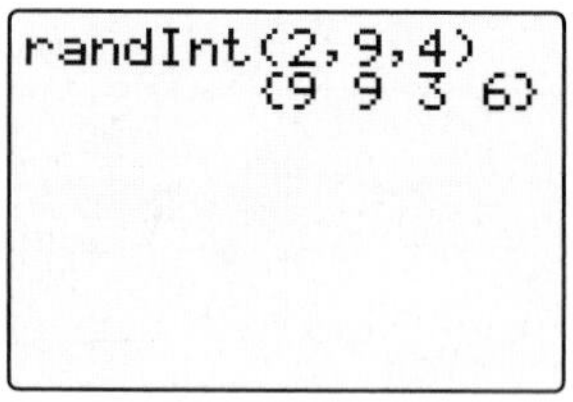

Press MATH, then scroll over to PRB, choose option `5:randInt(` and press ENTER. Type `2,9,4)` and press ENTER. You will then have something similar to the screenshot in the margin.

This screenshot shows a random sample of 9, 9, 3, and 6. Your random sample will obviously be different.

Duplicate numbers are a possibility, as shown in our example. If distinct numbers are required, you might need to increase the number of random digits you ask the calculator to generate.

Microsoft Excel

Generating a random number in Microsoft Excel is very easy. You can use the function command =RANDBETWEEN(*bottom*, *top*), which returns a random integer between the numbers you specify. For example, if we want to generate a random number between 0 and 9, inclusive, we would type the formula **=RANDBETWEEN(0, 9)** in one of the cells in the worksheet.

Example T.2

Using Microsoft Excel to Generate Random Integers

Use Microsoft Excel to create a list of five random numbers between 200 and 299, inclusive.

Solution

In cell A1, type **=RANDBETWEEN(200, 299)**. This will give you one random number. Copy that formula and paste it into four additional cells to get the remaining four random numbers. Your worksheet might look like the screenshot in the margin.

	A	B
1	213	
2	290	
3	250	
4	265	
5	261	

In this screenshot, the random numbers are 213, 290, 250, 265, and 261.

It is possible for Excel to duplicate random numbers here. If distinct random numbers are required, you will likely need to generate additional numbers.

MINITAB

Random numbers can also be generated using MINITAB, as shown in the following example.

Example T.3

Using MINITAB to Generate Random Integers

Generate 20 random integers between 1 and 100, inclusive, using MINITAB.

Solution

Go to **Calc ▶ Random Data ▶ Integer**. Enter the following parameters: Number of rows of data to generate: **20**; Store in column(s): **C1**; Minimum value: **1**; Maximum value: **100**; then click **OK**. The results will be produced in column C1.

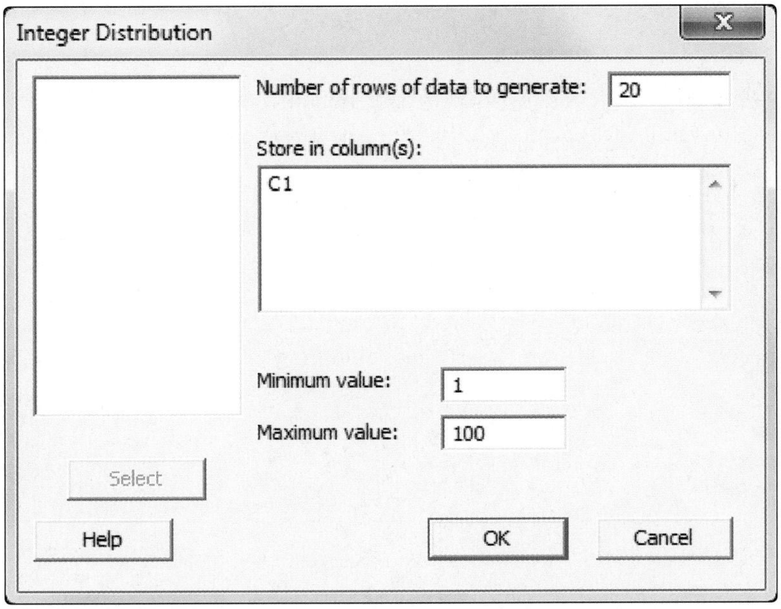

Chapter Two

Graphical Descriptions of Data

Sections

2.1 Frequency Distributions

2.2 Graphical Displays of Data

2.3 Analyzing Graphs

Objectives

1. Construct a frequency distribution.
2. Create and interpret the basic types of graphs used to display data.
3. Distinguish between the basic shapes of a distribution.
4. Identify misleading characteristics of a graph.

2

Introduction

In our modern, fast-paced society, graphs are useful for quickly providing a wealth of information to a reader. However, you must always use a discriminating eye and beware of graphs that are poorly constructed or misleading. For instance, Figure 2.1 illustrates the Google Voice international calling rates per minute for five selected countries.

Figure 2.1: Google Voice International Calling Rates

The picture looks nice, and the use of coins for bars is a clever idea. But take a closer look. Are the calling rates for France, Germany, and India really as similar as they appear? India, at 2¢ per minute, has a rate that is in reality considerably lower than the rates for France and Germany. The graph does not accurately depict the differences in calling rates.

Unfortunately, you can find misleading graphs everywhere. Sometimes these poorly constructed graphs occur by accident. At other times, they are purposefully misleading. Either way, it is essential that readers pay attention and be wary of misleading graphs.

2.1 Frequency Distributions

After collecting data as described in Chapter 1, you need to organize it so that inferences and conclusions can be drawn. A set of raw data is not very meaningful to an audience. Imagine looking up the prices of new 3-D TVs on an Internet search engine and getting the following dollar values.

Table 2.1: 3-D TV Prices (in Dollars)

1999	1599	1999	1899	1899
1699	1699	1899	1799	1685
1888	1787	1984	1699	1799
1699	1885	1999	1595	1757

This list is not very useful if you want to know the lowest price of a 3-D TV or the average price. If you create an **ordered array**, an ordered list of the data from largest to smallest or vice versa, then the lowest price and highest price are easy to see.

Table 2.2: 3-D TV Prices (in Dollars and in an Ordered Array)				
1595	1599	1685	1699	1699
1699	1699	1757	1787	1799
1799	1885	1888	1899	1899
1899	1984	1999	1999	1999

Now we can see that the lowest price for the TV is $1595, and the highest price is $1999. But which price is most common? Is there a pattern to the data? If we look at a distribution of the data, we can answer these questions. A **distribution** is a way to describe the structure of a particular data set or population. Throughout the book, we will look at the two types of statistical distributions: frequency distributions and probability distributions. A **frequency distribution** is a display of the values that occur in a data set and how often each value, or range of values, occurs. However, a **probability distribution** is a theoretical distribution used to predict the probabilities of particular data values occurring in a population. In Chapter 5 we begin our discussion of probability distributions, but for now, we'll focus only on frequency distributions.

There are two basic types of frequency distributions: grouped distributions and ungrouped distributions. If the data set is relatively small, or contains only a few possible values, then every value might be shown individually in the distribution. A frequency distribution such as this, where each category represents a single value, is called an **ungrouped frequency distribution**, and its **frequencies (f)**, or counts of data values, are listed for each category. For example, if you were looking at a distribution of qualitative data such as letter grades, there are only five possible choices for the data values: A, B, C, D, and F. In this case, it seems natural for each data value to have its own category, or **class**. It would be strange to group the letter grades, and since there are only five classes, an ungrouped frequency distribution is reasonable.

Table 2.3: Frequency Distribution of Grades

Class	Frequency
A	5
B	14
C	9
D	3
F	1

However, suppose we formed an ungrouped frequency distribution for the number grades of a test with a 100-point scale and not the letter grades. We would then have one hundred different classes. Not only would this be unreasonable, it probably would not depict a clear pattern to the data. Our example of 3-D TV prices would result in 12 different classes if we used each value as a separate class. (Look again at the ordered data set above and see if you can verify this.) In cases such as these, the data are often grouped into ranges of values, thus allowing for fewer classes and a more manageable frequency distribution. A frequency distribution of this type, where the classes are ranges of possible values, is called a **grouped frequency distribution**. Grouped distributions are more common and take more skill to create, so we will focus the remainder of our discussion on grouped frequency distributions. The procedure for creating a grouped frequency distribution (which from this point on we will refer to simply as a frequency distribution) is outlined on the following page.

Definition

A **distribution** is a way to describe the structure of a particular data set or population.

A **frequency distribution** is a display of the values that occur in a data set and how often each value, or range of values, occurs.

Frequencies (f) are the numbers of data values in the categories of a frequency distribution.

A **class** is a category of data in a frequency distribution.

Procedure

Constructing a Frequency Distribution

1. *Decide how many classes should be in the distribution.* There are typically between 5 and 20 classes in a frequency distribution. Several different methods can be used to determine the number of classes that will show the data most clearly, but in this textbook, the number of classes for a given data set will be suggested.
2. *Choose an appropriate class width.* In some cases, the data set easily lends itself to natural divisions, such as decades or years. At other times, we must choose divisions for ourselves. When starting a frequency distribution from scratch, one method of finding an appropriate class width is to begin by subtracting the lowest number in the data set from the highest number in the data set and dividing the difference by the number of classes. Rounding this number up gives a good starting point from which to choose the class width. You will want to choose a width so that the classes formed present a clear representation of the data and include all members of the data set, so make a sensible choice.
3. *Find the class limits.* The **lower class limit** is the smallest number that can belong to a particular class, and the **upper class limit** is the largest number that can belong to a class. Using the minimum data value, or a smaller number, as the lower limit of the first class is a good place to begin. However, judgment is required. You should choose the first lower limit so that reasonable classes will be produced, and it should have the same number of decimal places as the largest number of decimal places in the data. After choosing the lower limit of the first class, add the class width to it to find the lower limit of the second class. Continue this pattern until you have the desired number of lower class limits. The upper limit of each class is determined such that the classes do not overlap. If, after creating your classes, there are any data values that fall outside the class limits, you must adjust either the class width or the choice for the first lower class limit.
4. *Determine the frequency of each class.* Make a tally mark for each data value in the appropriate class. Count the marks to find the total frequency for each class.

Rounding Rule

Class limits should have the same number of decimal places as the largest number of decimal places in the data.

Definition

The **class width** is the difference between the lower limits or upper limits of two consecutive classes of a frequency distribution.

The **lower class limit** is the smallest number that can belong to a particular class.

The **upper class limit** is the largest number that can belong to a particular class.

Let's look at an example of creating a frequency distribution using these guidelines.

Example 2.1

Constructing a Frequency Distribution

Create a frequency distribution using five classes for the list of 3-D TV prices given earlier.

Solution

Because we were told how many classes to include, we will begin by deciding on a class width. Subtract the lowest data value from the highest and divide by the number of classes, as shown below.

$$\frac{1999-1595}{5}=80.8\approx 81$$

This would give us a class width of \$81. We will stop here and consider some options. Choosing a class width of \$81 does seem perfectly reasonable from a theoretical point of view. However, one should consider the impression created by having TV prices grouped in intervals of \$81. Can you imagine presenting this data to a client? Instead, it would be more reasonable to group TV prices by intervals of \$100. Therefore, we will choose our class width to be \$100.

Next, we need to choose a starting point for the classes, that is, the first lower class limit. One should always first consider using the smallest data value for the beginning point. In this case, if we choose the smallest TV price, we would be starting the first class at \$1595 with a width of \$100. However, given that we've chosen a class width of \$100, it is more natural to begin the first class at \$1500.

Now let's continue building the class limits. Adding the class width of \$100 to \$1500, we obtain a second lower class limit of \$1600. The next lower limit is found by adding \$100 to \$1600. We continue in this fashion until we have five lower class limits, one for each of our five classes.

Finally, we need to determine appropriate upper class limits. Again, be reasonable. Remember, too, that the classes are not allowed to overlap. Because the data are in whole dollar amounts, it makes sense to choose upper class limits that are one dollar less than the next lower limit. The classes we have come up with are as follows.

3-D TV Prices

Class	Frequency
\$1500–\$1599	
\$1600–\$1699	
\$1700–\$1799	
\$1800–\$1899	
\$1900–\$1999	

Note that the last upper class limit is also the maximum value in the data set. This will not necessarily occur in every frequency table. However, we have included all the data values in our range of classes, so no adjustments to the classes are necessary.

Tabulating the number of data values that occur in each class produces the following frequency table.

Math Symbols

≈: read as "approximately equals"

3-D TV Prices	
Class	**Frequency**
\$1500–\$1599	2
\$1600–\$1699	5
\$1700–\$1799	4
\$1800–\$1899	5
\$1900–\$1999	4

Note that the sum of the frequency column should equal the number of data values in the set. Check for yourself that this is true.

Characteristics of a Frequency Distribution

There are other characteristics of a frequency distribution that can be calculated once the basic frequency table has been constructed. Let's look at four of them.

The first calculation is that of **class boundaries**, which are similar to the class limits. The class boundaries split the difference in the gap between the upper limit of one class and the lower limit of the next class. To find a class boundary, add the upper limit of one class to the lower limit of the next class and divide by 2. For example, if an upper class limit is 10, and the next lower class limit is 11, the class boundary would be calculated as follows.

$$\frac{10+11}{2}=10.5$$

Thus 10.5 is the boundary between those two classes. You can use one class boundary to find all of the other class boundaries by adding (or subtracting) the class width. Each class has both an upper and a lower boundary, and the boundaries for a particular class are typically given in interval form, that is, lower boundary–upper boundary. Class boundaries are useful in constructing frequency histograms, which we will cover in the next section on graphing distributions.

Definition

A **class boundary** is the value that lies halfway between the upper limit of one class and the lower limit of the next class. After finding one class boundary, add (or subtract) the class width to find the next class boundary. The boundaries of a class are typically given in interval form: lower boundary–upper boundary.

Example 2.2

Calculating Class Boundaries

Calculate the class boundaries for each class in the frequency distribution from Example 2.1.

Solution

Look at the first and second classes. The upper limit of class one is 1599. The lower limit of class two is 1600. Thus, the class boundary between the first two classes is calculated as follows.

$$\frac{1599+1600}{2}=1599.5$$

Recall that the class width is 100. Adding 100 to 1599.5 gives the next class boundary. You can repeat this step to find the remaining class boundaries.

3-D TV Prices with Class Boundaries

Class	Frequency	Class Boundaries
\$1500–\$1599	2	1499.5–1599.5
\$1600–\$1699	5	1599.5–1699.5
\$1700–\$1799	4	1699.5–1799.5
\$1800–\$1899	5	1799.5–1899.5
\$1900–\$1999	4	1899.5–1999.5

The **midpoint**, or class mark, of a class is the sum of the lower and upper limits of the class divided by 2. The midpoints are often used for estimating the average value in each class.

Formula

Class Midpoint

$$\text{Class Midpoint} = \frac{\text{Lower Limit} + \text{Upper Limit}}{2}$$

Example 2.3

Calculating Class Midpoints

Calculate the midpoint of each class in the frequency distribution from Example 2.1.

Solution

The midpoint is the sum of the class limits divided by two. For the first class, the midpoint is calculated as follows.

$$\frac{1500+1599}{2} = 1549.5$$

We can use this same calculation to find the midpoints of the remaining classes. Another method is to add 100 (the class width) to the first midpoint, as we did with class boundaries.

3-D TV Prices with Class Midpoints

Class	Frequency	Midpoint
\$1500–\$1599	2	1549.5
\$1600–\$1699	5	1649.5
\$1700–\$1799	4	1749.5
\$1800–\$1899	5	1849.5
\$1900–\$1999	4	1949.5

The third calculation we will discuss, **relative frequency**, is the fraction or percentage of the data set that falls into a particular class. It is calculated by dividing the class frequency by the sample size. The **sample size**, n, for a frequency distribution can be found by adding all of the class frequencies together. Relative frequencies are useful because fractions or percentages make it easier to quickly analyze the data set as a whole.

Formula

Relative Frequency

The **relative frequency** is the fraction or percentage of the data set that falls into a particular class, given by

$$\text{Relative Frequency} = \frac{f}{n}$$

where f is the class frequency,

n is the sample size, given by $n = \sum f_i$, and

f_i is the frequency of the ith class.

Math Symbols

Σ: Greek letter, Sigma; indicates to take the sum of what follows

Example 2.4

Calculating Relative Frequencies

Calculate the relative frequency for each class in the frequency distribution from Example 2.1.

Solution

We first find the sample size by summing the class frequencies.

$$\begin{aligned} n &= \sum f_i \\ &= 2+5+4+5+4 \\ &= 20 \end{aligned}$$

Then divide each class frequency by 20.

3-D TV Prices with Relative Frequencies

Class	Frequency	Relative Frequency
\$1500–\$1599	2	$\frac{2}{20} = \frac{1}{10} = 0.1 = 10\%$
\$1600–\$1699	5	$\frac{5}{20} = \frac{1}{4} = 0.25 = 25\%$
\$1700–\$1799	4	$\frac{4}{20} = \frac{1}{5} = 0.2 = 20\%$
\$1800–\$1899	5	$\frac{5}{20} = \frac{1}{4} = 0.25 = 25\%$
\$1900–\$1999	4	$\frac{4}{20} = \frac{1}{5} = 0.2 = 20\%$

The final calculation we will look at, **cumulative frequency**, is the sum of the frequencies of a given class and all previous classes. The cumulative frequency of the last class equals the sample size.

Definition

The **cumulative frequency** is the sum of the frequencies of a given class and all previous classes. The cumulative frequency of the last class equals the sample size.

Example 2.5

Calculating Cumulative Frequencies

Calculate the cumulative frequency for each class in the frequency distribution from Example 2.1.

Solution

3-D TV Prices with Cumulative Frequencies

Class	Frequency	Cumulative Frequency
\$1500–\$1599	2	2
\$1600–\$1699	5	7 (2 + 5)
\$1700–\$1799	4	11 (2 + 5 + 4)
\$1800–\$1899	5	16 (2 + 5 + 4 + 5)
\$1900–\$1999	4	20 (2 + 5 + 4 + 5 + 4)

Now let's put all of these concepts together in a single example.

Example 2.6

Characteristics of a Frequency Distribution

Data collected on the numbers of miles that professors drive to work daily are listed below. Use these data to create a frequency distribution that includes the class boundaries, midpoint, relative frequency, and cumulative frequency of each class. Use six classes. Be sure that your class limits have the same number of decimal places as the largest number of decimal places in the data.

Numbers of Miles Professors Drive to Work Each Day

3.8	2.7	9.3	6.5	5.8	7
10.2	1	3.7	9.1	6.2	11
11.9	5.5	4.8	7.3	9.1	1.4

Solution

Since this example calls for six classes, a good starting point for the class width is calculated as follows.

$$\frac{11.9-1}{6}=1.81\overline{6}\approx 1.8$$

Because our data are in miles, a more sensible class width to use is 2.

Next, to choose the lower class limit of the first class, begin by considering the smallest data value, which is 1 mile. In this case, 1.0 is a reasonable place to begin our classes. Adding the class width of 2 gives us the following table.

Numbers of Miles Professors Drive to Work Each Day	
Class	**Frequency**
1.0–2.9	
3.0–4.9	
5.0–6.9	
7.0–8.9	
9.0–10.9	
11.0–12.9	

Once again, note that all of the data values fall within the range of the class limits. So, no adjustments in the classes are necessary.

The upper class boundary and midpoint of the first class are calculated as follows.

$$\text{Upper Boundary of Class 1: } \frac{2.9+3.0}{2}=2.95$$

$$\text{Midpoint of Class 1: } \frac{1.0+2.9}{2}=1.95$$

Memory Booster

Relative frequency is a ratio that relates a class to the whole.

Cumulative frequency is a running total of the number of data values.

Use the class width to find the other class boundaries and midpoints. The frequency, relative frequency, and cumulative frequency of each class are calculated as in previous examples.

Numbers of Miles Professors Drive to Work Each Day

Class	Frequency	Class Boundaries	Midpoint	Relative Frequency	Cumulative Frequency
1.0–2.9	3	0.95–2.95	1.95	$\frac{3}{18}=\frac{1}{6}=0.1\overline{6}\approx 17\%$	3
3.0–4.9	3	2.95–4.95	3.95	$\frac{3}{18}=\frac{1}{6}=0.1\overline{6}\approx 17\%$	6
5.0–6.9	4	4.95–6.95	5.95	$\frac{4}{18}=\frac{2}{9}=0.\overline{2}\approx 22\%$	10
7.0–8.9	2	6.95–8.95	7.95	$\frac{2}{18}=\frac{1}{9}=0.\overline{1}\approx 11\%$	12
9.0–10.9	4	8.95–10.95	9.95	$\frac{4}{18}=\frac{2}{9}=0.\overline{2}\approx 22\%$	16
11.0–12.9	2	10.95–12.95	11.95	$\frac{2}{18}=\frac{1}{9}=0.\overline{1}\approx 11\%$	18

2.1 Section Exercises

Characteristics of Frequency Distributions

Directions: For Exercises 1–10, find the following for each frequency distribution.

a. Class width

b. Class boundaries for each class

c. Midpoint of each class

d. Relative frequency for each class

e. Cumulative frequency for each class

1. Ages of Taste-Test Participants (in Years)

Class	Frequency
15–19	7
20–24	8
25–29	10
30–34	2
35–39	3

2. Braking Times for Vehicles at 60 mph (in Minutes)

Class	Frequency
0.05–0.07	12
0.08–0.10	15
0.11–0.13	14
0.14–0.16	15
0.17–0.19	14

3. Age at Time of First Marriage (in Years)

Class	Frequency
15–18	2
19–22	5
23–26	4
27–30	5
31–34	4

4.

Hourly Wage at First Job (in Dollars)	
Class	**Frequency**
7.50–8.49	12
8.50–9.49	50
9.50–10.49	48
10.50–11.49	45
11.50–12.49	34

5.

Ages of Survey Participants (in Years)	
Class	**Frequency**
15–24	9
25–34	8
35–44	12
45–54	1
55–64	3

6.

Cost of a 12 oz Soda (in Dollars)	
Class	**Frequency**
0.25–0.49	2
0.50–0.74	15
0.75–0.99	12
1.00–1.24	5
1.25–1.49	9

7.

Ages of First-Time Home Buyers (in Years)	
Class	**Frequency**
18–24	2
25–31	7
32–38	4
39–45	15
46–52	3

8.

Hourly Wages of Surveillance Operators (in Dollars)	
Class	**Frequency**
10.50–11.49	92
11.50–12.49	78
12.50–13.49	68
13.50–14.49	45
14.50–15.49	34

9.

Age at Time of First Car Purchase (in Years)	
Class	**Frequency**
16–19	12
20–23	8
24–27	15
28–31	12
32–35	9

10.

Grades on a Difficult Test	
Class	**Frequency**
A	2
B	5
C	7
D	13
F	10

Directions: Complete the frequency distribution that has been started for each set of data.

11. The following data describe the heights, in inches, of 30 men entering the military in April 2011.

72.8	71.2	70.3	73.4	72.6	74.1
70.9	71.6	72.1	74.6	75.0	72.0
69.1	69.5	72.6	72.4	73.6	75.1
71.8	71.6	71.9	70.9	70.2	69.3
72.1	72.3	72.5	73.4	74.0	75.0

Heights of Men (in Inches)	
Class	**Frequency**
69.0–69.9	3
70.0–70.9	
71.0–71.9	
72.0–72.9	
73.0–73.9	
74.0–74.9	
75.0–75.9	

12. The following data represent the number of exercises at the end of various sections in a traditional college algebra textbook.

145	137	138	112	137	100	78	127	97
70	143	133	150	124	115	110	45	141
119	92	84	94	105	71	95	117	104

Number of Section Exercises	
Class	**Frequency**
40–59	
60–79	
80–99	
100–119	
120–139	
140–159	

13. At a state fair, one game involves guessing the number of marbles in a glass jar. The following data represent the guesses that people made during one hour at the state fair.

1234	1645	1469	1467	1549	1348	1671	1300	1200	1199
1621	1547	1501	1410	1487	1299	1500	1688	1301	1399

Number of Marbles in the Jar	
Class	**Frequency**
1100–1199	
1200–	
1300–1399	4
	4
1500–1599	
1600–	4

14. The following data represent the amounts of pocket change, in dollars, in the pockets of a sample of men in an office building.

0.23	0.52	0.76	0.79	0.8	0.21	0.13
1.05	1.24	1.15	1.10	0.98	0.28	0.64
1.34	0.38	0.31	0.42	0.41	0.24	1.42

Amount of Pocket Change (in Dollars)	
Class	**Frequency**
–0.24	4
0.25–0.49	
0.50–	2
0.75–0.99	
–1.24	4
1.25–1.49	

Directions: **Create a frequency distribution with the indicated number of classes for each set of data. Include the frequency, class boundaries, midpoint, relative frequency, and cumulative frequency of each class.**

15. The following data represent the numbers of curl-ups completed in 60 seconds for a group of 16 eight-year-old boys. Use six classes that have a class width of 5. Begin with a lower limit of 15.

31	34	41	36	27	29	18	33
31	28	34	22	26	28	36	42

16. The following data represent times in minutes for completing a one mile run/walk from a group of 24 seventeen-year-old girls. Use six classes that have a class width of 2.00. Begin with a lower limit of 6.00.

15.23	13.52	11.35	11.15	12.20	9.90	10.37	14.05
10.02	17.35	8.33	8.05	9.87	9.28	10.62	6.65
9.55	10.23	13.93	10.97	9.75	12.85	12.82	10.93

17. The following data represent the caloric intakes in one day for a group of 15 men between the ages of 20 and 39. Use five classes that have a class width of 400. Begin with a lower limit of 1800.

2700	2200	2500	2800	2600
3000	2600	2200	3100	2800
1800	3500	2500	3000	2900

18. The following data represent the precipitation totals in inches for the month of September in 21 different towns in Alaska. Use six classes that have a class width of 3.50. Begin with a lower limit of 0.00.

2.7	1.72	1.39	6.88	2.59	2.04	2.43
9.28	1.06	3.29	1.57	4.23	0.95	8.37
0.6	4.41	6.73	1.92	2.28	2.74	18.65

Directions: Use the given frequency distribution to answer the questions.

19.

Amounts of Weight (in Pounds) Lost by Women Following a Low-Carbohydrate Diet	
Weight Lost	**Number of Women**
1–5	4
6–10	8
11–15	11
16–20	3
21–25	0
26–30	1

a. How many women were involved in the study?

b. How much weight did most participants lose?

c. What percentage of women lost more than 15 pounds?

d. How long did the study last?

e. What is the lower class limit of the 4th class?

f. What is the midpoint for the 1st class?

20.

Grams of Sugar per Serving in Children's Breakfast Cereal	
Class	**Frequency**
10.00–11.99	21
12.00–13.99	19
14.00–15.99	11
16.00–17.99	18
18.00–19.99	18

a. How big was the original data set?

b. What percentage of the data fell in the 3rd class?

c. Estimate the average value of the data in the highest class.

21.

Distribution of Attitudes Toward College Success					
Scale	Freshmen	Sophomores	Juniors	Seniors	Graduate Students
1.0–1.9	11	10	2	9	19
2.0–2.9	3	2	5	5	1
3.0–3.9	15	2	3	11	1
4.0–4.9	11	15	1	1	10
5.0–5.9	6	7	8	11	12
6.0–6.9	16	9	17	16	16
7.0–7.9	5	19	1	17	5
8.0–8.9	3	5	19	14	17
9.0–9.9	5	11	1	14	10
10.0–10.9	9	3	4	15	7

a. How many students were surveyed for the study?

b. What percentage of sophomores gave a response smaller than 5?

c. How many people gave a response in the 9th class?

d. What percentage of respondents were graduate students?

22. Below is an example of a frequency distribution produced by a statistical software package called SPSS. Use it to answer the following questions.

Statistics

Respondent's Ethnicity

N	Valid	936
	Missing	4

Respondent's Ethnicity

		Frequency	Percent	Valid Percent	Cumulative Percent
Valid	Caucasian	476	50.6	50.9	50.9
	African American	74	7.9	7.9	58.8
	Asian or Pacific Islander	123	13.1	13.1	71.9
	Hispanic	198	21.1	21.2	93.1
	Native American or Alaskan Native	4	.4	.4	93.5
	Other	61	6.5	6.5	100.0
	Total	936	99.6	100.0	
Missing	System	4	.4		
Total		940	100.0		

a. How many respondents gave their ethnicities?

b. What percentage of respondents who gave their ethnicities were Native American or Alaskan Natives?

c. What percentage of respondents who gave their ethnicities were either Caucasian or African American?

d. What percentage of respondents who gave their ethnicities were either Asian/Pacific Islanders or Hispanic?

2.2 Graphical Displays of Data

Sometimes it is easier to view the data we gather in a graph rather than a table. A graph is a snapshot that allows us to view patterns at a glance without undergoing lengthy analysis of the data. Also, graphs are much more visually appealing than a table or list.

Once you decide that a graph is the best way to present your data, you will need to decide which type of graph is most appropriate. It is important to know that not all graphs can be used for all types of data. As each new type of graph is presented, be sure to look for the type of data that you must have in order to use that type of graph. We will begin this section with graphs that use qualitative data and then discuss graphs that use quantitative data.

Before discussing *specific* types of graphs, let's briefly consider the information that every graph should contain. Every graph must be given a title as well as labels for both the horizontal and vertical axes. A footnote usually lists the source of the data, and the date should be included also since data can change dramatically over time. When appropriate, a legend should be provided that labels the different data sets being compared. As a general rule, a graph should be labeled well enough so that it is able to stand alone, without the original data. We will look at these details more in Section 2.3.

Pie Charts

We will begin with a graph that is used to display *qualitative* (or categorical) data. When you have nominal data where the categories of data are parts of a whole, such as parts of a budget, a *pie chart* can be created. A **pie chart** shows how large each category is in relation to the whole; that is, it uses the relative frequencies from the frequency distribution to divide the "pie" into different-sized wedges. The size, or central angle measure, of each wedge in the pie chart is calculated by multiplying 360° by the relative frequency of each class and rounding to the nearest whole degree.

Rounding Rule

When constructing a pie chart, round each angle measure to the nearest whole degree.

In the following example we will demonstrate how to calculate the central angle measures for a pie chart; however, pie charts can be created using available technology without having to calculate the central angle measures. The technology section at the end of this chapter shows you how to use a few different types of technology to create the graphs discussed in this section.

Example 2.7

Creating a Pie Chart

Create a pie chart from the following data describing the distribution of housing types for students in a statistics class. Calculate the size of each wedge in the pie chart to the nearest whole degree.

Housing Types for Students in a Statistics Class

Type of Housing	Number of Students
Apartment	20
Dorm	15
House	9
Sorority/Fraternity House	5
Total	49

Solution

	Relative Frequencies	**Central Angle Measures**
Apartment:	$\frac{f}{n}=\frac{20}{49}\approx 0.41=41\%$	$\frac{20}{49}\cdot 360°\approx 147°$
Dorm:	$\frac{f}{n}=\frac{15}{49}\approx 0.31=31\%$	$\frac{15}{49}\cdot 360°\approx 110°$
House:	$\frac{f}{n}=\frac{9}{49}\approx 0.18=18\%$	$\frac{9}{49}\cdot 360°\approx 66°$
Sorority/Fraternity House:	$\frac{f}{n}=\frac{5}{49}\approx 0.10=\underline{10\%}$	$\frac{5}{49}\cdot 360°\approx \underline{37°}$
	100%	360°

Housing Types for Students in a Statistics Class

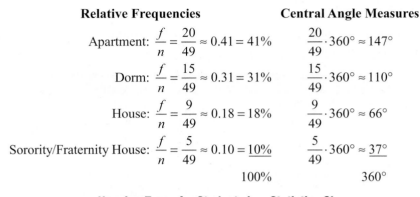

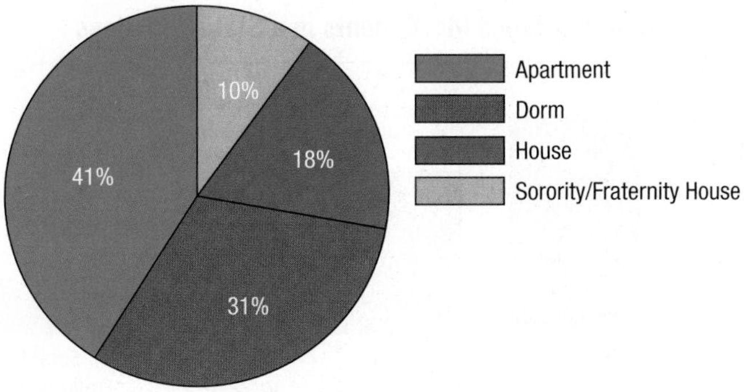

Bar Graphs

Pie charts are great if you have a specific type of nominal data, but what happens if your data are not nominal or do not represent parts of a whole? Graphs that use bars, instead of parts of a circle, to represent the amount of data in each category are called **bar graphs**. Bar graphs can be used for qualitative data, without restrictions on what types of qualitative data you must have. In addition, bar graphs can make it easier to see small differences in the amounts of data in different categories. Thus, we might choose to use a bar graph instead of a pie chart if the amounts of data in two or more categories are close to being equal.

In a bar graph, one axis displays the categories of data and the other axis displays the frequencies. In a traditional bar graph, the horizontal axis contains the qualitative categories. Because the bars represent categories, the width of each bar along the horizontal axis is meaningless. Thus, to avoid misrepresenting the data, the bars should be of uniform width. In addition, the bars usually do not touch since the categories are separate. In the traditional bar graph, the vertical axis represents the frequency of each category. The height of the bar should equal the frequency of the category. Some software packages, such as Microsoft Excel, call the traditional bar graph a "column graph" since the bars in the graph look like vertical columns. These software packages will create what they call "bar graphs" but the bars are horizontal. These two graphs are equivalent, and it is a matter of preference as to which you use.

Example 2.8

Creating a Bar Graph

Create a bar graph of the following data describing the distribution of housing types for students in a statistics class.

Housing Types for Students in a Statistics Class

Type of Housing	Number of Students
Apartment	20
Dorm	15
House	9
Sorority/Fraternity House	5

Solution

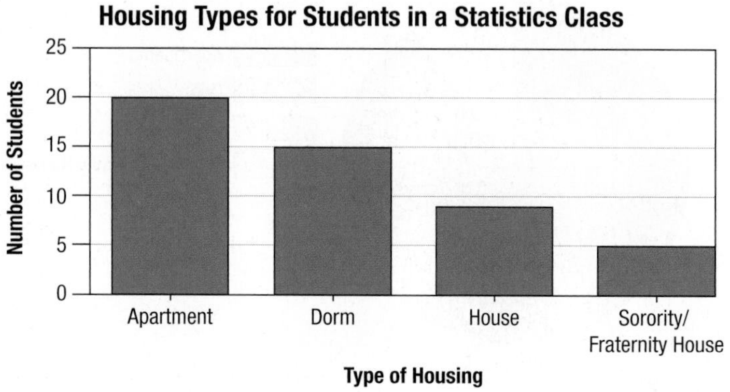

Notice that the bars in the previous example are in order from largest to smallest. This special type of bar graph with the bars in descending order is called a **Pareto chart**. Pareto charts are typically used with nominal data. The reason for this is that if a Pareto chart were created from ordinal data, the values on the x-axis might seem out of order after the bars were rearranged from largest to smallest based on frequency. To demonstrate this, consider the two bar graphs in Figures 2.2 and 2.3 showing the frequencies of various shoe sizes in a women's sorority. In Figure 2.2, the bar graph has the bars arranged according to shoe size, with the first bar representing the smallest shoe size and the last bar representing the largest shoe size. Figure 2.3 is a Pareto chart of the same data. Look at the labels on the x-axis. The shoe sizes seem out of order because they represent ordinal data, which have a natural order to them. Therefore, the data set does not lend itself to being rearranged in a Pareto chart.

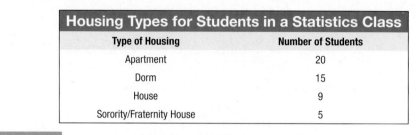

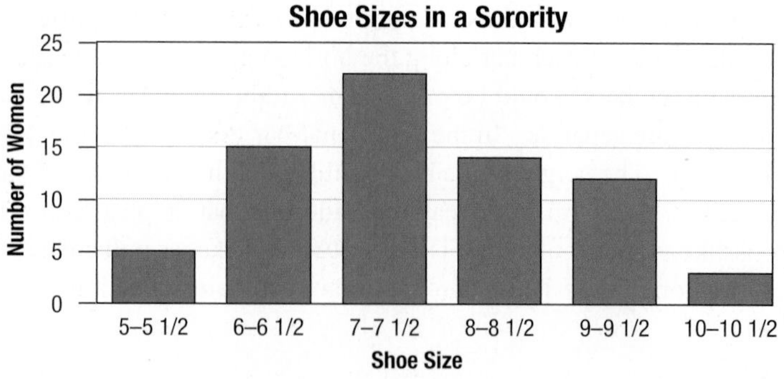

Appropriate

Figure 2.2: Bar Graph

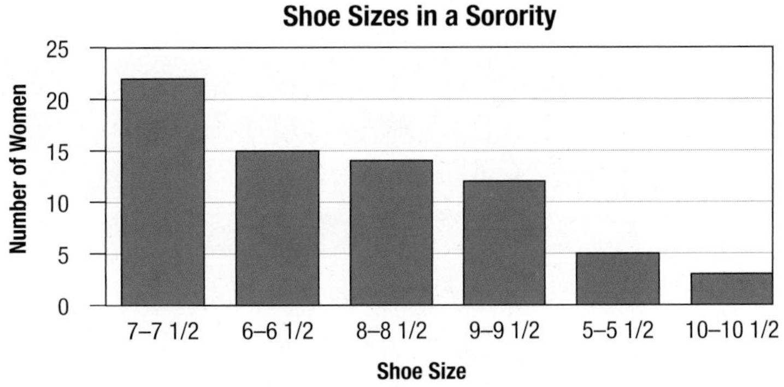

Not Appropriate

Figure 2.3: Pareto Chart

Example 2.9

Creating a Pareto Chart

Create a Pareto chart, if appropriate, for the following data.

Fundraising Results	
Class	**Number of Fundraiser Items Sold**
Ms. Boyd's class	200
Ms. Willard's class	157
Mr. Smith's class	176
Mr. Carter's class	181
Ms. Meredith's class	143

Solution

The data are nominal; therefore, it would be appropriate to create a Pareto chart from these data. Begin by rearranging the data from the largest frequency to the smallest frequency.

Ordered Fundraising Results	
Class	**Number of Fundraiser Items Sold**
Ms. Boyd's class	200
Mr. Carter's class	181
Mr. Smith's class	176
Ms. Willard's class	157
Ms. Meredith's class	143

Next, create a bar graph of the ordered data. The resulting Pareto chart is shown.

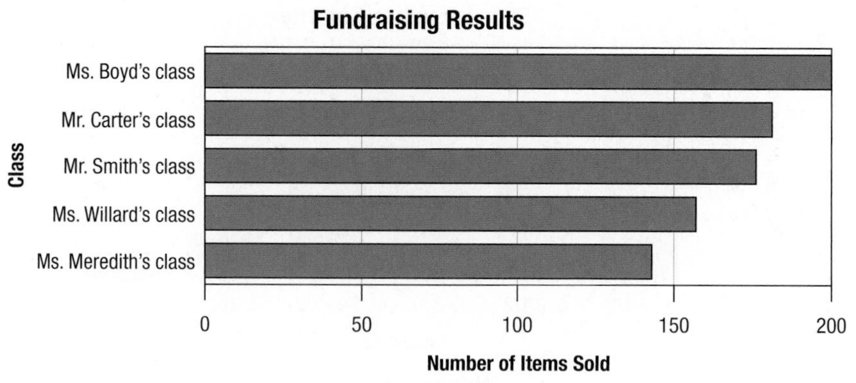

When we want to create a bar graph that compares the same categories for different groups, we use a **side-by-side bar graph**. In Example 2.8, we looked at the housing data from one sample. Let's compare those data to the data from another sample. To do so, create a bar from each sample for each category. Identify the bars in some way, such as different colors, to denote which bars represent a given sample. In this type of graph, it is important to include a legend that denotes which color represents which sample.

Example 2.10

Creating a Side-by-Side Bar Graph

Create a side-by-side bar graph of the following data describing the distribution of housing types for two different samples of students.

Housing Types for Two Samples of Students

Type of Housing	Number of Students from Sample A	Number of Students from Sample B
Apartment	20	13
Dorm	15	24
House	9	6
Sorority/Fraternity House	5	7

Solution

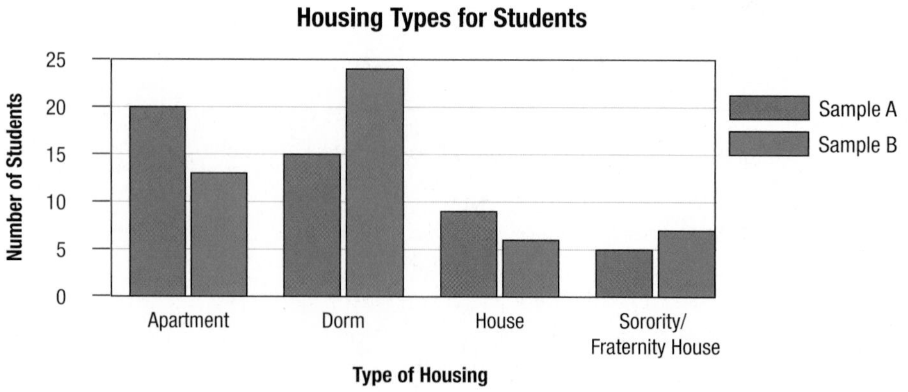

With a side-by-side bar graph, it is very easy to compare the housing characteristics of two samples. For instance, Sample A has more students who live in apartments, and Sample B has more students who live in dorms. Unfortunately, side-by-side bar graphs are not very convenient if you want to know how many students from both samples live in a certain type of housing. A more efficient graph for displaying the data for this purpose is the **stacked bar graph**. Imagine picking up each bar for Sample B and placing it on top of the bar for Sample A. The result would be a stacked bar graph. Here too, a legend is essential.

Example 2.11

Creating a Stacked Bar Graph

Create a stacked bar graph for the data in Example 2.10.

Solution

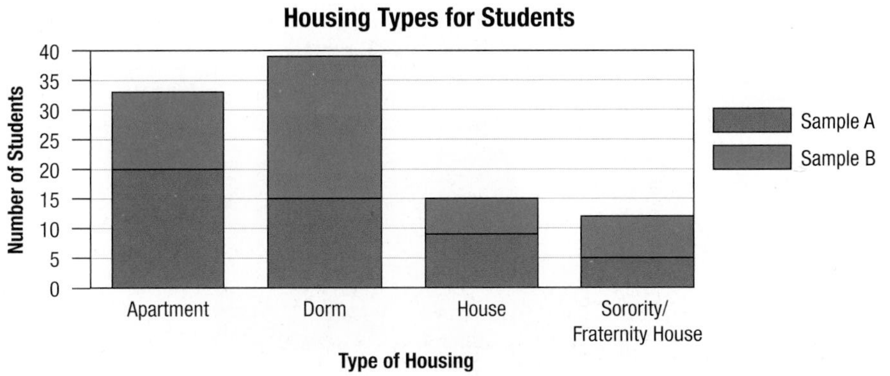

With the stacked bar graph, it is easier to see that more students live in the dorms than in apartments.

Histograms

Pie charts and bar graphs are types of graphs that are used for *qualitative* data. When we want to display *quantitative* data, we need other types of graphs. We have seen how to create bar graphs from frequency distributions of qualitative data, but we also know it is possible to create frequency distributions from quantitative data. When we create a bar graph of a frequency distribution of *quantitative* data, it is called a **frequency histogram**, which we will usually shorten to **histogram**.

We construct a histogram in a manner similar to how we constructed a bar graph. The classes are put along the horizontal axis and the frequencies are marked on the vertical axis. We begin constructing a histogram by finding the class boundaries for the frequency distribution. The horizontal axis of a histogram is a real number line on which you mark the class boundaries of every class. The width of each bar represents the width of each class. Since the upper class boundary of one class is the same as the lower class boundary of the next class, the bars in a histogram should touch (unlike in a bar graph). Since the classes in a frequency distribution are uniform in width, the bars should be uniform in width as well. Thus, histograms are not appropriate for frequency distributions that have classes with an undetermined width, such as "15 or greater." Although we use class boundaries to draw the histogram, it is appropriate to use either the class boundaries or the class midpoints when labeling the x-axis of a frequency histogram. (We will show both labeling methods in the next example.) The last step is to draw in the bars on the histogram. The height of each bar should be the frequency of the class it represents.

Example 2.12

Constructing a Histogram

Construct a histogram of the 3-D TV prices from the previous section. The frequency distribution of the data is restated here.

3-D TV Prices			
Class	Frequency	Midpoint	Class Boundaries
$1500–$1599	2	1549.5	1499.5–1599.5
$1600–$1699	5	1649.5	1599.5–1699.5
$1700–$1799	4	1749.5	1699.5–1799.5
$1800–$1899	5	1849.5	1799.5–1899.5
$1900–$1999	4	1949.5	1899.5–1999.5

Solution

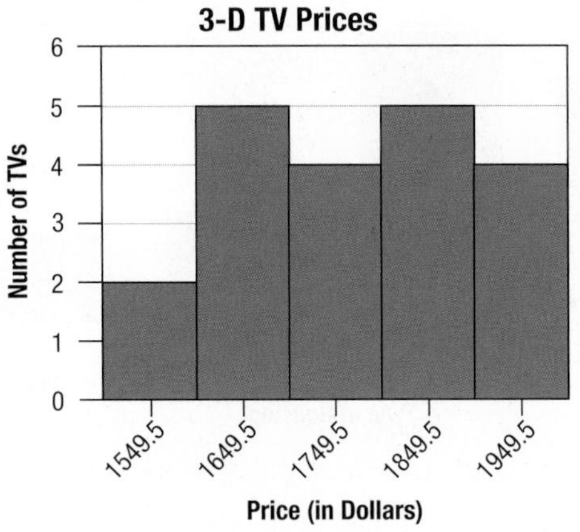

With Midpoints

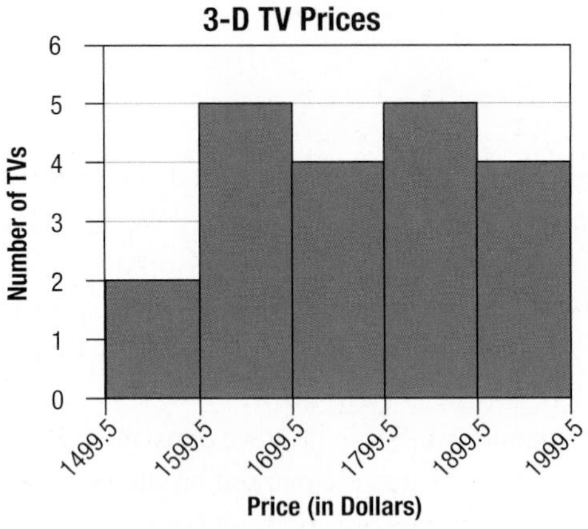

With Class Boundaries

Relative Frequency Histograms

There are situations in which it is beneficial to display the *relative* frequencies for a distribution. A **relative frequency histogram** is a histogram in which the heights of the bars represent the relative frequencies of each class rather than simply the frequencies. It is appropriate to label the *y*-axis of a relative frequency histogram with either decimals or percentages. The following example shows a relative frequency histogram drawn for the previous example about 3-D TVs. Notice that the only difference between the histogram and the relative frequency histogram is the labeling on the *y*-axis. In particular, note that the shapes are the same.

Example 2.13

Constructing a Relative Frequency Histogram

Construct a relative frequency histogram of the 3-D TV prices from the previous section. The frequency distribution of the data is reprinted here.

3-D TV Prices

Class	Frequency	Relative Frequency
\$1500–\$1599	2	$\frac{2}{20}=\frac{1}{10}=0.1=10\%$
\$1600–\$1699	5	$\frac{5}{20}=\frac{1}{4}=0.25=25\%$
\$1700–\$1799	4	$\frac{4}{20}=\frac{1}{5}=0.2=20\%$
\$1800–\$1899	5	$\frac{5}{20}=\frac{1}{4}=0.25=25\%$
\$1900–\$1999	4	$\frac{4}{20}=\frac{1}{5}=0.2=20\%$

Solution

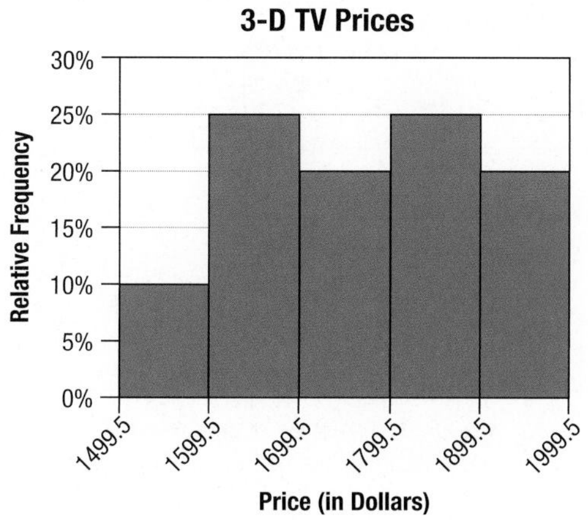

Frequency Polygons

Another way to graphically display the data from a quantitative frequency distribution is called a **frequency polygon**. A frequency polygon is a visual display created by plotting a point at the frequency of each class above each class midpoint and connecting the points using straight lines. The steps for creating a frequency polygon are shown in the following box.

Procedure

Constructing a Frequency Polygon

1. *Mark the class boundaries on the x-axis and the frequencies on the y-axis.* Note, extra classes at the lower and upper ends will be added, each having a frequency of 0. In our 3-D TV price example, these classes will be 1400–1499 at the lower end and 2000–2099 at the upper end.

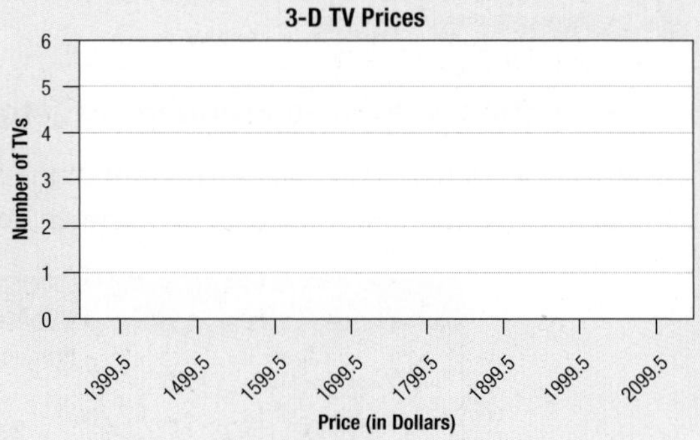

2. *Add the midpoints to the x-axis and plot a point at the frequency of each class directly above its midpoint.* Notice that the class boundaries on the *x*-axis have been lightened. This is due to the fact that frequency polygons represent the midpoints of the classes.

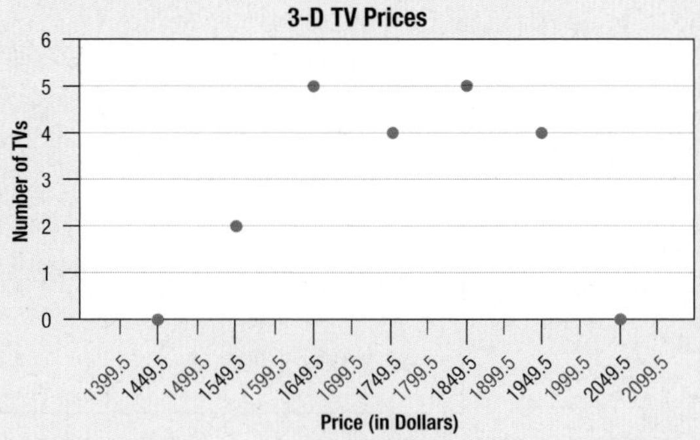

3. *Join each point to the next with a line segment.* Notice, this is not a smooth curve, but a polygon.

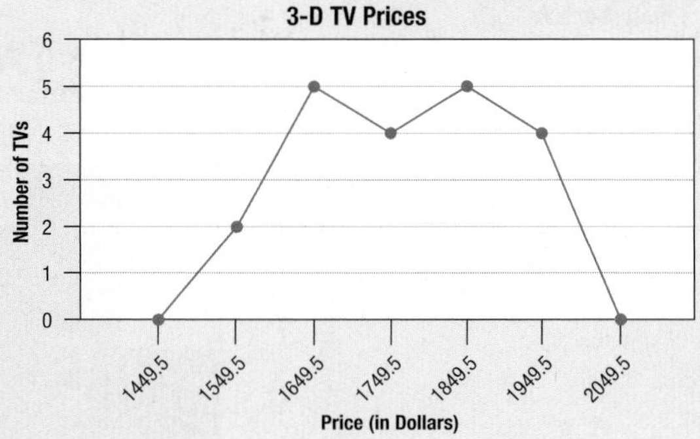

Ogives

An **ogive** (pronounced "oh-jive") is another type of graph that depicts information from a frequency distribution. An ogive shows the cumulative frequency of each class. Begin by tabulating the cumulative frequency for each class. Unlike when creating a frequency polygon, we only include an extra class at the lower end for this graph, giving it a frequency of 0. Next, plot a point at the cumulative frequency for each class directly above its *upper class boundary*. The ogive is created by joining the points together with line segments.

Example 2.14

Creating an Ogive

Below is the frequency distribution of the 3-D TV prices with the cumulative frequency column included. Create an ogive of the data.

3-D TV Prices			
Class	**Frequency**	**Cumulative Frequency**	**Class Boundaries**
\$1400–\$1499	0	0	1399.5–1499.5
\$1500–\$1599	2	2	1499.5–1599.5
\$1600–\$1699	5	7	1599.5–1699.5
\$1700–\$1799	4	11	1699.5–1799.5
\$1800–\$1899	5	16	1799.5–1899.5
\$1900–\$1999	4	20	1899.5–1999.5

Solution

Notice we have included the class 1400–1499 with a frequency of 0. The following is an ogive of the data.

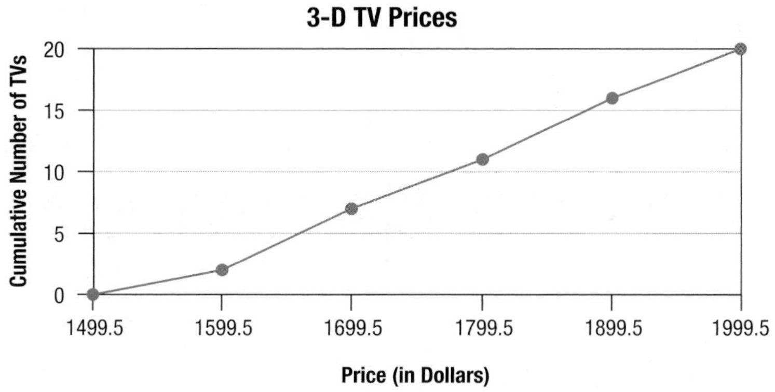

Stem-and-Leaf Plots

With many graphs, the original data are lost once the graph of the data is created, meaning that you cannot recreate the original data set using only the graph. However, one graph that retains the data is the **stem-and-leaf plot**. A stem-and-leaf plot is a graph of quantitative data that is similar to a histogram in the way that it visually displays the distribution. A stem-and-leaf plot is a quick way to depict the frequencies of the data without having to first create a frequency distribution. In fact, when turned on its side, this type of graph looks similar to a histogram.

In a stem-and-leaf plot, the leaves are usually the last digit in each data value and the stems are the remaining digits. For example, in the number 189, 9 is the last digit so it is the leaf, and the first part of the number, 18, is the stem. Be sure to include a legend for the stem-and-leaf plot, sometimes called a **key**, so that the reader can interpret the information. For example, 18 | 9 = 189.

Procedure

Constructing a Stem-and-Leaf Plot

1. *Create two columns, one on the left for stems and one on the right for leaves.*
2. *List each stem that occurs in the data set in numerical order.* Each stem is normally listed only once; however, the stems are sometimes listed two or more times if splitting the leaves would make the data set's features clearer.
3. *List each leaf next to its stem.* Each leaf will be listed as many times as it occurs in the original data set. There should be as many leaves as there are data values. Be sure to line up the leaves in straight columns so that the table is visually accurate.
4. *Create a key to guide interpretation of the stem-and-leaf plot.*
5. *If desired, put the leaves in numerical order to create an* ***ordered stem-and-leaf plot****.*

Example 2.15

Creating a Stem-and-Leaf Plot

Create a stem-and-leaf plot of the following ACT scores from a group of college freshmen.

ACT Scores				
18	23	24	31	19
27	26	22	32	18
35	27	29	24	20
18	17	21	25	26

Solution

ACT Scores

Stem	Leaves
1	8 9 8 8 7
2	3 4 7 6 2 7 9 4 0 1 5 6
3	1 2 5

Key: 1 | 8 = 18

The ordered stem-and-leaf plot from Example 2.15 is as follows.

ACT Scores

Stem	Leaves
1	7 8 8 8 9
2	0 1 2 3 4 4 5 6 6 7 7 9
3	1 2 5

Key: 1 | 8 = 18

Figure 2.4: Ordered Stem-and-Leaf Plot

Memory Booster

Although this is not a correct way to draw a stem-and-leaf plot, this figure will help you remember the concepts associated with a stem-and-leaf plot.

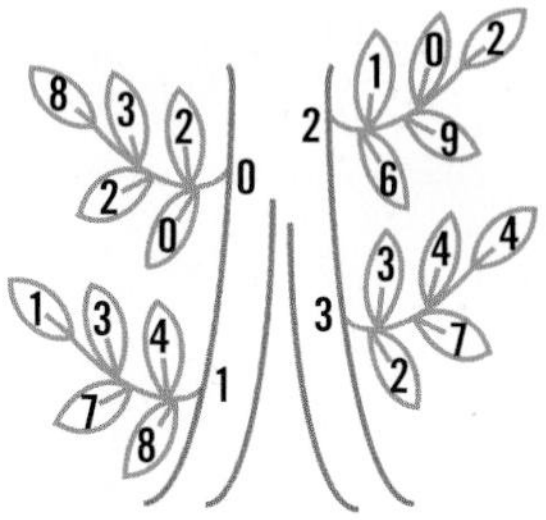

From this stem-and-leaf plot we can see that the majority of students scored in the 20s. But with only three categories, patterns may be difficult to see. We can divide the leaves into two groups, 0–4 and 5–9. To do this, write each stem *twice* instead of just once. Write any leaves from 0 to 4 with the first stem and any leaves 5 to 9 with the second stem. The ordered stem-and-leaf plot in Figure 2.4 would then appear as follows. Although no obvious pattern emerges in this example, splitting stems can be a helpful way to divide the data into more manageable chunks.

ACT Scores

Stem	Leaves
1	
1	7 8 8 8 9
2	0 1 2 3 4 4
2	5 6 6 7 7 9
3	1 2
3	5

Key: 1 | 8 = 18

Figure 2.5: Stem-and-Leaf Plot with Split Stems

Example 2.16

Creating and Interpreting a Stem-and-Leaf Plot

Create a stem-and-leaf plot for the following starting salaries for entry-level accountants at public accounting firms. Use the stem-and-leaf plot that you create to answer the following questions.

Starting Salaries for Entry-Level Accountants				
$51,500	$48,300	$40,900	$40,700	$48,200
$45,500	$42,500	$44,200	$46,400	$48,600
$45,800	$46,300	$50,000	$50,700	$44,300
$43,000	$42,700	$49,000	$46,700	$43,200
$42,900	$46,500	$47,700	$48,000	$46,300
$44,500	$47,900	$45,300	$46,100	$45,000

a. What were the smallest and largest salaries recorded?

b. Which salary appears the most often?

c. How many salaries were in the range $41,000–$41,900?

d. In which salary range did the most salaries lie: $40,000–$44,900, $45,000–$49,900, or $50,000 and above?

Solution

This example is different than Example 2.15 in that the data have more than two digits and every data value ends in two zeros. It is important to choose the stems of the numbers so that the last significant digit in each data point will be the leaf in the chart. Listing the zeros as the leaves for each different stem would be of little use to anyone interpreting the stem-and-leaf plot. Instead, the key will denote that the salaries listed are in hundreds. Using the first two digits of each salary as a stem, we have the following plot. (**Note:** It is helpful to first list the salaries in numerical order.)

Starting Salaries for Entry-Level Accountants

Stem	Leaves
40	7 9
41	
42	5 7 9
43	0 2
44	2 3 5
45	0 3 5 8
46	1 3 3 4 5 7
47	7 9
48	0 2 3 6
49	0
50	0 7
51	5

Key: 40 | 7 = $40,700

a. The smallest salary is $40,700; the largest salary is $51,500.

b. $46,300 appears twice, which is more than any other salary.

c. No salaries are in the range $41,000–$41,900 because there are no leaves listed for the stem 41.

d. By counting the leaves in the given groups, we see that there are more salaries in the range $45,000–$49,900.

As we saw in the last example, it is often the case that carefully choosing the stems will provide a clear picture for others to interpret the data without lots of excess digits. You can see how important the key is in situations like this. Consider for a moment what you would use for stems if all the data had decimals. Would a stem-and-leaf plot be appropriate with decimals? The answer is yes! There is no need to list the decimals in either the stems or leaves, but instead, treat the numbers as if they did not contain decimals, making sure to clearly note the decimal in the key, as shown in Figure 2.6.

US Retail Gas Prices from May 4, 2012

Stem	Leaves
32	0
33	7 7 7 8 9 9 9 9
34	0 0 0 0
35	4 8
36	0
37	9 9 9 9
38	4 5 5 5 5 5 6 9 9
39	9
40	0

Key: 32 | 0 = $3.20 per gallon

Figure 2.6: Stem-and-Leaf Plot with Decimals

Note that, although it may appear that the gas prices are hundreds of dollars per gallon, the key tells us that 32 | 0 is actually only $3.20 per gallon.

Dot Plots

As we mentioned before, stem-and-leaf plots are desirable because they retain the original data in the chart. A **dot plot** is similar to a stem-and-leaf plot in that respect. However, a dot plot displays the data without grouping certain points together like the stem-and-leaf plot does. Instead, only data that are exactly the same appear together. Therefore, it is useful for identifying extreme values as well as clusters in data sets. Because a dot plot is just what its name suggests—a dot representing each data value on a number line—it provides an easy way to visually spot data trends. Of course, various forms of statistical technology will make dot plots, but to construct them by hand, "dots" are stacked on top of each other at the appropriate places along a number line. Here's an example of what one looks like.

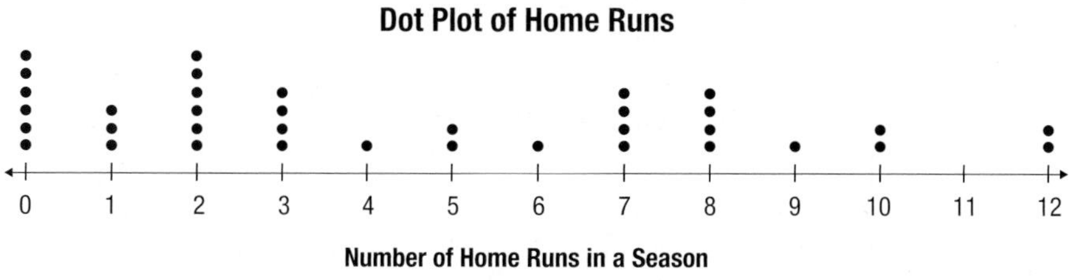

Figure 2.7: Dot Plot

Line Graphs

All of the previous graphs depict data at one point in time. However, not all data come from a single moment in time. As we learned in Chapter 1, some studies gather measurements over a period of time. When your data are measurements over time, it is best to display the data using a **line graph**. In a line graph, the horizontal axis represents time. The vertical axis represents the variable being measured. A line graph is constructed by plotting each value of the measurement above the time it was taken, and then joining the data points in order with line segments. It is these line segments that give the graph its name.

Example 2.17

Constructing a Line Graph

The Consumer Price Index (CPI) is a measure of the average change in value over time for a basket of goods and services. It is an index calculated by the US Bureau of Labor and Statistics. The table below shows the values of the CPI from several years. Construct a line graph of these data.

Consumer Price Index

Year	CPI
1920	20.0
1930	16.7
1940	14.0
1950	24.1
1960	29.6
1970	38.8
1980	82.4
1990	130.7
2000	172.2
2010	218.1

Source: US Bureau of Labor Statistics. "Consumer Price Index History Table." 19 Jan. 2012. ftp://ftp.bls.gov/pub/special.requests/cpi/cpiai.txt (24 Jan. 2012).

Solution

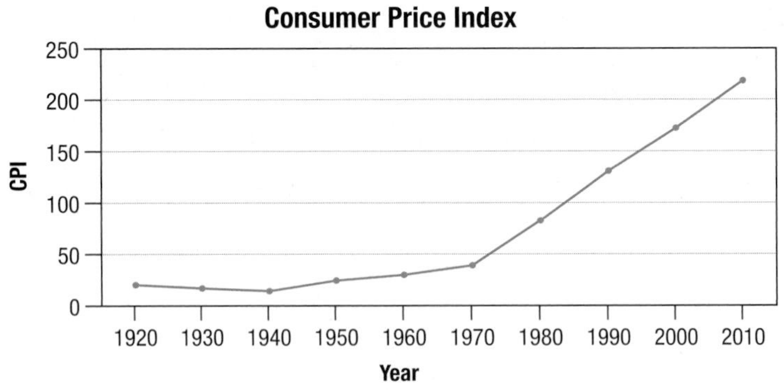

Summary

The following table is a summary of the types of graphs we have discussed and when to use them.

	Type of Graph		Definition
Qualitative Data	**Pie Chart**	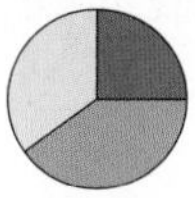	A pie chart shows how large each category is in relation to the whole; that is, it uses the relative frequencies from the frequency distribution to divide the "pie" into different-sized wedges. It can only be used to display qualitative data.
Qualitative Data	**Bar Graph**	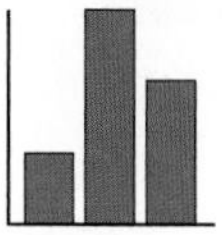	In a bar graph, bars are used to represent the amount of data in each category; one axis displays the categories of qualitative data and the other axis displays the frequencies.
Qualitative Data	**Pareto Chart**	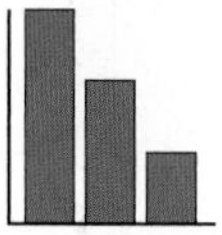	A Pareto chart is a bar graph with the bars in descending order of frequency. Pareto charts are typically used with nominal data.
Qualitative Data	**Side-by-Side Bar Graph**	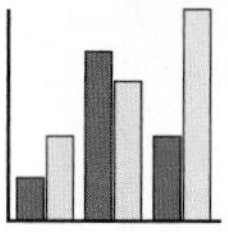	A side-by-side bar graph is a bar graph that compares the same categories for different groups.
Qualitative Data	**Stacked Bar Graph**	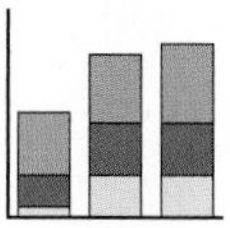	A stacked bar graph is a bar graph that compares the same categories for different groups and shows category totals.
Quantitative Data	**Histogram**	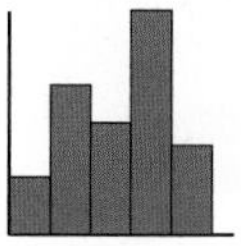	A histogram is a bar graph of a frequency distribution of quantitative data; the horizontal axis is a number line.
Quantitative Data	**Frequency Polygon**	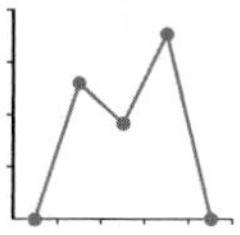	A frequency polygon is a visual display of the frequency of each class of quantitative data that uses straight lines to connect points plotted above the class midpoints.
Quantitative Data	**Ogive**	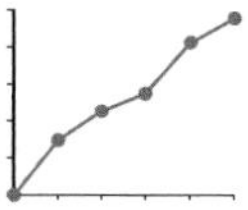	An ogive displays the cumulative frequency of each class of quantitative data by using straight lines to connect points plotted above the upper class boundaries.
Quantitative Data	**Stem-and-Leaf Plot**	Stem | Leaves 32 | 0 33 | 7 7 7 8 34 | 0 0 0 0	A stem-and-leaf plot retains the original data; the leaves are the last significant digit in each data value and the stems are the remaining digits.
Quantitative Data	**Dot Plot**	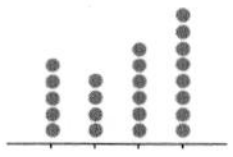	A dot plot retains the original data by plotting a dot above each data value on a number line.
Quantitative Data	**Line Graph**	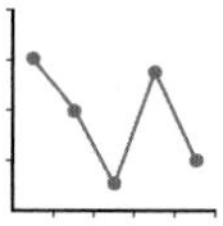	A line graph uses straight lines to connect points plotted at the value of each measurement above the time it was taken.

2.2 Section Exercises

Bar Graphs

Directions: Construct a bar graph for the given data. If it is appropriate, make the bar graph a Pareto chart.

1.

Math Grades on Test 1	
Grade	**Number of Students**
A	30
B	56
C	47
D or F	12

2.

2008 US Presidential Election	
Candidate	**Percentage of Popular Vote**
Obama	52.9%
McCain	45.5%
Nader	0.5%
Other	1.2%

Source: US National Archives and Records Administration: Office of the Federal Register. "2008 Presidential Election: Popular Vote Totals." http://www.archives.gov/federal-register/electoral-college/2008/popular-vote.html (24 Jan. 2012).

3.

Value-Added Tax	
Country	**Tax Percentage**
Spain	16%
Canada	7%
Norway	25%
Japan	5%
United Kingdom	17.5%

Source: Wikipedia contributors. "Value added tax." *Wikipedia, The Free Encyclopedia.* 24 Jan. 2012. http://en.wikipedia.org/wiki/Value_added_tax (24 Jan. 2012).

Directions: Construct a side-by-side bar graph and a stacked bar graph for the given data. Then answer the questions.

4.

Math Enrollment		
Math Class	**Freshmen**	**Sophomores**
Statistics	147	45
Algebra	160	73
Calculus	23	92
Quantitative Reasoning	12	120

a. Which course has the most students enrolled? Which graph did you use to answer this question?

b. Which course has the most sophomores enrolled? Which graph did you use to answer this question?

c. Which course has the most freshmen enrolled? Which graph did you use to answer this question?

d. Which course has the least students enrolled? Which graph did you use to answer this question?

5.

Pets Seen by the Vet in One Month		
Type of Animal	**Number Seen by Dr. Warren**	**Number Seen by Dr. Campbell**
Cats	47	59
Dogs	56	37
Reptiles	13	6
Birds	28	30

a. Which veterinarian saw more dogs this month? Which graph did you use to answer this question?

b. Which type of animal was seen the least this month? Which graph did you use to answer this question?

c. Which veterinarian saw more cats this month? Which graph did you use to answer this question?

d. Which animal did Dr. Warren see the least this month? Which graph did you use to answer this question?

Pie Charts

Directions: Calculate the central angle measures needed to construct a pie chart for the given data. Round your answers to the nearest whole degree.

6.

Attitude Toward Math	
Attitude	**Number of Students**
Love	23
Like	46
Indifferent	9
Hate	12

7.

College Majors	
Major	**Number of Students**
English	171
Business	569
Education	346
Science	285

Histograms

Directions: Do the following for each frequency distribution.

a. Construct a histogram.

b. Calculate the relative frequency for each class.

c. Construct a relative frequency histogram.

8.

Ages of Taste-Test Participants (in Years)	
Class	**Frequency**
15–19	7
20–24	8
25–29	10
30–34	2
35–39	3

9.

Braking Times for Vehicles at 60 mph (in Minutes)	
Class	**Frequency**
0.05–0.07	12
0.08–0.10	15
0.11–0.13	14
0.14–0.16	15
0.17–0.19	14

10.

Age at Time of First Marriage (in Years)	
Class	**Frequency**
15–18	2
19–22	5
23–26	4
27–30	5
31–34	4

11.

Hourly Wage at First Job (in Dollars)	
Class	**Frequency**
7.50–8.49	12
8.50–9.49	50
9.50–10.49	48
10.50–11.49	45
11.50–12.49	34

Frequency Polygons

12. Use the data from Exercise 8 to create a frequency polygon.

13. Use the data from Exercise 9 to create a frequency polygon.

Ogives

14. Use the data from Exercise 10 to create an ogive.

15. Use the data from Exercise 11 to create an ogive.

Stem-and-Leaf Plots

Directions: Create a stem-and-leaf plot for the given data.

16. The following data represent the numbers of curl-ups completed in 60 seconds for a group of sixteen 8-year-old boys.

31	34	41	36	27	29	18	33
31	28	34	22	26	28	36	42

17. The following data represent the caloric intakes in one day for a group of fifteen men between the ages of 20 and 39.

2700	2200	2500	2800	2600
3000	2600	2200	3100	2800
1800	3500	2500	3000	2900

Directions: Create an ordered stem-and-leaf plot for the given data.

18. The following data represent times in minutes for completing a one mile run from a group of twenty-four 17-year-old girls.

12.4	12.3	11.1	11.9	12.1	9.5	11.6	10.8
10.2	9.3	10.1	11.2	8.2	9.3	9.5	12.5
9.4	9.7	10.7	10.9	9.3	10.4	12.9	10.6

19. The following data represent the precipitation totals in inches for the month of September in 21 different towns in Alaska.

2.73	2.81	2.54	2.59	2.70	2.88	2.64
2.55	2.86	2.68	2.77	2.61	2.56	2.62
2.78	2.64	2.50	2.67	2.89	2.74	2.81

Dot Plots

Directions: Create a dot plot for the given data.

20. The following data represent the numbers of visitors, in thousands, which various online clothing retailers had on their websites during the month of February.

13	11	19	15	11	11	17
3	10	14	14	9	12	15
12	16	12	13	15	12	10

21. The following data represent the numbers of plastic shopping bags that customers used in a single shopping trip at a local grocery store one afternoon.

3	2	5	5	7	7	3	10	2	3	9	2	6	3	4
2	1	3	5	1	5	0	4	2	8	2	6	2	1	6
2	8	10	4	5	3	8	0	3	2	2	7	11	1	3
4	4	3	10	4	2	6	14	5	3	4	4	10	3	1
3	19	1	15	2	4	1	1	4	4	7	3	2	1	1
4	8	8	6	8	3	1	0	30	1	4	1	11	5	6
19	1	9	0	5	1	5	7	5	1	9	2	10	5	8
3	3	4	4	5	2	2	4	9	1	2	2	0	2	7
7	13	3	1	7										

Directions: Use the given dot plot to answer the questions.

22. A local fast food restaurant collected the following data while checking the weights of their quarter-pound burgers, which should be approximately 4 ounces.

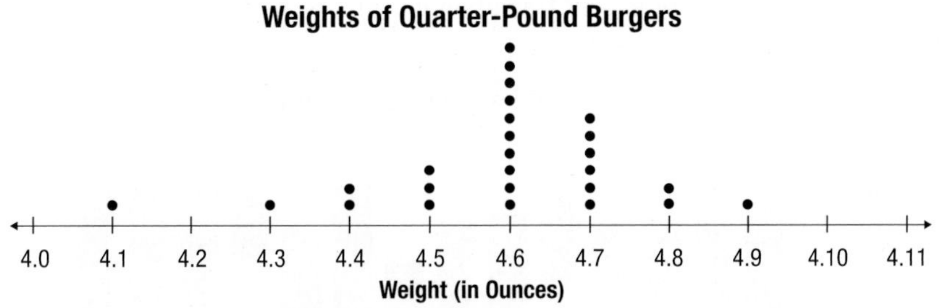

a. How many burgers were checked for weight?

b. What was the most common weight?

c. If you were describing the average weight for this sample rounded to the nearest whole ounce, would you say 4 ounces or 5 ounces? Explain your reasoning.

d. If you were a customer at this restaurant, would you feel satisfied that you were always getting your money's worth for a quarter-pound burger? Why or why not?

23. A quality control manager for a manufacturer of household products collected the following data while measuring the diameters of stainless steel cylinders that the company uses to make large trash cans.

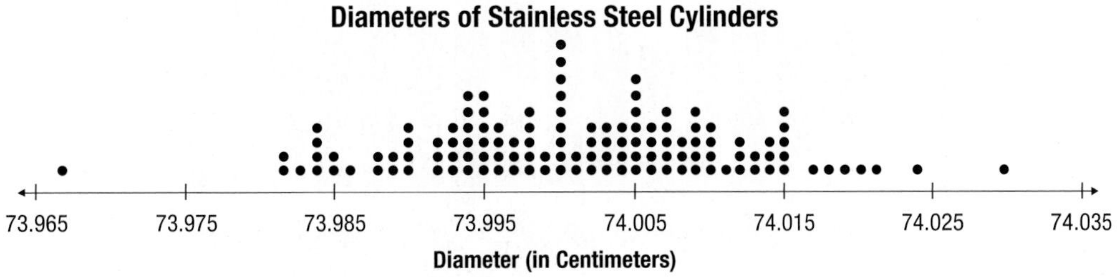

a. Which diameter is the most common?

b. What is the largest diameter in the data?

c. Discuss the pros and cons of displaying this data set in this type of dot plot, keeping in mind any difficulties you had in answering part b.

Interpreting Graphs

Directions: **Use the given graphs to answer the questions.**

24. Suppose the graphs below represent Canadian gas prices in 2005 (adjusted to US dollars per gallon from Canadian dollars per liter).

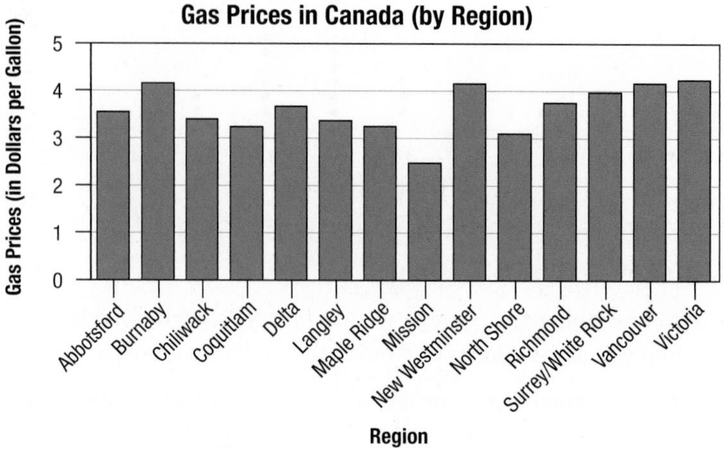

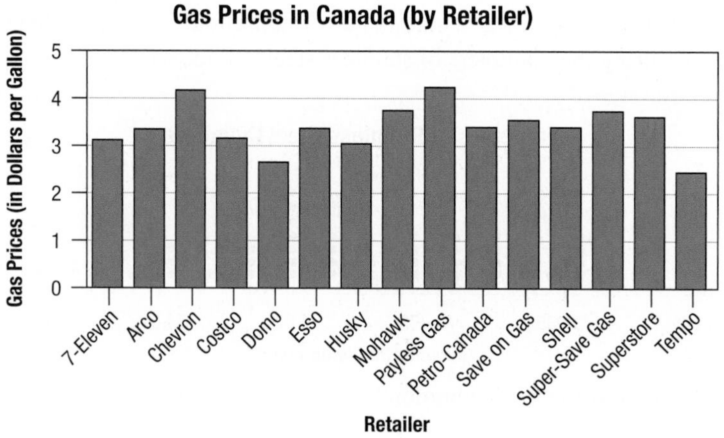

a. In what region(s) was gas the most expensive?

b. What retailer(s) offered the cheapest price for gas?

c. What was the cheapest price for gas in any given region?

d. What was the most expensive price any retailer charged for gas?

e. Would a Pareto chart be an appropriate way to display the data? Why or why not?

25. Below is a line graph depicting the spot and retail prices for regular gasoline in the United States for the year 2010.

Source: US Energy Information Administration. "Petroleum & Other Liquids - Data." 19 Jan. 2012. http://www.eia.gov/petroleum/data.cfm (24 Jan. 2012).

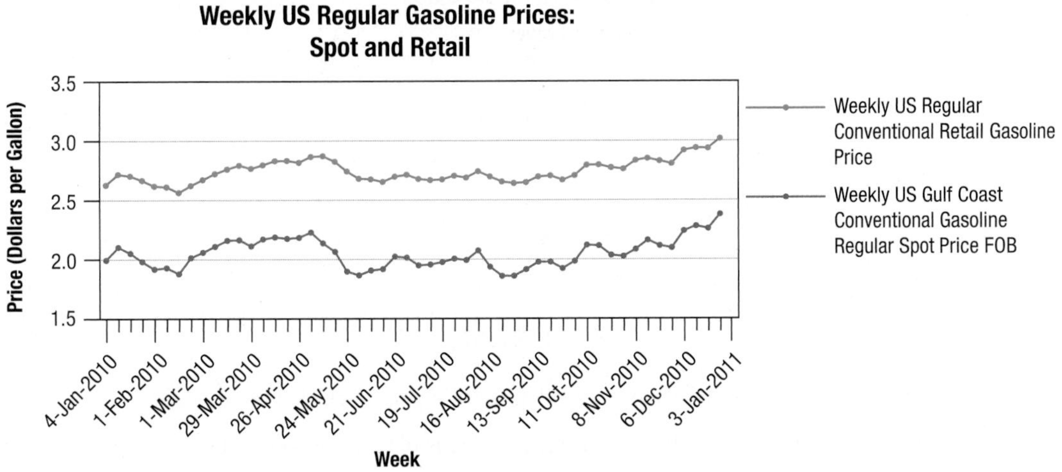

a. Approximately, what was the highest spot price for gasoline?

b. Approximately, what was the lowest retail price for gasoline?

c. Approximately, when did the lowest spot price for gasoline occur?

d. Approximately, when did the highest retail price for gasoline occur?

26. The following pie chart depicts Harrison's monthly budget.

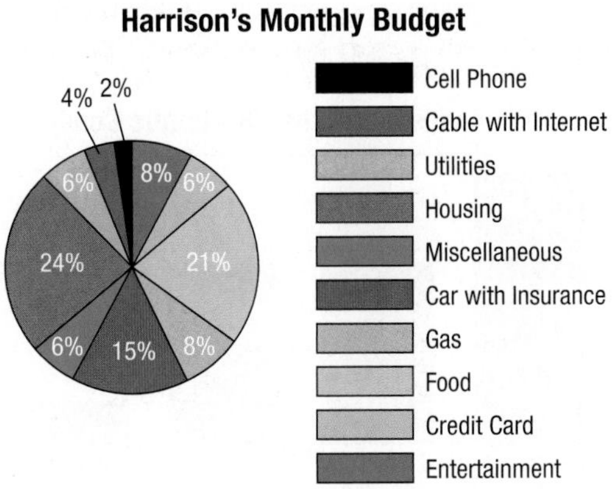

a. What does Harrison spend the most money on each month?

b. What does Harrison spend the least on each month?

c. What percentage of Harrison's budget is spent on household bills? (Housing, Utilities, Cable with Internet, and Cell Phone)

d. What percentage of his monthly budget does Harrison spend on his car? (Car with Insurance and Gas)

e. Disposable income is the money remaining after all essentials have been paid. What percentage of Harrison's budget is disposable income? (Entertainment, Cable with Internet, Miscellaneous, Cell Phone)

27. Below is a stem-and-leaf plot of ACT math scores for a group of college music students.

ACT Math Scores of Students in a College Music Class

Stem	Leaves
1	2 3 3 4
1	5 7 8 8 8 8 9 9
2	0 0 0 1 1 2 2 2 3
2	5 5 5 6 7
3	2 3 4
3	

Key: 1 | 2 = 12

a. What was the lowest math score a student in this class received on the ACT?

b. What was the highest math score a student in this class received on the ACT?

c. Which math score occurred most often?

d. How many students are represented by this information?

28. Meteorologists categorize hurricanes according to their maximum sustained wind speed using the Saffir-Simpson scale. This scale divides hurricanes into five categories, with Category 1 hurricanes having maximum sustained winds between 74 and 95 miles per hour (mph). The bar graph below depicts the number of hurricanes along the Atlantic coast of the United States from 1851 to 2010. Each bar in the graph represents one of the categories used to classify hurricanes.

Source: National Weather Service: National Hurricane Center. "Best Track Data (HURDAT): Atlantic Tracks File 1851–2010." 18 Aug. 2011. http://www.nhc.noaa.gov/data/hurdat/tracks1851to2010_atl_2011rev.txt (24 Jan. 2012).

Hurricanes along the US Atlantic Coast

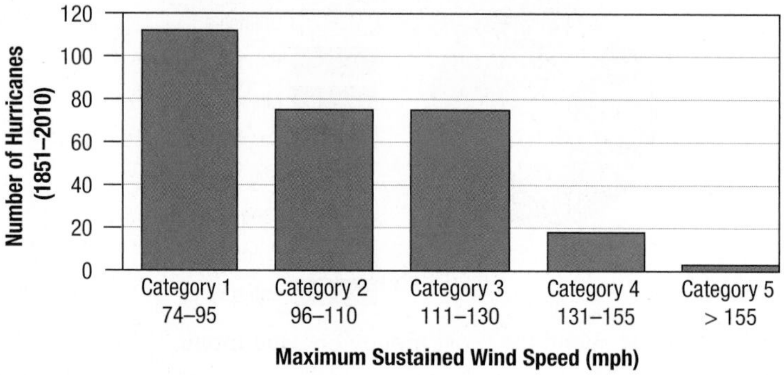

a. How many Category 3 hurricanes, with wind speeds of 111 to 130 mph, have hit the US Atlantic coast between 1851 and 2010?

b. A major hurricane is considered a hurricane in categories 3, 4, or 5 (that is, wind speeds greater than 111 mph). How many major hurricanes have hit the US Atlantic coast between 1851 and 2010?

c. What is the total number of hurricanes to hit the US Atlantic coast between 1851 and 2010?

29. Consider the following stem-and-leaf plot from a study published in an academic journal, which displays the measurements of the anterior radius of the left otolith for female stone flounder obtained from commercial capture in Tokyo Bay. The numbers to the left of the stems in this graph represent the cumulative frequencies of the data to the closest end. The middle stem is denoted by parentheses around the frequency of data in that category.

Source: Salgado Ugarte, Isaías H. "Exploratory Analysis of Some Measures of the Asymmetric Otoliths of Stone Flounder *Kareius bicoloratus* (Pisces: Pleuronectidae) in Tokyo Bay." *Anales del Instituto de Ciencias del Mar y Limnología*, Vol. 18, 1991 No. 2.

Anterior Radius of the Left Otolith for Female Stone Flounder

Unit = 0.01   10 | 2 represents 1.02 mm

N = 85

1	18	2
	19	
5	20	0 5 6 8
	21	
8	22	1 3 8
11	23	2 6 7
15	24	2 3 4 6
21	25	0 4 4 5 6 7
27	26	0 2 2 7 8 8
32	27	1 2 4 8 8
36	28	3 6 8 9
(7)	29	0 4 7 7 9 9 9
42	30	0 1 2 2 3 4 5 6 6 8 8 8 9
29	31	0 0 2 4 5 5 7 8 9
20	32	0 0 2 4 5 5 7
13	33	3 3 5 6 6 9
7	34	0 3
5	35	3 6
3	36	1 9
1	37	6

a. How many fish were in the study?

b. What were the smallest and largest measurements recorded?

c. Which length(s) appeared most often?

d. Did more fish have an otolith radius shorter or longer than the middle length category?

30. Consider the pie chart below, which shows the highest level of education attained by Americans age 30 and over in 2007. Assume that people who earn master's, professional, and doctoral degrees must first earn a bachelor's degree, and people who earn a bachelor's degree must first complete high school.

Source: US Department of Commerce, Census Bureau. "Current Population Survey (CPS)." March 2007.

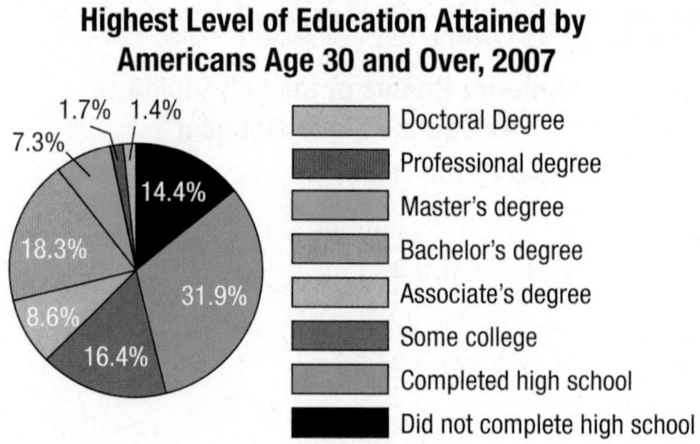

a. What percentage of all the people surveyed had earned only a bachelor's degree?

b. What percentage of respondents had completed some college but not a bachelor's degree?

c. What percentage of respondents had completed a degree higher than a bachelor's degree?

d. What percentage of respondents had not completed a bachelor's degree?

e. What percentage of respondents with bachelor's degrees had earned professional degrees?

31. For a school project, you need to graphically display data that you have collected. Determine the best way to display the following data. Explain your choice.

Number of Social Network Friends for Adults Over the Age of 40											
213	583	114	317	80	497	524	352	126	627	250	710
528	438	347	944	721	753	302	349	101	812	74	849
314	841	411	160	174	439	435	569	323	444	529	556

2.3 Analyzing Graphs

Let's now consider various ways to analyze graphs. Graphs can come in any number of forms, a few of which we looked at in depth in the previous section. But how should a reader approach a graph or chart that is shown in the media? What types of information should we expect to find? In what ways might we be misled? Here are a few pointers and key ideas to pay attention to when interpreting information presented in a graph.

Title, Axes, Source

First and foremost, every graph should be properly labeled. The title of the graph should be clearly displayed and should succinctly tell what type of information is being given. For instance, in the graph shown below, the title, "Generational Divide Over American Exceptionalism," quickly and accurately tells what is going on in the graph. Also notice that each of the bars shown in this bar graph is labeled both according to generation name and percentage. The title subheading explains that the percentages represent "% saying U.S. is 'the greatest country in the world.'"

Next, it is of utmost importance to consider the source of the information. What good is the best of graphs if it is not backed up by a reliable source? This is where we, as busy consumers, should not be in so much of a hurry that we forget to pay attention to who is telling the story. It certainly makes a difference whether the information comes from a reputable news outlet or simply the high school newspaper, doesn't it? In the case of our chosen example, the source is the Pew Research Center, which is a reputable organization.

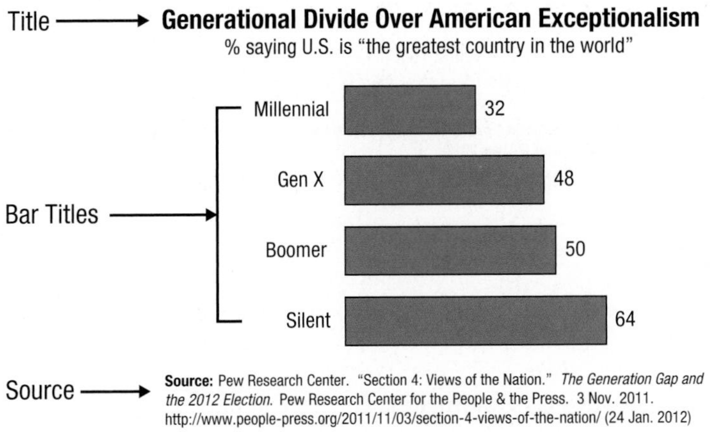

Figure 2.8: Properly Labeled Bar Graph

Appropriateness of the Graph

In Section 2.2 we looked at a number of different types of graphs. What we want to consider now is the *appropriateness* of the graph. That is, we want to be able to determine whether the type of graph is correct for the data being displayed. For instance, let's contrast the different uses of line graphs and cross-sectional graphs. A line graph should only be used to display a variable whose values change over time. Because time is continuous in nature, it makes sense to "connect the dots" and form the trademark line segments of a line graph. This type of graph is sometimes referred to as a **time-series graph**. A **cross-sectional graph**, on the other hand, is just the opposite. A cross-sectional graph displays information collected at only one point in time. In Figures 2.9 and 2.10 you will see correct representations of both a time-series graph and cross-sectional graph.

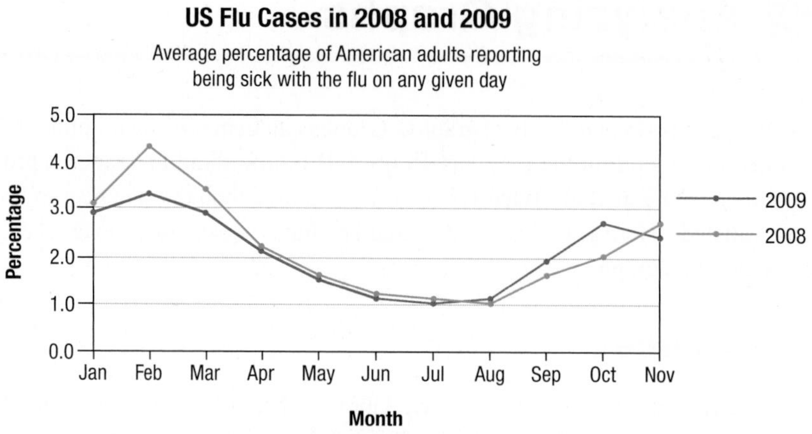

Source: Witters, Dan. "U.S. Flu and Cold Cases Decline in November." Gallup Wellbeing. 18 Dec. 2009. http://www.gallup.com/poll/124748/Flu-Cold-Cases-Decline-November.aspx (24 Jan. 2012).

Figure 2.9: Time-Series Graph

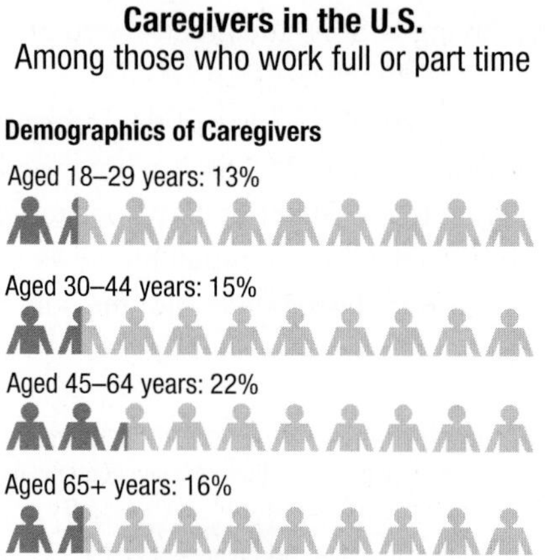

Source: Cynkar, Peter and Mendes, Elizabeth. "More Than One in Six American Workers Also Act as Caregivers." Gallup Wellbeing. 26 July 2011. http://www.gallup.com/poll/148640/One-Six-American-Workers-Act-Caregivers.aspx (24 Jan. 2012).

Figure 2.10: Cross-Sectional Graph

Definition

A **time-series graph** is a line graph that is used to display a variable whose values change over time.

A **cross-sectional graph** displays information collected at only one point in time.

Another often-used type of graph, which can be misleading, is the pictograph. A **pictograph** is a bar graph that uses pictures of objects instead of bars. These graphs tend to be visually appealing and simple to understand, so it's easy to see why they are frequently used in the media. For example, Figure 2.10 shows an accurate pictograph from a report by Gallup Wellbeing. However, you must be cautious in interpreting a pictograph because looks can be deceiving. The problem with a pictograph is that when the size of the pictured object is increased, the change is not simply one-dimensional. In Figure 2.11, the cost of housing doubled between 1994 and 2006. However, when both the length and width of the house portraying housing costs are doubled, the resulting area is not double but *four times* the original area. Just imagine what a large distortion there would be if the houses were drawn in 3-D rather than 2-D! For this reason, it is essential that the artist be aware of the overall area that is being displayed in order for the pictograph to be truly accurate.

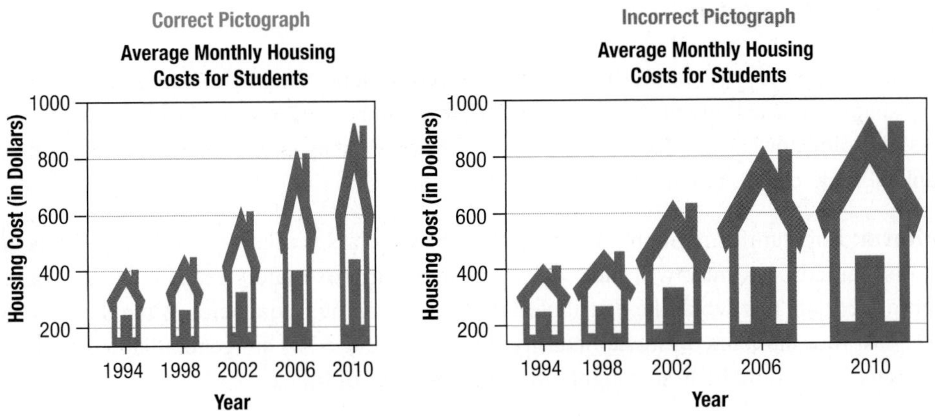

Figure 2.11: Pictographs

Scaling of Graphs

Another important feature to look out for is that the graph is scaled properly. If you stretch or shrink the scale on the y-axis, the shape of the graph may change dramatically. A line that rises gently on one scale might look very steep with a different scale. When analyzing a graph, check to make sure that the scale represents the data well. If there are large differences between data, then the graph should accurately reflect those differences. On the other hand, if the differences between the data values are small, then the graph should reflect this as well.

Example 2.18

Scaling of Graphs

Consider the graph below on US federal minimum hourly wage rates, unadjusted for inflation. What errors can you find in the graph? How should they be fixed?

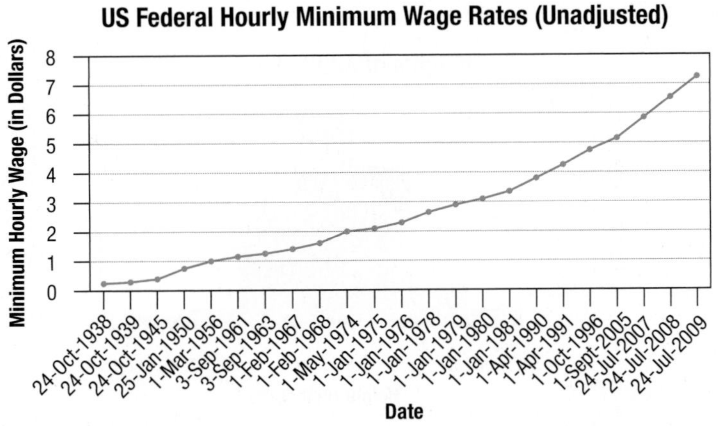

Source: US Department of Labor, Wage and Hour Division (WHD). "History of Federal Minimum Wage Rates Under the Fair Labor Standards Act, 1938-2009." http://www.dol.gov/whd/minwage/chart.htm (24 Jan. 2012).

Solution

Notice that the x-axis does not have a consistent scale. The years are as few as one year apart and as many as nine years apart, so the shape of the graph is distorted. To correct this graph, the x-axis needs to be changed to use a consistent scale. The corrected graph can be found in Exercise 10 in the Chapter 2 Exercises.

Shapes of Graphs

It is sometimes helpful to also be able to classify the overall shape of a graph. Here are a few commonly occurring shapes of distributions. Note that these shapes can be used to describe histograms, frequency polygons, dot plots, or bar graphs of *ordinal* data. However, they cannot be used to describe ogives, line graphs, Pareto charts, or bar graphs of *nominal* data.

1. **Uniform:** For a uniform graph, the frequency of each class is relatively the same. Data sets that are distributed uniformly will have graphs that are approximately rectangular in shape. One example of data that would be distributed uniformly are the frequencies of the digits (0–9) used in the last four numbers of Social Security IDs. Since these IDs are generated randomly, each of the digits 0–9 should be used roughly the same number of times.

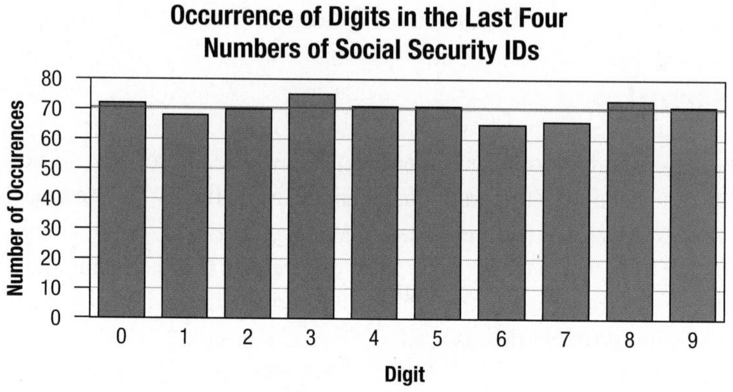

Figure 2.12: Uniform

2. **Symmetric:** For a symmetric graph, the data lie evenly on both sides of the distribution. That is, a vertical line drawn through the center of the graph would give mirror images of the data on either side. An example of data distributed symmetrically would be heights of adult men. Note that although the sketch below shows a bell-shaped distribution, a symmetric graph does not have to be bell-shaped in order to be symmetric.

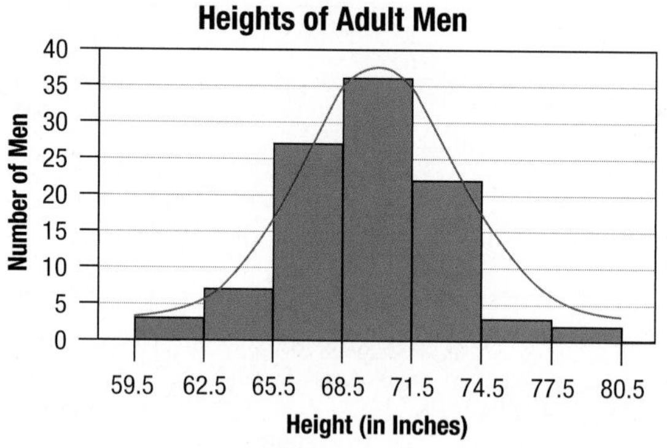

Figure 2.13: Symmetric

Memory Booster

Skewed distributions are shaped like a whale, and they are skewed to the whale's tail!

3. **Skewed to the Right:** We say that a graph is skewed to the right when the majority of the data fall on the left side of the distribution. The "tail" of the distribution, so to speak, is on the right. When you see a graph that is skewed to the right, this may indicate that there is an outlier on the right side of the graph. An **outlier** is a data value that falls outside the shape of the distribution. In this case the outlier is larger than the rest of the values and is "skewing" the data towards itself. For example, suppose that we graphed a group of test scores for which everyone in the class scored really low except for one student that made a 100. Everyone in the class would say that this student "blew the curve," wouldn't they? Indeed, the graph created for this scenario would be skewed to the right, pulled towards that high grade of 100.

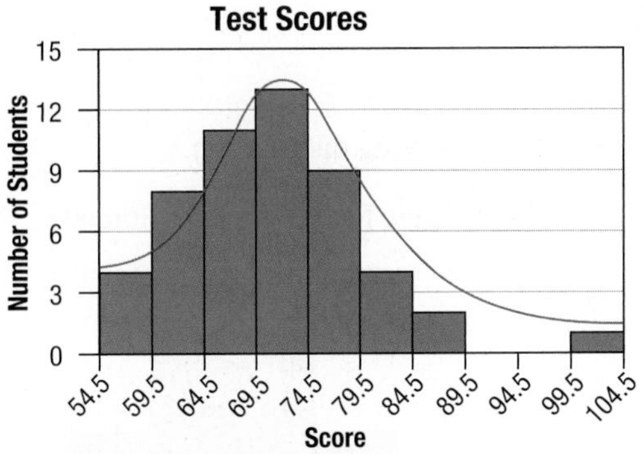

Figure 2.14: Skewed to the Right

4. **Skewed to the Left:** We say that a graph is skewed to the left when the majority of the data fall on the right side of the distribution. This is just the opposite of the previous example. So then, a graph that is skewed to the left may indicate that there is an outlier on the left side of the graph, that is, a number smaller than the rest of the data values. An example of a graph that is skewed to the left might be the graph of finishing times if an Olympic runner participated in his or her hometown July 4th 10K race one summer. A graph of finishing times would likely show one runner with a much faster (so smaller) finishing time than everyone else.

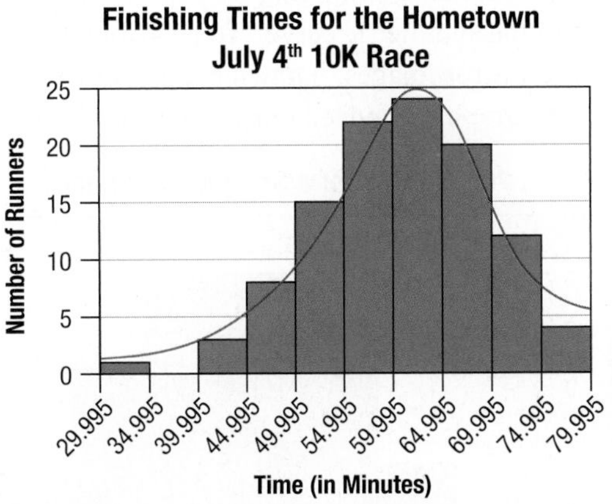

Figure 2.15: Skewed to the Left

Example 2.19

Shapes of Graphs

Describe the overall shape of the following distribution.

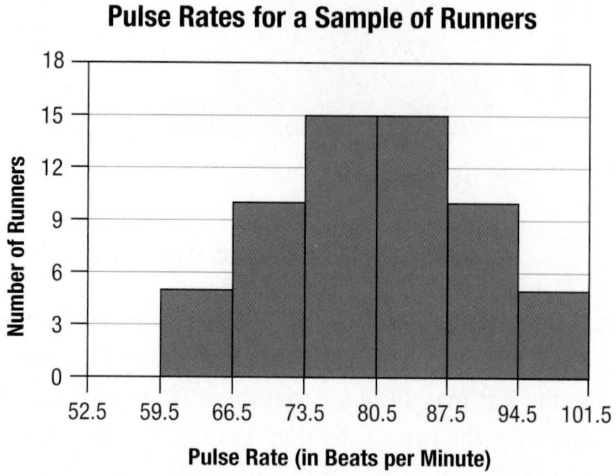

Solution

Notice that if we draw a smooth curve skimming the top of the histogram, we begin to see a curve similar to the shape of the symmetric curve. To be symmetric, the left and right sides of the graph should be close to mirror images. Drawing a line down the center of the graph, we can see that both sides of the graph are indeed mirror images of each other.

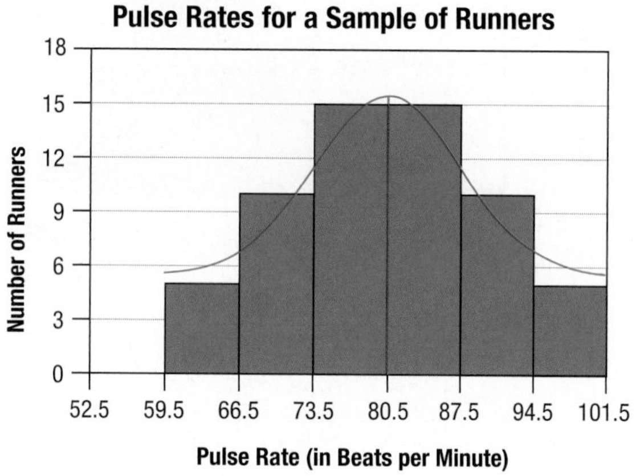

Thus, this histogram has a symmetric shape.

2.3 Section Exercises

Time-Series vs. Cross-Sectional Graph

Directions: For each graph described below, decide whether it is a time-series or a cross-sectional graph.

1. A map of Louisiana has the individual parishes shaded to represent the average income level for each parish.

2. A line graph depicts the change in the minimum wage since 1965.

3. A bar graph shows the average tax rates for different countries around the world.

4. A stem-and-leaf plot displays the number of home runs hit by each player on the Yankees ball club.

5. A bar graph displays the average SAT scores each year for the last five years.

Analyzing Graphs

Directions: Use the given graphs to answer the questions.

6. Consider the following pictograph used to display the increase in funds donated to one university's scholarship program.

   a. By what percentage did funds to the university scholarship program increase between 2005 and 2010?

   b. Does the graph shown accurately depict the change in scholarship funds? Explain your answer.

   c. What changes could be made to better display the given information?

7. Does the scale on the following graph depict the situation accurately? Why or why not?

Source: US Energy Information Administration. "Petroleum & Other Liquids - Data." 19 Jan. 2012. http://www.eia.gov/petroleum/data.cfm (24 Jan. 2012).

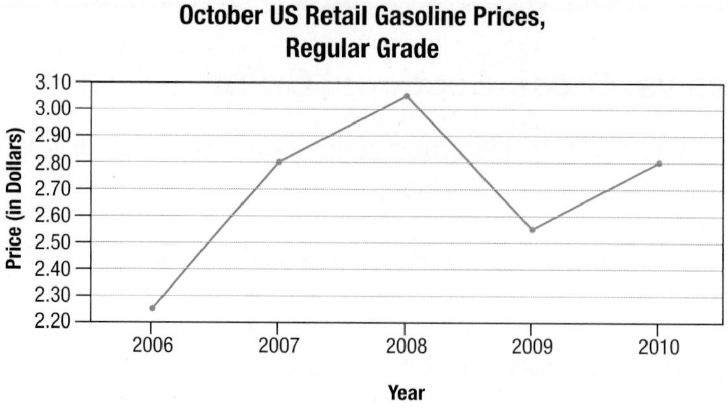

8. Look at the two graphs shown below depicting the *same* data on people's overall satisfaction level with the care they received at their local hospital. Which of these two graphs shows the more accurate picture of hospital satisfaction? Has hospital satisfaction increased? Are people satisfied with the care at their local hospital according to these graphs? How do you know?

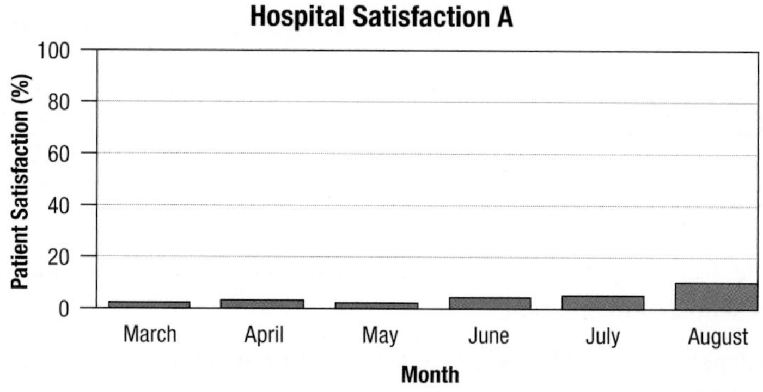

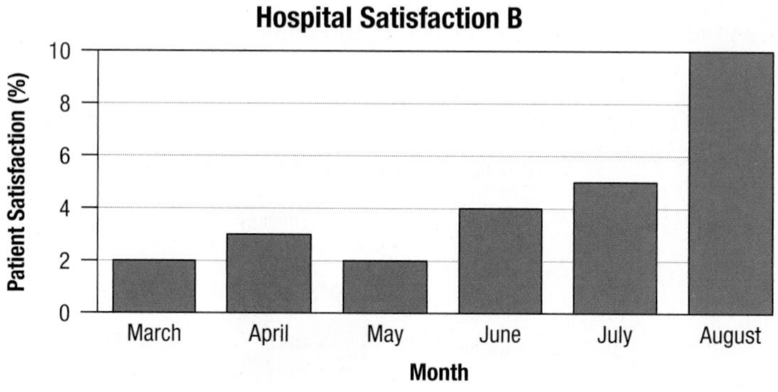

9. What errors occur in the following histogram?

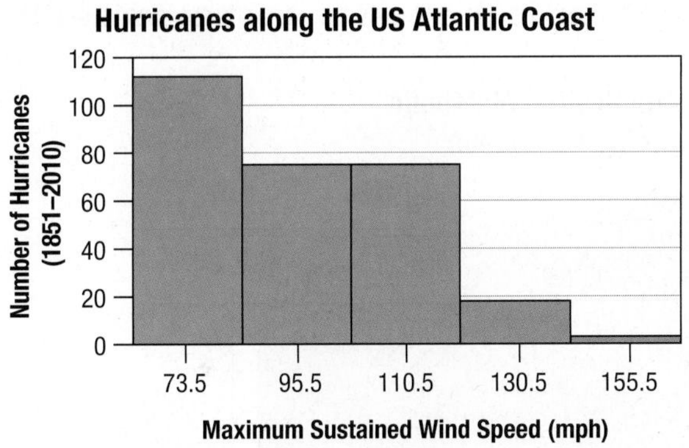

10. Consider the following excerpt from an online publication. Is the graph correctly labeled? If not, identify the corrections needed.

Source: NHHealthCost.org. "Health Costs for Consumers - Methodology." 16 Feb. 2007. http://www.nhhealthcost.org/method.aspx (24 Jan. 2012).

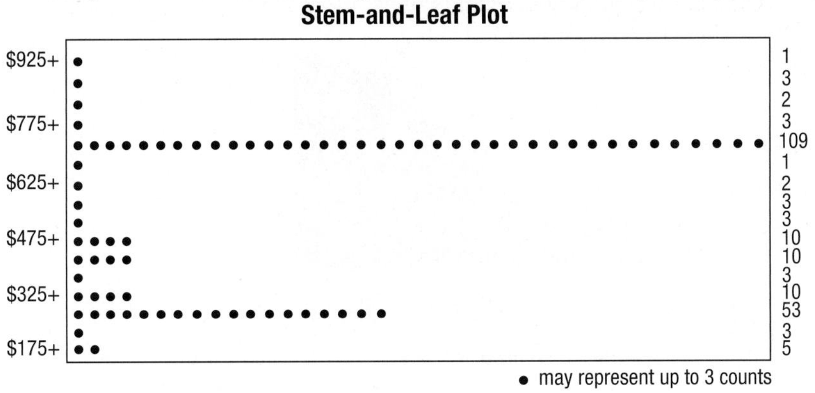

Shapes of Graphs

Directions: For each set of data described below, discuss the most likely shape of its distribution.

11. The weights of the defensive linemen on football teams in the Big Ten Conference.

12. The math scores on the ACT of a large randomly selected group of high school seniors.

13. The lengths of the pregnancies of a group of gorillas being studied in the wild.

14. The last four digits used to generate telephone numbers.

15. The income levels of a group of professional baseball players.

Directions: For each set of data displayed below, choose which of the four shapes defined in this section best describes the distribution.

16.

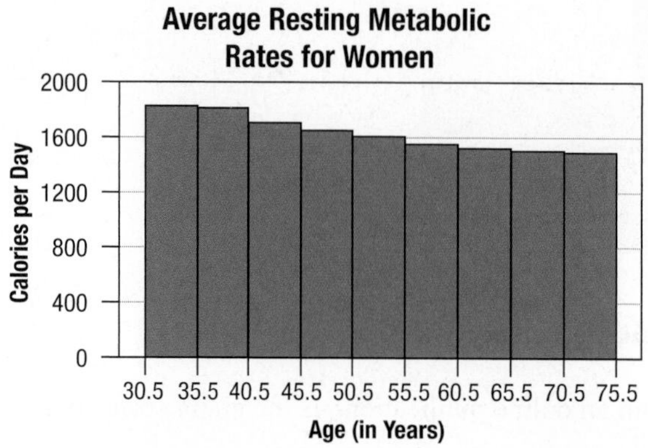

17.

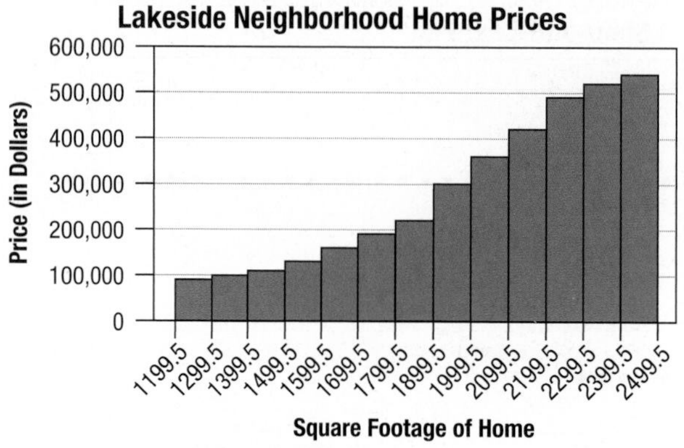

18.

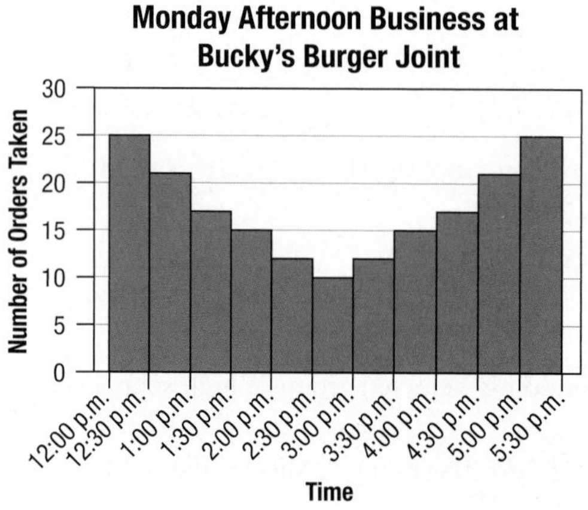

19.

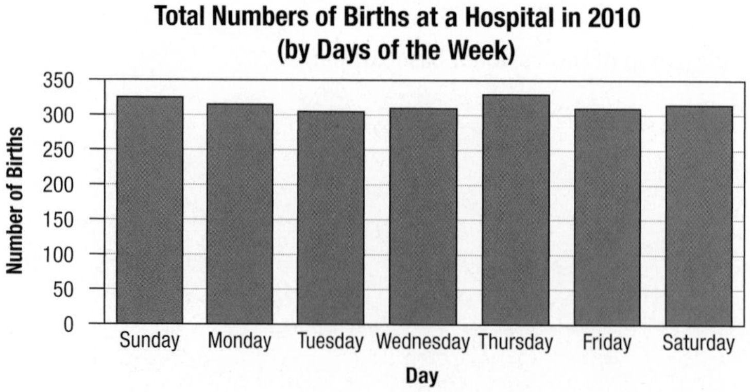

R Chapter 2 Review

Section 2.1: Frequency Distributions

Definitions

Ordered array
An ordered list of data from largest to smallest or vice versa

Distribution
A way to describe the structure of a particular data set or population

Frequency distribution
A display of the values that occur in a data set and how often each value, or range of values, occurs

Probability distribution
A theoretical distribution used to predict the probabilities of particular data values occurring in a population

Ungrouped frequency distribution
A frequency distribution in which each category represents a single value

Frequency, *f*
The number of data values in a category of a frequency distribution

Class
A category of data in a frequency distribution

Grouped frequency distribution
A frequency distribution in which the classes are ranges of possible values

Class width
The difference between the lower limits or upper limits of two consecutive classes of a frequency distribution

Lower class limit
The smallest number that can belong to a particular class

Upper class limit
The largest number that can belong to a particular class

Sample size, *n*
The sum of all the frequencies in a frequency distribution

Constructing a Frequency Distribution

1. Decide how many classes should be in the distribution.
2. Choose an appropriate class width.
3. Find the class limits.
4. Determine the frequency of each class.

Section 2.1: Frequency Distributions (cont.)

Characteristics of a Frequency Distribution

1. **Class boundary:** The value that lies halfway between the upper limit of one class and the lower limit of the next class
2. **Midpoint:** $\text{Class Midpoint} = \dfrac{\text{Lower Limit} + \text{Upper Limit}}{2}$
3. **Relative frequency:** The fraction or percentage of the data set that falls into a particular class, given by

$$\text{Relative Frequency} = \frac{f}{n}$$

4. **Cumulative frequency:** The sum of the frequencies of a given class and all previous classes

Section 2.2: Graphical Displays of Data

Definitions

Pie chart

Shows how large each category of qualitative data is in relation to the whole; uses relative frequencies to divide the "pie" into wedges

Bar graph

Bars are used to represent the amount of data in each catagory; one axis displays the categories of qualitative data and the other axis displays the frequencies

Pareto chart

A bar graph with the bars in descending order of frequency; typically used with nominal data

Side-by-side bar graph

A bar graph that compares the same categories for different groups

Stacked bar graph

A bar graph that compares the same categories for different groups and shows category totals

Histogram

A bar graph of a frequency distribution of quantitative data; the horizontal axis is a number line

- **Frequency histogram:** A histogram in which the heights of the bars represent frequencies
- **Relative frequency histogram:** A histogram in which the heights of the bars represent relative frequencies in either decimals or percentages

Frequency polygon

A visual display of the frequency of each class of quantitative data that uses straight lines to connect points plotted above the class midpoints

Ogive

A graph that displays the cumulative frequency of each class of quantitative data by using straight lines to connect points plotted above the upper class boundaries

Stem-and-leaf plot

Retains the original data; the leaves are the last significant digit in each data value and the stems are the remaining digits

Section 2.2: Graphical Displays of Data (cont.)

Dot plot

Retains the original data by plotting a dot above each data value on a number line

Line graph

Uses straight lines to connect points plotted at the value of each measurement above the time it was taken

Section 2.3: Analyzing Graphs

Definitions

Time-series graph

A line graph that is used to display a variable whose values change over time

Cross-sectional graph

Displays information collected at only one point in time

Pictograph

A bar graph that uses pictures of objects instead of bars

Outlier

A data value that falls outside the shape of the distribution

Shapes of Graphs

1. **Uniform:** The frequency of each class is relatively the same; graph has a rectangular shape
2. **Symmetric:** The data lie evenly on both sides of the distribution
3. **Skewed to the right:** The majority of the data fall on the left side of the distribution
4. **Skewed to the left:** The majority of the data fall on the right side of the distribution

E Chapter 2 Exercises

Directions: Create a frequency distribution with the indicated number of classes for each set of data. Include the frequency, class boundaries, midpoint, relative frequency, and cumulative frequency of each class.

1. The following data represent the numbers of days absent from school in one school year for each of the 24 students in Ms. Jinn's fourth grade class. Use six classes that have a class width of 5 days. Begin with a lower limit of 0.

17	8	12	3	0	5	13	12
25	10	6	8	11	0	1	4
19	21	22	9	16	9	3	2

2. The following data represent the weights in pounds of 24 collegiate offensive linemen in a particular state. Use six classes that have a class width of 25 pounds. Begin with a lower limit of 175.

195	210	255	267	231	229	301	199
178	281	245	256	278	205	217	223
279	196	235	248	262	291	302	189

Directions: Create a stem-and-leaf plot for the given data. Be sure to include a key to your stem-and-leaf plot.

3. The following data represent the saline concentrations (in terms of specific gravity) in a saltwater aquarium on various days. Order your leaves, but do not split the stems.

1.022	1.021	1.022	1.023	1.019	1.021	1.022	1.017	1.024
1.021	1.022	1.022	1.023	1.020	1.019	1.023	1.025	1.018

4. The following data represent the numbers of grams of fat per serving for a representative sample of various foods found in someone's kitchen pantry. Split the stems in the stem-and-leaf plot to better reveal the patterns in the data. Do you believe that this person is on a low-fat diet? Explain your reasoning using the data as your evidence.

0	2	0	6	8	1	10	2
3	14	4	21	13	7	9	17

Directions: Determine which type of graph would most clearly depict the data described.

5. **a.** The number of tickets sold at one theater over the course of a year

b. The number of tickets sold at one theater for each movie showing during one week

c. The number of tickets sold at one theater for each movie this week, specifically comparing the movie choices of patrons aged 18–35 to the choices of patrons aged 36 and older

d. Ticket prices for all theaters across the country

Directions: Use the given frequency distribution to answer the questions.

6. The frequency distribution shows the ages of students volunteering at a local animal shelter.

Ages of Student Volunteers (in Years)	
Class	**Frequency**
10–12	5
13–15	9
16–18	2
19–21	11
22–24	6

a. What is the relative frequency for the 4th class?

b. What is the cumulative frequency for ages 18 and under?

c. What is the upper class boundary for the 2nd class?

d. What is the class width?

e. What percentage of the student volunteers were between the ages 19 and 24?

f. How many students were surveyed?

g. What was the youngest age of the students surveyed?

Directions: Use the given graphs to answer the questions.

7. The stacked bar graph shows the average number of hours that married people in Japan spent each day doing various activities.

Source: Statistics Bureau (Japan). "2001 Survey on Time Use and Leisure Activities." 30 Sept. 2002. http://www.stat.go.jp/english/data/shakai/2001/jikan/yoyakuj.htm (24 Jan. 2012).

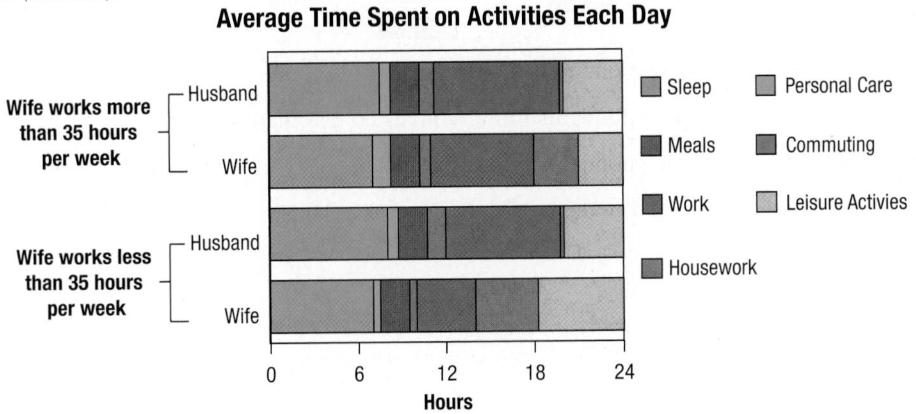

a. For wives who spend more than 35 hours per week working, how many hours on average are spent each day sleeping?

b. For husbands whose wives work fewer than 35 hours per week, approximately how many hours on average are spent each day on personal care?

c. Compare the number of hours spent on leisure activities for wives in each category.

d. What type of graph could be used to represent these data in a clearer fashion?

8. The side-by-side bar graph shows the percentages of American adults who use various wireless devices to access the Internet.

Source: Smith, Aaron. "Mobile Access 2010." Pew Internet & American Life Project. 7 July 2010. http://www.pewinternet.org/Reports/2010/Mobile-Access-2010.aspx (24 Jan. 2012).

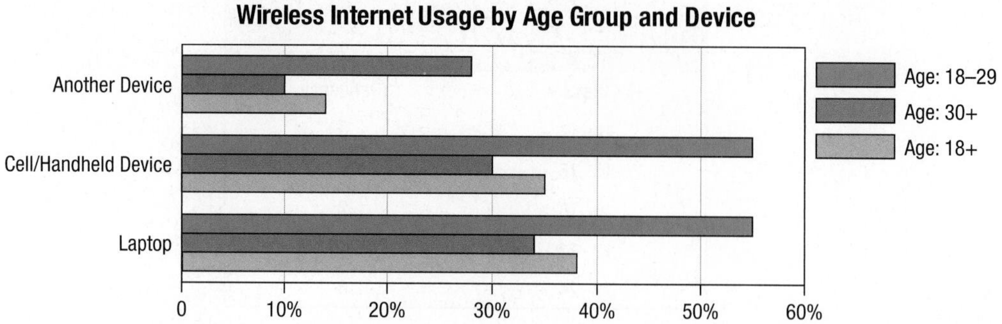

a. What percentage of people ages 18–29 use a cell/handheld device for wireless Internet access?

b. What device is used most for wireless Internet access by people in the 30+ age category?

c. What age group is most likely to use a device other than a cell phone or laptop to wirelessly access the Internet?

9. Consider the pie chart below.

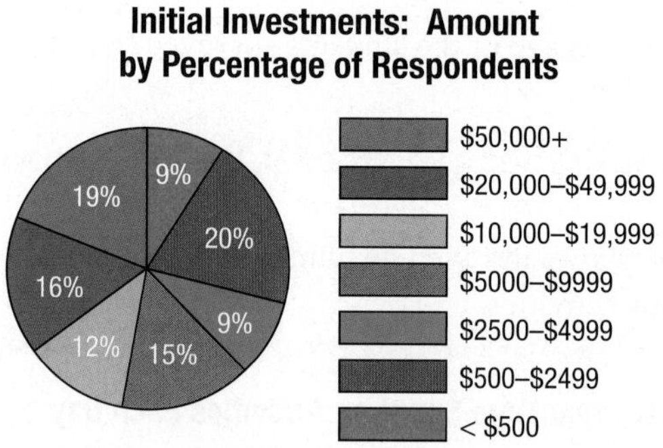

a. What percentage of respondents initially invested $50,000 or more?

b. What percentage of respondents initially invested less than $10,000?

c. What percentage of respondents initially invested between $5000 and $19,999?

d. What percentage of respondents initially invested $10,000 or more?

10. The line graph shows the actual dollar amounts of the federal hourly minimum wage over time, unadjusted for inflation.

Source: US Department of Labor, Wage and Hour Division (WHD). "History of Federal Minimum Wage Rates Under the Fair Labor Standards Act, 1938-2009." http://www.dol.gov/whd/minwage/chart.htm (24 Jan. 2012).

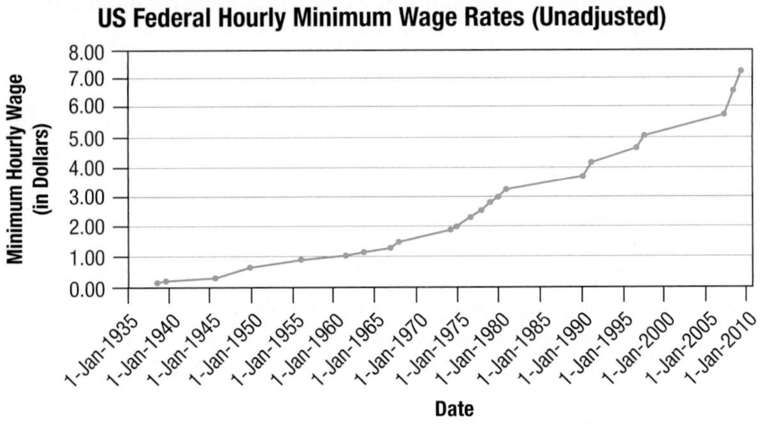

a. What is the overall trend in the unadjusted minimum wage?

b. When did the highest unadjusted minimum wage occur?

c. Has there been any time period when the unadjusted minimum wage decreased?

11. A pretest and a posttest were administered to one class at the beginning and end of two weeks of classes. The scores for each of the tests are shown graphically below.

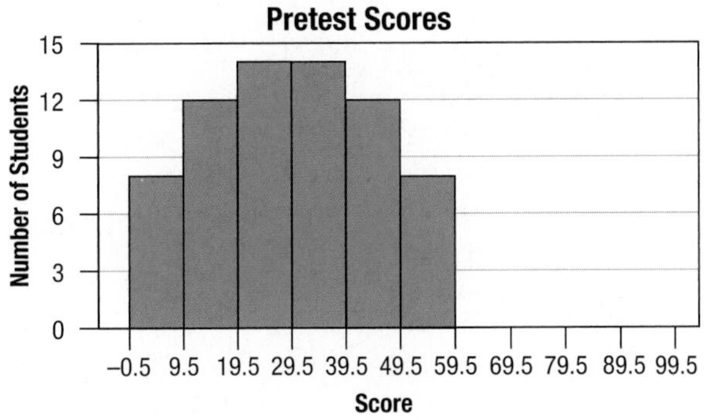

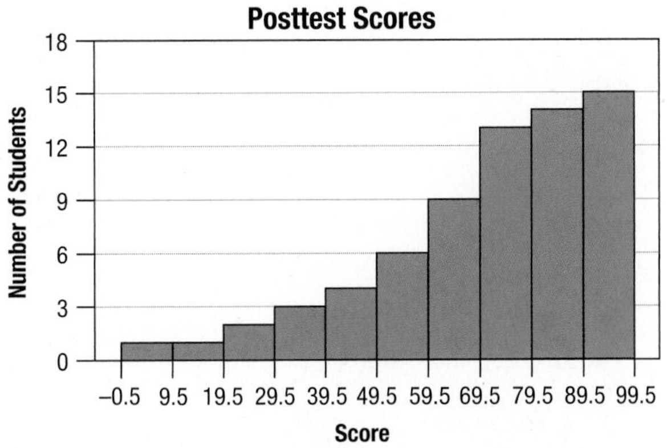

a. Identify the general shape of the distribution of pretest scores.

b. Identify the general shape of the distribution of posttest scores.

c. Why do you think that the pretest scores are distributed the way they are?

d. Why do you think that the posttest scores are distributed the way they are?

12. The line graph shows the federal hourly minimum wage over time, adjusted for inflation.

Source: US Department of Labor, Wage and Hour Division (WHD). "History of Federal Minimum Wage Rates Under the Fair Labor Standards Act, 1938-2009." http://www.dol.gov/whd/minwage/chart.htm (24 Jan. 2012).
Source: US Department of Labor, Bureau of Labor Statistics (BLS). "CPI Inflation Calculator." http://data.bls.gov/cgi-bin/cpicalc.pl (18 Jun. 2012).

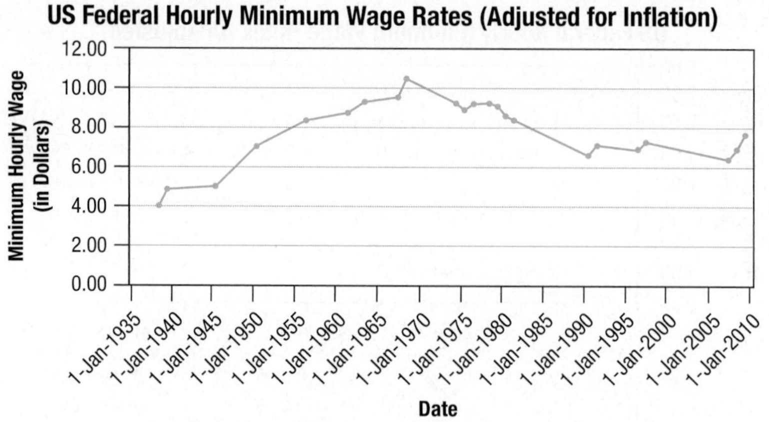

a. What is the overall trend in the value of minimum wage? How does this trend compare to the trend in the unadjusted minimum wage graph?

b. When did the highest adjusted minimum wage occur?

c. Has there been any time period when the adjusted minimum wage decreased?

P Chapter 2 Project

Fast Food Comparison

For three fast-food restaurants, a random sample of service times (in seconds) was taken. The results are to the right.

1. For each restaurant's service times, do the following.

a. Create a frequency distribution with the given classes. Include the frequency, class boundaries, relative frequency, and cumulative frequency of each class.

Class
30–49
50–69
70–89
90–109
110–129

b. Use the frequency distribution to construct a histogram of the data.

2. Based on the graphical descriptions of the data you created in part 1, which fast-food restaurant do you believe consistently has the shortest service times? Explain why.

Service Times (in Seconds)

Kim's Kajun Kitchen	Chez Carolyn	Emily's Eatery
111	109	99
94	84	63
57	93	53
80	123	82
78	97	75
109	56	112
92	79	65
34	57	55
67	32	65
122	68	86
95	45	94
46	99	87
103	82	62
110	76	68
93	54	61
86	84	49
75	86	37
49	73	46
61	79	65
82	67	86
76	96	48
92	112	95
65	94	33
92	82	49
94	96	76
106	125	35
112	98	68
83	120	57
72	78	41
119	63	49

T Chapter 2 Technology

Creating Graphs

TI-83/84 Plus

The graphing function on the TI-83/84 Plus calculator will allow you to create many of the graphs we have discussed in this chapter. We will look at two of these graphs, the histogram and the line graph.

Histograms

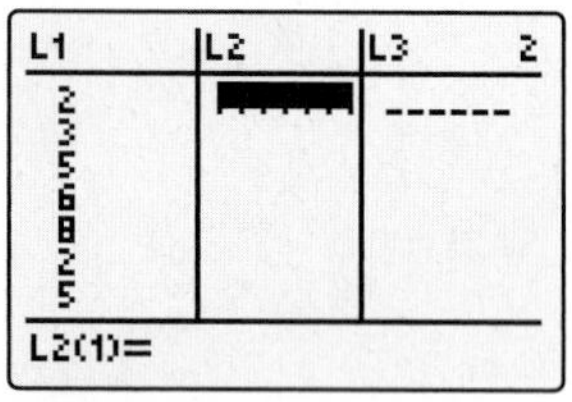

Let's begin by creating a histogram. First, you must enter your data into the calculator. To do so, press STAT, then choose `1:Edit`, and type your data in the first list, `L1`. Your screen should now look similar to the screenshot in the margin.

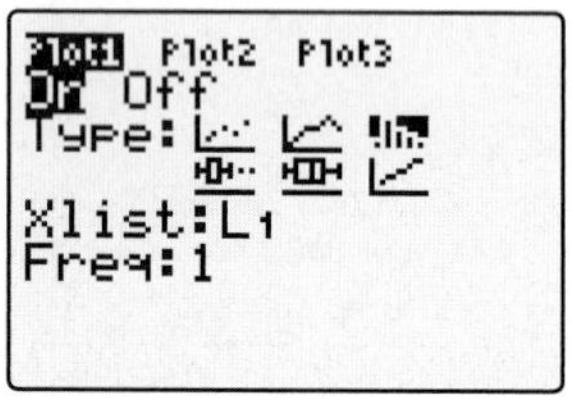

Once your data are entered, you need to set the specifications for the graph. Press 2ND and then press Y=. This displays the STAT PLOTS menu. Press ENTER. This brings up the menu of graph settings shown in the second screenshot in the margin.

Choose `On`, highlight the type of graph you want to create, and press ENTER. The histogram is the third option. Now, when you press GRAPH, you should see a histogram of your data. If your graph does not appear or it seems strange, you can modify the screen window by pressing ZOOM and then choosing option `9:ZoomStat`. This will resize the screen so that it is appropriate for a statistical graph.

Example T.1

Using a TI-83/84 Plus Calculator to Create a Histogram

Use a TI-83/84 Plus calculator to recreate the histogram of our 3-D TV prices. The prices are reprinted below for your convenience.

3-D TV Prices (in Dollars)				
1595	1599	1685	1699	1699
1699	1699	1757	1787	1799
1799	1885	1888	1899	1899
1899	1984	1999	1999	1999

Solution

Enter your data into the list editor by pressing STAT and then choosing `1:Edit`.

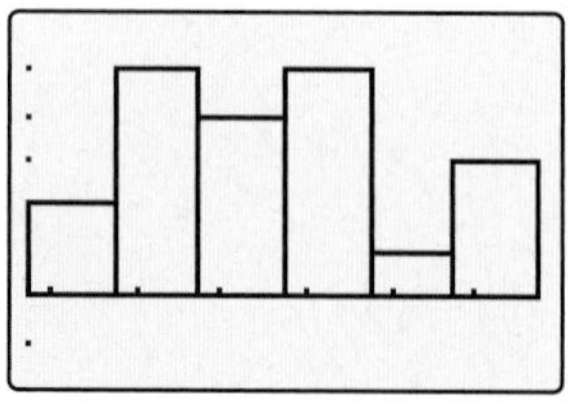

Once your data have been entered, press 2ND and then press Y= to bring up the STAT PLOTS menu.

Press ENTER. Set the menu so that the graph will be a histogram. Press GRAPH. Your screen should look like the screenshot in the margin. If it does not appear that way at first, pressing ZOOM and then choosing option `9:ZoomStat` should correct the problem.

Line Graphs

The process for creating a line graph is similar to constructing a histogram; however, to create a line graph we need both x- and y- values. This means we will need to enter data into `L1` and `L2`. If your data are not in these lists, you can change the `Xlist` and/or `Ylist` in the `STAT PLOTS` menu.

Example T.2

Using a TI-83/84 Plus Calculator to Create a Line Graph

Use a TI-83/84 Plus calculator to recreate the line graph of CPI values from Section 2.2. The data are reproduced below.

Consumer Price Index

Year	CPI
1920	20.0
1930	16.7
1940	14.0
1950	24.1
1960	29.6
1970	38.8
1980	82.4
1990	130.7
2000	172.2
2010	218.1

Source: US Bureau of Labor Statistics. "Consumer Price Index History Table." 19 Jan. 2012. ftp://ftp.bls.gov/pub/special.requests/cpi/cpiai.txt (24 Jan. 2012).

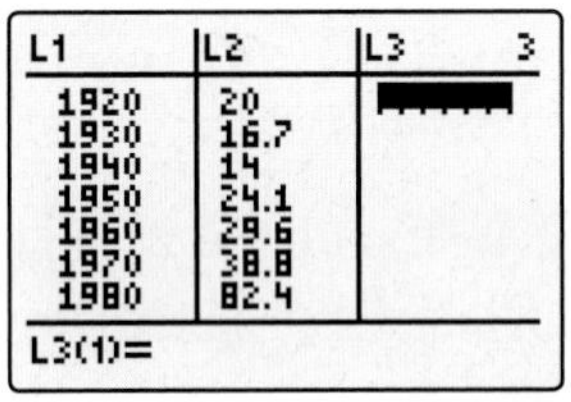

Solution

Press STAT, choose `1:Edit`, and enter the years in `L1` and the CPI values in `L2`. You should have the screen like the screenshot in the margin.

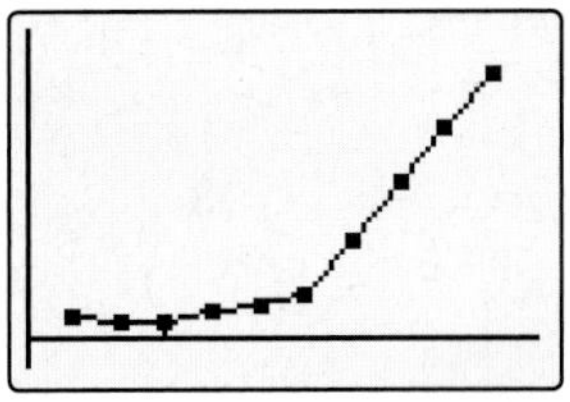

Now set up the `STAT PLOTS` menu to produce a line graph by selecting the second graph type in the menu. Press GRAPH. If you do not see a line graph at first, press ZOOM and choose option `9:ZoomStat`. You should then see the line graph shown like the screenshot in the margin.

Microsoft Excel

Basic Charts

You can create many more graphs in Microsoft Excel than you can on a TI-83/84 Plus calculator. The chart options under the Insert tab can be used to create bar graphs, pie charts, and line graphs, to name a few. Begin by entering your data into the worksheet. Then click on the button in the tool bar for the type of chart that you would like to create.

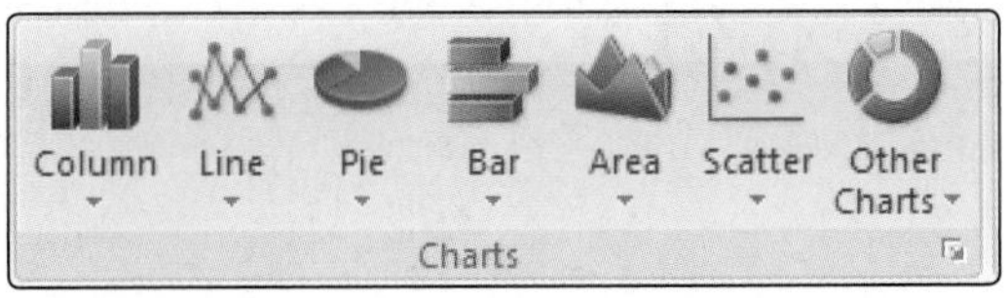

Figure T.1: Microsoft Excel Chart Options

By clicking on the parts of the resulting graph, you can change the various options, such as the title, the values shown on the axes, and so forth.

Example T.3

Using Microsoft Excel to Create a Pie Chart

Use Microsoft Excel to recreate the pie chart of the housing data from Section 2.2. The data are reproduced below.

Types of Housing

Type of Housing	Number of Students
Apartment	20
Dorm	15
House	9
Sorority/Fraternity House	5

Solution

Begin by entering the types of housing in cells A1 through A4. Then enter the numbers of students in cells B1 through B4. Your Excel worksheet should look like the one below.

	A	B	C
1	Apartment	20	
2	Dorm	15	
3	House	9	
4	Sorority/Fraternity House	5	
5			

Now click on the **Insert** tab in the tool bar. Click on the **Pie Chart** button and choose the first 2-D pie chart option. The pie chart is then displayed within the worksheet. By clicking on the **Layout** tab under Chart Tools, you can add a title to the chart, change the legend, or add data labels to the chart using the various options provided.

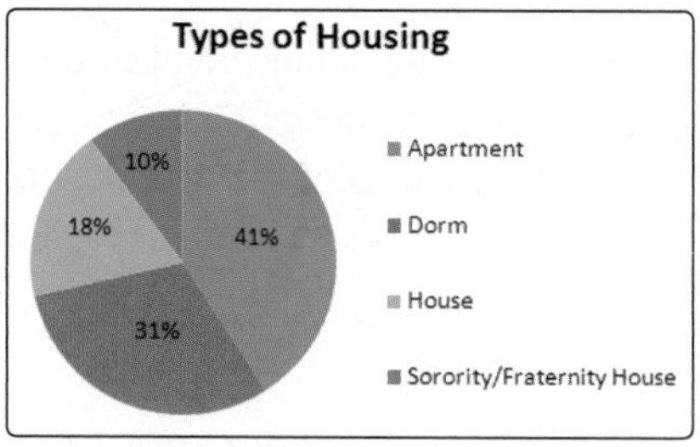

Histograms and Frequency Distributions

The charts under the Insert tab are wonderful for creating a wide variety of very nice graphs; however, if you want to create a histogram, you would not use these options. You do not need to create a frequency distribution first either, as Excel does both at the same time. The process is described in the following example.

Example T.4

Using Microsoft Excel to Create a Histogram

Let's recreate the histogram of 3-D TV prices using Microsoft Excel. The data are reprinted below.

3-D TV Prices (in Dollars)				
1595	1599	1685	1699	1699
1699	1699	1757	1787	1799
1799	1885	1888	1899	1899
1899	1984	1999	1999	1999

Solution

Begin by entering the data in column A in Excel. Next, enter the upper boundaries of the classes you want to use in Column B. Otherwise, the class boundaries will be chosen for you and may not be the best choice. Once your data are entered, go to the **Data** tab, then **Data Analysis**. Choose **Histogram** from the menu and click **OK**. The *Input Range* is where your data are located. The *Bin Range* is where your upper class boundaries have been entered. Click on **Labels** if the first cells in your input range and bin range are data labels. Next, click on **Chart Output** to create a histogram instead of just a frequency distribution. Finally, click **OK** to generate the output.

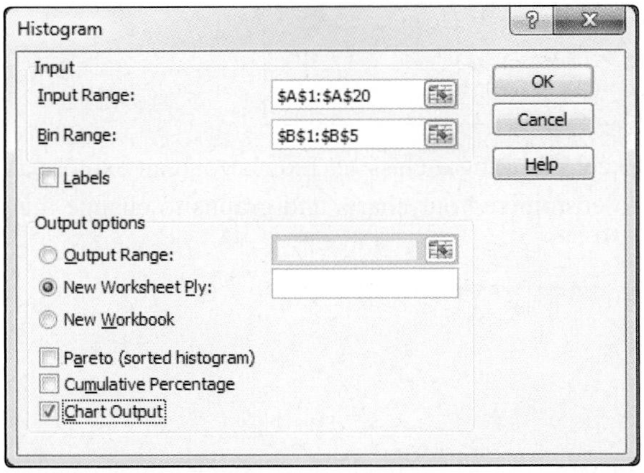

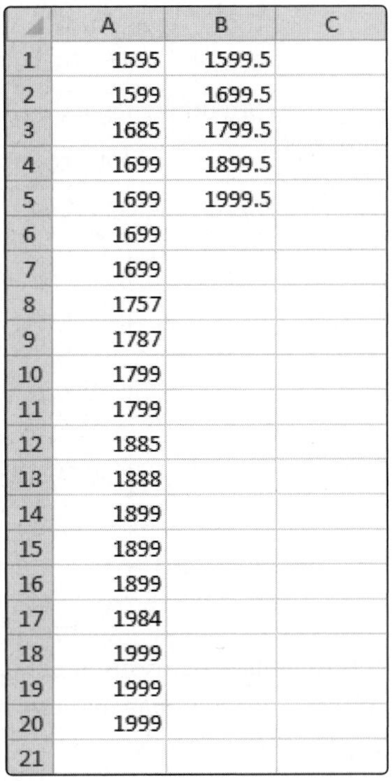

	A	B	C
1	1595	1599.5	
2	1599	1699.5	
3	1685	1799.5	
4	1699	1899.5	
5	1699	1999.5	
6	1699		
7	1699		
8	1757		
9	1787		
10	1799		
11	1799		
12	1885		
13	1888		
14	1899		
15	1899		
16	1899		
17	1984		
18	1999		
19	1999		
20	1999		
21			

Side Note

If Data Analysis does not appear in your Excel menu under the Data tab, you can easily add this function. See the directions in Appendix B: Getting Started with Microsoft Excel.

When your histogram has been created, you will notice that the bars do not touch. To correct this error, click on one of the bars in the histogram. Click on the **Layout** tab under Chart Tools. Click on the **Format Selection** button on the far left. The first option is Series Options. Move the **Gap Width** slider to **No Gap** and click on **Close**. You will also notice that the class boundaries are appearing on the *x*-axis where the midpoints should be, and there is a "More" category on the graph. To correct these errors, go to the Excel worksheet where the histogram's frequency distribution is located. Change the bins to be the midpoints and delete the "More" entry from the spreadsheet. Once these adjustments are made, the graph will be a histogram like the one shown in the following screenshot.

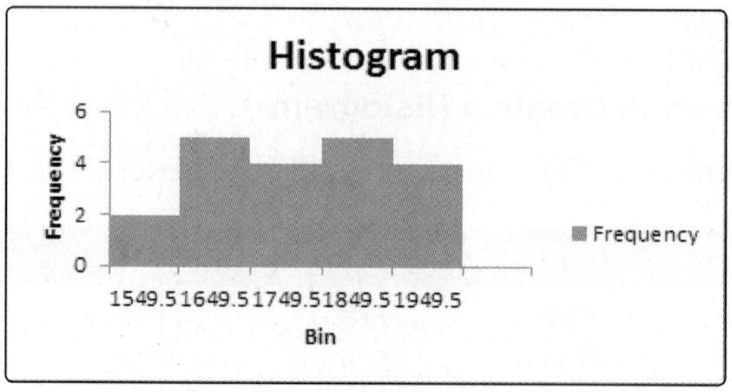

Notice that this histogram does not look like the one you saw earlier in the chapter. In order to change the histogram you just created to look like the one shown at the end of this example, several additional steps have to be taken to tweak the options of the chart. You must:

- Delete the whole row of the spreadsheet that contained the "More" category.
- Change the size, title, and axis labels.
- Turn off the legend.
- Format the Plot Area to have a gray background and darker gray border.
- Format the Data Series to add the solid line border around the bars.

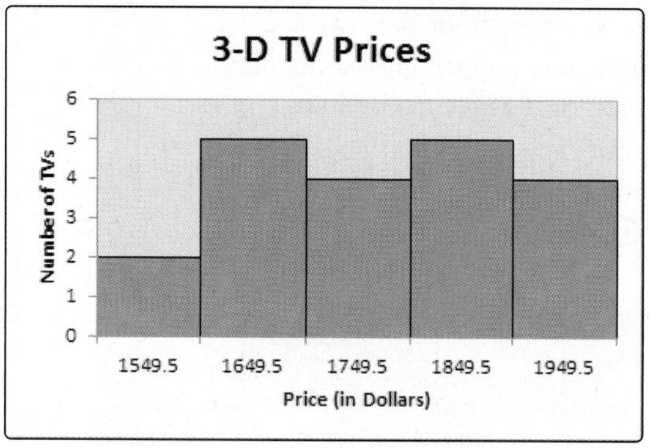

By using the Microsoft online help for Excel or taking a class on Excel, you can explore all of the various options that can be used to customize your charts and graphs to change their appearance in countless ways

Minitab

You can also create many statistical graphs in Minitab. All of the graphs shown in this chapter can be made in Minitab. A few examples are shown below.

Example T.5

Using Minitab to Create a Histogram

Create a histogram in Minitab using the data set containing the heart rates of 50 students, given below.

77	84	79	80	67	84	82	84	69	81
94	68	65	86	78	79	83	83	84	82
93	80	81	80	62	98	77	93	82	80
82	73	77	79	81	70	72	85	84	80
83	77	80	70	75	74	85	87	79	88

Solution

Begin by entering the data in column 1, C1, in MINITAB. Once your data are entered, go to the **Graph** menu and choose **Histogram**. Select **Simple** from the types of histograms and click **OK**. When the Histogram - Simple menu appears, select **C1** as the Graph variables. Then click **OK** to generate the graph.

To change the way the classes are displayed in the histogram, double-click on one of the numbers shown on the x-axis of the histogram. Select the **Binning** tab on the Edit Scale menu. If you want the histogram to only mark the midpoints of the classes, choose **Midpoint** for the Interval Type, select **Midpoint/Cutpoint positions** under Interval Definition, and enter the midpoints, **63 68 73 78 83 88 93 98**, in the box. To display the class boundaries as shown in the histogram in the following screenshot, first choose **Cutpoint** for the Interval Type, select **Midpoint/Cutpoint positions** under Interval Definition, and enter the class boundaries, **60.5 65.5 70.5 75.5 80.5 85.5 90.5 95.5 100.5**, in the box. Then select the **Scale** tab on the Edit Scale menu, choose **Position of ticks** under Major Tick Positions, and enter the class boundaries, **60.5 65.5 70.5 75.5 80.5 85.5 90.5 95.5 100.5**, in the box. Then click **OK** to see the updated histogram. You can change the title, change the axis labels, and add color by double-clicking on the various parts of the graph shown in the following screenshot.

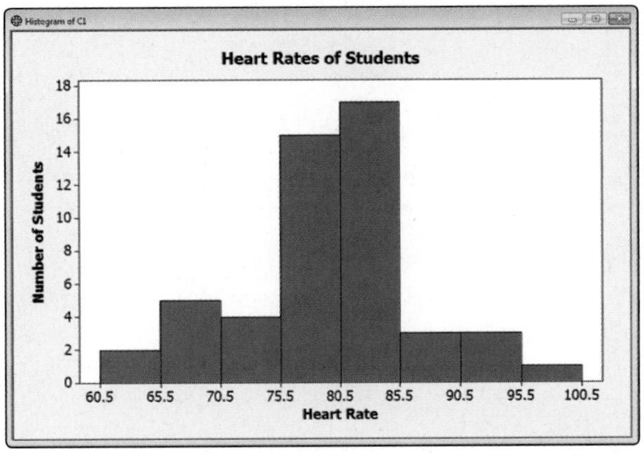

Example T.6

Using MINITAB to Create a Stem-and-Leaf Plot

Using the data from Example T.5, create a stem-and-leaf plot in MINITAB.

Solution

Begin by entering the data in column 1, C1, in MINITAB. Next, choose **Graph** and select **Stem-and-Leaf**. Select **C1** for the Graph variables and enter **10** for the Increment. Click **OK** to display the following stem-and-leaf plot.

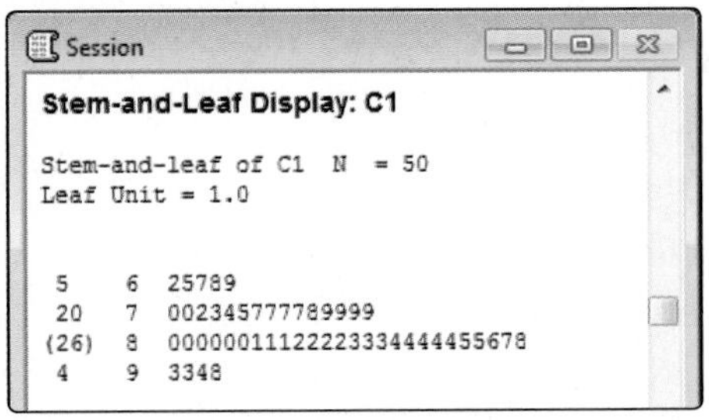

Example T.7

Using MINITAB to Create a Dot Plot

Using the data from Example T.5, create a dot plot in MINITAB.

Solution

Begin by entering the data in column 1, C1, in MINITAB. Next, choose **Graph** and select **Dotplot**. Select **Simple** from the types of dot plots under One Y and click **OK**. Select **C1** for the Graph variables and click **OK** to display the following dot plot. You can change the title, change the *x*-axis label, and add color by double-clicking on the various parts of the graph.

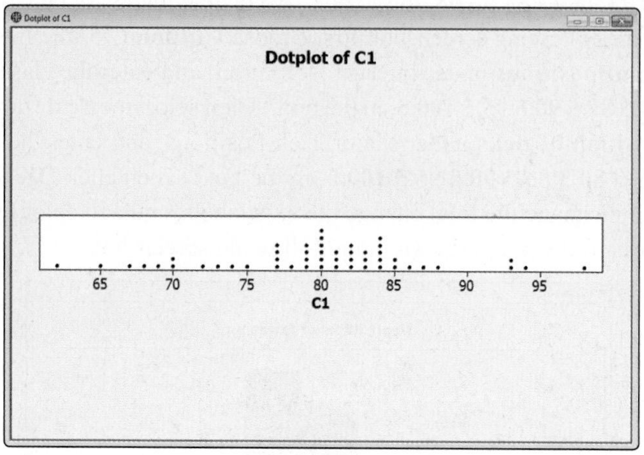

Chapter Three

Numerical Descriptions of Data

Sections

Objectives

1. Calculate the mean, median, and mode.
2. Determine the most appropriate measure of center.
3. Compute the range, variance, and standard deviation.
4. Determine the percentiles, quartiles, and five-number summary of a data set.
5. Construct a box plot.

3

Introduction

Many children regard professional athletes as heroes, faithfully watching every game and memorizing every last statistic of their favorite players. This is especially true for fans of Major League Baseball, since the statistics are such an important part of the sport. Little boys dreaming of the major leagues memorize batting averages, runs earned, and home runs scored—not to mention the salaries—of their favorite superstars. A player who makes it to the big leagues not only gets to play his beloved sport professionally, but he is automatically a millionaire, right?

Not necessarily. Take for instance, the salary statistics for the members of the 2012 Atlanta Braves. The average salary for these players is reported to be \$2,776,998. However, upon closer inspection of the raw data, the salaries range from \$480,000 for the lowest-paid player to \$14,000,000 for Chipper Jones, who currently has the highest salary on the team. In fact, the majority of players actually earn less than a million dollars each year. Though a salary of \$480,000 per year is certainly not shabby, it is far from the nearly \$3 million reported average.

This is just one example of why it is important to look at more than one characteristic of a data set. In this chapter we will learn how to use measures of center, dispersion, and relative position to analyze data.

3.1 Measures of Center

We have seen how to describe data using tables and graphs; now we will learn how to summarize data using numbers. In order to summarize data numerically, we will use the center, dispersion, and shape of the data. Let's begin with the center of the data.

Memory Booster

Quantitative data are counts and measurements.

To describe the numerical center of a data set, a number (or numbers) that best represents a typical value in the data set is used. For quantitative data, it also describes the location of the data on the real number line. Thus the center of the data is the "average" value of the data. In statistics, however, "average" can be calculated many ways.

Mean

The general usage of the word average indicates the **arithmetic mean**, often simply called the *mean*. You can calculate the means of both sample data and population data. The mean is the sum of all of the data values divided by the number of data values.

Formula

Sample Mean

The **sample mean** is the arithmetic mean of a set of sample data, given by

$$\bar{x} = \frac{x_1 + x_2 + \cdots + x_n}{n} = \frac{\sum x_i}{n}$$

where x_i is the ith data value and

n is the number of data values in the sample.

This formula is actually the **sample mean**, $\bar{x}$, because we are calculating the mean of a set of sample values. If we are calculating the mean of all the values in a population, we call it the **population mean**, μ. The procedure is the same: divide the sum of all the population values by the number of values in the population, N. Remember that it is most often the case that we are calculating the sample mean and not the population mean because that would require a census, a collection of data from all members of the population.

Math Symbols

$\bar{x}$**:** sample mean; read as "x-bar"

μ**:** population mean; Greek letter, mu

Rounding Rule

When calculating the mean, round to one more decimal place than the largest number of decimal places given in the data. Occasional exceptions to this rule can be made when the type of data lends itself to a more natural rounding scheme, such as rounding values of currency to two decimal places.

Formula

Population Mean

The **population mean** is the arithmetic mean of all the values in a population, given by

$$\mu = \frac{x_1 + x_2 + \cdots + x_N}{N} = \frac{\sum x_i}{N}$$

where x_i is the ith data value in the population and

N is the number of values in the population.

Remember that this value only makes sense when the data are quantitative! It is meaningless to find the mean of qualitative data such as the ratings of a hotel: "very satisfactory," "satisfactory," "needs improvement," or "don't plan to visit again." Because there is not a measurable difference between each of the ratings, assigning a number value to each and taking the mean carries little meaning. However, you might often see such calculations being exhibited. Later in this section, we'll show you a more appropriate "average" to calculate in this case.

When we do calculate the mean of quantitative data, often it's the case that the value we get is not actually a member of the data set—it is simply a descriptive value of the entire set. It is an especially useful summary statistic when data sets are large and you may not be able to examine all the data values individually. Let's look at a simple example of finding the mean. When calculating a mean, round to one more decimal place than the largest number of decimal places given in the data. Occasional exceptions to this rule can be made when the type of data lends itself to a more natural rounding scheme, such as rounding values of currency to two decimal places.

Example 3.1

Calculating the Sample Mean

Students were surveyed to find out the number of hours they sleep per night during the semester. Here is a sample of their self-reported responses. Calculate the mean.

$$5, 6, 8, 10, 4, 6, 9$$

Solution

Because we are given a sample of the student responses, we are calculating the *sample mean*. Add the hours together and then divide by 7, which is the number of students in the sample. At the end of the calculation, round to one decimal place since the data values are whole numbers.

$$\begin{aligned} \bar{x} &= \frac{\sum x_i}{n} \\ &= \frac{5+6+8+10+4+6+9}{7} \\ &= \frac{48}{7} \\ &\approx 6.857143 \\ &\approx 6.9 \end{aligned}$$

The sample mean for the number of hours that students reported sleeping per night during the semester is 6.9.

Alternate Calculator Method

To find the sample mean on a TI-83/84 Plus calculator, follow the steps below.

Side Note

The most recent TI-84 Plus calculators (Jan. 2011 and later) contain a new feature called STAT WIZARDS. This feature is not used in the directions in this text. By default, this feature is turned ON. STAT WIZARDS can be turned OFF under the second page of MODE options.

- Press STAT.
- Choose option 1:Edit and press ENTER.
- Enter the data in L1.
- Press STAT again.
- Choose CALC.
- Choose option 1:1-Var Stats.
- Press ENTER twice. (Note: If your data are not in L1, before pressing ENTER the second time, enter the list where your data are located, such as L3 or L5.)

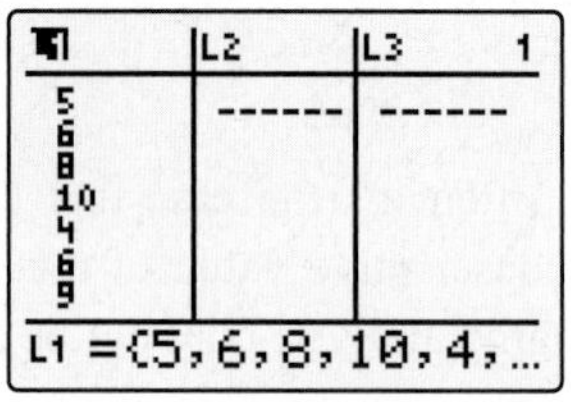

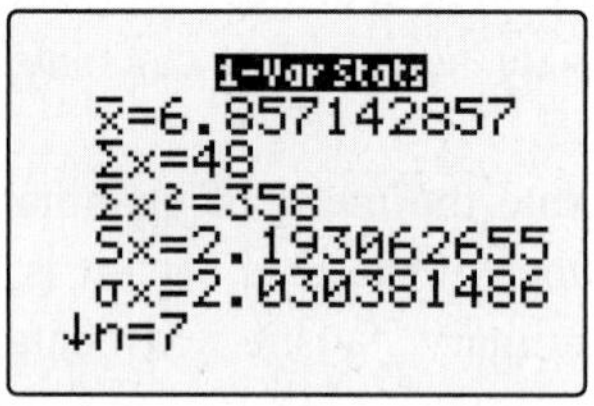

The first value in the output, seen in the screenshot on the right, shows the value of $\bar{x} = 6.857142857$. In addition, the calculator displays many other descriptive statistics, not just the sample mean. We will use the above procedure repeatedly to find various descriptive statistics throughout this chapter.

Example 3.2

Using the Mean to Find a Data Value

Rutherford downloaded five new songs from the Internet. He knows that, on average (mean), the songs cost \$1.23. If four of the songs cost \$1.29 each, what was the price of the fifth song he downloaded?

Solution

Since we are given the value of the mean, we can use algebra on the formula to find the missing value. Substituting all the values we are given into the formula for the mean, we have the following.

$$\mu = \frac{x_1 + x_2 + x_3 + x_4 + x_5}{N}$$

$$\$1.23 = \frac{\$1.29 + \$1.29 + \$1.29 + \$1.29 + x_5}{5}$$

$$5 \cdot \$1.23 = \frac{\$5.16 + x_5}{\cancel{5}} \cdot \cancel{5}$$

$$\$6.15 = \$5.16 + x_5$$

$$\$6.15 - \$5.16 = \$5.16 + x_5 - \$5.16$$

$$\$0.99 = x_5$$

So the cost of Rutherford's fifth song was \$0.99.

For certain data sets, the formula for calculating a mean value is not sufficient. For example, suppose that Walter wants to calculate his overall average in his US history course. If the course syllabus states that the final grade is determined by tests (40%), homework (20%), quizzes (10%) and a final exam (30%), how should Walter calculate his grade? He would have to "weight" each grade he received based on the percentages given in the syllabus. This type of average is called the *weighted mean.* We will calculate Walter's grade in Example 3.3 after we discuss the details of the formula for the weighted mean.

When each data value in the set does not hold the same relative importance, a **weighted mean** must be used. To calculate a weighted mean for a sample, first multiply each value by its respective weight. Then divide the sum of these products by the sum of the weights to obtain the mean.

Formula

Weighted Mean

The **weighted mean** is the mean of a data set in which each data value in the set does not hold the same relative importance, given by

$$\bar{x} = \frac{\sum (x_i \cdot w_i)}{\sum w_i}$$

where x_i is the ith data value and

w_i is the weight of the ith data value.

The procedure for calculating the weighted mean for a population is the same, but with the notation μ rather than $\bar{x}$.

Example 3.3

Calculating a Weighted Mean

The syllabus in Walter's US history class states that the final grade is determined by tests (40%), homework (20%), quizzes (10%), and a final exam (30%). Two students in the class, Walter and Virginia, want to calculate their final grades. Below are their average grades in each of the categories for tests, homework, and quizzes. They have also individually guessed at what they might score on their final exam.

a. Calculate what Walter's final grade would be if these were his ultimate scores.

Tests: 83

Homework: 98

Quizzes: 90

Final Exam: 87

b. Calculate Virginia's final grade, given her scores below, using a TI-83/84 Plus calculator.

Tests: 95

Homework: 45

Quizzes: 66

Final Exam: 90

Solution

a. First, let's determine which numbers are values of x and which are weights. The grade earned in each category is weighted by the percentage for that category in the syllabus. For instance, Walter's test average of 83 gets a weight of 40%. Thus, in this case the weights are the percentages for the categories. The values for x are then Walter's grades in the categories. The weighted mean is calculated as follows.

$$\begin{aligned}\bar{x} &= \frac{\sum(x_i \cdot w_i)}{\sum w_i}\\ &= \frac{83(0.4)+98(0.2)+90(0.1)+87(0.3)}{0.4+0.2+0.1+0.3}\\ &= \frac{87.9}{1}\\ &= 87.9\end{aligned}$$

Therefore, Walter's final grade for the class would be 87.9.

b. As noted when calculating Walter's grade, the weights are the percentages for the categories and the values for x are Virginia's grade in each category.

To calculate a weighted mean using a TI-83/84 Plus calculator, we will follow steps similar to those used for calculating a sample mean. However, we will need to enter two lists of data instead of just one.

- Press STAT.
- Choose option 1:Edit and press ENTER.
- Enter the values of x for your data in L1, and enter the weights for your data in L2.
- Press STAT again.
- Choose CALC.
- Choose option 1:1-Var Stats and press ENTER.
- Press 2ND 1 , 2ND 2 to enter L1,L2 on the screen. This tells the calculator that this is a weighted mean and that the data are in two lists.
- Press ENTER.

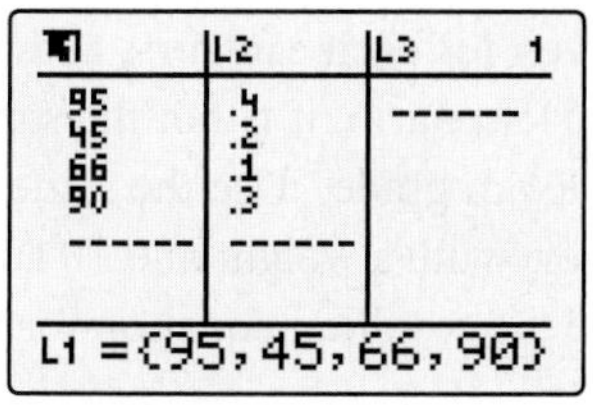

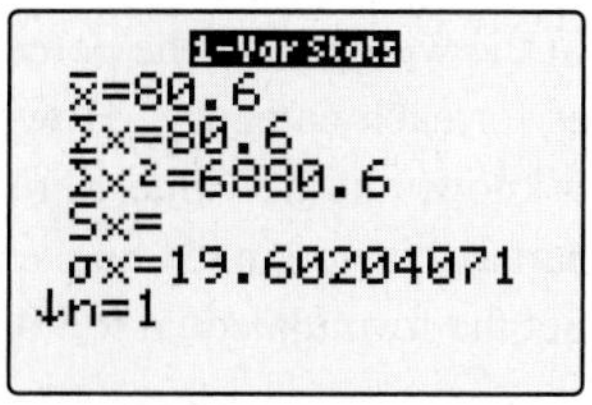

The output, seen in the screenshot on the right, shows the value of $\bar{x} = 80.6$, along with the other descriptive statistics for this data set.

Although Virginia didn't do as well on her homework and quizzes, if she manages to get a 90 on the final exam like she predicts, she'll end up with an 80.6 for her final grade. Take a moment and consider the effect on her grade if her test score was low, but the homework and quiz scores were high. Would she come out with the same grade?

Example 3.4

Calculating a Weighted Mean

At the end of the semester, Heather knows all of her grades in her sculpting class except for the final exam. Here's a breakdown of her points and how much each category counts toward the final course grade.

Tests (35%): 78

Class Assignments (20%): 92

Semester Project (35%): 93

Final Exam (10%): ?

a. Calculate the weighted mean for the portion of the grade that she has completed.

b. What grade must Heather make on her final exam in order to have a final grade of 90%, which would be the A that she desires? Assume that the final exam is worth 100 points.

c. What grade must Heather make on her final exam in order to have a final grade of 80%, which would give her a B for the course? Again, assume that the final exam is worth 100 points.

Solution

a. Just like in the previous example, the weights are the percentages given for each category, and the values for x are the scores she has so far. We will only include the portions of the final grade in which she has scores, not all of the categories. The weighted mean is then calculated as follows. (Remember to change the percentages to decimals for the formula.)

$$\begin{aligned}\bar{x} &= \frac{\sum (x_i \cdot w_i)}{\sum w_i} \\ &= \frac{78(0.35)+92(0.2)+93(0.35)}{0.35+0.2+0.35} \\ &= \frac{78.25}{0.9} \\ &= 86.9\overline{4} \\ &\approx 86.9\end{aligned}$$

Notice that the total in the denominator no longer equals 1 since we are not including all of the components for the final grade.

So, Heather has 86.9 for her sculpting grade so far without the final exam.

b. We know that the weights are the percentages given for each category, and the values for x are the scores for each category. However, in this scenario, it is not the sample mean, $\bar{x}$, that we don't know, but the value, x, of the final exam grade. Use the same formula as the one used in part a., but add in the final exam category with a weight of 0.10 and the unknown value of x. Set the formula equal to 90, and solve the resulting equation for x as follows.

$$\begin{aligned}\bar{x} &= \frac{\sum(x_i \cdot w_i)}{\sum w_i}\\ 90 &= \frac{78(0.35)+92(0.2)+93(0.35)+x\cdot(0.1)}{0.35+0.2+0.35+0.1}\\ 90 &= \frac{78.25+0.1x}{1}\\ 90 &= 78.25+0.1x\\ 11.75 &= 0.1x\\ x &= 117.5\end{aligned}$$

Since $x > 100$, it appears as though it is mathematically impossible for Heather to make an A in the course. The highest final grade that Heather could achieve in the class is 88.25. We will leave this for you to verify on your own.

c. To find this solution, we will set up the problem just the same as in part b., except now we will set the formula equal to 80 rather than 90. We will again solve for x, which represents the unknown final exam score. We then have the following.

$$\begin{aligned}\bar{x} &= \frac{\sum(x_i \cdot w_i)}{\sum w_i}\\ 80 &= \frac{78(0.35)+92(0.2)+93(0.35)+x\cdot(0.1)}{0.35+0.2+0.35+0.1}\\ 80 &= \frac{78.25+0.1x}{1}\\ 80 &= 78.25+0.1x\\ 1.75 &= 0.1x\\ x &= 17.5\end{aligned}$$

Hence, we see that although an A is not possible, Heather must score only 17.5 or higher on the final exam in order to secure a B in the course.

Median

Another measure of center is the median. The **median** is the middle value in an ordered array of data. For this calculation to make sense, the data must be interval or ratio. The median would not make sense for nominal or ordinal data, despite the fact that it technically can be calculated for ordinal data.

Procedure

Finding the Median of a Data Set

1. List the data in ascending (or descending) order, making an ordered array.
2. If the data set contains an ODD number of values, the median is the middle value in the ordered array.
3. If the data set contains an EVEN number of values, the median is the arithmetic mean of the two middle values in the ordered array. Note that this implies that the median may not be a value in the data set.

Example 3.5

Finding the Median

Given the numbers of absences for samples of students in two different classes, find the median for each sample.

a. 3, 4, 6, 7, 2, 8, 9

b. 5, 7, 8, 1, 4, 9, 8, 9

Solution

a. First, put the data in order: 2, 3, 4, 6, 7, 8, 9. Since there are an odd number of values, the median is the number in the middle: 2, 3, 4, 6, 7, 8, 9. Thus the median for this sample is 6 absences.

b. First, put the data in order: 1, 4, 5, 7, 8, 8, 9, 9. Since there are an even number of values, the median is the mean of the two numbers in the middle.

1, 4, 5, 7, 8, 8, 9, 9

$$\frac{7+8}{2}=7.5$$

Thus the median for this sample is 7.5 absences.

Alternate Calculator Method

The median is one of the descriptive statistics that the TI-83/84 Plus calculator displays when you choose the `1-Var Stats` option from the STAT > `CALC` menu. Recall from Example 3.1 that the steps to find the descriptive statistics are as follows.

- Press STAT.
- Choose option `1:Edit` and press ENTER.
- Enter the data in `L1`.
- Press STAT again.
- Choose `CALC`.
- Choose option `1:1-Var Stats`.
- Press ENTER twice.

Do you see an output value for the median? Probably not. That is because the median is actually on the second "page" of the output. Use the down arrow to scroll down to the other descriptive statistics. The one labeled "`Med=7.5`" is the median.

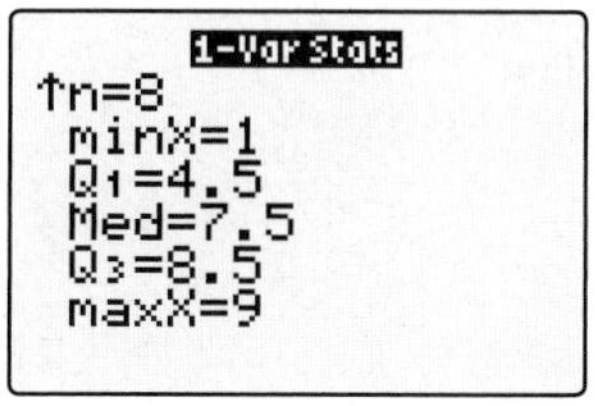

Mode

The last measure of center we will discuss is the mode. This measurement can be calculated for data of any level of measurement. The **mode** is the value in the data set that occurs most frequently. If all of the data values occur only once, or they each occur an equal number of times, we say that there is **no mode**. If only one value occurs most often, then the data set is said to be **unimodal**. If exactly two values occur equally often, then the data set is said to be **bimodal**. If more than two values occur equally often, the data set is **multimodal**. Note that if there is a mode, it will always be a value in the data set. To find the mode of a data set, it is helpful to first put the data in an ordered array, as demonstrated in the following example.

Example 3.6

Finding the Mode

Given the number of phone calls received each hour during business hours for four different companies, find the mode of each, and state if the data set is unimodal, bimodal, or neither.

a. 6, 4, 6, 1, 7, 8, 7, 2, 5, 7

b. 3, 4, 7, 8, 1, 6, 9

c. 2, 5, 7, 2, 8, 7, 9, 3

d. 2, 2, 3, 3, 4, 4, 5, 5

Solution

a. To make it easier to see which value(s) occur most often, begin by putting the data in numerical order. The ordered data set is: 1, 2, 4, 5, 6, 6, 7, 7, 7, 8. The number 7 occurs more than any other value, so the mode is 7. This data set is unimodal.

b. Begin by sorting the data as follows: 1, 3, 4, 6, 7, 8, 9. The values all occur only once, so there is no mode. This data set is neither unimodal nor bimodal.

c. As before, sort the data: 2, 2, 3, 5, 7, 7, 8, 9. The values 2 and 7 both occur an equal number of times; thus they are both modes. This data set is bimodal.

d. Note that this data set is already sorted for us. All of the data values occur the same number of times, so there is no mode. This data set is neither unimodal nor bimodal.

Note that, although some statistical software packages will identify the mode of a data set, the mode is not one of the descriptive statistics listed by a TI-83/84 Plus calculator.

Choosing an Appropriate Measure of Center

We now know how to calculate three measures of center: mean, median, and mode. For a given data set, which measure of center is best? Well, it depends on the data set. In order to answer that question, we must consider the type of data in the set. Let's first consider what we might mean by the "average" college major. The "average" in this sense would refer to the most typical major chosen by college students. Thus, it describes the most *frequently occurring* college major, which is the mode of that nominal data set. Thus for nominal data, the mode should be used. The mode is also the best choice for ordinal data. Therefore, we can say that the mode is the best measure of center for qualitative data.

This answers the question for qualitative data, but what about quantitative data? Recall from Chapter 2 that an outlier is an extreme data value. Because it is much larger or much smaller than the rest of the data, it can affect the center. Let's consider the following example to see how an outlier affects the various measures of center in a quantitative data set.

Example 3.7

Calculating Measures of Center—Mean, Median, and Mode

Given the recent economy and change of attitude in society, many people chose to take on another job after retiring from one. Below is a sample of ages at which people truly retired; that is, they stopped working for pay. Calculate the mean, median, and mode for the data.

84, 80, 82, 77, 78, 80, 79, 42

Solution

Mean: Remember, the mean is the sum of all the data points divided by the number of points.

$$\begin{aligned} \bar{x} &= \frac{\sum x_i}{n} \\ &= \frac{84+80+82+77+78+80+79+42}{8} \\ &= \frac{602}{8} \\ &= 75.25 \\ &\approx 75.3 \end{aligned}$$

Median: We have an even number of values, so we will need the mean of the middle two values in the ordered array.

42, 77, 78, 79, 80, 80, 82, 84

$$\frac{79+80}{2} = 79.5$$

Mode: The number 80 occurs more than any other number, so it is the mode.

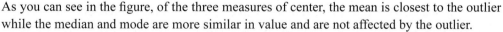

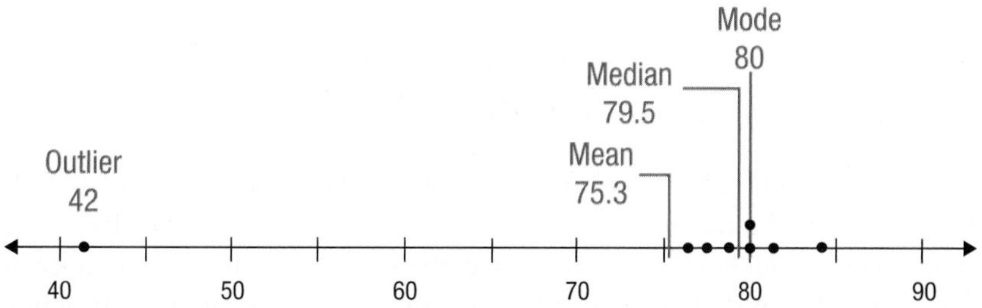

As you can see in the figure, of the three measures of center, the mean is closest to the outlier while the median and mode are more similar in value and are not affected by the outlier.

Note that the median and mode of the previous example are close to the same value, but the mean is smaller. This is because the outlier, 42, is much smaller than the rest of the data. The value for the mean will be pulled toward any outlier; thus the mean is pulled towards the tail of a skewed distribution. For this reason, if a quantitative data set has an outlier or is skewed, you should use the median. In summary, here are a few guidelines for choosing the most appropriate measure of center for a given data set.

Procedure

Determining the Most Appropriate Measure of Center

1. For qualitative data, the mode should be used.
2. For quantitative data, the mean should be used, unless the data set contains outliers or is skewed.
3. For quantitative data sets that are skewed or contain outliers, the median should be used.

Side Note

Inappropriate Average

In the late 1980s, the University of North Carolina reported an unusually high average starting salary for graduates of their geography program. What could have been the reason for UNC's superior average? Were UNC instructors better qualified? Did the program attract higher quality students? Or was there some other reason for this exceptional mean?

It turns out that the source of this inflated mean was the inclusion of one famous graduate of UNC's geography program: Michael Jordan. Though he earned his salary from playing professional basketball rather than from the field of geography, by keeping his multimillion dollar salary in the mix the value of the mean was significantly increased. Needless to say, Michael Jordan's salary would be considered an outlier. In this case, the median salary would more accurately represent the true center of the data.

Example 3.8

Choosing the Most Appropriate Measure of Center

Choose the best measure of center for the following data sets.

a. T-shirt sizes (S, M, L, XL) of American women

b. Salaries for a professional team of baseball players

c. Prices of homes in a subdivision of similar homes

d. Professor rankings from student evaluations on a scale of *best*, *average*, and *worst*

Solution

a. T-shirt sizes are ordinal data; since they are qualitative, the mode is the best measure of center.

b. The players' salaries are quantitative data with outliers, since the superstars on the team make substantially more than the typical players. Therefore, the median is the best choice.

c. The home prices are quantitative data with no outliers, since the homes are similar. Therefore, the mean is the best choice.

d. The rankings are ordinal data; since they are qualitative, it's best to use the mode as a measure of center.

Graphs and Measures of Center

When looking at the graph of a distribution, it is often possible to estimate certain measures of center. For example, the mode is the number that occurs most frequently. Graphically, the data value at which a distribution has its highest peak is the mode. Since the median is the number in the middle, on a graph of a distribution, it will be the number that divides the area of the distribution in half. As we previously stated, the mean of a distribution will be pulled toward any outliers. For example, if the distribution is skewed to the left, the mean of the distribution will be shifted to the left. When all three measures of center are plotted on a distribution that is skewed to the left, a good rule of thumb to remember is that the mean is most likely the measure of center farthest to the left. In a unimodal data set, if the mean, median, and mode are all equal, the shape of the distribution will be bell-shaped. Using this information, we can determine the general location (and therefore an approximate value) of each measure of center.

Properties

Graphs and Measures of Center

1. The mode is the data value at which a distribution has its highest peak.
2. The median is the number that divides the area of the distribution in half.
3. The mean of a distribution will be pulled toward any outliers.

Example 3.9

Determining Mean, Median, and Mode from a Graph

Determine which letter represents the mean, the median, and the mode in the graph to the right.

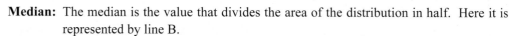

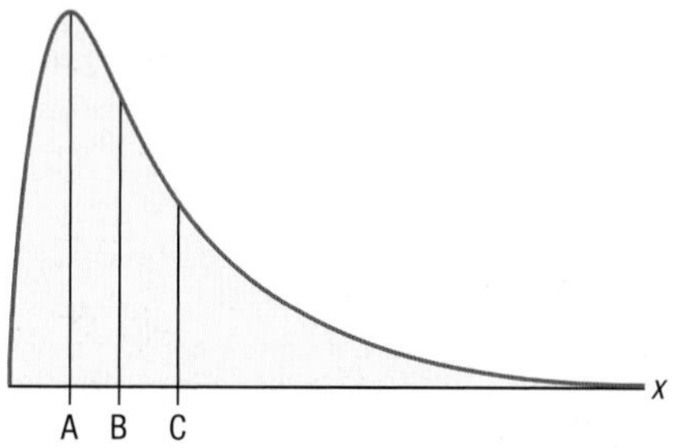

Solution

Mode: The mode is always located at the peak of a distribution, so it is line A.

Median: The median is the value that divides the area of the distribution in half. Here it is represented by line B.

Mean: This distribution is skewed to the right, so the mean will be the measure of center farthest to the right. Here it is represented by line C.

Table 3.1: Properties of Mean, Median, and Mode

Mean	Median	Mode
"Average"	"Middle value"	"Most frequent value"
May not be a data value	May not be a data value	Must be a data value
Single value	Single value	Could be one value, multiple values, or not exist
Affected by outliers	Not affected by outliers	Not affected by outliers
Use for quantitative data with *no* outliers	Use for quantitative data with outliers	Use for qualitative data

3.1 Section Exercises

Mean, Median, and Mode

Directions: Find the mean, median, and mode for each of the given data sets, and state whether the data set is unimodal, bimodal, multimodal or has no mode.

1. 10, 1, 5, 9, 4, 1

2. 190, 219, 160, 250, 175, 180, 240, 290

3. 8, 9, 10, 12, 4, 3, 5, 7, 10

4. 5, 5, 5, 5, 5, 5, 5

5. 4.25, 8.3, 6.0, 3.2, 1.9, 4.4, 4.9, 5.9, 5.1

6. $25.00, $42.00, $15.00, $6.16, $194.73, $60.00, $16.00, $6.40, $1.10

7. 0.7941, 0.8333, 0.6667, 0.8750, 0.7222, 0.9412, 0.8421, 0.7333, 0.6176, 0.7353, 0.5833

8. 36, –76, 36, –20, 100, 98, 18

9. –22, –83, 77, –79, –2, –42, 98, –66, –42

10. 0.5, –0.6, 1.7, 0.5, 1.3, 1.0, 1.3, 0.5

11. The following data represent weights of newborn babies (in pounds).

Weights of Newborn Babies

Stem	Leaves
5	4 9
6	5 6 8
7	3 3 5 5 8
8	2 5 7
9	1 3

Key: 5 | 4 = 5.4 pounds

12. The following data represent batting averages for a sample of professional baseball players.

.275	.265	.333	.244	.279
.250	.288	.292	.370	.243
.236	.321	.305	.292	.250

13. The following data represent average numbers of Tweets per day posted on Twitter for 16 high school students.

0.8	42.2	20.6	2.8
36.7	18.6	23.3	11.5
3.7	14.9	9.4	1.5
14.9	31.1	23.5	9.5

14. The following data represent high temperatures for cities in the Southeast (in degrees Fahrenheit).

High Temperatures for Cities in the Southeast

Stem	Leaves
7	
7	7 9
8	2 3 4
8	5 5 7 8 8 9 9
9	0 0 1 1 2 2 3
9	5

Key: 7 | 7 = 77 °F

15. The following data represent the ages of 20 American entrepreneurs (in years).

28	39	43	53
35	32	34	29
33	31	32	31
25	22	30	29
41	36	23	47

Using the Mean to Find a Data Value

Directions: Use the given information to determine the unknown value.

16. The mean cost for items in a bag of groceries is \$1.96. There are 12 items in the bag, and the following are the prices for 11 of those items. Determine the price of the 12th item in the bag.

\$2.69, \$1.88, \$2.18, \$1.99, \$0.99, \$1.99, \$0.97, \$3.49, \$1.97, \$2.48, \$0.52

17. A plane that ferries visitors to a small resort island has strict guidelines on the weight allowed for passenger luggage. Consequently, the six passengers are limited to a maximum average luggage weight of 35 pounds (lb). The following are the weights of four out of six pieces of luggage: 39 lb, 22 lb, 35 lb, and 37 lb. The two pieces of luggage that haven't been weighed will have to split the remaining weight allowance. If the remaining weight allowance is split evenly between the bags, determine the maximum possible weight allowance for each remaining bag.

Weighted Mean

Directions: Calculate the weighted mean as described in each exercise.

18. Marquis is calculating his cumulative GPA. His grades are as follows: A (15 hours), B (18 hours), C (8 hours), and D (3 hours). Note that an A is equivalent to a 4.0, a B is equivalent to a 3.0, a C is equivalent to a 2.0, and a D is equivalent to a 1.0. What is Marquis' GPA?

19. Beth is calculating her cumulative GPA. Her grades are as follows: A (12 hours), B (22 hours), C (14 hours), and F (3 hours). Note that an A is equivalent to a 4.0, a B is equivalent to a 3.0, a C is equivalent to a 2.0, a D is equivalent to a 1.0, and an F is equivalent to 0. What is Beth's GPA?

20. The following table gives the average balances for one bank customer for the months of October through December.

Average Monthly Balances for a Bank Customer (October through December)	
Month	**Average Balance**
October	$2251.33
November	$2490.51
December	$1478.27

Calculate the average monthly balance for the three-month period of October through December. Note that, since each month contains a different number of days, the average balance for each month must be weighted by the number of days in that month.

21. The following table gives the average balances for one bank customer for the months of July through September.

Average Monthly Balances for a Bank Customer (July through September)	
Month	**Average Balance**
July	$402.45
August	$322.97
September	$298.64

Calculate the average monthly balance for the three-month period of July through September. Note that, since each month contains a different number of days, the average balance for each month must be weighted by the number of days in that month.

22. Susan is calculating her final grade in biology. The grade is broken down as follows: tests (50%), lab (30%), and final exam (20%). Susan's category averages are the following.

Tests:	82
Lab:	78
Final Exam:	86

What is her final grade?

23. If I survey 20% of 50 people and 80% of 500 people, what percentage of the total population of 550 people have I surveyed?

24. Maeve wants to calculate the average final exam grade of the students that she teaches. Her first class, which had 42 students in it, averaged an 88 on their final exam. Her second class of 55 students averaged an 81 and her last class of 48 students averaged an 84. What is the average final exam grade of her students?

25. Let's extend the concept of weighted mean to include the mean for a frequency distribution. The following frequency distribution gives the hourly wage for a person's first job.

Hourly Wage at First Job (in Dollars)

Class	Frequency
7.50–8.49	12
8.50–9.49	50
9.50–10.49	48
10.50–11.49	45
11.50–12.49	34

Since we do not know the exact value of each person's hourly wage, we will estimate that each value in a class is equal to the midpoint of that class. To calculate the mean of the frequency distribution by hand, we can use the formula for a weighted mean, with the data values equal to the class midpoints and the weights equal to the class frequencies. To use a TI-83/84 Plus calculator to find the mean of the frequency distribution, we can use the same procedure outlined in Example 3.3, again with the data values equal to the class midpoints and the weights equal to the class frequencies. Estimate the mean hourly wage for the given data.

Determining the Most Appropriate Measure of Center

Directions: Determine the most appropriate measure of center for the described data set.

26. Hairstyles of female students on a given college campus (length, color, straight/curly, and so forth)

27. Prices of used cars on a lot including the following models: Buick LaCrosse, Toyota 4Runner, Honda Prelude, Ford Escape, Dodge Grand Caravan, Ford Explorer, and Ferrari F430

28. Number of minutes students spend completing a homework assignment for an honors class

29. Prices of similar sofas at different furniture stores

30. B, C, D, B, D, B, D, A, B, A, D, C, A, B

31. The ratings below for a popular vacation spot in Florida, which have the following scale.

***** = Excellent

**** = Above average

*** = Average

** = Below average

* = Terrible

, *, ***, ***, **, ***, ***

Graphs and Measures of Center

Directions: For each graph, determine which letter represents the mean, the median, and the mode.

32.

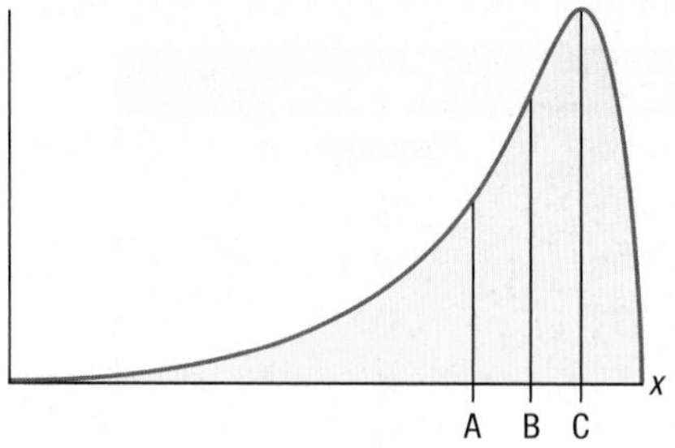

33.

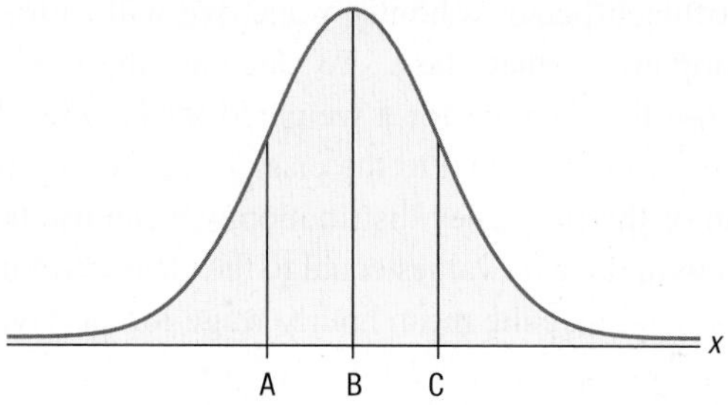

34. Given the following measures of center for a data set, describe the most likely shape of the distribution.

Mean = 121.89, Median = 163.5, Mode = 165.2

3.2 Measures of Dispersion

In the previous section, we calculated various measures of center. These numerical descriptions tell us where the data set is centered on a number line, but we still need more information to determine the shape of the distribution of the data. For instance, consider the average wait time for patients in the waiting room at a doctor's office. If patients are told that the average wait time is 20 minutes, does this mean that patients will always wait for about 20 minutes? Or is it possible that although the average is 20 minutes, some patients get right in to see the doctor and others wait for much longer than 20 minutes? Knowing the true variation in wait times might affect the way in which patients schedule their days. At the very least, it might allay frustration! To better analyze a scenario such as this, we need to learn about several other numerical descriptions called *measures of dispersion.*

Range

The easiest measure of dispersion to calculate is the **range**. The range of a data set is the difference between the largest and smallest values in the data set.

Formula

Range

$$\text{Range} = \text{Maximum Data Value} - \text{Minimum Data Value}$$

Example 3.10

Calculating the Range

The following data were collected from samples of call lengths (in minutes) observed for two different mobile phone users. Calculate the range of each data set.

a. 2, 25, 31, 44, 29, 14, 22, 11, 40

b. 2, 2, 44, 2, 2, 2, 2, 2

Solution

a. The maximum value is 44 minutes and the minimum value is 2 minutes, so the range is as follows.

$$\begin{aligned}\text{Range} &= \text{Maximum Data Value} - \text{Minimum Data Value}\\ &= 44 - 2\\ &= 42 \text{ minutes}\end{aligned}$$

b. The maximum value for the second data set is also 44 minutes and the minimum value is also 2 minutes, so the range is calculated in the same way.

$$\begin{aligned}\text{Range} &= \text{Maximum Data Value} - \text{Minimum Data Value}\\ &= 44 - 2\\ &= 42 \text{ minutes}\end{aligned}$$

Calculating the range is very easy; however, the range is not as descriptive as other measures of dispersion. Consider the two data sets in the previous example. Notice that both data sets have the same range. However, almost all of the values in the second data set are the same, while the values in the first data set are more spread out. To distinguish between these two situations, we must use other measures of dispersion.

Standard Deviation

The most commonly used measure of dispersion is **standard deviation**. The standard deviation provides a measure of how much we might expect a typical member of the data set to differ from the mean. Thus, the units of the standard deviation are the same as the units in which the data are measured. The greater the standard deviation, the more the data values are spread out. Similarly, a smaller standard deviation indicates that the data values lie closer together. If the standard deviation equals 0, then none of the values differ from the mean, and therefore must all be the same as the mean. We also know that the standard deviation could never be a negative number, since measures of distances are always positive. Let's take a look at the formulas for calculating standard deviation.

Rounding Rule

When calculating the standard deviation, round to one more decimal place than the largest number of decimal places given in the data. Occasional exceptions to this rule can be made when the type of data lends itself to a more natural rounding scheme, such as rounding values of currency to two decimal places.

Formula

Standard Deviation

The **standard deviation** is a measure of how much we might expect a typical member of the data set to differ from the mean.

The **population standard deviation** is given by

$$\sigma = \sqrt{\frac{\sum (x_i - \mu)^2}{N}}$$

where x_i is the ith value in the population,

μ is the population mean, and

N is the number of values in the population.

The **sample standard deviation** is given by

$$s = \sqrt{\frac{\sum (x_i - \bar{x})^2}{n-1}}$$

where x_i is the ith data value,

$\bar{x}$ is the sample mean, and

n is the number of data values in the sample.

Math Symbols

σ: population standard deviation; Greek letter, sigma

s: sample standard deviation

Notice that there is a difference between the formulas for population and sample standard deviation, unlike in the case of population and sample means. The reason for this is that adjustments must be made in the formula for sample standard deviation to account for the fact that the sample standard deviation is a biased estimator of the population standard deviation. We'll explore biased estimators more in later chapters.

Should you desire to calculate standard deviation by hand, especially with large data sets, it is helpful to use a table as illustrated in the following example. However, using a calculator or other technology can greatly simplify this task as further demonstrated in the example.

Example 3.11

Calculating Standard Deviation

Calculate the sample standard deviation of the following data collected regarding the numbers of hours students studied for a physics exam.

5, 8, 7, 6, 9

Solution

Let's calculate the sample standard deviation by hand using the following formula.

$$s=\sqrt{\frac{\sum(x_i-\bar{x})^2}{n-1}}$$

To start, we need the mean and sample size of the data set. We calculate that $\bar{x}=7$ and $n=5$. Next, we need to subtract the mean from each number in the sample, and then square each of these differences. Let's use a chart to keep everything organized.

Deviations and Squared Deviations of the Data

x_i	$(x_i-\bar{x})$	$(x_i-\bar{x})^2$
5	$5-7=-2$	4
8	$8-7=1$	1
7	$7-7=0$	0
6	$6-7=-1$	1
9	$9-7=2$	4

Next, find the sum of the squared deviations by adding up the values in the last column.

$$\begin{aligned}\sum(x_i-\bar{x})^2&=4+1+0+1+4\\&=10\end{aligned}$$

Finally, substitute the appropriate values into the sample standard deviation formula as follows.

$$\begin{aligned}s&=\sqrt{\frac{\sum(x_i-\bar{x})^2}{n-1}}\\&=\sqrt{\frac{10}{5-1}}\\&=\sqrt{2.5}\\&\approx1.581139\\&\approx1.6\end{aligned}$$

Alternate Calculator Method

To find the sample standard deviation on a TI-83/84 Plus calculator, follow the steps below.

- Press **STAT**.
- Choose option `1:Edit` and press **ENTER**.
- Enter the data in L1.
- Press **STAT** again.
- Choose CALC.
- Choose option `1:1-Var Stats`.
- Press **ENTER** twice. (Note: If your data are not in L1, before pressing **ENTER** the second time, enter the list where your data are located, such as L3 or L5.)

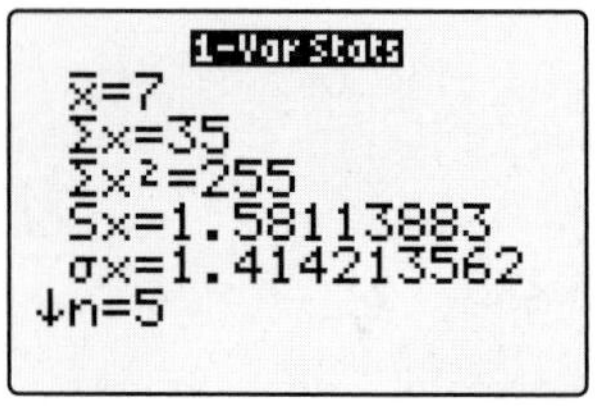

The fourth value in the output, seen in the screenshot in the margin, gives the sample standard deviation, which is Sx = 1.58113883 ≈ 1.6.

Formula

Where Does the Formula Come From?

You can think of the standard deviation as a sort-of average distance that values in a set lie from the mean. While not the actual average, the standard deviation is usually very close to the average and, conceptually, that is a good way to think about standard deviation as we discuss how the formula is derived. To derive the formula for standard deviation, we will use a method similar to finding average distance, or deviation, from the mean. First, we must know the actual deviation from the mean for every data value in the set. The deviation is simply the difference between each value and the mean.

Next, we need to find the sum of all the deviations. Here we run into a problem because the sum is zero. In fact, the sum of the deviations from the mean for any data set is always equal to zero because the positive deviations cancel out the negative deviations. To help eliminate the problem with the negative values canceling the positive values, we can make all the deviations positive by squaring each one.

Now, find the average by summing up all the squared deviations and dividing by the population size just as you would find a traditional mean. To find the standard deviation, we must take one additional step and "undo" the previous squaring by taking the square root. The completed population standard deviation formula is as follows.

$$\sigma = \sqrt{\frac{\sum (x_i - \mu)^2}{N}}$$

Example 3.12

Using a TI-83/84 Plus Calculator to Find Standard Deviation

Use a TI-83/84 Plus calculator to find the standard deviation of the data shown below, given the following conditions.

11, 18, 25, 51, 44, 29, 30, 17, 29, 47, 52, 60

a. Assume that the values represent the ages (in years) of patients randomly sampled from an urgent care clinic.

b. Assume that the values represent the ages (in years) of all patients seen by Dr. Dabbs one afternoon.

Solution

- Press **STAT**.
- Choose `1:Edit`.
- Enter the data in `L1`.
- Press **STAT** again.
- Choose `CALC`.
- Choose option `1:1-Var Stats`.
- Press **ENTER** twice.

A list of numerical summaries will be generated for the data. The beginning of the list is shown in the margin. Use these values to find the answers for this example.

```
1-Var Stats
x̄=34.41666667
Σx=413
Σx²=16991
Sx=15.88857985
σx=15.21215705
↓n=12
```

a. We are told that the values in this case represent a random sample of patients. We will then need to use the sample standard deviation, denoted on the calculator by `Sx`. From the list, we see that $s \approx 15.9$ years.

b. In this scenario, the values represent all patients seen in one afternoon. The population standard deviation is most appropriate here, and it is denoted on the calculator by `σx`. From the list, we see that $\sigma \approx 15.2$ years.

Now that we know how to calculate the standard deviation, let's look at an example that highlights one way in which the standard deviation of a data set might be useful.

Example 3.13

Interpreting Standard Deviations

Mark is looking into investing a portion of his recent bonus into the stock market. While researching different companies, he discovers the following standard deviations of one year of daily stock closing prices.

Profacto Corporation: Standard deviation of stock prices = \$1.02

Yardsmoth Company: Standard deviation of stock prices = \$9.67

What do these two standard deviations tell you about the stock prices of these companies?

Solution

A smaller standard deviation indicates that the data values are closer together, while a larger standard deviation indicates that the data values are more spread out. In this example, the standard deviation of stock prices for the Profacto Corporation is considerably smaller than that of the Yardsmoth Company. Hence, there is less variability in the daily closing prices of the Profacto stock than in the Yardsmoth stock prices. If Mark wants a stable long-term investment, then Profacto appears to be the better choice. If, however, Mark is looking to make a quick profit and is willing to take the risk, then the Yardsmoth stock would seem to better suit his purposes. Note that looking at the standard deviations is just one component of evaluating market prices.

The standard deviation allows us to interpret differences from the mean with some sense of scale. For instance, if a data set consisted of house prices, then differences of thousands of dollars might be considered small compared to differences of tens of thousands of dollars. However, if a data set consisted of gas prices in various towns, a difference of even a single dollar would be considered very large. As we shall see, the standard deviation allows us to make such judgments of whether a difference is large or small, in a systematic way.

Which of the two graphs has the larger standard deviation?

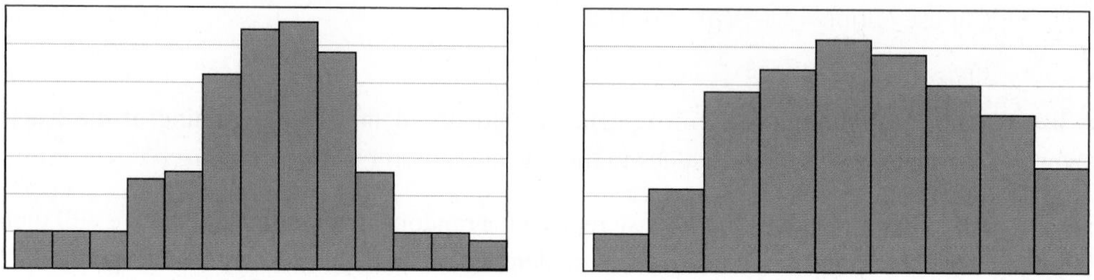

A **B**

Figure 3.1

We know that the graph with the larger standard deviation should appear more spread out; but without the numerical labeling for these graphs, one cannot tell which graph has the largest spread. In fact, if these graphs are from two different types of data, they may not have the same units of measurement, and so comparison is meaningless. However, we can do this comparison mathematically by using the **coefficient of variation, CV**. The CV is the ratio of the standard deviation to the mean as a percentage.

Formula

Coefficient of Variation

The **coefficient of variation** for a set of data is the ratio of the standard deviation to the mean as a percentage. For a population, it is given by

$$\mathrm{CV} = \frac{\sigma}{\mu} \cdot 100\%$$

where σ is the population standard deviation and

μ is the population mean.

For a sample, it is given by

$$\mathrm{CV} = \frac{s}{\bar{x}} \cdot 100\%$$

where s is the sample standard deviation and

$\bar{x}$ is the sample mean.

The coefficient of variation allows us to compare the spreads of data from two different sources, as shown in the following example.

Example 3.14

Calculating and Interpreting Coefficient of Variation

Suppose that Graph A from Figure 3.1 represents average amounts of annual rainfall for a sample of farms in the United States and Graph B represents prices per 20 acres of farmland for the same farms. The mean and standard deviation of Data Set A are 26.08 inches and 7.55 inches, respectively, whereas the mean and standard deviation of Data Set B are \$117,000 and \$42,931, respectively.

Which of the two graphs has the larger standard deviation relative to its mean?

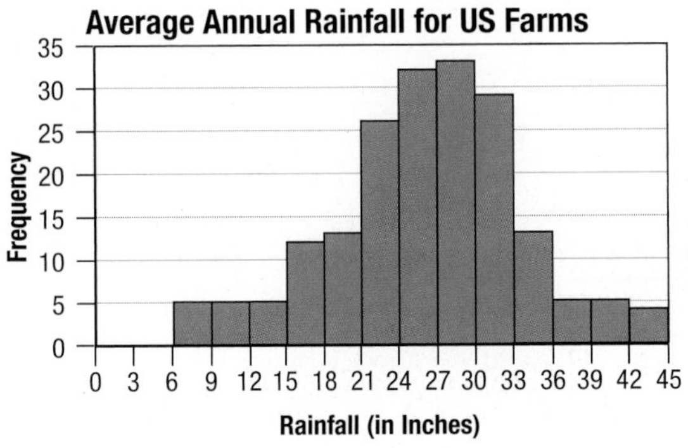

A

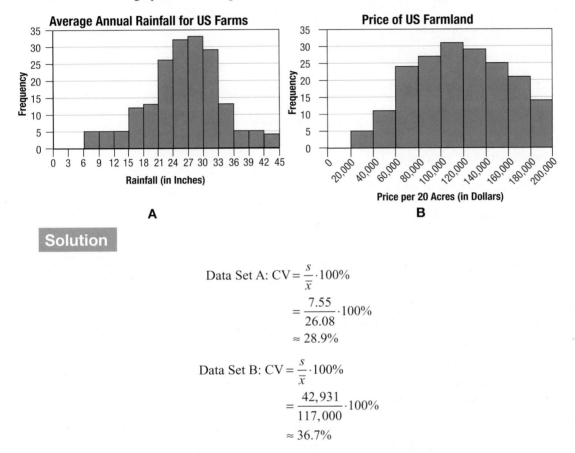

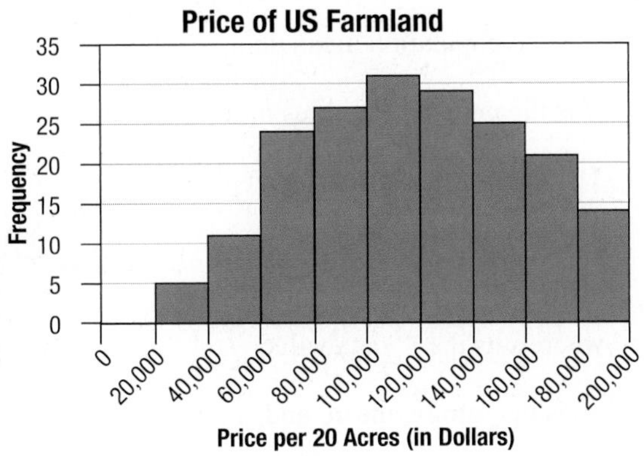

B

Solution

$$\text{Data Set A: CV} = \frac{s}{\bar{x}} \cdot 100\%$$
$$= \frac{7.55}{26.08} \cdot 100\%$$
$$\approx 28.9\%$$

$$\text{Data Set B: CV} = \frac{s}{\bar{x}} \cdot 100\%$$
$$= \frac{42{,}931}{117{,}000} \cdot 100\%$$
$$\approx 36.7\%$$

Now, it is clear to see that Data Set B has the larger standard deviation relative to its own mean. Thus, our comparison of CV shows that there is more variability in Data Set B than in Data Set A.

Variance

Variance and standard deviation are very closely related. In fact, **variance** is simply the square of the standard deviation. That's right; if you know the standard deviation of a data set, then to get the variance, all you have to do is square the standard deviation. So simple! However, when calculating the standard deviation by hand, you actually calculate the variance first. The variance is the value you obtain before you take the square root in the standard deviation formula. It is important to note that, since the variance is the square of the standard deviation, the units of the variance are the square of the units in which the data are measured.

Rounding Rule

When calculating the variance, round to one more decimal place than the largest number of decimal places given in the data.

Math Symbols

σ^2: population variance; read as "sigma squared"

s^2: sample variance

Formula

Variance

The **variance** is the square of the standard deviation.

The **population variance** is given by

$$\sigma^2 = \frac{\sum (x_i - \mu)^2}{N}$$

where x_i is the ith value in the population,

μ is the population mean, and

N is the number of values in the population.

The **sample variance** is given by

$$s^2 = \frac{\sum (x_i - \bar{x})^2}{n-1}$$

where x_i is the ith data value,

$\bar{x}$ is the sample mean, and

n is the number of data values in the sample.

Example 3.15

Calculating Variance

Calculate the variance of the data shown below, given the following conditions.

3, 2, 5, 6, 4

a. Assume that the data represent the actual weight changes (in pounds) for a sample of fitness club members during the month of April.

b. Assume that the data represent the actual weight changes (in pounds) of every member of a book club during the month of April. Use a TI-83/84 Plus calculator to perform the calculation.

Solution

a. As these data represent a sample, we need to calculate the sample variance. The formula that we need is given below.

$$s^2 = \frac{\sum (x_i - \bar{x})^2}{n-1}$$

To start, we need the mean and sample size of the data set. We calculate that $\bar{x} = 4$ and $n = 5$. Next, we need to subtract the mean from each number in the sample, and then square each of these differences. Let's use a chart to keep everything organized.

Deviations and Squared Deviations of the Data		
x_i	$(x_i - \bar{x})$	$(x_i - \bar{x})^2$
3	$3-4=-1$	1
2	$2-4=-2$	4
5	$5-4=1$	1
6	$6-4=2$	4
4	$4-4=0$	0

Next, find the sum of the squared deviations by adding up the values in the last column.

$$\begin{aligned}\sum(x_i-\bar{x})^2 &= 1+4+1+4+0\\ &= 10\end{aligned}$$

Finally, substitute the appropriate values into the formula for sample variance as follows.

$$\begin{aligned}s^2 &= \frac{\sum(x_i-\bar{x})^2}{n-1}\\ &= \frac{10}{5-1}\\ &= 2.5\end{aligned}$$

b. To calculate a variance on a TI-83/84 Plus calculator, you must actually calculate the standard deviation and then square that value to get the variance. The steps for calculating the standard deviation of a data set were presented in Example 3.12, and are repeated below.

- Press STAT.
- Choose 1:Edit.
- Enter the data in L1.
- Press STAT again.
- Choose CALC.
- Choose option 1:1-Var Stats.
- Press ENTER twice.

The calculator actually presents the values for both the population and sample standard deviations. Since the data set in this scenario represents a population (all members of the book club), we are looking for the population variance. Thus, we need to square the population standard deviation, given by $\sigma = 1.414213562$.

$$\begin{aligned}\sigma^2 &\approx (1.414214)^2\\ &\approx 2.0\end{aligned}$$

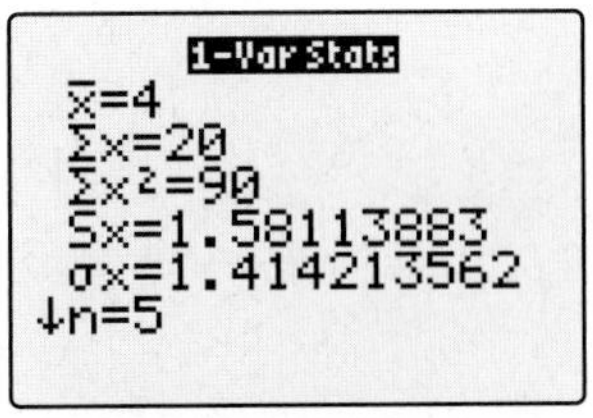

Rounding Rule

Very important! DO NOT round the standard deviation to fewer than six decimal places before you square it to get the variance. If you round too early, then your answer may not be accurate!

If standard deviation is so useful, then what about the variance? Since variance is a squared unit, we cannot use it to compare distances between data values. So for the layperson, variance really is not all that useful in analyzing data. However, in many of the statistical tests that we will learn about later in this book, the variance is often preferred. This makes sense when you look back at the formulas for standard deviation and variance. Since the formula for the variance does not contain a square root, it is much easier to work with than the standard deviation.

Empirical Rule

Now that we know how to calculate standard deviation, we can use the standard deviation to help us determine how spread out the data values are in relation to the mean. When the distribution of a set of data is *bell-shaped*, the **Empirical Rule** can be used to estimate the percentage of values within a few standard deviations of the mean. Recall that describing a curve as bell-shaped implies that there is symmetry about the middle value, which is the mean. This fact will help us when working with the Empirical Rule. The Empirical Rule is as follows.

Theorem

Empirical Rule for Bell-Shaped Distributions

- Approximately 68% of the data values lie within one standard deviation of the mean.
- Approximately 95% of the data values lie within two standard deviations of the mean.
- Approximately 99.7% of the data values lie within three standard deviations of the mean.

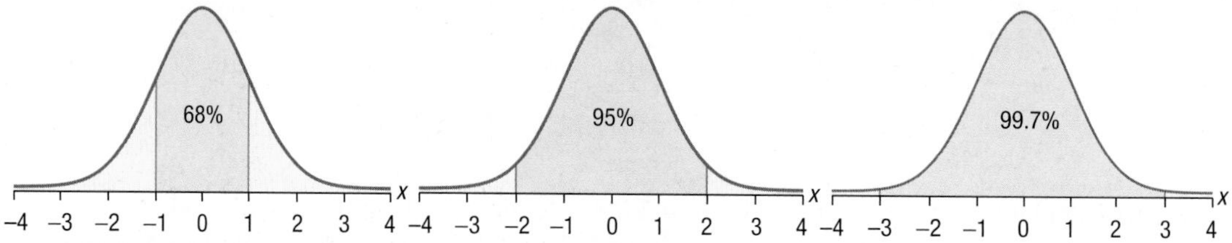

Figure 3.2: Graphical Depiction of the Empirical Rule

Note: The units on the *x*-axis in the graphs above are distances from the mean in terms of numbers of standard deviations.

Example 3.16

Applying the Empirical Rule for Bell-Shaped Distributions

The distribution of weights of newborn babies is bell-shaped with a mean of 3000 grams and standard deviation of 500 grams.

a. What percentage of newborn babies weigh between 2000 and 4000 grams?

b. What percentage of newborn babies weigh less than 3500 grams?

c. Calculate the range of birth weights that would contain the middle 68% of newborn babies' weights.

Solution

a. Since we know that the distribution of the data is bell-shaped, we can apply the Empirical Rule. We need to know how many standard deviations 2000 grams and 4000 grams are from the mean. By subtracting, we can find how far each of these figures is from the mean. Then, dividing by the standard deviation, we can convert these differences into numbers of standard deviations. Here are the calculations.

$$2000 - 3000 = -1000 \quad \text{and} \quad 4000 - 3000 = 1000$$

$$\frac{-1000}{500} = -2 \qquad \frac{1000}{500} = 2$$

Thus these weights lie two standard deviations above and below the mean. According to the Empirical Rule, approximately 95% of values lie within two standard deviations of the mean. Therefore, we can say that approximately 95% of newborn babies weigh between 2000 and 4000 grams.

b. To begin, let's find out how many standard deviations a weight of 3500 grams is away from the mean by performing the same calculation as before.

$$3500 - 3000 = 500$$

$$\frac{500}{500} = 1$$

Thus it is one standard deviation above the mean. The Empirical Rule says that 68% of data values lie within one standard deviation of the mean. Because of the symmetry of the distribution, half of this 68% is above the mean and half is below. Putting the upper 34% together with the 50% of data that is below the mean, we have that approximately

$$50\% + 34\% = 84\%$$

of newborn babies weigh less than 3500 grams.

Weights of Newborn Babies

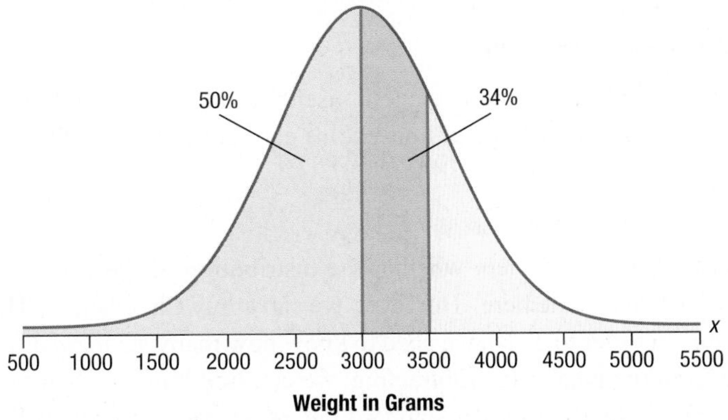

c. From the Empirical Rule, we know that 68% of the data values lie within one standard deviation of the mean for bell-shaped distributions. The standard deviation of this distribution is 500; thus, by adding 500 to and subtracting 500 from the mean of the distribution, we will get the range of birth weights that contain the middle 68% of newborn babies' weights.

$$\text{Upper end: } 3000 + 500 = 3500$$

$$\text{Lower end: } 3000 - 500 = 2500$$

Thus, 68% of newborn babies weigh between 2500 and 3500 grams.

Although the Empirical Rule is a good rule of thumb, in later chapters we will learn techniques for calculating areas under bell-shaped curves more precisely.

Chebyshev's Theorem

Although the Empirical Rule is handy for bell-shaped distributions of data, it cannot be applied to other distributions. Chebyshev's Theorem is helpful when the Empirical Rule cannot be used. However, Chebyshev's Theorem simply gives a minimum estimate; it is a less accurate estimate than given by the Empirical Rule for bell-shaped distributions.

Theorem

Chebyshev's Theorem

The proportion of data that lie within K standard deviations of the mean is at least $1-\frac{1}{K^2}$, for $K>1$. When $K=2$ and $K=3$, Chebyshev's Theorem says the following.

- $K=2$: At least $1-\frac{1}{2^2}=\frac{3}{4}=75\%$ of the data lie within two standard deviations of the mean.
- $K=3$: At least $1-\frac{1}{3^2}=\frac{8}{9}\approx 88.9\%$ of the data lie within three standard deviations of the mean.

Example 3.17

Applying Chebyshev's Theorem

Suppose that in one small town, the average household income is \$34,200, with a standard deviation of \$2200. What percentage of households earn between \$27,600 and \$40,800?

Solution

Since we are not told in the problem whether the distribution of the data is bell-shaped, we cannot apply the Empirical Rule here. However, we can apply Chebyshev's Theorem to find a minimum estimate. In order to do so, we need to know how many standard deviations \$27,600 and \$40,800 are from the mean. By subtracting, we can find how far each of these figures is from the mean. Then, dividing by the standard deviation, we can convert these differences into numbers of standard deviations. Here are the calculations.

$$\$27,600-\$34,200=-\$6600 \quad \text{and} \quad \$40,800-\$34,200=\$6600$$

$$\frac{-\$6600}{\$2200}=-3 \qquad \frac{\$6600}{\$2200}=3$$

Thus these incomes lie three standard deviations above and below the mean. Chebyshev's Theorem can then be applied for $K=3$. Using the calculation previously shown in the box with the theorem, we can say that at least 88.9% of the household incomes lie within this range.

3.2 Section Exercises

Range, Population Standard Deviation, and Population Variance

Directions: Calculate the range, population standard deviation, and population variance of the given data set.

1. 3, 5, 7, 8

2. 8, 12, 10, 12, 4, 13, 5, 17, 10

3. 5, 5, 5, 5, 5, 5, 5

4. –5, 6, 3, –5, –7, –7, 8, –25

5. 17, 19, 21, 16, 20, 19, 17, 13

6. 1.6, 2.2, 1.1, 3.0, 4.9, 2.8, 5.7, 4.5, 6.0

Range, Sample Standard Deviation, and Sample Variance

Directions: Calculate the range, sample standard deviation, and sample variance of the given data set.

7. 2, 2, 3, 4, 7

8. 15, 12, 13, 14, 15, 17, 18, 11, 12, 15

9. 4, 4, 4, 4, 4, 4, 4

10. –1, 2, 3, –2, –4, 4, 5, –35, 2

11. 2.0, 1.7, 1.9, 1.5, 1.6, 1.9, 1.7, 3.0

12. 8.1, 9.5, 10.7, 12.3, 4.4, 3.9, 5.1, 7.3, 10.1

Sample Standard Deviation

Directions: Calculate the sample standard deviation for the given data set.

13. The following data represent weights of newborn babies (in pounds).

6.8	9.1	8.7	7.5	8.2
5.4	6.5	8.5	7.3	6.6
5.9	7.3	9.3	7.5	7.8

14. The following data represent batting averages for a sample of professional baseball players.

.275	.265	.333	.244	.279
.250	.288	.292	.370	.243
.236	.321	.305	.292	.250

15. The following data represent average numbers of Tweets per day posted on Twitter for 16 high school students.

0.8	42.2	20.6	2.8
36.7	18.6	23.3	11.5
3.7	14.9	9.4	1.5
14.9	31.1	23.5	9.5

16. The following data represent high temperatures for cities in the Southeast (in degrees Fahrenheit).

High Temperatures for Cities in the Southeast

Stem	Leaves
7	
7	7 9
8	2 3 4
8	5 5 7 8 8 9 9
9	0 0 1 1 2 2 3
9	5

Key: 7 | 7 = 77 °F

17. The following data represent ages of 20 American entrepreneurs (in years).

Ages of American Entrepreneurs

Stem	Leaves
2	2 3
2	5 8 9 9
3	0 1 1 2 2 3 4
3	5 6 9
4	1 3
4	7
5	3
5	

Key: 2 | 2 = 22 years old

Standard Deviation and Variance

Directions: Decide if each statement is true or false. Explain why.

18. If the standard deviation of a data set is zero, then all entries in the data must equal zero.

19. The population variance and sample variance are the same value for the same set of data.

20. It is possible to have a standard deviation of –3 for some data set.

21. It is possible to have a standard deviation of 435,000 for some data set.

Coefficient of Variation

Directions: Calculate the coefficient of variation, CV, for each data set and then answer the question.

22. The data in set A represent numbers of hours worked in one week for a sample of employees of a fast food restaurant. The data in set B represent numbers of minutes spent waiting for food for a sample of customers at the same restaurant. Which of the two data sets has the *larger* spread relative to its own mean?

A: 31, 33, 35, 39, 31, 32, 30, 40, 13, 41, 38, 32, 37, 33

B: 2.3, 3.5, 4.3, 2.1, 4.8, 3.9, 2.0, 3.3, 4.0, 1.6, 2.2

23. The data in set A represent prices (with tax included) of a large cup of regular coffee for a sample of coffee shops. The data in set B represent numbers of large cups of regular coffee sold in one weekend for the same sample of coffee shops. Which of the two data sets has the *larger* spread relative to its own mean?

A: $1.23, $1.55, $1.01, $1.89, $2.35, $2.56, $2.71, $1.75, $2.01, $1.59

B: 119, 145, 97, 121, 118, 98, 102, 114, 118, 99

24. The data in set A represent numbers of orders received by an online retailer for a random sample of months from the past two years. The data in set B represent package weights (in pounds) for a random sample of customer orders from the same online retailer in the past two years. Which of the two data sets has the *smaller* spread relative to its own mean?

A: 21,568 20,888 20,037 21,932 22,000 21,123 21,567 22,298

B: 4.5, 4.3, 6.5, 3.3, 4.7, 3.67, 4.01, 3.89, 4.4, 2.99, 4.88, 3.77

25. The data in set A represent numbers of graduates from a high school for a random sample of years since the school opened. The data in set B represent the numbers of graduates who started college during the year after graduating from the same high school for the same sample of years. Which of the two data sets has the *smaller* spread relative to its own mean?

A: 328, 444, 283, 289, 345, 327, 298, 277, 419, 402, 399, 418, 401

B: 78, 73, 92, 89, 74, 88, 91, 71, 70, 89, 81, 83, 84

Empirical Rule

Directions: Use the Empirical Rule to answer the questions.

26. Suppose that starting salaries for graduates of one university have a bell-shaped distribution with a mean of $25,400 and a standard deviation of $1300. Approximately what percentage of graduates have starting salaries between $24,100 and $26,700?

27. Suppose that starting salaries for graduates of one university have a bell-shaped distribution with a mean of $25,400 and a standard deviation of $1300. Approximately what percentage of graduates have starting salaries between $22,800 and $28,000?

28. Suppose that electric bills for the month of May in one city have a bell-shaped distribution with a mean of $119 and a standard deviation of $22. Approximately what percentage of electric bills are greater than $97?

29. Suppose that electric bills for the month of May in one city have a bell-shaped distribution with a mean of $119 and a standard deviation of $22. Approximately what percentage of electric bills are less than $163?

30. Suppose it is known that verbal SAT scores have a bell-shaped distribution with a mean of 500 and a standard deviation of 100. Approximately what percentage of verbal SAT scores are no more than 600?

31. Suppose it is known that verbal SAT scores have a bell-shaped distribution with a mean of 500 and a standard deviation of 100. Approximately what percentage of verbal SAT scores are at least 300?

Chebyshev's Theorem

Directions: Use Chebyshev's Theorem to answer the questions.

32. Suppose that salaries for associate mathematics professors at one university have a mean of $64,900 and a standard deviation of $9400. What is the minimum percentage of associate professors with salaries between $46,100 and $83,700?

33. Suppose that household electric bills for the months of May through August in a city in Florida have a mean of $230 and a standard deviation of $58. What is the minimum percentage of electric bills between $56 and $404?

34. Car insurance premiums in one region have a quarterly mean of $246 and a standard deviation of $31. What is the minimum percentage of car insurance premiums between $184 and $308?

Standard Deviation and Variance of Grouped Data

Directions: Estimate the sample standard deviation or variance of the data in each frequency distribution using the given formula.

35. Let's extend the concept of standard deviation to include the standard deviation for a frequency distribution. The following frequency distribution gives the final grades for students in a statistics class.

Final Grades	
Grade	**Frequency**
66–72	4
73–79	7
80–86	12
87–93	8
94–100	5

Since we do not know the exact value of each final grade, we will estimate that each value in a class is equal to the midpoint of that class. Use the following formula to estimate the sample standard deviation of the data in the frequency distribution, if you calculate this value by hand. To calculate this estimate using a TI-83/84 Plus calculator, use the same directions given in Section 3.1 for calculating a weighted mean, entering the midpoints in L1 and the frequencies in L2.

$$s=\sqrt{\frac{n\left[\sum\left(f_i\cdot x_i^2\right)\right]-\left[\sum\left(f_i\cdot x_i\right)\right]^2}{n(n-1)}},$$

where n = sample size,

f_i = frequency of class i, and

x_i = midpoint of class i.

36. Use the formula given in Exercise 35 to estimate the sample standard deviation of the gas prices in the following frequency distribution.

Gas Prices	
Price in Dollars per Gallon	**Frequency**
3.55–3.59	1
3.60–3.64	3
3.65–3.69	5
3.70–3.74	6
3.75–3.79	2
3.80–3.84	1

37. What is the approximate sample variance of the gas prices in the frequency distribution given in Exercise 36?

3.3 Measures of Relative Position

In addition to measuring the center of a data set and how spread out the data values are, we can also measure the relative positions of values in the data set, that is, how they compare to the other data points. Often, it is useful to know exactly where a particular value is located within a set, or the number of values that are above or below it. One example of data values for which relative positions are commonly measured are standardized test scores. A standardized test score is often reported in terms of a *percentile*, which is one measure of relative position.

Percentiles

In order to calculate a value's relative position, we can divide the data into equal parts and state in which part the value lies. We may choose to divide the data up into any number of parts. For example, data divided into 10 parts are called *deciles* and data divided into 8 parts are called *octiles*. Let's consider dividing the data into *100* parts. These divisions are called **percentiles**. As we already stated, percentiles are commonly seen in standardized test scores; they are also often used in medical records. They tell you *approximately* what percentage of the data lie *at or below* a given value. For instance, if your ACT score of 24 falls in the 74th percentile, this indicates that 74% of ACT scores are less than or equal to your score.

One difficulty when it comes to calculating percentiles is that there is little agreement on exactly how to do it. Various statistical software packages, such as SAS, actually allow you to choose the method you want to use to calculate the value of a given percentile. Each of the methods could result in a different answer, depending on the size and variation of your data set. Remember then that these calculations are approximations and, as always, larger data sets will result in a better approximation. In particular, since we are dividing the data set into 100 parts, these calculations give the best approximation for data sets with over 100 values.

Despite the controversy surrounding how best to calculate percentiles, time and space limitations require us to choose a single method in order to demonstrate how to find percentiles in this text. Therefore, we will use the following formula to approximate the value of a certain percentile in a given data set.

Formula

Location of Data Value for the Pth Percentile

To find the *data value* for the Pth percentile, the location of the data value in the data set is given by

$$l = n \cdot \frac{P}{100}$$

where l is the location of the Pth percentile in the *ordered array* of data values,

n is the number of data values in the sample, and

P stands for the Pth percentile.

When using this formula to find the location of the percentile's value in the data set, you must make sure to follow these two rules.

1. If the formula results in a decimal value for l, the location is the next *larger* whole number.
2. If the formula results in a whole number, the percentile's value is the arithmetic mean of the data value in that location and the data value in the next *larger* location.

Example 3.18

Finding Data Values Given the Percentiles

A car manufacturer is studying the highway miles per gallon (mpg) for a wide range of makes and models of vehicles. The stem-and-leaf plot to the right contains the average highway mpg for each of the 135 different vehicles the manufacturer tested.

a. Find the value of the 10th percentile.

b. Find the value of the 20th percentile.

Solution

First, it is important to notice that the data values are presented in an *ordered* stem-and-leaf plot, as it is essential that the data values be in numerical order. This is an important first step, since the location of the percentile refers to the location in the *ordered array* of values.

Highway Gas Mileage for Various Vehicles

Stem	Leaves
12	1
13	3
14	1
15	5 6
16	1 1 7 8
17	0 0 1 2 3 4 4 5 6 9
18	2 3 4 5
19	1 2 2 2 3 3 4 6 6 7 8 9
20	1 2 3 3 3 4 5 6 6 7 8
21	0 1 1 2 3 5 7 8 9
22	2 3 4 7 8 9
23	1 1 1 4 4 5 6 6 6 6 7 8 9 9
24	0 1 2 3 4 4 4 5 5 5 5 6 7 8 8 8 9 9
25	0 0 1 1 1 2 3 3 3 3 4 4 5 6 6 7 8 9
26	0 0 0 1 2 5 5 6 7 9
27	1 4 7
28	3 5
29	2 4 9
30	0 7
31	3
32	7
33	
34	5
35	9

Key: 12 | 1 = 12.1 mpg

a. There are 135 values in this data set, thus $n = 135$. We want the 10th percentile, so $P = 10$. Substituting these values into the formula for the location of a percentile gives us the following.

$$l = n \cdot \frac{P}{100} = 135 \cdot \frac{10}{100} = 13.5$$

Since the formula resulted in a decimal value for l, we round the number 13.5 to the next larger whole number, 14, to determine the location. Thus, the 10th percentile is approximately the value in the 14th spot in the data set. Counting data values, we find that the 14th value is 17.3. Thus, the value of the 10th percentile of this data set is 17.3 mpg. This means that approximately 10% of the values in the data set are less than or equal to 17.3 mpg.

b. We still have $n = 135$, but to find the value of the 20th percentile, $P = 20$. Substituting these new values into the formula, we get the following.

$$l = n \cdot \frac{P}{100}$$
$$= 135 \cdot \frac{20}{100}$$
$$= 27$$

Since the value calculated for l is a whole number, we must find the mean of the data value in that location and the one in the next larger location. Thus, the 20th percentile is the arithmetic mean of the 27th and 28th values in the data set, which are 19.2 and 19.3, respectively. Hence the value of the 20th percentile is 19.25 mpg. This means that approximately 20% of the values in the data set are less than or equal to 19.25 mpg.

Instead of just finding the value that represents a given percentile, we can also find the approximate percentile of a specific value in a data set. We can use the same formula and solve for P, which gives us the following formula.

Rounding Rule

When calculating the percentile of a data value, round to the nearest whole number.

Formula

P^{th} Percentile of a Data Value

The Pth **percentile** of a particular value in a data set is given by

$$P = \frac{l}{n} \cdot 100$$

where P is the percentile rounded to the nearest whole number,

l is the number of values in the data set *less than or equal to* the given value, and

n is the number of data values in the sample.

When using this formula, begin as you did when finding the value of a given percentile by ordering the data set from smallest to largest. Note that when you have repeated values in the data set, l represents the highest location of the given data value. Always round the percentile, P, to the nearest whole number; for example, $P = 36.1$ rounds down to the 36th percentile and $P = 36.57$ rounds up to the 37th percentile.

Example 3.19

Finding the Percentile of a Given Data Value

In the data set from the previous example, the Nissan Xterra averaged 21.1 mpg. In what percentile is this value?

Solution

We begin by making sure that the data are in order from smallest to largest. We know from the previous example that they are, so we can proceed with the next step.

The Xterra's value of 21.1 mpg is repeated in the data set, in both the 48th and 49th positions, so we will pick the one with the largest location value, which is the 49th. Using a sample size of $n = 135$ and a location of $l = 49$, we can substitute these values into the formula for the percentile of a given data value, which gives us the following.

$$\begin{aligned} P &= \frac{l}{n} \cdot 100 \\ &= \frac{49}{135} \cdot 100 \\ &\approx 36.296 \end{aligned}$$

Since we always need to round a percentile to a whole number, we round 36.296 to 36. Thus, approximately 36% of the data values are less than or equal to the Xterra's mpg rating. That is, 21.1 mpg is in the 36th percentile of the data set.

Quartiles

Percentiles divide the data into 100 equal parts, but we mentioned at the beginning of the section that we can divide the data into any number of equal parts. If we divide a data set into four parts, the numbers that form the divisions are called **quartiles**. If we wanted to divide a line segment into four parts, we would draw three dividing lines. Similarly, when dividing a data set into four parts, we use three quartiles.

Definition

Q_1 = First Quartile: 25% of the data are less than or equal to this value.

Q_2 = Second Quartile: 50% of the data are less than or equal to this value.

Q_3 = Third Quartile: 75% of the data are less than or equal to this value.

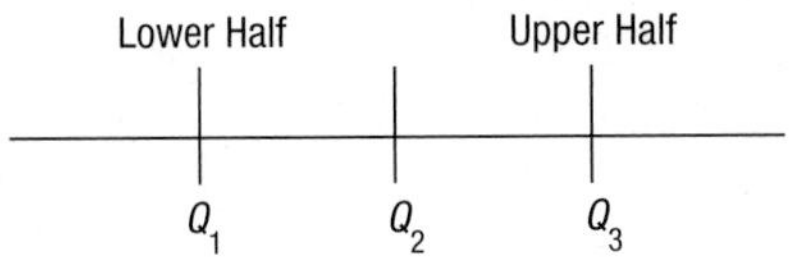

Figure 3.3: Quartiles

To find the quartiles, first note that they are equivalent to percentiles. The first quartile is equivalent to the 25th percentile, the second quartile is equivalent to the 50th percentile, and the third quartile is equivalent to the 75th percentile. Thus, to find the first quartile of a data set, we can use the method described previously to find the 25th percentile of the data set.

An alternative method for finding a rough approximation for the quartiles is as follows. Begin by ordering the data set and finding the median, which is Q_2. Next, use the median to divide the data set into an upper half and a lower half. For an odd number of data values, do not include the median in each half. If there are an even number of data values, then the median would not be a value in the data set, so we would not include it in either the upper or lower half of the data anyway. That is, never include the median in either half of the data. Next, find the median of the lower half of the data, which is the first quartile. The third quartile is the median of the upper half of the data. This approximation method results in the values `Q1`, `Med`, and `Q3` given on the TI- 83/84 Plus calculator when calculating `1-Var Stats` as discussed in previous sections.

Side Note

Hinges vs. Quartiles

If you do not include the median when you are approximating the first and third quartiles, you are actually calculating a completely different value, created by statistician John Tukey, called a **hinge**. The lower hinge is a rough approximation of the first quartile, and the upper hinge is a rough approximation of the third quartile. To complicate matters further, some textbooks (and the TI-83/84 Plus calculator) do not differentiate between quartiles and hinges. The solutions given in this textbook will actually be hinges, so as to match the values given by the TI-83/84 Plus calculator.

Example 3.20

Finding the Quartiles of a Given Data Set

Using the following set of mpg data from the previous examples, find the quartiles.

a. Use the percentile method to find the quartiles.

b. Use the approximation method to find the quartiles.

c. How do these values compare?

Solution

The data are already in order from smallest to largest. We also know that $n = 135$.

a. Percentile Method

To find the first quartile, we want to find the 25th percentile, so $P = 25$. Substituting the values into the formula for the location of a percentile, we get the following.

$$l = n \cdot \frac{P}{100}$$

$$= 135 \cdot \frac{25}{100}$$

$$= 33.75$$

Rounding up to the next whole number, we can say that the 34th value, which is 19.8 mpg, is the first quartile.

The second quartile is the median, or the 50th percentile. Thus, $n = 135$ and $P = 50$. Substituting these values into the formula for the location of a percentile, we get the following.

$$l = n \cdot \frac{P}{100}$$

$$= 135 \cdot \frac{50}{100}$$

$$= 67.5$$

Once again we round up, so the second quartile is the 68th value of 23.6 mpg. This is also the median.

The third quartile is the 75th percentile, so $n = 135$ and $P = 75$. Substituting these values into the formula, we get the following.

Highway Gas Mileage for Various Vehicles

Stem	Leaves
12	1
13	3
14	1
15	5 6
16	1 1 7 8
17	0 0 1 2 3 4 4 5 6 9
18	2 3 4 5
19	1 2 2 2 3 3 4 6 6 7 8 9
20	1 2 3 3 3 4 5 6 6 7 8
21	0 1 1 2 3 5 7 8 9
22	2 3 4 7 8 9
23	1 1 1 4 4 5 6 6 6 6 7 8 9 9
24	0 1 2 3 4 4 4 5 5 5 5 6 7 8 8 8 9 9
25	0 0 1 1 1 2 3 3 3 3 4 4 5 6 6 7 8 9
26	0 0 0 1 2 5 5 6 7 9
27	1 4 7
28	3 5
29	2 4 9
30	0 7
31	3
32	7
33	
34	5
35	9

Key: 12 | 1 = 12.1 mpg

$$
\begin{aligned}
l &= n \cdot \frac{P}{100} \\
&= 135 \cdot \frac{75}{100} \\
&= 101.25
\end{aligned}
$$

Again, we round the decimal value for the location up to the next whole number; thus the third quartile is the number in the 102nd position, which is 25.3 mpg.

b. Approximation Method

To begin, divide the data in half. There are an odd number of data values, so the median is the number exactly in the middle of the data set. Thus, the median is the number in the 68th position (halfway), which is 23.6 mpg. This also means that the second quartile is 23.6 mpg.

The first quartile is then approximately the median of the lower half of the data. Look at the data from the 1st position to the 67th position, since we do not include the median in the lower half of the data. The middle value is in the 34th position. So the first quartile is the value of 19.8 mpg.

The third quartile is the median of the upper half of the data. Look at the data from the 69th to the 135th positions. The data value in the middle is the value in the 102nd position. This value is 25.3. Thus, the third quartile is the value of 25.3 mpg.

c. These two methods result in the same values, which are also the values given by a TI-83/84 Plus calculator, as shown in the margin. This will always be true for any data set with an even number of data values. For a data set with an odd number of data values (like this one), the larger the data set, the closer the approximations will be to the percentile method's values.

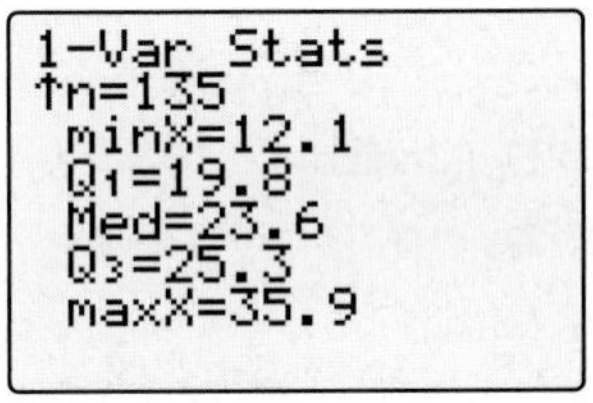

Example 3.21

Finding the Quartiles of a Given Data Set

The following speeds of motorists (in mph) were obtained by a Highway Patrol officer on duty one weekend. Determine the quartiles of each data set using the approximation method.

a. 60, 62, 63, 65, 65, 67, 70, 71, 71, 75, 78, 79, 80, 81

b. 59, 66, 67, 67, 72, 74, 75, 75, 75, 76, 78, 79, 80, 81, 85

Solution

a. Using the approximation method, the first step in calculating quartiles is to find the median. Note that the data set is already ordered. Since $n = 14$, the median is the arithmetic mean of the values in the 7th and 8th positions, which is calculated as follows.

$$
\begin{aligned}
Q_2 &= \frac{70+71}{2} \\
&= 70.5
\end{aligned}
$$

Since the data set contains an even number of values, to find Q_1 we will take the median of the lower half of data. Q_1, then, is 65. Finally, to find Q_3 take the median of the upper half of the data, which is 78. The quartiles, then, are as follows.

$$Q_1 = 65,\ Q_2 = 70.5,\ \text{and } Q_3 = 78$$

b. To find the quartiles of the second set of data using the approximation method, again start with the median. Note again that the data set is already ordered. Since $n = 15$, an odd number of values, the median is the value located at the middle, 75. Remember, when there are an odd number of values in the data set, the median is not included in either the lower or upper half of the data when finding Q_1 and Q_3. Hence the median of the resulting lower group is 67. The median of the resulting upper group is 79. The quartiles, then, are as follows.

$$Q_1 = 67,\ Q_2 = 75,\ \text{and } Q_3 = 79$$

Five-Number Summary and Box Plots

Memory Booster

The **five-number summary** contains:

1. Minimum
2. Q_1
3. Q_2 (Median)
4. Q_3
5. Maximum

Quartiles are used in a numerical description, aptly called the **five-number summary** because it contains five numbers: the minimum value; the first quartile, Q_1; the median or second quartile, Q_2; the third quartile, Q_3; and the maximum value. The five-number summary is made up of these five numbers listed in order from smallest to largest.

Example 3.22

Writing the Five-Number Summary of a Given Data Set

Write the five-number summary for the data from Example 3.20.

Solution

The minimum value is 12.1 mpg, the maximum value is 35.9 mpg, and we have previously determined that the quartiles are 19.8 mpg, 23.6 mpg, and 25.3 mpg. Thus the five-number summary is 12.1, 19.8, 23.6, 25.3, 35.9.

If we want to represent a five-number summary graphically, we can use a graph called a **box plot**, as shown in Figure 3.4. A box plot is sometimes also referred to as a "box-and-whisker plot."

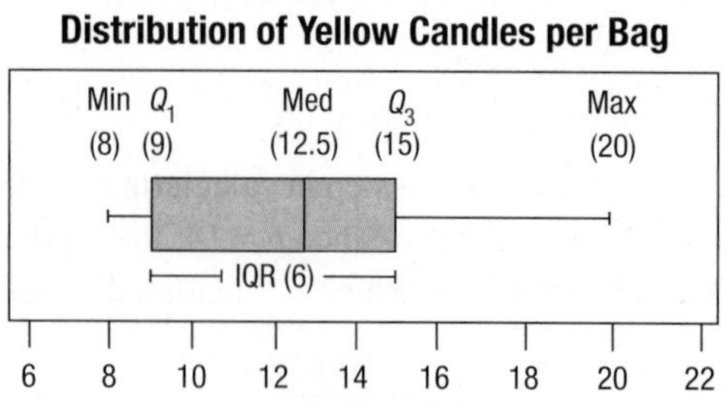

Figure 3.4: Box Plot

A vertical line represents each of the five-number summary values. The "box" refers to the rectangle that is created from joining the lines representing the first and third quartiles. This box represents the **interquartile range**, or **IQR**. The interquartile range is the difference between the third quartile and the first quartile, and is seen as the size of the box in a box-and-whisker plot. It is the range of the middle 50% of the data. The "whiskers" are the lines that extend to reach the minimum and maximum values.

Formula

Interquartile Range (IQR)

The **interquartile range** is the range of the middle 50% of the data, given by

$$\text{IQR} = Q_3 - Q_1$$

where Q_3 is the third quartile and

Q_1 is the first quartile.

In Figure 3.5, five box plots are drawn together for the sake of comparison. Each box plot shown represents the distribution of a different color of candy found in a sample of randomly chosen bags of colored candies. Notice that the median for the red-colored candies is higher than the median for the other colors. Further, the interquartile range for the red-colored candies is also larger and higher than the other boxes, meaning that the middle 50% of the data is more spread out and has larger values than those of the other colors. In contrast, the interquartile range for the green color is small, and its median is one of the smallest as well, indicating that fewer green candies are likely to appear in a bag. In summary, the box plots shown in the graphic below are good news for those who love red-colored candy, but for those who prefer green, there may not be as many as they would hope!

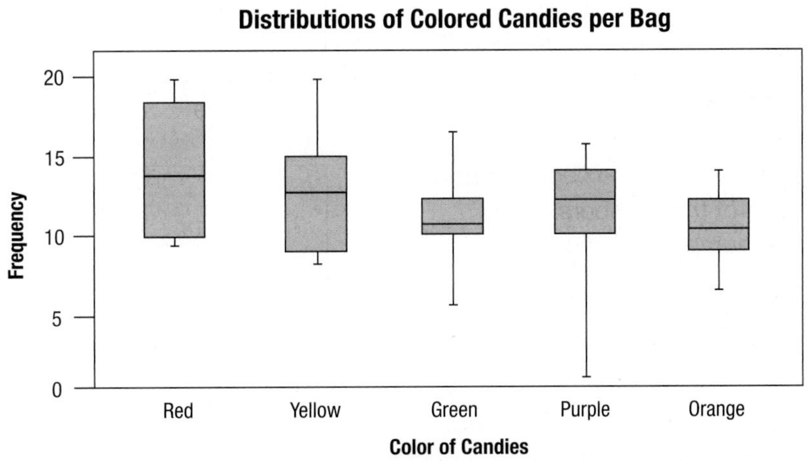

Source: Data obtained from a class experiment in Carolyn Warren's statistics class, March 2011.

Figure 3.5: Comparing Box Plots

To create a box plot, follow the steps outlined in the following box. To compare several data sets at once, as in Figure 3.5, draw multiple box plots on the same horizontal and vertical axes.

Procedure

Creating a Box Plot

1. Begin with a horizontal (or vertical) number line that contains the five-number summary.
2. Draw a small line segment above (or next to) the number line to represent each of the numbers in the five-number summary.
3. Connect the line segment that represents the first quartile to the line segment representing the third quartile, forming a box with the median's line segment in between.
4. Connect the "box" to the line segments representing the minimum and maximum values to form the "whiskers."

Example 3.23

Creating a Box Plot

Draw a box plot to represent the five-number summary from the previous example. Recall that the five-number summary was 12.1, 19.8, 23.6, 25.3, 35.9.

Solution

Step 1: Label the horizontal axis at even intervals.

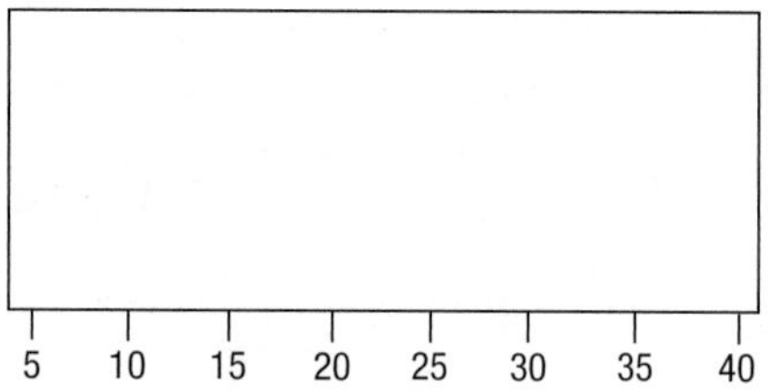

Step 2: Place a small line segment above each of the numbers in the five-number summary.

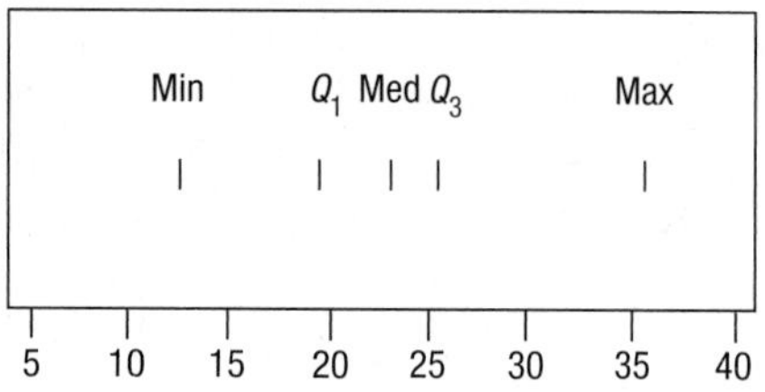

Step 3: Connect the line segment that represents Q_1 to the line segment that represents Q_3, forming a box with the median's line segment in between.

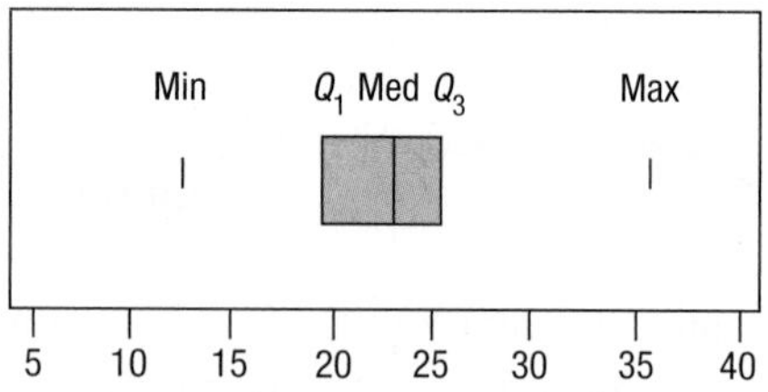

Step 4: Connect the "box" to the line segments representing the minimum and maximum to form the "whiskers."

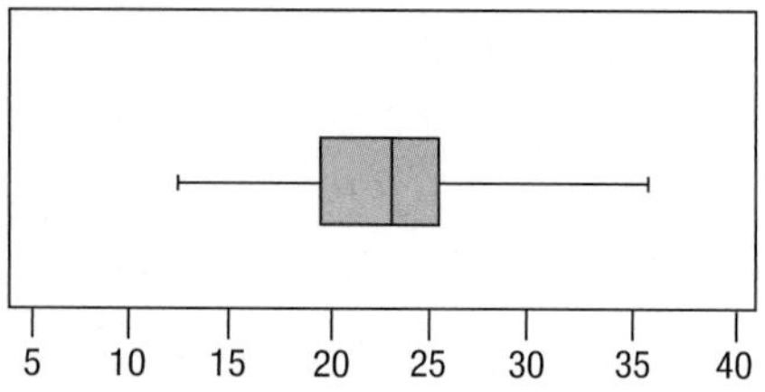

Example 3.24

Interpreting Box Plots

The box plots below are from the US Geological Survey website. Use them to answer the following questions.

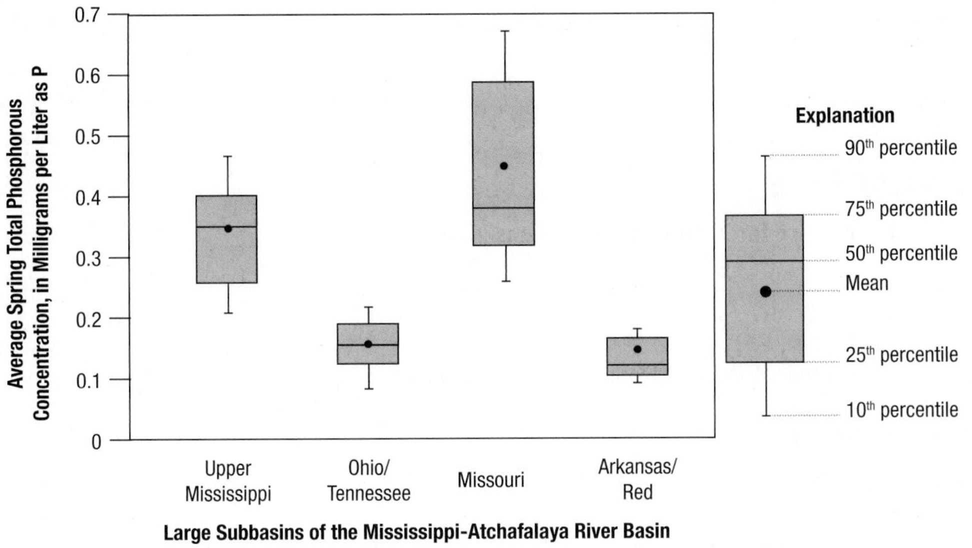

Note: Box plots showing the distribution of average Spring (April and May) total phosphorous concentrations, for the years 1979 to 2008, for four of the five large subbasins that comprise the Mississippi-Atchafalaya River Basin. (The Lower Mississippi River subbasin was excluded due to the large errors in estimating the average concentrations.)

Source: US Geological Survey. "2009 Preliminary Mississippi-Atchafalaya River Basin Flux Estimate." US Department of the Interior. 2009. http://toxics.usgs.gov/hypoxia/mississippi/oct_jun/images/figure9.png (9 Aug. 2010).

a. What do the top and bottom bars represent in these box plots according to the key?

b. Which subbasin had the highest median average spring total phosphorus concentration?

c. Which subbasin had the lowest average spring total phosphorus concentration? (**Note:** Each data value is an average of April's and May's totals, and the lowest average shown for each subbasin is the *10th percentile*.)

d. Which subbasin had the largest interquartile range?

Solution

a. In each box plot, the top bar represents the 90th percentile of average spring total phosphorous concentration, and the bottom bar represents the 10th percentile.

b. The subbasin with the highest median average spring total phosphorus concentration was the Missouri.

c. The subbasin with the lowest average spring total phosphorus concentration was the Ohio/Tennessee.

d. The subbasin with the largest interquartile range was the Missouri.

Standard Scores

We can compare two data values from different populations by comparing their respective percentiles. For example, an SAT score in the 56th percentile is not as high as an ACT score in the 87th percentile. We could also determine how the values relate to the respective means of their data sets. This measure is called a **standard score**, or ***z*-score**. A standard score tells us how far a value is from the mean, specifically, how many standard deviations it is away from the mean. The formula is as follows.

Rounding Rule

When calculating a *z*-score, round to two decimal places.

Formula

Standard Score

The **standard score for a population** value is given by

$$z = \frac{x - \mu}{\sigma}$$

where x is the value of interest from the population,

μ is the population mean, and

σ is the population standard deviation.

The **standard score for a sample** value is given by

$$z = \frac{x - \bar{x}}{s}$$

where x is the value of interest from the sample,

$\bar{x}$ is the sample mean, and

s is the sample standard deviation.

Standard scores are useful in comparing data values from populations with different means and standard deviations. For example, they could be used to determine if you scored better on the ACT exam or the SAT exam (assuming you took both).

Example 3.25

Calculating a Standard Score

If the mean score on the math section of the SAT test is 500 with a standard deviation of 150 points, what is the standard score for a student who scored a 630?

Solution

$\mu = 500$ and $\sigma = 150$. The value of interest is $x = 630$, so we have the following.

$$\begin{aligned} z &= \frac{x - \mu}{\sigma} \\ &= \frac{630 - 500}{150} \\ &\approx 0.87 \end{aligned}$$

Thus the student's math SAT score of 630 is approximately 0.87 standard deviations above the mean.

Example 3.26

Comparing Standard Scores

Jodi scored an 87 on her calculus test and was bragging to her best friend about how well she had done. She said that her class had a mean of 80 with a standard deviation of 5; therefore, she had done better than the class average. Her best friend, Ashley, was disappointed. She had scored only an 82 on her calculus test. The mean for her class was 73 with a standard deviation of 6.

Who *really* did better on her test, compared to the rest of her class, Jodi or Ashley?

Solution

Let's calculate each student's standard score.

Jodi's standard score can be calculated as follows.

$$\begin{aligned} z &= \frac{x-\mu}{\sigma} \\ &= \frac{87-80}{5} \\ &= 1.4 \end{aligned}$$

Ashley's standard score can be calculated in a similar fashion.

$$\begin{aligned} z &= \frac{x-\mu}{\sigma} \\ &= \frac{82-73}{6} \\ &= 1.5 \end{aligned}$$

Thus Ashley actually did better on her calculus test with respect to her class, despite the fact that Jodi had the higher score, because Ashley's score was more standard deviations above her class mean.

3.3 Section Exercises

Percentiles

Directions: Answer the questions that follow each set of data.

1. The following data represent weights of newborn babies (in pounds).

6.8	9.1	8.7	7.5	8.2
5.4	6.5	8.5	7.3	6.6
5.9	7.3	9.3	7.4	7.8

   a. Which weight represents the 50th percentile?

   b. What is the percentile of a weight of 8.2 pounds?

2. The following data represent high temperatures for cities in the Southeast (in degrees Fahrenheit).

85	82	93	88
92	79	84	90
77	83	91	89
90	85	87	91
89	92	95	88

   a. Which temperature represents the 75th percentile?

   b. What is the percentile of a temperature of 93 °F?

3. The following data represent average numbers of Tweets per day posted on Twitter for 16 high school students.

0.8	42.2	20.6	2.8
36.7	18.6	23.3	11.5
3.7	14.9	9.4	1.5
14.9	31.1	23.5	9.5

   a. Which number represents the 25th percentile?

   b. What is the percentile of an average of 11.5 Tweets per day?

Directions: Exercises 4–8 reference data sets that are available for download from **http://www.hawkeslearning.com/support/downloads.htm** in Excel or Minitab format. The patient data are medical statistics for a sample of patients from a hospital. The state data are measurements collected by the US Census Bureau for each of the 50 states. Use the data sets to answer the following questions.

4. Refer to the Patient Data.

   a. Which patient weight represents the 35th percentile?

   b. What is the percentile of a weight of 189 pounds?

5. Refer to the State Data.

   a. Which population represents the 82nd percentile?

   b. What is the percentile of a mean travel time of 27 minutes?

6. What is the 85th percentile for the patients' systolic blood pressure readings from the Patient Data?

7. In the State Data, what is the percentile of Oregon's mean travel time to work, 22.1 minutes?

8. What is the 50th percentile of the patients' ages in the Patient Data?

Five-Number Summary

Directions: **Find the five-number summary for each data set. Use the approximation method to calculate quartiles. These values will match those produced by a TI-83/84 Plus calculator.**

9. The following data represent prices (in dollars) of used cars listed on www.autotrader.com for one zip code.

18,865	11,442	15,750	10,960	15,635
15,963	13,702	14,788	15,495	8250
	12,900	14,850	6450	

10. The following data represent weights of dimes, measured in grams.

2.268	2.267	2.269	2.268	2.271
2.266	2.267	2.268	2.270	2.272
	2.265	2.269	2.268	

11. The following data represent INR (International Normalized Ratio) readings of patients with blood clotting disorders.

1.5	2.1	1.7	3.5
4.1	1.2	1.7	1.8

12. The following data represent SAT Critical Reading scores for a randomly selected group of high school seniors.

520	750	620	470
520	660	780	580
390	460	660	570
290	500	690	540

13. The following data represent birth weights (in pounds) of eight newborn babies born on the same day at a local hospital.

5.4	6.5	7.8	9.1
9.3	10.1	7.8	9.0

14. The following data represent times, in minutes, taken by students in a physical education class to run/walk two miles.

12.22	12.35	13.45	16.78	19.01
21.34	24.87	25.10	26.93	29.81

15. The following data represent changes in weight, measured in pounds, from the beginning of a new diet to one month later.

–12	–11	–10
–8	–7	–5
–2	0	1
3	4	

16. The following data represent differences in high temperatures, in degrees Fahrenheit, for the same day from one year to the next in nine metropolitan areas.

9	–2	–8
12	15	19
21	–14	–11

Box Plots

Directions: Draw a box plot for each set of data on the same graph. Use the approximation method to calculate quartiles. These values will match those produced by a TI-83/84 calculator. Use your box plots to answer the following questions.

a. Which data set has the smallest value?

b. Which data set has the larger median?

c. Which data set has the larger interquartile range?

17. Weight loss (in pounds) from diet A: 2, 3, 5, 5, 5, 6, 6, 6, 7, 7, 8

Weight loss (in pounds) from diet B: 3, 3, 4, 4, 4, 5, 6, 6, 9, 12

18. Respiratory rates at rest (in breaths per minute) of adults in group A:

10, 12, 13, 14, 15, 16, 17, 18

Respiratory rates at rest (in breaths per minute) of adults in group B:

11, 15, 17, 18, 19, 19, 20

19. Test scores for class A: 45, 60, 57, 83, 72, 93, 87, 73, 92

Test scores for class B: 23, 88, 67, 89, 91, 76, 72, 100, 95, 35

20. Diameters of cans (in cm) from assembly line A:

5.6, 5.7, 5.1, 5.7, 5.5, 5.9, 5.7, 5.5, 5.6, 5.6

Diameters of cans (in cm) from assembly line B:

5.4, 5.7, 5.6, 5.5, 5.6, 5.7, 5.7, 5.8, 5.6, 5.5

Standard Scores

Directions: Calculate the standard score using the given values. Round your answer to two decimal places.

21. $\mu = 25, \quad \sigma = 3, \quad x = 27$

22. $\bar{x} = 37, \quad s = 8, \quad x = 34$

23. $\mu = 0.32, \quad \sigma = 0.01, \quad x = 0.29$

24. $\mu = 2, \quad \sigma = 0.5, \quad x = 1.8$

25. $\bar{x} = 180, \quad s = 10, \quad x = 210$

26. Carlita scored 32 on the ACT Mathematics Test and 730 on the mathematics section of the SAT. If the ACT Mathematics Test had a mean score of 21.0 with a standard deviation of 5.3, and the mathematics section of the SAT had a mean score of 516 with a standard deviation of 116, on which exam did Carlita earn a better math score with respect to her peers?

27. A manufacturer makes aluminum cans and longneck bottles. The average diameter of an aluminum can is supposed to be 4.2 inches, with an allowable standard deviation of 0.01 inches. The average diameter on a longneck bottle is supposed to be 3.8 inches, with an allowable standard deviation of 0.02 inches. A factory worker randomly selects a can from the assembly line and it has a diameter of 4.3 inches. The worker then selects a bottle from the assembly line and it has a diameter of 3.75 inches. Which assembly line is closest to specifications?

28. Don played in a local golf tournament for charity and scored a round of 63 while the average round for the day was a 74 with a standard deviation of 3 strokes. Later that week, Don played in a Pro-Am tournament and scored a 65 while the average score for the day was a 79 with a standard deviation of 4 strokes. Which was Don's better round of golf in comparison to the competition? (Remember, in golf, lower scores are better!)

Directions: For each graph, where the mean is marked by the dotted line, which is a likely *z*-score for the indicated value of *x*? Choose from the following *z*-scores:

a. –1.3

b. 0

c. 1.7

29.

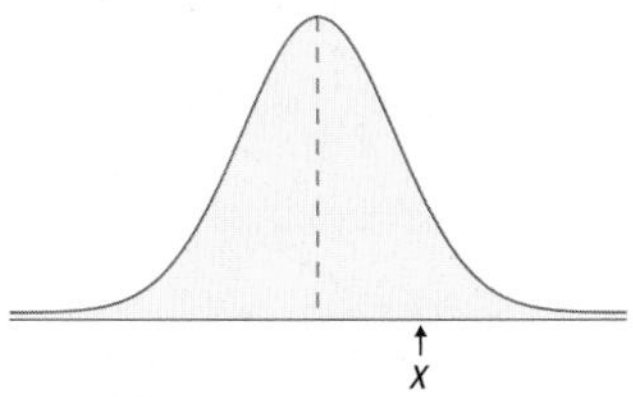

30.

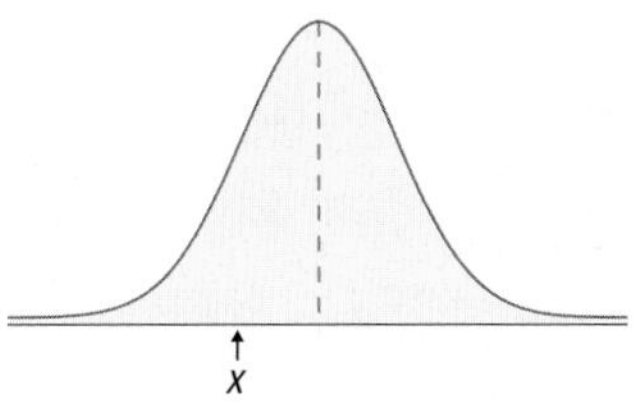

31.

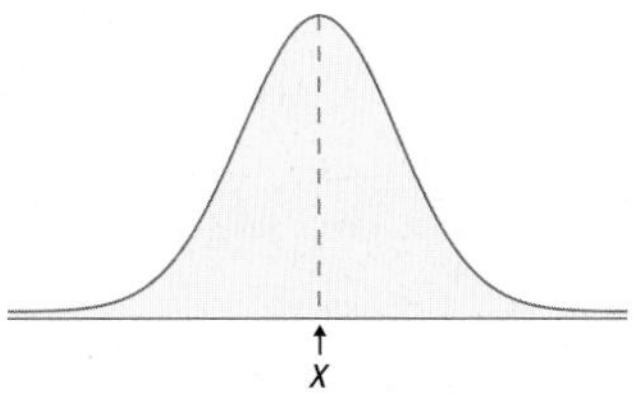

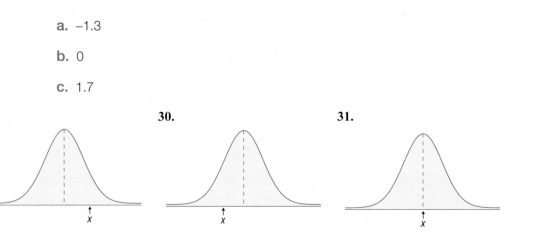

R Chapter 3 Review

Section 3.1: Measures of Center

Definitions

Arithmetic mean

The sum of all of the data values divided by the number of data values, often simply called the mean

Median

The middle value in an ordered array of data

Mode

The value in a data set that occurs most frequently

No mode

Describes a data set in which all of the data values occur only once or each value occurs an equal number of times

Unimodal

Describes a data set in which only one data value occurs most often

Bimodal

Describes a data set in which exactly two data values occur equally often

Multimodal

Describes a data set in which more than two data values occur equally often

Sample Mean

The arithmetic mean of a set of sample data, given by

$$\bar{x} = \frac{x_1 + x_2 + \cdots + x_n}{n} = \frac{\sum x_i}{n}$$

Population Mean

The arithmetic mean of all the values in a population, given by

$$\mu = \frac{x_1 + x_2 + \cdots + x_N}{N} = \frac{\sum x_i}{N}$$

Section 3.1: Measures of Center (cont.)

Weighted Mean

The mean of a data set in which each data value in the set does not hold the same relative importance, given by

$$\bar{x} = \frac{\sum (x_i \cdot w_i)}{\sum w_i}$$

Properties of the Mean

1. Most familiar and widely used measure of center
2. Its value is affected by EVERY value in the data set.
3. May not be a value in the data set
4. Appropriate measure of center for quantitative data with no outliers

Finding the Median of a Data Set

1. List the data in ascending (or descending) order, making an ordered array.
2. If the data set contains an ODD number of values, the median is the middle value in the ordered array.
3. If the data set contains an EVEN number of values, the median is the arithmetic mean of the two middle values in the ordered array.

Properties of the Median

1. Easy to compute by hand
2. Only determined by middle values of a data set, not affected by outliers
3. May not be a value in the data set if there are an even number of values in the set
4. Useful measure of center for skewed distributions

Properties of the Mode

1. A data set may not have a mode.
2. A data set may have one or more than one mode.
3. If a mode exists for a data set, the mode is a value in the data set.
4. Not affected by outliers in the data set
5. Only measure of center appropriate for qualitative data

Determining the Most Appropriate Measure of Center

1. For qualitative data, the mode should be used.
2. For quantitative data, the mean should be used, unless the data set contains outliers or is skewed.
3. For quantitative data sets that are skewed or contain outliers, the median should be used.

Graphs and Measures of Center

1. The mode is the data value at which a distribution has its highest peak.
2. The median is the number that divides the area of the distribution in half.
3. The mean of a distribution will be pulled toward any outliers.

Section 3.2: Measures of Dispersion

Definitions

Standard deviation

A measure of how much we might expect a typical member of the data set to differ from the mean

Coefficient of variation, CV

The ratio of the standard deviation to the mean as a percentage; allows comparison of the spreads of data from different sources, regardless of differences in units of measurement

Variance

The square of the standard deviation

Empirical Rule

Used with bell-shaped distributions of data to estimate the percentage of values within a few standard deviations of the mean

Chebyshev's Theorem

Gives a minimum estimate of the percentage of data within a few standard deviations of the mean for any distribution

Range

The difference between the largest and smallest values in the data set, given by

$$\text{Range} = \text{Maximum Data Value} - \text{Minimum Data Value}$$

Properties of the Range

1. Easiest measure of dispersion to calculate
2. Only affected by the largest and smallest values in the data set, so it can be misleading

Population Standard Deviation

The standard deviation of a population data set, given by

$$\sigma = \sqrt{\frac{\sum (x_i - \mu)^2}{N}}$$

Sample Standard Deviation

The standard deviation of a set of sample data, given by

$$s = \sqrt{\frac{\sum (x_i - \bar{x})^2}{n-1}}$$

Section 3.2: Measures of Dispersion (cont.)

Properties of the Standard Deviation

1. Easily computed using a calculator or computer
2. Affected by every value in the data set
3. Population standard deviation and sample standard deviation formulas yield different results
4. Interpreted as the average distance a data value is from the mean; thus it cannot take on negative values
5. Same units as the units of the data
6. Larger standard deviation indicates that data values are more spread out, smaller standard deviation indicates that data values lie closer together
7. If it equals 0, then all of the data values are equal to the mean.
8. Equal to the square root of the variance

Population Coefficient of Variation

$$\text{CV} = \frac{\sigma}{\mu} \cdot 100\%$$

Sample Coefficient of Variation

$$\text{CV} = \frac{s}{\bar{x}} \cdot 100\%$$

Population Variance

The variance of a population data set, given by

$$\sigma^2 = \frac{\sum (x_i - \mu)^2}{N}$$

Sample Variance

The variance of a set of sample data, given by

$$s^2 = \frac{\sum (x_i - \bar{x})^2}{n-1}$$

Properties of the Variance

1. Easily computed using a calculator or computer
2. Affected by every value in the data set
3. Population variance and sample variance formulas yield different results
4. Difficult to interpret because of its unusual squared units
5. Equal to the square of the standard deviation
6. Preferred over the standard deviation in many statistical tests because of its simpler formula

Section 3.2: Measures of Dispersion (cont.)

Empirical Rule for Bell-Shaped Distributions

- Approximately 68% of the data values lie within one standard deviation of the mean.
- Approximately 95% of the data values lie within two standard deviations of the mean.
- Approximately 99.7% of the data values lie within three standard deviations of the mean.

Chebyshev's Theorem

The proportion of data that lie within K standard deviations of the mean is at least $1-\frac{1}{K^2}$ for $K>1$. When $K=2$ and $K=3$, Chebyshev's Theorem says the following.

- $K=2$: At least $1-\frac{1}{2^2}=\frac{3}{4}=75\%$ of the data values lie within two standard deviations of the mean.
- $K=3$: At least $1-\frac{1}{3^2}=\frac{8}{9}\approx 88.9\%$ of the data values lie within three standard deviations of the mean.

Section 3.3: Measures of Relative Position

Definitions

Percentiles
Values that divide the data into 100 equal parts; each percentile indicates approximately what percentage of the data lie at or below a given value

Quartiles
Values that divide the data into four equal parts; equivalent to the 25th, 50th, and 75th percentiles

- Q_1 = First Quartile: 25% of the data are less than or equal to this value.
- Q_2 = Second Quartile: 50% of the data are less than or equal to this value.
- Q_3 = Third Quartile: 75% of the data are less than or equal to this value.

Hinge
An approximation of the first or third quartile, found by using the median to divide the data set into an upper half and a lower half (without including the median in either half), and then finding the median of either half of the data set

Five-number summary
A numerical description of a data set that lists the minimum value; the first quartile, Q_1; the median or second quartile, Q_2; the third quartile, Q_3; and the maximum value in order from smallest to largest

Box plot
A graphical representation of a five-number summary, sometimes referred to as a "box-and-whisker plot"

Standard score (or *z*-score)
Indicates how many standard deviations from the mean a particular data value lies

Section 3.3: Measures of Relative Position (cont.)

Location of Data Value for the Pth Percentile

The location of the Pth percentile in an ordered array of data values, given by

$$l = n \cdot \frac{P}{100}$$

1. If the formula results in a decimal value for l, the location is the next *larger* whole number.
2. If the formula results in a whole number, the percentile's value is the arithmetic mean of the data value in that location and the data value in the next *larger* location.

P^{th} Percentile of a Data Value

The Pth percentile of a particular value in a data set is given by

$$P = \frac{l}{n} \cdot 100$$

where P is rounded to the nearest whole number.

Creating a Box Plot

1. Begin with a horizontal (or vertical) number line that contains the five-number summary.
2. Draw a small line segment above (or next to) the number line to represent each of the numbers in the five-number summary.
3. Connect the line segment that represents the first quartile to the line segment representing the third quartile, forming a box with the median's line segment in the middle.
4. Connect the "box" to the line segments representing the minimum and maximum values to form the "whiskers."

Interquartile Range (IQR)

The range of the middle 50% of the data, given by

$$\text{IQR} = Q_3 - Q_1$$

Standard Score for a Population

$$z = \frac{x - \mu}{\sigma}$$

Standard Score for a Sample

$$z = \frac{x - \overline{x}}{s}$$

E Chapter 3 Exercises

Directions: Respond thoughtfully to the following exercises.

1. Four friends went out to eat one evening, and the mean price of their dinners was \$10.54. If the prices of the first three meals were \$9.62, \$11.59, and \$10.03, what was the cost of the fourth meal?

2. Suppose that a list of company CEOs is compiled, and 37% are over the age of 45. True or false: Q_3 must be greater than 45.

3. Suppose that 200 employees at a major theme park are surveyed, and 54% are under the age of 25.

   a. True or false: Of those surveyed, the mean age must be under 25.

   b. True or false: Of those surveyed, the median age must be under 25.

4. Give one example of a data set with a large variation and one example of a data set with a small variation.

5. The mean grade for Test 2 in Marquetta's biology class is 73, and the standard deviation is 8 points. If the standard score for Marquetta's test is 1.25, what is her grade?

6. Neikia wants to determine the average amount of money moviegoers typically spend on snacks at the theater. He surveys 150 people leaving the theater one evening, and finds that 79 did not spend any money on refreshments. The maximum amount spent was \$22.50.

   a. What is the largest possible value for the mean?

   b. What is the median amount spent on refreshments?

7. Suppose that Julie's height has a standard score of 0.44. True or false: Julie is taller than average.

8. The mean cost of individual health insurance in one region of the country is \$421 per month with a standard deviation of \$78. Of those in the region who have individual health insurance, approximately what percentage spend between \$343 and \$499 on their insurance each month? Assume that individual health insurance costs in the region have a bell-shaped distribution.

9. The mean upload speed for a particular Internet provider is 500 KBps with a standard deviation of 60 KBps. Use Chebyshev's Theorem to estimate the percentage of customers who should be able to upload data at speeds between 320 and 680 KBps.

10. True or false: If the standard deviation of a data set is zero, then all of the values in the set must be the same number.

11. Calculate the sample standard deviation and sample variance of the following monthly prices of cellular phone plans.

$44.99, $59.99, $34.49, $89.99, $54.99, $65.99, $49.99

12. Find the GPA of a student who received the following grades. Note that an A is equivalent to a 4.0, a B is equivalent to a 3.0, and a C is equivalent to a 2.0.

Biology II (3-hour class): B

Biology II Lab (1-hour class): A

English 212 (3-hour class): B

German 101 (3-hour class): C

Precalculus (3-hour class): C

Scuba Diving (2-hour class): A

13. Find the approximate mean of the following data, which represent the number of debit card purchases in one month for a random sample of college students.

Number of Debit Card Purchases in One Month	
Class	**Frequency**
0–7	8
8–15	3
16–23	9
24–31	5

14. If we know that an ACT score was 21 and that score was in the 67th percentile, can we determine how many students were in the sample? Why or why not?

15. The following numbers represent the numbers from a five-number summary for the number of child abuse cases reported per county during a one-week period. Unfortunately, the numbers have been shuffled and are no longer in the right order. Can you still determine which number is the first quartile, Q_1? Explain your answer.

14, 17.5, 8, 12.5, 21

P Chapter 3 Project

Project A: Olympic Gold

The following tables give the years of the Winter Olympics and the number of gold medals won in each year by the United States.

USA Gold Medal Count											
Year	**1924**	**1928**	**1932**	**1936**	**1948**	**1952**	**1956**	**1960**	**1964**	**1968**	**1972**
Number of Gold Medals	1	2	6	1	3	4	2	3	1	1	3
Number of Events	16	14	14	17	22	22	24	27	34	35	35
Year	**1976**	**1980**	**1984**	**1988**	**1992**	**1994**	**1998**	**2002**	**2006**	**2010**	
Number of Gold Medals	3	6	4	2	5	6	6	10	9	?	
Number of Events	37	38	39	46	57	61	68	78	84	86	

Source: Olympic.org. "Olympic Games." http://www.olympic.org/olympic-games (13 Jun. 2011).

Note that the number of events is not the same for every year. When analyzing the data, it will become necessary to take into account the number of gold medals as a percentage of the number of events. Now we are ready to estimate the expected number of gold medals for the United States in the 2010 Winter Games.

1. Begin by calculating the number of gold medals won each year as a percentage of the number of events. To do this, divide the number of gold medals by the number of events, and then multiply by 100. Round your answers to the nearest whole percentage. Create a table similar to the one above with the percentage of gold medals won in each year.

2. Calculate the range of the number of gold medals and the range of the percentage of gold medals.

3. Calculate the median number of gold medals and the median percentage of gold medals.

4. Calculate the mode of the number of gold medals and the mode of the percentage of gold medals.

5. Calculate the mean number of gold medals won.

6. Calculate the mean percentage of gold medals won by adding to get the total number of gold medals won in all years, dividing by the total number of events in all years and then multiplying by 100. Round your answer to the nearest whole percentage. (Note that you cannot average percentages. This is why we have to go back to the numbers of gold medals and the numbers of events.)

7. Calculate the five-number summary for the number of gold medals won.

8. Draw a box plot for the number of gold medals won.

9. Using the fact that there were 86 events in the 2010 Olympics, estimate the expected number of gold medals by multiplying 86 by the median percentage of gold medals won by the United States.

10. The United States actually won 9 gold medals in the 2010 Winter Games in Vancouver, British Columbia. How does your calculation for the expected number of gold medals compare to the actual number?

Project B: Where Would You Invest Your Money?

Let's look at several sets of stock prices. The following prices were obtained from historical records from 2009, and are listed in dollars per share.

Stock Prices		
Coca-Cola (KO)	**Bank of America (BAC)**	**General Electric (GE)**
45	7	12
44	4	9
52	7	10
52	9	13
49	11	13
55	13	12
57	15	13
54	18	14
48	17	16
45	15	14
47	16	16
54	15	15

Source: Yahoo! Finance. http://finance.yahoo.com (5 Aug. 2010).

1. Find the mean of each set of stock prices.

2. Find the median of each set of stock prices.

3. Calculate the variance and standard deviation of each set of stock prices.

4. If you have $10,000 to invest, what stock would you buy under the following circumstances? Justify your reasoning.

a. You are nearing retirement and need a stable investment for the future.

b. You are a wealthy entrepreneur hoping to make a large profit in a short amount of time.

T Chapter 3 Technology

Calculating Descriptive Statistics

TI-83/84 Plus

One-Variable Statistics

Side Note

The most recent TI-84 Plus calculators (Jan. 2011 and later) contain a new feature called STAT WIZARDS. This feature is not used in the directions in this text. By default this feature is turned ON. STAT WIZARDS can be turned OFF under the second page of MODE options.

As shown in the text, you can use a TI-83/84 Plus calculator to generate all of the numerical summaries at once using the CALC menu. Press STAT, then choose 1:Edit, and enter your data in L1. Then press STAT again, and now choose CALC (calculate). Choose option 1:1-Var Stats and press ENTER. If your data are in L1 press ENTER again, since L1 is the default list. If you did not type your data in L1, enter the list where your data are located, such as L3 or L5. (These list names are in blue, above the numeric keys.)

The values shown on the screen will be listed, in order, as follows.

Table T.1: TI-83/84 Plus 1-Var Stats

Value	Definition
$\bar{x}$	Sample mean
Σx	Sum of the values
Σx^2	Sum of the squares of the values
Sx	Sample standard deviation
σx	Population standard deviation
n	Number of data values
minX	Minimum value
Q_1	Lower hinge (approximation of the first quartile)
Med	Median
Q_3	Upper hinge (approximation of the third quartile)
maxX	Maximum value

Example T.1

Using a TI-83/84 Plus Calculator to Calculate One-Variable Statistics

Use a TI-83/84 Plus calculator to calculate the list of numerical summaries for the following data values, which represent the breathing rates for a sample of adults (in breaths per minute).

12, 14, 15, 19, 12, 10, 13, 19, 20, 12, 23

Solution

Begin by entering your data in the first list, L1. Press STAT, then choose CALC, and option 1:1-Var Stats. Press ENTER. The beginning of the list is shown in the margin.

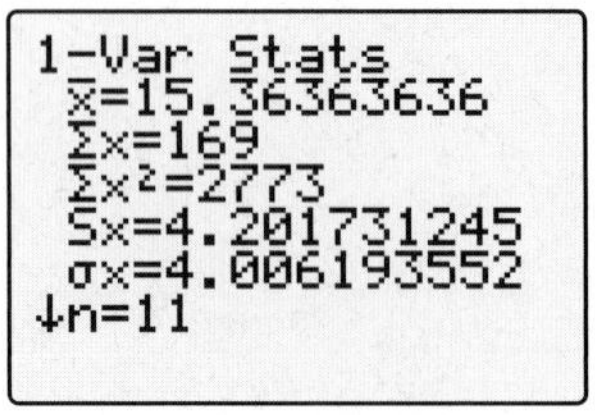

Box Plots

The TI-83/84 Plus calculator can also produce box plots, as shown in the following example.

Example T.2

Using a TI-83/84 Plus Calculator to Create a Box Plot

Create a box plot, using a TI-83/84 Plus calculator, for the following data values, which represent the breathing rates for a sample of adults (in breaths per minute).

12, 14, 15, 19, 12, 10, 13, 19, 20, 12, 23

Solution

Begin by entering the data in the first list, L1. Next, go to the STAT PLOTS menu by pressing 2ND Y=. Select option 1:Plot1; turn on Plot1 by highlighting On and pressing ENTER. Then choose the second of the two box plot options. Next, enter L1 in the Xlist. Then press GRAPH. You should see the box plot shown in the screenshot in the margin. If it does not appear that way at first, pressing ZOOM and choosing option 9:ZoomStat should correct the problem.

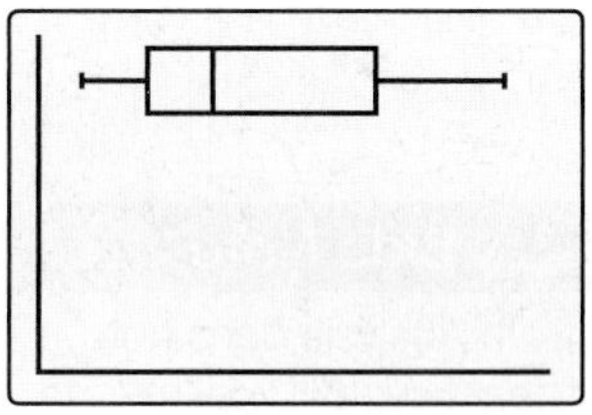

Microsoft Excel

Descriptive Statistics

Excel can calculate the values of the descriptive statistics individually or collectively. Individually, the functions are as follows.

Table T.2: Excel Functions

Calculation	Excel 2010 Function	Excel 2007 Function
Mean	=AVERAGE(*cell range*)	=AVERAGE(*cell range*)
Median	=MEDIAN(*cell range*)	=MEDIAN(*cell range*)
Single Mode	=MODE.SING(*cell range*)	=MODE(*cell range*)
Multiple Modes	=MODE.MULT(*cell range*)	
Population Variance	=VAR.P(*cell range*)	=VARP(*cell range*)
Sample Variance	=VAR.S(*cell range*)	=VAR(*cell range*)
Population Standard Deviation	=STDEV.P(*cell range*)	=STDEVP(*cell range*)
Sample Standard Deviation	=STDEV.S(*cell range*)	=STDEV(*cell range*)
Minimum Value	=MIN(*cell range*)	=MIN(*cell range*)
First Quartile	=QUARTILE.INC(*cell range*, 1)	=QUARTILE(*cell range*, 1)
Third Quartile	=QUARTILE.INC(*cell range*, 3)	=QUARTILE(*cell range*, 3)
Maximum Value	=MAX(*cell range*)	=MAX(*cell range*)

Note that the Excel 2007 functions in Table T.2 can also be used in Excel 2010 or earlier versions of Excel. Also note that the method used by Microsoft Excel to calculate the first and third quartiles is different from the methods discussed in this book, and the method used by the TI-83/84 Plus calculator, so the values given by Microsoft Excel may differ slightly from the provided textbook answers depending on the data set.

You can also use the Data Analysis menu under the Data tab to create a list of numerical summaries. The output will produce some values we have not discussed, as well as many we have studied. The values you should recognize are the mean, median, mode, standard deviation (for a sample), sample variance, range, minimum, maximum, sum (which is the sum of all the values), and count (which is the number of data values).

Example T.3

Using Microsoft Excel to Calculate Descriptive Statistics

Use the Data Analysis menu in Microsoft Excel to calculate the list of descriptive statistics for the following data values, which represent the breathing rates for a sample of adults (in breaths per minute).

12, 14, 15, 19, 12, 10, 13, 19, 20, 12, 23

Solution

Side Note

If Data Analysis does not appear in your Excel menu under the Data tab, you can easily add this function. See the directions in Appendix B: Getting Started with Microsoft Excel.

Begin by typing the data in column A. Go to the **Data** tab, then **Data Analysis**, then choose **Descriptive Statistics**, and click **OK**. In the *Input Range* box, enter the cells where your data are located. Select **New Worksheet Ply** and type **Descriptive Statistics** in the box. Click on the box in front of **Summary Statistics**. The Descriptive Statistics menu should look similar to the following.

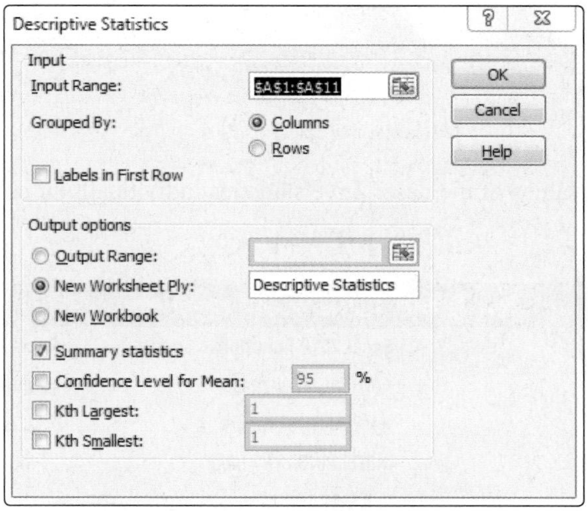

Clicking **OK** will produce the following list of descriptive statistics.

	A	B
1	Column1	
2		
3	Mean	15.36363636
4	Standard Error	1.266869637
5	Median	14
6	Mode	12
7	Standard Deviation	4.201731245
8	Sample Variance	17.65454545
9	Kurtosis	-0.955866366
10	Skewness	0.588007341
11	Range	13
12	Minimum	10
13	Maximum	23
14	Sum	169
15	Count	11

MINITAB

Descriptive Statistics

MINITAB can be used to find the mean, median, and sample standard deviation of a data set. It can also give the values for the five-number summary. There is not a consensus among statisticians nor software packages on how to calculate the first and third quartiles; thus, as with Microsoft Excel, the method used by MINITAB to calculate the first and third quartiles is different from the other methods previously discussed, so the values given by MINITAB may differ slightly from the provided textbook answers depending on the data set.

Example T.4

Using MINITAB to Calculate Descriptive Statistics

Suppose six people participated in a 1000-meter run. Their times, measured in minutes, are given below.

4, 10, 9, 11, 9, 7

Use the Display Descriptive Statistics option in MINITAB to calculate the list of descriptive statistics for the data values.

Solution

Begin by typing the data in column C1. Go to **Stat ▶ Basic Statistics ▶ Display Descriptive Statistics**. In the dialog box, input **C1** under Variables. If you want to choose which descriptive statistics are shown, click on **Statistics**, select or deselect the various options, and click **OK**. Once this is done, click **OK** in the main dialog window. Observe the Session window for the descriptive statistics.

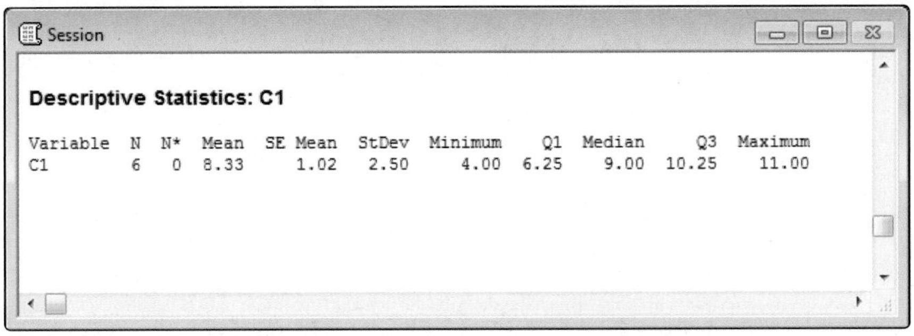
Session

Descriptive Statistics: C1

Variable	N	N*	Mean	SE Mean	StDev	Minimum	Q1	Median	Q3	Maximum
C1	6	0	8.33	1.02	2.50	4.00	6.25	9.00	10.25	11.00

Chapter Four

Probability, Randomness, and Uncertainty

Sections

4.1 Introduction to Probability
4.2 Addition Rules for Probability
4.3 Multiplication Rules for Probability
4.4 Combinations and Permutations
4.5 Combining Probability and Counting Techniques

Objectives

1. Identify the sample space of a probability event.
2. Calculate basic probabilities.
3. Determine if two events are mutually exclusive.
4. Determine if two events are independent.
5. Use the addition rules to calculate probability.
6. Use the multiplication rules to calculate probability.
7. Calculate numbers of permutations and combinations.
8. Use basic counting rules to calculate probability.

4

Introduction

Assessing how likely things are to happen is a useful tool in everyday life. In statistics, a measure of this likelihood is called probability. How would it affect your wager if you could *know* that you have only a 6% chance of winning in a game of cards? Or, what if instead you could know your chance of winning the hand is 58%? In both cases, in order to calculate your probabilities of winning, you have to know which cards are left in the deck. That is, you have to be able to count all of the possible outcomes that could be dealt.

Casinos know all too well what happens when players use statistical techniques to help them beat the odds. These players are called card counters, and casinos do not like them. By keeping track of the cards that have already been dealt, card counters are able to determine which cards are left in the deck, and therefore can predict when the high cards are more likely to turn up. Casinos make money because the odds of each game are stacked slightly in their favor, and the element of chance is essential.

By using probability to help them decide how to place their bets, card counters are able to walk away with winnings that are unlikely to occur simply by chance. For example, in the early 1990s a group of MIT students studied card counting techniques in order to win big in blackjack at the Las Vegas casinos. Win big they did, as their winnings peaked at $4 million before the casinos caught on. Unfortunately for the students, not only were they unable to hold onto their money, but they were also quickly blacklisted from every casino in the state!

4.1 Introduction to Probability

In this chapter, our primary goal is to introduce the basic techniques and terminology that will allow us to calculate probabilities. First, a **probability experiment**, or *trial*, is any process with a result determined by chance. Examples of probability experiments include flipping a coin, tossing a pair of dice, or drawing a raffle ticket. In each of these examples, there is more than one possible result and that result is determined at random. In a given probability experiment, each individual result that is possible is called an **outcome**. The set of all possible outcomes for a given probability experiment is called the **sample space**. An **event** is a subset of outcomes from the sample space.

Definition

A **probability experiment** (or *trial*) is any process with a result determined by chance.

Each individual result that is possible for a probability experiment is an **outcome**.

The **sample space** is the set of all possible outcomes for a given probability experiment.

An **event** is a subset of outcomes from the sample space.

Note that it is possible for an event to include the entire sample space. For example, consider the experiment of rolling a die. There are six possible outcomes, namely the numbers 1 through 6. The sample space is the set of all outcomes, which in this case is simply $\{1, 2, 3, 4, 5, 6\}$. The event "rolling an even number" is the subset of outcomes $\{2, 4, 6\}$. On the other hand, the event "rolling a number less than 10" is the set $\{1, 2, 3, 4, 5, 6\}$.

Example 4.1

Identifying Outcomes in a Sample Space or Event

Consider an experiment in which a coin is tossed and then a six-sided die is rolled.

a. List the outcomes in the sample space for the experiment.

b. List the outcomes in the event "tossing a tail then rolling an odd number."

Solution

a. Each outcome consists of a coin toss and a die roll. For example, heads and a 3 could be denoted as H3. Using this notation, the sample space can be written as follows.

$$\text{Sample space} = \begin{Bmatrix} \text{H1} & \text{T1} \\ \text{H2} & \text{T2} \\ \text{H3} & \text{T3} \\ \text{H4} & \text{T4} \\ \text{H5} & \text{T5} \\ \text{H6} & \text{T6} \end{Bmatrix}$$

b. Choosing the members of the sample space that fit the event "tossing a tail then rolling an odd number" gives the following.

$$\{\text{T1}, \text{T3}, \text{T5}\}$$

Sometimes it is easy to list the outcomes for an experiment, as in the previous example. However, at times, the sample space is large, and techniques should be used to ensure that no outcomes are omitted. An experiment may have a particular pattern that enables you to more easily organize the outcomes. For example, a pattern was used to create the sample space in Example 4.1. The first column shows each possibility beginning with heads, and the second column shows the possibilities beginning with tails. By using this pattern, we can easily see that every possible outcome is accounted for. The following is another example of using patterns to exhaustively list the outcomes in the sample space.

Example 4.2

Using a Pattern to List All Outcomes in a Sample Space

Consider the experiment in which a red six-sided die and a blue six-sided die are rolled together.

a. Use a pattern to help list the outcomes in the sample space.

b. List the outcomes in the event "the sum of the numbers rolled on the two dice equals 6."

Solution

a. Although there are many patterns that could help us list all outcomes in the sample space, our pattern will be to start with keeping the red die at 1, and allowing the blue die to vary from 1 to 6.

There are six of these outcomes: ⚀ ⚀, ⚀ ⚁, ⚀ ⚂, ⚀ ⚃, ⚀ ⚄, and ⚀ ⚅. Next we list all outcomes in which a 2 is rolled on the red die. Again, there are six outcomes of this form. The pattern continues until all 36 outcomes are listed.

Sample space =

⚀ ⚀ = 2	⚀ ⚁ = 3	⚀ ⚂ = 4	⚀ ⚃ = 5	⚀ ⚄ = 6	⚀ ⚅ = 7
⚁ ⚀ = 3	⚁ ⚁ = 4	⚁ ⚂ = 5	⚁ ⚃ = 6	⚁ ⚄ = 7	⚁ ⚅ = 8
⚂ ⚀ = 4	⚂ ⚁ = 5	⚂ ⚂ = 6	⚂ ⚃ = 7	⚂ ⚄ = 8	⚂ ⚅ = 9
⚃ ⚀ = 5	⚃ ⚁ = 6	⚃ ⚂ = 7	⚃ ⚃ = 8	⚃ ⚄ = 9	⚃ ⚅ = 10
⚄ ⚀ = 6	⚄ ⚁ = 7	⚄ ⚂ = 8	⚄ ⚃ = 9	⚄ ⚄ = 10	⚄ ⚅ = 11
⚅ ⚀ = 7	⚅ ⚁ = 8	⚅ ⚂ = 9	⚅ ⚃ = 10	⚅ ⚄ = 11	⚅ ⚅ = 12

Note that rolling a 1 on the red die and a 2 on the blue die, denoted ⚀ ⚁, is a different outcome than rolling a 2 on the red die and a 1 on the blue die, denoted ⚁ ⚀.

b. To list the outcomes in the event "the sum of the numbers rolled on the two dice equals 6," look through the list above for each individual outcome where the sum equals 6. Note that $1 + 5 = 6$, so the outcome of rolling a 1 on the red die and a 5 on the blue die is in the event. The reverse would be rolling a 5 on the red die and a 1 on the blue die. These are two separate possible outcomes that are both in the event. The event consists of the following outcomes.

{⚀ ⚄ = 6, ⚄ ⚀ = 6, ⚁ ⚃ = 6, ⚃ ⚁ = 6, ⚂ ⚂ = 6}

Caution

The sample space is the *set* of outcomes for an experiment, not the *number* of outcomes.

Sometimes a pattern is more easily displayed by other means. For example, when an experiment consists of several stages, a **tree diagram** allows the outcomes to be organized in a systematic manner. The tree begins with the possible outcomes for the first stage and then branches for each additional possibility. Each of the elements of the last row in the tree diagram represents a unique outcome in the sample space. To identify a specific outcome, list the sequence of events that describes a path from the top of the tree to the bottom of the tree. Notice that the number of possibilities in the bottom row of the tree is equal to the *number* of outcomes in the sample space. The tree diagram method is demonstrated in the following example.

Example 4.3

Using a Tree Diagram to List All Outcomes in a Sample Space

Consider a family with three children. Use a tree diagram to find the sample space for the gender of each child in regard to birth order.

Solution

The tree begins with the two possibilities for the first child—girl or boy. It then branches for each of the other two births in the family as shown.

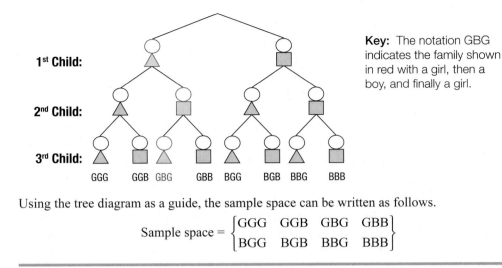

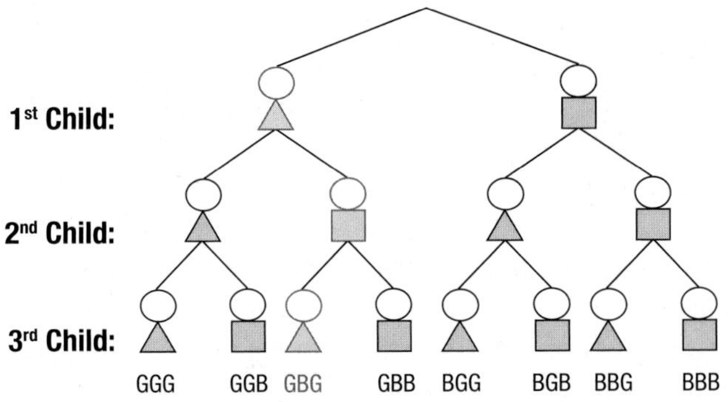

Key: The notation GBG indicates the family shown in red with a girl, then a boy, and finally a girl.

Using the tree diagram as a guide, the sample space can be written as follows.

$$\text{Sample space} = \begin{Bmatrix} \text{GGG} & \text{GGB} & \text{GBG} & \text{GBB} \\ \text{BGG} & \text{BGB} & \text{BBG} & \text{BBB} \end{Bmatrix}$$

Now that we have established ways to identify all of the possible outcomes we'll look at three methods for calculating the probability of each outcome: subjective, experimental, and classical probability.

The least precise type of probability is subjective probability. **Subjective probability** is simply an educated guess regarding the chance that an event will occur. For example, the local meteorologist gives a subjective probability when he predicts the chance of rain. Clearly, subjective probability is only as good as the expertise of the person giving that probability. If the weatherman says there is a 90% chance of rain, you will probably take your umbrella with you to work. However, if your roommate looks out the window and guesses a 90% chance of rain, you won't be so quick to believe him or her. For this reason, you should always be cautious with subjective probabilities.

Experimental probability, on the other hand, is found, obviously enough, by performing an experiment. Specifically, the experimental probability of an event is calculated by dividing the number of times an event occurs by the total number of trials performed. This is also known as *empirical probability*.

Formula

Experimental Probability

In **experimental probability**, if E is an event, then $P(E)$, read "the probability that E occurs," is given by

$$P(E) = \frac{f}{n}$$

where f is the frequency of event E and

n is the total number of times the experiment is performed.

Rounding Rule

When calculating probability, either give the exact fraction or a decimal rounded to four decimal places. If the probability is extremely small, it is permissible to round the decimal to the first nonzero digit.

For example, suppose a fisherman wants to know the probability that a fish he catches in his favorite fishing hole is a catfish. For his experiment, he records the numbers and types of fish he catches one week. Suppose that during this week he catches 92 fish, 43 of which are catfish. He can then calculate the experimental probability of catching a catfish as follows.

$$\begin{aligned} P(\text{catfish}) &= \frac{\text{number of catfish caught}}{\text{total number of fish caught}} \\ &= \frac{43}{92} \\ &\approx 0.4674 \end{aligned}$$

Side Note

Expressing Probability

A probability may be written as a fraction, decimal, or percentage. For instance, we may say that the probability of the fisherman catching a catfish is $\frac{43}{92}$, 0.4674, or 46.74%.

The fisherman has obtained a reasonable estimate based on one week's worth of fishing. How do you think his estimate of the probability of catching a catfish would change if instead he recorded his catches for a month? Two months? A year? The **Law of Large Numbers** says that the greater the number of trials, the closer the experimental probability will be to the *true probability*.

What, then, is the true probability? **Classical probability**, also called *theoretical probability*, is the most precise type of probability. This is because classical probability is calculated by taking *all* possible outcomes for an experiment into account. In the previous example about fishing, the only way that the classical probability of catching a catfish could be calculated is if the exact numbers of catfish and all fish living in the pond were both known.

Let's give a proper definition for classical probability. Classical probability states that if all outcomes are equally likely, the probability of an event is equal to the number of outcomes included in the event divided by the total number of outcomes in the sample space. Classical probability is written mathematically in the following box.

Formula

Classical Probability

In **classical probability**, if all outcomes are equally likely to occur, then $P(E)$, read "the probability that E occurs," is given by

$$P(E) = \frac{n(E)}{n(S)}$$

where $n(E)$ is the number of outcomes in the event and

$n(S)$ is the number of outcomes in the sample space.

So back at our local fishing hole, a sudden inexplicable drought causes all of the water from the pond to evaporate, leaving the fish in the pond exposed and vulnerable. Wildlife Management is able to step in and save the fish, relocating them to other ponds nearby. During this relocation process, the exact count of all fish in the pond, along with their species, is recorded. There are 1235 fish in the pond, 541 of which are catfish. With this new information, what is the classical probability of the fisherman catching a catfish, assuming that it is equally likely to catch any given fish in the pond?

$$\begin{aligned} P(\text{catfish}) &= \frac{n(E)}{n(S)} \\ &= \frac{541}{1235} \\ &\approx 0.4381 \end{aligned}$$

As an alternate example of classical probability, consider the probability of rolling a six-sided die and obtaining an even number. There are three outcomes in the event: rolling a 2, 4, or 6, so $n(E)=3$. There are six outcomes in the sample space (and all outcomes are equally likely); thus $n(S)=6$. So we calculate the probability as follows.

$$\begin{aligned} P(\text{even}) &= \frac{n(E)}{n(S)} \\ &= \frac{3}{6} \\ &= \frac{1}{2} \\ &= 0.5 \end{aligned}$$

It's important to note that this form of classical probability applies only when the outcomes are equally likely. That is, each outcome has the same probability of occurring. There is an obvious fallacy in the statement "There is a 50% chance I will be hit by lightning tomorrow." Even though there are only two outcomes in the sample space, {hit by lightning, not hit by lightning}, the probability of being struck by lightning is *not* $\frac{1}{2}=50\%$ because the outcomes are not equally likely. (Thank goodness!) In fact, the National Weather Service estimates that the probability of being struck by lightning in a lifetime is $\frac{1}{10,000}$.

Definition

Subjective probability is an educated guess regarding the chance that an event will occur.

Experimental probability (or *empirical probability*) uses the outcomes obtained by repeatedly performing an experiment to calculate the probability.

Classical probability (or *theoretical probability*) is the most precise type of probability and can only be calculated when all possible outcomes in the sample space are known and equally likely to occur.

Let's look at an example of how to distinguish between the three different types of probability.

Example 4.4

Identifying Types of Probability

Determine whether each probability is subjective, experimental, or classical.

a. The probability of selecting the queen of spades out of a well-shuffled standard deck of cards is $\frac{1}{52}$.

b. An economist predicts a 20% chance that technology stocks will decrease in value over the next year.

c. A police officer wishes to know the probability that a driver chosen at random will be driving under the influence of alcohol on a Friday night. At a roadblock, he records the number of drivers stopped and the number of drivers driving with a blood alcohol level over the legal limit. He determines that the probability is 3%.

Solution

a. All 52 outcomes are known and are equally likely to occur, so this is an example of classical probability.

b. The economist is giving an educated guess; therefore, this is an example of subjective probability.

c. The police officer's probability is based on the evidence gathered from the roadblock. Because not *all* drivers were evaluated at that roadblock, the probability is experimental.

We'll spend the remainder of this section finding classical probabilities for different scenarios.

Example 4.5

Calculating Classical Probability

Beck is allergic to peanuts. At a large dinner party one evening, he notices that the cheesecake options on the dessert table contain the following flavors: 10 slices of chocolate, 12 slices of caramel, 12 slices of peanut butter chocolate, and 8 slices of strawberry. Assume that the desserts are served to guests at random.

a. What is the probability that Beck's cheesecake contains peanuts?

b. What is the probability that Beck's dessert does not contain chocolate?

Solution

a. There are 12 slices of chocolate peanut butter cheesecake, and $10 + 12 + 12 + 8 = 42$ pieces of cheesecake total. The probability is then calculated as follows.

$$\begin{aligned} P(\text{peanut butter chocolate}) &= \frac{n(E)}{n(S)} \\ &= \frac{12}{42} \\ &= \frac{2}{7} \\ &\approx 0.2857 \end{aligned}$$

b. There are $12 + 8 = 20$ slices of cheesecake that do not contain chocolate. The probability of being served one of these desserts is calculated as follows.

$$\begin{aligned} P(\text{not chocolate}) &= \frac{n(E)}{n(S)} \\ &= \frac{20}{42} \\ &= \frac{10}{21} \\ &\approx 0.4762 \end{aligned}$$

Example 4.6

Calculating Classical Probability

Consider a beginning archer who only manages to hit the target 50% of the time. What is the probability that in three shots, the archer will hit the target all three times?

Solution

For each shot, the arrow will either hit the target or miss the target. Note that these two outcomes are equally likely here since his chance of hitting the target is 50%. To determine how many outcomes are in the sample space, we can use either a pattern or a tree diagram. Let's use a tree diagram.

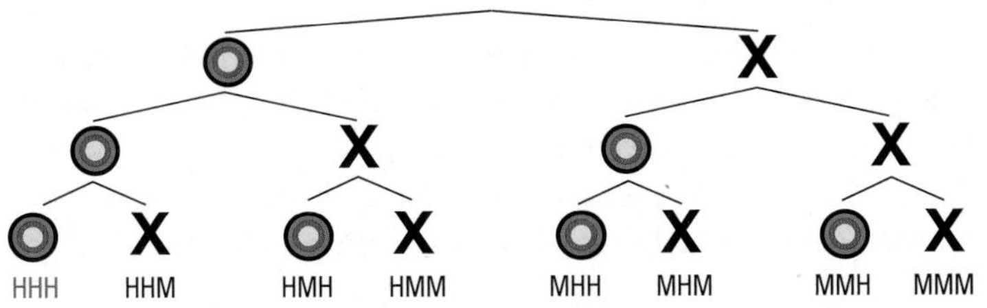

Notice that this gives 8 possible outcomes for the three shots, only 1 of which consists of hitting the target all three times. Thus, the probability of the novice archer hitting the target three times in a row is calculated as follows.

$$\begin{aligned} P(\text{all three hits}) &= \frac{n(E)}{n(S)} \\ &= \frac{1}{8} \\ &= 0.125 \end{aligned}$$

Example 4.7

Calculating Classical Probability

Consider a family with six boys. What is the probability that the seventh child will also be a boy?

Solution

Many people would say that this family is due to have a girl, but let's consider the mathematics. Suppose we took the time to write down every possible combination for birth orders of seven children. Because the family already has six boys, we would have to mark out all combinations that contain a girl in the first six children. The only two outcomes left in the sample space would be BBBBBBG and BBBBBBB. The probability of baby number seven being a boy is then

$$\begin{aligned} P(\text{all seven boys}) &= \frac{n(E)}{n(S)} \\ &= \frac{1}{2} \\ &= 0.5, \end{aligned}$$

just as with any other pregnancy!

Example 4.8

Calculating Classical Probability

In biology, we learn that many diseases are genetic. One example of such a disease is Huntington's disease, which causes neurological disorders as a person ages. Each person has two huntingtin genes—one inherited from each parent. If an individual inherits a mutated huntingtin gene from either of his or her parents, he or she will develop the disease. On the TV show *House*, the character who Dr. House calls "13" inherited this disease from her mother. Assume for a moment that "13" has a child with a person who has two healthy huntingtin genes. What is the probability that her child will develop Huntington's disease?

Solution

Side Note

Punnett Square

Reginald Punnett (1875–1967) was a British geneticist who developed the method discussed in Example 4.8, which is named after him, called the Punnett square. He is also known for cofounding the *Journal of Genetics* and writing one of the first textbooks on genetics.

So far, we have listed the outcomes of an experiment in an orderly fashion and by using a tree diagram. In biology, we commonly use a Punnett square to help list the outcomes of a genetic experiment. The mother's two alleles of a specific gene are listed along one side of a square, and the father's two alleles of the gene are listed along an adjacent side. We then fill in the square by writing in the four possible combinations of alleles inherited—one from each parent.

For this experiment, we know that each person has two huntingtin genes. We will label a mutated huntingtin gene with an uppercase "H" and an unaffected huntingtin gene with a lowercase "h." For this experiment, the mother has at least one mutated gene. Let's assume it is just one, so her genetic makeup would be written as Hh. She has a child with someone who has no affected huntingtin genes, so his genetic makeup can be written as hh. Let's write these gene combinations on the sides of a Punnett square.

		Mother	
		H	h
Father	h	Hh	hh
	h	Hh	hh

So the four possibilities for the child's genes are $\{\text{Hh, Hh, hh, hh}\}$. Someone only has to have one mutated gene to develop the disease, so that means two of the four combinations would result in a child having the disease, $\{\text{Hh, Hh}\}$. Thus, if E = the event of the child inheriting a mutated huntingtin gene, then the probability that E occurs is calculated as follows.

$$\begin{aligned} P(E) &= \frac{n(E)}{n(S)} \\ &= \frac{2}{4} \\ &= \frac{1}{2} \\ &= 0.5 \end{aligned}$$

This means that "13" has a 50% chance of passing on the disease to a child. This is true for anyone with Huntington's disease. The probability goes up if the other parent also has the disease.

4.1 Section Exercises

Sample Spaces

Directions: Find the sample space for the given experiment.

1. Two coins are tossed.

2. A child's board game contains a spinner with three colors: orange, yellow, and green. Give the sample space for two consecutive spins.

3. A family has four children. Give the sample space in regard to the genders of the children.

4. A bag contains four marbles: one each of green, red, blue, and violet. Two marbles are drawn from the bag. (Assume that the first marble is not put back in the bag before drawing the second marble.)

5. When buying a new car, you've narrowed your choices to three colors: red, black, or silver. You also need to decide whether to have a sunroof or not, and whether you want leather or cloth interior.

6. When ordering a pizza with a coupon, you have a choice of crusts: thin, hand-tossed, or stuffed. You can also choose one topping from the following: pepperoni, ham, sausage, onion, bell pepper, or olives.

7. When choosing your seat at the opera, you can choose from three levels and then whether you want an aisle seat or not.

8. When building your new house, you have a choice of flooring for the kitchen: tile, concrete, or wood. You must also choose the counter tops from the following: granite, concrete, wood, or stainless steel.

Types of Probability

Directions: Determine whether each probability is subjective, experimental, or classical.

9. Jeff wants to know whether a certain coin is fair or not. He flips the coin 100 times and obtains tails 61 times. He calculates that the probability of obtaining a tail with his coin is 61%.

10. Caroline estimates that there is only a 10% chance that they will have a quiz in biology.

11. Mr. Dorrough's 18 students have dropped their names in the hat for a prize drawing. Stephanie calculates that she has a 1/18 chance of winning.

12. On a game show, the contestant must choose one of three doors, behind one of which is a new car. He has a one-in-three chance of winning the car.

13. A computer manufacturer brags that there is less than a 5% chance that its customers will want to switch brands within the first year.

Experimental Probability

Directions: Calculate each experimental probability.

14. A very large bag contains more coins than you are willing to count. Instead, you draw a random sample of coins from the bag and record the following numbers of each type of coin in the sample before returning the sampled coins to the bag.

Coins in a Bag			
Quarters	**Dimes**	**Nickels**	**Pennies**
23	29	17	38

If you randomly draw a single coin out of the bag, what is the probability that you will obtain:

a. A nickel?

b. A penny?

c. Either a quarter or a dime?

15. A telemarketer's computer selects phone numbers at random. The telemarketer has recorded the number of respondents in each age bracket for one evening in the following table.

Number of Respondents by Age			
18–25	**26–35**	**36–45**	**Over 45**
29	40	55	51

What is the probability that the next respondent will be:

a. Over 45?

b. Between 26 and 35?

c. At least 36?

Classical Probability

Directions: Calculate each classical probability. Assume that individual outcomes are equally likely.

16. Martha has a box full of 17 different CDs: 5 rock, 3 blues, 6 pop, and 3 R&B. If she randomly pulls out a CD, what is the probability that it is a blues CD?

17. Chloë puts a coin into a gumball machine that contains 12 blue, 15 pink, 9 orange, 16 yellow, and 14 white gumballs. What is the probability that Chloë gets a yellow gumball?

18. At one hospital, there were 796 boys and 764 girls born during one particular year. If a delivery from that year at the hospital is chosen at random, what is the probability that the baby is a girl?

19. What is the probability that a person selected at random will have a March birthday? (Assume that every day of the year contains an equal number of birthdays, and the person was not born in a leap year.)

20. If Mark grabs a utensil out of a drawer without looking, what is the probability that he grabs a fork if there are 7 knives, 9 spoons, and 6 forks in the drawer?

21. If Timmy is fishing in his newly stocked pond and knows that there are 200 bream and 150 bass in it, what is the probability that the first fish he catches will be a bass?

22. John is at a cookout and wants to get a drink from the cooler. If there are 12 colas, 10 bottles of water, and 5 root beers in the cooler, what is the probability that he randomly grabs a root beer?

23. A college algebra class has 14 freshmen, 21 sophomores, 9 juniors, and a senior enrolled. What is the probability that the professor randomly selects the senior to answer a question?

24. Mary Ann is sewing and needs the spool of white thread. Her basket of sewing supplies is sitting next to her, and it contains 26 different colors of thread, including the white spool she needs. If she grabs one spool without looking, what is the probability that she has chosen the white spool of thread?

25. For a school fundraiser, 1000 raffle tickets are sold for $5 each. Each ticket is assigned a three-digit number using the digits 0–9. What is the probability that the winning ticket will be one with three repeating digits?

26. If you roll one six-sided die twice, what is the probability that the second number rolled is at least as large as the first?

4.2 Addition Rules for Probability

In the previous section we introduced the concept of probability, discussed basic terminology, and looked at how to compute simple probabilities. We will now begin learning various tools for calculating more complex probabilities, the first of which is the addition rule. But before we are ready to calculate more involved probabilities, let's first cover some basic properties of probability.

Properties of Probability

First, probability is always a number between 0 and 1. That is, $0 \le P(E) \le 1$. Why is this the case? Well, using what we have learned from the previous section, what is the probability of an event that will never occur? Consider the probability of rolling a number greater than a six on a standard six-sided die. The numerator in this probability would be 0 since there are 0 outcomes for an event that cannot happen, and 0 divided by any number is 0, so the probability is 0. Now, what would be the probability of an event that will occur every time, or is certain to occur? Consider rolling a number less than seven on a standard six sided die. In this probability, the numerator and denominator would be the same, so the probability would be 1. These are the two extremes, so every other probability would have to fall between these values of 0 and 1. The closer a probability is to 0, the less likely the event is to happen; and the closer a probability is to 1, the more likely it is to happen.

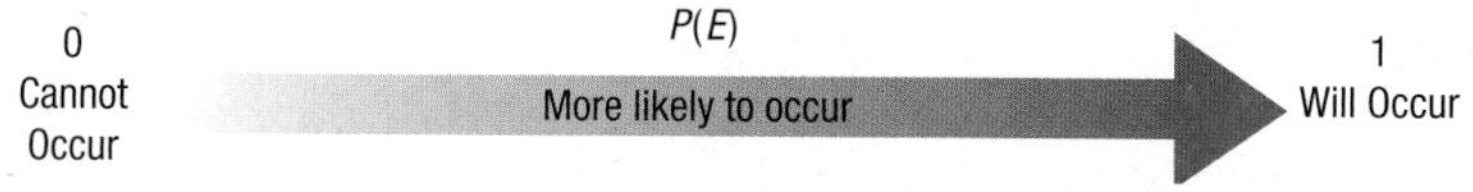

Figure 4.1: Properties of Probability

Memory Booster

When an event includes the entire sample space, the probability is equal to 1.

Another absolute regarding probability is that for a given experiment, one of the outcomes in the sample space has to occur. If the sample space, S, is the set of all possible outcomes, then when the experiment is performed, one of them must occur. Mathematically, we write this fact as $P(S) = 1$. Along those same lines, it is impossible for none of the outcomes in the sample space to occur. We can write this second fact as $P(\varnothing) = 0$, where $\varnothing$ is the "empty set," which is the set of no outcomes.

Properties

Properties of Probability

1. For any event, E, $0 \le P(E) \le 1$.
2. For any sample space, S, $P(S) = 1$.
3. For the empty set, $\varnothing$, $P(\varnothing) = 0$.

The Complement

Math Symbols

The complement of an event, E, can be denoted in several different ways, such as E^c, $\bar{E}$, or E'.

Recall that an event, E, is defined to be a subset of outcomes from a sample space. The **complement** for E, denoted E^c, consists of all the outcomes in the sample space that are not in E. This means that if you combine the outcomes from E and the outcomes from E^c, you will have the entire sample space. That is, $E + E^c = S$.

As an example, consider the experiment of rolling a six-sided die. If E is the event of rolling an even number, $E = \{2, 4, 6\}$, then its complement is rolling an odd number, $E^c = \{1, 3, 5\}$. Notice that $E + E^c = S$; that is, $\{2, 4, 6\} + \{1, 3, 5\} = \{1, 2, 3, 4, 5, 6\}$. An illustration of this die example is given in Figure 4.2.

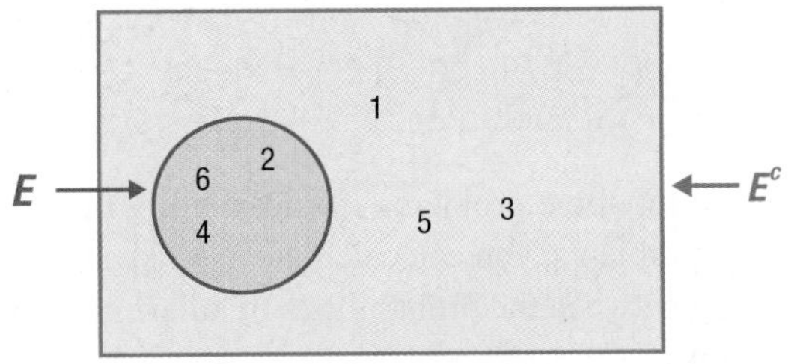

Figure 4.2: Venn Diagram Illustrating the Complement of an Event

Definition

The **complement** of an event E, denoted E^c, is the set of all outcomes in the sample space that are not in E.

Example 4.9

Describing the Complement of an Event

Describe the complement of each of the following events.

a. Choose a red card out of a standard deck of cards.

b. Out of 31 students in your statistics class, 15 are out sick with the flu.

c. In your area, 91% of phone customers use PhoneSouth.

Solution

a. A standard deck of cards contains 26 red and 26 black cards. Since all 26 red cards are contained in the event, the complement contains all 26 black cards.

b. Since 15 students are out with the flu, the complement contains the other 16 students who are not sick.

c. Since the event contains all PhoneSouth customers, everyone who is not a PhoneSouth customer is in the complement. Thus, the complement contains the other 9% of phone customers in your area.

Memory Booster

$E + E^c = S$

That is, the outcomes in the event, E, plus the outcomes in the complement, E^c, equal the entire possible set of outcomes, which is the sample space, S.

Because combining the two subsets, E and E^c, gives the entire sample space, it is true that the sum of their probabilities is always one. This property is written mathematically as follows.

Formula

Complement Rule for Probability

The sum of the probabilities of an event, E, and its complement, E^c, is equal to one.

$$P(E)+P(E^c)=1$$

This fact is convenient because for some problems the probability of E^c is easier to calculate than the probability of E. For these problems, you can calculate $P(E^c)$ and subtract that value from 1 to obtain $P(E)$. This relationship between the probabilities of an event, E, and its complement, E^c, is given by the following equation.

$$P(E)=1-P(E^c)$$

Example 4.10

Using the Complement Rule for Probability

a. If you are worried that there is a 35% chance that you will fail your upcoming test, what is the probability that you will pass the test?

b. If there is a 5% chance that none of the items on a scratch-off lottery ticket will be a winner, what is the probability that at least one of the scratch-off items will win?

Solution

a. The complement to the outcome of failing your upcoming test is passing it. Thus, the probability of passing is calculated as follows.

$$\begin{aligned} P(\text{passing}) &= 1-P(\text{failing}) \\ &= 1-0.35 \\ &= 0.65 \\ &= 65\% \end{aligned}$$

So the good news is that you have more than a 50% chance of passing the test.

b. The complement to having none of something is having at least one of that thing. Thus, the probability is calculated as follows.

$$\begin{aligned} P(\text{at least one winner}) &= 1-P(\text{no winners}) \\ &= 1-0.05 \\ &= 0.95 \\ &= 95\% \end{aligned}$$

Thus, there is a 95% chance of having at least one winning scratch-off item.

Example 4.11

Using the Complement Rule for Probability

Roll a pair of standard six-sided dice. What is the probability that neither die is a three?

Solution

We could list the outcomes in E, every combination of the dice that does not have a three. It is much easier to count the outcomes in the complement, E^c. The complement of this event contains the outcomes in which either die is a three. (Check for yourself by making sure that adding the event and its complement covers the entire sample space.) Let's list these outcomes.

$E^c =$

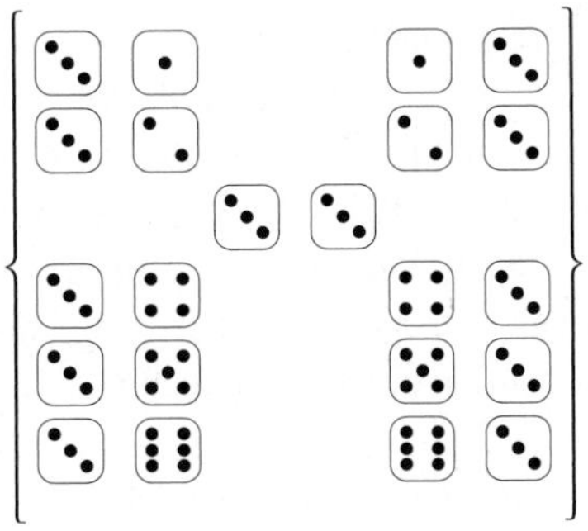

There are 11 outcomes where at least one of the dice is a three. Since we have already seen that there are 36 possible ways to roll a pair of dice (Section 4.1 Example 4.2), we have that $P(E^c) = \frac{n(E^c)}{n(S)} = \frac{11}{36}$. Subtracting this probability from 1 gives us the following.

$$\begin{aligned} P(E) &= 1 - P(E^c) \\ &= 1 - \frac{11}{36} \\ &= \frac{25}{36} \\ &\approx 0.6944 \end{aligned}$$

Therefore, the probability that neither die is a three is approximately 0.6944.

Addition Rules for Probability

So far in this chapter we have discussed probabilities that involve a single event, such as drawing a heart from a deck of cards or rolling seven with a pair of dice, but now we want to discuss how to find probabilities that involve combinations of events. The first type of combination we will consider is the probability that one event *or* another event occurs. For example, what is the probability of drawing a heart *or* a queen from a deck of cards? There are 52 cards in a deck and 13 of these cards are hearts, so the probability of drawing a heart is $\frac{13}{52}$. Similarly, there are 4 queens in a deck, so the probability of drawing a queen is $\frac{4}{52}$. But what is the probability of drawing one or the other? Wouldn't it be convenient if we could add the previous two probabilities together? Let's look at a picture illustrating this situation to help us see if adding these probabilities will give us the correct solution.

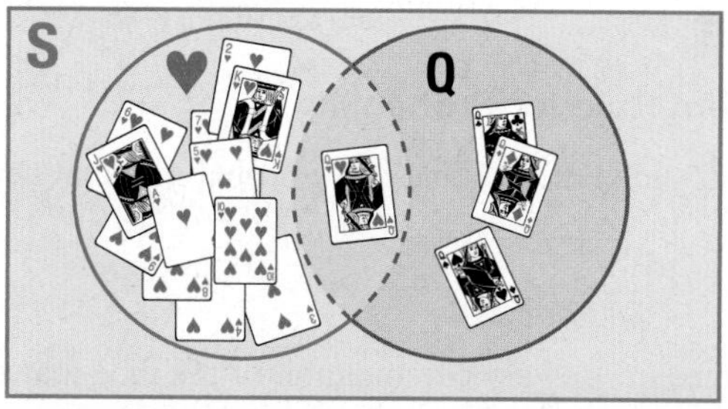

Figure 4.3: Venn Diagram Illustrating the Probability of Choosing Either a Heart or a Queen

The rectangle represents the entire sample space, all 52 cards in the deck. The circle on the left represents the 13 hearts, and the circle on the right represents the four queens. The shaded area in the middle represents any cards that are both hearts and queens, namely the queen of hearts. Figure 4.3 shows a problem with our proposed solution—the overlapping region has been counted twice! The queen of hearts was counted as a heart and again as a queen. Because its probability was counted both for the hearts and the queens, we must subtract it in order to obtain the correct probability. Therefore, the probability of choosing a heart or a queen is calculated as follows.

$$\begin{aligned}P(\text{heart or queen}) &= P(\text{heart}) + P(\text{queen}) - P(\text{heart and queen})\\ &= \frac{13}{52} + \frac{4}{52} - \frac{1}{52}\\ &= \frac{16}{52}\\ &= \frac{4}{13}\\ &\approx 0.3077\end{aligned}$$

We can generalize this concept to any probability of one event or another occurring as shown in the formula, often called the **Addition Rule for Probability**, in the following box.

Formula

Addition Rule for Probability

For two events, E and F, the probability that E or F occurs is given by the following formula.

$$P(E \textbf{ or } F) = P(E) + P(F) - P(E \text{ and } F)$$

Example 4.12

Using the Addition Rule for Probability

Ceresa is looking for a new condo to rent. Ceresa's realtor has provided her with the following list of amenities for 17 available properties. The list contains the following.

- Close to the subway: 6 properties
- Low maintenance fee: 7 properties
- Green space: 5 properties
- Newly renovated: 2 properties

- Close to the subway and low maintenance fee: 2 properties
- Green space and newly renovated: 1 property

If Ceresa's realtor selects the first condo they visit at random, what is the probability that the property is either close to the subway or has a low maintenance fee?

Solution

Before we calculate the probability, let's verify that Ceresa's realtor has accurately counted the total number of properties. At first glance, it might seem that there are more than 17 properties if you simply add all the numbers in the list the realtor gave. However, there are $6+7+5+2=20$ properties that have single characteristics, and $2+1=3$ properties containing two characteristics each. So, there are in fact only $20-3=17$ individual properties.

To calculate Ceresa's probability, the key word here is "or," which tells us we will be using the Addition Rule for Probability to find the solution. Using this formula gives the following.

$$\begin{aligned}
&P(\text{close to subway or low fee}) \\
&\quad= P(\text{close to subway}) + P(\text{low fee}) - P(\text{close to subway and low fee}) \\
&\quad= \frac{6}{17} + \frac{7}{17} - \frac{2}{17} \\
&\quad= \frac{11}{17} \\
&\quad\approx 0.6471
\end{aligned}$$

Therefore, the probability that the first condo Ceresa sees is either close to the subway or has a low maintenance fee is approximately 64.71%.

Example 4.13

Using the Addition Rule for Probability

Suppose that after a vote in the US Senate on a proposed health care bill, the following table shows the breakdown of the votes by party.

Votes on Health Care Bill

	Voted in Favor	Voted Against
Democrat	23	21
Republican	43	7
Independent	2	4

If a lobbyist stops a random senator after the vote, what is the probability that this senator will either be a Republican or have voted against the bill?

Solution

The key word "or" tells us to use the Addition Rule once again. There are a total of 100 US senators, all of whom voted on this bill according to the table. Of these senators, $43+7=50$ are Republicans and $21+7+4=32$ voted against the bill. However, 7 of the senators are both Republican and voted against the bill. Thus, the Addition Rule would apply as follows.

$$\begin{aligned}
&P(\text{Republican or against}) \\
&\quad = P(\text{Republican}) + P(\text{against}) - P(\text{Republican and against}) \\
&\quad = \frac{50}{100} + \frac{32}{100} - \frac{7}{100} \\
&\quad = \frac{75}{100} \\
&\quad = \frac{3}{4} \\
&\quad = 0.75
\end{aligned}$$

So the probability that the senator the lobbyist stops will either be a Republican or have voted against the bill is 75%.

Example 4.14

Using the Addition Rule for Probability

Roll a pair of dice. What is the probability of rolling either a total less than four or a total equal to ten?

Solution

The key word "or" tells us to use the Addition Rule once again, which we apply as follows.

$$P(\text{less than four or ten}) = P(\text{less than four}) + P(\text{ten}) - P(\text{less than four and ten})$$

We know from previous examples that there are 36 outcomes in the sample space for rolling a pair of dice. We just need to determine the number of outcomes that give a total less than four and the number of outcomes that give a total of ten. Let's list these outcomes in a table.

Totals Less Than Four	Totals of Ten
2: 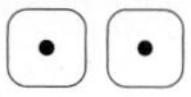	10: 
3: 	

By counting the outcomes, we see that there are 3 outcomes that have totals less than four and 3 outcomes that have totals of exactly ten. Note that there are no outcomes that are both less than four and exactly ten. Hence we fill in the probabilities as follows.

$$\begin{aligned}
P(\text{less than four or ten}) &= P(\text{less than four}) + P(\text{ten}) - P(\text{less than four and ten}) \\
&= \frac{3}{36} + \frac{3}{36} - \frac{0}{36} \\
&= \frac{6}{36} \\
&= \frac{1}{6} \\
&\approx 0.1667
\end{aligned}$$

We saw in the previous example that the events "roll a total less than four" and "roll a total equal to ten" did not have any outcomes in common. This made the calculation of their combined probability even easier than previous examples for the Addition Rule. Events that share no outcomes are called **mutually exclusive** events. For example, consider the events of choosing a seven or a face card out of a standard deck of cards. Is it possible to choose a card that is both a face card and a seven? No. These two events have no outcomes in common, so they are mutually exclusive. Figure 4.4 illustrates this relationship.

Memory Booster

Mutually exclusive events have no outcomes in common.

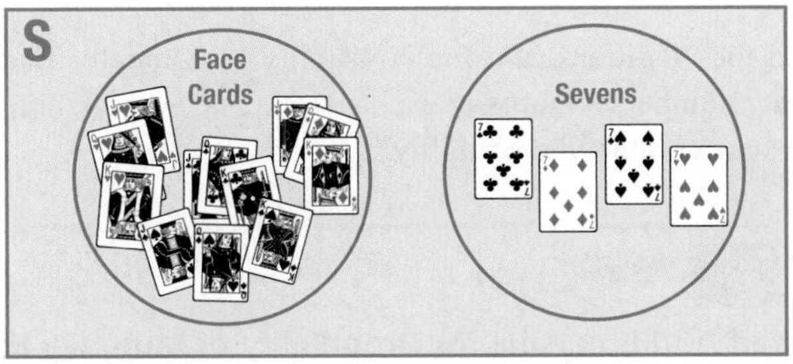

Figure 4.4: Venn Diagram Illustrating Mutually Exclusive Events

Notice that since these two events have no outcomes in common, the probability of both events occurring at the same time is zero. In fact, this will be true for any two events that are mutually exclusive. We can therefore shorten the formula for the Addition Rule for Probability of Mutually Exclusive Events to the following.

Formula

Addition Rule for Probability of Mutually Exclusive Events

If two events, E and F, are **mutually exclusive**, then the probability that E or F occurs is given by the following formula.

$$P(E\ \textbf{or}\ F) = P(E) + P(F)$$

Example 4.15

Using the Addition Rule for Probability of Mutually Exclusive Events

Caleb is very excited that it's finally time to purchase his first new car. After much thought, he has narrowed his choices down to four. Because it has taken him so long to make up his mind, his friends have started to bet on which car he will choose. They have given each car a probability based on how likely they think Caleb is to choose that car. Devin is betting that Caleb will choose either the Toyota or the Jeep. Find the probability that Devin is right.

Toyota: 0.40

Honda: 0.10

Ford: 0.10

Jeep: 0.35

Solution

Because Caleb can only choose one car, these two events are mutually exclusive. So, we can use the shortened Addition Rule for Probability of Mutually Exclusive Events. Let's assume that Caleb's friends have accurately determined Caleb's interest in each car with their probability predictions. Then the probability that Caleb chooses either the Toyota or the Jeep is calculated as follows.

$$\begin{aligned}P(\text{Toyota or Jeep}) &= P(\text{Toyota}) + P(\text{Jeep})\\ &= 0.40 + 0.35\\ &= 0.75\end{aligned}$$

Thus, Devin has a 75% chance of correctly picking which car Caleb will buy.

It is possible to extend the Addition Rule for Probability of Mutually Exclusive Events to more than two events. For any number of mutually exclusive events, you can add the probabilities of the individual events to find the combined probability.

Example 4.16

Using the Extended Addition Rule for Probability of Mutually Exclusive Events

At a certain major exit on the interstate, past experience tells us that the probabilities of a truck driver refueling at each of the five possible gas stations are given in the table below. Assuming that the truck driver will refuel at only one of the gas stations (thus making the events mutually exclusive), what is the probability that the driver will refuel at Shell, Exxon, or Chevron?

Probabilities of Refueling

Gas Station	Probability
BP	0.0351
Chevron	0.1539
Exxon	0.2793
Shell	0.3207
Texaco	0.2110

Solution

Since these three events are mutually exclusive, we can add the individual probabilities together.

$$\begin{aligned}P(\text{Shell or Exxon or Chevron}) &= P(\text{Shell}) + P(\text{Exxon}) + P(\text{Chevron})\\ &= 0.3207 + 0.2793 + 0.1539\\ &= 0.7539\end{aligned}$$

4.2 Section Exercises

Complement

Directions: Describe the complement for the given event.

1. Out of 253 apple trees in the orchard, just 42 are ready for harvesting.

2. In the graduating class at one large high school, 39% of the graduates plan to attend college out-of-state.

3. Out of 21 players on the baseball team, 4 are left-handed.

4. The adult literacy rate for the United States is 97%.

5. A local news station claims that 70% of its viewers are over the age of 30.

Complement Rule

Directions: Use the Complement Rule to find each probability.

6. A food distributor estimates that 2% of the eggs it delivers to grocery stores are cracked. What is the probability that an egg selected at random will not be cracked?

7. Every eleventh box of Blue Zinga cereal that is produced contains a special toy. If you buy a box of Blue Zinga, what is the probability that your box does not contain a toy?

8. Find the probability of rolling two dice and not getting doubles.

9. Suppose that a family has five children. What is the probability that the family has at least one girl?

10. Given that every fifth person in line will get a coupon for a free box of popcorn at the movies, what is the probability that you don't get a coupon when you're in line?

Mutually Exclusive Events

Directions: Determine whether the events are mutually exclusive.

11. Out of a standard deck, choose a jack or choose a diamond.

12. Out of a standard deck, choose a red card or a club.

13. For a survey on campus, a sophomore or a history major is chosen.

14. When determining his schedule for the spring semester, Troy must decide whether to take a history class that meets at 9:00 a.m. on Mondays, Wednesdays, and Fridays or a science class that meets Mondays and Wednesdays at 9:30 a.m. (Assume that each class meets for 50 minutes.)

15. When assigning parts in a high school play, a senior or someone who's been in at least two other plays can play the lead role.

16. Choose a diet cola or a bottle of water out of the refrigerator.

17. Rent a movie that is rated G, or rent a movie that is rated R.

Probability

Directions: Calculate each probability.

18. Find the probability of choosing either a club or a ten out of a standard deck of cards.

19. Find the probability of choosing either a face card (king, queen, jack) or a diamond out of a standard deck of cards.

20. Suppose that 4 out of the 15 doctors in a small hospital are female, 11 out of the 15 are under the age of 45, and 2 are both female and under the age of 45. What is the probability that you are randomly assigned a female doctor or a doctor under the age of 45?

21. Nineteen college graduates are applying for internships at a Wall Street financial institution. Of these applicants, 12 graduated from Ivy League universities, 8 are women, 9 graduated summa cum laude, and 5 graduated summa cum laude from Ivy League universities. If one resume is selected at random for the first interview, what is the probability that the applicant is either an Ivy League graduate or a summa cum laude graduate?

22. The local arbor society is giving out free trees to the public. They have 200 pine trees, 100 dogwoods, 125 oaks, and 200 birch trees. If you are the first in line to receive your free tree and you cannot choose what type of tree you are given, what is the probability that you randomly get either an oak or a dogwood?

23. A flight attendant for USA Airlines is given an assignment that is randomly chosen from the following available flights: 15 European destinations, 6 Caribbean destinations, 8 South American destinations, 5 Asian destinations, and 11 destinations in the continental United States. What is the probability that her next assignment takes her either to Europe or Asia?

24. Consider the experiment in which two six-sided dice are tossed. What is the probability that the total is not four?

25. Consider the experiment in which two six-sided dice are tossed. What is the probability that either both dice are sixes or neither die is a six?

26. What's the probability of rolling at most one even number with two six-sided dice?

27. Jacki has a sizable collection of music downloaded to her MP3 player. The relative frequencies for each genre of music she has stored are given in the table below.

Jacki's Music Collection

Genre	Relative Frequency
Rock	0.3256
Country	0.1181
R&B	0.2068
Oldies	0.0843
Show Tunes	0.1956
Bluegrass	0.0696

If Jacki chooses the shuffle option and one song is randomly chosen for play, what is the probability that it is either country or bluegrass?

28. Ana is looking forward to starting kindergarten in her Texas hometown. Of 14 kindergarten teachers, 3 are new this year, 11 are bilingual, and 1 is both new and bilingual. If class rosters are generated at random, what is the probability that Ana's teacher is either new or bilingual?

29. What is the probability of rolling two six-sided dice and obtaining either an odd total or a total less than six?

30. In the teacher's pencil box, 13 out of the 20 pencils have no eraser, and 12 out of the 20 are not mechanical pencils. There are 4 pencils that are both mechanical and have erasers on them. What is the probability that Ava chooses a pencil at random from the box and gets either a mechanical pencil or a pencil with an eraser?

31. For a story she is writing in her high school newspaper, Grace surveys moviegoers selected at random as they leave the new feature *Mystery on Juniper Island.* She simply asks each moviegoer to rate the show using a thumbs-up or thumbs-down. The results of her survey are given in the table below.

Survey Results

	Men	Women
Thumbs-Up	23	40
Thumbs-Down	19	11

What is the probability that one of Grace's survey respondents has either given a thumbs-up rating or is a man?

4.3 Multiplication Rules for Probability

In the previous section, we were asked about the probability of a single event, such as "drawing a queen or a king from a deck of cards." Note that only one card was drawn, and we were interested in the probability that it was one thing "or" another. For these situations, we used the Addition Rule. In this section we will consider the probability of two events happening in two trials. For example, what is the probability of first drawing a queen *and* then drawing a king from a deck of cards? The key word in these problems is the word "and."

Before we can calculate these types of probabilities, we must carefully consider how the experiment is to be performed. For the example of "first drawing a queen and then drawing a king," it is important to know if we are drawing two cards at the same time or perhaps one card and then another. If we are drawing one card first and then a second card, do we put the first card back before drawing a second card? These various scenarios have different probabilities, as we shall see, so we must know the details of the experiment.

Memory Booster

Mutually Exclusive
Consider whether the events can happen at the same time.

Independent
Consider whether one event occurring affects the probability of the other event occurring.

Let's begin with the scenario of drawing one card from a standard 52-card deck, replacing the card and shuffling the deck, then drawing a second card from the same deck. Since we have two steps in this experiment, it can be referred to as a **multistage experiment**. Secondly, since we are replacing the first card before drawing the second card, we say we are performing the experiment **with replacement**. The phrase, "with replacement," refers to placing objects back into consideration before performing the next stage of the experiment.

When considering the probability of two events occurring, it is important to know if the two events are independent. Two events are **independent** if one event happening does not influence the probability of the other event happening. For example, if after drawing a card you replace the card drawn—and shuffle the deck—then the probability for the next card drawn is not affected by what was picked first. Therefore, these two selections of a card are independent events. When two events are independent, the probability that both events occur can easily be calculated by multiplying their respective probabilities together. Written mathematically, the formula referred to as the **Multiplication Rule for Probability of Independent Events** is shown in the following box.

Formula

Multiplication Rule for Probability of Independent Events

For two independent events, E and F, the probability that E and F occur is given by the following formula.

$$P(E \textbf{ and } F) = P(E) \cdot P(F)$$

Note that if $P(E \text{ and } F) = P(E) \cdot P(F)$, then we also know that E and F are independent events.

Example 4.17

Using the Multiplication Rule for Probability of Independent Events

Choose two cards from a standard deck, with replacement. What is the probability of choosing a king and then a queen?

Solution

Because the cards are replaced after each draw, the probability of the second event occurring is not affected by the outcome of the first event. Thus, these two events are independent. Using the Multiplication Rule for Probability of Independent Events, we have the following.

$$\begin{aligned} P(\text{king and queen, with replacement}) &= P(\text{king}) \cdot P(\text{queen}) \\ &= \frac{\overset{1}{\cancel{4}}}{\underset{13}{\cancel{52}}} \cdot \frac{\overset{1}{\cancel{4}}}{\underset{13}{\cancel{52}}} \\ &= \frac{1}{169} \\ &\approx 0.0059 \end{aligned}$$

Just as we can extend the Addition Rule for Probability of Mutually Exclusive Events, we can extend the Multiplication Rule for Probability of Independent Events as well. It doesn't matter how many events are in question; if each one is independent of the others, then we can multiply the individual probabilities.

Example 4.18

Using the Extended Multiplication Rule for Probability of Independent Events

Assume that a study by Human Resources has found that the probabilities of an employee being written up for the following infractions are the values shown in the following table. Assume that each infraction is independent of the others. What is the probability that a given employee will be written up for being late for work, taking unauthorized breaks, and leaving early?

Solution

Probabilities of Being Written Up at Work

Infraction	Probability
Insubordination	0.1356
Late for work	0.2478
Failure to show up for work	0.1026
Leaving early	0.1954
Taking unauthorized breaks	0.3186

Since these three events are independent, we can apply the Multiplication Rule for Probability of Independent Events to this situation.

$$\begin{aligned}
&P(\text{late for work and taking unauthorized breaks and leaving early})\\
&\quad = P(\text{late for work})\cdot P(\text{taking unauthorized breaks})\cdot P(\text{leaving early})\\
&\quad = 0.2478\cdot 0.3186\cdot 0.1954\\
&\quad \approx 0.0154
\end{aligned}$$

When two events are not independent, it means that the occurrence of one influences the probability that the other occurs. For example, consider drawing two cards from a standard deck *without replacement*. The phrase **without replacement** means that your first choice is not put back in for consideration before performing the next stage of the experiment. When a multistage experiment is performed without replacement, the outcome of the first part affects the probability of the second event occurring; therefore, the events are not independent. Instead, they are said to be **dependent**. To calculate the combined probability, we can still multiply the probability of the first event occurring in the first stage of the experiment by the probability of the second event occurring in the second stage of the experiment. The catch is that we must take into account the outcome of the first stage when calculating the probability of the second event occurring.

Example 4.19

Calculating Probability of Dependent Events

What is the probability of drawing a king and then a queen from a standard deck if the cards are drawn without replacement?

Solution

This situation is essentially the same as drawing two cards at the same time from a standard deck. We think of the experiment in two stages, though, for ease in calculation. Begin by determining the probability of drawing a king from a standard deck of cards. Since there are 4 kings in a deck of 52 cards, the probability is $P(\text{king}) = \frac{4}{52} = \frac{1}{13}$. Next, assume that when you drew the first card, it was a king. What is the probability that you now draw a queen? Well, since we are holding a king in our hand, there are still 4 queens left in the deck of cards. But, there are no longer 52 cards total, only 51. Thus, $P(\text{queen, given king drawn first, without replacement}) = \frac{4}{51}$.

We can now find the probability of the multistage experiment by multiplying the two individual probabilities.

$$\begin{aligned}
P(\text{king and queen, without replacement}) &= P(\text{king})\cdot P(\text{queen, given king drawn first})\\
&= \frac{\overset{1}{\cancel{4}}}{\underset{13}{\cancel{52}}}\cdot\frac{4}{51}\\
&= \frac{4}{663}\\
&\approx 0.0060
\end{aligned}$$

Notice that in the previous example, the second probability was not just $P(\text{queen})$. It was the probability of drawing a queen, given that some other condition had already occurred. In general, this type of probability is called **conditional probability**, denoted as $P(F|E)$ and read "the probability of *F* given *E*." For our previous example, we would write $P(\text{queen}|\text{king})$. This notation is read "the probability of drawing a queen, given that a king was drawn first." This type of conditional probability is necessary when performing multistage experiments without replacement.

Definition

Conditional probability, denoted $P(F|E)$ and read "the probability of *F* given *E*," is the probability of event *F* occurring given that event *E* occurs first.

Example 4.20

Calculating Conditional Probability

One card has already been chosen from a standard deck without replacement. What is the probability of now choosing a second card from the deck and it being red, given that the first card was a diamond?

Solution

First, we must determine how many red cards are left in the deck after the first pick. Because the first card was a diamond (which is a red card), there are only 25 red cards left in the deck instead of 26. There are also only 51 cards total left in the deck. Thus the conditional probability is calculated as follows.

$$\begin{aligned} P(\text{red}|\text{diamond}) &= \frac{25}{51} \\ &\approx 0.4902 \end{aligned}$$

Conditional probability is necessary in order to find the probability of two events happening consecutively in a multistage experiment when the events are not independent. As we noted before, in this case, we say that the two events are dependent. As we saw in the example, for these types of problems we multiply the probability of one event by the conditional probability of the other event as shown in the following formula, which is the **Multiplication Rule for Probability of Dependent Events**.

Formula

Multiplication Rule for Probability of Dependent Events

For two dependent events, *E* and *F*, the probability that *E* and *F* occur is given by the following formula.

$$\begin{aligned} P(E \textbf{ and } F) &= P(E) \cdot P(F|E) \\ &= P(F) \cdot P(E|F) \end{aligned}$$

Example 4.21

Using the Multiplication Rule for Probability of Dependent Events

What is the probability of choosing two face cards in a row? Assume that the cards are chosen without replacement.

Solution

We are dealing with dependent events, so we must use the Multiplication Rule for Probability of Dependent Events. When the first card is picked, all 12 face cards are available out of 52 cards. When the second card is drawn, there are only 11 face cards left out of the 51 cards remaining in the deck. Thus, we have the following.

$$\begin{aligned} P(\text{face card and face card}) &= P(\text{face card})\cdot P(\text{face card}|\text{face card}) \\ &= \frac{\overset{1}{\cancel{\overset{\cancel{3}}{\cancel{12}}}}}{\underset{13}{\cancel{52}}}\cdot\frac{11}{\underset{17}{\cancel{51}}} \\ &= \frac{11}{221} \\ &\approx 0.0498 \end{aligned}$$

Example 4.22

Using the Multiplication Rule for Probability of Dependent Events

Assume that there are 17 men and 24 women in the Rotary Club. Two members are chosen at random each year to serve on the scholarship committee. What is the probability of choosing two members at random and the first being a man and the second being a woman?

Solution

Note that since we are choosing two members, the first choice will influence the probability for the second choice, assuming we do not want to choose the same member twice. This means we are dealing with dependent events.

We want to find $P(\text{man and woman})$, which according to the Multiplication Rule for Probability of Dependent Events, equals $P(\text{man})\cdot P(\text{woman}|\text{man})$. When the first member is picked, there are 17 men out of 41 members. When the second member is picked, we assume that we have already picked a man, so that leaves all 24 women, but only 40 remaining members. The calculation is as follows.

$$\begin{aligned} P(\text{man and woman}) &= P(\text{man})\cdot P(\text{woman}|\text{man}) \\ &= \frac{17}{41}\cdot\frac{\overset{3}{\cancel{24}}}{\underset{5}{\cancel{40}}} \\ &= \frac{51}{205} \\ &\approx 0.2488 \end{aligned}$$

The Multiplication Rule for Probability of Dependent Events can be rewritten as a rule for finding the conditional probability of an event.

Formula

Conditional Probability

For two dependent events, E and F, the probability that F occurs given that E occurs is given by the following formula.

$$P(F|E) = \frac{P(E \textbf{ and } F)}{P(E)}$$

Note, if E and F are *independent* events, then the conditional probability of F given E is the same as the probability of F.

Example 4.23

Using the Rule for Conditional Probability

Out of 300 applicants for a job, 212 are female and 110 are female and have a graduate degree.

a. What is the probability that a randomly chosen applicant has a graduate degree, given that she is female?

b. If 152 of the applicants have graduate degrees, what is the probability that a randomly chosen applicant is female, given that the applicant has a graduate degree?

Solution

a. The question asks for $P(\text{graduate degree}|\text{female})$. Thus, in order to use the formula for conditional probability, we need to find $P(\text{female and graduate degree})$ as well as $P(\text{female})$. Since only one applicant is chosen, this scenario cannot technically be classified as a multistage experiment. However, you can think of this problem in stages because the field of applicants is narrowed first by "females" and then by "graduate degree." Hence, the formula for conditional probability is appropriate to use here. There are 300 applicants total, so the probability of choosing an applicant who is female and has a graduate degree is $\frac{110}{300}$. Similarly, the probability of choosing a female applicant is $\frac{212}{300}$. Thus, we calculate the conditional probability as follows.

$$
\begin{aligned}
P(\text{graduate degree}|\text{female}) &= \frac{P(\text{female and graduate degree})}{P(\text{female})} \\
&= \frac{\frac{110}{300}}{\frac{212}{300}} \\
&= \frac{110}{300} \div \frac{212}{300} \\
&= \frac{\overset{55}{\cancel{110}}}{\cancel{300}} \cdot \frac{\cancel{300}}{\underset{106}{\cancel{212}}} \\
&= \frac{55}{106} \\
&\approx 0.5189
\end{aligned}
$$

b. This question asks for the reverse of the probability calculated in part a. Let's compare and see if we get a similar result. We want to calculate $P(\text{female}|\text{graduate degree})$. We know from part a. that the probability of choosing one of the 300 applicants who is female and has a graduate degree is $\frac{110}{300}$.

Also, we know from the information given in part b. that the probability that a randomly chosen applicant has a graduate degree is $\frac{152}{300}$. Using the formula for conditional probability, we have the following.

$$
\begin{aligned}
P(\text{female}|\text{graduate degree}) &= \frac{P(\text{female and graduate degree})}{P(\text{graduate degree})} \\
&= \frac{\frac{110}{300}}{\frac{152}{300}} \\
&= \frac{110}{300} \div \frac{152}{300} \\
&= \frac{\overset{55}{\cancel{110}}}{\cancel{300}} \cdot \frac{\cancel{300}}{\underset{76}{\cancel{152}}} \\
&= \frac{55}{76} \\
&\approx 0.7237
\end{aligned}
$$

As you can see, $P(F|E)$ is not necessarily the same as $P(E|F)$.

Fundamental Counting Principle

We have learned that we can multiply probabilities together to extend the calculation from a single trial to multiple trials. We also know that when calculating probabilities, it is important to be able to count the number of outcomes in each stage of the experiment. One method for counting the total number of possible outcomes for a multistage experiment is called the **Fundamental Counting Principle**. Simply put, it states that one can multiply together the number of possible outcomes for each stage in a multistage experiment in order to obtain the total number of outcomes for that experiment.

As an example of how this rule works, let us revisit the example of rolling a red die and then a blue die, which we discussed in Section 4.1. We could call the first stage of the experiment "rolling the red die" and the second stage "rolling the blue die." There are 6 possible outcomes for stage one, and for each of the 6 outcomes in stage one, there are 6 possible outcomes for stage two. We can multiply the number of outcomes from stage one by the number of outcomes from stage two in order to obtain the total number of outcomes for our experiment. In this case, there are $6 \cdot 6 = 36$ total outcomes for the die-rolling experiment. We know that this is the correct number because we methodically listed all 36 outcomes in Example 4.2 from Section 4.1.

Theorem

Fundamental Counting Principle

For a multistage experiment with n stages where the first stage has k_1 outcomes, the second stage has k_2 outcomes, the third stage has k_3 outcomes, and so forth, the total number of possible outcomes for the sequence of stages that make up the multistage experiment is $k_1 \cdot k_2 \cdot k_3 \cdot \cdots \cdot k_n$.

A helpful method for setting up a Fundamental Counting Principle problem is to think of each stage as a "slot." First, determine the number of slots that must be filled. Then decide how many outcomes are possible for each slot. For example, suppose one university issues student identification codes that consist of two letters followed by three digits (0–9). How can we determine the number of unique ID codes the school can assign, if letters and digits can be repeated? The first step is to notice that there are five ID code "slots" to be filled.

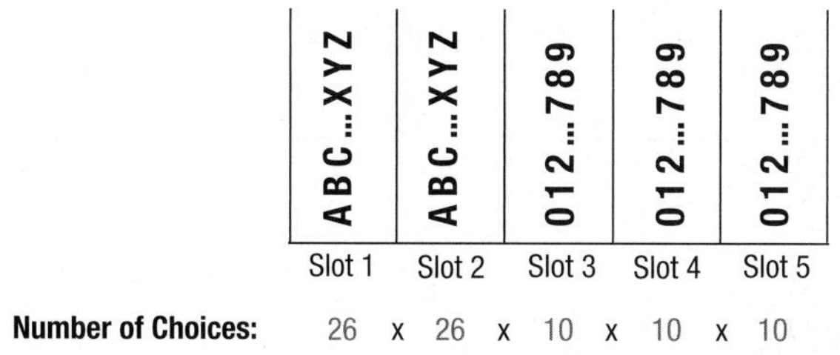

Figure 4.5: Applying the Fundamental Counting Principle

The first two slots contain 26 outcomes each, one outcome for each letter of the alphabet. The other three slots contain 10 possibilities each, one for each digit 0–9. After determining this information, we simply multiply together the number of possibilities for each slot and obtain $26 \cdot 26 \cdot 10 \cdot 10 \cdot 10 = 676{,}000$ unique ID codes.

Example 4.24

Using the Fundamental Counting Principle

Kilby begins her first year in an online degree program in July. The first semester she will randomly be assigned to one section for each of four different core courses. If there are 8 English I sections, 12 College Algebra sections, 11 American History sections, and 5 Physical Science sections, how many different options are there for Kilby's schedule for her first semester?

Solution

There are four slots to fill—one for each of the first four courses in Kilby's schedule. For the first slot, any of the 8 English I sections may be assigned to Kilby. For the second slot, any of the 12 College Algebra sections can be chosen. There are 11 American History sections for slot three and 5 Physical Science sections for slot four. Using the Fundamental Counting Principle, we then multiply. There are $8 \cdot 12 \cdot 11 \cdot 5 = 5280$ possible options for Kilby's schedule.

Example 4.25

Using the Fundamental Counting Principle (Without Replacement)

The governing board of the local charity, Mission Stateville, is electing a new vice president and secretary to replace outgoing board members. If the board consists of 11 members who don't already hold an office, in how many different ways can the two positions be filled if no one may hold more than one office?

Solution

There are two slots to fill in this example. However, this time, our choice to fill the second slot is made *without replacement*. Once someone is chosen for one office, they cannot be chosen for the other. The first slot may be filled with any of the 11 board members. The second position has one fewer to choose from, so there are only 10 choices for it. Using the Fundamental Counting Principle, we then multiply. There are $11 \cdot 10 = 110$ possible ways to elect the new officers.

Example 4.26

Using the Fundamental Counting Principle to Calculate Probability

Robin is preparing an afternoon snack for her twins, Matthew and Lainey. She wants to give each child one item. She has the following snacks on hand: carrots, raisins, crackers, grapes, apples, yogurt, and granola bars. If she randomly chooses one snack for Matthew and one snack for Lainey, what is the probability that each child gets the same snack as yesterday?

Solution

To begin, we need to count the number of ways in which Robin can randomly choose a snack for her twins. To do this, think of there being two slots to fill—one for each twin. Because there is no requirement that the twins have different snacks or the same snack, there are 7 possibilities for each child. That is, the choices are made *with replacement*. Therefore, there are $7 \cdot 7 = 49$ possible ways she can prepare the snacks.

Next, we need to count the number of ways that she can choose the same afternoon snack as yesterday for each child. Let's assume that the twins get only one snack each afternoon, (probably a safe assumption); then there is only 1 way to choose the same snack as yesterday for each child.

Putting this information together, we can calculate the probability.

$$
\begin{aligned}
P(\text{choose same snack}) &= \frac{\text{number of ways to choose same snack}}{\text{total number of snacks}} \\
&= \frac{1}{49} \\
&\approx 0.0204
\end{aligned}
$$

Thus, there is about a 2% chance that each child will eat the same thing two days in a row.

4.3 Section Exercises

Independent Events

Directions: Determine whether the events are independent.

1. Finding the batteries in your calculator dead. Finding the battery in your car dead.

2. Getting a letter from your aunt. Getting a bank statement in the mail.

3. Claire's name being drawn in the school raffle, without replacement. Ben's name being drawn in the school raffle.

4. Buying a new shirt on sale. Having enough money for lunch that day.

5. Lois ordering a steak. Ann ordering a salad.

6. Winning the Mississippi state lottery. Winning the Florida state lottery.

Multiplication Rule for Independent Events

Directions: Calculate each probability.

7. What is the probability of choosing a jack and then a club out of a standard deck of cards with replacement?

8. There are 15 different crayons in a new box. What is the probability that the orange and then the green will be chosen at random, without replacement?

9. Suppose you draw two cards out of a standard deck with replacement. What is the probability that one of the cards is a jack and the other is a club?

10. Ashley's Internet service is terribly unreliable. In fact, on any given day, there is a 15% chance that her Internet connection will be lost at some point that day. What is the probability that her Internet service is not broken for five days in a row?

11. On a five-day vacation, the forecast is a 50% chance of rain every day. What's the probability that it rains every day?

12. On awards day at the end of the year, Jasmine has an 85% chance of winning the top award in English and a 4 out of 5 chance of winning an award for athletics. What's the probability that Jasmine wins both awards?

Conditional Probability

Directions: Calculate each conditional probability.

13. What is the probability of drawing a queen from a standard deck of cards, given that you drew a red card?

14. What is the probability of drawing a face card from a standard deck of cards, given that you drew a spade?

15. A six-sided die is rolled. What is the probability of rolling a two, assuming that you rolled an even number?

16. A six-sided die is rolled. What is the probability of rolling a five, assuming you roll a number greater than or equal to three?

17. Mrs. Harvey's algebra class has 42 students, classified by academic year and gender as follows.

Mrs. Harvey's Algebra Class

	Male	Female
Freshman	9	13
Sophomore	4	5
Junior	4	2
Senior	2	3

Mrs. Harvey randomly chooses one student to collect yesterday's homework.

a. What is the probability that she selects a male, given that she chooses from only the sophomores?

b. What is the probability that she selects a junior, given that she chooses a female?

c. What is the probability that she selects a female, given that she chooses a junior?

Multiplication Rule for Dependent Events

Directions: Calculate each probability.

18. Suppose you draw two cards out of a standard deck without replacement. What is the probability that you draw the queen of hearts and then another heart?

19. Suppose you draw two cards out of a standard deck without replacement. What is the probability that you draw a jack and then a face card?

20. Mrs. Harvey's algebra class has 42 students, classified by academic year and gender as follows.

Mrs. Harvey's Algebra Class		
	Male	**Female**
Freshman	9	13
Sophomore	4	5
Junior	4	2
Senior	2	3

Mrs. Harvey must choose two students at random for a special project.

a. What is the probability that a sophomore boy and then a freshman girl are chosen?

b. What is the probability that two boys are chosen?

21. David likes to keep a jar of change on his desk. Right now his jar contains 26 pennies, 19 nickels, 11 dimes, and 16 quarters. What is the probability that David reaches in and randomly grabs a quarter and then a nickel?

Fundamental Counting Principle

Directions: Use the Fundamental Counting Principle to determine the total number of outcomes.

22. Determine the number of five-digit ZIP codes that can be made from the digits 0–9. (Assume that digits may repeat.)

23. Henry is setting a six-character password on his computer. He is told that the first two characters must be letters and the last four must be numbers. No character may be used twice. How many choices does Henry have for his password?

24. There are six children, three girls and three boys, singing in the school program. As they line up to perform, the choral director insists that the first person be a girl and the last a boy. How many ways can the children line up?

25. How many even four-digit pin numbers can be created from the digits 0–9?

26. How many odd six-digit pin numbers can be created from the digits 0–9?

27. As Candy is trying to decide what to wear, she can choose from eight blouses, three skirts and four pairs of shoes. Assuming everything coordinates, how many different possible outfits does she have?

Fundamental Counting Principle and Probability

Directions: **Use the Fundamental Counting Principle to determine the total number of outcomes and then calculate the probability.**

28. In the game of Clue, the guilty person can be chosen from 6 people, and there are 9 different possible weapons and 9 possible rooms. What is the probability of making a random guess of the guilty person, location, and murder weapon, and the guess being correct?

29. When ordering his new computer, Joe must choose one of 5 monitors, one of 4 printers, and one of 6 scanners. Joe likes 2 of the monitors, 1 of the printers, and 3 of the scanners. If his wife randomly chooses his computer system for him, what is the probability that she will choose a system that makes him happy?

30. A couple having twins is deciding on names. They narrowed their choices to 5 boy names and 7 girl names. The new father's parents like only 1 of the boy names and 2 of the girl names. Assuming that the new parents have one child of each gender, what is the probability that the names they choose will make the new grandparents happy?

31. Stephanie has 11 different outfits in her closet that she wears on Sundays. Ann has 14 different outfits, of which 3 are the same as Stephanie's. What is the probability that they wear the same outfit on a particular Sunday?

4.4 Combinations and Permutations

When you need to count the number of ways objects can be chosen out of a group of distinct objects, without replacement, then the problem you are dealing with involves either *combinations* or *permutations*. This is different than counting the number of outcomes when objects are chosen with replacement, which uses the Fundamental Counting Principle covered in Section 4.3. In this section, we will discuss these two counting methods and look at how to use them to solve probability problems. Before we can do that, we need to first discuss the mathematical functions called *factorials*.

Factorials

A **factorial** is the product of all positive integers less than or equal to a given positive integer, n. Symbolically, a factorial is written as $n!$, which is read "n factorial."

Formula

Factorial

A factorial is given by

$$n! = n(n-1)(n-2)\cdots(2)(1)$$

where n is a positive integer.

Thus 5! is the product of all positive integers less than or equal to 5, which equals $5 \cdot 4 \cdot 3 \cdot 2 \cdot 1 = 120$. One important factorial expression to memorize is 0!, which appears in the denominator of many calculations involving combinations or permutations. By special definition, the value of 0! is defined as 1.

Math Symbols

$0! = 1$

Since factorials equal the products of strings of positive numbers, their values get very big, very quickly. While it was easy to calculate 5! without a calculator, expressions such as 100! are much too large even for ordinary calculators to compute. (Check to see how large a factorial your calculator can compute!) Hence, it's convenient to use some methods for reducing factorials at the start of a calculation if at all possible. Let's take a look at some common ways to compute factorials.

Example 4.27

Calculating Factorial Expressions

Calculate the following factorial expressions.

a. $7!$ **b.** $\frac{4!}{0!}$ **c.** $\frac{95!}{93!}$ **d.** $\frac{5!}{(5-3)!}$ **e.** $\frac{6!}{2!(6-2)!}$

Solution

a. Multiply each positive integer less than or equal to 7.

$$7! = 7 \cdot 6 \cdot 5 \cdot 4 \cdot 3 \cdot 2 \cdot 1 = 5040$$

b. Calculate each factorial and then divide. Remember that 0! = 1.

$$\begin{aligned}\frac{4!}{0!} &= \frac{4\cdot 3\cdot 2\cdot 1}{1}\\ &= \frac{24}{1}\\ &= 24\end{aligned}$$

c. It would be very cumbersome to multiply out 95! and 93! and then divide. Instead, we will cancel first.

$$\begin{aligned}\frac{95!}{93!} &= \frac{95\cdot 94\cdot \cancel{93\cdot 92\cdots 2\cdot 1}}{\cancel{93\cdot 92\cdots 2\cdot 1}}\\ &= 95\cdot 94\\ &= 8930\end{aligned}$$

d. You must begin by evaluating the expression in parentheses. Next, cancel common factors in the numerator and denominator and then simplify.

$$\begin{aligned}\frac{5!}{(5-3)!} &= \frac{5!}{2!}\\ &= \frac{5\cdot 4\cdot 3\cdot \cancel{2\cdot 1}}{\cancel{2\cdot 1}}\\ &= 5\cdot 4\cdot 3\\ &= 60\end{aligned}$$

e. Make sure that you begin by subtracting to simplify the expression in parentheses.

$$\begin{aligned}\frac{6!}{2!(6-2)!} &= \frac{6!}{2!4!}\\ &= \frac{6\cdot 5\cdot \cancel{4\cdot 3\cdot 2\cdot 1}}{(2\cdot 1)(\cancel{4\cdot 3\cdot 2\cdot 1})}\\ &= \frac{30}{2}\\ &= 15\end{aligned}$$

Combinations and Permutations

Imagine you were given the responsibility of handing out door prizes at the first meeting of the education conference you are helping to plan. You must choose two random attendees out of the 50 present to be given a \$50 gift card. How many ways do you think there are for you to choose the lucky winners? Choosing two random people from a group of 50 is called a combination if the order that you chose them in is *not* important. Formally, a mathematical **combination** is a selection of elements chosen from a group without regard to their order. Because both the door prizes were the same in this scenario, and therefore the order of choosing the winners is not important, we are interested in counting the number of combinations of two things from a set of 50.

However, what if the door prizes weren't the same? What if, instead, you needed to choose two winners to receive \$100 and \$50 gift certificates, respectively? Would this change the number of ways you could choose the winners? Yes, it would! There would be more ways to give out the prizes since giving the first person chosen the \$100 gift certificate and the second person the \$50 gift certificate is not the same as giving the \$50 certificate out first. When you choose elements from a group and their order is important, the ordered subgroup is called a **permutation**.

Definition

A **combination** is a selection of objects from a group without regard to their arrangement.

A **permutation** is a selection of objects from a group where the arrangement is specific.

The first step in counting the number of ways subgroups of objects can be chosen is to decide whether you are choosing combinations or permutations of things, based on whether or not order is important. Once you have determined whether the order in which objects are chosen is important in your problem, you apply one of the following rules to calculate the total number of possibilities.

Memory Booster

Combinations
order does not matter

Permutations
order matters

Formula

Combinations and Permutations

When *order is not important*, the following formula is used to calculate the number of **combinations**.

$$_nC_r = \frac{n!}{r!(n-r)!}$$

When *order is important*, the following formula is used to calculate the number of **permutations**.

$$_nP_r = \frac{n!}{(n-r)!}$$

In both of these formulas, r objects are selected from a group of n distinct objects, so r and n are both positive integers with $r \le n$.

Math Symbols

Combinations

$$_nC_r = C(n,r) = C_{n,r} = {}^nC_r = \binom{n}{r}$$

Permutations

$$_nP_r = P(n,r) = P_{n,r} = {}^nP_r$$

The number of combinations might also be written as $\binom{n}{r}$. A permutation is read "n things permuted r at a time," whereas a combination can be read in the shorter form "n choose r."

Example 4.28

Calculating Numbers of Combinations and Permutations by Hand and by Using Formulas

Given a group of three friends, Emre, Heather, and Sheena:

a. How many ways can you arrange the way they stand in line for the movies?

b. How many ways can you choose two of them to ride in a car together?

Solution

a. First, notice that the arrangement of the three friends is important for counting purposes. To count how they can line up for a movie, the order in which they line up does make a difference, so this is a permutation. We are choosing to arrange three objects, so $r = 3$, from a group of three objects, so n also equals three, $n = 3$. Therefore, the number of permutations of 3 things permuted 3 at a time is calculated as follows.

Side Note

Special Case

Note that

$$_nP_n = \frac{n!}{(n-n)!} = \frac{n!}{0!} = \frac{n!}{1} = n!$$

Thus, $n!$ gives the number of different sequences (or permutations) for a group of n distinct objects.

$$\begin{aligned} {}_3P_3 &= \frac{3!}{(3-3)!} \\ &= \frac{3\cdot 2\cdot 1}{0!} \\ &= \frac{6}{1} \\ &= 6 \end{aligned}$$

So, there are six unique ways for the friends to line up for the movie. Let's check this by listing out all the possible arrangements. We'll use a pattern to help us.

EHS	HSE	SHE
ESH	HES	SEH

b. In the second part, we are interested in choosing two of the three friends to ride in a car together, where the arrangement is not specified to be important. Therefore, we'd like to count the number of combinations of two things from a group of three.

$$\begin{aligned} {}_3C_2 &= \frac{3!}{2!(3-2)!} \\ &= \frac{3!}{2!1!} \\ &= \frac{3\cdot \cancel{2\cdot 1}}{\cancel{2\cdot 1}\cdot 1} \\ &= \frac{3}{1} \\ &= 3 \end{aligned}$$

From this, we know there are three distinct ways that we can choose two of the friends to ride in the car. Let's list out all of the possibilities.

EH ES HS

In the previous example, we were able to list out all of the possible combinations and permutations that we counted. However, as we saw earlier, factorials can quickly become very large numbers, and hence the number of arrangements of things can also grow quickly. This is why the formulas for counting the number of possible combinations and permutations are so valuable. In the remaining examples in this section, we will not list out all of the possible choices for each solution.

Example 4.29

Calculating the Number of Permutations

A class of 18 fifth graders is holding elections for class president, vice president, and secretary. In how many different ways can the officers be elected?

Solution

First, note that the order of the students chosen is important; that is, it is different if someone is elected for president rather than vice president. Therefore, we are counting permutations. There are 18 students in the class, so $n = 18$. Because 3 students are to be elected, $r = 3$.

$$\begin{aligned} {}_{18}P_3 &= \frac{18!}{(18-3)!} \\ &= \frac{18!}{15!} \\ &= \frac{18\cdot 17\cdot 16\cdot \cancel{15\cdot 14\cdots 2\cdot 1}}{\cancel{15\cdot 14\cdots 2\cdot 1}} \\ &= 18\cdot 17\cdot 16 \\ &= 4896 \end{aligned}$$

So there are 4896 ways the class officers can be elected.

Example 4.30

Calculating the Number of Combinations

Consider that a cafeteria is serving the following vegetables for lunch one weekday: carrots, green beans, lima beans, celery, corn, broccoli, and spinach. Suppose now that Bill wishes to order a vegetable plate with 3 different vegetables. How many ways can his plate be prepared?

Solution

For this problem, we are choosing 3 vegetables out of a list of 7. The order in which the vegetables are placed on Bill's plate makes no difference, so we are counting combinations. Fill in the combination formula using $n = 7$ and $r = 3$.

$$\begin{aligned} {}_7C_3 &= \frac{7!}{3!(7-3)!} \\ &= \frac{7!}{3!4!} \\ &= \frac{7\cdot \cancel{6}\cdot 5\cdot \cancel{4\cdot 3\cdot 2\cdot 1}}{\cancel{(3\cdot 2\cdot 1)}\cancel{(4\cdot 3\cdot 2\cdot 1)}} \\ &= 7\cdot 5 \\ &= 35 \end{aligned}$$

Therefore, Bill has 35 different ways to order his vegetable plate. Let's hope he doesn't hold up the line making his choice!

Example 4.31

Calculating Probability Using Permutations

Suppose that a little league baseball coach is randomly listing the 9 starting baseball players in a batting order for their second game. At this level, the batting order is randomly chosen to give all players the opportunity to experience different batting positions. What is the probability that the order chosen for the second game is exactly the same as that of the first game?

Solution

The answer to our probability question will be the following fraction.

$$\frac{\text{Number of batting orders that are the same as that of the first game}}{\text{Number of possible batting orders of the 9 players}}$$

Before we can calculate the probability, we must first know how many possible ways there are to make a batting order from the 9 players. In baseball, the batting order is important, so we know that we are counting permutations. Because there are 9 players, $n = 9$. However, for this problem, r is also 9, because all 9 players are to be chosen for the lineup. Substituting these values into the permutation formula gives us the following.

$$\begin{aligned} {}_9P_9 &= \frac{9!}{(9-9)!} \\ &= \frac{9!}{0!} \\ &= \frac{9\cdot 8\cdot 7\cdot 6\cdot 5\cdot 4\cdot 3\cdot 2\cdot 1}{1} \\ &= 362{,}880 \end{aligned}$$

There are 362,880 possible batting orders from which the coach can choose.

For the numerator of the fraction, we know that there is only 1 batting order that is exactly like the order used in the first game, so the probability of randomly choosing that same order is calculated as follows.

$$\begin{aligned} P(\text{same batting order as first game}) &= \frac{1}{362{,}880} \\ &\approx 0.000003 \end{aligned}$$

In other words, players can feel sure that there will be a difference of some sort in the lineup!

Example 4.32

Calculating Probability Using Combinations

Maya has a bag of 15 blocks, each of which is a different color including red, blue, and yellow. Maya reaches into the bag and pulls out 3 blocks. What is the probability that the blocks she has chosen are red, blue, and yellow?

Solution

Let's first determine the number of outcomes in the sample space. We want to count the number of ways that 3 blocks can be drawn. This is a combination because the order of the colors is not important. The number of ways to choose 3 blocks out of a group of 15 is calculated as follows.

$$
\begin{aligned}
{}_{15}C_3 &= \frac{15!}{3!(15-3)!} \\
&= \frac{15!}{3!12!} \\
&= \frac{\overset{5}{\cancel{15}}\cdot \overset{7}{\cancel{14}}\cdot 13\cdot \cancel{12\cdot 11\cdots 2\cdot 1}}{(\cancel{3}\cdot \cancel{2}\cdot 1)(\cancel{12\cdot 11\cdots 2\cdot 1})} \\
&= 5\cdot 7\cdot 13 \\
&= 455
\end{aligned}
$$

How many combinations contain the red, blue, and yellow blocks? Order does not matter, so there is only one way to choose these colors. Thus, the probability that Maya chooses red, blue, and yellow is calculated as follows.

$$
\begin{aligned}
P(\text{red, blue, yellow}) &= \frac{1}{455} \\
&\approx 0.0022
\end{aligned}
$$

Special Permutations

Finally, let's consider the special permutations in which some of the objects being permuted are identical. For example, consider the permutation on the letters in the word MISSISSIPPI. Because the letters I, S, and P are *repeated*, we must use a different permutation formula that takes into consideration the repeated objects, or in this case, the repeated letters.

Formula

Special Permutations

Special permutations involve objects that are identical. The number of distinguishable permutations of n objects, of which k_1 are all alike, k_2 are all alike, and so forth, is given by

$$\frac{n!}{k_1!k_2!\cdots k_p!}$$

where $k_1 + k_2 + \cdots + k_p = n$.

Example 4.33

Calculating the Number of Special Permutations

How many different ways can you arrange the letters in the word TENNESSEE?

Solution

Because we can make no distinction between each E, N, or S in the word, we need to group the letters together. The letters in TENNESSEE are grouped as follows.

T: 1

E: 4

N: 2

S: 2

Since there are a total of 9 letters in TENNESSEE, substituting these values into the special permutation formula gives the following.

$$\begin{aligned}\frac{9!}{1!4!2!2!} &= \frac{9\cdot \overset{4}{\cancel{8}}\cdot 7\cdot \overset{3}{\cancel{6}}\cdot 5\cdot \cancel{4\cdot 3\cdot 2\cdot 1}}{(1)(\cancel{4\cdot 3\cdot 2\cdot 1})(\cancel{2\cdot 1})(\cancel{2\cdot 1})}\\ &= 9\cdot 4\cdot 7\cdot 3\cdot 5\\ &= 3780\end{aligned}$$

Thus, there are 3780 ways to arrange the letters in the word TENNESSEE.

4.4 Section Exercises

Factorials

Directions: Evaluate each factorial expression.

1. $6!$	**2.** $8!$	**3.** $\frac{6!}{4!}$	**4.** $\frac{8!}{5!}$
5. $\frac{6!}{4!2!}$	**6.** $\frac{8!}{5!3!}$	**7.** $\frac{6!}{4!(6-4)!}$	**8.** $\frac{8!}{5!(8-5)!}$
9. $0!$	**10.** $\frac{7!}{0!}$		

Combinations and Permutations

Directions: Evaluate each combination or permutation expression.

11. ${}_5C_2$	**12.** ${}_8C_5$	**13.** ${}_4C_4$	**14.** ${}_5C_1$
15. ${}_{12}C_1$	**16.** ${}_7P_4$	**17.** ${}_5P_3$	**18.** ${}_7P_6$
19. ${}_8P_1$	**20.** ${}_3P_3$	**21.** $\frac{{}_3C_2}{{}_3P_1}$	**22.** $\frac{{}_{10}P_2}{{}_{10}C_2}$

23. ${}_6C_4 + {}_6C_3 + {}_6C_2 + {}_6C_1$

24. ${}_5P_4 + {}_5P_3 + {}_5P_2 + {}_5P_1$

Directions: Simplify the formula for each expression.

25. ${}_{n}C_{n}$ **26.** ${}_{n}C_{1}$ **27.** ${}_{n}P_{1}$ **28.** ${}_{n}P_{n}$

29. ${}_{n}P_{n-1}$

Directions: Use a combination or permutation expression to determine the total number of outcomes.

30. Farmer John has nine prize-winning cows. How many ways can he choose three of his cows to show at the state fair?

31. There are 12 members in the local garden club. In how many ways can a president and secretary be chosen? (Assume that no member can hold both positions at the same time.)

32. A teacher must choose parts for the upcoming Thanksgiving play from her class of 17 students. How many ways can she choose the parts of pilgrim, Native American, and turkey?

33. A teacher must choose parts for the upcoming Thanksgiving play from her class of 18 students. She needs a group of four students to serve as program attendants before the start of the play. In how many ways can this group be chosen?

34. There are eight people hosting a party. Three people are needed to stay and clean up after the party is over. How many ways can the clean-up crew be chosen?

35. There are eight people hosting a party. One person must set up the catering, another must bring flowers, and someone else needs to bring drinks. In how many ways can these tasks be assigned?

36. In assigning seats for a classroom, how many ways can a teacher place 8 students in the front row from her roll of 35?

37. In how many ways can a graduate student fulfill the master's degree requirements in mathematics if 10 classes are needed from a choice of 15 classes?

38. In how many ways can 1st, 2nd, and 3rd place prizes be awarded in a local science fair, if there are 25 participants?

39. In how many ways can a task force of 4 people be chosen from a group of 12 employees?

40. If 3 people need to serve as chaperones on a school trip, in how many ways can they be chosen from the parents of the 20 students? (Assume that each child has two parents available.)

41. The Seago family is planning their vacation. Each of the five family members is allowed to nominate three places they would like to visit. If they want to visit four different places during the trip, in how many ways can they plan their road trip, assuming that no family members choose the same place?

42. When painting a plate at a paint-your-own-pottery studio, Sara wants to use 3 different colors from the palette of 30 colors. How many different color combinations can she choose from?

43. In how many ways can the letters in the word STATISTICS be arranged?

44. In how many ways can the letters in the word PROBABILITY be arranged?

45. Karran was born on 11/21/1992. He would like to make an eight-digit code from all of the digits in his birth date. How many different eight-digit codes could he create?

46. Employees at a local factory need a unique seven-digit code to access the building. The manager wants to make each person's code from the factory's phone number, 555-9313.

a. If there are 509 employees who need codes, will the manager have enough unique codes using only the digits in the phone number?

b. Would there be enough ten-digit codes if he used the area code, 514, as well?

47. Which of the following words would produce the greatest number of different five-letter arrangements? (**Hint:** Think before you calculate!)

a. BEAST

b. ORDER

c. TESTS

d. GOING

Counting Techniques and Probability

Directions: Use counting techniques to compute each classical probability.

48. At a carnival entrance, tickets are assigned five-digit numbers using the digits 0–9. If one ticket number is chosen randomly for a prize, what is the probability that every number on the ticket is even? (Assume that all possible ticket numbers are eligible to be chosen.)

49. Suppose that your boss must choose three employees in your office to attend a conference in Jamaica. Because all 20 of you want to go, he decides that the only fair way is to draw names out of a hat. What is the probability that you, Suzanne, and Alex are chosen?

50. Rhonda and Laura are planning to watch two movies over the weekend from Laura's collection of 24 DVDs. Rhonda has two favorites among the collection. What is the probability that the girls would randomly choose those two movies to watch?

51. Bill is planting tulip bulbs in the front yard. There are two white bulbs and two red bulbs mixed together in a bucket. What is the probability that Bill plants the four bulbs in a row so that they are alternating in color?

52. Kim is playing Scrabble. What is the probability that she chooses the tiles with the letters of her name, in order, when she draws three tiles from the bag? (Assume that when she begins there is one tile of each letter in the alphabet in the bag.)

4.5 Combining Probability and Counting Techniques

In previous sections, we have looked at techniques for solving simple counting problems. In this section, we will now introduce an assortment of counting problems that have more complicated solutions. A helpful trick for these examples is to look for certain **key terms**. The key terms are important because they define the method that must be used in order to obtain the correct answer. Words to look for include *at least*, *at most*, *greater than*, *less than*, *between*, and so forth. A few examples of the many different types of problems you may encounter are given below.

Example 4.34

Calculating Probability Using Combinations

A group of 12 tourists is visiting London. At one particular museum, a discounted admission is given to groups of at least ten.

a. How many combinations of tourists can be made for the museum visit so that the group receives the discounted rate?

b. Suppose that a group of the tourists does get the discount. What's the probability that it was made up of 11 of the tourists?

Solution

a. The key words for this problem are *at least*. If at least 10 are required, then the group will get the discount rate if 10, 11, or 12 tourists go to the museum. We must then calculate the number of combinations for each of the three possibilities. Note that we are counting combinations, not permutations, because the order of tourists chosen to go to the museum is irrelevant.

Number of groups of 10 tourists:

$$\begin{aligned}
{}_{12}C_{10} &= \frac{12!}{10!(12-10)!} \\
&= \frac{12!}{10!2!} \\
&= \frac{\cancelto{6}{12}\cdot 11\cdot \cancel{10\cdot 9\cdots 2\cdot 1}}{\cancel{(10\cdot 9\cdots 2\cdot 1)}(\cancel{2}\cdot 1)} \\
&= 6\cdot 11 \\
&= 66
\end{aligned}$$

Number of groups of 11 tourists:

$$\begin{aligned}{}_{12}C_{11} &= \frac{12!}{11!(12-11)!} \\ &= \frac{12!}{11!1!} \\ &= \frac{12\cdot \cancel{11\cdot 10 \cdots 2\cdot 1}}{(\cancel{11\cdot 10 \cdots 2\cdot 1})(1)} \\ &= 12\end{aligned}$$

Number of groups of 12 tourists:

$$\begin{aligned}{}_{12}C_{12} &= \frac{12!}{12!(12-12)!} \\ &= \frac{\cancel{12!}}{\cancel{12!}0!} \\ &= \frac{1}{1} \\ &= 1\end{aligned}$$

As we noted, the problem implies that the group will get the discount rate if 10, 11, *or* 12 tourists attend. Remember, the word "or" tells us to add the results together. So, to find the total number of groups, we add the numbers of combinations together.

$$66+12+1=79$$

There are 79 different groups that can be formed to tour the museum at the discounted rate.

b. To calculate this probability, we need to think of this as a conditional probability. Think of the sentence as reading, "Given that a group of tourists received the discount, what's the probability that it was a group of 11?" So, to find the probability, we need the total number of ways a group of 11 can be chosen from the group of 12 divided by the number of ways that a group could receive the discount. Since both of these were calculated in part a., we can simply substitute these numbers in our fraction. Therefore, we calculate the probability as follows.

$$\begin{aligned}P(11\text{ tourists}|\text{discount}) &= \frac{12}{79} \\ &\approx 0.1519\end{aligned}$$

Example 4.35

Calculating Probability Using Permutations

Jack is setting a password on his computer. He is told that his password must contain at least three, but no more than five, characters. He may use either letters or numbers (0–9).

a. How many different possibilities are there for his password if each character can only be used once?

b. Suppose that Jack's computer randomly sets his password using all of the restrictions given above. What is the probability that this password would contain only the letters in his name?

Solution

a. First, notice the key words *at least* and *no more than*. These words tell us that Jack's password can be 3, 4, *or* 5 characters in length. Also note that for a password, the order of characters is important, so we are counting permutations. There are 26 letters and 10 digits to choose from, so we must calculate the number of permutations possible from taking groups of 3, 4, or 5 characters out of 36 possibilities.

Number of permutations of 3 characters:

$$\begin{aligned} {}_{36}P_3 &= \frac{36!}{(36-3)!} \\ &= \frac{36!}{33!} \\ &= \frac{36\cdot35\cdot34\cdot\cancel{33\cdot32\cdots2\cdot1}}{\cancel{33\cdot32\cdots2\cdot1}} \\ &= 36\cdot35\cdot34 \\ &= 42{,}840 \end{aligned}$$

Number of permutations of 4 characters:

$$\begin{aligned} {}_{36}P_4 &= \frac{36!}{(36-4)!} \\ &= \frac{36!}{32!} \\ &= \frac{36\cdot35\cdot34\cdot33\cdot\cancel{32\cdot31\cdots2\cdot1}}{\cancel{32\cdot31\cdots2\cdot1}} \\ &= 36\cdot35\cdot34\cdot33 \\ &= 1{,}413{,}720 \end{aligned}$$

Number of permutations of 5 characters:

$$\begin{aligned} {}_{36}P_5 &= \frac{36!}{(36-5)!} \\ &= \frac{36!}{31!} \\ &= \frac{36\cdot35\cdot34\cdot33\cdot32\cdot\cancel{31\cdot30\cdots2\cdot1}}{\cancel{31\cdot30\cdots2\cdot1}} \\ &= 36\cdot35\cdot34\cdot33\cdot32 \\ &= 45{,}239{,}040 \end{aligned}$$

To find the total number of possible passwords, once again we need to add all three of the above numbers of permutations together since the solution contains an "or" statement. Therefore, we obtain 46,695,600 passwords.

$$42{,}840 + 1{,}413{,}720 + 45{,}239{,}040 = 46{,}695{,}600$$

(Imagine how many possibilities there would be if he could use up to eight or nine characters!)

b. To find the probability that a randomly chosen password would include only the four letters from Jack's name, we need the total number of possible passwords calculated in part a. as well as the number of possible passwords made from the four letters in Jack's name. To find this second number, we need to calculate the number of permutations of 4 things from a set of 4.

$$\begin{aligned} {}_4P_4 &= \frac{4!}{(4-4)!} \\ &= \frac{4 \cdot 3 \cdot 2 \cdot 1}{0!} \\ &= \frac{24}{1} \\ &= 24 \end{aligned}$$

So, the probability is calculated as follows.

$$\begin{aligned} P(\text{password containing only the letters J, A, C, K}) &= \frac{24}{46{,}695{,}600} \\ &\approx 0.0000005 \end{aligned}$$

Combining Counting Techniques

At times it is necessary to *combine* counting techniques in order to find all possibilities for a given problem. A common scenario that occurs is when a combination or permutation formula must be used in order to determine the number of outcomes for each slot of a Fundamental Counting Principle problem. Solving this type of problem is best learned by example.

Example 4.36

Using Numbers of Combinations with the Fundamental Counting Principle

Tina is packing her suitcase to go on a weekend trip. She wants to pack 3 shirts, 2 pairs of pants, and 2 pairs of shoes. She has 9 shirts, 5 pairs of pants, and 4 pairs of shoes to choose from. How many ways can Tina pack her suitcase? (We will assume that everything matches.)

Solution

Begin by thinking of this problem as a Fundamental Counting Principle problem. Consider the act of packing the suitcase to be a multistage experiment. There are three stages—packing the shirts, packing the pants, and packing the shoes. Thus, there are three suitcase "slots" for Tina to fill—a slot for shirts, a slot for pants, and a slot for shoes. To fill these slots, the combination formula must be used because we are choosing some items out of her closet, and the order of the choosing does not matter. Let's first calculate these individual numbers of combinations.

Number of combinations of shirts:

$$\begin{aligned} {}_9C_3 &= \frac{9!}{3!(9-3)!} \\ &= \frac{9!}{3!6!} \\ &= \frac{\overset{3}{\cancel{9}} \cdot \overset{4}{\cancel{8}} \cdot 7 \cdot \cancel{6 \cdot 5 \cdots 2 \cdot 1}}{(\cancel{3} \cdot \cancel{2} \cdot 1)(\cancel{6 \cdot 5 \cdots 2 \cdot 1})} \\ &= 3 \cdot 4 \cdot 7 \\ &= 84 \end{aligned}$$

Number of combinations of pants:

$$\begin{aligned} {}_5C_2 &= \frac{5!}{2!(5-2)!} \\ &= \frac{5!}{2!3!} \\ &= \frac{5 \cdot \overset{2}{\cancel{4}} \cdot \cancel{3 \cdot 2 \cdot 1}}{(\cancel{2} \cdot 1)(\cancel{3 \cdot 2 \cdot 1})} \\ &= 5 \cdot 2 \\ &= 10 \end{aligned}$$

Number of combinations of shoes:

$$\begin{aligned} {}_4C_2 &= \frac{4!}{2!(4-2)!} \\ &= \frac{4!}{2!2!} \\ &= \frac{\overset{2}{\cancel{4}} \cdot 3 \cdot \cancel{2 \cdot 1}}{(\cancel{2} \cdot 1)(\cancel{2 \cdot 1})} \\ &= 2 \cdot 3 \\ &= 6 \end{aligned}$$

Shirts		Pants		Shoes
${}_9C_3$	×	${}_5C_2$	×	${}_4C_2$
84	×	10	×	6
		5040		

The final step in this problem is different than the final step in the last two examples, because of the word "and." The Fundamental Counting Principle tells us we have to *multiply*, rather than add, the numbers of combinations together to get the final result. There are then

$$84 \cdot 10 \cdot 6 = 5040$$

different ways that Tina can pack for the weekend.

Memory Booster

And vs. Or in Probability

"And" ⇔ Multiply

"Or" ⇔ Add

Example 4.37

Using Numbers of Combinations with the Fundamental Counting Principle

An elementary school principal is putting together a committee of 6 teachers to head up the spring festival. There are 8 first-grade, 9 second-grade, and 7 third-grade teachers at the school.

a. In how many ways can the committee be formed?

b. In how many ways can the committee be formed if there must be 2 teachers chosen from each grade?

c. Suppose the committee is chosen at random and with no restrictions. What is the probability that 2 teachers from each grade are represented?

Solution

a. There are no stipulations as to what grades the teachers are from and the order in which they are chosen does not matter, so we simply need to calculate the number of combinations by choosing 6 committee members from the $8 + 9 + 7 = 24$ teachers.

$$\begin{aligned}
{}_{24}C_6 &= \frac{24!}{6!(24-6)!} \\
&= \frac{24!}{6!18!} \\
&= \frac{\overset{4}{\cancel{24}}\cdot 23\cdot \overset{11}{\cancel{22}}\cdot \overset{7}{\cancel{21}}\cdot \overset{4}{\cancel{20}}\cdot 19\cdot \cancel{18\cdot 17\cdots 2\cdot 1}}{\left(\cancel{6}\cdot\cancel{5}\cdot\cancel{4}\cdot\cancel{3}\cdot\cancel{2}\cdot 1\right)\cancel{\left(18\cdot 17\cdots 2\cdot 1\right)}} \\
&= 23\cdot 11\cdot 7\cdot 4\cdot 19 \\
&= 134{,}596
\end{aligned}$$

b. For this problem we need to combine the Fundamental Counting Principle and the combination formula. First, we see that there are three slots to fill: one first-grade slot, one second-grade slot, and one third-grade slot. Each slot is made up of 2 teachers, and we must use the combination formula to determine how many ways these slots can be filled.

Number of first-grade combinations:

$$\begin{aligned}
{}_8C_2 &= \frac{8!}{2!(8-2)!} \\
&= \frac{8!}{2!6!} \\
&= \frac{\overset{4}{\cancel{8}}\cdot 7\cdot \cancel{6\cdot 5\cdots 2\cdot 1}}{\left(\cancel{2}\cdot 1\right)\cancel{\left(6\cdot 5\cdots 2\cdot 1\right)}} \\
&= 4\cdot 7 \\
&= 28
\end{aligned}$$

Number of second-grade combinations:

$$
\begin{aligned}
{}_{9}C_{2} &= \frac{9!}{2!(9-2)!} \\
&= \frac{9!}{2!7!} \\
&= \frac{9\cdot \overset{4}{\cancel{8}}\cdot \cancel{7\cdot 6\cdot 5\cdots 2\cdot 1}}{(\cancel{2}\cdot 1)\cancel{(7\cdot 6\cdot 5\cdots 2\cdot 1)}} \\
&= 9\cdot 4 \\
&= 36
\end{aligned}
$$

Number of third-grade combinations:

$$
\begin{aligned}
{}_{7}C_{2} &= \frac{7!}{2!(7-2)!} \\
&= \frac{7!}{2!5!} \\
&= \frac{7\cdot \overset{3}{\cancel{6}}\cdot \cancel{5\cdot 4\cdot 3\cdot 2\cdot 1}}{(\cancel{2}\cdot 1)\cancel{(5\cdot 4\cdot 3\cdot 2\cdot 1)}} \\
&= 7\cdot 3 \\
&= 21
\end{aligned}
$$

Finally, the Fundamental Counting Principle says that we have to multiply together the outcomes for all three slots in order to obtain the total number of ways to form the committee. Thus, there are

$$28\cdot 36\cdot 21 = 21{,}168$$

ways to choose the committee.

1st Grade		2nd Grade		3rd Grade
${}_{8}C_{2}$	×	${}_{9}C_{2}$	×	${}_{7}C_{2}$
28	×	36	×	21
		21,168		

c. The final part of this question asks us to find the probability that if a committee was chosen at random, it would meet the requirements given in part b. That is, each grade would have two teachers on the committee. We can combine our answers from parts a. and b. to find the answer to this probability question.

$$
\begin{aligned}
&P(\text{committee has 2 teachers from each grade}) \\
&\quad = \frac{\text{Number of possible committees with 2 teachers from each grade}}{\text{Number of possible committees of 6 teachers}} \\
&\quad = \frac{21{,}168}{134{,}596} \\
&\quad \approx 0.1573
\end{aligned}
$$

In other words, if the members of the committee are randomly selected, there is a 15.73% chance that the committee will be made up of 2 teachers from each grade.

4.5 Section Exercises

Combining Probability and Counting Techniques

Directions: Use the Fundamental Counting Principle, combination formula, permutation formula, or a combination of the methods to solve each problem.

1. Suppose that license tags in a particular area of the state must begin with one of the following letters: L, D, Y, or K. The rest of the tag must contain two letters followed by three digits (0–9).

a. If characters cannot repeat, how many unique tags can be made?

b. What's the probability that you are randomly assigned a license tag that begins with the letter K? (Assume that all possible license tags are available to be assigned.)

2. Suppose that license tags in another state are made up of three letters followed by three digits (0–9), none of which can repeat. Additionally, each tag number may be assigned to a regular tag, a wildlife-conservation tag, or a veterans tag.

a. How many unique tags can be made?

b. What's the probability that you are parked next to a car from this state that has a tag that ends in a 9?

3. **a.** How many four-digit numbers can be created from the digits 0–9 if the first and last digits must be odd and no digit can repeat?

b. What's the probability that a number that starts with 5 is randomly chosen from this group?

4. How many four-digit numbers can be made from the digits 0–9 if each number created must be greater than 5000?

5. **a.** How many five-card hands containing exactly two hearts and three clubs can be drawn without replacement from a standard deck of cards?

b. What's the probability that you are dealt one of these hands from a standard deck of cards? (Assume that you are randomly dealt a five-card hand.)

6. **a.** How many five-card hands containing at least four face cards can be drawn without replacement from a standard deck of cards?

b. What's the probability that you are not dealt one of these hands from a standard deck of cards? (Assume that you are randomly dealt a five-card hand.)

7. Virginia's Veggie Café offers 5 types of homemade bread, 10 toppings, and 6 different condiments. How many different super sandwiches can be made if a super sandwich consists of 6 different toppings and 2 different condiments?

8. Mysti is picking out material for her new quilt. There are 12 possible plaids, 8 different solids, and 4 floral prints that she can choose from. If she needs 3 plaids, 2 solids, and 2 floral prints for her quilt, how many different ways can she choose the materials?

9. Because Tristan has diabetes, he must make sure that his daily diet includes 2 vegetables, 3 fruits, and 2 breads. At the grocery store, he has a choice of 20 vegetables, 8 fruits, and 5 breads.

a. In how many ways can he make up his daily requirements if he doesn't like to eat 2 helpings of the same thing in one day?

b. What's the probability that a random choice from his possibilities would yield either carrots or spinach in its menu, given that carrots and spinach are both vegetable choices at the grocery store?

10. How many different teams of 4 can be chosen from a group of 20 girls and 15 boys if each team must have at least 1 boy on it?

11. A football coach needs to choose 11 players to start on offense. There are 6 freshmen, 6 sophomores, 8 juniors, and 7 seniors on the team. In how many ways can the starting 11 be chosen if the coach wants all 7 seniors to play?

12. Lindsay is checking out books at the library, and she is primarily interested in mysteries and nonfiction. She has narrowed her selections down to seven mysteries and eight nonfiction books.

a. How many different combinations of books can she check out if she is only allowed three books at a time?

b. How many different combinations of books can she check out if she is only allowed three books at a time, and she wants at least one mystery?

c. If she randomly chooses three books from her selections, what's the probability that they will all be mysteries?

13. In choosing what music to play at a charity fund-raising event, Marlow has 41 Mozart, 104 Haydn, and 8 Schubert symphonies from which to choose. He is setting up a schedule of the 12 songs to be played during the show.

a. How many different schedules are possible if he needs to have an equal number of symphonies from Mozart, Haydn, and Schubert?

b. If the songs are chosen randomly, what's the probability that all 12 symphonies are by Mozart?

14. A baseball coach needs to choose 9 players to be in the batting lineup for the first game of the season. There are 5 freshmen, 4 sophomores, 7 juniors, and 4 seniors on the team.

a. In how many ways can the batting order be chosen if the coach wants no more than 2 freshmen to play?

b. In how many ways can the batting order be chosen if the coach wants all 4 seniors to play?

R Chapter 4 Review

Section 4.1: Introduction to Probability

Definitions

Probability experiment (or trial)
Any process with a result determined by chance

Outcome
Each individual result that is possible for a probability experiment

Sample space
Set of all possible outcomes for a given probability experiment

Event
Subset of outcomes from the sample space

Tree diagram
Diagram that provides a systematic way of listing all the outcomes in a sample space for a probability experiment consisting of several stages

Subjective probability
Educated guess regarding the chance that an event will occur

Experimental Probability (or Empirical Probability)

Found by experiment; probability that an event E occurs is calculated by dividing the number of times E occurs by the total number of trials performed, given by

$$P(E) = \frac{f}{n}$$

Classical Probability (or Theoretical Probability)

Most precise type of probability; if all outcomes are equally likely to occur, then the probability that an event E occurs is calculated by dividing the number of outcomes included in the event by the total number of outcomes in the sample space, given by

$$P(E) = \frac{n(E)}{n(S)}$$

Law of Large Numbers

The greater the number of trials, the closer the experimental probability will be to the true probability.

Section 4.2: Addition Rules for Probability

Definitions

Complement
Set of all outcomes in the sample space that are not in the event

Mutually exclusive
Describes events that have no outcomes in common

Properties of Probability

1. For any event, E, $0 \le P(E) \le 1$.
2. For any sample space, S, $P(S) = 1$.
3. For the empty set, $\varnothing$, $P(\varnothing) = 0$.

Probability Rules

Complement Rule for Probability
The sum of the probabilities of an event, E, and its complement, E^c, is equal to one.

$$P(E) + P(E^c) = 1$$

Addition Rule for Probability
Probability that event E or event F occurs, given by

$$P(E \textbf{ or } F) = P(E) + P(F) - P(E \text{ and } F)$$

Addition Rule for Probability of Mutually Exclusive Events
For two mutually exclusive events, E and F, the probability that E or F occurs is given by the following formula.

$$P(E \textbf{ or } F) = P(E) + P(F)$$

This rule can also be extended to more than two mutually exclusive events.

Section 4.3: Multiplication Rules for Probability

Definitions

Multistage experiment
Experiment with more than one step

With replacement
Objects are placed back into consideration before performing the next stage of a multistage experiment

Without replacement
Objects are not placed back into consideration before performing the next stage of a multistage experiment

Independent
Describes two events if one event happening does not influence the probability of the other event happening

Section 4.3: Multiplication Rules for Probability (cont.)

Dependent

Describes two events if one event happening influences the probability of the other event happening

Conditional probability

Probability of an event occurring given that another event occurs first, used when the two events are dependent

Fundamental Counting Principle

Method for counting the total number of possible outcomes for a multistage experiment

Probability Rules

Multiplication Rule for Probability of Independent Events

For two independent events, E and F, the probability that E and F occur is given by the following formula.

$$P(E \textbf{ and } F) = P(E) \cdot P(F)$$

This rule can also be extended to more than two independent events.

Multiplication Rule for Probability of Dependent Events

For two dependent events, E and F, the probability that E and F occur is given by the following formula.

$$\begin{aligned} P(E \textbf{ and } F) &= P(E) \cdot P(F|E) \\ &= P(F) \cdot P(E|F) \end{aligned}$$

Conditional Probability

For two dependent events, E and F, the probability that F occurs given that E occurs is given by the following formula.

$$P(F|E) = \frac{P(E \textbf{ and } F)}{P(E)}$$

Fundamental Counting Principle

For a multistage experiment with n stages where the first stage has k_1 outcomes, the second stage has k_2 outcomes, the third stage has k_3 outcomes, and so forth, the total number of possible outcomes for the sequence of stages that make up the multistage experiment is $k_1 \cdot k_2 \cdot k_3 \cdot \dots \cdot k_n$.

Section 4.4: Combinations and Permutations

Definitions

Combination

Selection of objects from a group without regard to their arrangement

Permutation

Selection of objects from a group where the arrangement is specific

Section 4.4: Combinations and Permutations (cont.)

Counting Formulas

Factorial

Product of all positive integers less than or equal to a given positive integer, n, given by

$$n! = n(n-1)(n-2) \cdots (2)(1)$$

By definition, $0! = 1$.

Combinations

When order is not important, the number of combinations of r objects chosen from n distinct objects is given by the following formula.

$${}_nC_r = \frac{n!}{r!(n-r)!}$$

Permutations

When order is important, the number of permutations of r objects chosen from n distinct objects is given by the following formula.

$${}_nP_r = \frac{n!}{(n-r)!}$$

Special Permutations

Involve objects that are identical; the number of distinguishable permutations of n objects, of which k_1 are all alike, k_2 are all alike, and so forth, is given by

$$\frac{n!}{k_1!k_2!\cdots k_p!}$$

where $k_1 + k_2 + \cdots + k_p = n$.

Section 4.5: Combining Probability and Counting Techniques

Key Terms

Words to look for in counting problems, including *at least, at most, greater than, less than, between*, and so forth

Combining Counting Techniques

Occurs primarily when the formula for counting either combinations or permutations is used in conjunction with the Fundamental Counting Principle

E Chapter 4 Exercises

Counting Techniques

Directions: Use counting techniques to answer each question.

1. Suppose you have money to buy 5 new downloads for your iPod. If there are 12 new downloads you'd like to have, in how many different ways could you purchase your 5? (Assume that the order of songs does not matter.)

2. In planning the teaching assignments for next semester, Mr. Hinton must have a teacher in each of the 7 grades during each of the 6 periods of the day. If he has 10 teachers to choose from, which allows 3 teachers to be on break at a time, how many different teaching schedules could he come up with?

3. The student council at one high school must choose 2 representatives from each of the sophomore, junior, and senior classes to attend the annual student council convention. If there are 6 sophomores, 5 juniors, and 7 seniors on the student council, in how many ways can the group be chosen for the convention?

4. The Pancake House offers hash browns with up to five different toppings: cheese, ham, bacon, onions, and mushrooms. In how many different ways can the hash browns be served?

5. Richard is assigned the task of setting the passwords for every computer at his office. Different computers have different guidelines as to how the password may be set. How many different four-digit passwords using the digits 0–9 can he create under each of the following guidelines?
   - **a.** The passwords must be odd and greater than 3000.
   - **b.** The passwords must be even and greater than 4000.
   - **c.** The passwords must be even and less than 5000.

6. Suppose that there are eight employees at a café in downtown Jackson, MS. The boss needs one employee to serve as host or hostess, one to clean tables, and one to supervise the kitchen. In how many ways can these tasks be assigned?

7. Chandra has ten pieces of jewelry in her jewelry case, and she wants to take four pieces with her on vacation. In how many ways can she select the jewelry for her trip?

8. A homeowner wants to plant some new trees in her yard. If she has 6 pink crape myrtles, 4 red oak trees, and 5 yellow poplar trees to plant in one row down the drive, in how many different ways can she plant them?

9. In planning the room assignments for the overnight field trip, Mrs. Viant needs to put 3 girls and 1 chaperone in each room. If there are 12 girls and 4 chaperones on the trip, how many room assignments can she possibly have for the rooms 100, 101, 102, and 103?

Counting Techniques and Probability

Directions: Use the most appropriate method to find the probability for each scenario.

10. If 56% of adults attend church services in a typical month, 46% listen to a Christian radio broadcast in a typical month, and 39% do both, what's the probability that a randomly selected adult will neither attend church nor listen to a Christian radio broadcast in a typical month?

11. Find the probability of being dealt five hearts from a standard deck of playing cards. (Assume that you are randomly dealt a five-card hand.)

12. Find the probability that out of the 1319 college freshmen on campus, you are randomly chosen by the yearbook staff to give a quote about your first year at college, which will be published in the yearbook.

13. In the dice game Yahtzee you achieve the most points by throwing a Yahtzee, which is throwing the same number on 5 different dice. Find the probability that you don't roll a Yahtzee in one roll.

14. In a large doll store, 47% of the dolls blink, 56% of the dolls have movable legs and arms, and 23% of the dolls neither blink nor have movable limbs. Find the probability that a doll chosen at random will both blink and have movable limbs.

15. In a class of 87 people, 27 wear glasses, 32 are blonde, and 38 are neither blonde nor wear glasses. Find the probability that a student chosen at random will have blonde hair and wear glasses.

16. Joe is eating colored candies, and he has 5 orange candies and 5 yellow candies left. If he randomly selects one candy at a time to eat, what is the probability that he eats the rest of the candies so that the colors alternate?

17. Emily has a bad habit of losing her keys. This time, she estimates that there is a 0.2 chance that she left them in her office, a 0.35 chance that she left them in her bag, and a 0.15 chance that she left them on the kitchen counter. What is the probability that her keys will be in none of these places?

18. Robert is ordering dessert, and he wants the fudge pie. The waitress tells him that he can have up to four toppings: ice cream, chocolate sauce, whipped cream, and a cherry. Since he cannot decide how many of the toppings he wants, he tells the waitress to surprise him. If the waitress randomly chooses which toppings to add, what is the probability that Robert gets just chocolate sauce and a cherry?

19. Out of 62 people surveyed, 22 own a laptop computer, 39 own a desktop computer, and 9 own no computer at all. What is the probability that a person selected at random from those surveyed owns both a laptop and a desktop computer?

20. Find the probability of drawing four cards from a standard deck of cards and not getting all aces.

21. What's the probability of rolling a total of at least 11 with two dice?

22. Raffle tickets for a trip to Miami are assigned three-digit numbers using the digits 0–9. What is the probability that the number on the winning ticket is either even or less than 300? (Assume that all possible ticket numbers are eligible to win.)

23. Calculate the probability of choosing, without replacement, a spade and then a heart out of a standard deck of cards.

24. Calculate the probability of drawing two cards from a standard deck, with replacement, and obtaining a five and an ace.

25. A gumball machine at the local pizza place is filled with plastic toys. The machine is filled with 28 rings, 35 bouncy balls, 18 rubber spiders, and 41 tattoos. Suppose you and three friends want to get four toys from the machine, and you would like 2 rings and 2 tattoos. Find the probability that you get what you want on the first 4 tries. (This requires some thought. Don't undercount!)

P Chapter 4 Project

Law of Large Numbers

Directions: This project is designed to be completed in groups, but it may be completed by an individual. Each group is to follow the steps below using a standard six-sided die.

Step 1: Calculate the probability of rolling a single die and getting a four. Round the probability to four decimal places.

$$\text{Probability} = \frac{\#\text{ of ways of getting a four}}{\#\text{ of ways to roll a die}} = \boxed{}$$

Step 2: Each group member is to roll a die ten times and record the outcomes. Compile the outcomes for all group members, and compute the proportion of the group's rolls that were fours. Round the proportion to four decimal places.

$$\text{Proportion of group rolls that were fours} = \frac{\text{Total }\#\text{ of group rolls that were fours}}{\text{Total }\#\text{ of group rolls}}$$

$$= \boxed{}$$

Step 3: This time, each group member is to roll a die an additional 40 times, for a total of 50 rolls per person. Again, combine the outcomes for all group members, and compute the proportion of the group's rolls that were fours. Round the proportion to four decimal places.

$$\text{Proportion of group rolls that were fours} = \frac{\text{Total }\#\text{ of group rolls that were fours}}{\text{Total }\#\text{ of group rolls}}$$

$$= \boxed{}$$

Step 4: Let's combine the information from all groups in order to look at the results for the class as a whole. On the board, make a chart with one column for "number of fours" and one for "number of rolls." Fill in the chart with the information from each group.

Number of Fours	Number of Rolls

Using this information, calculate the proportion of all rolls for the class that were fours. Round the proportion to four decimal places.

$$\text{Proportion of class rolls that were fours} = \frac{\text{Total \# of class rolls that were fours}}{\text{Total \# of class rolls}}$$

$$= \boxed{}$$

Step 5: Let's evaluate the results of this experiment. Compare the proportions calculated in Steps 2, 3, and 4. You should see that as the number of times the dice were tossed increases, the proportion of fours rolled becomes closer to the probability calculated in Step 1. This is precisely what the Law of Large Numbers says: the greater the number of trials, the closer the experimental probability comes to the classical probability. In fact, if it were possible to complete an infinite number of die tosses, the proportion of all rolls that were fours would indeed be equal to the classical probability.

T Chapter 4 Technology

Counting Techniques and Probability

TI-83/84 Plus

Factorials

To calculate a factorial of the form $n!$ using a TI-83/84 Plus calculator, enter the n-value in the calculator, press MATH, and then scroll over to the probability menu, PRB. The factorial is option 4:!. Select option 4 and ! will appear on the screen after the value you entered. To obtain the answer, press ENTER.

Example T.1

Using a TI-83/84 Plus Calculator to Calculate a Factorial

Use a TI-83/84 Plus calculator to calculate 9!

Solution

Press 9, then MATH, and then select PRB and option 4:!. Press ENTER.

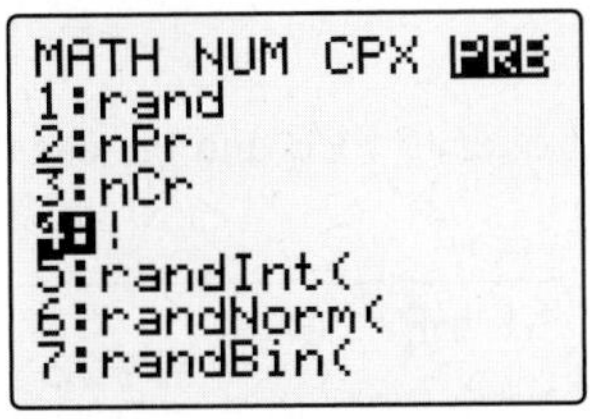

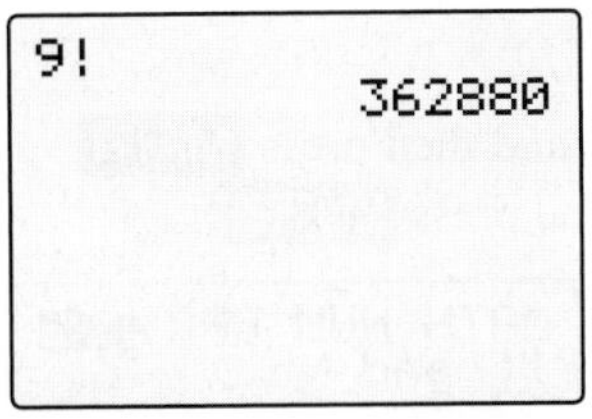

As you can see in the screenshot on the right, 9! = 362,880.

Combinations

To calculate a combination of the form ${}_nC_r$ using a TI-83/84 Plus calculator, enter the n-value in the calculator, press MATH, and then scroll over to the probability menu, PRB. The combination is option 3:nCr. Select option 3 and nCr will appear on the screen after the value you entered. Enter the value for r and press ENTER.

Example T.2

Using a TI-83/84 Plus Calculator to Calculate the Number of Combinations

Use a TI-83/84 Plus calculator to calculate ${}_{15}C_9$.

Solution

Enter 15, and then press MATH. Next, scroll over to PRB and choose option 3:nCr. Then enter 9 and press ENTER.

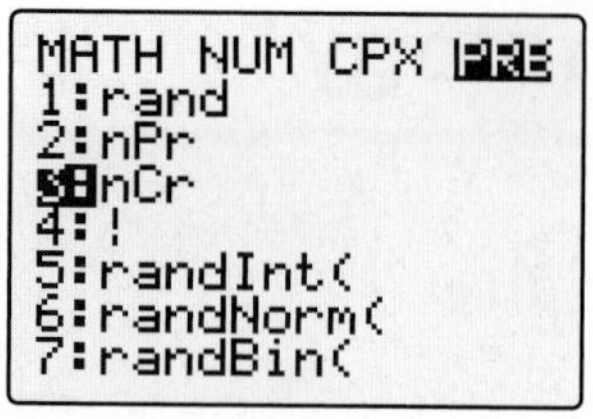

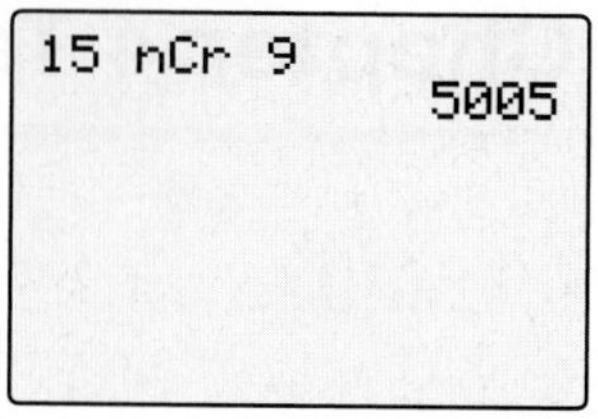

The screenshot on the right shows that ${}_{15}C_9 = 5005$.

Permutations

To calculate a permutation of the form ${}_nP_r$ using a TI-83/84 Plus calculator, enter the n-value in the calculator, press MATH, and then scroll over to the probability menu, PRB. The permutation is option 2:nPr. Select option 2 and nPr will appear on the screen after the value you entered. Enter the value for r and press ENTER.

Example T.3

Using a TI-83/84 Plus Calculator to Calculate the Number of Permutations

Use a TI-83/84 Plus calculator to calculate ${}_{11}P_4$.

Solution

First, enter 11 and then press MATH. Next, scroll over to PRB and choose option 2:nPr. Then enter 4 and press ENTER.

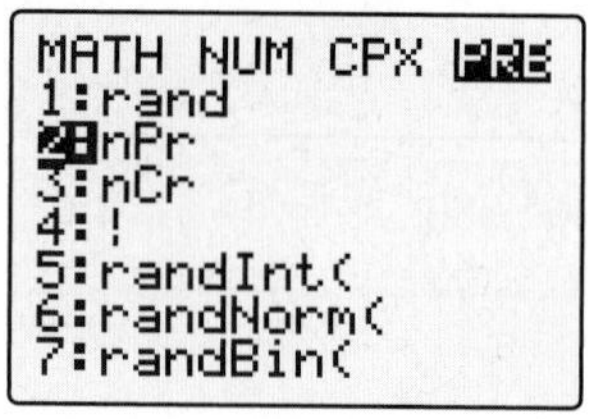

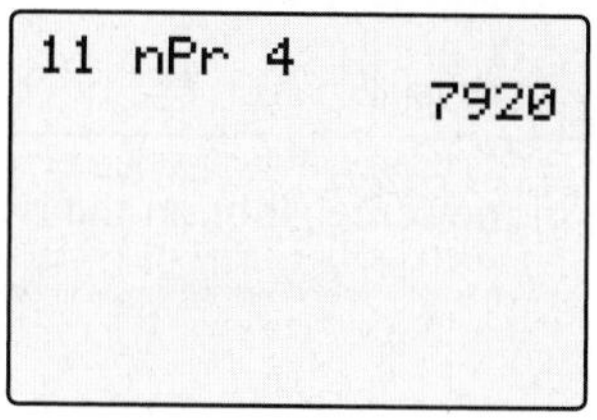

As shown in the screenshot on the right, ${}_{11}P_4 = 7920$.

Microsoft Excel

Counting Techniques

Microsoft Excel can calculate many mathematical functions used in probability. Individually, the commands are as follows.

Table T.1: Excel Commands

Calculation	Command
Factorial	=FACT(*number*)
Permutation	=PERMUT(*number, number_chosen*)
Combination	=COMBIN(*number, number_chosen*)

Probability

Excel is also a great tool to use when trying to calculate the probability of several independent events as shown in the following example.

Example T.4

Using Microsoft Excel to Calculate the Probability of Several Independent Events

A coin is flipped, a die is rolled, and a card is drawn from a deck. Use Microsoft Excel to find the probability of getting a tail on the coin, rolling a 5 on the die, and drawing a heart from the deck of cards.

Solution

First enter the data in an Excel worksheet as shown below.

	A	B	C	D	E
1		Coin	Die	Card	
2	Chance of Success	1	1	13	
3	Possible Outcomes	2	6	52	
4					
5					

Then in cell B4, divide B2 by B3. To do this, type the formula **=B2/B3**. Copy the formula from cell B4 into C4 and D4. Now the probabilities for each event are listed in row 4. Remember, when events are independent we multiply the probabilities of each event occurring with each other. Therefore, in cell E4, multiply the cells B4 through D4 by typing **=B4*C4*D4**. The result, shown in the following screenshot, will be the probability of getting a tail on the coin, a 5 on the die, and drawing a heart from the deck of cards, which is approximately 0.0208.

	A	B	C	D	E
1		Coin	Die	Card	
2	Chance of Success	1	1	13	
3	Possible Outcomes	2	6	52	
4		0.5	0.166667	0.25	0.020833
5					

Minitab

Factorials, Combinations, and Permutations

Minitab can be also be used to calculate factorials and find numbers of combinations or permutations, as shown in the following examples.

Example T.5

Using Minitab to Calculate a Factorial

Use Minitab to calculate 10!

Solution

Go to **Calc ▶ Calculator**. Type **C1** in the box after Store result in variable. Select **Factorial** under All functions, and then type **10** to replace "number of items" in the expression. Then click **OK**; the result, 3628800, will be displayed in row 1 of column C1. Thus, 10! = 3,628,800.

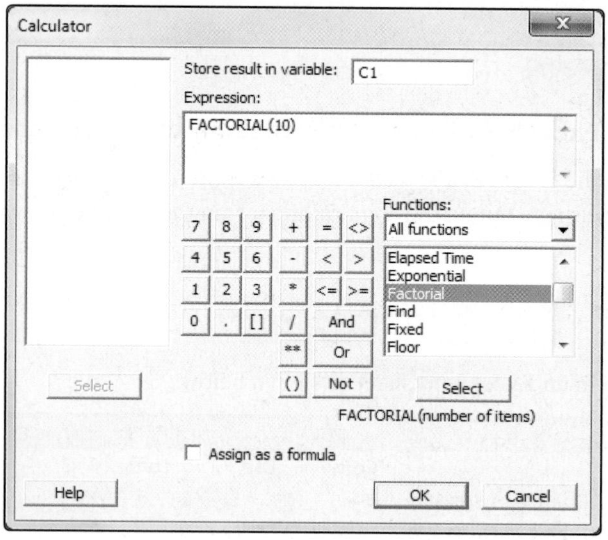

Example T.6

Using Minitab to Calculate the Number of Combinations

Use Minitab to calculate ${}_{15}C_{13}$.

Solution

Go to **Calc ▶ Calculator**. Type **C1** in the box after Store result in variable. Select **Combinations** under All functions. Type **15** to replace "number of items" and type **13** to replace "number to choose" in the expression. Then click **OK**; the result, 105, will be displayed in row 1 of column C1. Thus, ${}_{15}C_{13} = 105$.

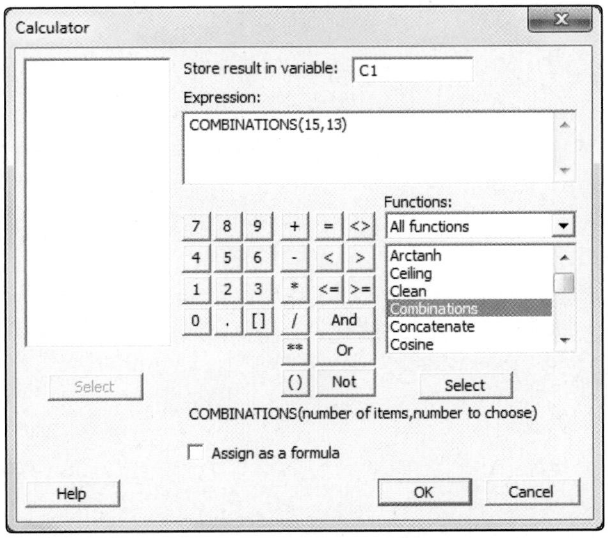

Example T.7

Using MINITAB to Calculate the Number of Permutations

Use MINITAB to calculate ${}_{18}P_7$.

Solution

Go to **Calc ▶ Calculator**. Type **C1** in the box after Store result in variable. Select **Permutations** under All functions. Type **18** to replace "number of items" and type **7** to replace "number to choose" in the expression. Then click **OK**; the result, 160392960, will be displayed in row 1 of column C1. Thus, ${}_{18}P_7 = 160{,}392{,}960$.

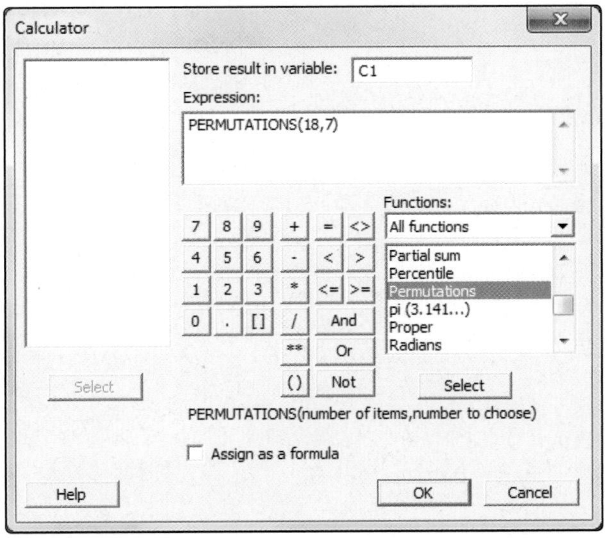

Chapter Five

Discrete Probability Distributions

Sections

Objectives

1. Calculate the expected value of a probability distribution.
2. Calculate the variance and the standard deviation of a probability distribution.
3. Identify a distribution as binomial, Poisson, or hypergeometric.
4. Calculate probabilities using a binomial, Poisson, or hypergeometric distribution.

5

Introduction

In the previous chapter we looked at methods for computing simple probabilities. For example, we learned how to find the probability of events such as winning the local school raffle or the probability that a student randomly—and unfortunately—sits down at one of the two broken computers in the computer lab on campus.

Suppose instead that you wanted to know a more complicated probability, one that cannot easily be solved by using classical probability techniques alone. For instance, consider the scenario in which a university randomly selects sixty students to complete a survey regarding the cost of tuition. What is the probability that all sixty students chosen are freshmen and male? Or, suppose that an investor wants to compare the risks of two different groups of stocks for his portfolio. Is there a way to estimate the long-term benefit for each investment option? Let's look at some additional concepts needed to solve these types of problems.

5.1 Discrete Random Variables

Side Note

Possible Outcomes for the Coin Toss Challenge

HHH, HHT, HTH, THH, TTH, THT, HTT, TTT

Let's suppose that you and some friends go to a fall carnival. One game at the carnival is a coin toss challenge where you get a chance to flip three coins. If all three coins are the same (heads or tails), then you win three dollars. If any other combination of heads and tails turns up, you don't win anything. Let's consider for a moment the probabilities associated with this carnival game.

The number of coins that turn up heads each time you play the game will be some number between 0 and 3, which is completely determined by chance. A variable like this is called a **random variable** and is usually denoted by a capital letter, such as X, Y, or Z. For example, if $X =$ the number of coins showing heads, then an outcome of three coins showing two heads and one tail gives X the value of 2, while an outcome of all three coins showing heads would make $X = 3$.

2 Heads, 1 Tail
$X = 2$

3 Heads
$X = 3$

Figure 5.1

In our example above, notice that, although the values of X are determined by chance, the values of X are not all equally likely. Of the eight possible outcomes, only one gives a value of $X = 3$. Therefore, the probability of getting three heads is

$$P(X = 3) = \frac{1}{8}.$$

On the other hand, there are three throws that give $X = 2$, so

$$P(X=2)=\frac{3}{8}.$$

A table or formula that gives the probabilities for every value of the random variable X is called a **probability distribution**. Consider the probability distribution for our example.

Memory Booster

Uppercase or lowercase?

X = Random Variable

x = specific value of the random variable

Table 5.1: Number of Heads in Three Coin Tosses

x	0	1	2	3
$P(X=x)$	$\frac{1}{8}$	$\frac{3}{8}$	$\frac{3}{8}$	$\frac{1}{8}$

Notice that the following properties are true about the above distribution. In fact, these two conditions hold true for all probability distributions. A list of probabilities for the values of a random variable is only a probability distribution if the following conditions are met.

Properties

Properties of a Probability Distribution

1. All of the probabilities are between 0 and 1, inclusive. That is, $0 \le P(X=x) \le 1$.
2. The sum of the probabilities is 1. That is, $\sum P(X=x_i)=1$.

Definition

A **random variable** is a variable whose numeric value is determined by the outcome of a probability experiment.

A **probability distribution** is a table or formula that gives the probabilities for every value of the random variable X, where $0 \le P(X=x) \le 1$ and $\sum P(X=x_i)=1$.

Recall from Chapter 1 that data may be discrete or continuous. Similarly, random variables may also be discrete or continuous. A **discrete random variable** may have either finitely many possible values or infinitely many possible values that are determined by a counting process. Our example of counting heads in a coin toss is an example of a discrete random variable, and its table is called a **discrete probability distribution**. The remainder of this chapter will be focused on several types of discrete probability distributions. We'll then turn our attention to continuous probability distributions in Chapter 6.

Example 5.1

Creating a Discrete Probability Distribution

Create a discrete probability distribution for X, the sum of two rolled dice.

Solution

To begin, let's list all of the possible values for X.

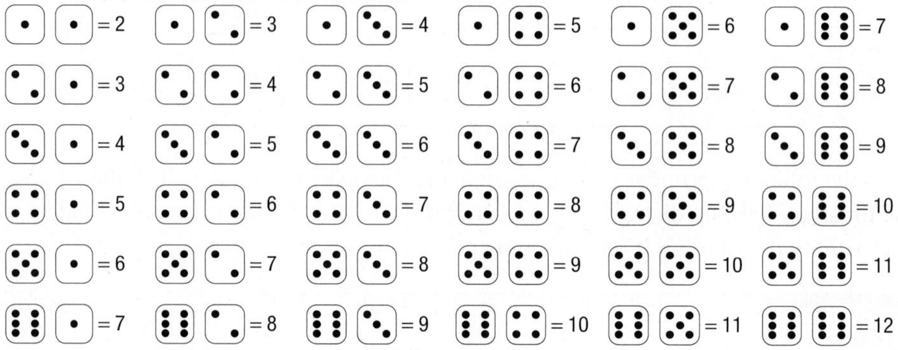

When rolling two dice, there are 36 possible rolls, each giving a sum between 2 and 12, inclusive. To find the probability distribution, we need to calculate the probability for each value.

$P(X=2)=\frac{1}{36}$ because there is only one way to get a sum of 2: 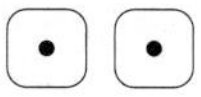

$P(X=3)=\frac{2}{36}$ because you may get the sum of 3 in two ways:

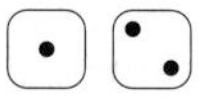 or 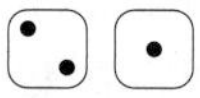

Continuing this process will give us the following probability distribution.

Sum of Two Rolled Dice											
x	2	3	4	5	6	7	8	9	10	11	12
P(X = x)	$\frac{1}{36}$	$\frac{2}{36}$	$\frac{3}{36}$	$\frac{4}{36}$	$\frac{5}{36}$	$\frac{6}{36}$	$\frac{5}{36}$	$\frac{4}{36}$	$\frac{3}{36}$	$\frac{2}{36}$	$\frac{1}{36}$

Check for yourself that the probabilities listed are the true values for the probability distribution of X, the sum of two rolled dice. Note that all of the probabilities are numbers between 0 and 1, inclusive, and that the sum of the probabilities is equal to 1. Check this for yourself.

Memory Booster

Properties of a Probability Distribution

$0 \le P(X=x) \le 1$

$\sum P(X=x_i)=1$

Probability distributions are important when calculating the average value for a random variable. Let's go back to our original example of the coin toss challenge. Now, if you play the game a number of times, are you likely to come out winning money? The solution to this problem cannot be found simply by using one of the patterns we have seen in the previous sections. In order to calculate this probability, you need to be able to find something called the *expected value* of a random variable. Also, the random variable that we need to use is not X, the number of heads, but a new random variable W = the amount of money won playing a single game.

There are two possible results for the coin toss game—you win \$3.00 or walk away with nothing. So, the possible values for the discrete random variable W are \$3.00 and \$0.00. To calculate the expected value of W for our example, we first need to determine the discrete probability distribution for W. Thus, we need to know the probability for each value of W. There are two ways for all three coins to be the same—all heads or all tails—and eight possible outcomes for flipping three coins. The probability of all three coins being the same is then $\frac{2}{8}=\frac{1}{4}$. By the Complement Rule, we know that the probability of not getting all three alike must be $1-\frac{1}{4}=\frac{3}{4}$. Let's look at the discrete probability distribution of W.

Table 5.2: Coin Toss Game Winnings

w	\$3.00	\$0.00
$P(W=w)$	$\frac{1}{4}$	$\frac{3}{4}$

What this table shows us is that we should anticipate winning \$3.00 about a fourth of the time and winning nothing about three-fourths of the time. So what would our average winnings be? If we calculate a weighted mean here, we can find the value we should *expect* to win on average. The **expected value**, $E(X)$, is exactly that—the value that we expect the random variable to have on average. Formally, it is the mean of the probability distribution, that is, $E(X)=\mu$. To calculate it, as we just said, find the weighted mean of the probability distribution by multiplying each value of the random variable in the distribution by its probability and then summing these products. The formula for expected value is as follows.

Formula

Expected Value

The **expected value** for a discrete random variable X is equal to the mean of the probability distribution of X and is given by

$$E(X)=\mu=\sum\left[x_i\cdot P(X=x_i)\right]$$

where x_i is the ith value of the random variable X.

Rounding Rule

When calculating the expected value, round to one more decimal place than the largest number of decimal places given in the values of the random variable. Occasional exceptions to this rule can be made when the type of data lends itself to a more natural rounding scheme, such as rounding values of currency to two decimal places.

Let's calculate the expected value of the amount of money won for the coin toss game by calculating the mean of the probability distribution for W.

$$\begin{aligned}E(W)&=\sum\left[w_i\cdot P(W=w_i)\right]\\&=3.00\cdot P(W=3.00)+0.00\cdot P(W=0.00)\\&=(3.00)\left(\frac{1}{4}\right)+(0.00)\left(\frac{3}{4}\right)\\&=\frac{3}{4}+0\\&=\frac{3}{4}\\&=\$0.75\end{aligned}$$

Now that we know the expected value of the random variable W, we can say whether we are likely to come out winning money after having played the game a number of times. The expected value of money won in the coin toss challenge is \$0.75. This tells us that if you were to play the game a large number of times, you would expect an average gain of 75¢ for each game you play. Notice that the expected value gives us a *long-term average*. For any given try, a gain of 75¢ is not an option. (Remember, you either win \$3.00 or nothing at all.) However, if you repeat the game many times, an average gain of only 75¢ per game is to be expected. So, what is the answer to our original question? Are you likely to come out winning money after playing the game over a period of time? Although it is possible for you to win \$3.00 every time, the expected value tells us that over time you are more likely to win a small amount of money per game.

Now, before you throw on your lucky shoes and head to the nearest carnival in search of the winning coin toss game, let's stop and think about the reality of a carnival giving away an average of 75¢ every time a person plays the game. The carnival would likely be broke after one weekend once word got around. What we failed to take into account is that carnivals make you pay for each entertaining game you play. Suppose we bring our coin toss game into reality and adjust for the fact that the carnival charges \$1.00 every time you play the game. How does that affect our winnings? Well, if we expected to win 75¢ on average for each game we play, but we have to spend \$1.00 to play it, we should actually expect to *lose* 25¢ per game on average. This changes things quite considerably when we figure in the money we must pay to play the game, and it explains why the game is profitable for the carnival.

Instead of waiting until the end to account for the cost of the game, we can adjust the original probability distribution to begin with. Let's revisit the coin toss challenge from this perspective. For each individual game, since you must pay \$1.00 to play, the net amount of money won for the game is either \$2.00 if you win, or –\$1.00 if you lose (the original dollar you paid to play the game). So, the possible values for W are \$2.00 and –\$1.00. (A loss is considered a negative value.) The probability of winning the coin toss game remains the same, so our distribution looks like the following.

Table 5.3: Coin Toss Game Winnings

w	\$2.00	–\$1.00
$P(W = w)$	$\frac{1}{4}$	$\frac{3}{4}$

If we calculate the expected value of the amount of money won using the formula, we have the following.

$$\begin{aligned} E(W) &= \sum\left[w_i \cdot P(W = w_i)\right] \\ &= 2.00 \cdot P(W = 2.00) + (-1.00) \cdot P(W = -1.00) \\ &= (2.00)\left(\frac{1}{4}\right) + (-1.00)\left(\frac{3}{4}\right) \\ &= \frac{2}{4} - \frac{3}{4} \\ &= -\frac{1}{4} \\ &= -\$0.25 \end{aligned}$$

Just as we thought—on average, we would lose 25¢ per game.

Let's take another look at calculating expected values.

Example 5.2

Calculating Expected Values

Suppose that Randall and Blake decide to make a friendly wager on the football game they are watching one afternoon. For every kick the kicker makes, Blake has to pay Randall $30.00. For every kick the kicker misses, Randall has to pay Blake $40.00. Prior to this game, the kicker has made 18 of his past 23 kicks this season.

a. What is the expected value of Randall's bet for one kick?

b. Suppose that the kicker attempts four kicks during the game. How much should Randall expect to win in total?

Solution

a. There are two possible outcomes for this bet: Randall wins $30.00 ($x = 30.00$) or Randall loses $40.00 ($x = -40.00$). If the kicker has made 18 of his past 23 kicks, then we assume that the probability that he will make a kick—and that Randall will win the bet—is $\frac{18}{23}$. By the Complement Rule, the probability that the kicker will miss—and Randall will lose the bet—is $1-\frac{18}{23}=\frac{5}{23}$.

Randall's Bet for One Kick

x	$30.00	–$40.00
$P(X=x)$	$\frac{18}{23}$	$\frac{5}{23}$

Then we calculate the expected value as follows.

$$\begin{aligned} E(X) &= \sum\left[x_i \cdot P(X=x_i)\right] \\ &= (30.00)\left(\frac{18}{23}\right)+(-40.00)\left(\frac{5}{23}\right) \\ &= \frac{540}{23}-\frac{200}{23} \\ &= \frac{340}{23} \\ &\approx \$14.78 \end{aligned}$$

We see that the expected value of the wager is $14.78. Randall should expect that if the same bet were made many times, he would win an average of $14.78 per bet.

b. We know that over the long term Randall would win an average of $14.78 per bet. So for four attempted kicks, we multiply the expected value for one bet by four: $14.78 \cdot 4 = 59.12$. If he and Blake place four bets, then Randall can expect to win approximately $59.12.

Now let's look at a situation where we can use expected values to help us make an informed decision.

Example 5.3

Calculating Expected Values

Peyton is trying to decide between two different investment opportunities. The two plans are summarized in the table below. The left column for each plan gives the potential earnings, and the right columns give their respective probabilities. Which plan should he choose?

Investment Plans			
Plan A		Plan B	
Earnings	Probability	Earnings	Probability
\$1200	0.1	\$1500	0.3
\$950	0.2	\$800	0.1
\$130	0.4	–\$100	0.2
–\$575	0.1	–\$250	0.2
–\$1400	0.2	–\$690	0.2

Solution

It is difficult to determine which plan will yield the higher return simply by looking at the probability distributions. Let's use the expected values to compare the plans. Let the random variable X_{A} be the earnings for Plan A, and let the random variable X_{B} be the earnings for Plan B.

For Investment Plan A:

$$\begin{aligned} E(X_{\text{A}}) &= \sum\left[x_i \cdot P(X_{\text{A}} = x_i)\right] \\ &= (1200)(0.1)+(950)(0.2)+(130)(0.4)+(-575)(0.1)+(-1400)(0.2) \\ &= 120+190+52-57.5-280 \\ &= \$24.50 \end{aligned}$$

For Investment Plan B:

$$\begin{aligned} E(X_{\text{B}}) &= \sum\left[x_i \cdot P(X_{\text{B}} = x_i)\right] \\ &= (1500)(0.3)+(800)(0.1)+(-100)(0.2)+(-250)(0.2)+(-690)(0.2) \\ &= 450+80-20-50-138 \\ &= \$322.00 \end{aligned}$$

From these calculations, we see that the expected value of Plan A is \$24.50, and the expected value of Plan B is \$322.00. Therefore, Plan B appears to be the wiser investment option for Peyton.

At times, it is also important to have a measure of the variation of a discrete probability distribution. This is especially important for problems such as the previous investment example, where a high variation in profit would make for a riskier investment. The variance and standard deviation of a discrete probability distribution are found using the following formulas. Note that the two formulas given for the variance are equivalent. The second formula given for the variance can be interpreted as the weighted mean of the squared deviations, and it is easier to use when doing calculations by hand. The two formulas given for the standard deviation are simply the square roots of the formulas for the variance.

Formula

Variance and Standard Deviation for a Discrete Probability Distribution

The **variance for a discrete probability distribution** of a random variable X is given by

$$\sigma^2 = \sum\left[x_i^2 \cdot P(X = x_i)\right] - \mu^2$$
$$= \sum\left[(x_i - \mu)^2 \cdot P(X = x_i)\right]$$

where x_i is the ith value of the random variable X and

μ is the mean of the probability distribution.

The **standard deviation for a discrete probability distribution** of a random variable X is the square root of the variance, given by the following formulas.

$$\sigma = \sqrt{\sigma^2}$$
$$= \sqrt{\sum\left[x_i^2 \cdot P(X = x_i)\right] - \mu^2}$$
$$= \sqrt{\sum\left[(x_i - \mu)^2 \cdot P(X = x_i)\right]}$$

Rounding Rule

When calculating the variance or standard deviation for a discrete probability distribution, round to one more decimal place than the largest number of decimal places given in the values of the random variable. Occasional exceptions to this rule can be made when the type of data lends itself to a more natural rounding scheme, such as rounding values of currency to two decimal places.

When calculating the standard deviation, do not round the value of the variance before taking the square root.

Example 5.4

Calculating the Variances and Standard Deviations for Discrete Probability Distributions

Which of the investment plans in the previous example carries more risk, Plan A or Plan B?

Solution

To decide which plan carries more risk, we need to look at their standard deviations, which requires that we first calculate their variances. Let's calculate the variance separately for each investment plan. To do this, we will use a table to organize our calculations as we compute the variance. We will use the expected values that we calculated in the previous example as the means.

For Investment Plan A, $\mu = E(X_A) = \$24.50$.

Investment Plan A				
x	$P(X = x)$	$x - \mu$	$(x - \mu)^2$	$(x - \mu)^2 \cdot P(X = x)$
\$1200	0.1	1175.50	1,381,800.25	138,180.025
\$950	0.2	925.50	856,550.25	171,310.05
\$130	0.4	105.50	11,130.25	4452.1
−\$575	0.1	−599.50	359,400.25	35,940.025
−\$1400	0.2	−1424.50	2,029,200.25	405,840.05

$$\begin{aligned}\sigma^2 &= \sum\left[(x_i-\mu)^2 \cdot P(X_A = x_i)\right]\\ &= 138{,}180.025 + 171{,}310.05 + 4452.1 + 35{,}940.025 + 405{,}840.05\\ &= 755{,}722.25\end{aligned}$$

$$\begin{aligned}\sigma &= \sqrt{\sigma^2}\\ &= \sqrt{755{,}722.25}\\ &\approx \$869.32\end{aligned}$$

For Investment Plan B, $\mu = E(X_B) = \$322.00$.

Investment Plan B				
x	$P(X=x)$	$x-\mu$	$(x-\mu)^2$	$(x-\mu)^2 \cdot P(X=x)$
\$1500	0.3	1178	1,387,684	416,305.2
\$800	0.1	478	228,484	22,848.4
–\$100	0.2	422	178,084	35,616.8
–\$250	0.2	572	327,184	65,436.8
–\$690	0.2	1012	1,024,144	204,828.8

$$\begin{aligned}\sigma^2 &= \sum\left[(x_i-\mu)^2 \cdot P(X_B = x_i)\right]\\ &= 416{,}305.2 + 22{,}848.4 + 35{,}616.8 + 65{,}436.8 + 204{,}828.8\\ &= 745{,}036\end{aligned}$$

$$\begin{aligned}\sigma &= \sqrt{\sigma^2}\\ &= \sqrt{745{,}036}\\ &\approx \$863.15\end{aligned}$$

What do these results tell us? Comparing the standard deviations, we see that not only does Plan B have a higher expected value, but its profits vary slightly less than those of Plan A. We may conclude that Plan B carries a slightly lower amount of risk than Plan A.

5.1 Section Exercises

Properties of a Probability Distribution

Directions: For each table, determine whether it could represent a valid discrete probability distribution. If not, explain why.

1.

x	$P(X=x)$
1	0.2
2	0.6
3	0.05
4	0.15
5	0.0

2.

x	$P(X=x)$
–22	0.4
53	1.05
–15	0.05

3.

x	$P(X=x)$
0.5	0.4
2.5	0.7
4.5	–0.3
6.5	0.2

4.

x	$P(X=x)$
–15	0.4
–20	0.3
–25	0.4

Discrete Probability Distributions

Directions: Create the probability distribution for each random variable described.

5. The number of tails showing when flipping four coins.

6. The number of even numbers showing when a pair of standard six-sided dice are rolled.

7. The difference between the two numbers showing when a pair of standard six-sided dice are rolled (largest value – smallest value).

8. The number of heads showing in five tosses of a coin.

Mean and Standard Deviation for Discrete Probability Distributions

Directions: For each discrete probability distribution, find the mean and the standard deviation.

9.

x	$P(X=x)$
15	0.6
22	0.4

10.

x	$P(X=x)$
–55	0.45
30	0.55

11.

x	$P(X = x)$
14	0.3
21.5	0.4
−2	0.3

12.

x	$P(X = x)$
−$1.50	0.3
$0.00	0.5
$2.75	0.1
$5.00	0.1

Expected Values

Directions: Determine the expected values for each scenario.

13. Suppose that you and a friend are playing cards and you decide to make a friendly wager. The bet is that you will draw two cards without replacement from a standard deck. If both cards are hearts, your friend will pay you $25.00. Otherwise, you have to pay your friend $5.00.

a. What is the expected value of your bet?

b. If this same bet is made 100 times, how much would you expect to win or lose?

14. Scott likes to day-trade on the Internet. On a good day, he averages a $2200.00 gain. On a bad day, he averages a $1600.00 loss. Suppose that he has good days 25% of the time, bad days 35% of the time, and the rest of the time he breaks even.

a. What is the expected value for one day of Scott's day-trading hobby?

b. If Scott day-trades every weekday for three weeks, how much money should he expect to gain or lose?

c. What is the variance for one day of Scott's hobby?

15. Mike's older brother, Jeff, bets him that he can't roll two dice and get doubles three times in a row. If Mike does it, Jeff will give him $100.00. Otherwise, Mike has to give Jeff $5.00.

a. What is the expected value of Mike's bet?

b. What is the expected value of Jeff's bet?

c. If Mike and Jeff make the same bet 30 times, how much can Mike expect to win or lose?

16. An insurance company offers Mississippi adults between the ages of 25 and 34 a $100,000 life insurance policy for $18 a month. They use the fact that Mississippi has a yearly death rate of 172.8 per 100,000 residents aged 25–34 years.

a. Find the expected value per customer for the insurance company at the end of one year for the policy described.

b. If the insurance company has 10,000 customers with these life insurance policies in Mississippi, what is its profit at the end of the year?

17. Suppose the same insurance company as in the previous question insures adults ages 25 to 34 in California for the same amount of money per month, but offers a $175,000 policy for that amount of money. The reason for the difference in the payout is that the death rate in California for that age group is 81.6 per 100,000 residents.

   a. Find the expected value per customer for the insurance company at the end of one year for the policy described in California.

   b. If the insurance company has 10,000 customers with these life insurance policies in California, what is its profit at the end of the year?

   c. Which state is more profitable for the insurance company (as compared to Mississippi in the previous problem)?

18. Taranique loves to play the floating duck game at the carnival. For $2.00 per try, she gets to choose one duck out of the 50 swimming in the water. If Taranique is lucky, she will pick one of the 8 winning ducks and go home with a pink teddy bear.

   a. What is the expected value of the game for Taranique if the value of the prize is $5.00?

   b. If Taranique plays the game 10 times, how much can she expect to win or lose?

19. A church in town is raffling off $50.00. You can buy one ticket for $1.00, three tickets for $2.50, or five tickets for $4.00. Assume that the church sells 100 tickets.

   a. Find the expected value for each of the three ticket options.

   b. Should you buy one, three, or five tickets in order to maximize your expected winnings?

20. A department store is running a promotion one Saturday by giving out coupons for $10 of free merchandise. Based on data collected in the past, only one-fourth of customers who shop on that Saturday use the coupon but do not purchase any other merchandise. However, one-third of customers purchase $40 in merchandise and then use the coupon. Another one-third of customers use the coupon after ringing up a total of $75 in merchandise. The remainder of customers who come in the store do not take advantage of the promotion at all.

   a. Find the expected value of the promotion per customer for the department store.

   b. If the store has 720 customers on the promotional Saturday, what is its expected revenue for the day?

21. A car dealership is offering an interesting incentive in order to get people to come and test drive its new sports cars. Everyone who agrees to a test drive gets to choose a key. There are 75 car keys in the bag, and 4 of them unlock a sports car. For the customers who choose a winning key, the dealership agrees to knock $1000 off of the price if they buy a new car. (Assume that each key is returned after being drawn.)

   a. From the perspective of the car dealership, what is the expected value of the incentive for one customer who chooses a key and buys a new car?

   b. If 90 customers come in and choose a key and all of them buy new cars, how much can the dealership expect to give up in sales?

22. a. In the following probability distribution for the cost of textbooks in a fall semester for various liberal arts majors, each probability represents the chance that the total cost of a student's books will be the given amount. Find the expected value for the cost of books for the semester.

Cost of Textbooks for Liberal Arts Majors	
x	$P(X = x)$
\$262	0.19
\$410	0.21
\$590	0.17
\$653	0.43

b. In the following probability distribution for the cost of textbooks in a fall semester for various business majors, each probability represents the chance that the total cost of a student's books will be the given amount. Find the expected value for the cost of books for the semester.

Cost of Textbooks for Business Majors	
x	$P(X = x)$
\$378	0.35
\$389	0.14
\$392	0.25
\$401	0.26

c. Which of the groups of majors would you guess is more likely to feel that their textbooks are priced fairly? Explain your answer. (**Hint:** Use the standard deviations to help you make your *informed* decision.)

23. The managing director of a traveling carnival needs to add a new game to the carnival lineup. Given below are the probability distributions for his top two choices. The values of the random variable are the amounts the carnival would either gain (positive values) or have to pay out (negative values). Which game would you advise the director to choose? Why?

Carnival Game 1	
x	$P(X = x)$
\$1.00	0.8
–\$5.00	0.135
–\$10.00	0.06
–\$100.00	0.005

Carnival Game 2	
x	$P(X = x)$
\$1.00	0.65
–\$1.00	0.25
–\$2.00	0.1

5.2 Binomial Distribution

Now that the groundwork has been laid for how discrete probability distributions work, in the remainder of the chapter we are going to learn about some specific types of discrete probability distributions. The distributions we will learn about include the binomial distribution, Poisson distribution, and the hypergeometric distribution. These are not the only discrete probability distributions; they are just the ones we will be learning about in this chapter. Continuous probability distributions will be discussed in the next chapter.

The **binomial distribution** is a special discrete probability function for problems with a fixed number of independent trials, where each trial has only two possible outcomes and one of these outcomes is counted. For example, a common binomial experiment is that of flipping a coin. The coin is flipped a fixed number of times, and the only possible outcomes are heads and tails. We can then choose to count either the number of heads or the number of tails obtained for an experiment. The outcome that is counted is called a **success**.

Let's say that we want to determine the probability of getting fewer than two heads when a coin is tossed three times. We want to count the number of heads obtained in three coin tosses, so getting a head will be considered the success. We can calculate this probability by using classical techniques from Section 4.1. First, let's find the sample space.

HHH	HHT	HTH	THH
TTT	TTH	THT	HTT

If we want the probability of getting fewer than two heads, then we need to count the outcomes that have either exactly one head or no heads. There are four outcomes that meet these criteria, so the probability of getting fewer than two heads is $\frac{4}{8}=\frac{1}{2}$.

That's easy enough. But what if we want to find the probability of getting exactly six heads in ten trials? Or fifteen heads in twenty trials? The sample spaces for these problems are much too large to list by hand. We must find a better way to calculate these types of probabilities.

The binomial distribution can be used to calculate more complicated binomial probabilities. First, let's see when it would be appropriate to use the binomial distribution. All of the following guidelines must be met for an experiment to have a binomial distribution.

Properties

Properties of a Binomial Distribution

1. The experiment consists of a fixed number, n, of identical trials.
2. Each trial is independent of the others.
3. For each trial, there are only two possible outcomes. For counting purposes, one outcome is labeled a success, and the other a failure.
4. For every trial, the probability of getting a success is called p. The probability of getting a failure is then $1-p$.
5. The binomial random variable, X, counts the number of successes in n trials.
6. For a binomial distribution, the mean is given by $\mu = np$ and the variance is given by $\sigma^2 = np(1-p)$.

Memory Booster

$E(X)=\mu$
for a random variable X.

When all of these rules are satisfied, the following formula can be used to determine the probability of obtaining x successes out of n trials.

Rounding Rule

When calculating probabilities for binomial distributions, round to four decimal places. This follows the convention used in the binomial probability distribution tables in Appendix A.

Formula

Probability for a Binomial Distribution

For a binomial random variable X, the probability of obtaining x successes in n independent trials is given by

$$P(X = x) = {}_nC_x \cdot p^x (1-p)^{(n-x)}$$

where x is the number of successes,

n is the number of trials, and

p is the probability of getting a success on any trial.

Example 5.5

Calculating a Binomial Probability Using the Formula

What is the probability of getting exactly six heads in ten coin tosses?

Solution

We showed earlier that coin tosses meet the criteria of the binomial distribution. For this problem, let X = the number of heads obtained out of ten coin tosses. There are ten coin tosses, so $n = 10$. We will say that a success is getting a head. We want the probability of *exactly six* successes, so $x = 6$. The probability of flipping a head in one coin toss is 0.5, which means that $p = 0.5$. Substituting these values into the binomial probability formula gives us the following.

$$\begin{aligned}
P(X = x) &= {}_nC_x \cdot p^x (1-p)^{(n-x)} \\
P(X = 6) &= {}_{10}C_6 \cdot (0.5)^6 (1-0.5)^{(10-6)} \\
&= \frac{10!}{6!4!} \cdot (0.5)^6 (0.5)^4 \\
&= (210)(0.015625)(0.0625) \\
&\approx 0.2051
\end{aligned}$$

Thus, the probability of getting exactly six heads in ten coin tosses is approximately 0.2051.

Though the binomial formula is not an overly complex formula to use, many graphing calculators can calculate a binomial probability directly. In this text, we will use a TI-83/84 Plus calculator to calculate the necessary probabilities. The following examples continue to demonstrate the use of the binomial formula and also describe the two different types of binomial calculations that can be performed on a TI-83/84 Plus. For specific directions on how to use a TI-83/84 Plus, or other available technology including Microsoft Excel and Minitab, see the technology section at the end of this chapter.

Example 5.6

Calculating a Binomial Probability Using the Formula or a TI-83/84 Plus Calculator

A quality control expert at a large factory estimates that 10% of all batteries produced are defective. If he takes a random sample of fifteen batteries, what is the probability that exactly two are defective?

Solution

First, let's verify that this process meets the criteria of a binomial distribution. Since the batteries are randomly sampled and, presumably, more batteries continue to be produced by the factory while the sampling takes place, we can consider the selection of the batteries to be identical, independent trials. Since we are testing fifteen batteries, the number of trials is $n = 15$. For each trial, there are two possible outcomes: either the battery is defective or it is not. We will consider a defective battery to be a success and 10% of all batteries produced are defective, so the probability of getting an individual success is $p = 0.1$. Let X = the number of defective batteries found in a sample of 15 batteries. We are looking for the probability that *exactly two* are defective, so we want the binomial probability, $P(X = 2)$.

Using the binomial probability formula for our solution would require us to calculate the following expression.

$$P(X = x) = {}_nC_x \cdot p^x (1-p)^{(n-x)}$$
$$P(X = 2) = {}_{15}C_2 \cdot (0.1)^2 (0.9)^{13}$$
$$\approx 0.2669$$

However, the TI-83/84 Plus calculator can calculate $P(X = x)$ directly using the following procedure.

- Press **2ND** and then **VARS** to access the `DISTR` menu.
- Choose option `A:binompdf(`.
- Enter n, p, and x in the parentheses as: `binompdf(`n, p, x`)`.

Thus, using a TI-83/84 Plus, we would calculate the probability as shown below and in the screenshot in the margin.

$$P(X = x) = \texttt{binompdf}(n, p, x)$$
$$P(X = 2) = \texttt{binompdf(15,0.1,2)}$$
$$\approx 0.2669$$

Therefore, the probability that exactly two out of the fifteen batteries are defective is approximately 0.2669.

Side Note

On TI-83 Plus calculators and older TI-84 Plus calculators, the function `binompdf(` is option `Ø` on the `DISTR` menu.

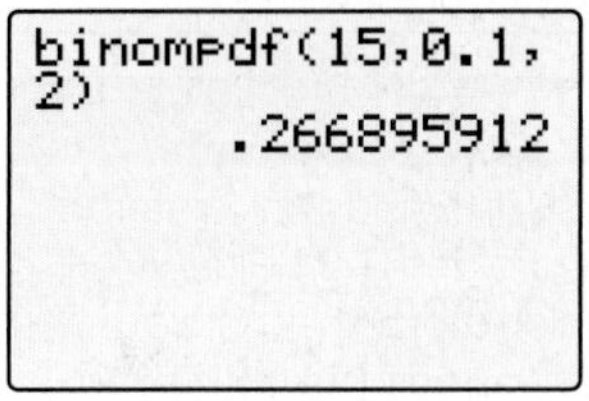

Let's change up the last example a little and instead of looking for the probability of getting exactly two defective batteries, find the probability of having at most two defective batteries in the sample.

Example 5.7

Calculating Binomial Probabilities Using the Formula or a TI-83/84 Plus Calculator

A quality control expert at a large factory estimates that 10% of all the batteries produced at the factory are defective. If he takes a random sample of fifteen batteries, what is the probability that no more than two are defective?

Solution

This scenario is the same as in the previous example; therefore, we know that we have a binomial distribution with $n = 15$ and $p = 0.1$. This time we want the probability that *no more than two* are defective, which is $P(X \le 2)$. Thus we are looking for the probability that $X = 0$, or $X = 1$, or $X = 2$. We can find $P(X \le 2)$ by adding these three individual probabilities.

$$P(X \le 2) = P(X = 0) + P(X = 1) + P(X = 2)$$

Using the binomial probability formula for our solution would require us to calculate the following expression.

$$\begin{aligned} P(X \le 2) &= P(X = 0) + P(X = 1) + P(X = 2) \\ &= {}_{15}C_0 (0.1)^0 (0.9)^{15} + {}_{15}C_1 (0.1)^1 (0.9)^{14} + {}_{15}C_2 (0.1)^2 (0.9)^{13} \\ &\approx 0.8159 \end{aligned}$$

A TI-83/84 Plus calculator can calculate $P(X = x)$ directly as seen in the previous example. Thus, using a TI-83/84 Plus, we would calculate the probability as shown below and in the screenshot in the margin.

$P(X \le 2) = P(X = 0) + P(X = 1) + P(X = 2)$

= `binompdf(15,Ø.1,Ø)+binompdf(15,Ø.1,1)+binompdf(15,Ø.1,2)`

≈ 0.8159

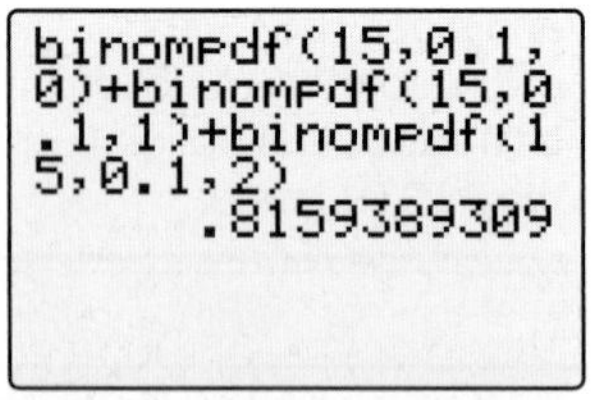

However, the TI-83/84 Plus calculator allows us to use an even more efficient method for this particular problem, as it will also directly calculate a cumulative probability, $P(X \le x)$.

- Press 2ND and then VARS to access the `DISTR` menu.
- Choose option `B:binomcdf(`.
- Enter n, p, and x in the parentheses as: `binomcdf(`n, p, x`)`.

Side Note

On TI-83 Plus calculators and older TI-84 Plus calculators, the function `binomcdf(` is option A on the `DISTR` menu.

So, using this more efficient method, we would calculate the probability as shown below and in the screenshot in the margin.

$P(X \le x) =$ `binomcdf(`n, p, x`)`

$P(X \le 2) =$ `binomcdf(15,Ø.1,2)`

≈ 0.8159

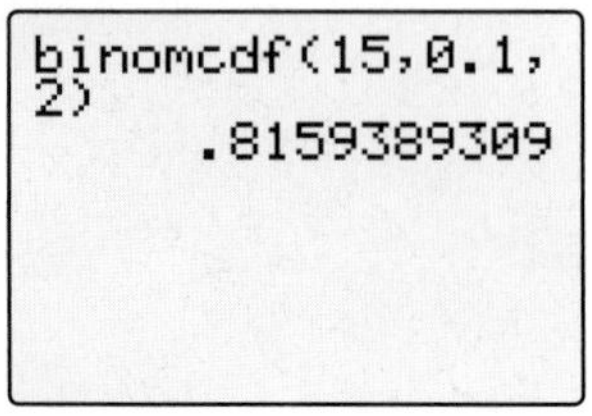

Therefore, the probability that no more than two out of the fifteen batteries are defective is approximately 0.8159.

Example 5.8

Calculating a Cumulative Binomial Probability Using a TI-83/84 Plus Calculator

What is the probability that a family with six children has more than two girls? Assume that the gender of one child is independent of the gender of any of the other children.

Memory Booster

Note the difference in commands on a TI-83/84 Plus calculator.

`binompdf(` is for a **single** binomial probability: $P(X = x)$

`binomcdf(` is for a **cumulative** binomial probability: $P(X \le x)$

Solution

First, let's verify that this scenario meets the criteria of a binomial distribution. We are told that the gender of each child is independent, so we can consider the births of the children to be identical, independent trials. The family has six children, so the number of trials is $n = 6$. For each trial, there are two possible outcomes: the child is either a girl or a boy. Let's define a success as having a girl. We can assume that both genders are equally likely; thus the probability of obtaining a success is $p = 0.5$. Let X = the number of girls out of the six children. We are considering the event of having *more than two* girls, $X > 2$. This is the complement to the event of having no more than two girls, $X \le 2$. Thus we can calculate the binomial probability by using the Complement Rule, as follows.

$$P(X > 2) = 1 - P(X \le 2)$$

Using the binomial formula for this problem would be cumbersome, so let's use a TI-83/84 Plus calculator. We would enter the probability expression as shown below and in the screenshot in the margin.

$$\begin{aligned} P(X > 2) &= 1 - P(X \le 2) \\ &= \texttt{1-binomcdf(6,0.5,2)} \\ &\approx 0.6563 \end{aligned}$$

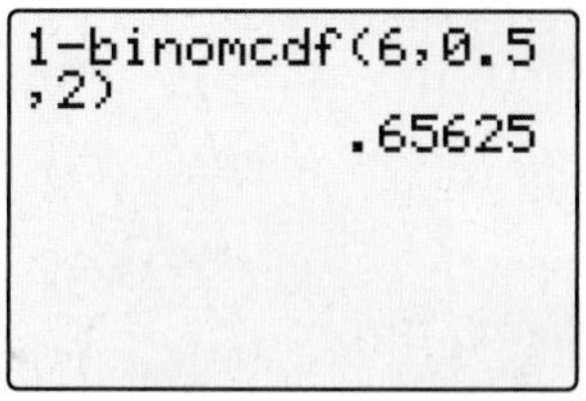

Therefore, the probability that a family with six children has more than two girls is approximately 0.6563.

Example 5.9

Calculating a Cumulative Binomial Probability Using a TI-83/84 Plus Calculator

Suppose that 20% of the programs sold at the home games of a professional sports team during the course of one season contain a special discount coupon. If all eight friends in your group bought programs at one game, what is the probability that at least half of your friends received the discount coupon?

Solution

Since there are a significant number of programs sold at the home games of a professional sports team throughout one season, and the number of trials we are considering is relatively small in comparison, we can model this situation as if the trials are independent. The reasoning is that the precise probabilities would not actually change enough to affect the value of the answer. We are considering eight programs; thus $n = 8$. If we define a success to be receiving a discount coupon, then the probability of obtaining a success is $p = 0.2$. Let X = the number of discount coupons received in the eight programs bought by your friends. We are interested in the probability that *at least half of the eight* friends get a discount coupon, so at least four out of the eight, or $P(X \ge 4)$. As in the previous example, in order to use the cumulative binomial probability function on a TI-83/84 Plus calculator to solve this problem, we will need to use the Complement Rule.

$$P(X \geq 4) = 1 - P(X < 4)$$

This is still not exactly what we need because the TI-83/84 Plus calculator can only calculate cumulative probabilities of the form $P(X \leq x)$. Fortunately, this situation is not too difficult to deal with due to one of the characteristics of the binomial distribution. The value for x must be a whole number; therefore, $P(X < 4) = P(X \leq 3)$.

Using all of this information and a TI-83/84 Plus calculator, we calculate the probability as shown below and in the screenshot in the margin.

$$\begin{aligned} P(X \geq 4) &= 1 - P(X < 4) \\ &= 1 - P(X \leq 3) \\ &= \texttt{1-binomcdf(8,0.2,3)} \\ &\approx 0.0563 \end{aligned}$$

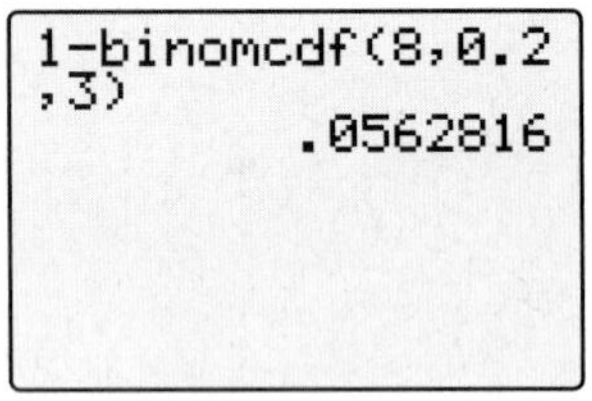

Therefore, the probability that at least half of the eight friends find discount coupons in their programs is approximately 0.0563, which indicates that it is not very likely.

While the binomial formula and different technologies are nice tools for finding binomial probabilities, binomial tables can be used as well. One drawback to these tables is that there are a limited number of probability values available. Thus, you must still use the formula or appropriate technology if the necessary probability is not in the table. We will look briefly at two binomial tables, the standard binomial table and the cumulative binomial table. The standard binomial table gives the probability for a specific number of successes. For example, if we want to know the probability of getting exactly six heads in ten coin tosses (as in Example 5.5), we could use the standard binomial table, Table D in Appendix A.

In contrast, the cumulative binomial table gives the cumulative probability up to and including a certain number of successes. That is, the cumulative binomial table gives probabilities of the form $P(X \leq x)$. For example, suppose we want to know the probability of getting six or fewer heads in ten coin tosses. For this problem, we need to use the cumulative binomial table, Table E in Appendix A, because it will give the cumulative probability up to and including six successes, $P(X \leq 6)$.

Both types of binomial tables are read the same way, so after you have decided which table is best for the problem, you need to know how to read the table. Within the binomial tables in Appendix A, there is a table for each value of n between 1 and 20, inclusive, so the first step is to choose the table corresponding to the number of trials, n, for your problem. Next, find the probability of success for any trial, p, in the top row of the table and the number of successes, x, in the second column. The cell where the row and column intersect gives the single or cumulative binomial probability. Let's walk through one complete example of using a binomial table.

Example 5.10

Finding a Binomial Probability Using a Table

What is the probability of rolling a die ten times and obtaining odd digits on eight of the rolls?

Solution

We know that this process can be modeled by a binomial distribution since the ten rolls of a die are identical, independent trials. In the context of this scenario, each roll has two possible outcomes: an odd number or an even number. Let X = the number of odd digits obtained in ten rolls of a die. We see for this problem that there are ten trials ($n = 10$) and exactly eight successes ($x = 8$), since we consider rolling an odd digit to be a success. Three of the six numbers on the die are odd, so the probability of rolling an odd digit for any one roll is $\frac{3}{6} = \frac{1}{2} = 0.5$. Thus $p = 0.5$. We have the three values we need, so we are now ready to find the probability in the table. Which table should we use? Because we are given a specific value for x, rather than a range of values, we need to use the standard binomial table. We choose the table for $n = 10$ and then look for the cell where the $p = 0.5$ column and $x = 8$ row meet. We then see that the probability of getting eight odd numbers in ten rolls is 0.0439.

						p				
n	***x***	**0.1**	**0.2**	**0.3**	**0.4**	**0.5**	**0.6**	**0.7**	**0.8**	**0.9**
10	**0**	0.3487	0.1074	0.0282	0.0060	0.0010	0.0001	0.0000	0.0000	0.0000
	1	0.3874	0.2684	0.1211	0.0403	0.0098	0.0016	0.0001	0.0000	0.0000
	2	0.1937	0.3020	0.2335	0.1209	0.0439	0.0106	0.0014	0.0001	0.0000
	3	0.0574	0.2013	0.2668	0.2150	0.1172	0.0425	0.0090	0.0008	0.0000
	4	0.0112	0.0881	0.2001	0.2508	0.2051	0.1115	0.0368	0.0055	0.0001
	5	0.0015	0.0264	0.1029	0.2007	0.2461	0.2007	0.1029	0.0264	0.0015
	6	0.0001	0.0055	0.0368	0.1115	0.2051	0.2508	0.2001	0.0881	0.0112
	7	0.0000	0.0008	0.0090	0.0425	0.1172	0.2150	0.2668	0.2013	0.0574
	8	0.0000	0.0001	0.0014	0.0106	0.0439	0.1209	0.2335	0.3020	0.1937
	9	0.0000	0.0000	0.0001	0.0016	0.0098	0.0403	0.1211	0.2684	0.3874
	10	0.0000	0.0000	0.0000	0.0001	0.0010	0.0060	0.0282	0.1074	0.3487

Memory Booster

When using the binomial tables, precision is limited to the choices for the probability, *p*, given in the tables.

5.2 Section Exercises

Properties of a Binomial Distribution

Directions: **Determine whether the given procedure meets the criteria of a binomial distribution. If not, identify at least one requirement that is not satisfied.**

1. A survey of college students rating the food in the campus dining hall on a scale from 1–10.

2. Spinning a roulette wheel nine times and recording the number the ball lands on.

3. Spinning a roulette wheel nine times and recording the number of times the ball lands on black.

4. Surveying 124 people living in the United States who use Internet service and recording their "no" responses to the question: "Do you think that Internet sites should be federally regulated?"

Probability for Binomial Distributions

Directions: Assume that the random variable X has a binomial distribution with the given probability of obtaining a success. Find each specified probability, given the number of trials.

5. $P(X=3),\quad n=5,\quad p=0.4$

6. $P(X=4),\quad n=10,\quad p=0.3$

7. $P(X\le 8),\quad n=12,\quad p=0.1$

8. $P(X\le 2),\quad n=3,\quad p=0.9$

9. $P(X<7),\quad n=18,\quad p=0.4$

10. $P(X<9),\quad n=17,\quad p=0.5$

11. $P(X>3),\quad n=4,\quad p=0.8$

12. $P(X>5),\quad n=10,\quad p=0.7$

13. $P(X\ge 6),\quad n=7,\quad p=0.2$

14. $P(X\ge 8),\quad n=15,\quad p=0.6$

Directions: Find each specified probability for the given scenario.

15. Suppose that the probability of Thad making a free throw in the championship basketball game is 60%, and each throw is independent of his last throw. Assume that Thad attempts seven free throws during the game.

- **a.** What is the probability that he will make more than four of his free throws?
- **b.** What is the probability that he will make all of his free throws?

16. At one large university, freshmen account for 30% of the student body.

- **a.** If a group of twelve students is randomly chosen by the school newspaper to comment on textbook prices, what is the probability that fewer than three of the students are freshmen? (Assume that this situation can be modeled using a binomial distribution.)
- **b.** If a group of ten students is randomly chosen by the school newspaper to comment on textbook prices, what is the probability that more than three of the students are freshmen? (Assume that this situation can be modeled using a binomial distribution.)

17. The SugarBear Candy Factory makes two types of chocolate candy bars—milk chocolate and milk chocolate with almonds. In a typical day, 60% of the candy bars being made are milk chocolate, and the rest are milk chocolate with almonds. At the end of the day, a quality control expert randomly chooses 14 chocolate bars for inspection.

- **a.** What is the probability that half of them contain almonds?
- **b.** What is the probability that at least half of them contain almonds?

18. Suppose that the probability of your favorite baseball player getting a hit at each at bat is 0.30. Assume that each at bat is independent of any other at bat.

- **a.** What is the probability that he bats six times and gets fewer than two hits?
- **b.** What is the probability that he bats nine times and gets at most four hits?

19. Suppose that Carlos is taking a multiple-choice test where there are five answer choices for each question, and he randomly guesses on four questions.

a. What is the probability that he gets exactly three of those questions correct?

b. What is the probability that he gets at least one out of the four questions correct?

c. What is the probability that he gets none of the four questions correct?

20. In a pediatrician's office, the probability of a "no show" for any checkup appointment on any given day is 1 out of 10. Suppose that there are 18 appointments scheduled for one day.

a. Find the probability that fewer than 4 don't show.

b. Find the probability that at least 2 don't show.

c. Find the probability that the doctor sees every patient scheduled.

21. The probability of any plant surviving in Kerry's garden is 0.8. Suppose she plants 19 new plants this year.

a. What is the probability that at least 5 of them survive?

b. What is the probability that more than $\frac{3}{4}$ of them survive?

c. What is the probability that she has a bad year and none of them survive?

22. In a national park in Alaska, there are 100 polar bears. As part of the monitoring of the park, the rangers caught 20 bears, tagged them and released them back into the park during the first year. A year later they caught 13 bears. (Assume that the bears are caught at different times, and the same bear could be caught more than once.)

a. What is the probability that 6 are tagged?

b. What is the probability that at least 10 are tagged?

c. What is the probability that none are tagged?

23. In a standard deck of 52 cards, 12 are face cards (king, queen, jack). Assume that 5 cards are selected with replacement out of a well-shuffled deck.

a. What is the probability of getting exactly 3 face cards?

b. What is the probability of getting at least 1 face card?

24. Underneath each cap of Brand X cola bottles is a chance to win a free cola. Suppose that the probability of winning is 1 out of 11 and you buy 16 colas.

a. What is the probability that you win at least once?

b. What is the probability that you don't win at all?

c. What is the probability that you win with half of the bottles?

25. Ronnie owns a fireworks stand and knows that in the fireworks business, 1 out of every 13 fireworks is a dud. Suppose that Juanita buys 10 firecrackers at Ronnie's stand.

a. What is the probability that no more than 3 are duds?

b. What is the probability that she has a perfect display with no duds?

c. What is the probability that more than half are duds?

5.3 Poisson Distribution

The next distribution that we will learn about is the **Poisson distribution**, pronounced "pwah-**sohn**." The Poisson distribution is a peculiar, though interesting, distribution. The Poisson distribution is similar to the binomial in that for the Poisson distribution, we are again looking to count the number of "successes" obtained. The major difference between the Poisson distribution and the binomial distribution is that for the Poisson distribution, there is not a set number of trials in which the successes must occur. Instead, the Poisson random variable, X, counts the number of successes in a given interval, such as a period of time. For example, a Poisson distribution could be used for problems involving such scenarios as the number of defects on a length of copper wiring or the number of calls into one company's computer tech support line one evening. Thus, a random variable distributed according to the Poisson distribution may take on infinitely many values, but these values are determined by counting. Therefore, the Poisson distribution is a discrete probability distribution.

Properties

Properties of a Poisson Distribution

1. Each success must be independent of any other successes.
2. The Poisson random variable, X, counts the number of successes in the given interval.
3. The mean number of successes in a given interval must remain constant.
4. For a Poisson distribution, the mean and variance are given by $\mu = \sigma^2 = \lambda$, where λ is the mean number of successes in a given interval.

Memory Booster

$E(X) = \mu$
for a random variable X.

If these conditions are met, then the following Poisson distribution formula can be used to find the probability of x successes occurring during the given interval, such as a period of time.

Math Symbols

λ: mean number of successes in one interval; Greek letter lambda

Formula

Probability for a Poisson Distribution

For a Poisson random variable X, the probability of obtaining x successes in any particular interval is given by

$$P(X = x) = \frac{e^{-\lambda}\lambda^x}{x!}$$

where x = the number of successes,

$e \approx 2.718282$, and

λ = the mean number of successes in each interval.

Rounding Rule

When calculating probabilities for Poisson distributions, round to four decimal places. This follows the convention used in the Poisson probability distribution tables in Appendix A.

Notice that there is only one parameter that must be determined for a Poisson distribution and it is λ (pronounced "lam-duh"). The parameter λ gives the mean number of successes for the given interval. One caution when you are working a Poisson problem is that you must make sure that your calculation for λ is in the same unit of measurement as the question. For example, suppose you are told that the local barber finishes one haircut every fifteen minutes. If you are asked to find the probability that he will finish six haircuts in one hour, you must calculate the average number of haircuts per hour, not per minute. In this case, you must convert 1 haircut per 15 minutes to the number of haircuts per hour as follows.

Memory Booster

Make sure that your calculation for λ is in the same unit of measurement as the interval in the question.

$$\frac{1 \text{ haircut}}{15 \text{ minutes}} \cdot \frac{60 \text{ minutes}}{1 \text{ hour}} = \frac{4 \text{ haircuts}}{1 \text{ hour}}$$

Thus $\lambda = 4$.

Example 5.11

Calculating a Poisson Probability Using the Formula

Calculate the probability that our barber is feeling extra-speedy one day and finishes six haircuts in one hour. Recall that he usually finishes one haircut every fifteen minutes.

Solution

First, we need to check that the conditions of the Poisson distribution are met, namely independence of successes and the mean number of successes per interval remaining constant. Finishing one haircut is independent of finishing any other haircut. We also know that the average is usually a constant value of one haircut every fifteen minutes. Thus, this scenario can be modeled by a Poisson distribution. Let X = the number of haircuts finished in one hour. We are looking for the probability of the barber finishing *exactly six* haircuts in one hour, so $x = 6$. We already determined that the barber averages four haircuts per hour and thus $\lambda = 4$. We have the two values needed for the Poisson probability formula, so we can substitute into the formula as follows.

$$\begin{aligned} P(X = x) &= \frac{e^{-\lambda}\lambda^x}{x!} \\ P(X = 6) &= \frac{e^{-4} \cdot 4^6}{6!} \\ &\approx \frac{(0.018316)(4096)}{720} \\ &\approx 0.1042 \end{aligned}$$

We see that the probability that he finishes six haircuts in one hour is 0.1042 or 10.42%.

Just as we learned in the previous section, using a TI-83/84 Plus calculator can simplify the process of calculating probabilities. The technology section at the end of this chapter describes how to use various technologies to calculate Poisson probabilities, but in these next few examples, we will demonstrate how to use a TI-83/84 Plus calculator to perform calculations, in addition to using the formula.

Example 5.12

Calculating a Poisson Probability Using the Formula or a TI-83/84 Plus Calculator

A popular accounting office takes in an average of three new tax returns per day during tax season. What is the probability that on a given day during tax season the firm will take in just one new tax return?

Solution

Again, to use the Poisson distribution we need to check for independence of successes and consistent mean number of successes per interval. One person bringing in a new tax return to the firm is independent of any other new tax return coming in. Also, we know that the average per day is three new tax returns, thus it is a constant. So this scenario can be modeled by a Poisson distribution. Let X = the number of new tax returns taken in by the firm in one day. We will consider obtaining a new tax return to be a success. Because we are looking for the probability that *exactly one* new return will come in, we are looking for one success, $x = 1$. The business averages three new tax returns each day, so $\lambda = 3$. Substituting the values of x and λ into the Poisson probability formula, we have the following.

$$P(X = x) = \frac{e^{-\lambda}\lambda^x}{x!}$$

$$P(X = 1) = \frac{e^{-3} \cdot 3^1}{1!}$$

$$\approx 0.1494$$

Use a TI-83/84 Plus calculator as follows.

- Press **2ND** and then **VARS** to access the `DISTR` menu.
- Choose option `C:poissonpdf(`.
- Enter λ and x in the parentheses as: `poissonpdf(`λ, x`)`.

Side Note

On TI-83 Plus calculators and older TI-84 Plus calculators, the function `poissonpdf(` is option `B` on the `DISTR` menu.

Thus, using a TI-83/84 Plus calculator, we would calculate the probability as shown below and in the screenshot in the margin.

$$P(X = x) = \texttt{poissonpdf}(\lambda, x)$$

$$P(X = 1) = \texttt{poissonpdf(3,1)}$$

$$\approx 0.1494$$

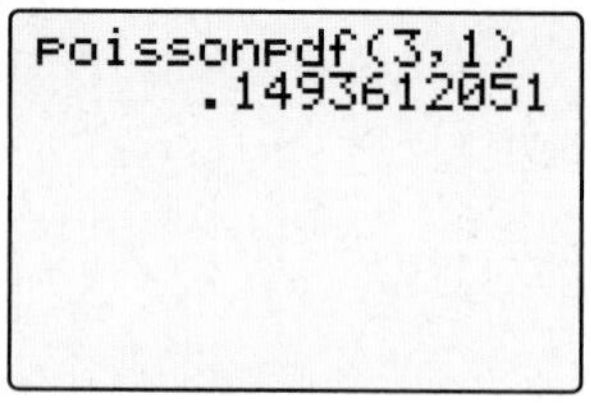
poissonpdf(3,1)
.1493612051

Therefore, the probability of the business getting just one new tax return on a given day during tax season is 0.1494.

Example 5.13

Calculating a Poisson Probability Using the Formula or a TI-83/84 Plus Calculator

Suppose that the dial-up Internet connection at your house goes out an average of 0.9 times every hour. If you plan to be connected to the Internet for three hours one afternoon, what is the probability that you will stay connected the entire time?

Solution

We will need to consider a disconnection to be a success for the purpose of this problem. With this in mind, note that each disconnection is independent of any other disconnection and the average rate of disconnections per hour is constant. Thus, this scenario can be modeled by a Poisson distribution. Let X = the number of Internet disconnections in the given three-hour period. We are looking for the probability that *no* successes occur over the course of three hours; thus $x = 0$. What is λ? The dial-up connection averages 0.9 disconnections each hour, so for three hours we multiply to get $\lambda = 0.9 \cdot 3 = 2.7$. We can substitute the values for x and λ into the Poisson probability formula as follows.

$$P(X = x) = \frac{e^{-\lambda}\lambda^x}{x!}$$

$$P(X = 0) = \frac{e^{-2.7}(2.7)^0}{0!}$$

$$\approx 0.0672$$

Using a TI-83/84 Plus calculator, we would enter the values as shown below and in the screenshot in the margin.

$P(X = x) =$ `poissonpdf(`λ, x`)`

$P(X = 0) =$ `poissonpdf(2.7,Ø)`

≈ 0.0672

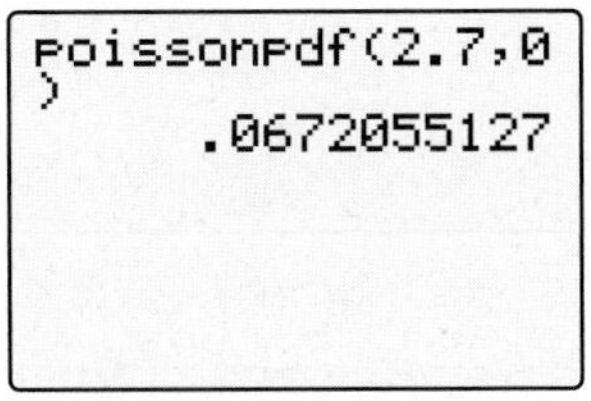
poissonpdf(2.7,0
)
.0672055127

Thus, the probability of staying connected for all three hours is 0.0672 or 6.72%.

Example 5.14

Calculating Poisson Probabilities Using the Formula or a TI-83/84 Plus Calculator

A fast-food restaurant averages one incorrect order every three hours. What is the probability that the restaurant will get no more than three orders wrong on any given day between 5 p.m. and 11 p.m.? Assume that the number of incorrect orders follows a Poisson distribution.

Solution

We will consider a wrong order a success for this example. Let X = the number of incorrect orders that occur between 5 p.m. and 11 p.m. on a given day. We are looking for the probability of getting *no more than three* successes, which we can write as $P(X \le 3)$. To find λ, we need to calculate the average number of incorrect orders that occur in a six-hour period. This is twice the length of a three-hour period, so we multiply the number of wrong orders for a three-hour period by two: $\lambda = 1 \cdot 2 = 2$. We must find the probability that X equals 0, 1, 2, *or* 3. We will use the Poisson formula to find each individual probability, and then add these probabilities together as shown below.

$$P(X \le 3) = P(X = 0) + P(X = 1) + P(X = 2) + P(X = 3)$$

$$= \frac{e^{-2} \cdot 2^0}{0!} + \frac{e^{-2} \cdot 2^1}{1!} + \frac{e^{-2} \cdot 2^2}{2!} + \frac{e^{-2} \cdot 2^3}{3!}$$

$$\approx 0.8571$$

Using a TI-83/84 Plus calculator, we would calculate the probability as shown below and in the screenshot in the margin.

$P(X \le 3) = P(X = 0) + P(X = 1) + P(X = 2) + P(X = 3)$

$=$ `poissonpdf(2,Ø)+poissonpdf(2,1)`

`+poissonpdf(2,2)+poissonpdf(2,3)`

≈ 0.8571

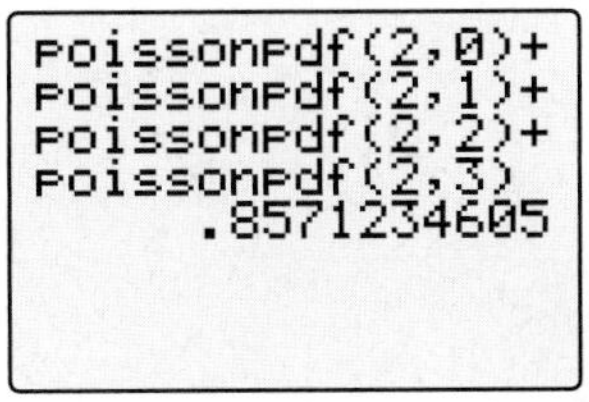
poissonpdf(2,0)+
poissonpdf(2,1)+
poissonpdf(2,2)+
poissonpdf(2,3)
.8571234605

However, just as the TI-83/84 Plus calculator can calculate a cumulative binomial probability, it can calculate a cumulative Poisson probability as well.

- Press 2ND and then VARS to access the `DISTR` menu.
- Choose option `D:poissoncdf(`.
- Enter λ and x in the parentheses as: `poissoncdf(`λ, x`)`.

Side Note

On TI-83 Plus calculators and older TI-84 Plus calculators, the function `poissoncdf(` is option `C` on the `DISTR` menu.

Using the cumulative Poisson distribution on the TI-83/84 Plus calculator simplifies our calculation as shown below and in the screenshot in the margin.

$$P(X \le x) = \texttt{poissoncdf}(\lambda, x)$$

$$P(X \le 3) = \texttt{poissoncdf}(2,3)$$

$$\approx 0.8571$$

poissoncdf(2,3)
.8571234606

Thus, the fast-food restaurant has an 85.71% chance of getting no more than three orders wrong on any given day between 5 p.m. and 11 p.m.

Memory Booster

Note the difference in commands on a TI-83/84 Plus calculator.

`poissonpdf(` is for a **single** Poisson probability: $P(X = x)$

`poissoncdf(` is for a **cumulative** Poisson probability: $P(X \le x)$

Example 5.15

Calculating a Cumulative Poisson Probability Using a TI-83/84 Plus Calculator

A math professor averages grading 20 exams per hour. What is the probability that she grades more than 35 of her 60 statistics exams during her uninterrupted hour and a half between classes? Assume that the number of exams graded follows a Poisson distribution.

Solution

Let's define a success to be grading an exam. Let X = the number of exams graded in 1.5 hours. We want to find the probability that *more than 35* successes occur. This probability can be written as $P(X > 35)$. We can use the Complement Rule to find the probability that we need: $P(X > 35) = 1 - P(X \le 35)$. Next, we need to find λ. We know that the average for 1 hour is 20, so the average for 1.5 hours is $\lambda = 20 \cdot 1.5 = 30$. Using the Poisson formula to calculate each of the necessary probabilities would be very time-consuming, so let's use a TI-83/84 Plus calculator as shown below and in the screenshot in the margin.

$$P(X > 35) = 1 - P(X \le 35)$$

$$= \texttt{1-poissoncdf(30,35)}$$

$$\approx 0.1574$$

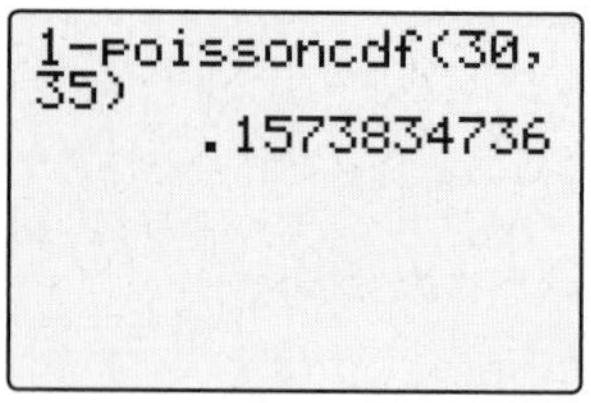
1-poissoncdf(30,
35)
.1573834736

Thus, the math professor has a 15.74% chance of getting more than 35 statistics exams graded between classes.

Example 5.16

Calculating a Cumulative Poisson Probability Using a TI-83/84 Plus Calculator

A typist averages four typographical errors per paragraph. If he is typing a five-paragraph document, what is the probability that he will make fewer than ten mistakes? Assume that the number of errors follows a Poisson distribution.

Solution

Let's define a success as making a mistake. (Yes, it does sound strange, but it is the best way to solve the problem!) Let X = the number of mistakes made in the document. If the typist averages four mistakes per paragraph, his average for five paragraphs is $4 \cdot 5 = 20$. Thus $\lambda = 20$. We are looking for the probability of *fewer than ten* mistakes, $P(X < 10)$. We need to rewrite the probability as $P(X < 10) = P(X \le 9)$ in order to use the cumulative Poisson distribution on our calculator. So we enter the probability expression as shown below and in the screenshot in the margin.

$$P(X<10) = P(X \le 9)$$
$$= \texttt{poissoncdf(20,9)}$$
$$\approx 0.0050$$

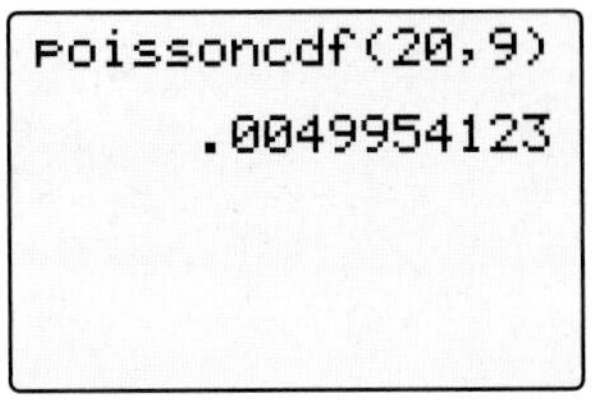
poissoncdf(20,9)
.0049954123

Thus the typist has a 0.0050 probability of making fewer than ten mistakes.

As with the binomial distribution, a Poisson distribution table can be used for certain values of λ. Otherwise, the Poisson distribution formula or technology must be used. To use the table, you must simply know the values of λ and x. The probability of obtaining x successes in the given interval will be located in the cell where the λ column and x row meet. The Poisson distribution table is Table F in Appendix A.

Example 5.17

Finding a Poisson Probability Using a Table

Suppose that a length of copper wiring averages one defect every 200 feet. What is the probability that a 300-foot stretch will have no defects?

Solution

Each defect in the wire is independent of any other defect, and the average number of defects in a given length of wire is constant. Thus, this scenario can be modeled by a Poisson distribution. Next, we must determine λ and x. If there is 1 defect on average every 200 feet, then we can expect 1.5 defects for a 300-foot stretch; thus $\lambda = 1.5$. Because we are looking for the probability of seeing *no* defects, $x = 0$. Using the Poisson table, we find that the probability that corresponds with these values of $\lambda = 1.5$ and $x = 0$ is 0.2231.

	λ										
x	**1.50**	**1.60**	**1.70**	**1.80**	**1.90**	**2.00**	**2.10**	**2.20**	**2.30**	**2.40**	**2.50**
0	0.2231	0.2019	0.1827	0.1653	0.1496	0.1353	0.1225	0.1108	0.1003	0.0907	0.0821
1	0.3347	0.3230	0.3106	0.2975	0.2842	0.2707	0.2572	0.2438	0.2306	0.2177	0.2052
2	0.2510	0.2584	0.2640	0.2678	0.2700	0.2707	0.2700	0.2681	0.2652	0.2613	0.2565
3	0.1255	0.1378	0.1496	0.1607	0.1710	0.1804	0.1890	0.1966	0.2033	0.2090	0.2138
4	0.0471	0.0551	0.0636	0.0723	0.0812	0.0902	0.0992	0.1082	0.1169	0.1254	0.1336
5	0.0141	0.0176	0.0216	0.0260	0.0309	0.0361	0.0417	0.0476	0.0538	0.0602	0.0668
6	0.0035	0.0047	0.0061	0.0078	0.0098	0.0120	0.0146	0.0174	0.0206	0.0241	0.0278
7	0.0008	0.0011	0.0015	0.0020	0.0027	0.0034	0.0044	0.0055	0.0068	0.0083	0.0099
8	0.0001	0.0002	0.0003	0.0005	0.0006	0.0009	0.0011	0.0015	0.0019	0.0025	0.0031
9	0.0000	0.0000	0.0001	0.0001	0.0001	0.0002	0.0003	0.0004	0.0005	0.0007	0.0009
10	0.0000	0.0000	0.0000	0.0000	0.0000	0.0000	0.0001	0.0001	0.0001	0.0002	0.0002

We can say that there is a 22.31% chance of finding no defects in a 300-foot section of wire.

Memory Booster

When using the Poisson tables, precision is limited to the choices given in the tables for the average number of successes in the given interval, λ, and the number of successes, x.

5.3 Section Exercises

Determining the Value of λ for a Poisson Distribution

Directions: Determine the value of λ for each scenario.

1. A fifth grader averages three grammatical errors per paragraph. What is his average for four paragraphs?

2. A surveillance officer reports two incidents of shoplifting per month, on average. How many incidents of shoplifting on average are reported per year?

3. Baggage handlers at a certain airport move an average of 1500 bags on an eight-hour shift. On average, how many bags are moved per hour?

4. An assembly line, on average, produces 1 defective part for every 100 parts that roll off the line. What is the average number of defects for a group of 20 parts?

5. You average 70 heartbeats per minute. What is the average number of heartbeats you have in 10 seconds?

Probability for Poisson Distributions

Directions: Find each specified probability. Assume that the random variable X has a Poisson distribution with the given value of λ.

6. $P(X=2), \quad \lambda=1.80$

7. $P(X=0), \quad \lambda=2.60$

8. $P(X=13), \quad \lambda=6.50$

9. $P(X=8), \quad \lambda=9.30$

10. $P(X\le 2), \quad \lambda=3.80$

11. $P(X\le 5), \quad \lambda=7.10$

12. $P(X<4), \quad \lambda=8.30$

13. $P(X<8), \quad \lambda=2.90$

14. $P(X\ge 3), \quad \lambda=4.90$

15. $P(X\ge 12), \quad \lambda=9.60$

Directions: Find each specified probability for the given scenario. Assume that each scenario can be modeled by a Poisson distribution.

16. The Oxford Gift Shop averages four sales each hour. Betty is scheduled to work the cash register from 1:00–3:00 on Saturday afternoon.

a. What is the probability that Betty rings up exactly ten customers?

b. What is the probability that Betty rings up more than ten customers?

17. The Pancake House is so popular that it boasts of selling a stack of pancakes every two minutes.

a. What is the probability that there is a ten-minute interval in which no pancakes are sold?

b. What is the probability that there is a five-minute interval in which fewer than three stacks of pancakes are sold?

18. The pizza place next to the local college receives an average of 20 pizza orders per hour during lunch.

a. In any given hour during lunch, what is the probability that the pizza place receives at least 22 pizza orders?

b. In any given hour during lunch, what is the probability that the pizza place receives more than 22 pizza orders?

19. Rob is a busy physician in the emergency room. He sees an average of four major trauma patients each night.

a. What is the probability that fewer than three major trauma patients will be admitted on any given night?

b. What is the probability that no more than five major trauma patients will be admitted on any given night?

20. Suppose that a bank drive-through serves customers at a rate of twelve cars every hour.

a. What is the probability that the bank drive-through will serve fewer than five customers in 30 minutes?

b. What is the probability that the bank drive-through will serve three customers in 15 minutes?

21. On average, Patrick sees a spider in his home once a month.

a. What is the probability that Patrick sees two spiders in a month and a half?

b. What is the probability that Patrick sees no more than one spider in a month and a half?

22. Jane cannot sell a finished piece of pottery if she discovers that the clay has a defect in it. Suppose that she has to discard 2 pieces for every 56 pieces she makes.

a. What is the probability that in 14 pieces of pottery, just 1 piece is defective?

b. What is the probability that in 28 pieces of pottery, at least 1 piece is defective?

23. Suppose that, on average, 45 books are checked out of the local public library per day.

a. What is the probability that 100 books are checked out in two days?

b. What is the probability that at most 200 books are checked out in four weeks?

24. A landscape architect knows that in the cable that he lays for landscape lighting he can expect one defect in 300 yards of cable.

a. What is the probability that in 100 yards of cable, he would find two defects?

b. What is the probability that in 200 yards of cable, he would find fewer than three defects?

5.4 Hypergeometric Distribution

As with the Poisson distribution, the hypergeometric distribution is also similar to the binomial distribution. The **hypergeometric distribution** is characterized by a given number of trials, n, and a specified number of countable successes, x, that occur within those n trials, in the same manner as the binomial distribution. However, the hypergeometric distribution is distinguished because it deals with *dependent* trials rather than independent trials.

Recall that two events, or trials, are dependent if one occurring affects the probability that the other will occur. A common example of dependent events is that of choosing cards without replacement. For example, suppose that a king is chosen out of a standard deck of cards and not replaced. The probability that a king will be chosen again on the second draw is affected by the first card choice. In fact, the chances of getting a king a second time are lowered because there is one less king in the stack. Choosing cards without replacement is just one example of dependent events. Any other experiment in which there is no replacement or repetition of objects will also be *dependent*.

What does it take for an experiment to have a hypergeometric distribution? The following properties are true for all hypergeometric distributions.

Properties

Properties of a Hypergeometric Distribution

1. Each trial consists of selecting one of the N items in the population and results in either a *success* or a *failure*. Note that this means that the population is of a *known* size.
2. The experiment consists of n trials.
3. The total number of possible successes in the entire population is k.
4. The trials are dependent (that is, selections are made without replacement).
5. The hypergeometric random variable, X, counts the number of successes in n trials.
6. For a hypergeometric distribution, the mean is given by $\mu = n \cdot \frac{k}{N}$ and the variance is given by $\sigma^2 = \frac{nk(N-k)(N-n)}{N^2(N-1)}$.

Memory Booster

$E(X) = \mu$
for a random variable X.

If a given scenario displays all of the properties above, then the following formula can be used to determine the probability of x successes occurring in n trials.

Formula

Probability for a Hypergeometric Distribution

For a hypergeometric random variable X, the probability of obtaining x successes in n dependent trials is given by

$$P(X=x)=\frac{\left({}_kC_x\right)\left({}_{N-k}C_{n-x}\right)}{\left({}_NC_n\right)}$$

where N is the number of items in the entire population,

n is the number of trials,

k is the number of successes in the entire population, and

x is the number of successes obtained in n trials.

Rounding Rule

When calculating probabilities for hypergeometric distributions, round to four decimal places.

For example, let's suppose that you bring a cooler full of soft drinks to a football tailgate party. You have packed 12 cans of regular soda and 6 cans of diet soda scattered randomly throughout the cooler. If you grab 3 drinks for your friends without looking, what is the probability that exactly 2 of the drinks will be regular sodas?

Notice that this is a hypergeometric problem because you are selecting from a finite population without replacement. That is, the drink you pull out first affects the probability of which drink you will pull out second. We now need to determine all of the necessary variables. There are $12+6=18$ drinks in the cooler, so the entire population size is $N=18$. Just 3 drinks are actually taken out of the cooler, making the number of trials $n=3$. Let $X=$ the number of regular sodas out of the 3 drinks chosen. For this example, we will consider a regular soda to be a success. Because there are 12 regular sodas in the cooler, there are 12 successes in the entire population, so $k=12$. We are looking for the probability of getting exactly 2 regular sodas, so $x=2$. Now we simply substitute these values in the hypergeometric probability formula as follows.

$$\begin{aligned}
P(X=x)&=\frac{\left({}_kC_x\right)\left({}_{N-k}C_{n-x}\right)}{{}_NC_n}\\
P(X=2)&=\frac{\left({}_{12}C_2\right)\left({}_{18-12}C_{3-2}\right)}{{}_{18}C_3}\\
&=\frac{\left({}_{12}C_2\right)\left({}_6C_1\right)}{{}_{18}C_3}\\
&=\frac{\left(\overset{33}{\cancel{66}}\right)\left(\cancel{6}\right)}{\underset{\underset{68}{\cancel{136}}}{\cancel{816}}}\\
&=\frac{33}{68}\\
&\approx 0.4853
\end{aligned}$$

We see that there is a 48.53% chance that you will pull exactly two regular sodas out of the cooler when you grab three drinks without looking.

Unlike the binomial and Poisson distributions, there is neither a table in the back of this textbook for easy reference nor a preprogrammed function on the TI-83/84 calculator for the hypergeometric distribution. This means that we will have to use the formula for calculating the probability for this distribution.

Example 5.18

Calculating a Hypergeometric Probability

At the local grocery store there are twenty boxes of cereal on one shelf, half of which contain a prize. Suppose that you buy three boxes of cereal. What is the probability that all three boxes contain a prize?

Solution

There are twenty boxes total, so the entire population contains $N = 20$ boxes. Three boxes are purchased and each is considered a trial, so there are a fixed number of trials ($n = 3$). Since we would not buy a box of cereal and then put it back on the shelf, we can assume that these trials are performed without replacement, and thus the trials are dependent. For each trial, there are two possible outcomes: the box of cereal either contains a prize or it does not. Let $X =$ the number of boxes with prizes out of the three boxes purchased. A box with a prize is considered a success, and we are looking for the probability that *all three* trials are successes, so $x = 3$. Because half of the 20 boxes on the shelf have prizes, we know that there are 10 possible successes in the entire population and $k = 10$. Therefore, we can conclude that this process meets the criteria of a hypergeometric distribution. Substituting into the hypergeometric probability formula, we have the following.

$$\begin{aligned}
P(X = x) &= \frac{\left({}_{k}C_{x}\right)\left({}_{N-k}C_{n-x}\right)}{{}_{N}C_{n}} \\
P(X = 3) &= \frac{\left({}_{10}C_{3}\right)\left({}_{20-10}C_{3-3}\right)}{{}_{20}C_{3}} \\
&= \frac{\left({}_{10}C_{3}\right)\left({}_{10}C_{0}\right)}{{}_{20}C_{3}} \\
&= \frac{(120)(1)}{1140} \\
&= \frac{2}{19} \\
&\approx 0.1053
\end{aligned}$$

Thus, $P(X = 3) \approx 0.1053$. This means that you have a 10.53% chance of getting a prize in all three boxes of cereal.

Example 5.19

Calculating a Hypergeometric Probability

There are seven yellow and nine green marbles in a bag. If five marbles are chosen at random without replacement, what is the probability that exactly three of the marbles chosen will be yellow?

Solution

There are 7 + 9 = 16 total marbles in the bag, so the entire population contains $N = 16$ marbles. Because five marbles are to be taken out and their colors noted, the number of trials is $n = 5$. The marbles are drawn without replacement, so these trials are dependent. For each trial, there are two possible outcomes: either the marble chosen is yellow or it is green. We will consider a success to be obtaining a yellow marble. Out of the entire population, seven marbles are yellow, so the total number of successes in the entire population is $k = 7$. Thus, we can conclude that this process follows a hypergeometric distribution. Let X = the number of yellow marbles out of the five marbles chosen. We are looking for the probability that *exactly three* trials are successes, so $x = 3$. We then have the following.

$$P(X = x) = \frac{\left({}_{k}C_{x}\right)\left({}_{N-k}C_{n-x}\right)}{{}_{N}C_{n}}$$

$$\begin{aligned} P(X = 3) &= \frac{\left({}_{7}C_{3}\right)\left({}_{16-7}C_{5-3}\right)}{{}_{16}C_{5}} \\ &= \frac{\left({}_{7}C_{3}\right)\left({}_{9}C_{2}\right)}{{}_{16}C_{5}} \\ &= \frac{\left(\overset{5}{\cancel{35}}\right)\left(\overset{3}{\cancel{36}}\right)}{\underset{\underset{52}{\cancel{364}}}{\cancel{4368}}} \\ &= \frac{15}{52} \\ &\approx 0.2885 \end{aligned}$$

The probability of getting exactly three yellow marbles when choosing five marbles without replacement is $P(X = 3) \approx 0.2885$, or 28.85%.

Example 5.20

Calculating a Cumulative Hypergeometric Probability

A shipment of 25 light bulbs contains 3 defective bulbs. If 5 bulbs are selected randomly without replacement, what is the probability that fewer than 2 of the bulbs chosen are defective?

Solution

There are 25 light bulbs in the entire population, so $N = 25$. Five bulbs are to be selected without replacement and checked for defects, so the trials are dependent and the number of trials is $n = 5$. For each trial, there are two possible outcomes: either the bulb chosen is defective or it is not. A success for this problem is a defective light bulb, and since there are 3 defective bulbs in the whole shipment, $k = 3$. Thus, we can conclude that this process follows a hypergeometric distribution. Let $X =$ the number of defective bulbs out of the 5 bulbs selected. Because we are looking for the probability that *fewer than 2* successes occur, we want to find $P(X < 2)$. This example is different from the others because we need to calculate the probability that fewer than 2 light bulbs chosen are defective. To do this, we must remember that the numbers of successes have to be whole numbers. This means that we can apply the Addition Rule for Probability of Mutually Exclusive Events and write the probability as $P(X < 2) = P(X = 0) + P(X = 1)$. We use the hypergeometric probability formula to obtain the following.

$$\begin{aligned} P(X<2) &= P(X=0)+P(X=1) \\ &= \frac{({}_3C_0)({}_{25-3}C_{5-0})}{{}_{25}C_5} + \frac{({}_3C_1)({}_{25-3}C_{5-1})}{{}_{25}C_5} \\ &= \frac{({}_3C_0)({}_{22}C_5)}{{}_{25}C_5} + \frac{({}_3C_1)({}_{22}C_4)}{{}_{25}C_5} \\ &= \frac{(1)(26{,}334)}{53{,}130} + \frac{(3)(7315)}{53{,}130} \\ &= \frac{48{,}279}{53{,}130} \\ &= \frac{209}{230} \\ &\approx 0.9087 \end{aligned}$$

The probability of getting fewer than 2 defective light bulbs when selecting 5 bulbs randomly without replacement is 90.87%.

Example 5.21

Calculating Hypergeometric Probabilities

A produce distributor is carrying nine boxes of Granny Smith apples and eight boxes of Golden Delicious apples. If four boxes are randomly delivered to a local market, what is the probability that at least three of the boxes delivered contain Golden Delicious apples?

Solution

The truck is carrying a total of $9 + 8 = 17$ boxes, which means that the entire population contains $N = 17$ boxes. Four of these boxes are actually delivered, so $n = 4$. Note that the boxes being delivered are chosen without replacement; thus these are dependent trials. For each trial, there are two possible outcomes: either the box delivered contains Granny Smith apples or it contains Golden Delicious apples. We will consider a box of Golden Delicious apples to be a success, and there are $k = 8$ successes in the entire population. Thus, we can conclude that this process meets the criteria for a hypergeometric distribution. Let $X =$ the number of boxes of Golden Delicious apples out of the four boxes delivered. We want to calculate the probability that *at least three* successes occur, so we want to find $P(X \geq 3)$. However, if at least three of the four boxes delivered are successes, then it is possible that either three or four successes occur. We then have the following.

$$
\begin{aligned}
P(X \geq 3) &= P(X = 3) + P(X = 4) \\
&= \frac{({}_8C_3)({}_{17-8}C_{4-3})}{{}_{17}C_4} + \frac{({}_8C_4)({}_{17-8}C_{4-4})}{{}_{17}C_4} \\
&= \frac{({}_8C_3)({}_9C_1)}{{}_{17}C_4} + \frac{({}_8C_4)({}_9C_0)}{{}_{17}C_4} \\
&= \frac{(56)(9)}{2380} + \frac{(70)(1)}{2380} \\
&= \frac{574}{2380} \\
&= \frac{41}{170} \\
&\approx 0.2412
\end{aligned}
$$

The probability that at least three boxes of Golden Delicious apples are delivered to the market is 24.12%.

Comparison of Several Discrete Probability Distributions

In this chapter, we have discussed three different discrete probability distributions. A summary of the three distributions is shown in Table 5.4. This is by no means an exhaustive list of the possible discrete probability distributions; it is simply a sample to demonstrate how to work with different types of discrete probability distributions.

Table 5.4: Comparing the Binomial, Poisson, and Hypergeometric Distributions

	Binomial	Poisson	Hypergeometric
Discrete Distribution	✓	✓	✓
Independent Trials	✓	✓	✗
Fixed Number of Trials	✓	✗	✓
Probability Formula	$P(X=x) = {}_nC_x \cdot p^x (1-p)^{(n-x)}$ x = the number of successes, n = the number of trials, and p = the probability of getting a success on any trial	$P(X=x) = \frac{e^{-\lambda}\lambda^x}{x!}$ x = the number of successes, $e \approx 2.718282$ and λ = the mean number of successes in each interval	$P(X=x) = \frac{({}_kC_x)({}_{N-k}C_{n-x})}{({}_NC_n)}$ N = the number of items in the entire population, n = the number of trials, k = the number of successes in the entire population, and x = the number of successes obtained in n trials.
Description	Use for problems with a fixed number of trials, where each trial is independent and only has two possible outcomes.	Use for problems with independent trials where successes occur in a given interval, such as a period of time.	Use for problems with a fixed number of trials, but unlike binomial distributions, each trial is dependent.

One of the most challenging parts of working with discrete probability distributions is deciding which distribution you should use to solve a particular problem. Table 5.4 should help you make this decision when working these types of problems. Let's practice choosing the correct distribution in the next example.

Example 5.22

Determining Which Discrete Probability Distribution to Use

For each scenario, determine the discrete probability distribution that should be used.

a. Based on data from an online search provider, a new company believes that its website is viewed 1500 times a week. What is the probability that in the next four weeks, the company's website will be viewed more than 5000 times?

b. Assume that the proportions of boys and girls born in Cincinnati, Ohio have remained constant for the last few years, and that 48% of babies born in Cincinnati, Ohio are girls. What is the probability that of the next 250 babies that are born, no more than 100 of them are girls?

c. Of the 500 programs printed for the University Opera, several contain coupons for the local coffee shop. Of these coupons, 100 are \$1-off coupons and 50 are for a free latte. Assuming that the different coupons are randomly disbursed throughout the programs, what is the probability that the first 50 programs that are handed out each contain a coupon for the coffee shop?

Solution

a. In this scenario, there is not a fixed number of trials. We are counting the number of times the website is viewed in a given period of time. In a fixed number of trials, a precise number of people would report either looking at the website or not looking at the website. Furthermore, the fact that one person chooses to access the website does not affect the probability of another person accessing the website. Therefore, this is a scenario with an unknown number of independent trials where successes occur in a given interval. A Poisson distribution should be used.

b. In this scenario, there is a fixed number of trials, namely $n = 250$. These trials are independent since the gender of one baby does not affect the probability for the gender of another baby born in the city. For each baby born, there are only two possible outcomes: a boy or a girl. Thus, a binomial distribution should be used.

c. In the last scenario, there is a fixed number of trials, namely $n = 50$. However, these trials are dependent since the probability of getting a coupon changes with each program that is handed out. With a fixed number of dependent trials, use a hypergeometric distribution.

5.4 Section Exercises

Probability for Hypergeometric Distributions

Directions: Find each specified probability for the given scenario. Assume that every scenario follows a hypergeometric distribution.

1. In a standard deck of 52 cards, 13 are hearts. Assume that 5 cards are selected without replacement out of a well-shuffled deck.

a. What is the probability of getting exactly 2 hearts?

b. What is the probability that all 5 cards will be hearts?

2. Suppose that one Christmas, Abby and Andrew's mother forgot to label their gifts. Out of ten wrapped presents, five are for Abby and five are for Andrew.

a. What is the probability that exactly one of the first four presents opened is for Abby?

b. What is the probability at most three of the first five gifts opened are for Andrew?

3. Suppose that 12 of the 20 azaleas for sale at a large nursery have pink flowers and the rest have red flowers. Because it is early in the season, they have not begun to bloom and you cannot yet tell what color each plant will be.

a. If eight azaleas are chosen at random without replacement, what is the probability that exactly 6 will be pink?

b. If five azaleas are chosen at random without replacement, what is the probability that none of the azaleas will be pink?

4. Eloise loves jelly beans, and the yellow ones are her favorite. One afternoon she is snacking on a bag of 18 jelly beans, 5 of which are yellow. She grabs a handful of 6 jelly beans.

a. What is the probability that more than half of the jelly beans are yellow?

b. What is the probability that fewer than 2 of the jelly beans are yellow?

5. The manager of a furniture store has just received a shipment of sofas and recliners. He knows that the order contains five sofas and nine recliners.

a. What is the probability that the first three items brought into the store are recliners?

b. What is the probability that out of the first seven items brought into the store, no more than two are sofas?

6. Suppose Audrey received a box of chocolates for Valentine's Day. Just after opening the box, she lost the paper which had the description of each chocolate on it. However, she knows that there were six truffles and five caramel candies left.

a. What is the probability that the first two chocolates she eats are both truffles?

b. What is the probability that at least one of the first three chocolates she eats is caramel?

7. Karen has 20 squares of material to use for her quilt; 8 are polka-dotted and the rest are floral.

a. What's the probability that she randomly uses all floral squares for the first 5 pieces of the quilt?

b. What's the probability that 2 out of the first 6 pieces she randomly chooses are polka-dotted?

8. Jay has ten pieces of mail to open, four of which are junk mail.

a. What is the probability that he randomly opens two pieces of mail and they are both junk mail?

b. What is the probability that he randomly opens three pieces of mail and at least two of them are junk mail?

9. Grab bags at the town festival are filled with either a coupon for a free hamburger or a coupon for a free order of french fries. Suppose there are 22 hamburger coupons left and 18 french-fry coupons left.

a. What is the probability that you and your friend both get bags with hamburger coupons in them?

b. Given that you both got hamburger coupons, what is the probability that if you each choose again, you both get french-fry coupons?

10. Suppose there are eight green tags, twelve white tags, and four red tags left to use as name tags at a conference. Tags are given out randomly at the registration desk.

a. What is the probability that the first two tags given out are red?

b. What is the probability that if part a. did happen, then more than three out of the next five tags would be white?

11. An antiques dealer has fifteen antique cedar chests for sale. Unknown to the dealer, one of these cedar chests is actually a modern reproduction. If the dealer randomly chooses three of these cedar chests for display, what is the probability that one of the cedar chests on display will be the reproduction?

12. Of the 30 pairs of size 8 jeans on the display table at a local retail store, 2 of the pairs are incorrectly labeled with the wrong size. A customer comes into the store and randomly chooses 5 pairs of jeans from the table, believing them all to be the same size.

a. What is the probability that all 5 pairs of jeans that the customer chose are the correct size?

b. What is the probability that 1 of the 5 pairs of jeans that the customer chose is incorrectly labeled?

c. What is the probability that both of the incorrectly labeled pairs of jeans are in the 5 pairs that the customer chose?

Determining Which Discrete Probability Distribution to Use

Directions: Each of the following exercises can be solved using the binomial distribution, the Poisson distribution, or the hypergeometric distribution. Begin by stating which distribution to use, and then solve the problem.

13. A bank receives an average of 8 bad checks per day. What is the probability that during a five-day bank week, the bank will receive 50 bad checks?

14. The IRS is considering auditing 250 tax returns within a given region, 15 of which have serious errors (unknown to the IRS, of course). If an IRS agent randomly chooses 25 of these 250 tax returns on which to perform an audit, what is the probability that 3 of these tax returns will have serious errors?

15. It is commonly stated that one out of every two marriages will end in divorce. Assuming this is true, what is the probability that of ten randomly selected married couples, seven of the couples' marriages will end in divorce?

16. According to records from a large public university, 85% of students who graduate from the university successfully find employment in their chosen field within six months of graduation. What is the probability that of eight randomly selected students who have graduated from this university, at least five of them find employment in their chosen field within six months?

17. Suppose that one of a local breeder's ten newborn poodles has a genetic birth defect that will only appear later in life. Due to the high expense of the testing, the breeder randomly chooses only three puppies to test for genetic birth defects. What is the probability that one of the puppies will test positive for a genetic birth defect?

18. If a child welfare office receives an average of five reports of child abuse per day, what is the probability that the child welfare office will have a slow day and receive no more than two reports?

19. The male-to-female ratio at one large university is 5:4. If eight students are selected at random for a survey regarding student housing, what is the probability that at least six of them are female? (**Hint:** Note that if the male-to-female ratio is 5:4, then out of nine students, five are male and four are female.)

R Chapter 5 Review

Section 5.1: Discrete Random Variables

Definitions

Random variable
A variable whose numeric value is determined by the outcome of a probability experiment

Probability distribution
A table or formula that gives the probabilities for every value of a random variable

Discrete random variable
A variable that may have either finitely many possible values or infinitely many possible values that are determined by a counting process

Discrete probability distribution
A table or formula that gives the probability for each outcome of a discrete random variable

Properties of a Probability Distribution

1. All of the probabilities are between 0 and 1, inclusive. That is, $0 \le P(X = x) \le 1$.
2. The sum of the probabilities is 1. That is, $\sum P(X = x_i) = 1$.

Expected Value

The expected value for a discrete random variable *X*, equal to the mean of the probability distribution of *X*, given by

$$E(X) = \mu = \sum \left[x_i \cdot P(X = x_i) \right]$$

Variance for a Discrete Probability Distribution

$$\begin{aligned}\sigma^2 &= \sum \left[x_i^2 \cdot P(X = x_i) \right] - \mu^2 \\ &= \sum \left[(x_i - \mu)^2 \cdot P(X = x_i) \right]\end{aligned}$$

Standard Deviation for a Discrete Probability Distribution

$$\begin{aligned}\sigma &= \sqrt{\sigma^2} \\ &= \sqrt{\sum \left[x_i^2 \cdot P(X = x_i) \right] - \mu^2} \\ &= \sqrt{\sum \left[(x_i - \mu)^2 \cdot P(X = x_i) \right]}\end{aligned}$$

Section 5.2: Binomial Distribution

Definitions

Binomial distribution

A special discrete probability function for problems with a fixed number of independent trials, where each trial has only two possible outcomes and one of these outcomes is counted

Success

The outcome that is counted in certain discrete probability distributions, including the binomial distribution

Properties of a Binomial Distribution

1. The experiment consists of a fixed number, n, of identical trials.
2. Each trial is independent of the others.
3. For each trial, there are only two possible outcomes. For counting purposes, one outcome is labeled a success, and the other a failure.
4. For every trial, the probability of getting a success is called p. The probability of getting a failure is then $1 - p$.
5. The binomial random variable, X, counts the number of successes in n trials.
6. For a binomial distribution, the mean is given by $\mu = np$ and the variance is given by $\sigma^2 = np(1-p)$.

Probability for a Binomial Distribution

For a binomial random variable X, the probability of obtaining x successes in n independent trials, given by

$$P(X = x) = {}_nC_x \cdot p^x (1-p)^{(n-x)}$$

Section 5.3: Poisson Distribution

Definition

Poisson distribution

A discrete probability distribution that counts the number of successes that occur in a given interval, such as a period of time

Properties of a Poisson Distribution

1. Each success must be independent of any other successes.
2. The Poisson random variable, X, counts the number of successes in the given interval.
3. The mean number of successes in a given interval must remain constant.
4. For a Poisson distribution, the mean and variance are given by $\mu = \sigma^2 = \lambda$, where λ is the mean number of successes in a given interval.

Section 5.3: Poisson Distribution (cont.)

Probability for a Poisson Distribution

For a Poisson random variable X, the probability of obtaining x successes in any particular interval, given by

$$P(X = x) = \frac{e^{-\lambda}\lambda^x}{x!}$$

Section 5.4: Hypergeometric Distribution

Definition

Hypergeometric distribution

A discrete probability distribution for problems with a fixed number of dependent trials and a specified number of countable successes

Properties of a Hypergeometric Distribution

1. Each trial consists of selecting one of the N items in the population and results in either a *success* or a *failure*. Note that this means that the population is of a *known* size.
2. The experiment consists of n trials.
3. The total number of possible successes in the entire population is k.
4. The trials are dependent (that is, selections are made without replacement).
5. The hypergeometric random variable, X, counts the number of successes in n trials.
6. For a hypergeometric distribution, the mean is given by $\mu = n \cdot \frac{k}{N}$ and the variance is given by $\sigma^2 = \frac{nk(N-k)(N-n)}{N^2(N-1)}$.

Probability for a Hypergeometric Distribution

For a hypergeometric random variable X, the probability of obtaining x successes in n dependent trials, given by

$$P(X = x) = \frac{\left({}_kC_x\right)\left({}_{N-k}C_{n-x}\right)}{\left({}_NC_n\right)}$$

E Chapter 5 Exercises

Properties of a Probability Distribution

Directions: Answer the question and explain your reasoning.

1. A poster advertising the latest game at a casino lists the probabilities of winning various amounts of money. Can this be a true probability distribution?

Mean and Standard Deviation for Discrete Probability Distributions

Directions: Answer the question thoughtfully.

2. A nationwide coffee chain is downsizing its branches in six states. It has already told the store managers in five of the six states the amount of downsizing that will occur. The store managers in the sixth state know that the downsizing that has occurred in the other states has the following distribution. If they assume that the probabilities in this distribution represent the likelihood that their stores will be downsized by the given percentages, what percentage of their employees could they expect to be downsized, and with how much variation?

Coffee Chain Downsizing	
Percentage Downsized	**Probability**
10%	0.25
15%	0.35
20%	0.25
25%	0.10
40%	0.05

Directions: Find each indicated expected value or standard deviation.

3. Suppose that a friend is helping put on a fundraiser for the local animal shelter. One activity is a game using a bowl that contains six green marbles and eight blue marbles. To play the game, each person draws two marbles without replacement (and without looking). If both marbles are green, the player wins \$25. If not, the player must donate \$10 to the animal shelter. The marbles are then replaced for the next player.

a. Calculate the expected value of the game for the player.

b. Suppose that the marble game is played 350 times. How much money would you expect to be donated to the animal shelter?

4. Begin by creating a probability distribution for the number of heads in five tosses of a coin.

a. Explain why this distribution meets the conditions for a binomial distribution.

b. Use the discrete probability distribution to calculate the expected value of the random variable.

c. The expected value of the random variable is equal to the mean of the distribution. As mentioned in the list of properties of the binomial distribution, the mean of a binomial distribution is given by $\mu = np$. Use this formula to calculate the mean of this probability distribution.

d. Does the value you calculated in part b. match the value you calculated in part c.? Explain why or why not.

5. Use the discrete probability distribution you created for Exercise 4.

a. Calculate the standard deviation of this distribution. Interpret the standard deviation.

b. For a binomial distribution, the formula for standard deviation is $\sigma = \sqrt{np(1-p)}$. Use this formula to calculate the standard deviation for the coin toss distribution. Does this number match the value you calculated in part a.? Explain why or why not.

Discrete Probability Distributions

Directions: For each scenario, state the appropriate probability distribution to use, and then find each specified probability.

6. If, in a typical work day, the president of a company is interrupted twice, what is the probability that in a five-day work week she will be interrupted more than 12 times?

7. One day while cleaning out your junk drawer at home, you put all of the loose batteries in a pile. Realizing later that two out of the eight were actually new batteries, you decide that it's necessary to test each battery before throwing it away. What is the probability that the two new batteries are within the first four batteries that you test?

8. Philip is running late for work, and he needs his black socks. Unfortunately, Philip never matches his socks, so there are ten loose white and two loose black socks in his drawer. If he reaches in and randomly grabs two socks at once, what is the probability that he gets the two black socks he needs?

9. Every time Brietta goes to the mall, she has a 0.6 chance of forgetting where she parked her car. If she goes to the mall five times this month, what is the probability that she will forget where she has parked fewer than two times?

10. Your boss has ordered lunch for everyone at the office. There are twelve chicken salad plates and six tuna salad plates to choose from. If you randomly select three plates for yourself and two coworkers, what is the probability that they all contain chicken salad?

11. You forgot to study for your statistics quiz, and you don't know the answers to any of the ten questions on the page. If each question has four multiple-choice options (one of which is correct), what is the probability that you will guess the right answer on at least one question?

12. One southern city experiences an ice storm on average once every eight years. Calculate the probability that there is an ice storm in the city twice in the next three years.

13. It is more common in southern states for high school students to take the ACT than the SAT. Assume that the probability that a southern high school student takes the ACT is 0.93 and the probability that a southern high school student takes the SAT is 0.35. Suppose that a group of 24 southern high school students is randomly selected.

- **a.** What is the probability that all 24 students have taken the ACT?
- **b.** What is the probability that half have taken the SAT?

P Chapter 5 Project

Playing Roulette

We all dream of winning big, becoming an instant millionaire; but how likely is that? Let's say we decide to pursue our goal of winning big money by going to a casino and continually playing what we think will be an easy game: roulette. Can we expect to win big in the long run? Is one bet better than another? How do the casinos make so much money anyway? If we are betting against the casino, how do they make sure that they always win? This project will help you answer these questions.

Let's begin with a lesson in roulette. Roulette is a casino game that involves spinning a ball on a wheel that is marked with numbered squares that are red, black, or green. Half of the numbers 1–36 are colored red and half are black and the numbers 0 and 00 are green. Each number occurs only once on the wheel.

We can make many different types of bets, but two of the most common are to bet on a single number (1–36) or to bet on a color (either red or black). These will be the two bets we will consider in this project. After all players place their bets on the table, the wheel is spun and the ball tossed onto the wheel. The pocket in which the ball lands on the wheel determines the winning number and color. The ball can land on only one color and number at a time.

We begin by placing a bet on a number between 1 and 36. This bet pays 36 to 1 in most casinos, which means we will be paid \$36 for each \$1 we bet on the winning number. If we lose, we simply lose whatever amount of money we bet.

1. Calculate the probability that we will win on a single spin of the wheel.

2. Calculate the probability that we will lose.

3. If we bet \$8 on the winning number, how much money will we win?

4. What is the expected value of a bet on a single number if we bet \$1?

5. For a \$5 bet, what is the expected value of a bet on a single number?

6. What is the expected value of a bet on a single number if we bet \$10?

7. Do you see a pattern in the answers to the last three questions?

We decide that we can certainly increase our chances of winning if we bet on a color instead of a number. Roulette allows us to bet on either red or black and if the number is that color, we win. This bet pays even money in most casinos. This means that for each dollar we bet, we will win $1 for choosing the winning color. So, if we bet $5 and win, we would keep our $5 and win $5 more. If we lose, we lose whatever amount of money we bet, just as before.

8. What is the probability that we will win on a single spin if we bet on red?

9. What is the probability that we will lose on a single spin if we bet on red?

10. If we bet $60 on the winning color, how much money will we win? Is this more or less than we will win by betting $8 on our favorite number? Explain why.

11. What is the expected value of a bet on red if we bet $1?

12. For a $5 bet, what is the expected value of a bet on red?

13. What is the expected value of a bet on red if we bet $10?

14. Do you see a pattern in the answers to the last three questions?

15. How does the expected value of betting on a number compare to the expected value of betting on a color? Is one bet more profitable than another?

16. If our goal was to play roulette so that we can "win it big," what does the expected value of a bet tell us about our chances of winning a large amount of money?

17. Are the casinos really gambling when we place a bet against them? Explain.

T Chapter 5 Technology

Discrete Probability Distributions

TI-83/84 Plus

Binomial Distribution

Side Note

On TI-83 Plus calculators and older TI-84 Plus calculators, the functions `binompdf(` and `binomcdf(` are options `Ø` and `A`, respectively, on the `DISTR` menu.

To calculate the probability $P(X = x)$ for x successes in n independent trials using a TI-83/84 Plus calculator, use option `A:binompdf(` under the `DISTR` menu. Press 2ND and then VARS to access the menu and then scroll down to option `A`. The function syntax is `binompdf(`*numtrials*, p, x), where *numtrials* is n, the number of trials, p is the probability of a success on each trial, and x is the number of successes.

Calculating the cumulative probability, $P(X \le x)$, using a TI-83/84 Plus calculator is similar. Use option `B:binomcdf(` under the `DISTR` menu. Press 2ND and then VARS to access the menu and then scroll down to option `B`. The function syntax is `binomcdf(`*numtrials*, p, x), where *numtrials*, p, and x are defined as before.

Poisson Distribution

Side Note

On TI-83 Plus calculators and older TI-84 Plus calculators, the functions `poissonpdf(` and `poissoncdf(` are options `B` and `C`, respectively, on the `DISTR` menu.

To use a TI-83/84 Plus calculator to calculate the value of a probability for a Poisson distribution, use option `C:poissonpdf(` under the `DISTR` menu. Press 2ND and then VARS to access the menu and then scroll down to option `C`. The function syntax is `poissonpdf(`μ, x`)`, where μ is the symbol that the TI-83/84 Plus Guidebook uses for the value of λ, and x represents the number of successes.

The TI-83/84 Plus calculator can also calculate the value of a cumulative probability for a Poisson distribution, $P(X \le x)$, which we do not calculate directly in the Poisson tables in this text. Use option `D:poissoncdf(` under the `DISTR` menu. Press 2ND and then VARS to access the menu and then scroll down to option `D`. The function syntax is `poissoncdf(`μ, x`)`. Again, μ is the value of λ, and x represents the number of successes.

Example T.1

Using a TI-83/84 Plus Calculator to Calculate a Poisson Probability

A popular accounting office takes in an average of 2.78 new tax returns per day during tax season. What is the probability that on a given day during tax season the firm will take in 4 new tax returns? Assume that the number of tax returns follows a Poisson distribution. Use a TI-83/84 Plus calculator to calculate this value.

Solution

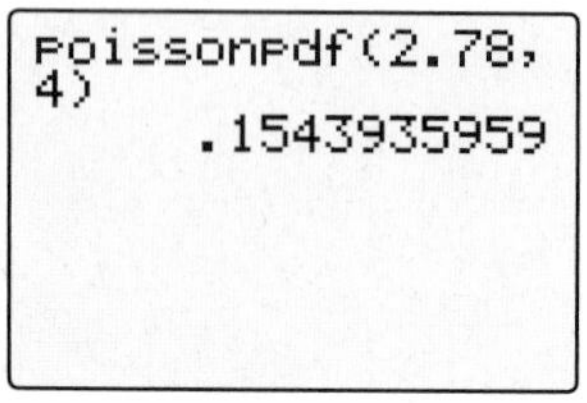
poissonpdf(2.78,
4)
.1543935959

We need the value of the Poisson probability for $\lambda = 2.78$ and $x = 4$. Since we want the probability of exactly four successes, we will use option `C:poissonpdf(`. Press 2ND and then VARS to access the `DISTR` menu and then scroll down to option `C`. Enter `2.78` for μ and `4` for x, as shown in the screenshot in the margin. Thus, $P(X = 4) \approx 0.1544$.

Microsoft Excel

Binomial Distribution

The formula for calculating the probability of getting x successes in n independent trials in Microsoft Excel 2010 is =BINOM.DIST(*number_s*, *trials, probability_s*, *cumulative*), where *number_s* is x, the number of successes; *trials* is n, the number of trials; *probability_s* is p, the probability of getting a success on any trial; and *cumulative* is TRUE if we want the probability of getting at most x successes and FALSE if we want the probability of getting exactly x successes. You may enter a 1 in place of TRUE or a 0 in place of FALSE. Note that the equivalent function for calculating a binomial probability in Excel 2007 and earlier versions of Excel, BINOMDIST, is still available in Excel 2010 to maintain compatibility.

Example T.2

Using Microsoft Excel to Calculate a Binomial Probability

What is the probability of getting exactly six heads in ten tosses of a fair coin?

Solution

If we define "getting a head" as a success, then we want the probability of 6 successes. There are 10 trials and the probability of getting a success on any trial is 0.5. Since we want the probability of exactly 6 successes, we do not want the cumulative probability, so we will let cumulative be FALSE. So we would type in the following formula: **=BINOM.DIST(6, 10, 0.5, FALSE)**.

This formula returns the value 0.205078125. Thus, the probability of getting exactly six heads in ten tosses of a fair coin is approximately 0.2051.

Poisson Distribution

The formula for calculating the value of a probability for a Poisson distribution in Microsoft Excel 2010 is =POISSON.DIST(x, *mean*, *cumulative*), where x is the number of successes, *mean* is the value of λ, and *cumulative* is TRUE if we want the probability of getting at most x successes and FALSE if we want the probability of getting exactly x successes. You may enter a 1 in place of TRUE or a 0 in place of FALSE. Note that the equivalent function for calculating a Poisson probability in Excel 2007 and earlier versions of Excel, POISSON, is still available in Excel 2010 to maintain compatibility.

Example T.3

Using Microsoft Excel to Calculate a Poisson Probability

Suppose that a length of copper wiring averages one defect every 200 feet. What is the probability that a 300-foot stretch will have no defects?

Solution

Each defect in the wire is independent of any other defect, and the average number of defects in a given length of wire is constant. Thus, this scenario can be modeled by a Poisson distribution. Because we are looking for the probability of seeing *no* defects, $x = 0$. If there is

1 defect on average every 200 feet, then we can expect 1.5 defects for a 300-foot stretch; thus $\lambda = 1.5$. Since we want the probability of exactly 0 defects, we do not want the cumulative probability, so we will let cumulative be FALSE. Thus, we would enter the following formula: **=POISSON.DIST(0, 1.5, FALSE)**.

This formula returns the value 0.22313016. Thus, the probability that a 300-foot section of wiring will have no defects is approximately 0.2231.

Hypergeometric Distribution

The formula for calculating the value of a probability for a hypergeometric distribution in Microsoft Excel is =HYPGEOM.DIST(*sample_s*, *number_sample*, *population_s*, *number_pop*, *cumulative*), where *sample_s* is x, the number of successes obtained in n trials; *number_sample* is n, the number of trials; *population_s* is k, the number of successes in the population; *number_pop* is N, the population size; and *cumulative* is TRUE if we want the probability of getting at most x successes and FALSE if we want the probability of getting exactly x successes. Note that the equivalent function for calculating a hypergeometric probability in Excel 2007 and earlier versions of Excel, HYPGEOMDIST, is still available in Excel 2010 to maintain compatibility.

Example T.4

Using Microsoft Excel to Calculate a Hypergeometric Probability

At the local grocery there are twenty boxes of cereal on one shelf, half of which contain a prize. Suppose that you buy three boxes of cereal. What is the probability that all three boxes contain a prize?

Solution

Each box purchased is considered a trial, so the number of trials is 3. A box with a prize is considered a success, and since we are looking for the probability that all 3 trials are successes, the number of successes obtained is also 3. The population size is 20. Half of the boxes in the population are successes, so there are 10 successes in the population. Since we want the probability of exactly 3 successes, we do not want the cumulative probability, so we will let cumulative be FALSE. Thus, we enter the following formula: **=HYPGEOM.DIST(3, 3, 10, 20, FALSE)**.

This formula returns the value 0.105263158. Thus, the probability that all three boxes contain a prize is approximately 0.1053.

Minitab

Discrete Probability Distributions

Minitab will calculate probabilities using a wide variety of distributions including binomial, Poisson, hypergeometric, and discrete distributions defined by the user. In fact, every probability distribution that you will encounter in this text is available by selecting **Calc** and then choosing **Probability Distributions**. Options for each type of distribution include *Probability*, *Cumulative probability*, and *Inverse cumulative probability* (introduced in Chapter 6).

Example T.5

Using Minitab to Calculate a Cumulative Binomial Probability

A local pizza place offers free large pizzas on Thursday nights. The offer is good provided that the customer correctly guesses the outcome of a coin toss when the pizza is ordered. Also, the offer can be used to receive a maximum of four free pizzas per customer. John arrives at the restaurant wanting to purchase at least three pizzas. What is the probability that he will be able to get at least three pizzas for free? Use Minitab to calculate the probability.

Solution

First, state the problem: we want to know the probability of at least 3 successful coin flips, where each trial (coin flip) has a 0.5 probability of success. The number of trials is 4 since the offer can be used for a maximum of 4 free pizzas per customer. This is the complement of the cumulative probability of 2 or fewer successes in 4 trials. Choose **Calc ► Probability Distributions ► Binomial**. Enter **4** for the number of trials and **0.5** for the event probability. Select **Input constant** and enter **2** in the box. Make sure **Cumulative probability** is selected and click **OK**. The Binomial Distribution dialog box is shown in the following screenshot.

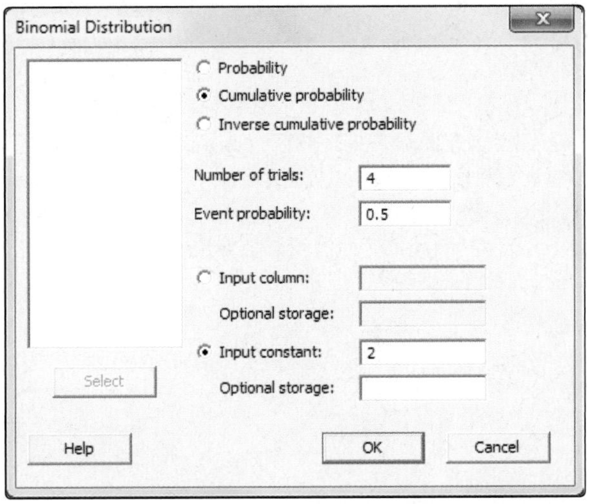

The probability, 0.6875, is displayed in the Session window. Since this is the probability of two or fewer successful coin flips, $1 - 0.6875 = 0.3125$ is the probability of getting at least three pizzas for free.

Chapter Six

Normal Probability Distributions

Sections

6.1 Introduction to the Normal Distribution
6.2 Finding Area under a Normal Distribution
6.3 Finding Probability Using a Normal Distribution
6.4 Finding Values of a Normally Distributed Random Variable
6.5 Approximating a Binomial Distribution Using a Normal Distribution

Objectives

1. Identify the properties of a normal distribution.
2. Find areas under the standard normal distribution.
3. Calculate probabilities for a normal distribution.
4. Identify *z*-values for given areas under the standard normal curve.
5. Calculate probabilities by using the normal distribution as an approximation to the binomial distribution.

6

Introduction

Suppose that a prize of $500 is being offered for the biggest pumpkin entered at the state fair. You have worked hard in the garden all summer long, and you are certain that your pumpkin will win the grand prize. Upon arriving at the fair, you are shocked to see a pumpkin that appears to be as big as yours. How can that be? You wait anxiously as the judges measure both pumpkins in order to determine the winner.

One thousand eight hundred ten pounds for both pumpkins, the measurements are identical! Determined to find a winner, the judges take a second look. This time they measure to the nearest half-pound, and find that both pumpkins are about 1810.5 pounds in weight. Because the measurements are still equal, a more precise measuring tool must be used. Now, the judges use a scale of 1/8-pound and find that the pumpkins both weigh 1810.375 pounds. The most precise measurement on the judges' scale has been used, but still there is no contest winner.

Imagine that the judges are so determined to find the winner that they continue to use more and more precise tools to weigh each of the pumpkins. In theory, how many measurements could be taken? The measuring process could theoretically go on forever, because the measurement of weight is continuous in nature. That is, for any two numbers in an interval, there is always a number between them. Unfortunately, this means that the state fair judges may be in for a very long day. And, if it were *my* pumpkin, I believe that I would just split the prize money and call it a day!

6.1 Introduction to the Normal Distribution

In Chapter 2, we discussed different types and shapes of distributions of data. In Chapter 5, we discussed probability distributions for discrete random variables. In this chapter, we will focus on a particular type of continuous probability distribution that occurs frequently, the normal distribution.

The most prevalent continuous distribution is the normal distribution. A **normal distribution** is a continuous probability distribution for a continuous random variable, X, with a very particular shape. In contrast to discrete probability distributions, which were covered in the last chapter, the possible values for a continuous random variable are not countable. Even if we had all the time in the world, we couldn't list out in table form all the possible values and their probabilities for a continuous distribution. Therefore, we define continuous probability distributions using other methods. A normal distribution, for example, is completely defined by its mean μ and standard deviation σ. A graphical representation of a normal distribution is a symmetric, bell-shaped curve centered above the mean of the distribution. Figure 6.1 shows an example of a normal distribution. Since normal distributions often appear as curves drawn above the x-axis, it is common to see normal distributions referred to as normal curves as well. This practice comes from the term **density curve**, which is the name of the line drawn to show the shape of the distribution.

Side Note

Prince of Mathematics

The Gaussian normal distribution was first developed by German mathematician Carl Friedrich Gauss (1777–1855). He is generally regarded as one of the greatest mathematicians of all time and is sometimes called "the prince of mathematics." As a child he often surprised both his parents and his teachers with the complicated arithmetic that he could do in his head.

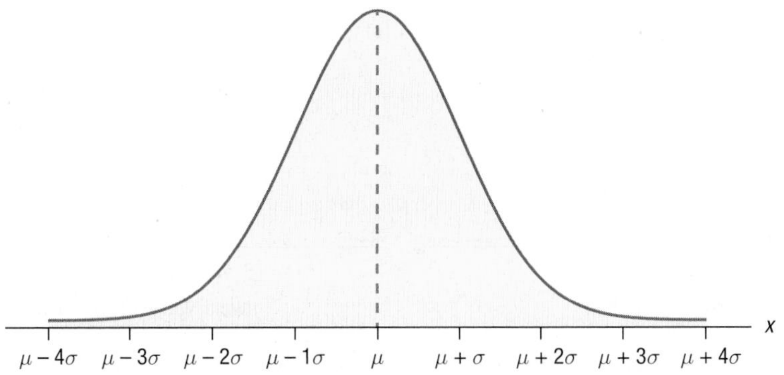

Figure 6.1: Normal Distribution

An example of a data set that would produce an approximately normal distribution is the set of heights of 500 randomly selected men in the United States, such as the distribution in Figure 6.2. Recall from Chapter 5 that a random variable is a variable whose numeric value is determined by the outcome of a probability experiment. The values of the random variable in this example are men's heights. The "probability experiment" is simply choosing a man from the population and measuring his height. The heights are approximately normally distributed with a mean close to 69.2 inches. Heights of men produce a normal distribution because most men are fairly close to the same height, give or take a few inches. Very tall and very short men are rare. Some other examples of data that are normally distributed over a large randomly selected sample are body temperature, weight, and pregnancy duration.

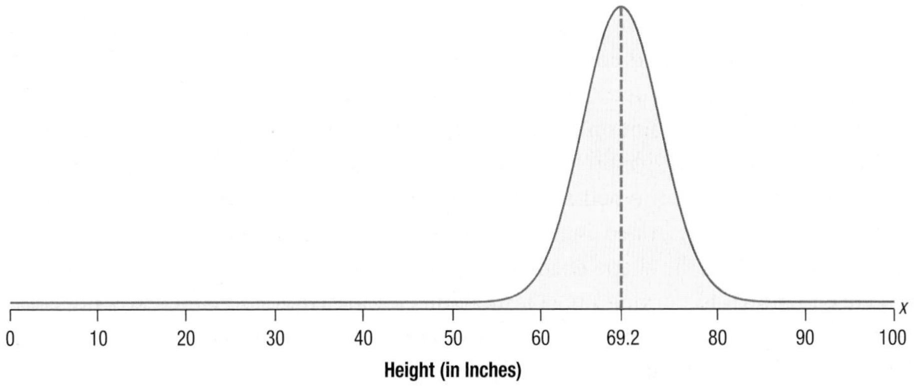

Figure 6.2: Example of Normally Distributed Data

Properties of a Normal Distribution

We have already mentioned that a normal distribution is completely defined by its mean and standard deviation, but it has other properties as well. Let's look at each of the four properties that characterize a normal distribution.

1. Symmetric and Bell-Shaped

First, a normal distribution is symmetric and bell-shaped. The symmetry of the curve means that if you cut the curve in half, the left and right sides are mirror images. We say that the line of symmetry is the vertical line $x = \mu$. The bell shape of the curve implies that the majority of the data are in the middle of the distribution, and the amount of data tapers off evenly in both directions from the center.

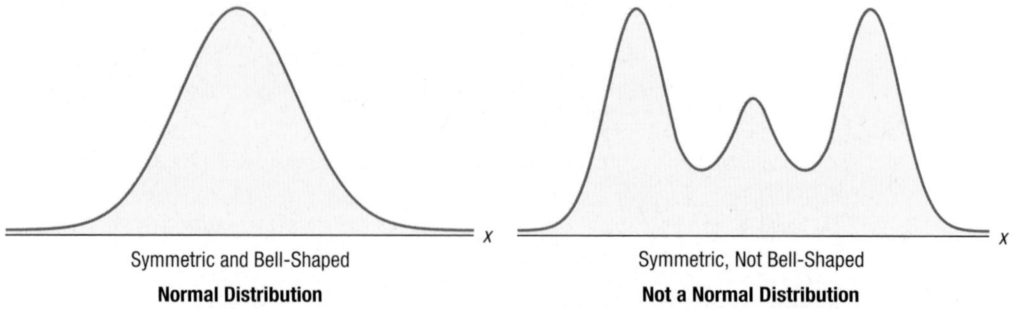

Figure 6.3: General Shape of a Normal Distribution

The fact that a normal distribution is symmetric and bell-shaped tells us many other facts about the distribution. First, the bell-shape indicates that there is only one mode, and it is located at the center of the distribution. Second, recall from Chapter 3 that a symmetric bell-shaped curve has the property that its mean equals its median and its mode. For a normal distribution, the mean, median, and mode are all the same value.

2. Completely Defined by Its Mean and Standard Deviation

As we stated previously, a normal distribution is completely defined by its mean and standard deviation. In order to sketch a normal distribution, you first need to know where to place the center of the distribution on the x-axis. We were just reminded that a bell-shaped, symmetrical distribution such as a normal distribution has a mean equal to its mode, or peak. Thus, the center (or peak) of the distribution will be located on the x-axis directly above the value of the mean of the distribution.

The second thing you need to know about the distribution is its shape. We already know it should be symmetric about the mean, and bell-shaped, but there are many such curves. The distance of the inflection points from the mean determines the shape of a specific curve. An **inflection point** is a calculus term that indicates a point on a curve where the curve changes from curving up to curving down, or vice versa. However, we do not need to know calculus to "feel" this change in curvature. Practice drawing a few symmetric bell-shaped curves similar to the examples shown thus far. Start on the left and draw the curve in one continuous stroke ending on the bottom right. You should feel a point in the process when the shape changes from curving upward to curving downward and then again from curving down to curving up. Those points on the curve are called the inflection points. The distance along the x-axis from one inflection point to the mean is equal to the value of the standard deviation of the particular normal distribution.

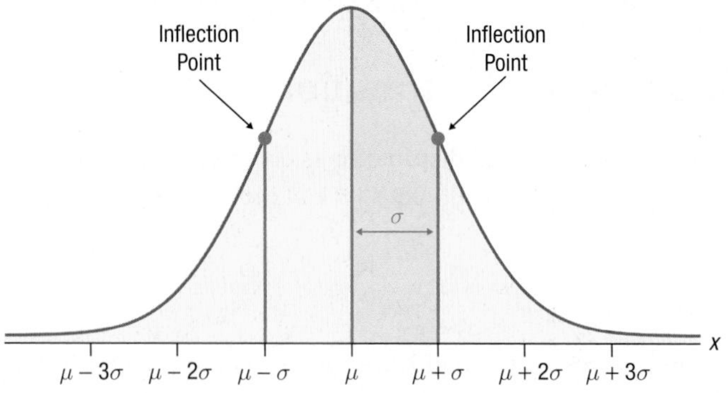

Figure 6.4: Shape of a Normal Distribution

3. Total Area under the Curve Equals One

An important property of any continuous probability distribution is that the total area under the curve of the distribution is equal to 1. We have already seen in Chapter 5 that the probabilities in any discrete probability distribution must add up to 1. All of the possible outcomes in the sample space should be listed in a discrete probability distribution, and the probability for each outcome should be listed, too. The collection of all the possible outcomes should form the sample space, and thus the sum of the probabilities should equal 100%, or 1. The equivalent property for continuous distributions is the fact that the total area under the curve of a continuous distribution equals 1. This property can be demonstrated using calculus, but we'll save that for a more advanced statistics course.

For a continuous probability distribution such as the normal distribution, the area under the curve to the left of a specific value x of the continuous random variable equals the probability that the random variable will take on a value less than or equal to x; that is, $P(X \leq x)$ = area under the curve to the left of x. This is a very important concept that we will continue to discuss and use for the next several chapters. For the moment though, it tells us that the *total* area under the curve is equivalent to the probability of the random variable taking on *any* possible value, that is, any value in the sample space. This probability certainly equals 1, just as it did for discrete distributions; therefore, the total area under the curve equals 1.

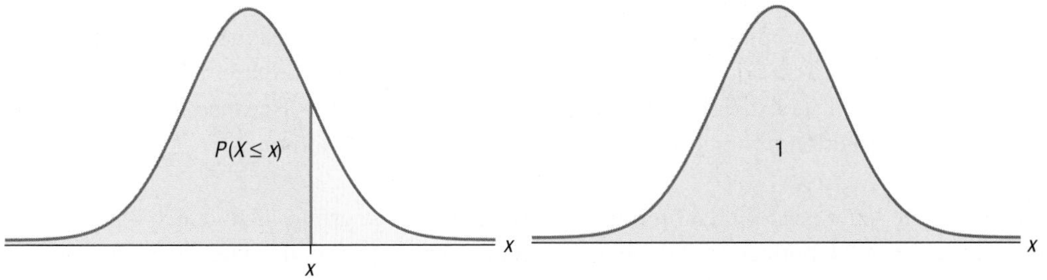

Figure 6.5: Area under a Normal Distribution Curve

4. *x*-Axis Is a Horizontal Asymptote

The last property we will discuss is that the x-axis is a horizontal asymptote for a normal distribution. A **horizontal asymptote** is another calculus term that indicates in this situation that the curve of a normal distribution will approach the x-axis on both ends, but will never touch or cross it. The farther away from the mean (center) of the distribution we travel in either direction along the x-axis, the closer the curve will come to touching the x-axis. The curve will never touch the axis though because that would indicate a point at which no further values are possible. Though further values might be so unlikely as to seem impossible, this would not be the case theoretically.

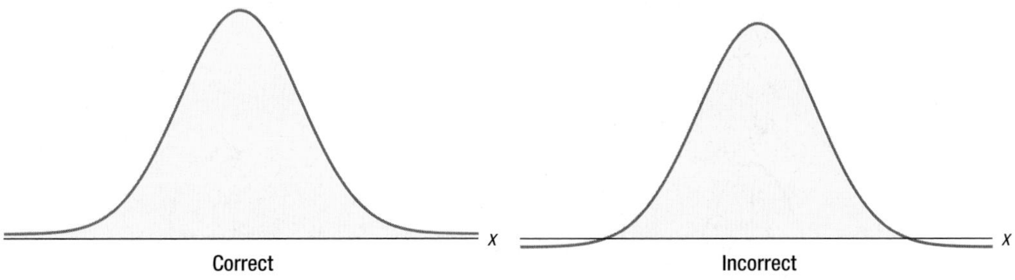

Figure 6.6: Horizontal Axis of a Normal Distribution Curve

Properties

Properties of a Normal Distribution

1. A normal distribution is bell-shaped and symmetric about its mean.
2. A normal distribution is completely defined by its mean, μ, and standard deviation, σ.
3. The total area under a normal distribution curve equals 1.
4. The x-axis is a horizontal asymptote for a normal distribution curve.

Determining Whether a Distribution Is Normal

Now that we've looked at properties of normal distributions, let's consider how to determine if a distribution could be classified as normal. Several obvious conditions must be met first. The data must be continuous, not discrete. In addition, most data values must lie close to the mean of the distribution, while data values much smaller or larger than the mean are less likely to occur (that is, the distribution must be bell-shaped). We will leave more sophisticated tests of normality for a more advanced statistics course.

Consider the heights of the players in the men's collegiate basketball league. First of all, the data are continuous in nature. In addition, there are both shorter people and extremely tall people in the set, but the majority of their heights cluster around the mean height. Because the data meet these conditions, the distribution of heights is likely to be approximately normal.

Now consider the letter grades on a biology final given to a large group of students. Although there are likely to be grades at both ends of the scale, the data are not continuous. They are, in fact, not even quantitative data. Therefore, one would be less likely to call the distribution normal. However, the *shape* of the grade distribution may be similar to that of a normal distribution. In fact, it is possible to use the normal distribution to approximate values of discrete probability distributions under certain conditions. For example, we will talk about how to use the normal distribution to approximate the binomial distribution in the last section of this chapter.

The Standard Normal Distribution

How many normal curves are there? Look again for a moment at the second property of normal curves, which says that a normal curve is completely defined by its mean and standard deviation. Because there are an infinite number of possibilities for μ and σ, there are an infinite number of normal curves.

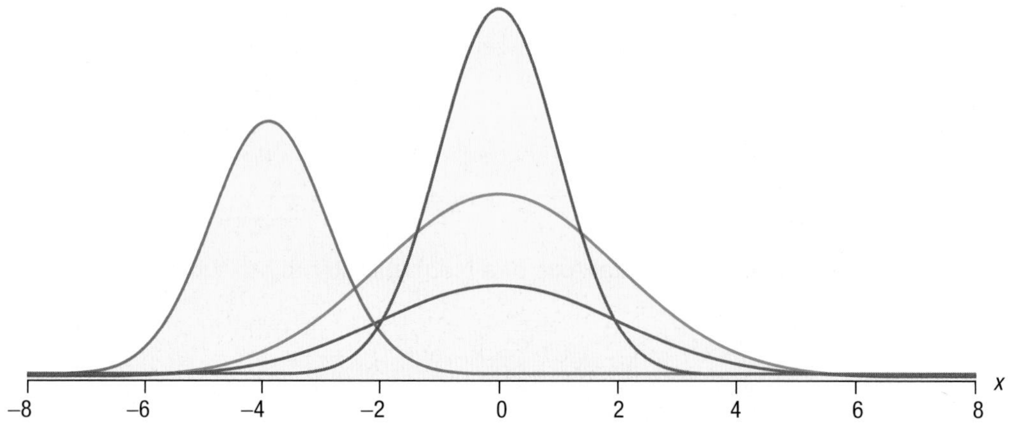

Figure 6.7: Various Normal Curves

Since there are infinitely many normal curves, each with a different mean and standard deviation, we would have to perform separate probability calculations for each normal distribution in order to determine probabilities for each normally distributed random variable. Without the use of technology, calculating the area under a normal curve would be a tedious process that would have to be repeated for every new situation. To avoid these repeated calculations, we can standardize normal curves using standard scores, which we covered in Chapter 3. We can then reference this **standard normal distribution** rather than having to compute the area under each different normal curve. In addition to the properties that any normal distribution has, a standard normal distribution has the two additional properties that the mean is always 0 and the standard deviation is always 1.

Properties

Properties of the Standard Normal Distribution

1. The standard normal distribution is bell-shaped and symmetric about its mean.
2. The standard normal distribution is completely defined by its mean, $\mu = 0$, and standard deviation, $\sigma = 1$.
3. The total area under the standard normal distribution curve equals 1.
4. The x-axis is a horizontal asymptote for the standard normal distribution curve.

Standardizing a Normal Distribution

To standardize any normal distribution to the standard normal distribution, we convert each x-value to a standard score, z, using the following formula, which we first saw in Section 3.3.

$$z = \frac{x - \mu}{\sigma}$$

For the remainder of this book, we will reference a standard score in several different ways, as is the common practice among statisticians. A standard score may also be referred to as a z-score or a z-value.

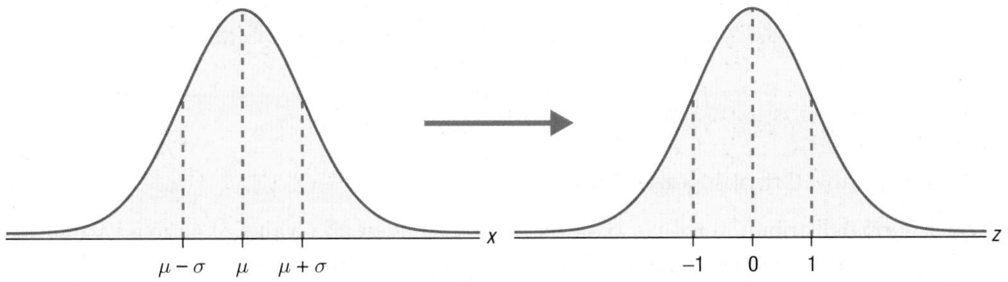

Figure 6.8: Converting to the Standard Normal Distribution

Let's look at how we convert data values to z-scores. Below is a normal curve with a mean of 25 and a standard deviation of 3.2, where an x-value of 27 is indicated.

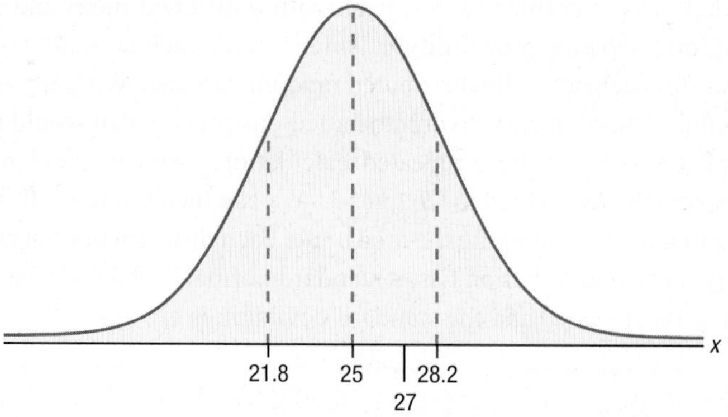

Figure 6.9: Normal Distribution

Rounding Rule

When calculating a z-score, round to two decimal places.

The procedure for standardizing a normal distribution is to obtain the corresponding z-value for the given x-value, which in this example is 27. To convert from an x-value to a z-value, we substitute the mean and standard deviation of the normal curve into the standard score formula as follows.

$$\begin{aligned} z &= \frac{x-\mu}{\sigma} \\ &= \frac{27-25}{3.2} \\ &\approx 0.63 \end{aligned}$$

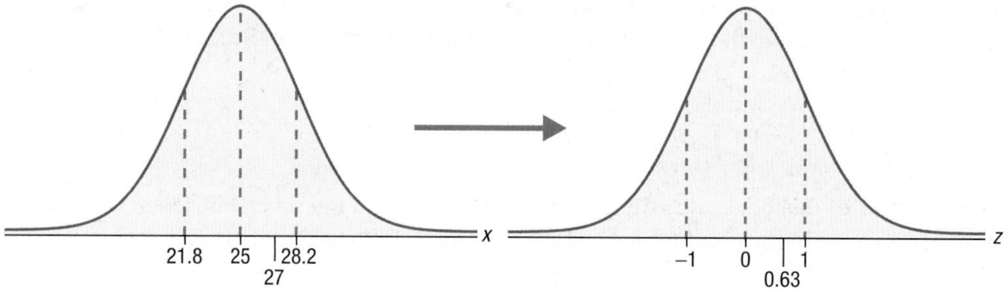

Figure 6.10: Converting to the Standard Normal Distribution

Example 6.1

Calculating and Graphing z-Values

Given a normal distribution with $\mu = 48$ and $\sigma = 5$, convert an x-value of 45 to a z-value and indicate where this z-value would be on the standard normal distribution.

Solution

Begin by finding the z-score for $x = 45$ as follows.

$$\begin{aligned} z &= \frac{x-\mu}{\sigma} \\ &= \frac{45-48}{5} \\ &= -0.60 \end{aligned}$$

Now draw each of the distributions, marking a standard score of $z = -0.60$ on the standard normal distribution.

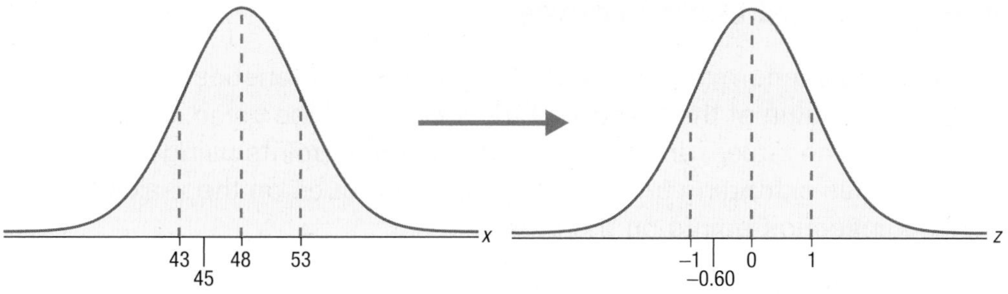

The distribution on the left is a normal distribution with a mean of 48 and a standard deviation of 5. The distribution on the right is a standard normal distribution with a standard score of $z = -0.60$ indicated.

6.1 Section Exercises

Properties of Normal Distributions

Directions: Decide if each statement is true or false. Explain why.

1. There are a limited number of normal distributions.

2. There is only one standard normal distribution.

3. The mean of a normal distribution is always 0.

4. The mean of the standard normal distribution is always 0.

5. The standard deviation of the standard normal distribution is always 0.

6. For any normal distribution, the mean, median, and mode are equal.

7. The line of symmetry for a normal distribution is $x = \mu$.

8. The y-axis is a vertical asymptote for all normal distributions.

9. The inflection points for any normal distribution are one standard deviation on either side of the mean.

10. Normal distributions are symmetric, but they do not have to be bell-shaped.

Drawing Normal Distributions

Directions: Draw a normal curve with the given characteristics. Indicate the value of the mean and the location of the given *x*-value on the *x*-axis, and indicate the inflection points using vertical lines extending from the appropriate values on the *x*-axis to the inflection points on the curve.

11. $\mu = 65$ and $\sigma = 20$; $x = 40$

12. $\mu = 5.00$ and $\sigma = 0.25$; $x = 4.80$

13. $\mu = 15$ and $\sigma = 2$; $x = 19$

14. $\mu = 0.023$ and $\sigma = 0.001$; $x = 0.020$

15. $\mu = 12,000$ and $\sigma = 2000$; $x = 10,750$

Converting to the Standard Normal Distribution

Directions: Calculate the standard score of the given *x*-value. Indicate where the *z*-value would be on the standard normal distribution.

16. $\mu = 65$ and $\sigma = 20$; $x = 40$

17. $\mu = 5.00$ and $\sigma = 0.25$; $x = 4.80$

18. $\mu = 15$ and $\sigma = 2$; $x = 19$

19. $\mu = 0.023$ and $\sigma = 0.001$; $x = 0.020$

20. $\mu = 12,000$ and $\sigma = 2000$; $x = 10,750$

Using the *z*-Score Formula

Directions: Use the *z*-score formula to complete each exercise.

21. Find the missing value in each row of the table.

	z	x	μ	σ
a.		35.0	37.0	1.6
b.	1.75	12.3		2.8
c.	–3.10		3.40	0.20
d.	2.15	479	436	

22. Find the missing value in each row of the table.

	z	x	μ	σ
a.		2.87	2.45	0.21
b.	–2.80	579.0		8.5
c.	1.40		89.10	7.45
d.	–3.20	13.67	15.11	

23. Write a formula for x in terms of the population mean, population standard deviation, and z-score.

Comparing Normal Distributions

Directions: Draw two normal distributions, on the same *x*-axis, that have the given characteristics.

24. Population means that differ by 5 units.

25. The same mean but different standard deviations.

6.2 Finding Area under a Normal Distribution

Many of the distributions in which statisticians are interested are approximately normal distributions. One important application of a normal distribution is that the *area* under any part of the normal curve is equal to the *probability* of the random variable falling within that region. Recall from the previous section that the area under a normal curve to the left of a specific value x of the random variable is equal to $P(X \le x)$. Similarly, the area to the right of x equals $P(X \ge x)$. (Notice that the inequality symbol points in the direction of the area!) Furthermore, we do not talk about finding the probability that x is a specific value because of the fact that the normal curve is a continuous probability distribution. Instead, we refer to the probability that x is within a range of values. Thus, choosing to include the endpoint of our range does not change the value of the probability or the area. So, we can say that $P(X \le x) = P(X < x)$.

To demonstrate how area and probability are related, let's look at an example. Suppose that the average weight of adult Labrador Retrievers seen by Dr. Sullivan's animal clinic is 75 pounds with a standard deviation of 5 pounds and that these weights are approximately normally distributed. What is the probability that an adult Labrador Retriever chosen at random from the clinic weighs more than 80 pounds? Look at the illustration of the distribution shown in the figure below. Notice that the mean is 75 and, because the standard deviation is 5, a weight of 80 is one standard deviation above the mean. Suppose you were told that the *area* represented in the region shaded to the right of 80 is equal to 0.1587. If the area to the right of 80 is equal to 0.1587, then that means that 15.87% of all the weights fall in that region. Hence, we are able to say that the *probability* is 15.87% that a dog chosen at random from all the adult Labrador Retrievers seen by the clinic weighs more than 80 pounds.

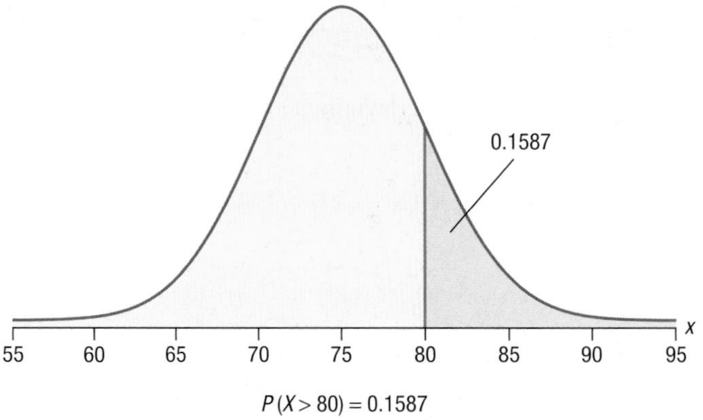

Figure 6.11: Normal Distribution of Labrador Retriever Weights (in Pounds)

Now that we have demonstrated the usefulness of the area under a normal curve, we need to be able to calculate the area under a normal curve in any given region. The area under a curve can be found using *integration*, which is a calculus technique beyond the scope of this course. Also complicating the matter is the fact that there are an infinite number of normal curves, meaning that the area would need to be calculated for each unique curve. Thankfully, the area under any given part of the *standard* normal curve is fixed, and tables have been created so that one can look up the area under the standard normal curve without relying on calculus. In addition, area under any normal curve can also be found using calculators or statistical software. We will focus on the use of tables and graphing calculators to find area under the standard normal curve in this section. We do, however, include instructions for using Microsoft Excel and Minitab in the technology section at the end of the chapter.

Let's first look at the most traditional method for finding area under the standard normal curve: the standard normal distribution tables. There are several different standard normal distribution tables that statisticians use, each one giving us a different method for finding the same area under the normal curve. Although you can use different tables to produce the same results, we will use the **cumulative normal distribution tables**, Table A and Table B in Appendix A, which give the area under the standard normal curve to the left of the given z-value. A small excerpt of Table B from Appendix A is shown below.

z	0.00	0.01	0.02	0.03	0.04
0.0	0.5000	0.5040	0.5080	0.5120	0.5160
0.1	0.5398	0.5438	0.5478	0.5517	0.5557
0.2	0.5793	0.5832	0.5871	0.5910	0.5948
0.3	0.6179	0.6217	0.6255	0.6293	0.6331
0.4	0.6554	0.6591	0.6628	0.6664	0.6700
0.5	0.6915	0.6950	0.6985	0.7019	0.7054
0.6	0.7257	0.7291	0.7324	0.7357	0.7389
0.7	0.7580	0.7611	0.7642	0.7673	0.7704
0.8	0.7881	0.7910	0.7939	0.7967	0.7995

Figure 6.12: Excerpt from Table B: Standard Normal Distribution

Notice that the z-values given in the table are rounded to two decimal places. The first decimal place of each z-value is listed in the left column, with the second decimal place in the top row. Where the appropriate row and column intersect, we find the amount of area under the standard normal curve to the *left* of that particular z-value. Let's now look at an example of finding the area to the left of a particular value of z.

Example 6.2

Finding Area to the Left of a Positive z-Value Using a Cumulative Normal Table

Find the area under the standard normal curve to the left of $z = 1.37$.

Solution

To read the table, we must break the given z-value (1.37) into two parts: one containing the first decimal place (1.3) and the other containing the second decimal place (0.07). So, in Table B from Appendix A, look across the row labeled 1.3 and down the column labeled 0.07. The row and column intersect at 0.9147. Thus, the area under the standard normal curve to the left of $z = 1.37$ is 0.9147.

z	0.05	0.06	0.07	0.08	0.09
1.0	0.8531	0.8554	0.8577	0.8599	0.8621
1.1	0.8749	0.8770	0.8790	0.8810	0.8830
1.2	0.8944	0.8962	0.8980	0.8997	0.9015
1.3	0.9115	0.9131	0.9147	0.9162	0.9177
1.4	0.9265	0.9279	0.9292	0.9306	0.9319
1.5	0.9394	0.9406	0.9418	0.9429	0.9441

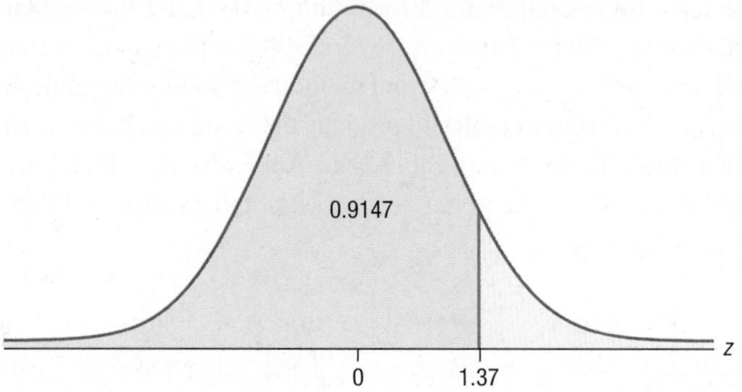

Example 6.3

Finding Area to the Left of a Negative *z*-Value Using a Table or a TI-83/84 Plus Calculator

Find the area under the standard normal curve to the left of $z = -2.03$.

Solution

The first part of the *z*-value is –2.0 and the second part is 0.03. This time, use Table A from Appendix A since the *z*-value is negative; look across the row labeled –2.0 and down the column labeled 0.03. The row and column intersect at 0.0212. Thus, the area under the normal curve to the left of $z = -2.03$ is 0.0212.

z	**0.04**	**0.03**	**0.02**	**0.01**	**0.00**
–2.2	0.0125	0.0129	0.0132	0.0136	0.0139
–2.1	0.0162	0.0166	0.0170	0.0174	0.0179
–2.0	0.0207	0.0212	0.0217	0.0222	0.0228
–1.9	0.0262	0.0268	0.0274	0.0281	0.0287
–1.8	0.0329	0.0336	0.0344	0.0351	0.0359
–1.7	0.0409	0.0418	0.0427	0.0436	0.0446

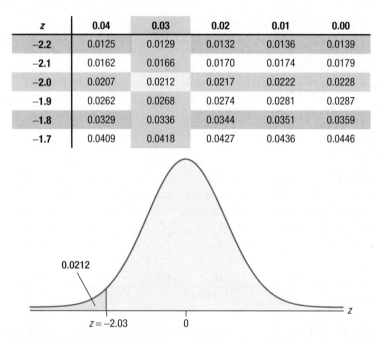

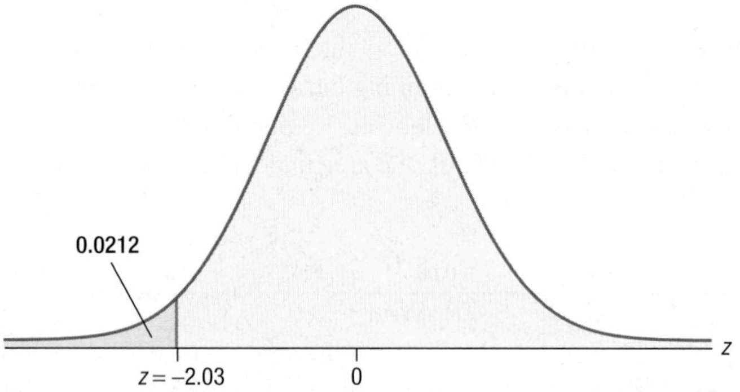

To obtain the solution using a TI-83/84 Plus calculator, perform the following steps.

- Press 2ND and then VARS to access the `DISTR` menu.
- Choose option `2:normalcdf(`.
- Enter *lower bound, upper bound,* μ, σ. **Note:** If you want to find area under the standard normal curve, as in this example, then you do not need to enter μ or σ.

- Since we are asked to find the area to the left of z, the lower bound is $-\infty$. We cannot enter $-\infty$ into the calculator, so we will enter a very small value for the lower endpoint, such as -10^{99}. This number appears as `-1E99` when entered correctly into the calculator. To enter `-1E99`, press [(−)] [1] [2ND] [,] [9] [9].

Enter `normalcdf(-1E99,-2.03)`, as shown in the screenshot in the margin. The area is approximately 0.0212.

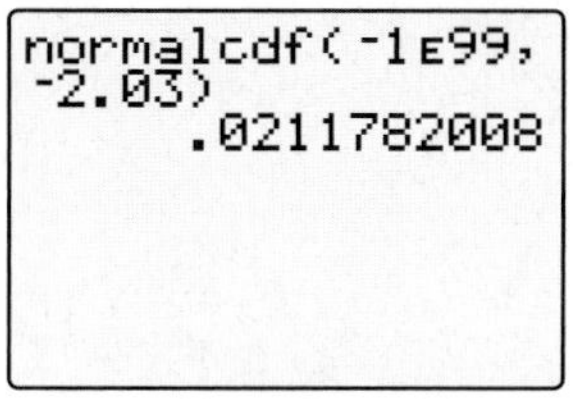

The cumulative normal curve tables we are using only give the area to the *left* of a given z-value, but we can use the tables, along with the properties of the standard normal distribution, to find other areas as well. First, let's consider the area to the right of a given z-score. To obtain this area, remember that the total area under the standard normal curve is 1. So, if the table gives us the area to the left of z, then subtracting that area from 1 gives us the area to the *right* of z.

Memory Booster

Total area under a normal curve = 1.

A shortcut to finding the area to the right of a given value of z is to use the symmetry of the curve. In terms of area under the curve, this means that the area to the right of z is equal to the area to the left of $-z$. To find the area to the right of z, instead of looking up the area to the left of z and subtracting that area from 1, you can simply look up the area to the left of $-z$.

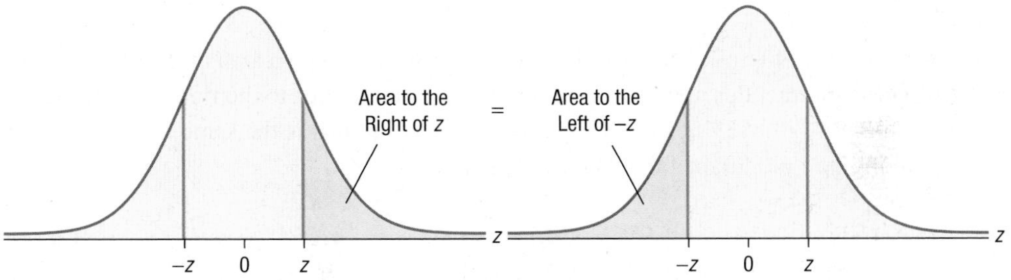

Figure 6.13: Finding Area to the Right of z

Example 6.4

Finding Area to the Right of a Positive *z*-Value Using a Cumulative Normal Table

Find the area under the standard normal curve to the right of $z = 1.37$.

Solution

- Method 1: From Example 6.2, we know that the area under the standard normal curve to the left of $z = 1.37$ is 0.9147. So, the area under the standard normal curve to the right of $z = 1.37$ is $1 - 0.9147 = 0.0853$.

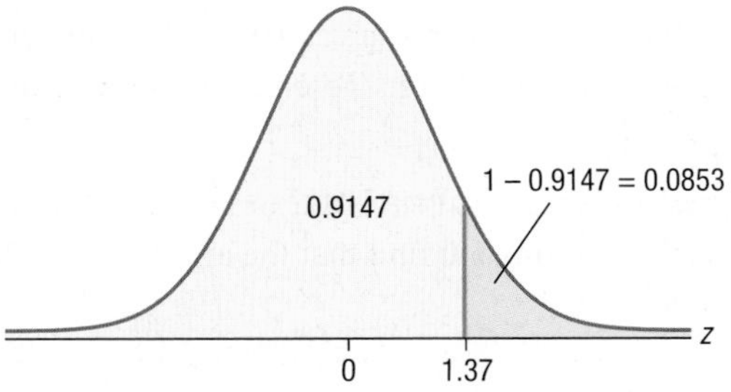

- Method 2: We can look up $z = -1.37$ in Table A from Appendix A, which also gives us 0.0853.

z	0.09	0.08	0.07	0.06	0.05
–1.6	0.0455	0.0465	0.0475	0.0485	0.0495
–1.5	0.0559	0.0571	0.0582	0.0594	0.0606
–1.4	0.0681	0.0694	0.0708	0.0721	0.0735
–1.3	0.0823	0.0838	0.0853	0.0869	0.0885
–1.2	0.0985	0.1003	0.1020	0.1038	0.1056
–1.1	0.1170	0.1190	0.1210	0.1230	0.1251

Example 6.5

Finding Area to the Right of a Negative *z*-Value Using a Table or a TI-83/84 Plus Calculator

Find the area under the standard normal curve to the right of $z = -0.90$.

Solution

If we look up $z = -0.90$ in Table A from Appendix A, we see that an area of 0.1841 lies to the left of z. Then, subtract this area from 1 to find the amount of area to the right of z. Hence, an area of $1 - 0.1841 = 0.8159$ lies to the right of z. Note that this is the same area you find by simply looking up $z = 0.90$ in Table B from Appendix A.

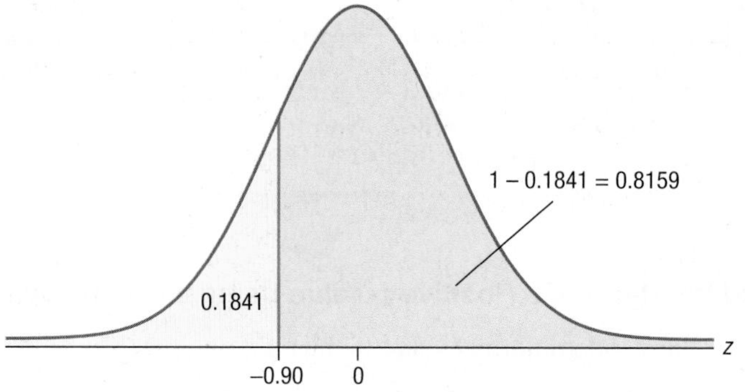

To obtain the solution using a TI-83/84 Plus calculator, perform the following steps.

- Press 2ND and then VARS to access the DISTR menu.
- Choose option 2:normalcdf(.
- Because the calculator will not allow us to enter ∞ for the upper endpoint, we will choose a sufficiently large number, such as 10^{99}. This number appears as 1E99 when entered correctly into the calculator. To enter 1E99 into the calculator, press 1 2ND , 9 9.

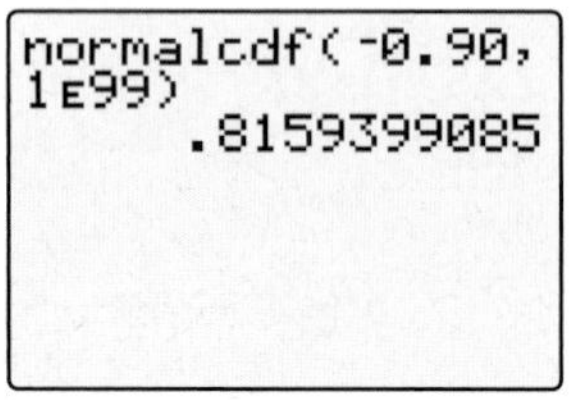

Thus, to obtain the answer, we enter normalcdf(-Ø.9Ø,1E99) into the calculator, as shown in the screenshot in the margin, and find that the area to the right of $z = -0.90$ is indeed approximately 0.8159.

Let's consider the area under the standard normal curve that is between two z-scores. To find the area between two values of z, use the table to look up the area to the left of each z-value and then subtract the smaller area from the larger area.

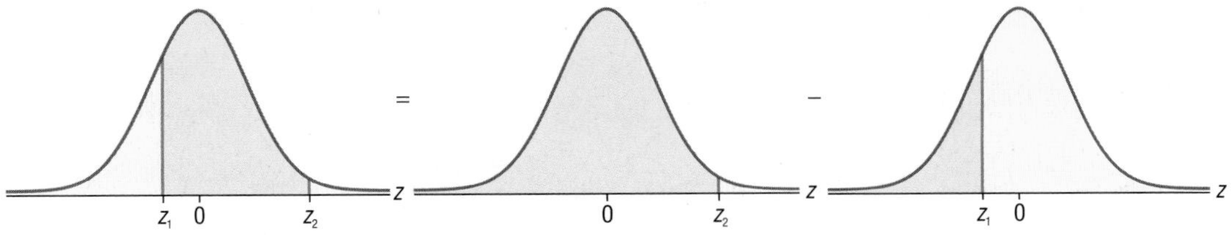

Figure 6.14: Calculating Area between Two z-Values

Example 6.6

Finding Area between Two z-Values Using Tables or a TI-83/84 Plus Calculator

Find the area under the standard normal curve between $z_1 = -1.68$ and $z_2 = 2.00$.

Solution

First, look up the area to the left of $z_1 = -1.68$, which is 0.0465. Second, look up the area to the left of $z_2 = 2.00$, which is 0.9772. Finally, subtract: $0.9772 - 0.0465 = 0.9307$. Thus, the area between the two z-values is 0.9307.

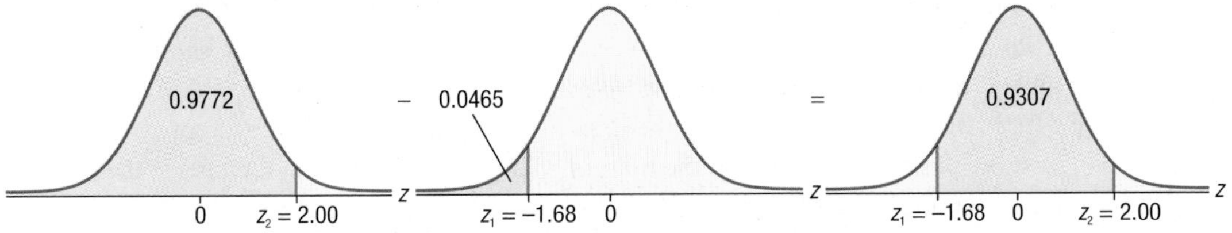

Enter `normalcdf(-1.68,2.ØØ)`, as shown in the screenshot in the margin, to find that the area between $z_1 = -1.68$ and $z_2 = 2.00$ is approximately 0.9308.

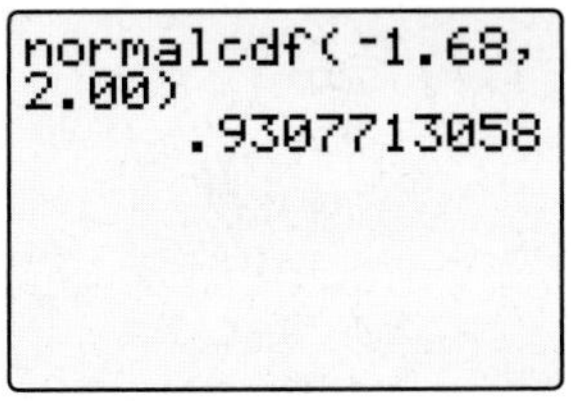

Notice that this is a slightly different answer than what we obtained using the tables. The reason is that the table values have been rounded in the intermediate steps and the calculator values have not. For this reason, calculator solutions may vary slightly from those obtained by using the tables.

Example 6.7

Finding Area between Two z-Values Using a TI-83/84 Plus Calculator

Find the area under the standard normal curve between $z_1 = 1.50$ and $z_2 = 2.75$.

Solution

Let's use a TI-83/84 Plus calculator to find this area. We want to use a lower bound of 1.50 and an upper bound of 2.75. Enter `normalcdf(1.5Ø,2.75)` into the calculator, which gives a value of approximately 0.0638 as shown in the screenshot in the margin.

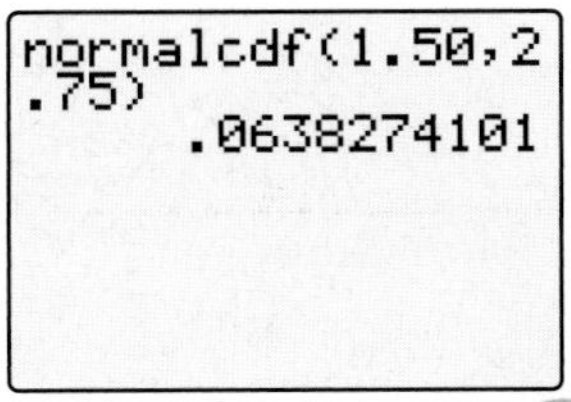

Therefore, we see that an area of approximately 0.0638 lies between $z_1 = 1.50$ and $z_2 = 2.75$.

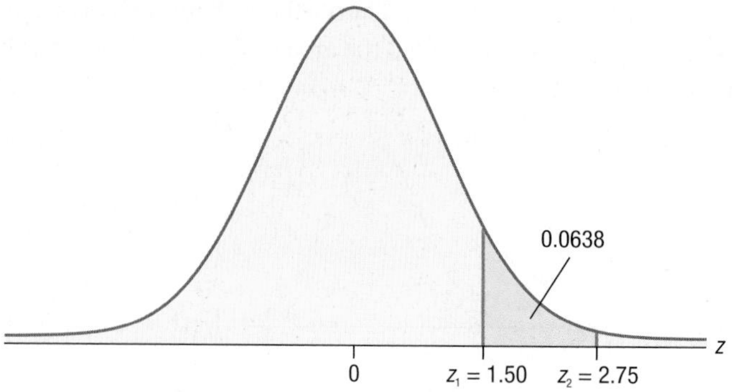

Example 6.8

Finding Area in the Tails for Two *z*-Values Using a TI-83/84 Plus Calculator

Find the total of the areas under the standard normal curve to the left of $z_1 = -2.50$ and to the right of $z_2 = 3.00$.

Solution

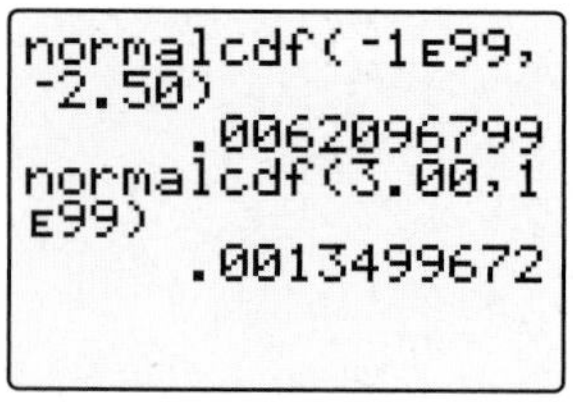

There are two areas that we must find. The total we are interested in is the sum of these two areas. Let's begin by finding the area to the left of $z_1 = -2.50$. Enter `normalcdf(-1E99,-2.50)` to find that the area to the left of $z_1 = -2.50$ is approximately 0.006210, as shown in the screenshot in the margin. Next, we need to find the area to the right of $z_2 = 3.00$. Enter `normalcdf(3.00,1E99)` to find that the area to the right of $z_2 = 3.00$ is approximately 0.001350, as shown in the screenshot in the margin. Thus, the total area in the tails is the sum of the two areas, $0.006210 + 0.001350 \approx 0.0076$.

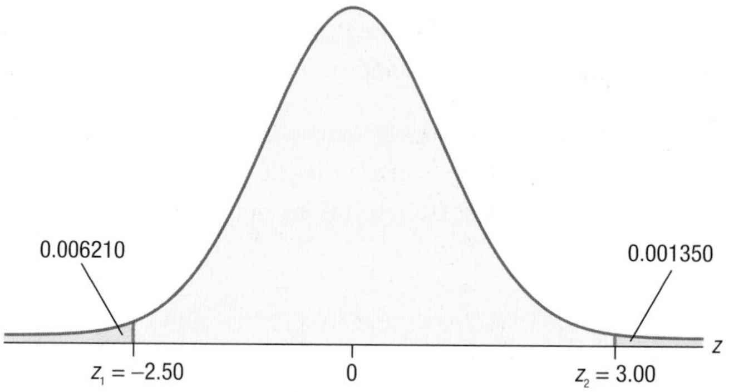

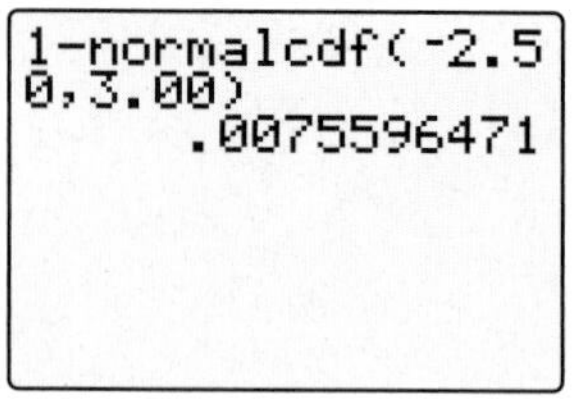

Note an alternative method for finding this area that is particularly clever. By definition, we know that the total area under the curve equals 1. Using this fact, the area in the tails can be obtained by finding the area between $z_1 = -2.50$ and $z_2 = 3.00$ and then subtracting that area from 1. This method can even be entered in one step into your calculator as shown below and in the screenshot in the margin.

`1-normalcdf(-2.50,3.00)` ≈ 0.0076

Example 6.9

Finding Area in the Tails for Two *z*-Values Using a TI-83/84 Plus Calculator

Find the total of the areas under the standard normal curve to the left of $z_1 = -1.23$ and to the right of $z_2 = 1.23$.

Solution

Notice that the absolute values of z_1 and z_2 are the same. Thus, the areas in the two tails of the distribution will be the same because of the symmetric property of the standard normal curve. So, to find the total area in the two tails, we only need to look up the area in the left tail and multiply that by 2.

By entering `normalcdf(-1E99,-1.23)`, we find that the area to the left of $z_1 = -1.23$ is approximately 0.109349, as shown in the screenshot in the margin. Multiply this area by 2 in order to obtain the combined area in the tails.

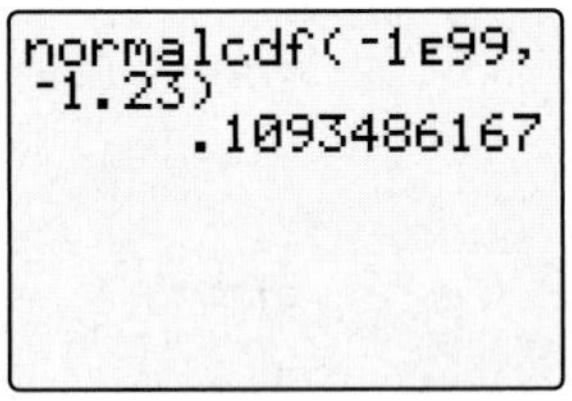

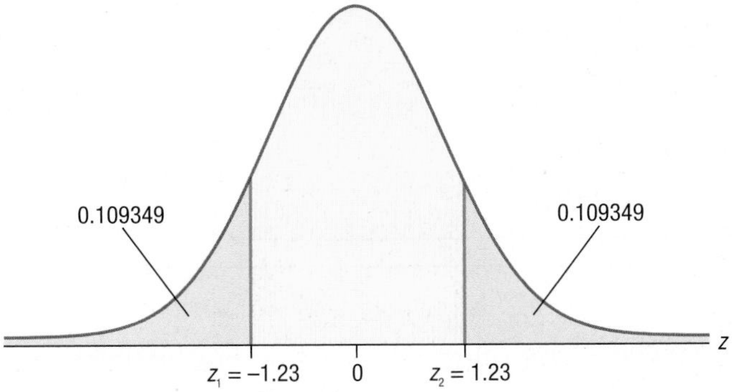

Thus, $(0.109349)(2) \approx 0.2187$. So, the total area in the two tails is approximately 0.2187.

Now that we know how to find areas under the standard normal curve, we know how to find the probability that a value of a normally distributed random variable will occur within a specific range of values. For example, finding the area to the left of $z = 1.37$ is the same as calculating $P(z \le 1.37)$. Notice that the inequality sign points the same direction in which we shaded the area under the curve in Example 6.2. Also recall that $P(z \le 1.37) = P(z < 1.37)$ because whether or not we include the endpoint does not change the value of the probability. The following procedures summarize the basic rules we use for each of the four general types of probability problems for the standard normal distribution.

Procedure

Using the *Cumulative Normal Distribution Tables* to Find Areas under the Standard Normal Curve

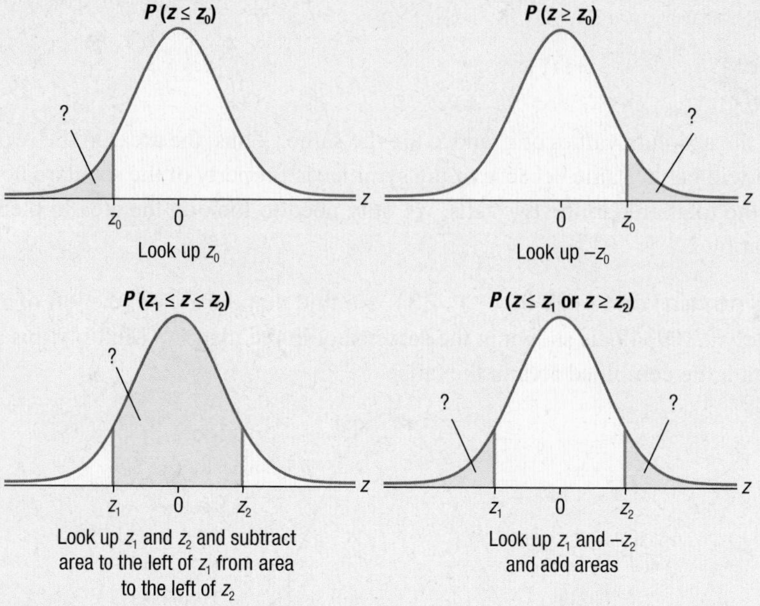

Procedure

Using a *TI-83/84 Plus Calculator* to Find Areas under the Standard Normal Curve

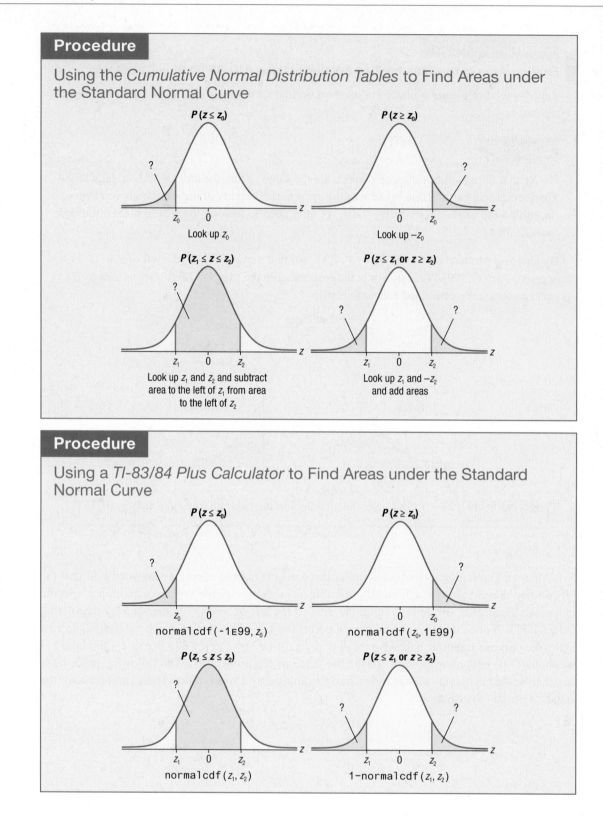

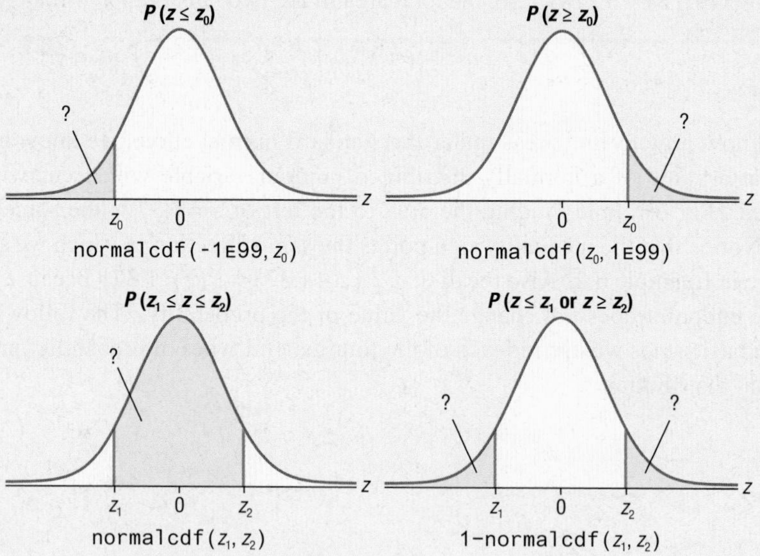

Example 6.10

Interpreting Probability for the Standard Normal Distribution as an Area under the Curve

Interpret $P(z \le -2.67)$.

Solution

$P(z \le -2.67)$ stands for the probability that z is less than or equal to -2.67. This is equal to the area under the standard normal curve to the left of $z = -2.67$.

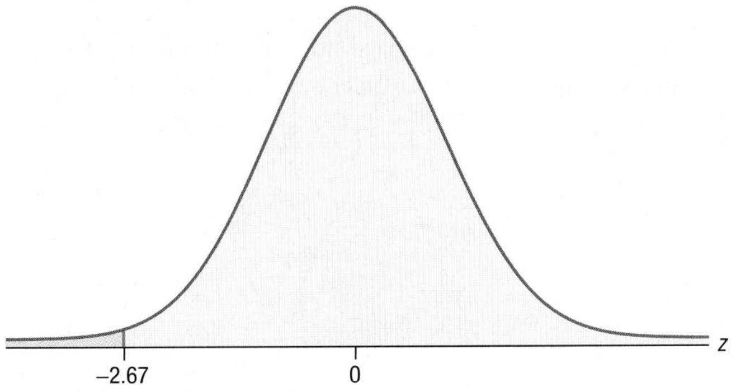

Note that the probability that z is less than a value is the same as the probability that z is less than or equal to that value, or symbolically, $P(z \le -2.67) = P(z < -2.67)$.

Example 6.11

Finding Probabilities for the Standard Normal Distribution Using Tables or a TI-83/84 Plus Calculator

Find the following probabilities using the cumulative normal distribution tables or a TI-83/84 Plus calculator.

a. $P(z < 1.45)$

b. $P(z \ge -1.37)$

c. $P(1.25 < z < 2.31)$

d. $P(z < -2.5 \text{ or } z > 2.5)$

e. $P(z < -4.01)$

f. $P(z \le 3.98)$

Solution

For each part of this example, the solution first describes the method for finding the answer using the cumulative normal distribution tables and then explains how to find the answer using a TI-83/84 Plus calculator.

a. $P(z < 1.45)$ is the area under the standard normal curve to the left of $z = 1.45$. Look up $z = 1.45$ in the cumulative normal table for positive z-values, Table B. The area is 0.9265.

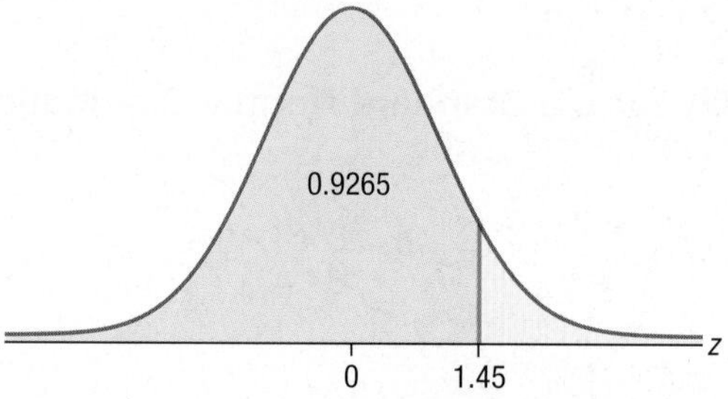

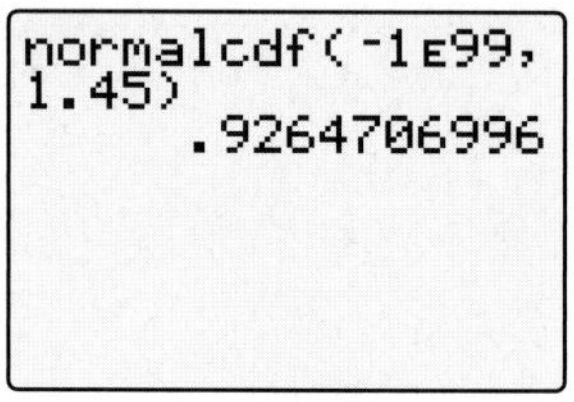

By entering `normalcdf(-1E99,1.45)`, we find that the area to the left of $z = 1.45$ is approximately 0.9265, as shown in the screenshot in the margin.

b. $P(z \geq -1.37)$ is the area under the standard normal curve to the right of $z = -1.37$. Use the symmetry property of the standard normal curve and look up $z = 1.37$ in the cumulative normal table for positive z-values, Table B. The area to the right of $z = -1.37$ is 0.9147.

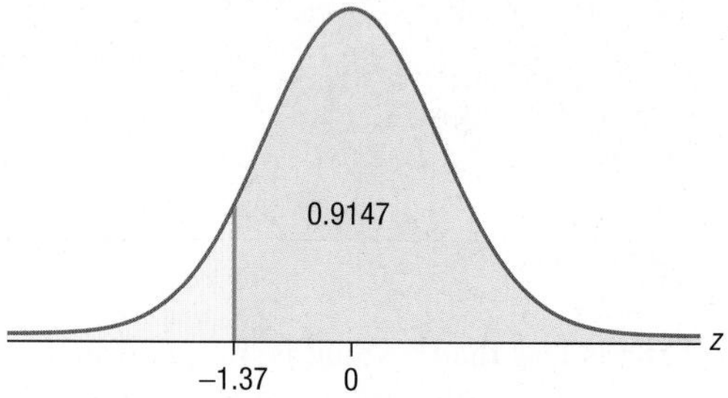

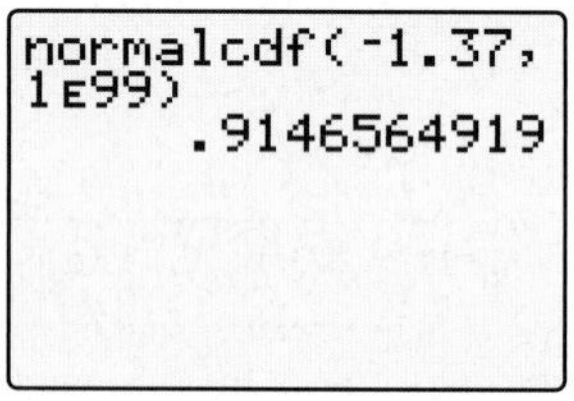

By entering `normalcdf(-1.37,1E99)`, we find that the area to the right of $z = -1.37$ is approximately 0.9147, as shown in the screenshot in the margin.

c. $P(1.25 < z < 2.31)$ is the area under the standard normal curve between $z_1 = 1.25$ and $z_2 = 2.31$. Look up each value in the cumulative normal table for positive z-values, Table B. The area to the left of $z_1 = 1.25$ is 0.8944. The area to the left of $z_2 = 2.31$ is 0.9896. The area between $z_1 = 1.25$ and $z_2 = 2.31$ is the difference between these two areas. Thus, the area is $0.9896 - 0.8944 = 0.0952$.

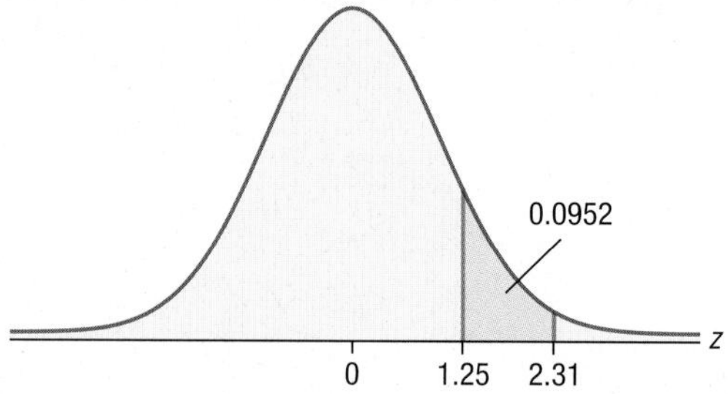

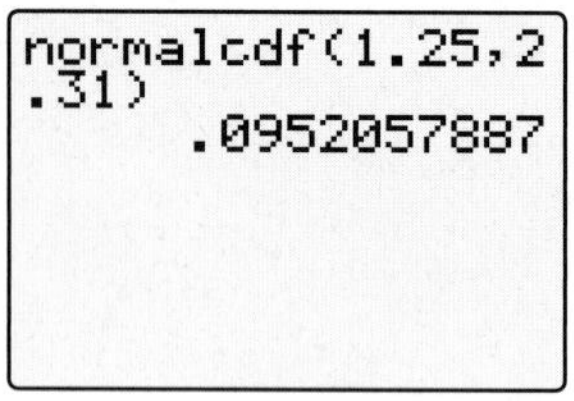

By entering `normalcdf(1.25,2.31)`, we find that the area between $z_1 = 1.25$ and $z_2 = 2.31$ is approximately 0.0952, as shown in the screenshot in the margin.

d. $P(z < -2.5 \text{ or } z > 2.5)$ is the sum of the areas to the left of $z_1 = -2.5$ and to the right of $z_2 = 2.5$. Since the normal curve is symmetric, these two areas are the same; therefore, look up the area to the left of $z = -2.50$ in the cumulative normal table for negative z-values, Table A, and multiply that area by 2. The area to the left of $z = -2.50$ is 0.0062. Thus, the area we are interested in is $(0.0062)(2) = 0.0124$.

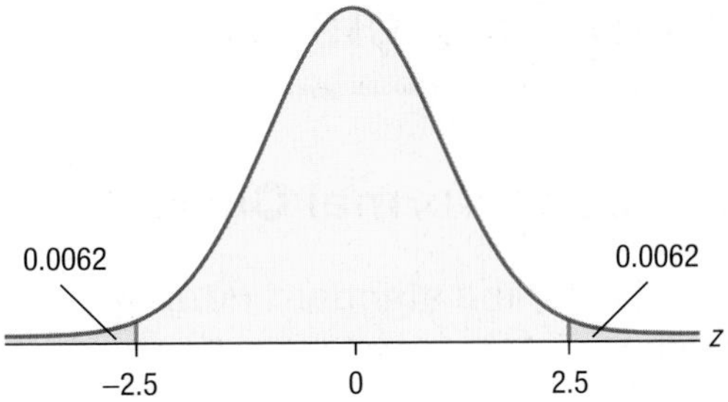

Enter `1-normalcdf(-2.5,2.5)`. We find that the sum of the areas to the left of $z_1 = -2.5$ and to the right of $z_2 = 2.5$ is approximately 0.0124, as shown in the screenshot in the margin.

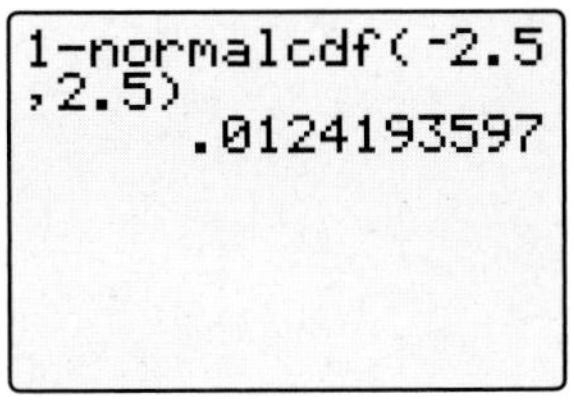

e. $P(z < -4.01)$ is the area under the standard normal curve to the left of $z = -4.01$. Notice that $z = -4.01$ is not in the cumulative normal table for negative z-values, Table A. It is smaller than the z-values in the table, which means that it is further to the left. Thus, the area under the curve is smaller than all of the areas listed in the table. Therefore, $P(z < -4.01)$ is approximately 0.0000.

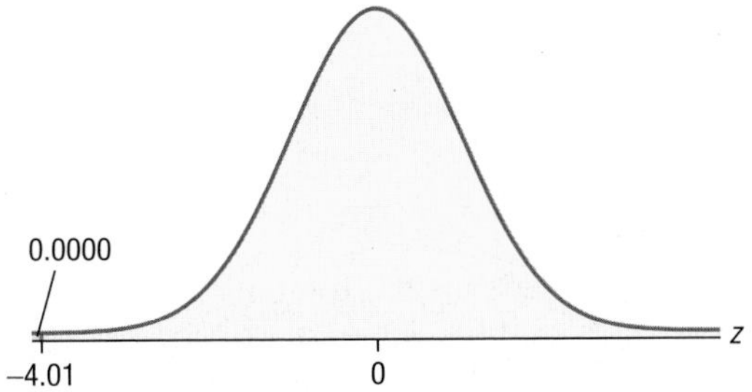

By entering `normalcdf(-1E99,-4.01)`, we find that the area to the left of $z = -4.01$ is approximately 0.00003, as shown in the screenshot in the margin.

f. $P(z \le 3.98)$ is the area under the standard normal curve to the left of $z = 3.98$. Notice that $z = 3.98$ is not in the cumulative normal table for positive z-values, Table B. It is larger than the z-values in the table, which means that it is further to the right. Thus, the area under the curve is larger than all of the areas listed in the table. Therefore, $P(z \le 3.98)$ is approximately 1.0000.

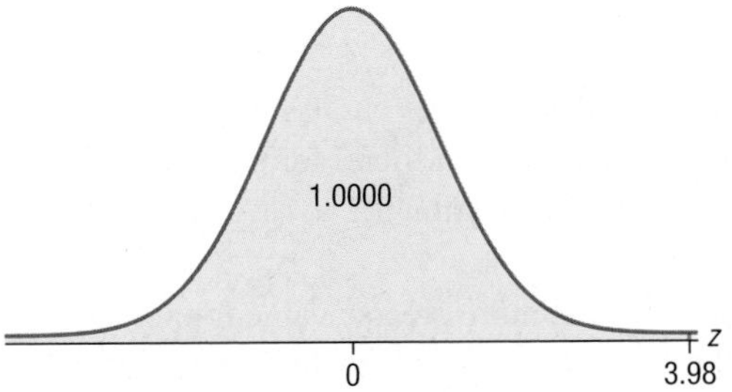

By entering `normalcdf(-1E99,3.98)`, we find that the area to the left of $z = 3.98$ is approximately 1.0000, as shown in the screenshot in the margin.

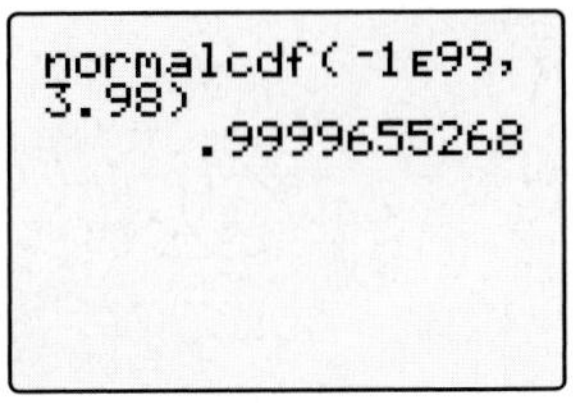

6.2 Section Exercises

Area under the Standard Normal Curve

Directions: Find the area under the standard normal curve to the left of the given *z*-value.

1. $z = 2.35$

2. $z = 1.78$

3. $z = -1.25$

4. $z = -0.19$

5. $z = 2$

6. $z = 1.3$

7. $z = 3.57$

8. $z = -4.12$

Directions: Find the area under the standard normal curve to the right of the given *z*-value.

9. $z = 1.35$

10. $z = 1.7$

11. $z = -2.51$

12. $z = -0.39$

13. $z = -1$

14. $z = 3.68$

Directions: Find the area under the standard normal curve between the given *z*-values.

15. $z_1 = 0.35, \quad z_2 = 1.85$

16. $z_1 = -1.25, \quad z_2 = 2.16$

17. $z_1 = -1.78, \quad z_2 = -0.95$

18. $z_1 = -0.19, \quad z_2 = 1$

19. $z_1 = -3.57, \quad z_2 = 1.85$

20. Find the area under the standard normal curve between $z_1 = -1.00$ and $z_2 = 1.00$. What is the difference between the amount you found for this area and the amount the Empirical Rule approximates for this area? (See Section 3.2 to review the Empirical Rule.)

21. Find the difference between the more precise value found using the normal distribution tables or technology and the approximation given by the Empirical Rule for the area under the standard normal curve within two standard deviations of the mean. Do the same for the area within three standard deviations of the mean. (See Section 3.2 to review the Empirical Rule.)

Directions: Find the total of the areas under the standard normal curve to the left of z_1 and to the right of z_2.

22. $z_1 = -1.46, \quad z_2 = 1.46$

23. $z_1 = -2.11, \quad z_2 = 2.11$

24. $z_1 = -3.05, \quad z_2 = 3.05$

25. $z_1 = -2.31, \quad z_2 = 1.67$

26. $z_1 = -1.75, \quad z_2 = 1.89$

27. $z_1 = -2, \quad z_2 = 1$

28. $z_1 = -3.81, \quad z_2 = 2.37$

29. $z_1 = 1.31, \quad z_2 = 1.93$

30. $z_1 = 0.35, \quad z_2 = 1.75$

31. $z_1 = -2.31, \quad z_2 = -1.67$

32. $z_1 = -3, \quad z_2 = -2$

Probability for the Standard Normal Distribution

Directions: Find each specified probability.

33. $P(z < -3.14)$

34. $P(z < 1.43)$

35. $P(z > 2.72)$

36. $P(z > -0.81)$

37. $P(-1.86 < z < 3.14)$

38. $P(0.78 < z < 2.64)$

39. $P(0 < z < 2.78)$

40. $P(-2.81 < z < -1.14)$

41. $P(z < -1.26 \text{ or } z > 1.26)$

42. $P(z < -2.39 \text{ or } z > 2.39)$

43. Find the probability that z differs from the mean by more than one standard deviation.

44. Find the probability that z differs from the mean by more than two standard deviations.

45. Find the probability that z differs from the mean by less than two standard deviations.

46. Find the probability that z differs from the mean by less than 1.5 standard deviations.

47. $P(z < 0.85 \text{ or } z > -2.34)$

48. $P(-4.92 < z < 3.68)$

49. $P(z > 4.08)$

6.3 Finding Probability Using a Normal Distribution

Recall that one of the key properties of any normal curve is that the area under the curve for any region is equivalent to the probability of the random variable falling within that region. In the previous section, we learned how to find areas under the standard normal curve. But what about finding areas under other normal curves? We can find the area under any normal curve and, therefore, the probability for any normally distributed random variable by first converting the x-values to standard scores and then using the standard normal curve.

Recall from the previous section that there are four basic types of probability problems:

1. Probability that the random variable will be less than some value
2. Probability that the random variable will be greater than some value
3. Probability that the random variable will be between two values
4. Probability that the random variable will be less than one value or greater than another value

Let's look at an example of each type of probability problem.

Example 6.12

Finding the Probability that a Normally Distributed Random Variable Will Be *Less Than* a Given Value

Body temperatures of adults are normally distributed with a mean of 98.60 °F and a standard deviation of 0.73 °F. What is the probability of a healthy adult having a body temperature less than 96.9 °F?

Solution

We are interested in finding the probability that a person's body temperature X is *less than* 96.9 °F, written mathematically as $P(X < 96.9)$. The following graphs illustrate this. First, we need to convert 96.9 to a standard score.

$$\begin{aligned} z &= \frac{x-\mu}{\sigma} \\ &= \frac{96.9-98.60}{0.73} \\ &\approx -2.33 \end{aligned}$$

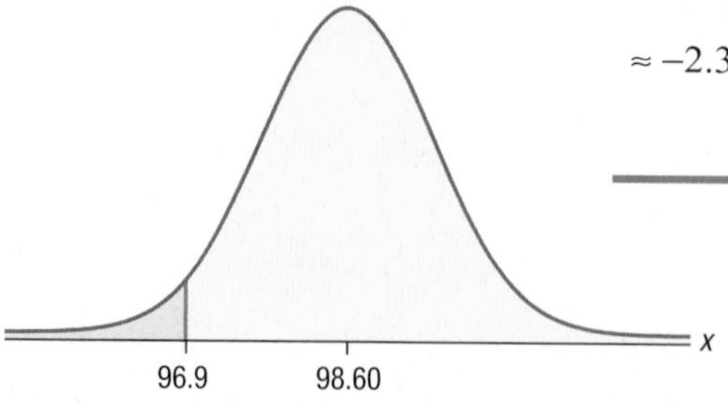

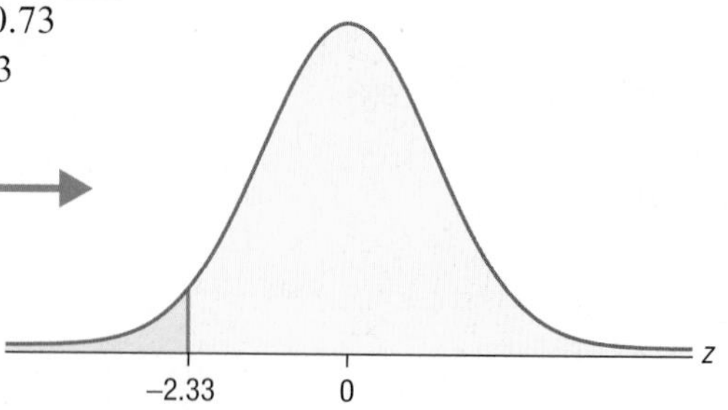

Second, we need to find the area to the left of $z \approx -2.33$, written mathematically as $P(z < -2.33)$. The picture on the right shows this portion under the standard normal curve. Either looking up −2.33 in the cumulative normal table or using a calculator, we see that the shaded region has an area of approximately 0.0099. Thus, the probability of a healthy adult having a body temperature less than 96.9 °F is approximately 0.0099.

$$P(X < 96.9) \approx P(z < -2.33)$$
$$\approx 0.0099$$

Alternate Calculator Method

Note that when using a TI-83/84 Plus calculator to solve this problem, it is not necessary to first convert 96.9 to a standard score. The probability can be found in one step by entering `normalcdf(-1E99,96.9,98.60,0.73)`, as shown in the screenshot in the margin.

Memory Booster

To find an area under any normal curve when using a TI-83/84 Plus calculator, enter `normalcdf(`*lower bound, upper bound,* μ, σ`)`.

normalcdf(-1E99,
96.9,98.60,0.73)
.0099356802

Example 6.13

Finding the Probability that a Normally Distributed Random Variable Will Be *Greater Than* a Given Value

Body temperatures of adults are normally distributed with a mean of 98.60 °F and a standard deviation of 0.73 °F. What is the probability of a healthy adult having a body temperature greater than 97.6 °F?

Solution

This time, we are interested in finding the probability that a person's body temperature X is *greater than* 97.6 °F, written mathematically as $P(X > 97.6)$. The following graph illustrates this. First, convert 97.6 to a standard score.

$$z = \frac{x-\mu}{\sigma}$$
$$= \frac{97.6-98.60}{0.73}$$
$$\approx -1.37$$

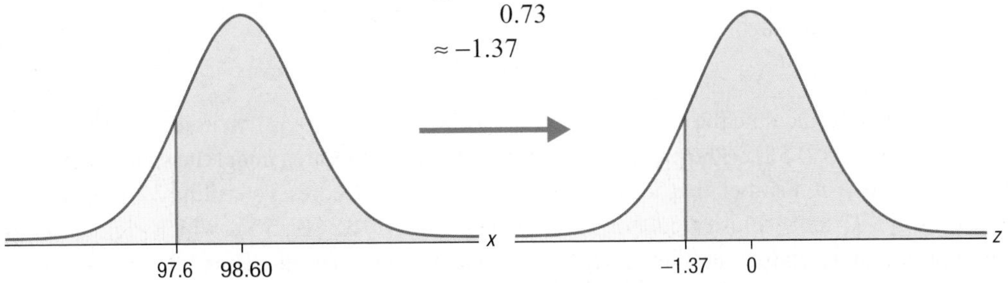
x
97.6
98.60
z
−1.37
0

Second, we need to find the area under the standard normal curve to the right of $z \approx -1.37$, which we denote mathematically as $P(z > -1.37)$. The right graph in the previous figure shows this region under the curve. Using a calculator or the cumulative normal tables, we find that this area is approximately 0.9147. Hence, the probability of a healthy adult having a body temperature greater than 97.6 °F is approximately 0.9147.

$$P(X > 97.6) \approx P(z > -1.37)$$
$$\approx 0.9147$$

Alternate Calculator Method

When using a TI-83/84 Plus calculator to solve this problem, it is not necessary to first convert 97.6 to a standard score. The probability can be found in one step by entering `normalcdf(97.6,1E99,98.60,0.73)`, as shown in the screenshot in the margin. Note that by solving for the probability in one step on the calculator, we eliminate the rounding error that is introduced when we first round the standard score to $z \approx -1.37$. Thus, the calculator gives the more accurate value of approximately 0.9146 for the probability when this method is used.

normalcdf(97.6,1
E99,98.60,0.73)
.9146351091

Example 6.14

Finding the Probability that a Normally Distributed Random Variable Will Be *between* Two Given Values

Body temperatures of adults are normally distributed with a mean of 98.60 °F and a standard deviation of 0.73 °F. What is the probability of a healthy adult having a body temperature between 98 °F and 99 °F?

Solution

We are now interested in finding the probability that a person's body temperature X is *between* 98 °F and 99 °F, written mathematically as $P(98 < X < 99)$. The left graph in the following figure illustrates this region under the normal curve. First, convert both 98 and 99 to standard scores.

For 98 °F:
$$z_1 = \frac{x-\mu}{\sigma} = \frac{98-98.60}{0.73} \approx -0.82$$

For 99 °F:
$$z_2 = \frac{x-\mu}{\sigma} = \frac{99-98.60}{0.73} \approx 0.55$$

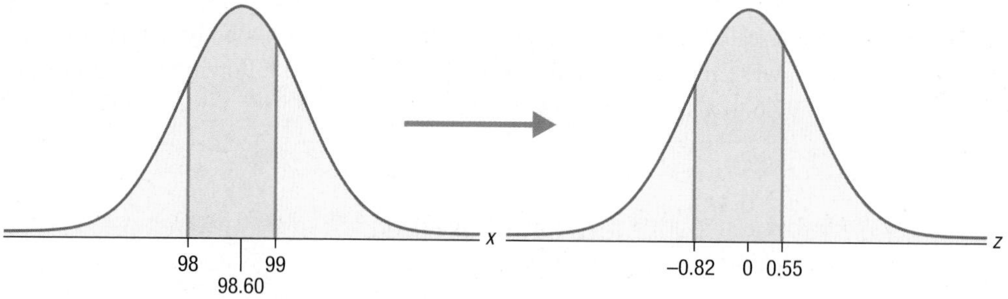

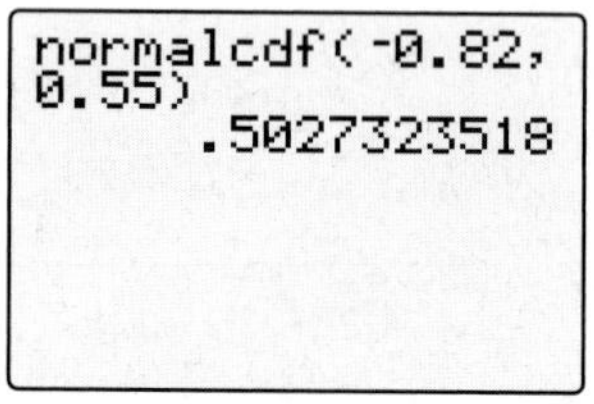

We now need to calculate the area between $z_1 \approx -0.82$ and $z_2 \approx 0.55$, written mathematically as $P(-0.82 < z < 0.55)$. The picture on the right illustrates this area under the standard normal curve. Finally, remember that to find area between two z-scores you simply enter the two z-scores in a TI-83/84 Plus calculator as `normalcdf(-Ø.82,Ø.55)`, which yields a value of approximately 0.5027, as shown in the screenshot in the margin. Thus, the probability of a healthy adult having a body temperature between 98 °F and 99 °F is approximately 0.5027.

$$P(98 < X < 99) \approx P(-0.82 < z < 0.55) \approx 0.5027$$

Alternate Calculator Method

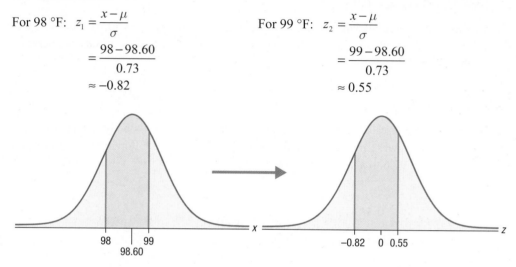

When using a TI-83/84 Plus calculator to solve this problem, it is not necessary to first convert 98 and 99 to standard scores. The probability can be found in one step by entering `normalcdf(98,99,98.6Ø,Ø.73)`, as shown in the screenshot in the margin. Note that by solving for the probability in one step on the calculator, we eliminate the rounding error that is introduced when we first round the standard scores to $z_1 \approx -0.82$ and $z_2 \approx 0.55$. Thus, the calculator gives the more accurate value of approximately 0.5026 for the probability when this method is used.

Example 6.15

Finding the Probability that a Normally Distributed Random Variable Will Be *in the Tails* Defined by Two Given Values

Body temperatures of adults are normally distributed with a mean of 98.60 °F and a standard deviation of 0.73 °F. What is the probability of a healthy adult having a body temperature less than 99.6 °F or greater than 100.6 °F?

Solution

We are interested in finding the probability that a person's body temperature X is either *less than* 99.6 °F or *greater than* 100.6 °F. We can denote this mathematically as $P(X < 99.6 \text{ or } X > 100.6)$. The picture on the left below shows the area under the normal curve in which we are interested. First, convert each temperature to a standard score.

For 99.6 °F: $$z_1 = \frac{x-\mu}{\sigma} = \frac{99.6-98.60}{0.73} \approx 1.37$$

For 100.6 °F: $$z_2 = \frac{x-\mu}{\sigma} = \frac{100.6-98.60}{0.73} \approx 2.74$$

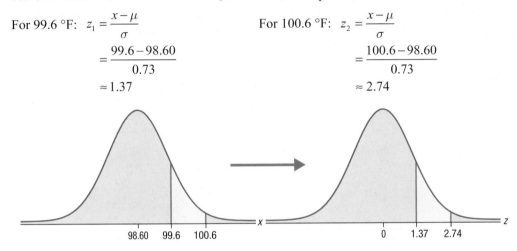

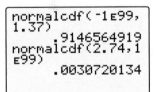

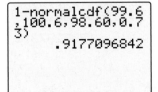

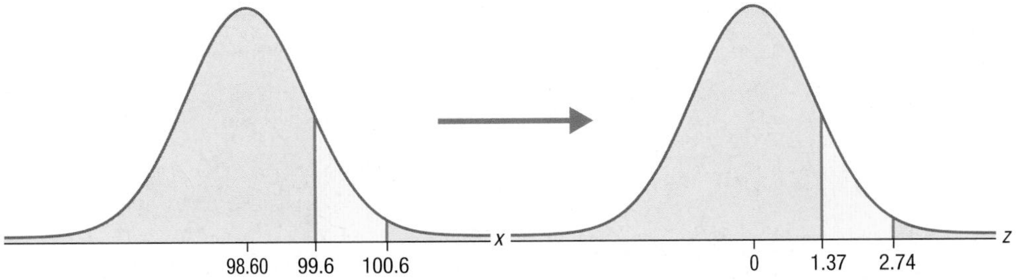

We now need to calculate the sum of the areas to the left of $z_1 \approx 1.37$ and to the right of $z_2 \approx 2.74$. This may be denoted mathematically as $P(z < 1.37 \text{ or } z > 2.74)$. The right graph in the previous figure illustrates these areas under the standard normal curve.

Using a TI-83/84 Plus calculator, we find each area required as shown below and in the screenshot in the margin.

normalcdf(-1E99,1.37) ≈ 0.914656

normalcdf(2.74,1E99) ≈ 0.003072

```
normalcdf(-1E99,
1.37)
         .9146564919
normalcdf(2.74,1
E99)
        .0030720134
```

Adding these together, we have $0.914656 + 0.003072 \approx 0.9177$. Thus, the probability of a healthy adult having a body temperature either less than 99.6 °F or greater than 100.6 °F is approximately 0.9177.

$$P(X < 99.6 \text{ or } X > 100.6) \approx P(z < 1.37 \text{ or } z > 2.74) \approx 0.9177$$

Alternate Calculator Method

Note that when using a TI-83/84 Plus calculator to solve this problem, it is not necessary to first convert 99.6 and 100.6 to standard scores. The probability can be found in one step by entering 1-normalcdf(99.6,1ØØ.6,98.6Ø,Ø.73), as shown in the screenshot in the margin.

```
1-normalcdf(99.6
,100.6,98.60,0.7
3)
         .9177096842
```

Example 6.16

Finding the Probability that a Normally Distributed Random Variable Will *Differ* from the Mean *by More Than* a Given Value

Body temperatures of adults are normally distributed with a mean of 98.60 °F and a standard deviation of 0.73 °F. What is the probability of a healthy adult having a body temperature that differs from the population mean by more than 1 °F?

Solution

Memory Booster

"Differs by less": between

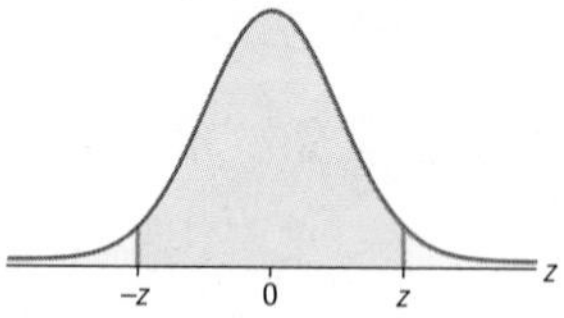

"Differs by more": two tails

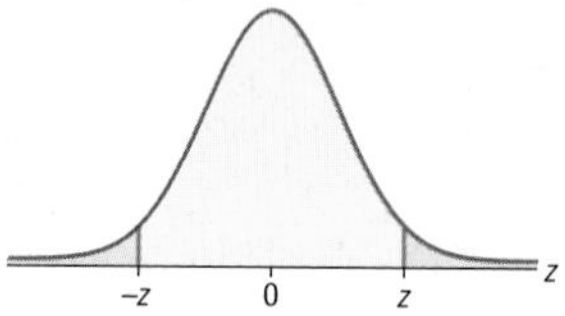

Because we want the temperatures to *differ from the population mean by more than* 1 °F, we need to add 1 °F to and subtract 1 °F from the mean of 98.6 °F to get the temperatures in which we are interested. Doing this tells us we are interested in finding the probability that a person's body temperature X is less than 97.6 °F or greater than 99.6 °F. We can denote this mathematically as $P(X < 97.6 \text{ or } X > 99.6)$. The left graph in the following figure shows the area under the normal curve in which we are interested. Just as we did in the previous example, we first convert each temperature to a standard score.

For 97.6 °F:
$$z_1 = \frac{x-\mu}{\sigma} = \frac{97.6-98.60}{0.73} \approx -1.37$$

For 99.6 °F:
$$z_2 = \frac{x-\mu}{\sigma} = \frac{99.6-98.60}{0.73} \approx 1.37$$

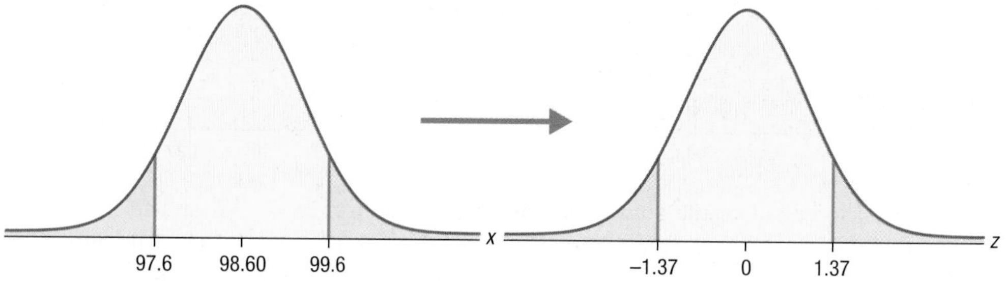

We now need to calculate the sum of the areas to the left of $z_1 \approx -1.37$ and to the right of $z_2 \approx 1.37$. This may be denoted mathematically as $P(z < -1.37 \text{ or } z > 1.37)$. The graph on the right in the previous figure illustrates these areas under the standard normal curve. Notice that these z-scores are equal distances from the mean. Because of the symmetry of the curve, the shaded areas must then be equal. The easiest way to calculate the total area is to find the area to the left of $z_1 \approx -1.37$ and simply double it. We then have $(0.0853)(2) = 0.1706$. Thus, the probability of a healthy adult having a body temperature either less than 97.6 °F or greater than 99.6 °F is approximately 0.1706.

$$P(X < 97.6 \text{ or } X > 99.6) \approx P(z < -1.37 \text{ or } z > 1.37) \approx 0.1706$$

Alternate Calculator Method

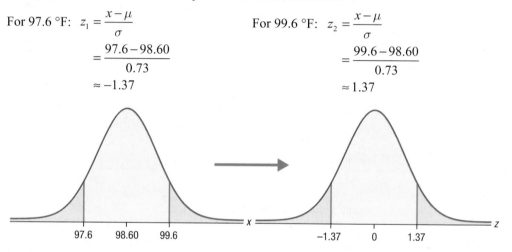

When using a TI-83/84 Plus calculator to solve this problem, it is not necessary to first convert 97.6 and 99.6 to standard scores. The probability can be found in one step by entering `1-normalcdf(97.6,99.6,98.6Ø,Ø.73)`, as shown in the screenshot in the margin. Note that by solving for the probability in one step on the calculator, we eliminate the rounding error that is introduced when we first round the standard scores to $z_1 \approx -1.37$ and $z_2 \approx 1.37$. Thus, when this one-step method is used, the calculator gives the more accurate value of approximately 0.1707 for the probability.

6.3 Section Exercises

Probability for Normal Distributions

Directions: Complete the exercises for the given scenario. Round your answer to two decimal places for a percentage and four decimal places for a probability.

1. Recently, birth weights of Norwegians were reported to have a mean of 3668 grams (g) and a standard deviation of 511 g. Suppose that a Norwegian baby was chosen at random.

a. Find the probability that the baby's birth weight was less than 4000 g.

b. Find the probability that the baby's birth weight was greater than 3750 g.

c. Find the probability that the baby's birth weight was between 3000 g and 4000 g.

d. Find the probability that the baby's birth weight was less than 2650 g or greater than 4650 g.

2. According to a recent report, replacement times for CD players are normally distributed with a mean of 7.1 years and a standard deviation of 1.4 years.

a. Find the probability that a randomly selected CD player will have a replacement time of less than 8.0 years.

b. Find the probability that a randomly selected CD player will have a replacement time of more than 9.0 years.

c. Find the probability that a randomly selected CD player will have a replacement time between 5 and 8 years.

d. Find the probability that a randomly selected CD player will have a replacement time of less than 5.1 or greater than 9.1 years.

3. Data collected on the total number of minutes people spent in the local emergency room revealed that patients spent on average 166.9 minutes in the ER, with a standard deviation of 55.4 minutes. Assume that the times collected follow a normal distribution.

a. What percentage of the ER patient population spent less than 3 hours there?

b. What percentage of the ER patient population spent more than 2 hours there?

c. What percentage of the ER patient population spent between 2.5 hours and 3.5 hours there?

d. What percentage of the ER patient population spent either less than 2.5 hours or more than 3.5 hours there?

4. The heights of American women ages 18 to 29 are normally distributed with a mean of 64.3 inches and a standard deviation of 3.8 inches. An American woman in this age bracket is chosen at random.

a. What is the probability that she is more than 68 inches tall?

b. What is the probability that she is less than 70 inches tall?

c. What is the probability that she is between 64 and 69 inches tall?

d. What is the probability that she is less than 59 or greater than 71 inches tall?

5. Suppose that the mean calorie intake for women is 2050 calories per day, with a standard deviation of 175 calories. Assume that calorie intakes follow a normal distribution.

   a. Find the probability that a female in your class consumes more than 1800 calories per day.

   b. Find the probability that a randomly selected woman consumes fewer than 1500 calories per day.

   c. Even though the "freshman fifteen" is a common occurrence, for a woman to put on 15 pounds in one semester she would have to consume between 2100 and 2600 calories per day. Find the probability that a randomly chosen woman would consume this many calories.

6. The total blood cholesterol levels in a certain Mediterranean population are found to be normally distributed with a mean of 160 milligrams/deciliter (mg/dL) and a standard deviation of 50 mg/dL. Researchers at the National Heart, Lung, and Blood Institute consider this pattern ideal for a minimal risk of heart attacks.

   a. Find the percentage of this population who have blood cholesterol levels less than 150 mg/dL.

   b. Find the percentage of this population who have blood cholesterol levels which exceed the ideal level by at least 10 mg/dL.

   c. Find the percentage of this population who have blood cholesterol levels between 150 and 200 mg/dL.

   d. Find the percentage of this population who have blood cholesterol levels less than 100 mg/dL or greater than 220 mg/dL.

7. Suppose that motorists in the southeastern United States use a mean of 8.20 gallons of gasoline per week with a standard deviation of 0.47 gallons. Assume that gasoline consumption levels are approximately normally distributed.

   a. What percentage of southeastern drivers use more than 9.0 gallons of gasoline per week?

   b. What percentage of southeastern drivers use no more than 7.0 gallons of gasoline per week?

   c. What percentage of southeastern drivers use an amount of gasoline that differs from the mean by more than 0.5 gallons per week?

8. Assume for the moment that the distribution of weights of adult males is approximately normal with a mean of 179.8 lb and a standard deviation of 93.0 lb.

   a. Find the probability that a randomly selected adult male would have a weight at least two standard deviations above the mean.

   b. Find the probability that a randomly selected adult male would have a weight at least two standard deviations below the mean.

   c. Given your answers in parts a. and b., do you think it is reasonable to assume that weights of adult males follow this normal distribution? Explain your answer.

9. Suppose that lifetimes for a particular car battery are normally distributed with a mean of 148 weeks and a standard deviation of 8 weeks.

a. If the company guarantees its battery for 3 years, what percentage of the batteries sold would you expect to be returned before the end of the warranty period? Assume that there are 52 weeks in a year.

b. Imagine you were the CEO of the battery company. Evaluate the warranty offer and list any changes you would make as the CEO.

10. Ella refuses to tell you her weight in her ninth month of pregnancy. However, she does tell you that her weight is above the mean. Which of the following z-scores is possible for her weight?

a. −0.08

b. 1.43

c. 0

d. Not enough information

11. The mean distance a new sales representative travels per week in the first year is 300 miles. At Olivia's job interview, she found out that she would travel at most the mean distance per week for first-year sales reps. Which of the following z-scores are possible for her mean weekly traveling distance during her first year?

a. 0

b. −1.42

c. 0.78

d. Not enough information

6.4 Finding Values of a Normally Distributed Random Variable

In the previous section, we used the properties of the standard normal distribution to solve probability problems. In this section we will use these same properties but in reverse. For example, instead of trying to find the probability of someone having a cholesterol reading of greater than 200 mg/dL, we might want to know what cholesterol reading would represent the 95th percentile. In this case we know the probability; the 95th percentile represents an area equal to 0.9500 under the normal curve to the left of some value of the random variable. We now want to know the value of the random variable, which is the cholesterol reading. Specifically, if we know a probability, we can use the standard normal distribution to find the corresponding z-value and then we can convert the z-score to the corresponding value of the random variable.

Memory Booster

Not sure about your answer? Draw a sketch to make sure your answer makes sense.

In this section, since we will just work the problems from the previous section "in reverse," we're still essentially working with the same four basic types of problems: area to the left, area to the right, area in between, and area in the tails. This time though, we'll be given a particular amount of area and asked to identify the z-score that marks the cutoff point for that area under the normal curve. We will look at examples of each type of problem along with how to find solutions using both the cumulative normal distribution tables from the appendix as well as a TI-83/84 Plus calculator. Instructions on using Microsoft Excel and Minitab can be found in the technology section at the end of the chapter.

Example 6.17

Finding the z-Value with a Given Area to Its Left

What z-value has an area of 0.7357 to its left?

Solution

Caution

Careful! If the problem asks for a z-score, then the answer should come from the edges of the chart, not the middle.

In order to sketch a graph of where the area under the curve lies, we need to shade under more than half of the curve because $0.7357 > 0.5$. Therefore, we know that the z-score will be a positive number.

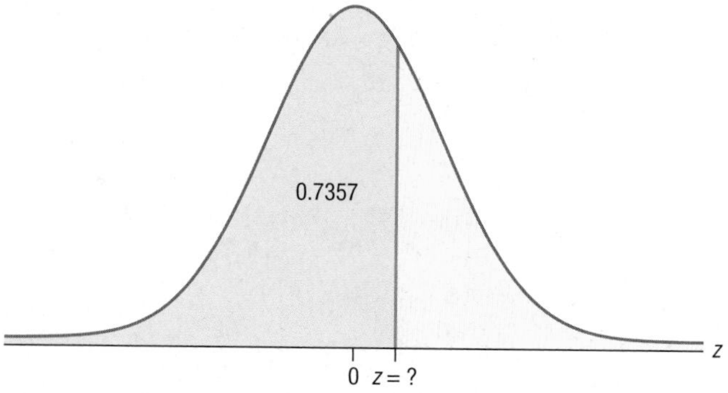

Scan through the *interior* of Table B from Appendix A for an area of 0.7357. Look to the left and up to find the corresponding z-value. Looking at the row and column titles, you will see that $z = 0.63$.

z	0.00	0.01	0.02	0.03	0.04
0.0	0.5000	0.5040	0.5080	0.5120	0.5160
0.1	0.5398	0.5438	0.5478	0.5517	0.5557
0.2	0.5793	0.5832	0.5871	0.5910	0.5948
0.3	0.6179	0.6217	0.6255	0.6293	0.6331
0.4	0.6554	0.6591	0.6628	0.6664	0.6700
0.5	0.6915	0.6950	0.6985	0.7019	0.7054
0.6	0.7257	0.7291	0.7324	0.7357	0.7389
0.7	0.7580	0.7611	0.7642	0.7673	0.7704

Let's also demonstrate how to find the solution by using a TI-83/84 Plus calculator. Begin by pressing 2ND and then VARS. Choose option `3:invNorm(`. This option requires you to enter the area to the left of the z-value that you are interested in finding. For problems where the area to the left is given, such as this one, knowing the value to enter in the calculator is easy. This will become more challenging with the other types of problems. The area to the left of the z-score we are interested in is 0.7357. Enter `invNorm(Ø.7357)`, as shown in the screenshot in the margin, and press ENTER. The answer is $z \approx 0.63$.

Memory Booster

When finding area under a normal curve, use `normalcdf(`.

When finding a z-score or raw score, use `invNorm(`.

Example 6.18

Finding the z-Value with a Given Area to Its Left

Find the value of z such that the area to the left of z is 0.2000.

Solution

First, sketching a graph of the curve shows us that since $0.2000 < 0.5$, our z-score will be a negative value.

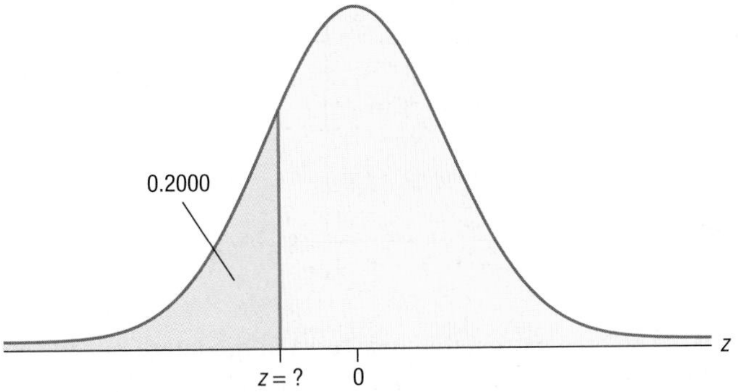

We can then begin to find 0.2000 in the body of the cumulative normal distribution table for negative z-values, Table A from Appendix A. Notice that none of the areas is exactly 0.2000. Our area falls between the areas of 0.2005 and 0.1977 in the table. One method for finding the z-value for an area not in the table is *interpolation*; however, for our purposes, we will use the area closest to the one we want. The value of 0.2005 is the closest one to 0.2000. Its z-value is -0.84. So, the approximate z-score for an area of 0.2000 is $z = -0.84$.

z	0.07	0.06	0.05	0.04	0.03
–1.2	0.1020	0.1038	0.1056	0.1075	0.1093
–1.1	0.1210	0.1230	0.1251	0.1271	0.1292
–1.0	0.1423	0.1446	0.1469	0.1492	0.1515
–0.9	0.1660	0.1685	0.1711	0.1736	0.1762
–0.8	0.1922	0.1949	0.1977	0.2005	0.2033
–0.7	0.2206	0.2236	0.2266	0.2296	0.2327
–0.6	0.2514	0.2546	0.2578	0.2611	0.2643

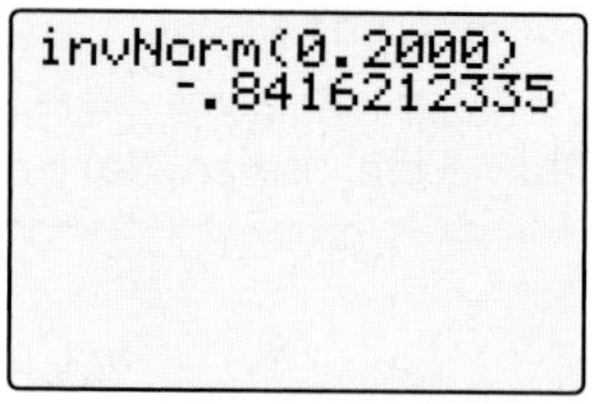

Since the area we are given is the area to the left of the unknown z-value, using a TI-83/84 Plus calculator is straightforward. Begin by pressing 2ND and then VARS. Choose option `3:invNorm(`. The area to the left of the z-score we are interested in is 0.2000. Enter `invNorm(0.2000)`, as shown in the screenshot in the margin, and press ENTER. The answer is $z \approx -0.84$.

Example 6.19

Finding the z-Value That Represents a Given Percentile

What z-value represents the 90th percentile?

Solution

The 90th percentile is the z-value for which 90% of the area under the standard normal curve is to the left of z. So, we need to find the value of z that has an area of 0.9000 to its left.

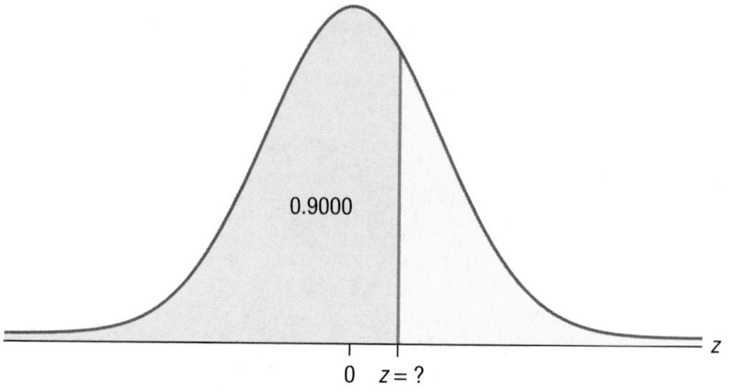

Looking for 0.9000 (or an area extremely close to it) in the interior of the cumulative normal tables, we find 0.8997, which corresponds to a z-value of 1.28. Thus $z \approx 1.28$ represents the 90th percentile.

z	0.05	0.06	0.07	0.08	0.09
1.0	0.8531	0.8554	0.8577	0.8599	0.8621
1.1	0.8749	0.8770	0.8790	0.8810	0.8830
1.2	0.8944	0.8962	0.8980	0.8997	0.9015
1.3	0.9115	0.9131	0.9147	0.9162	0.9177
1.4	0.9265	0.9279	0.9292	0.9306	0.9319
1.5	0.9394	0.9406	0.9418	0.9429	0.9441

Using a TI-83/84 Plus calculator, we can find a value of z with a given area to its left. As noted previously, the 90th percentile is the z-value that has an area of 0.9000 to its left. Enter invNorm(Ø.9ØØØ), as shown in the screenshot in the margin, and press ENTER. The answer is $z \approx 1.28$.

Next, let's consider a problem in which the area given lies to the *right* of z. The method for finding these z-scores using a cumulative normal distribution table is to first find the area to the left of z, since the area to the right is not given in the table. Then, look up the area to the left in the table and find its corresponding z-value.

Example 6.20

Finding the *z*-Value with a Given Area to Its Right

What z-value has an area of 0.0096 to its right?

Solution

First, sketch a graph of the curve. Because 0.0096 is such a small area on the right, we know we should sketch the z-score on the positive side of the graph. Since we need the area to the left, not the area to the right, in order to use the tables, subtract the area to the right from 1. This gives us an area to the left of z equal to 0.9904.

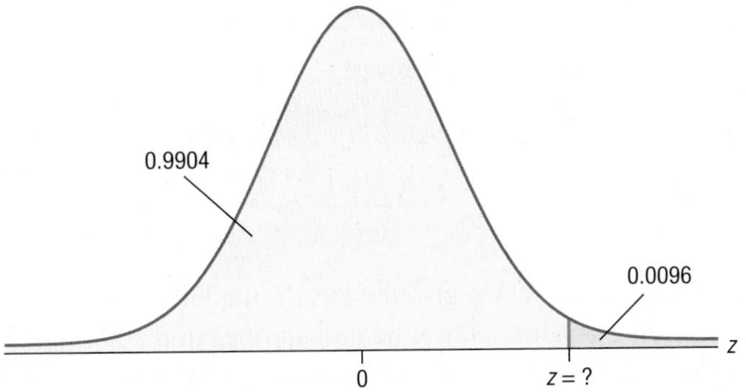

Scan through the interior of the table for an area of 0.9904. Look to the left and up to find the corresponding z-value. What we find is that $z = 2.34$.

z	0.00	0.01	0.02	0.03	0.04
2.0	0.9772	0.9778	0.9783	0.9788	0.9793
2.1	0.9821	0.9826	0.9830	0.9834	0.9838
2.2	0.9861	0.9864	0.9868	0.9871	0.9875
2.3	0.9893	0.9896	0.9898	0.9901	0.9904
2.4	0.9918	0.9920	0.9922	0.9925	0.9927
2.5	0.9938	0.9940	0.9941	0.9943	0.9945

Memory Booster

The area entered must be to the *left* of the unknown z-value when using invNorm(.

Since the area we are given is the area to the right of the unknown z-value, using a TI-83/84 Plus calculator requires that we first find the area to the left of z. We have already calculated that the area to the left is 0.9904, but we can have the calculator find this area at the same time that it finds the z-score. Enter invNorm(1−Ø.ØØ96), as shown in the screenshot in the margin, and press ENTER. The answer is $z \approx 2.34$.

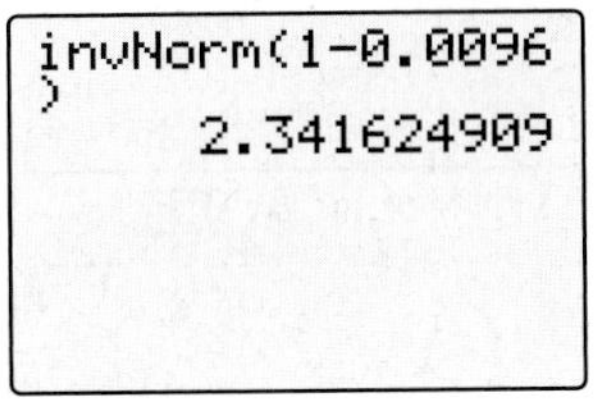

The third problem type, in which the area given lies between $-z$ and z, requires more work than the previous examples. First, find the area remaining in the two tails by subtracting the given area from the total area of one. Next, divide the difference by two to find the area in just the left tail. Use this area to obtain $-z$ from the table or using technology. Finally, because the question asks for the positive value of z, simply switch the sign of z to obtain the correct answer.

Example 6.21

Finding the z-Value with a Given Area between −z and z

Find the value of z such that the area between $-z$ and z is 0.90.

Solution

If the area between $-z$ and z is 0.90, then the total area in the two tails must be $1 - 0.90 = 0.10$. In our sketch, the shaded area will take up most of the area under the curve. Because the curve is symmetric, the area left after subtracting from 1 is divided equally between the two tails. So, half of that area, which is $\frac{0.10}{2} = 0.05$, is in the left tail.

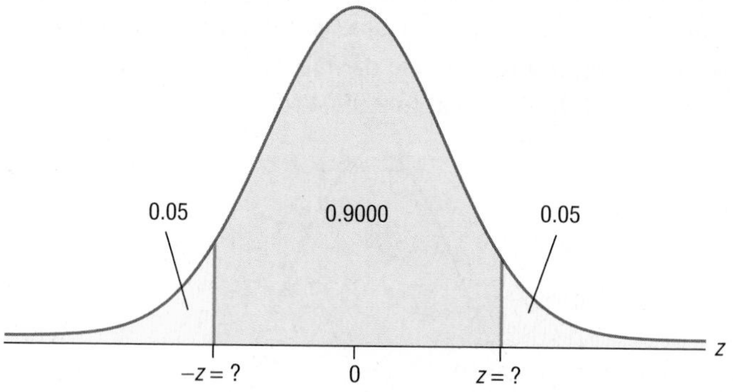

Because the cumulative normal tables give the area to the left, we can look up this tail area and find the corresponding z-value. Since the probabilities in the normal tables are given to four decimal places, we look for the probability that is closest to the left tail area of 0.0500.

z	0.07	0.06	0.05	0.04	0.03
−2.0	0.0192	0.0197	0.0202	0.0207	0.0212
−1.9	0.0244	0.0250	0.0256	0.0262	0.0268
−1.8	0.0307	0.0314	0.0322	0.0329	0.0336
−1.7	0.0384	0.0392	0.0401	0.0409	0.0418
−1.6	0.0475	0.0485	0.0495	0.0505	0.0516
−1.5	0.0582	0.0594	0.0606	0.0618	0.0630

In the interior of the table, we find two areas that are equally close to 0.0500, so use the z-value exactly halfway between the z-scores corresponding to those two areas. The z-value halfway between $z = -1.64$ and $z = -1.65$ is -1.645. Thus the positive z-value such that the area between $-z$ and z is 0.90 is $z = 1.645$.

Since the area we are given is the area between the unknown z-values, using a TI-83/84 calculator requires that we first find the area to the left of $-z$. We have already calculated that the area to the left of $-z$ is 0.05. Enter `invNorm(Ø.Ø5)`, as shown in the screenshot in the margin, and press ENTER. The answer is $-z \approx -1.644854$. Rounded to three decimal places, the positive z-value is 1.645.

The final problem type is a problem in which the area given is contained in the tails of the standard normal curve. Because the curve is symmetric, divide the area in half to find the amount of area in the left tail. Look in the table to find the z-value that corresponds with this area. Finally, as in the previous example, the question asks for the positive value of z. Simply switch the sign of z to obtain the correct answer.

Example 6.22

Finding the *z*-Value with a Given Area in the Tails to the Left of $-z$ and to the Right of z

Find the value of z such that the area to the left of $-z$ plus the area to the right of z is 0.1616.

Solution

If the total area in both tails is 0.1616, then the area in one of the tails must be half of that or

$$\frac{0.1616}{2}=0.0808.$$

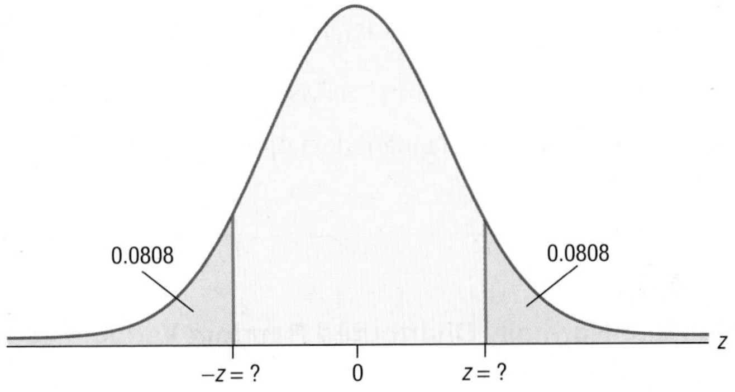

Scan through the interior of the cumulative normal table for an area of 0.0808. The corresponding z-value is $-z=-1.40$. Thus, the z-value such that the area to the left of $-z$ plus the area to the right of z is 0.1616 is $z=1.40$.

z	0.04	0.03	0.02	0.01	0.00
−1.6	0.0505	0.0516	0.0526	0.0537	0.0548
−1.5	0.0618	0.0630	0.0643	0.0655	0.0668
−1.4	0.0749	0.0764	0.0778	0.0793	0.0808
−1.3	0.0901	0.0918	0.0934	0.0951	0.0968
−1.2	0.1075	0.1093	0.1112	0.1131	0.1151
−1.1	0.1271	0.1292	0.1314	0.1335	0.1357

Since the area we are given is the total area in the two tails, we must first find the area to the left of $-z$. We have already calculated that the area to the left of $-z$ is 0.0808, but we can have the calculator find this area at the same time that it finds the z-score. Enter `invNorm(Ø.1616/2)`, as shown in the screenshot in the margin, and press ENTER. The answer is $-z\approx -1.399711$. Rounded to two decimal places, the positive value of z is 1.40.

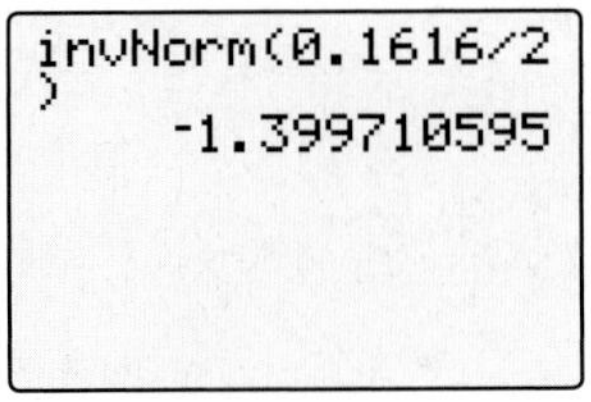

Applications

All of the previous examples have asked for the value of z, but a question could ask for the value of a normally distributed random variable, X. Recall that the steps for finding the area under a normal curve when given a value of X are as follows.

1. Use the values of the mean and standard deviation to convert the value of X to a z-value.
2. Use the z-value to find the area under the standard normal curve.

When we want to find the value of a normally distributed random variable X given the area under a normal curve, we use the steps in reverse. First, we will rearrange the z-score formula, using algebra to solve for x, as shown below.

$$z = \frac{x-\mu}{\sigma}$$
$$z \cdot \sigma = x - \mu$$
$$z \cdot \sigma + \mu = x$$

So $x = z \cdot \sigma + \mu$. Using this altered formula, the steps are as follows.

1. Use the given area under the standard normal curve to find the corresponding value of z.
2. Use the values of z, the mean, and the standard deviation to find the value of the random variable X by using the formula $x = z \cdot \sigma + \mu$.

Example 6.23

Finding the Value of a Normally Distributed Random Variable with a Given Area to Its Right

If a normal distribution has a mean of 28.0 and a standard deviation of 2.5, what is the value of the random variable X that has an area to its right equal to 0.6700?

Solution

Remember that we need to use the steps in reverse. Begin by using the given area to find the value of z. We will use a TI-83/84 Plus calculator to find the solution to this question.

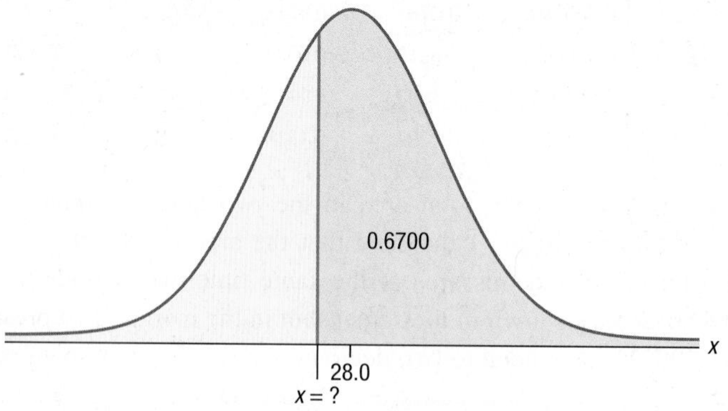

Since we know that the area to the right of x is 0.6700, enter `invNorm(1-Ø.67ØØ)`, as shown in the screenshot in the margin, and press ENTER. The answer is $z \approx -0.439913$. The final step is to use this z-score, along with the given values of the mean and standard deviation, to solve for x. Substituting these values into the formula for x we have the following.

$$\begin{aligned} x &= z \cdot \sigma + \mu \\ &= (-0.439913)(2.5) + 28.0 \\ &\approx 26.9 \end{aligned}$$

Alternate Calculator Method

It is also possible to have the calculator solve this problem directly without doing any intermediate calculations. Instead of finding the z-score first and then solving for x, we can have the calculator find the value of the normally distributed value in one step. The only difference when working with a normal distribution that is not the standard normal distribution is that we must enter the mean and standard deviation in addition to the area to the left of the value in which we are interested. The function syntax is `invNorm(`*area*, μ, σ`)`. Enter `invNorm(1-Ø.67ØØ,28.Ø,2.5)`, as shown in the screenshot in the margin, and press ENTER. The answer is $x \approx 26.9$.

```
invNorm(1-0.6700
,28.0,2.5)
        26.90021708
```

Now that we can find z for a given area, we can use this process to find particular values of a normally distributed random variable X, as we just saw in Example 6.23. This is extremely helpful when we want to know data values that represent cutoff points or mark certain intervals in our data set. A percentile is one type of "cutoff" for which we might want to find the value of the random variable.

Example 6.24

Finding the Value of a Normally Distributed Random Variable That Represents a Given Percentile

The body temperatures of adults are normally distributed with a mean of 98.60 °F and a standard deviation of 0.73 °F. What temperature represents the 90th percentile?

Solution

In order to determine the temperature that represents the 90th percentile, we first need to find the z-value that represents the 90th percentile. In Example 6.19, we found this value to be $z \approx 1.281552$. Once we have the value of z, we can substitute z, σ, and μ into the formula for x that we have from earlier in this section.

```
invNorm(0.9000)
        1.281551567
```

$$\begin{aligned} x &= z \cdot \sigma + \mu \\ &\approx (1.281552)(0.73) + 98.60 \\ &\approx 99.535533 \end{aligned}$$

Thus, a temperature of approximately 99.54 °F represents the 90th percentile.

Alternate Calculator Method

When using a TI-83/84 Plus calculator to solve this problem, it is not necessary to first find the z-value that represents the 90th percentile. The temperature that represents the 90th percentile can be found in one step by entering `invNorm(Ø.9Ø,98.6Ø,Ø.73)`, as shown in the screenshot in the margin.

```
invNorm(0.90,98.
60,0.73)
        99.53553264
```

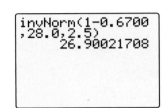

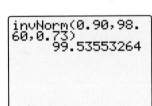

Example 6.25

Finding the Value of a Normally Distributed Random Variable That Represents a Given Quartile

Let's assume that the lengths of newborn full-term babies in the United States are normally distributed with a mean length of 20.0 inches and a standard deviation of 1.2 inches. What is the minimum length that a baby could be and still have a length that is amongst the top 25% of baby lengths?

Solution

If we want to find the minimum length for the top 25% of baby lengths, we can simply find the value of the 75th percentile, or Q_3.

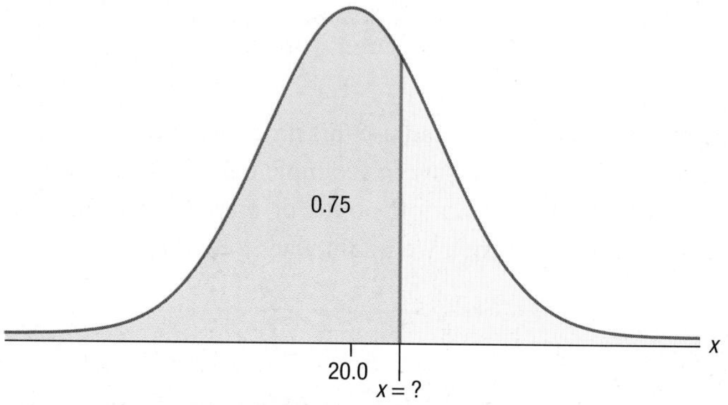

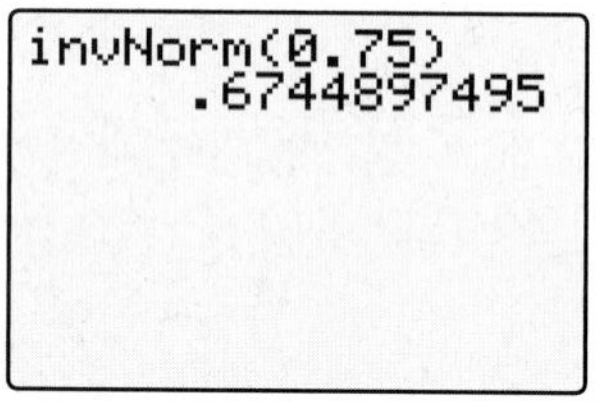

A percentage of 75% is the same as an area of 0.75 under the normal distribution. We will begin by finding the z-score that has an area of 0.75 to its left. Enter `invNorm(Ø.75)`, as shown in the screenshot in the margin, and press ENTER. The answer is $z \approx 0.674490$.

Now that we know the value of z, we must use it along with the mean and standard deviation to calculate the corresponding value of the random variable X that represents the length of the baby. We can use the formula for x that we saw previously.

$$\begin{aligned} x &= z \cdot \sigma + \mu \\ &\approx (0.674490)(1.2) + 20.0 \\ &= 20.809388 \end{aligned}$$

Rounding this to one decimal place (the same as in the standard deviation), we see that the minimum length a baby can be and still be in the top 25% of lengths of full-term newborn babies is approximately 20.8 inches.

Alternate Calculator Method

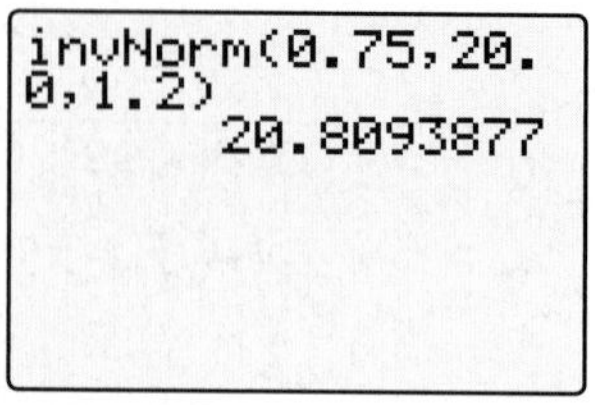

When using a TI-83/84 Plus calculator to solve this problem, it is not necessary to first find the z-score that has an area of 0.75 to its left. The length that represents the 75th percentile can be found in one step by entering `invNorm(Ø.75,2Ø.Ø,1.2)`, as shown in the screenshot in the margin.

6.4 Section Exercises

Finding the z-Value That Corresponds to a Given Area

Directions: Find the indicated value of z.

1. What z-value has an area of 0.0038 to its left?

2. What z-value has an area of 0.9803 to its left?

3. What z-value has an area of 0.0212 to its left?

4. What z-value represents the 95th percentile?

5. What z-value represents the 30th percentile?

6. What z-value represents the 75th percentile?

7. What z-value has an area of 0.0838 to its right?

8. What z-value has an area of 0.9706 to its right?

9. What z-value has an area of 0.5987 to its right?

10. Find the value of z such that the area between $-z$ and z is 0.99.

11. Find the value of z such that the area between $-z$ and z is 0.80.

12. Find the value of z such that the area between $-z$ and z is 0.95.

13. Find the value of z such that the area to the left of $-z$ plus the area to the right of z is 0.5686.

14. Find the value of z such that the area to the left of $-z$ plus the area to the right of z is 0.0286.

15. Find the value of z such that the area to the left of $-z$ plus the area to the right of z is 0.0040.

16. What z-value represents the third quartile?

17. What z-value represents the first quartile?

Applications

Directions: Answer each question for the given scenario.

18. If a normal distribution has a mean of 95.0 and a standard deviation of 8.8, what is the value of the random variable X that has an area to its right equal to 0.0526?

19. If a normal distribution has a mean of 33.7 and a standard deviation of 10.5, what is the value of the random variable X that has an area to its left equal to 0.9904?

20. The body temperatures of adults are normally distributed with a mean of 98.60 °F and a standard deviation of 0.73 °F. What temperature represents the 85th percentile?

21. Heights of river birch trees at a large nursery are approximately normally distributed with a mean of 92.3 inches and a standard deviation of 4.1 inches. What is the cutoff height for birch trees in the tallest 10%?

22. The weights of Jersey cows offered at auction in one region are normally distributed with a mean of 825.0 pounds and a standard deviation of 74.8 pounds. One cattleman endeavors to only bid on cows that are in the top 5% of weight. What is the lowest weight cow that the cattleman should bid on?

23. Suppose that the weights of female college students are normally distributed with a mean of 150 pounds and a standard deviation of 20 pounds. What weight represents the first quartile for female college students?

24. The cruising altitudes for the fleet of one commercial airliner are normally distributed with a mean of 34,950 feet and a standard deviation of 1830 feet. What cruising altitude represents the 80th percentile?

25. In one region of the Caribbean Sea, daily water temperatures are normally distributed with a mean of 77.9 °F and a standard deviation of 2.4 °F. What is the third quartile for water temperatures in this region?

26. During one season of racing at the Talladega Superspeedway, the mean speed of the cars racing there was found to be 158.900 mph with a standard deviation of 6.700 mph. What speed represents the 30th percentile for speeds of race cars at Talladega? Assume that the racing speeds are normally distributed.

27. Suppose that preschoolers spend a mean of 25 hours per week in day care with a standard deviation of 5 hours per week. A newspaper journalist wants to point out that preschoolers are staying in day cares too long, from his perspective. If he only looks at the extreme end of the distribution, that is, the top 3%, what's the minimum number of hours per week those preschoolers spend in day care? Assume that the numbers of hours per week that preschoolers spend in day care are normally distributed.

28. School-age children should drink approximately 40.5 oz of water per day according to a recent report. Suppose the amounts of water that schoolchildren actually consume in a day are approximately normally distributed with a mean of 32.0 oz and a standard deviation of 7.1 oz.

a. What is the probability that a randomly selected student will drink less than the suggested amount of water in a day?

b. If the standard deviation remained the same and the daily water intakes were still normally distributed, how much would the mean need to increase so that only 5% of students drink less than 36 oz per week?

29. Pop-It popcorn maker has a mean time before failure of 36 months with a standard deviation of 5 months, and the failure times are normally distributed. What should be the warranty period, in months, so that the manufacturer will not have more than 10% of the poppers returned?

6.5 Approximating a Binomial Distribution Using a Normal Distribution

In Section 5.2 we learned how to calculate probabilities using a binomial distribution. In this section, we will discuss a technique for approximating these probabilities using a normal distribution.

Let's begin by reviewing the following properties of a binomial distribution.

1. The experiment consists of a fixed number, n, of identical trials.
2. Each trial is independent of the others.
3. For each trial, there are only two possible outcomes. For counting purposes, one outcome is labeled a success, and the other a failure.
4. For every trial, the probability of getting a success is called p. The probability of getting a failure is then $1 - p$.
5. The binomial random variable, X, counts the number of successes in n trials.
6. For a binomial distribution, the mean is given by $\mu = np$ and the variance is given by $\sigma^2 = np(1-p)$.

In order to calculate the probability, $P(X = x)$, for a binomial random variable, we learned to use technology, the binomial table in the appendix (Table D), or the binomial formula, $P(X = x) = {}_nC_x \cdot p^x (1-p)^{(n-x)}$. However, we cannot use the table for values of n larger than 20. Calculators cannot compute combinations for large values of n either, thus making the formula impractical for very large values of n.

Memory Booster

In the binomial probability formula,

$${}_nC_x \cdot p^x (1-p)^{(n-x)},$$

x = the number of successes,

n = the number of trials, and

p = the probability of getting a success on any trial.

If a binomial distribution meets the conditions that $np \geq 5$ and $n(1-p) \geq 5$, then a new distribution can be created with a mean of $\mu = np$ and a standard deviation of $\sigma = \sqrt{np(1-p)}$ that will be approximately normal.

Formula

Normal Distribution Approximation of a Binomial Distribution

If the conditions that $np \geq 5$ and $n(1-p) \geq 5$ are met for a given binomial distribution, then a normal distribution can be used to approximate the binomial probability distribution with the mean and standard deviation given by

$$\mu = np$$

$$\sigma = \sqrt{np(1-p)}$$

where n is the number of trials and

p is the probability of getting a success on any trial.

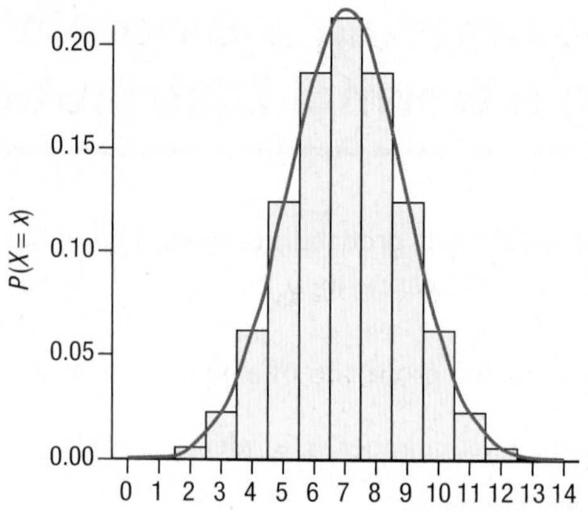

Figure 6.15: Example of Binomial Distribution with Normal Curve

Thus, when we need to calculate a probability for a binomial random variable with a large number of trials, n, we do not need to use Table D or the binomial formula. Instead, we can use the normal distribution to approximate the binomial probability.

Remember that the basic steps in finding a probability using the normal distribution are as follows.

1. Draw a normal curve labeled with the information given in the problem.
2. Convert each value of the random variable to a z-value.
3. Use the standard normal distribution tables, a calculator, or statistical software to find the appropriate area under the standard normal curve.

These steps do not change when we are approximating a binomial distribution; the first step is simply more involved. Let's now look at calculating the mean and standard deviation in more depth.

When using the normal distribution to approximate the binomial distribution, begin by determining the values of n and p. These values are necessary to verify that the conditions $np \geq 5$ and $n(1-p) \geq 5$ are met. Without these conditions being satisfied, the normal curve cannot be used as an appropriate approximation. Once the conditions have been verified, calculate the values of the mean and standard deviation of the underlying binomial random variable, using the formulas $\mu = np$ and $\sigma = \sqrt{np(1-p)}$. Draw a normal curve and mark the value of the mean in the center. Finding the value of x used in the probability calculation involves using a continuity correction, which we will discuss in detail next.

We are trying to use the normal distribution, which is continuous, to approximate the binomial distribution, which is *discrete*. Therefore, a **continuity correction** must be used to convert the whole number value of the discrete binomial random variable to an interval range of the continuous normal random variable. To make this correction, determine the value x of the binomial random variable and convert it to the interval from $(x-0.5)$ to $(x+0.5)$.

Definition

A **continuity correction** is a correction factor used to convert a value of a discrete random variable to an interval range of a continuous random variable when using a continuous distribution to approximate a discrete distribution.

Let's practice using the continuity correction before solving any probability problems.

Example 6.26

Using the Continuity Correction Factor with a Normal Distribution to Approximate a Binomial Probability

Use the continuity correction factor to describe the area under the normal curve that approximates the probability that at least 2 people in a statistics class of 50 students regularly cheat on their math tests. Assume that the number of people in a statistics class of 50 students who consistently cheat on their math tests has a binomial distribution with a mean of 5.00 and a standard deviation of approximately 2.12.

Solution

Begin by converting the discrete number 2 into an interval by adding 0.5 to and subtracting 0.5 from the number 2. The discrete number 2 is changed to the continuous interval from 1.5 to 2.5. Now, draw a normal curve with a mean of 5.00 and a standard deviation of 2.12, and indicate the interval from 1.5 to 2.5 to represent the number 2. Next, shade the area corresponding to the phrase *at least 2*. This would be the area under the curve for x-values greater than or equal to 2. Thus, the area corresponding to *at least 2* would include the interval from 1.5 to 2.5 and all x-values to the right of 2.5.

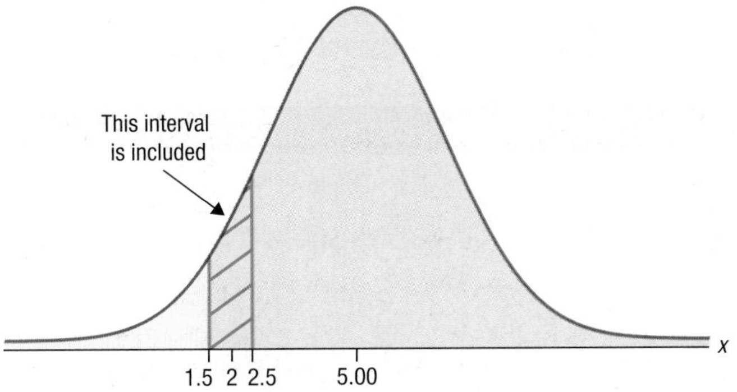

Thus, the area under the normal curve with a mean of 5.00 and a standard deviation of 2.12 that approximates the probability that at least 2 people in the statistics class regularly cheat on their math tests is the area to the right of 1.5.

Example 6.27

Using the Continuity Correction Factor with a Normal Distribution to Approximate a Binomial Probability

Use the continuity correction factor to describe the area under the normal curve that approximates the probability that fewer than 5 out of the 25 students on the 4th floor of a dorm stream TV programs online on their computers instead of using a cable television connection. Assume that the number of students who stream programs on their computers on any given dorm floor with 25 residents has a binomial distribution with a mean of 7.00 and a standard deviation of approximately 2.16.

Solution

To begin, we need to convert the discrete number 5 to the continuous interval from 4.5 to 5.5 by adding 0.5 to and subtracting 0.5 from 5. Draw a normal curve with a mean of 7.00 and a standard deviation of 2.16. Indicate the interval from 4.5 to 5.5 under the curve. Next, shade the area under the curve for x-values corresponding to the phrase *fewer than 5*. We need the area to the left of our interval. Since the phrase *fewer than 5* does not include the number 5, we do *not* shade the interval from 4.5 to 5.5.

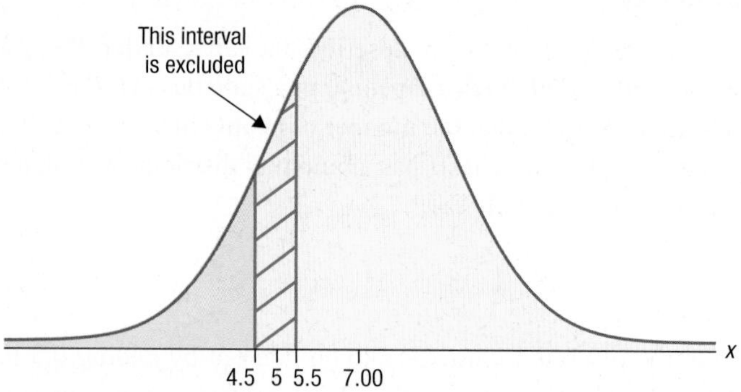

Thus, the area under the normal curve with a mean of 7.00 and a standard deviation of 2.16 that approximates the probability that fewer than 5 students on the 4th floor of the dorm stream TV programs online is the area to the left of 4.5.

To help you use the continuity correction properly, refer to the following table.

Table 6.1: Examples of the Continuity Correction

Statement	Symbolically	Graphically	Area
At least 45, 45 or more, No fewer than 45	≥ 45	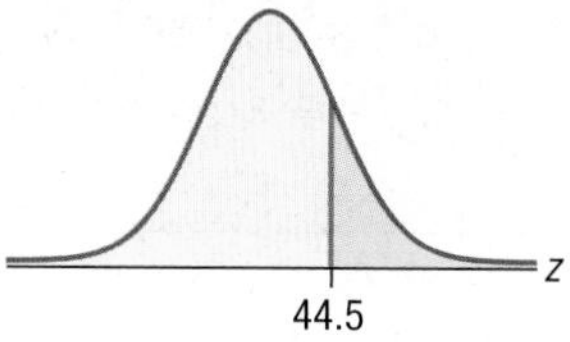	Area to the right of 44.5
More than 45, Greater than 45	> 45	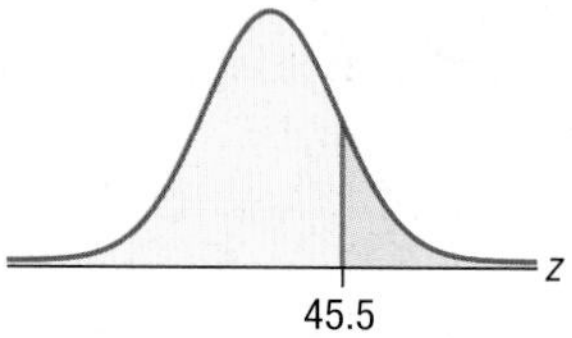	Area to the right of 45.5
At most 45, No more than 45	≤ 45	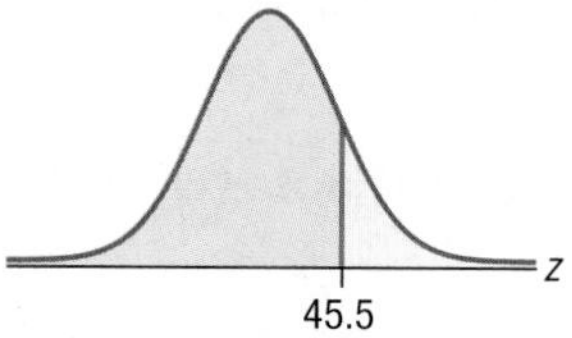	Area to the left of 45.5
Less than 45, Fewer than 45	< 45	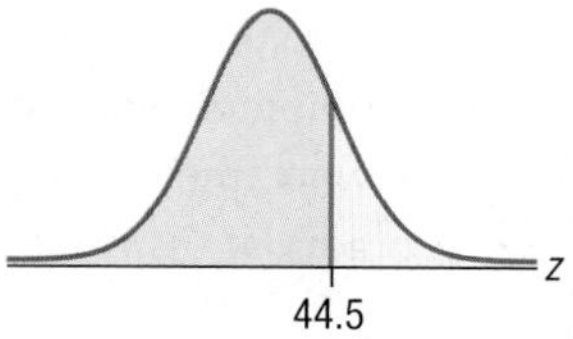	Area to the left of 44.5
Exactly 45, Equal to 45	$= 45$	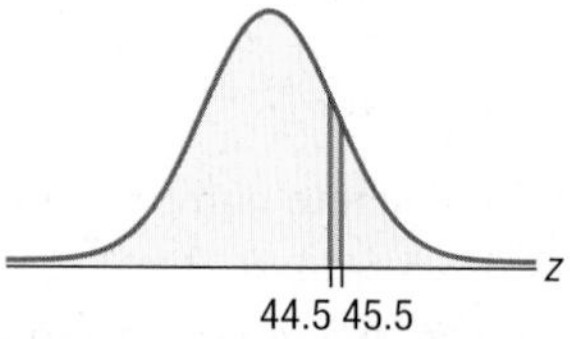	Area between 44.5 and 45.5

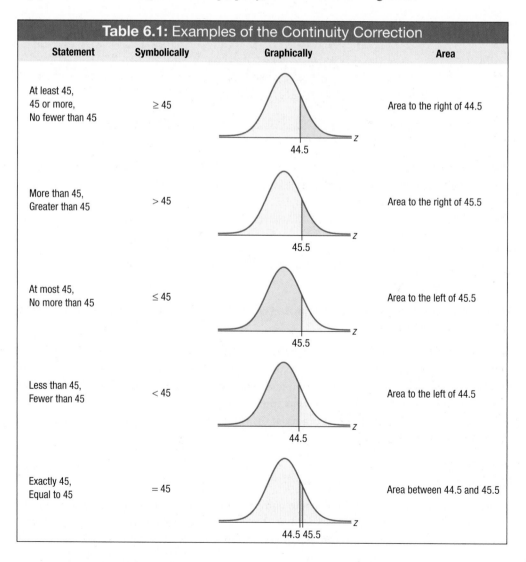

Procedure

Using a Normal Distribution to Approximate a Binomial Distribution

1. Determine the values of n and p.
2. Verify that the conditions $np \geq 5$ and $n(1-p) \geq 5$ are met.
3. Calculate the values of the mean and standard deviation of the binomial random variable using the formulas $\mu = np$ and $\sigma = \sqrt{np(1-p)}$.
4. Use a continuity correction to determine the interval corresponding to the given value of x.
5. Draw a normal curve using the mean and standard deviation calculated in Step 3, and label it with the information given in the problem.
6. Convert each value of the random variable to a z-value.
7. Use the standard normal distribution tables, a calculator, or statistical software to find the appropriate area under the standard normal curve.

Now that we have the tools to translate a discrete value into a continuous interval, let's look at some examples of probability questions.

Example 6.28

Using a Normal Distribution to Approximate a Binomial Probability of the Form $P(X > x)$

Use a normal distribution to estimate the probability of more than 55 girls being born in 100 births. Assume that the probability of a girl being born in an individual birth is 50%.

Solution

First, we are assuming that the probability of having a girl is 50%, so if we let having a girl represent a success, then $p = 0.50$. We also are considering 100 births. So, if each birth represents an individual trial, then $n = 100$. In order to use the normal distribution approximation, we must verify that $np \geq 5$ and $n(1-p) \geq 5$. Substituting the values for n and p into the conditions, we get

$$np = 100(0.50) = 50 \geq 5, \text{ as necessary,}$$

and

$$n(1-p) = 100(1-0.50) = 50 \geq 5, \text{ as necessary.}$$

Thus, the conditions are met and we can use the normal distribution approximation to estimate the binomial probability. Next, we must calculate the mean and standard deviation of the binomial random variable, which are also the mean and standard deviation of the normal distribution we will use to approximate the binomial probability. Substituting the values for n and p into the formulas, we get the following.

$$\begin{aligned}\mu &= np\\ &= 100(0.50)\\ &= 50\end{aligned}$$

$$\begin{aligned}\sigma &= \sqrt{np(1-p)}\\ &= \sqrt{100(0.50)(1-0.50)}\\ &= \sqrt{25}\\ &= 5\end{aligned}$$

Therefore, the mean is 50 and the standard deviation is 5.

Memory Booster

Using a normal distribution to *approximate* a binomial probability results in an APPROXIMATE answer.

We now need to use the continuity correction to determine the interval corresponding to our discrete x-value of 55. By adding and subtracting 0.5, we get the interval from 54.5 to 55.5. Now, we can draw the normal curve with a mean of 50 and a standard deviation of 5. Mark the interval for the area under the curve from 54.5 to 55.5. We are asked for the probability of obtaining *more than 55* girls, so we want the area to the *right* of the interval, but *not* including it. That is, we want the area under the curve to the right of 55.5.

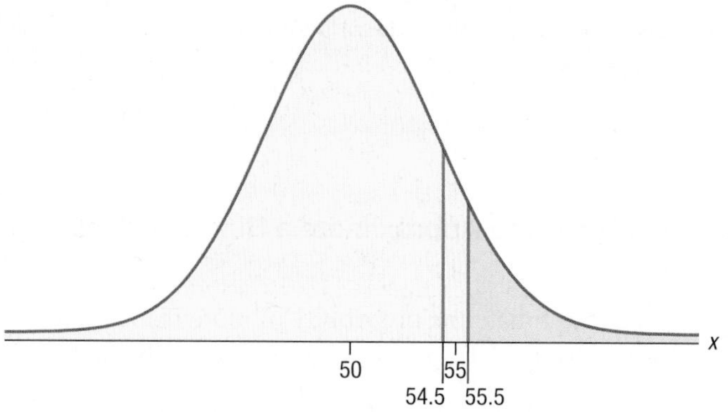

Next, convert the value 55.5 to a z-value using the z-score formula.

$$\begin{aligned}z &= \frac{x-\mu}{\sigma}\\ &= \frac{55.5-50}{5}\\ &= 1.10\end{aligned}$$

Lastly, use the cumulative normal distribution tables to find the area to the right of $z = 1.10$. Using the symmetry property of the standard normal curve, look up the area to the left of $z = -1.10$ instead. This gives us an area of 0.1357. Therefore, the probability of having more than 55 girls out of 100 births is approximately 0.1357.

```
normalcdf(1.10,1
E99)
        .1356661013
```

To use a TI-83/84 Plus calculator to find the area under the curve, you use option `2:normalcdf(` under the `DISTR` (distributions) menu. In this example, we want the area under the standard normal curve to the right of $z = 1.10$. Enter `normalcdf(1.10,1E99)`, as shown in the screenshot in the margin. The area is approximately 0.1357.

Alternate Calculator Method

```
normalcdf(55.5,1
E99,50,5)
        .1356661013
```

When using a TI-83/84 Plus calculator to solve this problem, it is not necessary to convert 55.5 to a z-score. After determining that we are looking for the area to the right of 55.5 under the normal distribution curve with a mean of 50 and a standard deviation of 5, the probability can be found by entering `normalcdf(55.5,1E99,50,5)`, as shown in the screenshot in the margin.

Example 6.29

Using a Normal Distribution to Approximate a Binomial Probability of the Form $P(X \leq x)$

After many hours of studying for your statistics test, you believe that you have a 90% probability of answering any given question correctly. Your test includes 50 true/false questions. Assuming that your estimate is the true probability that you will answer a question correctly, use a normal distribution to estimate the probability that you will miss no more than 4 questions.

Solution

If a trial is a single question on the test, then we have 50 trials, so $n = 50$. Since we are asked about missing questions on the test, we should define a success as *missing* a question. This definition might seem "backwards," but it will result in a more straightforward solution to the problem. If a success is missing a question, and you have a 90% chance of getting any question right, then you have a 10% chance of missing a question. Thus, $p = 0.10$.

Next, we need to verify the conditions that allow us to use the normal curve approximation. Substituting the values $n = 50$ and $p = 0.10$ into the conditions, we get

$$np = 50(0.10) = 5 \geq 5, \text{ as necessary,}$$

and

$$n(1-p) = 50(1-0.10) = 45 \geq 5, \text{ as necessary.}$$

Thus, both conditions are met.

Now, we need to calculate the mean and standard deviation of the binomial random variable, which are also the mean and standard deviation of the normal distribution we will use to approximate the binomial probability. Substituting the values for n and p into the formulas, we get the following.

$$\begin{aligned}\mu &= np \\ &= 50(0.10) \\ &= 5\end{aligned}$$

$$\begin{aligned}\sigma &= \sqrt{np(1-p)} \\ &= \sqrt{50(0.10)(1-0.10)} \\ &= \sqrt{4.5} \\ &\approx 2.121320\end{aligned}$$

Therefore, the mean is 5 and the standard deviation is approximately 2.121320.

Using a continuity correction to determine the interval corresponding to our discrete x-value of 4, add and subtract 0.5 to get the interval from 3.5 to 4.5. Draw a normal curve with a mean of 5 and a standard deviation of 2.121320 and mark the interval for the area under the curve from 3.5 to 4.5. We are asked for the probability of missing *no more than 4* questions, so we want the area to the *left* of the interval and *including* the interval. That is, we want the area under the curve to the left of 4.5.

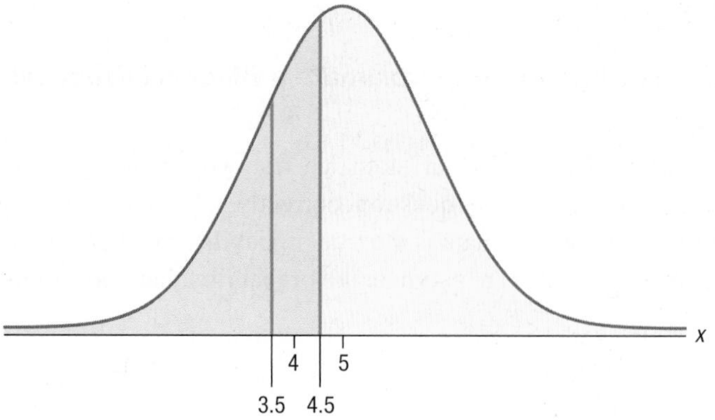

Next, convert the value 4.5 to a z-value using the z-score formula.

$$\begin{aligned} z &= \frac{x-\mu}{\sigma} \\ &= \frac{4.5-5}{2.121320} \\ &\approx -0.235702 \\ &\approx -0.24 \end{aligned}$$

Lastly, use a calculator or the cumulative normal distribution tables to find the area to the left of $z = -0.24$. This gives us an area of 0.4052. Therefore, the probability of missing no more than 4 questions out of 50 is approximately 0.4052.

Alternate Calculator Method

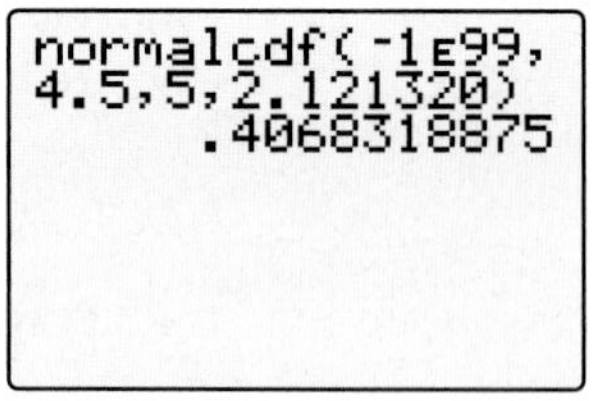

It is also possible to have a TI-83/84 Plus calculator find the area under the normal curve to the left of 4.5 without first calculating the z-score for 4.5 if we enter the mean and standard deviation of the distribution. The probability can be found by entering `normalcdf(1E99,4.5,5,2.12132Ø)`, as shown in the screenshot in the margin. Note that by solving for the probability in one step on the calculator, we eliminate the rounding error that is introduced when we first round the standard score to $z = -0.24$. Thus, the calculator gives the more accurate value of approximately 0.4068 for the probability when this method is used.

Example 6.30

Using a Normal Distribution to Approximate a Binomial Probability of the Form $P(X = x)$

Many toothpaste commercials advertise that 3 out of 4 dentists recommend their brand of toothpaste. Use a normal distribution to estimate the probability that in a random survey of 400 dentists, 300 will recommend Brand X toothpaste. Assume that the commercials are correct, and therefore, there is a 75% chance that any given dentist will recommend Brand X toothpaste.

Solution

Let's define a success to be a dentist who recommends Brand X toothpaste. Then, the probability of obtaining a success is $p = 0.75$. Since we are surveying 400 dentists, $n = 400$. Substituting these values into the conditions that $np \geq 5$ and $n(1-p) \geq 5$, we get

$$np = 400(0.75) = 300 \geq 5, \text{ as necessary,}$$

and

$$n(1-p) = 400(1-0.75) = 100 \geq 5, \text{ as necessary.}$$

Thus, the conditions are met and we can use the normal distribution approximation.

Using these values again, we calculate that the mean is 300 and the standard deviation is approximately 8.660254 as follows.

$$\begin{aligned}\mu &= np \\ &= 400(0.75) \\ &= 300\end{aligned}$$

$$\begin{aligned}\sigma &= \sqrt{np(1-p)} \\ &= \sqrt{400(0.75)(1-0.75)} \\ &= \sqrt{75} \\ &\approx 8.660254\end{aligned}$$

Using the continuity correction, we add and subtract 0.5 to determine the interval corresponding to our discrete x-value of 300. Thus, our interval is 299.5 to 300.5. Draw a normal curve with a mean of 300 and a standard deviation of 8.660254 and mark the interval for the area under the curve from 299.5 to 300.5.

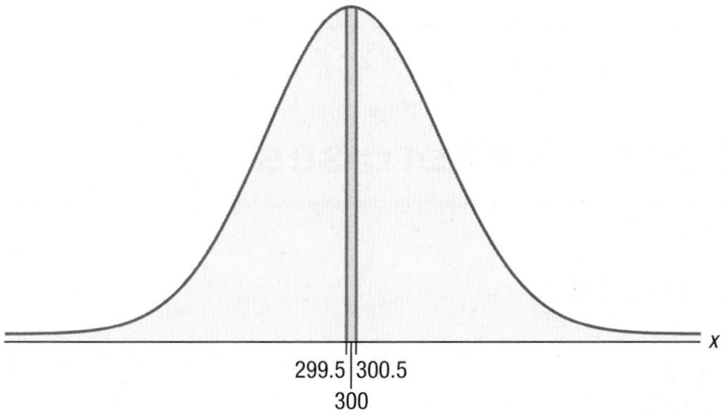

We are interested in the probability that $x = 300$, so we only want the area in our interval. We will need to convert both 299.5 and 300.5 to standard scores and then find the area between them. First, convert 299.5 to a standard score as follows.

$$\begin{aligned}z_1 &= \frac{x-\mu}{\sigma} \\ &= \frac{299.5-300}{8.660254} \\ &\approx -0.06\end{aligned}$$

Converting 300.5 to a standard score, we get the following.

$$\begin{aligned}z_2 &= \frac{x-\mu}{\sigma} \\ &= \frac{300.5-300}{8.660254} \\ &\approx 0.06\end{aligned}$$

Using the cumulative normal distribution tables to find the area to the left of each z-score, $z_1 \approx -0.06$ and $z_2 \approx 0.06$, we find that the areas are 0.4761 and 0.5239, respectively. Subtracting the smaller area from the larger area, we find that the area within our interval is

$$0.5239 - 0.4761 = 0.0478.$$

Thus, the probability of exactly 300 out of a sample of 400 dentists recommending Brand X toothpaste is approximately 0.0478.

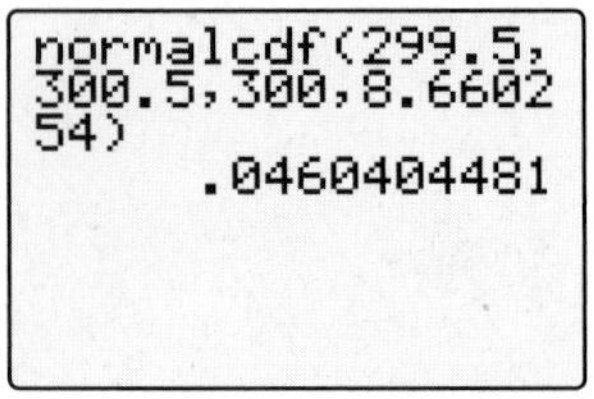
normalcdf(299.5,
300.5,300,8.6602
54)
.0460404481

We can use a TI-83/84 Plus calculator to find a more accurate value for the area under the normal curve between 299.5 and 300.5 if we do not first convert the values to standard scores. Enter `normalcdf(299.5,300.5,300,8.660254)`, which includes the mean and standard deviation of the distribution. As shown in the screenshot in the margin, this method gives the much more accurate estimate of approximately 0.0460 for the probability.

This section has demonstrated how to approximate probabilities for a binomial distribution using a normal distribution. This concept is not specific to the binomial distribution, however. Many distributions that are truly discrete are often estimated using the normal distribution if the shape of the discrete probability distribution closely matches that of a normal distribution. Some examples of random variables that are discrete but whose probabilities are often estimated using the normal distribution include scores on standardized exams and IQ scores. In order to use the normal distribution as an estimate, it is important that the graph of the probability distribution for the random variable be symmetric and bell-shaped. The Chapter 6 Project is an example of using the normal distribution to estimate probabilities for scores on a standardized exam.

6.5 Section Exercises

Continuity Correction

Directions: Describe the area under the normal curve that would be used to approximate the binomial probability.

1. Consider the probability that at least 40 out of 232 planes will malfunction on the runway.

2. Consider the probability that at least 100 out of 250 new mothers received prenatal care.

3. Consider the probability that more than 500 out of 12,000 tax returns were filed incorrectly.

4. Consider the probability that more than 180 out of a sample of 1200 elderly people in the United States will have the flu this winter.

5. Consider the probability that at most 30 out of 534 DVD players on the assembly line are defective.

6. Consider the probability that at most 15 out of 400 high school seniors will not graduate on time.

7. Consider the probability that fewer than 12 out of 367 fourth graders will not pass the state placement test.

8. Consider the probability that 70 out of 100 trees planted in an orchard will live to maturity.

9. Consider the probability that 35 out of 200 registered voters will not vote in the election.

10. Consider the probability that at least 225 out of a sample of 1200 people will use the emergency room at the hospital this year.

Conditions for Using the Normal Distribution Approximation

Directions: Verify that a normal distribution can be used to approximate the binomial probability, or show how the conditions have not been met.

11. Consider the probability that fewer than 15 out of the 123 people watching a movie have already read the book. Assume that the probability of a given person having read the book is 40%.

12. Consider the probability that more than 100 out of 238 fifth graders have seen all of the Harry Potter movies. Assume that the probability of a given fifth grader having seen all of the Harry Potter movies is 54%.

13. Consider the probability that at most 2 out of 30 television sets on an assembly line are defective. Assume that the probability of a given television set being defective is 5%.

14. Consider the probability that no more than 5 out of 120 teenage girls become pregnant before finishing high school. Assume that the teen pregnancy rate is 4%.

Normal Distribution Approximation of Binomial Probability

Directions: Approximate the binomial probability using the normal distribution. You may safely assume that the conditions for using the normal distribution approximation have been met for each scenario.

15. What is the probability that more than 150 out of 230 eighth graders at a local middle school have been exposed to drugs? Assume that a previous study at this school reported that the probability of an individual eighth-grade student having been exposed to drugs is 63%.

16. What is the probability that more than 100 out of 300 elections will contain voter fraud? One report suggests that there is a 32% chance of an individual election containing voter fraud.

17. What is the probability that more than 20 out of a class of 347 high school seniors will drive under the influence of alcohol on prom night? The local chapter of MADD fears that the probability of a high school senior drinking and driving on prom night is 38%.

18. What is the probability that more than 200 out of the 248 hunters staying at a hunting club this season will obtain their game of choice? Club records indicate that hunters have an 83% chance of obtaining their desired game animal.

19. What is the probability that at least 67 out of 100 cars stopped at a roadblock will not be given a ticket? Local authorities report that tickets usually are given to 23% of cars stopped.

20. What is the probability that at least 25 out of a survey of 200 Cajuns do not actually like Cajun food? A regional food critic believes that the probability of a Cajun not liking Cajun food is 14%.

21. What is the probability that at least 140 out of 200 drivers surveyed speed on a regular basis? The state's highway patrol estimates that 78% of drivers typically exceed the speed limit.

22. What is the probability that at least 130 out of 145 preschoolers watch more than four hours of television per night? One previous study indicates that the probability that a preschooler watches more than four hours of television per night is 84%.

23. What is the probability that no more than 32 out of 150 vehicles inspected at a service station will fail their yearly inspection and not receive an inspection sticker? One service station's records indicate that the probability of a car failing its inspection is 18%.

24. What is the probability that no more than 130 out of the 2300 tax returns filed at a local CPA's office will be inaccurate? Previous records indicate only a 7% probability that a given tax return from this office is incorrect.

25. What is the probability that no more than 5 out of 250 puppies born to a well-respected dog breeder will have birth defects? This dog breeder usually averages only 2 birth defects in 50 births.

26. What is the probability that no more than 50 out of 150 former smokers will resume smoking within six months of quitting? Assume that the probability of a former smoker resuming smoking is 35%.

27. What is the probability that fewer than 100 homes out of a sample of 1200 homes will not have a TV? A study indicates that 9 out of 10 homes have televisions.

28. What is the probability that fewer than 100 out of 320 high school graduates will not attend college? The registrar's office at a local college estimates that the probability of a high school graduate going on to college is 68.1%.

29. What is the probability that 125 out of 200 people are overweight? Estimates show that the probability of someone being overweight is 65%.

30. What is the probability that out of 350 entering college freshmen, 100 will need to take developmental mathematics? Research shows that there is a 36% chance of a college freshman needing to take developmental mathematics.

R Chapter 6 Review

Section 6.1: Introduction to the Normal Distribution

Definitions

Normal distribution
A continuous probability distribution for a continuous random variable X

Density curve
The line drawn to show the shape of a continuous distribution

Inflection point
A point on a curve where the curve changes from curving up to curving down, or vice versa

Horizontal asymptote
A horizontal line that a curve approaches as x approaches positive infinity or negative infinity

Standard normal distribution
A normal distribution with a mean of 0 and a standard deviation of 1

Properties of a Normal Distribution

1. A normal distribution is bell-shaped and symmetric about its mean.
2. A normal distribution is completely defined by its mean μ and standard deviation σ.
3. The total area under a normal distribution curve equals 1.
4. The x-axis is a horizontal asymptote for a normal distribution curve.

Standard Score

Also called a z-score or z-value, it indicates how many standard deviations a particular data value is away from the mean and is the corresponding value of the standard normal random variable for a particular value of a normally distributed random variable, given by

$$z = \frac{x - \mu}{\sigma}$$

Section 6.2: Finding Area under a Normal Distribution

Definition

Cumulative normal distribution table
Table that gives the area under the standard normal curve to the left of the given z-value

Using the Cumulative Normal Distribution Tables to Find Areas under the Standard Normal Curve

1. For area to the left of z_0, $P(z \le z_0) = P(z < z_0)$: Look up z_0
2. For area to the right of z_0, $P(z \ge z_0) = P(z > z_0)$: Look up $-z_0$
3. For area between z_1 and z_2,
   $P(z_1 \le z \le z_2) = P(z_1 \le z < z_2) = P(z_1 < z \le z_2) = P(z_1 < z < z_2)$:
   Look up z_1 and z_2 and subtract area to the left of z_1 from area to the left of z_2
4. For area to the left of z_1 plus area to the right of z_2,
   $P(z \le z_1 \text{ or } z \ge z_2) = P(z \le z_1 \text{ or } z > z_2) = P(z < z_1 \text{ or } z \ge z_2) = P(z < z_1 \text{ or } z > z_2)$:
   Look up z_1 and $-z_2$ and add areas

Using a TI-83/84 Plus Calculator to Find Areas under the Standard Normal Curve

1. For area to the left of z_0, $P(z \le z_0) = P(z < z_0)$: Enter `normalcdf(-1E99,`$z_0$`)`
2. For area to the right of z_0, $P(z \ge z_0) = P(z > z_0)$: Enter `normalcdf(`$z_0$`,1E99)`
3. For area between z_1 and z_2,
   $P(z_1 \le z \le z_2) = P(z_1 \le z < z_2) = P(z_1 < z \le z_2) = P(z_1 < z < z_2)$:
   Enter `normalcdf(`z_1`,`z_2`)`
4. For area to the left of z_1 plus area to the right of z_2,
   $P(z \le z_1 \text{ or } z \ge z_2) = P(z \le z_1 \text{ or } z > z_2) = P(z < z_1 \text{ or } z \ge z_2) = P(z < z_1 \text{ or } z > z_2)$:
   Enter `1-normalcdf(`z_1`,`z_2`)`

Section 6.3: Finding Probability Using a Normal Distribution

Relationship between Area and Probability for a Normal Distribution

The area under the normal curve for any region is equivalent to the probability of the normally distributed random variable falling within that region.

Finding Probability for Any Normal Curve

1. Convert each value of the random variable (x-value) to a standard score (z-value).
2. Use the standard normal distribution tables, a calculator, or statistical software to find the appropriate area under the standard normal curve.

Section 6.4: Finding Values of a Normally Distributed Random Variable

Finding the Value of a Normally Distributed Random Variable for a Given Probability

1. Use the standard normal distribution tables, a calculator, or statistical software to find the z-score corresponding to the given area under the standard normal curve.
2. Convert the z-score to the corresponding value of the random variable (x-value) using the formula $x = z \cdot \sigma + \mu$.

Section 6.5: Approximating a Binomial Distribution Using a Normal Distribution

Definition

Continuity correction

A correction factor used to convert a value of a discrete random variable to an interval range of a continuous random variable when using a continuous distribution to approximate a discrete distribution

Using a Normal Distribution to Approximate a Binomial Distribution

1. Determine the values of n and p.
2. Verify that the conditions $np \geq 5$ and $n(1-p) \geq 5$ are met.
3. Calculate the values of the mean and standard deviation of the binomial random variable using the formulas $\mu = np$ and $\sigma = \sqrt{np(1-p)}$.
4. Use a continuity correction to determine the interval corresponding to the given value of x.
5. Draw a normal curve using the mean and standard deviation calculated in Step 3, and label it with the information given in the problem.
6. Convert each value of the random variable to a z-value.
7. Use the standard normal distribution tables, a calculator, or statistical software to find the appropriate area under the standard normal curve.

E Chapter 6 Exercises

Properties of Normal Distributions

Directions: Complete each statement.

1. A normal curve is single-peaked, which tells us that it has __________ mode(s). (State the number of modes.)

2. A normal curve is symmetric about its mean, which tells us that the value of the mean equals the value of the __________ and equals the value of the __________.

3. A normal curve is __________-shaped.

4. You can completely describe a normal distribution by two numbers, the __________ and the __________.

5. The total area under a normal curve equals __________. (State the numerical value.)

Probability, Area, and Standard Scores for Normal Distributions

Directions: Complete each exercise. Round your answer to two decimal places for a percentage and four decimal places for a probability. (Hint: Recall from Chapter 3 that a *z*-score is the number of standard deviations above or below the mean.)

6. What percentage of the values in a normal distribution are more than one standard deviation away from the mean?

7. What percentage of the values in a normal distribution are greater than the mean but less than two standard deviations above the mean?

8. What is the probability that a single value from a normal distribution is greater than three standard deviations above the mean?

9. What is the area under the curve of a normal distribution between one and three standard deviations below the mean?

10. What is the relative frequency of the data greater than two standard deviations above the mean of a normal distribution?

11. What percentage of the data values in a normal distribution are less than the first quartile?

12. Kanuk, the polar bear, is pregnant. If gestation periods for polar bears are normally distributed with a mean of 32.0 weeks and a standard deviation of 1.5 weeks, then what is the likelihood that Kanuk's baby will be born before 30 weeks?

13. A local pizza parlor offers a free pizza if it is not delivered in 30 minutes or less. The mean time between ordering and delivery is 24.0 minutes with a standard deviation of 3.0 minutes, and the times are normally distributed. What percentage of the pizzas will be given away?

14. Candy was surprised when the doctor told her that her baby's weight was in the 85th percentile. If babies' weights are normally distributed, which of the following would be a reasonable standard score for her baby's weight?

a. 85

b. −2.75

c. 1.04

d. 0

e. Not enough information

15. The following table gives values for four normal distributions. Find the missing value in each row of the table.

	μ	σ	x	$P(X \leq x)$
a.	22.0	1.8	24.0	
b.		0.24	116.00	0.3228
c.	0.46		0.72	0.85
d.	1052	66		0.1131

16. The following table gives values for four normal distributions. Find the missing values in each row of the table.

	μ	σ	Q_1	Q_3
a.	15.0	2.1		
b.		200	3300	
c.			55.0	85.0
d.	1.8			2.2

17. The US Department of Labor reports that in 2009 self-employed individuals worked an average of 35.6 hours per week. Do you think that this variable could be represented by a normal distribution? If the distribution is not actually normal, how would you expect it to deviate from normal?

Source: Hipple, Steven F. "Self-employment in the United States." *Monthly Labor Review*, Vol. 133, No. 9. US Department of Labor, Bureau of Labor Statistics. September 2010. http://www.bls.gov/opub/mlr/2010/09/art2full.pdf (12 July 2012).

18. The mean demand for electricity in the average household is 936 kilowatt hours (kWh) in any given month. Suppose the usage is described by a normal random variable with a standard deviation of 110 kWh. If this description of the demand for electricity is correct, what generating capacity per household must be available in order for the load to be met 99.5% of the time?

19. In order to ride the Death Trap at the amusement park, guidelines say that participants must be between 60 and 74 inches tall. Given that the mean height of women is 64.00 inches with a standard deviation of 1.77 inches and the mean height of men is 69.50 inches with a standard deviation of 1.50 inches, give the percentage of each gender that would be able to ride the Death Trap. Assume that the heights of men and women are normally distributed.

Normal Distribution Approximation of Binomial Probability

Directions: Approximate the binomial probability using the normal distribution. You may safely assume that the conditions for using the normal distribution approximation have been met for each scenario.

20. A phone company reports that 51% of its customers subscribe to its high-speed Internet service. If 180 customers are randomly selected for a survey, what is the probability that more than 94 of them subscribe to the high-speed Internet service?

21. What is the probability that more than 35 out of 119 college algebra students will attend a late afternoon study session provided by their professor? Previous experience tells the professor that there is a 40% probability that a given student will attend the study session.

22. What is the probability that fewer than 45 calculus students will fail their calculus course this semester? There are 210 students currently enrolled in the class, and previous semesters indicate that there is a 28% chance of an individual student failing this course.

P Chapter 6 Project

Curving Grades Using a Normal Distribution

Dr. Smith, a biology professor at Bradford University, has decided to give his classes a standardized biology exam that is nationally normed. This indicates that the normal distribution is an appropriate approximation for the probability distribution of students' scores on this exam. The probability distribution of students' scores on this standardized exam can be estimated using the normal distribution shown below.

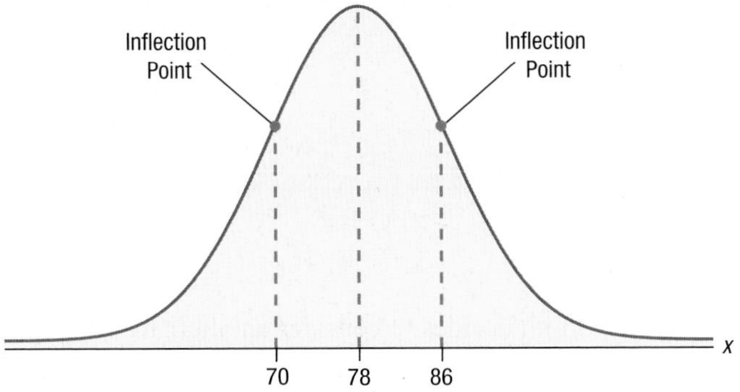

1. State the mean of the distribution of the biology exam scores.

2. State the standard deviation of the distribution of the biology exam scores.

Grading Curve Option I

Originally, Dr. Smith decides to curve his students' exam grades as follows.

- Students whose scores are at or above the 90th percentile will receive an A.
- Students whose scores are in the 80th–89th percentiles will receive a B.
- Students whose scores are in the 70th–79th percentiles will receive a C.
- Students whose scores are in the 60th–69th percentiles will receive a D.
- Students whose scores are below the 60th percentile will receive an F.

3. Find the z-scores that correspond to the following percentiles.

- 90th percentile
- 80th percentile
- 70th percentile
- 60th percentile

4. Using that information, find the exam scores that correspond to the curved grading scale. Assume that the exam scores range from 0 to 100. (Round to the nearest whole number.)

A: ___________–100

B: ___________–___________

C: ___________–___________

D: ___________–___________

F: 0–___________

5. The following is a partial list of grades for students in Dr. Smith's class. Using the grading scale you just created, find the new curved *letter grades* that the students will receive on their tests given their raw scores.

Biology Exam Grades	
Name	**Raw Score / Grade**
Adam	82 / B
Bill	77 / C
Susie	91 / A
Troy	86 / B
Sharon	75 / C
Laura	66 / D
Eric	88 / B
Marcus	69 / D
Stephanie	79 / C

Grading Curve Option II

After reviewing the results, Dr. Smith decides to consider an alternate curving method. He decides to assign exam grades as follows.

- A: Students whose scores are at least two standard deviations above the mean of the standardized test.
- B: Students whose scores are from one up to two standard deviations above the mean of the standardized test.
- C: Students whose scores are from one standard deviation below the mean up to one standard deviation above the mean of the standardized test.
- D: Students whose scores are from two standard deviations below the mean up to one standard deviation below the mean of the standardized test.
- F: Students whose scores are more than two standard deviations below the mean of the standardized test.

6. Using the previous information, create Dr. Smith's new grading scale. (Round to the nearest whole number.)

A: __________–100

B: __________–__________

C: __________–__________

D: __________–__________

F: 0–__________

7. Using the grading scale you just created, return to the partial list of grades and find the new curved *letter grades* that the students will receive on their tests given their raw scores.

8. Review the grades each student received using the two grading scales. Which grading scale do you feel is fairer? Explain why.

T Chapter 6 Technology

Normal Probability Distributions

Note: *When using technology to find an area under a normal curve, the value may be slightly different than the value found using the normal distribution tables. The value found using technology is the more accurate of the two values.*

TI-83/84 Plus

Finding Area under a Normal Curve

To use a TI-83/84 Plus calculator to find an area under any normal curve, use option `2:normalcdf(` under the `DISTR` menu. Press 2ND and then VARS to access the menu and then scroll down to option 2. The function syntax is `normalcdf(`*lower bound, upper bound,* μ, σ`)`. This means that you do not have to convert data values to z-scores before using a TI-83/84 Plus calculator. If you want to find an area under the standard normal curve, you do not need to enter μ or σ. It is sufficient to enter `normalcdf(`*lower bound, upper bound*`)`.

Side Note

The most recent TI-84 Plus calculators (Jan. 2011 and later) contain a new feature called `STAT WIZARDS`. This feature is not used in the directions in this text. By default this feature is turned `ON`. `STAT WIZARDS` can be turned `OFF` under the second page of MODE options.

Example T.1

Using a TI-83/84 Plus Calculator to Find Areas under Normal Curves

Use a TI-83/84 Plus calculator to find the indicated area under the appropriate normal curve.

a. Area between $z_1 = -1.23$ and $z_2 = 2.04$.

b. Area to the left of $z = 2.37$.

c. Area to the right of $z = -1.09$.

d. Area under a normal curve with a mean of 64.0 and a standard deviation of 5.3 to the left of $x = 58$.

Solution

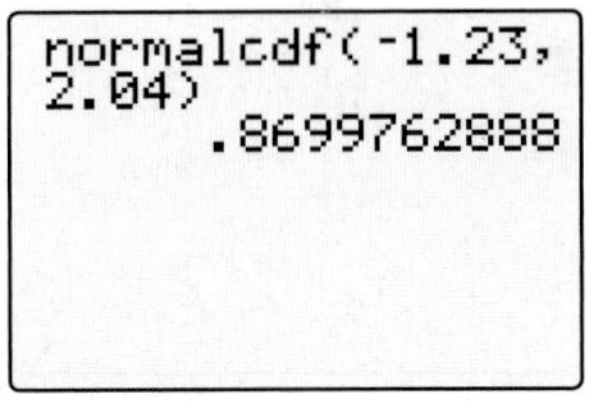

a. Press 2ND and then VARS to access the `DISTR` menu and then scroll down to option 2. Since this is the area under the standard normal curve, we do not need to enter μ or σ. Enter `normalcdf(-1.23,2.04)`, as shown in the screenshot in the margin. The area is approximately 0.8700.

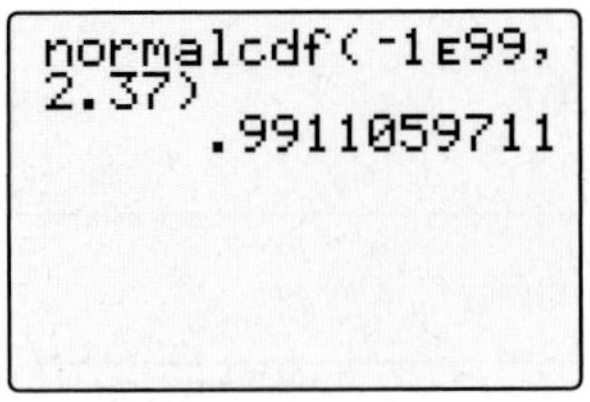

b. Press 2ND and then VARS to access the menu and then scroll down to option 2. Since this is the area under the standard normal curve, we do not need to enter μ or σ. We want the area to the left, so the lower bound would be $-\infty$. We cannot enter $-\infty$, so we will enter a very small value for the lower endpoint, such as -10^{99}. This number appears as `-1E99` when entered correctly into the calculator. To enter `-1E99`, press (−) 1 2ND , 9 9. Enter `normalcdf(-1E99,2.37)`, as shown in the screenshot in the margin. The area is approximately 0.9911.

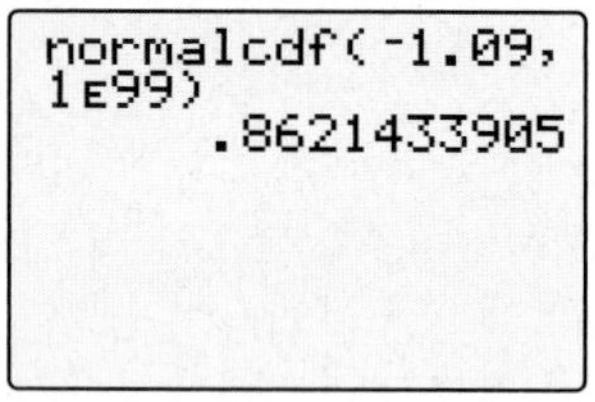
normalcdf(-1.09,
1E99)
.8621433905

c. Press 2ND and then VARS to access the menu and then scroll down to option 2. Again, this is the area under the standard normal curve, so we do not need to enter μ or σ. We want the area to the right, so the upper bound would be ∞. We cannot enter ∞, so we will enter a very large value for the upper endpoint, such as 10^{99}. This number appears as `1E99` when entered correctly into the calculator. To enter `1E99`, press 1 2ND , 9 9. Enter `normalcdf(-1.Ø9,1E99)`, as shown in the screenshot in the margin. The area is approximately 0.8621.

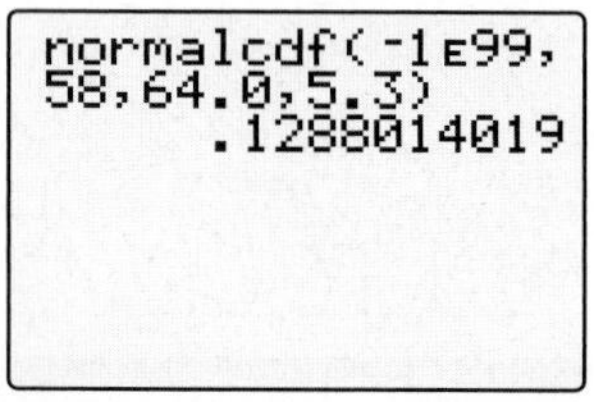
normalcdf(-1E99,
58,64.0,5.3)
.1288014019

d. Press 2ND and then VARS to access the menu and then scroll down to option 2. This is the area under a normal curve, but not the standard normal curve, so we will enter $\mu = 64.0$ and $\sigma = 5.3$. We want the area to the left, so we will enter a very small value for the lower endpoint, such as -10^{99}. Enter `normalcdf(-1E99,58,64.Ø,5.3)`, as shown in the screenshot in the margin. The area is approximately 0.1288.

Finding Values of Normally Distributed Random Variables

A TI-83/84 Plus calculator can find a z- or x-value for a given probability also. To find an x- or z-value, use option `3:invNorm(` under the `DISTR` menu. Press 2ND and then VARS to access the menu and then scroll down to option 3. The function syntax is `invNorm(`*area*, μ, σ`)`. If you want to find z instead of x, you do not need to enter μ or σ. It is sufficient to enter `invNorm(`*area*`)`. The *area* is the area to the left of x or z only. If you know the area to the right or the area between, you must first determine the corresponding area to the left.

Example T.2

Using a TI-83/84 Plus Calculator to Find Values of Normally Distributed Random Variables

Use a TI-83/84 Plus calculator to find the indicated value.

a. The value of z with an area of 0.6781 to its left.

b. The value of z with an area of 0.2500 to its right.

c. The value of z such that 90% of the area is between z and $-z$.

d. A normal distribution has a mean of 25.00 and a standard deviation of 2.45. What value of x has an area of 0.3761 to its left?

Solution

invNorm(0.6781)
.4623923306

a. Press 2ND and then VARS to access the `DISTR` menu and then scroll down to option 3. Enter `invNorm(Ø.6781)`, as shown in the screenshot in the margin. The value of z is approximately 0.46.

invNorm(0.7500)
.6744897495

b. Since the area given is the area to the right, we must first determine the area to the left. The area to the left is $1 - 0.2500 = 0.7500$. Press 2ND and then VARS to access the menu and then scroll down to option 3. Enter `invNorm(Ø.75ØØ)`, as shown in the screenshot in the margin. The value of z is approximately 0.67.

c. Since the area given is between two values of z, $1 - 0.90 = 0.10$ is the area in the two tails. Thus, half of that area, or 0.05, is the area to the left of $-z$. Press 2ND and then VARS to access the menu and then scroll down to option 3. Enter `invNorm(Ø.Ø5)`, as shown in the screenshot in the margin. The value returned by the calculator is approximately -1.644854. Note that the value we obtained earlier in this chapter using the normal distribution table was 1.645.

invNorm(0.05)
-1.644853626

d. This is not the standard normal distribution, so we will need to enter $\mu = 25.00$ and $\sigma = 2.45$. The area given is to the left, so we can enter it directly. Press 2ND and then VARS to access the menu and then scroll down to option 3. Enter `invNorm(Ø.3761,25.ØØ,2.45)`, as shown in the screenshot in the margin. The value of x is approximately 24.23.

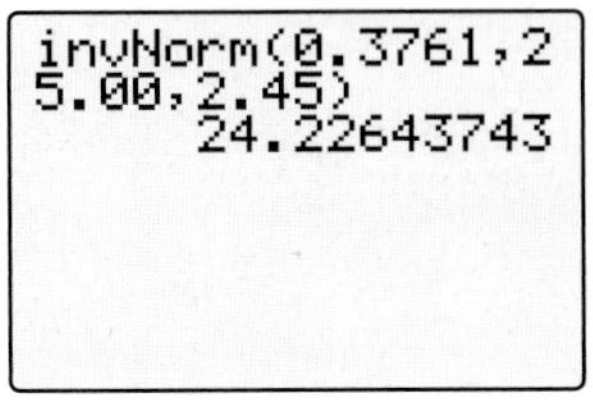
invNorm(0.3761,2
5.00,2.45)
24.22643743

Microsoft Excel

Finding Area under a Normal Curve

The formula that can be used to find an area under any normal curve in Microsoft Excel 2010 is =NORM.DIST(*x, mean, standard_dev, cumulative*). The value of the random variable in which you are interested is x, *mean* is μ, the mean of the distribution, *standard_dev* is σ, the standard deviation of the distribution, and *cumulative* is TRUE, which can be entered as the number 1, so that we will get the area to the left, just as in the cumulative normal distribution tables. Further, Microsoft Excel only calculates the area to the left of the given value of x or z. To find the area between two values or to the right of a value, you must use the methods discussed in this chapter. Note that the equivalent function for calculating an area under a normal curve in Excel 2007 and earlier versions of Excel, NORMDIST, is still available in Excel 2010 to maintain compatibility.

To find an area under the standard normal curve to the left of a particular z-value in Microsoft Excel 2010 without having to enter the mean of 0 and the standard deviation of 1, use the formula =NORM.S.DIST(*z, cumulative*). The z-score in which you are interested is z, and *cumulative* is TRUE, which can be entered as the number 1, so that we will get the area to the left, just as in the cumulative normal distribution tables. Note that the equivalent function for calculating an area under the standard normal curve in Excel 2007 and earlier versions of Excel, NORMSDIST, is still available in Excel 2010 to maintain compatibility.

Finding Values of a Normally Distributed Random Variable

The formula to find a z- or x-value of a normally distributed random variable for a given probability in Microsoft Excel 2010 is =NORM.INV(*probability, mean, standard_dev*). The *probability* is the area under the normal distribution to the left of the x- or z-value; thus if the area we are given is to the right or the area in between, we must first determine the area to the left, just as we did with the TI-83/84 Plus calculator. The *mean* is μ, the mean of the distribution, and *standard_dev* is σ, the standard deviation of the distribution. Note that the equivalent function for calculating a value of a normally distributed random variable in Excel 2007 and earlier versions of Excel, NORMINV, is still available in Excel 2010 to maintain compatibility.

To find a z-score for a given probability in Microsoft Excel 2010 without having to enter the mean of 0 and the standard deviation of 1, use the formula =NORM.S.INV(*probability*). The *probability* is the area under the standard normal curve to the left of the z-value you are calculating. Note that the equivalent function for calculating a z-score for a given probability in Excel 2007 and earlier versions of Excel, NORMSINV, is still available in Excel 2010 to maintain compatibility.

Example T.3

Using Microsoft Excel to Find a Value of a Normally Distributed Random Variable

The body temperatures of adults are normally distributed with a mean of 98.60 °F and a standard deviation of 0.73 °F. What temperature represents the 90th percentile?

Solution

The 90th percentile is the temperature x for which 90% of the area under the normal curve is to the left of x. So, we need to find the value of the normally distributed random variable with a mean of 98.60 and a standard deviation of 0.73 that has an area of 0.90 to its left. Thus, we enter the following formula: **=NORM.INV(0.90, 98.60, 0.73)**.

This formula returns the value 99.53553264. Thus, a temperature of approximately 99.54 °F represents the 90th percentile.

MINITAB

Finding Area under a Normal Curve

To find an area under a normal curve in MINITAB, enter the given x- or z-value in the first column and row. Once the value is entered, go to **Calc ► Probability Distributions ► Normal**. When the Normal Distribution menu appears, make sure **Cumulative probability** is selected and enter the **Mean** and **Standard deviation**. Select **C1** as the **Input column**. Once you are finished, click **OK**, and the probability will appear in the Session window. Like Excel, MINITAB only calculates the area to the left of the given value of x or z. To find the area between two values or to the right of a value, you must use the methods discussed in this chapter.

Finding Values of a Normally Distributed Random Variable

To find a z- or x-value for a given probability in MINITAB, enter the probability in the first column and row. Once the probability is entered, go to **Calc ► Probability Distributions ► Normal**. When the Normal Distribution menu appears, select **Inverse cumulative probability** and enter the **Mean** and **Standard deviation**. Select **C1** as the **Input column**. Once you are finished, click **OK**, and the z- or x-value will appear in the Session window. The probability is the area to the left of the z- or x-value that is calculated by MINITAB. Thus, if the area we are given is to the right or the area in between, we must first determine the area to the left, just as we did with the TI-83/84 Plus calculator.

Example T.4

Using MINITAB to Find the z-Value with a Given Area to Its Left

Find the value of z given that the area to the left of z is 0.9265.

Solution

Enter **0.9265** in the first column and row and then go to **Calc ► Probability Distributions ► Normal**. Select **Inverse cumulative probability**, and make sure the **Mean** is **0.0** and the **Standard deviation** is **1.0**. Enter **C1** as the **Input column** and click **OK**. The Normal Distribution dialog box is shown in the following screenshot.

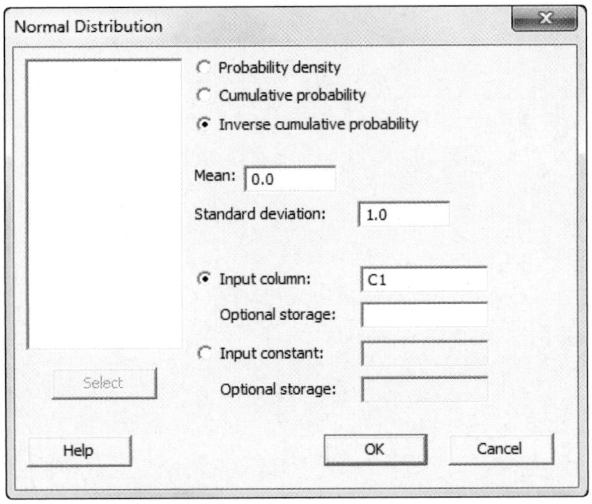

The answer, which appears in the Session window, is $z \approx 1.45$, as shown in the following screenshot.

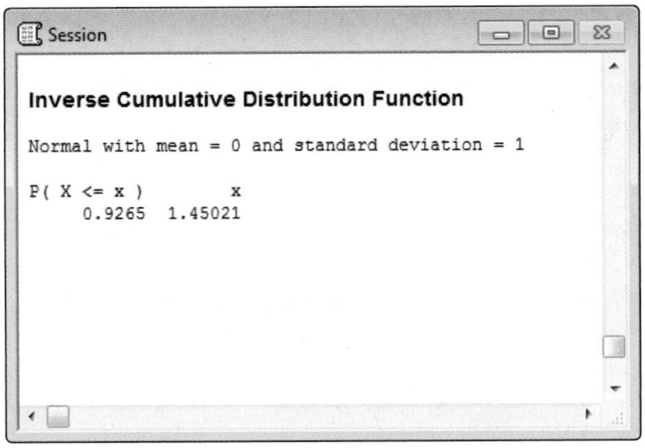

Chapter Seven

The Central Limit Theorem

Sections

Objectives

1. Use the Central Limit Theorem to identify characteristics of a sampling distribution.
2. Use the properties of the Central Limit Theorem to estimate the means and standard deviations of sampling distributions of sample means and sample proportions.
3. Use the Central Limit Theorem to find probabilities for sample means.
4. Use the Central Limit Theorem to find probabilities for sample proportions.

7

Introduction

Imagine that it is a cold, rainy day—the perfect kind of day for a bowl of hot vegetable soup. Fortunately, you have what you need on hand, and soon your soup is simmering away on the stove. The only ingredients left to add are the seasonings, and you want to make sure that the flavor is just right. How will you test to see if the seasonings are correct? Is it necessary to eat the entire pot of soup? Certainly not! Most likely, you would try just one or two spoonfuls in order to decide whether or not the soup is ready to eat.

Little did you know that something as simple as testing soup is actually an example of statistical sampling. Just as you estimate the seasoning of a whole batch of soup from a bite or two, often information about an entire population can be obtained from one well-chosen sample. On a small scale, you are using statistical sampling when you simply poll your friends about what TV shows they watch. On a large scale, The Nielsen Company is using statistical sampling when it surveys families across the country to find out the same information. Indeed, the statistical sampling process is everywhere.

7.1 Introduction to the Central Limit Theorem

We have established the fact that sampling is an essential part of the statistical process. Once an appropriate sample has been gathered, what do you do next? How does a statistician evaluate the significance of sample results? In this section we will introduce an important concept, the Central Limit Theorem, which allows us to use samples in order to make useful predictions about a population.

Let's begin by discussing the idea of a **sampling distribution**. Suppose we start with a population distribution of any shape—such as skewed, uniform, or symmetric. From this population we can take a sample of size n and find its sample mean. If we continue taking samples of size n until all possible samples have been selected, then the distribution created from the means of these samples would be called a **sampling distribution of sample means**. Sampling distributions are not limited to only means, however. Instead of the sample mean, we could also calculate some other statistic, such as the sample proportion, sample variance, or sample standard deviation. Each statistic would have its own sampling distribution.

To illustrate this complex idea, let's look at a simple example. Consider the outcomes obtained by rolling a six-sided die. Suppose that we could go to a casino and record the first 300 outcomes obtained from each die rolled at a craps table one Friday evening. Since the outcomes of rolling a fair die are all equally likely, we know that the frequencies for each of the outcomes 1–6 should be roughly equal. Recall that distributions of this sort are called uniform, and if we graph the outcomes the resulting picture should look somewhat like a box.

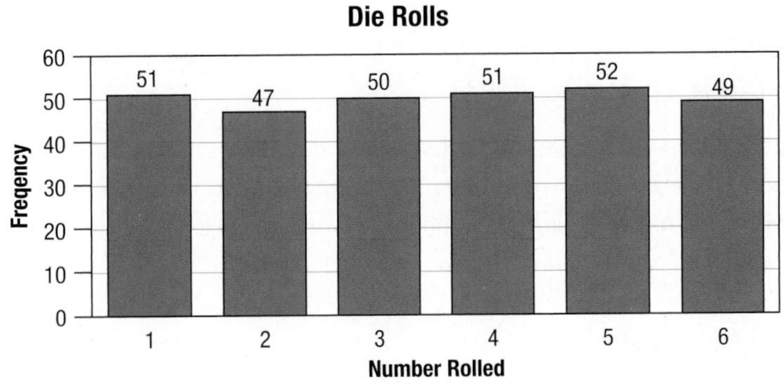

Figure 7.1: Population Distribution of 300 Die Rolls

The population mean and population standard deviation of the 300 die rolls are $\mu \approx 3.5$ and $\sigma \approx 1.7$. Using Figure 7.1, you can verify the values for the population mean and standard deviation using the various methods discussed in Chapter 3.

Here's where things get interesting. Let's now look at *groups* of rolls rather than simply individual rolls. For the sake of example, let's suppose that we want to consider groups of 40 rolls, where each group of 40 is randomly selected from the population of 300 die rolls that we observed at the casino. Then our sample size is $n = 40$, and for each sample we will calculate the sample mean, that is, the mean number rolled for the group. What kind of numbers would we expect for the sample means? The original distribution of single rolls was approximately uniform, but what would we expect the graph of the sample means to look like? Would it also be uniform, or some other shape? The remainder of this section will help you answer these questions.

Let's suppose, in theory of course, that we could take every possible sample of size 40 from the population of 300 die rolls. The resulting distribution of *all* the possible sample means makes up the sampling distribution of sample means for $n = 40$. Note that a sampling distribution is always complete in that it contains *all possible samples for the chosen sample size*. To clarify, you do not go out and collect samples in order to create a sampling distribution. Once the sample size is chosen, the sampling distribution is already determined based on the sample size and the features of the parent, that is, population, distribution.

Properties of the Central Limit Theorem

For sampling distributions of sample means in which the sample size is "large" or the population is normally distributed, a remarkable and immensely useful concept can be utilized. This concept is called the **Central Limit Theorem (CLT)**, and it states the conditions under which a sampling distribution will be approximately normal. It is the cornerstone of much of statistical theory, and it would be difficult to overstate its importance. In the chapters that follow, we will look at ways statisticians apply the CLT when estimating population parameters or performing statistical tests on data. For now, let's examine each of the three properties of the Central Limit Theorem. We will begin our discussion of the CLT in terms of a sampling distribution of sample means, but later in this chapter we will see that it can also be applied to a sampling distribution of sample proportions. Note again that either the sample size must be "large" or else the population must be normally distributed for these properties to hold true for a sampling distribution of sample means. For our purposes, "large" will be defined as $n \geq 30$, which is a common practice. This is not a magic number for n. If the parent distribution already closely resembles a normal distribution, then the value of n does not have to be as large.

Theorem

The Central Limit Theorem (CLT)

For any given population with mean, μ, and standard deviation, σ, a sampling distribution of sample means will have the following three characteristics if either the sample size, n, is at least 30 or the population is normally distributed.

1. The mean of a sampling distribution of sample means, $\mu_{\bar{x}}$, equals the mean of the population, μ.

$$\mu_{\bar{x}} = \mu$$

2. The standard deviation of a sampling distribution of sample means, $\sigma_{\bar{x}}$, equals the standard deviation of the population, σ, divided by the square root of the sample size, $\sqrt{n}$.

$$\sigma_{\bar{x}} = \frac{\sigma}{\sqrt{n}}$$

3. The shape of a sampling distribution of sample means will approach that of a normal distribution, regardless of the shape of the population distribution. The larger the sample size, the better the normal distribution approximation will be.

1. Mean of a Sampling Distribution of Sample Means Equals μ

The first property of the Central Limit Theorem states that the mean of a sampling distribution of sample means, denoted $\mu_{\bar{x}}$, is equal to the population mean, μ. Consider again the example of rolling a die. If we want to know the mean of the resulting sampling distribution, then the Central Limit Theorem tells us it is the value of the mean of the population distribution. We previously stated the mean of the population of die rolls, $\mu \approx 3.5$. Hence, the CLT tells us that the mean of the sampling distribution of sample means is also 3.5. We denote this value $\mu_{\bar{x}} \approx 3.5$.

2. Standard Deviation of a Sampling Distribution of Sample Means Equals $\frac{\sigma}{\sqrt{n}}$

The second property of the Central Limit Theorem provides a method for determining the standard deviation of a sampling distribution of sample means, which, for brevity, is also known as the *standard error of the mean*. To do this, we must first know the population standard deviation, σ. For our die-rolling example, we previously stated that the standard deviation of the population is $\sigma \approx 1.7$. Recall from our earlier discussion that we are using a sample size of $n = 40$. We can then use the Central Limit Theorem to find the standard error of the mean, denoted $\sigma_{\bar{x}}$, as follows.

$$\begin{aligned}\sigma_{\bar{x}} &= \frac{\sigma}{\sqrt{n}} \\ &\approx \frac{1.7}{\sqrt{40}} \\ &\approx 0.27\end{aligned}$$

Notice how the standard deviation differs between the parent distribution and the sampling distribution. The standard deviation of the population is 1.7, but the standard deviation of the sampling distribution is only 0.27. Why the change? Remember that standard deviation tells us how spread out the data values are. It is not unlikely to obtain an individual value that is very different from the mean. It is very unlikely, however, that a sample mean will be far from the population mean. Since the values contained in the sampling distribution do not vary as much from the mean, the standard deviation is therefore a smaller number.

Rounding Rule

When calculating the standard deviation of a sampling distribution of sample means, round the final answer to one more decimal place than what is given in the population standard deviation. Occasional exceptions to this rule can be made when the type of data lends itself to a more natural rounding scheme, such as rounding values of currency to two decimal places.

Let's consider a simple example that illustrates this last point. Suppose that a group of 100 women has a mean height of 67 inches. What is the likelihood that one woman from the population, chosen at random, is only 60 inches (5 feet) tall? Although 60 inches is pretty short for a woman, it is certainly possible to randomly choose a woman of this height. What if, instead, we consider the mean height of 40 women within the population? What is the likelihood that the sample mean is 60 inches? Intuition tells us that this is very unlikely, so we would expect the probability to be very close to zero. The reason this is true is that, although a distribution of individual heights would contain a wide range of values, the distribution of mean heights would contain values that hover closer to the center of the distribution.

Example 7.1

Calculating the Standard Deviation for a Sampling Distribution of Sample Means Using the Central Limit Theorem

Suppose that the standard deviation of movie ticket prices in the United States is \$0.72. Suppose sample means are calculated for samples of size 52; that is, ticket prices are recorded for different samples of 52 theaters and the sample means are calculated. What would be the standard deviation of the sampling distribution of the sample means, that is, the standard error of the mean?

Solution

$$\begin{aligned}\sigma_{\bar{x}} &= \frac{\sigma}{\sqrt{n}}\\ &= \frac{0.72}{\sqrt{52}}\\ &\approx 0.099846\\ &\approx 0.100\end{aligned}$$

So, the sampling distribution's standard deviation would be $\sigma_{\bar{x}} \approx \$0.10$.

3. Shape of the Sampling Distribution Approaches That of a Normal Distribution

The final property of the Central Limit Theorem states that, for sufficiently large samples ($n \geq 30$, for the purposes of this text) the shape of a sampling distribution of sample means will approach that of a normal distribution. In fact, the larger the value of n, the better the normal distribution approximation will be. Alternatively, if the population is normally distributed in the first place, then the sampling distribution is normal for any value of n. This one simple property of the Central Limit Theorem is *enormously* useful! Recall that in Chapter 6, we discussed at length the many problem types and applications associated with the normal distribution. This final property of the CLT allows for us to now apply those same techniques to sampling distributions. The ripple effect is that, since the use of samples is so prevalent in statistics, sampling distributions and the Central Limit Theorem are the foundation for much of what is covered in the remainder of this text.

In Sections 7.2 and 7.3 we will look in depth at applications of the Central Limit Theorem. For now, let's look at an illustration of the sampling distribution of sample means for our die-rolling example with $n = 40$. Note that although the population distribution was *uniform*, according to the Central Limit Theorem, the resulting sampling distribution is *approximately normal.*

In our die-rolling example, there are ${}_{300}C_{40} \approx 9.79 \times 10^{49}$ samples in the sampling distribution for $n = 40$. Because this is such a large number, something greater than the number of stars in the universe, we will illustrate this property of the Central Limit Theorem by showing a distribution of sample means for a large number of samples, specifically 1000 samples of size 40 from our original population of 300 die rolls. Note that even though the distribution shown in Figure 7.2 is only an empirical approximation of the theoretical sampling distribution of sample means, one can already see that the theoretical sampling distribution will be approximately normal.

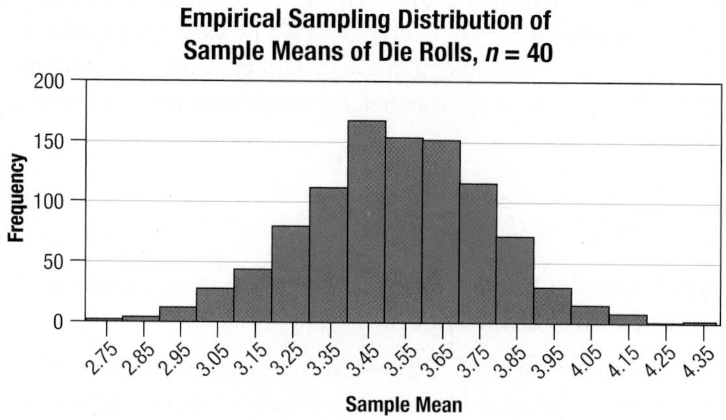

Figure 7.2: Sampling Distribution of Sample Means for $n = 40$

Example 7.2

Applying the Central Limit Theorem

The following histogram represents the population distribution of the weights of 600 horse jockeys. The mean of the population is 116.2 pounds and the standard deviation is 3.9 pounds.

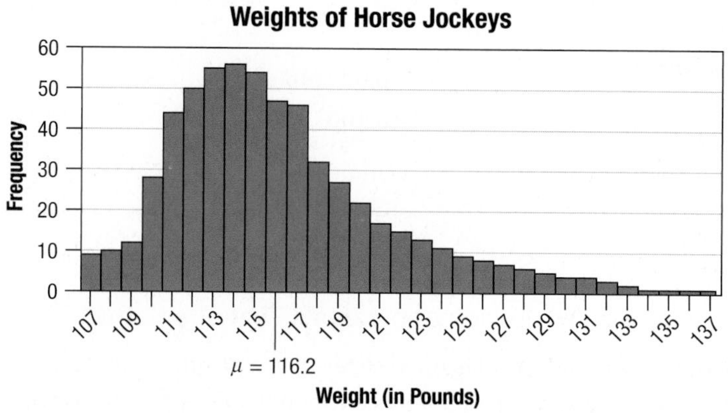

Let's now consider the sampling distribution of the sample means for samples of size $n = 45$.

a. Can the Central Limit Theorem be applied to this sampling distribution? If so, explain how the conditions are met.

b. What is the sampling distribution's mean?

c. What is the sampling distribution's standard deviation?

d. What is the sampling distribution's shape?

Solution

a. Yes, the Central Limit Theorem is applicable in this scenario because the sample size, $n = 45$, satisfies the criteria $n \geq 30$.

b. According to the CLT, the sampling distribution's mean is equal to the mean of the population. Therefore, $\mu_{\bar{x}} = \mu = 116.2$.

c. To find the standard deviation of the sampling distribution we will have to do a small calculation. The CLT says that $\sigma_{\bar{x}} = \frac{\sigma}{\sqrt{n}}$. Hence, we have $\sigma_{\bar{x}} = \frac{3.9}{\sqrt{45}} \approx 0.58$.

d. Although the original distribution was skewed to the right, the CLT tells us that the sampling distribution will be approximately normal, so it will be bell-shaped.

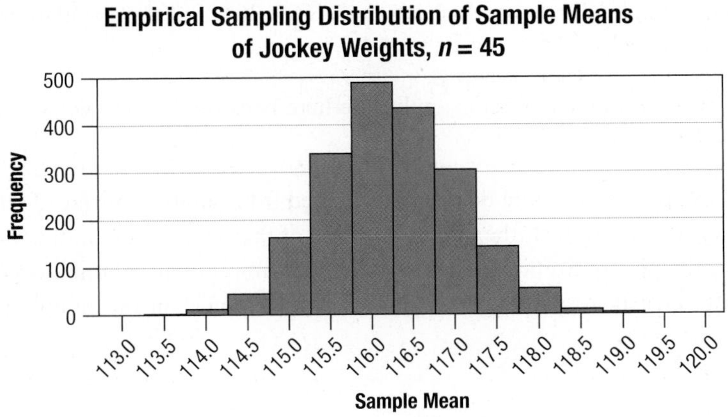

Example 7.3

Applying the Central Limit Theorem

The following graph displays the starting salaries for law school graduates entering the workforce in 2007.

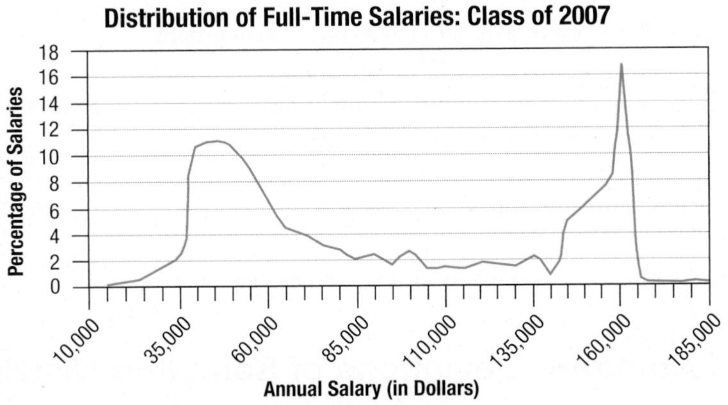

Note: The graph is based on 23,337 salaries. A few salaries above $200,000 are excluded for clarity.
Reproduced with permission of the National Association for Law Placement, Inc. (NALP).
Source: NALP: The Association for Legal Career Professionals. "Another Picture Worth 1,000 Words." July 2008. http://www.nalp.org/anotherpicture?s=Another%20Picture%20Worth%201%2C000%20Words (19 Dec. 2011).

a. Describe the shape of this distribution.

b. Consider the sampling distribution of the sample means created from this population for samples of size 25. Can the Central Limit Theorem be applied in this situation?

c. Consider the sampling distribution of the sample means created from this population for samples of size 50. Can the Central Limit Theorem be applied in this situation?

d. Consider the sampling distribution of the sample means created from this population for samples of size 200. What would you expect the shape of this distribution to be? How would it compare to the sampling distribution described in part c.?

Solution

a. The distribution is best classified as bimodal.

b. No, the Central Limit Theorem is not applicable in this situation because neither condition is met. The sample size is not large enough $(25 < 30)$, and the population data are not normally distributed.

c. Yes, the Central Limit Theorem is applicable here because the sample size is sufficiently large $(50 \geq 30)$.

d. We would expect the sampling distribution created from samples of size 200 to resemble a normal distribution. In fact, the larger the value of the sample size, n, the more "normal" the graph will appear. So then, it follows that the sampling distribution described in part d. will more closely resemble a true normal distribution than the sampling distribution described in part c.

7.1 Section Exercises

Properties of Sampling Distributions

Directions: Decide if each statement is true or false. Explain why.

1. A sampling distribution refers to individuals rather than groups.

2. The shape of a sampling distribution of sample means that follows the requirements of the Central Limit Theorem will be approximately bell-shaped.

3. A sampling distribution of sample means has a standard deviation equal to $\frac{\sigma}{\sqrt{n}}$.

4. A sampling distribution of sample means has a mean equal to $\frac{\mu}{\sqrt{n}}$.

Means and Standard Deviations of Sampling Distributions

Directions: Find the mean and standard deviation of the sampling distribution of sample means using the given information.

5. $\mu = 35$ and $\sigma = 9$; $n = 64$

6. $\mu = 28$ and $\sigma = 6$; $n = 81$

7. $\mu = 12.0$ and $\sigma = 2.3$; $n = 36$

8. $\mu = 52$ and $\sigma = 7$; $n = 100$

9. $\mu = 9.5$ and $\sigma = 10.0$; $n = 39$

10. $\mu = 582.0$ and $\sigma = 23.6$; $n = 1201$

Applying the Central Limit Theorem

Directions: Use the Central Limit Theorem to answer each question.

11. According to a report in the *Portland Press Herald*, the mean price of heating oil in Maine in December 2010 was \$2.98 per gallon. If 100 samples of 37 heating oil prices were collected from around Maine during that time, what would you expect to be the mean of the sampling distribution of the sample means?

Source: "Avg. Maine Heating Oil Price Rises to \$2.98." *Portland Press Herald.* 20 Dec. 2010. http://www.pressherald.com/news/maine-Heating-oil-energy-gallon.html (20 Dec. 2011).

12. According to a local school district, middle school students are assigned a mean of 2.5 hours of homework per night. If 144 samples of 50 students from this school district are collected and the amount of time spent per night on homework is recorded for each student, what would you expect to be the mean of the sampling distribution of the sample means?

13. Some health reports claim that the mean duration of a cold is seven days. If 120 samples of 100 people with colds are taken from across the United States and the duration of each person's cold is recorded, what would you expect to be the mean of the sampling distribution of the sample means?

14. Suppose that an Internet source shows that the mean fare for one-way flights for business travel is \$217, the lowest in five years. If 215 samples of 45 one-way fares for business travel are collected from across the United States, what would you expect to be the mean of the sampling distribution of the sample means?

15. The Federal Reserve Bank of New York conducted a study and claims that the inflation rates of American households had a population standard deviation of 0.2% in 1996. If a sampling distribution is created using samples of inflation rates from 1996 for 78 households, what would be the standard deviation of the sampling distribution of the sample means?

Source: Hobijn, Bart and David Lagakos. "Inflation Inequality in the United States." Federal Reserve Bank of New York Staff Reports, Number 173. October 2003. http://www.newyorkfed.org/research/staff_reports/sr173.pdf (5 Sept. 2011).

16. Suppose that a study of elementary school students reports that the mean age at which children begin reading is 5.7 years with a standard deviation of 1.1 years. If a sampling distribution is created using samples of the ages at which 55 children began reading, what would be the standard deviation of the sampling distribution of the sample means?

17. A study on the latest fad diet claimed that the amounts of weight lost by all people on this diet had a standard deviation of 5.8 pounds. If a sampling distribution is created using samples of the amounts of weight lost by 100 people on this diet, what would be the standard deviation of the sampling distribution of the sample means?

18. The Council of Christian Colleges and Universities Tuition Survey claims that in 2011–12 the population standard deviation of tuition and fees at CCCU-member colleges was \$4385. If a sampling distribution that has a standard error of the mean equal to \$750 is desired, how many CCCU-member colleges would need to be in each sample?

Source: Clark, Stanley A. "CCCU Tuition Survey: 2011–12 Update." 10 Nov. 2011. http://www.cccu.org/professional_development/resource_library/research_and_surveys/~/media/Resource%20Library/Documents/2011/CCCU_Tuition_Survey_2011-12.doc (8 Dec. 2011).

19. Researchers French and Jones reported that the standard deviation of health care costs (in 1998 dollars) was \$4271 per year for people aged 51 to 64 years. If a sampling distribution is created using samples of health care costs (in 1998 dollars) for 500 people aged 51 to 64 years, what would be the standard deviation of the sampling distribution of the sample means?

Source: French, Eric and John Bailey Jones. "On the Distributions and Dynamics of Health Care Costs: Extended Version." 18 Dec. 2004. http://www.albany.edu/~jbjones/healc15.pdf (20 Dec. 2011).

20. Using the report by researchers French and Jones, a medical analyst uses a value of \$324.56 per year as the standard error of the mean for a sampling distribution of health care costs (in 1998 dollars) for people aged 65 or older in a report to the state health care agency. If this sampling distribution was created using samples of 350 people aged 65 or older, could we determine the value of the population standard deviation originally reported by French and Jones?

Source: French, Eric and John Bailey Jones. "On the Distributions and Dynamics of Health Care Costs: Extended Version." 18 Dec. 2004. http://www.albany.edu/~jbjones/healc15.pdf (20 Dec. 2011).

7.2 Central Limit Theorem with Means

Recall that the procedure for solving an application problem using a normal probability distribution is as follows.

1. Draw a picture of a normal curve with the appropriate mean and standard deviation, and shade the region under the curve corresponding to the conditions of the problem.
2. Use the population mean, μ, and population standard deviation, σ, to convert each value of the random variable to a standard score, or z-value.
3. Use the normal distribution tables, a calculator, or statistical software to find the appropriate area under the standard normal curve.

All of the normal distribution application problems that we have studied thus far have involved randomly choosing a single individual from a population. Therefore, it was important that the population be normally distributed in order to use the normal distribution to solve the problem. We now know that if we look at the sampling distribution of the sample means for which the sample size is sufficiently large, the sampling distribution will be approximately normal despite the shape of the population distribution. Thus, the Central Limit Theorem allows us to use the same steps listed above, regardless of the shape of the original distribution, to solve application problems involving sample means. Applications of the CLT must be based on finding the probability of a *sample mean* being in a given interval, instead of the probability of an *individual value* being in a given interval.

Now that we know when we can use the Central Limit Theorem, we need to know how to change the standard score formula in order to use it with the CLT. We need to replace the values of the population mean and standard deviation with the values of the mean and standard deviation of the sampling distribution.

- The mean of a sampling distribution of sample means equals the mean of the population; that is, $\mu_{\bar{x}} = \mu$.
- The standard deviation of a sampling distribution of sample means, or standard error of the mean, equals the population standard deviation divided by the square root of the sample size; that is, $\sigma_{\bar{x}} = \dfrac{\sigma}{\sqrt{n}}$.

Further, we are looking at values of the sample mean, $\bar{x}$, not specific values of the random variable X.

For Individual Values	**For Sample Means**
$z = \dfrac{x-\mu}{\sigma}$	$z = \dfrac{\bar{x}-\mu}{\left(\dfrac{\sigma}{\sqrt{n}}\right)}$

Formula

Standard Score for a Sample Mean

According to the Central Limit Theorem, if either the sample size, n, is at least 30 or the population is normally distributed, then the standard score for a sample mean in a sampling distribution is given by

$$z = \frac{\bar{x} - \mu_{\bar{x}}}{\sigma_{\bar{x}}} = \frac{\bar{x} - \mu}{\left(\frac{\sigma}{\sqrt{n}}\right)}$$

where $\bar{x}$ is the given sample mean,

μ is the population mean,

σ is the population standard deviation, and

n is the sample size used to create the sampling distribution.

Example 7.4

Finding the Probability that a Sample Mean Will Be *Less Than* a Given Value

According to the Centers for Disease Control and Prevention (CDC), the mean total cholesterol level for persons 20 years of age and older in the United States from 2007–2010 was 197 mg/dL.

a. What is the probability that a randomly selected adult from the population will have a total cholesterol level of less than 183 mg/dL? Use a standard deviation of 35 mg/dL and assume that the cholesterol levels for the population are normally distributed.

b. What is the probability that a randomly selected sample of 150 adults from the population will have a mean total cholesterol level of less than 183 mg/dL? Use a standard deviation of 35 mg/dL.

Source: Centers for Disease Control and Prevention. "FastStats - Cholesterol." National Center for Health Statistics. 16 May 2012. http://www.cdc.gov/nchs/fastats/cholest.htm (6 July 2012).

Solution

a. Begin by sketching a normal curve with a mean of 197. We need to find the area under the curve to the left of 183. Since this value is smaller than the population mean of 197, we can mark a value to the left of the center, and shade from there to the left, as shown in the following picture.

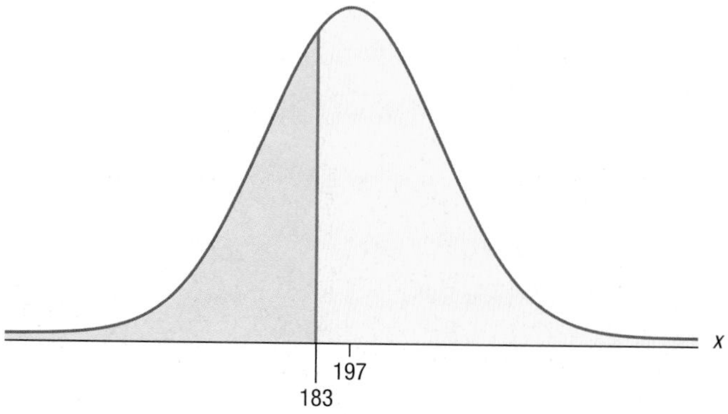

Next, we need to convert the value of 183 to a standard score. It is important to note that in this first part, we are referring to an *individual* and not a sample; thus, use the z-score formula as follows.

$$\begin{aligned} z &= \frac{x-\mu}{\sigma} \\ &= \frac{183-197}{35} \\ &= -0.40 \end{aligned}$$

Using the normal distribution tables or appropriate technology, we find that the area under the standard normal curve to the left of $z = -0.40$ is approximately 0.3446. Thus, the probability of one randomly selected adult in the United States having a total cholesterol level of *less than* 183 mg/dL is approximately 0.3446.

Memory Booster

To refresh your memory on how to find area under a normal curve, review Chapter 6.

Alternate Calculator Method

Alternately, you can find the area under the curve in one step without having to calculate the z-score by entering `normalcdf(-1E99,183,197,35)`, which gives approximately 0.3446.

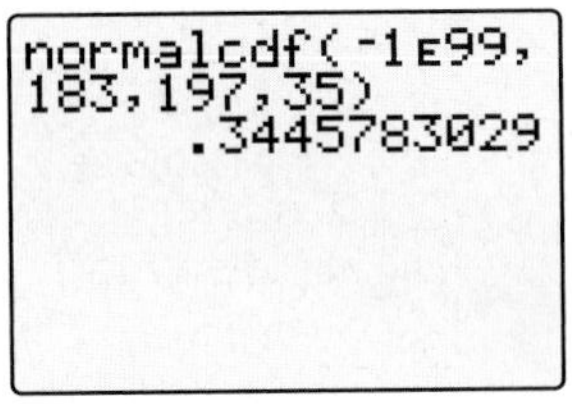

b. The sketch of the normal curve will be the same as shown in the first part of the solution, with a mean of 197 and the area to the left of 183 shaded, except now we are considering the sampling distribution of sample means rather than the distribution of individual total cholesterol levels.

Next, we need to convert the value of 183 to a standard score. Since we are referring to a *sample mean* and not an individual, we use the z-score formula that is given by the CLT as follows.

$$\begin{aligned} z &= \frac{\bar{x}-\mu}{\left(\frac{\sigma}{\sqrt{n}}\right)} \\ &= \frac{183-197}{\left(\frac{35}{\sqrt{150}}\right)} \\ &\approx -4.90 \end{aligned}$$

Using the normal distribution tables or appropriate technology, we find that the area under the standard normal curve to the left of $z \approx -4.90$ is approximately 0.0000. Thus, the probability of a randomly selected sample of 150 US adults having a mean total cholesterol level of *less than* 183 mg/dL is approximately 0.0000 (rounded to four decimal places), which means it is very unlikely, although not impossible.

Note how much more likely it would be to randomly select one person with a cholesterol level of less than 183 mg/dL (as found in part a.) than to get a sample of 150 adults with a mean that low.

Alternate Calculator Method

Alternately, you can find the area under the curve in one step without having to calculate the z-score by entering `normalcdf(-1E99,183,197,35/√(150))`, which gives approximately 0.0000005. Note that by solving for the area in one step on the calculator, we eliminate the rounding error from the intermediate calculations. Thus, the one-step solution of 0.0000005 is more accurate.

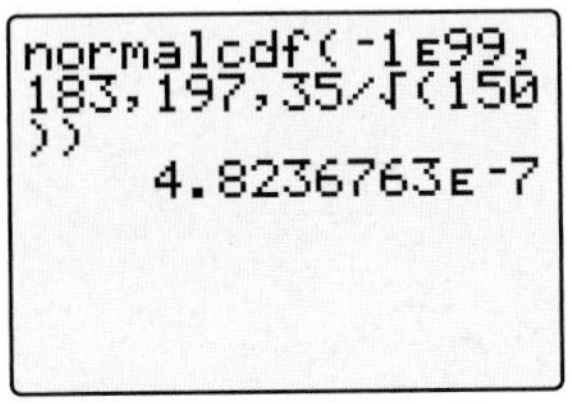

Example 7.5

Finding the Probability that a Sample Mean Will Be *Greater Than* a Given Value

Elevators are required to have weight capacities posted. If the weights of US men are normally distributed with a mean of 194.7 pounds and a standard deviation of 32.0 pounds, what is the probability that a random group of 10 men who enter an elevator will have a combined weight greater than the weight capacity of 2000 pounds that is posted in that elevator?

Solution

Let's begin by noting that, although the sample size is small ($n = 10$), the CLT can be applied because the population is normally distributed. This problem is worded differently than previous problems; note that we are given the combined weight instead of the mean. However, we can calculate the mean weight using the given information. If there are 10 men, and we are interested in a combined total of 2000 pounds, then the mean weight is $\frac{2000}{10} = 200$ pounds. So, we can restate the problem as: "What is the probability that a random group of 10 men have a mean weight *greater than* 200 pounds?"

Begin by drawing a normal curve with a mean of 194.7 (the mean of the sampling distribution) and shade the area *to the right* of 200.

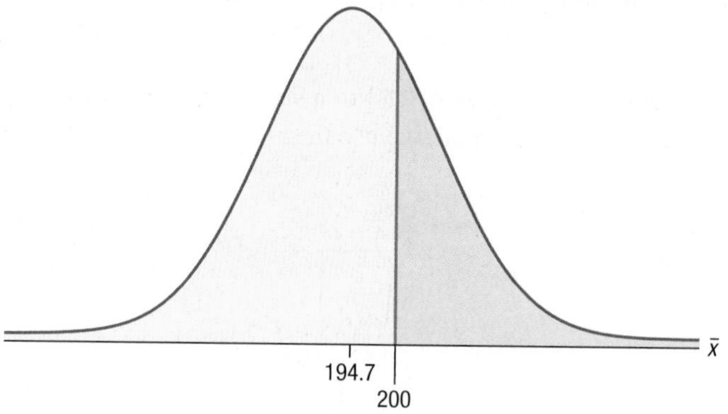

Next, convert the value of 200 to a standard score.

$$\begin{aligned} z &= \frac{\bar{x} - \mu}{\left(\frac{\sigma}{\sqrt{n}}\right)} \\ &= \frac{200 - 194.7}{\left(\frac{32.0}{\sqrt{10}}\right)} \\ &\approx 0.52 \end{aligned}$$

Use appropriate technology or the normal distribution tables to find the area under the standard normal curve to the right of $z \approx 0.52$. The area is approximately 0.3015. Thus, the probability that a random group of 10 men who enter the elevator will exceed the posted weight capacity is approximately 30.15%.

Alternate Calculator Method

Enter `normalcdf(200,1E99,194.7,32/√(10))`, as shown in the screenshot in the margin. Note that by solving for the probability in one step on the calculator, we eliminate the rounding error that is introduced when we first round the standard score to $z \approx 0.52$. Thus, the calculator gives the more accurate value of approximately 0.3002, or 30.02%, for the probability when the one-step method is used.

Example 7.6

Finding the Probability that a Sample Mean Will *Differ* from the Population Mean *by Less Than* a Given Amount

Suppose that prices of women's athletic shoes have a mean of \$75.15 and a standard deviation of \$17.89. What is the probability that the mean price of a random sample of 50 pairs of women's athletic shoes will differ from the population mean by less than \$5.00?

Solution

By subtracting \$5.00 from and adding \$5.00 to the population mean, we find that the area under the normal curve in which we are interested is *between* \$70.15 and \$80.15.

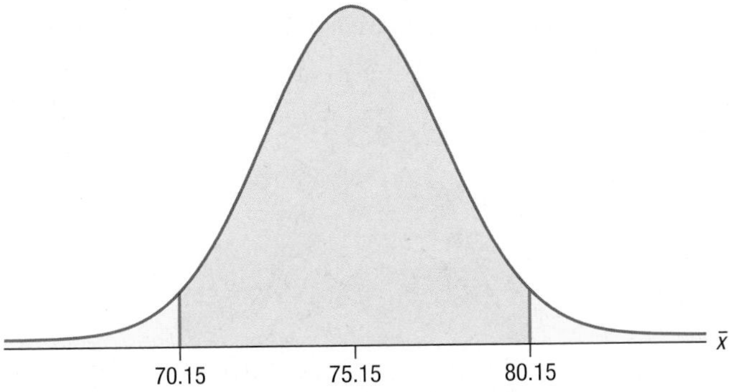

Memory Booster

"Differs by less": between

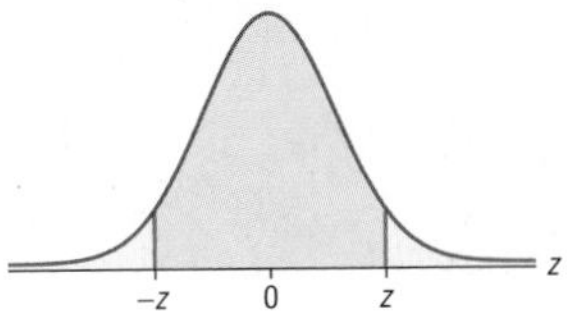

"Differs by more": two tails

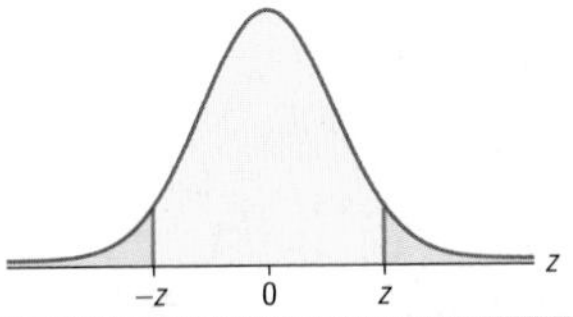

By converting to z-scores, we can standardize this normal distribution. First, let's find the z-score for the lower value of \$70.15 by substituting the appropriate values into the z-score formula.

$$z = \frac{\bar{x} - \mu}{\left(\frac{\sigma}{\sqrt{n}}\right)} = \frac{70.15 - 75.15}{\left(\frac{17.89}{\sqrt{50}}\right)} \approx -1.98$$

Notice that the absolute value of the numerator in our calculation equals the difference allowed from the population mean in the question. This is not a coincidence. Therefore, since the distance between both of the endpoints of interest and the population mean is 5.00, we do not need to find two distinct z-scores. They will both have the same absolute value, but one will be positive and the other negative. Thus, we need to find the area between the two z-scores, $z_1 \approx -1.98$ and $z_2 \approx 1.98$. Using technology, the area is found to be approximately 0.9523 (the tables give 0.9522).

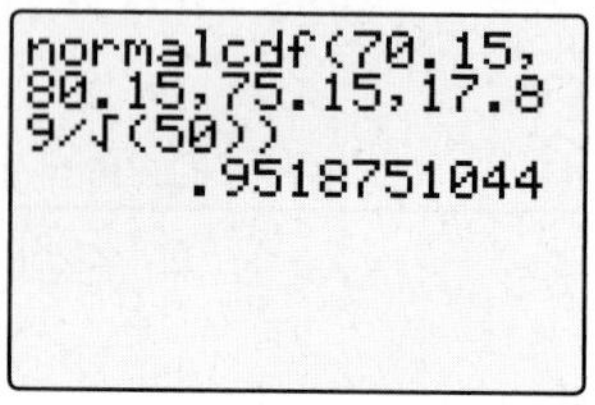
normalcdf(70.15,
80.15,75.15,17.8
9/√(50))
.9518751044

Thus, the probability of a random sample of 50 pairs of women's athletic shoes having a mean price that *differs* from the population mean *by less than* \$5.00 is approximately 0.9523.

Alternate Calculator Method

The one-step calculator method for this problem is `normalcdf(7Ø.15,8Ø.15,75.15,17.89/√(5Ø))`, as shown in the screenshot in the margin. The calculator gives the more accurate value of approximately 0.9519 for the probability when this method is used.

Example 7.7

Finding the Probability that a Sample Mean Will *Differ* from the Population Mean *by More Than* a Given Amount

At a local manufacturing plant, the screws being manufactured have a mean length of 1.625 inches, with a standard deviation of 0.010 inches. If the quality control director randomly chooses a batch of 50 screws, what is the probability that their mean length differs from the mean of the population by more than the allowed 0.004 inches?

Solution

By subtracting 0.004 inches from and adding 0.004 inches to the mean of 1.625 inches, we find that the area under the normal curve in which we are interested is the sum of the areas *to the left* of 1.621 inches *and to the right* of 1.629 inches.

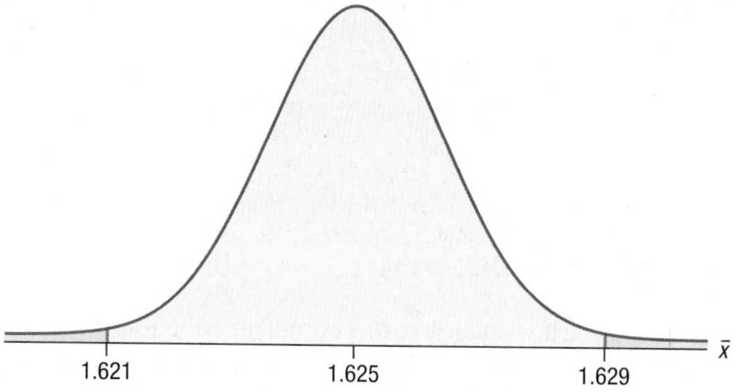
1.621
1.625
1.629
$\bar{x}$

By converting these values to z-scores, we can standardize the normal distribution. Notice that because the distance between both of the endpoints of interest and the population mean is 0.004, we do not need to find two distinct z-scores. They will both have the same absolute value, but one will be positive and the other negative. To calculate the positive z-score, the distance from the population mean is substituted into the numerator of the formula, as follows.

$$z = \frac{\bar{x} - \mu}{\left(\frac{\sigma}{\sqrt{n}}\right)}$$

$$= \frac{0.004}{\left(\frac{0.010}{\sqrt{50}}\right)}$$

$$\approx 2.83$$

Thus, the two z-scores are $z_1 \approx -2.83$ and $z_2 \approx 2.83$. Because we want to find the total area in the two tails, using the symmetry property of the normal distribution, we can find the area to the left of $z_1 \approx -2.83$ and double it. Using appropriate technology, or tables, we find that the area to the left of $z_1 \approx -2.83$ is approximately 0.0023, so the total area in the two tails is approximately $(0.0023)(2) = 0.0046$. Thus, the probability of the sample batch not meeting the requirements is very small, approximately 0.0046.

Alternate Calculator Method

The one-step calculator solution is `2*normalcdf(-1E99,1.621,1.625,Ø.Ø1Ø/√(5Ø))`, as shown in the screenshot in the margin. The calculator gives the more accurate value of approximately 0.0047 for the probability when this method is used.

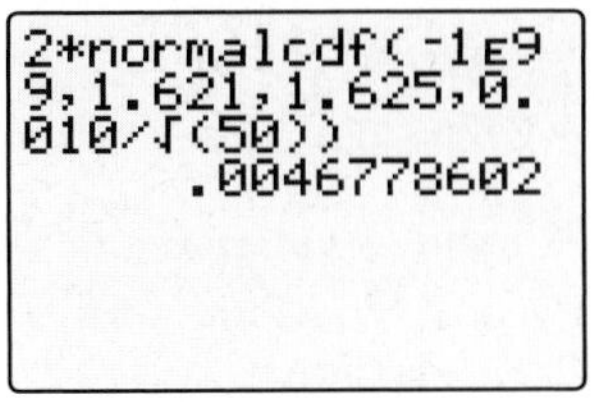

7.2 Section Exercises

Calculating Standard Scores for Sample Means Using the Central Limit Theorem

Directions: Calculate the standard score (*z*-score) of the given sample mean using the Central Limit Theorem. Round your answer to two decimal places.

1. $\mu = 36$ and $\sigma = 39$; $n = 73$; $\bar{x} = 38$

2. $\mu = 11.0$ and $\sigma = 10.2$; $n = 100$; $\bar{x} = 10.0$

3. $\mu = 1.10$ and $\sigma = 1.10$; $n = 17$; $\bar{x} = 0.55$

4. $\mu = 0.49$ and $\sigma = 0.60$; $n = 1500$; $\bar{x} = 0.52$

5. $\mu = 114,023$ and $\sigma = 30,000$; $n = 100$; $\bar{x} = 110,045$

6. $\mu = 514.2$ and $\sigma = 30.2$; $n = 1000$; $\bar{x} = 513.0$

Probability

Directions: Find each specified probability. Use the Central Limit Theorem only where appropriate.

7. Suppose that the walking step lengths of adult males are normally distributed with a mean of 2.4 feet and a standard deviation of 0.3 feet. A sample of 34 men's step lengths is taken.

a. Find the probability that an individual man's step length is less than 2.1 feet.

b. Find the probability that the mean of the sample taken is less than 2.1 feet.

c. Find the probability that the mean of the sample taken is more than 2.5 feet.

d. Find the probability that the sample mean differs from the population mean by more than 0.06 feet.

8. The distribution of the number of people in line at a grocery store checkout has a mean of 3 and a variance of 9. A sample of the numbers of people in 50 grocery checkout lines is taken.

   a. Calculate the probability that the sample mean is more than 4.

   b. Calculate the probability that the sample mean is less than 2.5.

   c. Calculate the probability that the sample mean differs from the population mean by less than 0.5.

9. Intelligence quotient (IQ) scores are often reported to be normally distributed with $\mu = 100.0$ and $\sigma = 15.0$.

   a. What is the probability of a random person on the street having an IQ score of less than 95?

   b. If a random sample of 50 people is taken, what is the probability that their mean IQ score will be less than 95?

   c. If a random sample of 50 people is taken, what is the probability that their mean IQ score will be more than 95?

   d. If a random sample of 50 people is taken, what is the probability that their mean IQ score will be more than 105?

   e. If a random sample of 50 people is taken, what is the probability that their mean IQ score will differ from the population mean by more than 5?

10. Suppose that the diameters of oak trees are normally distributed with a mean of 4.000 feet and a standard deviation of 0.375 feet.

   a. What is the probability of walking down the street and finding an oak tree with a diameter of more than 5 feet?

   b. What is the probability of sampling a set of 87 oak trees and finding their mean to be more than 4.1 feet in diameter?

   c. What is the probability of sampling a set of 87 oak trees and finding their mean to be less than 3.92 feet in diameter?

   d. What is the probability of sampling a set of 87 oak trees and finding their mean to differ from the population mean by less than 0.1 feet in diameter?

11. A glass tube maker claims that his tubes have lengths that are normally distributed with a mean of 9.00 cm and a variance of 0.25.

   a. What is the probability that a quality control regulator will pull a tube off the assembly line that has a length between 8.6 and 9 cm?

   b. What is the probability that a random sample of 40 tubes will have a mean of less than 8.8 cm?

   c. What is the probability that a random sample of 35 tubes will have a mean of more than 9.2 cm?

   d. What is the probability that a random sample of 75 tubes will have a mean that differs from the population mean by more than 0.1 cm?

12. A pencil manufacturer claims that its pencils have lengths that are normally distributed with a mean of 6.0 inches and a variance of 0.2. What is the probability that a randomly chosen pencil will have a length of more than 6.4 inches?

13. A vineyard claims that the mean berry count per grape cluster is 43 with a standard deviation of 3.6 berries. What is the probability that, in next year's crop, the mean berry count for a sample of 50 clusters will be between 42 and 44?

14. A book publisher claims that its mean book length is 250 pages with a standard deviation of 70 pages. As a reviewer, you get paid per book, and not per page, to read over a manuscript. What is the probability that for a sample of 45 randomly selected books, the mean length of a book is less than 230 pages?

15. Airlines predict that the mean number of "no shows" per 100 seats on each flight is 10.0 with a standard deviation of 3.4. What is the probability that a random sample of 45 flights has a mean of more than 12 "no shows" per 100 seats?

16. In a Scrabble tournament, the scores were normally distributed and the mean score was 420.2 points with a standard deviation of 105.0 points. What is the probability that the score of a randomly selected competitor differs from the mean score by less than 50 points?

17. A tea bag manufacturer needs to place 2 g of tea in each bag. If the machinery places a mean of 2.6 g of tea in each bag with a standard deviation of 0.3 g, what is the probability that a randomly chosen bag will have between 2 and 2.8 g of tea, assuming that the amounts of tea per bag are normally distributed?

18. A pollster claims that the mean amount spent on Christmas gifts by an American family is \$927 with a standard deviation of \$200. What is the probability that the mean amount spent on Christmas gifts for a sample of 500 families differs from the population mean by less than \$15?

19. A medical journal reports the mean fetal heart rate to be 140 beats per minute (bpm) with a standard deviation of 12 bpm. What is the probability that a fetal heart rate differs from the mean by more than 25 bpm, assuming that fetal heart rates are normally distributed?

20. A local sports magazine reports the mean length of a baseball game to be 175.9 minutes with a standard deviation of 27.0 minutes. For a random sample of 30 games, what is the probability that the mean game length is at most 170 minutes?

21. Teenagers spend a mean of 7.5 hours per week playing computer and video games. For a random sample of 110 teenagers, what is the probability that the mean amount of time spent playing games is more than 8 hours per week? Let $\sigma = 3.0$.

22. The mean wait time for a drive-through chain is 193.2 seconds with a standard deviation of 29.5 seconds. What is the probability that for a random sample of 45 wait times, the mean is between 185.7 and 206.5 seconds?

23. The mean amount spent per order at one fast food restaurant is \$8.43 with a standard deviation of \$1.52. What is the probability that for a random sample of 75 orders from this week's customers, the mean amount spent will be between \$8 and \$9?

24. The mean hourly rate for babysitters in one town is \$7.05 with a standard deviation of \$0.55. What is the probability that a babysitter chosen at random will charge an hourly rate that differs from the mean by less than \$1.00, assuming that the hourly rates for babysitters in this town are normally distributed?

25. The mean sale price for a piece of art from members of a large artists' guild is \$545 with a standard deviation of \$76. If, at one show, 45 pieces are for sale, what is the probability that the mean sale price for the show will be higher than \$575?

26. The mean cost for plumbing repairs in one area is \$208 with a standard deviation of \$64. If 31 homeowners in that area are surveyed, what is the probability that the mean cost for their plumbing repairs will be over \$200?

27. One pediatric clinic sees a mean of 42.0 patients per day with a standard deviation of 3.4 patients per day. What is the probability that the mean number of patients per day for March (31 days) will be between 41 and 44?

28. At a large university, the mean amount spent by students for cellular phone service is \$38.90 per month with a standard deviation of \$3.64 per month. Consider a group of 44 randomly chosen university students. What is the probability that the mean amount of their monthly cell phone bills differs from the mean for the university by more than \$1?

29. A large bakery sells a mean of 11.0 dozen cookies per day with a standard deviation of 1.1 dozen cookies per day. Consider the mean number of cookies sold per day in March (31 days) of one year. What is the probability that the mean differs from the population mean by less than 0.5 dozen?

30. According to data released by the New Town Chamber of Commerce, the weekly wages of office workers have a mean of \$723 and a standard deviation of \$151. If 57 office workers are chosen at random from New Town, what is the probability that the mean weekly wage of these workers will be greater than \$750?

7.3 Central Limit Theorem with Proportions

How often have you read in the newspapers that some percentage of the population would do "this" or favor "that"? Have you ever wondered how researchers come up with these figures? We're now in a position to explain perhaps the most commonly encountered use of statistics—proportions.

A **population proportion**, p, is the fraction or percentage of the population that has a certain characteristic. In other words, a proportion is the relationship of one part to the whole. In our text we will often write a proportion in the form of a fraction. However, it is common practice to use percentages or decimals to express proportions, depending on which form best suits the needs of your audience. For instance, if a census shows that 9 out of every 10 people carry a cell phone, then the population proportion is $\frac{9}{10}$, but it's more often written as 0.90 or 90%, as this is more easily understood by more people. All three are valid ways to denote the relationship between how many people carry a cell phone and the total population.

In the same manner, a **sample proportion** is the fraction or percentage of a sample that has a certain characteristic. We denote the sample proportion as $\hat{p}$, which is read "p-hat." For example, if we ask a sample of 20 people if they carry a cell phone and 16 of them say yes, then the sample proportion is $\hat{p} = \frac{16}{20} = \frac{4}{5} = 0.80 = 80\%$.

Side Note

Americans Prefer Smaller Families

A June 2011 article on Gallup.com claims that "58% of U.S. adults [are] now saying that having no more than two children is the ideal for a family."

Source: Saad, Lydia. "Americans' Preference for Smaller Families Edges Higher." Gallup. 30 June 2011. http://www.gallup.com/poll/148355/Americans-Preference-Smaller-Families-Edges-Higher.aspx (6 May 2012).

Formula

Proportion

A **population proportion** is given by

$$p = \frac{x}{N}$$

where x is the number of individuals in the population that have a certain characteristic and N is the size of the population.

A **sample proportion** is given by

$$\hat{p} = \frac{x}{n}$$

where x is the number of individuals in the sample that have a certain characteristic and n is the sample size.

Math Symbols

p: population proportion

$\hat{p}$: sample proportion; read as "p-hat"

Memory Booster

Properties of a Binomial Distribution

1. The experiment consists of a fixed number, *n*, of identical trials.
2. Each trial is independent of the others.
3. For each trial, there are only two possible outcomes. For counting purposes, one outcome is labeled a success, and the other a failure.
4. For every trial, the probability of getting a success is called *p*. The probability of getting a failure is then $1 - p$.
5. The binomial random variable, *X*, counts the number of successes in *n* trials.

Going back to our original question, can researchers really know that exactly 45% of the entire population has some opinion? It seems rather unlikely. Of course, we all understand that only by conducting a census could a researcher establish the true proportion of the population who favor a particular issue. In Chapter 8, we will see how we can use the Central Limit Theorem to estimate population parameters. Now, let's discuss how the CLT applies with population proportions.

A sampling distribution of sample proportions has a binomial distribution, not a normal distribution. However, as we learned in Section 6.5, we can use a normal distribution to approximate a binomial distribution when the following conditions are met.

- The samples are simple random samples.
- The conditions for a binomial distribution are met.
- The sample size is large enough to ensure that $np \geq 5$ and $n(1-p) \geq 5$.

The last condition ensures that the sample size is large enough for the normal distribution to be used as an approximation for the binomial distribution, just as for sample means we had to make sure that the sample size was at least 30 if the population was not known to be normally distributed. In this section you can safely assume that all of the necessary conditions have been met for each example and exercise. (You may certainly verify them for yourself, if you wish!)

In order to use a normal distribution and *z*-scores to calculate probability for sample proportions, we must know the mean and standard deviation for the normal distribution approximation to the sampling distribution of sample proportions. These formulas are derived from the formulas for the mean and standard deviation of a binomial distribution, since a sampling distribution of sample proportions is a binomial distribution. The formulas are as follows.

Formula

Mean and Standard Deviation of a Sampling Distribution of Sample Proportions

When the samples taken are simple random samples, the conditions for a binomial distribution are met, and the sample size is large enough to ensure that $np \geq 5$ and $n(1-p) \geq 5$, a normal distribution can be used to approximate the binomial sampling distribution of sample proportions with the mean and standard deviation given by

$$\mu_{\hat{p}} = p$$

$$\sigma_{\hat{p}} = \sqrt{\frac{p(1-p)}{n}}$$

where *p* is the population proportion and

n is the sample size.

Substituting the formulas just given into the generic *z*-score formula results in the following *z*-score formula, which we use for sample proportions in sampling distributions.

For Individual Values	**For Sample Proportions**
$z = \dfrac{x-\mu}{\sigma}$	$z = \dfrac{\hat{p}-p}{\sqrt{\dfrac{p(1-p)}{n}}}$

Formula

Standard Score for a Sample Proportion

According to the Central Limit Theorem, the standard score, or z-score, for a sample proportion in a sampling distribution is given by

$$z = \frac{\hat{p} - \mu_{\hat{p}}}{\sigma_{\hat{p}}} = \frac{\hat{p} - p}{\sqrt{\frac{p(1-p)}{n}}}$$

where $\hat{p}$ is the given sample proportion,

p is the population proportion, and

n is the sample size used to create the sampling distribution.

We can use this z-score formula along with our traditional procedures for using the normal distribution tables to calculate probabilities involving proportions.

Example 7.8

Finding the Probability that a Sample Proportion Will Be *At Least* a Given Value

In a certain conservative precinct, 79% of the voters are registered Republicans. What is the probability that in a random sample of 100 voters from this precinct at least 68 of the voters would be registered Republicans?

Solution

From the information given, we see that $\hat{p} = \frac{68}{100} = 0.68 = 68\%$. Now let's sketch a normal curve. Because we're looking for the probability that *at least* 68% of voters in the sample are registered Republicans, we need to find the area under the normal curve of the sampling distribution *to the right* of 68%. Since this value is smaller than the population proportion of 79%, we can mark a value to the left of the center, and shade from there to the right, as shown in the following picture.

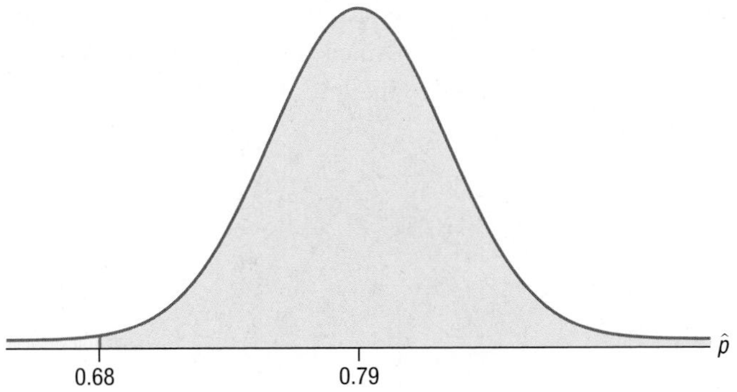

Next, we must calculate the z-score. To do this, we need to use $\hat{p} = 0.68$, $p = 0.79$, and $n = 100$ to calculate z as follows.

Rounding Rule

When calculations involve several steps, avoid rounding at intermediate calculations. If necessary, round intermediate calculations to at least six decimal places to avoid additional rounding error in subsequent calculations.

$$\begin{aligned} z &= \frac{\hat{p}-p}{\sqrt{\frac{p(1-p)}{n}}} \\ &= \frac{0.68-0.79}{\sqrt{\frac{0.79(1-0.79)}{100}}} \\ &\approx \frac{-0.11}{0.040731} \\ &\approx -2.70 \end{aligned}$$

Using the normal distribution tables or appropriate technology, we find that the area under the standard normal curve to the right of $z \approx -2.70$ is approximately 0.9965.

Alternate Calculator Method

normalcdf(0.68,1
E99,0.79,√(0.79(
1-0.79)/100))
.9965398214

The one-step calculator method is `normalcdf(Ø.68,1E99,Ø.79,√(Ø.79(1-Ø.79)/1ØØ))`, as shown in the screenshot in the margin.

Thus, the probability of at least 68 voters in a sample of 100 being registered Republicans is approximately 0.9965.

Example 7.9

Finding the Probability that a Sample Proportion Will Be *No More Than* a Given Value

In another precinct across town, the population is very different. In this precinct, 81% of the voters are registered Democrats. What is the probability that, in a random sample of 100 voters from this precinct, no more than 80 of the voters would be registered Democrats?

Solution

From the information given, we see that $\hat{p} = \frac{80}{100} = 0.80 = 80\%$. Now let's sketch a normal curve. This time we're interested in the probability that *no more than* 80% of voters in the sample are registered Democrats, so we need to find the area under the normal curve of the sampling distribution *to the left* of 80%. Since this value is smaller than the population proportion of 81%, we can mark a value to the left of the center, and shade from there to the left, as shown in the following picture.

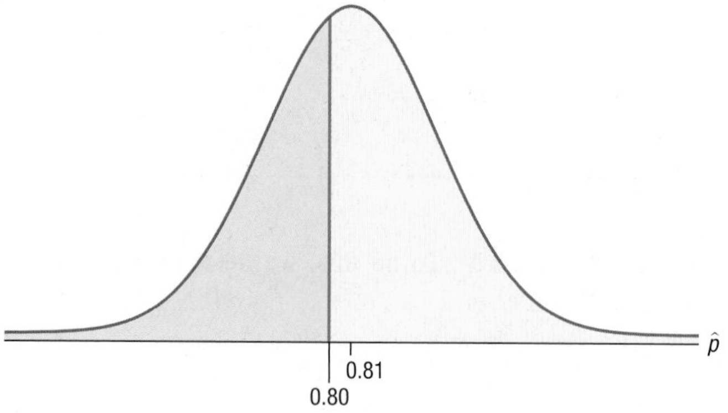
$\hat{p}$
0.81
0.80

Next, we need to calculate the z-score. To do this, we need to use $\hat{p} = 0.80$, $p = 0.81$, and $n = 100$ to calculate z as follows.

$$\begin{aligned} z &= \frac{\hat{p} - p}{\sqrt{\frac{p(1-p)}{n}}} \\ &= \frac{0.80 - 0.81}{\sqrt{\frac{0.81(1-0.81)}{100}}} \\ &\approx \frac{-0.01}{0.039230} \\ &\approx -0.25 \end{aligned}$$

Using the normal distribution tables or appropriate technology, we find that the area under the standard normal curve to the left of $z \approx -0.25$ is approximately 0.4013. Thus, the probability of no more than 80 voters in a randomly selected sample of 100 voters being registered Democrats is approximately 0.4013.

Alternate Calculator Method

The one-step calculator method is `normalcdf(-1E99,0.80,0.81,√(0.81(1-0.81)/100))`, as shown in the screenshot in the margin. The calculator gives the more accurate value of approximately 0.3994 for the probability when this method is used.

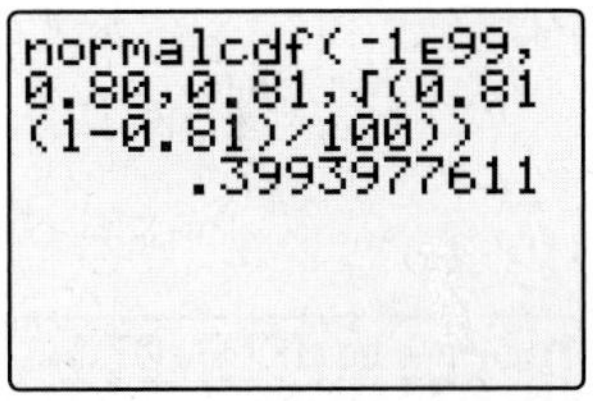

Example 7.10

Finding the Probability that a Sample Proportion Will *Differ* from the Population Proportion *by Less Than* a Given Amount

It is estimated that 8.3% of all Americans have diabetes. Suppose that a random sample of 74 Americans is taken. What is the probability that the proportion of people in the sample who are diabetic differs from the population proportion by less than 1%?

Solution

By subtracting 0.01 from and adding 0.01 to the population proportion, $p = 0.083$, we find that the area under the normal curve in which we are interested is *between* 0.073 and 0.093.

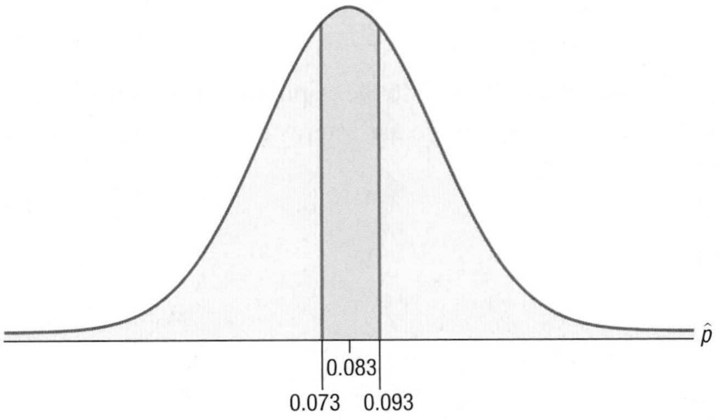

By converting these values to z-scores, we can standardize this normal distribution. Since the distance between both of the endpoints of interest and the population proportion is 0.01, we do not need to find two distinct z-scores. They will both have the same absolute value, but one will be positive and the other negative. To calculate the positive z-score, the difference allowed from the population proportion is substituted into the numerator of the z-score formula, as shown below.

$$\begin{aligned} z &= \frac{\hat{p}-p}{\sqrt{\frac{p(1-p)}{n}}} \\ &= \frac{0.01}{\sqrt{\frac{0.083(1-0.083)}{74}}} \\ &\approx \frac{0.01}{0.032071} \\ &\approx 0.31 \end{aligned}$$

So, we want to find the area between $z_1 \approx -0.31$ and $z_2 \approx 0.31$. Using the normal distribution tables or appropriate technology, we find that the area under the standard normal curve between the two z-scores is approximately 0.2434. Thus, the probability of the proportion of people in the sample who are diabetic *differing* from the population proportion *by less than* 1% is approximately 0.2434.

Alternate Calculator Method

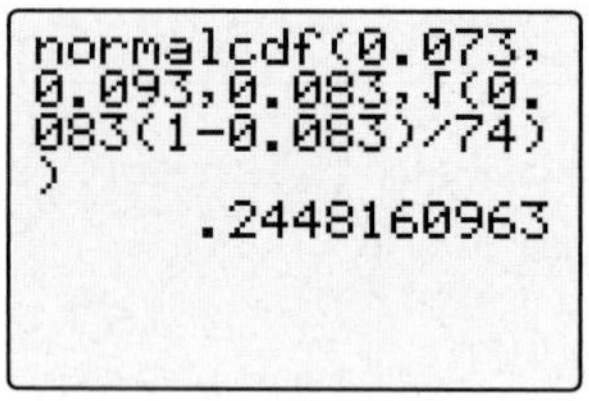

The one-step calculator method is `normalcdf(Ø.Ø73,Ø.Ø93,Ø.Ø83,√(Ø.Ø83(1-Ø.Ø83)/74))`, as shown in the screenshot in the margin. The calculator gives the more accurate value of approximately 0.2448 for the probability when this method is used.

Example 7.11

Finding the Probability that a Sample Proportion Will *Differ* from the Population Proportion *by More Than* a Given Amount

It is estimated that 8.3% of all Americans have diabetes. Suppose that a random sample of 91 Americans is taken. What is the probability that the proportion of people in the sample who are diabetic differs from the population proportion by more than 2%?

Solution

By subtracting 0.02 from and adding 0.02 to the population proportion, 0.083, we find that the area under the normal curve in which we are interested is the sum of the areas *to the left* of 0.063 *and to the right* of 0.103.

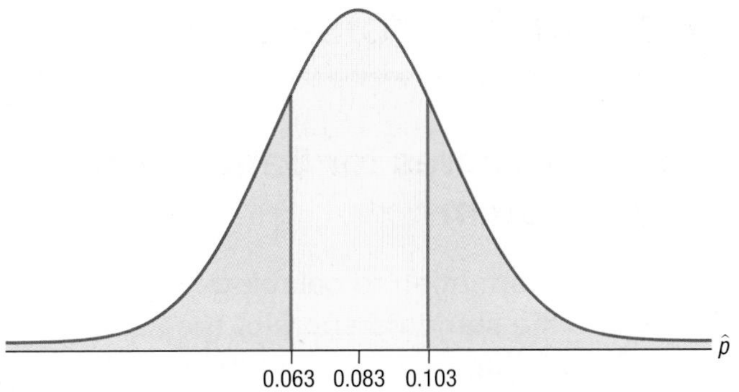

By converting to z-scores, we can standardize this normal distribution. Recall that we do not need to find two distinct z-scores. They will both have the same absolute value, but one will be positive and the other negative. To calculate the positive z-score, the given distance from the population proportion is substituted into the numerator of the formula, giving the following calculation.

$$\begin{aligned} z &= \frac{\hat{p}-p}{\sqrt{\frac{p(1-p)}{n}}} \\ &= \frac{0.02}{\sqrt{\frac{0.083(1-0.083)}{91}}} \\ &\approx \frac{0.02}{0.028920} \\ &\approx 0.69 \end{aligned}$$

Because we want to find the total area in the two tails, using the symmetry property of the normal distribution, we can find the area to the left of $z_1 \approx -0.69$ and double it. Using the normal distribution tables or appropriate technology, we find that the area to the left of $z_1 \approx -0.69$ is approximately 0.2451, so the total area in the two tails is approximately $(0.2451)(2) = 0.4902$. Thus, the probability that the sample proportion *differs* from the population proportion *by more than* 2% is approximately 0.4902.

Alternate Calculator Method

The one-step calculator method is `2*normalcdf(-1E99,0.063,0.083,√(0.083(1-0.083)/91))`, as shown in the screenshot in the margin. The calculator gives the more accurate value of approximately 0.4892 for the probability when this method is used.

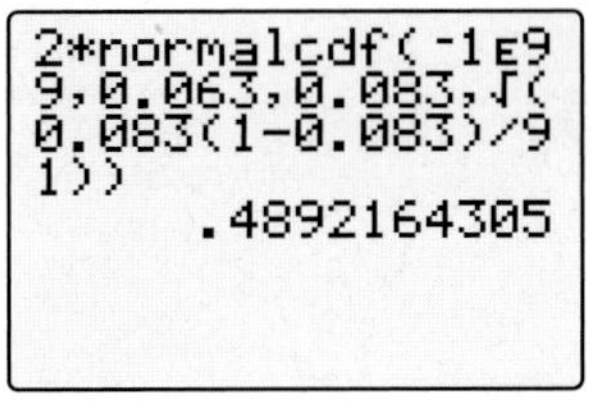

7.3 Section Exercises

Calculating Standard Scores for Sample Proportions Using the Central Limit Theorem

Directions: Use the given information to calculate the sample proportion, and then calculate the standard score of the sample proportion using the Central Limit Theorem. Round the standard score to two decimal places.

1. $p = 0.34, \quad x = 35, \quad n = 100$

2. $p = 0.54, \quad x = 563, \quad n = 1000$

3. $p = 0.8, \quad x = 24, \quad n = 29$

4. $p = 0.61, \quad x = 66, \quad n = 98$

5. $p = 0.15, \quad x = 1152, \quad n = 12{,}500$

6. $p = 0.93, \quad x = 987, \quad n = 1012$

Probability

Directions: Find each specified probability.

7. A large car dealership claims that 47% of its customers are looking to buy a sport utility vehicle (SUV). A random sample of 61 customers is surveyed. What is the probability that less than 40% are looking to buy an SUV?

8. The local nursery is waiting for its spring annuals to be delivered, and 20% of the plants ordered are petunias. If the first truck contains 120 plants packed at random, what is the probability that no more than 30 of the plants are petunias?

9. A car lot sells pre-owned cars. Its sales manager reports that 78% of the cars are sold for less than \$10,000. What is the probability that in a random sample of 45 cars, more than 75% are sold for less than \$10,000?

10. A news report stated that 65% of the vehicles sold nationally were SUVs. A random sample of 100 vehicles sold in the last month is taken. What is the probability that at least 70 of the vehicles in the sample were SUVs?

11. A local florist says that 85% of all flowers sold for Valentine's Day are roses. Consider a random sample of 150 orders for Valentine's Day flowers. What is the probability that less than 80% of the orders in the sample are for roses?

12. At one private college, 34% of students are business majors. Suppose that 260 students are randomly selected from a list in the registrar's office. What is the probability that the proportion of business students in the sample differs from the population proportion by less than 2%?

13. At a large grocery store, 72% of shoppers are women. In order to obtain information about spending habits, 40 shoppers are randomly chosen for a survey. What is the probability that the proportion of women in the sample differs from the population proportion by more than 3%?

14. A major appliance retailer claims that 18% of all appliances sold are dishwashers. If 112 appliances sold are randomly selected, what is the probability that the proportion of dishwashers sold differs from the population proportion by over 2%?

15. A popular restaurant downtown says that 20% of diners order the daily special. Consider a random sample of 128 people who dined at the restaurant. What is the probability that the proportion of diners who ordered the special differs from the population proportion by less than 1%?

R Chapter 7 Review

Section 7.1: Introduction to the Central Limit Theorem

Definitions

Sampling distribution

The distribution of the values of a particular sample statistic for all possible samples of a given size, n

Sampling distribution of sample means

The distribution of sample means for all possible samples of a given size, n

The Central Limit Theorem (CLT)

For any given population with mean, μ, and standard deviation, σ, a sampling distribution of sample means will have the following three characteristics if either the sample size, n, is at least 30 or the population is normally distributed.

1. The mean of a sampling distribution of sample means, $\mu_{\bar{x}}$, equals the mean of the population, μ; that is, $\mu_{\bar{x}} = \mu$.
2. The standard deviation of a sampling distribution of sample means, $\sigma_{\bar{x}}$, equals the standard deviation of the population, σ, divided by the square root of the sample size, $\sqrt{n}$; that is, $\sigma_{\bar{x}} = \dfrac{\sigma}{\sqrt{n}}$.
3. The shape of a sampling distribution of sample means will approach that of a normal distribution, regardless of the shape of the population distribution. The larger the sample size, the better the normal distribution approximation will be.

Section 7.2: Central Limit Theorem with Means

Standard Score for a Sample Mean

$$z = \frac{\bar{x} - \mu_{\bar{x}}}{\sigma_{\bar{x}}} = \frac{\bar{x} - \mu}{\left(\dfrac{\sigma}{\sqrt{n}}\right)}$$

Section 7.3: Central Limit Theorem with Proportions

Population Proportion

The fraction or percentage of a population that has a certain characteristic, given by

$$p = \frac{x}{N}$$

Sample Proportion

The fraction or percentage of a sample that has a certain characteristic, given by

$$\hat{p} = \frac{x}{n}$$

Mean of a Sampling Distribution of Sample Proportions

$$\mu_{\hat{p}} = p$$

Standard Deviation of a Sampling Distribution of Sample Proportions

$$\sigma_{\hat{p}} = \sqrt{\frac{p(1-p)}{n}}$$

Standard Score for a Sample Proportion

$$z = \frac{\hat{p} - \mu_{\hat{p}}}{\sigma_{\hat{p}}} = \frac{\hat{p} - p}{\sqrt{\frac{p(1-p)}{n}}}$$

E Chapter 7 Exercises

Sampling Distributions and Probability

Directions: Answer each question.

1. The mean lifetime of laptop computers produced by one company is 4.6 years with a standard deviation of 1.1 years. Consider random samples of 55 laptop computers. What is the standard deviation of the sampling distribution of the sample means?

2. Suppose that for 18-to-25-year-old men, the mean number of football games watched each football season is 19.0 games with a standard deviation of 3.7 games. If 31 randomly selected men in this age range are surveyed, what is the probability that the mean number of football games watched by this group is no more than 17 games per season?

3. In one city, 83% of eligible voters are registered to vote. If 425 randomly selected eligible voters are surveyed, what is the probability that at least 82% are registered to vote?

4. A local fast-food restaurant reports that the mean number of hamburgers it sells per day is 235.0 with a standard deviation of 11.3 hamburgers. Consider a random sample of 31 days. What is the probability that the mean number of hamburgers sold per day for the sample will differ from the population mean by more than 3 hamburgers?

5. The mean size of candy bars from one candy maker is 39.0 grams with a standard deviation of 2.3 grams. If you produce a sampling distribution of sample means using samples of size 40 from the candy production line, what would be the mean of the sampling distribution?

6. The mean blade length of a rapier, which is a fencing sword, is 90.0 cm with a standard deviation of 4.2 cm. If 38 rapiers are randomly sampled, what is the probability that their mean blade length is less than 88 cm?

7. The mean length of an adult blue whale is 30.0 meters with a standard deviation of 5.7 meters. If a random sample of 31 whales is taken, what is the probability that their mean length differs from the population mean by less than 2 meters?

8. Suppose that, for a particular make and model of car, the mean sale price for car dealerships nationwide is $24,560 with a standard deviation of $390. What is the probability that the sale price for this car at the car dealership in your area will be no more than $24,000? Assume that the car prices are normally distributed.

9. One brand of light bulbs has a mean lifetime of 400 hours with a standard deviation of 28 hours. What is the probability that the lifetime of one randomly chosen bulb will differ from the mean by over 50 hours? Assume that the population is normally distributed.

10. At one regional grocery store chain, a customer spends a mean of $7.29 on produce in each visit with a standard deviation of $2.33. Find the probability that the mean amount spent on produce per visit for 55 randomly selected shoppers will be less than $7.

11. At a local elementary school, students miss a mean of 4.1 days of school each year with a standard deviation of 1.1 days. What is the probability that the mean number of days missed for a random sample of 30 students will be no more than 3.5?

12. At one theater, moviegoers spend a mean of \$6.32 on refreshments with a standard deviation of \$2.55. What is the probability that the mean amount spent on refreshments by a random sample of 50 moviegoers will be between \$6 and \$7?

13. At a local accounting firm, the mean fee for completing an individual's tax return is \$225 with a standard deviation of \$28. Consider a random sample of 33 individual tax returns from this firm. What is the probability that the mean amount billed for the sample differs from the population mean by more than \$5?

14. At one supermarket, patrons purchase a mean of 17.0 items each visit with a standard deviation of 8.2 items. Suppose that 53 customers are randomly selected. What is the probability that the mean number of items purchased by customers in the sample differs from the population mean by less than 2 items?

15. Salaries of employees at a large factory are normally distributed with a mean of \$27,500 and a standard deviation of \$2900. Consider the mean salary of a group of 8 employees and the mean salary of a group of 15 employees. Which group's mean would you expect to be closer to \$27,500?

P Chapter 7 Project

Project A: Central Limit Theorem Experiment

Directions: You will need a standard six-sided die and at least six sets of data to complete this project.

Consider the distribution of the possible outcomes from rolling a single die; that is, 1, 2, 3, 4, 5, and 6. Let's use this distribution as our theoretical population distribution. We want to use this population distribution to explore the properties of the Central Limit Theorem. Let's begin by determining the shape, center, and dispersion of the population distribution.

1. What would you expect the distribution of the outcomes from repeated rolls of a single die to look like; in other words, what is its shape? (**Hint:** What is the probability of getting each value?)

   Shape: ___________

2. Calculate the mean of the population. (**Hint:** What is the mean outcome for rolling a single die?)

   $\mu =$ _______________

3. Calculate the standard deviation of the population. (**Hint:** What is the standard deviation of all possible outcomes from rolling a single die?)

   $\sigma =$ _______________

Let's continue by exploring the distribution of the original population empirically. To do so, follow these steps.

Step 1: Roll your die 60 times and record each outcome.

Step 2: Combine your results with at least two other students and tally the frequency of each roll of the die from the combined results. Record your results in a table similar to the following.

Outcome	Frequency
1	
2	
3	
4	
5	
6	

Step 3: Draw a bar graph of these frequencies.

Step 4: Does the distribution appear to be a normal distribution? Is this what you expected from question 1?

The Central Limit Theorem is not about individual rolls like we just looked at, but is about the averages of sample rolls. Thus we need to create samples in order to explore the properties of the Central Limit Theorem.

Step 5: Return to your original data from Step 1. To create samples from your data you can group the rolls into sets of 10. For each sequence of 10 rolls, calculate the mean of that sample. Round your answers to one decimal place. (You should have six sample means.)

Step 6: Combine your sample means with those of as many of your classmates as you can. Record the sample means of each of your classmates' six samples.

Step 7: Tally the frequencies of the sample means from your combined results in a table like the one that follows.

Sample Mean	Frequency
1.0–1.2	
1.3–1.5	
1.6–1.8	
1.9–2.1	
2.2–2.4	
2.5–2.7	
2.8–3.0	
3.1–3.3	
3.4–3.6	
3.7–3.9	
4.0–4.2	
4.3–4.5	
4.6–4.8	
4.9–5.1	
5.2–5.4	
5.5–5.7	
5.8–6.0	

Step 8: Draw a histogram of the sample means.

Step 9: What is the shape of this distribution?

Step 10: What is the mean of your sample means? (**Hint:** Use the sample means you collected in Step 6.)

$\mu_{\bar{x}} =$ __________

How does $\mu_{\bar{x}}$ compare to μ from question 2?

Step 11: What is the standard deviation of the sample means? (Again, go back to the sample means you collected in Step 6 and use a calculator or statistical software.)

$\sigma_{\bar{x}} =$ __________

How does $\sigma_{\bar{x}}$ compare to σ from question 3?

Since our samples were groups of 10 rolls, $n = 10$. Using σ from question 3, calculate $\frac{\sigma}{\sqrt{n}}$.

$\frac{\sigma}{\sqrt{n}} =$ __________

How does $\sigma_{\bar{x}}$ compare to $\frac{\sigma}{\sqrt{n}}$?

The Central Limit Theorem says that the distribution of the sample means should be closer to a normal distribution when the sample size becomes larger. To see this effect, group your original data from Step 1 into two samples of 30 rolls instead of six sets of 10.

Repeat Steps 5–11 using the new sample size of $n = 30$.

Step 12: Do your results seem to verify the three properties of the Central Limit Theorem?

Project B: Sampling Distribution Simulation

In the Hawkes Learning Systems software, *Beginning Statistics*, open Lesson 7.1, Central Limit Theorem. This lesson is a simulation designed to help you better understand sampling distributions as well as the Central Limit Theorem. Begin the simulation by choosing a parent distribution from the **Distribution** menu at the top of the screen. If you do not have a preference, the computer will automatically begin with a uniform parent distribution. Click **Simulate** to generate the first iteration of data from that distribution. For each iteration, the computer randomly chooses 30 numbers from the parent distribution and displays them in the parent histogram. At the same time, the computer chooses samples of size 5, 15, and 30 from those same 30 numbers. The mean of each sample is calculated and then displayed in its respective histogram. Answer the following questions.

1. How many numbers are displayed in the parent histogram?
2. How many numbers are displayed in each of the sampling distribution histograms?

Click **Next** to obtain a second iteration.

3. After the second iteration, how many numbers are displayed in the parent histogram?
4. After the second iteration, how many numbers are displayed in each of the sampling distribution histograms?

Now let's see what happens after many iterations. Click **Auto** and let the simulation run until about 100 iterations have passed. Click the same button, which now says **Stop**, to stop the process.

5. How many numbers are displayed in the parent histogram? (This number will vary, depending on how many iterations have passed.)
6. How many numbers are displayed in each of the sampling distributions?
7. Which, if any, of the sampling distributions appear to have a normal shape?

Finally, click **Auto** again and allow the program to process at least 1500 iterations before clicking **Stop**.

8. Which of the sampling distributions appear to have a normal shape?
9. Compare the means of sample means listed in the table. Do these numbers behave as you would expect according to the Central Limit Theorem?
10. Compare the standard deviations of sample means listed in the table. Do these numbers behave as you would expect according to the Central Limit Theorem?

Take some time to explore the other parent distributions available. Although the shape of the parent histogram will differ with each parent used, the Central Limit Theorem principles will always remain constant.

T Chapter 7 Technology

Calculating Standard Scores Using the Central Limit Theorem

TI-83/84 Plus

This chapter extends the concepts discussed in the previous chapter. In this chapter, the methods for calculating probabilities using normal distributions are the same as those discussed in Chapter 6. The difference is that we now have introduced new formulas for z that are needed when calculating a probability for a sample mean or sample proportion in a sampling distribution. To learn how to use MINITAB, Microsoft Excel, or a TI-83/84 Plus calculator to find a given probability, refer to the Chapter 6 Technology section. In this section, we will focus on how to use a TI-83/84 Plus calculator to find the value of z for a sample mean or sample proportion in a sampling distribution.

Side Note

The most recent TI-84 Plus calculators (Jan. 2011 and later) contain a new feature called `STAT WIZARDS`. This feature is not used in the directions in this text. By default this feature is turned `ON`. `STAT WIZARDS` can be turned `OFF` under the second page of MODE options.

Example T.1

Using a TI-83/84 Plus Calculator to Calculate the Standard Score for a Sample Mean

Find the value of z for the sample mean using the formula from the Central Limit Theorem given that $\bar{x} = 34$, $\mu = 35$, $\sigma = 5$, and $n = 100$.

Solution

The formula we need is the equation of the z-value for a sample mean in a sampling distribution from the Central Limit Theorem. Let's begin by substituting the given values into the equation.

$$z = \frac{\bar{x} - \mu}{\left(\frac{\sigma}{\sqrt{n}}\right)} = \frac{34 - 35}{\left(\frac{5}{\sqrt{100}}\right)}$$

Now we need to enter this into the calculator. We must make sure that we put parentheses around the numerator and the denominator. Enter the following into the calculator: `(34-35)/(5/√(1ØØ))`. Press ENTER. Thus, $z = -2$.

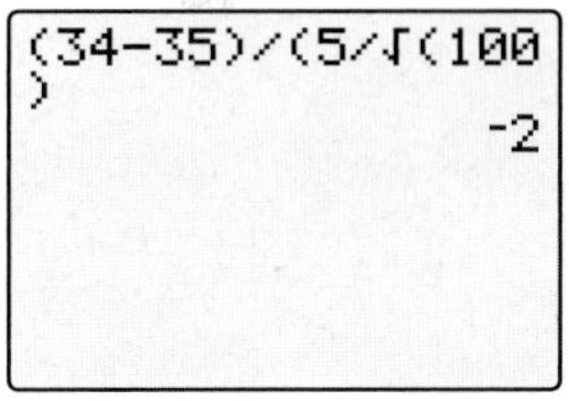

Example T.2

Using a TI-83/84 Plus Calculator to Calculate the Standard Score for a Sample Proportion

Find the value of z for the sample proportion using the formula from the Central Limit Theorem given that $\hat{p} = 0.56$, $p = 0.54$, and $n = 81$.

Solution

The formula we need is the equation of the z-value for a sample proportion in a sampling distribution from the Central Limit Theorem. Let's begin by substituting the given values into the equation.

$$z = \frac{\hat{p} - p}{\sqrt{\frac{p(1-p)}{n}}} = \frac{0.56 - 0.54}{\sqrt{\frac{0.54(1-0.54)}{81}}}$$

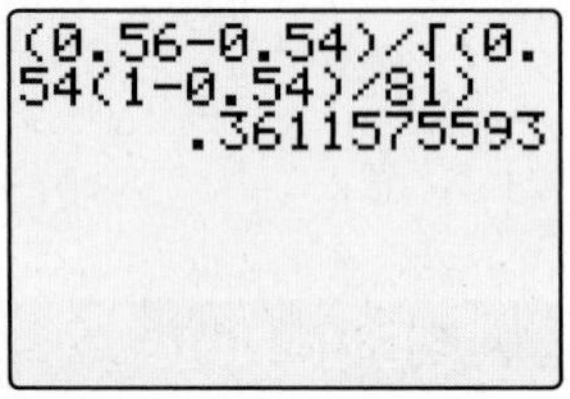

Now we need to enter this into the calculator. We must make sure that we put parentheses around the numerator and the denominator. Enter the following into the calculator: (Ø.56-Ø.54)/√(Ø.54(1-Ø.54)/81). Press ENTER. Thus, $z \approx 0.36$.

Microsoft Excel

We can also find the values of z from the previous examples using Microsoft Excel.

Example T.3

Using Microsoft Excel to Calculate the Standard Score for a Sample Mean

Find the value of z for the sample mean using the formula from the Central Limit Theorem given that $\bar{x} = 34$, $\mu = 35$, $\sigma = 5$, and $n = 100$.

Solution

Recall that the Central Limit Theorem states that the standard deviation of a sampling distribution of sample means, $\sigma_{\bar{x}}$, equals the standard deviation of the population divided by the square root of the sample size. That is, $\sigma_{\bar{x}} = \frac{\sigma}{\sqrt{n}}$. The formula for calculating the value of z in Microsoft Excel is =STANDARDIZE(x, *mean*, *standard_dev*). Applying the Central Limit Theorem, we input **=STANDARDIZE(34, 35, 5/SQRT(100))**. Just as we found in Example T.1, $z = -2$.

Example T.4

Using Microsoft Excel to Calculate the Standard Score for a Sample Proportion

Find the value of z for the sample proportion using the formula from the Central Limit Theorem given that $\hat{p} = 0.56$, $p = 0.54$, and $n = 81$.

Solution

Here we will use $\sigma_{\hat{p}} = \sqrt{\frac{p(1-p)}{n}}$. Enter

=STANDARDIZE(0.56, 0.54, SQRT(0.54*(1-0.54)/81))

into Excel. The answer, 0.361158, is displayed. Thus, $z \approx 0.36$.

Minitab

Minitab can calculate a column of z-scores with a single operation.

Example T.5

Using Minitab to Calculate Standard Scores for Sample Means

The call processing times at an emergency dispatch center have a population mean of 45 seconds and a standard deviation of 50 seconds. Five operators are evaluated using random samples of the calls they have handled. The total number of calls sampled and the corresponding mean processing time for each operator are displayed in the table below. Find the z-score for each sample mean.

Call Processing Times (in Seconds)					
n	98	85	105	110	91
Sample Mean	52	48	63	45	55

Solution

First, enter the data into columns C1 and C2 in the worksheet. The first column is n, the number of calls sampled, and the second column contains each operator's sample mean. Go to **Calc ► Calculator** and enter the following expression: **(C2-45)/(50/SQRT(C1))**. Choose to store the result in column C3 and click **OK**. The dialog box appears as follows.

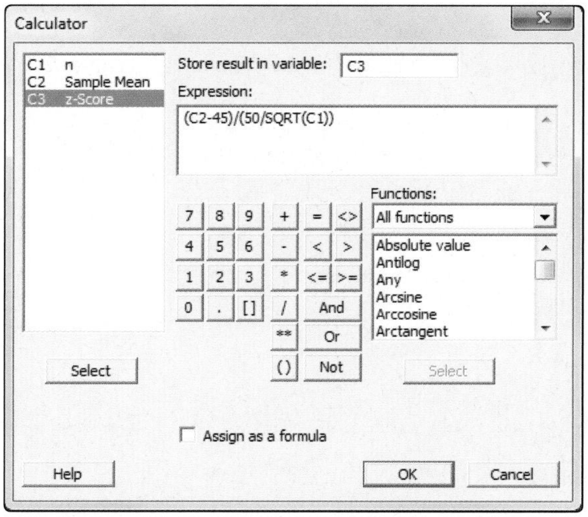

The column produced contains the z-score for each operator.

Worksheet 1 ***

↓	C1	C2	C3
	n	Sample Mean	z-Score
1	98	52	1.38593
2	85	48	0.55317
3	105	63	3.68890
4	110	45	0.00000
5	91	55	1.90788

Chapter Eight
Confidence Intervals

Sections

Objectives

1. Determine the best point estimates for population parameters.
2. Calculate margins of error for confidence intervals for population parameters.
3. Construct confidence intervals for population parameters.
4. Determine the minimum sample size required to obtain the desired level of precision for estimation.

8

Introduction

One of the key concepts in statistical analysis is that of estimating population parameters. Researchers and statisticians estimate a wide variety of parameters: anything from the average price of a flight to Europe to the proportion of voters who favor one candidate over another. Some of the predictions made with these estimates have important ramifications.

One important prediction is the estimate of risk assumed each day by traders at JPMorgan Chase & Co., one of the world's largest banks. For the first quarter of 2012, JPMorgan released to its investors that the average value-at-risk factor for its Chief Investment Office (CIO) was \$67 million. The average value-at-risk factor is described by the company as the maximum potential for loss during a single trading day, assuming ordinary market conditions and a 95% level of confidence. The first quarter average value-at-risk factor was later revised in May 2012 to be \$129 million.

However, on May 11, 2012, JPMorgan reported an *actual* first-quarter trading loss of *\$2 billion*. In other words, their estimate was off by over \$1.8 billion. The prediction model used by JPMorgan to assess risk was sorely inaccurate. The company has explained the discrepancy by saying that a new—and inadequate—model was used to predict value-at-risk for the first quarter of 2012. To put things in perspective, a quarterly loss of \$2 billion for a company with over \$17 billion in annual profit is not as disastrous as it appears at first glance. Unfortunately, the real damage done by the inaccurate prediction was the public relations problem, which ultimately cost JPMorgan's CIO her job.

The example of JPMorgan highlights the need for the best possible methods for estimating population parameters. In Chapter 8, we will put together concepts of descriptive statistics and probabilities in order to estimate unknown population parameters.

8.1 Estimating Population Means (σ Known)

The major purpose of statistics is to provide information so that informed decisions can be made. Statisticians are asked to provide information about a variety of population parameters. For example, a television network may want to know how many Americans watch its TV shows each week. It is impossible to survey each and every household, yet the network needs the best information available in order to make programming decisions. This is where inferential statistics plays an important role.

Recall that inferential statistics is the branch that uses sample statistics to provide estimates for population parameters. For a particular parameter, there may be several sample statistics that could be used to provide estimates. When the estimate for a population is a single value, it is called a **point estimate**.

Definition

A **point estimate** is a single-number estimate of a population parameter.

The best point estimate for a population parameter is one that is *unbiased*—that is, a sample statistic that does not consistently underestimate or overestimate the population parameter. Let's consider trying to estimate the population mean with an unbiased point estimate. Three obvious statistics that could be used to estimate the population mean are the mean, median, and mode of a sample. It is not necessarily true of sample modes or medians that their values will be centered about the mean of the population. However, recall from our discussion of the Central Limit Theorem in Section 7.1 that the mean of a sampling distribution of sample means *is* equal to the mean of the population. Thus, the sample mean does not consistently underestimate or overestimate the population mean and is actually the best point estimate of the population mean. Therefore, we refer to the sample mean as an **unbiased estimator** of the population mean. Similar unbiased point estimates exist for other population parameters, such as population proportion and population variance. We will discuss the unbiased estimators for each of these parameters in later sections of this chapter.

Definition

An **unbiased estimator** is a point estimate that does not consistently underestimate or overestimate the population parameter.

Example 8.1

Finding a Point Estimate for a Population Mean

Find the best point estimate for the population mean of test scores on a standardized biology final exam. The following is a simple random sample taken from the population of test scores.

45	68	72	91	100	71
69	83	86	55	89	97
76	68	92	75	84	70
81	90	85	74	88	99
76	91	93	85	96	100

Solution

The best point estimate for the population mean is a sample mean because it is an unbiased estimator. The sample mean for the given sample of test scores is $\bar{x} = \frac{\sum x_i}{n} \approx 81.6$. Thus, the best point estimate for the population mean of test scores on this standardized exam is 81.6.

As you might imagine, using only one value for estimating the population mean, or any other population parameter, leaves much room for error. In fact, we have no way of knowing the likelihood that we have chosen the exact value of the parameter, or that we are even close. It would be much better to provide a range of values rather than just one number to estimate the parameter. This range of possible values is called an **interval estimate**.

Notice that an interval estimate is still an *estimate*. Just as we couldn't be certain that the point estimate was exact, we also cannot be certain that the interval estimate contains the true population parameter. Although not certain, how confident can we be? The probability that the interval actually contains the true population parameter is called the **level of confidence**, and is represented by the symbol c. An interval estimate associated with a certain level of confidence is called a **confidence interval**. Thus, if we are 95% confident that the interval estimate contains the population parameter, then $c = 0.95$ and the interval estimate is a 95% confidence interval.

Definition

An **interval estimate** is a range of possible values for a population parameter.

The **level of confidence** is the probability that the interval estimate contains the population parameter.

A **confidence interval** is an interval estimate associated with a certain level of confidence.

In general, how do we begin to find a confidence interval for a population parameter? Obviously, the confidence interval should be constructed around the best point estimate we can find, which would be an unbiased estimator as we discussed earlier. But, how big do we make the interval? The largest possible distance away from the point estimate that the confidence interval will cover is called the **margin of error**, or *maximum error of estimate, E*. To construct confidence intervals for the most common population parameters, simply subtract the margin of error from and add the margin of error to the point estimate.

Definition

The **margin of error**, or *maximum error of estimate, E*, is the largest possible distance from the point estimate that a confidence interval will cover.

Figure 8.1: Constructing a Confidence Interval

Let us stop at this point and emphasize how to interpret confidence intervals. We have no way of being certain, short of knowing the actual parameter, if a particular interval estimate actually contains the population parameter—just as we had no way of knowing how close the point estimate was to the true population parameter. Furthermore, the population parameter is a fixed value. It is already determined—we just don't know its value. So when we interpret a confidence interval, we do *not* want to say that there is a 95% chance of the population parameter falling within the confidence interval. We *can* say that we are 95% confident that the confidence interval contains the population parameter, because this statement correctly implies that the population mean is fixed and it is either in the interval or not.

Example 8.2

Constructing a Confidence Interval with a Given Margin of Error

A college student researching study habits collects data from a random sample of 250 college students on her campus and calculates that the sample mean is $\bar{x} = 15.7$ hours per week. If the margin of error for her data using a 95% level of confidence is $E = 0.6$ hours, construct a 95% confidence interval for her data. Interpret your results.

Solution

The best point estimate for the population mean is a sample mean, so use $\bar{x} = 15.7$ as the point estimate for this population parameter. We are given the value of the margin of error for our confidence interval, $E = 0.6$. As indicated in Figure 8.1, the endpoints of a confidence interval are found by subtracting E from and adding E to the point estimate.

To find the lower endpoint, subtract the margin of error from the sample mean; that is, calculate $\bar{x} - E$. Thus, the lower endpoint is calculated as follows.

$$\begin{aligned}\text{Lower endpoint: } \bar{x} - E &= 15.7 - 0.6 \\ &= 15.1 \text{ hours per week}\end{aligned}$$

To find the upper endpoint, add the margin of error to the sample mean; that is, calculate $\bar{x} + E$. Thus, the upper endpoint is calculated as follows.

$$\begin{aligned}\text{Upper endpoint: } \bar{x} + E &= 15.7 + 0.6 \\ &= 16.3 \text{ hours per week}\end{aligned}$$

Therefore, the confidence interval ranges from 15.1 to 16.3. The confidence interval can be written mathematically using either inequality symbols or interval notation, as shown below.

$$15.1 < \mu < 16.3$$

or

$$(15.1, 16.3)$$

The interpretation of our confidence interval is that we are 95% confident that the true population mean for the number of hours per week that students on this campus spend studying is between 15.1 and 16.3 hours.

So far, constructing a confidence interval for a population mean does not seem much more complicated than finding a point estimate for the population mean, and then adding the margin of error to and subtracting the margin of error from that point estimate. In the previous example, the margin of error was given to us since we did not yet have the tools to calculate this value. In the remainder of this chapter, we will see how to calculate the margin of error for various population parameters. Calculating the margin of error requires choosing the right distribution or using advanced statistical techniques. Since we are familiar with the standard normal distribution, we will begin by calculating the margin of error in situations where the standard normal distribution is appropriate.

Using the Standard Normal Distribution to Estimate a Population Mean

When the following conditions are met, the distribution used to calculate the margin of error of a confidence interval for the population mean is the standard normal distribution.

- All possible samples of a given size have an equal probability of being chosen; that is, a simple random sample is used.
- The population standard deviation, σ, is *known*.
- Either the sample size is at least 30 $(n \geq 30)$ *or* the population distribution is approximately normal.

These conditions come from the Central Limit Theorem (discussed in Chapter 7), which allows us to use the normal distribution when estimating population means.

Notice that the second condition requires that the population standard deviation, σ, is known. If σ is known, then either the standard normal distribution or more advanced statistical techniques must be used to calculate the margin of error. However, essentially the only way you could calculate the population standard deviation is to know every value in the population. With this knowledge you could calculate *any* parameter you needed—there would be no need to estimate anything! However, since we are familiar with the standard normal distribution, we can begin by learning how to calculate a margin of error using this distribution, and then move on to learn about other distributions in order to calculate the margin of error in other situations and for other population parameters.

Remember that we want to be confident that the population mean is within the interval that we are creating. The measure of how confident we are that the interval contains this parameter is the level of confidence, c. This level of confidence corresponds to the area under the normal curve centered at the estimate for the population mean, $\bar{x}$. Thus, if we want a 95% level of confidence, we need to find the z-values that correspond to the middle 0.95 of the area under the standard normal distribution. These z-values that mark the boundaries for the area of c under the middle of the standard normal curve are called the **critical z-values**, and are denoted as $-z_{\alpha/2}$ and $z_{\alpha/2}$.

Memory Booster

c is the complement of α.

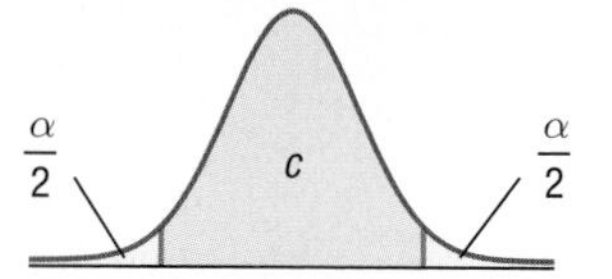

If the area under the middle of the curve equals c, then the total area in the two tails equals $1-c$ by the Complement Rule. This value is called α (pronounced "alpha"), and $\alpha = 1-c$. Since α equals the combined area in the tails, $\frac{\alpha}{2}$ equals the area in one tail. Thus, the notation $z_{\alpha/2}$ indicates the critical z-value corresponding to an area of $\frac{\alpha}{2}$ in the right tail of the standard normal distribution. Furthermore, as the normal distribution is symmetric, we do not have to find each critical value separately. The critical value at the left end of the area of c under the curve is simply the negative of the critical value at the right end. For this reason, only the positive value is reported as the critical z-value.

For example, to find the critical value for a 95% confidence interval, we need to find the values of z that correspond to an area in the middle of the normal distribution equal to 0.95. This is actually one type of problem that we learned how to solve in Section 6.4. If the area in the middle of the distribution equals 0.95, then the area in the tails of the distribution equals 0.05. Using our knowledge of the symmetry of the normal curve, we know that an area of 0.05 is split equally between the two tails, so each tail has an area of 0.025. The standard normal distribution tables tell us that the z-value with an area of 0.025 under the curve to its left is $z = -1.96$. By the symmetry property, we can then say that the critical value is $z_{0.025} = 1.96$.

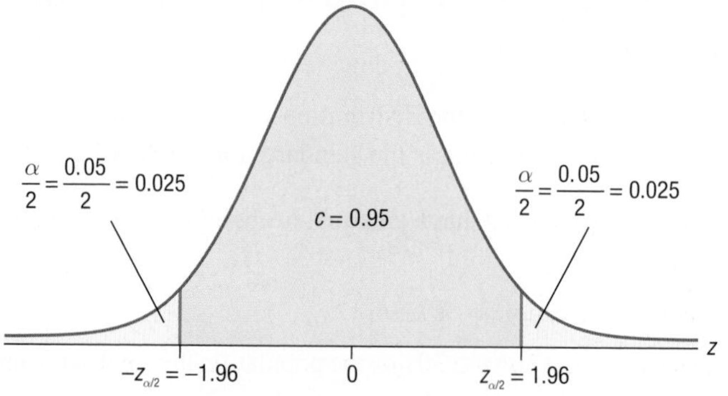

Figure 8.2: Critical z-Value for a 95% Confidence Interval

The critical value for any given level of confidence may be found using the same method. Notice that the critical value is based only on the level of confidence and not on the size of the sample or any of the population's parameters. Hence, the critical z-value for a 95% confidence interval remains the same for every problem for which it is appropriate to use the standard normal distribution. Commonly used levels of confidence include 90%, 95%, and 99%, although the critical value for any level of confidence can be calculated. The table of critical z-values shown below may be used as a reference.

Table 8.1: Critical z-Values for Confidence Intervals

Level of Confidence, c	$\alpha = 1 - c$	$z_{\alpha/2}$
0.80	0.20	1.28
0.85	0.15	1.44
0.90	0.10	1.645
0.95	0.05	1.96
0.98	0.02	2.33
0.99	0.01	2.575

The formula for the margin of error of a confidence interval for a population mean when the population standard deviation (σ) is known is as follows.

$$E = \left(z_{\alpha/2}\right)\left(\sigma_{\bar{x}}\right)$$

By applying the Central Limit Theorem, this formula can be rewritten as shown below.

$$E = \left(z_{\alpha/2}\right)\left(\frac{\sigma}{\sqrt{n}}\right)$$

We leave the details of how this formula is rewritten for the exercises. Note that to use this formula, you need to know the critical z-value, the population standard deviation, and the sample size for the sample you are using to estimate the population mean.

Formula

Margin of Error of a Confidence Interval for a Population Mean (σ Known)

When the population standard deviation is known, the sample taken is a simple random sample, and either the sample size is at least 30 or the population distribution is approximately normal, the margin of error of a confidence interval for a population mean is given by

$$E = \left(z_{\alpha/2}\right)\left(\sigma_{\bar{x}}\right) = \left(z_{\alpha/2}\right)\left(\frac{\sigma}{\sqrt{n}}\right)$$

where $z_{\alpha/2}$ is the critical value for the level of confidence, $c = 1 - \alpha$, such that the area under the standard normal distribution to the right of $z_{\alpha/2}$ is equal to $\frac{\alpha}{2}$,

σ is the population standard deviation, and

n is the sample size.

Rounding Rule

When calculating a margin of error for a confidence interval, round to at least six decimal places to avoid additional rounding errors in the subsequent calculations of the endpoints of the confidence interval.

Example 8.3

Finding the Margin of Error of a Confidence Interval for a Population Mean (σ Known)

Researchers want to estimate the mean monthly electricity bill in a large urban area using a simple random sample of 100 households. Assume that the population standard deviation is known to be \$15.50. Find the margin of error for a 99% confidence interval. Round your answer to two decimal places.

Solution

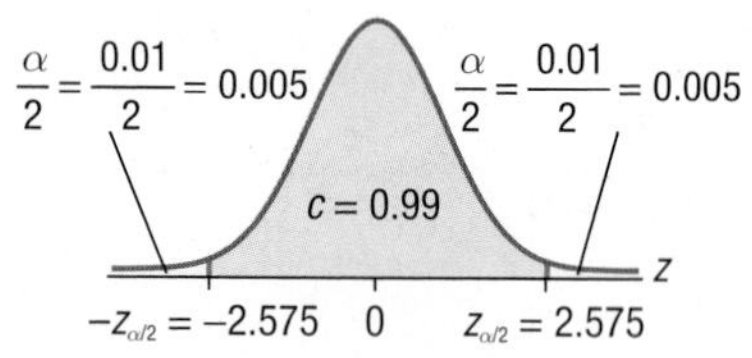

First, refer to the table of critical z-values to find the critical value for $c = 0.99$. In the table, we see that for $c = 0.99$, the critical value is $z_{\alpha/2} = z_{0.01/2} = z_{0.005} = 2.575$. The population standard deviation is $\sigma = 15.50$, and the sample size is $n = 100$. Substituting these values into the formula for the margin of error, we get the following.

$$
\begin{aligned}
E &= \left(z_{\alpha/2}\right)\left(\frac{\sigma}{\sqrt{n}}\right) \\
&= (2.575)\left(\frac{15.50}{\sqrt{100}}\right) \\
&= 3.99125
\end{aligned}
$$

So the margin of error for this 99% confidence interval is approximately \$3.99.

Recall that once the value of the margin of error for a particular confidence interval has been calculated, we simply subtract the margin of error from and add the margin of error to the point estimate to construct the confidence interval.

Rounding Rule

Round the endpoints of a confidence interval for a population mean as follows:

- If sample data are given, round to one more decimal place than the largest number of decimal places in the given data.
- If statistics are given, round to the same number of decimal places as given in the standard deviation or variance.

Formula

Confidence Interval for a Population Mean

The confidence interval for a population mean is given by

$$\bar{x} - E < \mu < \bar{x} + E$$

or

$$(\bar{x} - E, \bar{x} + E)$$

where $\bar{x}$ is the sample mean, which is the point estimate for the population mean, and

E is the margin of error.

Example 8.4

Constructing a Confidence Interval for a Population Mean (σ Known)

In order to estimate the number of calls to expect at a new suicide hotline, volunteers contact a random sample of 35 similar hotlines across the nation and find that the sample mean is 42.0 calls per month. Construct a 95% confidence interval for the mean number of calls per month. Assume that the population standard deviation is known to be 6.5 calls per month.

Solution

Step 1: Find the point estimate.

To find a confidence interval, you need to know a point estimate and a margin of error. The point estimate for a population mean is the sample mean. In this example, the sample mean is given to us as 42.0 calls per month. The margin of error must be calculated using the formula for population means with σ known.

Step 2: Find the margin of error.

To calculate the margin of error, refer to the table of critical z-values to find the critical value for $c = 0.95$. Note that $z_{\alpha/2} = z_{0.05/2} = z_{0.025} = 1.96$ is the critical value. The population standard deviation is $\sigma = 6.5$, and the sample size is $n = 35$. Substituting these values into the formula for the margin of error, we get the following.

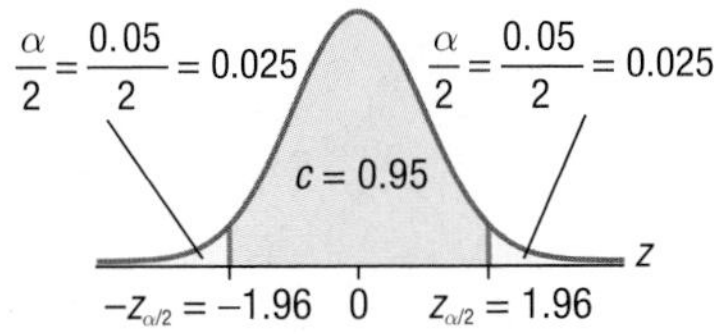

$$E = \left(z_{\alpha/2}\right)\left(\frac{\sigma}{\sqrt{n}}\right)$$
$$= (1.96)\left(\frac{6.5}{\sqrt{35}}\right)$$
$$\approx 2.153453$$

Step 3: Subtract the margin of error from and add the margin of error to the point estimate.

To find the lower endpoint, subtract the margin of error from the sample mean. Thus, the lower endpoint is calculated as follows.

$$\text{Lower endpoint: } \bar{x} - E = 42.0 - 2.153453$$
$$\approx 39.8 \text{ calls per month}$$

To find the upper endpoint, add the margin of error to the sample mean. Thus, the upper endpoint is calculated as follows.

$$\text{Upper endpoint: } \bar{x} + E = 42.0 + 2.153453$$
$$\approx 44.2 \text{ calls per month}$$

Therefore, the 95% confidence interval ranges from 39.8 to 44.2 calls per month. The confidence interval can be written mathematically using either inequality symbols or interval notation, as shown below.

$$39.8 < \mu < 44.2$$

or

$$(39.8,\ 44.2)$$

The interpretation of our confidence interval is that we are 95% confident that the true population mean for the numbers of calls to suicide hotlines across the nation is between 39.8 and 44.2 calls per month.

Example 8.5

Constructing a Confidence Interval for a Population Mean (σ Known)

A toy company wants to know the mean number of new toys per child bought each year. Marketing strategists at the toy company collect data from the parents of 1842 randomly selected children. The sample mean is found to be 4.7 toys per child. Construct a 90% confidence interval for the mean number of new toys per child purchased each year. Assume that the population standard deviation is known to be 1.9 toys per child per year.

Solution

Step 1: **Find the point estimate.**

The point estimate for the population mean is the sample mean, which we are told is 4.7 toys per child.

Step 2: **Find the margin of error.**

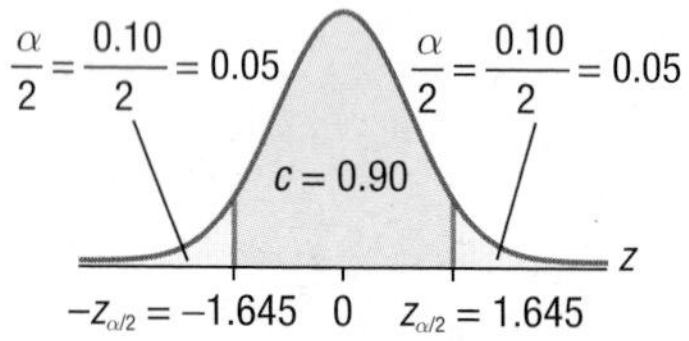

Next, we need to calculate the margin of error. Since we want a 90% confidence interval, we can use the table to find the critical z-value, which is $z_{\alpha/2}=z_{0.10/2}=z_{0.05}=1.645$. Substituting the values we have into the formula for margin of error, we get the following.

$$E=\left(z_{\alpha/2}\right)\left(\frac{\sigma}{\sqrt{n}}\right)$$
$$=(1.645)\left(\frac{1.9}{\sqrt{1842}}\right)$$
$$\approx 0.072824$$

Step 3: **Subtract the margin of error from and add the margin of error to the point estimate.**

To find the lower endpoint, subtract the margin of error from the best point estimate, that is, the sample mean.

$$\text{Lower endpoint: } \bar{x}-E=4.7-0.072824$$
$$\approx 4.6 \text{ toys per child}$$

To find the upper endpoint, add the margin of error to the point estimate.

$$\text{Upper endpoint: } \bar{x}+E=4.7+0.072824$$
$$\approx 4.8 \text{ toys per child}$$

Thus, the 90% confidence interval ranges from 4.6 to 4.8 new toys per child per year. The confidence interval can be written mathematically using either inequality symbols or interval notation, as shown below.

$$4.6<\mu<4.8$$

or

$$(4.6,\ 4.8)$$

Therefore, we are 90% confident that the true population mean for the number of new toys per child bought each year is between 4.6 and 4.8 toys per child.

Using a TI-83/84 Plus Calculator to Find a Confidence Interval for a Population Mean (σ Known)

You can also use technology to calculate a confidence interval for a population mean with a known population standard deviation. Using a TI-83/84 Plus calculator, press STAT, scroll to TESTS, and then select option `7:ZInterval`. You then have the option of selecting the raw data from an existing list, or entering the known population standard deviation and sample statistics. When we know the statistics, as in the previous examples, we highlight the `Stats` option, and enter the values for σ, x̄, and n. `C-Level` is the confidence level, which should be entered as a decimal. Highlight `Calculate` and press ENTER. The calculator screen then gives the confidence interval in interval form and also reiterates the sample mean and sample size. For example, the following screenshots show how a TI-83/84 Plus calculator can be used to find the confidence interval we constructed in Example 8.5.

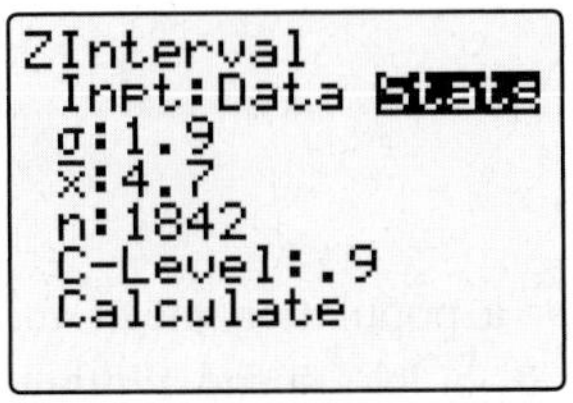

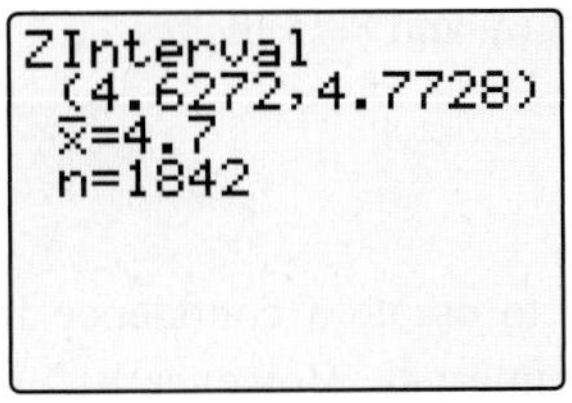

There are a few things to note here. First, the margin of error is not reported when using a TI-83/84 Plus calculator to find a confidence interval. If you want to know the margin of error, you will have to calculate the margin of error by finding how large the interval is and then dividing that number in half. Second, we have been rounding the endpoints of each confidence interval to the same number of decimal places as the given standard deviation, but you will see many more decimal places in the confidence interval given by the calculator. Continue to round the endpoints of the confidence interval using the same rounding rule we have been using throughout this section.

Example 8.6

Using a TI-83/84 Plus Calculator to Find a Confidence Interval for a Population Mean (σ Known)

The owners of a local company that produces hand-knitted socks want to know, for women in their area, the average length of a woman's foot, from toe to heel. They collect data from 431 randomly selected women. The sample mean is found to be 8.72 inches. Construct a 95% confidence interval for the mean length of a woman's foot in the company's area. Assume the owners know that the population standard deviation is 0.36 inches.

Solution

Using a TI-83/84 Plus calculator, press STAT, scroll to TESTS, and then select option `7:ZInterval`, since the population standard deviation is known. We are given the sample statistics, so we highlight the `Stats` option, and enter the values for σ, x̄, and n. For this example, $\sigma = 0.36$, $\bar{x} = 8.72$, and $n = 431$. `C-Level` is the confidence level, which should be entered as a decimal. The confidence level for this example is 95%, so we enter 0.95.

Highlight `Calculate` and press ENTER. The calculator screen then gives the confidence interval in interval form and also reiterates the sample mean and sample size.

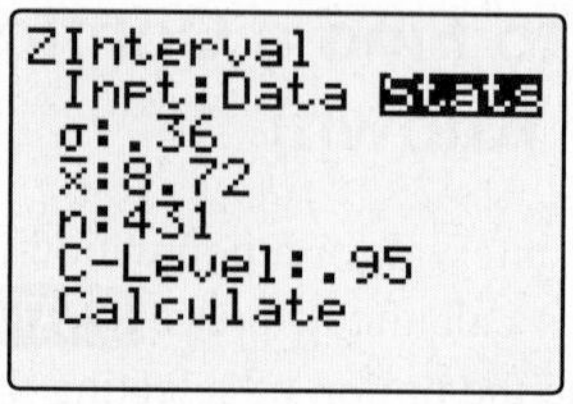

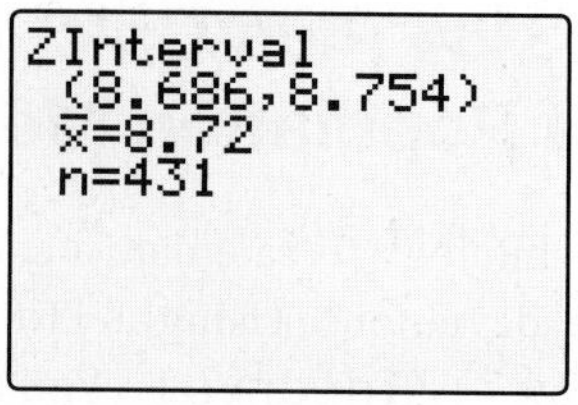

Thus, the 95% confidence interval ranges from 8.69 to 8.75 inches. The confidence interval can be written mathematically using either inequality symbols or interval notation, as shown below.

$$8.69 < \mu < 8.75$$

or

$$(8.69, 8.75)$$

Therefore, we are 95% confident that the true population mean for the length of a woman's foot is between 8.69 and 8.75 inches.

We are now armed to create a confidence interval for a population mean and provide a general interpretation of the interval. However, before we move on, let's spend a little time delving into the more practical side of using confidence intervals to make inferences about populations. Remember that when we say we have a 95% confidence interval, we mean that we are 95% confident that the true population mean falls within that interval. The important key here is that the population value could fall *anywhere* in the interval. We cannot say that it is more likely to fall in the middle of the interval than it is to fall near one of the endpoints. This is a distinction that is sometimes misrepresented at best. Depending on your agenda, it may be more convenient to focus your attention at one end of the interval over the other. However, doing so provides a biased view of the true meaning of the interval.

Example 8.7

Interpreting a Confidence Interval

A new study compared the sensitivity of HIV home tests to that of professionally administered tests that used the same testing procedure. Measures of test sensitivity represent the percentages of results that are correctly labeled positive when HIV is actually present. The table shows the reported 95% confidence intervals for the sensitivity of both tests.

Sensitivity of HIV Tests

Method	Sensitivity (95% Confidence Interval)
Home Test	86.6%–96.9%
Professionally Administered Test	95.85%–99.08%

Source: Pebody, Roger. "US regulators set to approve HIV home-testing kit." NAM Publications. 16 May 2012. http://www.aidsmap.com/US-regulators-set-to-approve-HIV-home-testing-kit/page/2356465/ (20 May 2012).

Which of the following statements provides the most unbiased interpretation of the confidence intervals?

i. The professionally administered tests were found to be far more reliable, with 99.08% sensitivity, than the home tests, with only 86.6% sensitivity.

ii. Home testing methods for HIV are just as sensitive as those administered professionally. In fact, they are even more sensitive since they have a possible 96.9% sensitivity rating compared with a possible 95.85% rating for the professional ones.

iii. The results provide reasonable assurance that the home test is as effective as the professionally administered test.

Solution

Option iii. gives the best interpretation of the confidence intervals given. Because the confidence intervals overlap, it is possible that both tests have the same sensitivity rating, and therefore can provide a reasonable assurance that they are similarly accurate. This was, in fact, a part of the findings that the expert panel reported to the Food and Drug Administration (FDA) in the United States.

Option i. does not give a fair picture of the results by only providing the upper endpoint of the confidence interval for the professional tests compared to the lower endpoint of the interval for the home tests. This analysis might lead you to believe that there was definitely a large difference between the sensitivity ratings for the two tests. It does not reflect the possibility that the sensitivity ratings did not differ at all.

Option ii. is similar to option i. in that, by giving the upper endpoint of the home test interval compared to the lower endpoint of the professional test interval, one is led to believe that the home tests are out-performing the professional tests. It also does not account for the fact that it is possible for the professional tests to have a much higher sensitivity than the home tests.

Minimum Sample Size for Estimating a Population Mean

Suppose that a researcher wants to conduct a statistical study with a high level of confidence and a reasonably small margin of error. By using a larger sample size, his estimates are based on more data. This results in an increase in the level of confidence without increasing the margin of error. A logical question at this point might be, "How large a sample size does he need?" The answer is that it depends on the level of confidence and margin of error that he plans to use. It would be nice to have very large samples with thousands of data values, but time and money will usually prevent one from such luxuries. So, what is the fewest number of data values that he must collect to obtain the desired level of confidence with a particular margin of error? That is, what is the minimum sample size he must have to obtain his requirements?

Rather than simply using a very large sample, we can calculate the minimum sample size necessary using the following formula. This formula for the minimum sample size is derived from the formula for the margin of error by algebraically solving for n.

Margins of Error of Confidence Intervals for Population Means (σ Known)

Directions: Calculate the margin of error of a confidence interval for the population mean at the given level of confidence.

5. $n = 56, \quad \sigma = 3.14, \quad c = 0.90$

6. $n = 81, \quad \sigma = 2.45, \quad c = 0.95$

7. $n = 93, \quad \sigma = 1.25, \quad c = 0.95$

8. $n = 134, \quad \sigma = 0.27, \quad c = 0.99$

Confidence Intervals for Population Means (σ Known)

Directions: Construct a confidence interval for the population mean at the given level of confidence.

9. $n = 89, \quad \sigma = 2.01, \quad c = 0.95, \quad \bar{x} = 45.00$

10. $n = 64, \quad \sigma = 8.01, \quad c = 0.90, \quad \bar{x} = 90.40$

11. $n = 607, \quad \sigma = 1.92, \quad c = 0.99, \quad \bar{x} = 18.45$

12. $n = 1123, \quad \sigma = 7.31, \quad c = 0.95, \quad \bar{x} = 87.12$

Minimum Sample Sizes for Estimating Population Means

Directions: Calculate the minimum sample size needed to construct a confidence interval with the desired characteristics. The value of the population standard deviation is an estimate based on a previous reliable study.

13. $E = 0.5, \quad \sigma = 5.25, \quad c = 0.95$

14. $E = 2, \quad \sigma = 12.10, \quad c = 0.90$

15. $E = 1.5, \quad \sigma = 4.75, \quad c = 0.99$

16. $E = 3, \quad \sigma = 15.03, \quad c = 0.95$

Confidence Intervals for Population Means (σ Known)

Directions: Construct and interpret each specified confidence interval.

17. A professor wants to estimate how many hours per week her students study. A simple random sample of 78 students had a mean of 15.0 hours of studying per week. Construct and interpret a 90% confidence interval for the mean number of hours a student studies per week. Assume that the population standard deviation is known to be 2.3 hours per week.

18. A faculty advocacy group is concerned about the amount of time teachers spend each week doing schoolwork at home. A simple random sample of 56 teachers had a mean of 8.0 hours per week working at home after school. Construct and interpret a 95% confidence interval for the mean number of hours per week a teacher spends working at home. Assume that the population standard deviation is 1.5 hours per week.

19. A writer for a computer magazine is working on an article about computer usage in American households. A simple random sample of 120 American households has a mean computer usage time of 19.2 hours per week. Construct and interpret a 95% confidence interval for the mean computer usage time per week for all American households. Assume that the population standard deviation is 3.3 hours per week.

20. A survey of 85 randomly selected homeowners finds that they spend a mean of \$67 per month on home maintenance. Construct and interpret a 99% confidence interval for the mean amount of money spent per month on home maintenance by all homeowners. Assume that the population standard deviation is \$14 per month.

21. A survey of 97 randomly selected homeowners found that the mean amount spent on lawn service was \$720 per year. Construct and interpret a 98% confidence interval for the mean amount of money spent on lawn service per household each year. Assume that the population standard deviation is \$123 per year.

22. A survey of a simple random sample of 140 dieters revealed that the numbers of times they "cheated" on their diets had a mean of 7.0 times per week. Construct and interpret a 99% confidence interval for the mean number of times dieters "cheat" on their diets each week. Assume that the population standard deviation is 1.5 times per week.

23. A physical therapist is investigating the mean recovery time after ACL surgery for patients involved in a new therapy regimen. For the purpose of the study, a successful recovery was defined to be the ability to walk without crutches. For 38 randomly selected patients, the mean recovery time after ACL surgery was found to be 22.6 days. Assume that the population standard deviation is 3.7 days. Find and interpret a 99% confidence interval for the mean recovery time for all ACL surgery patients undergoing the same new therapy.

24. The manufacturers of Caudill automotive oil wish to estimate the mean number of miles that motorists drive between oil changes. A random sample of 54 motorists has a mean of 5900 miles driven between oil changes. Assume that the population standard deviation is 1350 miles. Construct and interpret a 95% confidence interval for the mean number of miles driven between oil changes for all motorists.

Minimum Sample Sizes for Estimating Population Means

Directions: Calculate the minimum sample size needed to construct a confidence interval with the desired characteristics. The value of the population standard deviation is an estimate based on a previous reliable study.

25. The upper management at a bank would like to estimate the mean number of credit cards college students have in their wallets. They would like to create a 98% confidence interval with a maximum error of 1 card. Assuming a standard deviation of 3.25 cards, what is the minimum number of college students they must include in their sample?

26. A social worker is concerned about the number of prescriptions her elderly clients have. She would like to create a 99% confidence interval for the mean number of prescriptions per client with a maximum error of 2 prescriptions. Assuming a standard deviation of 5.2 prescriptions, what is the minimum number of clients she must sample?

27. Suppose you are interested in determining the mean number of hours students spend studying each week. You want a 95% level of confidence and a maximum error of 0.5 hours. Assuming the standard deviation is 2.5 hours, what is the minimum number of students you must include in your sample?

28. For a psychology experiment, Emma is assigned the task of finding the mean number of hours students sleep per night. Her results must be at the 99% level of confidence with a maximum error of 0.25 hours. Assuming the standard deviation is 1.4 hours, how many students must Emma survey?

Directions: Respond thoughtfully to the following exercises.

29. Given a 99% confidence interval for a population mean of $(1.01, 1.97)$, is it possible to determine the original point estimate for the interval? Explain your answer.

30. Two individual researchers report a confidence interval for the mean number of unreported domestic violence incidents per month in one small county over the past year. One gives a 95% confidence interval of $(4.93, 6.14)$ while the other gives a 98% confidence interval of $(3.72, 5.01)$. Is it possible for both of the intervals to contain the true mean number of unreported domestic violence incidents? Explain your answer.

31. Last quarter, your sales team reported a mean amount of revenue of $131,540 per salesperson. Based on a survey of a portion of the sales team and last quarter's data, you estimate that this quarter, the mean revenue will be between $129,915 and $154,798 per salesperson, with a 95% level of confidence. Would you report to your supervisor that this quarter's sales will be up from last quarter? Why or why not?

32. The formula for the margin of error for a confidence interval for a population mean (when the population standard deviation is known) is given as $E=\left(z_{\alpha/2}\right)\left(\sigma_{\bar{x}}\right)$ or $E=\left(z_{\alpha/2}\right)\left(\frac{\sigma}{\sqrt{n}}\right)$. Explain how the Central Limit Theorem can be used to rewrite the first formula as the second one.

8.2 Student's *t*-Distribution

In this chapter we are learning how to estimate population parameters. We began with learning how to estimate a population mean when the population standard deviation, σ, is known because it allowed us to use a distribution with which we were familiar, the standard normal distribution. However, essentially the only way you could know the population standard deviation is to know every value in the population. With this knowledge you could calculate *any* parameter you needed—there would be no need to estimate anything! Thus, we now need to learn about a new distribution in order to learn how to estimate a population mean when σ is unknown.

Rarely do we actually know the population standard deviation. As a result, two facts are crucial: the distribution of the population and the size of the sample. If the population is not normally distributed and the sample size is small (generally thought of as less than 30), then advanced statistical techniques that are beyond the scope of this book are required to estimate the population mean. However, if either the population is *normally distributed* or the sample size is *sufficiently large*, then another distribution called the **Student's *t*-distribution** may be used to calculate the margin of error for a population mean when the population standard deviation, σ, is unknown.

Side Note

William Gossett published his work regarding the *t*-distribution in 1908 while working at the Guinness Brewery in Dublin, Ireland. Because his employer required him to publish under a pseudonym, Gossett chose the name "Student." Over a century later, Gossett's work lives on, as does the pseudonym: Student's *t*-distribution.

The Student's *t*-distribution is another important continuous probability distribution. Let's begin our discussion of this distribution by looking at its properties. First, the *t*-distribution is similar to the standard normal distribution. It, too, is bell-shaped and centered at 0, but with more area under the tails, or "fatter" tails, on each end. Also, whereas the standard normal distribution has two parameters, the mean ($\mu = 0$) and standard deviation ($\sigma = 1$), the *t*-distribution has only one parameter, the number of degrees of freedom (*df*). There are infinitely many *t*-distribution curves, one for each number of degrees of freedom. A *t*-distribution with fewer degrees of freedom has more area under the tails of the curve. As the numbers of degrees of freedom increase, the *t*-distribution curves approach the shape of the standard normal curve. Figure 8.3 is a diagram comparing the two distributions.

Memory Booster

df = degrees of freedom

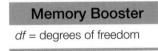

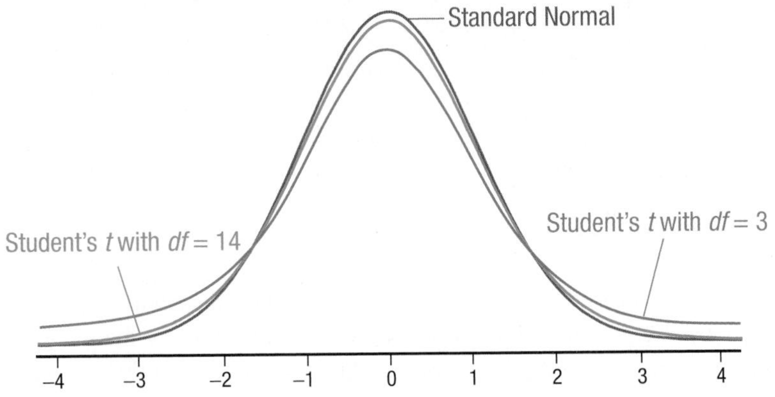

Figure 8.3: Comparison of the Standard Normal Distribution and Student's *t*-Distributions

Properties

Properties of a *t*-Distribution

1. A *t*-distribution curve is symmetric and bell-shaped, centered about 0.
2. A *t*-distribution curve is completely defined by its number of degrees of freedom, *df*.
3. The total area under a *t*-distribution curve equals 1.
4. The *x*-axis is a horizontal asymptote for a *t*-distribution curve.

Just as there are infinitely many normal curves, there are also infinitely many *t*-distribution curves. In contrast to the normal distribution, there is no "standard" *t*-distribution that we can use to calculate probabilities. Instead, each *t*-distribution curve must be considered individually. Tables can be used to look up a critical *t*-value corresponding to a given probability value and number of degrees of freedom. In this text, we combine these tables into one, which contains the most commonly used probability values. In general, if the number of degrees of freedom you desire is not listed in the *t*-table you are using, simply use the entry for the next smaller number of degrees of freedom. This guarantees a conservative estimate. Figure 8.4 shows a small portion of the Student's *t*-distribution table. Note that the interior of the Student's *t*-distribution table gives the *t*-values associated with particular areas under the curve. This is the reverse of a standard normal distribution table, which gives areas in the interior of the table.

			Area in One Tail		
	0.100	0.050	0.025	0.010	0.005
			Area in Two Tails		
df	0.200	0.100	0.050	0.020	0.010
1	3.078	6.314	12.706	31.821	63.657
2	1.886	2.920	4.303	6.965	9.925
3	1.638	2.353	3.182	4.541	5.841
4	1.533	2.132	2.776	3.747	4.604
5	1.476	2.015	2.571	3.365	4.032
6	1.440	1.943	2.447	3.143	3.707
7	1.415	1.895	2.365	2.998	3.499
8	1.397	1.860	2.306	2.896	3.355

Figure 8.4: Excerpt from Table C: Critical Values of *t*

As the number of degrees of freedom increases, the tails of the *t*-distribution become thinner. They begin to appear more like the tails of a normal curve until, at infinitely many degrees of freedom, the *t*-distribution matches the standard normal distribution. Consequently, as you look down each column in the table the *t*-values decrease. A full *t*-distribution table can be found in Appendix A. Notice that the last row of that table has some familiar numbers in it. Can you figure out why?

In general, the value of t such that an area of α is to the right of t is denoted by t_α, as shown in Figure 8.5.

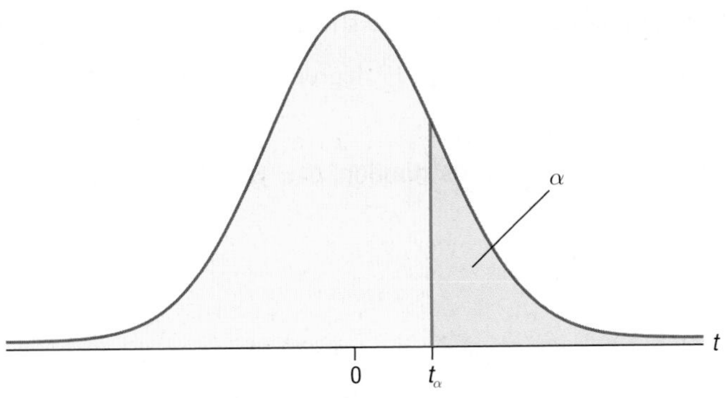

Figure 8.5: Area to the Right of t_α

Example 8.9

Finding the Value of t_α

Find the value of $t_{0.025}$ for the *t*-distribution with 25 degrees of freedom.

Solution

The number of degrees of freedom is listed in the first column of the *t*-distribution table. Since the *t*-distribution in our example has 25 degrees of freedom, the value we need lies on the row corresponding to $df = 25$. Looking at the value of the subscript on *t*, which is the area in the right tail, 0.025, tells us to use the column for an area of 0.025 in one tail. This row and column intersect at 2.060. Thus, $t_{0.025} = 2.060$.

	Area in One Tail				
	0.100	0.050	0.025	0.010	0.005
	Area in Two Tails				
df	**0.200**	**0.100**	**0.050**	**0.020**	**0.010**
23	1.319	1.714	2.069	2.500	2.807
24	1.318	1.711	2.064	2.492	2.797
25	1.316	1.708	2.060	2.485	2.787
26	1.315	1.706	2.056	2.479	2.779
27	1.314	1.703	2.052	2.473	2.771
28	1.313	1.701	2.048	2.467	2.763

Example 8.10

Finding the Value of *t* Given the Area to the Right

Find the value of t for a t-distribution with 17 degrees of freedom such that the area under the curve to the right of t is 0.10.

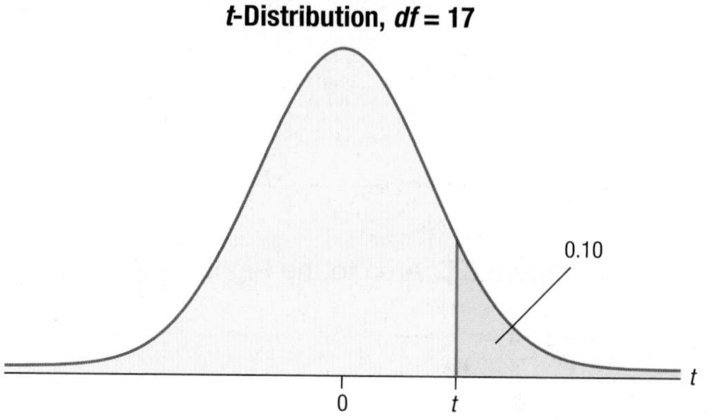

Solution

Note that according to the picture, the area under the curve to the right of t is 0.10. This means that $\alpha = 0.10$. We are told that the distribution has 17 degrees of freedom. Looking across the row for $df = 17$ and down the column for an area in one tail of 0.100 we see that $t_{0.10} = 1.333$.

	Area in One Tail				
	0.100	0.050	0.025	0.010	0.005
	Area in Two Tails				
df	**0.200**	**0.100**	**0.050**	**0.020**	**0.010**
15	1.341	1.753	2.131	2.602	2.947
16	1.337	1.746	2.120	2.583	2.921
17	1.333	1.740	2.110	2.567	2.898
18	1.330	1.734	2.101	2.552	2.878
19	1.328	1.729	2.093	2.539	2.861
20	1.325	1.725	2.086	2.528	2.845

Example 8.11

Finding the Value of *t* Given the Area to the Left

Find the value of t for a t-distribution with 11 degrees of freedom such that the area under the curve to the left of t is 0.05.

t-Distribution, $df = 11$

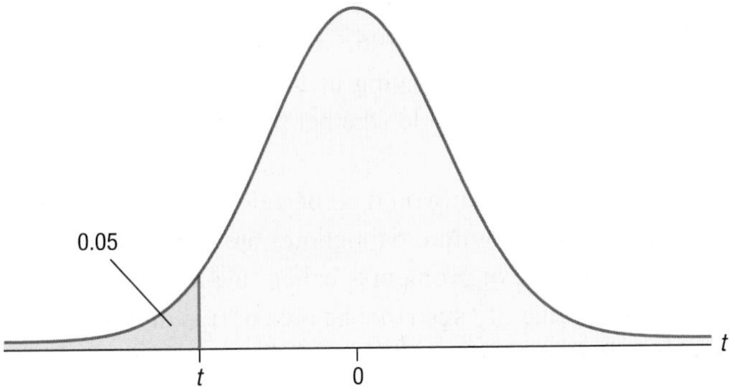

Solution

Because the t-distribution is symmetric, we can look up the t-value for an area of 0.05 under the curve to the right of t. Using the table, we get $t_{0.05} = 1.796$.

	Area in One Tail				
	0.100	0.050	0.025	0.010	0.005
	Area in Two Tails				
df	0.200	0.100	0.050	0.020	0.010
9	1.383	1.833	2.262	2.821	3.250
10	1.372	1.812	2.228	2.764	3.169
11	1.363	1.796	2.201	2.718	3.106
12	1.356	1.782	2.179	2.681	3.055
13	1.350	1.771	2.160	2.650	3.012
14	1.345	1.761	2.145	2.624	2.977

However, since the given area is to the left of t, the t-value needs to be negative. So, for this example, $-t_{0.05} = -1.796$.

Some TI-84 Plus Silver Edition calculators can also be used to find the t-value.

- Press 2ND and then VARS to go to the `DISTR` menu.
- Choose option `4:invT(`.
- Enter the area to the left of t and df in the parentheses as: `invT(`*area to the left of t, df*`)`.
- Enter `invT(Ø.Ø5,11)`.

The answer given by the calculator is $t \approx -1.796$.

Side Note

The `invT` option is available on TI-84 Plus Silver Edition calculators with OS 2.30 or higher and TI-Nspire calculators.

Rounding Rule

When calculating a t-value, round to three decimal places. This follows the convention used in the t-distribution table in Appendix A.

Note that we had to use the symmetry property of the Student's t-distribution in Example 8.11 to find the t-value with a given area to its left using the table. We must use this symmetry property when using the calculator to find the t-value with a given area to its right, as well. You could enter the area to the right of t, but the calculator will return the t-value with the wrong sign because it will assume that the area you enter is the area to the left as required. The "fix" is simple. Switch the sign! Make a negative calculator value positive, and vice versa.

For instance, if Example 8.11 had asked for the value of t, given the same parameters, such that the area under the curve to the *right* of t is 0.05, we could proceed as follows.

- Press **2ND** and then **VARS** to go to the `DISTR` menu.
- Choose option `4:invT(`.
- Enter `invT(Ø.Ø5,11)`, just as if we had been given the area to the left.

The answer given by the calculator is $t \approx -1.796$. Using the symmetry property, we can simply switch the negative sign to a positive sign, giving us the answer $t \approx 1.796$. As always, it's best to sketch a picture for yourself to help you decide whether the answer you get is in the right ballpark!

Memory Booster

One-Tailed α

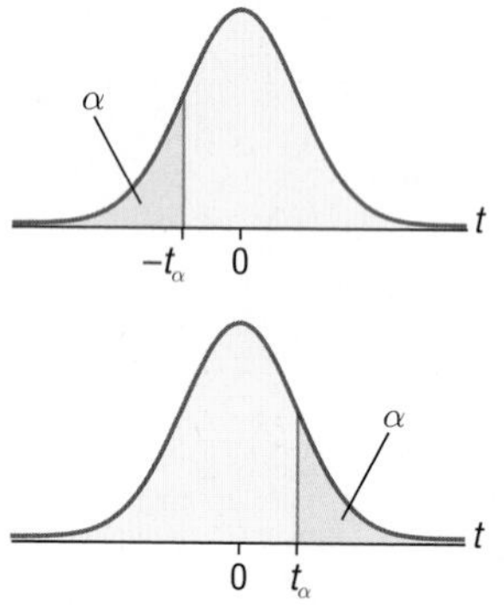

The three examples we have just looked at would all be called *one-tailed*, because the area that was shaded was in only one tail of the distribution. Sometimes the area that we are concerned with lies in both the right and left tails. This type of problem is called *two-tailed* since the area is in two tails of the distribution. In general, the value of t such that an area of α is divided equally between both tails is denoted by $t_{\alpha/2}$, as shown in Figure 8.6.

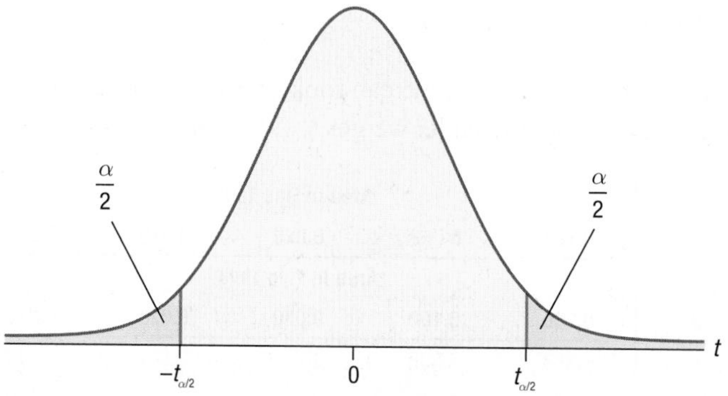

Figure 8.6: Two-Tailed α

Example 8.12

Finding the Value of *t* Given the Area in Two Tails

Find the value of t for a t-distribution with 7 degrees of freedom such that the area to the left of $-t$ plus the area to the right of t is 0.02, as shown in the picture.

***t*-Distribution, *df* = 7**

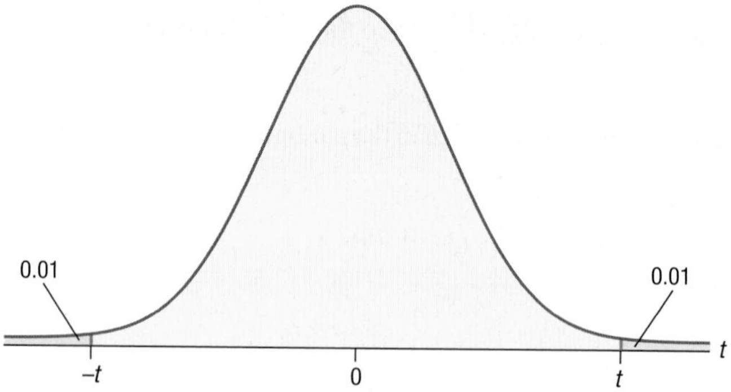

Solution

This is a two-tailed problem because the given area, 0.02, is divided between both sides of the distribution. Therefore, when looking up the t-value in the table, we simply find the given area in the row labeled "Area in Two Tails" as shown in the following excerpt from the table. So the value of t for a t-distribution with 7 degrees of freedom such that the total area in the two tails is 0.02 is $t = 2.998$.

	Area in One Tail				
	0.100	**0.050**	**0.025**	**0.010**	**0.005**
	Area in Two Tails				
df	**0.200**	**0.100**	**0.050**	**0.020**	**0.010**
5	1.476	2.015	2.571	3.365	4.032
6	1.440	1.943	2.447	3.143	3.707
7	1.415	1.895	2.365	2.998	3.499
8	1.397	1.860	2.306	2.896	3.355
9	1.383	1.833	2.262	2.821	3.250
10	1.372	1.812	2.228	2.764	3.169

To use a TI-84 Plus calculator to find t given the area in two tails, you need to enter the area in the left tail only. Since the problem indicates that the area is divided between both ends, we must divide the area in half before we use the calculator. Therefore, we calculate the area in one tail as follows: $\frac{\alpha}{2} = \frac{0.02}{2} = 0.01$.

- Press 2ND and then VARS to go to the `DISTR` menu.
- Choose option `4:invT(`.
- Enter `invT(Ø.Ø1,7)`.

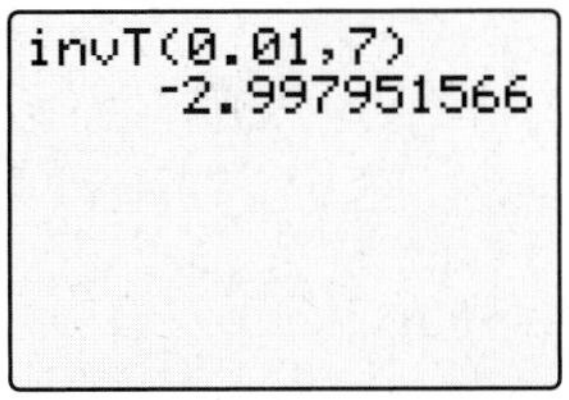

Notice that the value of t that is returned is negative, $-t_{\alpha/2} = -t_{0.01} \approx -2.998$. If you want the positive value of t, just ignore the negative sign since the t-distribution is symmetric.

Another time we use the concept of two-tailed is when the area in which we are interested lies between two t-values. In this case, we need to do a little more manipulation with the given area in order to know which column of the t-distribution table to use. Consider the picture in Figure 8.7.

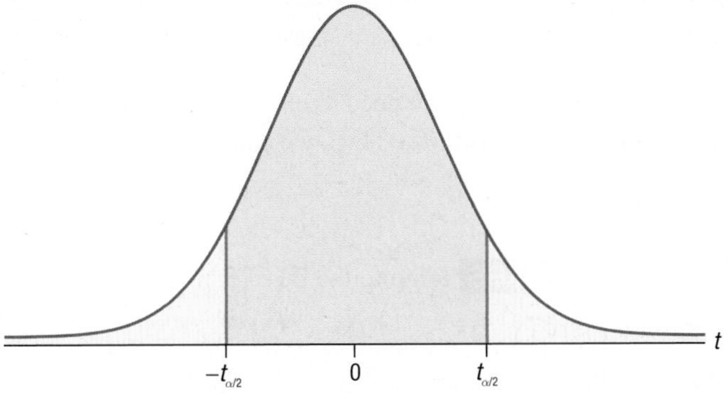

Figure 8.7: Area between Two *t*-Values

Suppose we are looking for the t-value that has 95% of the area under the curve between the positive and negative values of t. Since our table of critical t-values deals only with the area in one or two tails, what we need to consider is the amount of area that remains in the tails. If 95% of the area is on the inside, then 5% remains in the two tails. So the column we would use is the column for an area in two tails of 0.050. Since the t-distribution is symmetric, this column also represents an area of 0.025 (half of 0.050) in each one of the tails.

Example 8.13

Finding the Value of *t* Given Area between $-t$ and *t*

Find the critical value of t for a t-distribution with 29 degrees of freedom such that the area between $-t$ and t is 99%.

t-Distribution, $df = 29$

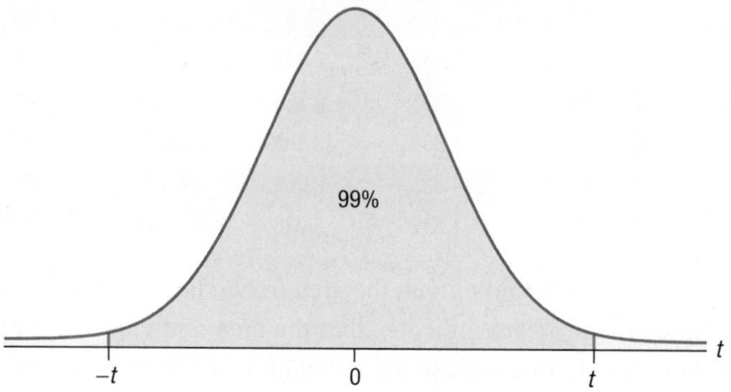

Solution

Since 99% of the area under the curve is in the middle, that leaves 1%, or 0.01 of the area in the two tails. Since the t-distribution has 29 degrees of freedom, look across the row for $df = 29$ and down the column for an area in two tails of 0.010. Thus, $t = 2.756$.

	Area in One Tail				
	0.100	0.050	0.025	0.010	0.005
	Area in Two Tails				
df	**0.200**	**0.100**	**0.050**	**0.020**	**0.010**
27	1.314	1.703	2.052	2.473	2.771
28	1.313	1.701	2.048	2.467	2.763
29	1.311	1.699	2.045	2.462	2.756
30	1.310	1.697	2.042	2.457	2.750
31	1.309	1.696	2.040	2.453	2.744
32	1.309	1.694	2.037	2.449	2.738

To use a TI-84 Plus calculator to find t, you need to enter the area in the left tail only. We have determined that the area in two tails is 0.01. Thus, we calculate the area in one tail as follows: $\frac{\alpha}{2} = \frac{0.01}{2} = 0.005.$

- Press 2ND and then VARS to go to the `DISTR` menu.
- Choose option `4:invT(`.
- Enter `invT(Ø.ØØ5,29)`.

Notice that the value of t that is returned is negative, $-t_{\alpha/2} = -t_{0.005} \approx -2.756$. If you want the positive value of t, just ignore the negative sign since the t-distribution is symmetric.

Hopefully when you saw the notation $t_{\alpha/2}$ for the first time, it looked familiar. It should have resembled the notation $z_{\alpha/2}$, and for good reason. The value of $z_{\alpha/2}$ is the critical value of the standard normal distribution for a certain confidence level, *c*. That confidence level is the area under the standard normal curve between $-z_{\alpha/2}$ and $z_{\alpha/2}$ where $c = 1 - \alpha$. So for example, for a 95% confidence level, $c = 0.95$ and $\alpha = 0.05$ and 95% of the area under the standard normal curve is between $-z_{0.025} = -1.96$ and $z_{0.025} = 1.96$.

The same thing is true for critical *t*-values as for critical *z*-values. Thus, to find the critical *t*-value for a 95% confidence interval, find the *t*-value such that the area between $-t_{\alpha/2}$ and $t_{\alpha/2}$ is 0.95. However, we will have a different critical *t*-value for each 95% confidence interval, depending on the number of degrees of freedom for the *t*-distribution in that problem.

Memory Booster

Unlike critical *z*-values, to find critical *t*-values you must know the number of degrees of freedom in addition to the level of confidence.

Example 8.14

Finding the Critical *t*-Value for a Confidence Interval

Find the critical *t*-value for a 95% confidence interval using a *t*-distribution with 24 degrees of freedom.

Solution

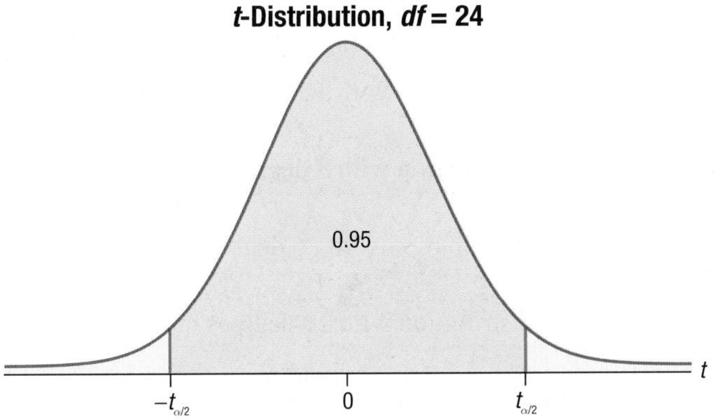

Since we are looking for the critical value for a 95% confidence interval, we want to find the value of *t* such that the area between $-t_{\alpha/2}$ and $t_{\alpha/2}$ is 0.95. If the area under the curve between the two *t*-values is $c = 0.95$, then $\alpha = 1 - c = 1 - 0.95 = 0.05$ is the area in the two tails. Since the *t*-distribution has 24 degrees of freedom and the area in two tails is 0.05, looking across the row for $df = 24$ and down the column for an area in two tails of 0.050, we find a critical *t*-value of $t_{\alpha/2} = t_{0.025} = 2.064$.

	Area in One Tail				
	0.100	0.050	0.025	0.010	0.005
	Area in Two Tails				
df	**0.200**	**0.100**	**0.050**	**0.020**	**0.010**
21	1.323	1.721	2.080	2.518	2.831
22	1.321	1.717	2.074	2.508	2.819
23	1.319	1.714	2.069	2.500	2.807
24	1.318	1.711	2.064	2.492	2.797
25	1.316	1.708	2.060	2.485	2.787
26	1.315	1.706	2.056	2.479	2.779

To use a TI-84 Plus calculator to find t, you need to enter the area in the left tail only. We have determined that the area in two tails is 0.05. Thus, we calculate the area in one tail as follows: $\frac{\alpha}{2} = \frac{0.05}{2} = 0.025$.

- Press 2ND and then VARS to go to the `DISTR` menu.
- Choose option `4:invT(`.
- Enter `invT(0.025,24)`.

Notice that the value of t that is returned is negative, $-t_{\alpha/2} = -t_{0.025} \approx -2.064$. If you want the positive value of t, just ignore the negative sign since the t-distribution is symmetric.

8.2 Section Exercises

Finding *t*-Values

Directions: Find each specified t-value.

1. Find the value of $t_{0.05}$ for a t-distribution with 15 degrees of freedom.

2. Find the value of $t_{0.01}$ for a t-distribution with 8 degrees of freedom.

3. Find the value of $t_{0.005}$ for a t-distribution with 26 degrees of freedom.

4. Find the value of $t_{0.10}$ for a t-distribution with 20 degrees of freedom.

5. Find the value of $t_{0.01}$ for a t-distribution with 29 degrees of freedom.

6. Find the value of $t_{0.025}$ for a t-distribution with 6 degrees of freedom.

7. Find the value of t for a t-distribution with 9 degrees of freedom such that the area to the right of t equals 0.05.

8. Find the value of t for a t-distribution with 11 degrees of freedom such that the area to the right of t equals 0.025.

9. Find the value of t for a t-distribution with 5 degrees of freedom such that the area to the right of t equals 0.01.

10. Find the value of t for a t-distribution with 14 degrees of freedom such that the area to the right of t equals 0.005.

11. Find the value of t for a t-distribution with 10 degrees of freedom such that the area to the left of t equals 0.05.

12. Find the value of t for a t-distribution with 3 degrees of freedom such that the area to the left of t equals 0.10.

13. Find the value of t for a t-distribution with 12 degrees of freedom such that the area to the left of t equals 0.005.

14. Find the value of t for a t-distribution with 27 degrees of freedom such that the area to the left of t equals 0.025.

15. Find the value of t for a t-distribution with 13 degrees of freedom such that the area to the left of $-t$ plus the area to the right of t equals 0.01.

16. Find the value of t for a t-distribution with 23 degrees of freedom such that the area to the left of $-t$ plus the area to the right of t equals 0.20.

17. Find the value of t for a t-distribution with 12 degrees of freedom such that the area to the left of $-t$ plus the area to the right of t equals 0.02.

18. Find the value of t for a t-distribution with 18 degrees of freedom such that the area to the left of $-t$ plus the area to the right of t equals 0.05.

19. Find the value of t for a t-distribution with 18 degrees of freedom such that the area between $-t$ and t equals 95%.

20. Find the value of t for a t-distribution with 4 degrees of freedom such that the area between $-t$ and t equals 98%.

21. Find the value of t for a t-distribution with 25 degrees of freedom such that the area between $-t$ and t equals 90%.

22. Find the critical t-value for a 90% confidence interval using a t-distribution with 25 degrees of freedom.

23. Find the critical t-value for a 99% confidence interval using a t-distribution with 18 degrees of freedom.

24. Compare and contrast Exercises 21 and 22.

25. Compare and contrast Exercise 15 and Example 8.13.

26. In the paragraph after Figure 8.4 in this section, we mention that the last row of the t-distribution table has some familiar numbers in it. Why should these numbers be familiar?

8.3 Estimating Population Means (σ Unknown)

As we saw in the first section of this chapter, a confidence interval is a valuable tool in inferential statistics used to estimate a population parameter. We learned how to construct confidence intervals for population means for populations where the standard deviation, σ, was known. Now that we have looked at the Student's t-distributions in Section 8.2, we are equipped to discuss how to estimate the mean of a population when σ is *not* known. In this section we will discuss how to construct confidence intervals in situations where the standard normal distribution cannot be applied and a Student's t-distribution must be used instead.

The basic principles for finding a confidence interval are the same as in Section 8.1—calculating a margin of error and then adding it to and subtracting it from a point estimate—the only change is in the distribution that we will use to calculate the margin of error.

In this section we will focus on estimating the population mean when the following criteria are true.

- All possible samples of a given size have an equal probability of being chosen; that is, a simple random sample is used.
- The population standard deviation, σ, is *unknown*.
- Either the sample size is at least 30 $(n \geq 30)$ *or* the population distribution is approximately normal.

The first step in constructing a confidence interval for any population parameter is to identify the best point estimate. To build the confidence interval for the population mean, we use an unbiased estimator, the sample mean, as the point estimate to construct the interval around (just as we did in Section 8.1). This leaves us with the second step, calculating the margin of error.

When all of the above conditions are met, the distribution used to calculate the margin of error for a confidence interval for the population mean is the Student's t-distribution. The formula for the margin of error using the Student's t-distribution is as follows.

Rounding Rule

When calculating a margin of error for a confidence interval, round to at least six decimal places to avoid additional rounding errors in the subsequent calculations of the endpoints of the confidence interval.

Formula

Margin of Error of a Confidence Interval for a Population Mean (σ Unknown)

When the population standard deviation is unknown, the sample taken is a simple random sample, and either the sample size is at least 30 or the population distribution is approximately normal, the margin of error of a confidence interval for a population mean is given by

$$E = \left(t_{\alpha/2}\right)\left(\frac{s}{\sqrt{n}}\right)$$

where $t_{\alpha/2}$ is the critical value for the level of confidence, $c = 1-\alpha$, such that the area under the t-distribution with $n-1$ degrees of freedom to the right of $t_{\alpha/2}$ is equal to $\frac{\alpha}{2}$,

s is the sample standard deviation, and

n is the sample size.

Notice that the formula for the margin of error uses $t_{\alpha/2}$ and not t_{α}. This is because when we construct an interval estimate using a t-distribution, the level of confidence, c, is the area between $-t_{\alpha/2}$ and $t_{\alpha/2}$. The remaining area under the curve, α, is split between the two tails. Recall from the previous section that to find the value of $t_{\alpha/2}$, you may use appropriate technology such as a TI-84 Plus calculator or Table C: Critical Values of t in Appendix A.

Example 8.15

Finding the Margin of Error of a Confidence Interval for a Population Mean (σ Unknown)

Dental researchers want to estimate the mean leakage, measured in nanometers (nm), of a new filling material for cavities using a simple random sample of 10 trials. Assuming that the population distribution is approximately normal and the population standard deviation is unknown, find the margin of error for a 95% confidence interval for the population mean given that the sample standard deviation is 15.5 nm.

Solution

Since we know that the population distribution is approximately normal and the population standard deviation is unknown, we are able to use the t-distribution to calculate the margin of error. The problem tells us the values for s and n ($s = 15.5$, $n = 10$), so the only missing value in the calculation of E is $t_{\alpha/2}$. Since the level of confidence is 95%, $\alpha = 1 - 0.95 = 0.05$. Therefore, $t_{\alpha/2} = t_{0.05/2} = t_{0.025}$. A sample size of 10 means that there are 9 degrees of freedom, $df = 9$.

Memory Booster

df = degrees of freedom

To find this value by hand using the t-distribution table, look across the row for 9 degrees of freedom and down the column for an area in one tail of 0.025. This shows a critical t-value of $t_{0.025} = 2.262$. Notice that, using our table, we could also have looked up the area in two tails, $\alpha = 0.05$, instead of the area in one tail, $\frac{\alpha}{2} = 0.025$. Both give the same answer.

	Area in One Tail				
	0.100	0.050	0.025	0.010	0.005
	Area in Two Tails				
df	**0.200**	**0.100**	**0.050**	**0.020**	**0.010**
7	1.415	1.895	2.365	2.998	3.499
8	1.397	1.860	2.306	2.896	3.355
9	1.383	1.833	2.262	2.821	3.250
10	1.372	1.812	2.228	2.764	3.169
11	1.363	1.796	2.201	2.718	3.106
12	1.356	1.782	2.179	2.681	3.055

We can also use a TI-84 Plus calculator to find the critical t-value. Recall that you need to enter the area in the left tail only, so remember to divide α by 2 when using the calculator: $t_{\alpha/2} = t_{0.05/2} = t_{0.025}$.

- Press 2ND and then VARS to go to the DISTR menu.
- Choose option `4:invT(`.
- Enter `invT(Ø.Ø25,9)`.

Notice that the value of t that is returned is negative, $-t_{\alpha/2} = -t_{0.025} \approx -2.262$. Because we want the positive value of t, we can just ignore the negative sign since the t-distribution is symmetric.

Substituting these values into the formula for the margin of error, we get the following.

$$\begin{aligned} E &= \left(t_{\alpha/2}\right)\left(\frac{s}{\sqrt{n}}\right) \\ &= (2.262)\left(\frac{15.5}{\sqrt{10}}\right) \\ &\approx 11.087262 \end{aligned}$$

Although we are not calculating the endpoints of the confidence interval in this example, we will round the margin of error to six decimal places. So the margin of error for this 95% confidence interval is approximately 11.087262 nm.

Once we have calculated E, the confidence interval is constructed in the same way as before. We simply subtract the margin of error from and add the margin of error to the point estimate, that is, the sample mean, to find the endpoints of the interval.

Rounding Rule

Round the endpoints of a confidence interval for a population mean as follows:

- If sample data are given, round to one more decimal place than the largest number of decimal places in the given data.
- If statistics are given, round to the same number of decimal places as given in the standard deviation or variance.

Formula

Confidence Interval for a Population Mean

The confidence interval for a population mean is given by

$$\bar{x} - E < \mu < \bar{x} + E$$

or

$$(\bar{x} - E, \bar{x} + E)$$

where $\bar{x}$ is the sample mean, which is the point estimate for the population mean, and

E is the margin of error.

Example 8.16

Constructing a Confidence Interval for a Population Mean (σ Unknown)

A marketing company wants to know the mean price of new vehicles sold in an up-and-coming area of town. Marketing strategists collected data over the past two years from all of the dealerships in the new area of town. From previous studies about new car sales, they believe that the population distribution looks somewhat like the following graph.

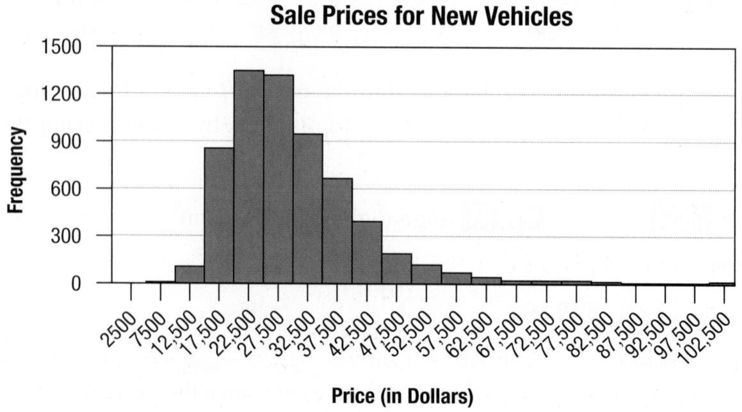

The simple random sample of 756 cars has a mean of \$27,400 with a standard deviation of \$1300. Construct a 95% confidence interval for the mean price of new cars sold in this area.

Solution

Step 1: **Find the point estimate.**

The point estimate for the population mean is the sample mean, which we are told is \$27,400.

Step 2: **Find the margin of error.**

Next, we need to calculate the margin of error. From the graph, we know that the population distribution is not guaranteed to be normal, but instead is considered skewed to the right. However, since the sample size, $n = 756$, is sufficiently large, we use the Student's t-distribution to calculate the margin of error. We are told that the standard deviation of the sample is \$1300, that is, $s = 1300$, so all that remains is to find the critical t-value.

Since our sample is so large, we will use the calculator method instead of the table to find $t_{\alpha/2}$. As we saw in the previous example, to find $t_{0.025}$ for the t-distribution with $df = n - 1 = 756 - 1 = 755$ degrees of freedom, enter `invT(Ø.Ø25,755)` and find that the critical value we need is $t_{0.025} \approx 1.963$. Notice that this is the same value that you get if you look in the table of critical t-values for $df = 750$ with a 95% level of confidence. With such a large sample, there is no difference in the first three decimal places of the critical value.

```
invT(0.025,755)
   -1.963110982
```

Substituting into the margin of error formula, we have the following.

$$\begin{aligned} E &= \left(t_{\alpha/2}\right)\left(\frac{s}{\sqrt{n}}\right) \\ &= (1.963)\left(\frac{1300}{\sqrt{756}}\right) \\ &\approx 92.811706 \end{aligned}$$

Memory Booster

If the number of degrees of freedom needed is not in the t-table, use the entry for the next smaller number of degrees of freedom.

Step 3: **Subtract the margin of error from and add the margin of error to the point estimate.**

The third step is to subtract the margin of error that we just calculated from the point estimate we were given in the problem, and then add the margin of error to the point estimate to get the lower and upper endpoints of the confidence interval.

$$\text{Lower endpoint: } \bar{x} - E = 27{,}400 - 92.811706 \approx \$27{,}307$$

$$\text{Upper endpoint: } \bar{x} + E = 27{,}400 + 92.811706 \approx \$27{,}493$$

Thus, the 95% confidence interval ranges from \$27,307 to \$27,493. The confidence interval can be written mathematically using either inequality symbols or interval notation, as shown below.

$$27{,}307 < \mu < 27{,}493$$

or

$$(27{,}307,\ 27{,}493)$$

So with 95% confidence, we can say that the mean price of new cars sold in the area is between \$27,307 and \$27,493.

Example 8.17

Constructing a Confidence Interval for a Population Mean (σ Unknown)

A student records the repair costs for 20 randomly selected computers from a local repair shop where he works. A sample mean of \$216.53 and standard deviation of \$15.86 are subsequently computed. Assume that the population distribution is approximately normal and σ is unknown.

a. Determine the 98% confidence interval for the mean repair cost for all computers repaired at the local shop by first calculating the margin of error, E.

b. Use a TI-83/84 Plus calculator to determine the 98% confidence interval from the given statistics.

Solution

Because σ is unknown, we need to ensure that either the sample size is large enough $(n \geq 30)$ or the population is normally distributed in order to use the t-distribution for the confidence interval. We are given the assumption of normality, so we can proceed.

a.

Step 1: Find the point estimate.

The point estimate for the population mean is the sample mean, which we are told is \$216.53.

Step 2: Find the margin of error.

Calculating the margin of error first requires that we find $t_{\alpha/2}$. The level of confidence is $c = 0.98$, so $\frac{\alpha}{2} = 0.01$. There are 20 computers in the sample, so $df = 19$. Using either the table of critical t-values or technology, we find that $t_{0.01} = 2.539$.

Next, substitute these values into the formula for the margin of error.

$$\begin{aligned} E &= \left(t_{\alpha/2}\right)\left(\frac{s}{\sqrt{n}}\right) \\ &= (2.539)\left(\frac{15.86}{\sqrt{20}}\right) \\ &\approx 9.004319 \end{aligned}$$

Step 3: Subtract the margin of error from and add the margin of error to the point estimate.

Subtracting the margin of error that we just calculated from the point estimate we were given in the problem and then adding the margin of error to the point estimate gives us the following endpoints of the confidence interval.

$$\begin{aligned} \text{Lower endpoint: } \bar{x} - E &= 216.53 - 9.004319 \\ &\approx \$207.53 \end{aligned}$$

$$\text{Upper endpoint: } \bar{x} + E = 216.53 + 9.004319$$
$$\approx \$225.53$$

Thus, the 98% confidence interval ranges from \$207.53 to \$225.53. The confidence interval can be written mathematically using either inequality symbols or interval notation, as shown below.

$$207.53 < \mu < 225.53$$
$$\text{or}$$
$$(207.53,\ 225.53)$$

Therefore, the student can be 98% confident that the mean repair cost for all computers repaired at the local shop is between \$207.53 and \$225.53.

b. Now we'll show you how to do the same interval calculation using a TI-83/84 Plus calculator.

- Press STAT.
- Scroll over and choose TESTS.
- Choose option 8:TInterval.
- Choose the Stats option because we were given sample statistics.

We'll need to enter the following sample statistics, as shown in the calculator screenshot in the margin.

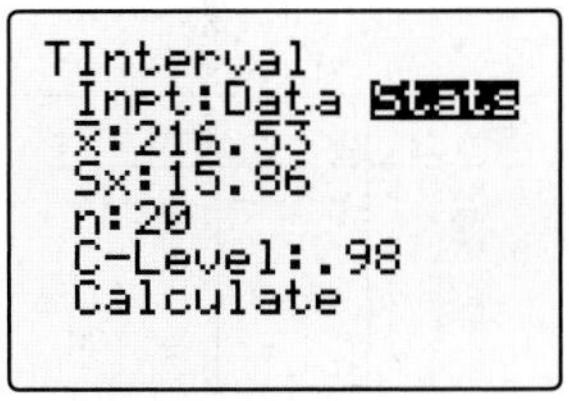

- Sample mean, x̄:216.53
- Sample standard deviation, Sx:15.86
- Sample size, n:20
- Level of confidence, C-Level:.98

After highlighting Calculate and pressing ENTER, we get the results shown in the second screenshot in the margin. Notice that the last digits of the endpoints in the interval given by the calculator, (207.52, 225.54), are different from those in our hand-calculated interval. This is because we used a rounded value of $t_{\alpha/2}$ to find the margin of error in our first method of calculation.

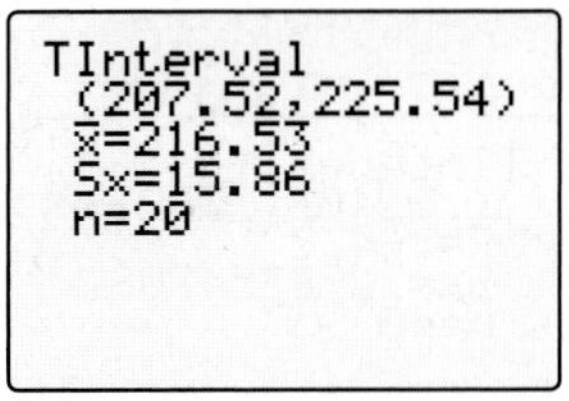

Let's look at another example using a TI-83/84 Plus calculator to calculate a confidence interval when we are given the actual sample data and not just the sample statistics.

Example 8.18

Constructing a Confidence Interval for a Population Mean (σ Unknown) from Original Data

Given the following sample data from a study on the average amount of water used per day by members of a household while brushing their teeth, calculate the 99% confidence interval for the population mean using a TI-83/84 Plus calculator. Assume that the sample used in the study was a simple random sample.

Household Water Used for Brushing Teeth (in Gallons per Day)				
0.485	0.428	0.390	0.308	0.231
0.587	0.516	0.465	0.370	0.282
0.412	0.367	0.336	0.269	0.198
0.942	0.943	0.940	0.941	0.946
0.868	0.898	0.889	0.910	0.927
0.925	0.950	0.959	0.948	0.956
0.805	0.810	0.839	0.860	0.861
0.515	0.463	0.420	0.326	0.243

Solution

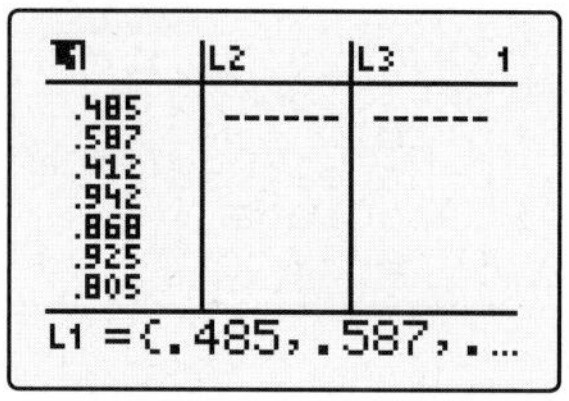

Since we are not told any population parameters for the study, we cannot assume that σ is known or that the population distribution is approximately normal. However, since the sample size ($n = 40$) is large enough ($n \geq 30$), we can use the t-distribution to construct a confidence interval for the population mean.

To begin with, since we are given the raw data and not the sample statistics, we need to enter the data in the calculator list. Recall, to enter data in a TI-83/84 Plus calculator, press STAT, choose EDIT, select 1:Edit, and then enter the data in L1. (Remember to clear the list before entering the data.)

Once the data are entered, press STAT, choose TESTS, and select 8:TInterval on the calculator. This time, however, choose the Data option. You'll need to specify which list your data are in, which is L1, and the confidence level (C-Level), which is Ø.99 in this example. The value of Freq should be left as the default value of 1. After you select Calculate, you should see the results shown in the second screenshot in the margin.

Thus, the 99% confidence interval ranges from 0.5240 to 0.7624. The confidence interval can be written mathematically using either inequality symbols or interval notation, as shown below.

$$0.5240 < \mu < 0.7624$$

or

$$(0.5240, 0.7624)$$

We are 99% confident that the mean amount of water used per household for brushing teeth is between 0.5240 and 0.7624 gallons per day.

8.3 Section Exercises

Directions: For all exercises in this section, you may assume that the requirements mentioned in this section are met; namely, the population standard deviation is unknown, all samples are simple random samples, and either the sample size is at least 30 or the population distribution is approximately normal.

Confidence Intervals for Population Means (σ Unknown)

Directions: Construct a confidence interval for the population mean at the given level of confidence.

1. $n = 14$, $\bar{x} = 95.0$, $s = 4.8$, level of confidence is 95%

2. $n = 25$, $\bar{x} = 56$, $s = 8$, level of confidence is 90%

3. $n = 8$, $\bar{x} = 7.0$, $s = 1.2$, level of confidence is 99%

4. $n = 13$, $\bar{x} = 1.97$, $s = 0.03$, level of confidence is 98%

5. $c = 0.98$

192	465	321	299	516	256	339	311	407

6. $c = 0.95$

Stem-and-Leaf Plot

Stem	Leaves
2	2 3 3 8 9 9
3	0 1 1 4
4	5 6 7 7 8 8

Key: 2 | 2 = 22

7. Level of confidence is 90%

11	47	95	54	33	64	4	8	57	9	80	32	19
8	90	3	49	4	44	79	80	48	16	64	55	68
31	7	15	21	52	6	78	109	40	50	12	29	22

8. Level of confidence is 80%

Ages of Participants in a Study

Age	Frequency
16	9
17	15
18	16
19	5
20	3

Margins of Error of Confidence Intervals for Population Means (σ Unknown)

Directions: Calculate each specified margin of error.

9. The mean distance commuters drove to work each day was estimated to be 40.8 miles from a sample of 45 commuters. The sample standard deviation was 5.8 miles. Calculate the margin of error for a 95% confidence interval.

10. The mean amount of money spent per week on gas by a sample of 25 drivers was found to be \$57.00 with a standard deviation of \$2.36. Calculate the margin of error for a 90% confidence interval. Assume that the population distribution is approximately normal.

Confidence Intervals for Population Means (σ Unknown)

Directions: Construct and interpret each specified confidence interval.

11. Wildlife conservationists studying grizzly bears in the United States found that the mean weight of 25 adult males was 600 pounds with a standard deviation of 90 pounds. Construct and interpret a 98% confidence interval for the mean weight of all adult male grizzly bears in the United States. Assume that the weights of all adult male grizzly bears in the United States are normally distributed.

12. The mean length of 12 newly hatched iguanas is 7.00 inches with a standard deviation of 0.75 inches. Construct and interpret a 90% confidence interval for the mean length of all newly hatched iguanas. Assume that the lengths of all newly hatched iguanas are normally distributed.

13. Suppose that you sample 59 high school baseball pitchers in one county and find that they have a mean fastball pitching speed of 80.00 miles per hour (mph) with a standard deviation of 4.98 mph. Find a 95% confidence interval for the mean fastball pitching speed of all high school baseball pitchers in the county. Interpret the interval.

14. Given the following data, construct and interpret a 99% confidence interval for the mean face amount of an individual life insurance policy.

Face Amounts of Individual Life Insurance Policies

Stem	Leaves
15	0 0 0 0 0 0 1 2 2 3 3 4 5 8 9 9
16	0 0 0 2 3 3 3 3 5 5 8 8 8
17	1 1 2 2 2 3 4 5 5 5 6
18	2 2 3 5 5
19	0 0 5 5 5 6
20	0 0 0 0 0 2 2

Key: 15 | 0 = \$150,000

15. The attendance records for a random sample of 28 men's basketball games at one university revealed that the mean number of fans at each game was 4125.0 with a standard deviation of 741.0. In order to predict ticket sales for next year, the athletic office needs a confidence interval for attendance at these games. Using a confidence level of 95%, construct and interpret a confidence interval for the mean number of fans at men's basketball games at this university. Assume that the population distribution is approximately normal.

16. Suppose you are thinking about getting a puppy and want to know the amount of time people spend caring for puppies. You survey 31 puppy owners and find that the mean amount of time they spend caring for their puppies is 108.0 minutes per day. If the standard deviation is 17.0 minutes, construct a 98% confidence interval for the mean amount of time puppy owners spend on their puppies per day. Interpret the interval.

17. The following is a random sample of the annual salaries of high school counselors in the United States. Assuming that the distribution of salaries is approximately normal, construct a 90% confidence interval for the mean salary of high school counselors across the United States. Interpret the interval.

\$51,050	\$38,740	\$65,360	\$42,640	\$55,340	\$32,980	\$49,540

Directions: Respond thoughtfully to the following exercises.

18. Suppose you were told that the confidence interval for a population mean is $(16.30, 19.70)$. Is it possible for you to determine what the margin of error, *E,* is? If so, what is it?

19. Suppose you are told that a 95% confidence interval for the mean monthly household electric bill in the county where you live is $(205.56, 253.90)$. Is it possible for you to determine how many electric bills were sampled to construct the interval?

20. You are presented with the following reports estimating the mean increase in monthly household spending on gasoline over the past six months. Determine which report you find most convincing and explain your reasoning.

i. Based on a simple random sample of 15 households, we are 98% confident that the mean increase in monthly household spending on gasoline over the past six months is between \$43.65 and \$58.93.

ii. Based on a simple random sample of 52 households, we are 95% confident that the mean increase in monthly household spending on gasoline over the past six months is between \$48.82 and \$52.10.

21. If you were presented with a margin of error of 15,642, would you believe that the margin of error was reported correctly?

Discussion Questions

Directions: **Discuss each question with your classmates. Focus on the relationships between the parameters in each question.**

22. Lisa sets out to survey 500 people; however, she only receives responses from 387 people.

a. How will this decrease in her sample size affect the margin of error for her confidence interval for a population mean?

b. How will this decrease in her sample size affect the width of her confidence interval for a population mean?

23. How will increasing the level of confidence without changing the sample size affect the width of a confidence interval for a population mean?

24. How will increasing the level of confidence without changing the sample size affect the margin of error for a confidence interval for a population mean?

25. Which level of confidence will produce a wider confidence interval for a population mean: a 95% level of confidence or a 99% level of confidence?

26. If you decrease the sample size while keeping the margin of error for a confidence interval for a population mean constant, what effect will this have on the level of confidence?

27. How will the width of a confidence interval for a population mean change if you increase the sample size and keep the same level of confidence?

8.4 Estimating Population Proportions

Another parameter often estimated is a population proportion. Recall that a population proportion, denoted p, is the fraction or percentage of a population that displays a certain characteristic. For example, the proportion of women around the world is usually thought to be $p \approx \frac{1}{2} = 0.5$. Of course, when a population is very large, it is impractical or impossible to find p. A sample must be taken and the sample information is then used to analyze the population. The sample proportion, denoted $\hat{p}$, is the fractional part of a sample that displays a certain characteristic. For example, if in a sample of 135 teachers, 100 of the teachers are women, then the sample proportion of women teachers is $\hat{p} = \frac{x}{n} = \frac{100}{135} \approx 0.741 = 74.1\%$.

The best point estimate used for estimating the population proportion is the sample proportion, $\hat{p}$. Let's look at an example of finding the best point estimate for a population proportion.

Example 8.19

Finding a Point Estimate for a Population Proportion

A graduate student wishes to know the proportion of American adults who speak two or more languages. He surveys 565 randomly selected American adults and finds that 226 speak two or more languages. Estimate the proportion of all American adults who speak two or more languages.

Solution

Use $\hat{p}$ as a point estimate for p.

$$\begin{aligned}\hat{p} &= \frac{x}{n}\\ &= \frac{226}{565}\\ &= 0.4\end{aligned}$$

We estimate the proportion of all American adults who speak two or more languages to be 40%.

As we've noted earlier, using a range of values rather than a point estimate increases the likelihood of estimating the true population proportion. In order to construct a confidence interval for estimating a population proportion, the following conditions must be satisfied.

- All possible samples of a given size have an equal probability of being chosen; that is, a simple random sample is used.
- The conditions for a binomial distribution are met.
- The sample size is large enough to ensure that $n\hat{p} \geq 5$ and $n(1-\hat{p}) \geq 5$.

When these conditions are met, we can apply the Central Limit Theorem to the sampling distribution of sample proportions.

Memory Booster

Properties of a Binomial Distribution

1. The experiment consists of a fixed number, n, of identical trials.
2. Each trial is independent of the others.
3. For each trial, there are only two possible outcomes. For counting purposes, one outcome is labeled a success, and the other a failure.
4. For every trial, the probability of getting a success is called p. The probability of getting a failure is then $1 - p$.
5. The binomial random variable, X, counts the number of successes in n trials.

Constructing a confidence interval for the population proportion is similar to constructing a confidence interval for the population mean. The first step in constructing a confidence interval is to find the point estimate, and the second step is to calculate the margin of error. The formula for the margin of error of a confidence interval for a population proportion is as follows.

Rounding Rule

When calculating a margin of error for a confidence interval, round to at least six decimal places to avoid additional rounding errors in the subsequent calculations of the endpoints of the confidence interval.

Formula

Margin of Error of a Confidence Interval for a Population Proportion

When the sample taken is a simple random sample, the conditions for a binomial distribution are met, and the sample size is large enough to ensure that $n\hat{p} \geq 5$ and $n(1-\hat{p}) \geq 5$, the margin of error of a confidence interval for a population proportion is given by

$$E = z_{\alpha/2}\sqrt{\frac{\hat{p}(1-\hat{p})}{n}}$$

where $z_{\alpha/2}$ is the critical value for the level of confidence, $c = 1-\alpha$, such that the area under the standard normal distribution to the right of $z_{\alpha/2}$ is equal to $\frac{\alpha}{2}$,

$\hat{p}$ is the sample proportion, and

n is the sample size.

The margin of error formula uses the critical value of $z_{\alpha/2}$, which can be found using the same table of critical z-values that we used when calculating margins of error for population means (with σ known) in Section 8.1. The table of critical values is given below.

Table 8.2: Critical z-Values for Confidence Intervals

Level of Confidence, c	$\alpha = 1-c$	$z_{\alpha/2}$
0.80	0.20	1.28
0.85	0.15	1.44
0.90	0.10	1.645
0.95	0.05	1.96
0.98	0.02	2.33
0.99	0.01	2.575

Once the margin of error is calculated, find the confidence interval for p by subtracting E from $\hat{p}$ and adding E to $\hat{p}$. Thus, a confidence interval for a population proportion can be written as shown in the following box.

Rounding Rule

Round the endpoints of a confidence interval for a population proportion to three decimal places.

Formula

Confidence Interval for a Population Proportion

The confidence interval for a population proportion is given by

$$\hat{p} - E < p < \hat{p} + E$$

or

$$(\hat{p} - E,\ \hat{p} + E)$$

where $\hat{p}$ is the sample proportion, which is the point estimate for the population proportion, and

E is the margin of error.

Note: In this textbook we will assume that the conditions for a binomial distribution are met for all examples and exercises involving confidence intervals for population proportions.

Example 8.20

Constructing a Confidence Interval for a Population Proportion

A survey of 345 randomly selected students at one university found that 301 students think that there is not enough parking on campus. Find the 90% confidence interval for the proportion of all students at this university who think that there is not enough parking on campus.

Solution

Step 1: **Find the point estimate.**

First calculate the point estimate, $\hat{p}$.

$$\hat{p} = \frac{x}{n} = \frac{301}{345} \approx 0.872464$$

Step 2: **Find the margin of error.**

Refer to the table of critical z-values to find $z_{\alpha/2} = z_{0.10/2} = z_{0.05} = 1.645$. Use the formula for the margin of error for proportions to calculate E.

$$\begin{aligned} E &= z_{\alpha/2}\sqrt{\frac{\hat{p}(1-\hat{p})}{n}} \\ &= 1.645\sqrt{\frac{0.872464(1-0.872464)}{345}} \\ &\approx 0.029542 \end{aligned}$$

Step 3: **Subtract the margin of error from and add the margin of error to the point estimate.**

Subtracting the margin of error from the sample proportion and then adding the margin of error to the sample proportion gives us the following endpoints of the confidence interval.

$$\text{Lower endpoint: } \hat{p} - E = 0.872464 - 0.029542 \approx 0.843$$

$$\text{Upper endpoint: } \hat{p} + E = 0.872464 + 0.029542 \approx 0.902$$

Thus, the 90% confidence interval for the population proportion ranges from 0.843 to 0.902. The confidence interval can be written mathematically using either inequality symbols or interval notation, as shown below.

$$0.843 < p < 0.902$$

or

$$(0.843,\ 0.902)$$

Note that these proportions are expressed as decimals, which is typical for confidence intervals. In this example, we can interpret the confidence interval $(0.843, 0.902)$ to mean that, with 90% confidence, we estimate that between 84.3% and 90.2% of all students at this university think there is not enough parking on campus.

Rounding Rule

When calculations involve several steps, avoid rounding at intermediate calculations. If necessary, round intermediate calculations to at least six decimal places to avoid additional rounding errors in subsequent calculations.

Using a TI-83/84 Plus Calculator to Find a Confidence Interval for a Population Proportion

When we use a calculator to find a confidence interval for a population proportion, we actually need less information than when we find it by hand. We no longer need to calculate the sample proportion first; the calculator does that for us. In fact, to find the interval estimate for a population proportion, the only values that are required are n, the sample size; x, the number of members of the sample that have the characteristic in which we are interested; and c, the level of confidence.

Let's use the information from the last example and illustrate the process on the calculator. Here's the problem that was given:

"A survey of 345 randomly selected students at one university found that 301 students think that there is not enough parking on campus. Find the 90% confidence interval for the proportion of all students at this university who think that there is not enough parking on campus."

The sample size, n, is 345 and x is 301 because that is the number of students sampled who have the characteristic in which we are interested, that is, the number who believe there is not enough parking on campus. To put these in the calculator, press **STAT** and scroll over to choose `TESTS`. Choose option `A:1-PropZInt`. Enter `3Ø1` for `x`, `345` for `n`, and `Ø.9Ø` for `C-Level`. Highlight `Calculate` and press **ENTER**. The screen will display the results shown in the screenshot on the right below.

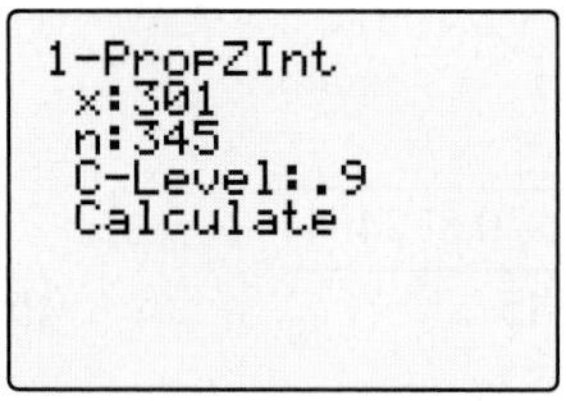

Notice that the confidence interval is displayed, as well as the value of the sample proportion, $\hat{p}$. Note that we rounded the endpoints of the confidence interval to three decimal places in Example 8.20, but in the interval given by the calculator, $(0.84292, 0.902)$, five decimal places are shown for the lower endpoint. We then follow the rounding rule and round to three decimals places, which gives us the same interval that we found by hand in Example 8.20, $(0.843, 0.902)$. Regardless of the method used to calculate the endpoints of a confidence interval for a population proportion, we follow the rounding rule and round the endpoints to three decimal places.

Example 8.21

Using a TI-83/84 Plus Calculator to Find a Confidence Interval for a Population Proportion

In a recent study, 820 American adults were surveyed and 61% said they would travel less this year for nonbusiness trips than last year. Find the 95% confidence interval for the true proportion of adults who said that they would travel less this year for nonbusiness trips.

Solution

We need three pieces of information to use the calculator to construct the confidence interval for the population proportion: n, x, and *level of confidence*. From the problem we know that $n = 820$ and $c = 0.95$. We only need to find x, or the number of adults who said they would travel less this year. You should note that the calculator wants a whole number for x and will return an error if you enter either a decimal or the percentage of the sample. In general, since

$\hat{p}$ is equivalent to the fraction of the sample that has the characteristic of interest, $\hat{p} = \frac{x}{n}$, we can multiply both sides of this equation by n to get a formula for x: $x = \hat{p} \cdot n$. We were told that 61% of the adults surveyed said they would travel less, so we can find x by multiplying 0.61 by the sample size, 820, as follows.

$$\begin{aligned} x &= \hat{p} \cdot n \\ &= 0.61 \cdot 820 \\ &= 500.2 \end{aligned}$$

Because we cannot enter a decimal, we can round this value to the nearest whole number. Approximately 500 adults have the characteristic of interest, so $x = 500$. To input these values into the calculator, press STAT, select TESTS, and choose option A:1-PropZInt. Once entered, the calculator will give the results shown below.

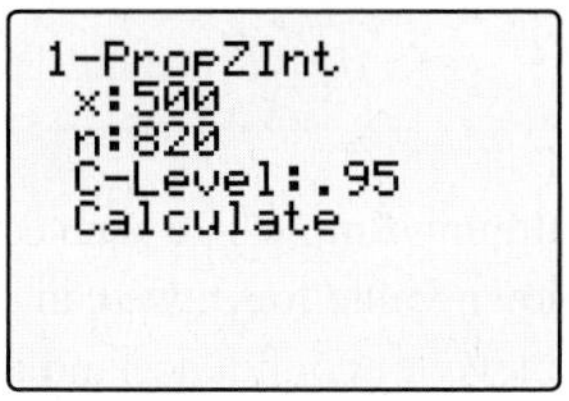

Thus, the 95% confidence interval for the population proportion ranges from 0.576 to 0.643. The confidence interval can be written mathematically using either inequality symbols or interval notation, as follows.

$$0.576 < p < 0.643$$

or

$$(0.576, 0.643)$$

We are 95% confident that the true proportion of American adults who say they will travel less this year for nonbusiness trips is between 57.6% and 64.3%.

Minimum Sample Size for Estimating a Population Proportion

As we saw when estimating population means, it is often desirable to know the fewest number of data values required to obtain a certain level of confidence for our interval estimate. The minimum sample size required for estimating a population proportion may be calculated using the following formula. This formula for the minimum sample size is derived from the formula for the margin of error by algebraically solving for n.

Formula

Minimum Sample Size for Estimating a Population Proportion

The minimum sample size required for estimating a population proportion at a given level of confidence with a particular margin of error is given by

$$n = p(1-p)\left(\frac{z_{\alpha/2}}{E}\right)^2$$

where p is the population proportion,

$z_{\alpha/2}$ is the critical value for the level of confidence, $c = 1-\alpha$, such that the area under the standard normal distribution to the right of $z_{\alpha/2}$ is equal to $\frac{\alpha}{2}$, and

E is the desired maximum margin of error.

Typically, the value used for p in the formula for the minimum sample size is an estimate or a value of $\hat{p}$ from a previously conducted study. If there is no known value for p, we can still use the formula to calculate a sample size that will guarantee our desired level of confidence and maximum margin of error. If we let $p = 0.5$, we will find a sample size that is large enough to give us a confidence interval with our requirements regardless of the actual value of p. Note that because we are calculating the number of data values needed, we must always *round up* to the next *larger* whole number. Again, this is a *minimum* number of data values we need in our sample. It is always a good idea to sample as many as possible.

Example 8.22

Finding the Minimum Sample Size Needed for a Confidence Interval for a Population Proportion

What is the minimum sample size needed for a 99% confidence interval for the population proportion if previous studies indicated $p \approx 0.54$ and we desire no more than a 2% margin of error?

Solution

To find the minimum sample size, we need values for p, E, and $z_{\alpha/2}$. One way to estimate p is from previous studies. Therefore, we will use $p \approx 0.54$ as given. Since we'd like the margin of error to be at most 2%, we convert to a decimal to have $E = 0.02$. Finally, we can look up the critical value in the table to find $z_{\alpha/2} = z_{0.01/2} = z_{0.005} = 2.575$.

$$\begin{aligned} n &= p(1-p)\left(\frac{z_{\alpha/2}}{E}\right)^2 \\ &= 0.54(1-0.54)\left(\frac{2.575}{0.02}\right)^2 \\ &= 4117.618125 \\ &\approx 4118 \end{aligned}$$

Remembering to always round a minimum sample size up to the next larger whole number, we then need a sample size of 4118 or larger.

When using a TI-83/84 Plus calculator, a more accurate value for $z_{\alpha/2}$ can be calculated using the function, `invNorm`, under the `DISTR` menu.

- Press 2ND and then VARS.
- Choose option `3:invNorm(`. The function syntax for this option, when using the standard normal curve, is `invNorm(`*area to the left of z*`)`.
- The area to the left of a critical *z*-value, $z_{\alpha/2}$, is equal to $1-\alpha/2$. For this problem, $\alpha = 1-c = 1-0.99 = 0.01$. Thus, $1-\alpha/2 = 1-0.01/2 = 0.995$, so we would enter `invNorm(Ø.995)`.

The calculator returns a value of approximately 2.575829. Using this more accurate value for $z_{\alpha/2}$ in the formula results in a minimum sample size of 4121. Notice that this value is close to the previous answer we calculated, but not the same.

Example 8.23

Finding the Minimum Sample Size, Point Estimate, and Confidence Interval for a Population Proportion

The state education commission wants to estimate the proportion of tenth-grade students in the state who read at or below the eighth-grade level. The commissioner of education believes that the proportion is around 0.21.

a. How large a sample would be required to estimate the proportion of all tenth graders in the state who read at or below the eighth-grade level with an 85% level of confidence and an error of at most 0.03?

b. Suppose a sample of minimum size is chosen. Of these students, 84 read at or below the eighth-grade level. Using this information, compute the sample proportion of all tenth graders reading at or below the eighth-grade level.

c. Construct the 85% confidence interval for the population proportion of tenth graders in the state reading at or below the eighth-grade level.

Solution

a. From the problem, we know that $E = 0.03$ and $p = 0.21$. Using the table of critical *z*-values, we find that $z_{\alpha/2} = z_{0.15/2} = z_{0.075} = 1.44$. We can now use this information to calculate the minimum sample size.

$$\begin{aligned} n &= p(1-p)\left(\frac{z_{\alpha/2}}{E}\right)^2 \\ &= 0.21(1-0.21)\left(\frac{1.44}{0.03}\right)^2 \\ &= 382.2336 \end{aligned}$$

Thus, the minimum sample size required is 383.

(**Note:** If you use the invNorm function on a TI-83/84 Plus calculator to find the value of $z_{\alpha/2}$, it will result in $z_{\alpha/2} \approx 1.439531$. Using this value in the above formula will result in a minimum sample size of 382, when rounded up to the next whole number.)

b. To compute the sample proportion, we divide the number reading at or below the eighth-grade level by the sample size.

$$\begin{aligned}\hat{p} &= \frac{x}{n}\\ &= \frac{84}{383}\\ &\approx 0.219321\end{aligned}$$

(**Note:** If the value found using the critical value of z from a TI-83/84 Plus calculator, 382, is used for the minimum sample size in this calculation, the resulting sample proportion is $\hat{p} \approx 0.219895$.)

c. To construct the confidence interval, we need to calculate the margin of error. Substituting the appropriate values into the formula, we have the following.

$$\begin{aligned}E &= z_{\alpha/2}\sqrt{\frac{\hat{p}\left(1-\hat{p}\right)}{n}}\\ &= 1.44\sqrt{\frac{0.219321(1-0.219321)}{383}}\\ &\approx 0.030447\end{aligned}$$

Notice that the margin of error we calculated is very close to the value of E that we were willing to accept in part a. Since we used the minimum sample size calculated in part a., this is to be expected.

Therefore, subtracting the margin of error from the sample proportion and then adding the margin of error to the sample proportion gives us the following endpoints of the confidence interval.

$$\text{Lower endpoint: } \hat{p} - E = 0.219321 - 0.030447 \approx 0.189$$

$$\text{Upper endpoint: } \hat{p} + E = 0.219321 + 0.030447 \approx 0.250$$

Thus, the 85% confidence interval for the population proportion ranges from 0.189 to 0.250. The confidence interval can be written mathematically using either inequality symbols or interval notation, as follows.

$$0.189 < p < 0.250$$

or

$$(0.189, 0.250)$$

(**Note:** Using the values obtained from a TI-83/84 Plus calculator in parts a. and b. also results in a confidence interval of $0.189 < p < 0.250$.)

In other words, the state education commission can be 85% confident that the proportion of tenth-grade students in the state who read at or below the eighth-grade level is between 18.9% and 25.0%.

The previous example points out that using traditional calculations, versus using technology, will produce slightly different answers. The technology can provide more accurate answers, as the critical values are no longer limited to those shown in the tables. When appropriate, we have tried to provide both answers in examples and exercise solutions.

8.4 Section Exercises

Directions: For all exercises in this section, you may assume that the requirements mentioned in this section are met; namely, all samples are simple random samples, the conditions for a binomial distribution are met, and the sample size is large enough to ensure that $n\hat{p} \geq 5$ and $n(1-\hat{p}) \geq 5$.

Point Estimates and Confidence Intervals for Population Proportions

Directions: Find each specified point estimate or confidence interval.

1. Out of 50 randomly selected students who were surveyed, 41 felt that dorm renovation should be the administration's top priority. What is the best point estimate for the proportion of all students who feel that dorm renovation should be the administration's top priority?

2. In a random sample of 130 students, only 7 had been placed in the wrong math class. What is the best point estimate for the population proportion of all students who have been placed in the wrong math class?

3. A survey revealed that 48 out of 112 randomly selected homeowners in one neighborhood pay for lawn service every month. What is the best point estimate for the population proportion of all homeowners in that neighborhood who pay for lawn service every month?

4. Out of 210 randomly selected registered voters who were polled, 107 declared that they intended to vote for the incumbent in the upcoming election. What is the best point estimate for the proportion of all registered voters who intend to vote for the incumbent?

5. Out of 50 randomly selected students who were surveyed, 41 felt that dorm renovation should be the administration's top priority. Using a margin of error of 8.9%, give the interval estimate for the proportion of all students who feel that dorm renovation should be the administration's top priority.

6. Out of 210 randomly selected registered voters who were polled, 107 declared that they intended to vote for the incumbent in the upcoming election. Using a margin of error of 4.4%, give the interval estimate for the proportion of all registered voters who intend to vote for the incumbent.

Confidence Intervals for Population Proportions

Directions: Construct and interpret each specified confidence interval.

7. Thirteen out of 147 randomly selected faculty members who were surveyed at a community college know sign language. Construct and interpret a 90% confidence interval for the proportion of all faculty members at the community college who know sign language.

8. A random sample of 200 computer chips is obtained from one factory and 4% are found to be defective. Construct and interpret a 95% confidence interval for the proportion of all computer chips from that factory that are defective.

9. Out of 140 randomly selected kindergartners who were surveyed, 32 said that pancakes are their favorite breakfast food. Construct and interpret a 90% confidence interval for the percentage of all kindergartners who say pancakes are their favorite breakfast food.

10. Out of 54 randomly selected patients of a local hospital who were surveyed, 49 reported that they were satisfied with the care they received. Construct and interpret a 95% confidence interval for the percentage of all patients satisfied with their care at that hospital.

11. A survey of 145 randomly selected students at one college showed that only 87 checked their campus e-mail account on a regular basis. Construct and interpret a 90% confidence interval for the percentage of students at that college who do not check their e-mail account on a regular basis.

12. A random sample of 200 computer chips is obtained from one factory and 4% are found to be defective. Construct and interpret a 99% confidence interval for the percentage of all computer chips from that factory that are *not* defective.

13. Out of 190 randomly selected adults in the United States who were surveyed, 71 exercise on a regular basis. Construct and interpret a 95% confidence interval for the proportion of all adults in the United States who exercise on a regular basis.

14. A survey of 47 randomly selected members of a health club reported that 68.1% prefer walking or running on a treadmill to any other type of aerobic exercise. Construct and interpret a 95% confidence interval for the proportion of health club members who prefer walking or running on a treadmill.

15. Suppose that a national wireless phone company is interested in determining how many of its customers who use smartphones might switch to an iPhone if it began to carry this phone. In a survey of 730 customers who use smartphones, 323 said they would be likely to switch to the iPhone if it is offered. Construct and interpret a 95% confidence interval for the percentage of customers who would be likely to switch to the new iPhone if it is offered by the company.

Minimum Sample Sizes for Estimating Population Proportions

Directions: **Calculate the minimum sample size needed to construct a confidence interval with the desired characteristics.**

16. Suppose you wish to determine the proportion of college students in your state who receive some form of financial aid. You want to be 98% confident of your results and have a maximum error of 5%. Calculate the minimum sample size that you must have to meet these requirements, given that the financial aid office at a local institution estimates the percentage to be 78%.

17. Suppose that you wish to determine the proportion of college students in the United States who own a smartphone. You want to be 95% confident in your results and have a maximum error of 3%. Calculate the minimum number of college students that you must survey to meet these requirements, given that a previous study estimated that 68% of all college students in the United States own a smartphone.

18. A market researcher wishes to determine the proportion of American women who shop online. The results must be accurate at the 90% level of confidence with a maximum error of 2%. Calculate the minimum sample size needed to meet these requirements, given that a previous study estimated that 59% of American women shop online.

19. Suppose a researcher needs to verify the proportion of patients who will have mild side effects when taking a recently approved drug. The researcher must be 99% sure of her results with a maximum error of 1%. Calculate the minimum sample size the researcher needs to meet her requirements, given that the studies used to get the drug approved reported that only 12% of patients experienced mild side effects.

20. Solve the formula for the margin of error of a confidence interval for a proportion, $E = z_{\alpha/2}\sqrt{\dfrac{\hat{p}\left(1-\hat{p}\right)}{n}}$, for n, to show that it is equivalent to the formula for the minimum sample size for estimating a population proportion, $n = \hat{p}\left(1-\hat{p}\right)\left(\dfrac{z_{\alpha/2}}{E}\right)^2$. (Recall that a value of $\hat{p}$ from a previous study is typically used as an estimate for p in the formula for the minimum sample size.)

Directions: Respond thoughtfully to the following exercise.

21. The Deseret News in Salt Lake City, Utah reported in April 2012 that the Zions Bank Consumer Attitude Index for Utah increased to 85.3, which was 5.4 points higher than the month prior. It stated that the index was "based on a representative sample of 500 Utah households … and has a confidence interval of plus or minus 4.38 percent at a 95 percent confidence level." Based on the information given in the article, is it possible that the index reported in April actually did not increase over the previous month's index? Why or why not?

Source: "Utah Consumers Remain Confident, New Report States." *Deseret News.* 24 Apr. 2012. http://www.deseretnews.com/article/765571132/Utah-consumers-remain-confident-new-report-states.html (20 May 2012).

8.5 Estimating Population Variances

The final parameter that we will estimate is the population variance. Up to this point, the applications for creating confidence intervals for parameters such as the mean and proportion were obvious. Less obvious, perhaps, is the need to estimate the *variance* of a population. However, there are times when it is important for products or items to be identical. For example, an auto parts manufacturer must test to ensure that each auto part that comes off the assembly line matches the necessary specifications. The consistency of quality in the auto parts is essential. For quality control, the manufacturer would likely estimate the variation of all parts by using a sample of parts on the assembly line. If the estimate of the variance is too high, then adjustments in the manufacturing process must be made.

Let's begin as before with simply estimating the parameter from a single source. Once again, the best single estimate for a population parameter is a sample statistic. In this case the unbiased estimator, and hence the best point estimate, for the population variance, σ^2, is the sample variance, s^2. The most commonly used point estimate for the population standard deviation is the sample standard deviation. The following example illustrates how point estimates may be used to estimate the variance and standard deviation of a population.

Example 8.24

Finding Point Estimates for the Population Standard Deviation and Variance

General Auto is testing the variance in the lengths of its windshield wiper blades. A sample of 12 windshield wiper blades is randomly selected, and the following lengths are measured in inches.

22.1	22.0	22.1	22.4	22.3	22.5
22.3	22.1	22.2	22.6	22.5	22.7

a. Find a point estimate for the population standard deviation.

b. Find a point estimate for the population variance.

Rounding Rule

Recall that standard deviation and variance are rounded to one more decimal place than the largest number of decimal places given in the data.

Solution

a. The sample standard deviation, s, is the most common point estimate for σ. According to the calculator, $s \approx 0.224958 \approx 0.22$. For review on how to calculate s, see Section 3.2. The point estimate for the population standard deviation is then 0.22 inches.

b. The sample variance, s^2, is the best point estimate for σ^2. Using a calculator, we calculate $s^2 \approx (0.224958)^2 \approx 0.05$. The point estimate for the population variance is then 0.05.

As we noted in previous sections, using an interval estimate, or confidence interval, rather than a point estimate will increase our likelihood of estimating the true population variance. However, the necessary conditions and critical values for constructing confidence intervals for variances and standard deviations are different than those we used for confidence intervals for means and proportions. In order to construct a confidence interval for a variance or standard deviation, the following conditions must be met.

- All possible samples of a given size have an equal probability of being chosen; that is, a simple random sample is used.
- The population distribution is approximately normal.

We will assume that all examples and exercises meet these requirements so that we may concentrate on the applications here, not on verifying normality. The formula for calculating the lower and upper endpoints of the confidence interval requires that we use critical values of the chi-square random variable, denoted χ^2. Like the Student's *t*-distribution, the χ^2-distribution is one that is completely defined by the number of degrees of freedom, $df = n - 1$. However, unlike the normal and Student's *t*-distributions, the χ^2-distribution is not symmetric, as shown in Figure 8.8.

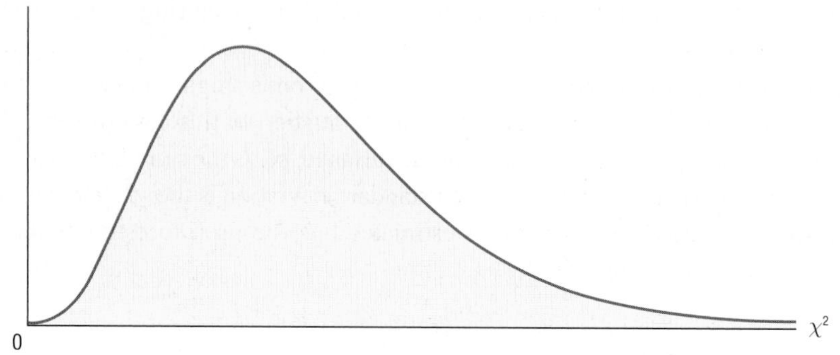

Figure 8.8: χ^2-Distribution

We will discuss the χ^2-distribution in more detail in Chapter 10. For now, a basic knowledge of how to read the χ^2-distribution table will be sufficient. Let's look at how to read the table. The table of critical values for the chi-square distribution is Table G in Appendix A. Notice that the first column contains the number of degrees of freedom, *df*. When constructing an interval estimate for a population variance, the number of degrees of freedom is simply $df = n - 1$. Across the top row of the table, you will see ten commonly used values for the area in the right tail of the χ^2-distribution. The column you will use depends on the level of confidence given in the problem, which dictates the corresponding level of α. (Remember that $\alpha = 1 - c$.) The critical value is found in the cell where the row for the number of degrees of freedom and the column for the appropriate area intersect.

	Area to the Right of the Critical Value of χ^2									
df	**0.995**	**0.990**	**0.975**	**0.950**	**0.900**	**0.100**	**0.050**	**0.025**	**0.010**	**0.005**
1	0.000	0.000	0.001	0.004	0.016	2.706	3.841	5.024	6.635	7.879
2	0.010	0.020	0.051	0.103	0.211	4.605	5.991	7.378	9.210	10.597
3	0.072	0.115	0.216	0.352	0.584	6.251	7.815	9.348	11.345	12.838
4	0.207	0.297	0.484	0.711	1.064	7.779	9.488	11.143	13.277	14.860
5	0.412	0.554	0.831	1.145	1.610	9.236	11.070	12.833	15.086	16.750
6	0.676	0.872	1.237	1.635	2.204	10.645	12.592	14.449	16.812	18.548
7	0.989	1.239	1.690	2.167	2.833	12.017	14.067	16.013	18.475	20.278
8	1.344	1.646	2.180	2.733	3.490	13.362	15.507	17.535	20.090	21.955

Figure 8.9: Excerpt from Table G: Critical Values of χ^2

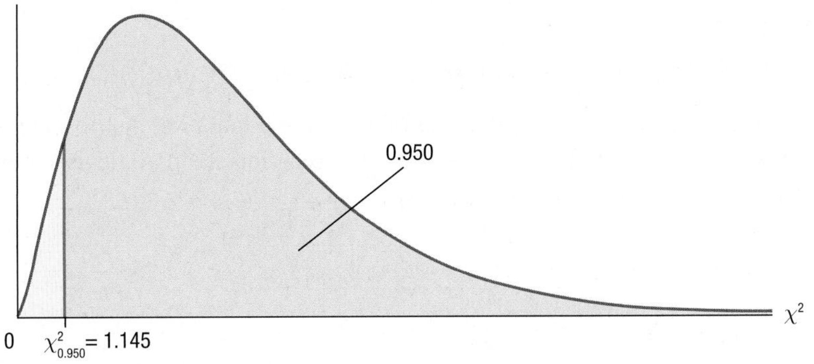

Figure 8.10: Critical Value for χ^2-Distribution with $df = 5$

Another way in which confidence intervals for variance and standard deviation are unique is the way in which the interval estimate is created. Rather than adding and subtracting a margin of error, we have formulas for calculating the endpoints. Because the chi-square distribution is not symmetric, the left and right critical values must be determined separately, and the endpoints calculated individually. The formulas for the endpoints of a confidence interval for a population variance are as follows.

Formula

Confidence Interval for a Population Variance

When the sample taken is a simple random sample and the population distribution is approximately normal, the confidence interval for a population variance is given by

$$\frac{(n-1)s^2}{\chi^2_{\alpha/2}} < \sigma^2 < \frac{(n-1)s^2}{\chi^2_{(1-\alpha/2)}}$$

where n is the sample size,

s^2 is the sample variance, and

$\chi^2_{\alpha/2}$ and $\chi^2_{(1-\alpha/2)}$ are the critical values for the level of confidence, $c = 1-\alpha$, such that the area under the χ^2-distribution with $n - 1$ degrees of freedom to the right of $\chi^2_{\alpha/2}$ is equal to $\frac{\alpha}{2}$, and the area to the right of $\chi^2_{(1-\alpha/2)}$ is equal to $1-\frac{\alpha}{2}$.

Rounding Rule

Round the endpoints of a confidence interval for a population variance or standard deviation as follows:

- If sample data are given, round to one more decimal place than the largest number of decimal places in the given data.
- If statistics are given, round to the same number of decimal places as given in the point estimate for the standard deviation or variance.

The formula for creating a confidence interval for a population standard deviation follows easily from this by simply taking the square root of each side to give the following.

Formula

Confidence Interval for a Population Standard Deviation

When the sample taken is a simple random sample and the population distribution is approximately normal, the confidence interval for a population standard deviation is given by

$$\sqrt{\frac{(n-1)s^2}{\chi^2_{\alpha/2}}} < \sigma < \sqrt{\frac{(n-1)s^2}{\chi^2_{(1-\alpha/2)}}}$$

where n is the sample size,

s^2 is the sample variance, and

$\chi^2_{\alpha/2}$ and $\chi^2_{(1-\alpha/2)}$ are the critical values for the level of confidence, $c = 1-\alpha$, such that the area under the χ^2-distribution with $n - 1$ degrees of freedom to the right of $\chi^2_{\alpha/2}$ is equal to $\frac{\alpha}{2}$, and the area to the right of $\chi^2_{(1-\alpha/2)}$ is equal to $1-\frac{\alpha}{2}$.

Note that when we construct an interval estimate using a χ^2-distribution, the level of confidence, c, is the area between the critical values $\chi^2_{(1-\alpha/2)}$ and $\chi^2_{\alpha/2}$. The remaining area under the curve, α, is split between the two tails.

Procedure

Constructing a Confidence Interval for a Population Variance (or Standard Deviation)

1. Find the point estimate, s^2 (or s).
2. Based on the level of confidence given, calculate $\frac{\alpha}{2}$ and $1-\frac{\alpha}{2}$.
3. Use the χ^2-distribution table to find the critical values, $\chi^2_{\alpha/2}$ and $\chi^2_{(1-\alpha/2)}$, for a distribution with $n-1$ degrees of freedom.
4. Substitute the necessary values into the formula for the confidence interval.

Note that for a chi-square distribution, when the number of degrees of freedom needed is not in the tables, use the closest value given. If the number of degrees of freedom is exactly halfway between the two closest values given in the table, find the mean of the critical values of χ^2 for those two values of *df*. Let's look at an example of how this type of confidence interval is created.

Example 8.25

Constructing a Confidence Interval for a Population Variance

A commercial bakery is testing the variance in the weights of the cookies it produces. A random sample of 15 cookies is chosen; the weights of the cookies are measured in grams and found to have a variance of 3.4. Build a 95% confidence interval for the variance of the weights of all cookies produced by the company.

Solution

Step 1: **Find the point estimate, s^2 (or s).**

We are given the sample variance in the problem: $s^2 = 3.4$.

Step 2: **Based on the level of confidence given, calculate $\frac{\alpha}{2}$ and $1-\frac{\alpha}{2}$.**

Because $c = 0.95$, we know that $\alpha = 1-0.95 = 0.05$. Then $\frac{\alpha}{2} = \frac{0.05}{2} = 0.025$. Also, $1-\frac{\alpha}{2} = 1-0.025 = 0.975$.

Step 3: **Find the critical values, $\chi^2_{\alpha/2}$ and $\chi^2_{(1-\alpha/2)}$.**

Using the χ^2-distribution table with $df = n-1 = 15-1 = 14$ degrees of freedom, we see that $\chi^2_{\alpha/2} = \chi^2_{0.025} = 26.119$ and $\chi^2_{(1-\alpha/2)} = \chi^2_{0.975} = 5.629$.

Area to the Right of the Critical Value of χ^2

df	0.995	0.990	0.975	0.950	0.900	0.100	0.050	0.025	0.010	0.005
11	2.603	3.053	3.816	4.575	5.578	17.275	19.675	21.920	24.725	26.757
12	3.074	3.571	4.404	5.226	6.304	18.549	21.026	23.337	26.217	28.300
13	3.565	4.107	5.009	5.892	7.042	19.812	22.362	24.736	27.688	29.819
14	4.075	4.660	5.629	6.571	7.790	21.064	23.685	26.119	29.141	31.319
15	4.601	5.229	6.262	7.261	8.547	22.307	24.996	27.488	30.578	32.801
16	5.142	5.812	6.908	7.962	9.312	23.542	26.296	28.845	32.000	34.267

Memory Booster

Since the χ^2-distribution is not symmetric, the left and right critical values must be determined separately.

Step 4: **Substitute the necessary values into the formula for the confidence interval.**

Substituting into the formula for a confidence interval for a population variance gives us the following.

$$\frac{(n-1)s^2}{\chi^2_{\alpha/2}} < \sigma^2 < \frac{(n-1)s^2}{\chi^2_{(1-\alpha/2)}}$$

$$\frac{(15-1)3.4}{26.119} < \sigma^2 < \frac{(15-1)3.4}{5.629}$$

$$1.8 < \sigma^2 < 8.5$$

Using interval notation, the confidence interval can also be written as $(1.8, 8.5)$.

The bakery estimates with 95% confidence that the variance in the weights of their cookies is between 1.8 and 8.5.

It is worth pointing out that the critical values of the chi-square distribution behave in a way that is backwards from what you might expect. Take, for instance, the critical values found in the previous example. The critical value associated with the lower endpoint, $\chi^2_{0.025} = 26.119$, is a larger number than the critical value associated with the upper endpoint, $\chi^2_{0.975} = 5.629$. The reason why is that in the χ^2-distribution table, the area associated with each critical value is the area under the curve in the right tail of the distribution with the given number of degrees of freedom. Hence, a critical value associated with a larger area would lie further to the left of a critical value with a smaller area. Also, since the critical values are substituted into the denominators of the confidence interval formula, the final result does indeed reflect the lower and upper endpoints as expected.

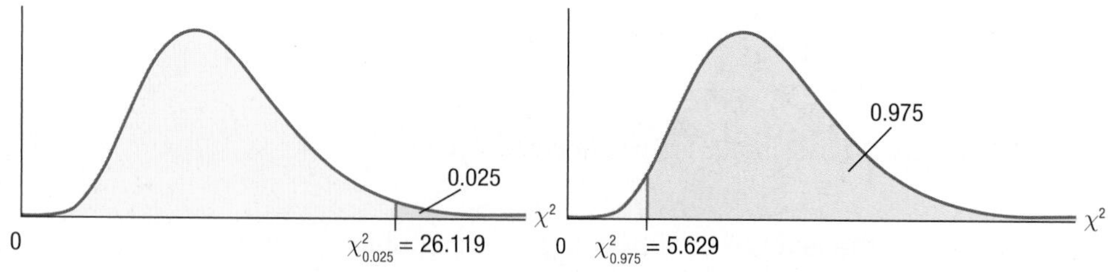

Figure 8.11: Critical Values for χ^2 -Distribution with *df* = 14

Example 8.26

Constructing a Confidence Interval for a Population Standard Deviation

Consider again the bakery in the previous example. Suppose that the bakery needs to estimate the standard deviation of the weights of their cookies as well. Construct a 95% confidence interval for the standard deviation of the weights of all cookies produced at the bakery.

Solution

Remember the relationship between standard deviation and variance. To find the 95% confidence interval for standard deviation, simply take the *square roots* of the expressions used to find the endpoints of the 95% confidence interval for the variance. This gives us the following.

$$\sqrt{\frac{(n-1)s^2}{\chi^2_{\alpha/2}}} < \sigma < \sqrt{\frac{(n-1)s^2}{\chi^2_{(1-\alpha/2)}}}$$

$$\sqrt{\frac{(15-1)3.4}{26.119}} < \sigma < \sqrt{\frac{(15-1)3.4}{5.629}}$$

$$1.3 < \sigma < 2.9$$

Using interval notation, the confidence interval can also be written as $(1.3, 2.9)$.

The bakery can be 95% confident that the standard deviation of the weights of all cookies produced is between 1.3 and 2.9 grams.

Example 8.27

Constructing a Confidence Interval for a Population Variance

A seed company is researching the consistency in the output of a new hybrid tomato plant that was recently developed. The weights of a random sample of 75 tomatoes produced by the hybrid plant are measured in pounds, and the sample has a variance of 0.0225. Construct a 90% confidence interval for the variance in weights for all tomatoes produced by the new hybrid plant.

Solution

Step 1: Find the point estimate, s^2 (or s).

We are given the sample variance, $s^2 = 0.0225$.

Step 2: Based on the level of confidence given, calculate $\frac{\alpha}{2}$ and $1-\frac{\alpha}{2}$.

Because $c = 0.90$, we know that $\alpha = 1 - 0.90 = 0.10$. Then $\frac{\alpha}{2} = \frac{0.10}{2} = 0.05$. Also, $1-\frac{\alpha}{2} = 1 - 0.05 = 0.95$.

Step 3: Find the critical values, $\chi^2_{\alpha/2}$ and $\chi^2_{(1-\alpha/2)}$.

We need to find the critical values for the χ^2-distribution with $df = n - 1 = 75 - 1 = 74$ degrees of freedom, but 74 is not one of the numbers of degrees of freedom listed in the table. Using the closest value available in the χ^2-distribution table, $df = 70$, we see that $\chi^2_{\alpha/2} = \chi^2_{0.05} = 90.531$ and $\chi^2_{(1-\alpha/2)} = \chi^2_{0.95} = 51.739$.

	Area to the Right of the Critical Value of χ^2									
df	**0.995**	**0.990**	**0.975**	**0.950**	**0.900**	**0.100**	**0.050**	**0.025**	**0.010**	**0.005**
40	20.707	22.164	24.433	26.509	29.051	51.805	55.758	59.342	63.691	66.766
50	27.991	29.707	32.357	34.764	37.689	63.167	67.505	71.420	76.154	79.490
60	35.534	37.485	40.482	43.188	46.459	74.397	79.082	83.298	88.379	91.952
70	43.275	45.442	48.758	51.739	55.329	85.527	90.531	95.023	100.425	104.215
80	51.172	53.540	57.153	60.391	64.278	96.578	101.879	106.629	112.329	116.321
90	59.196	61.754	65.647	69.126	73.291	107.565	113.145	118.136	124.116	128.299

Step 4: Substitute the necessary values into the formula for the confidence interval.

Substituting into the formula for a confidence interval for a population variance gives us the following.

$$\frac{(n-1)s^2}{\chi^2_{\alpha/2}} < \sigma^2 < \frac{(n-1)s^2}{\chi^2_{(1-\alpha/2)}}$$

$$\frac{(75-1)0.0225}{90.531} < \sigma^2 < \frac{(75-1)0.0225}{51.739}$$

$$0.0184 < \sigma^2 < 0.0322$$

Using interval notation, the confidence interval can also be written as $(0.0184, 0.0322)$.

Therefore, the seed company can estimate with 90% confidence that the true variance in weights of the new hybrid tomatoes is between 0.0184 and 0.0322.

Example 8.28

Constructing a Confidence Interval for a Population Standard Deviation

Let's consider again the scenario of the seed company given in Example 8.27. A new random sample of 86 tomatoes is taken from the hybrid plants and found to have a standard deviation of 0.1400 pounds. Construct a 99% confidence interval to estimate the standard deviation in the weights of all tomatoes produced by the new hybrid plant.

Solution

Step 1: Find the point estimate, s^2 (or s).

We are given the sample standard deviation, $s = 0.1400$.

Step 2: **Based on the level of confidence given, calculate $\frac{\alpha}{2}$ and $1-\frac{\alpha}{2}$.**

Because $c = 0.99$, we know that $\alpha = 1-0.99 = 0.01$. Then $\frac{\alpha}{2} = \frac{0.01}{2} = 0.005$. Also, $1-\frac{\alpha}{2} = 1-0.005 = 0.995$.

Step 3: **Find the critical values, $\chi^2_{\alpha/2}$ and $\chi^2_{(1-\alpha/2)}$.**

We need to find the critical values for the χ^2-distribution with $df = n-1 = 86-1 = 85$ degrees of freedom, but we see that not only is 85 degrees of freedom not listed in the table, but it is exactly halfway between the two closest values available, 80 and 90 degrees of freedom. Therefore for each critical value, we will need to find the mean of the values listed for 80 and 90 degrees of freedom in order to obtain the value we need.

Area to the Right of the Critical Value of χ^2

df	0.995	0.990	0.975	0.950	0.900	0.100	0.050	0.025	0.010	0.005
40	20.707	22.164	24.433	26.509	29.051	51.805	55.758	59.342	63.691	66.766
50	27.991	29.707	32.357	34.764	37.689	63.167	67.505	71.420	76.154	79.490
60	35.534	37.485	40.482	43.188	46.459	74.397	79.082	83.298	88.379	91.952
70	43.275	45.442	48.758	51.739	55.329	85.527	90.531	95.023	100.425	104.215
80	51.172	53.540	57.153	60.391	64.278	96.578	101.879	106.629	112.329	116.321
90	59.196	61.754	65.647	69.126	73.291	107.565	113.145	118.136	124.116	128.299
100	67.328	70.065	74.222	77.929	82.358	118.498	124.342	129.561	135.807	140.169

Taking the mean of 116.321 and 128.299, we obtain a critical value of

$\chi^2_{\alpha/2} = \chi^2_{0.005} = \frac{116.321+128.299}{2} = 122.310$ for 85 degrees of freedom.

Taking the mean of 51.172 and 59.196, we obtain a critical value of

$\chi^2_{(1-\alpha/2)} = \chi^2_{0.995} = \frac{51.172+59.196}{2} = 55.184.$

Step 4: **Substitute the necessary values into the formula for the confidence interval.**

Substituting into the formula for a confidence interval for a population standard deviation gives us the following.

$$\sqrt{\frac{(n-1)s^2}{\chi^2_{\alpha/2}}} < \sigma < \sqrt{\frac{(n-1)s^2}{\chi^2_{(1-\alpha/2)}}}$$

$$\sqrt{\frac{(86-1)(0.1400)^2}{122.310}} < \sigma < \sqrt{\frac{(86-1)(0.1400)^2}{55.184}}$$

$$0.1167 < \sigma < 0.1738$$

Using interval notation, the confidence interval can also be written as $(0.1167, 0.1738)$.

Therefore, the seed company can estimate with 99% confidence that the true standard deviation in weights of all tomatoes produced by the new hybrid plant is between 0.1167 and 0.1738 pounds.

Minimum Sample Size for Estimating a Population Variance or Population Standard Deviation

As we have discussed in previous sections, there are times at which it is desirable to estimate how large a sample must be chosen in order for the results of a study to have a certain level of precision. Unfortunately, calculating a minimum sample size for estimating the variance of a population is not as simple as we have seen before. In fact, the best way to determine the minimum sample size for these types of problems is to use statistical software. For our purposes, we will use the following two examples of tables created using statistical software. These tables give the minimum sample size needed to estimate a population variance or population standard deviation in order to be 95% or 99% confident that the value of the sample statistic will be within a certain percentage of the actual value of the population parameter.

Table 8.3: Minimum Sample Sizes for Estimating Variance

s^2 Is within This Percentage of the Value of σ^2	Minimum Sample Size Needed for 95% Level of Confidence	Minimum Sample Size Needed for 99% Level of Confidence
1%	77,209	133,450
5%	3150	5459
10%	807	1403
20%	212	370
30%	99	173
40%	58	102
50%	39	69

Table 8.4: Minimum Sample Sizes for Estimating Standard Deviation

s Is within This Percentage of the Value of σ	Minimum Sample Size Needed for 95% Level of Confidence	Minimum Sample Size Needed for 99% Level of Confidence
1%	19,206	33,220
5%	769	1337
10%	193	337
20%	49	86
30%	22	39
40%	13	23
50%	9	15

Let's look at an example of how to use these tables to find the minimum sample size necessary for a particular study.

Example 8.29

Finding the Minimum Sample Size Needed for a Confidence Interval for a Population Standard Deviation

A market researcher wants to estimate the standard deviation of home prices in a metropolitan area in the Northeast. She needs to be 99% confident that her sample standard deviation is within 5% of the true population standard deviation. Assuming that the home prices in that area are normally distributed, what is the minimum number of home prices she must acquire?

Solution

According to the table, we see that, in order to be 99% confident that the sample standard deviation will be within 5% of the true population standard deviation, the minimum sample size required is 1337.

Minimum Sample Sizes for Estimating Standard Deviation

s Is within This Percentage of the Value of σ	Minimum Sample Size Needed for 95% Level of Confidence	Minimum Sample Size Needed for 99% Level of Confidence
1%	19,206	33,220
5%	769	1337
10%	193	337
20%	49	86
30%	22	39
40%	13	23
50%	9	15

Therefore, the market researcher must include at least 1337 home prices in order for her study to have the level of precision that she needs.

8.5 Section Exercises

Note: For all exercises in this section, you may assume that the requirements mentioned in this section are met; namely, all samples of a given size have an equal probability of being chosen and the population distribution is approximately normal.

Point Estimates for Population Variances and Population Standard Deviations

Directions: Find each specified point estimate.

1. What is the best point estimate for the population standard deviation if the sample standard deviation is 3.5?

2. What is the best point estimate for the population standard deviation if the sample variance is 25?

3. Consider a sample of caps for three-inch pipes. Their diameters are measured and found to have a variance of 0.12. Give a point estimate for the population variance in diameter lengths of the caps.

4. For a random sample of 112 men, the mean height was found to be 72 inches with a standard deviation of 2.2 inches. Give a point estimate for the population variance of heights for men.

Critical Values for Confidence Intervals for Population Variances

Directions: Determine the critical values for the left and right endpoints of a confidence interval for the population variance using the given information.

5. $n = 25, \quad \alpha = 0.05$

6. $n = 17, \quad \alpha = 0.10$

7. $n = 22, \quad c = 0.90$

8. $n = 10, \quad c = 0.95$

Confidence Intervals for Population Variances and Population Standard Deviations

Directions: Construct a confidence interval for the population variance at the given level of confidence.

9. $n = 12, \quad s^2 = 19.2, \quad c = 0.95$

10. $n = 20, \quad s^2 = 14.2, \quad c = 0.99$

11. $n = 23, \quad s^2 = 11.9, \quad c = 0.99$

12. $n = 29, \quad s^2 = 26.5, \quad c = 0.98$

Directions: Construct a confidence interval for the population standard deviation at the given level of confidence.

13. $n = 63, \quad s = 2.4, \quad c = 0.99$

14. $n = 41, \quad s = 1.2, \quad c = 0.95$

15. $n = 50, \quad s = 6.8, \quad c = 0.98$

16. $n = 56, \quad s = 3.5, \quad c = 0.99$

Directions: Construct and interpret each specified confidence interval.

17. A commercial grocer is testing the variance in the weights of packages of strawberries. The weights of the packages are measured in ounces, and a random sample of 15 packages of strawberries has a variance of 3.80. Construct and interpret a 90% confidence interval for the variance in weights of all packages of strawberries.

18. A tire manufacturer is testing the tire pressure in its new line of SUV tires. A random sample of 10 tire pressure readings, measured in pounds per square inch (psi), yields a variance of 31.8. Construct and interpret a 95% confidence interval for the variance in tire pressures for all new SUV tires produced by the manufacturer.

19. Speeds were measured in miles per hour (mph) for a random sample of 26 fastballs thrown by major league pitchers, and the variance in speed was 21.5. Construct and interpret a 98% confidence interval for the variance in speeds of all fastballs thrown by major league pitchers.

20. A company is testing 63 randomly selected compact fluorescent light bulbs to see how long they last. The study results show a standard deviation of 121.4 hours. Construct and interpret a 99% confidence interval for the true standard deviation for the lifetimes of the compact fluorescent light bulbs.

21. A transit system at a large theme park is testing how long it takes for a driver to complete one circuit. A sample of 40 completed circuits shows a standard deviation of 7.7 minutes. Construct and interpret a 98% confidence interval for the standard deviation in completion times for all circuits driven.

22. After testing 31 pairs of noise-reducing earmuffs, the standard deviation of the reductions in noise levels is calculated to be 0.9 decibels. Construct and interpret a 95% confidence interval for the standard deviation of noise level reductions produced by all pairs of this type of noise-reducing earmuffs.

23. Consider the following sample data.

3.2	3.6	2.9	3.0	3.0
3.1	3.2	3.3	2.9	3.3
2.9	3.1	3.4	3.3	3.0

Build and interpret a 99% confidence interval for the population variance.

24. The following sample of weights (in ounces) was taken from 14 boxes of crackers randomly selected from the assembly line.

16.87	16.92	17.01	16.98	16.99	16.92	16.91
17.00	17.01	16.96	16.95	16.94	17.00	16.92

Build and interpret a 98% confidence interval for the population variance for the weights of all boxes of crackers that come off the assembly line.

25. The weights of 89 randomly selected new truck engines from one factory were found to have a standard deviation of 1.59 pounds. Construct and interpret a 95% confidence interval for the population standard deviation of the weights of all new truck engines in this particular factory.

26. A butcher uses a machine that packages ground beef in one-pound portions. A sample of 52 packages of ground beef has a standard deviation of 0.2 pounds. Construct and interpret a 99% confidence interval to estimate the standard deviation of the weights of all packages prepared by the machine.

Minimum Sample Sizes for Estimating Population Variances and Population Standard Deviations

Directions: **Find the minimum sample size needed to construct a confidence interval with the desired characteristics.**

27. Find the minimum sample size needed to be 95% confident that the sample variance is within 20% of the population variance.

28. Find the minimum sample size needed to be 99% confident that the sample variance is within 5% of the population variance.

29. Find the minimum sample size needed to be 99% confident that the sample standard deviation is within 10% of the population standard deviation.

30. Find the minimum sample size needed to be 95% confident that the sample standard deviation is within 1% of the population standard deviation.

R Chapter 8 Review

Section 8.1: Estimating Population Means (σ Known)

Definitions

Point estimate
A single-number estimate of a population parameter

Unbiased estimator
A point estimate that does not consistently underestimate or overestimate the population parameter

Interval estimate
A range of possible values for a population parameter

Level of confidence
The probability that the interval estimate contains the population parameter

Confidence interval
An interval estimate associated with a certain level of confidence

Margin of error
Denoted by E and also called the *maximum error of estimate*, it is the largest possible distance from the point estimate that a confidence interval will cover

Critical z-Values for Confidence Intervals

Level of Confidence, c	$\alpha = 1 - c$	$z_{\alpha/2}$
0.80	0.20	1.28
0.85	0.15	1.44
0.90	0.10	1.645
0.95	0.05	1.96
0.98	0.02	2.33
0.99	0.01	2.575

Margin of Error of a Confidence Interval for a Population Mean (σ Known)

Used when a simple random sample of size n is taken from a population with a known standard deviation, σ, where either $n \geq 30$ or the population distribution is approximately normal, given by

$$E = (z_{\alpha/2})(\sigma_{\bar{x}}) = (z_{\alpha/2})\left(\frac{\sigma}{\sqrt{n}}\right)$$

Section 8.1: Estimating Population Means (σ Known) (cont.)

Confidence Interval for a Population Mean

Can be written mathematically using either inequality symbols or interval notation, given by

$$\bar{x} - E < \mu < \bar{x} + E$$

or

$$(\bar{x} - E, \bar{x} + E)$$

Minimum Sample Size for Estimating a Population Mean

Minimum sample size required for estimating a population mean at a given level of confidence with a particular margin of error, given by

$$n = \left(\frac{z_{\alpha/2} \cdot \sigma}{E}\right)^2$$

Section 8.2: Student's t-Distribution

Properties of a t-Distribution

1. A t-distribution curve is symmetric and bell-shaped, centered about 0.

2. A t-distribution curve is completely defined by its number of degrees of freedom, df.

3. The total area under a t-distribution curve equals 1.

4. The x-axis is a horizontal asymptote for a t-distribution curve.

Section 8.3: Estimating Population Means (σ Unknown)

Margin of Error of a Confidence Interval for a Population Mean (σ Unknown)

Used when a simple random sample of size n is taken from a population with an unknown standard deviation, where either $n \geq 30$ or the population distribution is approximately normal, given by

$$E = \left(t_{\alpha/2}\right)\left(\frac{s}{\sqrt{n}}\right) \text{ with } df = n - 1$$

Section 8.4: Estimating Population Proportions

Margin of Error of a Confidence Interval for a Population Proportion

Used when a simple random sample of size n is large enough to ensure that $n\hat{p} \geq 5$ and $n(1-\hat{p}) \geq 5$, where the conditions for a binomial distribution are met, given by

$$E = z_{\alpha/2}\sqrt{\frac{\hat{p}(1-\hat{p})}{n}}$$

Confidence Interval for a Population Proportion

Can be written mathematically using either inequality symbols or interval notation, given by

$$\hat{p} - E < p < \hat{p} + E$$

or

$$(\hat{p} - E, \hat{p} + E)$$

Minimum Sample Size for Estimating a Population Proportion

Minimum sample size required for estimating a population proportion at a given level of confidence with a particular margin of error, given by

$$n = p(1-p)\left(\frac{z_{\alpha/2}}{E}\right)^2$$

Section 8.5: Estimating Population Variances

Confidence Interval for a Population Variance

Used when a simple random sample of size n is taken from a population whose distribution is approximately normal population, given by

$$\frac{(n-1)s^2}{\chi^2_{\alpha/2}} < \sigma^2 < \frac{(n-1)s^2}{\chi^2_{(1-\alpha/2)}} \text{ with } df = n-1$$

Confidence Interval for a Population Standard Deviation

Used when a simple random sample of size n is taken from a population whose distribution is approximately normal population, given by

$$\sqrt{\frac{(n-1)s^2}{\chi^2_{\alpha/2}}} < \sigma < \sqrt{\frac{(n-1)s^2}{\chi^2_{(1-\alpha/2)}}} \text{ with } df = n-1$$

Section 8.5: Estimating Population Variances (cont.)

Constructing a Confidence Interval for a Population Variance (or Standard Deviation)

1. Find the point estimate, s^2 (or s).
2. Based on the level of confidence given, calculate $\frac{\alpha}{2}$ and $1-\frac{\alpha}{2}$.
3. Use the χ^2-distribution table to find the critical values, $\chi^2_{\alpha/2}$ and $\chi^2_{(1-\alpha/2)}$, for a distribution with $n-1$ degrees of freedom.
4. Substitute the necessary values into the formula for the confidence interval.

Minimum Sample Size for Estimating a Population Variance (or Standard Deviation)

Minimum Sample Sizes for Estimating Variance

s^2 Is within This Percentage of the Value of σ^2	Minimum Sample Size Needed for 95% Level of Confidence	Minimum Sample Size Needed for 99% Level of Confidence
1%	77,209	133,450
5%	3150	5459
10%	807	1403
20%	212	370
30%	99	173
40%	58	102
50%	39	69

Minimum Sample Sizes for Estimating Standard Deviation

s Is within This Percentage of the Value of σ	Minimum Sample Size Needed for 95% Level of Confidence	Minimum Sample Size Needed for 99% Level of Confidence
1%	19,206	33,220
5%	769	1337
10%	193	337
20%	49	86
30%	22	39
40%	13	23
50%	9	15

E Chapter 8 Exercises

Note: For all exercises in this section, you may assume that all samples of a given size have an equal probability of being chosen. Unless otherwise indicated, assume that the population standard deviation is unknown.

Confidence Intervals and Margins of Error

Directions: Find each specified confidence interval or margin of error.

1. A survey of 21 accountants finds that they spend a mean of \$45.00 each week on lunch with a standard deviation of \$3.63. Using a margin of error of \$2.00, give an interval estimate for the mean amount of money accountants spend on lunch each week, and interpret the interval. Assume that the population is normally distributed.

2. In a random sample of 35 students at a local college, 19 regularly carry a laptop computer to class. Using this sample and a margin of error of 12.1%, construct and interpret an interval estimate for the proportion of all students at the college who carry a laptop computer to class.

3. Out of 300 randomly selected DVDs, a manufacturer found that only 2% were defective. Using a 95% level of confidence, calculate the margin of error for the proportion of all DVDs that are defective.

4. A random sample of 253 working mothers was surveyed regarding the cost of health care. The mean amount of money spent on health care by these working mothers each year was \$1300. Suppose that the population standard deviation is known to be \$289 per year. Calculate the margin of error for the population mean using a 98% level of confidence.

5. Fifteen randomly selected dentists were timed while brushing their own teeth for a recent study. The mean amount of time spent on brushing their teeth was 2.13 minutes with a standard deviation of 0.45 minutes. Calculate the margin of error for the mean amount of time all dentists spend brushing their teeth. Use a 99% level of confidence. Assume that the population is normally distributed.

6. Of 600 bags of rice randomly selected from a train after a wreck, 230 were damaged in the train wreck. The insurance company needs to estimate the proportion of bags of rice on the entire train that were damaged in the wreck. Construct and interpret a 99% confidence interval for the proportion of all bags of rice that were damaged.

7. From a random sample of 49 stray dogs, a veterinarian finds that their mean weight is 41.0 pounds with a standard deviation of 8.5 pounds. Construct and interpret a 98% confidence interval for the mean weight of all stray dogs in this area.

8. Although male polar bears weigh only about one pound at birth, the mean weight of a random sample of 10 adult male polar bears is 1200 pounds with a standard deviation of 100 pounds. Construct and interpret a 98% confidence interval for the mean weight of all adult male polar bears. Assume that the population is normally distributed.

9. The following sample of volumes of soda (in ounces) was taken from 14 cans randomly selected from an assembly line.

12.01	12.02	11.95	11.99	11.94	12.01	12.03
11.98	12.00	12.03	11.98	12.05	11.93	11.98

Build and interpret a 95% confidence interval for the population variance for the volumes of soda in all the soda cans that come off that particular assembly line.

10. The following is a random sample of the numbers of daily hits a website receives. Construct and interpret a 99% confidence interval for the mean number of hits the website gets if the population standard deviation is known to be 3822.1.

7881	10,101	16,157	17,377	16,690	16,363	13,282	6882	8207	12,098	13,971
17,968	16,615	14,310	8401	10,689	19,011	18,222	17,538	16,005	13,192	7344

Minimum Sample Sizes for Interval Estimates

Directions: Calculate the minimum sample size needed to construct a confidence interval with the desired characteristics.

11. A researcher wants to interview military personnel returning from war regarding their experiences overseas. The researcher wants to know the proportion of soldiers who would not choose to return to combat duty overseas if given the choice. If the researcher desires to construct a 90% confidence interval with a 5% margin of error, how many soldiers must the researcher interview? An estimate for the proportion of soldiers who would not return to combat duty overseas if given the choice is 76%.

12. Juan is interning for the summer at a marketing firm. One of his assignments is to conduct research to estimate the percentage of Americans who shop at after-Christmas sales. He needs his results to have a 99% level of confidence and a maximum error of 2%. If previous estimates have shown the percentage to be 20%, how many people must Juan survey?

13. A researcher is looking to estimate the mean LSAT score of law school applicants who completed their undergraduate degrees at small liberal arts colleges. She wants a 99% level of confidence and a maximum margin of error of 2 points. How many applicants must she survey if previous studies show the standard deviation to be approximately 9 points?

P Chapter 8 Project

Project A: Constructing a Confidence Interval for a Population Mean

Directions: Choose a study question and follow the given steps to construct a confidence interval for a population mean.

Choose one of the study questions below, or write your own study question for which you could collect data from students on your campus in order to estimate a population mean.

- How many hours per week do college students on your campus study?
- What is the mean number of parking tickets college students on your campus receive per semester?
- What is the mean price of rent paid per month by college students on your campus?

For the study question you chose, construct a confidence interval for that parameter by doing each of the following.

1. Collect data on your study question from at least 30 students on your campus.
2. Calculate the sample mean and sample standard deviation of your sample data.
3. Calculate the margin of error for a 95% confidence interval using your sample statistics.
4. Construct a 95% confidence interval using your sample statistics.
5. Write a summary of your results.

Project B: Constructing a Confidence Interval for a Population Proportion

Directions: Choose a study question and follow the given steps to construct a confidence interval for a population proportion.

Choose one of the study questions below, or write your own study question for which you could collect data from students on your campus in order to estimate a population proportion.

- What percentage of college students on your campus own a smartphone?
- What percentage of college students on your campus are seeking a degree in your field of study?
- What percentage of college students on your campus live in on-campus housing?

For the study question you chose, construct a confidence interval for that parameter by doing each of the following.

1. Collect data on your study question from at least 30 students on your campus.
2. Calculate the sample proportion from your sample data.
3. Calculate the margin of error for a 90% confidence interval using your sample statistics.
4. Construct a 90% confidence interval using your sample statistics.
5. Write a summary of your results.

T Chapter 8 Technology

Confidence Intervals

TI-83/84 Plus

Finding *t*-Values

Side Note

The most recent TI-84 Plus calculators (Jan. 2011 and later) contain a new feature called `STAT WIZARDS`. This feature is not used in the directions in this text. By default this feature is turned `ON`. `STAT WIZARDS` can be turned `OFF` under the second page of **MODE** options.

Some TI-84 Plus Silver Edition calculators can be used to find the value of t given the number of degrees of freedom and a specified area under the curve. The `invT` option is available on TI-84 Plus Silver Edition calculators with OS 2.30 or higher and TI-Nspire calculators.

Example T.1

Using a TI-84 Plus Calculator to Find the Value of *t* Given the Area to the Left

Find the value of t for a t-distribution with 17 degrees of freedom such that the area under the curve to the left of t is 0.10.

Solution

Press **2ND** and then **VARS** to go to the `DISTR` menu. Choose option `4:invT(`. The function syntax for this option is `invT(`*area to the left of t, df*`)`. Since the area to the left is the given area in this example, we would enter `invT(Ø.1Ø,17)`. Press **ENTER**. The answer given by the calculator is $t \approx -1.333$.

Finding Confidence Intervals

Calculating a confidence interval on a TI-83/84 Plus calculator is a straightforward process. Of the confidence intervals that we studied in this chapter, the following three types of confidence intervals can be created using the calculator.

1. A confidence interval for a population mean constructed using a critical value of z.
2. A confidence interval for a population mean constructed using a critical value of t.
3. A confidence interval for a population proportion.

Confidence Intervals for Population Means (σ Known)

When the population standard deviation, σ, is known, the endpoints of a confidence interval for a population mean are calculated using a critical value of z. You need to know σ, $\bar{x}$, n, and your level of confidence, c. Now, let's calculate a confidence interval using the `ZInterval` option on a TI-83/84 Plus calculator.

Example T.2

Using a TI-83/84 Plus Calculator to Find a Confidence Interval for a Population Mean (σ Known)

Use a TI-83/84 Plus calculator to find an 85% confidence interval for the population mean amount of money that teachers spend on lunch each week. A survey of 42 randomly selected teachers found that they spend a mean of $18.00 per week on lunch. Assume that the population standard deviation is known to be $2.00 per week.

Solution

To begin, press STAT; then scroll over and choose TESTS. From the TESTS menu, choose option 7:ZInterval. Choose Stats, and then press ENTER. Enter 2 for σ, 18 for x̄, a sample size of 42 for n, and 0.85 for the level of confidence, C-Level. Highlight Calculate, and then press ENTER. The screen will display the results shown in the screenshot in the margin.

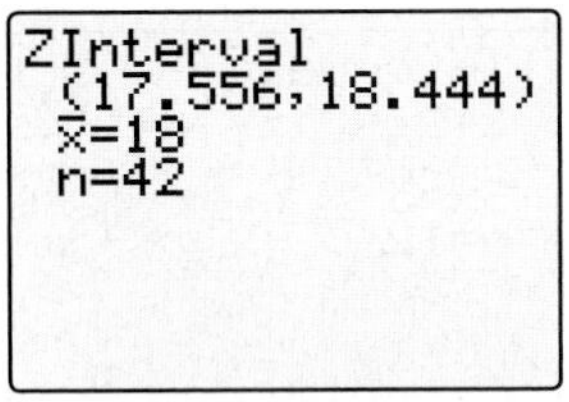

Thus we are 85% confident that the mean amount of money that teachers spend on lunch each week is between $17.56 and $18.44.

Confidence Intervals for Population Means (σ Unknown)

When the population standard deviation, σ, is unknown, you need to use a critical value of t when calculating the endpoints of a confidence interval for a population mean. The procedure on a TI-83/84 Plus calculator is basically the same as for calculating a confidence interval for a population mean with σ known, except you choose option 8:TInterval instead of option 7:ZInterval. The statistics needed are a little different as well. For example, if you were given a sample standard deviation of $2.33 instead of a population standard deviation of $2.00 for the previous example, the data would be entered as follows.

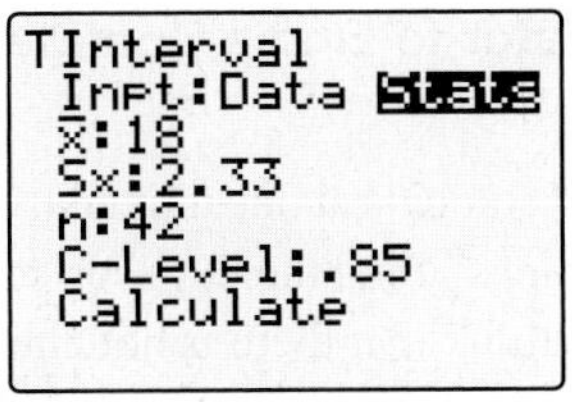

As you can see, you would need to enter the sample mean, x̄, the sample standard deviation, Sx, the sample size, n, and the level of confidence, C-Level. Highlight Calculate, and then press ENTER. The screen should display the following results.

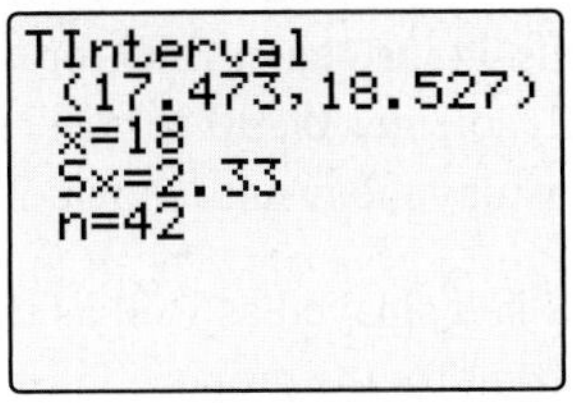

Example T.3

Using a TI-83/84 Plus Calculator to Find a Confidence Interval for a Population Mean (σ Unknown)

Use a TI-83/84 Plus calculator to find a 95% confidence interval for the population mean amount of money that teachers spend on lunch each week. A survey of 12 randomly selected teachers found that they spend a mean of $20.00 per week on lunch with a sample standard deviation of $3.00 per week.

Solution

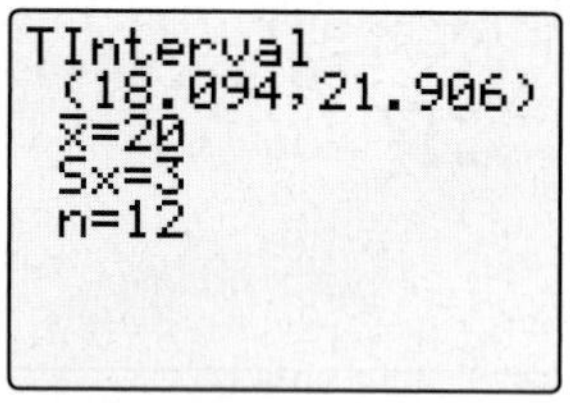

To begin, press STAT; then scroll over and choose `TESTS`. From the `TESTS` menu, choose option `8:TInterval`. Choose `Stats`, and then press ENTER. Enter `20` for `x̄`, `3` for `Sx`, a sample size of `12` for `n`, and `0.95` for the level of confidence, `C-Level`. Highlight `Calculate`, and then press ENTER. The screen will display the results shown in the screenshot in the margin.

Thus, we are 95% confident that the mean amount of money that teachers spend on lunch each week is between $18.09 and $21.91.

Confidence Intervals for Population Proportions

A TI-83/84 Plus calculator can calculate confidence intervals for population proportions as well. To begin, press STAT and then scroll over and choose `TESTS`. Choose option `A:1-PropZInt`. The only values that are required to calculate this confidence interval are the sample size, `n`, the number in the sample that have the characteristic in question, `x`, and the desired level of confidence, `C-Level`.

Example T.4

Using a TI-83/84 Plus Calculator to Find a Confidence Interval for a Population Proportion

A survey of 345 randomly selected students at one university found that 301 students think that there is not enough parking on campus. Find the 90% confidence interval for the proportion of all students at this university who think that there is not enough parking on campus.

Solution

Press STAT; then scroll over and choose `TESTS`. From the `TESTS` menu, choose option `A:1-PropZInt`. Of those surveyed, 301 students think that there is not enough parking, so enter `301` for `x`. Our sample size is 345, so enter `345` for `n`. Our level of confidence is 90%. This must be entered as a decimal, so enter `0.90` for `C-level`. Choose `Calculate` and press ENTER. The screen will display the results shown in the screenshot in the margin.

Notice that the confidence interval is displayed as well as the value of the sample proportion, $\hat{p}$. Thus, the 90% confidence interval for the proportion of all students at this university who think that there is not enough parking on campus is $(0.843, 0.902)$.

Microsoft Excel

Finding *t*-Values

In this chapter, we actually calculated the inverse *t*-distribution, which gives the value of *t* for a given area under the *t*-distribution. To calculate the inverse *t*, we can use Microsoft Excel. The formula for versions older than Excel 2010, which is still available in Excel 2010, is =TINV(*probability*, *deg_freedom*). This formula is actually for the area in two tails, so if we know the area in one tail, the probability we would enter would be the area times 2.

In Excel 2010, however, there now exists the ability to calculate the value of *t* given the area in one *or* two tails. For the area in the left tail of the distribution *only*, the formula is =T.INV(*probability*, *deg_freedom*). Since the *t*-distribution is symmetrical, to find the value of *t* for the area in the right tail, one easy solution is to simply change the sign of *t*. For finding the value of *t* for the area in both tails, there is now a new formula, =T.INV.2T(*probability*, *deg_freedom*), which is equivalent to =TINV(*probability*, *deg_freedom*) in earlier versions of Excel. The following examples demonstrate how to use each of these formulas.

Example T.5

Using Microsoft Excel to Find the Value of t_{α}

Find the value of $t_{0.05}$ for the *t*-distribution with 15 degrees of freedom.

Solution

We know the number of degrees of freedom is 15. The area, 0.05, is the area to the right of *t*, which is the area in one tail of the distribution. Using Microsoft Excel 2010, enter **=T.INV(0.05, 15)**. The value displayed will be approximately −1.753, since the formula assumes that we are entering the area in the left tail. By changing the resulting value of *t* to a positive number, we obtain $t_{0.05} \approx 1.753$.

Example T.6

Using Microsoft Excel to Find the Value of *t* Given Area between −*t* and *t*

Find the critical *t*-value for a *t*-distribution with 29 degrees of freedom such that the area between −*t* and *t* is 99%.

Solution

We know the number of degrees of freedom is 29. Since 99% of the area of the curve is in the middle, that leaves 1%, or 0.01 of the area in the two tails. Thus, we need to use the formula for area in two tails, with an area, or probability, of 0.01. Unlike other formulas we have used that required us to divide that area by 2, this formula has that calculation built in. Using Microsoft Excel 2010, enter **=T.INV.2T(0.01, 29)**. The value displayed will be approximately 2.756. This is the positive *t*-value. Thus, 99% of the area lies between $-t \approx -2.756$ and $t \approx 2.756$.

Finding Confidence Intervals

Microsoft Excel can calculate the margin of error of a confidence interval for a population mean directly. To calculate the margin of error using older versions of Excel, use the formula =CONFIDENCE(*alpha*, *standard_dev*, *size*), where *alpha* is equal to $1-c$. An issue with this older formula is that it uses the normal distribution to calculate the margin of error. The assumption is that *standard_dev* is the standard deviation of the population, which is assumed to be known. The value of *size* is the sample size. This is fine if the population standard deviation is known, but what if it isn't? In Excel 2010, this older formula is split into two new formulas to accommodate using either the normal distribution or the *t*-distribution. For the margin of error calculated using a normal distribution, use =CONFIDENCE.NORM(*alpha*, *standard_dev*, *size*), where *standard_dev* is the population standard deviation. To calculate the margin of error using a *t*-distribution, use =CONFIDENCE.T(*alpha*, *standard_dev*, *size*), where *standard_dev* is the sample standard deviation.

Important warning on using Microsoft Excel to calculate confidence intervals and margins of error: The input template for these functions says that they produce the confidence interval, but they only produce the value of the margin of error. Since the sample mean is not entered into the formula, there is no way for these formulas to produce the endpoints of the interval. Thus unless you are very familiar with the limitations of these functions in Microsoft Excel, it is best to avoid using them.

Example T.7

Using Microsoft Excel to Find the Margin of Error for a Confidence Interval for a Population Mean (σ Known)

Calculate the margin of error for a 95% confidence interval for a population mean given that $\sigma = 8.5$ and $n = 47$.

Solution

Since the value for the population standard deviation is given, we want to use a normal distribution to calculate the margin of error. Remember that $\alpha = 1 - c$, so $\alpha = 1 - 0.95 = 0.05$. In Excel 2010, enter **=CONFIDENCE.NORM(0.05, 8.5, 47)**. Thus, the margin of error is approximately 2.430066.

Example T.8

Using Microsoft Excel to Find the Margin of Error for a Confidence Interval for a Population Mean (σ Unknown)

Calculate the margin of error for a 95% confidence interval for a population mean given that $s = 8.5$ and $n = 47$. Use a *t*-distribution.

Solution

Since the value for the population standard deviation is unknown, it is appropriate to use a *t*-distribution. We are told to use this distribution as well. Remember that $\alpha = 1 - c$, so $\alpha = 1 - 0.95 = 0.05$. In Excel 2010, enter **=CONFIDENCE.T(0.05, 8.5, 47)**. Thus, the margin of error is approximately 2.495693.

Example T.9

Using Microsoft Excel to Construct a Confidence Interval for a Population Proportion

During the 2010–2011 NBA season, Kobe Bryant attempted 583 free throws and made 483 of these. Construct a 99% confidence interval for his true free throw percentage.

Source: FOX Sports on MSN. "NBA Sortable Stats." 2012. http://msn.foxsports.com/nba/sortableStats?league=NBA&dir=descending&stat=&low=1&high=50&table=free-throws&position=all&showPlayers=min&year=2010&seasonState=regular&Go=Go (22 Jan. 2012).

Solution

The first step is calculating a point estimate for the interval. Enter the number of successes, **483**, into the cell B1 and the total attempts, **583**, into B2. We divide the number of successes by the total attempts. Enter **=B1/B2** to calculate $\hat{p}$. The margin of error is calculated by entering **=NORM.S.INV(0.995)*SQRT(B3*(1-B3)/B2)**. Finally, subtracting the margin of error from $\hat{p}$ yields the lower endpoint of the interval and adding the margin of error to $\hat{p}$ yields the upper endpoint.

	A	B
1	*x*	483
2	*n*	583
3	*p-hat*	0.828473
4	*E*	0.040215
5	lower endpoint	0.788258
6	upper endpoint	0.868688

Thus, based on his performance during the 2010–2011 NBA season, we are 99% confident that Kobe Bryant's true free throw percentage is between 78.8% and 86.9%.

MINITAB

Finding t-Values

To find the t-value for a given area in MINITAB, enter the area to the left of the desired t-value in the first row of column C1. Once the value of the area is entered, go to **Calc ► Probability Distributions ► t**. When the t Distribution menu appears, select **Inverse cumulative probability** and enter the number of degrees of freedom. Select **C1** as the input column. Once you are finished, click **OK**, and the t-value will appear in the Session window. If the area we are given is to the right or the area in between, we must first determine the area to the left, just as we did when using a TI-84 Plus calculator.

Finding Confidence Intervals

MINITAB can produce a confidence interval for a population proportion or mean in a single step. The following examples demonstrate the procedures for calculating confidence intervals for population means and population proportions.

Example T.10

Using Minitab to Find a Confidence Interval for a Population Mean (σ Unknown)

Given the following sample data from a study on the average amount of water used per day by members of a household while brushing their teeth, calculate the 99% confidence interval for the population mean using Minitab. Assume that the sample used in the study was a simple random sample.

Household Water Used for Brushing Teeth (in Gallons per Day)				
0.485	0.428	0.390	0.308	0.231
0.587	0.516	0.465	0.370	0.282
0.412	0.367	0.336	0.269	0.198
0.942	0.943	0.940	0.941	0.946
0.868	0.898	0.889	0.910	0.927
0.925	0.950	0.959	0.948	0.956
0.805	0.810	0.839	0.860	0.861
0.515	0.463	0.420	0.326	0.243

Solution

Since we are not told any population parameters for the study, we cannot assume that σ is known or that the population distribution is approximately normal. However, since the sample is large enough, we can use the t-distribution to construct a confidence interval for the population mean. Begin by typing the sample data into the first column, C1. Next, to produce the dialog window for finding a confidence interval for a population mean using a t-distribution, go to **Stat ▶ Basic Statistics ▶ 1-Sample t**. Next, select **Samples in columns** and enter **C1** in the box. Then select **Options** to enter **99** for the confidence level. Click **OK** on the options window and **OK** on the main dialog window and your results will be displayed in the Session window.

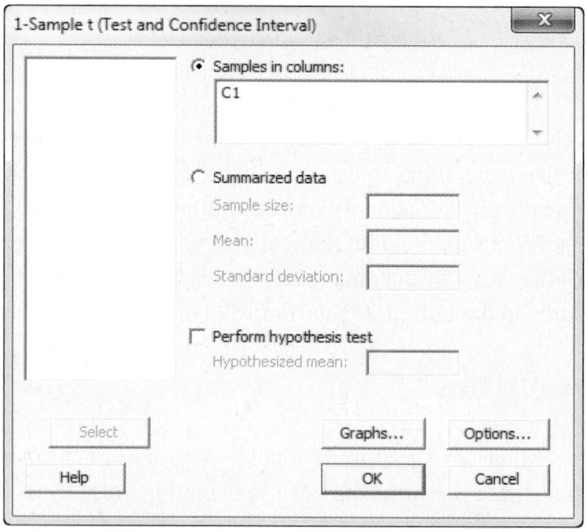

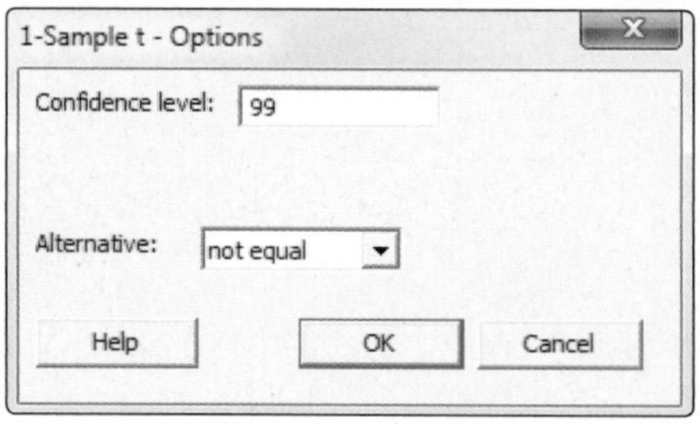

The results in the Session window should appear as follows.

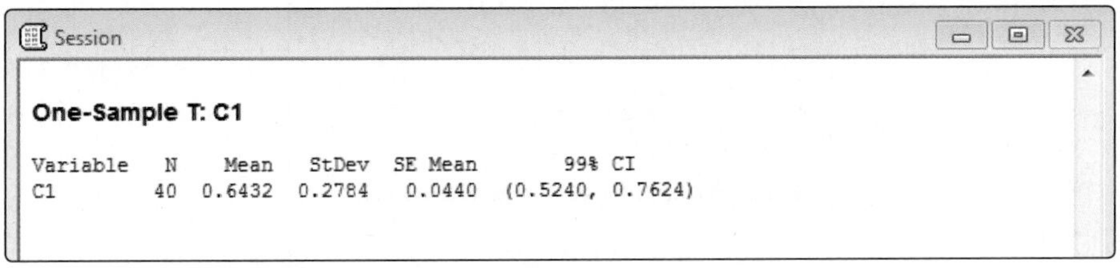

Session
One-Sample T: C1
Variable N Mean StDev SE Mean 99% CI
C1 40 0.6432 0.2784 0.0440 (0.5240, 0.7624)

The third column of the results displays the point estimate, $\bar{x}$, and the last column contains the confidence interval. The 99% confidence interval is $(0.5240, 0.7624)$. Therefore, we are 99% confident that the mean amount of water used per household for brushing teeth is between 0.5240 and 0.7624 gallons per day.

Example T.11

Using Minitab to Find a Confidence Interval for a Population Proportion

A certain dosage of insecticide is applied to a sample of 200 cockroaches in a laboratory. The scientists have predicted that this dose should kill more than 50% of the cockroaches. When the results are tallied, 83 cockroaches are dead. Construct a 95% confidence interval for the percentage of cockroaches killed by the insecticide and use the interval to determine whether the evidence supports the scientists' claim.

Solution

To produce the dialog window for a confidence interval for a population proportion, go to **Stat ► Basic Statistics ► 1 Proportion**. Next, select **Summarized data** and enter **83** for the number of events and **200** for the number of trials. Then select **Options** to enter **95** for the confidence level and select the option, **Use test and interval based on normal distribution**. Click **OK** on the options window and **OK** on the main dialog window and your results will be displayed in the Session window.

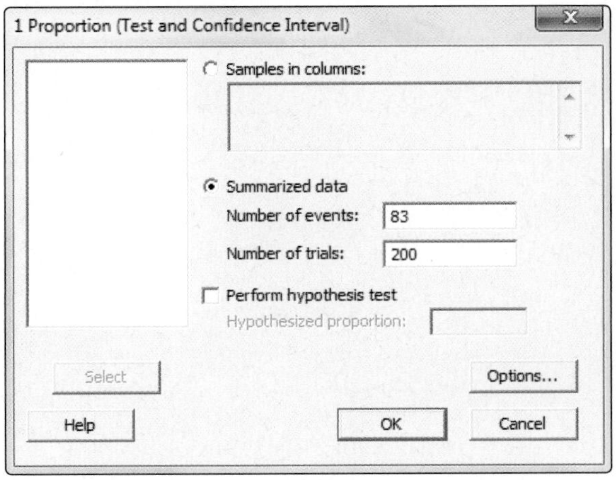

1 Proportion - Options

Confidence level: | 95

Alternative: | not equal ▼

☑ Use test and interval based on normal distribution

Help OK Cancel

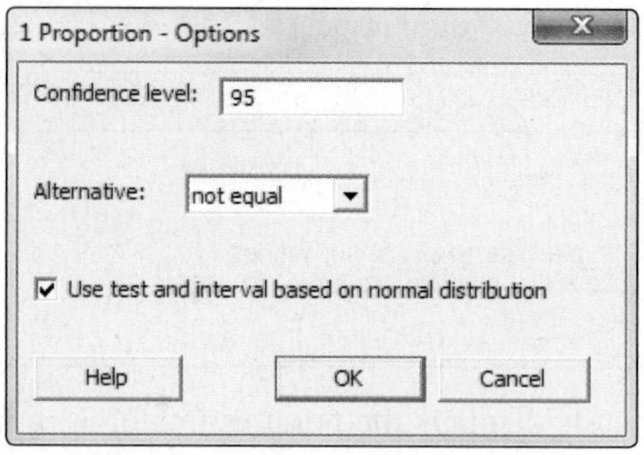

The results in the Session window should appear as follows.

Session

Test and CI for One Proportion

```
Sample   X    N   Sample p          95% CI
1       83  200  0.415000   (0.346713, 0.483287)

Using the normal approximation.
```

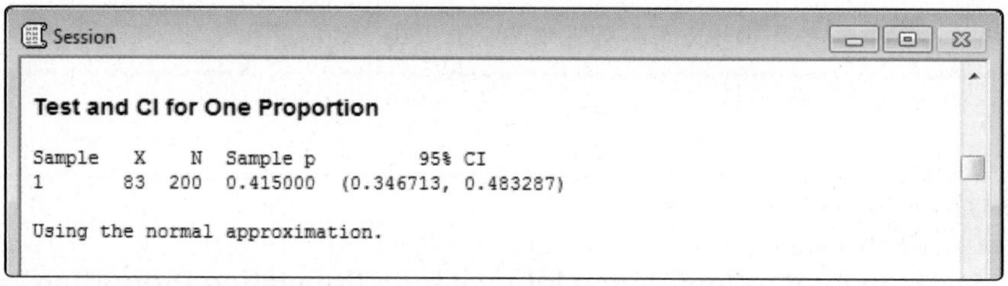

The fourth column displays the point estimate, \hat{p}, and the fifth contains the confidence interval. The 95% confidence interval, $(0.346713, 0.483287)$, does not contain any values greater than 0.50. Therefore, the scientists' claim that the tested dosage of insecticide should kill more than 50% of the cockroaches is not supported at a confidence level of 95%.

Chapter Nine

Confidence Intervals for Two Samples

Sections

9.1 Comparing Two Population Means (σ Known)

9.2 Comparing Two Population Means (σ Unknown)

9.3 Comparing Two Population Means (σ Unknown, Dependent Samples)

9.4 Comparing Two Population Proportions

9.5 Comparing Two Population Variances

Objectives

1. Construct a confidence interval for the difference between two population means using independent samples when σ is known for both populations.
2. Construct a confidence interval for the difference between two population means using independent samples when σ is unknown for both populations.
3. Determine when sample variances can be pooled to estimate the common population variance when constructing a confidence interval for the difference between two population means.
4. Construct a confidence interval for the mean of the paired differences for two populations of dependent data.
5. Construct a confidence interval for the difference between two population proportions.
6. Construct a confidence interval for the ratio of two population variances.

9

Introduction

It seems these days that everyone is trying to do things better, faster—you name it. Each company wants to outdo the next in order to win not only the market, but bragging rights to being the best. Let's face it: our world is competitive, to say the least.

Take, for instance, the competition between cellular phone companies. Each company claims that its phone service is superior to the others'. That is, each company wants to have the most reliable service covering the broadest territory. Is it possible to know exactly what percentage of the time *all* callers are able to connect for *each* phone company? Certainly not—especially when you consider the enormous volume of calls that are attempted even in a single hour.

Even though parameters such as these cannot be pinpointed, it is still possible to *compare* the percentage of calls completed for each company. Hey, each company doesn't really care what percentage of calls are going through anyway, only if its percentage is higher than the competitors', right? This is just one example of many cases in which knowing how to calculate a confidence interval comparing two population parameters comes in handy.

9.1 Comparing Two Population Means (σ Known)

In the previous chapter, we studied how to construct a confidence interval to estimate a single population parameter. However, it is often desirable or even necessary to compare two populations. In this section, we'll look at comparing two population means in order to identify if they are significantly different. One method of comparison is to look at the difference between the means of the two populations. Suppose you thought that there was no difference between two population means, that is, they were the same. Then subtracting the two means should give you zero. However, if you thought one was larger than the other, when you subtract the smaller from the larger, you would expect to get a positive number. This is the method that we will use to compare two population means; we will construct a confidence interval for the difference between the two means. First let's note what's different when we study two populations rather than just one.

If we are considering two population parameters, sample data must be drawn from each population. There are then two sets of sample statistics instead of the single set used in the previous formulas for confidence intervals. Hence, the formulas from the previous chapter must be modified to include a second set of sample statistics, namely, that from the second population.

One important distinction must be made regarding how the two populations are related. We must consider how the data from the two populations are connected, if at all. Two samples are said to be **independent** if the data from the first sample are not connected to the data from the second sample. For example, if we compare the mean ACT score of high school seniors entering college to the mean ACT score of adults returning to college, then the data sets would be independent. If the two data sets are systematically connected, then the samples are said to be **dependent**. Dependent data sets are also referred to as *matched data* or *paired data*. Examples of a pair of dependent data sets are the sets of grades on a pretest and posttest administered to the same group of students.

Definition

Two samples are **independent** if the data from the first sample are not connected to the data from the second sample.

Two samples are **dependent** if the data from both samples are systematically connected, or *paired.*

Whether or not the two data sets under consideration are independent or dependent is one factor in determining the appropriate formulas to use for calculating a confidence interval. Not only must we know whether the data sets are independent or not, but once again we must also consider whether the standard deviations of the populations are known values before choosing the appropriate method to use. As we stated in Chapter 8, although σ is rarely known for any population, we will start with this scenario because it requires the use of the standard normal distribution for calculating the margin of error. In this section, we will consider the case in which we have two populations whose standard deviations are known and from which independent samples have been chosen. That is, the following criteria must be met. Notice that these conditions are similar to those discussed in Section 8.1 for estimating a single population mean when σ is known.

- All possible samples of a given size have an equal probability of being chosen; that is, simple random samples are used.
- The samples are independent.
- Both population standard deviations, σ_1 and σ_2, are *known*.
- Either both sample sizes are at least 30 ($n_1 \geq 30$ and $n_2 \geq 30$) *or* both population distributions are approximately normal.

Recall from the last chapter the general process for creating a confidence interval for a population mean, which contains three basic steps.

1. Find the point estimate.

2. Find the margin of error.

3. Subtract the margin of error from and add the margin of error to the point estimate.

This process often can be streamlined into one step with the aid of technology, such as a TI-83/84 Plus calculator, as shown in the previous chapter's examples. However, for the sake of clarity, we will start by demonstrating each step in the process.

The process for constructing a confidence interval with two samples is the same as the one given above for one sample. Let's determine the point estimate to use when comparing two means first. Since we are actually considering two populations, we need to consider a different point estimate than simply $\bar{x}$. Remember, we are not constructing a separate confidence interval for each population mean; we are trying to *compare* the two means. As a result, we use the difference between the sample means, namely $\bar{x}_1 - \bar{x}_2$, as the point estimate for the difference between the population means. Looking at the value of the difference will tell us if the means are the same, or if one is estimated to be greater than the other. We'll discuss this more within each of the examples.

Memory Booster

The best point estimate for the difference between two population means is $\bar{x}_1 - \bar{x}_2$.

As you might imagine, the formula for the margin of error will be different as well. Just as in Chapter 8, when σ is known, the normal distribution is used in the formula. When we have two independent samples and σ is known for both populations, the formula for the margin of error of a confidence interval for the difference between two population means is as follows.

Rounding Rule

When calculating a margin of error for a confidence interval, round to at least six decimal places to avoid additional rounding errors in the subsequent calculations of the endpoints of the confidence interval.

Formula

Margin of Error of a Confidence Interval for the Difference between Two Population Means (σ Known)

When both population standard deviations are known, the samples taken are independent, simple random samples, and either both sample sizes are at least 30 or both population distributions are approximately normal, the margin of error of a confidence interval for the difference between two population means is given by

$$E = z_{\alpha/2}\sqrt{\frac{\sigma_1^2}{n_1}+\frac{\sigma_2^2}{n_2}}$$

where $z_{\alpha/2}$ is the critical value for the level of confidence, $c = 1 - \alpha$, such that the area under the standard normal distribution to the right of $z_{\alpha/2}$ is equal to $\frac{\alpha}{2}$,

σ_1 and σ_2 are the two population standard deviations, and

n_1 and n_2 are the two sample sizes.

The critical z-value, $z_{\alpha/2}$, that is used depends on the level of confidence, c, that is desired. The table of critical values is given below. It is the same table used in the previous chapter.

Table 9.1: Critical z-Values for Confidence Intervals

Level of Confidence, c	$\alpha = 1 - c$	$z_{\alpha/2}$
0.80	0.20	1.28
0.85	0.15	1.44
0.90	0.10	1.645
0.95	0.05	1.96
0.98	0.02	2.33
0.99	0.01	2.575

Let's stop here for a moment and practice calculating the point estimate and margin of error before moving on to constructing the entire confidence interval.

Example 9.1

Finding the Point Estimate and Margin of Error of a Confidence Interval for the Difference between Two Population Means (σ Known)

Use the given population standard deviations and statistics from two independent, simple random samples to calculate both the point estimate and margin of error that would be used to estimate the true difference between the population means with an 85% level of confidence.

$$\sigma_1 = 2.1 \qquad \sigma_2 = 1.8$$
$$n_1 = 48 \qquad n_2 = 30$$
$$\bar{x}_1 = 7.01 \qquad \bar{x}_2 = 6.93$$

Solution

The point estimate for the difference between the population means is found by simply subtracting one sample mean from the other. Therefore, the point estimate for $\mu_1 - \mu_2$ is calculated as follows.

$$\bar{x}_1 - \bar{x}_2 = 7.01 - 6.93$$
$$= 0.08$$

Since the population standard deviations, σ_1 and σ_2, are known; both samples are independent, simple random samples; and both sample sizes are at least 30, we can use the formula given on the previous page. To find the margin of error for an 85% level of confidence, we will use the critical value $z_{\alpha/2} = z_{0.15/2} = z_{0.075} = 1.44$. Substituting all known values into the formula for the margin of error, we have the following.

$$E = z_{\alpha/2}\sqrt{\frac{\sigma_1^2}{n_1} + \frac{\sigma_2^2}{n_2}}$$
$$= 1.44\sqrt{\frac{(2.1)^2}{48} + \frac{(1.8)^2}{30}}$$
$$\approx 0.643786$$

So the margin of error for an 85% confidence interval for the difference between the two population means is approximately 0.643786.

Finally, the last step in the process of constructing a confidence interval for the difference between two population means is simply subtracting the margin of error from and adding the margin of error to the point estimate to find the endpoints of the interval.

Formula

Confidence Interval for the Difference between Two Population Means (Independent Samples)

The confidence interval for the difference between two population means for independent data sets is given by

$$(\bar{x}_1 - \bar{x}_2) - E < \mu_1 - \mu_2 < (\bar{x}_1 - \bar{x}_2) + E$$

or

$$((\bar{x}_1 - \bar{x}_2) - E, (\bar{x}_1 - \bar{x}_2) + E)$$

where $\bar{x}_1$ and $\bar{x}_2$ are the two sample means,

$(\bar{x}_1 - \bar{x}_2)$ is the point estimate for the difference between the population means, $\mu_1 - \mu_2$, and

E is the margin of error.

Rounding Rule

Round the endpoints of a confidence interval for the difference between two population means as follows:

- If sample data are given, round to one more decimal place than the largest number of decimal places in the given data.
- If statistics are given, round to the same number of decimal places as given in the standard deviation or variance.

Interpreting the Confidence Interval

Remember that we are comparing the means of two populations. Our confidence interval is an interval estimate of the difference between the two means. As we stated earlier, if there is no difference between the two, then subtracting one from the other will give you 0. Therefore, if the confidence interval we calculate contains 0, we have to conclude that there is a possibility that the two population means are the same. If, however, the interval estimate lies completely on the positive side of the number line, this implies that the first mean is larger than the second. The reverse is true if the interval is completely on the negative side of the number line; that is, the first mean is smaller than the second. If the interval is either all positive or all negative, as just described, it is unlikely that the two population means are equal.

Let's pull it all together now and build a confidence interval for the difference between two population means using independent samples when the population standard deviations are known.

Example 9.2

Constructing a Confidence Interval for the Difference between Two Population Means (σ Known)

A researcher is looking at the study habits of college students. A random sample of 42 freshmen reported a mean study time of 15.0 hours per week. A random sample of 39 seniors reported a mean study time of 23.0 hours per week. Construct a 95% confidence interval to estimate the true difference between the amounts of time per week that freshmen and seniors spend studying. Assume that study times for freshmen have a population standard deviation of 4.7 hours per week, while the study times for seniors have a population standard deviation of 5.2 hours per week.

Solution

Step 1: Find the point estimate.

Let Population 1 be freshmen and Population 2 be seniors. Begin by calculating the point estimate for the difference between the population means.

$$\begin{aligned}\bar{x}_1 - \bar{x}_2 &= 15.0 - 23.0 \\ &= -8.0 \text{ hours per week}\end{aligned}$$

Step 2: Find the margin of error.

Because the samples are from two different classes of students who are not related in any way, we can say that the samples are independent. In addition, we are given both population standard deviations and the sample sizes are both at least 30. So, we will calculate the margin of error using the formula for independent samples, σ known. We need a 95% level of confidence, so $c = 0.95$. In the table of critical z-values, we see that $z_{\alpha/2} = z_{0.05/2} = z_{0.025} = 1.96$. Thus, the margin of error is calculated as follows.

$$\begin{aligned}E &= z_{\alpha/2}\sqrt{\frac{\sigma_1^2}{n_1} + \frac{\sigma_2^2}{n_2}} \\ &= 1.96\sqrt{\frac{(4.7)^2}{42} + \frac{(5.2)^2}{39}} \\ &\approx 2.164257\end{aligned}$$

Step 3: **Subtract the margin of error from and add the margin of error to the point estimate.**

Subtracting the margin of error from the point estimate gives a lower endpoint for the confidence interval, as shown below.

$$\text{Lower endpoint: } (\bar{x}_1 - \bar{x}_2) - E = -8.0 - 2.164257$$
$$\approx -10.2 \text{ hours per week}$$

By adding the margin of error to the point estimate, we obtain the upper endpoint of the confidence interval as follows.

$$\text{Upper endpoint: } (\bar{x}_1 - \bar{x}_2) + E = -8.0 + 2.164257$$
$$\approx -5.8 \text{ hours per week}$$

Thus, the 95% confidence interval for the difference between the two means ranges from −10.2 to −5.8 hours per week. The confidence interval can be written mathematically using either inequality symbols or interval notation, as shown below.

$$-10.2 < \mu_1 - \mu_2 < -5.8$$

or

$$(-10.2,\ -5.8)$$

The negative signs before both endpoints indicate that the difference between the means is negative, and that the mean of Population 1 is less than the mean of Population 2. Therefore, we are 95% confident that freshmen study between 5.8 and 10.2 hours per week *less* than seniors.

Note that in the previous example, we arbitrarily assigned freshmen as Population 1 and seniors as Population 2. If we had reversed the groups, the confidence interval would have positive endpoints, indicating that students in Population 1 (seniors) study more than students in Population 2 (freshmen). The result is the same, just worded differently. Therefore, it does not matter how the population numbers, 1 and 2, are assigned to the two groups. What matters is that you are consistent with the naming throughout the solution.

As we've seen throughout the text, using technology can streamline steps in the processes we've studied. The same is true in this section. Our next example illustrates how to use a TI-83/84 Plus calculator to find a confidence interval for the difference between two population means when σ is known.

Example 9.3

Using a TI-83/84 Plus Calculator to Find a Confidence Interval for the Difference between Two Population Means (σ Known)

An automotive repair shop advertises that using a fuel additive will decrease the wear and tear on an engine. An independent researcher collected data from 50 randomly selected cars that used the fuel additive and found that the mean cost of engine repairs was \$3250. He then collected data from 55 randomly selected cars that did not use the fuel additive and found that the mean cost of engine repairs was \$3445. The population standard deviation of engine-repair costs for cars that do not use the fuel additive is known to be \$812. Assume that the population standard deviation of engine-repair costs for cars that do use the additive is \$748. Construct an 85% confidence interval to estimate the true difference between the mean costs of engine repairs for cars that use the fuel additive and those that do not.

Solution

Once again, the first step is to assign each population a number. Let Population 1 consist of the cars that use the fuel additive and Population 2 consist of the cars that do not. Since we wish to construct a confidence interval for the difference between two population means using two independent samples where σ is known for both populations and both sample sizes are at least 30, we will use z to calculate the endpoints of the interval.

Using a TI-83/84 Plus calculator to find the interval is very similar to techniques discussed in Chapter 8. Press STAT; then scroll to TESTS and choose option 9:2-SampZInt. Since we are working with statistics and not the original data, select the Stats option and then enter the given values of σ1, σ2, x̄1, n1, x̄2, and n2. Notice that the calculator asks for the two population standard deviations first, and then the sample mean and size for the sample from Population 1, followed by the information for the sample from Population 2. Since we want the 85% confidence interval, enter 0.85 for C-Level. Be careful to be consistent with the population numbering.

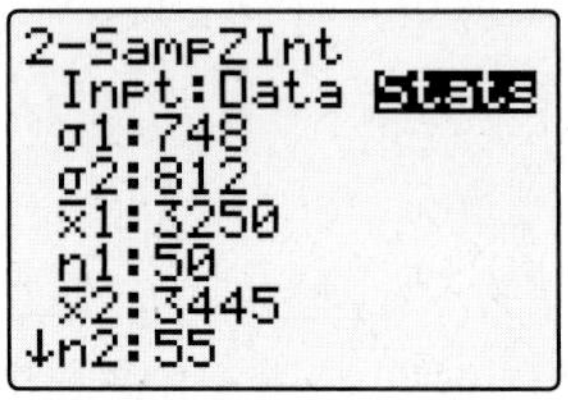

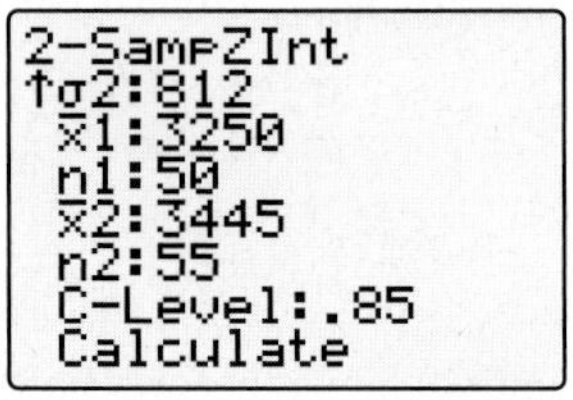

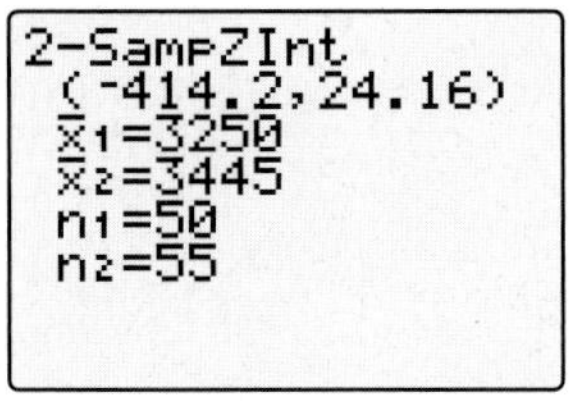

Once you have entered the values, highlight Calculate and press ENTER. The confidence interval is displayed along with the sample means and sizes, as shown in the screenshot in the margin.

Thus, the 85% confidence interval for the difference between the two means ranges from –\$414 to \$24. The confidence interval can be written mathematically using either inequality symbols or interval notation, as shown below.

$$-414 < \mu_1 - \mu_2 < 24$$

or

$$(-414, 24)$$

Because the interval ranges from a negative number to a positive one, the interval contains the number 0. In other words, the data do not provide evidence that the two population means are unequal at this level of confidence. Therefore, with 85% confidence, we can say that there is not sufficient evidence to support the claim that using the fuel additive results in a decrease in repair costs.

9.1 Section Exercises

Note: For all exercises in this section, you may assume that the requirements mentioned in this section are met; namely, all samples are independent, simple random samples, both population standard deviations are known, and either both sample sizes are at least 30 or both population distributions are approximately normal.

Point Estimates for Differences between Two Population Means

Directions: Find the point estimate for the true difference between the population means.

1. $\bar{x}_1 = 14$ and $\bar{x}_2 = 11$

2. Sample 1's mean is 45 and Sample 2's mean is 56.

3. One-mile running times for a sample of 65 participants had a mean of 8.45 minutes. The mean time for a sample of 77 participants is 8.25 minutes.

4. The mean temperature for Group A is 72.4 °F and the mean temperature for Group B is 77.3 °F.

5. The following data represent body temperatures in degrees Celsius for random samples of men and women.

Body Temperatures of Men

Stem	Leaves
34	6
35	2 4 4 5 5 6 7 7 7 7
36	1 2 3 4 4 5 5 5 6 7 8 8 9 9
37	0 0 0 0 1 1 2 2 2 2 5 6 6 9
38	0 0 1 1 2

Key: 34 | 6 = 34.6 °C

Body Temperatures of Women

Stem	Leaves
33	3 3 4 4 5 5 6 7 8 8 9 9
34	0 0 1 1 2 2 2 3 4 5 5 7 9 9
35	2 3 3 3 6 6 8 8 8
36	5
37	2 2 4 4 5 8

Key: 33 | 3 = 33.3 °C

6. Weights (in Grams) of Candy Bar A: 93, 92, 93, 94, 92, 93, 95, 94, 93, 93, 92

Weights (in Grams) of Candy Bar B: 93, 94, 95, 92, 93, 91, 96, 92, 93, 93, 94, 95, 92

Margins of Error of Confidence Intervals for Differences between Two Population Means (σ Known)

Directions: Calculate the margin of error of a confidence interval for the difference between the two population means using the given information.

7. $\sigma_1 = 2.88, \quad n_1 = 73, \quad \sigma_2 = 3.01, \quad n_2 = 99, \quad c = 0.99$

8. $\sigma_1 = 5.3, \quad n_1 = 31, \quad \sigma_2 = 4.9, \quad n_2 = 40, \quad \alpha = 0.02$

9. $\sigma_1 = 0.36, \quad n_1 = 88, \quad \sigma_2 = 1.09, \quad n_2 = 83, \quad c = 0.95$

10. $\sigma_1 = 11.54, \quad n_1 = 115, \quad \sigma_2 = 10.65, \quad n_2 = 122, \quad \alpha = 0.01$

Confidence Intervals for Differences between Two Population Means (σ Known)

Directions: Construct and interpret each specified confidence interval.

11. A local pizza shop claims to have a shorter delivery time than its competitor, a national chain. A local reporter collects data on random samples of deliveries from the local shop and from the competitor. Based on 48 deliveries, the local store has a mean delivery time of 29 minutes. Assume that the population standard deviation of the local shop's delivery times is 8 minutes. Based on 52 deliveries, the national chain store has an average delivery time of 31 minutes. Assume that the population standard deviation of the chain store's delivery times is 9 minutes. Construct and interpret a 90% confidence interval for the true difference between the mean delivery times for the two restaurants.

12. Dogsled drivers, known as mushers, use several different breeds of dogs to pull their sleds. One proponent of Siberian Huskies believes that sleds pulled by Siberian Huskies are faster than sleds pulled by other breeds. He times 35 teams of Siberian Huskies on a particular short course, and they have a mean time of 5.3 minutes. The mean time on the same course for 32 teams of other breeds of sled dogs is 6.1 minutes. Assume that the times on this course have a population standard deviation of 1.1 minutes for teams of Siberian Huskies and 1.9 minutes for teams of other breeds of sled dogs. Construct and interpret a 99% confidence interval for the true difference between the mean times on this course for teams of Siberian Huskies and teams of other breeds of sled dogs.

13. There are two elementary schools in Caleb's community. Third graders in both schools recently took the same standardized exam. A random sample of 45 third graders from the first school has a mean exam score of 75. A random sample of 39 third graders from the second school has a mean score of 78. Assume that the population standard deviations of the third graders' scores are known to be 10 for the first school and 5 for the second school. Construct and interpret a 95% confidence interval for the true difference between the mean exam scores for the two schools.

14. A contributor for the local newspaper is writing an article for the weekly fitness section. To prepare for the story, she conducts a study to compare the exercise habits of people who exercise in the morning to the exercise habits of people who work out in the afternoon or evening. She selects three different health centers from which to draw her samples. The 49 people she sampled who work out in the morning have a mean of 4.1 hours of exercise each week. The 54 people surveyed who exercise in the afternoon or evening have a mean of 3.7 hours of exercise each week. Construct and interpret a 95% confidence interval for the true difference between the mean amounts of time spent exercising each week by people who work out in the morning and those who work out in the afternoon or evening at the three health centers. Assume that the weekly exercise times have a population standard deviation of 0.7 hours for people who exercise in the morning and 0.5 hours for people who exercise in the afternoon or evening.

15. Russell is doing some research before buying his first house. He is looking at two different areas of the city, and he wants to know if there is a significant difference between the mean prices of homes in the two areas. For the 34 homes he samples in the first area, the mean home price is \$168,500. Public records indicate that home prices in the first area have a population standard deviation of \$22,950. For the 39 homes he samples in the second area, the mean home price is \$171,800. Again, public records show that home prices in the second area have a population standard deviation of \$30,995. Construct and interpret a 90% confidence interval for the true difference between the mean home prices in the two areas.

16. A sports news station wanted to know whether people who live in the North or the South are bigger sports fans. For its study, 121 randomly selected Southerners were surveyed and found to watch a mean of 4.1 hours of sports per week. In the North, 137 randomly selected people were surveyed and found to watch a mean of 3.9 hours of sports per week. Find and interpret a 99% confidence interval for the true difference between the mean numbers of hours of sports watched per week for the two regions if the South has a population standard deviation of 1.8 hours per week and the North has a population standard deviation of 1.9 hours per week.

9.2 Comparing Two Population Means (σ Unknown)

In Chapter 8, we discussed the fact that rarely is a population standard deviation known. Hence, we will now turn to the Student's t-distribution to help us compare two population means when σ is unknown. Now that we have the basic skills for building a confidence interval for the difference between two population means when σ is known, we'll use similar methods to evaluate means for populations where the standard deviation is unknown, but whose samples are still independent.

Our restrictions for this section, listed below, are much like those previously discussed for estimating a single population mean when σ is unknown.

- All possible samples of a given size have an equal probability of being chosen; that is, simple random samples are used.
- The samples are independent.
- Both population standard deviations, σ_1 and σ_2, are *unknown*.
- Either both sample sizes are at least 30 $(n_1 \geq 30$ and $n_2 \geq 30)$ *or* both population distributions are approximately normal.

Once again, we'll follow the same three steps for constructing a confidence interval involving population means.

1. Find the point estimate.
2. Find the margin of error.
3. Subtract the margin of error from and add the margin of error to the point estimate.

It still remains true that the best point estimate for the difference between two population means is the difference between the sample means, $\bar{x}_1 - \bar{x}_2$. So there is nothing new in this first step. However, now that σ is not known for either population, we need to consider what effect that might have on the formula for the margin of error. Rather than simply shooting in the dark for estimates of the two population variances (or standard deviations) since they are unknown, it is prudent to consider whether or not they are likely to be equal. Although neither population variance is actually known, in some cases it is reasonable to assume that they are the same. For instance, a golf ball manufacturer claims that its ball travels further off the tee than other brands. If a single player tests the new balls against other brands with the same club, it is reasonable to assume that the variances of the driving distances with the different balls will be equal, since the variances depend primarily on how consistently the player hits the ball with the same club, not on the balls themselves. If, instead, different players were to drive the different balls, it would no longer be reasonable to assume a common variance.

When the two population variances are assumed to be equal, the sample variances can be **pooled**, or *combined*, to give as good an estimate as possible for the common population variance. If the variances are not assumed to be equal, then the sample variances *cannot be pooled.* It is this second scenario that we will consider first in this section.

Unequal Variances

When we have two independent samples and the population variances are not assumed to be equal, the sample variances are not pooled. In this case, the formula for the margin of error of a confidence interval for the difference between two population means looks very similar to the one from the previous section that we use when σ is known for both populations. The only difference is that this formula uses a critical value of t instead of z to account for the error that arises from using the sample variances to estimate the unknown population variances.

Formula

Margin of Error of a Confidence Interval for the Difference between Two Population Means (σ Unknown, Unequal Variances)

When both population standard deviations are unknown and assumed to be unequal, the samples taken are independent, simple random samples, and either both sample sizes are at least 30 or both population distributions are approximately normal, the margin of error of a confidence interval for the difference between two population means is given by

$$E = t_{\alpha/2}\sqrt{\frac{s_1^2}{n_1} + \frac{s_2^2}{n_2}}$$

where $t_{\alpha/2}$ is the critical value for the level of confidence, $c = 1 - \alpha$, such that the area under the t-distribution to the right of $t_{\alpha/2}$ is equal to $\frac{\alpha}{2}$,

s_1 and s_2 are the two sample standard deviations, and

n_1 and n_2 are the two sample sizes.

The number of degrees of freedom for the t-distribution is given by the smaller of the values $n_1 - 1$ and $n_2 - 1$.

Side Note

Many calculators and statistical software packages use the following alternate formula for the number of degrees of freedom for the t-distribution when the population variances are assumed to be unequal.

$$df = \frac{\left(\frac{s_1^2}{n_1} + \frac{s_2^2}{n_2}\right)^2}{\frac{1}{n_1 - 1}\left(\frac{s_1^2}{n_1}\right)^2 + \frac{1}{n_2 - 1}\left(\frac{s_2^2}{n_2}\right)^2}$$

Let's get our feet wet first by just calculating the margin of error from some given sample data.

Example 9.4

Finding the Margin of Error of a Confidence Interval for the Difference between Two Population Means (σ Unknown, Unequal Variances)

Given the following data from two independent samples, calculate the margin of error of the 95% confidence interval for the difference between the two population means, assuming that the population variances are not equal and both population distributions are approximately normal.

Sample 1: 12, 9, 14, 16, 8, 9, 10, 11, 13, 13, 10

Sample 2: 14, 14, 15, 8, 11, 11, 12, 10, 9, 9

Solution

To use the formula for the margin of error, we need to know the critical value of t and the sample standard deviation and sample size for each sample. Finding the sample standard deviation for each sample is easily done using technology. Hence we have the following.

$$s_1 \approx 2.460599 \qquad s_2 \approx 2.406011$$
$$n_1 = 11 \qquad n_2 = 10$$

In order to find the critical t-value, we need the number of degrees of freedom, which is the smaller of the values $n_1 - 1$ and $n_2 - 1$.

$$n_1 - 1 = 11 - 1 \qquad n_2 - 1 = 10 - 1$$
$$= 10 \qquad\qquad = 9$$

Thus, $df = 9$. Since the level of confidence is 95%, $\alpha = 1 - 0.95 = 0.05$. So, $t_{\alpha/2} = t_{0.05/2} = t_{0.025}$. For a t-distribution with 9 degrees of freedom, $t_{0.025} = 2.262$. Substituting into the formula for the margin of error gives us the following.

$$\begin{aligned} E &= t_{\alpha/2}\sqrt{\frac{s_1^2}{n_1} + \frac{s_2^2}{n_2}} \\ &= 2.262\sqrt{\frac{(2.460599)^2}{11} + \frac{(2.406011)^2}{10}} \\ &\approx 2.403797 \end{aligned}$$

Just as before, the final step in the process of constructing a confidence interval for the difference between two population means is simply subtracting the margin of error from and adding the margin of error to the point estimate to find the lower and upper bounds of the interval.

Rounding Rule

Round the endpoints of a confidence interval for the difference between two population means as follows:

- If sample data are given, round to one more decimal place than the largest number of decimal places in the given data.
- If statistics are given, round to the same number of decimal places as given in the standard deviation or variance.

Formula

Confidence Interval for the Difference between Two Population Means (Independent Samples)

The confidence interval for the difference between two population means for independent data sets is given by

$$(\bar{x}_1 - \bar{x}_2) - E < \mu_1 - \mu_2 < (\bar{x}_1 - \bar{x}_2) + E$$

or

$$\left((\bar{x}_1 - \bar{x}_2) - E,\ (\bar{x}_1 - \bar{x}_2) + E\right)$$

where $\bar{x}_1$ and $\bar{x}_2$ are the two sample means,

$(\bar{x}_1 - \bar{x}_2)$ is the point estimate for the difference between the population means, $\mu_1 - \mu_2$, and

E is the margin of error.

The following example illustrates how these types of confidence intervals are constructed.

Example 9.5

Constructing a Confidence Interval for the Difference between Two Population Means (σ Unknown, Unequal Variances)

Unhappy with the class she is taking, Misty believes that the difficulty she is having in the class is the result of an inexperienced teacher. She believes that students in her particular class are receiving lower scores on their exams than students in another class with a more experienced teacher. She collects exam scores from a random sample of 11 of her classmates and calculates a mean exam score of 75.0 with a standard deviation of 8.0. Misty then collects data from a random sample of 9 students in the other class and calculates that the mean exam score is 82.0 with a standard deviation of 5.0. Assume that the population distributions of exam scores are approximately normal for both classes. Construct a 90% confidence interval for the true difference between the mean exam scores for the two classes. Does this confidence interval suggest that the inexperienced teacher's class is receiving lower scores than the experienced teacher's class?

Solution

Step 1: Find the point estimate.

Let Population 1 be the class with the inexperienced teacher and Population 2 be the class with the experienced teacher. Begin by calculating the point estimate for the difference between the population means.

$$\begin{aligned}\bar{x}_1 - \bar{x}_2 &= 75.0 - 82.0 \\ &= -7.0\end{aligned}$$

Step 2: Find the margin of error.

Notice the fact that we were not given the population standard deviation for either population. We must then decide if it's reasonable to assume that the variances for the two different populations are equal. Since we are not given details about the two classes, that is, if they are the same course, if they have the same prerequisites, and so forth, we cannot assume the variances to be equal. Therefore, we will use the formula for the margin of error for unequal variances.

$$E = t_{\alpha/2}\sqrt{\frac{s_1^2}{n_1} + \frac{s_2^2}{n_2}}$$

The sample sizes and standard deviations for both samples are given, so we only need to determine the critical t-value before substituting the values into the formula to calculate the margin of error.

For populations where the variances are assumed to be different, the number of degrees of freedom for the critical t-value is the smaller of the values $n_1 - 1$ and $n_2 - 1$.

$$\begin{aligned}n_1 - 1 &= 11 - 1 \qquad & n_2 - 1 &= 9 - 1 \\ &= 10 & &= 8\end{aligned}$$

Thus, $df = 8$. Since the level of confidence is 90%, $\alpha = 1 - 0.90 = 0.10$. Then $t_{\alpha/2} = t_{0.10/2} = t_{0.05}$. For the t-distribution with 8 degrees of freedom, $t_{0.05} = 1.860$. Substituting into the formula for the margin of error gives us the following.

$$\begin{aligned}E &= t_{\alpha/2}\sqrt{\frac{s_1^2}{n_1} + \frac{s_2^2}{n_2}} \\ &= 1.860\sqrt{\frac{(8.0)^2}{11} + \frac{(5.0)^2}{9}} \\ &\approx 5.453309\end{aligned}$$

Step 3: Subtract the margin of error from and add the margin of error to the point estimate.

Subtracting the margin of error from the point estimate and then adding the margin of error to the point estimate gives us the following endpoints of the confidence interval.

$$\begin{aligned}\text{Lower endpoint: } (\bar{x}_1 - \bar{x}_2) - E &= -7.0 - 5.453309 \\ &\approx -12.5\end{aligned}$$

$$\text{Upper endpoint: } (\bar{x}_1 - \bar{x}_2) + E = -7.0 + 5.453309$$
$$\approx -1.5$$

Thus, the 90% confidence interval for the difference between the two population means ranges from –12.5 to –1.5. The confidence interval can be written mathematically using either inequality symbols or interval notation, as shown below.

$$-12.5 < \mu_1 - \mu_2 < -1.5$$
or
$$(-12.5,\ -1.5)$$

Because both endpoints are negative numbers, it appears that the true difference between the class means is a negative number. In this scenario, a negative difference indicates that the second class performed better than the first class. Therefore, we can be 90% confident that the mean exam score for students in the inexperienced teacher's class is between 1.5 and 12.5 points lower than the mean exam score for students in the experienced teacher's class.

Equal Variances

Remember, in this section we are working with populations whose standard deviations are not known. Even though we don't know exactly what the standard deviation is, in some cases we can still assume that it is the same for both populations. For instance, the samples might come from populations with similar characteristics. However, we should be very careful when we assume that the variances are equal. If you are unsure of whether to assume the variances are equal, then you should err on the side of caution and assume that they are not equal.

When we assume that the population variances are equal, the sample variances can be pooled, or combined, to give a weighted estimate for the common population variance. When this is the case, the formula for the margin of error changes as follows.

Formula

Margin of Error of a Confidence Interval for the Difference between Two Population Means (σ Unknown, Equal Variances)

When both population standard deviations are unknown and assumed to be equal, the samples taken are independent, simple random samples, and either both sample sizes are at least 30 or both population distributions are approximately normal, the margin of error of a confidence interval for the difference between two population means is given by

$$E = t_{\alpha/2}\sqrt{\frac{(n_1-1)s_1^2+(n_2-1)s_2^2}{n_1+n_2-2}}\sqrt{\frac{1}{n_1}+\frac{1}{n_2}}$$

where $t_{\alpha/2}$ is the critical value for the level of confidence, $c = 1 - \alpha$, such that the area under the t-distribution with $n_1 + n_2 - 2$ degrees of freedom to the right of $t_{\alpha/2}$ is equal to $\frac{\alpha}{2}$,

s_1 and s_2 are the two sample standard deviations, and

n_1 and n_2 are the two sample sizes.

In the following examples, we will first construct this type of confidence interval without the use of technology, and then we will show how to use a TI-83/84 Plus calculator to perform interval estimation on this type of data.

Example 9.6

Constructing a Confidence Interval for the Difference between Two Population Means (σ Unknown, Equal Variances)

Obstetricians are concerned that a certain pain reliever may be causing lower birth weights. A medical researcher collects data from a random sample of 12 mothers who took the pain reliever while pregnant and calculates that the mean birth weight of their babies was 5.6 pounds with a standard deviation of 1.8 pounds. She also collects data from a random sample of 20 mothers who did not take the pain reliever while pregnant and calculates that the mean birth weight of their babies was 6.3 pounds with a standard deviation of 2.1 pounds. Assume that the distributions of birth weights are approximately normal for both populations. Construct a 99% confidence interval for the true difference between the mean birth weights. Does the confidence interval suggest that there is a difference in the birth weights?

Solution

Step 1: Find the point estimate.

Let Population 1 be the birth weights of babies whose mothers took the pain reliever while pregnant and Population 2 be the birth weights of babies whose mothers did not take the pain reliever while pregnant. Begin by calculating the point estimate for the difference between the population means.

$$\begin{aligned}\bar{x}_1 - \bar{x}_2 &= 5.6 - 6.3 \\ &= -0.7 \text{ pounds}\end{aligned}$$

Step 2: Find the margin of error.

Because the only difference in the two populations was the administration of the pain reliever, it's reasonable to assume that the population variances are the same. Therefore, we will use the following formula for the margin of error.

$$E = t_{\alpha/2}\sqrt{\frac{(n_1-1)s_1^2+(n_2-1)s_2^2}{n_1+n_2-2}}\sqrt{\frac{1}{n_1}+\frac{1}{n_2}}$$

The sample sizes and standard deviations for both samples are given, so we only need to determine the critical t-value before substituting the values into the formula to calculate the margin of error.

For populations where the variances are assumed to be equal, the number of degrees of freedom is calculated as follows.

$$\begin{aligned}df &= n_1 + n_2 - 2 \\ &= 12 + 20 - 2 \\ &= 30\end{aligned}$$

Since the level of confidence is 99%, $\alpha = 1 - 0.99 = 0.01$. Then $t_{\alpha/2} = t_{0.01/2} = t_{0.005}$. For the t-distribution with 30 degrees of freedom, $t_{0.005} = 2.750$.

Substituting into the formula for the margin of error gives us the following. Because this is such a large formula, we will calculate it in stages, rounding each intermediate value to six decimal places in order to keep the final calculation as exact as possible.

$$\begin{aligned}E &= t_{\alpha/2}\sqrt{\frac{(n_1-1)s_1^2+(n_2-1)s_2^2}{n_1+n_2-2}}\sqrt{\frac{1}{n_1}+\frac{1}{n_2}}\\ &= 2.750\sqrt{\frac{(12-1)(1.8)^2+(20-1)(2.1)^2}{12+20-2}}\sqrt{\frac{1}{12}+\frac{1}{20}}\\ &\approx 2.750(1.995244)(0.365148)\\ &\approx 2.003538\end{aligned}$$

Step 3: Subtract the margin of error from and add the margin of error to the point estimate.

Subtracting the margin of error from the point estimate and then adding the margin of error to the point estimate gives us the following endpoints of the confidence interval.

$$\text{Lower endpoint: } (\bar{x}_1-\bar{x}_2)-E = -0.7-2.003538 \approx -2.7$$

$$\text{Upper endpoint: } (\bar{x}_1-\bar{x}_2)+E = -0.7+2.003538 \approx 1.3$$

Thus, the 99% confidence interval for the difference between the two population means ranges from −2.7 to 1.3 pounds. The confidence interval can be written mathematically using either inequality symbols or interval notation, as shown below.

$$-2.7 < \mu_1-\mu_2 < 1.3$$

or

$$(-2.7, 1.3)$$

Therefore, we are 99% confident that the population mean birth weight for babies whose mothers took the pain reliever while pregnant is between 2.7 pounds less than and 1.3 pounds more than the population mean birth weight for babies whose mothers didn't take the pain reliever. Since the confidence interval contains 0, there is not sufficient statistical evidence to indicate that the mean birth weights are different for the two populations. Thus, we cannot conclude that the pain reliever causes lower birth weights.

Using a TI-83/84 Plus Calculator to Find a Confidence Interval for the Difference between Two Population Means (σ Unknown)

When we have independent samples from populations whose standard deviations are not known and we want to construct a confidence interval to estimate the difference between the population means, the formula used to calculate the margin of error depends on whether or not we can assume that the variances are equal. As we saw in the previous examples, we are often not told one way or another and must decide for ourselves. In the next example, we will see how to use a TI-83/84 Plus calculator to find a confidence interval for the difference between two population means in either of these scenarios. Using the same sample data, we will find two confidence intervals, first using unpooled sample variances under the assumption that the population variances are not equal, and then using pooled sample variances under the assumption that the population variances are equal.

Example 9.7

Using a TI-83/84 Plus Calculator to Find a Confidence Interval for the Difference between Two Population Means (σ Unknown)

Numbers of hours per week spent studying for a particular mathematics class were gathered from randomly selected students in a study conducted by sociology majors. One of the questions on the survey asked if the respondent usually chooses a seat at the front of the classroom or at the back. The results were divided into samples of students who usually choose to sit in the front and students who prefer to sit in the back of the classroom. The following data represent the numbers of hours per week that students in the samples reportedly spent studying math.

Front of classroom preferred:

12, 14, 17, 9, 7, 11, 12, 10, 8, 4, 5

Back of classroom preferred:

4, 11, 5, 8, 4, 5, 10, 7, 8, 7, 5, 2, 4, 6, 8

Use a TI-83/84 Plus calculator to find the 90% confidence interval for the true difference between the mean amounts of time per week spent studying for this particular math class by students who prefer to sit in the front of the classroom and those who prefer to sit in the back. Assume that both population distributions are approximately normal.

a. Assume that the population variances are not equal.

b. Assume that the population variances are equal.

Solution

Since the confidence intervals involve normally distributed populations where σ is not known, we will use the Student's t-distribution to calculate the endpoints of the intervals. Thus, we will use the option for a two-sample t-interval on the calculator. Since we are given the original data, we must first enter our data into the calculator as usual by pressing STAT, choosing option 1:Edit, and then entering the data in L1 and L2. Note that which data set we put in which list guides our interpretation of the interval, but the data sets can be entered here in either order. Once the data are entered, press STAT, scroll to TESTS, and then choose option Ø:2-SampTInt. Next we are prompted to enter either the data or the statistics. Choose Data and confirm that the names of the lists shown are correct. The values of Freq1 and Freq2 should both be left as the default value of 1. Since we want the 90% confidence interval, enter Ø.9Ø for C-Level.

a. We are told to assume that the variances are not equal, and therefore we cannot pool the sample variances. After choosing No for Pooled, select Calculate and press ENTER. The results are shown in the following screenshots.

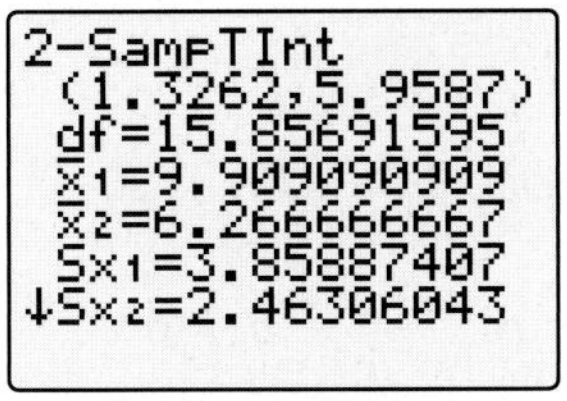

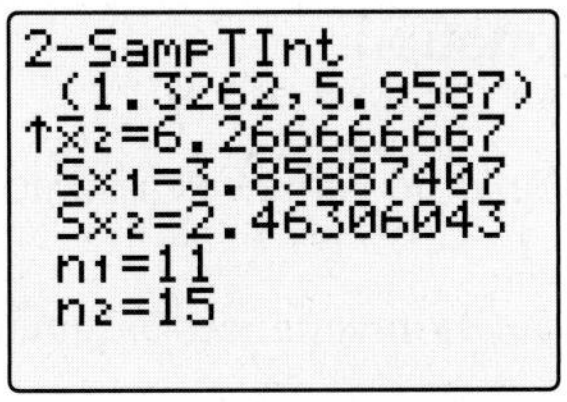

Notice that the number of degrees of freedom given by the calculator for this example is approximately 15.8569. In part b., we will see that the number of degrees of freedom is higher when the population variances are assumed to be equal. Since the data are whole numbers, we round the endpoints of the interval to one decimal place. Thus, the 90%

Memory Booster

Unequal Variances
Not Pooled

Equal Variances
Pooled

confidence interval is $(1.3, 6.0)$. Because the interval's endpoints are both positive, we can be 90% confident that the students who prefer to sit in the front of the classroom study for this particular math class between 1.3 and 6.0 hours per week more than their peers who choose to sit in the back.

b. For this part, we are told to assume that the variances are equal, and thus the sample variances can be pooled. We can use almost exactly the same settings in the calculator for this scenario. The only difference is that we choose `Yes` for `Pooled`. The results are shown in the following screenshots.

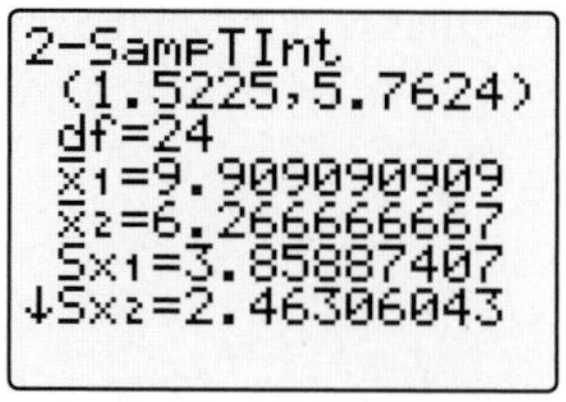

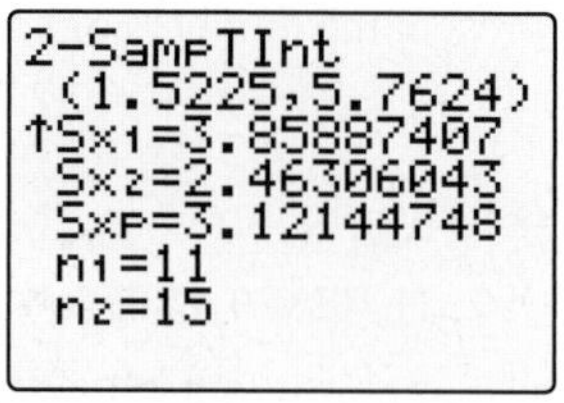

Notice that the number of degrees of freedom has increased now to 24. Although the endpoints of the confidence interval changed slightly to be $(1.5, 5.8)$, the conclusion that those students who prefer to sit in the front of the classroom study more than those who choose to sit in the back remains the same since the interval is still clearly on the positive side of the number line. If the population variances are equal, we can be 90% confident that students who choose to sit in the front study for this particular math class between 1.5 and 5.8 hours per week more than their classmates who choose seats in the back of the room.

9.2 Section Exercises

Note: For all exercises in this section, you may assume that the requirements mentioned in this section are met; namely, all samples are independent, simple random samples, both population standard deviations are unknown, and either both sample sizes are at least 30 or both population distributions are approximately normal.

Critical Values for Confidence Intervals for Differences between Two Population Means (σ Unknown)

Directions: Determine the critical value for a confidence interval for the difference between the two population means using the given information.

1. $c = 0.95$, population variances assumed to be equal, $n_1 = 11$, $n_2 = 20$

2. $c = 0.99$, population variances assumed to be equal, $n_1 = 9$, $n_2 = 6$

3. $c = 0.90$, population variances assumed to be unequal, $n_1 = 14$, $n_2 = 12$

4. $c = 0.95$, population variances assumed to be unequal, $n_1 = 25$, $n_2 = 26$

5. $c = 0.95$, population variances assumed to be unequal, $n_1 = 4$, $n_2 = 7$

6. $c = 0.99$, population variances assumed to be equal, $n_1 = 22$, $n_2 = 18$

7. $c = 0.98$, population variances assumed to be equal, $n_1 = 13$, $n_2 = 8$

Margins of Error of Confidence Intervals for Differences between Two Population Means (σ Unknown)

Directions: Calculate the margin of error of a confidence interval for the difference between the two population means using the given information.

8. Population variances assumed to be unequal, $t_{0.01} = 2.764$, $s_1 = 6$, $n_1 = 16$, $s_2 = 5$, $n_2 = 11$

9. Population variances assumed to be equal, $t_{0.025} = 2.021$, $s_1 = 2$, $n_1 = 22$, $s_2 = 3$, $n_2 = 20$

10. Population variances assumed to be equal, $\alpha = 0.01$, $s_1 = 10$, $n_1 = 15$, $s_2 = 13$, $n_2 = 10$

11. Population variances assumed to be unequal, $c = 0.90$, $s_1 = 2$, $n_1 = 6$, $s_2 = 5$, $n_2 = 5$

Confidence Intervals for Differences between Two Population Means (σ Unknown)

Directions: Construct and interpret each specified confidence interval.

12. Different breeds of dogs produce different-sized litters. Wesley believes that his Coonhounds will have litters that are twice as big as his Schnauzers. He has ten female Coonhounds and the mean size of their most recent litters was 11.0 puppies with a standard deviation of 3.0 puppies. Wesley also has nine female Schnauzers, whose most recent litters had a mean litter size of 5.0 puppies with a standard deviation of 2.0 puppies. Assume that the population variances are not the same. Create and interpret a 90% confidence interval to estimate the true difference between the mean sizes of the litters of Wesley's Coonhounds and Schnauzers.

13. Midwives claim that a water birth reduces the amount of pain the mother perceives. To test this theory, one midwife asks six of her clients who had a water birth and eight of her clients who did not have a water birth to rate their pain on a scale of 1 to 10. The mean pain score of mothers who had a water birth was a 7.0 with a standard deviation of 1.2. The mean pain score of mothers who did not have a water birth was 9.0 with a standard deviation of 0.9. Construct and interpret an 80% confidence interval for the true difference between the pain scores of mothers who had a water birth and mothers who did not. Assume that the variances of the two populations are not equal.

14. Steve believes that his wife's cell phone battery does not last as long as his cell phone battery. On eight different occasions, he measured the length of time his cell phone battery lasted, and calculated that the mean was 24.3 hours with a standard deviation of 6.1 hours. He measured the length of time his wife's cell phone battery lasted on nine different occasions and calculated a mean of 22.8 hours with a standard deviation of 8.3 hours. Construct and interpret a 95% confidence interval for the true difference in battery life between Steve's cell phone and his wife's. Assume that the population variances are the same.

15. Dentists believe that a diet low in sugary foods can reduce the number of cavities in children. Ten children whose diets are believed to be high in sugar are examined and the mean number of cavities is 3.0 with a standard deviation of 0.7. Twelve children whose diets are believed to be low in sugar are examined and the mean number of cavities is 1.8 with a standard deviation of 0.6. Construct and interpret a 99% confidence interval for the true difference between the mean numbers of cavities for children whose diets are high in sugar and those whose diets are low in sugar. Assume that the variances of the two populations are the same.

16. A company that manufactures baseball bats believes that its new bat will allow players to hit the ball 30 feet farther than its current model. The owner hires a professional baseball player known for hitting home runs to hit ten balls with each bat and he measure the distance each ball is hit to test the company's claim. The results of the batting experiment are shown in the following table. Construct and interpret a 90% confidence interval for the true difference between the mean distance hit with the new model and the mean distance hit with the older model. Assume that the variances of the two populations are the same.

Hitting Distance (in Feet)

New Model	Old Model
235	200
240	210
253	231
267	218
243	210
237	209
250	210
241	229
251	234
248	231

17. A marketing firm is doing research for an Internet-based company. It wants to appeal to the age group of people who spend the most money online. The company wants to know if there is a difference in the mean amount of money people spend per month on Internet purchases depending on their age bracket. The marketing firm looked at two age groups, 18–24 years and 25–30 years, and collected the data shown in the following table. Assume that the population variances are not the same. Construct and interpret an 80% confidence interval to estimate the true difference between the mean amounts of money per month that people in these two age groups spend on Internet purchases.

Internet Spending per Month

	18–24 Years	25–30 Years
Mean Amount Spent	\$52.00	\$45.49
Standard Deviation	\$21.30	\$13.19
Sample Size	18	25

18. Is it worth pursuing a doctoral degree in education if you already have an undergraduate degree? One way to help make this decision is to look at the mean incomes of these two groups. Suppose that 19 people with bachelor's degrees in education were surveyed. Their mean annual salary was \$28,500 with a standard deviation of \$6800. Eleven people with doctoral degrees in education were found to have a mean annual salary of \$55,500 with a standard deviation of \$10,200. Assume that the population variances are not the same. Construct and interpret a 90% confidence interval to estimate the true difference between the mean salaries for people with doctoral degrees and undergraduate degrees in education.

19. In evaluating his teaching, a literature professor decides to have a random sample of 15 students rate him on a scale of 1–10 with 10 being excellent. After getting the results, a mean of 6.2 with a standard deviation of 1.5, he decides to make an effort to improve his teaching. Next semester, for a random sample of 13 students, the results are a mean of 7.5 with a standard deviation of 1.6. Construct and interpret a 95% confidence interval to estimate the true difference between the literature professor's mean ratings from semester to semester. You can assume that the population variances are the same.

20. To conform to market trends, a prominent syrup manufacturer is changing the design of its syrup bottle from a tall slender bottle to a shorter rounder bottle that fits more easily in a microwave oven. The shareholders are concerned that the mean amount of syrup in each bottle remains the same. A sample of 16 bottles of the older design has a mean capacity of 36.20 fluid ounces (fl oz) with a standard deviation of 0.90 fl oz, and a sample of 20 bottles from the new design has a mean capacity of 35.90 fl oz with a standard deviation of 0.73 fl oz. Because the same machine is being used to fill the bottles, you may assume that the population variances are the same. Construct and interpret a 90% confidence interval to estimate the difference between the mean volumes for the bottles.

21. A researcher is interested in exploring the relationship between calcium intake and weight loss. Two different groups, each with 25 dieters, are chosen for the study. Group A is required to follow a specific diet and exercise regimen, and also take a 500-mg supplement of calcium each day. Group B is required to follow the same diet and exercise regimen, but with no supplemental calcium. After six months on the program, the members of Group A had lost a mean of 12.7 pounds with a standard deviation of 2.2 pounds. The members of Group B had lost a mean of 10.8 pounds with a standard deviation of 2.0 pounds during the same time period. Assume that the population variances are not the same. Create and interpret a 95% confidence interval to estimate the true difference between the mean amounts of weight lost by dieters who supplement with calcium and those who do not.

22. Gavin likes to grow tomatoes in the summer, and each year he experiments with ways to improve his crop. This summer, he wants to determine whether the new fertilizer he has seen advertised will increase the mean number of tomatoes produced per plant. Being careful to control the conditions in his garden, he uses the old fertilizer on 12 tomato plants and the new fertilizer on the other 8 plants. Over the course of the growing season, he calculates a mean of 18.2 tomatoes per plant for the old fertilizer, with a standard deviation of 1.9 tomatoes. He calculates a mean of 21.4 tomatoes per plant for the new fertilizer, with a standard deviation of 2.6 tomatoes. Create and interpret a 90% confidence interval for the true difference in tomato production. Assume that the population variances are the same.

23. A retailer is interested in comparing the shopping habits of customers as different types of music play throughout the store. One day when the store played slow instrumental music, the 17 customers who made purchases spent a mean of \$84 with a standard deviation of \$20. The next day, the store played upbeat instrumental music, and the 13 customers who made purchases spent a mean of \$93 with a standard deviation of \$22. Construct and interpret a 99% confidence interval for the true difference between the mean amounts spent by customers while listening to different types of music in the store. Assume that the population standard deviations are the same.

24. A researcher wants to know who reads more, women or men. She surveys 28 women and finds that they read a mean of 2.5 books a month, with a standard deviation of 0.9 books. For the 25 men she surveys, she finds that they read a mean of 2.1 books per month with a standard deviation of 0.8 books. Assuming that the population standard deviations are different, construct and interpret a 90% confidence interval for the true difference between the mean numbers of books read each month by women and men.

9.3 Comparing Two Population Means (σ Unknown, Dependent Samples)

So far in this chapter, we have looked at creating a confidence interval for the difference between two population means using independent samples, meaning that the data from the two samples have no influence on each other. However, sometimes situations arise when the data sets are dependent. In this section, we will discuss how to construct a confidence interval for the difference between two population means using dependent samples where the observations in one sample uniquely correspond with observations in the second sample. Two dependent data sets, in which the observations from one data set are matched directly to the observations from the other data set, are called **paired data**.

So how do you decide when to design an experiment that will give you paired data? In general, you should select to use paired data when you want to compare two subgroups of a population that are logically connected. Each member of the first subgroup is systematically paired with a single member of the second subgroup either by matching characteristics or by using a preexisting connection, for example, twins. Here are some specific situations in which paired data would be used.

Pretest/posttest studies on the same subjects: For instance, suppose researchers wanted to study whether a person's sleeping habits changed when taking a new drug. Data would be taken from a number of participants both before the drug was administered and after. The data from each participant would then be paired together.

Pairing subjects with similar characteristics: The same research on sleep could occur by recruiting subjects as pairs by matching variables such as age, ethnicity, work environment, and so forth, and then giving one group a treatment (that is, the new drug) and the other a placebo.

Pairing subjects who have a specific connection that is of interest: For instance, parent/child pairings or sibling/twin pairings could reveal how certain genetic traits are related to patients' responses to the new drug.

All of these methods allow the researcher to study the effects of the new drug on sleeping habits while eliminating as many confounding variables as possible. As you might imagine, the way in which data are paired is also a critical part of any study. In our sleep research example, if subjects are paired based on their similar characteristics, although age, sex, ethnicity, or work environment might be reasonable variables to consider, pairing subjects by height would be unnecessary. It is important to remember that all decisions about the design of the study, such as whether to pair the data, should be made in the planning stage of the study, not after data are collected.

When working with paired data, we actually construct a confidence interval as if there were only one set of data values. The set of data that we use contains the differences between each pair of values. Thus, the first step in constructing a confidence interval for the difference between the population means of two sets of paired data is to calculate the **paired differences**, d, for the sample data. The paired differences are found by simply subtracting each pair of values. Thus, the formula for a paired difference for any pair of values from the sample data is as follows.

Formula

Paired Difference

When two dependent samples consist of paired data, the paired difference for any pair of data values is given by

$$d = x_2 - x_1$$

where x_2 is a data value from the second sample and

x_1 is the data value from the first sample that is paired with x_2.

Example 9.8

Calculating Paired Differences

The amounts of home utility bills are given for two consecutive months for 5 different homes in a neighborhood. For each home in the sample, we can pair the billing amounts for March and April. Given the following data, calculate the paired differences for the five homes.

Utility Bills for Homes

March	April
\$119.75	\$121.06
\$68.43	\$79.04
\$202.39	\$189.55
\$47.88	\$49.64
\$66.01	\$68.52

Solution

Calculate each paired difference by subtracting the value of the March bill from the value of the April bill. By subtracting in this manner, the paired difference will reflect the increase (or decrease) in the utility bill from March to April.

Utility Bills for Homes

March, x_1	April, x_2	Paired Difference, $d = x_2 - x_1$
\$119.75	\$121.06	\$1.31
\$68.43	\$79.04	\$10.61
\$202.39	\$189.55	–\$12.84
\$47.88	\$49.64	\$1.76
\$66.01	\$68.52	\$2.51

To use paired data to construct a confidence interval, the following conditions must be met.

- All possible samples of a given size have an equal probability of being chosen; that is, simple random samples are used.
- The samples are dependent.
- Both population standard deviations, σ_1 and σ_2, are *unknown.*
- Either the number of pairs of data values in the sample data is at least 30 ($n \geq 30$) *or* the population distribution of the paired differences is approximately normal.

In this textbook, you may assume that these conditions are met for all examples and exercises involving paired data.

The value that we want to estimate is the mean of the paired differences for the two populations of dependent data, μ_d. Recall that the first step in constructing a confidence interval is to find the point estimate, and the best point estimate for a population mean is a sample mean. Therefore, the mean of the paired differences for the sample data, denoted $\bar{d}$, is the point estimate used here.

Formula

Mean of Paired Differences

When two dependent samples consist of paired data, the mean of the paired differences for the sample data is given by

$$\bar{d} = \frac{\sum d_i}{n}$$

where d_i is the paired difference for the ith pair of data values and

n is the number of paired differences in the sample data.

Example 9.9

Finding a Point Estimate for the Mean of the Paired Differences for Two Populations (σ Unknown, Dependent Samples)

Calculate the best point estimate for the mean of the paired differences for the two populations using the sample data given in Example 9.8.

Solution

Since the best point estimate for a population mean is the sample mean, we need to calculate the mean of the paired differences for the sample data, $\bar{d}$. The mean of the paired differences found in Example 9.8 is calculated as follows.

$$\begin{aligned}\bar{d} &= \frac{\sum d_i}{n} \\ &= \frac{1.31+10.61+(-12.84)+1.76+2.51}{5} \\ &= \frac{3.35}{5} \\ &= 0.67\end{aligned}$$

After we find the point estimate for the mean of the paired differences for the two populations, the next step in constructing a confidence interval is to calculate the margin of error. In order to calculate the margin of error, we need to know the standard deviation of the paired differences for the sample data, which is given by the following formula.

Formula

Sample Standard Deviation of Paired Differences

When two dependent samples consist of paired data, the sample standard deviation of the paired differences for the sample data is given by

$$s_d = \sqrt{\frac{\sum\left(d_i - \bar{d}\right)^2}{n-1}}$$

where d_i is the paired difference for the ith pair of data values,

$\bar{d}$ is the mean of the paired differences for the sample data, and

n is the number of paired differences in the sample data.

If there are n pairs of data values and the population distribution of the paired differences is approximately normal, then the sampling distribution for the sample statistic $\bar{d}$ follows a t-distribution with $n - 1$ degrees of freedom. Hence, the formula for the margin of error is as follows. Notice that this is the same formula that we used in Section 8.3 when estimating a single population mean when σ is unknown. This is because we use the paired differences as a single set of sample data rather than using the data from the two samples separately when working with paired data.

Rounding Rule

When calculating a margin of error for a confidence interval, round to at least six decimal places to avoid additional rounding errors in the subsequent calculations of the endpoints of the confidence interval.

Formula

Margin of Error of a Confidence Interval for the Mean of the Paired Differences for Two Populations (σ Unknown, Dependent Samples)

When both population standard deviations are unknown, the samples taken are dependent, simple random samples of paired data, and either the number of pairs of data values in the sample data is at least 30 or the population distribution of the paired differences is approximately normal, the margin of error of a confidence interval for the mean of the paired differences for two populations is given by

$$E = \left(t_{\alpha/2}\right)\left(\frac{s_d}{\sqrt{n}}\right)$$

where $t_{\alpha/2}$ is the critical value for the level of confidence, $c = 1 - \alpha$, such that the area under the t-distribution with $n - 1$ degrees of freedom to the right of $t_{\alpha/2}$ is equal to $\frac{\alpha}{2}$,

s_d is the sample standard deviation of the paired differences for the sample data, and

n is the number of paired differences in the sample data.

As usual, once the margin of error is calculated, subtracting E from and adding E to the point estimate gives the lower and upper endpoints of the confidence interval.

Formula

Confidence Interval for the Mean of the Paired Differences for Two Populations (Dependent Samples)

The confidence interval for the mean of the paired differences for two dependent population data sets is given by

$$\bar{d} - E < \mu_d < \bar{d} + E$$

or

$$\left(\bar{d} - E, \bar{d} + E\right)$$

where $\bar{d}$ is the mean of the paired differences for the sample data, which is the point estimate for the population mean of the paired differences, μ_d, and

E is the margin of error.

Rounding Rule

Round the endpoints of a confidence interval for the mean of the paired differences for two populations as follows:

- If sample data are given, round to one more decimal place than the largest number of decimal places in the given data
- If statistics are given, round to the same number of decimal places as given in the standard deviation or variance of the paired differences.

Example 9.10

Constructing a Confidence Interval for the Mean of the Paired Differences for Two Populations (σ Unknown, Dependent Samples)

In a CPR class, students are given a pretest at the beginning of the class to determine their initial knowledge of CPR and then a posttest at the end of the class to determine their new knowledge of CPR. Educational researchers are interested in whether the class increases a student's knowledge of CPR. Construct a 95% confidence interval for the mean increase in the students' CPR test scores. Data from a random sample of 10 students who took the CPR class are listed in the table below.

Students' CPR Test Scores										
Pretest Score	60	63	68	70	71	68	72	80	83	79
Posttest Score	80	79	83	90	83	89	94	95	96	93

Solution

Step 1: Find the point estimate.

The point estimate for the population mean of the paired differences is the mean of the paired differences for the sample data. Because these are pretest and posttest scores, we can easily pair the data to determine if the mean test score either increased or decreased. The first step is to calculate the paired difference for each pair of data. Subtract the pretest score from the posttest score.

Students' CPR Test Scores										
Pretest Score, x_1	60	63	68	70	71	68	72	80	83	79
Posttest Score, x_2	80	79	83	90	83	89	94	95	96	93
$d = x_2 - x_1$	20	16	15	20	12	21	22	15	13	14

Next, calculate the mean of the paired differences. The `1-Var Stats` option on a TI-83/84 Plus calculator can calculate this mean. It's worth noting that we will also need the sample standard deviation of the paired differences for Step 2 of the process, and the calculator can compute both of these values at once.

For a review of the calculator instructions as well as the methods for calculating these values by hand, see Chapter 3.

$$\bar{d} = 16.8$$
$$s_d \approx 3.614784$$

Step 2: **Find the margin of error.**

To calculate the margin of error, we need the critical value, $t_{\alpha/2}$. There are 10 students in the sample, so $n = 10$ and the number of degrees of freedom is $df = n - 1 = 10 - 1 = 9$. Because $c = 0.95$, we know that $\alpha = 1 - 0.95 = 0.05$. Then $t_{\alpha/2} = t_{0.05/2} = t_{0.025}$. For the t-distribution with 9 degrees of freedom, $t_{0.025} = 2.262$. Substituting into the formula for the margin of error gives us the following.

$$\begin{aligned} E &= \left(t_{\alpha/2}\right)\left(\frac{s_d}{\sqrt{n}}\right) \\ &= (2.262)\left(\frac{3.614784}{\sqrt{10}}\right) \\ &\approx 2.585681 \end{aligned}$$

Step 3: **Subtract the margin of error from and add the margin of error to the point estimate.**

Therefore, subtracting the margin of error from the point estimate and then adding the margin of error to the point estimate gives us the following endpoints of the confidence interval.

$$\begin{aligned} \text{Lower endpoint: } \bar{d} - E &= 16.8 - 2.585681 \\ &\approx 14.2 \\ \text{Upper endpoint: } \bar{d} + E &= 16.8 + 2.585681 \\ &\approx 19.4 \end{aligned}$$

Thus, the 95% confidence interval for the population mean of the paired differences ranges from 14.2 to 19.4. The confidence interval can be written mathematically using either inequality symbols or interval notation, as shown below.

$$14.2 < \mu_d < 19.4$$

or

$$(14.2, 19.4)$$

Because both endpoints are positive, we can say with 95% confidence that the mean difference between the posttest and pretest scores was positive and between 14.2 and 19.4 points. Therefore, the research suggests that the CPR class increases student knowledge of CPR, as evidenced by their exam grades, an average of 14.2 to 19.4 points.

Example 9.11

Using a TI-83/84 Plus Calculator to Find a Confidence Interval for the Mean of the Paired Differences for Two Populations (σ Unknown, Dependent Samples)

In a recent study, cholesterol levels were measured before and after participants took a new drug designed to lower their total cholesterol for four weeks. The results are shown in the following table.

Total Cholesterol Levels (in mg/dL)	
Before	**After**
238	235
240	241
220	219
246	235
202	198
222	208
210	202
233	211
204	188
229	201
244	235
220	211
219	207

Use a TI-83/84 Plus calculator to find the 92% confidence interval for the true mean of the differences between the cholesterol levels for the population from which the participants were sampled.

Solution

Because these are pretest/posttest results on the same sample of participants, we can pair the results for each participant to find a confidence interval for the mean difference.

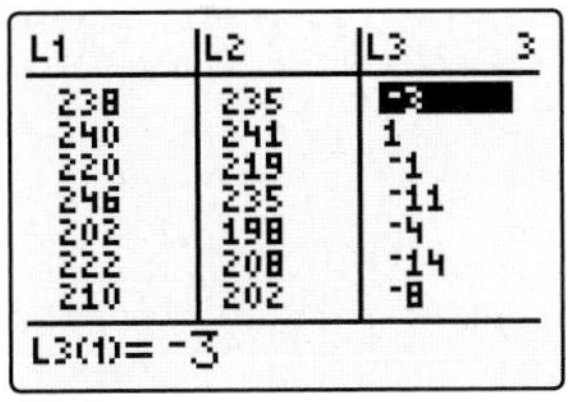

The first step in finding the interval when using a TI-83/84 Plus calculator is to enter the data in the lists L1 and L2. Let L1 be the "before" levels and L2 be the "after" levels. We need to perform one extra step that we haven't had to do before when looking at two sets of data. Since we also need the differences between the pairs of data, we can set L3 to calculate those differences for us. To do so, highlight L3 and enter the formula to subtract the "before" levels from the "after" levels (L2−L1) by pressing 2ND 2 − 2ND 1 and then ENTER. The screenshot in the margin shows how the data and the paired differences will appear in the calculator.

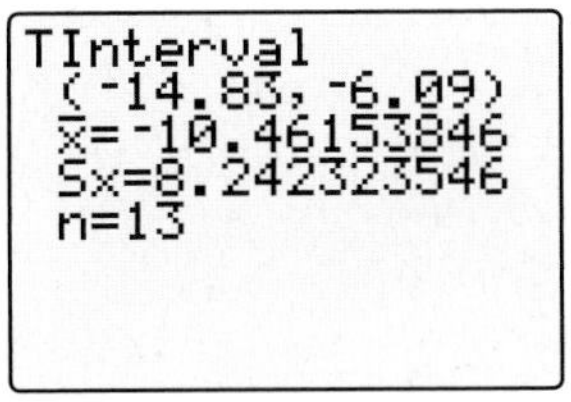

Now that we have the paired differences in L3, we can create a one-sample t-interval using those paired differences as our raw data. Press STAT, scroll to TESTS, and choose option `8:TInterval`. We want to calculate the confidence interval from the data, so choose the `Data` option. Our data are in List 3, so enter L3 by pressing 2ND and then 3. The frequency of the data (`Freq`) is the default value, which is 1. Also, we want a 92% confidence interval, so enter `Ø.92` for `C-Level`. Highlight `Calculate` and press ENTER. The results are shown in the screenshot in the margin.

Note that both endpoints of the confidence interval are negative, which indicates that the mean cholesterol level decreased significantly. We are 92% confident that, after taking the new cholesterol-lowering drug for four weeks, the mean decrease in the total cholesterol levels for the population from which the participants were sampled is between 6.1 and 14.8 mg/dL.

9.3 Section Exercises

Note: For all exercises in this section, you may assume that the requirements mentioned in this section are met; namely, the samples are dependent samples of paired data, and the population distributions of the paired differences are approximately normal.

Means and Standard Deviations of Paired Differences

Directions: Calculate $\bar{d}$ and s_d for each set of paired data.

1.

Sample A	22	21	19	20	17	18	20	19	22
Sample B	24	23	23	19	20	21	23	19	23

2.

Sample 1	2	1	1	2	1	1	2	1	2
Sample 2	4	3	2	1	2	2	2	1	3

3.

Red Group	78	79	91	83	79	92	45	68
Blue Group	67	85	90	84	86	96	51	66

4.

Group A	18	23	24	19	17	22	18	16	20
Group B	24	23	23	19	20	21	23	19	23

Confidence Intervals for Means of Paired Differences for Two Populations (σ Unknown, Dependent Samples)

Directions: Construct a confidence interval for the mean of the paired differences for the two populations using the given information.

5. $n = 41,\quad \bar{d} = 2.230,\quad \alpha = 0.01,\quad s_d = 0.567$

6. $n = 30,\quad \bar{d} = 12.0,\quad \alpha = 0.05,\quad s_d = 2.7$

7. $n = 5,\quad \bar{d} = 1.14,\quad \alpha = 0.10,\quad s_d = 1.30$

8. $n = 25,\quad \bar{d} = 3.40,\quad \alpha = 0.05,\quad s_d = 1.08$

Directions: **Construct and interpret each specified confidence interval.**

9. To determine if a new cold medicine works better than the traditional cold medicine, 16 people who had a cold volunteered for a study. The volunteers were matched based on age to create eight pairs. A double-blind study was constructed where one of the volunteers in the pair was given the new cold medicine and the other member of the pair was given the traditional medicine. The duration of the cold (in days) was measured for each person and the results are shown in the following table. Construct and interpret a 99% confidence interval for the true mean difference between the durations of a cold for those taking the traditional medication and those taking the new medication.

Duration of Cold (in Days)								
New Medicine	4	5	3	6	3	4	5	7
Traditional Medicine	6	5	4	8	4	5	7	7

10. To determine if his teaching method increases students' learning, a professor administers a pretest to his class at the beginning of the semester and then a posttest at the end of the semester. The results from 15 randomly chosen students are given below. Construct and interpret a 95% confidence interval for the true mean difference between the scores to determine if the teaching method increases students' knowledge of the course material.

Test Scores	
Pretest	**Posttest**
60	80
61	87
65	91
71	97
68	89
67	86
65	85
62	83
63	89
68	93
69	94
70	99
65	92
62	91
57	83

11. A dietician wants to see how much weight a certain diet can help patients lose. Nine people agree to participate in her study. She weighs each patient at the beginning of the study and after 30 days of being on the diet. Her results are shown in the table below. Construct and interpret a 99% confidence interval for the true mean difference between the weights to determine the mean amount of weight lost by going on this diet.

Patients' Weights (in Pounds)									
Starting Weight	180	176	152	230	183	214	250	197	201
Ending Weight	172	167	144	219	172	201	239	186	195

12. A workout coach believes that walking every day can produce the same health benefits as jogging. Ten volunteers are paired based on significant characteristics, and half of the group is asked to walk everyday and half of the group is asked to jog every day. The amounts of weight lost (in pounds) over a 30-day period are recorded in the following table. Construct and interpret a 95% confidence interval for the true mean difference between the amount of weight lost by jogging and the amount of weight lost by walking.

Weight Loss (in Pounds)					
Walking	8	9	10	7	9
Jogging	10	12	14	9	12

13. Researchers have developed a method to improve memory. To test their method, 12 participants are asked to memorize a list of words and the number of words remembered correctly is recorded. The participants are then taught the method to improve their memory and are asked to memorize another list of words and the number of words remembered correctly is again recorded. The results are shown in the following table. Construct and interpret a 98% confidence interval to estimate the true mean increase in the number of words that people can memorize after learning the memory method.

Number of Words Memorized	
Before	**After**
8	13
5	16
6	12
7	16
6	15
3	13
8	17
10	20
12	19
6	17
8	18
5	13

14. A pharmaceutical company is running tests to see how well its new drug lowers cholesterol. Ten adults volunteer to participate in the study. The total cholesterol level of each participant (in mg/dL) is recorded once at the start of the study and then again after three months of taking the drug. The results are given in the following table. Construct and interpret a 99% confidence interval for the true mean difference between the cholesterol levels for people who take the new drug.

Total Cholesterol Levels (in mg/dL)	
Initial Level	**Level after Three Months**
210	201
200	195
215	208
194	197
206	200
221	203
203	190
189	188
208	210
211	210

15. An infomercial claims that its new cooking device will dramatically reduce the time you spend preparing meals. Wondering if the claim is true, you set out to determine how much time the new cooking device will really save someone on average. Eight people who have purchased the item agree to participate in your study, and they each estimate the time they spent cooking dinner before they bought the new device and then after they began using it. Use the table of results to create and interpret a 90% confidence interval for the true mean change in the amount of time spent cooking dinner by using the infomercial's item.

Time Spent Preparing Dinner (in Minutes)	
Without Device	**With Device**
50	45
60	50
45	30
30	30
45	60
50	40
20	15
25	30

16. Philip wants to take a speed-reading course, but his wife thinks that it is a waste of time. To convince her that the course will really change the way that he reads, Philip decides to conduct an informal study. He polls seven people, asking them to tell the number of pages they were able to read in an hour before and then after they took the course. The results he obtained are found in the following table. Construct and interpret a 95% confidence interval for the true mean increase in reading speeds for people who have taken the speed-reading course.

Number of Pages Read in One Hour							
Before Course	35	50	45	50	60	70	65
After Course	50	60	80	70	85	100	90

17. A home improvement show gives tips on ways to improve your house before it is listed for sale. The show claims that their tips will help your house sell faster. To test the show's claim, a researcher found five pairs of houses that were similar in condition, area, and asking price. The homeowners from one house in each pair were asked to follow the tips given in the show before putting their houses on the market. (No major renovations were allowed.) The researchers then kept tabs on the subsequent length of time that it took for each house to sell. Use the following results to find and interpret a 90% confidence interval for the true mean difference between the numbers of weeks required to sell a house for homeowners who follow the show's tips and those who do not follow the tips.

Number of Weeks to Sell a House					
Without Tips	9	10.5	3	5.5	14.5
With Tips	7	6	5	1	4

18. During the fall semester of biology, the instructor teaching the class became ill and had to have another instructor stand in for him. After the students had taken a test under both instructors, the chairman of the biology department wanted to know if there was a significant difference in the performance of students under the different instructors. The results from each test are given in the table below. Construct and interpret a 90% confidence interval for the true mean difference between students' test scores under the two instructors.

Students' Test Scores												
Instructor A	58	82	78	91	83	77	45	87	94	92	68	77
Instructor B	52	60	83	90	87	70	48	81	90	90	71	70

19. At the beginning of an elementary physical education class, students are asked to do as many sit-ups as possible in a one-minute period. After practicing the proper technique for doing sit-ups, the students are once again timed to see if they increased the number of sit-ups they can do in one minute. Use the table of results to create and interpret a 95% confidence interval for the true mean change in the number of sit-ups that students were able to complete in one minute after the training.

Number of Sit-Ups								
Before	10	20	15	30	31	40	20	25
After	28	30	27	42	50	45	19	36

20. A learn-to-type software program claims that it can improve your typing skills. To test the claim and possibly help yourself out, you and three of your friends decide to try the program and see what happens. Use the table below to construct and interpret an 80% confidence interval for the true mean change in the typing speeds for people who have completed the typing program.

Typing Speeds (in Words per Minute)	
Before	**After**
42	54
50	52
37	41
22	30

21. After students were not doing so well in her math class, Ms. Comeaux decided to try a different approach and use verbal positive reinforcement at least once every hour. Use the following results to find and interpret a 99% confidence interval for the true mean change in the students' test scores after Ms. Comeaux started using positive reinforcement.

Students' Test Scores	
Without Reinforcement	**With Reinforcement**
66	72
57	55
61	80
72	79
38	49
47	56

9.4 Comparing Two Population Proportions

Memory Booster

Population Proportion

$$p = \frac{x}{N} = \frac{\text{\# of successes}}{\text{population size}}$$

Sample Proportion

$$\hat{p} = \frac{x}{n} = \frac{\text{\# of successes}}{\text{sample size}}$$

Memory Booster

Properties of a Binomial Distribution

1. The experiment consists of a fixed number, n, of identical trials.
2. Each trial is independent of the others.
3. For each trial, there are only two possible outcomes. For counting purposes, one outcome is labeled a success, and the other a failure.
4. For every trial, the probability of getting a success is called p. The probability of getting a failure is then $1 - p$.
5. The binomial random variable, X, counts the number of successes in n trials.

In this chapter we have seen how to compare population means using both independent and dependent samples. Now we will turn our attention to comparing two population proportions. Once again, there are times when we aren't necessarily focused on the exact proportion, but rather how proportions from two populations compare, that is, if they are equal, or if one is larger than the other.

When we were comparing population means, we constructed a confidence interval for the difference between the two population means. Similarly, when comparing two population proportions, we use a confidence interval for the difference between the population proportions. The best point estimate for the difference is $\hat{p}_1 - \hat{p}_2$. In this section we will restrict our discussion to comparing two population proportions when the following conditions are met. Notice that the conditions are similar to those discussed in Section 8.4 for estimating a single population proportion.

- All possible samples of a given size have an equal probability of being chosen; that is, simple random samples are used.
- The samples are independent.
- The conditions for a binomial distribution are met for both samples.
- The sample sizes are large enough to ensure that $n_1\hat{p}_1 \ge 5$, $n_1\left(1-\hat{p}_1\right) \ge 5$, $n_2\hat{p}_2 \ge 5$, and $n_2\left(1-\hat{p}_2\right) \ge 5$.

When these conditions are met, we can apply the Central Limit Theorem to the sampling distribution of the differences between the sample proportions for two independent samples. This means that we will use the standard normal distribution to calculate the margin of error of a confidence interval for the difference between two population proportions. You can assume that the necessary criteria are met for all examples and exercises in this section.

Example 9.12

Calculating Sample Proportions and Verifying Sample Size Conditions

Suppose that you wish to compare the proportions of registered voters who are women in two different towns. You take a random sample of registered voters from each town and record the numbers of men and women in each sample. The following table summarizes your results.

Registered Voters

	Women	Men
Town A	45	37
Town B	32	28

Calculate the sample proportion for each town and verify that the samples are large enough to use the normal distribution to compare the population proportions.

Solution

Let Population 1 be registered voters in Town A and Population 2 be registered voters in Town B. Begin by finding the sample size for each town. The size of each sample is the sum of the numbers of men and women.

For the first sample, from Town A, the sample size is calculated as follows.

$$\begin{aligned} n_1 &= 45 + 37 \\ &= 82 \end{aligned}$$

For the second sample, from Town B, the sample size is found as follows.

$$\begin{aligned} n_2 &= 32 + 28 \\ &= 60 \end{aligned}$$

The sample proportions of registered voters who are women are found by dividing the number of women in each sample by the sample size. Thus, the sample proportions are calculated as follows.

Sample 1:

$$\begin{aligned} \hat{p}_1 &= \frac{x_1}{n_1} \\ &= \frac{45}{82} \\ &\approx 0.548780 \end{aligned}$$

Sample 2:

$$\begin{aligned} \hat{p}_2 &= \frac{x_2}{n_2} \\ &= \frac{32}{60} \\ &\approx 0.533333 \end{aligned}$$

Next, verify that these samples are large enough to use the normal distribution to compare the population proportions by making sure that the products of the sample sizes and the proportions are all greater than or equal to five. The products are as follows.

Sample 1:

$$n_1\hat{p}_1 = 82\left(\frac{45}{82}\right) = 45 \ge 5$$

$$n_1\left(1-\hat{p}_1\right) = 82\left(1-\frac{45}{82}\right) = 37 \ge 5$$

Sample 2:

$$n_2\hat{p}_2 = 60\left(\frac{32}{60}\right) = 32 \ge 5$$

$$n_2\left(1-\hat{p}_2\right) = 60\left(1-\frac{32}{60}\right) = 28 \ge 5$$

Look back at the data table. The products we just calculated are the number of "successes," that is, the number of women, and the number of "failures," that is, the number of men. Therefore, when the data are given, to verify that the sample size is large enough, we simply need to check that we have at least five in each group.

Because these are independent samples and all four products are greater than or equal to five, the normal distribution can be used to compare the population proportions.

Memory Booster

The best point estimate for the difference between two population proportions is $\hat{p}_1 - \hat{p}_2$.

When comparing population proportions, the first step is the same as when building any other confidence interval. It is to calculate the point estimate. The best point estimate for the difference between two population proportions is the difference between the sample proportions, $\hat{p}_1 - \hat{p}_2$.

The second step is to calculate the margin of error. Since the sampling distribution of the differences between the sample proportions is approximately normal, we will use the critical value of z for the desired level of confidence to calculate the margin of error. These are the same critical z-values that we have used throughout this text. The margin of error is found by multiplying the appropriate critical z-value for the desired level of confidence by the standard deviation of the sampling distribution.

Table 9.2: Critical z-Values for Confidence Intervals

Level of Confidence, c	$\alpha = 1 - c$	$z_{\alpha/2}$
0.80	0.20	1.28
0.85	0.15	1.44
0.90	0.10	1.645
0.95	0.05	1.96
0.98	0.02	2.33
0.99	0.01	2.575

The margin of error of a confidence interval for the difference between two population proportions is given by the following formula.

Rounding Rule

When calculating a margin of error for a confidence interval, round to at least six decimal places to avoid additional rounding errors in the subsequent calculations of the endpoints of the confidence interval.

Formula

Margin of Error of a Confidence Interval for the Difference between Two Population Proportions

When the samples taken are independent, simple random samples, the conditions for a binomial distribution are met for both samples, and the sample sizes are large enough to ensure that $n_1\hat{p}_1 \ge 5$, $n_1\left(1-\hat{p}_1\right) \ge 5$, $n_2\hat{p}_2 \ge 5$, and $n_2\left(1-\hat{p}_2\right) \ge 5$, the margin of error of a confidence interval for the difference between two population proportions is given by

$$E = z_{\alpha/2}\sqrt{\frac{\hat{p}_1\left(1-\hat{p}_1\right)}{n_1} + \frac{\hat{p}_2\left(1-\hat{p}_2\right)}{n_2}}$$

where $z_{\alpha/2}$ is the critical value for the level of confidence, $c = 1 - \alpha$, such that the area under the standard normal distribution to the right of $z_{\alpha/2}$ is equal to $\frac{\alpha}{2}$,

$\hat{p}_1$ and $\hat{p}_2$ are the two sample proportions, and

n_1 and n_2 are the two sample sizes.

Finally, to construct a confidence interval for the true difference between the population proportions, subtract the margin of error from and add the margin of error to the point estimate.

Formula

Confidence Interval for the Difference between Two Population Proportions

The confidence interval for the difference between two population proportions is given by

$$\left(\hat{p}_1 - \hat{p}_2\right) - E < p_1 - p_2 < \left(\hat{p}_1 - \hat{p}_2\right) + E$$

or

$$\left(\left(\hat{p}_1 - \hat{p}_2\right) - E, \left(\hat{p}_1 - \hat{p}_2\right) + E\right)$$

where $\hat{p}_1$ and $\hat{p}_2$ are the two sample proportions,

$\left(\hat{p}_1 - \hat{p}_2\right)$ is the point estimate for the difference between population proportions, $p_1 - p_2$, and

E is the margin of error.

Rounding Rule

Round the endpoints of a confidence interval for the difference between two population proportions to three decimal places.

Example 9.13

Constructing a Confidence Interval for the Difference between Two Population Proportions

School administrators want to know if there is a smaller percentage of students at School A in a lower economic district who carry cell phones than the percentage of students who carry cell phones at School B in an upper economic district. A survey of students is conducted at each school and the following results are tabulated.

Responses to Survey

	School A	School B
Carry a Cell Phone	45	53
Do Not Carry a Cell Phone	31	25

Construct a 90% confidence interval for the true difference between the proportions of students who carry a cell phone at the two schools.

Solution

Step 1: Find the point estimate.

Let Population 1 be students at School A and Population 2 be students at School B. Note that in order to calculate the sample proportions, we first need to know the total sizes of the samples. Here, we must calculate the sample sizes by adding the numbers of yes and no responses for each school. Thus, the sample size for the first sample (School A) is calculated as follows.

$$\begin{aligned} n_1 &= 45 + 31 \\ &= 76 \end{aligned}$$

The sample size for the second sample (School B) is found as follows.

$$\begin{aligned} n_2 &= 53 + 25 \\ &= 78 \end{aligned}$$

Using these sample sizes to calculate the sample proportions of students who carry cell phones gives us the following.

$$\begin{aligned}\hat{p}_1 &= \frac{x_1}{n_1}\\ &= \frac{45}{76}\\ &\approx 0.592105\end{aligned}$$

$$\begin{aligned}\hat{p}_2 &= \frac{x_2}{n_2}\\ &= \frac{53}{78}\\ &\approx 0.679487\end{aligned}$$

Subtracting these two sample proportions produces the point estimate.

$$\begin{aligned}\hat{p}_1 - \hat{p}_2 &= 0.592105 - 0.679487\\ &= -0.087382\end{aligned}$$

Rounding Rule

When calculations involve several steps, avoid rounding at intermediate calculations. If necessary, round intermediate calculations to at least six decimal places to avoid additional rounding errors in subsequent calculations.

Step 2: **Find the margin of error.**

Notice that the samples are independent of one another because the responses from the two groups do not correspond with one another. We can assume that the other necessary conditions are met to allow us to use the standard normal distribution to calculate the margin of error. Since we want to create a 90% confidence interval, we need to use $z_{\alpha/2} = z_{0.10/2} = z_{0.05} = 1.645$ as the critical value. Thus, the margin of error is calculated as follows.

$$\begin{aligned}E &= z_{\alpha/2}\sqrt{\frac{\hat{p}_1\left(1-\hat{p}_1\right)}{n_1}+\frac{\hat{p}_2\left(1-\hat{p}_2\right)}{n_2}}\\ &= 1.645\sqrt{\frac{0.592105(1-0.592105)}{76}+\frac{0.679487(1-0.679487)}{78}}\\ &\approx 0.127102\end{aligned}$$

Step 3: **Subtract the margin of error from and add the margin of error to the point estimate.**

Subtracting the margin of error from the point estimate and then adding the margin of error to the point estimate gives us the following endpoints of the confidence interval.

$$\begin{aligned}\text{Lower endpoint: } \left(\hat{p}_1 - \hat{p}_2\right) - E &= -0.087382 - 0.127102\\ &\approx -0.214\end{aligned}$$

$$\begin{aligned}\text{Upper endpoint: } \left(\hat{p}_1 - \hat{p}_2\right) + E &= -0.087382 + 0.127102\\ &\approx 0.040\end{aligned}$$

Thus, the 90% confidence interval for the difference between the two population proportions ranges from −0.214 to 0.040. The confidence interval can be written mathematically using either inequality symbols or interval notation, as follows.

$$-0.214 < p_1 - p_2 < 0.040$$

or

$$(-0.214,\ 0.040)$$

Therefore, we are 90% confident that the difference between the population proportions of students who carry cell phones at School A and School B is between –0.214 and 0.040. Since the confidence interval contains 0, there is not sufficient statistical evidence to indicate that the population proportions are different for the two schools. Thus, we cannot conclude that a smaller percentage of students at School A carry cell phones.

Example 9.14

Constructing a Confidence Interval for the Difference between Two Population Proportions

In order to determine if a new instructional technology improves students' scores, a professor wants to know if a larger percentage of students using the instructional technology passed the class than the percentage of students who did not use the new technology. Records show that 45 out of 50 randomly selected students who were in classes that used the instructional technology passed the class and 38 out of 51 randomly selected students who were in classes that did not use the instructional technology passed the class. Construct a 95% confidence interval for the true difference between the proportion of students using the technology who passed and the proportion of students not using the technology who passed.

Solution

We are going to show how to construct the confidence interval first without a TI-83/84 Plus calculator and then with one.

Step 1: **Find the point estimate.**

First, we'll let Population 1 be those students who used the new technology and Population 2 be those students who did not. Next, we need to calculate the sample proportions. The sample proportion for Sample 1 (using instructional technology) is calculated as follows.

$$\begin{aligned}\hat{p}_1 &= \frac{x_1}{n_1}\\ &= \frac{45}{50}\\ &= 0.9\end{aligned}$$

The sample proportion for Sample 2 (without the instructional technology) is found as follows.

$$\begin{aligned}\hat{p}_2 &= \frac{x_2}{n_2}\\ &= \frac{38}{51}\\ &\approx 0.745098\end{aligned}$$

Now that we have the sample proportions, we can calculate the point estimate.

$$\begin{aligned}\hat{p}_1 - \hat{p}_2 &= 0.9 - 0.745098\\ &= 0.154902\end{aligned}$$

Step 2: **Find the margin of error.**

Notice that the samples are indeed independent of one another. Because they are two separate groups of students, they are not connected in any way. We can assume that the other necessary conditions are met to allow us to use the standard normal distribution to calculate the margin of error. The level of confidence is $c = 0.95$, so the critical value is $z_{\alpha/2} = z_{0.05/2} = z_{0.025} = 1.96$. Substituting the values into the formula gives us the following.

$$\begin{aligned} E &= z_{\alpha/2}\sqrt{\frac{\hat{p}_1\left(1-\hat{p}_1\right)}{n_1}+\frac{\hat{p}_2\left(1-\hat{p}_2\right)}{n_2}} \\ &= 1.96\sqrt{\frac{0.9(1-0.9)}{50}+\frac{0.745098(1-0.745098)}{51}} \\ &\approx 0.145675 \end{aligned}$$

Step 3: **Subtract the margin of error from and add the margin of error to the point estimate.**

Subtracting the margin of error from the point estimate and then adding the margin of error to the point estimate gives us the following endpoints of the confidence interval.

$$\begin{aligned} \text{Lower endpoint: } \left(\hat{p}_1-\hat{p}_2\right)-E &= 0.154902-0.145675 \\ &\approx 0.009 \\ \text{Upper endpoint: } \left(\hat{p}_1-\hat{p}_2\right)+E &= 0.154902+0.145675 \\ &\approx 0.301 \end{aligned}$$

Thus, the 95% confidence interval for the difference between the two population proportions ranges from 0.009 to 0.301. The confidence interval can be written mathematically using either inequality symbols or interval notation, as shown below.

$$0.009 < p_1 - p_2 < 0.301$$

or

$$(0.009,\ 0.301)$$

Therefore, we are 95% confident that the percentage of students who passed the class is between 0.9% and 30.1% higher for the population of students who used the new instructional technology (Population 1) than for the population of students who did not use the technology (Population 2). Thus, with 95% confidence, the professor can conclude that the new instructional technology improves students' scores.

To calculate the confidence interval for the difference between two proportions on the calculator, we don't need to find the individual sample proportions; we just need to enter the number of successes and the sample size for each sample, as well as the level of confidence. Press STAT, scroll to `TESTS`, and then choose option `B:2-PropZInt.` `x1` is the number of successes from the first sample and `n1` is the first sample's size. Similarly, `x2` is the number of successes from the second sample and `n2` is the second sample's size. As usual, `C-Level` is the confidence level, which must be entered as a decimal. The data should be entered as shown in the following screenshot on the left. After you select `Calculate` and press ENTER, the results will be displayed on the screen as shown in the screenshot on the right.

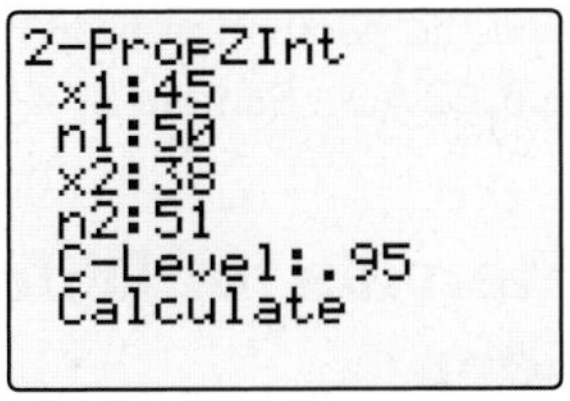

Notice that the calculator gives the same interval but with more decimal places. The interpretation of the confidence interval is still the same. The proportion of students passing the class was higher for the population of students who used the new instructional technology than for the population of students who did not use the technology.

9.4 Section Exercises

Necessary Conditions for Using the Normal Distribution to Compare Two Population Proportions

Directions: Verify that the normal distribution can be used to compare the population proportions using the given information, or show how the conditions have not been met.

1. $n_1 = 130, \quad n_2 = 200, \quad \hat{p}_1 \approx 0.869, \quad \hat{p}_2 = 0.72$

2. $n_1 = 1100, \quad n_2 = 1200, \quad \hat{p}_1 = 0.05, \quad \hat{p}_2 = 0.01$

3. $n_1 = 23, \quad n_2 = 14, \quad \hat{p}_1 \approx 0.435, \quad \hat{p}_2 \approx 0.714$

4. $n_1 = 19, \quad n_2 = 12, \quad \hat{p}_1 \approx 0.158, \quad \hat{p}_2 \approx 0.333$

Point Estimates for Differences between Two Population Proportions

Directions: Find each specified point estimate.

5. Sample 1 has 17 "yes" responses out of 97 responses in the sample, and Sample 2 has 46 "yes" responses out of 131 responses in the sample. Calculate the point estimate for the difference between the population proportions of "yes" responses.

6. A random sample of records from Public School A shows that 61 out of 156 students started first grade having already lost their first tooth, while at Public School B, 46 out of a random sample of 121 students had lost their first tooth before entering first grade. Find the point estimate for the difference between the population proportions of students who have lost their first tooth before entering first grade at the two schools.

7. Given that the first sample had 15 broken eggs out of 360 eggs in the sample and that the second sample had 12 broken eggs out of 540 eggs, find the point estimate for the difference between the population proportions of broken eggs.

8. Find the point estimate for the difference between the population proportions of "no" votes given the following information. Sample A: 16 no, 32 yes. Sample B: 43 no, 55 yes.

Margins of Error of Confidence Intervals for Differences between Two Population Proportions

Directions: Calculate the margin of error of a confidence interval for the difference between the two population proportions using the given information.

9. $n_1 = 130,\quad n_2 = 200,\quad \hat{p}_1 \approx 0.869231,\quad \hat{p}_2 = 0.72,$ 95% level of confidence

10. $n_1 = 1100,\quad n_2 = 1200,\quad \hat{p}_1 = 0.05,\quad \hat{p}_2 = 0.01,\quad \alpha = 0.05$

11. $n_1 = 88,\quad n_2 = 74,\quad \hat{p}_1 \approx 0.431818,\quad \hat{p}_2 \approx 0.608108,\quad c = 0.99$

12. The following table shows a random sample of the data collected by a state highway patrolman on whether the driver in the vehicle in a traffic stop was wearing a seat belt. Group A consists of vehicles that were stopped for what the officer considered serious offenses, and Group B consists of vehicles that were stopped for minor offenses. Using a 99% level of confidence, calculate the margin of error for the true difference between the proportions of drivers who do not wear their seat belts for the populations of drivers stopped for major offenses and those stopped for minor offenses.

Seat Belt Use		
	No	**Yes**
Group A	73	486
Group B	159	453

13. The data below represent a random sample of students from a community college (which would only have freshmen and sophomores), showing how many students in the sample are attending school without financial aid and with financial aid. Calculate the margin of error of the 90% confidence interval for the true difference between the proportions of freshmen and sophomores who are attending school without financial aid.

Financial Aid		
	Freshmen	**Sophomores**
Without Aid	36	68
With Aid	72	36

14. The data below are the proportions of respondents from surveys conducted in January and May who said they feel strongly that the government is doing a good job.

January: 365 out of 500

May: 402 out of 600

Using a 90% level of confidence, calculate the margin of error for the true difference between the population proportions of people who feel strongly that the government is doing a good job.

Confidence Intervals for Differences between Two Population Proportions

Directions: Construct and interpret each specified confidence interval. You may assume the requirements mentioned in this section are met; namely, all samples are independent, simple random samples, the conditions for a binomial distribution are met, and the sample sizes are large enough to ensure that $n_1\hat{p}_1 \geq 5$, $n_1(1-\hat{p}_1) \geq 5$, $n_2\hat{p}_2 \geq 5$, and $n_2(1-\hat{p}_2) \geq 5$.

15. Doctors at a fertility clinic wish to determine if taking fertility drugs increases the chances of a multiple birth (having two or more babies at once). Doctors record the number of multiple births and the number of single births for a sample of patients taking fertility drugs and a sample of patients not taking fertility drugs. The data are shown below. Construct and interpret a 99% confidence interval for the true difference between the proportions of multiple births for women taking fertility drugs and those who are not.

Fertility Drugs		
	With Drugs	**Without Drugs**
Single Birth	32	43
Multiple Birth	12	13

16. Psychiatrists wish to determine if there is a higher incidence of divorce among couples in which one of the spouses has suffered a serious head injury than among the general population. Of 51 couples who were married at the time one of the spouses had a serious head injury, 20 are still married. Of 50 randomly selected couples in which no head injury occurred, 24 have remained married during this same time period. Construct and interpret a 95% confidence interval for the true difference between the divorce rates of couples in which a head injury occurred and the general population.

17. Do a larger percentage of Southerners than Northerners attend church on a weekly basis? Random samples of Northerners and Southerners are interviewed about their church attendance and the results of the survey are shown below. Construct and interpret a 90% confidence interval for the true difference between the percentages of Northerners and Southerners who attend church on a weekly basis.

Church Attendance		
	Northerners	**Southerners**
Attend Church Weekly	12	35
Do Not Attend Church Weekly	34	16

18. Do more women than men exercise on a regular basis? To help answer this question, a survey was conducted asking men and women if they exercised on a regular basis, meaning three or more times a week. In response to the survey, 49 of the 100 women in the sample said yes, and 52 of the 100 men in the sample said yes. Construct and interpret a 95% confidence interval for the true difference between the proportions of men and women who exercise on a regular basis.

19. For his senior psychology project, Robert wants to explore which gender is more honest, men or women. To do this, he leaves a wallet filled with money and identification cards in a local store and hides a video camera to record the results. He repeats the experiment 80 times. Of the 37 women who picked up the wallet, 30 returned it to Robert. Of the 43 men who found the wallet, 29 returned it to him. Construct and interpret a 90% confidence interval for the true difference between the proportions of men and women who will return a lost wallet.

20. Ann, a teacher, is concerned with the obsession many of her students have with video games. She is afraid that the video games have a negative impact on her students' performance in the classroom. To test her theory, Ann sends home a letter explaining the study, and 82 parents agree to let their children participate. Ann then randomly divides the students into two equal groups. Group A is required to play video games for two hours one evening, while Group B is not allowed any time to play video games. The following day, the students are given a review test over previously learned material. From Group A, 29 students pass the test. From Group B, 34 students pass the test. Construct and interpret a 95% confidence interval for the true difference between the proportions of students who pass the test and play video games for two hours on the night before the test and those who pass the test but do not play games. Use the confidence interval to determine whether the evidence suggests that playing video games negatively impacts students' performance.

21. Frank owns a nursery, and he has had trouble in the past getting his gerbera daisies to bloom. This spring, he has decided to experiment with a new, more expensive fertilizer called FertiGro. Of the 40 plants that he treats with his regular fertilizer, 27 bloom. Of the 35 that he treats with FertiGro, 29 bloom. Construct and interpret a 99% confidence interval for the true difference between the proportions of gerbera daisy plants that bloom for the populations of plants treated with FertiGro and those treated with regular fertilizer.

22. A state politician is interested in knowing how voters in rural areas and cities differ in their opinions about gun control. For his study, 75 rural voters were surveyed, and 41 were found to support gun control. Also included in the study were 75 voters from cities, and 53 of these voters were found to support gun control. Construct and interpret a 90% confidence interval for the true difference between the proportions of rural and city voters who favor gun control.

23. Aaliyah wants to know if there is a difference between the proportions of customers who just order water to drink at two popular restaurants in town. She collected the data in the following table. Construct and interpret a 95% confidence interval for the true difference between the proportions of customers who just order water to drink at Restaurants A and B.

Drink Orders		
	Water	**Other Beverage**
Restaurant A	61	52
Restaurant B	68	72

24. A local city government is trying to promote road safety by encouraging drivers to buckle up. Its campaign director is trying to decide to which age group she should direct most of the promotions. She believes that fewer older adults buckle up, as they came from a generation who grew up without seat belts. She surveyed 49 senior adults and found that 30 of them buckle up on a regular basis. She then surveyed 52 middle-aged adults and found that 42 of them buckle up on a regular basis. Construct and interpret a 90% confidence interval to estimate the true difference between the proportions of senior adults and middle-aged adults who wear seat belts on a regular basis.

25. Early-childhood-development studies indicate that the more often a child is read to from birth, the earlier the child begins to read. A local parents' group wants to test this theory and samples families with young children. They find the following results. Construct and interpret a 98% confidence interval to estimate the true difference between the proportions of children who read at an early age when they are read to frequently compared to those who were read to less often, as described in the table of results.

Ages When Children Begin to Read

	Read to at Least Three Times per Week	Read to Fewer than Three Times per Week
Started Reading by Age 4	46	31
Started Reading after Age 4	40	57

9.5 Comparing Two Population Variances

So far in this chapter, we have discussed comparing two population means and two population proportions. Now let's turn our attention to the scenario of comparing two population variances.

To compare two population variances, we must ensure that the following conditions are met.

- All possible samples of a given size have an equal probability of being chosen; that is, simple random samples are used.
- The samples are independent.
- Both population distributions are approximately normal.

These conditions are very strict, much more so than the conditions for other confidence intervals, as the methods used to compare population variances produce very poor results when these conditions have been violated, even to a small degree. You can safely assume that these conditions are met for all examples and exercises in this section.

To compare population means and proportions, we used the differences between the sample statistics as our point estimates. To compare population variances, we will use the ratio of the sample statistics as our point estimate. Thus, the best point estimate for comparing two population variances is the ratio of the sample variances, $\frac{s_1^2}{s_2^2}$.

Memory Booster

Use the larger of the two sample variances as s_1^2 to simplify calculations.

To simplify the remainder of the calculations, let us impose the rule that s_1^2 is greater than or equal to s_2^2. This rule is not mathematically necessary; however, the calculations are more complicated if the ratio is less than 1, so we will consistently put the larger sample variance in the numerator of the point estimate.

Let's specify a few other notations that we will use in this section. First, remember that the level of confidence is c and $\alpha = 1 - c$. Second, if s_1^2 is the larger sample variance, then n_1 is the size of the sample with the larger variance. The value s_2^2 is then the smaller sample variance, so n_2 is the size of the sample with the smaller variance.

Formula

Point Estimate for Comparing Two Population Variances

The point estimate that is used to construct a confidence interval for comparing two population variances is the ratio of the two sample variances given by

$$\frac{s_1^2}{s_2^2}$$

where s_1^2 and s_2^2 are the two sample variances, with $s_1^2 \geq s_2^2$.

Calculating the endpoints of a confidence interval for the ratio of two population variances requires that we use critical values of a distribution that we have not discussed previously. When independent samples are drawn from normal distributions, the sampling distribution for the ratio of the sample variances is an F-distribution, named for Sir Ronald A. Fisher who first developed the F-distribution. For now, a few comments on the properties of the F-distribution and a brief discussion on how to read the F-distribution table should suffice.

Properties

Properties of an *F*-Distribution

1. The F-distribution is skewed to the right.
2. The values of F are always greater than or equal to 0.
3. The shape of an F-distribution is completely determined by its two parameters, the degrees of freedom for the numerator and the degrees of freedom for the denominator.

In general, the value of F such that an area of α *is to the right of F is denoted by* F_α, *as shown in Figure 9.1.*

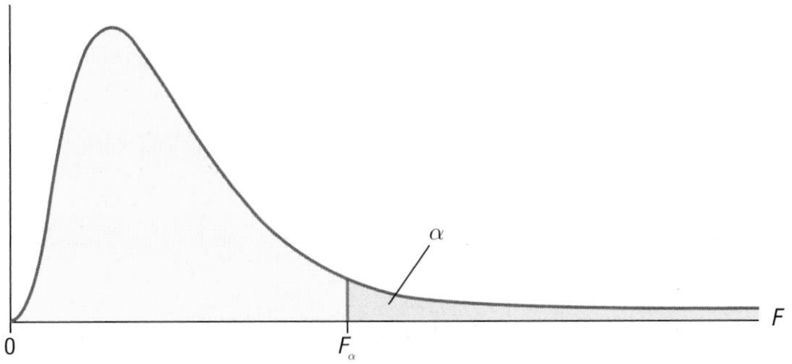

Figure 9.1: *F*-Distribution

Table H in Appendix A lists the critical values of the F-distributions for various numbers of degrees of freedom and areas in the right tails. In general, the number of degrees of freedom for the F-distribution is one less than the sample size, so the number of degrees of freedom for the numerator is $df_1 = n_1 - 1$ and the number of degrees of freedom for the denominator is $df_2 = n_2 - 1$. If the number of degrees of freedom needed is not in the F-distribution table, use the closest value given. To find the critical values using the F-distribution table, first find the section of the table for the appropriate area in the right tail of the distribution. The critical values we will need to construct a confidence interval to compare two population variances are $F_{\alpha/2}$ and $F_{(1-\alpha/2)}$, where the values of the subscripts, $\alpha/2$ and $(1-\alpha/2)$, are the areas under the F-distribution curve to the right of the critical values and $\alpha = 1 - c$ for the given level of confidence, c. Next, find the value in the table where the column for the number of degrees of freedom for the numerator (found across the top) and the row for the number of degrees of freedom for the denominator (found down the left side) intersect.

Rounding Rule

When calculating a value of *F*, round to four decimal places. This follows the convention used in the *F*-distribution table in Appendix A.

Denominator Degrees of Freedom	Numerator Degrees of Freedom: 1	2	3	4	5	6	7	8	9
1	0.0001	0.0050	0.0180	0.0319	0.0439	0.0537	0.0616	0.0681	0.0735
2	0.0001	0.0050	0.0201	0.0380	0.0546	0.0688	0.0806	0.0906	0.0989
3	0.0000	0.0050	0.0211	0.0412	0.0605	0.0774	0.0919	0.1042	0.1147
4	0.0000	0.0050	0.0216	0.0432	0.0643	0.0831	0.0995	0.1136	0.1257
5	0.0000	0.0050	0.0220	0.0445	0.0669	0.0872	0.1050	0.1205	0.1338
6	0.0000	0.0050	0.0223	0.0455	0.0689	0.0903	0.1092	0.1258	0.1402
7	0.0000	0.0050	0.0225	0.0462	0.0704	0.0927	0.1125	0.1300	0.1452
8	0.0000	0.0050	0.0227	0.0468	0.0716	0.0946	0.1152	0.1334	0.1494
9	0.0000	0.0050	0.0228	0.0473	0.0726	0.0962	0.1175	0.1363	0.1529
10	0.0000	0.0050	0.0229	0.0477	0.0734	0.0976	0.1193	0.1387	0.1558

Figure 9.2: Excerpt from Table H: Critical Values of *F* (Area = 0.995)

Example 9.15

Calculating the Point Estimate for Comparing Two Population Variances and Finding Critical *F*-Values

Consider the following information from two independent samples.

$$n_1 = 18 \qquad n_2 = 25$$
$$s_1^2 = 1.23 \qquad s_2^2 = 1.19$$

Using a 95% level of confidence, calculate the point estimate for comparing the population variances, $\frac{s_1^2}{s_2^2}$. Also find the critical values $F_{\alpha/2}$ and $F_{(1-\alpha/2)}$.

Solution

Let's begin by calculating the point estimate.

$$\frac{s_1^2}{s_2^2} = \frac{1.23}{1.19} \approx 1.033613$$

Next, remember that $\alpha = 1 - c$. The level of confidence is $c = 0.95$, so $\alpha = 0.05$. To find the F-critical values, we also need the numbers of degrees of freedom for the numerator and denominator.

$$df_1 = n_1 - 1 = 18 - 1 = 17$$

$$df_2 = n_2 - 1 = 25 - 1 = 24$$

Using the two numbers of degrees of freedom and $\alpha = 0.05$, find $F_{\alpha/2}$. Use the section of the table for an area of $\frac{\alpha}{2} = 0.025$, and find the value where the row for the denominator degrees of freedom, $df_2 = 24$, intersects with the column for the numerator degrees of freedom, $df_1 = 17$. Thus, $F_{\alpha/2} = F_{0.025} = 2.3865$.

Critical Values of *F* (Area = 0.025)

Denominator Degrees of Freedom	Numerator Degrees of Freedom: 10	11	12	13	14	15	16	17	18
21	2.7348	2.6819	2.6368	2.5978	2.5638	2.5338	2.5071	2.4833	2.4618
22	2.6998	2.6469	2.6017	2.5626	2.5285	2.4984	2.4717	2.4478	2.4262
23	2.6682	2.6152	2.5699	2.5308	2.4966	2.4665	2.4396	2.4157	2.3940
24	2.6396	2.5865	2.5411	2.5019	2.4677	2.4374	2.4105	2.3865	2.3648
25	2.6135	2.5603	2.5149	2.4756	2.4413	2.4110	2.3840	2.3599	2.3381
26	2.5896	2.5363	2.4908	2.4515	2.4171	2.3867	2.3597	2.3355	2.3137
27	2.5676	2.5143	2.4688	2.4293	2.3949	2.3644	2.3373	2.3131	2.2912
28	2.5473	2.4940	2.4484	2.4089	2.3743	2.3438	2.3167	2.2924	2.2704
29	2.5286	2.4752	2.4295	2.3900	2.3554	2.3248	2.2976	2.2732	2.2512

To find the critical value $F_{(1-\alpha/2)}$ we use similar methods. Thus, $F_{(1-\alpha/2)} = F_{0.975} = 0.3907$.

Critical Values of *F* (Area = 0.975)

Numerator Degrees of Freedom

Denominator Degrees of Freedom	10	11	12	13	14	15	16	17	18
21	0.2938	0.3114	0.3271	0.3410	0.3536	0.3649	0.3752	0.3846	0.3932
22	0.2950	0.3128	0.3286	0.3427	0.3554	0.3668	0.3773	0.3868	0.3955
23	0.2961	0.3140	0.3300	0.3442	0.3570	0.3686	0.3792	0.3888	0.3976
24	0.2971	0.3152	0.3313	0.3456	0.3586	0.3703	0.3809	0.3907	0.3996
25	0.2981	0.3163	0.3325	0.3470	0.3600	0.3718	0.3826	0.3924	0.4014
26	0.2990	0.3173	0.3336	0.3482	0.3614	0.3733	0.3841	0.3940	0.4031
27	0.2998	0.3183	0.3347	0.3494	0.3626	0.3746	0.3856	0.3956	0.4048
28	0.3006	0.3191	0.3357	0.3505	0.3638	0.3759	0.3869	0.3970	0.4063
29	0.3013	0.3200	0.3366	0.3515	0.3649	0.3771	0.3882	0.3984	0.4077

Now that we can calculate the point estimate and find the critical values, let's construct the confidence interval. Just as when constructing a confidence interval for a population variance with one sample, we cannot calculate a margin of error for a confidence interval for the ratio of two population variances. We must calculate the upper and lower endpoints for the confidence interval separately. The formulas for the endpoints of a confidence interval for a ratio of two population variances are given in the following box.

Formula

Confidence Interval for the Ratio of Two Population Variances

When the samples taken are independent, simple random samples and both population distributions are approximately normal, a confidence interval for the ratio of two population variances is given by

$$\left(\frac{s_1^2}{s_2^2}\cdot\frac{1}{F_{\alpha/2}}\right) < \frac{\sigma_1^2}{\sigma_2^2} < \left(\frac{s_1^2}{s_2^2}\cdot\frac{1}{F_{(1-\alpha/2)}}\right)$$

where s_1^2 and s_2^2 are the two sample variances, with $s_1^2 \ge s_2^2$,

$\frac{s_1^2}{s_2^2}$ is the point estimate for the ratio of the population variances, $\frac{\sigma_1^2}{\sigma_2^2}$, and

$F_{\alpha/2}$ and $F_{(1-\alpha/2)}$ are the critical values for the level of confidence, $c = 1 - \alpha$, such that the area under the F-distribution to the right of $F_{\alpha/2}$ is equal to $\frac{\alpha}{2}$, and the area to the right of $F_{(1-\alpha/2)}$ is equal to $1-\frac{\alpha}{2}$.

The F-distribution has $df_1 = n_1 - 1$ degrees of freedom for the numerator and $df_2 = n_2 - 1$ degrees of freedom for the denominator,

where n_1 and n_2 are the two sample sizes.

Rounding Rule

Round the endpoints of a confidence interval for the ratio of two population variances or standard deviations to four decimal places.

Notice that the larger critical value, $F_{\alpha/2}$, is actually used to calculate the left endpoint, and the smaller critical value, $F_{(1-\alpha/2)}$, is used to calculate the right endpoint. The reason for this is the same as we discussed for the chi-square distribution in Chapter 8.

The following formula for a confidence interval for the ratio of two population standard deviations is found by simply taking the square root of each side of the formula we just saw.

Formula

Confidence Interval for the Ratio of Two Population Standard Deviations

When the samples taken are independent, simple random samples and both population distributions are approximately normal, a confidence interval for the ratio of two population standard deviations is given by

$$\left(\frac{s_1}{s_2} \cdot \frac{1}{\sqrt{F_{\alpha/2}}}\right) < \frac{\sigma_1}{\sigma_2} < \left(\frac{s_1}{s_2} \cdot \frac{1}{\sqrt{F_{(1-\alpha/2)}}}\right)$$

where s_1 and s_2 are the two sample standard deviations, with $s_1 \geq s_2$,

$\frac{s_1}{s_2}$ is the point estimate for the ratio of the population standard deviations, $\frac{\sigma_1}{\sigma_2}$, and

$F_{\alpha/2}$ and $F_{(1-\alpha/2)}$ are the critical values for the level of confidence, $c = 1 - \alpha$, such that the area under the F-distribution to the right of $F_{\alpha/2}$ is equal to $\frac{\alpha}{2}$, and the area to the right of $F_{(1-\alpha/2)}$ is equal to $1-\frac{\alpha}{2}$.

The F-distribution has $df_1 = n_1 - 1$ degrees of freedom for the numerator and $df_2 = n_2 - 1$ degrees of freedom for the denominator,

where n_1 and n_2 are the two sample sizes.

Procedure

Constructing a Confidence Interval for the Ratio of Two Population Variances (or Standard Deviations)

1. Find the point estimate, $\frac{s_1^2}{s_2^2}\left(\text{or } \frac{s_1}{s_2}\right)$.
2. Based on the level of confidence given, calculate $\frac{\alpha}{2}$ and $1-\frac{\alpha}{2}$.
3. Use the F-distribution table to find the critical values, $F_{\alpha/2}$ and $F_{(1-\alpha/2)}$, for a distribution with $df_1 = n_1 - 1$ degrees of freedom for the numerator and $df_2 = n_2 - 1$ degrees of freedom for the denominator.
4. Substitute the necessary values into the formula for the confidence interval.

Interpreting the Confidence Interval

When we create a confidence interval to compare two population variances, we are estimating the value of the ratio, or fraction, $\frac{\sigma_1^2}{\sigma_2^2}$. Since a fraction equals 1 only if the numerator and denominator are equal, if the confidence interval contains 1, then we can conclude that the data do not provide evidence that the two population variances are unequal at a particular level of confidence. If a confidence interval for the ratio of two population variances does not contain 1, then we can determine which variance is most likely larger than the other. When the numerator of a fraction is larger than the denominator, the value of the fraction is greater than 1. Thus, if both endpoints of the interval are greater than 1, we can conclude that σ_1^2 is probably larger than σ_2^2. On the other hand, if both endpoints of the interval are less than 1, this implies that σ_1^2 is most likely smaller than σ_2^2, since a fraction is less than 1 only if the numerator is smaller than the denominator.

Example 9.16

Constructing a Confidence Interval for the Ratio of Two Population Variances

Suppose that a quality control inspector wants to make sure that two different machines are filling soda cans with the same level of precision. Specifically, she wants to ensure that the population variances of the amounts of soda per can are the same for the two machines. Random samples of cans filled by the two machines are taken, the amount of soda in each can is measured in milliliters, and the variance of the amounts of soda is calculated for each sample. A sample of 16 cans from Machine A has a sample variance of 0.93, while a sample of 18 cans from Machine B has a sample variance of 0.36. Assuming that both populations are normally distributed, construct a 90% confidence interval for the ratio of the population variances.

Solution

To begin, list the information that was given in the problem. We will call the sample from Machine A Sample 1 since its variance is larger. We can also determine the degrees of freedom for each sample by subtracting 1 from the sample size.

Machine A:	Machine B:
$n_1 = 16$	$n_2 = 18$
$s_1^2 = 0.93$	$s_2^2 = 0.36$
$df_1 = 15$	$df_2 = 17$

Step 1: Find the point estimate, $\frac{s_1^2}{s_2^2}$ $\left(\text{or } \frac{s_1}{s_2}\right)$.

The point estimate is the ratio of the sample variances.

$$\frac{s_1^2}{s_2^2} = \frac{0.93}{0.36} \approx 2.583333$$

Step 2: Based on the level of confidence given, calculate $\frac{\alpha}{2}$ and $1-\frac{\alpha}{2}$.

We know that $c = 0.90$; therefore $\alpha = 1 - 0.90 = 0.10$. Then $\frac{\alpha}{2} = \frac{0.10}{2} = 0.05$. Also, $1-\frac{\alpha}{2} = 1-0.05 = 0.95$.

Step 3: **Use the F-distribution table to find the critical values, $F_{\alpha/2}$ and $F_{(1-\alpha/2)}$.**

We need to find the critical values, $F_{\alpha/2}$ and $F_{(1-\alpha/2)}$, for the F-distribution with $df_1 = 15$ numerator degrees of freedom and $df_2 = 17$ denominator degrees of freedom. Using the row for $df_2 = 17$ and the column for $df_1 = 15$ in the table for an area of $\frac{\alpha}{2} = 0.05$ in the right tail gives us $F_{\alpha/2} = F_{0.05} = 2.3077$. Using the row for $df_2 = 17$ and the column for $df_1 = 15$ in the table for an area of $1 - \frac{\alpha}{2} = 0.95$ gives us $F_{(1-\alpha/2)} = F_{0.95} = 0.4222$.

Step 4: **Substitute the necessary values into the formula for the confidence interval.**

Substituting the values into the formula for a confidence interval for the ratio of two population variances gives us the following.

$$\left(\frac{s_1^2}{s_2^2} \cdot \frac{1}{F_{\alpha/2}}\right) < \frac{\sigma_1^2}{\sigma_2^2} < \left(\frac{s_1^2}{s_2^2} \cdot \frac{1}{F_{(1-\alpha/2)}}\right)$$

$$\left(2.583333 \cdot \frac{1}{2.3077}\right) < \frac{\sigma_1^2}{\sigma_2^2} < \left(2.583333 \cdot \frac{1}{0.4222}\right)$$

$$1.1194 < \frac{\sigma_1^2}{\sigma_2^2} < 6.1187$$

Using interval notation, the 90% confidence interval for the ratio of the population variances can also be written as $(1.1194, 6.1187)$.

We are comparing the two population variances to see if they are equal. If they are equal, their ratio would equal 1. Since both endpoints of the confidence interval are greater than 1, the sample data provide evidence at the 90% level of confidence that the population variance of the amounts of soda per can is greater for cans filled by Machine A than for cans filled by Machine B.

Example 9.17

Constructing a Confidence Interval for the Ratio of Two Population Standard Deviations

A Family Nurse Practitioner (FNP) is comparing the variability of normal body temperatures for infants and adults. Specifically, she wants to know if the population standard deviation of normal body temperatures is the same for infants as for adults. A simple random sample of 28 of her infant patients' normal body temperatures, measured in degrees Fahrenheit, reveals that they have a sample standard deviation of 0.567 °F. A simple random sample of 25 of her adult patients' normal body temperatures, also measured in degrees Fahrenheit, have a sample standard deviation of 0.663 °F. Assuming that both populations are normally distributed, construct a 95% confidence interval for the ratio of the population standard deviations.

Solution

To begin, let's list the information that is given in the problem, just as we did in the previous example. Let's call the sample of adults Sample 1 since its standard deviation is larger. Thus, we will label the sample of infants Sample 2.

Adults:	Infants:
$n_1 = 25$	$n_2 = 28$
$s_1 = 0.663$	$s_2 = 0.567$
$df_1 = 24$	$df_2 = 27$

Step 1: Find the point estimate, $\frac{s_1^2}{s_2^2}\left(\text{or } \frac{s_1}{s_2}\right)$.

The point estimate is the ratio of the sample standard deviations, which is calculated as follows.

$$\frac{s_1}{s_2} = \frac{0.663}{0.567} \approx 1.169312$$

Step 2: Based on the level of confidence given, calculate $\frac{\alpha}{2}$ and $1-\frac{\alpha}{2}$.

We know that $c = 0.95$; therefore $\alpha = 1 - 0.95 = 0.05$. Then $\frac{\alpha}{2} = \frac{0.05}{2} = 0.025$. Also, $1-\frac{\alpha}{2} = 1 - 0.025 = 0.975$.

Step 3: Use the *F*-distribution table to find the critical values, $F_{\alpha/2}$ and $F_{(1-\alpha/2)}$.

Next, we need to find the critical *F*-values for the *F*-distribution with $df_1 = 24$ numerator degrees of freedom and $df_2 = 27$ denominator degrees of freedom. Using the row for $df_2 = 27$ and the column for $df_1 = 24$ in the table for an area of $\frac{\alpha}{2} = 0.025$ in the right tail gives us $F_{\alpha/2} = F_{0.025} = 2.1946$. Using the row for $df_2 = 27$ and the column for $df_1 = 24$ in the table for an area of $1-\frac{\alpha}{2} = 0.975$ gives us $F_{(1-\alpha/2)} = F_{0.975} = 0.4472$.

Step 4: Substitute the necessary values into the formula for the confidence interval.

Substituting the values into the formula for a confidence interval for the ratio of two population standard deviations gives us the following.

$$\left(\frac{s_1}{s_2}\cdot\frac{1}{\sqrt{F_{\alpha/2}}}\right) < \frac{\sigma_1}{\sigma_2} < \left(\frac{s_1}{s_2}\cdot\frac{1}{\sqrt{F_{(1-\alpha/2)}}}\right)$$

$$\left(1.169312\cdot\frac{1}{\sqrt{2.1946}}\right) < \frac{\sigma_1}{\sigma_2} < \left(1.169312\cdot\frac{1}{\sqrt{0.4472}}\right)$$

$$0.7893 < \frac{\sigma_1}{\sigma_2} < 1.7486$$

Using interval notation, the 95% confidence interval for the ratio of the population standard deviations can also be written as $(0.7893, 1.7486)$.

Since a ratio of 1 would mean that the numerator and the denominator are equal, and the value of 1 is in this interval, the data do not provide evidence at the 95% level of confidence that the population variances of normal body temperatures of adults and infants are unequal.

9.5 Section Exercises

Note: For all exercises in this section, you may assume that the requirements mentioned in this section are met; namely, the samples are independent, simple random samples, and both population distributions are approximately normal.

Point Estimates for Ratios of Two Population Variances

Directions: Calculate the point estimate for the ratio of the population variances.

1. $n_1 = 13,\quad n_2 = 17,\quad s_1^2 = 3.467,\quad s_2^2 = 2.903$

2. $n_1 = 23,\quad n_2 = 20,\quad s_1^2 = 0.689,\quad s_2^2 = 0.542$

3. $n_1 = 8,\quad n_2 = 10,\quad s_1^2 = 12.103,\quad s_2^2 = 10.874$

4. $n_1 = 26,\quad n_2 = 28,\quad s_1^2 = 1.472,\quad s_2^2 = 1.327$

Critical Values for Confidence Intervals for Ratios of Two Population Variances

Directions: Determine the critical values for the left and right endpoints of a confidence interval for the ratio of the two population variances using the given information.

5. $n_1 = 18,\quad n_2 = 23,\quad s_1^2 = 12.470,\quad s_2^2 = 12.205,$ 95% level of confidence

6. $n_1 = 9,\quad n_2 = 9,\quad s_1^2 = 0.974,\quad s_2^2 = 0.903,\quad c = 0.95$

7. $n_1 = 21,\quad n_2 = 20,\quad s_1^2 = 134.943,\quad s_2^2 = 125.908,\quad \alpha = 0.10$

8. $n_1 = 18,\quad n_2 = 19,\quad s_1^2 = 20.461,\quad s_2^2 = 18.321,\quad c = 0.90$

9. $n_1 = 12,\quad n_2 = 13,\quad s_1^2 = 5.100,\quad s_2^2 = 5.098,\quad \alpha = 0.01$

10. $n_1 = 7,\quad n_2 = 6,\quad s_1^2 = 4.110,\quad s_2^2 = 3.873,$ 99% level of confidence

Confidence Intervals for Ratios of Two Population Variances

Directions: Construct a confidence interval for the ratio of the two population variances using the given information.

11. $\frac{s_1^2}{s_2^2} = 1.23, \quad F_{\alpha/2} = 2.4499, \quad F_{(1-\alpha/2)} = 0.3556$

12. $\frac{s_1^2}{s_2^2} = 6.31, \quad F_{\alpha/2} = 5.4160, \quad F_{(1-\alpha/2)} = 0.0688$

13. $n_1 = 13, \quad n_2 = 12, \quad \frac{s_1^2}{s_2^2} = 1.14, \quad \alpha = 0.05$

14. $n_1 = 14, \quad n_2 = 18, \quad \frac{s_1^2}{s_2^2} = 2.84, \quad \alpha = 0.05$

15. $n_1 = 9, \quad n_2 = 12, \quad \frac{s_1^2}{s_2^2} = 1.23, \quad \alpha = 0.01$

16. $n_1 = 20, \quad n_2 = 20, \quad \frac{s_1^2}{s_2^2} = 1.87, \quad \alpha = 0.10$

17. $n_1 = 15, \quad n_2 = 16, \quad s_1^2 = 8.455, \quad s_2^2 = 2.897,$ 95% level of confidence

18. $n_1 = 18, \quad n_2 = 17, \quad s_1^2 = 4.067, \quad s_2^2 = 3.903,$ 95% level of confidence

Directions: Construct each specified confidence interval and interpret the interval.

19. A medical researcher is trying to determine whether the population variances of systolic blood pressure levels are the same for patients who take a new medication for high blood pressure and patients who do not take the new medication. The control group consists of 20 patients with a sample variance in systolic blood pressure of 124.940. The treatment group, who is taking the new medication, consists of 23 patients with a sample variance in systolic blood pressure of 123.980. Construct and interpret a 90% confidence interval for the ratio of the population variances of systolic blood pressure levels for the two groups.

20. A track coach wants to make sure that two of her star runners are equally consistent in their times for the race. She times 16 of Amy's practice runs and calculates that the sample variance in times (measured in seconds) is 2.560. She then times 15 of Veronika's practice runs and calculates that the sample variance in times is 2.519. Construct and interpret a 90% confidence interval for the ratio of the population variances of Amy's and Veronika's run times.

21. A quality control inspector is testing two machines that are used to fill bags of flour to determine if they are running consistently. In particular, he wants to know whether the population variances of the amounts of flour per bag are equal for the two machines. He measures the weights, in pounds, of 11 bags of flour filled by Machine A and calculates a sample variance of 0.584. He then measures the weights of 10 bags of flour filled by Machine B and calculates a sample variance of 0.499. Construct and interpret a 99% confidence interval for the ratio of the population variances of the weights of the bags of flour for the two machines. If it is important that the two machines have the same variance, should the inspector adjust one of the machines, and if so, which machine needs adjusting?

22. A nutritionist is comparing two new diets, and after running previous tests to determine if the same amount of weight can be lost using both diets, he now wants to know if the amounts of weight lost are consistent for the two diets. In other words, he wants to estimate the ratio of the variances of the amounts of weight lost by people on the two diets. Group A consists of six people on Diet A, and after two months on the diet, the sample variance of the amounts of weight lost (measured in pounds) is 26.041. Group B consists of seven people on Diet B, and after the same two months on their diet, the sample variance of the amounts of weight lost by this group is 25.084. Construct and interpret a 99% confidence interval for the ratio of the population variances of the amounts of weight lost by people on these two diets.

23. A new brand of golf balls claims that using these golf balls can increase your precision. A golf pro tests this claim by hitting 45 golf balls of the old brand and 30 golf balls of the new brand from the same tee on the driving range and measuring the distance, in yards, that each ball travels. The sample variance of the driving distances for the old brand was 29.752 and the sample variance for the new brand was 15.910. Construct and interpret a 90% confidence interval for the ratio of the population variances.

24. A professor wants to make sure that two different versions of a test are equivalent. He decides to compare the variances of the test scores from each version. A sample of 19 scores on Version A has a sample variance of 2.450, while a sample of 20 scores from Version B has a sample variance of 2.391. Construct and interpret a 95% confidence interval for the ratio of the population variances of the scores on the two versions of the test.

25. A graduate student walking around a large college campus notices that students, not faculty, often own the more expensive cars. He collects data from 25 professors and 28 students on the prices of their cars and calculates the variance of the car prices from each sample. The sample variance for the sample of professors' cars is 7849.38 and the sample variance for the sample of students' cars is 3567.90. Construct and interpret a 95% confidence interval for the ratio of the population variances of prices of professors' and students' cars for this college.

26. A researcher wants to compare the consistency with which two marksmen hit the bull's-eye of a target. The first marksman hits an average of 4.5 bull's-eyes per session with a variance of 2.78. The second marksman hits an average of 5.3 bull's-eyes per session with a variance of 2.34. If 20 sessions were recorded for each marksman, construct and interpret a 95% confidence interval for the ratio of the population variances of the numbers of times per session that these two marksmen hit the bull's-eye.

27. A paint technician for an oil company has to make sure that the painted coatings on the insides of the oil tankers have a consistent thickness. He measures the thickness of the coating (in millimeters) at 15 spots inside the first tanker and calculates a sample variance of 0.3918. He then measures the thickness of the coating at 15 spots inside a second tanker sprayed using the same equipment and calculates a sample variance of 0.4231. Construct and interpret a 99% confidence interval for the ratio of the population variances to help determine if there was a significant difference in how consistently the equipment applied the coating in one tanker versus the other.

28. When shopping for a new car, Emily became interested in how consistently cars are priced in her area; she was planning to buy either a Honda Accord or a Toyota Camry. She priced ten different Accords with similar features at several dealerships in her area. The sample variance of the prices for the Accords was 273,529. Emily then priced nine different Camrys with similar features at several different dealerships in her area. The sample variance of the prices for the Camrys was 231,361. Construct and interpret a 99% confidence interval for the ratio of the population variances of the prices for the two different car models.

R Chapter 9 Review

Section 9.1: Comparing Two Population Means (σ Known)

Definitions

Independent

Describes two samples when the data from the first sample are not connected to the data from the second sample

Dependent

Describes two samples when the two data sets are systematically connected; data from dependent samples are referred to as *paired data*

Margin of Error of a Confidence Interval for the Difference between Two Population Means (σ Known)

Used when independent, simple random samples of sizes n_1 and n_2 are taken from populations with known standard deviations, σ_1 and σ_2, where either $n_1 \geq 30$ and $n_2 \geq 30$ or both population distributions are approximately normal, given by

$$E = z_{\alpha/2}\sqrt{\frac{\sigma_1^2}{n_1} + \frac{\sigma_2^2}{n_2}}$$

Critical z-Values for Confidence Intervals

Critical z-Values for Confidence Intervals

Level of Confidence, c	$\alpha = 1 - c$	$z_{\alpha/2}$
0.80	0.20	1.28
0.85	0.15	1.44
0.90	0.10	1.645
0.95	0.05	1.96
0.98	0.02	2.33
0.99	0.01	2.575

Confidence Interval for the Difference between Two Population Means

Can be written mathematically using either inequality symbols or interval notation, given by

$$(\bar{x}_1 - \bar{x}_2) - E < \mu_1 - \mu_2 < (\bar{x}_1 - \bar{x}_2) + E$$

or

$$((\bar{x}_1 - \bar{x}_2) - E,\ (\bar{x}_1 - \bar{x}_2) + E)$$

Section 9.2: Comparing Two Population Means (σ Unknown)

Definition

Pooled

Describes two sample variances that are combined to give a weighted estimate for the common population variance when the two population variances are assumed to be equal

Margin of Error of a Confidence Interval for the Difference between Two Population Means (σ Unknown, Unequal Variances)

Used when independent, simple random samples of sizes n_1 and n_2 are taken from populations with standard deviations that are unknown and assumed to be unequal, where either $n_1 \geq 30$ and $n_2 \geq 30$ or both population distributions are approximately normal, given by

$$E = t_{\alpha/2}\sqrt{\frac{s_1^2}{n_1}+\frac{s_2^2}{n_2}} \text{ with } df = \text{smaller of the values } n_1 - 1 \text{ and } n_2 - 1$$

Margin of Error of a Confidence Interval for the Difference between Two Population Means (σ Unknown, Equal Variances)

Used when independent, simple random samples of sizes n_1 and n_2 are taken from populations with standard deviations that are unknown and assumed to be equal, where either $n_1 \geq 30$ and $n_2 \geq 30$ or both population distributions are approximately normal, given by

$$E = t_{\alpha/2}\sqrt{\frac{(n_1-1)s_1^2+(n_2-1)s_2^2}{n_1+n_2-2}}\sqrt{\frac{1}{n_1}+\frac{1}{n_2}} \text{ with } df = n_1 + n_2 - 2$$

Section 9.3: Comparing Two Population Means (σ Unknown, Dependent Samples)

Definition

Paired data

Two data sets in which the observations from one data set are matched directly to the observations from the other data set

Paired Difference

Used for paired data from two dependent samples, the difference between any pair of data values, given by

$$d = x_2 - x_1$$

Mean of Paired Differences

Used for paired data from two dependent samples, given by

$$\bar{d} = \frac{\sum d_i}{n}$$

Section 9.3: Comparing Two Population Means (σ Unknown, Dependent Samples) (cont.)

Sample Standard Deviation of Paired Differences

Used for paired data from two dependent samples, given by

$$s_d = \sqrt{\frac{\sum\left(d_i - \bar{d}\right)^2}{n-1}}$$

Margin of Error of a Confidence Interval for the Mean of the Paired Differences for Two Populations (σ Unknown, Dependent Samples)

Used when paired data from two dependent, simple random samples of size n are taken from populations with unknown standard deviations, where either $n \geq 30$ or the population distribution of the paired differences is approximately normal, given by

$$E = \left(t_{\alpha/2}\right)\left(\frac{s_d}{\sqrt{n}}\right) \text{ with } df = n-1$$

Confidence Interval for the Mean of the Paired Differences for Two Populations (Dependent Samples)

Can be written mathematically using either inequality symbols or interval notation, given by

$$\bar{d} - E < \mu_d < \bar{d} + E$$

or

$$\left(\bar{d} - E, \bar{d} + E\right)$$

Section 9.4: Comparing Two Population Proportions

Margin of Error of a Confidence Interval for the Difference between Two Population Proportions

Used when independent, simple random samples of sizes n_1 and n_2 are large enough to ensure that $n_1\hat{p}_1 \geq 5$, $n_1\left(1-\hat{p}_1\right) \geq 5$, $n_2\hat{p}_2 \geq 5$, and $n_2\left(1-\hat{p}_2\right) \geq 5$, where the conditions for a binomial distribution are met for both samples, given by

$$E = z_{\alpha/2}\sqrt{\frac{\hat{p}_1\left(1-\hat{p}_1\right)}{n_1} + \frac{\hat{p}_2\left(1-\hat{p}_2\right)}{n_2}}$$

Confidence Interval for the Difference between Two Population Proportions

Can be written mathematically using either inequality symbols or interval notation, given by

$$\left(\hat{p}_1 - \hat{p}_2\right) - E < p_1 - p_2 < \left(\hat{p}_1 - \hat{p}_2\right) + E$$

or

$$\left(\left(\hat{p}_1 - \hat{p}_2\right) - E, \left(\hat{p}_1 - \hat{p}_2\right) + E\right)$$

Section 9.5: Comparing Two Population Variances

Point Estimate for Comparing Two Population Variances

The ratio of the two sample variances, given by

$$\frac{s_1^2}{s_2^2} \text{ with } s_1^2 \geq s_2^2$$

Properties of an *F*-Distribution

1. The *F*-distribution is skewed to the right.
2. The values of *F* are always greater than or equal to 0.
3. The shape of the *F*-distribution is completely determined by its two parameters, the degrees of freedom for the numerator and the degrees of freedom for the denominator.

Confidence Interval for the Ratio of Two Population Variances

Used when independent, simple random samples of sizes n_1 and n_2 are taken, and both population distributions are approximately normal, given by

$$\left(\frac{s_1^2}{s_2^2} \cdot \frac{1}{F_{\alpha/2}}\right) < \frac{\sigma_1^2}{\sigma_2^2} < \left(\frac{s_1^2}{s_2^2} \cdot \frac{1}{F_{(1-\alpha/2)}}\right) \text{ with } df_1 = n_1 - 1 \text{ and } df_2 = n_2 - 1$$

Confidence Interval for the Ratio of Two Population Standard Deviations

Used when independent, simple random samples of sizes n_1 and n_2 are taken, and both population distributions are approximately normal, given by

$$\left(\frac{s_1}{s_2} \cdot \frac{1}{\sqrt{F_{\alpha/2}}}\right) < \frac{\sigma_1}{\sigma_2} < \left(\frac{s_1}{s_2} \cdot \frac{1}{\sqrt{F_{(1-\alpha/2)}}}\right) \text{ with } df_1 = n_1 - 1 \text{ and } df_2 = n_2 - 1$$

Constructing a Confidence Interval for the Ratio of Two Population Variances (or Standard Deviations)

1. Find the point estimate, $\frac{s_1^2}{s_2^2}\left(\text{or } \frac{s_1}{s_2}\right)$.
2. Based on the level of confidence given, calculate $\frac{\alpha}{2}$ and $1-\frac{\alpha}{2}$.
3. Use the *F*-distribution table to find the critical values, $F_{\alpha/2}$ and $F_{(1-\alpha/2)}$, for a distribution with $df_1 = n_1 - 1$ degrees of freedom for the numerator and $df_2 = n_2 - 1$ degrees of freedom for the denominator.
4. Substitute the necessary values into the formula for the confidence interval.

E Chapter 9 Exercises

Directions: Construct and interpret each specified confidence interval.

1. Two math teachers each believe that his or her own very large class is doing better than the other class. The teachers decide to determine whose class is doing better by looking at scores from a recent exam. A random sample of 40 scores from Class A yields a mean exam score of 88.5. Assume that the population standard deviation of the exam scores for Class A is 5.7. A random sample of 37 scores from Class B yields a mean exam score of 85.3. Assume that the population standard deviation of the exam scores for Class B is 6.2. Construct and interpret a 95% confidence interval for the true difference between the mean exam scores for Classes A and B.

2. Insurance companies assert that more accidents are caused by people who talk on the phone while driving than by people who do not. Cellular phone companies dispute this assertion. Independent studies were done by an insurance company and a cell phone company. The results are in the following table. Construct and interpret a 95% confidence interval for the true difference between the proportions of accidents occurring while the driver was talking on the phone for the populations from which the samples were drawn for the two studies.

Number of Accidents		
	Driver on Phone	**Driver Not on Phone**
Cell Phone Company Study	44	50
Insurance Company Study	71	63

3. To determine how well a new drug reduces a fever, 10 adults running a fever in a hospital agree to take the new drug instead of the traditional treatment. Their temperatures are taken before the drug is administered and 30 minutes after taking the drug. The results are shown in the following table. Construct and interpret a 99% confidence interval for the mean change in temperatures of patients after taking this new drug.

Patient Temperatures (in °F)										
First Temperature	100.1	101.3	102.1	102.7	101.9	100.8	103.1	102.5	103.5	101.7
Second Temperature	98.9	99.1	99.2	99.0	98.7	98.6	99.4	99.2	100.1	99.2

4. Voters in a dry county (meaning no alcohol is sold in the county) will soon decide whether the county should remain dry. Local businesspeople fear that more people in the food industry than in the general population favor the sale of alcohol. A survey of local restaurant and convenience store owners is conducted, as well as a survey of the general population. The results of the surveys are shown in the following table. Construct and interpret a 90% confidence interval for the true difference between the population proportions of the food industry owners and the general population who favor the sale of alcohol.

Results of Survey		
	Food Industry	**General Population**
Favor Alcohol Sale	54	32
Oppose Alcohol Sale	19	20

5. Does an apple a day keep the doctor away? To test this theory, a pediatrician asks 12 children to be a part of a study. With parental consent, the number of doctor's visits from the previous six months is recorded for each child. The children are then asked to eat an apple each day for the next six months, after which the number of doctor's visits for that time period is again recorded for each child. Use the following table to construct and interpret a 95% confidence interval for the true mean change in the number of doctor's visits during a six-month period for children who eat an apple a day for six months.

Number of Doctor's Visits												
Without Apples	2	1	2	2	3	0	4	1	2	1	2	0
With Apples	1	1	0	2	2	1	1	2	0	0	2	1

6. Who makes better grades, men or women? To answer this question, suppose you survey male and female students on your campus, asking each to confidentially give you his or her GPA. For the 28 men surveyed, you calculate their mean GPA to be 3.15 with a standard deviation of 0.76 points. For the 26 women surveyed, you calculate their mean GPA to be 3.32 with a standard deviation of 0.68 points. Assuming that both population distributions are approximately normal and the population variances are equal, construct and interpret a 90% confidence interval for the true difference between the mean GPAs of men and women.

7. A popular magazine is conducting a study in order to rank each state according to the overall health of its citizens. One measure that is used is the mean number of times each citizen exercises per week. For the 85 people surveyed in State A, the mean exercise rate was 1.9 times per week. For the 91 people surveyed in State B, the mean exercise rate was 2.3 times per week. Assume that the population standard deviations of the citizens' exercise rates are 0.5 times per week for State A and 1.1 times per week for State B. Construct and interpret a 95% confidence interval for the true difference between the mean exercise rates for these two states.

8. A research firm is hired by a snack food company to determine where it should focus its resources for a new advertising campaign. Specifically, it is interested in comparing the snack food preferences of adults and children. To do this, 100 adults and 100 children are surveyed. Among the adults, 62% say that they prefer salty over sweet snack food. Among the children, 54% say that they prefer salty over sweet snack food. Construct and interpret a 95% confidence interval for the true difference between the population proportions of adults and children who prefer salty over sweet snack food.

9. Comparing results on standardized tests is a very common way to evaluate programs within certain university settings. The Board of Trustees of State Institutions for Higher Learning (IHL) in one state often evaluates programs based on how students perform on these tests. Suppose that in 2010, University A reported that a random sample of 101 students who took the GRE had a mean verbal score of 464.3. Also in 2010, University B reported that a random sample of 216 students who took the GRE had a mean verbal score of 455.8. Assume that the population standard deviations of students' GRE verbal scores are 5.1 for University A and 7.4 for University B. Construct and interpret a 90% confidence interval for the true difference between the population means of students' GRE verbal scores for University A and University B.

10. A television station is interested in the number of televisions that the average family has in its home. The station executives performed a survey in two different areas of town to see if there was a difference by area. After surveying 20 homes in Area A, they found that the mean number of televisions found in each home was 2.4 with a standard deviation of 0.5. In Area B, they found that after sampling 15 homes, the mean number of televisions was 1.9 with a standard deviation of 0.7. Assume that both population distributions are approximately normal and the population variances are not the same. Construct and interpret a 90% confidence interval to estimate the true difference between the mean numbers of TVs per home in the two areas.

11. One petroleum company claims that using its engine-cleaning product will improve a car's gas mileage. It was added to six cars, and here are the before and after gas mileages (in miles per gallon). Construct and interpret a 95% confidence interval for the true mean improvement in gas mileage for cars that use the engine-cleaning product.

Gas Mileage (in Miles per Gallon)

Without Engine Cleaner	29	48	43	35	21	19
With Engine Cleaner	31	50	46	43	29	20

12. Have you ever noticed that air currents can speed up air travel in certain directions? With this in mind, a businessperson wants to determine how much faster the return trip from Dallas to Atlanta is than the initial trip from Atlanta to Dallas. He records the times for his next seven flights to and from Dallas and obtains the following data. Assuming that both population distributions are approximately normal and the variances of the two populations are not the same, construct and interpret a 90% confidence interval for the true difference between the mean times for flights from Atlanta to Dallas and flights from Dallas to Atlanta. (**Hint:** Convert all times to minutes.)

Flight Times

Atlanta to Dallas	Dallas to Atlanta
1 h 49 min	1 h 32 min
1 h 51 min	1 h 23 min
2 h 03 min	1 h 38 min
1 h 47 min	1 h 26 min
1 h 53 min	1 h 30 min
2 h 01 min	1 h 19 min
1 h 45 min	1 h 28 min

13. A coach wants to compare the consistency of the two best batters on his baseball team. The coach considers the number of times per game that each player gets on base. For a sample of 15 games, John's sample variance is 4.389 and Mark's sample variance is 3.981. Assume that the population distributions of the numbers of times per game that the players get on base are both approximately normal. Construct and interpret a 95% confidence interval for the ratio of the population variances for the two different batters.

14. To ensure that the braking time of a client's car meets the manufacturer's specifications, a mechanic decides to compare the consistency of its braking distances to a car that is known to meet the manufacturer's specifications. The mechanic measures the braking distance, in feet, of the client's car 12 times and calculates a sample variance of 313.6. He then measures the braking distance of the control car 12 times and calculates a sample variance of 304.8. Assume that the population distributions of the braking distances are approximately normal for both cars. Construct and interpret a 99% confidence interval for the ratio of the population variances for the two different cars.

15. A professor decides to compare the variance in the annual salaries of associate professors with the variance in the annual salaries of full professors. A sample of 15 associate professors reveals a sample variance of 609,961. A sample of 14 full professors reveals a sample variance of 674,041. Construct and interpret a 90% confidence interval for the ratio of the population variances for the different ranks of professors.

P Chapter 9 Project

Project A: The Battle of the Sexes

The age-old battle of the sexes... Which gender is smarter, better, and so forth? Well, let's see if we can answer that question once and for all!

Step 1: The next page contains a short ten-question test. Feel free to photocopy that page for the purposes of this experiment. You will need to find 30 willing volunteers, approximately 15 male and 15 female (though it does not have to be evenly split), to take this short test. The answers to the test are provided for your benefit. Grade each person's test and record the number of answers correct.

Answers to Intelligence Test:

1. Eight
2. Benjamin Franklin
3. Red
4. 1440
5. A musical instrument
6. Six-cylinder engine in the shape of a "V"
7. Aurora
8. British Petroleum
9. 206
10. Ruby

Step 2: Divide the results into two groups, male and female; then calculate the mean number of correct answers and the sample standard deviation for each group. Record your statistics below.

$n_{\text{male}} =$ ______________ $n_{\text{female}} =$ ______________

$\bar{x}_{\text{male}} =$ ______________ $\bar{x}_{\text{female}} =$ ______________

$s_{\text{male}} =$ ______________ $s_{\text{female}} =$ ______________

Step 3: Construct a 95% confidence interval for the true difference between the mean numbers of correct answers for males and females. Assume that the population variances are not the same and the population distributions of the numbers of correct answers are approximately normal for both males and females.

Step 4: Write up your conclusion. Is one gender "smarter" than the other? Consider how you choose your sample, the validity of the intelligence test, and the limits of interpreting the confidence interval when drawing your conclusion.

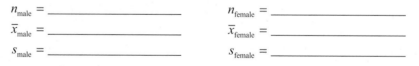

Intelligence Test for the Battle of the Sexes

Directions: Thank you for agreeing to help with this educational project. Please read and answer each of the 10 questions below. Do your best! Remember, you are representing your entire gender in this heated battle.

1. How many sides does a stop sign have?

____

2. Whose face is on the $100 bill?

____

3. Which color is on top on a traffic light?

____

4. How many minutes are in a day?

____

5. What is a viola?

____

6. What does V6 stand for when referring to a car?

____

7. What is the name of the princess in the Disney movie *Sleeping Beauty*?

____

8. What does BP stand for in the name of the chain of gas stations?

____

9. How many bones are in the human body?

____

10. What is the birthstone for July?

____

Project B: Knowledge of Current Events

Many adults assume that college students do not pay attention to current events. Let's do an experiment to see if a simple crash course can improve college students' knowledge of current events. This experiment will involve some research and preparation on your part.

Step 1: Begin by preparing two posters with pictures of the following people. Only label the posters with a letter next to each person's picture (A through F). Make sure that the pictures are not in the same order on the two posters and that the same letter is not used for the same person on the two posters. Using different pictures on the two posters is also a good idea.

President of the United States

Vice President of the United States

Secretary of State of the United States

Chief Justice of the Supreme Court of the United States

Prime Minister of the United Kingdom

Pope

Step 2: Next, you will need to find 10 willing college student volunteers. You will give each volunteer a pretest and a posttest using the posters. For each volunteer, show him the first poster and ask him to name all the people and state their titles. Record how many answers he gets correct out of 12 (6 names and 6 titles). After the pretest, give him the answers. Coach him for five minutes on who the people are and their titles. Next, visit with him for a few minutes about the class, what you are doing in this project, the weather, your favorite sports team, and so forth. Then, show him the other poster and ask him to name all the people and their titles. Record how many answers he gets correct out of 12. Thank him for his help with the project.

Step 3: Once data have been collected from 10 college students, calculate the paired difference for each student by subtracting the pretest score from the posttest score.

Pretest Score, x_1										
Posttest Score, x_2										
$d = x_2 - x_1$										

Step 4: Calculate each of the following sample statistics.

$\bar{d} =$ ____________________ $s_d =$ ____________________

Step 5: Construct a 95% confidence interval for the true mean difference between the students' scores on the posttest and pretest. Assume that the population distribution of the paired differences is approximately normal.

Step 6: Did your "crash course" on people in the news improve the students' knowledge? Consider how you chose the participants in your sample and whether their initial knowledge of the people on your posters impacted your results.

T Chapter 9 Technology

Confidence Intervals for Two Samples

TI-83/84 Plus

Confidence Intervals for Differences between Two Population Means (σ Known)

Using a TI-83/84 Plus calculator to calculate the endpoints of a confidence interval can be done from data or from statistics that have already been calculated. The process is similar to the way we calculated confidence intervals for one sample in Chapter 8. For the following example, we will be constructing a confidence interval from given statistics.

Side Note

The most recent TI-84 Plus calculators (Jan. 2011 and later) contain a new feature called `STAT WIZARDS`. This feature is not used in the directions in this text. By default this feature is turned `ON`. `STAT WIZARDS` can be turned `OFF` under the second page of MODE options.

Example T.1

Using a TI-83/84 Plus Calculator to Find a Confidence Interval for the Difference between Two Population Means (σ Known)

A nursing instructor has created a prep course for a standardized comprehensive exam used statewide at community colleges that she believes will result in students performing better on the exam than if they had not taken the course. She collects data from 40 students who took her course and calculates that they had a mean score of 88.1. She also collects data from 35 students who did not take her course and calculates that they had a mean score of 80.2. Assume that the population standard deviation of the scores for students who took the new prep course is known to be 4.1, and the population standard deviation of the scores for students who did not take the new prep course is 6.3. Construct a 90% confidence interval to estimate the true difference between the mean scores of students who took her course and those who did not.

Solution

Since this confidence interval involves known population standard deviations, we will use z to calculate the endpoints of the interval. Press STAT; then scroll to `TESTS` and choose option `9:2-SampZInt`. Next, select the `Stats` option, and enter the given values. Using the results from students who took the prep course as Group 1 and the results from students who did not take the prep course as Group 2, the values would be entered as shown in the following screenshots.

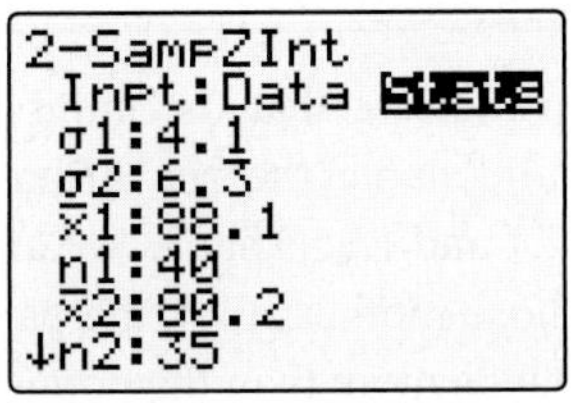

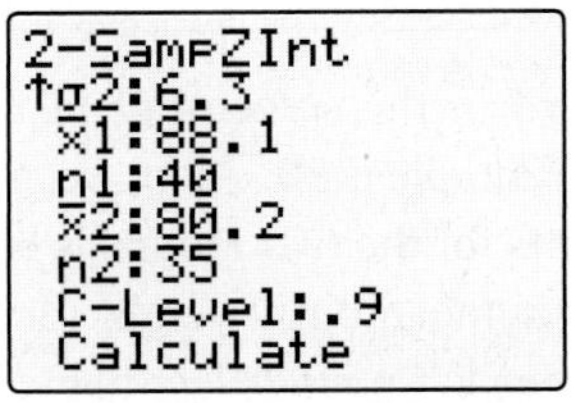

Once you have entered the values, highlight `Calculate` and press ENTER. The confidence interval is displayed along with several of the entered statistics, as shown in the screenshot in the margin.

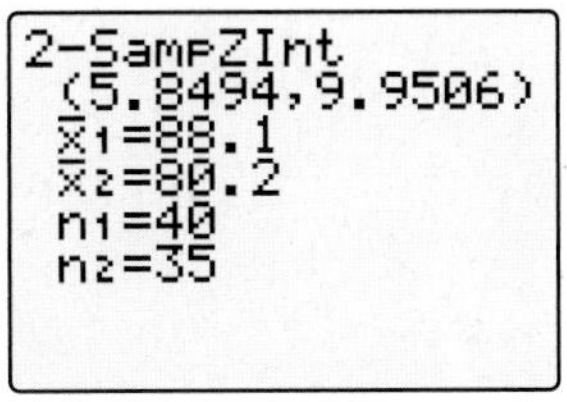

Thus, we are 90% confident that the true difference between Group 1's and Group 2's mean scores is between 5.8 and 10.0. That is, the instructor can conclude with 90% confidence that the mean score for students who take her prep course is between 5.8 and 10.0 points higher than the mean score for students who do not take her course.

Confidence Intervals for Differences between Two Population Means (σ Unknown)

For confidence intervals for differences between two population means where σ is unknown, we need to know if the population variances are assumed to be equal. The TI-83/84 Plus uses the same test for both scenarios by allowing you to set whether the sample variances are pooled (the variances are assumed to be equal) or not pooled (the variances are assumed to be unequal).

Example T.2

Using a TI-83/84 Plus Calculator to Find a Confidence Interval for the Difference between Two Population Means (σ Unknown, Equal Variances)

Obstetricians are concerned that a certain pain reliever may be causing lower birth weights. A medical researcher collects data from a random sample of 12 mothers who took the pain reliever while pregnant and calculates that the mean birth weight of their babies was 5.6 pounds with a standard deviation of 1.8 pounds. She also collects data from a random sample of 20 mothers who did not take the pain reliever while pregnant and calculates that the mean birth weight of their babies was 6.3 pounds with a standard deviation of 2.1 pounds. Assume that the distributions of birth weights are approximately normal for both populations. Assuming that the population variances are equal, construct a 99% confidence interval for the true difference between the mean birth weights for the two groups.

Solution

Since σ is unknown, we will use t to calculate the endpoints of the interval. So press **STAT**; then scroll to `TESTS` and choose option `Ø:2-SampTInt`. Next, select `Stats` and then enter the given statistics. Using the results from the mothers taking the pain reliever as Group 1 and the results from the mothers not using the pain reliever as Group 2, the statistics would be entered as shown in the following screenshots. Then enter the confidence level, which is $c = 0.99$. Since the population variances are assumed to be equal, we want to pool the sample variances, so we choose `Yes` for `Pooled`. Finally, select `Calculate`.

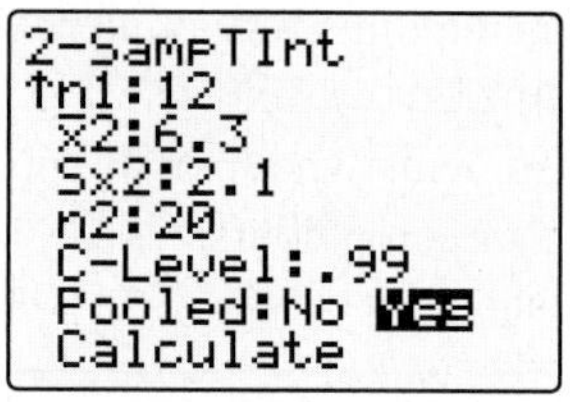

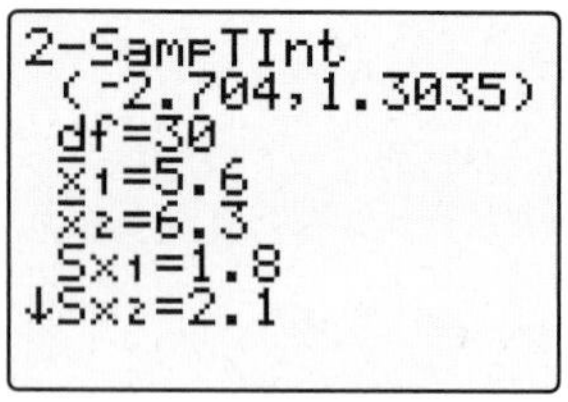

The results shown in the screenshot in the margin indicate that the number of degrees of freedom for this example is 30. Also, we are 99% confident that the true difference between the mean birth weights for the two groups is between –2.7 and 1.3. That is, we are 99% confident that the population mean birth weight for babies whose mothers took the pain reliever while pregnant is between 2.7 pounds less than and 1.3 pounds more than the population mean birth weight for babies whose mothers didn't take the pain reliever. Since the confidence interval contains 0, we cannot conclude that the pain reliever causes lower birth weights.

Confidence Intervals for Means of Paired Differences for Two Populations (σ Unknown, Dependent Samples)

A TI-83/84 Plus does not directly calculate a confidence interval for dependent samples, but we can create a one sample *t*-interval using the paired differences as our raw data.

Example T.3

Using a TI-83/84 Plus Calculator to Find a Confidence Interval for the Mean of the Paired Differences for Two Populations (σ Unknown, Dependent Samples)

In a CPR class, students are given a pretest at the beginning of the class to determine their initial knowledge of CPR and then a posttest at the end of the class to determine their new knowledge of CPR. Educational researchers are interested in whether the class increases a student's knowledge of CPR. Create a 95% confidence interval for the mean increase in the students' CPR test scores. Data from a random sample of 10 students who took the CPR class are listed in the following table.

Students' CPR Test Scores

Pretest Score	60	63	68	70	71	68	72	80	83	79
Posttest Score	80	79	83	90	83	89	94	95	96	93

Solution

First, enter the given data into the calculator. Press STAT, choose option 1:Edit, and enter the pretest scores in L1 and posttest scores in L2. Next, we want to calculate the paired differences in L3. To do so, highlight L3 and enter the formula to subtract the pretest scores from the posttest scores by pressing 2ND 2 − 2ND 1 and then ENTER. This will produce the formula L2−L1 and calculate the paired difference for each pair of data.

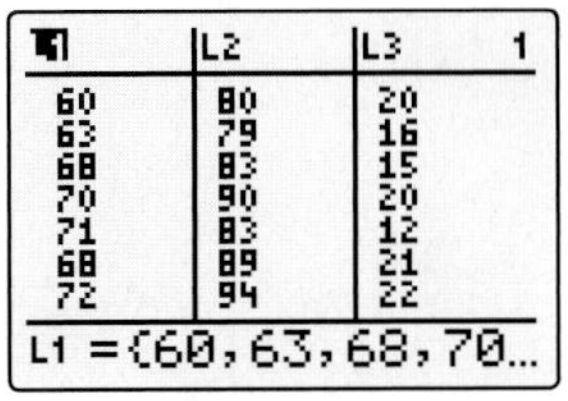

We now have the paired differences in L3 and can create a one-sample *t* confidence interval from the raw data. Press STAT, scroll to TESTS, and choose option 8:TInterval. We want to calculate the confidence interval from the data, so choose the Data option. Our data are in List 3, so enter L3 by pressing 2ND and then 3. The frequency of the data, Freq, is the default value, which is 1. Also, we want a 95% confidence interval, so enter Ø.95 for C-Level. The menu should then appear as it does in the following screenshot on the left. Choosing Calculate produces the results shown in the screenshot on the right.

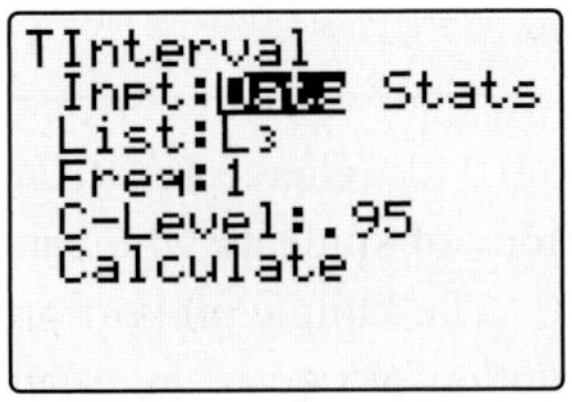

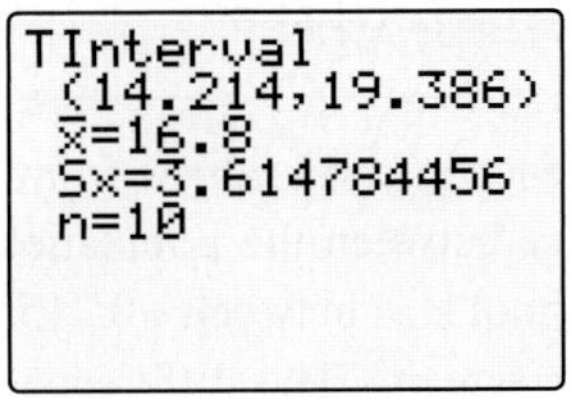

These results show a confidence interval of (14.2, 19.4). Thus, we are 95% confident that the population mean increase of the students' CPR test scores is between 14.2 and 19.4 points. This is the same confidence interval we got when solving this problem by hand earlier in the chapter. Included with the confidence interval on the calculator screen are also the mean of the paired differences, the sample standard deviation of the paired differences, and the sample size.

Confidence Intervals for Differences between Two Population Proportions

Using a TI-83/84 Plus to construct a confidence interval to compare two proportions is very simple. We simply enter the number of successes and the sample size for each sample, as well as the level of confidence.

Example T.4

Using a TI-83/84 Plus Calculator to Find a Confidence Interval for the Difference between Two Population Proportions

School administrators want to know if the percentage of students at School A in a lower economic district who carry cell phones is smaller than the percentage of students who carry cell phones at School B in an upper economic district. A survey of students is conducted at each school and the following results are tabulated.

Responses to Survey

	School A	School B
Carry a Cell Phone	45	53
Do Not Carry a Cell Phone	31	25

Construct a 90% confidence interval for the true difference between the proportions of students who carry a cell phone at the two schools. Does this confidence interval suggest that fewer students carry a cell phone at School A than School B?

Solution

Press STAT and scroll over to TESTS. Choose option B:2-PropZInt. x1 is the number of successes from the first sample, and n1 is the first sample's size. Similarly, x2 is the number of successes from the second sample, and n2 is the second sample's size. C-Level is the confidence level, which must be entered as a decimal. The data should be entered as shown in the screenshot on the left. Select Calculate. The results are shown in the screenshot on the right.

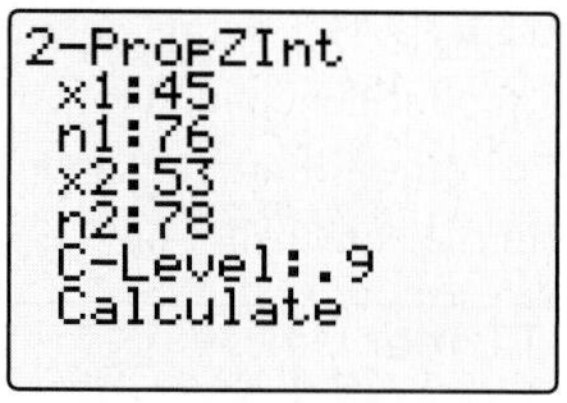

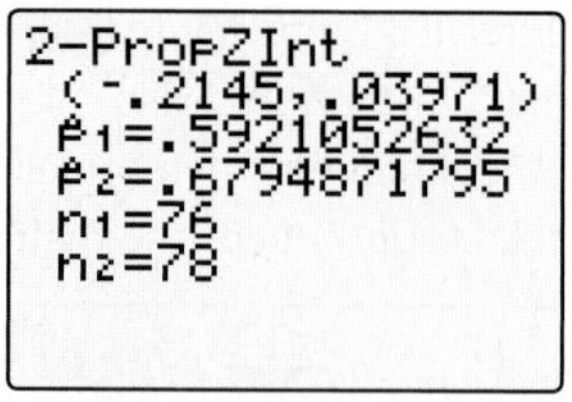

Thus, the confidence interval ranges from −0.215 to 0.040. Therefore, we are 90% confident that the difference between the population proportions of students who carry cell phones at School A and School B is between −0.215 and 0.040. The sample proportions are also shown on the calculator screen. The difference in the solution presented here versus the solution presented earlier in this chapter is due to rounding.

Microsoft Excel

Example T.5

Using Microsoft Excel to Construct a Confidence Interval for the Difference between Two Population Means (σ Known)

Francisco owns a small apple orchard and believes that the mild winter he experienced last year decreased his trees' production. He harvested, weighed, and recorded his trees' production on several randomly selected days during the harvest—35 days last year and 40 days this year. Last year the sample mean was 240.6 pounds of apples per day. Assume that the population standard deviation of the daily production levels last year was 5.2 pounds. This year the sample mean was 238.2 pounds of apples per day. Assume that the population standard deviation of the daily production levels this year was 6.8 pounds. Construct a 99% confidence interval to estimate the true difference between the mean daily yields from last year and this year.

Solution

Begin by entering the given information in cells A1 through C4 as shown in the following screenshot. Then calculate the point estimate, $\bar{x}_1 - \bar{x}_2$, which is $240.6 - 238.2 = 2.4$, by entering **=B2-C2** in cell B6.

Recall that the margin of error is given by the following formula.

$$E = z_{\alpha/2}\sqrt{\frac{\sigma_1^2}{n_1} + \frac{\sigma_2^2}{n_2}}$$

Next, we calculate the margin of error using the critical value obtained by using the NORM.S.INV function with a probability equal to the area under the standard normal distribution to the left of the critical value, which is $1 - \frac{\alpha}{2} = 1 - \frac{0.01}{2} = 0.995$. Enter the following formula in cell B7 to produce the margin of error, $E \approx 3.577132$.

=NORM.S.INV(0.995)*SQRT(B3^2/B4+C3^2/C4)

Then, subtract the margin of error from and add the margin of error to the point estimate to find the endpoints of the confidence interval. To calculate the endpoints, enter **=B6-B7** in cell B8 and **=B6+B7** in cell B9. We are 99% confident that the interval $(-1.2, 6.0)$ contains the true difference between the mean daily yields for the two years. Since the confidence interval contains 0, the data do not provide evidence that the mean daily yields for the two years are unequal. The Excel spreadsheet is shown below.

	A	B	C
1		Year 1	Year 2
2	sample mean	240.6	238.2
3	population standard deviation	5.2	6.8
4	*n*	35	40
5			
6	point estimate	2.4	
7	*E*	3.577132	
8	lower endpoint	-1.177132	
9	upper endpoint	5.977132	

Example T.6

Using Microsoft Excel to Construct a Confidence Interval for the Difference between Two Population Means (σ Unknown, Equal Variances)

Beth is a big hockey fan. One day she hears that the teams in the Eastern Conference score more than their counterparts in the Western Conference. For the games played during the 2011–2012 NHL regular season, a random sample of 10 teams in the East has a mean of 1.145 goals per game with a standard deviation of 0.147 goals per game. For the same season, a random sample of 10 Western teams has a mean of 1.122 goals per game with a standard deviation of 0.124 goals per game. Assume that the population variances are equal in order to construct a 90% confidence interval to estimate the true difference between the mean numbers of goals per game for Eastern teams and Western teams.

Source: NHL.com. "2011-2012 Conference Standings." 10 May 2012. http://www.nhl.com/ice/standings.htm?type=con#&navid=nav-stn-conf (10 May 2012).

Solution

First, enter the given information in cells A1 through C4 as shown in the following screenshot. Then calculate the point estimate, $\bar{x}_1 - \bar{x}_2$, which is $1.145 - 1.122 = 0.023$, by entering **=B2-C2** in cell B6.

Since the population variances are assumed to be equal, the margin of error is given by the following formula.

$$E = t_{\alpha/2}\sqrt{\frac{(n_1-1)s_1^2+(n_2-1)s_2^2}{n_1+n_2-2}}\sqrt{\frac{1}{n_1}+\frac{1}{n_2}}$$

Next, calculate the margin of error by using the critical value of t for a t-distribution with 18 degrees of freedom. Enter the following formula in cell B7 to calculate E, the margin of error.

=T.INV.2T(0.1, 18)*SQRT(((B4-1)*B3^2+(C4-1)*C3^2)/(B4+C4-2))*SQRT(1/B4+1/C4)

Finally, subtract the margin of error from the point estimate to determine the lower endpoint and add the margin of error to the point estimate to determine the upper endpoint. To calculate the endpoints, enter **=B6-B7** in cell B8 and **=B6+B7** in cell B9. Since the interval, (–0.082, 0.128), contains 0, the claim of Eastern scoring superiority is unsubstantiated at the 90% confidence level.

	A	B	C
1		Eastern Conf.	Western Conf.
2	sample mean	1.145	1.122
3	sample standard deviation	0.147	0.124
4	*n*	10	10
5			
6	point estimate	0.023	
7	*E*	0.105458	
8	lower endpoint	-0.082458	
9	upper endpoint	0.128458	

Example T.7

Using Microsoft Excel to Construct a Confidence Interval for the Mean of the Paired Differences for Two Populations (σ Unknown, Dependent Samples)

Abby thinks traffic is worse in the afternoon. She asks her coworkers to record their travel times, in minutes, to and from work one day. The following table lists the times they recorded.

Travel Times (in Minutes)								
	Darren	**Jane**	**Justin**	**Kyia**	**Libby**	**Ron**	**Rusty**	**Siani**
Morning	25	15	48	5	9	55	35	17
Afternoon	31	14	55	7	12	56	32	15

Construct a 95% confidence interval for the mean difference between afternoon and morning travel times.

Solution

First, we must choose the correct procedure. Since the data are paired, we will construct an interval centered at the mean of the paired differences using the Student's t-distribution. We enter the given data in cells A1 through I3, as shown in the following screenshot. Then, calculate the paired differences in row 4 by entering **=B3-B2** in cell B4 and copying and pasting that formula into cells C4–I4. Then we determine the sample size and calculate the mean and standard deviation of the paired differences—n, $\bar{d}$, and s_d, respectively. Enter **=COUNT(B4:I4)**, **=AVERAGE(B4:I4)**, and **=STDEV.S(B4:I4)** in cells B6, B7, and B8, respectively, to produce the count, mean, and standard deviation of the paired differences.

Recall that the margin of error is given by the following formula.

$$E = \left(t_{\alpha/2}\right)\left(\frac{s_d}{\sqrt{n}}\right)$$

The margin of error, $E \approx 3.028774$, is calculated by entering the following formula in cell B9.

=T.INV.2T(0.05, B6-1)*B8/SQRT(B6)

Finally, subtracting the margin of error from and adding the margin of error to the mean of the paired differences yields the lower and upper endpoints. To calculate the endpoints, enter **=B7-B9** in cell B10 and **=B7+B9** in cell B11. Since the interval, (−1.4, 4.7), contains 0, there is not sufficient evidence at the 95% confidence level to conclude that afternoon travel times are significantly longer than morning travel times for Abby's coworkers. The Excel spreadsheet is shown below.

	A	B	C	D	E	F	G	H	I
1		Darren	Jane	Justin	Kyia	Libby	Ron	Rusty	Siani
2	Morning, *x*_1	25	15	48	5	9	55	35	17
3	Afternoon, *x*_2	31	14	55	7	12	56	32	15
4	paired difference, *d* = *x*_2 - *x*_1	6	-1	7	2	3	1	-3	-2
5									
6	*n*	8							
7	mean of paired differences	1.625							
8	standard deviation of paired differences	3.622844							
9	*E*	3.028774							
10	lower endpoint	-1.403774							
11	upper endpoint	4.653774							

Example T.8

Using Microsoft Excel to Construct a Confidence Interval for the Difference between Two Population Proportions

A city planner believes that two separate districts have significantly different proportions of people living below the poverty line. She takes a sample from each district to justify her claim. In a sample of 252 people from District 1, 63 are living below the poverty line. On the other hand, District 2 reports 45 people living below the poverty line with a sample size of 256. Construct a 95% confidence interval for the difference between the proportions of people living below the poverty line in Districts 1 and 2.

Solution

First, we enter the given information in cells A1 through C3 as shown in the following screenshot. Then we calculate the sample proportions and the point estimate. Enter **=B2/C2** in cell D2 and **=B3/C3** in cell D3 to produce the values of $\hat{p}_1$ and $\hat{p}_2$, respectively. Next, enter **=D2-D3** in cell B5 to calculate the point estimate, $\hat{p}_1 - \hat{p}_2 \approx 0.074219$.

Recall that the margin of error is given by the following formula.

$$E = z_{\alpha/2}\sqrt{\frac{\hat{p}_1\left(1-\hat{p}_1\right)}{n_1}+\frac{\hat{p}_2\left(1-\hat{p}_2\right)}{n_2}}$$

Again, we find the critical value of the normal distribution by using the NORM.S.INV function with a probability equal to the area under the standard normal distribution to the left of the critical value, which is $1-\frac{\alpha}{2}=1-\frac{0.05}{2}=0.975$. The margin of error, E, is calculated by entering the following formula.

=NORM.S.INV(0.975)*SQRT((D2*(1-D2))/C2+D3*(1-D3)/C3)

Subtracting the margin of error from and adding the margin of error to the point estimate yields the lower and upper endpoints. To calculate the endpoints, enter **=B5-B6** in cell B7 and **=B5+B6** in cell B8. Since both endpoints of the interval, (0.003, 0.145), are positive, the claim that the two districts have significantly different proportions of people living below the poverty line is supported at the 95% confidence level. Specifically, we are 95% confident that the proportion of people living below the poverty line is between 0.003 and 0.145 higher in District 1 than in District 2. The Excel spreadsheet is shown in the following screenshot.

	A	B	C	D
1		*x*	*n*	*p-hat*
2	District 1	63	252	0.25
3	District 2	45	256	0.175781
4				
5	point estimate	0.074219		
6	*E*	0.070939		
7	lower endpoint	0.00328		
8	upper endpoint	0.145157		

Minitab

Example T.9

Using Minitab to Construct a Confidence Interval for the Mean of the Paired Differences for Two Populations (σ Unknown, Dependent Samples)

Geraldo's Tire Store claims to have better prices than a rival store. To prove this claim, the statistician at Geraldo's prices a random sample of 10 tire models at both stores. The data are in the table below. Construct a 95% confidence interval for the true mean difference between the tire prices at the two stores.

Tire Prices (in Dollars)										
Geraldo's	45	75	80	65	73	109	125	78	152	73
Rick's	47	80	85	62	80	105	130	80	155	78

Solution

First, enter the tire prices in the first two columns in the worksheet, C1 and C2. Next, go to **Stat ▶ Basic Statistics ▶ Paired t**. Next, select **Samples in columns**; enter **C1** for the first sample and **C2** for the second sample. Then click **Options** in order to specify a confidence level of **95**. Click **OK** on the options window and **OK** on the main dialog window and the confidence interval is produced in the Session window. The dialog boxes are shown in the following screenshots.

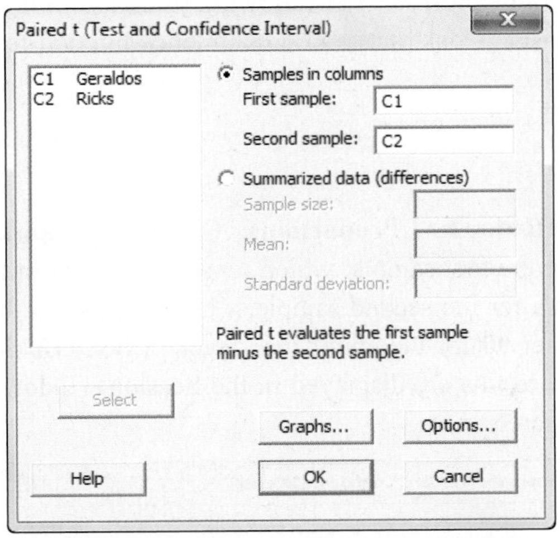

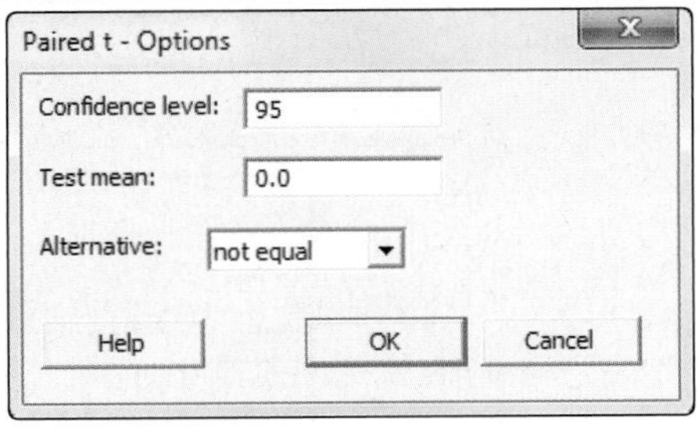

The results in the Session window should appear as follows.

```
Session

Paired T-Test and CI: Geraldos, Ricks

Paired T for Geraldos - Ricks

             N   Mean  StDev  SE Mean
Geraldos    10   87.5   31.7     10.0
Ricks       10   90.2   31.9     10.1
Difference  10  -2.70   3.62     1.15

95% CI for mean difference: (-5.29, -0.11)
T-Test of mean difference = 0 (vs not = 0): T-Value = -2.36  P-Value = 0.043
```

The interval produced, $(-5.3, -0.1)$, contains only negative values. Therefore, Geraldo's Tire Store can be 95% confident that it has lower prices than Rick's.

Example T.10

Using Minitab to Construct a Confidence Interval for the Difference between Two Population Proportions

A local tourism board wants to know if the percentage of people who are tourists at a particular site on Labor Day increased from one year to the next. The first year, a sample of 150 people was taken of which 115 were tourists. The following year, a sample of 174 people was taken of which 120 were tourists. Construct a 90% confidence interval for the true change in the percentage of tourists.

Solution

Go to **Stat ▶ Basic Statistics ▶ 2 Proportions**. Choose **Summarized data** and enter **120** events and **174** trials for the first sample, which represents the most recent year. Then enter **115** events and **150** trials for the second sample, which represents the previous year. Next, choose **Options** and enter **90** for the confidence level. Click **OK** on the options and main dialog windows and the results are displayed in the Session window. The dialog boxes are shown in the following screenshots.

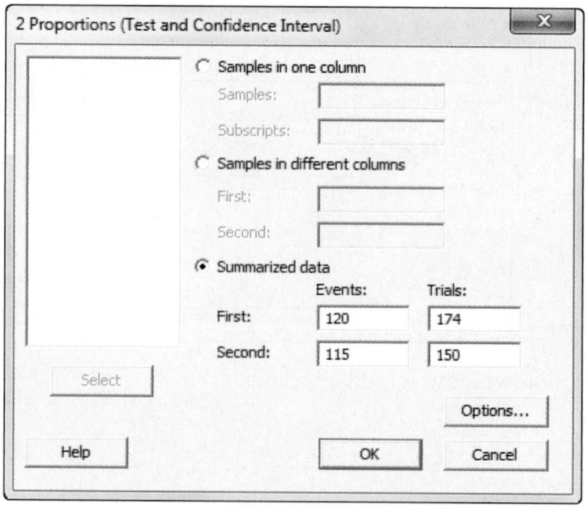

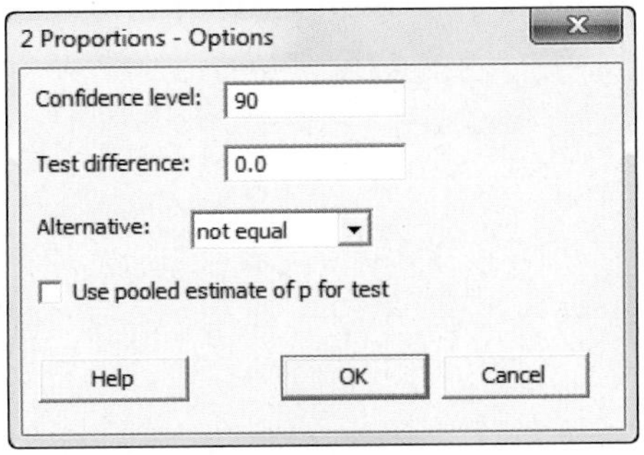

The results in the Session window should appear as follows.

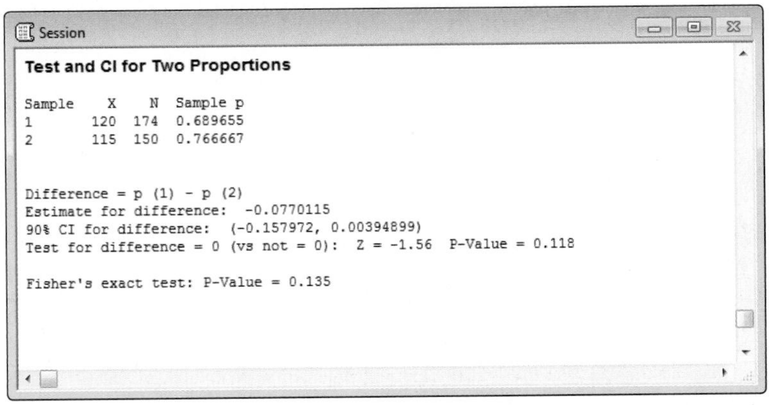

Session
Test and CI for Two Proportions
Sample X N Sample p
1 120 174 0.689655
2 115 150 0.766667
Difference = p (1) - p (2)
Estimate for difference: -0.0770115
90% CI for difference: (-0.157972, 0.00394899)
Test for difference = 0 (vs not = 0): Z = -1.56 P-Value = 0.118
Fisher's exact test: P-Value = 0.135

The confidence interval produced is (–0.158, 0.004). Since the confidence interval contains 0, there is not sufficient evidence at the 90% confidence level to conclude that the percentage of tourists either increased or decreased. Since the interval estimates the proportion for the most recent year minus the proportion for the previous year, with 90% confidence, we can say that the true percentage of people at the site who are tourists may have decreased by as much as 15.8% or increased slightly by as much as 0.4%.

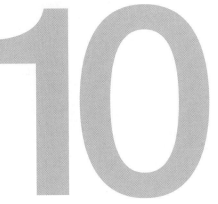

Chapter Ten

Hypothesis Testing

Sections

Objectives

1. State the null and alternative hypotheses.
2. Determine the distribution to use for the test statistic.
3. Use the rejection region to draw a conclusion.
4. Use the *p*-value to draw a conclusion.
5. Interpret the conclusion to a hypothesis test.
6. Describe Type I and Type II errors and recognize when they occur.
7. Perform hypothesis tests for means, proportions, and variances.
8. Perform a chi-square test for goodness of fit.
9. Perform a chi-square test for association.

10

Introduction

Anyone can make a claim about a population parameter. You have probably heard claims similar to the following: "Four out of five dentists prefer Star Brite toothpaste" and "Less than 1% of dogs attack people unprovoked." People make decisions every day based on claims such as these. Sometimes the decisions are relatively insignificant, like what type of toothpaste to buy. Other times, the decisions we make based on these claims are very important, like whether or not to use a new prescription drug that has come on the market. Or perhaps you may choose your next car based on reported safety ratings. In these cases, it is expected that extensive testing will be done to ensure that the information given to the public is as accurate as possible.

For example, the Food and Drug Administration has the daunting task of determining which drugs are safe and effective enough to be sold to the public. The process through which a drug obtains approval is grueling, and usually it takes years for the few drugs that pass inspection to be released. Even then, some drugs are later found to be dangerous enough to be taken back off the market. For example, recently a popular arthritis drug was found to increase the risk of heart attacks. For this reason, the FDA decided that this drug's risks outweighed its benefits, and it was subsequently pulled from the market.

How do you begin to test claims made about population parameters? Where do we draw the line? Hypothesis testing is one statistical process that sets a uniform standard for evaluating these types of claims.

10.1 Fundamentals of Hypothesis Testing

You probably remember the definition of a hypothesis from a science class. A **hypothesis** is a theory or premise—often it is a claim that someone is making that must be investigated. In science, a hypothesis is usually a claim about how the world works. In statistics, a hypothesis involves a claim about the numerical value of a population parameter, such as the population mean, proportion, or variance. Consider the following statements.

"In any given one-year period, 9.5% of the population, or about 18.8 million Americans suffer from a depressive illness."

Source: Russell, June. "What You Need to Know About Alcohol and Depression: Depression Often Goes Hand-in-Hand With Alcohol Use." About.com. 9 Dec. 2011. http://depression.about.com/od/drugsalcohol/a/alcoholanddep.htm (7 Feb. 2012).

"For cars, this Corporate Average Fuel Economy (CAFE) has been 27.5 miles per gallon since 1990. Truck fuel economy has gradually increased to 22.2 mpg for the 2007 model year."

Source: O'Donnell, Jayne. "Politicians Call for New Fuel Economy Rules." *USA Today.* 3 May 2006. http://www.usatoday.com/money/autos/2006-05-02-mileage-usat_x.htm (5 Nov. 2011).

"In 1967, Americans ate 114 pounds of sugar and sweeteners a year per capita.... In 2003, each person consumed about 142 pounds of sugar per year."

Source: "One Sweet Nation." *U.S. News and World Report.* 28 Mar. 2005.

Notice that each of these statements makes a claim about a population parameter. You can find statements like these in almost every newspaper article today. However, these claims are not limited only to journalism. They appear in any number of research studies—studies that decide whether new drugs are safe, whether new public safety campaigns are working, or whether one or another candidate is likely to win an election. How, then, do we test these claims using statistical principles? In short, they all depend on a technique appropriately called **hypothesis testing**.

Before we can work with the numbers in a hypothesis test, we must start with clearly stating what it is we want to investigate. For instance, suppose you wanted to know if the average person actually consumes 142 pounds of sugar per year, as reported above. After collecting data, the sample statistics that you calculate provide evidence that supports this claim. Is that strong enough evidence to say that people *always* consume 142 pounds of sugar per year? In statistics, unless you are able to ask every person in the population, your statement is only as strong as the amount of data that you were able to collect. On the other hand, it is a more powerful statement to be able to say that you found data that *contradicts* the believed value. Thus, when we set up our study, we want to phrase the research hypothesis in such a way that if evidence were found in support of it, it would contradict the currently held parameter. In our sugar-consumption example, it would be a stronger study to research whether people do not consume 142 pounds of sugar per year. Or, if there is an indication that perhaps the true parameter is greater than or less than 142 pounds, the research hypothesis could be phrased to reflect this inclination.

Null and Alternative Hypotheses

The first step in a hypothesis test, then, is to clearly state the *research hypothesis*. This hypothesis is also called the **alternative hypothesis**, denoted by H_a. It is a mathematical statement that describes a population parameter, and it is the hypothesis that the researcher is aiming to gather evidence in favor of. The mathematical opposite of this hypothesis is called the **null hypothesis**, denoted by H_0. These two hypotheses form a pair of logically opposite possibilities.

Math Symbols

H_0: null hypothesis

H_a: alternative hypothesis

Throughout a hypothesis test, the null hypothesis is assumed to be true. Note that equality must be included in the null hypothesis to establish the presumed value of the population parameter; hence the null hypothesis will always include equality. The data must then provide overwhelming evidence to contradict the null hypothesis for the alternative hypothesis to be supported at the conclusion of the test. Therefore, rejecting the null hypothesis is a stronger conclusion than failing to reject it, which is why the hypothesis that the researcher aims to gather evidence in favor of is the alternative hypothesis.

Let's now look at an example of how to set up the null and alternative hypotheses. Consider a claim made by the manager of a candy manufacturing plant. Ever since his machinery was overhauled, he has claimed that the mean weight of the chocolate bars produced is smaller than the correct mean weight of 4 ounces. Since his claim refers to the value of the population mean of the weights of the chocolate bars, μ, it is the research hypothesis and is expressed mathematically as $H_a: \mu < 4$. The mathematical opposite of this claim is $\mu \geq 4$. Hence, the null hypothesis is that the population mean is greater than or equal to 4 ounces and the alternative hypothesis is that it is less than 4 ounces. Mathematically, these hypotheses are written as follows.

Memory Booster

Begin by writing the research hypothesis as a mathematical statement for H_a, and then write the logical opposite, including equality, for H_0.

$$H_0: \mu \geq 4$$
$$H_a: \mu < 4$$

Definition

Hypothesis testing is a technique for testing a claim about a population parameter using statistical principles.

The **alternative hypothesis**, denoted by H_a, is a mathematical statement that describes a population parameter, and it is the hypothesis that the researcher is aiming to gather evidence in favor of; it is also referred to as the *research hypothesis*.

The **null hypothesis**, denoted by H_0, is the mathematical opposite of the alternative hypothesis; it will always include equality.

Example 10.1

Determining the Null and Alternative Hypotheses

Determine the null and alternative hypotheses for the following scenario.

At Northwest Mississippi Community College, syllabi for online classes state that students should expect to spend 10 hours per week doing coursework for each three-credit-hour class. The director of eLearning for the community college is concerned that students are being required to spend more than 10 hours per week doing work for each three-credit course.

Solution

In order to determine the hypotheses, we must first ask ourselves "What does the director want to gather evidence for?" The eLearning director's concern is that students are being required to spend more than the 10 hours per week on coursework that the syllabi state will be required. Hence, she is investigating whether the mean is greater than 10 hours. Therefore, the research hypothesis, H_a, is $\mu > 10$. The logical opposite of this hypothesis is $\mu \le 10$, which is the null hypothesis, H_0. Note that the null hypothesis contains a statement about what is currently believed to be true regarding the population mean, that is, $\mu = 10$. Thus, the two hypotheses are written as follows.

$$H_0: \mu \le 10$$
$$H_a: \mu > 10$$

Example 10.2

Determining the Null and Alternative Hypotheses

Determine the null and alternative hypotheses for the following scenario.

After learning that Peter Colat of Switzerland could hold his breath for 19 minutes and 21 seconds under water, a group of students decided to conduct their science fair project around the amount of time an average adult can hold his or her breath. Part of the group found research to suggest that adults can hold their breath for 30.0 seconds when submerged in lukewarm water. The other half of the group found evidence to suggest that the mean time is different if the water temperature is colder. The group would like to test this claim about adults holding their breath in cold water.

Solution

We begin by determining what the group hopes to gather evidence in favor of. They are investigating whether the mean amount of time that an adult can hold his or her breath underwater when the temperature is cold is different from the reported time of 30.0 seconds when the water is lukewarm. Note that they are simply interested in whether the time is different than 30.0 seconds, not necessarily whether it is longer or shorter than 30.0 seconds. Therefore, we write the research hypothesis mathematically as $H_a\colon \mu \neq 30.0$ The logical opposite is $\mu = 30.0$ and is the null hypothesis, H_0. Therefore, the two hypotheses are written as follows.

$$H_0\colon \mu = 30.0$$
$$H_a\colon \mu \neq 30.0$$

Example 10.3

Determining the Null and Alternative Hypotheses

Determine the null and alternative hypotheses for the following scenario.

A leading news authority claims that the President's job approval rating has dropped over the past three months. Previous polls put the President's approval rating at a minimum rate of 56%. The president's chief of staff is concerned about this news report since it is an election year, and he wants to run a test on the claim.

Solution

The news authority's claim refers to a population proportion, which we write symbolically as p. The claim being tested is that the President's approval rating has dropped since a previous poll estimated it to be at least 56%. Thus, the research hypothesis is that the President's approval rating is *less than* 56%, written mathematically as $H_a\colon p < 0.56$. The logical opposite of the research hypothesis is the null hypothesis, $H_0\colon p \geq 0.56$. Therefore, we have the following hypotheses.

$$H_0\colon p \geq 0.56$$
$$H_a\colon p < 0.56$$

Example 10.4

Determining the Null and Alternative Hypotheses

Determine the null and alternative hypotheses for the following scenario.

A survey of Super Bowl viewers found that 8.4% said that they were influenced to buy products from companies that advertise during the game. Because of rising costs, one company president has threatened the advertising team with not renewing the Super Bowl ad spot unless they get the results they want. He says that unless their company's ad stimulates at least 8.4% of viewers to buy products, then they will not be airing an ad during the Super Bowl next year.

Solution

In order to determine the hypotheses, we need to ask, "What is the company president researching?" He does not want to cancel next year's Super Bowl ad unless there is overwhelming evidence that the percentage of viewers persuaded to buy the company's products is too low. So the research hypothesis is that the percentage is less than 8.4%, written mathematically as $H_a: p < 0.084$. The mathematical opposite of $p < 0.084$ is $p \geq 0.084$. Hence, the null and alternative hypotheses are written as follows.

$$H_0: p \geq 0.084$$
$$H_a: p < 0.084$$

Memory Booster

Throughout a hypothesis test, the null hypothesis is assumed to be true.

Once the hypotheses have been clearly stated, we can begin to decide which of the two hypotheses the data support. Thus, a hypothesis test compares the merit of these two competing hypotheses by examining the data that are collected. However, it does not treat the two hypotheses equally. The null hypothesis describes the currently accepted value for the population parameter, which is why it is the hypothesis that contains equality. It states what the status quo is, that is, what value the parameter is assumed to *equal*. Throughout a hypothesis test, the null hypothesis is assumed to be true. A hypothesis test therefore examines whether or not the data collected are consistent with this belief that the null hypothesis is true. If the data disagree "sufficiently" with the null hypothesis, then the null is *rejected* in favor of the alternative hypothesis.

Let's consider the classic analogy of the United States criminal justice system to help us further understand how to perform a hypothesis test. First, the starting assumption in any trial is that the defendant is not guilty. This is to be accepted as true throughout the trial. Therefore, the null hypothesis is that the defendant is not guilty. The prosecutor, however, makes a claim that the defendant *is* guilty. It is then the prosecutor's responsibility to provide enough evidence *against* the assumption of innocence so that the jury will be persuaded beyond a reasonable doubt that the defendant is guilty of the crime. In just the same way, the null hypothesis is believed to be true unless the evidence against it is overwhelming.

Test Statistic

In a criminal court, the jurors listen to the evidence and have to decide whether the evidence provides proof of guilt *beyond a reasonable doubt*. Often this burden of proof depends on the expectation of the jurors themselves and is somewhat subjective. However, the *evidence* in a hypothesis test is provided by the data in the form of a statistic, such as a sample mean, which must be converted to a test statistic. A **test statistic** is the value used to make a decision about the null hypothesis and is derived from the sample statistic. In contrast to the criminal justice system, there is no subjective jury in hypothesis testing. Instead, a hypothesis test depends on the fact that sample statistics are unlikely to take values far away from the value of the corresponding population parameter.

Recall from the earlier example about the chocolate bars that the null hypothesis is $\mu \geq 4$. What kind of data would convince us that the mean weight of the chocolate bars is actually less than 4 ounces? If the null hypothesis were true, we would expect the mean weight of a sample of bars to be somewhere close to or greater than 4 ounces. If the sample mean were much smaller than 4 ounces, this evidence would certainly persuade us that there is a problem with the machinery. But how far away from 4 ounces does the sample mean need to be in order for us to be persuaded? Would a mean weight of 3.9 ounces be far enough away from the presumed population mean of 4 ounces? How about 3.01 ounces? Where do we draw the line that distinguishes "far enough" from the population parameter?

To answer this question, we need to introduce the term *statistically significant*. A sample statistic is said to be **statistically significant** if it is far enough away from the presumed value of the population parameter to conclude that it would be unlikely for that sample statistic to occur by chance if the null hypothesis is true. Hence, the test statistic calculated from a statistically significant sample statistic is also statistically significant. For example, if you toss a fair coin 100 times, it is very likely to land on heads 49 times. However, if you tossed a coin 100 times and it landed on heads only 20 times, you would probably say that this sample proportion was unlikely to occur by chance if the coin was fair, and that perhaps the coin is not fair after all. Thus, we might say that the coin landing on heads 20 times, assuming that the coin is fair, is statistically significant.

To determine if the sample statistic is statistically significant, a confidence interval for the population parameter can be constructed using the methods we saw in Chapter 8. The level of confidence used for this confidence interval is analogous to the concept of *reasonable doubt*. The evidence is considered statistically significant when the sample statistic is so inconsistent with the null hypothesis that the presumed value of the population parameter from the null hypothesis falls outside the confidence interval, that is, *beyond a reasonable doubt*.

Recall that the level of confidence, c, is a measure of how certain we are that the confidence interval contains the population parameter. If the level of confidence we choose for a particular hypothesis test is c, then the probability of making the error of rejecting a true null hypothesis in our testing is $1 - c$, which we denote as the Greek letter "alpha," α. We call this value, α, the **level of significance**. When constructing confidence intervals in the previous chapters, the level of confidence was always given in the problems. Throughout this chapter, we will state either the level of confidence or the level of significance when performing a hypothesis test. In the process of a statistical study, any level of significance used should be specified before collecting data. Alternatively, the level of significance can be left to the consumer of your research to decide for himself or herself if the test statistic is statistically significant. This is commonplace in academic research.

Math Symbols

α**:** level of significance; Greek letter, alpha

Memory Booster

c = level of confidence

$\alpha = 1 - c$
= level of significance

Definition

A **test statistic** is the value used to make a decision about the null hypothesis and is derived from the sample statistic.

A sample statistic is said to be **statistically significant** if it is far enough away from the presumed value of the population parameter to conclude that it would be unlikely for the sample statistic to occur by chance if the null hypothesis is true.

The **level of significance**, denoted by α, is the probability of making the error of rejecting a true null hypothesis in a hypothesis test; $\alpha = 1 - c$.

As we mentioned earlier in this section, the test statistic is the value used to test the null hypothesis. When we constructed confidence intervals in Chapter 8, we introduced all of the test statistics we will use for hypothesis testing now. They are the sample statistics, $\bar{x}$, $\hat{p}$, and s^2, which are used as point estimates for the population parameters when constructing confidence intervals and converted to scores of z, t, or χ^2 for use in hypothesis testing, as we will see in this chapter. After the hypotheses have been clearly stated, the next step in a hypothesis test is to determine which probability distribution to use for the test statistic and state the level of significance. In the following sections we will look at how to determine which distribution to use for the test statistic and practice calculating the test statistic, given data from a representative sample.

Gathering the data that are used to calculate the test statistic is Step 3 in a hypothesis test. Appropriate sampling techniques, as discussed in Chapter 1, should be used when gathering your own data. For the purposes of our textbook examples and exercises, appropriate data will be provided to you in the form of raw data or sample statistics.

The last step in a hypothesis test is to draw a conclusion based on the calculations performed in Step 3. That conclusion can take one of two forms: 1) reject the null hypothesis, or 2) fail to reject the null hypothesis. If the test statistic calculated from the data collected in a hypothesis test is found to be statistically significant, then we *reject the null hypothesis* in favor of the alternative hypothesis. This is the stronger of the two conclusions that can be made. However, if the test statistic is not statistically significant, then the decision is to *fail to reject the null hypothesis*. This is the weaker of the two conclusions. Notice that, in just the same sense that innocence is not proved in a trial, a hypothesis test never *proves* that either hypothesis is true. It can only say that either the evidence is strong enough to contradict the null hypothesis in favor of the alternative or else it is not strong enough to contradict the null hypothesis. The null hypothesis is presumed to be true unless there is statistically significant evidence to the contrary. Thus, there are only two possible conclusions that can be drawn from a statistical hypothesis test, and these are restated in the following box.

Properties

Conclusions for a Hypothesis Test

- Reject the null hypothesis.
- Fail to reject the null hypothesis.

Once you have decided upon a conclusion for the hypothesis test, a discussion of the meaning of this conclusion in terms of the original claim is appropriate. Because it is not necessarily obvious what each conclusion implies, a well-worded statement of your findings that references the claim, and not just mathematical symbols, is a useful summary of your results. Let's look at two examples.

Example 10.5

Interpreting the Conclusion to a Hypothesis Test

At Northwest Mississippi Community College, syllabi for online classes state that students should expect to spend 10 hours per week doing coursework for each three-credit-hour class. The director of eLearning for the community college is concerned that students are being required to spend more than 10 hours per week doing work for each three-credit course. A hypothesis test with a 5% level of significance is performed on the director's claim. The result is to reject the null hypothesis. Does the conclusion support the director's claim?

Solution

We must first review the null and alternative hypotheses in order to evaluate the conclusion. Recall from Example 10.1 that the two hypotheses for this hypothesis test are as follows.

$$H_0: \mu \le 10$$
$$H_a: \mu > 10$$

Since the null hypothesis was rejected, the evidence supports the alternative hypothesis, that is, the research hypothesis. Thus, the eLearning director can be confident in her assessment that students study more per week, on average, than the time stated on the syllabi, with a 5% level of significance.

Example 10.6

Interpreting the Conclusion to a Hypothesis Test

Based on historical data, the Board of Education for one large school district believes that the percentage of high school sophomores considering dropping out of school is at least 10%. A high school counselor in the district claims that this percentage is too high. A hypothesis test with $\alpha = 0.02$ is performed on the counselor's claim. The result is to fail to reject the null hypothesis. Do the findings support the counselor's claim?

Solution

Once again we must write down the null and alternative hypotheses to evaluate the conclusion. We need to ask, "What does the counselor hope that the data will show evidence for?" The school counselor claims that the percentage of high school sophomores in this district that are considering dropping out of school is less than the assumed rate of at least 10%. Therefore, the research hypothesis is written mathematically as $H_a: p < 0.10$, and its logical opposite is $H_0: p \geq 0.10$, so we write the hypotheses as follows.

$$H_0: p \geq 0.10$$
$$H_a: p < 0.10$$

Since the test resulted in a failure to reject the null hypothesis, the evidence is not strong enough at this level of significance to support the counselor's claim that the presumed rate is too high. Remember that performing a hypothesis test does not prove the null hypothesis to be true. Therefore, we can only say that there is not sufficient evidence to reject it.

Now that we understand the basis for performing a hypothesis test, let's outline the general steps of any hypothesis test. They are given in the box below.

Procedure

Performing a Hypothesis Test

1. State the null and alternative hypotheses.
2. Determine which distribution to use for the test statistic, and state the level of significance.
3. Gather data and calculate the necessary sample statistics.
4. Draw a conclusion and interpret the decision.

Types of Errors in Hypothesis Testing

As stated previously, there are only two possible conclusions to any hypothesis test: reject the null hypothesis or fail to reject the null hypothesis. When wording the conclusion, you must be careful to state only the findings and not be tempted to say more than the evidence supports. Remember the analogy of the United States judicial system? In a US trial, a finding of not guilty does not indicate that the defendant is innocent. It only means that the evidence presented in the trial did not support a finding of guilty. Similarly, if we fail to reject the null hypothesis, then we cannot say that the null hypothesis is supported. We can only say that the evidence does not support the alternative hypothesis. It is important to remember that no matter what the conclusion is, neither hypothesis has been proven.

Memory Booster

No matter what the conclusion of a hypothesis test is, neither hypothesis has been proven.

Let's consider an example of a student scheduling classes for the next semester. Suppose this student is currently enrolled in English Composition I (Comp I). Next semester, the student will need to take English Composition II (Comp II), unless he fails Comp I. A potential problem arises during preregistration. The student has not finished Comp I; therefore, he needs to use the evidence he has, that is, his grades Thus, far, to draw a conclusion about whether or not he will pass Comp I. Let's give the student the benefit of the doubt, and assume that he will pass Comp I. Thus, the null and alternative hypotheses can be written as follows.

H_0: The student passes Comp I.

H_a: The student does not pass Comp I.

If the student's grades are mostly Fs, then we might conclude that the evidence is sufficient to reject the null hypothesis, supporting the alternative hypothesis that the student will not pass Comp I and therefore must repeat Comp I the following semester. If the final grade in the class is an F, then a correct decision was made. If the final grade is a C, then an error was made, and the student would need to correct his preregistration schedule.

On the other hand, the student's grades could be mostly passing grades. In this case, we might conclude that the evidence is not sufficient to reject the null hypothesis. Thus, we would fail to reject the null hypothesis and the student would schedule Comp II for the following semester. If the final grade in the class turns out to be a passing grade, then a correct decision was made. If the student's grade is an F, then an error was made, and the schedule would need to be corrected. The following table summarizes these four possible situations.

		Reality	
		Student passes Comp I	**Student does not pass Comp I**
Decision	**Comp I scheduled for next semester**	Wrong schedule	Schedule correct Student retakes Comp I
	Comp II scheduled for next semester	Schedule correct Student takes Comp II	Wrong schedule

Figure 10.1: Possible Scheduling Errors

Notice that there are two types of errors possible: 1) rejecting a true null hypothesis, or 2) failing to reject a false null hypothesis. We call rejecting a true null hypothesis a **Type I error**, and failing to reject a false null hypothesis is called a **Type II error**.

Replacing the specific statements in the table with general statements about the null hypothesis, the table becomes the following.

		Reality	
		H_0 is true	H_0 is false
Decision	Reject H_0	Type I error	Correct decision
	Fail to reject H_0	Correct decision	Type II error

Figure 10.2: Types of Errors

The decision criteria used in a hypothesis test are usually based on the idea of minimizing the chance of committing a Type I error. The probability of committing a Type I error, in other words the probability of rejecting a true null hypothesis, is called the level of significance, which we denote with the Greek letter, α. We have already seen this value in this section. This is the same level of significance that is used to draw the conclusions in hypothesis tests.

In our earlier example about a student choosing whether to preregister for Comp I or Comp II, there are two different ways that scheduling errors can occur. The level of significance, α, specifies how willing the student is to make a Type I error, which in this scenario would be scheduling Comp I for next semester and then passing Comp I this semester. Would he be willing to accept a 5% chance of making this error? In this situation, it might depend on the amount of time and effort required to change a class schedule at his institution as well as the likelihood of a seat being available in Comp II. If a computerized system allows changes to be made with minimal effort and Comp II classes do not fill up during preregistration, he might be willing to make a Type I error 10% of the time. If changing his schedule requires trips to several professors' offices and the registrar, or if all Comp II classes are usually filled during preregistration, then he may be willing to accept only a 5%, or even a 1%, chance of making a Type I error. The smaller the level of significance, the more evidence required to reject the null hypothesis. The most common practice in academic research is to use a level of significance of 0.05, if one is stated at all. Many times, as we noted earlier, research studies do not provide a value for the level of significance, but simply state the probability that the results obtained would occur by chance under the assumption that the null hypothesis is true. Thus, readers can determine for themselves whether the null hypothesis should be rejected.

Technically, we can choose the level of significance, α, to be any value we wish. So why wouldn't we choose the level of significance to be as low as possible? The answer lies in the inverse relationship between the probability of making a Type I error and the probability of making a Type II error. The probability of making a Type II error is represented by the Greek letter "beta," β. The smaller the probability of making a Type I error, the larger the probability of making a Type II error will be. Thus, you must choose a level of significance small enough to control the risk of a Type I error while not making it too small, which would increase the chance of making a Type II error.

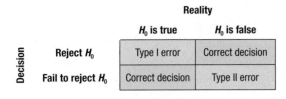

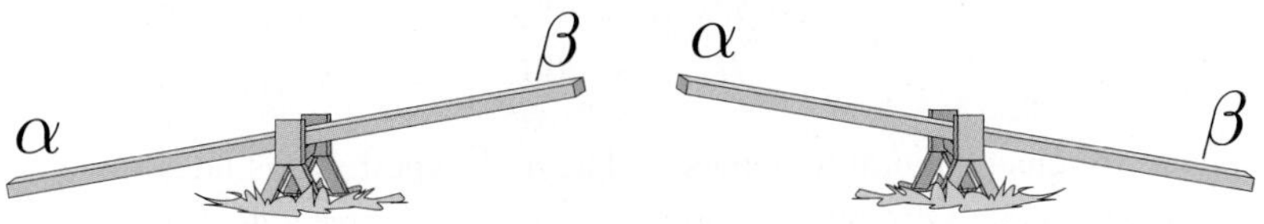

Figure 10.3: Inverse Relationship of α and β

Math Symbols

β: probability of failing to reject a false null hypothesis; Greek letter, beta

Memory Booster

α = probability of a Type I error

β = probability of a Type II error

Example 10.7

Determining the Type of Error

A television executive believes that at least 99% of households in the United States have at least one television. An intern at the executive's company is given the task of using a hypothesis test to determine whether the percentage is actually less than 99%. The hypothesis test is completed, and based on the sample collected, the intern decides to fail to reject the null hypothesis. If, in reality, 96.7% of households own a television set, was an error made? If so, what type?

Solution

Let's begin by writing the null and alternative hypotheses by asking "What is the intern gathering data for?" Because a hypothesis test is being used to determine if the percentage is *less than* 99%, the research hypothesis is $H_a: p < 0.99$. The logical opposite is $p \geq 0.99$, and this is the null hypothesis. So the two hypotheses are written as follows.

$$H_0: p \geq 0.99$$
$$H_a: p < 0.99$$

The hypothesis test fails to reject the null hypothesis. However, since the reality is that $p = 0.967$ and thus, the null hypothesis is false, the intern failed to reject a false null hypothesis. This is a Type II error.

Example 10.8

Determining the Type of Error

Insurance companies commonly use 1000 miles as the mean number of miles a car is driven per month. One insurance company claims that, due to our more mobile society, the mean is more than 1000 miles per month. The insurance company tests its claim with a hypothesis test and decides to reject the null hypothesis. Assume that in reality, the mean number of miles a car is driven per month is 1250 miles. Was an error made? If so, what type?

Solution

Begin by writing the null and alternative hypotheses. The insurance company wishes to gather data in support of the statement that the mean is *more than* 1000 miles. Therefore, the research hypothesis is $H_a: \mu > 1000$. The opposite is $\mu \leq 1000$. Thus, the null and alternative hypotheses are written as follows.

$$H_0: \mu \leq 1000$$
$$H_a: \mu > 1000$$

The decision was to reject the null hypothesis. The null hypothesis is false since $\mu = 1250$, so the decision was to reject a false null hypothesis, which is a correct decision.

Example 10.9

Determining the Type of Error

A study on the effects of television-viewing on children reports that children watch a mean of 4.0 hours of television per night. Kiko believes that the mean number of hours children in her neighborhood watch television per night is not 4.0. She performs a hypothesis test and rejects the null hypothesis. Assume that in reality, children in her neighborhood do watch a mean of 4.0 hours of television per night. Did she make an error? If so, what type?

Solution

Begin by writing the null and alternative hypotheses. Kiko wishes to gather data in support of her belief that the mean is not 4.0 hours per night. Therefore, her research hypothesis is written mathematically as H_a: $\mu \neq 4.0$. The logical opposite is H_0: $\mu = 4.0$. Thus, the null and alternative hypotheses are written as follows.

$$H_0: \mu = 4.0$$
$$H_a: \mu \neq 4.0$$

The decision was to reject the null hypothesis, when in reality, $\mu = 4.0$, so Kiko rejected a true null hypothesis. This is a Type I error.

Now that you have had practice distinguishing between Type I and Type II errors, it is important to note that in almost every case, we would not know "the reality" of a situation because population parameters are usually unknown. Thus, we would usually have no basis for determining if a Type I error or a Type II error has been made. Knowing what types of errors exist and how they are related to each other, as well as being able to distinguish between them, allows us to interpret decisions in the context of possible realities.

Memory Booster

It is important to note that we would usually have no basis for determining if a Type I error or a Type II error has been made because population parameters are usually unknown.

In this section, we discussed in detail the process of setting up the null and alternative hypotheses, as well as the types of conclusions and errors that can be made once the calculations in a hypothesis test have been completed. In the next several sections, we will discuss the details of determining the appropriate probability distribution to use for the test statistic for a given population parameter and the various methods for determining the values of that test statistic that result in rejecting the null hypothesis.

10.1 Section Exercises

Fundamentals of Hypothesis Testing

Directions: Decide if each statement is true or false. Explain why.

1. When we reject a null hypothesis, we have proven the alternative hypothesis to be true.

2. There are only two possible conclusions in a hypothesis test: reject or fail to reject the null hypothesis.

3. The null and alternative hypotheses form a pair of logically opposite statements.

4. A Type I error is made when a true null hypothesis is rejected.

5. A Type II error is made when we fail to reject a true null hypothesis.

6. The level of significance is the probability of making a Type II error.

7. The probability of making a Type I error is inversely related to the probability of making a Type II error.

Null and Alternative Hypotheses

Directions: State the null and alternative hypotheses for each scenario.

8. A sports analyst is testing his claim that the mean weight of this year's NFL linemen is heavier than in past years. Suppose that over the past five years, the mean weight of NFL linemen was 320.0 pounds.

9. Based on past sales, a shoe manufacturer considers the mean shoe size of women to be 7.50. The manufacturer would like to test if this is still the case.

10. Austin's company manufactures test tubes. He needs his tubes to be exactly 4.0 mm in diameter. If they are too narrow or too wide, he must recalibrate his machine. Austin randomly measured 150 tubes off the production line in order to perform a hypothesis test.

11. A nationwide study shows that children watch a mean of 3.0 hours of television per day. A mother in Louisiana believes that this is an underestimate for children in her area. She conducts a local survey to test her belief.

12. While continuing to keep abreast of local trends in education, a school administrator read a journal article that reported only 42% of high school students study on a regular basis. He claims that this percentage is too low for his district.

13. Guidelines for cereal manufacturers say that to claim a cereal is "mostly fruit," at least 45% of the cereal must be fruit. A local consumer watch group tests 15 randomly selected boxes of a cereal claiming to be mostly fruit to see if it meets the guidelines.

14. As a general guideline, tap water should contain 3 ppm (parts per million) chlorine. A local council wants to test the chlorine levels on 15 randomly selected days over the next two months to see if the guidelines are not being met.

15. Madison, an ambulance driver in a small town, believes that on a typical 24-hour shift, she receives fewer than 5 emergency calls on average. To test this belief, she records the number of calls she receives each shift for 10 randomly selected shifts.

16. Seawater is believed to have a mean fluoride concentration of 1.3 mg/L (milligrams per liter). A marine biologist is concerned that the level of fluoride is too high in a particular area and is killing the ocean life.

Source: "Fluoride in Drinking-Water: Background Document for Development of WHO *Guidelines for Drinking-Water Quality.*" World Health Organization. 2004. http://www.who.int/water_sanitation_health/dwq/chemicals/fluoride.pdf (9 Nov. 2011).

17. The city council of Oxford is thinking of building a road around the city. The council wants to know if the majority of the residents are in favor of the new road before pursuing the issue further.

Interpreting Conclusions of Hypothesis Tests

Directions: Answer each question.

18. A television network has believed for many years that at least 40% of its viewers are below the age of 22. For marketing purposes, a potential advertiser wishes to test the claim that the percentage is actually less than 40%. After performing the test at the 0.05 level of significance, the advertiser decides to reject the null hypothesis. What does this conclusion lead us to believe about the potential advertiser's claim?

19. A pharmaceutical company has publicized that no more than 4% of people who take a particular drug experience significant side effects. A researcher is concerned that the percentage is more than 4%, and she decides to test her claim with a hypothesis test. Based on the sample she collects, she decides to fail to reject the null hypothesis at the 0.01 level of significance. What does this conclusion tell us about the researcher's claim?

20. A radio station has always believed that the mean age of its listeners is 26. Because their ratings are slipping, the executives need to know if the mean age has changed, so that they can alter their programs accordingly. After information is collected from 329 listeners and a hypothesis test is completed, the radio station executives decide that they should reject the null hypothesis with a 95% level of confidence. Based on this conclusion, should the radio station look into changing its programming?

21. A cellular phone company promotes the claim that its customers spend a mean of no more than $60 per month on cell phone service. One skeptical college student is convinced that the mean is higher than $60 per month. He surveys a random sample of the company's customers and performs a hypothesis test to test his claim. In the end, he decides to fail to reject the null hypothesis with $\alpha = 0.01$. What does this conclusion tell us about the college student's claim?

22. A prominent travel agency claims that the mean cost of a four-day theme park vacation is no more than \$600 per person. A researcher believes that the mean cost is much higher, and decides to test her theory with a hypothesis test. The conclusion of the hypothesis test is to fail to reject the null hypothesis with $\alpha = 0.01$. What does this conclusion say about the researcher's claim?

Types of Errors in Hypothesis Testing

Directions: For each scenario, determine the type of error that was made, if any. (Hint: Begin by determining the null and alternative hypotheses.)

23. The percentage of children who leave foster care due to adoption is generally accepted to be 15%. A social worker claims that this percentage is incorrect. After performing a hypothesis test on his claim using sample data from 2009, he fails to reject the null hypothesis. According to a report by the US Department of Health and Human Services, the actual percentage of children who left foster care due to adoption was 20% in 2009. Was an error made? If so, what type?

Source: Child Welfare Information Gateway. "Outcomes." *Foster Care Statistics 2009.* US Department of Health and Human Services, Children's Bureau. May 2011. http://www.childwelfare.gov/pubs/factsheets/foster.pdf#page=5 (9 Nov. 2011).

24. A software company advertises that its software can improve students' grades by at least 15%. One student conducts a hypothesis test to see if the percentage increase is less than 15%. The conclusion of the hypothesis test is to reject the null hypothesis. If the true percentage increase in grades for all students using the software is 13%, was an error committed? If so, what type?

25. A new whitening toothpaste advertises that it whitens teeth up to three shades whiter. The product has been so successful that the company wants to change its slogan to say "more than three shades whiter." Before changing the slogan, the company's executives want to test this new claim with a hypothesis test. According to the sample obtained, the decision is to fail to reject the null hypothesis. If, in reality, the mean number of shades that the toothpaste whitens teeth is three shades whiter, was an error made with the hypothesis test? If so, what type?

26. Suppose that the mean price for a gallon of unleaded gasoline in the United States was considered by some economists to be \$3.510 in 2011. One consumer advocacy group claimed that the mean price was more than \$3.510 per gallon. A hypothesis test of the advocacy group's claim resulted in rejecting the null hypothesis. If the true mean price for a gallon of unleaded gasoline in the United States in 2011 was \$3.576, was an error made? If so, what type?

Source: US Energy Information Administration. "US Gasoline and Diesel Retail Prices." US Department of Energy. 6 Feb. 2012. http://www.eia.gov/dnav/pet/pet_pri_gnd_dcus_nus_a.htm (9 Feb. 2012).

27. The mathematics department at one university reports that the failure rate for College Algebra is no more than 35% in any given semester. A group of students believes the rate is higher, and they decided to use a hypothesis test to see if they are correct. Based on the sample they obtain, they decided to reject the null hypothesis. If, in reality, the failure rate is 34%, did the students make an error? If so, what type?

28. Dreamfilms Studios boasts that its summer features are so good that at least half of all tickets sold one summer are for its films. A competing studio claims that less than half of all tickets sold are for Dreamfilms' movies. After the hypothesis test is completed, the conclusion is to fail to reject the null hypothesis. If the true percentage of tickets sold for Dreamfilms Studios' movies that summer was only 48%, was an error committed? If so, what type?

29. An ice cream company must keep the temperature inside its delivery trucks at 30 degrees Fahrenheit. If the temperature is too warm, the ice cream will melt. The company also does not want to allow the temperature to be too low, because it does not want to waste the cost and energy it takes to keep the trucks cool. Periodically the company runs a hypothesis test to ensure that the temperature is as claimed. Suppose that after one hypothesis test, the company fails to reject the null hypothesis. If, in reality, the mean temperature of its trucks is 29 degrees Fahrenheit, was an error committed? If so, what type?

30. In 2003, the National Assessment of Adult Literacy (NAAL), conducted by the National Center for Education Statistics (NCES), found that 14.5% of American adults lack *Basic Prose Literacy Skills (BPLS)*; that is, they are functionally illiterate. The directors of a local organization promoting literacy claim that they have impacted their community and the illiteracy rate in their community is lower than the national rate. A hypothesis test of their claim results in rejecting the null hypothesis. If, in reality, the illiteracy rate in their community is 13%, was an error made? If so, what type?

Source: National Center for Education Statistics. "National Assessment of Adult Literacy: State and County Literacy Estimates – Overview." US Department of Education, Institute of Education Sciences. 2003. http://nces.ed.gov/naal/estimates/Overview.aspx (9 Nov. 2011).

10.2 Hypothesis Testing for Population Means (σ Known)

Following the guidelines presented in the last section, we will now look at performing a complete hypothesis test for a population mean. To choose which probability distribution to use for testing means, we use the same criteria that we used to determine which distribution to use for the critical values when constructing confidence intervals. Once again, we first consider whether the population standard deviation, σ, is known or unknown. If σ is known, which is rare in reality, we either use the z-test statistic or other more advanced methods. However, just as with confidence intervals, we will use this rare situation to explore how a basic hypothesis test is performed, and then move to the more practical case where σ is unknown in the next section.

Recall the general steps in a hypothesis test, which are outlined below.

1. State the null and alternative hypotheses.
2. Determine which distribution to use for the test statistic, and state the level of significance.
3. Gather data and calculate the necessary sample statistics.
4. Draw a conclusion and interpret the decision.

We know how to state the null and alternate hypotheses from the previous section, so now we must determine which distribution to use for calculating the test statistic. When the following conditions are met, the standard normal distribution should be used when testing the mean of the population.

- All possible samples of a given size have an equal probability of being chosen; that is, a simple random sample is used.
- The population standard deviation, σ, is *known*.
- Either the sample size is at least 30 ($n \geq 30$) *or* the population distribution is approximately normal.

The formula for the test statistic for testing a population mean with σ known is as follows.

Formula

Test Statistic for a Hypothesis Test for a Population Mean (σ Known)

When the population standard deviation is known, the sample taken is a simple random sample, and either the sample size is at least 30 or the population distribution is approximately normal, the test statistic for a hypothesis test for a population mean is given by

$$z = \frac{\bar{x} - \mu}{\left(\dfrac{\sigma}{\sqrt{n}}\right)}$$

where $\bar{x}$ is the sample mean,

μ is the presumed value of the population mean from the null hypothesis,

σ is the population standard deviation, and

n is the sample size.

Note that this is the formula for the standard score for a sample mean in a sampling distribution, which we first learned in Section 7.2. This application of the Central Limit Theorem is useful here because hypothesis testing involves gathering samples. As we learned before, the Central Limit Theorem can only be applied to a sampling distribution of sample means if the sample size taken is at least 30 or if the population distribution is approximately normal.

Rejection Regions

Before the data values are gathered from a representative sample of the population, there must be a standard in place for determining whether or not the test statistic calculated from the sample data is statistically significant. The first method we will use to evaluate z-scores in hypothesis testing is to use rejection regions. Rejection regions are determined by two things: 1) the type of hypothesis test, and 2) the level of significance, α. To determine the type of test, simply look at the symbol contained in the alternative hypothesis. There are then three different types of tests: left-tailed tests (<), right-tailed tests (>), and two-tailed tests (≠). Left- and right-tailed tests are both considered *one-tailed* tests.

Definition

Alternative Hypothesis, H_a	Type of Hypothesis Test
< Value	Left-tailed test
> Value	Right-tailed test
≠ Value	Two-tailed test

Before we look at each of these three cases individually, let's state the rule that will be used to determine the conclusion. The decision rule for rejection regions is that we reject the null hypothesis if the test statistic calculated from the sample data falls within the rejection region.

Procedure

Decision Rule for Rejection Regions

Reject the null hypothesis, H_0, if the test statistic calculated from the sample data falls within the rejection region.

Let's now look at how to construct rejection regions for the three types of hypothesis tests.

1. Rejection Region for a Left-Tailed Hypothesis Test (H_a contains the symbol <)

As you would expect, the area shaded for the rejection region is in the left tail of the distribution. This area is equal to the level of significance, α. The boundary of the rejection region is marked by a critical z-score, z_α. We will look at how to calculate the critical z-score next. Notice though, that because the critical z-score lies on the left side of the curve, it must be negative. Thus, the decision rule for a left-tailed test is to reject the null hypothesis if $z \le -z_\alpha$.

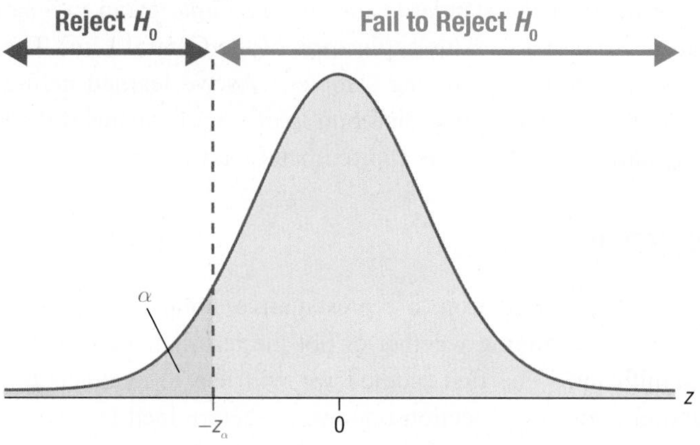

Figure 10.4: Rejection Region for a Left-Tailed Test

2. Rejection Region for a Right-Tailed Hypothesis Test (H_a contains the symbol >)

For a right-tailed test, an area of α is shaded in the right tail of the distribution. Again, a critical z-score marks the boundary of the rejection region. Because the critical z-score is on the right side of the curve, its value will be positive for these types of tests. Thus, the decision rule for a right-tailed test is to reject the null hypothesis if $z \geq z_\alpha$.

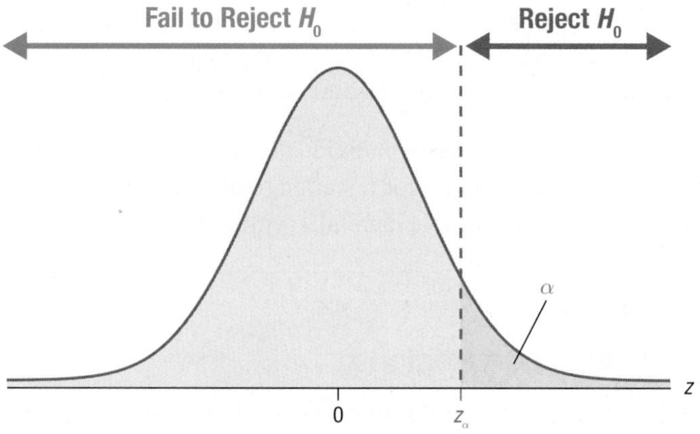

Figure 10.5: Rejection Region for a Right-Tailed Test

3. Rejection Region for a Two-Tailed Hypothesis Test (H_a contains the symbol ≠)

For a two-tailed test, a total area of α is divided equally between the two tails of the distribution. Notice in the following picture that the area shaded in each tail is $\frac{\alpha}{2}$. Thus, the decision rule for a two-tailed test is to reject the null hypothesis if $z \leq -z_{\alpha/2}$ or $z \geq z_{\alpha/2}$, that is, if $|z| \geq z_{\alpha/2}$.

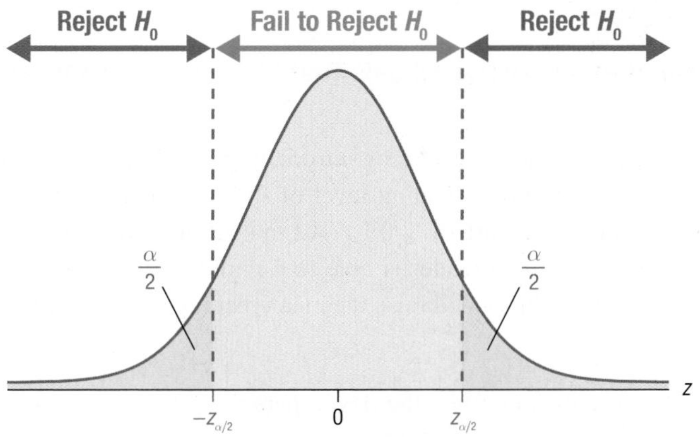

Figure 10.6: Rejection Region for a Two-Tailed Test

We can summarize these pictures with the following notation.

Procedure

Rejection Regions for Hypothesis Tests for Population Means (σ Known)

Reject the null hypothesis, H_0, if:

$z \leq -z_\alpha$ for a left-tailed test

$z \geq z_\alpha$ for a right-tailed test

$|z| \geq z_{\alpha/2}$ for a two-tailed test

The choice for the level of significance is usually one of only a few values of α. Thus, using the techniques we learned in Chapter 6 to find the value of z for an indicated area, we can easily create a table that allows us to look up critical z-values quickly for these choices of α for both one- and two-tailed tests.

Table 10.1: Critical z-Values for Rejection Regions

Level of Confidence c	Level of Significance $\alpha = 1 - c$	One-Tailed Test z_α	Two-Tailed Test $\pm z_{\alpha/2}$
0.90	0.10	1.28	±1.645
0.95	0.05	1.645	±1.96
0.98	0.02	2.05	±2.33
0.99	0.01	2.33	±2.575

Note that the one-tailed column contains critical values for right-tailed tests. Since the standard normal distribution is symmetrical about $z = 0$, for left-tailed tests we simply make these values negative. Let's look at a complete example of using a rejection region for conducting a hypothesis test. We will only look at an example of a right-tailed test using the z-score with a rejection region, and then we will turn our attention to another method for drawing the conclusion to a hypothesis test. We will see rejection regions again in the next section, in examples where the population standard deviation is unknown.

Example 10.10

Using a Rejection Region in a Hypothesis Test for a Population Mean (Right-Tailed, σ Known)

The state education department is considering introducing new initiatives to boost the reading levels of fourth graders. The mean reading level of fourth graders in the state over the last 5 years was a Lexile reader measure of 800 L. (A Lexile reader measure is a measure of the complexity of the language that a reader is able to comprehend.) The developers of a new program claim that their techniques will raise the mean reading level of fourth graders by more than 50 L.

To assess the impact of their initiative, the developers were given permission to implement their ideas in the classrooms. At the end of the pilot study, a simple random sample of 1000 fourth graders had a mean reading level of 856 L. It is assumed that the population standard deviation is 98 L. Using a 0.05 level of significance, should the findings of the study convince the education department of the validity of the developers' claim?

Solution

Memory Booster

Remember that the null hypothesis will contain equality.

Step 1: State the null and alternative hypotheses.

The developers want to show that their classroom techniques will raise the fourth graders' mean reading level to more than 850 L. This is written mathematically as $\mu > 850$, and since it is the research hypothesis, it will be H_a. The mathematical opposite is $\mu \le 850$. Thus, we have the following hypotheses.

$$H_0: \mu \le 850$$
$$H_a: \mu > 850$$

Step 2: Determine which distribution to use for the test statistic, and state the level of significance.

Note that the hypotheses are statements about the population mean, σ is known, the sample is a simple random sample, and the sample size is at least 30. Thus, we will use a normal distribution, which means we need to use the z-test statistic.

In addition to determining which distribution to use for the test statistic, we need to state the level of significance. The problem states that $\alpha = 0.05$.

Step 3: Gather data and calculate the necessary sample statistics.

At the end of the pilot study, a simple random sample of 1000 fourth graders had a mean reading level of 856 L. The population standard deviation is 98 L. Thus, the test statistic is calculated as follows.

$$\begin{aligned} z &= \frac{\bar{x} - \mu}{\left(\frac{\sigma}{\sqrt{n}}\right)} \\ &= \frac{856 - 850}{\left(\frac{98}{\sqrt{1000}}\right)} \\ &\approx 1.94 \end{aligned}$$

Step 4: Draw a conclusion and interpret the decision.

Remember that we determine the type of test based on the alternative hypothesis. In this case, the alternative hypothesis contains ">," which indicates that this is a right-tailed test. To determine the rejection region, we need a z-value so that 0.05 of the area under the standard normal curve is to its right. If 0.05 of the area is to the right then $1 - 0.05 = 0.95$ is the area to the left. If we look up 0.9500 in the body of the cumulative z-table, the corresponding critical z-value is 1.645. Alternately, we can look up $c = 0.95$ in the table of critical z-values for rejection regions. Either way, the rejection region is $z \geq 1.645$.

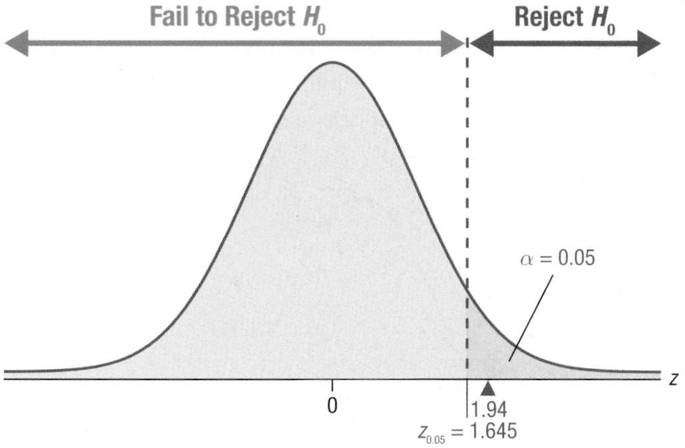

The z-value of 1.94 falls in the rejection region. So the conclusion is to reject the null hypothesis. Thus the evidence collected suggests that the education department can be 95% sure of the validity of the developers' claim that the mean Lexile reader measure of fourth graders will increase by more than 50 points.

p-Values

Although using rejection regions to draw a conclusion in a hypothesis test is a perfectly viable method, it is not the method most commonly found in research articles or journals. Instead, one is more likely to encounter the use of a *p-value*.

A ***p*-value** is the probability of obtaining a sample statistic as extreme or more extreme than the one observed in the data, when the null hypothesis, H_0, is assumed to be true. The definition of "more extreme" is based on the type of hypothesis test. For example, if testing a left-tailed claim, then "more extreme" would indicate farther to the left.

Definition

A ***p*-value** is the probability of obtaining a sample statistic as extreme or more extreme than the one observed in the data, when the null hypothesis, H_0, is assumed to be true.

For a left-tailed test, the probability of obtaining a sample statistic as extreme or more extreme than the one observed is the probability of getting a test statistic *less than or equal to* the calculated value. We can calculate these probabilities for z-test statistics using the cumulative normal distribution tables or available technology as we have done in previous chapters. Let's look at examples of calculating p-values for z-test statistics for each of the three types of hypothesis tests.

Example 10.11

Calculating the *p*-Value for a *z*-Test Statistic for a Left-Tailed Test

Calculate the *p*-value for a hypothesis test with the following hypotheses. Assume that data have been collected and the test statistic was calculated to be $z = -1.34$.

$$H_0\colon \mu \ge 0.15$$
$$H_a\colon \mu < 0.15$$

Solution

The alternative hypothesis tells us that this is a left-tailed test. Therefore, the *p*-value for this situation is the probability that z is less than or equal to -1.34, written $p\text{-value} = P(z \le -1.34)$. Use a table or appropriate technology to find the area under the standard normal curve to the left of $z = -1.34$. Thus, the $p\text{-value} \approx 0.0901$.

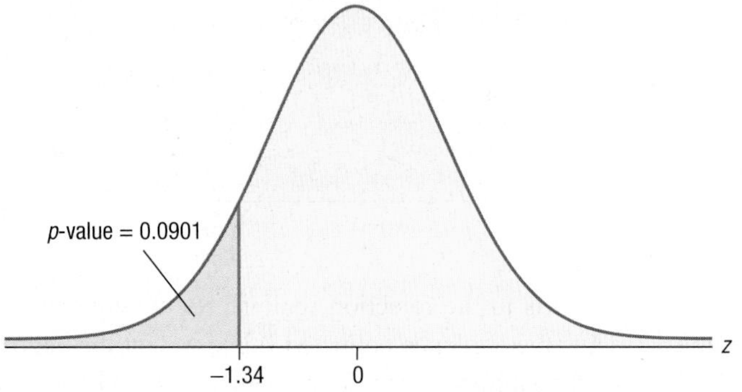

Example 10.12

Calculating the *p*-Value for a *z*-Test Statistic for a Right-Tailed Test

Calculate the *p*-value for a hypothesis test with the following hypotheses. Assume that data have been collected and the test statistic was calculated to be $z = 2.78$.

$$H_0\colon \mu \le 0.43$$
$$H_a\colon \mu > 0.43$$

Solution

The alternative hypothesis tells us that this is a right-tailed test. Therefore, the *p*-value for this situation is the probability that z is greater than or equal to 2.78, written $p\text{-value} = P(z \ge 2.78)$. Use a table or appropriate technology to find the area under the standard normal curve to the right of $z = 2.78$. Thus, the $p\text{-value} \approx 0.0027$.

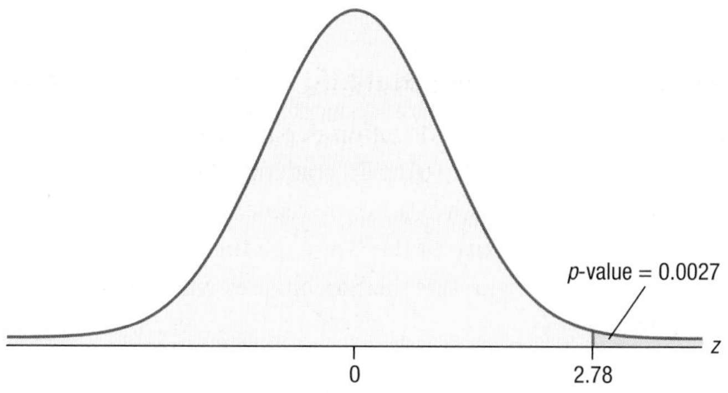

Example 10.13

Calculating the *p*-Value for a *z*-Test Statistic for a Two-Tailed Test

Calculate the *p*-value for a hypothesis test with the following hypotheses. Assume that data have been collected and the test statistic was calculated to be $z = -2.15$.

$$H_0: \mu = 0.78$$
$$H_a: \mu \neq 0.78$$

Solution

The alternative hypothesis tells us that this is a two-tailed test. Thus, the *p*-value for this situation is the probability that z is either less than or equal to -2.15 or greater than or equal to 2.15, which is written mathematically as $p\text{-value} = P(|z| \geq 2.15)$. Use a table or appropriate technology to find the area under the standard normal curve to the left of $z_1 = -2.15$. The area to the left of $z_1 = -2.15$ is 0.0158. Since the standard normal curve is symmetric about its mean, 0, the area to the right of $z_2 = 2.15$ is also 0.0158. Thus, the *p*-value is calculated as follows.

$$\begin{aligned} p\text{-value} &= (0.0158)(2) \\ &= 0.0316 \end{aligned}$$

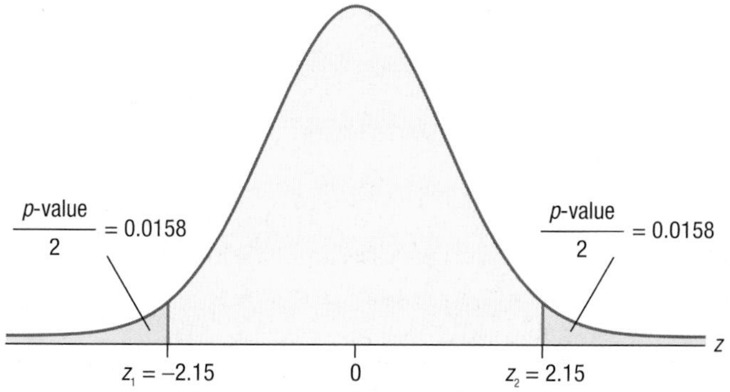

Example 10.14

Calculating the *p*-Value for a *z*-Test Statistic Using a TI-83/84 Plus Calculator

As we read in Example 10.10, the state education department is considering introducing new initiatives to boost the reading levels of fourth graders. The mean reading level of fourth graders in the state over the last 5 years was a Lexile reader measure of 800 L. (A Lexile reader measure is a measure of the complexity of the language that a reader is able to comprehend.) The developers of a new program claim that their techniques will raise the mean reading level of fourth graders by more than 50 L.

To assess the impact of their initiative, the developers were given permission to implement their ideas in the classrooms. At the end of the pilot study, a simple random sample of 1000 fourth graders had a mean reading level of 856 L. It is assumed that the population standard deviation is 98 L. Calculate the p-value for this hypothesis test.

Solution

First, recall from Example 10.10 that the hypotheses are as follows.

$$H_0: \mu \le 850$$
$$H_a: \mu > 850$$

The alternative hypothesis tells us that this is a right-tailed test. Thus, in terms of the sampling distribution of sample means for a sample size of $n = 1000$, the p-value for this situation is the probability of getting a sample mean greater than or equal to 856, written $p\text{-value} = P(\bar{x} \ge 856)$.

Memory Booster

Review the procedures for using a TI-83/84 Plus calculator to find areas under normal curves, such as *p*-values, in Section 6.2.

Press **2ND** and then **VARS** to get the `DISTR` (distributions) menu. Choose option `2:normalcdf(`. Remember that you must enter the lower and upper bounds of the area under the curve as well as the mean and standard deviation of the normal distribution. For this example, the lower bound of the area is the sample mean, $\bar{x} = 856$. Since we want the area to the right, the upper bound of the area is infinity, so we must enter a really large value of $\bar{x}$. Thus, we will enter `1E99` for the upper bound. The mean of the sampling distribution is the hypothesized value of the population mean, $\mu_{\bar{x}} = 850$. The standard deviation of the sampling distribution is the population standard deviation divided by the square root of the sample size, $\sigma_{\bar{x}} = \frac{98}{\sqrt{1000}}$. Enter `normalcdf(856,1E99,850,98/√(1000))` and press **ENTER**.

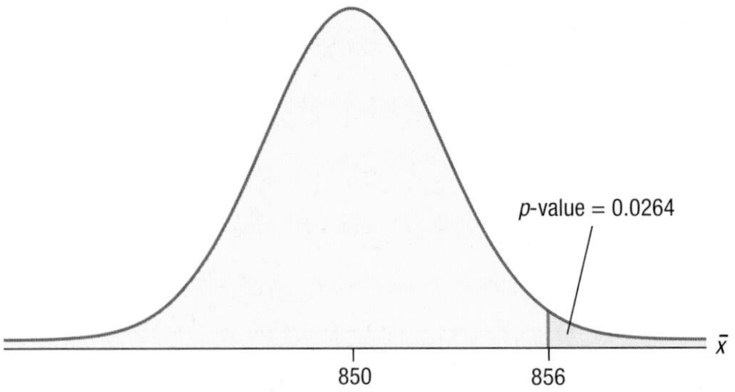

The p-value returned is approximately 0.0264, as shown in the screenshot in the margin. Note that by solving for the p-value in one step on the calculator, we eliminate any possible rounding errors. Thus, this method will always yield the most accurate p-value. Later in this chapter, we will see that the calculator can perform the entire hypothesis test in one step, and the output includes this p-value.

Recall from our discussion of statistical significance that if the test statistic calculated from the sample data is found to be statistically significant, then we *reject the null hypothesis* in favor of the alternative hypothesis. We can determine if the difference between the claimed value of the parameter and the data collected (that is, the sample statistic) is statistically significant by comparing the p-value of the test statistic to α, the level of significance. Throughout this text, we will state the necessary level of confidence or the level of significance when performing a hypothesis test. However, in practice it is common to simply state the p-value and allow the reader to draw conclusions based on his or her own desired level of significance.

Once the test statistic and the p-value have been calculated, we compare the p-value to the value of α, which was stated before the sample data were gathered, to draw a conclusion as described in the following box.

Procedure

Conclusions Using p-Values

- If p-value $\le \alpha$, then *reject* the null hypothesis.
- If p-value $> \alpha$, then *fail to reject* the null hypothesis.

Although we are using the p-value to make a decision when using a z-test statistic in this section, we can in fact use the same guidelines in future sections for rejecting a null hypothesis given α and the p-value for any test statistic.

Example 10.15

Determining the Conclusion to a Hypothesis Test Using the p-Value

For a certain hypothesis test, the p-value is calculated to be p-value $= 0.0146$.

a. If the stated level of significance is 0.05, what is the conclusion to the hypothesis test?

b. If the level of confidence is 99%, what is the conclusion to the hypothesis test?

Solution

a. The level of significance is $\alpha = 0.05$. Next, note that $0.0146 < 0.05$. Thus, p-value $\le \alpha$, so we *reject* the null hypothesis.

b. The level of confidence is 99%, so the level of significance is calculated as follows.

$$\begin{aligned}\alpha &= 1 - c \\ &= 1 - 0.99 \\ &= 0.01\end{aligned}$$

Next, note that $0.0146 > 0.01$. Thus, p-value $> \alpha$, so we *fail to reject* the null hypothesis.

Now that we are familiar with the necessary test statistic and the p-value method for determining a conclusion, let's perform complete hypothesis tests for population means with σ known.

Example 10.16

Performing a Hypothesis Test for a Population Mean (Right-Tailed, σ Known)

A researcher claims that the mean age of women in California at the time of a first marriage is higher than 26.5 years. Surveying a simple random sample of 213 newlywed women in California, the researcher found a mean age of 27.0 years. Assuming that the population standard deviation is 2.3 years and using a 95% level of confidence, determine if there is sufficient evidence to support the researcher's claim.

Solution

Step 1: State the null and alternative hypotheses.

The researcher's claim is investigating the mean age at first marriage for women in California. Therefore, the research hypothesis, H_a, is that the mean age is greater than 26.5, $\mu > 26.5$. The logical opposite is $\mu \le 26.5$. Thus, the null and alternative hypotheses are stated as follows.

$$H_0: \mu \le 26.5$$
$$H_a: \mu > 26.5$$

Step 2: Determine which distribution to use for the test statistic, and state the level of significance.

Note that the hypotheses are statements about the population mean, the population standard deviation is known, the sample is a simple random sample, and the sample size is at least 30. Thus, we will use a normal distribution and calculate the z-test statistic.

We will draw a conclusion by computing the p-value for the calculated test statistic and comparing the value to α. For this hypothesis test, the level of confidence is 95%, so the level of significance is calculated as follows.

$$\begin{aligned}\alpha &= 1 - c \\ &= 1 - 0.95 \\ &= 0.05\end{aligned}$$

Step 3: Gather data and calculate the necessary sample statistics.

From the information given, we know that the presumed value of the population mean is $\mu = 26.5$, the sample mean is $\bar{x} = 27.0$, the population standard deviation is $\sigma = 2.3$, and the sample size is $n = 213$. Thus, the test statistic is calculated as follows.

$$\begin{aligned}z &= \frac{\bar{x} - \mu}{\left(\frac{\sigma}{\sqrt{n}}\right)} \\ &= \frac{27.0 - 26.5}{\left(\frac{2.3}{\sqrt{213}}\right)} \\ &\approx 3.17\end{aligned}$$

Step 4: Draw a conclusion and interpret the decision.

The alternative hypothesis tells us that we have a right-tailed test. Therefore, the p-value for this test statistic is the probability of obtaining a test statistic greater than or equal to $z = 3.17$, written as $p\text{-value} = P(z \ge 3.17)$. To find the p-value, we need to find the area under the standard normal curve to the right of $z = 3.17$.

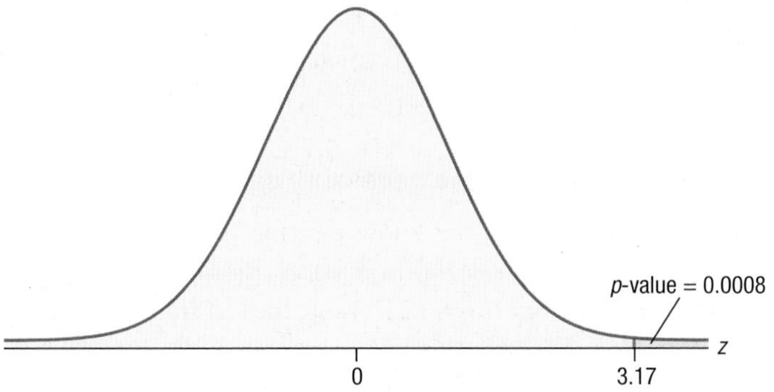

Using a normal distribution table or appropriate technology, we find that the area is $p\text{-value} \approx 0.0008$. Comparing this p-value to the level of significance, we see that $0.0008 < 0.05$, so $p\text{-value} \leq \alpha$. Thus, the conclusion is to reject the null hypothesis. Therefore, we can say with 95% confidence that there is sufficient evidence to support the researcher's claim that the mean age at first marriage for women in California is higher than 26.5 years.

Example 10.17

Performing a Hypothesis Test for a Population Mean (Two-Tailed, σ Known)

A recent study showed that the mean number of children for women in Europe is 1.5. A global watch group claims that German women have a mean fertility rate that is different from the mean for all of Europe. To test its claim, the group surveyed a simple random sample of 128 German women and found that they had a mean fertility rate of 1.4 children. The population standard deviation is assumed to be 0.8. Is there sufficient evidence to support the claim made by the global watch group at the 90% level of confidence?

Solution

Step 1: **State the null and alternative hypotheses.**

The watch group is investigating whether the mean fertility rate for German women is different from the mean for all of Europe. Thus, they need to find evidence that the mean fertility rate is not equal to 1.5. So the research hypothesis, H_a, is that the mean does not equal 1.5, $\mu \neq 1.5$. The logical opposite is $\mu = 1.5$. Thus, the null and alternative hypotheses are stated as follows.

$$H_0: \mu = 1.5$$
$$H_a: \mu \neq 1.5$$

Step 2: **Determine which distribution to use for the test statistic, and state the level of significance.**

Note that the hypotheses are statements about the population mean, σ is known, the sample is a simple random sample, and the sample size is at least 30. Thus, we will use a normal distribution and calculate the z-test statistic.

We will draw a conclusion by computing the p-value for the calculated test statistic and comparing the value to α. For this hypothesis test, the level of confidence is

90%, so the level of significance is calculated as follows.

$$\begin{aligned}\alpha &= 1 - c \\ &= 1 - 0.90 \\ &= 0.10\end{aligned}$$

Step 3: **Gather data and calculate the necessary sample statistics.**

From the information given, we know that the presumed value of the population mean is $\mu = 1.5$, the sample mean is $\bar{x} = 1.4$, the population standard deviation is $\sigma = 0.8$, and the sample size is $n = 128$. Thus, the test statistic is calculated as follows.

$$\begin{aligned}z &= \frac{\bar{x} - \mu}{\left(\frac{\sigma}{\sqrt{n}}\right)} \\ &= \frac{1.4 - 1.5}{\left(\frac{0.8}{\sqrt{128}}\right)} \\ &\approx -1.41\end{aligned}$$

Step 4: **Draw a conclusion and interpret the decision.**

The alternative hypothesis tells us that we have a two-tailed test. Therefore, the p-value for this test statistic is the probability of obtaining a test statistic that is either less than or equal to $z_1 = -1.41$ or greater than or equal to $z_2 = 1.41$, which is written mathematically as $p\text{-value} = P(|z| \geq 1.41)$. To find the p-value, we need to find the sum of the areas under the standard normal curve to the left of $z_1 = -1.41$ and to the right of $z_2 = 1.41$.

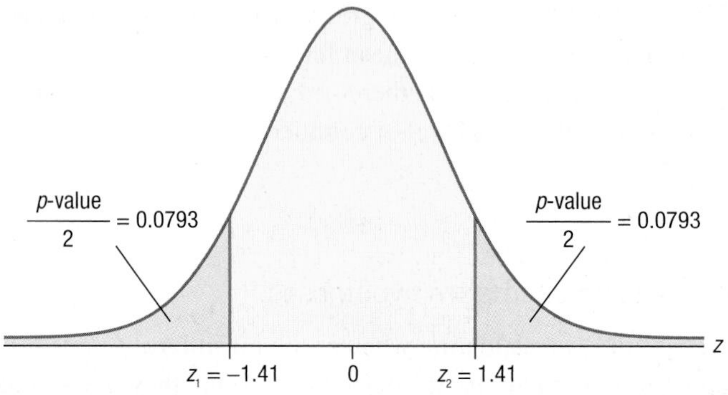

By looking up $z_1 = -1.41$ in the cumulative normal distribution table, we find that the area to the left is equal to 0.0793. Since the standard normal curve is symmetric about its mean, 0, the area to the right of $z_2 = 1.41$ is also 0.0793. Thus the p-value is calculated as follows.

$$\begin{aligned}p\text{-value} &= (0.0793)(2) \\ &= 0.1586\end{aligned}$$

Comparing the p-value to the level of significance, we see that $0.1586 > 0.10$, so $p\text{-value} > \alpha$. Thus, the conclusion is to fail to reject the null hypothesis. This means that, at a 90% level of confidence, the evidence does not support the watch group's claim that the fertility rate of German women is different from the mean for all of Europe.

The following example shows how the calculation of the test statistic in Step 3 can be performed by using a TI-83/84 Plus calculator.

Example 10.18

Performing a Hypothesis Test for a Population Mean Using a TI-83/84 Plus Calculator (Two-Tailed, σ Known)

A recent study showed that the mean number of children for women in Europe is 1.5. A global watch group claims that German women have a mean fertility rate that is different from the mean for all of Europe. To test its claim, the group surveyed a simple random sample of 128 German women and found that they had a mean fertility rate of 1.4 children. The population standard deviation is assumed to be 0.8. Is there sufficient evidence to support the claim made by the global watch group at the 90% level of confidence? Use a TI-83/84 Plus calculator to answer the question.

Solution

Even when working with technology, Steps 1 and 2 are the same, so we will just reiterate the important information for these two steps from Example 10.17.

Step 1: **State the null and alternative hypotheses.**

The hypotheses are stated as follows.

$$H_0: \mu = 1.5$$
$$H_a: \mu \neq 1.5$$

Step 2: **Determine which distribution to use for the test statistic, and state the level of significance.**

Since σ is known, the sample is a simple random sample, and the sample size is at least 30, we use the normal distribution and z-test statistic for this hypothesis test for the population mean. The level of significance is $\alpha = 0.10$.

Step 3: **Gather data and calculate the necessary sample statistics.**

This is where we begin to use a TI-83/84 Plus calculator. Let's start by writing down the information from the problem, as we will need to enter these values into the calculator. We know that the presumed value of the population mean is $\mu = 1.5$, the sample mean is $\bar{x} = 1.4$, the population standard deviation is $\sigma = 0.8$, and the sample size is $n = 128$.

Press STAT, scroll to TESTS, and choose option `1:Z-Test`. Since we know the sample statistics, choose `Stats` instead of `Data`. For `μo`, enter the value from the null hypothesis, thus enter `1.5` for `μo`. Enter the rest of the given values, as shown in the following screenshot on the left. Choose the alternative hypothesis `≠μo`. Highlight `Calculate` and press ENTER.

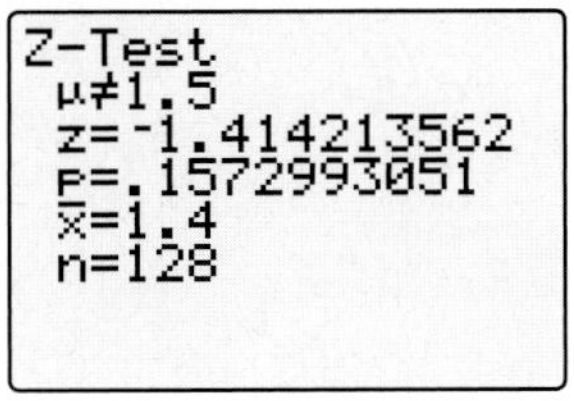

The output screen, shown on the right, displays the alternative hypothesis, calculates the z-test statistic and the p-value, and then reiterates the sample mean and sample size that were entered.

Step 4: Draw a conclusion and interpret the decision.

The p-value given by the calculator is approximately 0.1573. It is not identical to the one we found when using the table because there were several intermediate steps when we found the p-value by hand that reduced the accuracy of the p-value. However, the conclusion is the same since the two p-values are extremely close. Since $p\text{-value} > \alpha$, the conclusion is to fail to reject the null hypothesis. This means that at a 90% level of confidence, the evidence does not support the watch group's claim that the fertility rate of German women is different from the mean for all of Europe.

10.2 Section Exercises

Rejection Regions for Hypothesis Tests for Population Means (σ Known)

Directions: Draw the rejection region for the hypothesis test for the population mean with the given hypotheses. Assume that the population standard deviation is known and the sample size is at least 30.

1. H_0: $\mu \le 65$ and H_a: $\mu > 65$, $\alpha = 0.05$

2. H_0: $\mu = 5$ and H_a: $\mu \ne 5$, $c = 0.90$

3. H_0: $\mu \ge 3.2$ and H_a: $\mu < 3.2$, $\alpha = 0.01$

4. H_0: $\mu \le 700$ and H_a: $\mu > 700$, $c = 0.98$

5. H_0: $\mu = 0.19$ and H_a: $\mu \ne 0.19$, $c = 0.95$

6. H_0: $\mu \ge 0.166$ and H_a: $\mu < 0.166$, $\alpha = 0.10$

p-Values for Hypothesis Tests for Population Means (σ Known)

Directions: Calculate the p-value for the hypothesis test for the population mean with the given hypotheses and test statistic.

7. H_0: $\mu \geq 60$ and H_a: $\mu < 60$, $z = -1.48$

8. H_0: $\mu \leq 1.07$ and H_a: $\mu > 1.07$, $z = 2.27$

9. H_0: $\mu = 2.89$ and H_a: $\mu \neq 2.89$, $z = -2.21$

10. H_0: $\mu = 144$ and H_a: $\mu \neq 144$, $z = 1.65$

11. H_0: $\mu \leq 0.56$ and H_a: $\mu > 0.56$, $z = 1.61$

12. H_0: $\mu \geq 760$ and H_a: $\mu < 760$, $z = -2$

13. H_0: $\mu = 3$ and H_a: $\mu \neq 3$, $z = 2.04$

14. H_0: $\mu \leq 14.8$ and H_a: $\mu > 14.8$, $z = 3.01$

Conclusions of Hypothesis Tests for Population Means (σ Known)

Directions: Determine the appropriate conclusion for a hypothesis test with the given level of significance.

15. p-value $= 0.0166$, $\alpha = 0.01$

16. p-value $= 0.0197$, $\alpha = 0.02$

17. p-value $= 0.0465$, $\alpha = 0.05$

18. p-value $= 0.0485$, $\alpha = 0.02$

19. p-value $= 0.1190$, $\alpha = 0.10$

20. p-value $= 0.0094$, $\alpha = 0.01$

21. H_0: $\mu \geq 409$ and H_a: $\mu < 409$, $\alpha = 0.05$, $z = -1.87$

22. H_0: $\mu \leq 0.65$ and H_a: $\mu > 0.65$, $\alpha = 0.10$, $z = 1.22$

23. H_0: $\mu = 2$ and H_a: $\mu \neq 2$, $\alpha = 0.02$, $z = -2.28$

24. H_0: $\mu = 3010$ and H_a: $\mu \neq 3010$, $\alpha = 0.01$, $z = 2.69$

Hypothesis Tests for Population Means (σ Known)

Directions: **Perform each hypothesis test. For each exercise, complete the following steps.**

a. State the null and alternative hypotheses.

b. Determine which distribution to use for the test statistic, and state the level of significance.

c. Calculate the test statistic.

d. Draw a conclusion by comparing the *p*-value to the level of significance and interpret the decision.

25. A manufacturer must test that his bolts are 2.00 cm long when they come off the assembly line. He must recalibrate his machines if the bolts are too long or too short. After sampling 100 randomly selected bolts off the assembly line, he calculates the sample mean to be 1.90 cm. He knows that the population standard deviation is 0.50 cm. Assuming a level of significance of 0.05, is there sufficient evidence to show that the manufacturer needs to recalibrate the machines?

26. CNN/Money reports that the mean cost of a speeding ticket, including court fees, was \$150.00 in 2002. A local police department claims that this amount has increased. To test their claim, they collect data from a simple random sample of 160 drivers who have been fined for speeding in the last year, and find that they paid a mean of \$154.00 per ticket. Assuming that the population standard deviation is \$17.54, is there sufficient evidence to support the police department's claim at the 0.01 level of significance?

Source: Costello, Martine. "The Need for $peed -- It Will Cost You!" CNN/Money. 24 May 2002. http://money.cnn.com/2002/05/22/news/q_speed_cost (17 Nov. 2011).

27. BeachCalifornia.com reports that theme park travelers spend a mean of \$839.00 per trip. A consumer watch group states that this mean is too low. From a simple random sample of 104 people who recently visited theme parks, the group calculated a sample mean of \$851.33. Assuming that the population standard deviation is \$56.56, is there sufficient evidence at the 0.01 level of significance to support the group's statement?

Source: BeachCalifornia.com. "California Beaches Site Search." 1999-2011. http://www.beachcalifornia.com/california/beaches-search.html (17 Nov. 2011).

28. A national business magazine reports that the mean age of retirement for women executives is 61.0. A women's rights organization believes that this value does not accurately depict the current trend in retirement. To test this, the group polled a simple random sample of 95 recently retired women executives and found that they had a mean age of retirement of 61.5. Assuming the population standard deviation is 2.5 years, is there sufficient evidence to support the organization's belief at the 0.05 level of significance?

29. According to the National Center for Health Statistics, the mean weight for an adult female in the United States is 164.7 pounds. A group promoting healthier eating habits feels strongly that it has made an impact in one community by reducing the mean weight of women in the community to below the national mean. A simple random sample of 39 women in this community has a mean weight of 161.8 pounds. Assuming that the population standard deviation is 5.6 pounds, is there sufficient evidence at the 0.01 level of significance to say that the mean weight of women in this community is lower than 164.7 pounds?

Source: McDowell, M.A., C.D. Fryar, C.L. Ogden, and K.M. Flegal. "Anthropometric Reference Data for Children and Adults: United States, 2003-2006." *National Health Statistics Reports,* Number 10. National Center for Health Statistics. 22 Oct. 2008. http://www.cdc.gov/nchs/data/nhsr/nhsr010.pdf (17 Nov. 2011).

30. The board of a major credit card company requires that the mean wait time for customers when they call customer service is at most 3.00 minutes. To make sure that the mean wait time is not exceeding the requirement, an assistant manager tracks the wait times of 45 randomly selected calls. The mean wait time was calculated to be 3.40 minutes. Assuming the population standard deviation is 1.45 minutes, is there sufficient evidence to say that the mean wait time for customers is longer than 3.00 minutes with a 98% level of confidence?

10.3 Hypothesis Testing for Population Means (σ Unknown)

Following the guidelines presented in the last two sections, we will now look at performing a complete hypothesis test for a population mean when the population standard deviation, σ, is unknown. As we have mentioned several times, it is unlikely in practice that the population standard deviation would be known, so it is important to understand how to work with this more practical situation where σ is unknown.

In this section, we will focus on hypothesis tests for population means where the following conditions are met.

- All possible samples of a given size have an equal probability of being chosen; that is, a simple random sample is used.
- The population standard deviation, σ, is *unknown*.
- Either the sample size is at least 30 ($n \geq 30$) *or* the population distribution is approximately normal.

When the above conditions are met, a Student's t-distribution should be used when testing the mean. In this section, we will restrict our focus to these situations. Therefore, the test statistic for hypothesis testing involving a population mean with σ unknown is given by the following formula.

Rounding Rule

When calculating a t-value, round to three decimal places. This follows the convention used in the t-distribution table in Appendix A.

Formula

Test Statistic for a Hypothesis Test for a Population Mean (σ Unknown)

When the population standard deviation is unknown, the sample taken is a simple random sample, and either the sample size is at least 30 or the population distribution is approximately normal, the test statistic for a hypothesis test for a population mean is given by

$$t = \frac{\bar{x} - \mu}{\left(\frac{s}{\sqrt{n}}\right)}$$

where $\bar{x}$ is the sample mean,

μ is the presumed value of the population mean from the null hypothesis,

s is the sample standard deviation, and

n is the sample size.

Recall from Section 8.2 that a Student's t-distribution is completely defined by just one parameter, the number of degrees of freedom, which we abbreviate *df*. For the material covered in this section, the number of degrees of freedom for each hypothesis test will simply be $df = n - 1$.

Formula

Degrees of Freedom for t in a Hypothesis Test for a Population Mean (σ Unknown)

In a hypothesis test for a population mean where the population standard deviation is unknown, the number of degrees of freedom for the Student's t-distribution of the test statistic is given by

$$df = n - 1$$

where n is the sample size.

The number of degrees of freedom for the distribution of the t-test statistic is important to know when using either rejection regions or p-values for drawing conclusions. Let's review each method for drawing a conclusion in the context of using a Student's t-distribution.

Rejection Regions

Recall from the previous section that rejection regions are determined by two things: 1) the type of hypothesis test, and 2) the level of significance, α. Remember that to determine the type of test, simply look at the symbol contained in the alternative hypothesis. Lastly, recall that the decision rule for rejection regions is that we reject the null hypothesis if the test statistic calculated from the sample data falls *within* the rejection region.

Let's now look at how to construct rejection regions for each of the three types of hypothesis tests using a Student's t-distribution. The basic rules are the same as those for a normal distribution, except that instead of finding the critical z-score(s), you must find the critical t-score(s). We can summarize the decision rules for the t-distribution with the following notation.

Procedure

Rejection Regions for Hypothesis Tests for Population Means (σ Unknown)

Reject the null hypothesis, H_0, if:

$t \le -t_\alpha$ for a left-tailed test

$t \ge t_\alpha$ for a right-tailed test

$|t| \ge t_{\alpha/2}$ for a two-tailed test

The table that we will use when evaluating claims using a t-distribution is found in Table C of Appendix A. An excerpt from that table is shown in Figure 10.7. This table contains critical values of the t-distribution for certain numbers of degrees of freedom and commonly used levels of significance, α. To find the desired critical t-score, t_α or $t_{\alpha/2}$, first find the value of α at the top of the table to determine which column of the table to use. Look for the value of α under the label "Area in One Tail" for a one-tailed test, or find the value of α under the label "Area in Two Tails" for a two-tailed test. Then simply scan through the table and find the cell where the degrees of freedom row and the α column intersect. Note that for a two-tailed test, the value obtained from the chart is actually $t_{\alpha/2}$. The table of critical t-values used in this text is constructed in a simplified manner so that α is used regardless of what type of test is being performed. Let's now look at an example of how to use the table to find a critical t-score.

	Area in One Tail				
	0.100	0.050	0.025	0.010	0.005
	Area in Two Tails				
df	0.200	0.100	0.050	0.020	0.010
1	3.078	6.314	12.706	31.821	63.657
2	1.886	2.920	4.303	6.965	9.925
3	1.638	2.353	3.182	4.541	5.841
4	1.533	2.132	2.776	3.747	4.604
5	1.476	2.015	2.571	3.365	4.032
6	1.440	1.943	2.447	3.143	3.707

Figure 10.7: Excerpt from Table C: Critical Values of *t*

Example 10.19

Finding the Critical *t*-Value for a Right-Tailed Hypothesis Test

Find the critical *t*-score that corresponds with 14 degrees of freedom at the 0.025 level of significance for a right-tailed hypothesis test.

Solution

Scanning through the table, we see that the row for 14 degrees of freedom and column for a one-tailed area of $\alpha = 0.025$ intersect at a critical *t*-score of 2.145. Hence, $t_\alpha = t_{0.025} = 2.145$.

	Area in One Tail				
	0.100	0.050	0.025	0.010	0.005
	Area in Two Tails				
df	0.200	0.100	0.050	0.020	0.010
13	1.350	1.771	2.160	2.650	3.012
14	1.345	1.761	2.145	2.624	2.977
15	1.341	1.753	2.131	2.602	2.947
16	1.337	1.746	2.120	2.583	2.921
17	1.333	1.740	2.110	2.567	2.898
18	1.330	1.734	2.101	2.552	2.878

We now have all the tools we need to complete an entire hypothesis test using the *t*-score with a rejection region. Let's take a look at a full example.

Example 10.20

Using a Rejection Region in a Hypothesis Test for a Population Mean (Left-Tailed, σ Unknown)

Nurses in a large teaching hospital have complained for many years that they are overworked and understaffed. The consensus among the nursing staff is that the mean number of patients per nurse each shift is at least 8.0. The hospital administrators claim that the mean is lower than 8.0. In order to prove their point to the nursing staff, the administrators gather information from a simple random sample of 19 nurses' shifts. The sample mean is 7.5 patients per nurse with a standard deviation of 1.1 patients per nurse. Test the administrators' claim using

$\alpha = 0.025$, and assume that the number of patients per nurse has a normal distribution.

Solution

Step 1: **State the null and alternative hypotheses.**

In order to determine the hypotheses, we must first ask "What do the researchers want to gather evidence for?" In this scenario, the researchers are the hospital administrators, and they want to gather evidence in order to determine whether the mean number of patients per nurse is less than 8.0. We write this research hypothesis mathematically as H_a: $\mu < 8.0$. The logical opposite of $\mu < 8.0$ is $\mu \ge 8.0$. We then have the following hypotheses.

$$H_0: \mu \ge 8.0$$
$$H_a: \mu < 8.0$$

Memory Booster

Remember that the null hypothesis will contain equality.

Step 2: **Determine which distribution to use for the test statistic, and state the level of significance.**

The t-test statistic is appropriate to use in this case because the claim is about a population mean, the population is normally distributed, the population standard deviation is unknown, and the sample is a simple random sample.

In addition to determining which distribution to use for the test statistic, we need to state the level of significance. The problem states that $\alpha = 0.025$.

Step 3: **Gather data and calculate the necessary sample statistics.**

Substitute the information given in the scenario into the formula to obtain the t-test statistic.

$$\begin{aligned} t &= \frac{\bar{x} - \mu}{\left(\frac{s}{\sqrt{n}}\right)} \\ &= \frac{7.5 - 8.0}{\left(\frac{1.1}{\sqrt{19}}\right)} \\ &\approx -1.981 \end{aligned}$$

Step 4: **Draw a conclusion and interpret the decision.**

Next, notice from the symbol found in the alternative hypothesis that we are running a left-tailed test. We then need to obtain the critical t-score for the rejection region. From the table we see that a one-tailed t-test with $\alpha = 0.025$ and $n - 1 = 18$ degrees of freedom has a critical t-score of 2.101. Because the area in a left-tailed test is shaded in the left tail of the distribution, the critical t-score will be negative, $-t_\alpha = -t_{0.025} = -2.101$. The rejection region is shown in the following graph. Because the test statistic calculated from the sample, $t \approx -1.981$, does not fall in the rejection region, we must fail to reject the null hypothesis.

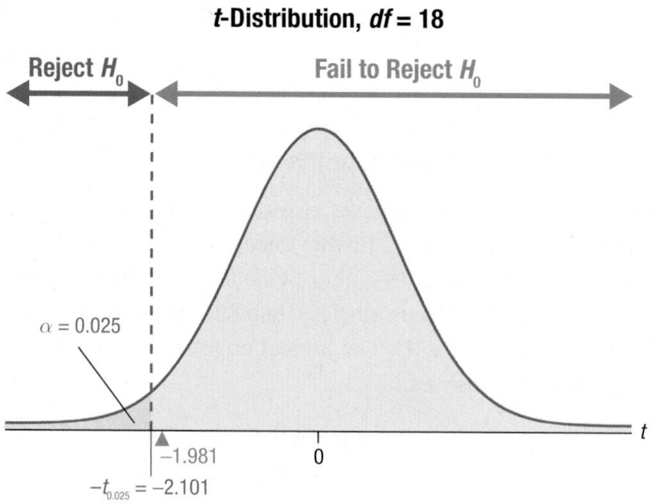

We can interpret this conclusion to mean that the evidence collected is not strong enough at the 0.025 level of significance to reject the null hypothesis in favor of the administrators' claim that the mean number of patients per nurse is less than 8.0.

p-Values

The *t*-distribution table in the back of the book cannot be used to calculate *p*-values in the same way that the normal distribution tables can; however, the *t*-table can be used to determine the range in which the *p*-value lies, such as $0.025 \le p\text{-value} \le 0.05$. When it is important to find an exact *p*-value for a *t*-distribution, we need to turn to technology, such as calculators or statistical software programs. In the following examples, we will use technology to calculate the *p*-value we need to draw a conclusion for each hypothesis test.

Recall from the previous section that the decision rules for *p*-values are as follows.

Procedure

Conclusions Using *p*-Values

- If $p\text{-value} \le \alpha$, then *reject* the null hypothesis.
- If $p\text{-value} > \alpha$, then *fail to reject* the null hypothesis.

Example 10.21

Performing a Hypothesis Test for a Population Mean Using a TI-83/84 Plus Calculator (Right-Tailed, σ Unknown)

A locally owned, independent department store has chosen its marketing strategies for many years under the assumption that the mean amount spent by each shopper in the store is no more than \$100.00. A newly hired store manager claims that the current mean is higher and wants to change the marketing scheme accordingly. A group of 27 shoppers is chosen at random and found to have spent a mean of \$104.93 with a standard deviation of \$9.07. Assume that the population distribution of amounts spent is approximately normal, and test the store manager's claim at the 0.05 level of significance.

Solution

Remember that using technology does not change the way in which a hypothesis test is performed. We will follow the steps as outlined in Section 10.1 to complete the hypothesis test.

Step 1: **State the null and alternative hypotheses.**

The store manager claims that the mean amount spent is more than \$100.00, which we write mathematically as $\mu > 100.00$. Since the manager is looking for evidence to support this statement, the research hypothesis is H_a: $\mu > 100.00$. The logical opposite of the claim is then $\mu \le 100.00$. Thus, we have the following hypotheses.

$$H_0: \mu \le 100.00$$
$$H_a: \mu > 100.00$$

Step 2: **Determine which distribution to use for the test statistic, and state the level of significance.**

A Student's t-distribution, and thus the t-test statistic for population means, is appropriate to use in this case because the claim is about a population mean, the population is normally distributed, the population standard deviation is unknown, and the sample is a simple random sample. The level of significance is stated in the problem to be $\alpha = 0.05$.

Step 3: **Gather data and calculate the necessary sample statistics.**

This is where we begin to use a TI-83/84 Plus calculator. Let's start by writing down the information from the problem, as we will need to enter these values into the calculator. We know that the presumed value of the population mean is $\mu = 100.00$, the sample mean is $\bar{x} = 104.93$, the sample standard deviation is $s = 9.07$, and the sample size is $n = 27$.

Press STAT, scroll to TESTS, and choose option `2:T-Test`. Since we know the sample statistics, choose `Stats` instead of `Data`. For `μ0`, enter the value from the null hypothesis; thus enter `100` for `μ0`. Enter the rest of the given values, as shown in the screenshot on the left below. Choose the alternative hypothesis `>μ0`. Select `Calculate`. Press ENTER.

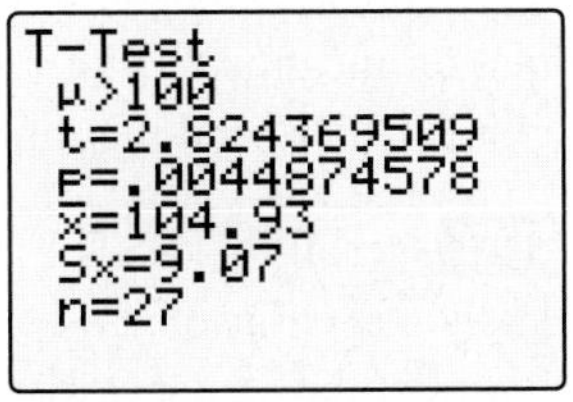

The output screen, shown above on the right, shows the alternative hypothesis, calculates the t-test statistic and the p-value, and then reiterates the sample mean, sample standard deviation, and sample size that were entered.

Step 4: **Draw a conclusion and interpret the decision.**

The p-value given by the calculator is approximately 0.0045. Since $0.0045 < 0.05$, we have $p\text{-value} \le \alpha$. Thus, the conclusion is to reject the null hypothesis. This means that, at the 0.05 level of significance, the evidence supports the manager's claim that the mean amount spent by each shopper in the department store is more than \$100.00.

Example 10.22

Performing a Hypothesis Test for a Population Mean Using a TI-83/84 Plus Calculator (Two-Tailed, σ Unknown)

The meat-packing department of a large grocery store packs ground beef in two-pound portions. Supervisors are concerned that their machine is no longer packaging the beef to the required specifications. To test the claim that the mean weight of ground beef portions is not 2.00 pounds, the supervisors calculate the mean weight for a simple random sample of 20 packages of beef. The sample mean is 2.10 pounds with a standard deviation of 0.33 pounds. Is there sufficient evidence at the 0.01 level of significance to show that the machine is not working properly? Assume that the weights are normally distributed.

Solution

Step 1: State the null and alternative hypotheses.

The supervisors wish to investigate whether the machinery is not working to specifications; that is, the mean weight is not 2.00 pounds per package. This research hypothesis, H_a, is written mathematically as $\mu \neq 2.00$. The logical opposite is then $\mu = 2.00$. We then have the following hypotheses.

$$H_0: \mu = 2.00$$
$$H_a: \mu \neq 2.00$$

Step 2: Determine which distribution to use for the test statistic, and state the level of significance.

The t-test statistic is appropriate to use in this case because the claim is about a population mean, the population is normally distributed, the population standard deviation is unknown, and the sample is a simple random sample. The level of significance is stated to be 0.01.

Step 3: Gather data and calculate the necessary sample statistics.

As stated previously in Example 10.21, this is where we begin to use a TI-83/84 Plus calculator. Let's start by writing down the information from the problem, as we will need to enter these values into the calculator. We know that the presumed value of the population mean is $\mu = 2.00$, the sample mean is $\bar{x} = 2.10$, the sample standard deviation is $s = 0.33$, and the sample size is $n = 20$.

Press STAT, scroll to TESTS, and choose option 2:T-Test. Since we know the sample statistics, choose Stats instead of Data. For μo, enter the value from the null hypothesis; thus enter 2 for μo.

Enter the rest of the given values, as shown in the screenshot on the left below. Choose the alternative hypothesis ≠μo. Select Calculate. Press ENTER.

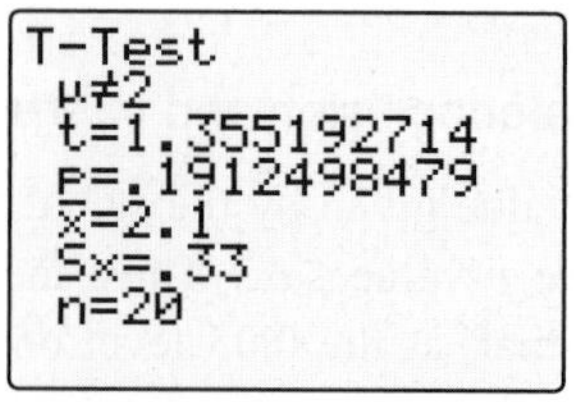

The output screen, shown above on the right, shows the alternative hypothesis, calculates the t-test statistic and the p-value, and then reiterates the sample mean, sample standard deviation, and sample size that were entered.

Step 4: Draw a conclusion and interpret the decision.

The p-value given by the calculator is approximately 0.1912. Since $0.1912 > 0.01$, we have p-value $> \alpha$. Thus, the conclusion is to fail to reject the null hypothesis. We interpret this conclusion to mean that the evidence collected is not strong enough at the 0.01 level of significance to say that the machine is working improperly.

10.3 Section Exercises

Rejection Regions for Hypothesis Tests for Population Means (σ Unknown)

Directions: Determine the rejection region for a hypothesis test for the population mean using the given information. Assume that the population standard deviation is unknown and the population distribution is approximately normal.

1. $\alpha = 0.05$, $df = 14$, left-tailed test

2. $\alpha = 0.01$, $df = 23$, right-tailed test

3. $c = 0.90$, $df = 18$, $H_a\text{: } \mu \neq 18$

4. $c = 0.95$, $df = 26$, $H_a\text{: } \mu < 0.18$

5. $\alpha = 0.10$, $n = 10$, $H_0\text{: } \mu = 102$

6. $\alpha = 0.005$, $n = 6$, $H_0\text{: } \mu \geq 203$

Directions: Draw the rejection region for the hypothesis test for the population mean and give the appropriate conclusion for the hypothesis test using the given information. Assume that the population distribution is approximately normal.

7. $H_0\text{: } \mu = 195$ and $H_a\text{: } \mu \neq 195$, $t = -3.11$, $n = 20$, $\alpha = 0.05$

8. $H_0\text{: } \mu \geq 17$ and $H_a\text{: } \mu < 17$, $t = -2.05$, $n = 6$, $\alpha = 0.05$

9. $H_0\text{: } \mu \leq 28$ and $H_a\text{: } \mu > 28$, $t = 1.99$, $n = 29$, $\alpha = 0.01$

10. $H_0\text{: } \mu = 7$ and $H_a\text{: } \mu \neq 7$, $t = -3$, $n = 15$, $\alpha = 0.01$

11. $H_0\text{: } \mu \geq 98$ and $H_a\text{: } \mu < 98$, $t = -2.73$, $n = 10$, $\alpha = 0.01$

12. H_0: $\mu \leq 39$ and H_a: $\mu > 39$, $t = 2.01$, $n = 8$, $\alpha = 0.05$

Hypothesis Tests for Population Means (σ Unknown)

Directions: **Perform each hypothesis test using the method of your choice or the one assigned by your instructor. For each exercise, complete the following steps. Assume that each population distribution is approximately normal.**

a. State the null and alternative hypotheses.

b. Determine which distribution to use for the test statistic, and state the level of significance.

c. Calculate the test statistic.

d. Draw a conclusion and interpret the decision.

13. The faculty at a large university are irritated by students' cell phones. They have been complaining for the past few years that a cell phone rings in each class at least 15.0 times per semester (which is about once a week). A reporter for the school newspaper claims that students are more courteous with their cell phones now than in past semesters, and that the mean is now lower than 15.0 cell phone disruptions per semester. The reporter asks instructors to keep track of the number of times a cell phone rings in a simple random sample of 12 different classes one semester. The sample mean is 14.8 calls with a standard deviation of 1.2 calls. Does the evidence support the reporter's claim at the 0.10 level of significance?

14. One cable company claims that it has excellent customer service. In fact, the company advertises that a technician will arrive within 30 minutes after a service call is placed. One frustrated customer believes this is not accurate, claiming that it takes over 30 minutes for the cable technician to arrive. The customer asks a simple random sample of 9 other cable customers how long it has taken for the cable technician to arrive when they have called for one. The sample mean for this group is 33.2 minutes with a standard deviation of 3.4 minutes. Test the customer's claim at the 0.025 level of significance.

15. A parenting magazine reports that the mean number of phone calls teenage girls make per night is at least 4. For her science fair project, Ella sets out to prove the magazine wrong. She claims that the mean among teenage girls in her area is less than reported. Ella collects information from a simple random sample of 25 teenage girls at her high school, and calculates a mean of 3.4 calls per night with a standard deviation of 0.9 calls per night. Test Ella's claim at the 0.01 level of significance.

16. A children's clothing company sells hand-smocked dresses for girls. The length of one particular size of dress is designed to be 26 inches. The company regularly tests the lengths of the garments to ensure quality control, and if the mean length is found to be significantly longer or shorter than 26 inches, the machines must be adjusted. The most recent simple random sample of 28 dresses had a mean length of 26.30 inches with a standard deviation of 0.77 inches. Perform a hypothesis test on the accuracy of the machines at the 0.01 level of significance.

17. A pizza delivery chain advertises that it will deliver your pizza in no more than 20 minutes from when the order is placed. Being a skeptic, you decide to test and see if the mean delivery time is actually more than 20 minutes. For the simple random sample of 7 customers who record the amount of time it takes for each of their pizzas to be delivered, the mean is 22.7 minutes with a standard deviation of 4.3 minutes. Perform a hypothesis test using a 0.05 level of significance.

18. Teachers' salaries in one state are very low, so low that educators in that state regularly complain about their compensation. The state mean is \$33,600, but teachers in one district claim that the mean in their district is significantly lower. They survey a simple random sample of 22 teachers in the district and calculate a mean salary of \$32,400 with a standard deviation of \$1520. Test the teachers' claim at the 0.01 level of significance.

19. It currently takes users a mean of 15 minutes to install the most popular computer program made by RodeTech, a software design company. After changes have been made to the program, the company executives want to know if the new mean is now different from 15 minutes so that they can change their advertising accordingly. A simple random sample of 20 new customers are asked to time how long it takes for them to install the software. The sample mean is 14.1 minutes with a standard deviation of 1.9 minutes. Perform a hypothesis test at the 0.05 level of significance to see if the mean installation time has changed.

10.4 Hypothesis Testing for Population Proportions

We have seen in the previous two sections how to apply the principles of hypothesis testing to questions regarding population means. In this section, we will turn our attention to questions about population proportions.

Recall the general steps in a hypothesis test, which are outlined below.

1. State the null and alternative hypotheses.
2. Determine which distribution to use for the test statistic, and state the level of significance.
3. Gather data and calculate the necessary sample statistics.
4. Draw a conclusion and interpret the decision.

At this point you should be very familiar with stating the null and alternative hypotheses, so we can focus our attention on determining which distribution to use for calculating the test statistic for population proportions and the methods for deciding when to reject the null hypothesis.

Recall from the discussion of confidence intervals in Chapter 8 that the best point estimate for a population proportion is a sample proportion, $\hat{p}$. In this section, we will perform hypothesis tests for population proportions in which the following conditions are satisfied.

- All possible samples of a given size have an equal probability of being chosen; that is, a simple random sample is used.
- The conditions for a binomial distribution are met.
- The sample size is large enough to ensure that $np \geq 5$ and $n(1-p) \geq 5$.

If these conditions are met, then the sampling distribution of sample proportions approximates a normal distribution. Thus, the test statistic for testing a population proportion is a *z*-score. The formula for this test statistic is as follows. Note that this is the formula for the standard score for a sample proportion in a sampling distribution, which we first learned in Section 7.3.

Memory Booster

Properties of a Binomial Distribution

1. The experiment consists of a fixed number, *n*, of identical trials.
2. Each trial is independent of the others.
3. For each trial, there are only two possible outcomes. For counting purposes, one outcome is labeled a success, and the other a failure.
4. For every trial, the probability of getting a success is called *p*. The probability of getting a failure is then $1-p$.
5. The binomial random variable, *X*, counts the number of successes in *n* trials.

Formula

Test Statistic for a Hypothesis Test for a Population Proportion

When the sample taken is a simple random sample, the conditions for a binomial distribution are met, and the sample size is large enough to ensure that $np \geq 5$ and $n(1-p) \geq 5$, the test statistic for a hypothesis test for a population proportion is given by

$$z = \frac{\hat{p}-p}{\sqrt{\frac{p(1-p)}{n}}}$$

where $\hat{p}$ is the sample proportion,

p is the presumed value of the population proportion from the null hypothesis, and

n is the sample size.

If the conditions for using the normal distribution are not met, then other methods that are beyond the scope of this text must be used to test a hypothesis about a population proportion or the sample size must be increased to allow for the use of the normal distribution. **Note:** In this textbook, we will assume that the conditions for a binomial distribution are met for all examples and exercises involving hypothesis tests for population proportions.

Example 10.23

Determining Which Distribution to Use for the Test Statistic in a Hypothesis Test

For each scenario, identify which distribution should be used to test the claim.

a. At least 15% of listeners of nonprofit radio stations generally favor commercials for other nonprofit organizations on the station. A local nonprofit radio station believes that less than 15% of its listeners favor such commercials. The station plans to survey a simple random sample of 50 of its listeners to test its claim.

b. The college paper reports that at least 10% of students turn off their cell phones during their classes. Aggravated with the number of cell phones that ring during his classes, a professor believes that fewer than 10% of his students turn off their cell phones during the class period. He plans on testing his claim by asking a simple random sample of 20 of his students to show him their cell phones one day during class.

c. The mean amount of rainfall in the southern part of Texas during the month of September is commonly believed to be no more than 3.02 inches. A meteorologist claims that the amount of rainfall this September has been higher than normal. He tests his claim by measuring the amounts of rainfall in a simple random sample of 12 locations across the region. Assume that the population standard deviation is unknown and the population distribution is approximately normal.

Solution

a. The claim refers to a population proportion, thus we must test the conditions necessary to use the normal distribution. First note that the sample will be a simple random sample. The station plans to survey 50 listeners, so $n = 50$, and the claim is referencing 15%, so $p = 0.15$. Therefore, we can calculate the following.

$$np = 50(0.15) = 7.5 \geq 5$$
$$n(1-p) = 50(1-0.15) = 42.5 \geq 5$$

All of the conditions are met, so the station should use the normal distribution to test its claim.

b. The claim refers to a population proportion, thus we must test the conditions necessary to use the normal distribution. The professor plans to survey 20 students, so $n = 20$, and the claim is referencing 10%, so $p = 0.10$. Therefore, we can calculate the following.

$$np = 20(0.10) = 2 < 5$$

The condition that the sample size must be large enough such that $np \geq 5$ is not met; therefore, the normal distribution cannot be used with the sample size given. If it is possible to increase the sample size to at least 50, then the normal distribution could be used. Otherwise, other methods are necessary.

c. The claim refers to a population mean, therefore we must use one of the distributions discussed in the previous sections. Note that the population standard deviation is unknown and the population is assumed to be normally distributed. Therefore, a Student's t-distribution should be used.

Once we have chosen the appropriate distribution, we must decide under what circumstances to reject the null hypothesis. In this section, we will continue to use the p-value method because it is the most popular method in research journals. However, if we wished to use rejection regions, the method is identical to using rejection regions in hypothesis tests for population means when σ is known since they also use a z-test statistic. Recall from the previous sections that the decision rules for p-values are as follows.

Procedure

Conclusions Using p-Values

- If p-value $\leq \alpha$, then *reject* the null hypothesis.
- If p-value $> \alpha$, then *fail to reject* the null hypothesis.

Caution

In standard practice, the symbol, p, can represent either *population proportion* or *p-value*. You must pay attention to the context in which the symbol is used in order to determine its meaning.

Memory Booster

Remember that the null hypothesis will contain equality.

Example 10.24

Performing a Hypothesis Test for a Population Proportion (Left-Tailed)

The local school board has been advertising that at least 65% of voters favor a tax increase to pay for a new school. A local politician believes that less than 65% of his constituents favor this tax increase. To test his belief, his staff asked a simple random sample of 50 of his constituents whether they favor the tax increase and 27 said that they would vote in favor of the tax increase. If the politician wishes to be 95% confident in his conclusion, does this information support his belief?

Solution

Step 1: **State the null and alternative hypotheses.**

The politician's belief is that less than 65% of the constituents favor a tax increase. Written mathematically, this claim is $p < 0.65$. The logical opposite of this claim is $p \geq 0.65$. Thus, the null and alternative hypotheses are stated as follows.

$$H_0: p \geq 0.65$$
$$H_a: p < 0.65$$

Step 2: **Determine which distribution to use for the test statistic, and state the level of significance.**

We are testing a population proportion, so we must check the necessary conditions to use the normal distribution and the z-test statistic. A simple random sample of 50 constituents were surveyed, giving us $n = 50$, and we know from Step 1 that $p = 0.65$. Therefore, we check the conditions for the sample size as follows.

$$np = 50(0.65) = 32.5 \geq 5$$
$$n(1-p) = 50(1-0.65) = 17.5 \geq 5$$

Since all of the conditions are satisfied, we can use the z-test statistic for the sample proportion.

For this hypothesis test, the level of confidence is 95%, so $c = 0.95$ and the level of significance is calculated as follows.

$$\begin{aligned}\alpha &= 1 - c \\ &= 1 - 0.95 \\ &= 0.05\end{aligned}$$

Step 3: Gather data and calculate the necessary sample statistics.

The sample data show that 27 out of 50 constituents favor the tax increase. Thus, the sample proportion is computed as follows.

$$\hat{p} = \frac{27}{50} = 0.54$$

Substituting the necessary values into the formula for the test statistic for the sample proportion, we obtain the following z-score.

$$\begin{aligned}z &= \frac{\hat{p} - p}{\sqrt{\frac{p(1-p)}{n}}} \\ &= \frac{0.54 - 0.65}{\sqrt{\frac{0.65(1-0.65)}{50}}} \\ &\approx -1.63\end{aligned}$$

Step 4: Draw a conclusion and interpret the decision.

The alternative hypothesis tells us that we are conducting a left-tailed test. Therefore, the p-value for this test statistic is the probability of obtaining a test statistic less than or equal to $z = -1.63$, written $p\text{-value} = P(z \le -1.63)$. To find the p-value, we need to find the area under the standard normal curve to the left of $z = -1.63$.

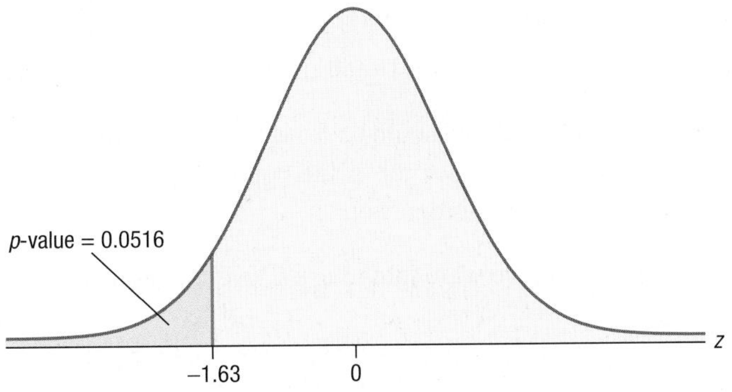

The p-value is 0.0516. Comparing this p-value to the level of significance, we see that $0.0516 > 0.05$, so $p\text{-value} > \alpha$. Thus, we fail to reject the null hypothesis. This means that, at the 0.05 level of significance, this evidence does not sufficiently support the politician's belief that less than 65% of his constituents favor a tax increase to pay for a new school.

Example 10.25

Performing a Hypothesis Test for a Population Proportion (Two-Tailed)

According to a report published by the Mathematical Association of America, 1% of bachelor's degrees in the United States are awarded in the fields of mathematics and statistics. One recent mathematics graduate believes that this number is incorrect. She has no indication if the actual percentage is more or less than 1%. To perform a hypothesis test, she uses a simple random sample of 12,317 recent college graduates for her sample. According to the graduation programs from the colleges, she notes that bachelor's degrees in mathematics or statistics were awarded to 148 graduates in her sample. Does this evidence support this graduate's belief that the percentage of bachelor's degrees awarded in the fields of mathematics and statistics is not 1%? Use a 0.10 level of significance.

Source: Bressoud, David. "Status of the Math-Intensive Majors." *Launchings.* Feb. 2011. http://www.maa.org/columns/launchings/launchings_02_11.html (1 Dec. 2011).

Solution

Step 1: State the null and alternative hypotheses.

The graduate is trying to show that the percentage of bachelor's degrees awarded in the fields of mathematics and statistics is not 1%. Written mathematically, this is $p \neq 0.01$. The logical opposite of this is $p = 0.01$. Thus, the null and alternative hypotheses are stated as follows.

$$H_0: p = 0.01$$
$$H_a: p \neq 0.01$$

Step 2: Determine which distribution to use for the test statistic, and state the level of significance.

We are testing a population proportion, so we must check the necessary conditions to use the normal distribution and the z-test statistic. Recall from the information given that the sample was a simple random sample and that $n = 12{,}317$ and $p = 0.01$. Therefore, we check the conditions for the sample size as follows.

$$np = 12{,}317(0.01) = 123.17 \geq 5$$
$$n(1-p) = 12{,}317(1-0.01) = 12{,}193.83 \geq 5$$

Due to the large sample size, it should have been obvious that both of these conditions would be easily satisfied. Since all of the conditions are satisfied, we can use the z-test statistic for the sample proportion.

For this hypothesis test, we were told to use a level of significance of $\alpha = 0.10$.

Step 3: Gather data and calculate the necessary sample statistics.

The sample data show that 148 out of 12,317 graduates obtained degrees in mathematics or statistics. Thus, the sample proportion is calculated as follows.

$$\hat{p} = \frac{148}{12{,}317} \approx 0.012016$$

Note that we rounded the sample proportion to six decimal places, rather than three decimal places, to avoid additional rounding error in the following calculation of the z-score.

Substituting the necessary values into the formula for the test statistic for the sample

Rounding Rule

When calculations involve several steps, avoid rounding at intermediate calculations. If necessary, round intermediate calculations to at least six decimal places to avoid additional rounding errors in subsequent calculations.

proportion, we obtain the following z-score.

$$\begin{aligned} z &= \frac{\hat{p}-p}{\sqrt{\frac{p(1-p)}{n}}} \\ &= \frac{0.012016-0.01}{\sqrt{\frac{0.01(1-0.01)}{12{,}317}}} \\ &\approx 2.25 \end{aligned}$$

Step 4: **Draw a conclusion and interpret the decision.**

The alternative hypothesis tells us that we are conducting a two-tailed test. Therefore, the p-value for this test statistic is the probability of obtaining a test statistic that is either less than or equal to $z_1 = -2.25$ or greater than or equal to $z_2 = 2.25$, which is written mathematically as $p\text{-value} = P(|z| \ge 2.25)$. To find the p-value, we need to find the area under the standard normal curve to the left of $z_1 = -2.25$. This area is 0.0122, which is just the area in the left tail. The total area is then twice this amount. Thus, the p-value is calculated as follows.

$$\begin{aligned} p\text{-value} &= (0.0122)(2) \\ &= 0.0244 \end{aligned}$$

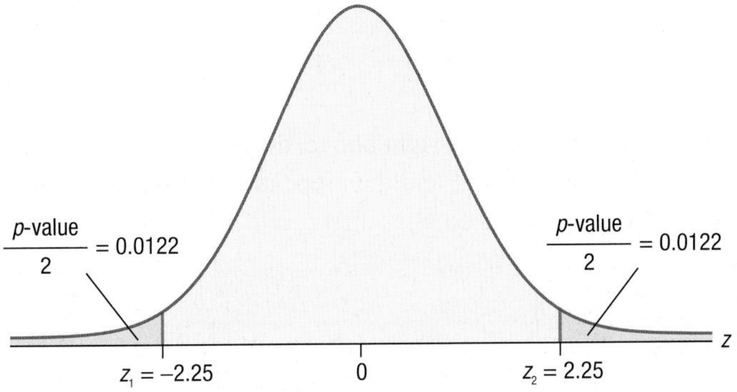

The p-value is 0.0244. Comparing this p-value to the level of significance, we see that $0.0244 \le 0.10$, so $p\text{-value} \le \alpha$. Thus, we reject the null hypothesis. This means that, at the 0.10 level of significance, this evidence supports the belief that the percentage of bachelor's degrees awarded in the fields of mathematics and statistics is not 1%. Notice that despite the sample proportion being larger than the assumed population proportion, our conclusion references only the original null and alternative hypotheses.

When we use a TI-83/84 Plus calculator to run a hypothesis test for a population proportion, we actually need less information than when we conduct the test by hand, just as we did with confidence intervals in Chapter 8. We no longer need to calculate the sample proportion first; the calculator does that for us. In fact, all that is required is the presumed value of the population proportion from the null hypothesis, which is denoted on the calculator by p_0; the sample size, n; and the number of members of the sample that have the characteristic in which you are interested, which is denoted by x.

Let's use the information from the last example and illustrate the process on the calculator. Here's a recap of the information that was given.

Report: 1% of bachelor's degrees in the United States are awarded in the fields of mathematics and statistics.

Claim: One recent mathematics graduate believes that this percentage is incorrect.

Sample: 12,317 recent college graduates

Number in sample who were awarded bachelor's degrees in mathematics or statistics: 148 graduates

Level of significance: 0.10

The sample size is $n = 12{,}317$ and $x = 148$ because that is the number of students sampled who have the characteristic in which we are interested, that is, the number of graduates in the sample who were awarded bachelor's degrees in mathematics or statistics. Finally, the presumed value of the population proportion from the null hypothesis is 1%; that is, $p_0 = 0.01$.

To put these in the calculator, press STAT and scroll over to choose TESTS. Choose option `5:1-PropZTest`. Enter the given values, as shown in the screenshot on the left below. From the research hypothesis, we know that this is a two-tailed test, so make sure that ≠p₀ is highlighted. Then select `Calculate` and press ENTER.

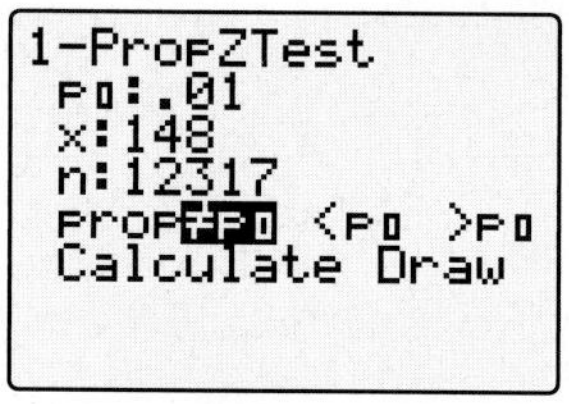

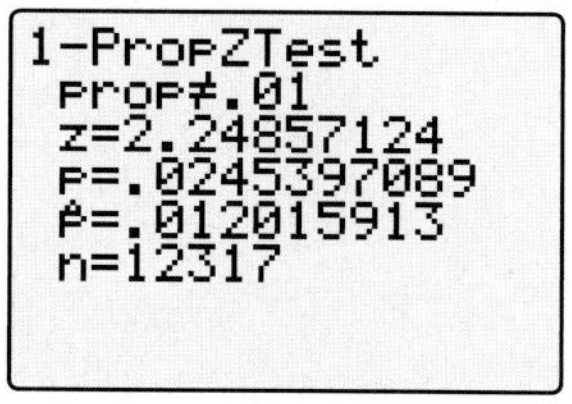

The output screen, which is the screen on the right above, displays the alternative hypothesis, the test statistic, the p-value, the sample proportion, and then repeats the sample size we entered. Since the p-value is approximately 0.0245, which is smaller than $\alpha = 0.10$, we reject the null hypothesis as we did in Example 10.25.

Example 10.26

Performing a Hypothesis Test for a Population Proportion Using a TI-83/84 Plus Calculator (Right-Tailed)

A crime watch group believes that the percentage of all inmates released from prison who were released because their sentences expired was higher in 2008–2009 for the whole United States than it was for the state of Florida. Between July 1, 2008 and June 30, 2009, 63.9% of all prison inmates released in Florida were released because their sentences expired. Suppose that, in a simple random sample of 36,463 prison inmates in the United States who were released during the same time period, 64.6% were released because their sentences expired. Use a TI-83/84 Plus calculator to run a hypothesis test to determine if the watch group's claim is correct at the 99% confidence level.

Source: Florida Department of Corrections. "Annual Statistics for Fiscal Year 2008–2009: Inmate Releases and Time Served." http://www.dc.state.fl.us/pub/annual/0809/stats/im_release.html (1 Dec. 2011).

Solution

Step 1: State the null and alternative hypotheses.

The watch group's belief is that the proportion of inmates released because of an expired sentence in 2008–2009 was higher for the whole United States than it was for the state of Florida. Since the proportion for Florida was 63.9% in 2008–2009, we write this belief mathematically as $p > 0.639$. The logical opposite is $p \le 0.639$. Thus the null and alternative hypotheses are stated as follows.

$$H_0: p \le 0.639$$
$$H_a: p > 0.639$$

Step 2: **Determine which distribution to use for the test statistic, and state the level of significance.**

Using the normal distribution, and thus the z-test statistic, in a hypothesis test for a population proportion requires first testing the necessary conditions. The sample is a simple random sample, the sample size is $n = 36{,}463$, and the presumed value of the population proportion from the null hypothesis is $p = 0.639$. Therefore, we check the conditions for the sample size as follows.

$$np = 36{,}463(0.639) = 23{,}299.857 \ge 5$$
$$n(1-p) = 36{,}463(1-0.639) = 13{,}163.143 \ge 5$$

Once again, the large sample size is an indication that both of these conditions are easily met. Therefore, we can use the z-test statistic for the sample proportion. For this hypothesis test, the confidence level is 99%, so $c = 0.99$ and the level of significance is calculated as follows.

$$\begin{aligned}\alpha &= 1-c\\ &= 1-0.99\\ &= 0.01\end{aligned}$$

Step 3: **Gather data and calculate the necessary sample statistics.**

Once again, this is where we begin to use a TI-83/84 Plus calculator. We need three pieces of information to use the calculator to run a hypothesis test for the population proportion: n, x, and p_0. From above we know that $n = 36{,}463$ and $p_0 = 0.639$. The calculator wants a whole number for x and will return an error if you enter either a decimal or the percentage of the sample. So, we need to find the number of inmates in the sample who were released from prison because their sentences expired. Here's where we actually have to work backwards a little bit. Since we were told that 64.6% of the inmates in the sample were released because of expired sentences, we can find x by multiplying 0.646 by the sample size, as follows.

$$\begin{aligned}x &= (0.646)(36{,}463)\\ &= 23{,}555.098\\ &\approx 23{,}555\end{aligned}$$

Because we cannot enter a decimal, we round to the nearest whole number. Approximately 23,555 inmates in the sample fit our characteristic, so $x = 23{,}555$.

The last bit of information we must input into the calculator is the type of test we are running: left-tailed, right-tailed, or two-tailed. Because the alternative hypothesis states that the percentage for the United States was higher than the percentage for Florida, this is a right-tailed test, represented by `>p0` on the calculator.

To input these values into the calculator, press STAT, scroll to TESTS, and choose option `5:1-PropZTest`. After entering the values, select `Calculate` and press ENTER.

Step 4: Draw a conclusion and interpret the decision.

The p-value given is approximately 0.0027, and our level of significance is $\alpha = 0.01$. Since $0.0027 < 0.01$, we have $p\text{-value} \le \alpha$. Thus, we reject the null hypothesis. This means that the evidence supports the watch group's claim that the percentage of inmates who were released from prison in 2008–2009 because their sentences expired was higher for the whole United States than it was for the state of Florida.

10.4 Section Exercises

Necessary Conditions for Using the Normal Distribution in Hypothesis Tests for Population Proportions

Directions: Determine whether the normal distribution can be used to perform a hypothesis test for the population proportion in each scenario.

1. After collecting data from a simple random sample of 43 townspeople on whether they approve of the local mayor, a statistics student wants to run a hypothesis test for the population proportion using a 95% level of confidence. Has he collected enough data to test the claim that the currently accepted approval rating is incorrect if the current belief is that only 5% of residents approve of the mayor?

2. An environmentalist wishes to conduct a hypothesis test on the percentage of cars driven in the city that are hybrids. Is it sufficient for him to use a simple random sample of 54 cars if hybrids currently account for 10% of the car sales in the country and he claims that the percentage of hybrids in the city is higher than that?

3. A scientist believes that 12 out of every 100 office workers suffer from extreme sensitivity to dust in the work place. Is it appropriate to conduct a hypothesis test at the 99% level of confidence if only 97 office workers are surveyed?

4. In a recent study at a local college, 48 students admitted that they had tried alcohol at least once while under the legal drinking age, while only 4 students said they had not. Is there enough data to conduct a hypothesis test to see if the percentage of students at the college who admit to drinking underage is the same as the national percentage of 95%?

Hypothesis Tests for Population Proportions

Directions: **Perform each hypothesis test. For each exercise, complete the following steps.**

a. State the null and alternative hypotheses.

b. Determine which distribution to use for the test statistic, and state the level of significance.

c. Calculate the test statistic.

d. Draw a conclusion by comparing the *p*-value to the level of significance and interpret the decision.

5. According to a report sponsored by the National Center for Health Statistics, 74% of American women have been married by the age of 30. Patrice, a single woman of 32 who has never been married, claims that this percentage is too high. To test her claim, she surveys a simple random sample of 125 American women between the ages of 30 and 44 and finds that 91 of them were married at least once by the age of 30. Does this evidence support Patrice's claim that less than 74% of American women have been married by the age of 30? Use a 0.10 level of significance.

Source: Goodwin, P., B. McGill, and A. Chandra. "Who Marries and When? Age at First Marriage in the United States: 2002." *NCHS Data Brief*, No. 19. National Center for Health Statistics. June 2009. http://www.cdc.gov/nchs/data/databriefs/db19.htm#probabilities (2 Dec. 2011).

6. The National Academy of Science reported in a 1997 study that 40% of research in mathematics is published by US authors. The mathematics chairperson of a prestigious university wishes to test the claim that this percentage is no longer 40%. He has no indication of whether the percentage has increased or decreased since that time. He surveys a simple random sample of 130 recent articles published by reputable mathematics research journals and finds that 62 of these articles have US authors. Does this evidence support the mathematics chairperson's claim that the percentage is no longer 40%? Use a 0.10 level of significance.

Source: Panel on International Benchmarking of US Mathematics Research and Committee on Science, Engineering, and Public Policy. *International Benchmarking of US Mathematics Research*. National Academy of Sciences, National Academy of Engineering, Institute of Medicine (SEM). 1997. http://www.nap.edu/openbook.php?record_id=9089&page=22 (2 Dec. 2011).

7. Sleep apnea is a condition in which the sufferers stop breathing momentarily while they are asleep. This condition results in lack of sleep and extreme fatigue during waking hours. A current estimate is that 18 million out of the 312.7 million Americans suffer from sleep apnea, or approximately 5.8%. A safety commission is concerned about the percentage of commercial truck drivers who suffer from sleep apnea. They do not have any reason to believe that it would be higher or lower than the population's percentage. To test the claim that the percentage of commercial truck drivers who suffer from sleep apnea is not 5.8%, a simple random sample of 350 commercial truck drivers is examined by a medical expert, who concludes that 30 suffer from sleep apnea. Does this evidence support the claim that the percentage of commercial truck drivers who suffer from sleep apnea is not 5.8%? Use a 0.01 level of significance.

Source: American Sleep Apnea Association. "Sleep Apnea." 2011. http://www.sleepapnea.org/learn/sleepapnea.html (2 Dec. 2011).
Source: US Census Bureau. "American FactFinder." http://factfinder2.census.gov (2 Dec. 2011).

8. Sleep experts believe that sleep apnea is more likely to occur in men than in the general population. In other words, they claim that the percentage of men who suffer from sleep apnea is greater than 5.8%. To test this claim, one sleep expert examines a simple random sample of 90 men and determines that 9 of these men suffer from sleep apnea. Does this evidence support the claim that the percentage of men who suffer from sleep apnea is greater than 5.8%? Use a 0.05 level of significance.

Source: American Sleep Apnea Association. "Sleep Apnea." 2011. http://www.sleepapnea.org/learn/sleepapnea.html (2 Dec. 2011).

9. A direct mail appeal for contributions from a university's alumni and supporters is considered cost effective if more than 15% of the alumni and supporters provide monetary contributions. To determine if a direct mail appeal is cost effective, the fundraising director sends the direct mail brochures to a simple random sample of 250 people on the alumni and supporters mailing lists. He receives monetary contributions from 40 people. Does this evidence support the cost effectiveness of the direct mail appeal? Use a 0.05 level of significance.

10. A study by a medical researcher of HIV infection rates among injection drug users reported that the rate of new infections steadily dropped over the last couple of decades. The latest report was that 2% of IV drug users are HIV positive. A drug counselor believes that this trend is continuing and that the percentage of IV drug users in his city who are HIV positive is now less than 2%. The counselor surveys a simple random sample of 415 IV drug users in the city and finds that 8 are HIV positive. Does this evidence support the counselor's belief? Use a 0.01 level of significance.

10.5 Hypothesis Testing for Population Variances

In the previous sections, we looked at testing claims involving either the population mean or the population proportion of certain sets of data. However, just as we stated in the beginning of the chapter, hypothesis testing is a procedure for evaluating claims about population parameters in general, not just means and proportions. Quite often, the variance of a population needs to be evaluated. Suppose for a moment that you are a prominent artist. Your work becomes more valuable if you produce one-of-a-kind pieces of art. In other words, you want each piece to vary from the others in some way, and variance is a good thing. If however, you are a pharmaceutical company executive monitoring the quality of each drug you produce, you want to reproduce the same product over and over with the smallest possible variance between the pills. In these cases, the variance of the population is of particular interest and worthy of being evaluated. In the same manner that we tested population means and proportions, we can perform hypothesis tests for population variances and population standard deviations. Note that throughout this section, we will speak generally in terms of variances, but these same techniques apply to standard deviations as well.

Caution

Read each example and exercise carefully. Some describe hypothesis tests for variance, σ^2, while others involve standard deviation, σ.

Recall the general steps in a hypothesis test, which are outlined below.

1. State the null and alternative hypotheses.
2. Determine which distribution to use for the test statistic, and state the level of significance.
3. Gather data and calculate the necessary sample statistics.
4. Draw a conclusion and interpret the decision.

Since the steps in hypothesis testing remain the same, let's look at what is different in hypothesis testing for a population variance or population standard deviation. First, the following criteria must be met before a hypothesis test for a variance or standard deviation can be performed.

- All possible samples of a given size have an equal probability of being chosen; that is, a simple random sample is used.
- The population distribution is approximately normal.

Stating the null and alternative hypotheses for the hypothesis tests in this section will require using either the symbol for population variance, σ^2, or the symbol for population standard deviation, σ. As discussed in Section 8.5, the chi-square test statistic is the appropriate choice when evaluating sample variances or sample standard deviations. This is because a sampling distribution of sample variances or sample standard deviations is a chi-square distribution with $n - 1$ degrees of freedom. Thus, the test statistic for a hypothesis test for a population variance or population standard deviation is the chi-square test statistic given by the following formula.

Formula

Test Statistic for a Hypothesis Test for a Population Variance or Population Standard Deviation

When the sample taken is a simple random sample and the population distribution is approximately normal, the test statistic for a hypothesis test for a population variance or population standard deviation is given by

$$\chi^2 = \frac{(n-1)s^2}{\sigma^2}$$

where n is the sample size,

s^2 is the sample variance, and

σ^2 is the presumed value of the population variance (or square of the presumed value of the population standard deviation) from the null hypothesis.

Formula

Degrees of Freedom for a Hypothesis Test for a Population Variance or Population Standard Deviation

In a hypothesis test for a population variance or population standard deviation, the number of degrees of freedom for the chi-square distribution of the test statistic is given by

$$df = n - 1$$

where n is the sample size.

When using the chi-square distribution, we will use a rejection region to draw a conclusion. We can compare the value of the test statistic to the critical value(s) of χ^2 found in Table G in Appendix A. Critical values for a chi-square distribution are found in much the same way as critical values for a Student's t-distribution. To find the critical value(s), we will need to know the number of degrees of freedom, which is given by $df = n - 1$. To determine the rejection region, we need to consider the type of test we are performing: left-tailed, right-tailed, or two-tailed.

Procedure

Rejection Regions for Hypothesis Tests for Population Variances and Standard Deviations

Reject the null hypothesis, H_0, if:

$\chi^2 \le \chi^2_{(1-\alpha)}$ for a left-tailed test

$\chi^2 \ge \chi^2_{\alpha}$ for a right-tailed test

$\chi^2 \le \chi^2_{(1-\alpha/2)}$ or $\chi^2 \ge \chi^2_{\alpha/2}$ for a two-tailed test

A critical value of χ^2 can be found in Table G in the cell where the row for the number of degrees of freedom and the column for the area in the right tail of the distribution intersect. For example, suppose you were running a right-tailed test for a population variance using a level of significance of $\alpha = 0.05$ and a sample size of $n = 7$. Then the critical value for the test, $\chi^2_{\alpha} = \chi^2_{0.05} = 12.592$, would be found in the cell where the row for $df = 7 - 1 = 6$ intersects the column for an area of 0.050, as shown in Figure 10.8.

	Area to the Right of the Critical Value of χ^2									
df	0.995	0.990	0.975	0.950	0.900	0.100	0.050	0.025	0.010	0.005
1	0.000	0.000	0.001	0.004	0.016	2.706	3.841	5.024	6.635	7.879
2	0.010	0.020	0.051	0.103	0.211	4.605	5.991	7.378	9.210	10.597
3	0.072	0.115	0.216	0.352	0.584	6.251	7.815	9.348	11.345	12.838
4	0.207	0.297	0.484	0.711	1.064	7.779	9.488	11.143	13.277	14.860
5	0.412	0.554	0.831	1.145	1.610	9.236	11.070	12.833	15.086	16.750
6	0.676	0.872	1.237	1.635	2.204	10.645	12.592	14.449	16.812	18.548
7	0.989	1.239	1.690	2.167	2.833	12.017	14.067	16.013	18.475	20.278
8	1.344	1.646	2.180	2.733	3.490	13.362	15.507	17.535	20.090	21.955

Figure 10.8: Excerpt from Table G: Critical Values of χ^2

Let's look at a few examples of hypothesis tests for population variances and population standard deviations. For each of these examples, you can safely assume that all of the necessary conditions are met.

Example 10.27

Performing a Hypothesis Test for a Population Variance (Left-Tailed)

Suppose that your pharmacy currently buys Drug A, a medication for high blood pressure. However, a new company says that it has a "better" drug for blood pressure than Drug A. Both drugs have exactly the same active ingredient in them. The new company claims that its pill is better because the variance in the amounts of the active ingredient in its pills is smaller than in Drug A, which is known to have a variance of 0.0009. To test the new company's claim, a simple random sample of 100 of the new pills are selected and the amounts of the active ingredient are found to have a mean of 2.470 mg and a standard deviation of 0.026 mg. Is this sufficient evidence, at the 0.01 level of significance, to support the new company's claim that its drug has a smaller variance in the amount of the active ingredient? Assume that the amounts of the active ingredient in the pills are normally distributed.

Solution

Step 1: State the null and alternative hypotheses.

The new drug company wants to show that its drug's variance is smaller than the variance, 0.0009, of the other drug. This research hypothesis can be written mathematically as $\sigma^2 < 0.0009$. The opposite of this claim is $\sigma^2 \geq 0.0009$. Thus, the null and alternative hypotheses are stated as follows.

$$H_0: \sigma^2 \geq 0.0009$$
$$H_a: \sigma^2 < 0.0009$$

Memory Booster

Remember that the null hypothesis will contain equality.

Step 2: Determine which distribution to use for the test statistic, and state the level of significance.

Since we are testing a population variance and we are told that we can safely assume that all necessary conditions are met, we must use the chi-square distribution and thus the χ^2-test statistic. The problem states that the level of significance is $\alpha = 0.01$.

Step 3: **Gather data and calculate the necessary sample statistics.**

Using the information provided in the problem, we know that $n = 100$, $s^2 = (0.026)^2 = 0.000676$, and $\sigma^2 = 0.0009$. Substituting these values into the formula, we obtain the χ^2-test statistic as shown below.

$$\begin{aligned}\chi^2 &= \frac{(n-1)s^2}{\sigma^2} \\ &= \frac{(100-1)(0.000676)}{0.0009} \\ &= 74.36\end{aligned}$$

Step 4: **Draw a conclusion and interpret the decision.**

This is a left-tailed test, so the rejection region is $\chi^2 \le \chi^2_{(1-\alpha)}$. The level of significance is $\alpha = 0.01$ and the number of degrees of freedom is $df = n - 1 = 100 - 1 = 99$. Since the chi-square table only gives multiples of 10 after 30 degrees of freedom, we will approximate the critical value using the closest number of degrees of freedom given. The closest value to 99 is 100, so we will use $df = 100$. Thus, the critical value is $\chi^2_{0.990} = 70.065$. Therefore, the decision rule is to reject H_0 if $\chi^2 \le 70.065$.

χ^2-Distribution, *df* = 100

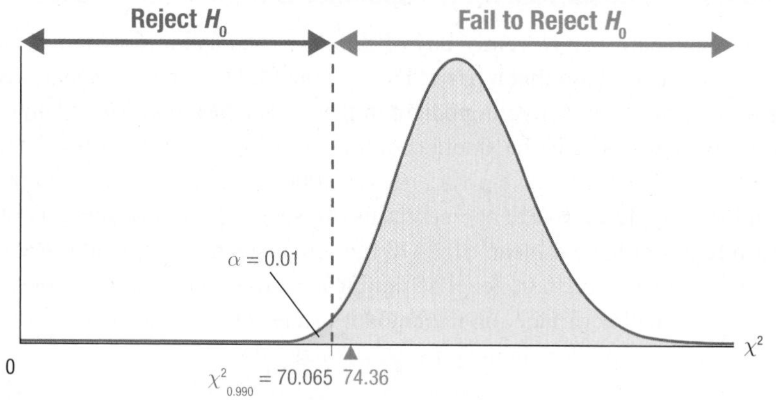

Since $\chi^2 = 74.36$, which is not in the rejection region, we fail to reject the null hypothesis. Thus, there is not sufficient evidence at the 0.01 level of significance to support the new drug company's claim that the variance in the amount of the active ingredient in its drug is less than 0.0009.

Example 10.28

Performing a Hypothesis Test for a Population Standard Deviation (Right-Tailed)

A manufacturer of golf balls requires that the weights of its golf balls have a standard deviation that does not exceed 0.08 ounces. One of the quality control inspectors says that the machines need to be recalibrated because he believes the standard deviation of the weights of the golf balls is more than 0.08 ounces. To test the machines, he selects a simple random sample of 30 golf balls off the assembly line and finds that they have a mean weight of 1.6200 ounces and a standard deviation of 0.0804 ounces. Does this evidence support the need to recalibrate the machines, at the 0.05 level of significance? Assume that the weights of the golf balls are normally distributed.

Solution

Step 1: State the null and alternative hypotheses.

The inspector believes that the machines need to be recalibrated because the standard deviation is greater than the allowed 0.08 ounces. Thus, the research hypothesis, H_a, is that the standard deviation of the weights of the golf balls is more than 0.08 ounces, which is written mathematically as $\sigma > 0.08$. Notice that we use σ and not σ^2 here because we are testing the standard deviation and not the variance. The logical opposite is $\sigma \le 0.08$. Thus, the hypotheses are stated as follows.

$$H_0: \sigma \le 0.08$$
$$H_a: \sigma > 0.08$$

Step 2: Determine which distribution to use for the test statistic, and state the level of significance.

Since we are testing a population standard deviation and we are told that we can safely assume that all necessary conditions are met, we can use the chi-square distribution just as we did for variance, and thus the χ^2-test statistic with the given level of significance of $\alpha = 0.05$.

Step 3: Gather data and calculate the necessary sample statistics.

Using the information provided in the problem, we know that $n = 30$, $s = 0.0804$, and $\sigma = 0.08$. Substituting these values into the formula, we obtain the χ^2-test statistic as shown below.

$$\begin{aligned}\chi^2 &= \frac{(n-1)s^2}{\sigma^2} \\ &= \frac{(30-1)(0.0804)^2}{(0.08)^2} \\ &\approx 29.291\end{aligned}$$

Step 4: Draw a conclusion and interpret the decision.

This is a right-tailed test, so the rejection region is $\chi^2 \ge \chi^2_\alpha$. The level of significance is $\alpha = 0.05$, and the number of degrees of freedom is $df = n - 1 = 30 - 1 = 29$. Thus, the critical value is $\chi^2_{0.050} = 42.557$. Therefore, the decision rule is to reject H_0 if $\chi^2 \ge 42.557$.

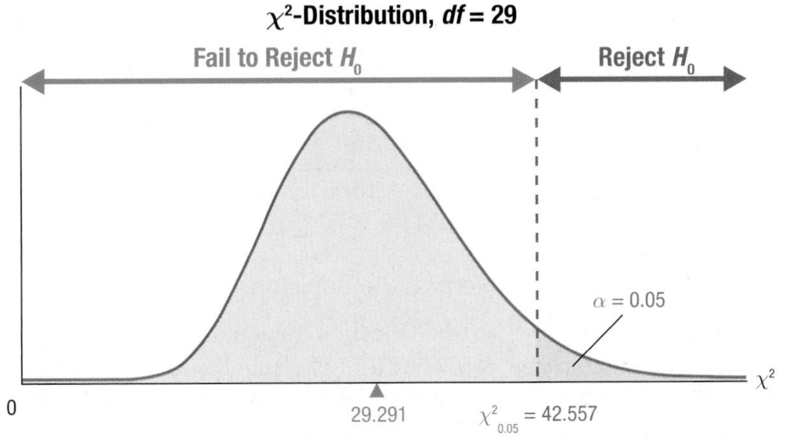

Since $\chi^2 = 29.291$, which is not in the rejection region, we fail to reject the null hypothesis. Thus, there is not sufficient evidence at the 0.05 level of significance to support the need to recalibrate the machines.

Example 10.29

Performing a Hypothesis Test for a Population Variance (Two-Tailed)

A tree farmer would like to know how much variance there is in the heights of two-year-old trees on his farm. It is generally accepted by botanists that after two years of growth, the heights of this variety of tree, measured in feet, should have a variance of 16. The farmer claims that after two years of growth, the variance in the heights of his trees is not 16. He selects a simple random sample of 40 two-year-old trees on his farm and finds that their heights have a standard deviation of 3.2 feet. Does this evidence, at the 0.10 level of significance, support the farmer's claim? Assume that the heights of the trees are normally distributed.

Solution

Step 1: State the null and alternative hypotheses.

The farmer's claim is that the variance of the heights of his trees is not 16. This claim is the research hypothesis, which is written mathematically as $H_a: \sigma^2 \neq 16$. The opposite of this claim is $\sigma^2 = 16$. Thus, the hypotheses are stated as follows.

$$H_0: \sigma^2 = 16$$
$$H_a: \sigma^2 \neq 16$$

Step 2: Determine which distribution to use for the test statistic, and state the level of significance.

Since we are testing a population variance and we are told that we can safely assume that all necessary conditions are met, we use the chi-square distribution and the χ^2-test statistic with the given level of significance of $\alpha = 0.10$.

Step 3: Gather data and calculate the necessary sample statistics.

Using the information provided in the problem, we know that $n = 40$, $s = 3.2$, and $\sigma^2 = 16$. Substituting these values into the formula, we obtain the χ^2-test statistic as shown below.

$$\begin{aligned}\chi^2 &= \frac{(n-1)s^2}{\sigma^2} \\ &= \frac{(40-1)(3.2)^2}{16} \\ &= 24.96\end{aligned}$$

Step 4: Draw a conclusion and interpret the decision.

This is a two-tailed test, so the rejection region is $\chi^2 \leq \chi^2_{(1-\alpha/2)}$ or $\chi^2 \geq \chi^2_{\alpha/2}$. The level of significance is $\alpha = 0.10$. The number of degrees of freedom is $df = n - 1 = 40 - 1 = 39$. Since the chi-square table only gives multiples of 10 after 30 degrees of freedom, we will approximate the critical value using the closest number of degrees of freedom given. The closest value to 39 is 40, so we will use

$df = 40$. The left-hand critical value is $\chi^2_{0.950} = 26.509$ and the right-hand critical value is $\chi^2_{0.050} = 55.758$. Thus, the decision rule is to reject H_0 if $\chi^2 \le 26.509$ or $\chi^2 \ge 55.758$.

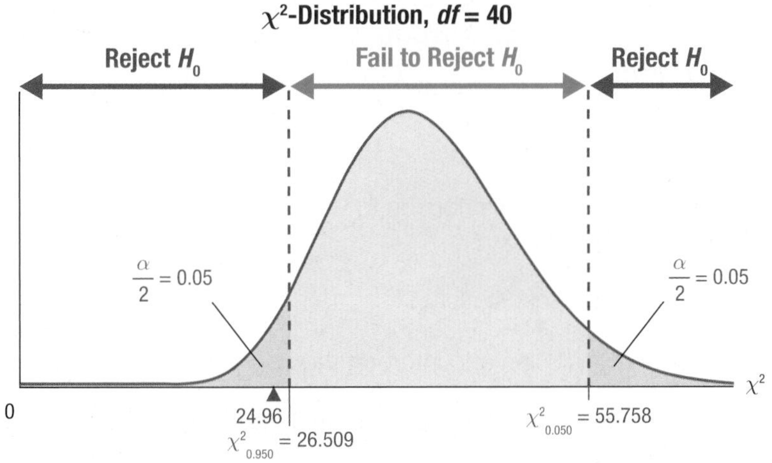

Since $\chi^2 = 24.96$, which is in the rejection region, we reject the null hypothesis. Thus, there is sufficient evidence at the 0.10 level of significance to support the farmer's claim that the variance of the heights of his two-year-old trees is not 16.

10.5 Section Exercises

Rejection Regions for Hypothesis Tests for Population Variances and Population Standard Deviations

Directions: State the critical value(s) of the test statistic, and determine the rejection region for a hypothesis test for the population variance or population standard deviation using the given information. Assume that the population distribution is approximately normal.

1. $\alpha = 0.05$, $df = 26$, left-tailed test

2. $\alpha = 0.10$, $df = 15$, left-tailed test

3. $\alpha = 0.05$, $n = 18$, $H_a\colon \sigma > 0.04$

4. $c = 0.90$, $n = 31$, $H_a\colon \sigma^2 > 0.65$

5. $c = 0.95$, $n = 25$, $H_0\colon \sigma^2 = 0.80$

6. $c = 0.99$, $n = 30$, $H_0\colon \sigma = 0.009$

Hypothesis Tests for Population Variances and Population Standard Deviations

Directions: **Perform each hypothesis test. For each exercise, complete the following steps. Assume that each population is normally distributed.**

a. State the null and alternative hypotheses.

b. Determine which distribution to use for the test statistic, and state the level of significance.

c. Calculate the test statistic.

d. Draw a conclusion and interpret the decision.

7. A dairy supplier fills hundreds of cartons with milk each day. He is contracted to fill each carton with exactly 1 gallon of milk. Because of the moving parts on the machine that fills the cartons, the amount of milk dispensed begins to vary slightly over time. When this happens, the machine must be serviced to realign it correctly. When testing the accuracy of the machine, the amount of milk dispensed into each carton sampled is measured in milliliters (mL). A variance of no more than 4 is acceptable. Servicing must occur when the variance in the amounts of milk in the cartons is more than 4 with a level of significance of 0.01. Twenty-five cartons are randomly chosen to be tested and the amounts of milk in the cartons are found to have a standard deviation of 2.5 mL. Perform a hypothesis test to determine if the machine needs servicing.

8. A potato chip manufacturer produces bags of potato chips that are supposed to have a net weight of 326 grams. Because the chips vary in size, it is difficult to fill the bags to the exact weight desired. However, the bags pass inspection so long as the standard deviation of their weights is no more than 3 grams. A quality control inspector wished to test the claim that one batch of bags has a standard deviation of more than 3 grams, and thus does not pass inspection. If a sample of 25 bags of potato chips is taken and the standard deviation is found to be 3.4 grams, does this evidence, at the 0.05 level of significance, support the claim that the bags should fail inspection?

9. At a shooting range, instructors can determine if a shooter is consistently missing the target because of his gun sight or because of his ability. If a gun's sight is off, the variance of the distances between the shots and the center of the shot pattern will be small (even if the shots are not in the center of the target). A student claims that it is the sight that is off, not his aim, and wants the instructor to confirm his claim. If a skilled shooter fires a gun at a target multiple times, the distances between the shots and the center of the shot pattern, measured in centimeters (cm), will have a variance of less than 0.33. After the student shoots 20 shots at the target, the instructor calculates that the distances between his shots and the center of the shot pattern, measured in cm, have a variance of 0.21. Does this evidence support the student's claim that the gun's sight is off? Use a 0.10 level of significance.

10. In order to ensure that there is not a large disparity in the quality of education at various schools in a school district, the school board wants to make sure that the variance of the mean standardized-test scores for all students at each school in the district is less than 0.05. To test this claim, they looked at the mean student scores for the standardized test from a random sample of 18 schools in the district. The results from the survey found that the overall mean was a score of 192.560 with a standard deviation of 0.162. With $\alpha = 0.10$, perform a hypothesis test to determine if the variance is less than 0.05.

11. The manufacturer of a popular antibiotic must ensure that each 5-mL dose contains 250 milligrams (mg) of the active ingredient. It is also essential that the variance of the amounts of active ingredient per dose be less than 0.1. For testing purposes, a random sample of 100 doses is taken, and the standard deviation of the amounts of active ingredient per dose in the sample is found to be 0.3 mg. Does the evidence support the claim that the variance is within the necessary bounds, at the 0.01 level of significance?

12. Claudia's bakery business has taken off. Her goods are so popular that she has decided to invest in kitchen equipment that will do some of the work for her. One machine in particular prepares bread dough, and it is set to measure 8 grams of yeast per loaf. Because baking requires precise measurements, she wants to test the new machine to make sure that the variance in the amounts of yeast per loaf is less than 0.30, at the 0.05 level of significance. A sample of the amounts of yeast added to the dough for 8 loaves has a variance of 0.25. Does this evidence support the claim that the new machine produces dough with a variance in the amounts of yeast per loaf of less than 0.30?

13. The temperatures in chicken incubators on a chicken farm, measured in degrees Fahrenheit (°F), are generally believed to have a variance of 0.50. The manager of the chicken farm claims that the variance has changed. He randomly tests 25 of his incubators and finds that their temperatures have a standard deviation of 0.55 °F. At the 0.05 level of significance, does this evidence support the manager's claim that the variance is not 0.50?

14. A grocery store needs the refrigeration section to have its coolers stay at the same temperature on a daily basis with little variance to help ensure quality. Daily temperatures are measured in degrees Fahrenheit (°F), and the manager of the store assumes that the variance in the daily temperatures is 3.8. The assistant manager claims that the variance is not 3.8 and decides to test his claim using a hypothesis test. For a random sample of 30 days, he finds that the standard deviation in the daily temperatures for one cooler is 2.9 °F. At the 0.01 level of significance, does this evidence support the claim that the variance in the daily temperatures for that cooler is not 3.8?

15. A health club needs to ensure that the temperature in its heated pool stays constant throughout the winter months. Otherwise it needs to invest in a new heater for the pool. It is assumed that the daily water temperatures, measured in degrees Fahrenheit (°F), have a variance of 2.25, which is considered to be within normal limits for a properly operating heater. The pool manager needs to determine if the variance is still 2.25, so she performs a hypothesis test to test the claim that the variance in the temperature is not 2.25. After testing the pool water for a random sample of 15 winter days, the pool manager finds a mean daily temperature of 78.60 °F with a variance of 3.81. At the 0.10 level of significance, does this evidence support the manager's claim that the variance in the water temperature is no longer 2.25?

10.6 Chi-Square Test for Goodness of Fit

In the previous section, we used chi-square as a test statistic when presented with a hypothesis test involving a population variance or population standard deviation. A chi-square test statistic can also be used in the following two cases: 1) comparing a frequency distribution that an experiment produces with what theoretically should be expected, and 2) assessing whether two traits or characteristics in a population are associated in some way. In Section 10.6 we tackle the first of these two by using the **chi-square test for goodness of fit**, and in Section 10.7 we compare traits in data with the **chi-square test for association**.

In both of these cases, the experiment must be set up in such a way that it satisfies the requirements of a *multinomial experiment.* In order for an experiment to be classified as multinomial, there must be a fixed number of identical independent trials, the outcome of each trial can be classified into exactly one of several different categories, and the probabilities for the different categories remain constant for each trial. In both Sections 10.6 and 10.7, we will focus on carrying out hypothesis tests for goodness of fit and association rather than emphasizing the conditions of a multinomial experiment. Therefore, you can safely assume that these multinomial conditions have been met for all examples and exercises in 10.6 and 10.7.

For a multinomial experiment, the chi-square test for goodness of fit evaluates how "good" the sample distribution and theoretical distribution fit each other. Consider the following scenario. Suppose you want to test if a die is actually fair (that is, not tampered with in any way). That is, you want to know whether the distribution of rolls obtained from the test die match the theoretical distribution of a fair die. If we record the outcomes of rolling the test die 100 times, we could compare them to the outcomes we would expect for a fair die. In other words, we would expect each number on the die to be rolled about $\frac{1}{6}$ of the time, so the expected value of the random variable X, the number of rolls for each value on the die, would be calculated as follows.

$$\begin{aligned} E(X) &= np \\ &= 100 \cdot \frac{1}{6} \\ &= \frac{100}{6} \\ &= \frac{50}{3} \\ &\approx 17 \end{aligned}$$

Thus, we would expect each number to be rolled about 17 times out of 100 rolls if the die is fair.

Let's say the following frequency distribution summarizes the results that our experiment produced.

Table 10.2: Results from Die Experiment

Number on Die	Number of Times It Appeared
1	17
2	18
3	8
4	14
5	25
6	18
Total	100

Now, although each number did not get rolled exactly 17 times as we predicted, are the results close enough for the die to be considered fair, or are the results far enough "off" to say the die is not fair? To answer our question, we will use a hypothesis test with a chi-square test statistic as we mentioned earlier. In a chi-square test for goodness of fit, the hypotheses we consider are as follows.

H_0: The observed values match the expected values.
H_a: The observed values do not match the expected values.

Recall that the null hypothesis always contains equality. Therefore, when we state the null hypothesis, the assumption is that the observed data values are equal to those of the theoretical distribution. Mathematically, the null and alternative hypotheses for a chi-square test for goodness of fit are written as follows.

Procedure

Null and Alternative Hypotheses for a Chi-Square Test for Goodness of Fit

When the theoretical probabilities for the various outcomes are all the same, the null and alternative hypotheses for a chi-square test for goodness of fit are as follows.

$H_0: p_1 = p_2 = \cdots = p_k$
H_a: There is some difference amongst the probabilities.

If the theoretical probabilities for the various outcomes are not all the same, each probability must be stated in the null hypothesis, so in that case, the hypotheses are as follows.

$H_0: p_1$ = probability of first outcome, p_2 = probability of second outcome, ..., p_k = probability of kth outcome
H_a: There is some difference from the stated probabilities.

k is the number of possible outcomes for each trial.

Note that the sum of the probabilities for all of the possible outcomes in a multinomial experiment must be 1, since it is certain that one of the outcomes will occur for each trial. That is, $\sum p_i = p_1 + p_2 + \cdots + p_k = 1$.

Let's now write the hypotheses for our die-rolling example. Since the theoretical probabilities are equal for all of the 6 possible outcomes when rolling a die, we can write the hypotheses as follows.

$H_0: p_1 = p_2 = p_3 = p_4 = p_5 = p_6$
H_a: There is some difference amongst the probabilities.

Now that we have set up the null and alternative hypotheses, let's consider how the test statistic is derived. Since we wish to know if there is a significant difference between the expected values and the actual values in our experiment, it seems reasonable to look at the differences between the observed frequencies and the expected frequencies. However, if we were to simply sum the differences, the negatives and positives would cancel each other out and the sum would be 0. In addition, we must take into account that in some scenarios, unlike our example of rolling a die, not all of the possible outcomes have the same probability. To adjust for both of these factors when calculating the test statistic, we will square the differences between the observed frequencies and the expected frequencies and divide each squared difference by its expected frequency.

Memory Booster

$E_i = np_i$

$$\chi^2 = \sum \frac{(O_i - E_i)^2}{E_i}$$

This test statistic will follow a chi-square distribution, given that the experiment is set up in such a way that it satisfies the requirements of a multinomial experiment, which we mentioned earlier. Furthermore, the following conditions must be met in order to test for goodness of fit.

- The sample data are randomly selected.
- Calculations use frequency counts of the data for each of the different categories.
- The expected frequency for each category is at least 5.

In this section you can safely assume that all of the necessary conditions have been met for each example and exercise.

Formula

Test Statistic for a Chi-Square Test for Goodness of Fit

The test statistic for a chi-square test for goodness of fit is given by

$$\chi^2 = \sum \frac{(O_i - E_i)^2}{E_i}$$

where O_i is the observed frequency for the ith possible outcome and

E_i is the expected frequency for the ith possible outcome.

As discussed earlier, this test statistic has a chi-square distribution. Recall that a chi-square distribution only has one parameter, the number of degrees of freedom. The number of degrees of freedom for a chi-square test for goodness of fit is given by the following formula.

Formula

Degrees of Freedom in a Chi-Square Test for Goodness of Fit

In a chi-square test for goodness of fit, the number of degrees of freedom for the chi-square distribution of the test statistic is given by

$$df = k - 1$$

where k is the number of possible outcomes for each trial.

For the die-rolling example, the formula for the test statistic would have six terms as shown in the following equation.

$$\begin{aligned}\chi^2 &= \sum \frac{(O_i - E_i)^2}{E_i} \\ &= \frac{(O_1 - E_1)^2}{E_1} + \frac{(O_2 - E_2)^2}{E_2} + \frac{(O_3 - E_3)^2}{E_3} + \frac{(O_4 - E_4)^2}{E_4} + \frac{(O_5 - E_5)^2}{E_5} + \frac{(O_6 - E_6)^2}{E_6}\end{aligned}$$

Let's now calculate the χ^2-test statistic for the die-rolling example. To be accurate, we will use $\frac{100}{6}$ in the formula for the expected frequency of each outcome, and not the approximate value of 17 that we calculated earlier for the value of E_i.

$$\chi^2 = \frac{\left(17-\frac{100}{6}\right)^2}{\frac{100}{6}} + \frac{\left(18-\frac{100}{6}\right)^2}{\frac{100}{6}} + \frac{\left(8-\frac{100}{6}\right)^2}{\frac{100}{6}} + \frac{\left(14-\frac{100}{6}\right)^2}{\frac{100}{6}} + \frac{\left(25-\frac{100}{6}\right)^2}{\frac{100}{6}} + \frac{\left(18-\frac{100}{6}\right)^2}{\frac{100}{6}}$$
$$= 9.32$$

As you can see from the formula, if the observed values are far from their expected values, the squared differences in the numerators of the terms will be large. Thus, when the terms are added, the result will be a large value of χ^2. That is, as the distances between the observed values and their expected values increase, so does the value of the test statistic. The question is, "How large is too large?" We can use critical values of the chi-square distribution to help us here. To do this, we will need the table of critical values of χ^2 along with the number of degrees of freedom, $df = k - 1$. In the die-rolling example, there are six possible outcomes when we roll the die, so the number of degrees of freedom is $df = 6 - 1 = 5$. We will compare the test statistic calculated from the sample to the critical value found in the chart. If the test statistic is as large or larger than the critical value, reject the null hypothesis that the actual frequencies are equal to the expected frequencies.

Procedure

Rejection Region for a Chi-Square Test for Goodness of Fit

Reject the null hypothesis, H_0, if:

$$\chi^2 \geq \chi^2_\alpha$$

Let's use a level of significance of $\alpha = 0.10$ and test the claim that the die is not fair. We have already calculated the χ^2-test statistic, so we just need to find the critical value, $\chi^2_{0.10}$, for the chi-square distribution with $df = 5$ in order to make a decision about the null hypothesis.

Area to the Right of the Critical Value of χ^2

df	0.995	0.990	0.975	0.950	0.900	0.100	0.050	0.025	0.010	0.005
1	0.000	0.000	0.001	0.004	0.016	2.706	3.841	5.024	6.635	7.879
2	0.010	0.020	0.051	0.103	0.211	4.605	5.991	7.378	9.210	10.597
3	0.072	0.115	0.216	0.352	0.584	6.251	7.815	9.348	11.345	12.838
4	0.207	0.297	0.484	0.711	1.064	7.779	9.488	11.143	13.277	14.860
5	0.412	0.554	0.831	1.145	1.610	9.236	11.070	12.833	15.086	16.750
6	0.676	0.872	1.237	1.635	2.204	10.645	12.592	14.449	16.812	18.548
7	0.989	1.239	1.690	2.167	2.833	12.017	14.067	16.013	18.475	20.278
8	1.344	1.646	2.180	2.733	3.490	13.362	15.507	17.535	20.090	21.955

Figure 10.9: Excerpt from Table G: Critical Values of χ^2

As shown in the table in Figure 10.9, the critical value is $\chi^2_{0.100} = 9.236$. Since the value of the test statistic is greater than the critical value, that is, $9.32 > 9.236$, we have $\chi^2 \geq \chi^2_\alpha$ and thus we reject the null hypothesis. We must conclude that our results were significantly "off" and therefore, at a confidence level of 90%, we can say that our die is not fair.

Now that we've discussed the concepts behind a chi-square test for goodness of fit, let's work through an example.

Example 10.30

Performing a Chi-Square Test for Goodness of Fit

A local bank wants to evaluate the usage of its ATM. Currently the bank manager assumes that the ATM is used consistently throughout the week, including weekends. She decides to use statistical inference with a 0.05 level of significance to test a customer's claim that the ATM is much busier on some days of the week than it is on other days. During a randomly selected week, the bank recorded the number of times the ATM was used on each day. The results are listed in the following table.

ATM Usage

	Number of Times Used
Monday	38
Tuesday	33
Wednesday	41
Thursday	25
Friday	22
Saturday	38
Sunday	45

Solution

Memory Booster

Remember that the null hypothesis will contain equality.

Step 1: **State the null and alternative hypotheses.**

When stating the hypotheses to be tested, we take the null hypothesis to be that the proportions of customers are the same for every day of the week.

H_0: The proportions of customers who use the ATM do not vary by the day of the week.

H_a: The proportions of customers who use the ATM do vary by the day of the week.

Mathematically, we can write the null and alternative hypotheses as follows.

$H_0: p_1 = p_2 = p_3 = p_4 = p_5 = p_6 = p_7$

H_a: There is some difference amongst the probabilities.

Step 2: **Determine which distribution to use for the test statistic, and state the level of significance.**

We are looking to see whether the observed values of ATM usage match the expected values for ATM usage at the bank. Remember that we can safely assume that the necessary conditions are met for examples in this section, so the chi-square test for goodness of fit is the appropriate choice for this scenario. Note that the level of significance given in the problem is $\alpha = 0.05$.

Step 3: **Gather data and calculate the necessary sample statistics.**

Before we begin to calculate the test statistic, let's calculate the expected value for each day of the week. Since we are assuming that the number of times the ATM is used does not vary for each day, then the probability will be the same for every day, so the expected number of customers for each day of the week is calculated as follows. Note that $n = 38 + 33 + 41 + 25 + 22 + 38 + 45 = 242$ (the total number of times the ATM was used).

$$E_i = np_i$$
$$= 242 \cdot \frac{1}{7}$$
$$= \frac{242}{7}$$

Let's calculate the χ^2-test statistic.

$$\chi^2 = \sum \frac{(O_i - E_i)^2}{E_i}$$
$$= \frac{\left(38 - \frac{242}{7}\right)^2}{\frac{242}{7}} + \frac{\left(33 - \frac{242}{7}\right)^2}{\frac{242}{7}} + \frac{\left(41 - \frac{242}{7}\right)^2}{\frac{242}{7}} + \frac{\left(25 - \frac{242}{7}\right)^2}{\frac{242}{7}}$$
$$+ \frac{\left(22 - \frac{242}{7}\right)^2}{\frac{242}{7}} + \frac{\left(38 - \frac{242}{7}\right)^2}{\frac{242}{7}} + \frac{\left(45 - \frac{242}{7}\right)^2}{\frac{242}{7}}$$
$$\approx 12.314$$

Step 4: **Draw a conclusion and interpret the decision.**

The number of degrees of freedom for the chi-square distribution for this test is $df = 7 - 1 = 6$, and $\alpha = 0.05$. Using the table, we find that the critical value is $\chi^2_{0.050} = 12.592$.

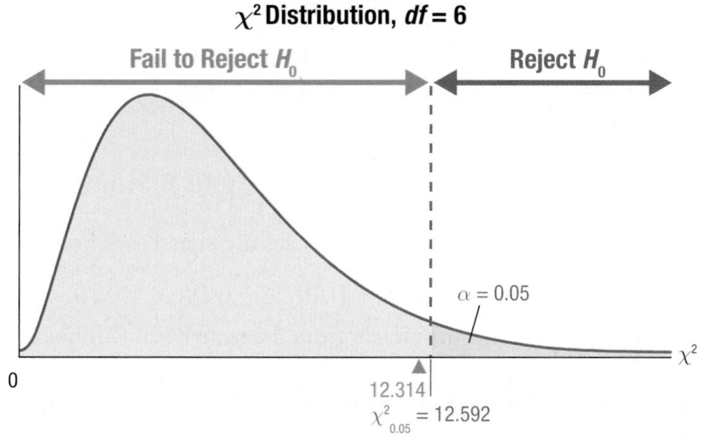

Comparing the test statistic to the critical value, we have 12.314 < 12.592 so $\chi^2 < \chi^2_{0.050}$, and thus we must fail to reject the null hypothesis. In other words, at the 0.05 level of significance, there is not sufficient evidence to support the customer's claim that the ATM is used significantly more on any particular day of the week.

Example 10.31

Performing a Chi-Square Test for Goodness of Fit Using a TI-84 Plus Calculator

A local fast-food restaurant serves buffalo wings. The restaurant's managers notice that they normally sell the following proportions of flavors for their wings: 20% Spicy Garlic, 50% Classic Medium, 10% Teriyaki, 10% Hot BBQ, and 10% Asian Zing. After running a campaign to promote their nontraditional specialty wings, they want to know if the campaign has made an impact. The results after 10 days are listed in the following table.

Buffalo Wing Sales

	Number Sold
Spicy Garlic	251
Classic Medium	630
Teriyaki	115
Hot BBQ	141
Asian Zing	121

Is there sufficient evidence at the 0.05 level of significance to say that the promotional campaign has made any difference in the proportions of flavors sold?

Solution

Step 1: State the null and alternative hypotheses.

The null hypothesis here is that the proportions of the flavors sold are the same as they were before the promotional campaign and the alternative is that they are different. To write this mathematically, we need to state the theoretical probabilities for the five different flavors sold. Let p_1, p_2, p_3, p_4, and p_5 be the probabilities for Spicy Garlic, Classic Medium, Teriyaki, Hot BBQ, and Asian Zing, respectively. Then we have the following.

$$p_1 = 0.20, p_2 = 0.50, p_3 = 0.10, p_4 = 0.10, p_5 = 0.10$$

Therefore, the null and alternative hypotheses are stated as follows.

H_0: $p_1 = 0.20, p_2 = 0.50, p_3 = 0.10, p_4 = 0.10, p_5 = 0.10$
H_a: There is some difference from the stated probabilities.

Step 2: Determine which distribution to use for the test statistic, and state the level of significance.

We are evaluating whether the observed proportions of wing flavors sold match the expected proportions for the five flavors. Remember that we can safely assume that the conditions are met for examples in this section, so the chi-square test for goodness of fit is again the appropriate choice. Note that the level of significance given in the problem is $\alpha = 0.05$.

Step 3: Gather data and calculate the necessary sample statistics.

Next, in order to calculate the χ^2-test statistic for the data given, let's begin by calculating the expected value of the number of orders for each flavor since they are not all the same. Here, $n = 251 + 630 + 115 + 141 + 121 = 1258$ (the total number of orders of wings sold).

$$E(\text{Spicy Garlic}) = E_1 = np_1 = (1258)(0.20) = 251.6$$
$$E(\text{Classic Medium}) = E_2 = np_2 = (1258)(0.50) = 629$$
$$E(\text{Teriyaki}) = E_3 = np_3 = (1258)(0.10) = 125.8$$
$$E(\text{Hot BBQ}) = E_4 = np_4 = (1258)(0.10) = 125.8$$
$$E(\text{Asian Zing}) = E_5 = np_5 = (1258)(0.10) = 125.8$$

Some TI-84 Plus calculators can calculate the χ^2-test statistic as well as the p-value for a chi-square test for goodness of fit. Begin by pressing STAT and then choose option `1:Edit`. Enter the observed values in L1 and the expected values in L2, as shown in the screenshot in the margin. Next, press STAT, scroll to TESTS, and then choose option `D:`χ^2`GOF-Test`. The calculator will prompt you for the following: `Observed`, `Expected`, and `df`, as shown in the screenshot on the left below. Enter L1 for `Observed` since that is the list where you entered the observed values. Enter L2 for `Expected` and enter the number of degrees of freedom for `df`. The number of degrees of freedom for this test is $df = 5 - 1 = 4$, so enter 4 for `df`. Finally, select `Calculate` and press ENTER.

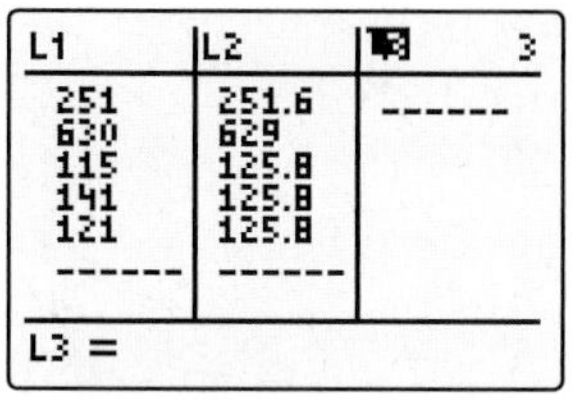

Side Note

The χ^2`GOF-Test` option is available on TI-84 Plus Silver Edition calculators with OS 2.30 or higher and TI-Nspire calculators.

The output screen, shown above on the right, displays the χ^2-test statistic and the p-value, and reiterates the number of degrees of freedom that was entered.

Step 4: Draw a conclusion and interpret the decision.

We see that $\chi^2 \approx 2.950$ and the p-value ≈ 0.5662. This p-value can be compared to the level of significance, $\alpha = 0.05$, to draw a conclusion. Remember that if p-value $\le \alpha$, then the conclusion is to reject the null hypothesis. In this case, p-value $> \alpha$, so the conclusion is to fail to reject the null hypothesis. In other words, the evidence does not support the claim that the proportions of wing flavors sold have changed. The restaurant cannot say that the promotional campaign has made any difference based on this evidence.

10.6 Section Exercises

Null and Alternative Hypotheses for Chi-Square Tests for Goodness of Fit

Directions: State the null and alternative hypotheses in words for each scenario.

1. A pharmacist believes that more people get their prescriptions filled on Fridays than on any other day of the week.

2. A game warden believes that the numbers of ducks, geese, and swans in the national park are not equal.

3. The hair stylists at a salon think that they are twice as busy on Thursdays and Fridays than the rest of the week.

4. A music producer believes that Southerners prefer country music over either pop music or R&B.

5. A cola company believes that the taste of its cola is preferred to those of its competitors.

Test Statistics for Chi-Square Tests for Goodness of Fit

Directions: Calculate the test statistic, χ^2, for a chi-square test for goodness of fit using the given information.

6. The following data represent numbers of drive-through customers at a fast-food restaurant on a weekday.

Drive-Through Customers				
	8:00–9:00 a.m.	12:00–1:00 p.m.	3:00–4:00 p.m.	5:00–6:00 p.m.
Observed Values, O_i	35	54	11	21
Expected Values, E_i	$\frac{121}{4}$	$\frac{121}{4}$	$\frac{121}{4}$	$\frac{121}{4}$

7. The following data represent numbers of four different candy bars sold at a concession stand in one day.

Candy Bar Sales				
	A	B	C	D
Observed Values, O_i	11	12	5	15
Expected Values, E_i	$\frac{43}{4}$	$\frac{43}{4}$	$\frac{43}{4}$	$\frac{43}{4}$

8. The following data represent job applicants' scoring levels on a company's entrance tests.

Applicants' Scoring Levels	Level 1	Level 2	Level 3
Observed Values, O_i	29	201	169
Expected Values, E_i	20.5	218.8	159.7

9. The following data represent numbers of hamburgers sold at three competing restaurants between 12 p.m. and 2 p.m. one weekday.

Hamburger Sales	Handy Andy	Philip's Grocery	Proud Larry
Observed Values, O_i	37	26	24
Expected Values, E_i	29	29	29

10. The following data represent numbers of ladies' jeans sold in various sizes at a department store in one day.

Ladies' Jeans Sales	Size 6	Size 8	Size 10	Size 12
Observed Values, O_i	4	7	5	5
Expected Values, E_i	5.25	5.25	5.25	5.25

11. The following data represent numbers of children at the local library on a Saturday.

Children at the Library	9:00–10:00 a.m.	10:00–11:00 a.m.	1:00–2:00 p.m.	3:00–4:00 p.m.
Observed Values, O_i	6	10	9	8
Expected Values, E_i	8.25	8.25	8.25	8.25

Critical Values for Chi-Square Tests for Goodness of Fit

Directions: Find the critical value of χ^2 for a chi-square test for goodness of fit using the given information.

12. $\alpha = 0.10, \quad k = 47$

13. $\alpha = 0.05, \quad k = 22$

14. $\alpha = 0.005, \quad k = 13$

15. $\alpha = 0.01, \quad k = 9$

16. $\alpha = 0.05, \quad k = 31$

Conclusions of Chi-Square Tests for Goodness of Fit

Directions: State the critical value of χ^2 and determine the appropriate conclusion for a chi-square test for goodness of fit using the given information.

17. $\alpha = 0.005, \quad k = 15, \quad \chi^2 = 29.674$

18. $\alpha = 0.10, \quad k = 8, \quad \chi^2 = 13.219$

19. $\alpha = 0.05, \quad k = 29, \quad \chi^2 = 43.222$

20. $\alpha = 0.025, \quad k = 17, \quad \chi^2 = 27.008$

Hypothesis Tests for Goodness of Fit

Directions: Perform each test for goodness of fit using the method of your choice or the one assigned by your instructor. For each exercise, complete the following steps.

a. State the null and alternative hypotheses.

b. Determine which distribution to use for the test statistic, and state the level of significance.

c. Find the expected value for each possible outcome, and calculate the test statistic.

d. Draw a conclusion and interpret the decision.

21. A school principal claims that the number of students who are tardy in her school does not vary from month to month. A survey over the school year produced the following results. Using a 0.05 level of significance, test a teacher's claim that the number of tardy students does vary by the month.

Tardy Students										
	Aug.	**Sept.**	**Oct.**	**Nov.**	**Dec.**	**Jan.**	**Feb.**	**Mar.**	**Apr.**	**May**
Number	7	18	16	5	8	12	15	18	11	15

22. A service station owner believes that equal numbers of customers prefer to buy gasoline on every day of the week. A manager at the service station disagrees with the owner and claims that the number of customers who prefer to buy gasoline on each day of the week varies. Test the manager's claim using $\alpha = 0.10$. The owner surveyed 739 customers over a period of time to record each customer's preferred day of the week. Here's what he found.

Preferred Day to Buy Gasoline							
	Mon.	**Tues.**	**Wed.**	**Thurs.**	**Fri.**	**Sat.**	**Sun.**
Number	103	103	126	103	111	96	97

23. At the emergency room at one hospital, the nurses are convinced that the number of patients that they see during the midnight shift is affected by the phases of the moon. The doctors think this is an old wives' tale. The nurses decide to test their theory at the 0.10 level of significance by recording the number of patients they see during the midnight shift for each moon phase over the course of one lunar cycle. The results are summarized in the table below.

ER Patients During the Midnight Shift				
	New Moon	**1st Quarter**	**Full Moon**	**3rd Quarter**
Number of Patients	85	66	97	68

24. The management of the local zoo wants to know if all of their animal exhibits are equally popular. If there is significant evidence that some of the exhibits are not being visited frequently enough, then changes may need to take place within the zoo. A tally of visitors is taken for each of the following animals throughout the course of a week, and the results are contained in the following table. At $\alpha = 0.05$, determine whether there is sufficient evidence to conclude that some exhibits are less popular than others.

Animal Exhibits at the Zoo

	Elephants	Lions/Tigers	Giraffes	Zebras	Monkeys	Birds	Reptiles
Number of Visitors	157	154	168	162	185	129	133

25. The manager of the city pool has scheduled extra lifeguards to be on staff for Saturdays. However, he suspects that Fridays may be more popular than the other weekdays as well. If so, he will hire extra lifeguards for Fridays, too. In order to test his theory that the daily number of swimmers varies on weekdays, he records the number of swimmers each day for the first week of summer. Test the manager's theory at the 0.01 level of significance.

Swimmers at the City Pool

	Monday	Tuesday	Wednesday	Thursday	Friday
Number	46	47	43	53	54

26. A manufacturer of children's vitamins claims that its vitamins are mixed so that each batch has exactly the following percentages of each color: 20% green, 40% yellow, 10% red, and 30% orange. To test the claim that these percentages are incorrect, 100 bottles of vitamins were pulled and the colors of the vitamins were tallied. The results are listed in the following table. At $\alpha = 0.05$, determine whether there is sufficient evidence to conclude that the percentages stated by the vitamin manufacturer are incorrect.

Children's Vitamins

	Green	Yellow	Red	Orange
Number	1149	1948	552	1401

10.7 Chi-Square Test for Association

Now we'll turn our attention to assessing whether two traits or characteristics in a population are associated in some way. When looking for a relationship between attributes, we will once again use the test statistic χ^2, so long as the scenario satisfies the properties of a multinomial experiment. Recall that in order for an experiment to be classified as multinomial, there must be a fixed number of identical independent trials, the outcome of each trial can be classified into exactly one of several different categories, and the probabilities for the different categories remain constant for each trial.

When using the chi-square test for association, the two variables in the population need not be numerical, nor does their relationship need to be linear. For instance, we could examine whether there is a dependent relationship between gender and choice of car color. However, in the calculation of the test statistic for a chi-square test for association, we use the observed frequencies for every possible outcome, which *are* numerical. Also, the sample data must be randomly selected and the expected frequency for each possible outcome must be at least 5. You can safely assume that all of the necessary conditions have been met for each example and exercise in this section. Let's use the example about the relationship between gender and choice of car color to walk through the steps in a chi-square test for association.

Step 1: **State the null and alternative hypotheses.**

To test whether there is dependence between two variables, such as gender and car color, we set up a hypothesis test. Let's start with the null hypothesis, which assumes that the two variables are not related; that is, the null hypothesis states that the variables are independent. The alternative hypothesis, then, states that the variables are not independent. We state these hypotheses, in general, as follows.

Procedure

Null and Alternative Hypotheses for a Chi-Square Test for Association

H_0: The two variables in the population are independent.

H_a: The two variables in the population are not independent.

For our example of gender and car color, we have the following null and alternative hypotheses.

H_0: Gender and car color are independent.

H_a: Gender and car color are not independent.

Step 2: **Determine which distribution to use for the test statistic, and state the level of significance.**

To determine whether gender and car color are related, we plan to randomly select 1000 cars and record the genders of the car owners along with the colors of their cars. Remember that we can safely assume that the conditions are met for examples in this section, so we will use the chi-square test statistic to test for association between the variables. We will use a level of significance of $\alpha = 0.01$ for this test.

Step 3: **Gather data and calculate the necessary sample statistics.**

Suppose that we took a random sample of 1000 cars on a university campus. We have summarized the data in the following table, which is called a contingency table.

Table 10.3: Observed Sample of 1000 Cars

	Silver	Black	Red	Other	Total
Male	190	117	215	53	575
Female	115	142	113	55	425
Total	305	259	328	108	1000

Once we have the sample data, the next thing we need to do is to calculate the expected value of each cell. The formula we will use to calculate the expected values is as follows.

Formula

Expected Value of a Frequency in a Contingency Table

The expected value of the frequency for the ith possible outcome in a contingency table is given by

$$E_i = \frac{(\text{row total})(\text{column total})}{n}$$

where n is the sample size.

For example, let's begin by calculating the expected value for the number of silver cars owned by men in our sample. The row total for males is 575, the column total for silver cars is 305, and the sample size for the contingency table is 1000. Substituting these values into the formula gives us the expected value as follows.

$$\begin{aligned} E_{\text{male and silver}} &= \frac{(575)(305)}{1000} \\ &= 175.375 \end{aligned}$$

We continue using this formula for each cell in the table until we have computed all of the required expected values. Note that all the expected frequencies are at least 5, which is one of the necessary conditions. The following table is the completed contingency table of expected values.

Table 10.4: Expected Values

	Silver	Black	Red	Other	Total
Male	175.375	148.925	188.6	62.1	575
Female	129.625	110.075	139.4	45.9	425
Total	305	259	328	108	1000

Now we are ready to calculate the chi-square test statistic. In the last section, we developed the formula for the χ^2-test statistic to determine how well data fit with theoretically expected values. We can use the same formula for the test statistic here.

Side Note

Recall that $E(X) = np$ for the binomial distribution. Furthermore, the multinomial distribution is just an extension of the binomial distribution where you have more than just two possible outcomes for each trial. For our example in the text, we have a random sample of 1000 cars; thus $n = 1000$. If we assume that gender and car color are independent, then we assume that the events of being male and having a silver car are independent events. Thus, we can use the formula for experimental probability and the Multiplication Rule for Independent Events to calculate the probability of being male and having a silver car. Therefore, we can calculate the expected value for the number of silver cars owned by men in our sample as follows.

$$\begin{aligned} E_{\text{male and silver}} &= np_{\text{male and silver}} \\ &= 1000 \cdot P\begin{pmatrix}\text{male} \\ \text{and} \\ \text{silver}\end{pmatrix} \\ &= 1000 \cdot \begin{bmatrix} P(\text{male}) \cdot \\ P(\text{silver}) \end{bmatrix} \\ &= \cancel{1000} \cdot \begin{bmatrix} \left(\frac{575}{\cancel{1000}}\right) \cdot \\ \left(\frac{305}{1000}\right) \end{bmatrix} \\ &= \frac{(575)(305)}{1000} \\ &= \frac{\begin{pmatrix}\text{row} \\ \text{total}\end{pmatrix}\begin{pmatrix}\text{column} \\ \text{total}\end{pmatrix}}{n} \end{aligned}$$

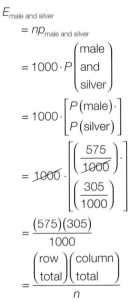

Formula

Test Statistic for a Chi-Square Test for Association

The test statistic for a chi-square test for association is given by

$$\chi^2 = \sum \frac{(O_i - E_i)^2}{E_i}$$

where O_i is the observed frequency for the ith possible outcome and

E_i is the expected frequency for the ith possible outcome.

For the data we have relating car colors and genders, the formula for the χ^2-test statistic is as follows.

$$\chi^2 = \frac{(O_{\text{male and silver}} - E_{\text{male and silver}})^2}{E_{\text{male and silver}}} + \cdots + \frac{(O_{\text{female and other}} - E_{\text{female and other}})^2}{E_{\text{female and other}}}$$

This test statistic has a chi-square distribution as we saw before. Let's calculate χ^2 for our particular example.

$$\begin{aligned}\chi^2 &= \frac{(190-175.375)^2}{175.375} + \frac{(117-148.925)^2}{148.925} + \frac{(215-188.6)^2}{188.6} + \frac{(53-62.1)^2}{62.1} \\ &\quad + \frac{(115-129.625)^2}{129.625} + \frac{(142-110.075)^2}{110.075} + \frac{(113-139.4)^2}{139.4} + \frac{(55-45.9)^2}{45.9} \\ &\approx 30.805\end{aligned}$$

To decide whether the χ^2-test statistic is significant or not, we need to know the number of degrees of freedom. In a chi-square test for association, the number of degrees of freedom is determined from the numbers of rows and columns in the contingency table.

Formula

Degrees of Freedom in a Chi-Square Test for Association

In a chi-square test for association, the number of degrees of freedom for the chi-square distribution of the test statistic is given by

$$df = (R-1)\cdot(C-1)$$

where R is the number of rows of data in the contingency table (not including the row of totals) and

C is the number of columns of data in the contingency table (not including the column of totals).

Therefore, the chi-square distribution for our example has $(2-1)\cdot(4-1)=3$ degrees of freedom.

Step 4: **Draw a conclusion and interpret the decision.**

As before, we will use the table of critical values for the chi-square distribution to draw our conclusion to the hypothesis test. If our test statistic is as large or larger than the critical value, we must reject the null hypothesis that the traits are independent.

Procedure

Rejection Region for Chi-Square Tests for Association

Reject the null hypothesis, H_0, if:

$$\chi^2 \geq \chi^2_\alpha$$

We will use a level of significance of $\alpha = 0.01$ to test our data. We have already calculated the χ^2-test statistic, so we just need to find the critical value, $\chi^2_{0.01}$, for the chi-square distribution with $df = 3$ in order to make a decision about the null hypothesis.

Area to the Right of the Critical Value of χ^2

df	0.995	0.990	0.975	0.950	0.900	0.100	0.050	0.025	0.010	0.005
1	0.000	0.000	0.001	0.004	0.016	2.706	3.841	5.024	6.635	7.879
2	0.010	0.020	0.051	0.103	0.211	4.605	5.991	7.378	9.210	10.597
3	0.072	0.115	0.216	0.352	0.584	6.251	7.815	9.348	11.345	12.838
4	0.207	0.297	0.484	0.711	1.064	7.779	9.488	11.143	13.277	14.860
5	0.412	0.554	0.831	1.145	1.610	9.236	11.070	12.833	15.086	16.750
6	0.676	0.872	1.237	1.635	2.204	10.645	12.592	14.449	16.812	18.548
7	0.989	1.239	1.690	2.167	2.833	12.017	14.067	16.013	18.475	20.278
8	1.344	1.646	2.180	2.733	3.490	13.362	15.507	17.535	20.090	21.955

Figure 10.10: Excerpt from Table G: Critical Values of χ^2

As shown in the table in Figure 10.10, the critical value is $\chi^2_{0.010} = 11.345$. Recall that the test statistic for our sample data is $\chi^2 \approx 30.805$. Since $\chi^2 > \chi^2_{0.010}$, we have $\chi^2 \geq \chi^2_\alpha$ and thus we reject the null hypothesis. We must conclude that the data do not fit with the assumption that gender and car color are independent. Therefore, with 99% confidence, we can say that gender and car color are related.

Now that we've walked carefully through a complete example of a chi-square test for association, let's work another example from start to finish.

Example 10.32

Performing a Chi-Square Test for Association

Suppose that the following data were collected in a poll of 13,660 randomly selected voters during the 2008 presidential election campaign.

Observed Sample of 13,660 Voters

	Obama	McCain	Other	Total
Male	3455	2764	65	6284
Female	3541	3762	73	7376
Total	6996	6526	138	13,660

Is there evidence at the 0.05 level to say that gender and voting choice were related for this election?

Solution

Step 1: State the null and alternative hypotheses.

We let the null hypothesis be that gender and voting preference are independent of one another.

H_0: Gender and voting preference are independent.

H_a: Gender and voting preference are not independent.

Step 2: Determine which distribution to use for the test statistic, and state the level of significance.

We wish to determine if there is an association between gender and voting preference. Since we are told that we can safely assume that the necessary conditions have been met for the examples in this section, we will use the chi-square test statistic to test for this association. We are told that the level of significance is $\alpha = 0.05$.

Step 3: Gather data and calculate the necessary sample statistics.

Before we begin to calculate the test statistic, we must calculate the expected value for each cell in the contingency table.

Expected Values

	Obama	McCain	Other	Total
Male	$\frac{(6284)(6996)}{13,660} \approx 3218.364861$	$\frac{(6284)(6526)}{13,660} \approx 3002.151098$	$\frac{(6284)(138)}{13,660} \approx 63.484041$	6284
Female	$\frac{(7376)(6996)}{13,660} \approx 3777.635139$	$\frac{(7376)(6526)}{13,660} \approx 3523.848902$	$\frac{(7376)(138)}{13,660} \approx 74.515959$	7376
Total	6996	6526	138	13,660

Memory Booster

Remember to round intermediate calculations to at least six decimal places.

Let's calculate the χ^2-test statistic.

$$\chi^2 = \sum \frac{(O_i - E_i)^2}{E_i}$$

$$= \frac{(3455-3218.364861)^2}{3218.364861} + \frac{(2764-3002.151098)^2}{3002.151098} + \frac{(65-63.484041)^2}{63.484041}$$

$$+ \frac{(3541-3777.635139)^2}{3777.635139} + \frac{(3762-3523.848902)^2}{3523.848902} + \frac{(73-74.515959)^2}{74.515959}$$

$$\approx 67.276$$

Step 4: Draw a conclusion and interpret the decision.

The number of degrees of freedom for this test is $df = (2-1)\cdot(3-1) = 2$, and $\alpha = 0.05$. Using the table, we find that the critical value is $\chi^2_{0.050} = 5.991$. Comparing the test statistic to the critical value, we have $67.276 > 5.991$, so $\chi^2 \geq \chi^2_{0.050}$, and thus we must reject the null hypothesis. In other words, at the 0.05 level of significance, we can conclude that gender and voting choice were related for the 2008 presidential election.

Example 10.33

Performing a Chi-Square Test for Association Using a TI-83/84 Plus Calculator

A local hairdresser is curious about whether there is a relationship between hair color and the combination of gender and marital status among his clients. He collects data from a random sample of his clients and records the data in the following contingency table.

Observed Sample of 232 Clients					
	Blonde	**Brown**	**Red**	**Black**	**Total**
Single Women	18	19	8	14	59
Married Women	20	18	9	17	64
Single Men	13	22	4	16	55
Married Men	12	24	3	15	54
Total	63	83	24	62	232

Based on these data, is there enough evidence at the 0.10 level of significance to say that there is a relationship between a person's hair color and the combination of gender and marital status for this hairdresser's clients?

Solution

For this example, we will use a TI-83/84 Plus calculator to calculate the test statistic and draw our conclusion. As we saw in previous sections when using technology, we must begin by doing the first few steps by hand.

Step 1: State the null and alternative hypotheses.

As always in a test for association, the null hypothesis is that the two variables, hair color and the combination of gender and marital status in this example, are independent.

H_0: Hair color and the combination of gender and marital status are independent.

H_a: Hair color and the combination of gender and marital status are not independent.

Step 2: Determine which distribution to use for the test statistic, and state the level of significance.

We wish to determine if there is an association between hair color and the combination of gender and marital status. Since we have been told that we can safely assume that the necessary criteria are met for the examples in this section, we will use the chi-square test statistic to test for this association. We are given a level of significance of $\alpha = 0.10$.

Step 3: Gather data and calculate the necessary sample statistics.

The data have been gathered and presented in the form of a contingency table. When using a TI-83/84 Plus calculator, you do not have to calculate the expected values as you would if you were performing a chi-square test for association by hand.

To use a TI-83/84 Plus calculator, start by entering the table of observed values into the calculator in the form of a matrix. Press 2ND and then x^{-1} to access the `MATRIX` menu. Scroll over to `EDIT` and choose option `1:[A]`, which is the

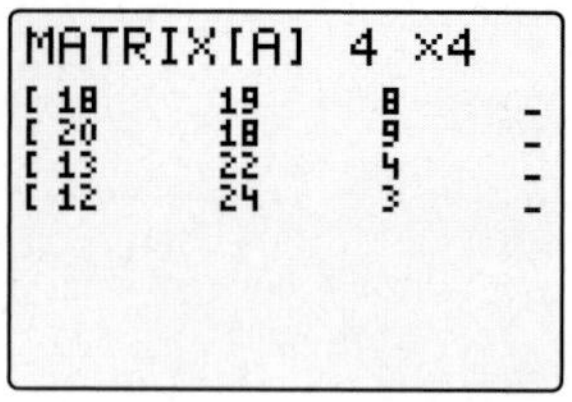

name of the first matrix. Now you need to enter the size of the matrix in the form of (Number of Rows)×(Number of Columns). *It is important to know that when using the calculator, you should not enter the total row or column!* For our table of observed values, there are 4 rows and 4 columns (omitting the totals), so the size of the matrix is 4 × 4. Now enter the data from the table, as shown in the screenshot in the margin.

Next press STAT and scroll to TESTS. Choose option C:χ^2-Test. We must then enter the name of the matrix containing the observed values and the name of the matrix where we want the expected values to be calculated. The TI-83/84 Plus calculates the expected values for the given matrix of observed values, and stores them in the matrix that you specify.

If the names of the matrices you wish to use are not the ones indicated, press 2ND and then x^{-1} to access the MATRIX menu. Then choose the name of the matrix needed. The default options of [A] for Observed and [B] for Expected are correct for our example. To run the test, scroll down to select Calculate and then press ENTER.

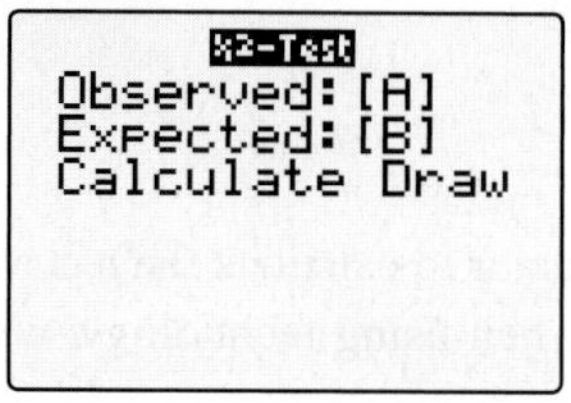

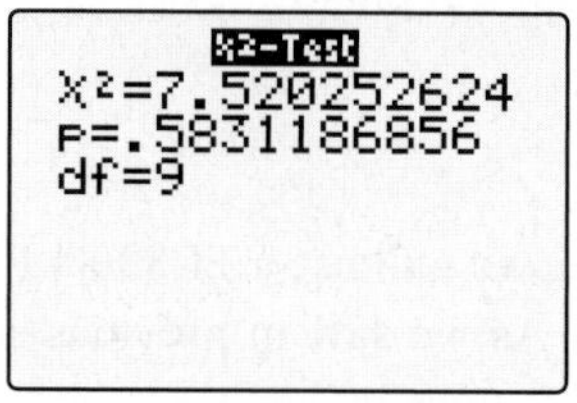

The output screen, shown above on the right, displays the test statistic, $\chi^2 \approx 7.520$, the p-value, which is approximately 0.5831, and the number of degrees of freedom, $df = 9$.

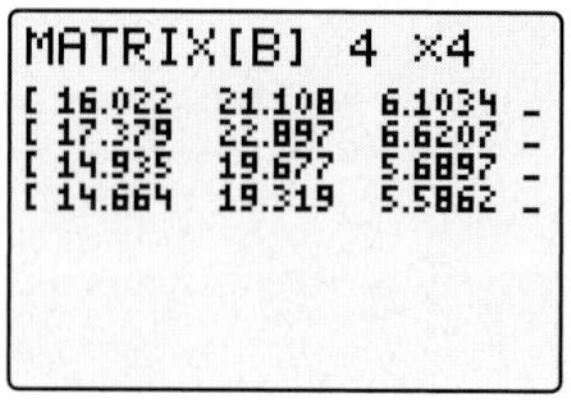

After running the test, we can view the matrix of expected values calculated by the TI-83/84 Plus. Press 2ND and then x^{-1} to access the MATRIX menu. Then scroll over to EDIT and choose option 2:[B]. The matrix of expected values is shown in the screenshot in the margin.

Step 4: Draw a conclusion and interpret the decision.

Instead of using a rejection region, the calculator reported a p-value of approximately 0.5831, so we can compare that to the level of significance, $\alpha = 0.10$. Since p-value $> \alpha$, we fail to reject the null hypothesis. Thus, there is not enough evidence at this level of significance to conclude that there is an association between hair color and the combination of gender and marital status for this hairdresser's clients.

10.7 Section Exercises

Contingency Tables of Expected Values

Directions: Use each contingency table of observed values to find the contingency table of expected values.

1. The following data represent preferred cereal brands for a random sample of adults.

Observed Sample of 146 Adults					
	Brand A	Brand B	Brand C	Brand D	Total
Male	28	14	11	27	80
Female	9	16	8	33	66
Total	37	30	19	60	146

2. The following data represent preferred writing hands for a random sample of adults.

Observed Sample of 237 Adults				
	Ages 20–29	Ages 30–39	Ages 40–49	Total
Left Hand	15	42	31	88
Right Hand	35	41	50	126
Ambidextrous	10	5	8	23
Total	60	88	89	237

3. The following data represent preferences for user interfaces of proposed tablet models, as chosen by participants of four different consumer focus groups.

Observed Sample of 119 Consumers					
	Group A	Group B	Group C	Group D	Total
Model 1	8	12	4	9	33
Model 2	11	15	6	3	35
Model 3	4	10	3	11	28
Model 4	6	11	4	2	23
Total	29	48	17	25	119

4. The following data represent the favorite sports of adults living in various regions of the United States for a random sample of adults.

Observed Sample of 336 Adults					
	Football	Basketball	Baseball	Soccer	Total
Northeast	27	25	23	20	95
Southeast	26	21	19	17	83
Midwest	24	25	18	11	78
West	20	22	17	21	80
Total	97	93	77	69	336

Test Statistics for Chi-Square Tests for Association

Directions: Calculate the test statistic, χ^2, for a chi-square test for association using the given contingency tables.

5. The following data represent the observed values and expected values of the eating habits of runners and swimmers for a random sample of adult athletes.

Observed Sample of 390 Adult Athletes			
	Prefer to Eat before a Workout	Prefer to Eat after a Workout	Total
Runner	68	121	189
Swimmer	73	128	201
Total	141	249	390

Expected Values			
	Prefer to Eat before a Workout	Prefer to Eat after a Workout	Total
Runner	68.330769	120.669231	189
Swimmer	72.669231	128.330769	201
Total	141	249	390

6. The following data represent the observed values and expected values of the song preferences for a random sample of young adults.

Observed Sample of 343 Young Adults					
	Song 1	Song 2	Song 3	Song 4	Total
Male 18–22	34	3	46	20	103
Male 23–27	33	25	48	17	123
Female 18–22	28	18	17	9	72
Female 23–27	3	15	20	7	45
Total	98	61	131	53	343

Expected Values					
	Song 1	Song 2	Song 3	Song 4	Total
Male 18–22	29.428571	18.317784	39.338192	15.915452	103
Male 23–27	35.142857	21.874636	46.976676	19.005831	123
Female 18–22	20.571429	12.804665	27.498542	11.125364	72
Female 23–27	12.857143	8.002915	17.186589	6.953353	45
Total	98	61	131	53	343

7. The following data represent the observed values and expected values of the fat contents and brands of microwaveable meals sold at a local grocery store for a random sample of meals sold.

Observed Sample of 3283 Microwaveable Meals Sold				
	Less Than 5 g Fat	**5–10 g Fat**	**More Than 10 g Fat**	**Total**
Brand A	160	27	1380	1567
Brand B	86	9	789	884
Brand C	55	7	770	832
Total	301	43	2939	3283

Expected Values				
	Less Than 5 g Fat	**5–10 g Fat**	**More Than 10 g Fat**	**Total**
Brand A	143.669510	20.524216	1402.806275	1567
Brand B	81.049041	11.578434	791.372525	884
Brand C	76.281450	10.897350	744.821200	832
Total	301	43	2939	3283

8. The following data represent the observed values and expected values of the movie preferences for a random sample of college students.

Observed Sample of 409 College Students					
	Suspense	**Drama**	**Comedy**	**Horror**	**Total**
Freshman	24	28	37	18	107
Sophomore	19	25	35	15	94
Junior	31	33	30	12	106
Senior	26	29	34	13	102
Total	100	115	136	58	409

Expected Values					
	Suspense	**Drama**	**Comedy**	**Horror**	**Total**
Freshman	26.161369	30.085575	35.579462	15.173594	107
Sophomore	22.982885	26.430318	31.256724	13.330073	94
Junior	25.916870	29.804401	35.246944	15.031785	106
Senior	24.938875	28.679707	33.916870	14.464548	102
Total	100	115	136	58	409

Conclusions of Chi-Square Tests for Association

Directions: State the critical value of χ^2 and determine the appropriate conclusion for a chi-square test for association using the given information.

9. $\alpha = 0.10$, Number of rows = 2, Number of columns = 3, $\chi^2 = 5.13$

10. $\alpha = 0.025$, Number of rows = 5, Number of columns = 5, $\chi^2 = 31.1$

11. $\alpha = 0.005$, Number of rows = 5, Number of columns = 7, $\chi^2 = 40.8$

12. $\alpha = 0.10$, Number of rows = 7, Number of columns = 6, $\chi^2 = 40.3$

Hypothesis Tests for Association

Directions: Perform each test for association using the method of your choice or the one assigned by your instructor. For each exercise, complete the following steps.

a. State the null and alternative hypotheses.

b. Determine which distribution to use for the test statistic, and state the level of significance.

c. Find the expected value for each possible outcome, and calculate the test statistic.

d. Draw a conclusion and interpret the decision.

13. One state's Department of Education wants to know if there is a relationship between grades and the combination of particular subject and gender for students in the state. A random sample of students in the state produces the following results. Is there sufficient evidence at the 0.005 level of significance to show that there is a relationship between grades and the combination of subject and gender?

Observed Sample of 440 Students						
	A	B	C	D	F	Total
Math, Female	12	30	44	15	9	110
Math, Male	5	24	37	36	8	110
English, Female	16	41	39	10	4	110
English, Male	10	40	35	19	6	110
Total	43	135	155	80	27	440

14. An insurance company wants to know if the color of an automobile has a relationship with the number of moving violations. The following contingency table gives the results of data collected from police reports across the nation. The columns list the numbers of reported moving violations in a year. Use a level of significance of $\alpha = 0.01$ to conduct this test.

Observed Sample of 443 Cars				
	0–1	2–3	More Than 3	Total
White	76	33	13	122
Black	44	21	8	73
Red	50	46	12	108
Silver	33	27	7	67
Other	28	34	11	73
Total	231	161	51	443

15. A soft drink company is interested in knowing whether there is a relationship between cola preference and age. A random sample of 800 people is chosen for a taste test. The results of the study are found in the following table. Is there sufficient evidence at the 0.005 level of significance to lead you to believe that there is an association between cola preference and age?

Observed Sample of 800 People				
	Cola A	Cola B	Cola C	Total
15–29	94	102	105	301
30–44	99	97	86	282
45–59	68	73	76	217
Total	261	272	267	800

16. An obstetrician has noticed that there seems to be a pattern to the genders of the babies that he delivers. That is, he tends to deliver boys and girls in clusters. In order to determine whether this pattern is true outside of his hospital, he calls four other hospitals in the region and obtains the following information regarding a random sample of births from the previous six months. Does the evidence gathered support the doctor's claim at $\alpha = 0.01$?

Observed Sample of 926 Births				
	Nov./Dec.	Jan./Feb.	Mar./Apr.	Total
Boy	158	165	153	476
Girl	149	147	154	450
Total	307	312	307	926

17. Suppose that a bookseller wants to study the relationship between book preference and gender. A random sample of readers is chosen for the study, and each participant is asked to choose his or her favorite genre out of the following choices: mystery, fiction, nonfiction, and self-help. The results are detailed below. Does the evidence gathered show a relationship between book preference and gender at $\alpha = 0.005$?

Observed Sample of 266 Readers					
	Mystery	**Fiction**	**Nonfiction**	**Self-Help**	**Total**
Male	28	30	41	22	121
Female	32	59	28	26	145
Total	60	89	69	48	266

18. A travel agency is interested in finding out if different age groups frequent different Spring Break destinations, in order to better target the appropriate audiences. A random sample of college Spring Break vacationers produces the results given in the table below. Is there enough evidence at the 0.05 level of significance to show that there is a relationship between age (by college classification) and destination?

Observed Sample of 192 College Students					
	Beach	**Mountains**	**City**	**Home**	**Total**
Freshman	19	2	7	24	52
Sophomore	15	4	3	20	42
Junior	18	1	9	19	47
Senior	21	6	4	20	51
Total	73	13	23	83	192

19. A marketing firm wants to know if there is an association between a person's age and gender, and his or her favorite Super Bowl commercial. A random sample of people is asked to choose between two commercials, and the results are in the following table. Given these results, is there enough evidence at the 0.10 level of significance to conclude that an association exists?

Observed Sample of 80 Adults			
	Commercial 1	**Commercial 2**	**Total**
Male 18–30	14	6	20
Female 18–30	9	11	20
Male 31–45	12	8	20
Female 31–45	6	14	20
Total	41	39	80

20. Pollsters want to test if an association exists between a person's profession and his or her political party. A random sample of 236 voters is polled, resulting in the data in the following table. Based on these results, is there enough evidence at the 0.025 level of significance to say that an association exists?

Observed Sample of 236 Voters			
	Democrat	**Republican**	**Total**
Doctor	13	35	48
Lawyer	33	19	52
Teacher	23	25	48
Farmer	39	11	50
Laborer	28	10	38
Total	136	100	236

R Chapter 10 Review

Section 10.1: Fundamentals of Hypothesis Testing

Definitions

Hypothesis
A theory or premise, a claim about the numerical value of a population parameter, such as the population mean, proportion, or variance

Hypothesis testing
A technique for testing a claim about a population parameter using statistical principles

Alternative hypothesis
Denoted by H_a, a mathematical statement that describes a population parameter, the hypothesis that the researcher is aiming to gather evidence in favor of; also referred to as the research hypothesis

Null hypothesis
Denoted by H_0, the mathematical opposite of the alternative hypothesis; will always include equality

Test statistic
The value used to make a decision about the null hypothesis; it is derived from the sample statistic

Statistically significant
Describes a sample statistic if it is far enough away from the presumed value of the population parameter to conclude that it would be unlikely for the sample statistic to occur by chance if the null hypothesis is true

Level of significance
Denoted by α, the probability of making the error of rejecting a true null hypothesis in a hypothesis test; $\alpha = 1 - c$

Type I error
Rejecting a true null hypothesis

Type II error
Failing to reject a false null hypothesis

Conclusions for a Hypothesis Test

- Reject the null hypothesis.
- Fail to reject the null hypothesis.

Performing a Hypothesis Test

1. State the null and alternative hypotheses.
2. Determine which distribution to use for the test statistic, and state the level of significance.
3. Gather data and calculate the necessary sample statistics.
4. Draw a conclusion and interpret the decision.

Section 10.1: Fundamentals of Hypothesis Testing (cont.)

Types of Errors in Hypothesis Testing

- The level of significance, α, is the probability of committing a Type I error.
- The probability of committing a Type II error is denoted by β.
- The values of α and β are inversely related.
- We would usually have no basis for determining if a Type I error or a Type II error has been made when performing actual research.

Decision	Reality: H_0 is true	Reality: H_0 is false
Reject H_0	Type I error	Correct decision
Fail to reject H_0	Correct decision	Type II error

Section 10.2: Hypothesis Testing for Population Means (σ Known)

Definition

***p*-value**

The probability of obtaining a sample statistic as extreme or more extreme than the one observed in the data, when the null hypothesis, H_0, is assumed to be true

Test Statistic for a Hypothesis Test for a Population Mean (σ Known)

Used when a simple random sample of size n is taken from a population with a known standard deviation, σ, where either $n \geq 30$ or the population distribution is approximately normal, given by

$$z = \frac{\bar{x} - \mu}{\left(\frac{\sigma}{\sqrt{n}}\right)}$$

Types of Hypothesis Tests

Alternative Hypothesis, H_a	Type of Test
< Value	Left-tailed test
> Value	Right-tailed test
≠ Value	Two-tailed test

Decision Rule for Rejection Regions

Reject the null hypothesis if the test statistic calculated from the sample data falls within the rejection region.

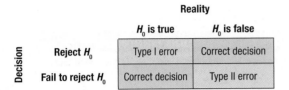

Section 10.2: Hypothesis Testing for Population Means (σ Known) (cont.)

Rejection Regions for Hypothesis Tests for Population Means (σ Known)

Reject the null hypothesis, H_0, if:

$z \le -z_\alpha$ for a left-tailed test

$z \ge z_\alpha$ for a right-tailed test

$|z| \ge z_{\alpha/2}$ for a two-tailed test

Critical z-Values for Rejection Regions

Critical z-Values for Rejection Regions			
Level of Confidence c	Level of Significance $\alpha = 1 - c$	One-Tailed Test z_α	Two-Tailed Test $\pm z_{\alpha/2}$
0.90	0.10	1.28	±1.645
0.95	0.05	1.645	±1.96
0.98	0.02	2.05	±2.33
0.99	0.01	2.33	±2.575

Conclusions Using p-Values

- If p-value $\le \alpha$, then *reject* the null hypothesis.
- If p-value $> \alpha$, then *fail to reject* the null hypothesis.

Section 10.3: Hypothesis Testing for Population Means (σ Unknown)

Test Statistic for a Hypothesis Test for a Population Mean (σ Unknown)

Used when a simple random sample of size n is taken from a population with an unknown standard deviation, where either $n \ge 30$ or the population distribution is approximately normal, given by

$$t = \frac{\bar{x} - \mu}{\left(\frac{s}{\sqrt{n}}\right)} \text{ with } df = n - 1$$

Rejection Regions for Hypothesis Tests for Population Means (σ Unknown)

Reject the null hypothesis, H_0, if:

$t \le -t_\alpha$ for a left-tailed test

$t \ge t_\alpha$ for a right-tailed test

$|t| \ge t_{\alpha/2}$ for a two-tailed test

Section 10.4: Hypothesis Testing for Population Proportions

Test Statistic for a Hypothesis Test for a Population Proportion

Used when a simple random sample of size n is large enough to ensure that $np \geq 5$ and $n(1-p) \geq 5$, where the conditions for a binomial distribution are met, given by

$$z = \frac{\hat{p} - p}{\sqrt{\frac{p(1-p)}{n}}}$$

Section 10.5: Hypothesis Testing for Population Variances

Test Statistic for a Hypothesis Test for a Population Variance or Population Standard Deviation

Used when a simple random sample of size n is taken from a population whose distribution is approximately normal, given by

$$\chi^2 = \frac{(n-1)s^2}{\sigma^2} \text{ with } df = n-1$$

Rejection Regions for Hypothesis Tests for Population Variances and Standard Deviations

Reject the null hypothesis, H_0, if:

$\chi^2 \leq \chi^2_{(1-\alpha)}$ for a left-tailed test

$\chi^2 \geq \chi^2_{\alpha}$ for a right-tailed test

$\chi^2 \leq \chi^2_{(1-\alpha/2)}$ or $\chi^2 \geq \chi^2_{\alpha/2}$ for a two-tailed test

Section 10.6: Chi-Square Test for Goodness of Fit

Null and Alternative Hypotheses for Chi-Square Tests for Goodness of Fit

- When the theoretical probabilities for the various outcomes are all the same:

  $H_0: p_1 = p_2 = \cdots = p_k$

  H_a: There is some difference amongst the probabilities.

- When the theoretical probabilities for the various outcomes differ:

  H_0: p_1 = probability of first outcome,

  p_2 = probability of second outcome, ...,

  p_k = probability of kth outcome

  H_a: There is some difference from the stated probabilities.

Section 10.6: Chi-Square Test for Goodness of Fit (cont.)

Test Statistic for a Chi-Square Test for Goodness of Fit

Used for a random sample where the conditions for a multinomial experiment are met, the data consist of frequency counts for each of the k different possible outcomes, and the expected frequency for each possible outcome is at least 5, given by

$$\chi^2 = \sum \frac{(O_i - E_i)^2}{E_i} \text{ with } df = k - 1$$

Rejection Region for a Chi-Square Test for Goodness of Fit

Reject the null hypothesis, H_0, if:

$$\chi^2 \geq \chi^2_\alpha$$

Section 10.7: Chi-Square Test for Association

Null and Alternative Hypotheses for Chi-Square Tests for Association

H_0: The two variables in the population are independent.

H_a: The two variables in the population are not independent.

Expected Value of a Frequency in a Contingency Table

$$E_i = \frac{(\text{row total})(\text{column total})}{n}$$

Test Statistic for a Chi-Square Test for Association

Used for a random sample where the conditions for a multinomial experiment are met, the data consist of frequency counts for each of the different possible outcomes in a contingency table, and the expected frequency for each possible outcome is at least 5, given by

$$\chi^2 = \sum \frac{(O_i - E_i)^2}{E_i} \text{ with } df = (R-1)\cdot(C-1)$$

Rejection Region for Chi-Square Tests for Association

Reject the null hypothesis, H_0, if:

$$\chi^2 \geq \chi^2_\alpha$$

E Chapter 10 Exercises

Hypothesis Tests

Directions: Perform each hypothesis test. Assume that all random samples had an equal chance of being chosen. For each exercise, complete the following steps.

a. State the null and alternative hypotheses.

b. Determine which distribution to use for the test statistic, and state the level of significance.

c. Calculate the test statistic.

d. Draw a conclusion and interpret the decision.

1. A manufacturer is responsible for making barrels used to store crude oil, which are designed to hold exactly 55 gallons of oil. As part of a routine quality check, a factory manager randomly tests 27 barrels off the assembly line and finds that their mean capacity is 54.7 gallons with a standard deviation of 0.8 gallons. Can the manager say with 99% confidence that the requirements are being met? Assume that the population distribution is approximately normal.

2. Sarah, a university junior, chose to live on the honors floor of her dorm because she was told that the girls on that floor would be quiet and studious. In fact, Sarah was told that students on the honors floor study at least 11 hours per week. After living on the honors floor for a few months, Sarah is convinced that the students around her study less than 11 hours a week. To test her claim, Sarah polls 10 girls on her floor regarding the number of hours per week that they spend studying. Use the results listed in the table below to test Sarah's claim at the 0.10 level of significance. Assume that the population distribution is approximately normal.

Hours of Study per Week										
Hours of Study	12	4	14	9	8	6	10	12	11	9

3. TransAir, an international airline company, has a new jumbo jet that it believes will decrease the mean transatlantic flight time from Newark to London Heathrow. In order to be faster than its competitors, the mean must be less than 6 h 50 min. Because of costs, the agency monitoring the flight times only requires that eight trials be run. The results are given in the table below. If the FAA requires that the agency be 90% confident of its findings, what should the agency report? Assume that the population distribution is approximately normal.

Flying Times from Newark to London

	Flying Time
Flight 1	6 h 32 min
Flight 2	6 h 28 min
Flight 3	6 h 51 min
Flight 4	6 h 2 min
Flight 5	6 h 51 min
Flight 6	7 h 2 min
Flight 7	6 h 36 min
Flight 8	6 h 42 min

4. A computer program designer has developed an online questionnaire that she believes will take a mean of 15 minutes to complete. Suppose that she samples 35 volunteers and finds that the mean time to complete the questionnaire is 14.0 minutes. Determine if there is sufficient evidence to conclude that the mean completion time for the online questionnaire differs from its intended duration. Use a 0.05 level of significance and assume that the population standard deviation is known to be 2.5 minutes.

5. A manufacturer of hearing-aid batteries has developed a new process that it believes will increase the usable life of each battery. Currently the usable lifetimes for these batteries have a mean of 264 hours. Test the hypothesis that the process increases the usable life of these batteries at the 0.01 level of significance. A sample of 50 batteries is tested and found to have a mean of 271.0 hours with a standard deviation of 16.9 hours. Assume that the population standard deviation is known to be 16.9 hours.

6. If it is true that fewer than 4 people per day use a particular public restroom at the state park, the park services will close the restroom. In order to determine the action the park should take, the number of uses is recorded each day for a 35-day period. The results are given in the following table. At the 5% level of significance, should the park close the restroom? Assume that the population standard deviation is known to be 2.1 uses per day.

Daily Restroom Uses						
3	3	3	1	2	6	9
0	2	4	3	3	5	3
7	3	0	2	1	4	1
0	0	6	3	5	10	5
3	4	0	5	2	4	6

Suppose that the actual mean usage rate for the restroom is 7 times a day. Did the park make the right decision? If not, what type of error was made?

7. While trying to convince her friends that more than half of all sophomores live off campus, Bren took a survey of a random sample of sophomores and collected the following data. Does this evidence support Bren's claim that more than half of all sophomores live off campus at the 0.10 level of significance?

Sophomores Living on Campus		
	Live On Campus	**Live Off Campus**
Male	25	47
Female	38	19

A friend of Bren's went to the registrar's office and found out that actually 975 out of 3788 sophomores lived off campus this year. Based on this information, did Bren make an error in her conclusion? If so, what type of error did she make?

8. Don believes that fewer than one-third of the students at his college hold part-time jobs while in school. To test his claim, he randomly selects 116 students from whom he collects the following data. Does this evidence support Don's claim at the 0.05 level of significance?

Students Holding Part-Time Jobs

	Part-Time Job Held
Yes	32
No	84

In a campus-wide end-of-year survey, the office of research was able to determine that 34% of students say they held a part-time job. Did Don draw the correct conclusion? If not, what type of error did he make?

9. A student organization that promotes diversity believes that the percentage of minority students is no longer 35%. In a random sample of 250 students, 97 of the students are minorities. Does this evidence support the organization's claim that the percentage of minority students is not 35% at the 0.01 level of significance?

The registrar reports that 38% of currently enrolled students are minorities according to their admissions records. Based on this new information, did the organization draw the correct conclusion? If not, what type of error was made?

10. A clothing designer produces thousands of pairs of blue jeans each week. The denim for each pair of jeans is cut by a machine, and if the cuts are not made properly, then the jeans will not be properly sized once they are sewn together. To ensure that the jeans are properly sized, the variance in the lengths of the denim, which are measured in millimeters, must be less than 0.25. To test the accuracy of the cuts, 20 pieces of denim are pulled from the line after being cut by the machine and their lengths are measured. The variance in these lengths is found to be 0.1369. Does this evidence support the claim that the variance in the lengths of the denim is less than 0.25, at the 0.10 level of significance?

11. From microminis to floor-length skirts, a historian believes that the variance in the lengths of ladies' hems on dresses and skirts is more now than at any other point in recorded history. Specifically, she claims that the variance in the lengths of skirts' hems, which are measured in inches, is more than 15.5. To test her claim, she randomly selects 25 dresses and skirts from a department store and measures the hem length of each. She finds that the variance in the lengths of the hems is 17.3. Does this evidence support the historian's claim, at the 0.05 level of significance?

12. A statistics professor is concerned that the variance in the grades on the final exam will not be 138, as it has been in previous semesters. She randomly selects 35 of her many exams and finds that the variance in the grades is 142. Does this evidence support the professor's claim at the 0.01 level of significance?

13. A city police department has always had an equal number of officers on patrol in each beat. However, each month the police chief reevaluates to make sure that he is utilizing the department's resources and manpower effectively. To do this, he records the number of calls received from each beat over the course of the month. The data for one month are summarized below. Does this evidence support the claim that there is a difference in the number of calls between the different areas of the city, at the 0.05 level of significance?

Calls Received in One Month				
	Beat 1	**Beat 2**	**Beat 3**	**Beat 4**
Number of Calls	28	19	31	38

14. A factory workers' union is appealing to the factory management for more vacation time for its employees. The union claims that workers are more productive when they come back from vacation than before. The following table contains the production rates, measured in units per hour, for a sample of employees. The union feels that it needs to be 90% confident of its claim before going into talks with the employers. Does the union have strong enough evidence to support its claim?

Production Rates				
	100–125 Units per Hour	**126–150 Units per Hour**	**More Than 150 Units per Hour**	**Total**
Before Vacation	210 workers	319 workers	58 workers	587
After Vacation	178 workers	387 workers	60 workers	625
Total	388	706	118	1212

15. A sociologist wants to study the education levels of people living in various regions of the United States. He surveys a random sample of 100 people in each of the following regions: Northeast, Southeast, Midwest, and West. The results obtained are found in the following table. Does this evidence support the sociologist's claim that the level of education differs by region, at the 0.05 level of significance?

Education Levels					
	Less Than a High School Diploma	**High School Graduate, No College**	**Some College or Associate Degree**	**College Graduate**	**Total**
Northeast	11	29	45	15	100
Southeast	19	28	44	9	100
Midwest	15	25	47	13	100
West	13	31	42	14	100
Total	58	113	178	51	400

P Chapter 10 Project

Project A: Hypothesis Testing for Population Means

Directions: Choose one of the three claims, collect data from members of the appropriate population, and perform a hypothesis test to determine if the evidence supports the claim or does not support the claim. After you have written your conclusion, look at the "real" value of the population mean and determine if your hypothesis test produced a correct decision, a Type I error, or a Type II error.

Pick one of the following claims to test:

- Parents of college freshmen believe that freshmen spend a mean of at most $50 per week on eating out.
- It is believed that college sophomores see at least 4 movies per month at the theater.
- The librarian claims that college juniors visit the library twice a week.

Step 1: State the null and alternative hypotheses.

Based on the claim you chose, what are the null and alternative hypotheses?

Step 2: Determine which distribution to use for the test statistic, and state the level of significance.

In the next step, you will collect data from 10 students. Assuming that the population distribution is approximately normal, what formula should be used for the test statistic? Also, choose a level of significance of 0.10, 0.05, or 0.01.

Step 3: Gather data and calculate the necessary sample statistics.

Collect data from 10 students who are in the appropriate population. Discuss which method of data collection you used. List any potential for bias. Calculate the sample mean and sample standard deviation.

Calculate the test statistic using the values you just calculated from your sample.

Step 4: Draw a conclusion and interpret the decision.

Determine the type of your hypothesis test: left-tailed, right-tailed, or two-tailed.

Draw a picture of your rejection region.

What is your conclusion?

Interpret your decision.

Types of Errors

Let's assume we find out that the truths are as follows.

- Freshmen spend a mean of $40 per week eating out.
- The mean number of movies seen by sophomores at the theater each month is more than 4.
- The mean number of times per week that juniors visit the library is 4.

Based on your conclusion, did you make a Type I error, a Type II error, or a correct decision? Explain.

Project B: Chi-Square Test for Goodness of Fit

In this project, we will look at whether the makeup of your institution has changed significantly from Year 1 to Year 2, for two nonconsecutive years. In other words, are the percentages of students in every classification in Year 2 equal to the percentages of students in every classification in Year 1? Let's begin by collecting some data from the earlier academic year in order to determine the hypotheses for the more recent academic year.

1. Choose two years that are not consecutive from which to collect data. (For instance, you may choose the current academic year for Year 2 and the academic year four years earlier for Year 1.) Find out the number of students who were enrolled at your institution for each classification during Year 1 and enter them in a table like the one below.

	Year 1
Number of Freshmen Enrolled	
Number of Sophomores Enrolled	
Number of Juniors Enrolled	
Number of Seniors Enrolled	
Total Number of Students Enrolled	

2. Now calculate the percentage of students in each category during Year 1 and enter them in a new table.

	Year 1
Percentage of Freshmen Enrolled	
Percentage of Sophomores Enrolled	
Percentage of Juniors Enrolled	
Percentage of Seniors Enrolled	

3. State the null and alternative hypotheses in words.

4. Specify the null and alternative hypotheses with mathematical symbols.

5. Now let's gather data for Year 2. Find out the number of students enrolled at your institution for each classification during Year 2 and enter them in a table like the one below.

	Year 2
Number of Freshmen Enrolled	
Number of Sophomores Enrolled	
Number of Juniors Enrolled	
Number of Seniors Enrolled	
Total Number of Students Enrolled	

6. Now calculate the percentage of students in each category during Year 2 and enter them in a new table.

	Year 2
Percentage of Freshmen Enrolled	
Percentage of Sophomores Enrolled	
Percentage of Juniors Enrolled	
Percentage of Seniors Enrolled	

7. Calculate the expected value for each classification.

8. Calculate the test statistic.

9. Determine your conclusion.

10. Is there sufficient evidence to conclude at $\alpha = 0.10$ that the makeup of your institution has significantly changed between Year 1 and Year 2? Explain.

T Chapter 10 Technology

Performing a Hypothesis Test

TI-83/84 Plus

Finding Area Under a Student's *t*-Distribution Curve

Side Note

The most recent TI-84 Plus calculators (Jan. 2011 and later) contain a new feature called `STAT WIZARDS`. This feature is not used in the directions in this text. By default this feature is turned `ON`. `STAT WIZARDS` can be turned `OFF` under the second page of MODE options.

To use a TI-83/84 Plus calculator to find the area under a Student's *t*-distribution curve, use option `6:tcdf(` under the `DISTR` menu. Press 2ND and then VARS to access the menu and then scroll down to option `6`. The function syntax is `tcdf(`*lower bound, upper bound, df*`)`. Just as we did with the normal distribution, if we want the area to the left of a given *t*-value, we will enter `-1E99` for the lower bound. If we want the area to the right, we will enter `1E99` for the upper bound. Remember that *df* stands for degrees of freedom.

Side Note

On TI-83 Plus calculators and older TI-84 Plus calculators, the function `tcdf(` is option `5` on the `DISTR` menu.

Example T.1

Using a TI-83/84 Plus Calculator to Find the Area to the Left of a Given *t*-Value

Find the area under the curve to the left of $t = -1.638$ for a *t*-distribution with 3 degrees of freedom.

Solution

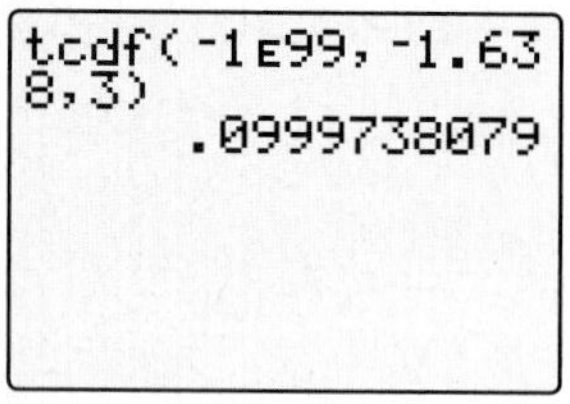

Press 2ND and then VARS to access the `DISTR` menu; then choose option `6:tcdf(`. Since we want the area to the left, the lower bound is $-\infty$, which we will enter as `-1E99`. Enter `tcdf(-1E99,-1.638,3)`, as shown in the screenshot in the margin. Thus, the area under the curve to the left of $t = -1.638$ for a *t*-distribution with 3 degrees of freedom is approximately 0.1000.

Hypothesis Testing

We can use a TI-83/84 Plus calculator to perform a hypothesis test by directly entering data or by entering the sample statistics. The process for performing a hypothesis test on a TI-83/84 Plus is basically the same for any type of test.

First, determine the type of test you need to perform and the null and alternative hypotheses. Then execute the following steps on the calculator.

1. Either enter your raw data in the `LIST` or `MATRIX` environments (whichever is appropriate) or calculate the necessary sample statistics for the type of test you are going to perform.
2. Press STAT.
3. Scroll over to the `TESTS` menu.
4. Scroll down the list and choose the type of test you need to perform.
5. Enter the appropriate information.
6. Choose `Calculate` and press ENTER.

Example T.2

Using a TI-83/84 Plus Calculator to Perform a Hypothesis Test for a Population Mean (Two-Tailed, σ Known)

The mean reading score for third graders in one state is believed to be 82. An educational researcher claims that this figure is wrong. A simple random sample of 45 third graders in the state is chosen and the mean reading score is calculated to be 81. The population standard deviation is believed to be 5. Do the data gathered support the researcher's claim? Use a 0.05 level of significance to draw your conclusion. Assume that the population distribution is approximately normal.

Solution

First, we notice that this will be a test for a population mean with σ known. This is a z-test, which is option 1 under the TESTS menu. We also need to state the null and alternative hypotheses.

$$H_0: \mu = 82$$
$$H_a: \mu \neq 82$$

Next, note that we have the following values from the problem heading.

$$\mu_0 = 82$$
$$\sigma = 5$$
$$\bar{x} = 81$$
$$n = 45$$
$$\alpha = 0.05$$

Since we have these values and not the raw data, we want to enter the values directly. Press STAT; then scroll to the TESTS menu and choose option `1:Z-Test`. Choose `Stats` and enter the values as shown in the screenshot on the left below. Choose the alternative hypothesis ≠μ0. Select `Calculate` and press ENTER.

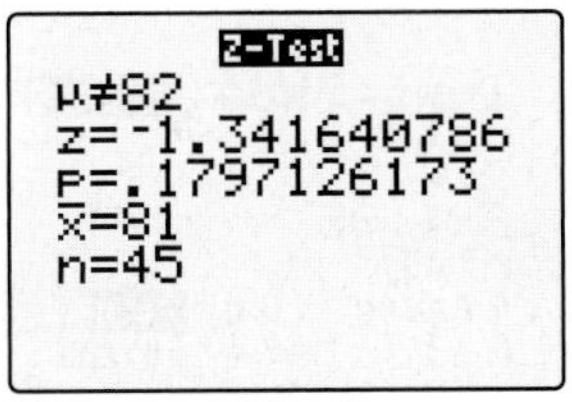

The output screen, shown above on the right, displays the alternative hypothesis, the z-test statistic, and the p-value, and then reiterates the sample mean and sample size that were entered.

The output screen for this test tells us that our alternative hypothesis is $\mu \neq 82$, our test statistic is $z \approx -1.34$, and our p-value is approximately 0.1797. Since our p-value is 0.1797 and it is greater than $\alpha = 0.05$, we fail to reject the null hypothesis. Therefore, the evidence does not support the educational researcher's claim that the mean reading score for third graders in the state is not 82.

Example T.3

Using a TI-83/84 Plus Calculator to Perform a Hypothesis Test for a Population Mean (Right-Tailed, σ Unknown)

The mean number of phone calls per night that teenage girls make is believed to be no more than 4. A concerned parent in one local neighborhood claims that the mean is really more than 4. A simple random sample of ten teenage girls in the neighborhood were asked how many phone calls they made last night. The mean number of phone calls they made was 4.50 with a standard deviation of 0.85. Does this information support the concerned parent's claim? Use a 0.10 level of significance to draw your conclusion. Assume that the population distribution is approximately normal.

Solution

The population standard deviation is unknown and the population distribution is approximately normal, so we will use a t-test. The t-test is option 2 under the TESTS menu. Next, we state the null and alternative hypotheses.

$$H_0: \mu \le 4$$
$$H_a: \mu > 4$$

So we have a right-tailed t-test. The following values are given to us in the problem heading.

$$\mu_0 = 4$$
$$\bar{x} = 4.50$$
$$s = 0.85$$
$$n = 10$$
$$\alpha = 0.10$$

Press STAT; then scroll to the TESTS menu and choose option `2:T-Test`. Choose `Stats` and enter the values as shown in the screenshot on the left below. Choose the alternative hypothesis `>μ0`. Select `Calculate` and press ENTER.

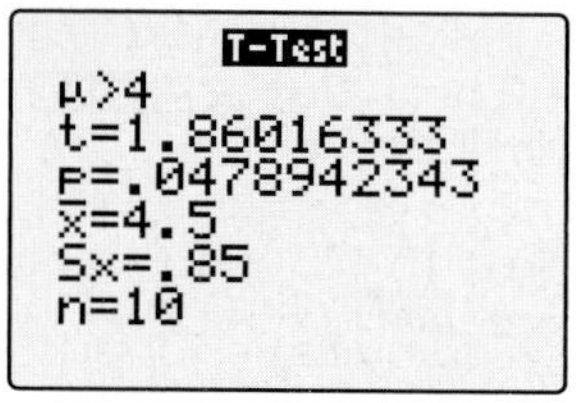

The output screen, shown above on the right, displays the alternative hypothesis, the t-test statistic, and the p-value, and then reiterates the sample mean, sample standard deviation, and sample size that were entered.

The output screen for this test tells us that our alternative hypothesis is $\mu > 4$, our test statistic is $t \approx 1.860$, and our p-value is approximately 0.0479. Since our p-value is 0.0479 and that is less than $\alpha = 0.10$, we reject the null hypothesis. Therefore, the evidence supports the parent's claim that the mean number of phone calls per night that teenage girls make is more than 4.

Example T.4

Using a TI-83/84 Plus Calculator to Perform a Hypothesis Test for a Population Proportion (Left-Tailed)

Let's consider Example 10.24 from Section 10.4. A local politician believes that less than 65% of his constituents will vote in favor of a tax increase to pay for a new school. His staff asked a simple random sample of 50 of his constituents whether they favor the tax increase and 27 said that they would vote in favor of the tax increase. Does this information support the politician's claim? Use a 0.01 level of significance to draw your conclusion.

Solution

This example deals with a population proportion, and we already verified that the necessary conditions are satisfied to allow us to use the normal distribution and the z-test statistic in Example 10.24. So we will use a z-test for a population proportion. This particular z-test is option `5` under the `TESTS` menu. Next, we state the null and alternative hypotheses.

$$H_0: p \geq 0.65$$
$$H_a: p < 0.65$$

So we have a left-tailed one-proportion z-test. The following values are given to us in the problem heading.

$$p_0 = 0.65$$
$$x = 27$$
$$n = 50$$
$$\alpha = 0.01$$

Press **STAT**; then scroll to the `TESTS` menu and choose option `5:1-PropZTest`. Enter the values as shown in the screenshot on the left below. Choose the alternative hypothesis `<p0`. Select `Calculate` and press **ENTER**.

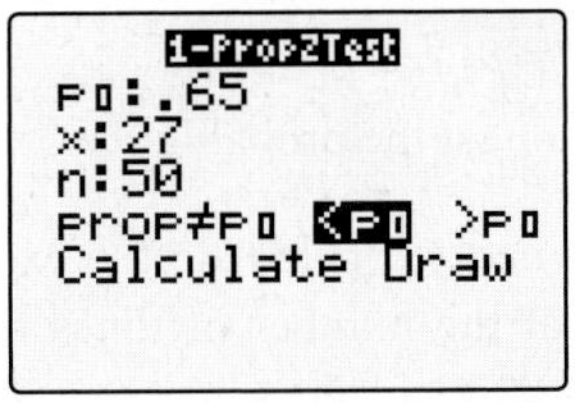

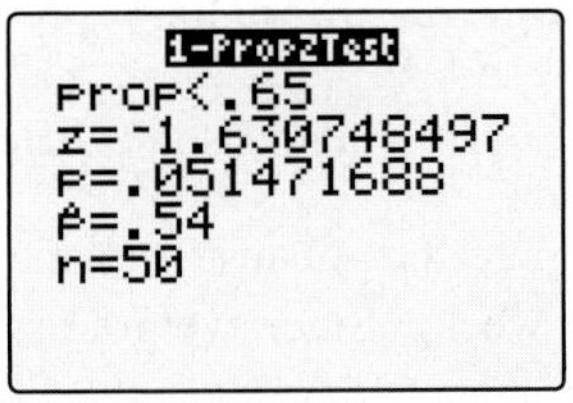

The output screen, shown above on the right, displays the alternative hypothesis, the z-test statistic, the p-value, and the sample proportion, and then repeats the sample size we entered.

The output screen for this test tells us that our alternative hypothesis is $p < 0.65$, our test statistic is $z \approx -1.63$, our p-value is approximately 0.0515, and our sample proportion is $\hat{p} = 0.54$. Since our p-value is 0.0515, which is greater than $\alpha = 0.01$, we fail to reject the null hypothesis. This means that, at the 0.01 level of significance, this evidence does not sufficiently support the politician's claim that less than 65% of his constituents favor a tax increase to pay for a new school.

Example T.5

Using a TI-84 Plus Calculator to Perform a Chi-Square Test for Goodness of Fit

Let's consider Example 10.30 from Section 10.6. A local bank wants to evaluate the usage of its ATM. Currently the bank manager assumes that the ATM is used consistently throughout the week, including weekends. She decides to use statistical inference with a 0.05 level of significance to test a customer's claim that the ATM is much busier on some days of the week than it is on other days. During a randomly selected week, the bank recorded the number of times the ATM was used on each day. The results are listed in the following table.

ATM Usage

	Number of Times Used
Monday	38
Tuesday	33
Wednesday	41
Thursday	25
Friday	22
Saturday	38
Sunday	45

Solution

Side Note

The χ^2`GOF-Test` option is available on TI-84 Plus Silver Edition calculators with OS 2.30 or higher and TI-Nspire calculators.

As we determined in Example 10.30, the chi-square test for goodness of fit is the appropriate choice for this scenario. On newer TI-84 Plus calculators, the chi-square test for goodness of fit is option `D:`χ^2`GOF-Test` under the `TESTS` menu. Before running the test on the calculator, we must determine the hypotheses and calculate the expected values.

When stating the hypotheses to be tested, we take the null hypothesis to be that the proportions of customers are the same for every day of the week. Thus, we can write the null and alternative hypotheses as follows.

$$H_0: p_1 = p_2 = p_3 = p_4 = p_5 = p_6 = p_7$$

H_a: There is some difference amongst the probabilities.

In order to compute the χ^2-test statistic, we first need to calculate the expected value for each day of the week. Since we are assuming that the number of times the ATM is used does not vary for each day, then the probability will be the same for every day, so the expected number of customers for each day of the week is calculated as follows. Note that $n = 38 + 33 + 41 + 25 + 22 + 38 + 45 = 242$ (the total number of times the ATM was used).

$$\begin{aligned} E_i &= np_i \\ &= 242 \cdot \frac{1}{7} \\ &= \frac{242}{7} \end{aligned}$$

Now we are ready to use a TI-84 Plus calculator to compute the χ^2-test statistic as well as the p-value. We will compare the p-value to the given level of significance, $\alpha = 0.05$, to make a decision. Begin by pressing STAT and then choose option `1:Edit`. Enter the observed values in `L1` and the expected values in `L2`. Next, press STAT; then scroll to the `TESTS` menu and choose option `D:`χ^2`GOF-Test`. The calculator will prompt you for the following: `Observed`, `Expected`, and `df`, as shown in the following screenshot on the left. Enter `L1` for `Observed` since that is the list where you entered the observed values. Enter `L2` for

Expected, and enter the number of degrees of freedom for df. The number of degrees of freedom for this test is $df = 7 - 1 = 6$, so enter 6 for df. Finally, select Calculate and press ENTER.

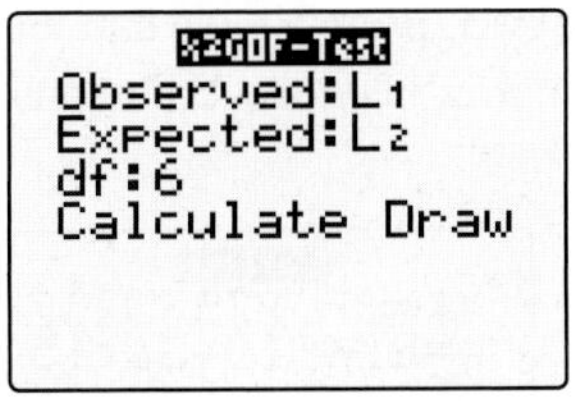

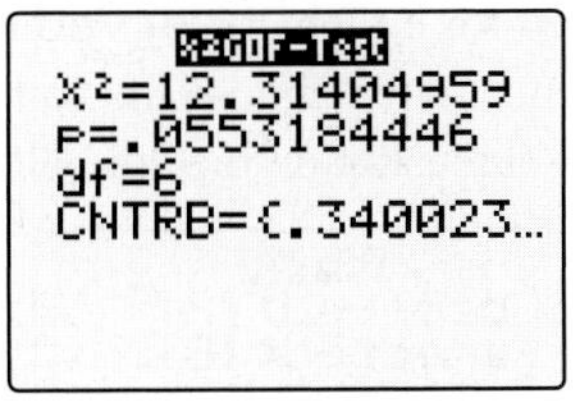

The output screen, shown above on the right, displays the χ^2-test statistic and the p-value and reiterates the number of degrees of freedom that was entered. We see that $\chi^2 \approx 12.314$ and the p-value ≈ 0.0553. Since $\alpha = 0.05$ and $0.0553 > 0.05$, we have p-value $> \alpha$, so the conclusion is to fail to reject the null hypothesis. Thus, there is not sufficient evidence at the 0.05 level of significance to support the customer's claim that the ATM is used significantly more on any particular day of the week.

Example T.6

Using a TI-83/84 Plus Calculator to Perform a Chi-Square Test for Association

Let's consider Example 10.32 from Section 10.7. Suppose that the following data were collected in a poll of 13,660 randomly selected voters during the 2008 presidential election campaign.

Observed Sample of 13,660 Voters

	Obama	McCain	Other	Total
Male	3455	2764	65	6284
Female	3541	3762	73	7376
Total	6996	6526	138	13,660

Is there evidence at the 0.05 level to say that gender and voting choice were related for this election?

Solution

Let's begin by stating the null and alternative hypotheses. We let the null hypothesis be that gender and voting preference are independent of one another.

H_0: Gender and voting preference are independent.

H_a: Gender and voting preference are not independent.

As noted in Example 10.32, we can safely assume that the necessary conditions have been met for this example, so we will perform a chi-square test for association.

To calculate χ^2 on a TI-83/84 Plus calculator, we need to enter the observed values in a matrix. The calculator will create and store the matrix of expected values as part of the χ^2-test. Begin by pressing 2ND and then x^{-1} to access the MATRIX menu. Scroll over to EDIT and choose the matrix you want to edit. Before entering values into the matrix, you must enter the number of rows and then the number of columns. We do not enter the "Total" row or column. Excluding the totals, the table of observed values has 2 rows and 3 columns, so the size of the matrix is 2×3. The matrix of observed values is shown in the screenshot in the margin.

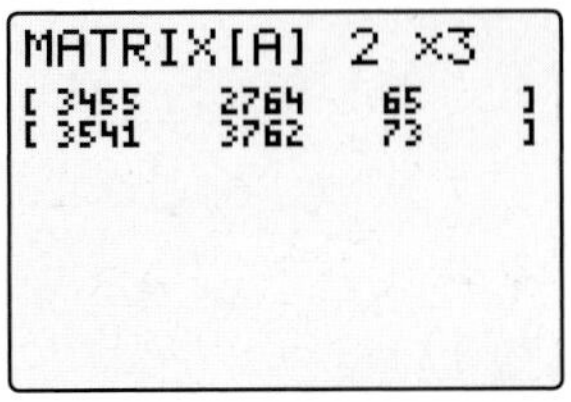

Next, press STAT and scroll over to the TESTS menu. Choose option C:χ^2-Test. If the observed values are in matrix [A] and we want the calculator to store the expected values in matrix [B], we do not need to change the default settings. If you need to change the name of the matrix entered for either Observed or Expected, press 2ND and then x^{-1} to access the MATRIX menu. Then choose the name of the matrix needed. Your screen should appear as shown in the screenshot on the left below.

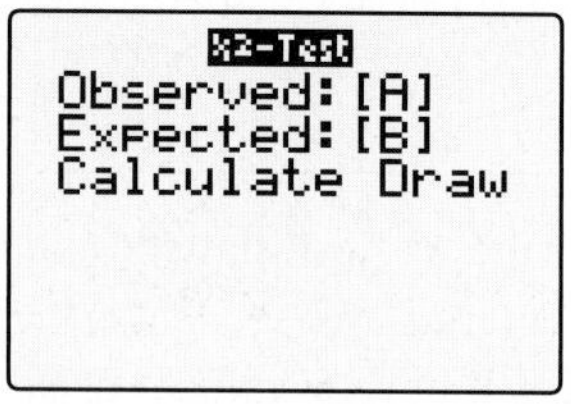

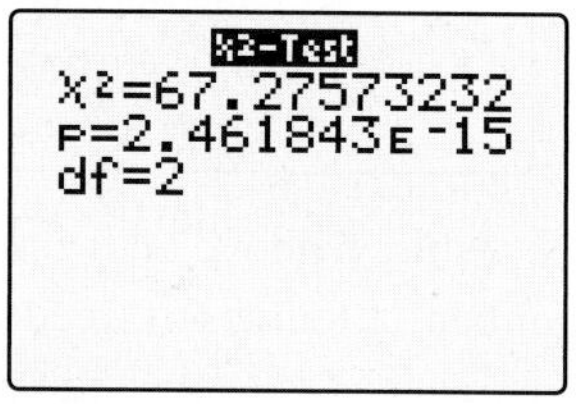

After selecting Calculate and pressing ENTER, the output screen, shown above on the right, displays the χ^2-test statistic, the p-value, and the number of degrees of freedom.

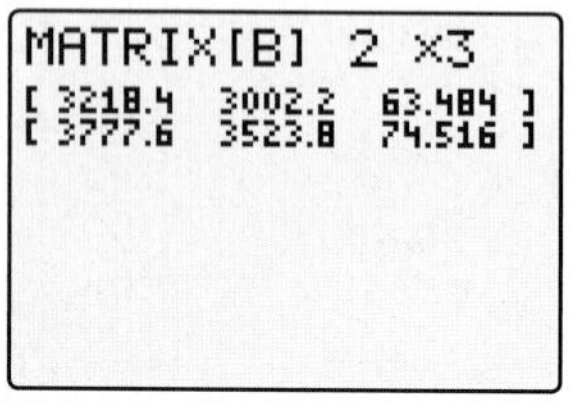

After running the test, we can view the matrix of expected values calculated by the TI-83/84 Plus. Press 2ND and then x^{-1} to access the MATRIX menu. Then scroll over to EDIT and choose option 2:[B]. The matrix of expected values for this test is shown in the screenshot in the margin.

The output screen for this test tells us that the value of our test statistic is $\chi^2 \approx 67.276$. It also tells us that the p-value is approximately 0.0000, which is less than $\alpha = 0.05$, so we reject the null hypothesis that the two variables are independent. Based on the given data, we can conclude that gender and voting choice were related for the 2008 presidential election.

Microsoft Excel

Excel can simplify hypothesis testing as shown in the following examples.

Example T.7

Using Microsoft Excel to Perform a Hypothesis Test for a Population Mean (Left-Tailed, σ Known)

Fred claims he has an 18-stroke golf handicap. This means he averages 18 strokes over par on a par-72 course. Ted thinks he is not telling the truth and collects a simple random sample of 33 of Fred's recent scorecards from par-72 courses. From these, he calculates the sample mean to be 88.1 strokes. Assume that the population standard deviation is known to be 3.5 strokes. Can he claim with 99% confidence that Fred has a handicap below 18 strokes?

Solution

First, we recognize that the desired method is a z-test for a population mean since the population standard deviation is known and the sample size is at least 30. If Ted has an 18-stroke handicap as he claims, then his mean score on a par-72 course is 72 + 18 = 90 strokes. Therefore, we state the null and alternative hypotheses as follows.

$$H_0: \mu \geq 90$$
$$H_a: \mu < 90$$

So we have a left-tailed z-test with the following values.

$\bar{x} = 88.1$ (Enter in cell B1.)

$\mu = 90$ (Enter in cell B2.)

$\sigma = 3.5$ (Enter in cell B3.)

$n = 33$ (Enter in cell B4.)

Now use Excel to calculate the test statistic and p-value. Remember that the test statistic is given by the following formula.

$$z = \frac{\bar{x} - \mu}{\left(\frac{\sigma}{\sqrt{n}}\right)}$$

Compute the test statistic by entering **=(B1-B2)/(B3/SQRT(B4))** in cell B6. Recall from Chapter 6 that the Excel formula, =NORM.S.DIST(z, TRUE), calculates the area under the standard normal distribution to the left of the z-value entered. Since this is a left-tailed test, the p-value is produced by entering **=NORM.S.DIST(B6, TRUE)**. The Excel spreadsheet is shown in the margin.

	A	B
1	sample mean	88.1
2	population mean	90
3	sigma	3.5
4	*n*	33
5		
6	*z*	-3.11848
7	*p*-value	0.000909

The p-value is approximately 0.0009, which is less than the level of significance, 0.01. Thus, we have p-value $\leq \alpha$, so the null hypothesis is rejected. With 99% confidence, Ted can conclude that Fred's handicap is below 18 strokes.

Example T.8

Using Microsoft Excel to Perform a Hypothesis Test for a Population Mean (Right-Tailed, σ Unknown)

A sanitation engineer believes that one waste management district is producing more garbage than usual. He believes the population mean is now more than 9.8 tons per truck. He chooses a simple random sample of 10 trucks that will be serving the district and weighs them before and after pickup. He calculates the sample mean to be 10.6 tons of garbage per truck with a standard deviation of 1.2 tons. Can his claim be substantiated at a confidence level of 95%? Assume that the population distribution of weights is approximately normal.

Solution

First, recognize that the desired test is a t-test for a population mean since the population standard deviation is unknown and the population distribution is approximately normal. Next, state the null and alternative hypotheses as follows.

$$H_0: \mu \leq 9.8$$
$$H_a: \mu > 9.8$$

So we have a right-tailed t-test with the following values.

$\bar{x} = 10.6$ (Enter in cell B1.)

$\mu = 9.8$ (Enter in cell B2.)

$s = 1.2$ (Enter in cell B3.)

$n = 10$ (Enter in cell B4.)

Next, use Excel to calculate the test statistic and critical value. The test statistic is given by the following formula.

$$t = \frac{\bar{x} - \mu}{\left(\frac{s}{\sqrt{n}}\right)}$$

	A	B
1	sample mean	10.6
2	population mean	9.8
3	*s*	1.2
4	*n*	10
5		
6	*t*	2.10818511
7	critical *t*	1.83311293

The test statistic is produced by entering **=(B1-B2)/(B3/SQRT(B4))** in cell B6. The Excel formula for the value of t corresponding to a given area in the left tail is =T.INV(*probability*, *deg_freedom*), where *probability* is the area in the left tail and *deg_freedom* is the number of degrees of freedom. Since this is a right-tailed test with a confidence level of 95%, we want the critical value of t with an area of $\alpha = 0.05$ in the right tail, which means that the area to the left of the critical value is $1 - 0.05 = 0.95$. Enter **=T.INV(0.95, 9)** to find the critical value for this test. Note that $n - 1$ degrees of freedom are used. The Excel spreadsheet is shown in the margin.

Since $t \approx 2.108$ and $t_\alpha = t_{0.05} \approx 1.833$, we have $t \geq t_\alpha$, so we reject the null hypothesis and conclude that the district produces more trash than is anticipated. At a confidence level of 95%, we can support the sanitation engineer's claim that the population mean for this district is more than 9.8 tons of garbage per truck.

Example T.9

Using Microsoft Excel to Perform a Hypothesis Test for a Population Proportion (Two-Tailed)

Randy has heard that 10% of the population is left-handed. To test this proposition, he takes a simple random sample and records the results. Out of the 65 people surveyed, 10 are left-handed. Using $\alpha = 0.10$, determine whether there is sufficient evidence to refute the claim that 10% of the population is left-handed.

Solution

Immediately, we recognize that we need to perform a z-test for a population proportion. Note that the sample size is $n = 65$, and the hypothesized value of the population proportion is $p = 0.10$. Since $np = 65(0.10) = 6.5 \geq 5$ and $n(1-p) = 65(1-0.10) = 58.5 \geq 5$, the conditions are met for utilizing this test. The null and alternative hypotheses are as follows.

$$H_0: p = 0.10$$
$$H_a: p \neq 0.10$$

First, note that $\hat{p} = \frac{x}{n} = \frac{10}{65}$, so enter **=10/65** in cell B1 to calculate the sample proportion. Then enter the hypothesized value of the population proportion, **0.10**, in cell B2 and the sample size, **65**, in cell B3. Next, use Excel to calculate the test statistic and p-value. Recall that the test statistic is given by the following formula.

$$z = \frac{\hat{p} - p}{\sqrt{\frac{p(1-p)}{n}}}$$

Enter **=(B1-B2)/SQRT(B2*(1-B2)/B3)** in cell B5 to calculate the test statistic, $z \approx 1.45$. Since the test is two-tailed, the p-value is the sum of the areas under the standard normal curve to the left of $z_1 \approx -1.45$ and to the right of $z_2 \approx 1.45$. The NORM.S.DIST function calculates the area to the left of the z-value entered, so enter **=2*(1-NORM.S.DIST(B5, TRUE))** in cell B6 to produce the p-value.

	A	B
1	sample proportion	0.153846
2	population proportion	0.1
3	*n*	65
4		
5	*z*	1.447072
6	*p*-value	0.147877

The p-value is approximately 0.1479, which is greater than $\alpha = 0.10$, so we fail to reject the null hypothesis. Thus, at the 0.10 level of significance, this evidence is not sufficient to refute the claim that 10% of the population is left-handed.

Example T.10

Using Microsoft Excel to Perform a Chi-Square Test for Goodness of Fit

A manager at a clothing store believes that pink shirts do not sell as well as other colors. Despite this observation, her buyer continues to supply the store with equal numbers of red, green, blue, and pink shirts. The manager tallies up last month's sales by shirt color. The results are shown below.

Last Month's Sales by Shirt Color

	Red	Green	Blue	Pink
Sales	27	25	26	17
Expected Sales	23.75	23.75	23.75	23.75

Can the manager prove with a confidence level of 90% that the numbers of red, green, blue, and pink shirts sold are not equal?

Solution

First, state the null and alternative hypotheses as follows.

$$H_0: p_1 = p_2 = p_3 = p_4$$

H_a: There is some difference amongst the probabilities.

Next, enter the data into cells A1 through E3, as shown in the screenshot at the end of this example, and then calculate the test statistic. The latter is actually a two-part task. The first part requires finding the addends that compose the test statistic using the calculation $\frac{(\text{Sales} - \text{Expected Sales})^2}{\text{Expected Sales}}$ for each shirt color. The second part is computing the sum of these four calculations. First, enter **=(B2-B3)^2/B3** in cell B4, and then copy this formula and paste it into cells C4, D4, and E4. Next, find the sum by entering **=SUM(B4:E4)** in cell B6. This sum is the test statistic, $\chi^2 \approx 2.642$.

The Excel formula for the value of χ^2 corresponding to a given area in the right tail is =CHISQ.INV.RT(*probability*, *deg_freedom*), where *probability* is the area in the right tail and *deg_freedom* is the number of degrees of freedom. Thus, the critical value for this test is calculated by entering **=CHISQ.INV.RT(0.10, 3)** into cell B7, where $\alpha = 1 - 0.90 = 0.10$, which is the area in the right tail of the chi-square distribution with $k - 1 = 3$ degrees of freedom. This critical value is $\chi^2_\alpha = \chi^2_{0.10} \approx 6.251$. Then, since $2.642 < 6.25$, we have $\chi^2 < \chi^2_\alpha$ and thus we fail to reject the null hypothesis. Therefore, at the 90% confidence level, the evidence does not support the manager's claim that the numbers of red, green, blue, and pink shirts sold are not equal.

	A	B	C	D	E
1		Red	Green	Blue	Pink
2	Sales	27	25	26	17
3	Expected Sales	23.75	23.75	23.75	23.75
4	(Sales - Expected Sales)^2/Expected Sales	0.44473684	0.06578947	0.21315789	1.91842105
5					
6	Test Statistic (Sum)	2.64210526			
7	Critical Value	6.25138863			

Example T.11

Using Microsoft Excel to Perform a Chi-Square Test for Association

The US Census Bureau reported that Charleston County, SC had the following population distribution in 2010.

Observed Sample of 350,209 Residents					
	Under 18	**18 to 44**	**45 to 64**	**65 and Over**	**Total**
Male	33,022	74,659	43,116	19,074	169,871
Female	35,722	71,044	47,925	25,647	180,338
Total	68,744	145,703	91,041	44,721	350,209

Source: American FactFinder. "Profile of General Population and Housing Characteristics: 2010 Demographic Profile Data." US Census Bureau. 2010. http://quickfacts.census.gov/qfd/states/45/45019lk.html (31 Jan. 2012).

Is there sufficient evidence at the 1% level to say that age group is related to gender?

Solution

Our null hypothesis is that age group and gender are independent of one another.

H_0: Gender and age group are independent.

H_a: Gender and age group are not independent.

After entering the given data into Excel, the next step is calculating the expected value for each cell in the contingency table. Enter the given contingency table of observed values in cells A1 through F4 as shown in the following screenshot. The expected values are calculated by using the formula $E_i = \frac{(\text{row total})(\text{column total})}{n}$. Enter **=B$4*$F2/$F$4** into cell B7 and then copy and paste this formula into cells B7 through E8. Note that the $ symbol is used to "lock" the column or row that follows the $ when a formula is copied from one cell and pasted into another. The only remaining step is calculating the p-value for the test. The Excel formula used to calculate the p-value for a chi-square test for association is =CHISQ.TEST(*actual_range*, *expected_range*), where *actual_range* is the range of cells containing the observed values and *expected_range* is the range of cells containing the expected values. Thus, the p-value for this test is found by entering **=CHISQ.TEST(B2:E3, B7:E8)** into cell B9.

	A	B	C	D	E	F
1		Under 18	18 to 44	45 to 64	65 and Over	Total
2	Male	33,022	74,659	43,116	19,074	169,871
3	Female	35,722	71,044	47,925	25,647	180,338
4	Total	68,744	145,703	91,041	44,721	350,209
5						
6						
7		33344.6942369	70674.12406020	44159.98935207	21692.19235085	
8		35399.3057631	75028.87593980	46881.01064793	23028.80764915	
9	*p*-value	4.9471E-239				

The p-value is approximately 0.0000. Since this p-value is less than $\alpha = 0.01$, we reject the null hypothesis. Therefore, at the 0.01 level of significance, the evidence supports the claim that age group is related to gender for residents of Charleston County, SC.

Minitab

Example T.12

Using Minitab to Perform a Hypothesis Test for a Population Proportion (Right-Tailed)

Alfredo thinks he is a better golfer than Tiger Woods was in 2008. He backs this up with the fact that in a simple random sample of 180 holes that he has played in the past year, he hit 133 greens in regulation. In 2008, Tiger had the highest greens in regulation percentage on the PGA Tour, 71.39%. Using $\alpha = 0.10$, test Alfredo's claim of golfing superiority.

Source: PGATOUR.com. "PGA TOUR – Performance Stats: Tiger Woods – 2008 Stats." 2010. http://www.pgatour.com/golfers/008793/tiger-woods/performance-stats/#uber (1 Feb. 2012).

Solution

First, we identify the test as a hypothesis test for a population proportion. Before we perform the hypothesis test, we must verify that the necessary conditions are satisfied. Note that the sample size is $n = 180$. Since Alfredo claims that his greens in regulation percentage is better than Tiger's, we will use Tiger's percentage in the hypotheses. Thus, the hypothesized value of the population proportion is $p = 0.7139$. Since $np = 180(0.7139) = 128.502 \geq 5$ and $n(1-p) = 180(1-0.7139) = 51.498 \geq 5$, the conditions are met for utilizing the z-test statistic. We have the following null and alternative hypotheses.

$$H_0: p \leq 0.7139$$
$$H_a: p > 0.7139$$

This test will use the following values.

$$x = 133$$
$$n = 180$$
$$p = 0.7139$$

In Minitab, go to **Stat ► Basic Statistics ► 1 Proportion**. Choose **Summarized data** and enter **133** for the number of events and **180** for the number of trials. Then select **Perform hypothesis test** and enter **0.7139** for the hypothesized proportion. Click on **Options** to enter **90** for the confidence level; then choose **greater than** for the alternative hypothesis. Since we are using the z-test statistic for this hypothesis test, select the option **Use test and interval based on normal distribution**.

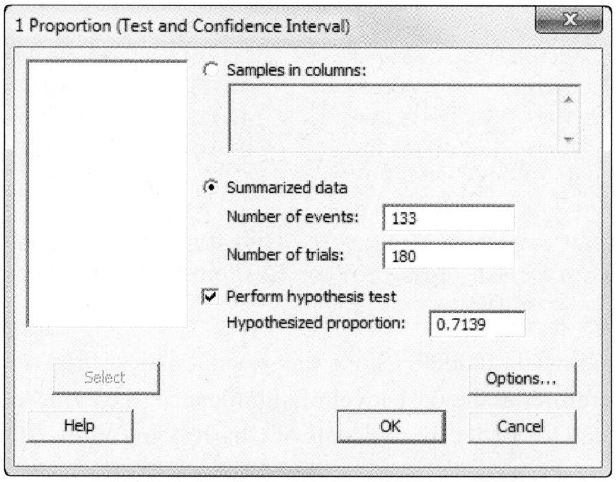

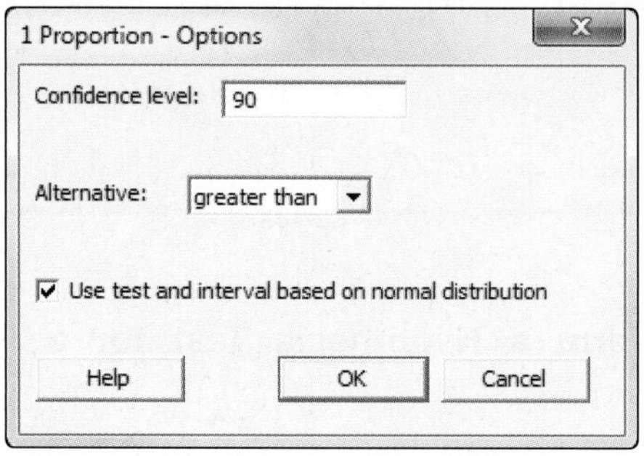

Click **OK** on the options window and **OK** on the main dialog window and the following results will be displayed in the Session window.

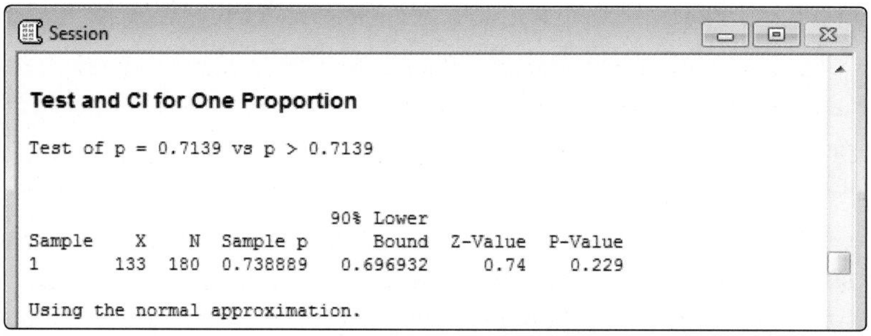

The last column in the results displays the p-value. With a p-value of 0.229, we fail to reject the null hypothesis, so the evidence does not support Alfredo's claim that he is a better golfer than Tiger Woods.

Example T.13

Using Minitab to Perform a Chi-Square Test for Goodness of Fit

A local library is open 20 hours per week. The librarian wants to know whether the proportions of books checked out are the same as the proportions of hours the library is open for every day of the week. She records the number of books checked out as well as the number of hours the library was open on each day of one randomly selected week. The results are as follows.

Library Books and Hours							
	Mon.	Tues.	Wed.	Thurs.	Fri.	Sat.	Sun.
Number of Books	101	165	162	180	117	82	closed
Number of Hours Open	2	4	4	5	3	2	closed

Calculate the χ^2-test statistic using MINITAB.

Solution

First, enter the given data into columns C1, C2, and C3 in the MINITAB worksheet, as shown in the following screenshot. Next, we need to calculate the expected number of books checked out for each day. We will calculate the expected number of books for each day using the formula $E = np_i$, where n is the total number of books checked out for the week, and p_i is the proportion of hours that the library is open on that day. Go to **Calc ▶ Calculator** and enter the following expression: **SUM(C2)*(C3/SUM(C3))**. Choose to store the result in column **C4** and click **OK**. The expected number of books checked out for each day will then be calculated and stored in column C4.

The next step is calculating the addends that compose the χ^2-test statistic using the formula $\frac{(O_i - E_i)^2}{E_i}$. Again, go to **Calc ▶ Calculator**. Enter the expression **(C2-C4)**2/C4**. This time you want to store the result in column **C5**. Click **OK** and the values will be calculated and stored in column C5. The completed worksheet will appear as shown in the following screenshot.

Worksheet 1 ***

↓	C1-T	C2	C3	C4	C5
	Day	Number of Books	Number of Hours Open	Expected Number of Books	Partial Chi-Square
1	Mon.	101	2	80.70	5.10644
2	Tues.	165	4	161.40	0.08030
3	Wed.	162	4	161.40	0.00223
4	Thurs.	180	5	201.75	2.34480
5	Fri.	117	3	121.05	0.13550
6	Sat.	82	2	80.70	0.02094

Finally, go to **Calc ▶ Column Statistics**. Select **Sum** and enter **C5** as the input variable; then click **OK**. The resulting sum that is shown in the Session window is our test statistic, $\chi^2 \approx 7.690$.

Example T.14

Using MINITAB to Perform a Chi-Square Test for Association

Gretchen wants to know whether gender has a relationship to beverage choice among her coworkers. She defines beverage choice as the first beverage consumed upon arrival at work and selects a random sample of her coworkers. She records her coworkers' consumption habits one morning and the results are shown in the following table. Is there sufficient evidence at the 0.05 level to say that gender and beverage choice are related?

Observed Sample of 90 Coworkers

	Coffee	Tea	Water	Other	Total
Male	10	7	18	9	44
Female	8	12	11	15	46
Total	18	19	29	24	90

Solution

First, we state our hypotheses, letting the null hypothesis be that gender and beverage choice are independent.

H_0: Gender and beverage choice are independent.

H_a: Gender and beverage choice are not independent.

Now enter the data into the worksheet as shown in the following screenshot. Do not enter the total column or the total row into the spreadsheet.

Worksheet 1 ***

↓	C1-T	C2	C3	C4	C5
		Coffee	Tea	Water	Other
1	Male	10	7	18	9
2	Female	8	12	11	15

Next, go to **Stat ► Tables ► Chi-Square Test (Table in Worksheet)**. Enter **C2-C5** as the columns containing the table, as shown in the following dialog box.

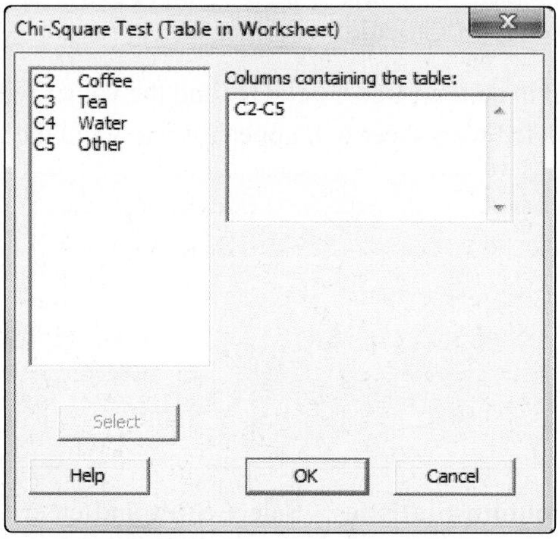

Click **OK** and the results in the Session window should appear as follows.

Session

Chi-Square Test: Coffee, Tea, Water, Other

```
Expected counts are printed below observed counts
Chi-Square contributions are printed below expected counts

        Coffee    Tea  Water  Other  Total
    1       10      7     18      9     44
          8.80   9.29  14.18  11.73
         0.164  0.564  1.030  0.637

    2        8     12     11     15     46
          9.20   9.71  14.82  12.27
         0.157  0.539  0.986  0.609

Total       18     19     29     24     90

Chi-Sq = 4.686, DF = 3, P-Value = 0.196
```

The expected values, χ^2-test statistic, number of degrees of freedom, and p-value are displayed in the results. Since the p-value $= 0.196$, we fail to reject the null hypothesis at $\alpha = 0.05$. Thus, there is not sufficient evidence at the 0.05 level to say that gender and beverage choice are related among Gretchen's coworkers.

Chapter Eleven

Hypothesis Testing (Two or More Populations)

Sections

Objectives

1. Perform a hypothesis test comparing two population means.
2. Perform a hypothesis test comparing two population proportions.
3. Perform a hypothesis test comparing two population variances.
4. Understand the basics of an ANOVA table.
5. Compare more than two population means using an ANOVA test.

11

Introduction

Consider the battle of the sexes, the age-old dispute about which gender is better. If we truly wanted to test the claim that one gender is better than the other, we could do this by performing a hypothesis test. Take, for example, IQ scores. Using our current knowledge of hypothesis testing, we could begin by testing the claim that the mean IQ of a man is greater than 100. Let's even assume that to be true! We could then test the claim that the mean IQ of a woman is greater than 100. Assuming this is true as well, we could raise the value of the claim. We could next test to see if men, and then women, had a mean IQ score greater than 105, and then 110, 115, and so on. We could continue in this manner until one gender's mean IQ was greater than our value and the other's was not. This tedious process would be similar to the high jump; the bar keeps being raised until only one person can successfully jump over the bar.

Wouldn't it be simpler to perform only one test? We are not interested in the particular value of the mean IQ for men or for women, only in whether one gender's mean IQ is higher than the other. Thus, what we really want is a test that compares two values, not one that identifies a particular value.

Just as we discussed creating a confidence interval for a single population parameter in Chapter 8 and then extended that process to creating a confidence interval to compare two population parameters in Chapter 9, we will extend the hypothesis testing procedure that we learned in the previous chapter so it is possible to test claims involving two or more populations. Then maybe we can settle the question of which gender is better once and for all.

11.1 Hypothesis Testing: Two Population Means (σ Known)

In this chapter we will be comparing two or more population parameters instead of testing a claim about a single parameter. However, the steps for a hypothesis test are the same with two or more parameters as they were before.

Procedure

Performing a Hypothesis Test

1. State the null and alternative hypotheses.
2. Determine which distribution to use for the test statistic, and state the level of significance.
3. Gather data and calculate the necessary sample statistics.
4. Draw a conclusion and interpret the decision.

Since we already know the hypothesis testing procedure, in order to perform a hypothesis test involving two populations, all we need are the relevant test statistics and a method of writing the null and alternative hypotheses. The ways that rejection regions are constructed, *p*-values are calculated, and conclusions are determined are the same as in the previous chapter.

First, let's look at how to write the null and alternative hypotheses for tests involving two population means. There are two different acceptable ways to write the null and alternative hypotheses. The first method directly compares the two populations. We can state that one population mean is greater than, less than, or not equal to the other population mean. For example, if we want to test the claim that the mean of Population 1 is greater than the mean of Population 2, then the hypotheses could be written as follows.

$$H_0: \mu_1 \leq \mu_2$$
$$H_a: \mu_1 > \mu_2$$

The second method of formulating the hypotheses considers the difference between the two population means and compares it to a specific value, such as zero. For example, by subtracting μ_2 from both sides of the inequality, the hypotheses from above can be restated as follows.

Memory Booster

The null hypothesis will contain equality.

$$H_0: \mu_1 - \mu_2 \leq 0$$
$$H_a: \mu_1 - \mu_2 > 0$$

This second method is the one we will use to write the null and alternative hypotheses for tests involving two population means since it directly relates to the test statistic we will use. The following examples illustrate how to determine the hypotheses.

Example 11.1

Determining the Null and Alternative Hypotheses for a Left-Tailed Test

Write the null and alternative hypotheses to test the claim that the mean of Population 1 is less than the mean of Population 2.

Solution

We can write the phrase "the mean of Population 1 is less than the mean of Population 2" symbolically as $\mu_1 < \mu_2$. Since we want to compare the difference between the two means to 0 as in the second method mentioned earlier, we can arrange the means so that they are both on the same side of the inequality. By subtracting μ_2 from both sides of the inequality, we have $\mu_1 - \mu_2 < 0$. Thus, the alternative hypothesis is $\mu_1 - \mu_2 < 0$, since that is what we are testing. The mathematical opposite of the claim is $\mu_1 - \mu_2 \geq 0$. Thus, we have the following hypotheses.

$$H_0: \mu_1 - \mu_2 \geq 0$$
$$H_a: \mu_1 - \mu_2 < 0$$

Example 11.2

Determining the Null and Alternative Hypotheses for a Right-Tailed Test

Write the null and alternative hypotheses to test the claim that the mean of Population 1 is greater than the mean of Population 2.

Solution

We can write the phrase "the mean of Population 1 is greater than the mean of Population

2" symbolically as $\mu_1 > \mu_2$. By subtracting μ_2 from both sides of the inequality, we have $\mu_1 - \mu_2 > 0$. Thus, the claim is $\mu_1 - \mu_2 > 0$. Since this is what we are testing, this is the alternative hypothesis. The mathematical opposite of the claim is $\mu_1 - \mu_2 \le 0$. Thus, we have the following hypotheses.

$$H_0: \mu_1 - \mu_2 \le 0$$
$$H_a: \mu_1 - \mu_2 > 0$$

Example 11.3

Determining the Null and Alternative Hypotheses for a Two-Tailed Test

Write the null and alternative hypotheses to test the claim that the mean of Population 1 is not equal to the mean of Population 2.

Solution

We can write the phrase "the mean of Population 1 is not equal to the mean of Population 2" symbolically as $\mu_1 \ne \mu_2$. By subtracting μ_2 from both sides, we have $\mu_1 - \mu_2 \ne 0$. Thus, the claim is $\mu_1 - \mu_2 \ne 0$. Once again, this would be what the researcher is looking to test; therefore, it is the alternative hypothesis. The mathematical opposite of the claim is $\mu_1 - \mu_2 = 0$. Thus, we have the following hypotheses.

$$H_0: \mu_1 - \mu_2 = 0$$
$$H_a: \mu_1 - \mu_2 \ne 0$$

Example 11.4

Determining the Null and Alternative Hypotheses

Write the null and alternative hypotheses to test the claim that the mean of Population 1 is more than 20 units above the mean of Population 2.

Solution

We can write the phrase "the mean of Population 1 is more than 20 units above the mean of Population 2" symbolically as $\mu_1 > \mu_2 + 20$. Subtracting μ_2 from both sides of the inequality, we have $\mu_1 - \mu_2 > 20$. Since the researcher would be looking for evidence to support this claim, this is the alternative hypothesis. The mathematical opposite is $\mu_1 - \mu_2 \le 20$. Thus, we have the following hypotheses.

$$H_0: \mu_1 - \mu_2 \le 20$$
$$H_a: \mu_1 - \mu_2 > 20$$

The test statistic for this type of hypothesis test is associated with the difference between the two population means. In this section, we will consider the case in which we have two populations whose standard deviations are known. As we've stated before, population standard deviations are rarely known, but we will start with this scenario, as we did in previous chapters, since it illustrates the use of the standard normal distribution. When the following criteria are met, the z-test statistic should be used.

- All possible samples of a given size have an equal probability of being chosen; that is, simple random samples are used.
- The samples are independent.
- Both population standard deviations, σ_1 and σ_2, are *known*.
- Either both sample sizes are at least 30 ($n_1 \geq 30$ and $n_2 \geq 30$) *or* both population distributions are approximately normal.

Formula

Test Statistic for a Hypothesis Test for Two Population Means (σ Known)

When both population standard deviations are known, the samples taken are independent, simple random samples, and either both sample sizes are at least 30 or both population distributions are approximately normal, the test statistic for a hypothesis test for two population means is given by

$$z = \frac{(\bar{x}_1 - \bar{x}_2) - (\mu_1 - \mu_2)}{\sqrt{\frac{\sigma_1^2}{n_1} + \frac{\sigma_2^2}{n_2}}}$$

where $\bar{x}_1$ and $\bar{x}_2$ are the two sample means,

$\mu_1 - \mu_2$ is the presumed value of the difference between the two population means from the null hypothesis,

σ_1 and σ_2 are the two population standard deviations, and

n_1 and n_2 are the two sample sizes.

Examples 11.5 and 11.6 illustrate hypothesis testing on the means of two populations when we simply want to know if there is any difference between the two means. In Example 11.7, we change the test slightly by considering the case where the amount by which the two population means differ is part of the claim.

Example 11.5

Performing a Hypothesis Test for Two Population Means (Right-Tailed, σ Known)

A researcher wants to test the claim that grocery shoppers spend more when shopping on an empty stomach than when shopping on a full stomach. One morning, 93 randomly selected shoppers were surveyed upon leaving a particular grocery store. Of those shoppers surveyed, 41 of the shoppers did not eat breakfast that morning. This group spent a mean of \$72.27 on groceries. The other 52 shoppers did eat breakfast that morning, and spent a mean of \$69.43 on groceries. Assume that the population standard deviation for shoppers shopping on an empty stomach is known to be \$8.05 and the population standard deviation for shoppers shopping on a full stomach is known to be \$9.22. Test the claim using a 0.05 level of significance.

Solution

Step 1: **State the null and alternative hypotheses.**

First, let Population 1 be the empty-stomach shoppers and Population 2 be the full-stomach shoppers. The researcher wants to know if grocery shoppers spend more when shopping on an empty stomach than when shopping on a full stomach. That is, the mean amount that shoppers spend when they are hungry is greater than the mean amount that shoppers spend when they are not hungry. Written mathematically, this is $\mu_1 > \mu_2$, which is the alternative hypothesis. By subtracting μ_2 from both sides of the inequality, we have $\mu_1 - \mu_2 > 0$. Thus, we have the following hypotheses.

$$H_0: \mu_1 - \mu_2 \le 0$$
$$H_a: \mu_1 - \mu_2 > 0$$

Step 2: **Determine which distribution to use for the test statistic, and state the level of significance.**

We are looking at the difference between two population means when both population standard deviations are known. We also know that independent random samples were drawn, and both sample sizes are at least 30. Therefore, the z-test statistic, which has a standard normal distribution, is appropriate.

We will draw a conclusion by computing the p-value for the calculated test statistic and comparing that value to the level of significance, α. For this hypothesis test, $\alpha = 0.05$.

Step 3: **Gather data and calculate the necessary sample statistics.**

The following values were given in the problem.

$$\bar{x}_1 = 72.27 \qquad \bar{x}_2 = 69.43$$
$$\sigma_1 = 8.05 \qquad \sigma_2 = 9.22$$
$$n_1 = 41 \qquad n_2 = 52$$

Substituting these values into the formula for the test statistic gives us the following.

$$\begin{aligned} z &= \frac{(\bar{x}_1 - \bar{x}_2) - (\mu_1 - \mu_2)}{\sqrt{\frac{\sigma_1^2}{n_1} + \frac{\sigma_2^2}{n_2}}} \\ &= \frac{(72.27 - 69.43) - 0}{\sqrt{\frac{8.05^2}{41} + \frac{9.22^2}{52}}} \\ &\approx 1.58 \end{aligned}$$

Note that, in the formula above, $\mu_1 - \mu_2 = 0$ because the hypothesized difference between the population means is zero. This is indicated in the statement of the null hypothesis, H_0.

Step 4: **Draw a conclusion and interpret the decision.**

Because this is a right-tailed test, the p-value for this test statistic is the probability of obtaining a test statistic greater than or equal to $z = 1.58$, written p-value $= P(z \geq 1.58)$. Thus, the p-value is equivalent to the area under the standard normal curve to the right of $z = 1.58$.

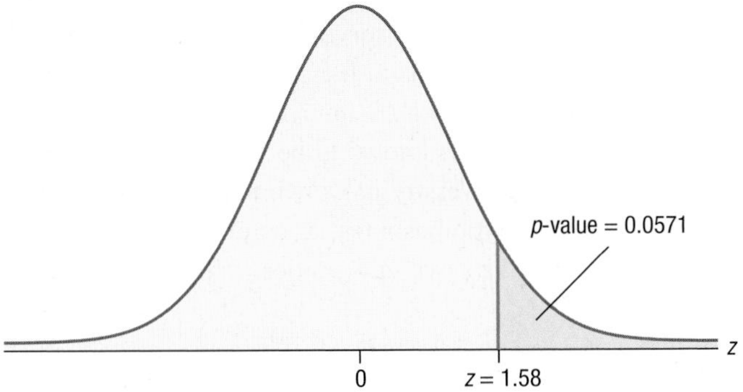

Using appropriate technology or a normal distribution table, we find a p-value of 0.0571. This p-value is greater than $\alpha = 0.05$, which means that we do not have enough evidence to reject the null hypothesis. Therefore, there is not sufficient evidence at the 0.05 level of significance to say that the mean amount spent on groceries by hungry shoppers is greater than the mean amount spent by those shopping on a full stomach.

Memory Booster

Conclusions Using p-Values

- If p-value $\leq \alpha$, then reject the null hypothesis.
- If p-value $> \alpha$, then fail to reject the null hypothesis.

We can also use a TI-83/84 Plus calculator to perform this hypothesis test by simply entering the necessary statistics into the calculator. To perform this test, press STAT, scroll over to TESTS, and choose option `3:2-SampZTest`. Choose `Stats`, since we know the sample statistics. (Choose the `Data` option if listing the original data in the Lists). After filling in the values as shown in the following screenshots, choose the type of test that the alternative hypothesis dictates, that is, a right-tailed test.

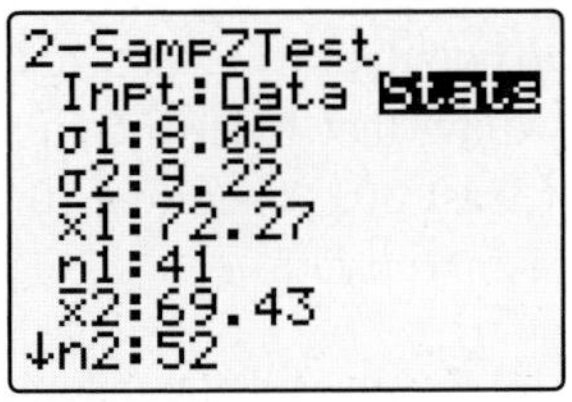

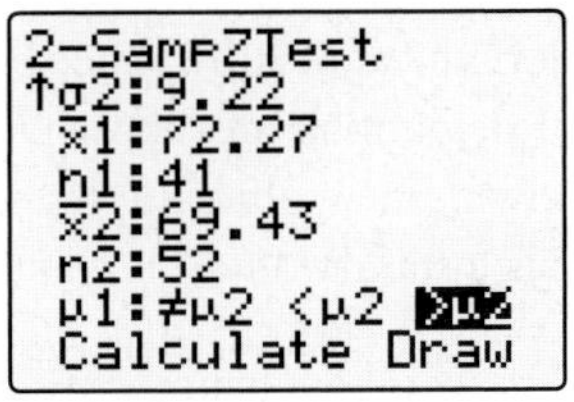

Choosing `Calculate` produces the following results.

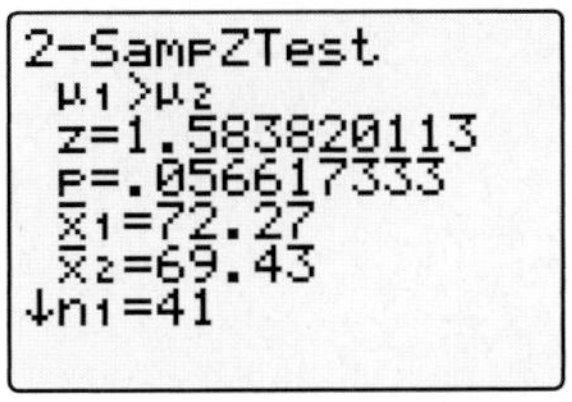

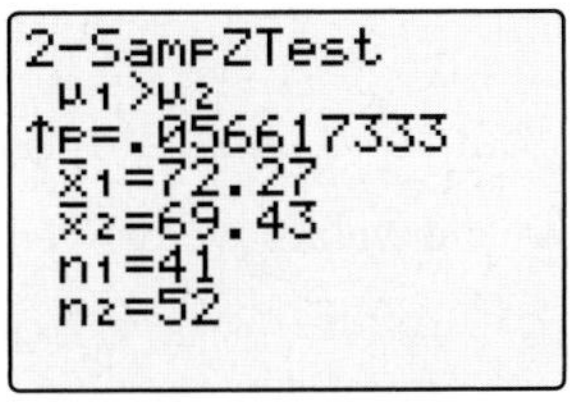

First, the alternative hypothesis is stated. Note that the calculator displays the alternative hypothesis formulated using the first method that we discussed earlier, but it is mathematically equivalent to the alternative hypothesis formulated using the second method, which we used in Step 1 of this example. As shown in the screenshots, the z-score is approximately 1.58 and the p-value is approximately 0.0566. The sample means and sample sizes are then repeated.

Example 11.6

Performing a Hypothesis Test for Two Population Means (Two-Tailed, σ Known)

Two universities in the same state are bitter rivals. Each university believes that its students are more physically fit than the students at the other university. To test the claim that there is a difference in the average fitness levels of students at the two universities, 36 randomly selected students at the first university were surveyed and exercised for a mean of 2.9 hours per week. A random sample of 38 students at the second university was also surveyed, and exercised for a mean of 2.7 hours per week. Assume that the population standard deviation for hours of exercise at the first university is known to be 1.1 hours per week and the population standard deviation for the second university is known to be 1.0 hour per week. Use a 0.05 level of significance to perform a hypothesis test to determine if there is a difference in the average fitness levels of students at the two universities.

Solution

Step 1: **State the null and alternative hypotheses.**

The claim is that there is a difference in the average physical fitness levels of students at the two universities. Thus, the universities want to determine if the mean numbers of hours per week spent exercising by students at the two universities are not equal. Written mathematically, this is $\mu_1 \neq \mu_2$. By subtracting μ_2 from both sides, we have $\mu_1 - \mu_2 \neq 0$. Thus, the hypotheses are as follows.

$$H_0: \mu_1 - \mu_2 = 0$$
$$H_a: \mu_1 - \mu_2 \neq 0$$

Step 2: **Determine which distribution to use for the test statistic, and state the level of significance.**

We are looking at the difference between two population means when both population standard deviations are known. We also know that independent random samples were drawn, and both sample sizes are at least 30. Therefore, the z-test statistic, which has a standard normal distribution, is appropriate.

We will draw a conclusion by computing the p-value for the calculated test statistic and comparing that value to the level of significance, α. For this hypothesis test, $\alpha = 0.05$.

Step 3: **Gather data and calculate the necessary sample statistics.**

The following values were given in the problem.

$$\bar{x}_1 = 2.9 \qquad \bar{x}_2 = 2.7$$
$$\sigma_1 = 1.1 \qquad \sigma_2 = 1.0$$
$$n_1 = 36 \qquad n_2 = 38$$

Substituting these values into the formula for the test statistic gives us the following.

$$
\begin{aligned}
z &= \frac{(\bar{x}_1 - \bar{x}_2) - (\mu_1 - \mu_2)}{\sqrt{\dfrac{\sigma_1^2}{n_1} + \dfrac{\sigma_2^2}{n_2}}} \\
&= \frac{(2.9 - 2.7) - 0}{\sqrt{\dfrac{1.1^2}{36} + \dfrac{1.0^2}{38}}} \\
&\approx 0.82
\end{aligned}
$$

Once again note that, in the formula above, $\mu_1 - \mu_2 = 0$ because the hypothesized difference between the population means is zero. This is indicated in the statement of the null hypothesis, H_0.

Step 4: **Draw a conclusion and interpret the decision.**

Because this is a two-tailed test, the p-value for this test statistic is the probability of obtaining a test statistic that is either less than or equal to $z_1 = -0.82$ or greater than or equal to $z_2 = 0.82$, which is written mathematically as $p\text{-value} = P(|z| \geq 0.82)$. Thus, the p-value is equivalent to the sum of the areas under the standard normal curve to the left of $z_1 = -0.82$ and to the right of $z_2 = 0.82$.

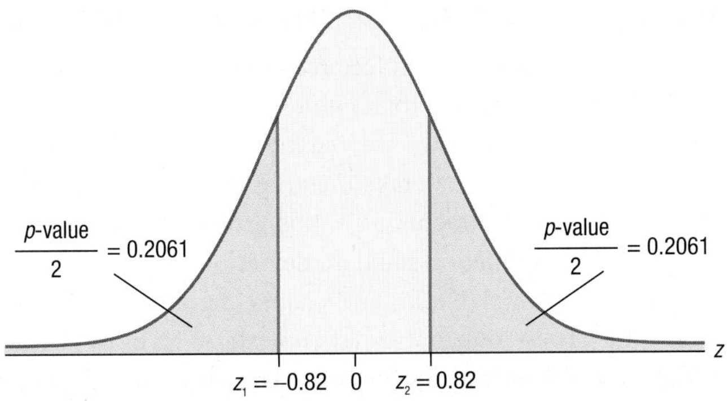

Using appropriate technology or a normal distribution table, we find a p-value of 0.4122. This p-value is greater than $\alpha = 0.05$, which means that we do not have enough evidence to reject the null hypothesis. Therefore, there is not sufficient evidence at the 0.05 level of significance to say that the mean numbers of hours per week spent exercising by students at the two universities are unequal. Thus, we cannot conclude that there is a difference in the average fitness levels of the students at the two universities.

The process for using a TI-83/84 Plus calculator to perform this test is the same as in Example 11.5. Press STAT, choose TESTS, and choose option `3:2-SampZTest`. Again, choose `Stats`, since we know the sample statistics. After filling in the required values, choose the symbol for a two-tailed test.

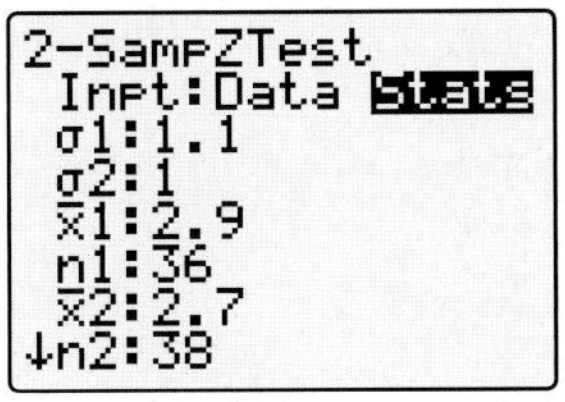

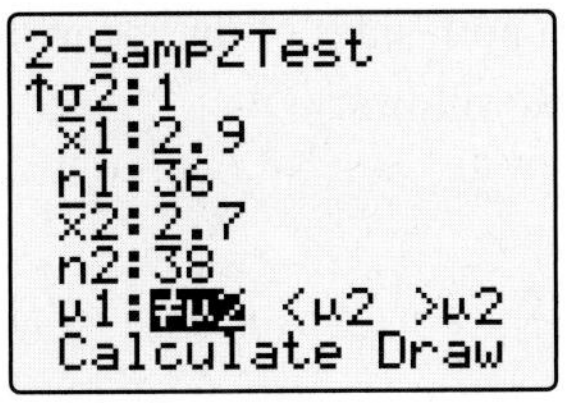

Choosing `Calculate` produces the following results.

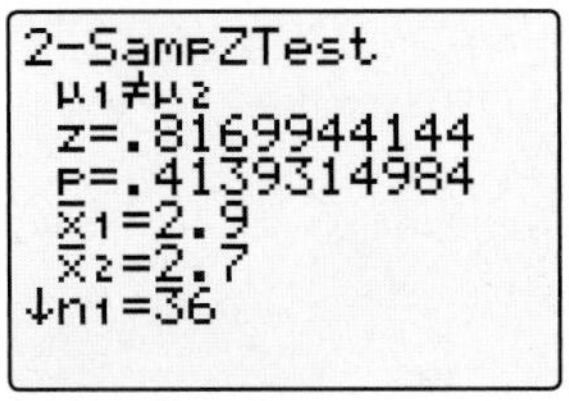

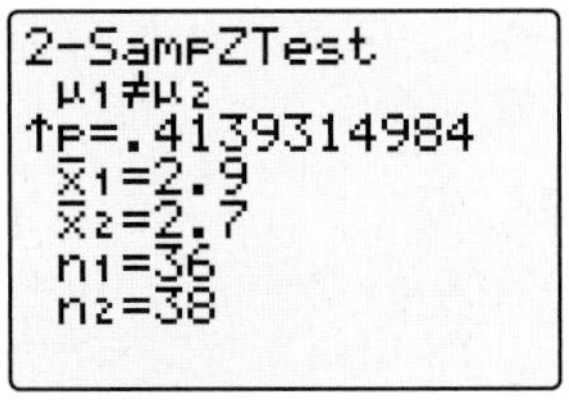

Again, the alternative hypothesis is stated first. As shown in the screenshots, the *z*-score is approximately 0.82 and the *p*-value is approximately 0.4139. The *p*-value given by the calculator is more accurate than the *p*-value we found using the rounded value of the test statistic because there are no intermediate steps to cause rounding errors. However, the conclusion of the test is the same since the two *p*-values are close. The sample means and sample sizes are also repeated in the results.

Now let's consider the case where the amount by which the population means differ is part of the claim.

Example 11.7

Performing a Hypothesis Test for Two Population Means (Right-Tailed, σ Known)

A drug manufacturer claims that its new cholesterol drug, when used together with a healthy diet and exercise plan, lowers a patient's total cholesterol level by over 20 points more than simply changing a patient's diet and exercise regimen. To test the claim, a sample of 55 patients with high cholesterol is chosen to take the drug in addition to changing their diet and exercise plans. Over the course of three months, this group lowers its total cholesterol level by a mean of 44.7 points. The population standard deviation for this group is known to be 6.8 points. Another 55 patients with high cholesterol change their diet and exercise regimens, but do not take the drug. This group lowers its total cholesterol level by a mean of 23.1 points. The population standard deviation for this group is known to be 5.3 points. Test the drug manufacturer's claim using a 0.01 level of significance.

Solution

Step 1: **State the null and alternative hypotheses.**

First, let Population 1 be patients who take the drug and Population 2 be patients who do not take the drug. The manufacturer's claim is that its drug lowers cholesterol by over 20 points more than diet and exercise alone. Notice that the means being tested represent the amounts by which the patients' total cholesterol levels are lowered. Hence, for each population, μ refers to the mean number of cholesterol points lost. Written mathematically, the claim is $\mu_1 > \mu_2 + 20$. Since this is what the manufacturer hopes to gather evidence in favor of, it is the alternative hypothesis. By subtracting μ_2 from both sides of the inequality, we have $\mu_1 - \mu_2 > 20$. Thus, the hypotheses are stated as follows.

$$H_0: \mu_1 - \mu_2 \le 20$$
$$H_a: \mu_1 - \mu_2 > 20$$

Step 2: **Determine which distribution to use for the test statistic, and state the level of significance.**

We are looking at the difference between two population means when both population standard deviations are known. We also know that independent random samples were drawn, and both sample sizes are at least 30. Therefore, the z-test statistic, which has a standard normal distribution, is appropriate.

We will draw a conclusion by computing the p-value for the calculated test statistic and comparing that value to the level of significance, α. For this hypothesis test, $\alpha = 0.01$.

Step 3: **Gather data and calculate the necessary sample statistics.**

The following values were given in the problem.

$$\bar{x}_1 = 44.7 \qquad \bar{x}_2 = 23.1$$
$$\sigma_1 = 6.8 \qquad \sigma_2 = 5.3$$
$$n_1 = 55 \qquad n_2 = 55$$

Substituting these values into the formula for the test statistic gives us the following.

$$\begin{aligned} z &= \frac{(\bar{x}_1 - \bar{x}_2) - (\mu_1 - \mu_2)}{\sqrt{\frac{\sigma_1^2}{n_1} + \frac{\sigma_2^2}{n_2}}} \\ &= \frac{(44.7 - 23.1) - 20}{\sqrt{\frac{6.8^2}{55} + \frac{5.3^2}{55}}} \\ &\approx 1.38 \end{aligned}$$

Note that, in the previous formula, $\mu_1 - \mu_2 = 20$ because the hypothesized difference between the population means is 20. This is indicated in the statement of the null hypothesis, H_0.

Step 4: **Draw a conclusion and interpret the decision.**

Because this is a right-tailed test, the p-value for this test statistic is the probability of obtaining a test statistic greater than or equal to $z = 1.38$, written $p\text{-value} = P(z \geq 1.38)$. Thus, the p-value is equivalent to the area under the standard normal curve to the right of $z = 1.38$.

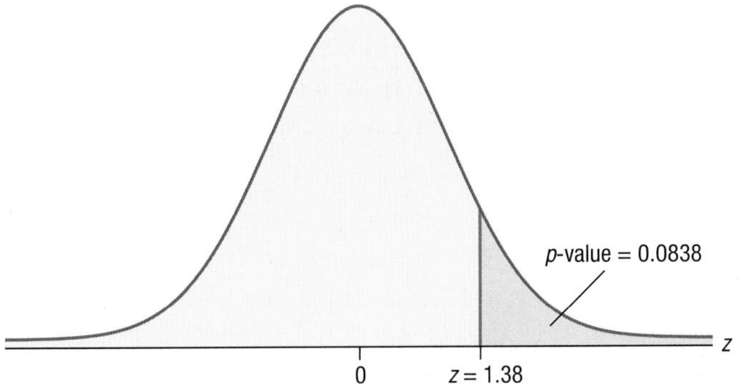

Using appropriate technology or a normal distribution table, we find a p-value of 0.0838. This p-value is greater than $\alpha = 0.01$, which means that we do not have enough evidence to reject the null hypothesis. Therefore, there is not sufficient evidence at the 0.01 level of significance to support the drug manufacturer's claim that the new drug, when used together with a healthy diet and exercise plan, lowers cholesterol by more than 20 points compared with changing diet and exercise alone.

Note that this test cannot be performed using a TI-83/84 Plus calculator because the hypothesized difference between the two population means is not zero.

11.1 Section Exercises

Test Statistics for Hypothesis Tests for Two Population Means

Directions: Calculate the test statistic for a hypothesis test for two population means using the given information.

1. $\bar{x}_1 = 9.21,\ \sigma_1 = 2.01,\ n_1 = 45,\ \bar{x}_2 = 8.76,\ \sigma_2 = 1.77,\ n_2 = 51,\ H_0: \mu_1 - \mu_2 \geq 0$

2. $\bar{x}_1 = 72.82,\ \sigma_1 = 7.90,\ n_1 = 31,\ \bar{x}_2 = 75.11,\ \sigma_2 = 6.54,\ n_2 = 39,\ H_0: \mu_1 - \mu_2 \leq 0$

3. $\bar{x}_1 = 118.4,\ \sigma_1 = 5.93,\ n_1 = 64,\ \bar{x}_2 = 104.3,\ \sigma_2 = 5.74,\ n_2 = 65,\ H_0: \mu_1 - \mu_2 \geq 15$

4. $\bar{x}_1 = 43.1,\ \sigma_1 = 2.33,\ n_1 = 71,\ \bar{x}_2 = 34.3,\ \sigma_2 = 2.96,\ n_2 = 70,\ H_0: \mu_1 - \mu_2 = 8$

Null and Alternative Hypotheses for Hypothesis Tests for Two Population Means

Directions: State the null and alternative hypotheses for each scenario.

5. The claim is that the mean of Population 1 is more than 30 units less than the mean of Population 2.

6. Ann claims that the mean drive time for her to get home from work on a Friday (Population 1) is less than her mean drive time from work to home on a Thursday (Population 2).

7. Carly claims that 6 months ago, her mean time to walk a mile (Population 1) was more than 4 minutes longer than it is currently (Population 2).

8. A newspaper claims that the mean age of its current readers (Population 1) has dropped 3.6 years over the last 10 years (Population 2). A market researcher believes that the newspaper's claim is incorrect and plans to conduct a hypothesis test to provide evidence against the newspaper's claim.

Hypothesis Tests for Two Population Means (σ Known)

Directions: **Perform each hypothesis test. For each exercise, complete the following steps.**

a. State the null and alternative hypotheses.

b. Determine which distribution to use for the test statistic, and state the level of significance.

c. Calculate the test statistic.

d. Draw a conclusion by comparing the *p*-value to the level of significance and interpret the decision.

9. A car company claims that its new SUV gets better gas mileage than its competitor's SUV. A random sample of 35 of its SUVs has a mean gas mileage of 12.6 miles per gallon (mpg). The population standard deviation is known to be 0.4 mpg. A random sample of 31 competitor's SUVs has a mean gas mileage of 12.4 mpg. The population standard deviation for the competitor is known to be 0.3 mpg. Test the company's claim at the 0.05 level of significance.

10. A college student is interested in investigating the claim that students who graduate with a master's degree earn higher salaries, on average, than those who finish with a bachelor's degree. She surveys, at random, 42 recent graduates who completed their master's degrees, and finds that their mean salary is $38,400 per year. The standard deviation of annual salaries for the population of recent graduates who have master's degrees is known to be $3100. She also surveys, at random, 45 recent graduates who completed their bachelor's degrees, and finds that their mean salary is $36,750 per year. The standard deviation of annual salaries for the population of recent graduates with only bachelor's degrees is known to be $3700. Test the claim at the 0.05 level of significance.

11. Fran is training for her first marathon, and she wants to know if there is a significant difference between the mean number of miles run each week by group runners and individual runners who are training for marathons. She interviews 32 randomly selected people who train in groups, and finds that they run a mean of 49.0 miles per week. Assume that the population standard deviation for group runners is known to be 4.2 miles per week. She also interviews a random sample of 30 people who train on their own and finds that they run a mean of 47.2 miles per week. Assume that the population standard deviation for people who run by themselves is 4.8 miles per week. Test the claim at the 0.05 level of significance.

12. Rob and Phil are both internal medicine residents in the Southeast, but they work at different hospitals. When they compare their schedules, Rob is convinced that, on average, residents at his hospital work for over 3 hours more per week than those at Phil's hospital. Rob asks a random sample of 30 residents at his hospital to record their hours for one week. He finds that they worked for a mean of 74.3 hours. Phil also asks 30 randomly selected residents at his hospital to record their hours for the same week. He calculates that they worked for a mean of 70.1 hours. Assume that the population standard deviation for the hospital where Rob works is known to be 2.6 hours and the population standard deviation for the hospital where Phil works is known to be 2.9 hours. Test Rob's claim at the 0.05 level of significance.

13. A weight-loss company wants to make sure that its clients lose more weight, on average, than they would without the company's help. An independent researcher collects data on the amount of weight lost in one month from 45 of the company's clients and finds a mean weight loss of 12 pounds. The population standard deviation for the company's clients is known to be 7 pounds per month. Data from 50 dieters not using the company's services reported a mean weight loss of 10 pounds in one month. The population standard deviation for dieters not using the company's services is known to be 6 pounds per month. Test the company's claim that using its services results in a greater mean weight loss at the 0.05 level of significance.

14. Two friends, Karen and Jodi, work different shifts for the same ambulance service. They wonder if the different shifts average different numbers of calls. Looking at past records, Karen determines from a random sample of 35 shifts that she had a mean of 5.2 calls per shift. She knows that the population standard deviation for her shift is 1.3 calls. Jodi calculates from a random sample of 34 shifts that her mean was 4.8 calls per shift. She knows that the population standard deviation for her shift is 1.2 calls. Test the claim that there is a difference between the mean numbers of calls for the two shifts at the 0.05 level of significance.

15. A professor believes that, for the introductory art history classes at his university, the mean test score of students in the evening classes is more than 5 points lower than the mean test score of students in the morning classes. He collects data from a random sample of 250 students in evening classes and finds that they have a mean test score of 80.2. He knows the population standard deviation for the evening classes to be 11.9 points. A random sample of 300 students from morning classes results in a mean test score of 86.8. He knows the population standard deviation for the morning classes to be 10.2 points. Test his claim with a 95% level of confidence.

16. A state board of directors for higher education is comparing the mean salaries of entry-level PhD positions at the state's two major universities to make sure there is not a difference in entry-level pay. Use the information given in the following table, which was collected by the board of directors for higher education, to perform the hypothesis test at the 0.05 level of significance.

Salaries of Entry-Level PhD Positions

	University A	University B
Sample Size	42	51
Mean Entry-Level Salary	\$58,500	\$60,200
Population Standard Deviation	\$3200	\$11,700

17. A parent interest group is looking at whether birth order affects scores on the ACT test. It was suggested that, on average, first-born children earn lower ACT scores than second-born children. After surveying a random sample of 100 first-born children, the parents' group found that they had a mean score of 20.9 on the ACT. A survey of 175 second-born children resulted in a mean ACT score of 21.1. Assume that the population standard deviation for first-born children is known to be 1.8 points and the population standard deviation for second-born children is known to be 2.3 points. Is there sufficient evidence at the 10% level of significance to say that the mean ACT score of first-born children is lower than the mean ACT score of second-born children?

18. Joe wants to get the best possible price on a used luxury car. He lives near the border of two states, and he believes that prices are better across the state line. For 31 randomly selected used luxury cars that were recently sold in his state, Joe finds a mean selling price of $62,065. He knows that the population standard deviation of selling prices for used luxury cars in his state is $1625. In the other state, he finds that a random sample of 40 recently sold used luxury cars has a mean selling price of $61,300. He also knows that the population standard deviation of selling prices for used luxury cars in the other state is $1475. Is there sufficient evidence at the 0.05 level to suggest that the mean selling price for used luxury cars is lower in the other state?

19. Lauren and Keri live in different states and disagree about who has higher electric bills. To settle their disagreement, the girls decide to sample electric bills in their area for the month of June and perform a hypothesis test. The electric company in Lauren's state reports that a random sample of 35 monthly residential electric bills has a mean of $104.53. For a random sample of 51 monthly residential electric bills in Keri's state, the mean is $101.48. Assume that the population standard deviation in Lauren's state is known to be $17.81, and the population standard deviation in Keri's state is known to be $25.30. Is there evidence at the 0.01 level to say that the mean monthly residential electric bill is higher for Lauren's state than for Keri's state?

20. A car servicing shop wants to use the best windshield wiper blades for its customers. It has kept track of the mean numbers of sets of blades needed per year for two different brands of blades. A random sample of 35 customers using Brand A needed a mean of 1.2 sets of blades per year, and 30 randomly selected customers using Brand B needed a mean of 1.3 sets of blades per year. Assume that the population standard deviations for Brand A and Brand B are 0.3 and 0.7 sets per year, respectively. Is there sufficient evidence at the 0.15 level to say that the mean number of sets of wiper blades needed per year is lower for Brand A than for Brand B?

21. Two college friends are big sports fans. While they are watching baseball one season, they think that there is probably a difference between the mean batting averages of players in the SEC East and SEC West divisions. To test their theory, they find the mean batting average of 40 randomly selected SEC West players to be .260. They also find the mean batting average of a random sample of 50 SEC East players to be .249. Assume that the population standard deviation for the SEC West division is known to be .026 and the population standard deviation for the SEC East division is known to be .051. Is there sufficient evidence at the 0.10 level of significance to say that there is a difference between the mean batting averages for the two SEC divisions?

11.2 Hypothesis Testing: Two Population Means (σ Unknown)

As we have seen in previous chapters, a population standard deviation (σ) is rarely known. When we wish to perform a hypothesis test that compares the means of two populations for which σ is unknown, a Student's t-distribution must be used, just as was the case for a test involving a single population mean with σ unknown, as discussed in Section 10.3.

As usual, the first step in a hypothesis test is to state the null and alternative hypotheses. For a test that compares the means of two populations for which σ is unknown, the hypotheses are written in the same manner as when σ is known.

In this section we will look at scenarios in which independent samples are drawn from two populations for which neither standard deviation is known. That is, the scenarios must meet the following criteria.

- All possible samples of a given size have an equal probability of being chosen; that is, simple random samples are used.
- The samples are independent.
- Both population standard deviations, σ_1 and σ_2, are *unknown*.
- Either both sample sizes are at least 30 ($n_1 \geq 30$ and $n_2 \geq 30$) *or* both population distributions are approximately normal.

When the above conditions are met, we must also consider whether or not the population variances are likely to be equal in order to determine the appropriate formula to use for the t-test statistic. Once again, just as we indicated in Section 9.2, although neither population variance is actually known, in some cases it is reasonable to assume that the population variances (or standard deviations) are the same.

When the two population variances are assumed to be equal, the sample variances can be *pooled*, or combined, to give as good an estimate as possible for the common population variance. On the other hand, it may not be appropriate to assume that the two samples are drawn from populations with the same variance. If the variances are *not assumed to be equal*, then the sample variances cannot be pooled.

In general, we can use either rejection regions or p-values to draw conclusions in hypothesis tests, and the examples in this section will demonstrate both of these methods. Recall from Chapter 10 that the rejection region for drawing a conclusion is based on the type of hypothesis test we are performing and the level of significance, α. The rejection regions for hypothesis tests that compare the means of two populations with unknown standard deviations are as follows. Note that these same decision rules are used in tests involving individual population means when σ is unknown, as we saw in Section 10.3.

Procedure

Rejection Regions for Hypothesis Tests for Two Population Means (σ Unknown)

Reject the null hypothesis, H_0, if:

$t \le -t_\alpha$ for a left-tailed test

$t \ge t_\alpha$ for a right-tailed test

$|t| \ge t_{\alpha/2}$ for a two-tailed test

Unequal Variances

Let's first consider the case when the population variances are not assumed to be equal. In this case, the sample variances are not pooled, and the formula for the test statistic looks very similar to the one from the previous section that we use when σ is known for both populations. The only difference is that the sample variances are used to estimate the unknown population variances. Thus the test statistic is a *t*-value, which has a Student's *t*-distribution, instead of a *z*-value, which has a standard normal distribution, in order to account for the error that arises from estimating the unknown population variances.

Formula

Test Statistic for a Hypothesis Test for Two Population Means (σ Unknown, Unequal Variances)

When both population standard deviations are unknown and assumed to be unequal, the samples taken are independent, simple random samples, and either both sample sizes are at least 30 or both population distributions are approximately normal, the test statistic for a hypothesis test for two population means is given by

$$t = \frac{(\bar{x}_1 - \bar{x}_2) - (\mu_1 - \mu_2)}{\sqrt{\frac{s_1^2}{n_1} + \frac{s_2^2}{n_2}}}$$

where $\bar{x}_1$ and $\bar{x}_2$ are the two sample means,

$\mu_1 - \mu_2$ is the presumed value of the difference between the two population means from the null hypothesis,

s_1 and s_2 are the two sample standard deviations, and

n_1 and n_2 are the two sample sizes.

The number of degrees of freedom for the *t*-distribution of the test statistic is given by the smaller of the values $n_1 - 1$ and $n_2 - 1$.

Side Note

Many calculators and statistical software packages use the following alternate formula for the number of degrees of freedom for the *t*-distribution of the test statistic when the population variances are assumed to be unequal.

$$df = \frac{\left(\frac{s_1^2}{n_1} + \frac{s_2^2}{n_2}\right)^2}{\frac{1}{n_1 - 1}\left(\frac{s_1^2}{n_1}\right)^2 + \frac{1}{n_2 - 1}\left(\frac{s_2^2}{n_2}\right)^2}$$

Let's look at an example using this test statistic for a scenario in which the two population variances are assumed to be unequal. In this example, we will first use a rejection region to draw the conclusion, and then show how to find the *p*-value using a TI-83/84 Plus calculator.

Example 11.8

Performing a Hypothesis Test for Two Population Means (Right-Tailed, σ Unknown, Unequal Variances)

Suppose that the Smith CPA firm claims that its clients receive larger tax refunds, on average, than clients of its competitor, Jones and Company CPA, located on the other side of Chicago. To test the claim, 15 clients from the Smith firm are randomly selected and found to have a mean tax refund of \$942 with a standard deviation of \$103. At Jones and Company, a random sample of 18 clients are surveyed and found to have a mean refund of \$898 with a standard deviation of \$95. Test the claim made by the Smith firm at the 0.05 level of significance. Assume that both population distributions are approximately normal.

Solution

Step 1: **State the null and alternative hypotheses.**

First, let the Smith clients be Population 1 and the Jones and Company clients be Population 2. The Smith CPA firm wants to show that Smith clients receive larger tax refunds, on average, than clients of Jones and Company. That is, the mean refund for Smith clients is greater than the mean refund for clients of Jones and Company. Written mathematically, this is $\mu_1 > \mu_2$. Thus, this is the alternative hypothesis. By subtracting μ_2 from both sides of the inequality, we have $\mu_1 - \mu_2 > 0$. Thus, the hypotheses are stated as follows.

$$H_0: \mu_1 - \mu_2 \le 0$$
$$H_a: \mu_1 - \mu_2 > 0$$

Step 2: **Determine which distribution to use for the test statistic, and state the level of significance.**

We are looking at the difference between two population means when both population standard deviations are unknown. Because the two firms are located in different parts of a large city, assume that the variances for the two populations are not equal. We also know that independent random samples were drawn, and both population distributions are approximately normal. Therefore, the t-test statistic for unequal variances is appropriate. We will test the Smith firm's claim at the 0.05 level of significance.

Step 3: **Gather data and calculate the necessary sample statistics.**

The following sample statistics were given in the problem.

$$\bar{x}_1 = 942 \qquad \bar{x}_2 = 898$$
$$s_1 = 103 \qquad s_2 = 95$$
$$n_1 = 15 \qquad n_2 = 18$$

We assume that the population variances are not equal, so we will use the test statistic for unequal variances. Substitute the sample statistics into the formula for the test statistic to obtain the following.

$$t = \frac{(\bar{x}_1 - \bar{x}_2) - (\mu_1 - \mu_2)}{\sqrt{\frac{s_1^2}{n_1} + \frac{s_2^2}{n_2}}} = \frac{(942 - 898) - 0}{\sqrt{\frac{103^2}{15} + \frac{95^2}{18}}} \approx 1.266$$

Step 4: Draw a conclusion and interpret the decision.

In order to draw a conclusion for a right-tailed test using a rejection region, we first need to determine the critical t-value. To do this, we need to know the number of degrees of freedom for the distribution of the t-test statistic. The number of degrees of freedom for populations with unequal variances is the smaller of $n_1 - 1 = 15 - 1 = 14$ and $n_2 - 1 = 18 - 1 = 17$. Thus, the number of degrees of freedom for this problem is 14. The critical t-value for a right-tailed test with 14 degrees of freedom and a 0.05 level of significance is $t_{0.05} = 1.761$. Because the calculated value of the test statistic, $t = 1.266$, is less than the critical value, it does not fall in the rejection region, so we must fail to reject the null hypothesis.

t-Distribution, $df = 14$

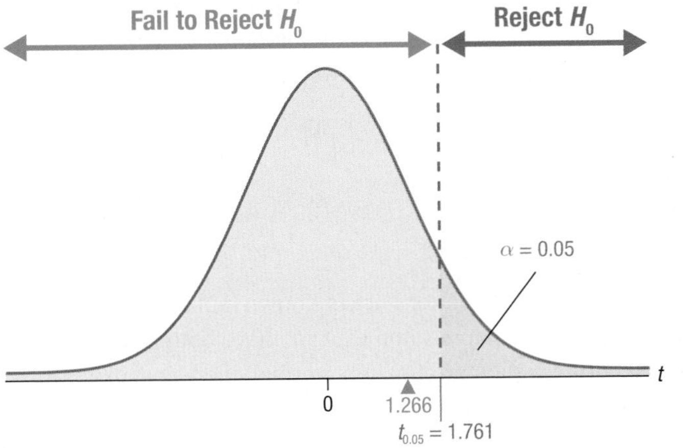

We interpret this conclusion to mean that there is not sufficient evidence at the 0.05 level of significance to support the Smith CPA firm's claim that its clients receive larger tax refunds, on average, than clients of Jones and Company.

To use a TI-83/84 Plus calculator to perform a hypothesis test for two population means where both population variances are unknown and assumed to be unequal, choose STAT, scroll over to TESTS, and select option `4:2-SampTTest`. Choose the `Stats` input option, since we have the sample statistics, and enter the requested values. This is a right-tailed test, so choose >μ2, for the alternative hypothesis. Since the population variances are assumed to be unequal, choose No next to `Pooled`.

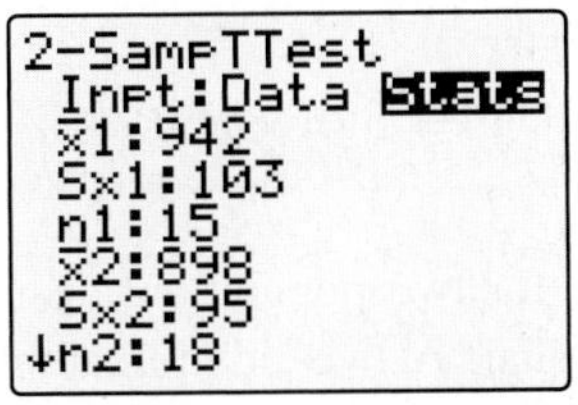

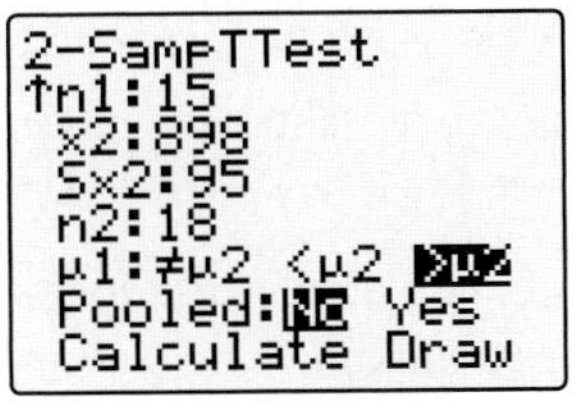

Highlight `Calculate` and press ENTER. The following results are displayed.

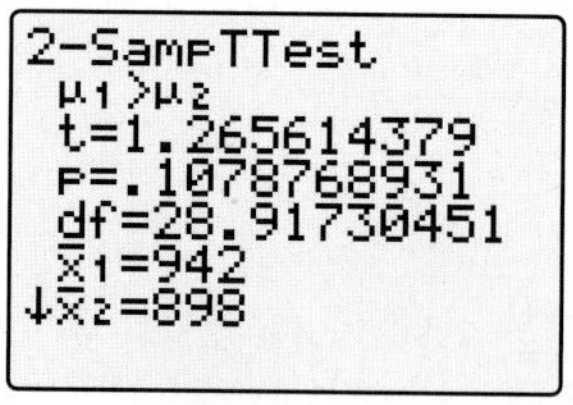

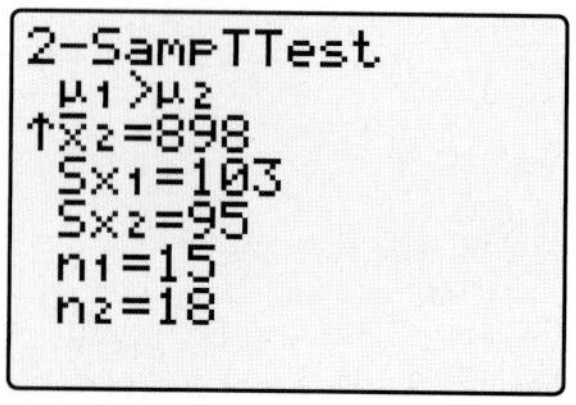

From the calculator output, we see that the p-value ≈ 0.1079. The sample means, sample standard deviations, and sample sizes are also repeated in the results. Since the p-value is greater than the level of significance, $\alpha = 0.05$, our conclusion is to fail to reject the null hypothesis.

Equal Variances

Remember, in this section we are working with populations whose standard deviations are not known. Even though we don't know exactly what the standard deviation is, in some cases we can still assume that it is the same for both populations. For instance, the samples might come from populations with similar characteristics. However, we should be very careful when we assume that the variances are equal. If you are unsure of whether to assume the variances are equal, then you should err on the side of caution and assume that they are not equal.

When we assume that the population variances are equal, the sample variances can be pooled, or combined, to give a weighted estimate for the common population variance. When this is the case, the formula for the t-test statistic changes as follows.

Formula

Test Statistic for a Hypothesis Test for Two Population Means (σ Unknown, Equal Variances)

When both population standard deviations are unknown and assumed to be equal, the samples taken are independent, simple random samples, and either both sample sizes are at least 30 or both population distributions are approximately normal, the test statistic for a hypothesis test for two population means is given by

$$t = \frac{(\bar{x}_1 - \bar{x}_2) - (\mu_1 - \mu_2)}{\sqrt{\frac{(n_1 - 1)s_1^2 + (n_2 - 1)s_2^2}{n_1 + n_2 - 2}}\sqrt{\frac{1}{n_1} + \frac{1}{n_2}}}$$

where $\bar{x}_1$ and $\bar{x}_2$ are the two sample means,

$\mu_1 - \mu_2$ is the presumed value of the difference between the two population means from the null hypothesis,

s_1 and s_2 are the two sample standard deviations, and

n_1 and n_2 are the two sample sizes.

The t-distribution of the test statistic has $n_1 + n_2 - 2$ degrees of freedom.

Now we will look at an example of a hypothesis test for two population means in which the two population variances are unknown but assumed to be equal. As we did in Example 11.8, we will first use a rejection region to draw the conclusion, and then show how to find the p-value using a TI-83/84 Plus calculator.

Example 11.9

Performing a Hypothesis Test for Two Population Means (Left-Tailed, σ Unknown, Equal Variances)

A home improvement warehouse claims that people interested in completing their own house improvement projects can save time by attending its workshops. Specifically, the warehouse claims that people who attend its tiling workshops take less time, on average, to complete comparable projects than those who do not. To test the claim, a randomly selected group of 10 people who attended the workshops is later surveyed about the time it took to finish their tiling projects. People in this group spent a mean of 14.1 hours completing their projects with a standard deviation of 2.3 hours. Another 10 people randomly chosen for the study did not attend the workshops prior to completing their tiling projects, and they spent a mean of 15.0 hours on their projects with a standard deviation of 2.4 hours. Test the home improvement warehouse's claim at the 0.01 level of significance. Assume that the variances of the populations are unknown but assumed to be equal and the distributions of times to complete the tiling projects are approximately normal for both populations.

Solution

Step 1: State the null and alternative hypotheses.

First, let Population 1 be those people who attend the workshops and Population 2 be those who do not attend the workshops before completing their projects. The home improvement warehouse's claim is that people who attend the workshops spend less time on their tiling projects, on average, than those who do not attend the workshops. That is, the mean time to complete a project is lower for people who attend the tiling workshops than for those who do not. When written mathematically, this is $\mu_1 < \mu_2$, and hence the alternative hypothesis. By subtracting μ_2 from both sides of the inequality, we have $\mu_1 - \mu_2 < 0$. Thus, the hypotheses are stated as follows.

$$H_0: \mu_1 - \mu_2 \geq 0$$
$$H_a: \mu_1 - \mu_2 < 0$$

Step 2: Determine which distribution to use for the test statistic, and state the level of significance.

We are looking at the difference between two population means when both population variances are unknown but assumed to be equal. We also know that independent random samples were drawn, and both population distributions are approximately normal. Therefore, the t-test statistic for equal variances is appropriate. The level of significance for this test is $\alpha = 0.01$.

Step 3: Gather data and calculate the necessary sample statistics.

The following sample statistics were given in the problem.

$$\bar{x}_1 = 14.1 \qquad \bar{x}_2 = 15.0$$
$$s_1 = 2.3 \qquad s_2 = 2.4$$
$$n_1 = 10 \qquad n_2 = 10$$

We are told that the population variances are assumed to be equal, so we will use the test statistic for equal variances. Substitute the sample statistics into the formula for the test statistic to obtain the following.

$$t = \frac{(\bar{x}_1 - \bar{x}_2) - (\mu_1 - \mu_2)}{\sqrt{\frac{(n_1 - 1)s_1^2 + (n_2 - 1)s_2^2}{n_1 + n_2 - 2}}\sqrt{\frac{1}{n_1} + \frac{1}{n_2}}}$$

$$= \frac{(14.1 - 15.0) - 0}{\sqrt{\frac{(10-1)2.3^2 + (10-1)2.4^2}{10 + 10 - 2}}\sqrt{\frac{1}{10} + \frac{1}{10}}}$$

$$\approx -0.856$$

Step 4: **Draw a conclusion and interpret the decision.**

Since the alternative hypothesis contains "<," we know that this is a left-tailed test. We need to determine the critical t-value to define the rejection region. To do this, we need to know the number of degrees of freedom for the t-distribution of the test statistic. Since the population variances are assumed to be equal, the number of degrees of freedom for this test is $df = n_1 + n_2 - 2 = 10 + 10 - 2 = 18$. The critical t-value for a left-tailed test with 18 degrees of freedom and a 0.01 level of significance is $-t_{0.01} = -2.552$. Because the calculated value of the test statistic, $t = -0.856$, is greater than the critical value, it does not fall in the rejection region, so we must fail to reject the null hypothesis.

***t*-Distribution, *df* = 18**

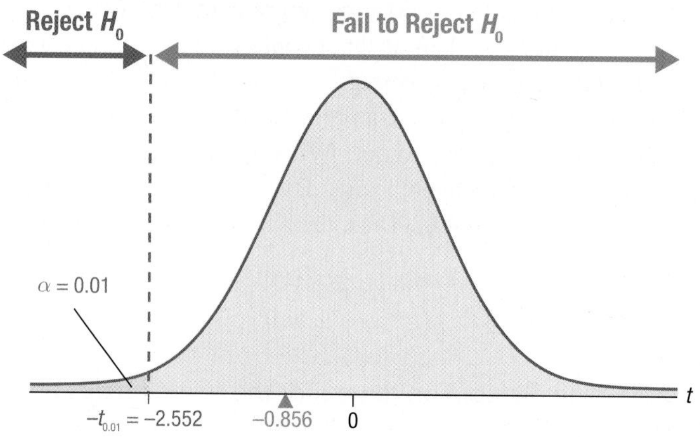

We interpret this conclusion to mean that there is not sufficient evidence at the 0.01 level to say that people who complete the tiling workshops take less time, on average, to complete their projects than those who do not attend the workshops.

The instructions for performing this test, using pooled sample variances, on a TI-83/84 Plus calculator are almost identical as in the last example where the sample variances were not pooled. Choose STAT, scroll over to TESTS, and select option 4:2-SampTTest. Use the Stats input option, since the sample statistics are given in the problem, and enter the requested values. This is a left-tailed test, so choose <μ2, for the alternative hypothesis. Since the population variances are assumed to be equal, choose Yes next to Pooled.

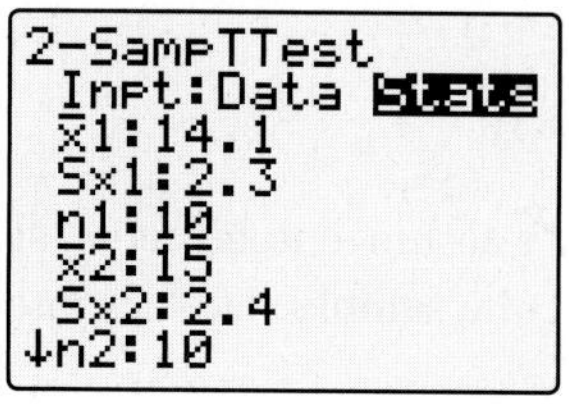

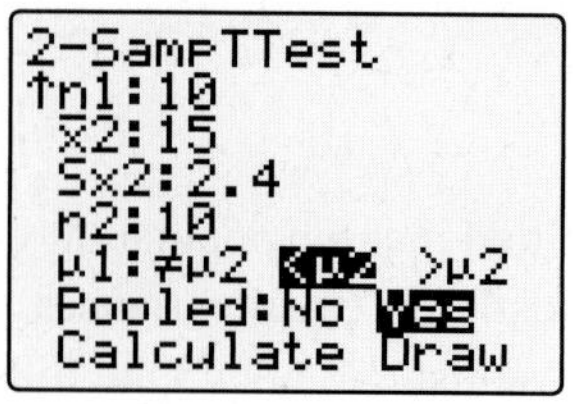

Highlight `Calculate`, and press ENTER. The following results are displayed.

```
2-SampTTest
 μ1<μ2
 t=-.8561726895
 p=.2015782063
 df=18
 x̄1=14.1
↓x̄2=15
```

```
2-SampTTest
 μ1<μ2
↑Sx1=2.3
 Sx2=2.4
 Sxp=2.35053185
 n1=10
 n2=10
```

From the calculator output, we see that the p-value ≈ 0.2016. Since the p-value is greater than the significance level, $\alpha = 0.01$, our conclusion is to fail to reject the null hypothesis.

11.2 Section Exercises

Test Statistics and Degrees of Freedom for Hypothesis Tests for Two Population Means (σ Unknown)

Directions: Calculate the test statistic and determine the number of degrees of freedom for a hypothesis test for two population means using the given information. Assume that both population distributions are approximately normal.

1. $\bar{x}_1 = 93.0, \quad s_1 = 10.4, \quad n_1 = 21, \quad \bar{x}_2 = 89.2, \quad s_2 = 9.5, \quad n_2 = 18, \quad H_0\colon \mu_1 - \mu_2 = 0$
Assume that the population variances are not the same.

2. $\bar{x}_1 = 3.4, \quad s_1 = 0.3, \quad n_1 = 5, \quad \bar{x}_2 = 3.7, \quad s_2 = 0.5, \quad n_2 = 7, \quad H_0\colon \mu_1 - \mu_2 \le 0$
Assume that the population variances are not the same.

3. $\bar{x}_1 = 33.5, \quad s_1 = 2.1, \quad n_1 = 14, \quad \bar{x}_2 = 31.1, \quad s_2 = 2.8, \quad n_2 = 11, \quad H_a\colon \mu_1 - \mu_2 > 0$
Assume that the population variances are equal.

4. $\bar{x}_1 = 24.1, \quad s_1 = 1.3, \quad n_1 = 19, \quad \bar{x}_2 = 23.0, \quad s_2 = 1.1, \quad n_2 = 22, \quad H_0\colon \mu_1 - \mu_2 = 0$
Assume that the population variances are equal.

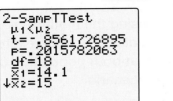

Hypothesis Tests for Two Population Means (σ Unknown)

Directions: **Perform each hypothesis test using the method of your choice or the one assigned by your instructor. For each exercise, complete the following steps. Assume that both population distributions are approximately normal in each scenario.**

a. State the null and alternative hypotheses.

b. Determine which distribution to use for the test statistic, and state the level of significance.

c. Calculate the test statistic.

d. Draw a conclusion and interpret the decision.

5. A professor is concerned that the two sections of college algebra that he teaches are not performing at the same level. To test his claim, he looks at the mean exam score for a random sample of students from each of his classes. In Class 1, the mean exam score for 12 students is 78.7 with a standard deviation of 6.5. In Class 2, the mean exam score for 15 students is 81.1 with a standard deviation of 7.4. Test the professor's claim at the 0.05 level of significance. Assume that the population variances are equal.

6. A manufacturer fills soda bottles. Periodically the company tests to see if there is a difference between the mean amounts of soda put in bottles of regular cola and diet cola. A random sample of 14 bottles of regular cola has a mean of 501.6 mL of soda with a standard deviation of 3.9 mL. A random sample of 16 bottles of diet cola has a mean of 498.9 mL of soda with a standard deviation of 5.3 mL. Test the claim that there is a difference between the mean fill levels for the two types of soda using a 0.01 level of significance. Assume that the population variances are not equal since different machines are used to fill bottles of regular cola and diet cola.

7. While shopping for a cookout, Ian notices that a particular brand of charcoal briquettes claims to burn longer because the briquettes are 60% thicker than the competitor's briquettes. Feeling tired of being taken for a shopper who just believes what the manufacturer wants him to believe, Ian decides to test the manufacturer's claim that the charcoal briquettes from Brand A are thicker than those from Brand B. He buys a bag of each kind of charcoal, randomly selects a few briquettes from each brand, and measures the thicknesses of the briquettes in his samples. His findings are given in the following table. Test the manufacturer's claim that the charcoal briquettes from Brand A are thicker than those from Brand B at the 0.01 level of significance. Assume that the population variances are different.

Thickness of Charcoal Briquettes (in cm)

	Brand A	Brand B
Sample Size	8	6
Mean Thickness	3.21	2.13
Standard Deviation	0.50	0.85

8. A new small business wants to know if its current radio advertising is effective. The owners decide to look at the mean number of customers who make a purchase in the store on days immediately following days when the radio ads are played as compared to the mean for those days following days when no radio advertisements are played. They found that for 11 days following no advertisements, the mean was 17.8 purchasing customers with a standard deviation of 3.5 customers. On 6 days following advertising, the mean was 22.8 purchasing customers with a standard deviation of 2.8 customers. Test the claim, at the 0.01 level, that the mean number of customers who make a purchase in the store is lower for days following no advertising compared to days following advertising. Assume that the population variances are equal.

9. Gary has discovered a new painting tool to help him in his work. If he can prove to himself that the painting tool reduces the amount of time it takes to paint a room, he has decided to invest in a tool for each of his helpers as well. From records of recent painting jobs that he completed before he got the new tool, Gary collected data for a random sample of 6 medium-sized rooms. He determined that the mean amount of time that it took him to paint each room was 4.2 hours with a standard deviation of 0.5 hours. For a random sample of 4 medium-sized rooms that he painted using the new tool, he found that it took him a mean of 3.9 hours to paint each room with a standard deviation of 0.7 hours. At the 0.10 level, can Gary conclude that his mean time for painting a medium-sized room without using the tool was greater than his mean time when using the tool? Assume that the population variances are equal.

10. Joyce is trying to convince her husband that, on average, she spends more money at the grocery store when she takes their children with her to shop, in hopes that she can make the trips alone in the future. She shows him that on 4 randomly selected trips with the children she spent a mean of \$122.56 with a standard deviation of \$13.12. But, on 5 trips without the children, she only spent a mean of \$108.31 with a standard deviation of \$17.06. Assume that the population variances are different. At the 0.10 level, can Joyce convince her husband that it's cheaper, on average, to not take her children with her to the grocery store?

11. While running some quality control tests, a manager at a factory that makes potato chips noticed a difference in the mean bag weights for chips coming from two different production lines. To see if the bags in Line A did actually have a mean weight lower than those in Line B, he randomly selected some of the bags from each line. His results are given in the following table. If he assumes that the population variances for the two lines are different, can he conclude at the 0.10 level that the mean weight of bags from Line A is lower than the mean weight of bags from Line B?

Weights of Bags of Chips (in g)

	Line A	Line B
Sample Size	20	15
Mean Weight	309.63	311.87
Standard Deviation	15.91	13.21

12. A physician wants to test the claim that the average adult height of premature baby boys is different from that of full-term baby boys. To do this, he finds a random sample of 18 men who were born prematurely and calculates that their mean height is 68.1 inches with a standard deviation of 2.3 inches. He also finds a random sample of 20 men who were carried full term and calculates their mean height to be 68.9 inches with a standard deviation of 2.0 inches. Assume that the population standard deviations are unequal, and test the claim at the 0.05 level of significance.

13. A pharmaceutical company needs to know if its new cholesterol drug, Praxor, is effective at lowering cholesterol levels. It believes that people who take Praxor will average a greater decrease in cholesterol level than people taking a placebo. After the experiment is complete, the researchers find that the 25 participants in the treatment group lowered their cholesterol levels by a mean of 23.5 points with a standard deviation of 5.8 points. The 25 participants in the control group lowered their cholesterol levels by a mean of 18.9 points with a standard deviation of 4.1 points. Assume that the population variances are not equal, and test the company's claim at the 0.01 level.

14. A speech pathology professor believes from experience that, on average, boys begin talking at a later age than girls. To test her theory, she gathers information from the parents of random samples of 12 boys and 14 girls. The boys began talking at a mean of 1.33 years of age with a standard deviation of 0.15 years. The girls began talking at a mean of 1.23 years of age with a standard deviation of 0.12 years. Assume that the population standard deviations are equal, and test the professor's claim at the 0.05 level of significance.

15. Insurance Company A claims that its customers pay less for car insurance, on average, than customers of its competitor, Company B. You wonder if this is true, so you decide to compare the average monthly costs of similar insurance policies from the two companies. For a random sample of 9 people who buy insurance from Company A, the mean cost is $152 per month with a standard deviation of $17. For 11 randomly selected customers of Company B, you find that they pay a mean of $155 per month with a standard deviation of $14. Assume that the population variances are equal, and test Company A's claim at the 0.01 level of significance.

16. A women's group believes that women are offered lower starting salaries than men applying for similar jobs. To test this claim, the group sends 10 women and 10 men to interviews for similar positions at various companies. The women were offered a mean starting salary of $29,500 with a standard deviation of $1100. The men were offered a mean starting salary of $30,500 with a standard deviation of $950. Assume that the population standard deviations are different, and test the group's claim at the 0.05 level of significance.

17. A psychologist wants to test the claim that the mean lengths of time spent to complete a sculpture are different for male and female artists. He randomly chooses 12 men and 11 women and asks them to estimate the amount of time it takes for them to complete a sculpture. According to the survey, the men spent a mean of 4.6 hours to complete a sculpture with a standard deviation of 1.3 hours. The women spent a mean of 5.4 hours to complete a sculpture with a standard deviation of 1.1 hours. Assume that the population variances are equal, and test the claim at the 0.05 level of significance.

11.3 Hypothesis Testing: Two Population Means (σ Unknown, Dependent Samples)

So far in this chapter, we have looked at hypothesis testing for two population means using data from independent samples. In this section, we will turn our attention to situations in which the samples are *dependent*. Recall from Section 9.3 that dependent data are connected in a specific manner, or *paired*. In other words, observations in one sample are matched directly to the observations in the second sample.

When performing a hypothesis test involving two population means using dependent samples, the first step is to calculate the paired difference for each pair of data values, which is given by the following formula.

Formula

Paired Difference

When two dependent samples consist of paired data, the paired difference for any pair of data values is given by

$$d = x_2 - x_1$$

where x_2 is a data value from the second sample and

x_1 is the data value from the first sample that is paired with x_2.

In this section, we will consider hypothesis tests involving two population means when the following conditions are met.

- All possible samples of a given size have an equal probability of being chosen; that is, simple random samples are used.
- The samples are dependent.
- Both population standard deviations, σ_1 and σ_2 are *unknown*.
- Either the number of pairs of data values in the sample data is at least 30 ($n \geq 30$) *or* the population distribution of the paired differences is approximately normal.

When these criteria are met, the test statistic will have a Student's *t*-distribution. In this textbook, you may assume that these conditions are met for all examples and exercises involving paired data.

The hypotheses for tests involving dependent samples are based on the *population mean of the paired differences*, μ_d, rather than the difference between the two population means, $\mu_1 - \mu_2$, like we used for hypothesis tests involving independent samples.

When calculating the paired differences for two dependent samples, the results are dependent on which population is designated as Population 1 and which is considered Population 2. For example, suppose 5 and 8 are paired data values where we have designated the population from which 5 was drawn as Population 1 and the population from which 8 was drawn as Population 2. Then the paired difference for these data values is $d = x_2 - x_1 = 8 - 5 = 3$. However, if we say that 5 is in the sample from Population 2 and 8 is in the sample from Population 1, then the paired difference is $d = x_2 - x_1 = 5 - 8 = -3$. So, for consistency, we need to have a standard way of deciding which population will be designated as Population 2. Let's impose the rule that in situations where measurements are taken before and after some treatment, each measurement that is taken after the treatment is contained in Population 2, and thus it will always be the first value in the subtraction. Therefore, if we expect the treatment to reduce the value of the measurement, then we would expect to get more negative differences than positive differences. On the other hand, if we expect the treatment to increase the value of the measurement, we would expect to get more positive differences than negative differences. This procedure for how to designate Population 1 and Population 2 should help us correctly set up our hypotheses. Let's look at an example of how to write the null and alternative hypotheses for a test involving paired differences.

Example 11.10

Determining the Null and Alternative Hypotheses for a Left-Tailed Test Using Dependent Samples

The manufacturer of a new diet pill claims that the pill helps a person lose, on average, more than 7 pounds in the first week. To test this claim, 15 people volunteer to take the diet pill for one week. Their weights are measured at the beginning and end of the week. State the null and alternative hypotheses for this hypothesis test.

Solution

For this example, Population 1 will be the starting weights and Population 2 will be the ending weights for people who take the new diet pill for one week. To calculate the paired differences for the population data, we would subtract each person's starting weight from his or her ending weight. If the diet pill helps people lose weight as intended, then the ending weights will be less than the starting weights and the paired differences will be negative values. The manufacturer's claim is that the mean amount of weight lost will be more than 7 pounds. Thus, the claim is that the population mean of the paired differences will be a negative value less than −7. Mathematically, this claim is written as $\mu_d < -7$. Since this is what the manufacturer seeks to gain evidence for, it is the alternative hypothesis. The mathematical opposite of this is $\mu_d \geq -7$. Therefore, the null and alternative hypotheses are as follows.

$$H_0: \mu_d \geq -7$$
$$H_a: \mu_d < -7$$

Memory Booster

When subtracting the before-treatment value from the after-treatment value, a *reduction* in the value would result in a negative number for the difference. This means that a "reduction of more than" would correspond to < and a "reduction of less than" would correspond to > in the alternative hypothesis.

Now that we know how to state the hypotheses, the next step is to determine the test statistic that will be used.

Formula

Test Statistic for a Hypothesis Test for the Mean of the Paired Differences for Two Populations (σ Unknown, Dependent Samples)

When both population standard deviations are unknown, the samples taken are dependent, simple random samples of paired data, and either the number of pairs of data values in the sample data is at least 30 or the population distribution of the paired differences is approximately normal, the test statistic for a hypothesis test for the mean of the paired differences for two populations is given by

$$t = \frac{\bar{d} - \mu_d}{\left(\frac{s_d}{\sqrt{n}}\right)}$$

where $\bar{d}$ is the mean of the paired differences for the sample data,

μ_d is the presumed value of the mean of the paired differences for the population data from the null hypothesis,

s_d is the sample standard deviation of the paired differences for the sample data, and

n is the number of paired differences in the sample data.

The t-distribution of the test statistic has $n - 1$ degrees of freedom.

Memory Booster

Recall that $\bar{d}$, the mean of the paired differences for the sample data, is given by the following formula.

$$\bar{d} = \frac{\sum d_i}{n}$$

Memory Booster

Recall that s_d, the sample standard deviation of the paired differences for the sample data, is given by the following formula.

$$s_d = \sqrt{\frac{\sum\left(d_i - \bar{d}\right)^2}{n-1}}$$

When drawing conclusions based on rejection regions, recall that the rejection region is based on the type of hypothesis test we are performing and the level of significance, α. Notice that these are the same decision rules that are used in other t-tests involving means of populations with unknown standard deviations, as we saw in Sections 10.3 and 11.2.

Procedure

Rejection Regions for Hypothesis Tests for the Mean of the Paired Differences for Two Populations (σ Unknown, Dependent Samples)

Reject the null hypothesis, H_0, if:

$t \le -t_\alpha$ for a left-tailed test

$t \ge t_\alpha$ for a right-tailed test

$|t| \ge t_{\alpha/2}$ for a two-tailed test

Example 11.11 illustrates a complete hypothesis test for the mean of the paired differences for two populations of dependent data using a rejection region to draw the conclusion, and Example 11.12 uses a TI-83/84 Plus calculator to draw a conclusion using the p-value.

Example 11.11

Using Rejection Regions in a Hypothesis Test for the Mean of the Paired Differences for Two Populations (Right-Tailed, σ Unknown, Dependent Samples)

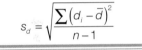

The standard course at a local defensive-driving school includes several films depicting violent car crashes and graphic pictures of injuries sustained in these crashes. The driving school has shown these videos for many years, believing that they reduce the students' average speeds on the highway. A group of concerned citizens, who feel that these videos are very disturbing, is not convinced that the videos reduce highway speeds enough to make a significant difference in highway safety. In fact, the group claims that these videos reduce a person's speed on the highway by less than 5 miles per hour, on average. To test the claim, the citizens install electronic data recorders (EDRs) on the vehicles of 10 volunteers, who agree to drive as they would normally for two weeks while the EDR records their vehicles' speeds. After the initial driving period, each volunteer watches the videos. Then the EDRs again record their vehicles' speeds for another two weeks. The following table contains the average highway speeds for each volunteer for the two-week periods before and after watching the videos. Use these data to test the concerned citizens' claim that these videos reduce a person's speed on the highway by less than 5 miles per hour, on average. Use a 0.05 level of significance.

Average Highway Speeds (in Miles per Hour)	
Before	**After**
75.83	72.13
80.12	73.87
65.41	66.09
70.03	68.43
73.91	71.45
76.02	73.67
75.10	70.19
67.89	65.34
81.12	75.31
77.67	70.92

Solution

Step 1: State the null and alternative hypotheses.

Let the average highway speeds after viewing the videos be Population 2 and the speeds before viewing the videos be Population 1. We want to subtract the average speed before viewing the videos from the average speed after viewing the videos to calculate each paired difference. The concerned citizens' claim is that viewing these videos reduces a person's speed on the highway by less than 5 miles per hour, on average. Thus, the claim is that the mean of the paired differences for the population data is greater than –5. (Note that a difference of –5 would represent a reduction of 5 mph.) Therefore, written symbolically, the citizens' claim is $\mu_d > -5$. Because the citizens are hoping to gather evidence that shows that the mean difference is greater than –5, this is the alternative hypothesis. The mathematical opposite of this is $\mu_d \le -5$. Thus, the hypotheses are stated as follows.

$$H_0: \mu_d \le -5$$
$$H_a: \mu_d > -5$$

Step 2: Determine which distribution to use for the test statistic, and state the level of significance.

Since we are performing a hypothesis test for the mean of the paired differences for the population data, and the samples are dependent samples of paired data, the test statistic will have a Student's t-distribution. The level of significance is given in the problem as $\alpha = 0.05$.

Step 3: **Gather data and calculate the necessary sample statistics.**

Since we were given raw data, we need to begin by calculating the paired differences, as well as the mean and sample standard deviation of these differences. The differences are listed in the following table.

Average Highway Speeds (in Miles per Hour)		
Before, x_1	**After, x_2**	**Paired Difference, $d = x_2 - x_1$**
75.83	72.13	–3.70
80.12	73.87	–6.25
65.41	66.09	0.68
70.03	68.43	–1.60
73.91	71.45	–2.46
76.02	73.67	–2.35
75.10	70.19	–4.91
67.89	65.34	–2.55
81.12	75.31	–5.81
77.67	70.92	–6.75

The mean of these paired differences is $\bar{d} = -3.57$ and the sample standard deviation, s_d, is approximately 2.352993. Substituting these values into the formula for the test statistic, we obtain the following.

$$t = \frac{\bar{d} - \mu_d}{\left(\frac{s_d}{\sqrt{n}}\right)} = \frac{-3.57 - (-5)}{\left(\frac{2.352993}{\sqrt{10}}\right)} \approx 1.922$$

Note that, in the previous formula, $\mu_d = -5$ because the hypothesized mean of the paired differences for the population data is –5. This is indicated in the statement of the null hypothesis, H_0.

Step 4: **Draw a conclusion and interpret the decision.**

For the rejection region, we have $df = n - 1 = 10 - 1 = 9$, which gives a critical value of $t_{0.05} = 1.833$. Since this is a right-tailed test, we will reject the null hypothesis if $t \geq 1.833$. Since the test statistic, $t \approx 1.922$, is greater than the critical value, it falls in the rejection region and we reject the null hypothesis.

Rounding Rule

When calculations involve several steps, avoid rounding at intermediate calculations. If necessary, round intermediate calculations to at least six decimal places to avoid additional rounding errors in subsequent calculations.

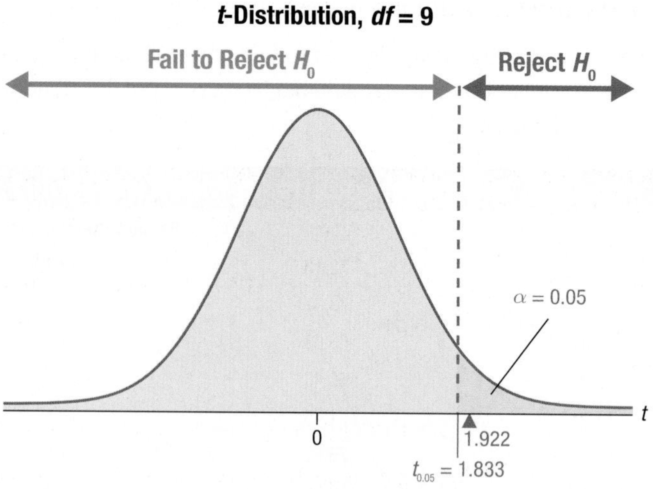

Thus, there is enough evidence, at the 0.05 level of significance, to say that, after watching the videos, the mean reduction in drivers' highway speeds is less than 5 miles per hour. That is, the evidence supports the concerned citizens' claim that watching these videos reduces a person's speed on the highway by less than 5 miles per hour, on average.

Example 11.12

Performing a Hypothesis Test for the Mean of the Paired Differences for Two Populations Using a TI-83/84 Plus Calculator (Two-Tailed, σ Unknown, Dependent Samples)

Dr. Xiong, a clinical psychologist, wishes to test the claim that there is a significant difference in a person's adult weight if he is raised by his father instead of his mother. Luckily, Dr. Xiong knows of five sets of identical twin boys who were raised separately, one twin by the mother and one twin by the father, and who are willing to participate in a study to help her test her claim. Each twin is weighed and identified as having been raised by his mother or his father. The following table lists the results. Do these data support Dr. Xiong's claim at the 0.01 level of significance?

Weights of Twins (in Pounds)					
Twin Raised by Father	143.67	235.91	156.34	187.21	129.81
Twin Raised by Mother	134.81	221.37	163.92	193.45	131.38

Solution

Step 1: State the null and alternative hypotheses.

In this example we are not comparing values before and after a treatment is applied, so it does not matter how we label Population 1 and Population 2. Let's identify the twins raised by their fathers as Population 1 and the twins raised by their mothers as Population 2. Dr. Xiong claims that there is a significant difference in a person's weight if he is raised by his father instead of his mother. If there is a significant difference in weight, then the mean of the paired differences is not zero. Written symbolically, the claim is $\mu_d \neq 0$. As this is what Dr. Xiong wishes to show with the data, it is the alternative hypothesis. The mathematical opposite of this is $\mu_d = 0$.

Thus, the hypotheses are stated as follows.

$$H_0\colon \mu_d = 0$$
$$H_a\colon \mu_d \neq 0$$

Step 2: **Determine which distribution to use for the test statistic, and state the level of significance.**

A Student's *t*-distribution, and thus the *t*-test statistic, is appropriate to use in this case because the claim is about the mean of the paired differences for the population data, and the samples are dependent samples of paired data. The level of significance is stated in the problem to be $\alpha = 0.01$.

Step 3: **Gather data and calculate the necessary sample statistics.**

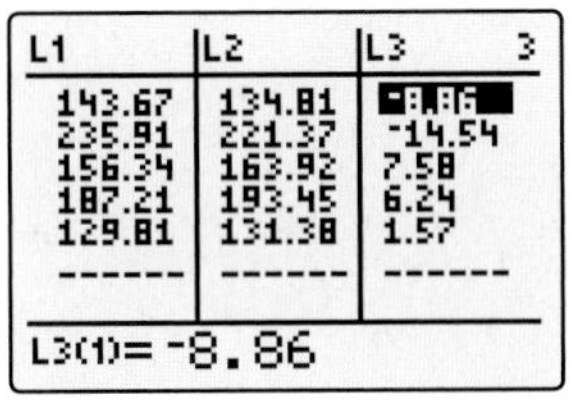

For this example, we are going to use the *p*-value to make our decision, so we will be using a TI-83/84 Plus calculator to perform the hypothesis test. Since we were given raw data, we need to begin by entering the data into the lists on the calculator. Enter the weights of the twins raised by their fathers in L1 and the weights of the twins raised by their mothers in L2. Next, we want to calculate the paired differences in L3. To do so, highlight L3 and enter the formula to subtract the values in L1 from the values in L2 (L2−L1) by pressing 2ND 2 − 2ND 1 and then ENTER.

We now have the paired differences in L3 and can perform a one-sample *t*-test using the paired differences as our raw data. Press STAT, scroll to TESTS, and choose option 2:T-Test. We want to perform the hypothesis test using raw data, so choose the Data option. The claim is that the mean of the paired differences is not zero, so enter Ø for μo. The data are in List 3, so enter L3 for List by pressing 2ND and then 3. The frequency of the data (Freq) is the default value, which is 1. We have a two-tailed test, so choose ≠μo for the alternative hypothesis. The menu should then appear as it does in the following screenshot on the left. Choosing Calculate produces the results shown in the screenshot on the right.

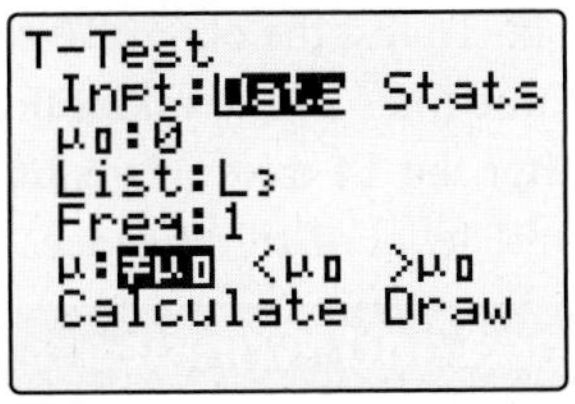

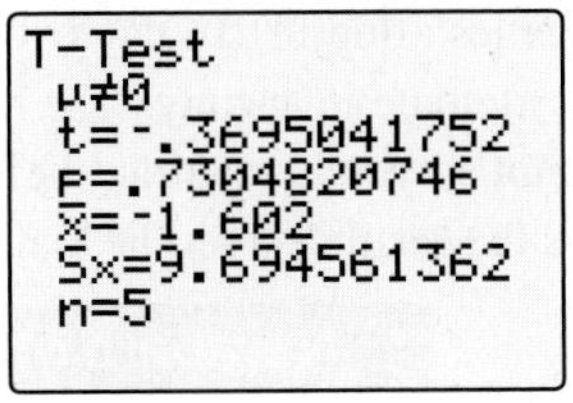

Step 4: **Draw a conclusion and interpret the decision.**

The *p*-value given by the calculator is approximately 0.7305. Since $0.7305 > 0.01$, we have *p*-value $> \alpha$, so the conclusion is to fail to reject the null hypothesis. Thus, there is not sufficient evidence at the 0.01 level of significance to say that the mean difference between the weights of male identical twins is not 0 for twins who are raised separately, one raised by each parent. That is, the evidence does not support Dr. Xiong's claim that there is a significant difference in a person's weight if he is raised by his father instead of his mother.

11.3 Section Exercises

Hypothesis Tests for the Mean of the Paired Differences for Two Populations (σ Unknown, Dependent Samples)

Directions: Perform each hypothesis test using the method of your choice or the one assigned by your instructor. For each exercise, complete the following steps.

a. State the null and alternative hypotheses.

b. Determine which distribution to use for the test statistic, and state the level of significance.

c. Calculate the necessary sample statistics then compute the test statistic.

d. Draw a conclusion and interpret the decision.

1. An anger-management course claims that, after completing its seminar, participants will lose their tempers less often. Always a skeptic, you decide to test this claim. A random sample of 12 seminar participants is chosen, and these participants are asked to record the number of times that they lost their tempers in the two weeks prior to the course. After the course is over, the same participants are asked to record the number of times that they lost their tempers in the next two weeks. The following table lists the results of the survey. Using these data, test the claim at the 0.05 level of significance.

Number of Times Temper Was Lost during a Two-Week Period												
Before	8	10	6	7	4	11	12	5	6	3	6	4
After	6	5	6	6	5	9	4	5	4	4	5	4

2. The manufacturer of a new eye cream claims that the cream reduces the appearance of fine lines and wrinkles after just 14 days of application. To test the claim, 10 women are randomly selected to participate in a study. The number of fine lines and wrinkles that are visible around each participant's eyes is recorded before and after the 14 days of treatment. The following table displays the results. Test the claim at the 0.01 level of significance.

Numbers of Fine Lines and Wrinkles										
Before	8	14	13	15	10	16	9	10	11	10
After	6	14	11	14	10	14	9	9	11	8

3. An SAT prep course claims to increase student scores by more than 60 points, on average. To test this claim, 9 students who have previously taken the SAT are randomly chosen to take the prep course. Their SAT scores before and after completing the prep course are listed in the following table. Test the claim at the 0.01 level of significance.

SAT Scores									
Before Prep Course	1010	980	1170	1200	1040	1280	1450	1470	1500
After Prep Course	1100	1260	1190	1280	1170	1370	1440	1500	1520

4. Sarah believes that completely cutting caffeine out of a person's diet will allow him or her more restful sleep at night. In fact, she believes that, on average, adults will have more than two additional nights of restful sleep in a four-week period after removing caffeine from their diets. She randomly selects 8 adults to help her test this theory. Each person is asked to consume two caffeinated beverages per day for 28 days, and then cut back to no caffeinated beverages for the following 28 days. During each period, the participants record the numbers of nights of restful sleep that they had. The following table gives the results of the study. Test Sarah's claim at the 0.05 level of significance.

Numbers of Nights of Restful Sleep in a Four-Week Period								
With Caffeine	21	20	26	20	24	21	18	15
Without Caffeine	22	24	27	23	21	26	22	23

5. To test the claim that children with the same parents do not have the same weights, some students in an upper-level statistics class surveyed nine families with at least two boys in the family having the same parents. The boys' weights, taken at the same age, are listed in the following table. Is there enough evidence, at the 0.10 level of significance, to support the claim that, on average, two boys with the same parents do not have equal weights?

Boys' Weights (in Pounds)									
Boy A	94	138	171	131	159	110	148	90	170
Boy B	93	140	176	130	173	112	145	90	178

6. One of the top golf camps in the country advertises that a week with its coaches will lower your average golf score by two points. One disgruntled customer claims that the camp does not live up to its advertised claim. To test the customer's claim, eight randomly selected men attending the camp agreed to participate in a study, and their pre-camp and post-camp average scores are listed in the following table. (These are their average scores on a par-72 golf course.) Test the customer's claim at the 0.10 level of significance.

Average Golf Scores								
Before Camp	75	74	76	75	76	76	74	77
After Camp	73	72	73	74	74	72	74	75

7. A violin teacher wants to convince parents of 5-year-olds that, after a year of lessons, the students will increase their stamina for standing in the correct position by more than 30 minutes, on average. She recorded the length of time that each student could hold the correct position while practicing at the beginning of the year, and then again at the end of the year. Her results are listed in the following table. Is there evidence, at the 0.10 level of significance, that the 5-year-olds' stamina times increase by a mean of more than 30 minutes after one year of violin lessons?

Stamina for Correct Position (in Minutes)								
Start of Lessons	5	7	5	4	10	6	8	5
After a Year of Lessons	36	39	40	35	41	37	38	37

8. A psychology graduate student wants to test the claim that there is a significant difference between the IQs of husbands and their wives. To test this claim, she measures the IQs of 9 married couples using a standard IQ test. The results of the IQ tests are listed in the following table. Using a 0.05 level of significance, test the claim that there is a significant difference between the IQs of husbands and their wives.

IQs of Married Couples

Husband	100	110	132	120	90	115	124	121	107
Wife	98	111	134	119	95	116	122	118	110

9. A local school district is looking at adopting a new textbook that, according to the publishers, will increase standardized test scores of second graders by more than 10 points, on average. Never willing to believe a publisher's claim without evidence to support it, the school board decides to test the claim. The school board chooses two second-grade classes for the study. One class was assigned the new textbook and the other class used the traditional textbook. Eight children from each class were then paired based on demographics and ability levels. The following table lists the standardized test scores for the pairs. Do the data support the company's claim at the 0.05 level of significance?

Standardized Test Scores of Second Graders

New Book	78	82	90	67	79	83	89	93
Old Book	67	70	79	54	68	71	78	82

10. An economist studying inflation in electricity prices in 2009 and 2010 believes that the average price of electricity, even after adjusting for inflation, changed between these two years. To test his claim, he samples 9 different counties and records the average price of electricity in each county from each year. He then adjusts the prices for inflation. His results are given in the following table. Test the economist's claim at the 0.01 level of significance.

Average Residential Retail Prices of Electricity ($/kWh)

2009	2010
20.37	19.30
15.44	15.48
17.45	15.29
16.47	16.06
15.73	15.99
14.82	15.36
16.63	16.55
17.66	18.63
11.67	12.79

11.4 Hypothesis Testing: Two Population Proportions

Let's now turn our attention to hypothesis tests that examine whether two population proportions are significantly different. In this section, we will restrict our discussion to hypothesis testing for two population proportions when the following conditions are met. Notice that the conditions are similar to those discussed in Section 10.4 for a hypothesis test for a single population proportion.

- All possible samples of a given size have an equal probability of being chosen; that is, simple random samples are used.
- The samples are independent.
- The conditions for a binomial distribution are met for both samples.
- The sample sizes are large enough to ensure that $n_1\hat{p}_1 \ge 5$, $n_1\left(1-\hat{p}_1\right) \ge 5$, $n_2\hat{p}_2 \ge 5$, and $n_2\left(1-\hat{p}_2\right) \ge 5$.

When these conditions are met, we can apply the Central Limit Theorem to the sampling distribution of the differences between the sample proportions for two independent samples. This means that the test statistic for a hypothesis test for two population proportions is a z-score, which has a standard normal distribution. You can assume that the necessary criteria are met for all examples and exercises in this chapter.

The steps in the hypothesis test are the same as in the previous sections. However, there is a new format required for the hypotheses, and we need to learn the new formula for the test statistic. When performing a hypothesis test for two population proportions using the method presented in this section, the null hypothesis must contain the assertion that the two population proportions are equal. We will write our null and alternative hypotheses by comparing the difference between the two population parameters to zero. For example:

$$H_0: p_1 - p_2 \ge 0$$
$$H_a: p_1 - p_2 < 0$$

Since H_0 always contains the assertion that the two population proportions are equal, we can pool the data from both samples to give a **weighted estimate of the common population proportion**, which we denote as $\bar{p}$. We will need this estimate to calculate the test statistic in these hypothesis tests.

Memory Booster

Properties of a Binomial Distribution

1. The experiment consists of a fixed number, n, of identical trials.
2. Each trial is independent of the others.
3. For each trial, there are only two possible outcomes. For counting purposes, one outcome is labeled a success, and the other a failure.
4. For every trial, the probability of getting a success is p. The probability of getting a failure is then $1 - p$.
5. The binomial random variable, X, counts the number of successes in n trials.

Formula

Test Statistic for a Hypothesis Test for Two Population Proportions

When the samples taken are independent, simple random samples, the conditions for a binomial distribution are met for both samples, and the sample sizes are large enough to ensure that $n_1\hat{p}_1 \ge 5$, $n_1\left(1-\hat{p}_1\right) \ge 5$, $n_2\hat{p}_2 \ge 5$, and $n_2\left(1-\hat{p}_2\right) \ge 5$, the test statistic for a hypothesis test for two population proportions is given by

$$z = \frac{\left(\hat{p}_1 - \hat{p}_2\right) - \left(p_1 - p_2\right)}{\sqrt{\bar{p}\left(1-\bar{p}\right)\left(\frac{1}{n_1}+\frac{1}{n_2}\right)}}$$

where $\hat{p}_1$ and $\hat{p}_2$ are the two sample proportions,

$p_1 - p_2$ is the presumed value of the difference between the two population proportions from the null hypothesis (thus, $p_1 - p_2 = 0$),

$\bar{p}$ is a weighted estimate of the common population proportion, $\bar{p} = \frac{x_1 + x_2}{n_1 + n_2}$,

x_1 and x_2 are the numbers of successes that occur in the two samples, and

n_1 and n_2 are the two sample sizes.

Memory Booster

x_i is the number of successes that occur in the ith sample.

We can use either rejection regions or the p-value method, as discussed earlier, to determine the conclusion of a hypothesis test for two population proportions. As in previous sections, we will continue to use the p-value method. The following examples illustrate how to use this new test statistic. Example 11.13 demonstrates how to calculate the test statistic by hand and Example 11.14 shows how to use a TI-83/84 Plus calculator to perform the test.

Example 11.13

Performing a Hypothesis Test for Two Population Proportions (Left-Tailed)

The mayor's chief of staff thinks that a local newspaper article has changed the way that the community thinks about the mayor. To test his theory, he finds a poll of the mayor's approval rating that was taken before the article came out and compares it to the mayor's approval rating after the article. Before the article ran in the paper 480 out of a simple random sample of 1200 voters thought the mayor was trustworthy. After the article, 550 out of a simple random sample of 1180 voters thought he was trustworthy. Based on these data, use a 5% level of significance to test the chief of staff's claim that the mayor's approval rating has increased.

Solution

Step 1: State the null and alternative hypotheses.

Let Population 1 be the voters' opinions of the mayor's trustworthiness before the article and Population 2 be the voters' opinions after the article. The chief of staff's claim is that the proportion of voters who thought the mayor was trustworthy before the newspaper article, p_1, was lower than the approval rating after the article, p_2. Written mathematically, this claim is $p_1 < p_2$. By subtracting p_2 from both sides of the inequality, we have $p_1 - p_2 < 0$. Thus, the hypotheses are stated as follows.

$$H_0: p_1 - p_2 \geq 0$$
$$H_a: p_1 - p_2 < 0$$

Step 2: Determine which distribution to use for the test statistic, and state the level of significance.

We are looking at the difference between two population proportions using large independent samples, so we use the z-test statistic.

We will draw a conclusion by computing the p-value for the calculated test statistic and comparing that value to the level of significance, α. For this hypothesis test, $\alpha = 0.05$.

Memory Booster

Conclusions Using *p*-Values

- If p-value $\leq \alpha$, then reject the null hypothesis.
- If p-value $> \alpha$, then fail to reject the null hypothesis.

Step 3: Gather data and calculate the necessary sample statistics.

Begin by calculating the sample proportions. The sample proportion for Sample 1, before the article, is $\hat{p}_1 = \frac{480}{1200} = 0.4$. The sample proportion for Sample 2, after the article, is $\hat{p}_2 = \frac{550}{1180} \approx 0.466102$. Lastly, we need to calculate $\bar{p}$ before substituting these values into the formula for the test statistic.

$$\begin{aligned}\bar{p} &= \frac{x_1 + x_2}{n_1 + n_2} \\ &= \frac{480 + 550}{1200 + 1180} \\ &\approx 0.432773\end{aligned}$$

Caution

In standard practice, the symbol p can represent either population proportion or p-value. You must pay attention to the context in which the symbol is used in order to determine its meaning.

Substituting these values into the formula for the z-test statistic, we obtain the following.

$$\begin{aligned}z &= \frac{(\hat{p}_1 - \hat{p}_2) - (p_1 - p_2)}{\sqrt{\bar{p}(1-\bar{p})\left(\frac{1}{n_1} + \frac{1}{n_2}\right)}} \\ &= \frac{(0.4 - 0.466102) - 0}{\sqrt{0.432773(1 - 0.432773)\left(\frac{1}{1200} + \frac{1}{1180}\right)}} \\ &\approx -3.25\end{aligned}$$

Rounding Rule

When calculations involve several steps, avoid rounding intermediate calculations. If necessary, round intermediate calculations to at least six decimal places to avoid additional rounding errors in subsequent calculations.

Note that, in the formula above, $p_1 - p_2 = 0$ because H_0 contains that assertion.

Step 4: Draw a conclusion and interpret the decision.

Because this is a left-tailed test, the p-value for this test statistic is the probability of obtaining a test statistic less than or equal to $z = -3.25$, written $p\text{-value} = P(z \leq -3.25)$. To find the p-value, we need the area under the standard normal curve to the left of $z = -3.25$.

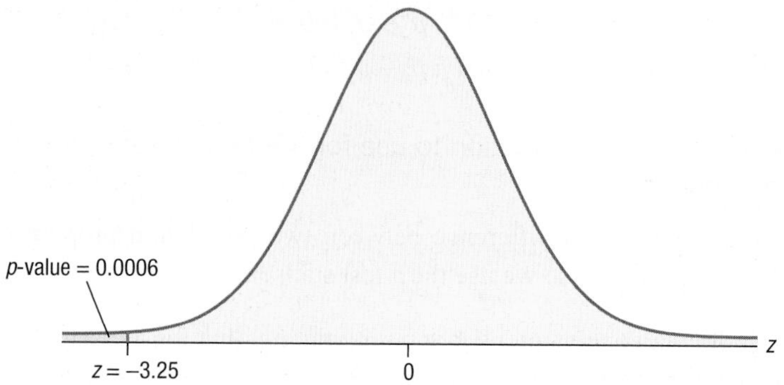

Using appropriate technology or a normal distribution table, we find a p-value of 0.0006. Since the p-value $\leq \alpha$, we reject the null hypothesis. Therefore, there is sufficient evidence at the 0.05 level of significance to support the chief of staff's claim that the mayor's approval rating has gone up.

Example 11.14

Performing a Hypothesis Test for Two Population Proportions Using a TI-83/84 Plus Calculator (Two-Tailed)

The numbers of students who passed and failed their driving tests at the DMV after completing courses at two driving schools in town, Flynt's School of Safe Driving and Pass with Pops, are listed in the following table.

Pass/Fail Records		
	Pass	**Fail**
Flynt's School of Safe Driving	17	8
Pass with Pops	23	6

Based on these data, use a 10% level of significance to test whether the passing proportions for the two driving schools are different.

Solution

Step 1: State the null and alternative hypotheses.

First, let Population 1 be students of Flynt's School of Safe Driving and Population 2 be students of Pass with Pops. The claim to be tested is that the passing proportion for Flynt's School, p_1, is different from the passing proportion for Pops, p_2. Written mathematically, this claim is $p_1 \neq p_2$, which can also be written as $p_1 - p_2 \neq 0$. Thus, the hypotheses are stated as follows.

$$H_0: p_1 - p_2 = 0$$
$$H_a: p_1 - p_2 \neq 0$$

Step 2: Determine which distribution to use for the test statistic, and state the level of significance.

We are looking at the difference between two population proportions using large independent samples, so we use the z-test statistic.

We will draw a conclusion by computing the p-value for the calculated test statistic and comparing that value to the level of significance, α. For this hypothesis test, $\alpha = 0.10$.

Step 3: **Gather data and calculate the necessary sample statistics.**

We need two pieces of information from each sample to use the calculator to run a hypothesis test for two population proportions: x, the number in the sample with the characteristic we are evaluating, and n, the sample size. The value for x in each of the two samples is the number of students who passed. To find the sample size for each school, we simply add the pass and fail totals together. So, we have $x_1 = 17$, $n_1 = 17 + 8 = 25$ and $x_2 = 23$, $n_2 = 23 + 6 = 29$.

Enter the values into the calculator by pressing STAT, scrolling over to TESTS, and then choosing option 6:2-PropZTest. First, enter the values of x and n for the two samples. Since this is a two-tailed test, choose ≠p2 for the alternative hypothesis. Then select Calculate and press ENTER to produce the results.

```
2-PropZTest
x1:17
n1:25
x2:23
n2:29
p1:≠p2 <p2 >p2
Calculate Draw
```

```
2-PropZTest
p1≠p2
z=-.9456876457
p=.3443078936
p̂1=.68
p̂2=.7931034483
↓p̂=.7407407407
```

Step 4: **Draw a conclusion and interpret the decision.**

The p-value given is approximately 0.3443. This value is greater than α, which means that we do not have enough evidence to reject the null hypothesis. Therefore, there is not sufficient evidence at the 0.10 level of significance to say that the passing proportions at the two driving schools are different.

11.4 Section Exercises

Test Statistics for Hypothesis Tests for Two Population Proportions

Directions: Calculate the test statistic for a hypothesis test for two population proportions using the given information.

1. $x_1 = 15,\quad n_1 = 28,\quad x_2 = 21,\quad n_2 = 33,\quad H_0: p_1 - p_2 = 0$

2. $x_1 = 139,\quad n_1 = 180,\quad x_2 = 178,\quad n_2 = 208,\quad H_0: p_1 - p_2 = 0$

3. $x_1 = 18,\quad n_1 = 67,\quad x_2 = 20,\quad n_2 = 59,\quad H_0: p_1 - p_2 = 0$

4. $x_1 = 62,\quad n_1 = 101,\quad x_2 = 71,\quad n_2 = 130,\quad H_0: p_1 - p_2 = 0$

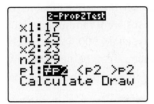

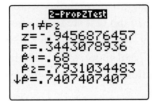

Hypothesis Tests for Two Population Proportions

Directions: Perform each hypothesis test. For each exercise, complete the following steps.

a. State the null and alternative hypotheses.

b. Determine which distribution to use for the test statistic, and state the level of significance.

c. Calculate the test statistic.

d. Draw a conclusion by comparing the *p*-value to the level of significance and interpret the decision.

5. A newspaper story claims that more houses are purchased by single women than single men. To test this claim, two studies were conducted on the buying habits of women and men, respectively. In the first study, 500 house purchases were randomly selected and 100 of those were made by single women. In the second study, again 500 house purchases were randomly selected and 72 of those were made by single men. Test the newspaper's claim using a 0.01 level of significance. Is there sufficient evidence to support the newspaper's claim?

6. Researchers claim that the birth rate in Bonn, Germany is higher than the national average. A random sample of 1200 Bonn residents produced 12 births in 2011, whereas a random sample of 1000 people from all over Germany had 8 births during the same year. Test the researchers' claim using a 0.05 level of significance.

7. University officials hope that changes they have made have improved the retention rate. In 2010, a sample of 1926 freshmen showed that 1400 returned as sophomores. In 2011, 1508 of 2011 freshmen sampled returned as sophomores. Determine if there is sufficient evidence at the 0.05 level to say that the retention rate has improved.

8. Adrian hopes that his new training methods have improved his batting average. Before starting his new regimen, he was batting .220 in a random sample of 50 at bats. For a random sample of 24 at bats since changing his training techniques, his batting average is .375. Determine if there is sufficient evidence to say that his batting average has improved at the 0.10 level of significance.

9. There is an old wives' tale that women who eat chocolate during pregnancy are more likely to have happy babies. A pregnancy magazine wants to test this claim, and it gathers 100 randomly selected pregnant women for its study. Half of the women sampled agree to eat chocolate at least once a day, while the other half agree to forego chocolate for the duration of their pregnancies. A year later, the ladies complete a survey regarding the overall happiness of their babies. The results are given in the following table. At the 0.01 level of significance, test the claim of the old wives' tale.

Numbers of Babies		
	Happy Babies	Unhappy Babies
With Chocolate	24	26
Without Chocolate	22	28

10. A new government program claims to lower high school dropout rates. In one school district, from a sample of 3400 students, the previous dropout rate was 4.5%. Two years after the start of the new program, the dropout rate has been lowered to 3.8% out of 1450 students sampled. Is there enough evidence to say that the government program is effective at lowering the high school dropout rate? Test the government's claim at the 0.10 level of significance.

11. To test the fairness of law enforcement in its area, a local citizens' group wants to know whether women and men are unequally likely to get speeding tickets. Two hundred randomly selected adults were phoned and asked whether or not they had been cited for speeding in the last year. Using the results in the following table and a 0.05 level of significance, test the claim of the citizens' group.

Speeding Tickets		
	Ticketed	**Not Ticketed**
Men	11	75
Women	12	102

11.5 Hypothesis Testing: Two Population Variances

The process for comparing two population variances using a hypothesis test is basically the same as we have used throughout this chapter. The main differences are the distribution and the formulas used to conduct the hypothesis test.

When comparing population means and proportions, we used either the standard normal distribution or a Student's t-distribution. When comparing population variances, we will use an F-distribution, named for Sir Ronald A. Fisher who first developed the F-distribution. The F-distribution is associated with test statistics that are ratios of two random variables. This is the same distribution that we used when constructing a confidence interval for the ratio of two population variances in Section 9.5. In order to compare two population variances, we must ensure that the following conditions are met.

- All possible samples of a given size have an equal probability of being chosen; that is, simple random samples are used.
- The samples are independent.
- Both population distributions are approximately normal.

These conditions are very strict, much more so than the conditions for other hypothesis tests, as the methods used to compare population variances produce very poor results when these conditions have been violated, even to a small degree. You can safely assume that these conditions are met for all examples and exercises in this section.

Let's review the basic properties of an F-distribution, which were introduced in Section 9.5.

Properties

Properties of an F-Distribution

1. The F-distribution is skewed to the right.
2. The values of F are always greater than or equal to 0.
3. The shape of the F-distribution is completely determined by its two parameters, the degrees of freedom for the numerator and the degrees of freedom for the denominator.

In general, the value of F such that an area of α is to the right of F is denoted by F_α, as shown in Figure 11.1.

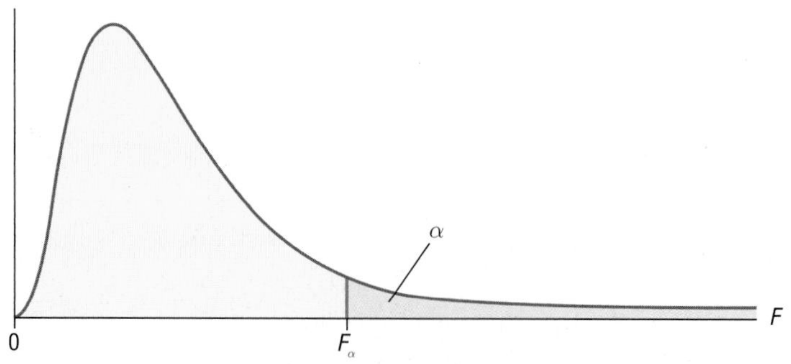

Figure 11.1: F-Distribution

Let's walk through the basic steps of a hypothesis test, focusing specifically on comparing two population variances.

Step 1: **State the null and alternative hypotheses.**

As always, the null and alternative hypotheses are logically opposite statements. In this section, the null and alternative hypotheses for comparing two population variances will take on one of the following three forms.

$H_0: \sigma_1^2 = \sigma_2^2$	$H_0: \sigma_1^2 \le \sigma_2^2$	$H_0: \sigma_1^2 \ge \sigma_2^2$
$H_a: \sigma_1^2 \ne \sigma_2^2$	$H_a: \sigma_1^2 > \sigma_2^2$	$H_a: \sigma_1^2 < \sigma_2^2$

Step 2: **Determine which distribution to use for the test statistic, and state the level of significance.**

To compare population variances, we will use the ratio of the sample variances, $\frac{s_1^2}{s_2^2}$, for the test statistic. This test statistic has an F-distribution with $df_1 = n_1 - 1$ numerator degrees of freedom and $df_2 = n_2 - 1$ denominator degrees of freedom.

We will no longer impose the rule that s_1^2 is greater than or equal to s_2^2. As discussed in Section 9.5, this rule is not mathematically necessary and was only used to simplify the calculations of the endpoints of confidence intervals. We do not want to impose this rule if we wish to consider cases where the researcher hopes to find evidence to support the claim that the population variance of Population 1 is less than the population variance of Population 2, that is, $\sigma_1^2 < \sigma_2^2$.

Note that the level of significance to be used, α, is stated in this step as well.

Step 3: **Gather data and calculate the necessary sample statistics.**

At this point, we would collect sample data from two independent populations with approximately normal distributions. Each sample should be chosen in such a way that all samples of the same size from the population have an equal chance of being chosen; that is, each sample must be a simple random sample. Once the data are collected, calculate the sample variance for each sample. The test statistic is then calculated by dividing the first sample variance by the second sample variance. That is, the test statistic is given by the following formula.

Formula

Test Statistic for a Hypothesis Test for Two Population Variances

When the samples taken are independent, simple random samples and both population distributions are approximately normal, the test statistic for a hypothesis test for two population variances is given by

$$F = \frac{s_1^2}{s_2^2}$$

where s_1^2 and s_2^2 are the two sample variances.

The F-distribution of the test statistic has $df_1 = n_1 - 1$ degrees of freedom for the numerator and $df_2 = n_2 - 1$ degrees of freedom for the denominator,

where n_1 and n_2 are the two sample sizes.

Rounding Rule

When calculating a value of *F*, round to four decimal places. This follows the convention used in the *F*-distribution table in Appendix A.

Step 4: **Draw a conclusion and interpret the decision.**

Remember that the decision rule for rejection regions is to reject the null hypothesis if the test statistic calculated from the sample data falls within the rejection region. Otherwise, fail to reject the null hypothesis. It is also important at the end of this step to interpret the meaning of the conclusion in reference to the question one is trying to answer.

To do this we must determine the values of the test statistic that would lead to a rejection of the null hypothesis. Recall that the size and location of the rejection region are determined by two things: 1) the type of hypothesis test being performed, and 2) the level of significance, α.

Procedure

Rejection Regions for Hypothesis Tests for Two Population Variances

Reject the null hypothesis, H_0, if:

$F \le F_{(1-\alpha)}$ for a left-tailed test

$F \ge F_{\alpha}$ for a right-tailed test

$F \le F_{(1-\alpha/2)}$ or $F \ge F_{\alpha/2}$ for a two-tailed test

We will use the F-distribution table, which is Table H in Appendix A, to determine the critical values for these rejection regions. The F-test statistic is the ratio of the sample variances. If the population variances are equal, then their ratio equals 1 and we would expect the ratio of the sample variances to be close to 1. Thus, the farther the F-test statistic is from 1, the less likely it is that the population variances are equal. Let's look at how to find the critical F-values and rejection regions for the three types of hypothesis tests.

1. Rejection Region for a Left-Tailed Test

For a left-tailed test, the critical value for the rejection region is $F_{(1-\alpha)}$, such that the area in the left tail of the distribution is equal to α. Thus, the rejection region is $F \le F_{(1-\alpha)}$.

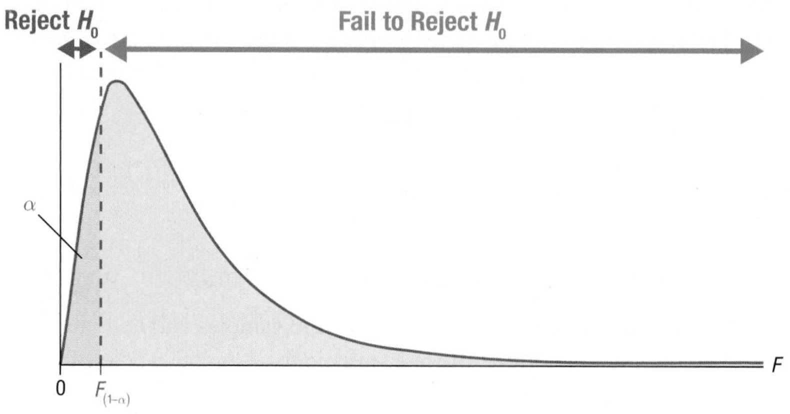

Figure 11.2: Rejection Region for a Left-Tailed Test

2. Rejection Region for a Right-Tailed Test

The critical value for a right-tailed test is F_α, such that the area in the right tail of the distribution is equal to α. Thus, the rejection region is $F \geq F_\alpha$.

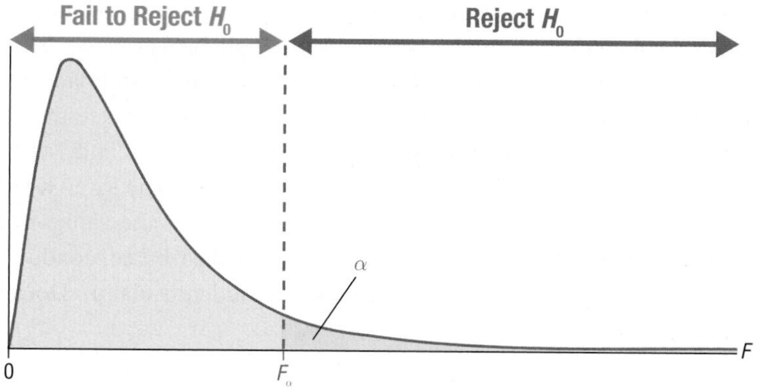

Figure 11.3: Rejection Region for a Right-Tailed Test

3. Rejection Region for a Two-Tailed Test

For a two-tailed test, a total area of α is divided equally between the two tails of the distribution. Thus, there are two critical values, $F_{(1-\alpha/2)}$ and $F_{\alpha/2}$. So the rejection region is $F \leq F_{(1-\alpha/2)}$ or $F \geq F_{\alpha/2}$.

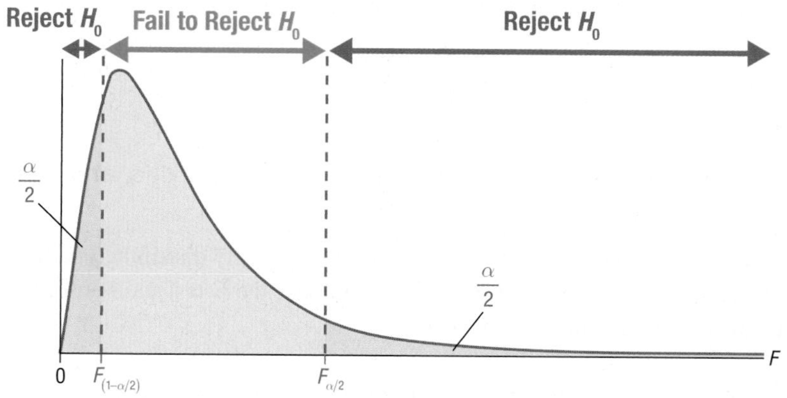

Figure 11.4: Rejection Region for a Two-Tailed Test

Example 11.15 illustrates calculating the F-test statistic by hand. Although a TI-83/84 Plus calculator cannot construct a confidence interval for comparing population variances, it can perform a hypothesis test for comparing two population variances using an F-test statistic. Therefore, Example 11.16 will illustrate a hypothesis test using a TI-83/84 Plus calculator. In Example 11.17, we will demonstrate both methods for a two-tailed hypothesis test.

Example 11.15

Performing a Hypothesis Test for Two Population Variances (Left-Tailed)

A graduate student walking around a large college campus notices that the values of professors' cars seem to have a much smaller variance than the values of the cars owned by students. Students, he notes, own vehicles ranging from small, broken-down trucks to Porsche convertibles, while the cars owned by faculty are more similar in style and value. To test his hypothesis that the values of professors' cars have a smaller variance than the values of students' cars, he collects data from random samples of 15 professors and 14 students on the values of their cars and calculates the variance for each sample. The sample variance of the values of the professors' cars is 34,057 and the sample variance of the values of the students' cars is 45,923. Assuming that both populations are normally distributed, conduct a hypothesis test using a 0.10 level of significance to test the graduate student's claim. Does the evidence support the student's claim?

Solution

Step 1: State the null and alternative hypotheses.

Let's represent the population variance of the values of professors' cars as σ_1^2 and the population variance of the values of students' cars as σ_2^2. Then the claim is that $\sigma_1^2 < \sigma_2^2$. The mathematical opposite of this claim is $\sigma_1^2 \geq \sigma_2^2$. The null and alternative hypotheses are stated as follows.

$$H_0: \sigma_1^2 \geq \sigma_2^2$$
$$H_a: \sigma_1^2 < \sigma_2^2$$

Step 2: Determine which distribution to use for the test statistic, and state the level of significance.

Since we are comparing the variances of two normally distributed populations using independent, simple random samples, we will use the F-test statistic. In this problem, the level of significance is $\alpha = 0.10$.

Step 3: Gather data and calculate the necessary sample statistics.

We have the sample variances, so let's calculate the test statistic.

$$\begin{aligned} F &= \frac{s_1^2}{s_2^2} \\ &= \frac{34{,}057}{45{,}923} \\ &\approx 0.7416 \end{aligned}$$

Step 4: Draw a conclusion and interpret the decision.

Next, we must determine the rejection region. Looking at the alternative hypothesis, we see that this is a left-tailed test. The critical F-value for a left-tailed test is $F_{(1-\alpha)}$, so we will use the section of Table H for an area of $1-\alpha = 0.90$ in the right tail of the distribution. Using $df_1 = n_1 - 1 = 14$ degrees of freedom for the numerator and $df_2 = n_2 - 1 = 13$ degrees of freedom for the denominator, we find that $F_{(1-\alpha)} = F_{0.900} = 0.4909$. Therefore, we will reject the null hypothesis if $F \leq 0.4909$.

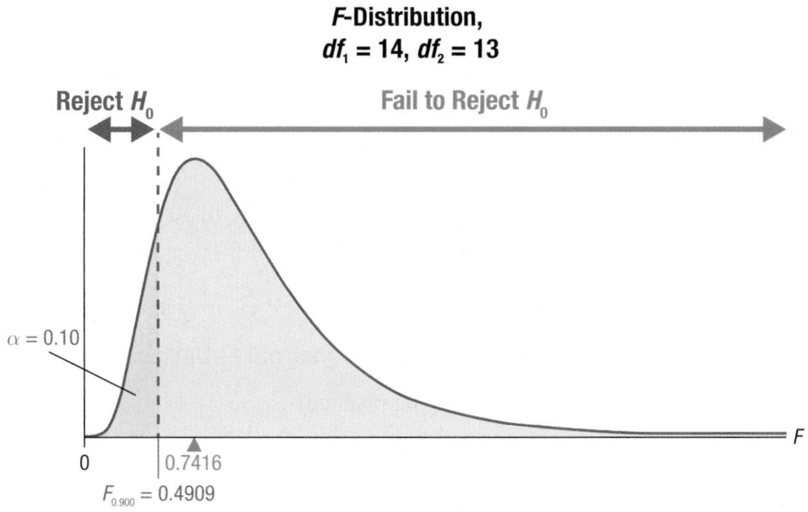

Because 0.7416 > 0.4909, the test statistic is not in the rejection region, so we must fail to reject the null hypothesis. This means that, at the 0.10 level of significance, the evidence does not support the graduate student's claim that the values of professors' cars have a smaller variance than the values of students' cars.

Example 11.16

Performing a Hypothesis Test for Two Population Variances Using a TI-83/84 Plus Calculator (Right-Tailed)

Suppose that a quality control inspector believes that the candy-making machines on Assembly Line A are not adjusted properly. She thinks that the variance in the sizes of the candy produced by Assembly Line A is greater than the variance in the sizes of the candy produced by Assembly Line B. A random sample of 20 pieces of candy is taken from each assembly line and measured. The sizes of the candy from Assembly Line A have a sample variance of 1.45, while the sizes of the candy from Assembly Line B have a sample variance of 0.47. Using a 0.01 level of significance, perform a hypothesis test to test the quality control inspector's claim. Assume that both populations are normally distributed.

Solution

Step 1: State the null and alternative hypotheses.

Let's represent the population variance for Assembly Line A as σ_1^2 and the population variance for Assembly Line B as σ_2^2. Then the quality control inspector's claim is that $\sigma_1^2 > \sigma_2^2$. The mathematical opposite of this claim is $\sigma_1^2 \le \sigma_2^2$. Thus, the hypotheses are stated as follows.

$$H_0: \sigma_1^2 \le \sigma_2^2$$
$$H_a: \sigma_1^2 > \sigma_2^2$$

However, a TI-83/84 Plus calculator will only perform an F-test to compare two population standard deviations. Since variance is necessarily a positive number, we can convert these hypotheses regarding variance to hypotheses regarding standard deviation as follows.

$$H_0: \sigma_1 \le \sigma_2$$
$$H_a: \sigma_1 > \sigma_2$$

Step 2: Determine which distribution to use for the test statistic, and state the level of significance.

Since we are comparing the variances of two normally distributed populations using independent, simple random samples, we will use the F-test statistic. In this problem, the level of significance is $\alpha = 0.01$.

Step 3: Gather data and calculate the necessary sample statistics.

The following values were given in the problem.

Assembly Line A:

$$s_1^2 = 1.45, \quad n_1 = 20$$

Assembly Line B:

$$s_2^2 = 0.47, \quad n_2 = 20$$

The calculator requires that we input the sample standard deviations and not the variances. Taking the square roots of the sample variances gives us the sample standard deviations.

Assembly Line A:

$$\begin{aligned} s_1 &= \sqrt{s_1^2} \\ &= \sqrt{1.45} \\ &\approx 1.204159 \end{aligned}$$

Assembly Line B:

$$\begin{aligned} s_2 &= \sqrt{s_2^2} \\ &= \sqrt{0.47} \\ &\approx 0.685565 \end{aligned}$$

Side Note

On TI-83 Plus calculators and older TI-84 Plus calculators, the function 2-SampFTest is option D on the TESTS menu.

Press STAT. Scroll over to TESTS and choose option E:2-SampFTest. Since we are entering statistics, choose Stats, and enter the values. Since this is a right-tailed test, choose >σ2 for the alternative hypothesis, as shown in the following screenshot. Then choose Calculate and press ENTER.

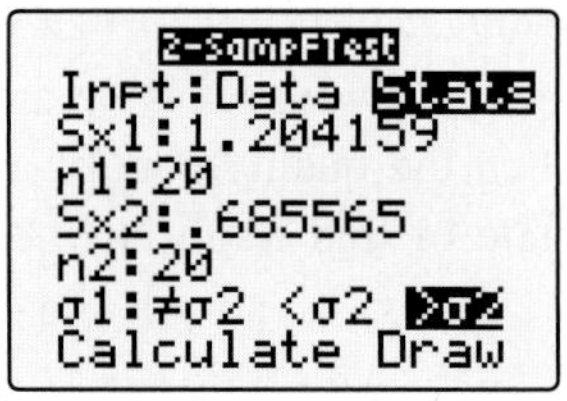

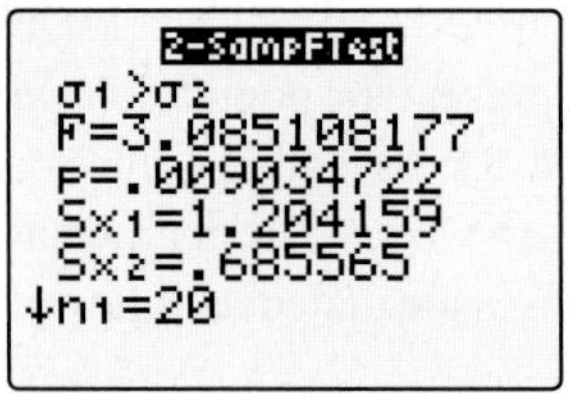

The results are shown in the previous screenshot on the right. The first line is the alternative hypothesis. The second line is the F-test statistic, and the third line is the p-value. The following lines repeat the statistics we entered.

Step 4: Draw a conclusion and interpret the decision.

The p-value is approximately 0.0090. Since 0.0090 is less than $\alpha = 0.01$, we have p-value $\le \alpha$, so the conclusion is to reject the null hypothesis. This means that, at the 0.01 level of significance, there is sufficient evidence to support the quality control inspector's claim that the variance of the sizes of the pieces of candy is greater for the candy produced by Assembly Line A than for the candy produced by Assembly Line B.

Example 11.17

Performing a Hypothesis Test for Two Population Variances (Two-Tailed)

A professor claims that the variances of the test scores on two different versions of an exam are not equal. To test his claim, he chooses a random sample of students' scores on each version. A sample of 20 scores on Version A has a sample variance of 3.8, while a sample of 20 scores on Version B has a sample variance of 3.3. Assuming that both populations of test scores are normally distributed, conduct a hypothesis test to test the professor's claim that the variances of the test scores are not equal. Use a 0.05 level of significance.

Solution

We will calculate the test statistic by hand first along with identifying the rejection region. Then we will show the calculator method at the end. Both methods produce the same results.

Step 1: State the null and alternative hypotheses.

Let the population variance for Version A be represented by σ_1^2 and the population variance for Version B be σ_2^2. Then the professor's claim is that $\sigma_1^2 \ne \sigma_2^2$. The mathematical opposite of this claim is $\sigma_1^2 = \sigma_2^2$. The null and alternative hypotheses are stated as follows.

$$H_0: \sigma_1^2 = \sigma_2^2$$
$$H_a: \sigma_1^2 \ne \sigma_2^2$$

Step 2: Determine which distribution to use for the test statistic, and state the level of significance.

Since we are comparing the variances of two normally distributed populations using independent, simple random samples, we will use the F-test statistic. In this problem, the level of significance is $\alpha = 0.05$.

Step 3: Gather data and calculate the necessary sample statistics.

We have the sample variances needed to calculate the F-test statistic by hand. Substituting these values into the formula, we have the following.

$$\begin{aligned} F &= \frac{s_1^2}{s_2^2} \\ &= \frac{3.8}{3.3} \\ &\approx 1.1515 \end{aligned}$$

Step 4: Draw a conclusion and interpret the decision.

If we are not using a calculator or other technology to calculate the p-value, we must determine the rejection region. Looking at the alternative hypothesis, we see that this is a two-tailed test. The critical F-values for a two-tailed test are $F_{(1-\alpha/2)}$ and $F_{\alpha/2}$. We need to find the critical values for the F-distribution with $df_1 = n_1 - 1 = 19$ numerator degrees of freedom and $df_2 = n_2 - 1 = 19$ denominator degrees of freedom. Using the section of Table H for an area of $1-\alpha/2 = 0.975$ in the right tail of the distribution, we find that $F_{(1-\alpha/2)} = F_{0.975} = 0.3958$. Using the section of Table H for an area of $\alpha/2 = 0.025$ gives us $F_{\alpha/2} = F_{0.025} = 2.5265$. Therefore, we will reject the null hypothesis if either $F \le 0.3958$ or $F \ge 2.5265$.

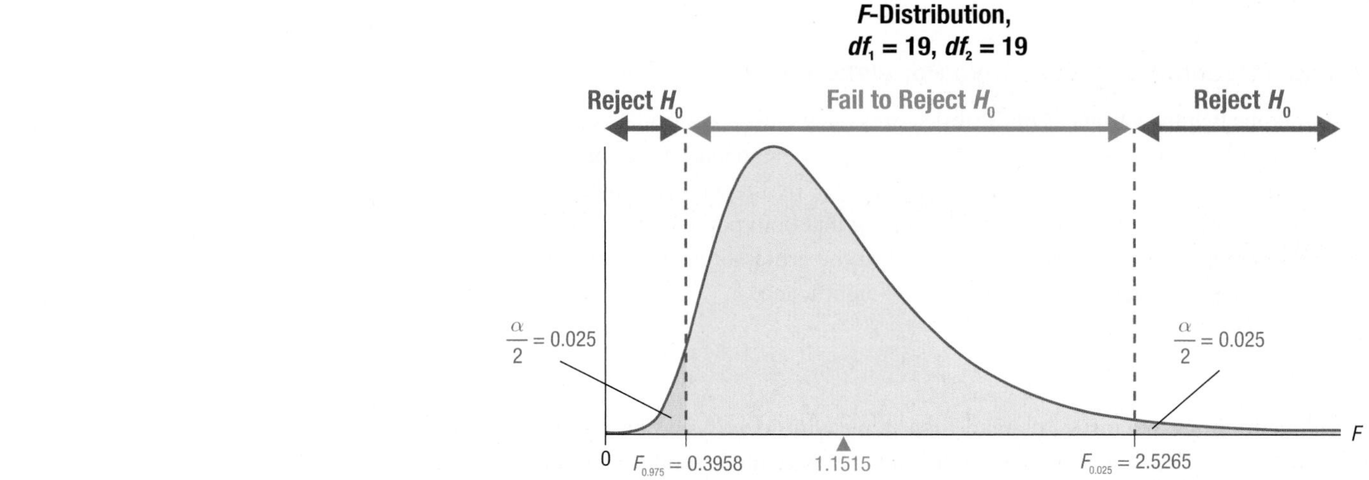

Because $F \approx 1.1515$ is not in the rejection region, we must fail to reject the null hypothesis. This means that, at the 0.05 level of significance, there is not sufficient evidence to say that the scores on the two versions of the test have different variances.

As stated in the previous example, the calculator requires that we input the standard deviation and sample size for each sample. Taking the square roots of the sample variances gives us the sample standard deviations.

Version A:

$$\begin{aligned} s_1 &= \sqrt{s_1^2} \\ &= \sqrt{3.8} \\ &\approx 1.949359 \end{aligned}$$

Version B:

$$\begin{aligned} s_2 &= \sqrt{s_2^2} \\ &= \sqrt{3.3} \\ &\approx 1.816590 \end{aligned}$$

The following screenshot displays the information as it should be entered, including the symbol that should be chosen for the alternative hypothesis for a two-tailed test. Choosing `Calculate` produces the results shown in the screenshot on the right. Remember, the p-value is given on the third line.

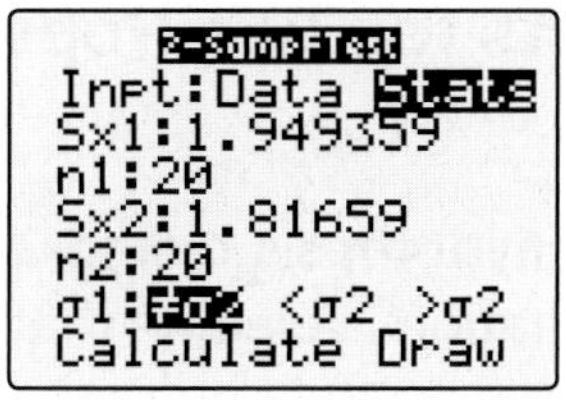

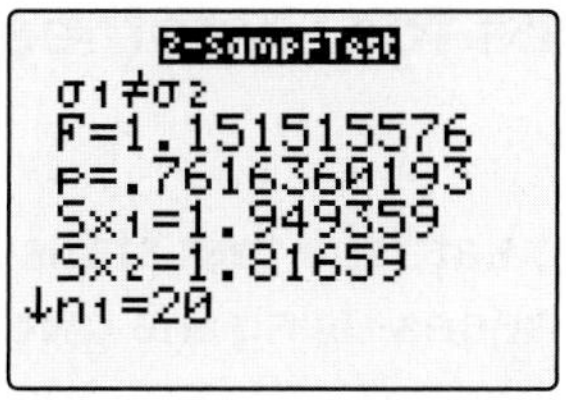

Since the p-value, 0.7616, is larger than $\alpha = 0.05$, we fail to reject the null hypothesis as we did when using the rejection region to draw our conclusion.

11.5 Section Exercises

Null and Alternative Hypotheses for Hypothesis Tests for Two Population Variances

Directions: State the null and alternative hypotheses for each scenario.

1. A medical researcher believes that the variance in the lung capacities of smokers is less than that of nonsmokers. Let σ_1^2 represent the population variance for smokers.

2. A professor believes that the variance of SAT scores of honor students is less than that of all students who take the SAT. Let σ_1^2 represent the population variance for honor students.

3. A quality control inspector believes that the variance in the diameters of soda cans produced by Machine 1 is greater than the variance in the diameters of soda cans produced by Machine 2. Let σ_1^2 represent the population variance for Machine 1.

4. A consumer protection agency believes that the prices of new SUVs have a greater variance than the prices of new luxury cars. Let σ_1^2 represent the population variance for new SUVs.

5. A golf pro believes that the variances of his driving distances are not the same when he uses different brands of golf balls. He is especially interested in comparing Titleist golf balls to a generic store brand. Let σ_1^2 represent the population variance for Titleist golf balls.

6. A paint technician believes that the variance of the thickness of the special coating in one tank is not the same as the variance of the thickness of the coating in another tank. Let σ_1^2 represent the population variance for Tank 1.

Test Statistics for Hypothesis Tests for Two Population Variances

Directions: Calculate the test statistic for a hypothesis test for two population variances using the given information. Assume that both population distributions are approximately normal.

7. $n_1 = 15,\quad s_1^2 = 3.007,\quad n_2 = 16,\quad s_2^2 = 2.897$

8. $n_1 = 4,\quad s_1^2 = 0.961,\quad n_2 = 6,\quad s_2^2 = 0.899$

9. $n_1 = 31,\quad s_1^2 = 46.821,\quad n_2 = 28,\quad s_2^2 = 57.024$

10. $n_1 = 23,\quad s_1^2 = 35{,}679,\quad n_2 = 24,\quad s_2^2 = 39{,}018$

Rejection Regions for Hypothesis Tests for Two Population Variances

Directions: State the critical value(s) of the test statistic, and determine the rejection region for the hypothesis test for the two population variances using the given information. Then give the appropriate conclusion for the hypothesis test. Assume that both population distributions are approximately normal.

11. $n_1 = 19,\quad s_1^2 = 0.3891,\quad n_2 = 24,\quad s_2^2 = 0.9579,\quad H_a\colon \sigma_1^2 < \sigma_2^2,\quad \alpha = 0.05$

12. $n_1 = 14,\quad s_1^2 = 3.152,\quad n_2 = 11,\quad s_2^2 = 9.300,\quad H_a\colon \sigma_1^2 < \sigma_2^2,\quad \alpha = 0.05$

13. $n_1 = 11,\quad s_1^2 = 31{,}207,\quad n_2 = 11,\quad s_2^2 = 38{,}916,\quad H_a\colon \sigma_1^2 < \sigma_2^2,\quad \alpha = 0.01$

14. $n_1 = 11,\quad s_1^2 = 3.007,\quad n_2 = 25,\quad s_2^2 = 2.897,\quad H_a\colon \sigma_1^2 > \sigma_2^2,\quad \alpha = 0.05$

15. $n_1 = 20,\quad s_1^2 = 10.453,\quad n_2 = 23,\quad s_2^2 = 3.199,\quad H_a\colon \sigma_1^2 > \sigma_2^2,\quad \alpha = 0.10$

16. $n_1 = 12,\quad s_1^2 = 1893,\quad n_2 = 26,\quad s_2^2 = 1066,\quad H_a\colon \sigma_1^2 > \sigma_2^2,\quad \alpha = 0.01$

17. $n_1 = 16,\quad s_1^2 = 18.01,\quad n_2 = 21,\quad s_2^2 = 17.07,\quad H_a\colon \sigma_1^2 \neq \sigma_2^2,\quad \alpha = 0.05$

18. $n_1 = 20,\quad s_1^2 = 27.08,\quad n_2 = 29,\quad s_2^2 = 11.77,\quad H_a\colon \sigma_1^2 \neq \sigma_2^2,\quad \alpha = 0.05$

19. $n_1 = 20,\quad s_1^2 = 8.12,\quad n_2 = 18,\quad s_2^2 = 16.78,\quad H_a\colon \sigma_1^2 \neq \sigma_2^2,\quad \alpha = 0.10$

20. $n_1 = 11,\quad s_1^2 = 12{,}047,\quad n_2 = 12,\quad s_2^2 = 18{,}019,\quad H_a\colon \sigma_1^2 \neq \sigma_2^2,\quad \alpha = 0.01$

Hypothesis Tests for Two Population Variances

Directions: **Perform each hypothesis test using the method of your choice or the one assigned by your instructor. For each exercise, complete the following steps. Assume that both population distributions are approximately normal in each scenario.**

a. State the null and alternative hypotheses.

b. Determine which distribution to use for the test statistic, and state the level of significance.

c. Calculate the test statistic.

d. Draw a conclusion and interpret the decision.

21. A golf pro believes that the variances of his driving distances are different for different brands of golf balls. In particular, he believes that his driving distances, measured in yards, have a smaller variance when he uses Titleist golf balls than when he uses a generic store brand. He hits 10 Titleist golf balls and records a sample variance of 201.65. He hits 10 generic golf balls and records a sample variance of 364.57. Test the golf pro's claim using a 0.05 level of significance. Does the evidence support the golf pro's claim?

22. A quality control inspector believes that the variance in the diameters of soda cans, measured in millimeters, is greater for soda cans produced by Machine A than for soda cans produced by Machine B. The sample variance of a random sample of 15 soda cans from Machine A is 2.788. The sample variance for a random sample of 17 soda cans from Machine B is 1.982. Test the inspector's claim using a 0.10 level of significance. Does the evidence support the inspector's claim?

23. A medical researcher believes that the variance of total cholesterol levels in men is greater than the variance of total cholesterol levels in women. The sample variance for a random sample of 8 men's cholesterol levels, measured in mg/dL, is 277. The sample variance for a random sample of 7 women is 89. Test the researcher's claim using a 0.10 level of significance. Does the evidence support the researcher's belief?

24. A basketball coach believes that the variance of the heights of adult male basketball players is different from the variance of heights for the general population of men. The sample variance of heights, measured in inches, for a random sample of 12 basketball players is 24.76. The sample variance for a random sample of 13 other men is 25.87. Test the coach's claim using a 0.01 level of significance. Does the evidence support the coach's claim?

25. One study claims that the variance in the resting heart rates of smokers is different than the variance in the resting heart rates of nonsmokers. A medical student decides to test this claim. The sample variance of resting heart rates, measured in beats per minute, for a random sample of 5 smokers is 545.1. The sample variance for a random sample of 5 nonsmokers is 103.7. Test the study's claim using a 0.01 level of significance. Does the evidence support the study's claim?

26. A professor believes that the variance of ACT composite scores of honor students is less than that of all students who take the ACT. The sample variance of the ACT composite scores for a random sample of 18 honors students is 12.1. The sample variance of the ACT composite scores for a random sample of 20 other students is 28.9. Test the professor's claim using a 0.05 level of significance. Does the evidence support the professor's claim?

11.6 ANOVA (Analysis of Variance)

Our discussion thus far in this chapter has shown us how to compare population parameters from two populations. But, what if you have four, eight, or even ninety-two? Consider the chair of an academic department who wants to know if the mean final grades are the same for the 12 sections of an introductory statistics class. If she has nothing better to do, she could compare the first class mean to each of the other 11, then the second class mean to each of the other 10, then the third class mean, and so on, using techniques that we have seen previously. However, this would result in her having to perform 66 different hypothesis tests! (Verify this yourself.) It would also reduce the level of confidence in the final result, because as we increase the number of tests performed, we increase the likelihood of finding a difference that occurs by chance as a result of random sampling. In order to avoid these issues, we need a hypothesis test that compares the means of more than two populations all at once. This test is called an ANOVA test, and will be the focus of this section.

ANOVA stands for **AN**alysis **O**f **VA**riance and is used to compare the means of three or more populations. When performing an ANOVA test, the null hypothesis states that all of the population means are equal, and the alternative hypothesis states that at least one of the population means is different.

Procedure

Null and Alternative Hypotheses for an ANOVA Test

The null and alternative hypotheses for an ANOVA test to compare the means of three or more populations are as follows.

$$H_0: \mu_1 = \mu_2 = \cdots = \mu_k$$

H_a: At least one mean differs from the others.

where $\mu_1, \mu_2, \ldots, \mu_k$ are the population means and

k is the number of populations being studied.

From our study of statistics so far, it should come as no great surprise that if we take samples from various populations, the sample means will not necessarily be equal, even if the population means are equal. Thus, an ANOVA test seeks to determine if the variation in the sample means from the different populations is due to random sampling or to a true difference in the population means. So, the reason for the test being called an analysis of variance instead of an analysis of means is that the test analyzes the variation between the sample means, rather than the means themselves.

In order for an ANOVA test to be used, the following assumptions must be met.

- All possible samples of a given size have an equal probability of being chosen; that is, simple random samples are used.
- The samples are independent.
- All of the population variances are equal. (If the sample sizes are nearly equal, this assumption is not essential.)
- All of the population distributions are approximately normal.

Before we look at an actual ANOVA table and learn how to determine if the data suggest that the population means are not all equal, we will begin by briefly introducing a few formulas. The purpose of showing these formulas is to better understand how the numbers in an ANOVA table are related;

however, using these formulas to calculate the values in an ANOVA table by hand would be tedious, at best. Therefore, we will use a TI-83/84 Plus calculator or Microsoft Excel to generate an ANOVA table, when necessary.

The first calculation is the **grand mean**, $\overline{\overline{x}}$, which is the weighted mean of the k sample means, one from each of the k populations.

Formula

Grand Mean

The **grand mean** is the weighted mean of the sample means from each of the populations, which is equivalent to the mean of all the sample data combined, given by

$$\overline{\overline{x}} = \frac{\sum_{i=1}^{k}\left(n_i\overline{x}_i\right)}{\sum_{i=1}^{k} n_i}$$

where k is the number of samples, one from each of the k populations,

n_i is the size of the sample drawn from the ith population, and

$\overline{x}_i$ is the mean of the sample drawn from the ith population.

The next two calculations that we need to look at are the values of two different types of variation, SST and SSE. First, **SST** stands for the **sum of squares among treatments** and measures the variation between the sample means and the grand mean. This value is sometimes referred to as SS(Between groups), SS(Factors), or SS(Treatments), where the treatments or factors are the characteristics that distinguish one population from another. For example, suppose that a farmer is interested in the effects that different kinds of fertilizers have on his tomato crop. He wants to know which plots produce the best yield of tomatoes, those in which he uses Fertilizer 1, those in which he uses Fertilizer 2, or those in which he does not use any fertilizer. In this situation, the type of fertilizer used (Fertilizer 1, Fertilizer 2, or none) is the treatment. SST would then refer to the variation in the sample data resulting from the differences between the sample mean yields of tomatoes for plots that use each type of fertilizer.

Formula

Sum of Squares among Treatments (SST)

The **sum of squares among treatments** (SST) measures the variation between the sample means and the grand mean and is given by

$$\text{SST} = \sum_{i=1}^{k} n_i\left(\overline{x}_i - \overline{\overline{x}}\right)^2$$

where k is the number of samples, one from each of the k populations,

n_i is the size of the sample drawn from the ith population,

$\overline{x}_i$ is the mean of the sample drawn from the ith population, and

$\overline{\overline{x}}$ is the grand mean.

SST measures the variation between the sample means. Thus, it makes sense that the size of this value would strongly influence whether we reject or fail to reject the null hypothesis. As the differences between the sample means increase, SST becomes larger. With a "large enough" value of SST, we will reject the null hypothesis and conclude that at least one of the population means is different from the others.

SSE stands for the **sum of squares for error** and is the variation in the sample data resulting from the variability *within* each sample.

Formula

Sum of Squares for Error (SSE)

The **sum of squares for error** (**SSE**) measures the variation in the sample data resulting from the variability within each sample, given by

$$\text{SSE} = \sum_{j=1}^{n_1}\left(x_{1j} - \bar{x}_1\right)^2 + \sum_{j=1}^{n_2}\left(x_{2j} - \bar{x}_2\right)^2 + \cdots + \sum_{j=1}^{n_k}\left(x_{kj} - \bar{x}_k\right)^2$$

where x_{ij} is the jth data value in the sample drawn from the ith population,

$\bar{x}_i$ is the mean of the sample drawn from the ith population,

n_i is the size of the sample drawn from the ith population, and

k is the number of samples, one from each of the k populations.

The **total variation** is the sum of the variations contributed by each sample. That is, the total variation is the sum of the squared deviations from the grand mean for all of the data values in each sample. Each summation in the formula for total variation is the amount of variation contributed by the sample taken from the ith population; thus, $\sum_{j=1}^{n_1}\left(x_{1j} - \bar{\bar{x}}\right)^2$ is the amount of total variation contributed by Sample 1. The total variation contains two types of variation; variation between the samples taken from the different populations that can be attributed to the treatments (SST), and variation within the samples themselves, which can be attributed to random error (SSE). Thus, we have the following relationship.

Formula

Total Variation

The **total variation**, also known as *total sum of squares*, is the sum of the variations contributed by each sample, given by

$$\text{Total Variation} = \sum_{j=1}^{n_1}\left(x_{1j} - \bar{\bar{x}}\right)^2 + \sum_{j=1}^{n_2}\left(x_{2j} - \bar{\bar{x}}\right)^2 + \cdots + \sum_{j=1}^{n_k}\left(x_{kj} - \bar{\bar{x}}\right)^2$$
$$= \text{SST} + \text{SSE}$$

where x_{ij} is the jth data value in the sample drawn from the ith population,

$\bar{\bar{x}}$ is the grand mean,

n_i is the size of the sample drawn from the ith population,

k is the number of samples, one from each of the k populations,

SST is the sum of squares among treatments, and

SSE is the sum of squares for error.

Two other calculations that are important to introduce at this point are **MST** and **MSE**, which stand for **mean square for treatments** and **mean square for error**, respectively. The mean square for treatments is found by dividing the sum of squares among treatments by its degrees of freedom (DFT), which is the number of samples being studied minus 1. The mean square for error is found by dividing the sum of squares for error by its degrees of freedom (DFE), which is the total number of data values minus the number of samples.

Formula

Mean Square for Treatments (MST)

The **mean square for treatments** (**MST**) is given by

$$\text{MST} = \frac{\text{SST}}{\text{DFT}}$$

where SST is the sum of squares among treatments,

$\text{DFT} = k - 1$ is the degrees of freedom for treatments, and

k is the number of samples, one from each of the k populations.

Formula

Mean Square for Error (MSE)

The **mean square for error (MSE)** is given by

$$\text{MSE} = \frac{\text{SSE}}{\text{DFE}}$$

where SSE is the sum of squares for error,

$\text{DFE} = n_T - k$ is the degrees of freedom for error,

$n_T = \sum_{i=1}^{k} n_i$ is the total number of data values in all samples combined,

k is the number of samples, one from each of the k populations, and

n_i is the size of the sample drawn from the ith population.

If SST is very large, then dividing that number by $k - 1$ will result in a large number as well. Thus, as SST increases, so does MST. Therefore, a large value of MST will cause us to doubt the null hypothesis that the population means are all equal.

The test statistic for an ANOVA test is an F-statistic. An ANOVA test uses this statistic instead of z, t, or χ^2 as some other tests do. We have seen the F-distribution before when we constructed confidence intervals for the ratio of two population variances (in Section 9.5) and performed hypothesis tests comparing two population variances (in Section 11.5). In an ANOVA test, the F-statistic is equal to the ratio of the mean squares, MST and MSE. Note that the mean squares are variances, so the F-statistic for an ANOVA test is also a ratio of variances (like F in a hypothesis test for two population variances).

Formula

Test Statistic for an ANOVA Test

The test statistic for an ANOVA test is the ratio of the mean squares, given by

$$F = \frac{\text{MST}}{\text{MSE}}$$

where MST is the mean square for treatments and

MSE is the mean square for error.

The F-distribution of the test statistic has $df_1 = \text{DFT} = k - 1$ degrees of freedom for the numerator and $df_2 = \text{DFE} = n_T - k$ degrees of freedom for the denominator,

where $n_T = \sum_{i=1}^{k} n_i$ is the total number of data values in all samples combined,

n_i is the size of the sample drawn from the ith population, and

k is the number of samples, one from each of the k populations.

The F-distribution is skewed to the right and is defined by only two parameters, df_1 and df_2. The first parameter, df_1, is the degrees of freedom for the numerator of F, which is given by $k - 1$ in an ANOVA test. Note that these are the degrees of freedom associated with SST, the sum of squares among treatments; that is, $df_1 = \text{DFT}$. The second parameter is df_2, which is the degrees of freedom for the denominator of F, given by $n_T - k$ for an ANOVA test statistic. Note that these are the degrees of freedom associated with SSE, the sum of squares of error; that is, $df_2 = \text{DFE}$. Notice that the total number of degrees of freedom is one less than the total number of data values: $\text{DFT} + \text{DFE} = n_T - 1$.

A picture of the F-distribution is shown in Figure 11.5. Notice that the critical F-value, denoted F_α, is the value of F such that the area to the right of F_α is equal to α.

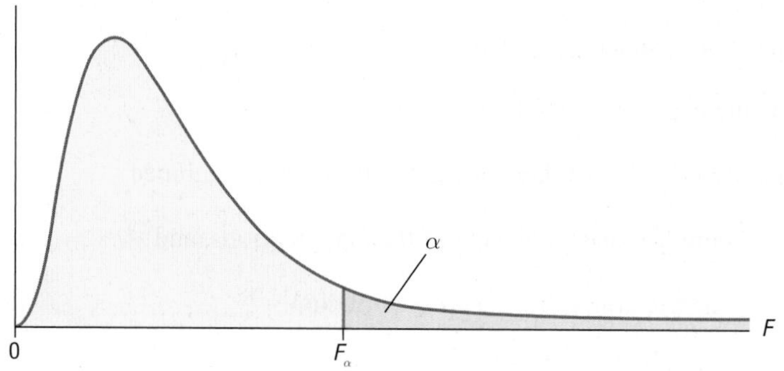

Figure 11.5: *F*-Distribution

The larger the differences between each pair of population means, the larger F will be. Thus, a large F-statistic supports the claim that at least one of the population means is different from the others. Therefore, a large F-value leads us to reject the null hypothesis that the k population means are all the same. This means that an ANOVA test is always a right-tailed test. The question naturally arises as this point, "How large is large enough?" In an ANOVA test, F is considered "large enough" to reject the null hypothesis if $F \geq F_\alpha$. If the null hypothesis is rejected in an ANOVA test, it does not indicate which of the population means is different from the others, or how many are different, or by how much they differ.

Procedure

Rejection Region for ANOVA Tests

Reject the null hypothesis, H_0, if:

$$F \geq F_\alpha$$

Now that we have discussed all of the elements of an ANOVA test, we can look at an ANOVA table, which is a summary of the calculations we have discussed in this section. The following table is an example of an ANOVA table.

	SS	*df*	*MS*	*F*
Treatments (T)	4035.51	9	448.39	0.9648
Error (E)	5112.14	11	464.74	
Total	9147.65	20		

Figure 11.6: Example of an ANOVA Table

Note that SST is the value in the cell where the *SS* column and the Treatments (T) row intersect. Thus, $\text{SST} = 4035.51$ in Figure 11.6. Similarly, MSE is the value in the cell where the *MS* column and the Error (E) row intersect. So $\text{MSE} = 464.74$ in Figure 11.6. Notice that the last row is the Total (or sum) of the previous two rows. Also notice that the formulas we discussed earlier in this section are evident in the table: $\text{MST} = \frac{\text{SST}}{\text{DFT}}$, $\text{MSE} = \frac{\text{SSE}}{\text{DFE}}$, and $F = \frac{\text{MST}}{\text{MSE}}$. (Verify that these formulas are applied correctly in this table.) The following table is a summary of the formulas we have discussed in this section, and how they are organized in an ANOVA table.

	SS	*df*	*MS*	*F*
Treatments (T)	SST	DFT	$\frac{\text{SST}}{\text{DFT}}$	$\frac{\text{MST}}{\text{MSE}}$
Error (E)	SSE	DFE	$\frac{\text{SSE}}{\text{DFE}}$	
Total	SST + SSE	DFT + DFE		

Figure 11.7: Formulas for the Values in an ANOVA Table

Example 11.18

Completing an ANOVA Table and Performing an ANOVA Test

Medical researchers studying the effect of vitamin C on the common cold grouped study participants into three groups: Group 1 took 1000 mg of vitamin C daily, Group 2 took 500 mg of vitamin C daily, and Group 3 took no vitamin C. The participants were observed for a year, during which time the number of colds that each participant suffered was recorded. Using these data, the researchers performed an ANOVA test, which resulted in the following ANOVA table. Note that this ANOVA table was produced by Microsoft Excel, and it contains two more values than the ANOVA tables shown in Figures 11.6 and 11.7. These additional values are the p-value for the F-statistic and the critical F-value for the ANOVA test with a 0.05 level of significance. Unfortunately, a few of the cells in the table have been lost.

	SS	*df*	*MS*	*F*	*P-value*	*F crit*
Treatments (T)		2	1.4288		0.4288	3.3158
Error (E)	49.2030					
Total	52.0606	32				

a. Determine the missing values in the ANOVA table using the given values.

b. Based on the completed ANOVA table, do the researchers have enough evidence, at the 0.05 level of significance, to support the claim that there is a significant difference between the mean numbers of colds suffered per year for patients who take 1000 mg of vitamin C daily, patients who take 500 mg of vitamin C daily, and patients who do not take vitamin C?

c. From the ANOVA table, can you conclude if one group had fewer colds than the others?

Solution

a. To complete the ANOVA table, we will work from left to right to find the missing values. That is, we will first compute the sum of squares among treatments, then the degrees of freedom for error, then the mean square for error, and finally the F-statistic.

To begin filling in the ANOVA table, we can use the fact that SST + SSE = Total Variation. Thus, by subtracting SSE from both sides of this equation, we have the following.

$$\begin{aligned} \text{SST} &= \text{Total Variation} - \text{SSE} \\ &= 52.0606 - 49.2030 \\ &= 2.8576 \end{aligned}$$

Similarly, DFT + DFE = DF (Total). Thus, by subtracting DFT from both sides of this equation, we have the following.

$$\begin{aligned} \text{DFE} &= \text{DF (Total)} - \text{DFT} \\ &= 32 - 2 \\ &= 30 \end{aligned}$$

To find the next missing value, we must divide.

$$\begin{aligned} \text{MSE} &= \frac{\text{SSE}}{\text{DFE}} \\ &= \frac{49.2030}{30} \\ &= 1.6401 \end{aligned}$$

Lastly, we must divide again to find the value of the F-test statistic.

$$\begin{aligned} F &= \frac{\text{MST}}{\text{MSE}} \\ &= \frac{1.4288}{1.6401} \\ &\approx 0.8712 \end{aligned}$$

We can now fill in all of the missing values to complete the ANOVA table.

	SS	*df*	*MS*	*F*	*P-value*	*F crit*
Treatments (T)	2.8576	2	1.4288	0.8712	0.4288	3.3158
Error (E)	49.2030	30	1.6401			
Total	52.0606	32				

b. First, let's state the null and alternative hypotheses for this test.

H_0: $\mu_1 = \mu_2 = \mu_3$

H_a: At least one mean differs from the others.

We have a level of significance of 0.05, thus $\alpha = 0.05$. We can draw our conclusion for an ANOVA test by using p-values, as we have in previous sections. Notice that the ANOVA table states that the p-value is 0.4288. Since $0.4288 > 0.05$, we have $p\text{-value} > \alpha$, so we fail to reject the null hypothesis. Thus, there is not sufficient evidence, at the 0.05 level of significance, to support the researcher's claim that there is a significant difference between the mean numbers of colds suffered per year for patients who take 1000 mg of vitamin C daily, patients who take 500 mg of vitamin C daily, and patients who do not take vitamin C.

c. Remember that rejecting the null hypothesis in an ANOVA test supports the claim that at least one of the population means is different from the others, but it does not indicate which one differs, or by how much. This test did not provide support for the claim that one group had fewer (or more) colds, on average, than another group; however, even if it had, we would not be able to tell from the ANOVA table which group had fewer colds.

Now that we are comfortable interpreting an ANOVA table, let's see how to use technology to generate the values in the table from raw data. We will use both Microsoft Excel and a TI-83/84 Plus calculator to generate the values in the next example. We will first use Excel, since the ANOVA table generated looks similar to those that we have seen so far in this section. A TI-83/84 Plus calculator can be used just as easily, but the output is in the form of a list, not a table.

Example 11.19

Performing an ANOVA Test Using Microsoft Excel and a TI-83/84 Plus Calculator

Researchers at a drug company wish to test whether their drug is the best on the market for lowering cholesterol levels in middle-aged women. To compare the drug with other drugs on the market, 30 women with high cholesterol volunteer to take part in a six-month study. The women are divided into three groups, and each group is given a different cholesterol-lowering drug, one of which is the drug made by the company. The women's cholesterol levels are measured at the beginning of the study, and again after six months. The difference between the before-treatment and after-treatment cholesterol levels for each woman is recorded in the following table. (Note that each study participant experienced a decrease in cholesterol level over the six-month period.) Use these data to perform an ANOVA test with a 0.10 level of significance. Assume that the three population distributions are all approximately normal with equal population variances. What conclusion can you draw from the ANOVA test? First, use Microsoft Excel and then use a TI-83/84 Plus calculator to perform the test.

Decreases in Cholesterol Levels (in mg/dL)		
Drug 1	**Drug 2**	**Drug 3**
18	18	21
19	20	22
20	16	17
21	20	18
22	21	22
23	20	19
18	18	21
19	19	20
20	17	18
21	13	23

Solution

To begin, we need to enter our data into Microsoft Excel. Enter the data from the table above, including the column headings, into columns A, B, and C. Under the **Data** tab, choose **Data Analysis**. In the Data Analysis menu, choose **Anova: Single Factor**. Next, fill out the ANOVA menu as shown in the following screenshot.

Side Note

If Data Analysis does not appear in your Excel menu under the Data tab, you can easily add this function. See the directions in Appendix B: Getting Started with Microsoft Excel.

Anova: Single Factor

Input			
Input Range:	A1:C11		OK
Grouped By:	⊙ Columns		Cancel
	○ Rows		
☑ Labels in first row			Help
Alpha: 0.1			

Output options

⊙ Output Range: A15
○ New Worksheet Ply:
○ New Workbook

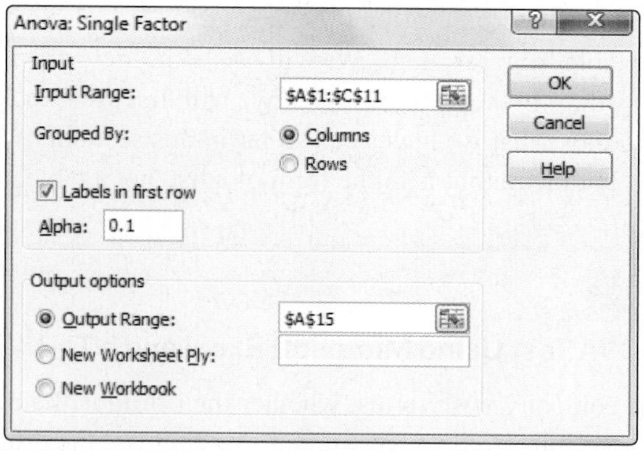

After clicking **OK**, the ANOVA table will be displayed along with a SUMMARY table, which contains the descriptive statistics for each sample.

Anova: Single Factor

SUMMARY

Groups	Count	Sum	Average	Variance		
Drug 1	10	201	20.1	2.766667		
Drug 2	10	182	18.2	5.733333		
Drug 3	10	201	20.1	4.1		

ANOVA

Source of Variation	SS	df	MS	F	P-value	F crit
Between Groups	24.06667	2	12.03333	2.865079	0.074407	2.510609
Within Groups	113.4	27	4.2			
Total	137.4667	29				

In the ANOVA table, note that Between Groups corresponds to Treatments (T) and Within Groups corresponds to Error (E). Notice that the p-value ≈ 0.0744, which is less than our level of significance, $\alpha = 0.10$, so we reject the null hypothesis. Thus, the researchers can conclude that there is sufficient evidence, at the 0.10 level of significance, to support the claim that at least one of the population means is different. That is, the evidence suggests that the mean decrease in a woman's cholesterol level for at least one of the drugs is different from the others. However, we cannot conclude, from the ANOVA test alone, how the population means differ. Looking at the descriptive statistics for the sample data, we see that Drugs 1 and 3 lowered the women's cholesterol levels by a mean of 20.1 points and Drug 2 lowered the women's cholesterol levels by a mean of 18.2 points, which suggests that the population mean for Drug 2 differs from the others, but we cannot make this conclusion from the ANOVA test alone.

Side Note

On TI-83 Plus calculators and older TI-84 Plus calculators, the function, ANOVA, is option **F** on the TESTS menu.

To use a TI-83/84 Plus calculator to perform an ANOVA test, begin by entering the sample data in the lists **L1**, **L2**, and **L3**. Next, press **STAT**, scroll to **TESTS**, and choose option **H:ANOVA(**. It will show **ANOVA(** on the screen. You need to enter the lists that contain the sample data for your ANOVA test. In our case we enter **ANOVA(L1,L2,L3)**. Press **ENTER**. The results will be displayed as shown in the following screenshots.

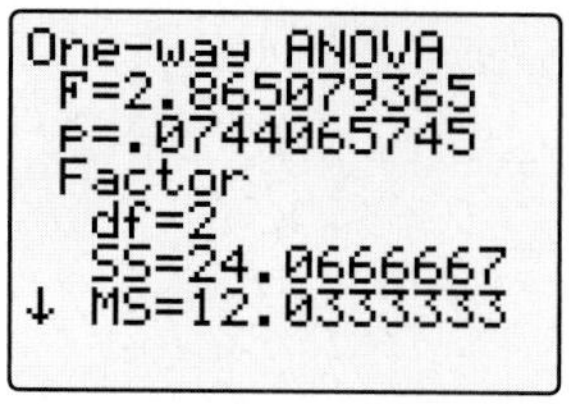

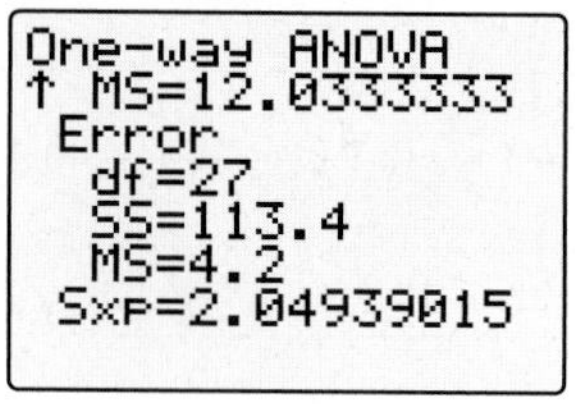

A traditional ANOVA table is not displayed; however, all of the values that would be in an ANOVA table are listed on the screen. The first value given is the F-test statistic, then the p-value. The values under `Factor` are the treatment values and the values under `Error` are the error values. Verify that the `Factor` and `Error` values are the same as those given in the ANOVA table generated by Microsoft Excel in the rows for Between Groups and Within Groups, respectively. Since the p-value, approximately 0.0744, is less than the level of significance, $\alpha = 0.10$, we will reject the null hypothesis, and again conclude that there is sufficient evidence to support the claim that at least one of the population means is different from the others.

11.6 Section Exercises

Critical Values for ANOVA Tests

Directions: Find the critical value for an ANOVA test using the given information.

1. $\alpha = 0.05, \quad df_1 = 10, \quad df_2 = 12$

2. $\alpha = 0.01, \quad df_1 = 9, \quad df_2 = 15$

3. $\alpha = 0.10, \quad df_1 = 7, \quad df_2 = 8$

4. $\alpha = 0.05, \quad df_1 = 3, \quad df_2 = 9$

5. $\alpha = 0.10, \quad df_1 = 1, \quad df_2 = 5$

6. $\alpha = 0.01, \quad df_1 = 6, \quad df_2 = 4$

ANOVA Tables

Directions: Complete each ANOVA table.

7.	*SS*	*df*	*MS*	*F*
Treatments (T)	4	5		
Error (E)	11	6		
Total				

8.	*SS*	*df*	*MS*	*F*
Treatments (T)	49.3	4		
Error (E)		3		
Total	142.5			

9.	*SS*	*df*	*MS*	*F*
Treatments (T)	50		12.5	
Error (E)		3		
Total	74			

10.	*SS*	*df*	*MS*	*F*
Treatments (T)		9		
Error (E)	21.8		5.45	
Total	101.9			

11.	*SS*	*df*	*MS*	*F*
Treatments (T)	313	6		
Error (E)		8		
Total	649			

ANOVA Tests

Directions: **Perform each ANOVA test. Assume that the population distributions for each test are all approximately normal with equal population variances. For each exercise, complete the following steps.**

a. Find all the values for the ANOVA table.

b. Draw a conclusion and interpret the decision.

12. A regional manager wants to know if there is a difference between the mean amounts of time that customers wait in line at the drive-through window for the three stores in her region. She samples the wait times at each store. Her data are given in the following table. Use an ANOVA test to determine if there is a difference between the mean wait times for the three stores, at the 0.05 level of significance.

Drive-Through Wait Times (in Minutes)		
Store 1	**Store 2**	**Store 3**
2.34	2.87	1.32
1.23	1.94	1.45
1.89	2.36	1.78
2.31	1.85	2.01
3.02	1.75	2.45
1.95	2.82	1.92
2.45	3.32	1.83

13. A manager is concerned that one of his workers is producing more defective parts, on average, than the other similarly skilled workers. He records the number of defective parts made by this worker and two others with similar experience during their shifts each day for one week. The results are provided in the following table. At the 0.10 level of significance, can you conclude that there is a difference between the mean numbers of defective parts produced each day by these three workers?

Number of Defective Parts per Day		
Worker A	**Worker B**	**Worker C**
3	2	1
2	2	2
2	1	1
4	0	1
2	1	2

14. The Panhellenic Council is comparing the mean GPAs of four sororities on campus. The GPAs for 10 girls in each sorority are given in the following table. Based on these data, can you conclude that there is a difference between the mean GPAs for these four sororities? Use a 0.05 level of significance.

Sorority GPAs			
ΛΧΒ	ΩΣ	ΘΠ	ΔΜΟ
2.5	3.1	2.7	3.1
3.4	3.4	2.9	3.2
4.0	3.6	2.8	3.1
3.8	3.6	2.8	2.8
2.7	3.7	2.9	2.9
2.8	3.0	3.1	3.1
2.8	4.0	2.9	3.0
3.2	3.9	2.7	3.3
3.1	4.0	2.6	2.7
3.9	3.4	3.3	2.6

15. A group of paramedics does not believe that the mean numbers of calls received in one shift are the same for the morning, afternoon, and night shifts. To test this claim, they record the number of calls received during each shift for seven days. Based on this evidence, can the paramedics conclude that the mean numbers of calls are different for the three shifts? Use a 0.01 level of significance.

Number of Calls per Shift		
Morning	Afternoon	Night
2	3	5
3	4	4
2	4	5
4	5	3
3	3	2
1	2	5
3	5	6

16. For those of Exercises 12–15 in which there was enough evidence to support the claim that at least one of the population means differs from the others, determine, if possible, which population mean(s) is (are) different.

R Chapter 11 Review

Section 11.1: Hypothesis Testing: Two Population Means (σ Known)

Test Statistic for a Hypothesis Test for Two Population Means (σ Known)

Used when independent, simple random samples of sizes n_1 and n_2 are taken from populations with known standard deviations, σ_1 and σ_2, where either $n_2 \geq 30$ and $n_2 \geq 30$ or both population distributions are approximately normal, given by

$$z = \frac{(\bar{x}_1 - \bar{x}_2) - (\mu_1 - \mu_2)}{\sqrt{\frac{\sigma_1^2}{n_1} + \frac{\sigma_2^2}{n_2}}}$$

Section 11.2: Hypothesis Testing: Two Population Means (σ Unknown)

Rejection Regions for Hypothesis Tests for Two Population Means (σ Unknown)

Reject the null hypothesis, H_0, if:

$t \leq -t_\alpha$ for a left-tailed test

$t \geq t_\alpha$ for a right-tailed test

$|t| \geq t_{\alpha/2}$ for a two-tailed test

Test Statistic for a Hypothesis Test for Two Population Means (σ Unknown, Unequal Variances)

Used when independent, simple random samples of sizes n_1 and n_2 are taken from populations with standard deviations that are unknown and assumed to be unequal, where either $n_1 \geq 30$ and $n_2 \geq 30$ or both population distributions are approximately normal, given by

$$t = \frac{(\bar{x}_1 - \bar{x}_2) - (\mu_1 - \mu_2)}{\sqrt{\frac{s_1^2}{n_1} + \frac{s_2^2}{n_2}}}$$

with df = smaller of the values $n_1 - 1$ and $n_2 - 1$

Section 11.2: Hypothesis Testing: Two Population Means (σ Unknown) (cont.)

Test Statistic for a Hypothesis Test for Two Population Means (σ Unknown, Equal Variances)

Used when independent, simple random samples of sizes n_1 and n_2 are taken from populations with standard deviations that are unknown and assumed to be equal, where either $n_1 \geq 30$ and $n_2 \geq 30$ or both population distributions are approximately normal, given by

$$t = \frac{(\bar{x}_1 - \bar{x}_2) - (\mu_1 - \mu_2)}{\sqrt{\frac{(n_1 - 1)s_1^2 + (n_2 - 1)s_2^2}{n_1 + n_2 - 2}}\sqrt{\frac{1}{n_1} + \frac{1}{n_2}}}$$

with $df = n_1 + n_2 - 2$

Section 11.3: Hypothesis Testing: Two Population Means (σ Unknown, Dependent Samples)

Paired Difference

Used for paired data from two dependent samples, the difference between any pair of data values, given by

$$d = x_2 - x_1$$

Test Statistic for a Hypothesis Test for the Mean of the Paired Differences for Two Populations (σ Unknown, Dependent Samples)

Used when paired data from two dependent, simple random samples of size n are taken from populations with unknown standard deviations, where either $n \geq 30$ or the population distribution of the paired differences is approximately normal, given by

$$t = \frac{\bar{d} - \mu_d}{\left(\frac{s_d}{\sqrt{n}}\right)} \text{ with } df = n - 1$$

Rejection Regions for Hypothesis Tests for the Mean of the Paired Differences for Two Populations (σ Unknown, Dependent Samples)

Reject the null hypothesis, H_0, if:

$t \leq -t_\alpha$ for a left-tailed test

$t \geq t_\alpha$ for a right-tailed test

$|t| \geq t_{\alpha/2}$ for a two-tailed test

Section 11.4: Hypothesis Testing: Two Population Proportions

Test Statistic for a Hypothesis Test for Two Population Proportions

Used when independent, simple random samples of sizes n_1 and n_2 are large enough to ensure that $n_1\hat{p}_1 \ge 5$, $n_1\left(1-\hat{p}_1\right) \ge 5$, $n_2\hat{p}_2 \ge 5$, and $n_2\left(1-\hat{p}_2\right) \ge 5$, where the conditions for a binomial distribution are met for both samples, given by

$$z = \frac{\left(\hat{p}_1 - \hat{p}_2\right) - \left(p_1 - p_2\right)}{\sqrt{\bar{p}\left(1-\bar{p}\right)\left(\frac{1}{n_1} + \frac{1}{n_2}\right)}}$$

Weighted Estimate of the Common Population Proportion

An estimate of the common value of the two population proportions, found using the pooled data from both samples since the null hypothesis contains the assertion that the two population proportions are equal, given by

$$\bar{p} = \frac{x_1 + x_2}{n_1 + n_2}$$

Section 11.5: Hypothesis Testing: Two Population Variances

Properties of an *F*-Distribution

1. The F-distribution is skewed to the right.
2. The values of F are always greater than or equal to 0.
3. The shape of the F-distribution is completely determined by its two parameters, the degrees of freedom for the numerator and the degrees of freedom for the denominator.

Test Statistic for a Hypothesis Test for Two Population Variances

Used when independent, simple random samples of sizes n_1 and n_2 are taken, and both population distributions are approximately normal, given by

$$F = \frac{s_1^2}{s_2^2} \text{ with } df_1 = n_1 - 1 \text{ and } df_2 = n_2 - 1$$

Rejection Regions for Hypothesis Tests for Two Population Variances

Reject the null hypothesis, H_0, if:

$F \le F_{(1-\alpha)}$ for a left-tailed test

$F \ge F_{\alpha}$ for a right-tailed test

$F \le F_{(1-\alpha/2)}$ or $F \ge F_{\alpha/2}$ for a two-tailed test

Section 11.6: ANOVA (Analysis of Variance)

Definition

ANOVA

ANalysis **O**f **VA**riance, used to compare the means of three or more populations

Null and Alternative Hypotheses for an ANOVA Test

$H_0: \mu_1 = \mu_2 = \cdots = \mu_k$

H_a: At least one mean differs from the others.

Grand Mean

Weighted mean of the k sample means, one from each of the k populations, equivalent to the mean of all the sample data combined, given by

$$\bar{\bar{x}} = \frac{\sum_{i=1}^{k} (n_i \bar{x}_i)}{\sum_{i=1}^{k} n_i}$$

Sum of Squares among Treatments (SST)

Measures the variation between the sample means and the grand mean, given by

$$\text{SST} = \sum_{i=1}^{k} n_i \left(\bar{x}_i - \bar{\bar{x}}\right)^2$$

Sum of Squares for Error (SSE)

Measures the variation in the sample data resulting from the variability within each sample, given by

$$\text{SSE} = \sum_{j=1}^{n_1} \left(x_{1j} - \bar{x}_1\right)^2 + \sum_{j=1}^{n_2} \left(x_{2j} - \bar{x}_2\right)^2 + \cdots + \sum_{j=1}^{n_k} \left(x_{kj} - \bar{x}_k\right)^2$$

Total Variation

Also known as total sum of squares, sum of the variations contributed by each sample; that is, the sum of the squared deviations from the grand mean for all of the data values in each sample, given by

$$\begin{aligned}\text{Total Variation} &= \sum_{j=1}^{n_1} \left(x_{1j} - \bar{\bar{x}}\right)^2 + \sum_{j=1}^{n_2} \left(x_{2j} - \bar{\bar{x}}\right)^2 + \cdots + \sum_{j=1}^{n_k} \left(x_{kj} - \bar{\bar{x}}\right)^2 \\ &= \text{SST} + \text{SSE}\end{aligned}$$

Mean Square for Treatments (MST)

Found by dividing the sum of squares among treatments by its degrees of freedom, given by

$$\text{MST} = \frac{\text{SST}}{\text{DFT}} \text{ with DFT} = k - 1$$

Section 11.6: ANOVA (Analysis of Variance) (cont.)

Mean Square for Error (MSE)

Found by dividing the sum of squares for error by its degrees of freedom, given by

$$\text{MSE} = \frac{\text{SSE}}{\text{DFE}} \text{ with DFE} = n_T - k$$

Test Statistic for an ANOVA Test

Used when independent, simple random samples are taken from populations with variances that are unknown and assumed to be equal, where all of the k population distributions are approximately normal, given by

$$F = \frac{\text{MST}}{\text{MSE}} \text{ with } df_1 = \text{DFT} = k - 1 \text{ and } df_2 = \text{DFE} = n_T - k$$

Rejection Region for ANOVA Tests

Reject the null hypothesis, H_0, if:

$$F \geq F_\alpha$$

ANOVA Table

	SS	*df*	*MS*	*F*
Treatments (T)	SST	DFT	$\frac{\text{SST}}{\text{DFT}}$	$\frac{\text{MST}}{\text{MSE}}$
Error (E)	SSE	DFE	$\frac{\text{SSE}}{\text{DFE}}$	
Total	SST + SSE	DFT + DFE		

E Chapter 11 Exercises

Directions: Perform each of the following tests. Assume that the necessary criteria are met for each scenario.

1. Psychologists believe that test anxiety can reduce a student's score on an exam. An educational psychologist administers a test to two groups of students, one group that has test anxiety and one group that does not. Use the information in the following table to test the psychologists' claim at the 0.05 level of significance.

Test Scores			
	n	$\bar{x}$	σ
With Test Anxiety	35	65.9	8.9
Without Test Anxiety	38	73.2	11.5

2. A company that manufactures baseball bats believes that its new bat will allow players to hit the ball more than 30 feet farther, on average, than its current model. The managers hire a professional baseball player known for hitting home runs to hit 10 balls with each bat and they measure the distance each ball is hit to test their claim. The results of the batting experiment are shown below. Test the company's claim using a 0.05 level of significance. Assume that the variances of the two populations are equal.

Baseball Hitting Distances (in Feet)	
New Model	**Old Model**
235	200
240	210
253	231
267	218
243	210
237	209
250	210
241	229
251	234
248	231

3. Do husbands gain more weight after marriage than their wives? Answer this question by performing a hypothesis test using a 0.05 level of significance. Data on the amount of weight gained during the first year of marriage were collected from 10 couples and are given in the following table. Negative values indicate weight lost.

Weight Gained (in Pounds)										
Husband	10	15	8	7	3	10	12	20	–5	4
Wife	11	8	4	2	–1	8	15	15	–10	3

4. A blackjack pit boss at a casino is concerned that one of the tables is not generating the same revenue, on average, as the other three tables. The amount of revenue (in whole dollars) received during one particular three-hour period is recorded at each table each night for five nights. The results are given in the table below. Based on these data, can the pit boss conclude that at least one of the tables is not generating the same mean revenue as the other tables? Use a 0.05 level of significance.

Revenue (in Dollars)			
Table 1	**Table 2**	**Table 3**	**Table 4**
8534	9821	10,542	7367
7845	8997	9982	8021
8901	7905	8934	9034
9371	8923	9344	6703
7782	6675	8745	8432

5. The company manufacturing a new antibiotic ointment claims that it speeds healing time for minor cuts and scrapes. To test this claim, a researcher conducts an experiment involving 30 participants who have recently suffered a minor cut or scrape. Half of the group is asked to apply the new ointment twice daily, and the other half of the group is not allowed to treat the cuts with any medication. The researcher tracks the length of time it takes for each cut to heal. The cuts treated with the new ointment took a mean of 4.6 days to heal, with a standard deviation of 1.3 days. The cuts that did not receive treatment took a mean of 6.1 days to heal, with a standard deviation of 1.6 days. Test the company's claim at the 0.01 level of significance. Assume that the population variances are not equal.

6. Are people with low incomes more likely to cheat on their taxes than people with high incomes? To test this claim, a researcher anonymously surveys 50 taxpayers with low annual incomes and 50 taxpayers with high annual incomes. Of those with low incomes, 7 admit that they were not completely honest on their most recent federal income tax returns. Of those with high incomes, 5 admit that they were not entirely honest on their most recent federal income tax returns. Evaluate the claim by performing a hypothesis test at the 0.05 level of significance.

7. An education professor believes that single-gender classes do not have the same number of discipline problems as mixed-gender classes. For her study, the professor randomly selects 10 students from one elementary school. For one month, the students are separated into boys' and girls' classes. The next month, the students are assigned to classrooms with boys and girls. For each month, the number of times that each student was disciplined is recorded. Test the professor's claim at the 0.01 level of significance.

Number of Times Disciplined										
Single-Gender Class	1	2	0	4	3	6	1	2	0	0
Mixed-Gender Class	2	2	1	6	3	9	0	4	0	1

8. Do more women read to their children than men? A survey of 50 mothers found that 37 of them read to their children. A survey of 55 fathers found that 35 of them read to their children. Answer the question by performing a hypothesis test using a 0.05 level of significance.

9. A psychologist is convinced that attitude plays a central role in health. Along those lines, he believes that people who attend church regularly are healthier. The psychologist decides to test his claim by comparing the mean number of doctor's visits per year, excluding routine checkups, for adults who attend church regularly and those who do not. To test his claim, he randomly selects 75 adults who regularly attend church and finds that in the past year they had a mean of 1.4 doctor's visits. Assume that the population standard deviation for adults who regularly attend church is known to be 0.4 doctor's visits per year. He also randomly selects 75 adults who do not regularly attend church, and he finds that in the past year they had a mean of 1.6 doctor's visits. Assume that the population standard deviation for adults who do not regularly attend church is known to be 0.5 doctor's visits per year. Use the 0.05 level of significance to test this claim.

10. A nutritionist believes that the variance of the weights, measured in pounds, of women on diets is less than the variance of the weights of women in the general population. The sample variance for a sample of 15 women on diets is 256.02. The sample variance for a sample of 15 other women is 367.91. Test the nutritionist's claim using a 0.05 level of significance. Does the evidence support the nutritionist's claim?

P Chapter 11 Project

Project A: Hypothesis Testing for Two Population Parameters

Directions: Choose one of the three questions to answer, collect data from members of the appropriate population, and perform a hypothesis test to answer the question. After you have written your conclusion, look at the "truth" and determine if your hypothesis test produced a correct decision, a Type I error, or a Type II error.

Pick one of the following questions to test:

- Do women in college spend more time studying than men in college? Ask at least 30 women and at least 30 men to estimate the amount of time they spend studying each week. Assume that the population variances are equal. Use a 0.10 level of significance.
- Do college freshmen spend less money each week eating out than seniors? Ask between 10 and 20 freshmen and between 10 and 20 seniors to estimate the amount of money they spend each week eating out. Assume that the population variances are different and both population distributions are approximately normal. Use a 0.05 level of significance.
- Are the percentages of men and women who exercise regularly the same? Ask at least 30 men and at least 30 women if they exercise at least three times per week. (**Note:** The participants do not need to be in college.) Record the number of men and the number of women who say "yes." Use a 0.01 level of significance.

Step 1: State the null and alternative hypotheses.

What are the null and alternative hypotheses?

Step 2: Determine which distribution to use for the test statistic, and state the level of significance.

Based on the description of the test you chose, what formula should be used for the test statistic? Also state the level of significance for your hypothesis test.

Step 3: Gather data and calculate the necessary sample statistics.

Collect data on the claim from the appropriate populations. Discuss which method of data collection you used. List any potential for bias. Calculate the sample statistics needed in order to compute the test statistic.

Calculate the test statistic using your sample statistics.

Step 4: Draw a conclusion and interpret the decision.

Determine the type of your hypothesis test: left-tailed, right-tailed, or two-tailed.

State the decision rule in terms of either the *p*-value or the rejection region for the test statistic.

What is your conclusion? Be sure to answer the original question.

Types of Errors

Suppose the "truths" are as follows.

- The mean amount of time spent studying each week is higher for women in college than for men in college.
- The mean amount of money spent eating out each week is higher for freshmen than for seniors.
- The percentage of women who exercise at least three times per week is the same as the percentage of men.

Based on your conclusion, did you make a Type I error, Type II error, or a correct decision? Explain.

Project B: ANOVA

A lack of adequate parking is one of the most common complaints of students on any college campus. Is there a difference between the mean numbers of parking tickets received in one semester by students who commute to campus, students who live in residential housing on campus, students who live in fraternity houses, and students who live in sorority houses? Let's perform an ANOVA test to help us answer this question.

To begin, label the populations as follows.

Population 1: Students who commute to campus

Population 2: Students living in residential housing on campus

Population 3: Students living in fraternity houses

Population 4: Students living in sorority houses

Step 1: State the null and alternative hypotheses.

What are the null and alternative hypotheses?

Step 2: Determine which distribution to use for the test statistic, and state the level of significance.

Assuming that the population distributions are all approximately normal and the population variances are all equal, what formula should be used for the test statistic? Also, choose a level of significance of 0.10, 0.05, or 0.01.

Step 3: Gather data and calculate the necessary sample statistics.

Collect data from five students in each population. Record the number of parking tickets that each student received last semester in a table similar to the one below.

Parking Tickets			
Commuters	**Residence Hall**	**Fraternity House**	**Sorority House**

Use the formulas given in Section 11.6, or available technology (as described in Example 11.19 and in the Chapter 11 Technology section), to complete an ANOVA table.

Step 4: Draw a conclusion and interpret the decision.

a. Draw a picture of your rejection region.

b. Based on the calculated value of F from the ANOVA table, should you reject the null hypothesis?

c. Is there a difference between the mean numbers of parking tickets received by students who commute to campus, students who live in residential housing on campus, students who live in fraternity houses, and students who live in sorority houses?

d. Does your hypothesis test provide any indication that one of the populations receives more parking tickets than another? Explain.

T Chapter 11 Technology

Performing a Hypothesis Test for Two or More Populations

TI-83/84 Plus

Comparing Two Population Means (σ Known)

If we have the raw data, we can use a TI-83/84 Plus calculator, Microsoft Excel, or MINITAB to perform a hypothesis test to compare two or more population parameters. If we only have the sample statistics, a TI-83/84 Plus calculator is much easier to use.

Side Note

The most recent TI-84 Plus calculators (Jan. 2011 and later) contain a new feature called `STAT WIZARDS`. This feature is not used in the directions in this text. By default this feature is turned `ON`. `STAT WIZARDS` can be turned `OFF` under the second page of MODE options.

Example T.1

Using a TI-83/84 Plus Calculator to Perform a Hypothesis Test for Two Population Means (Left-Tailed, σ Known)

A researcher is looking at the study habits of college students. She believes that freshmen study less than seniors. A random sample of 42 freshmen had a mean study time per week of 19.1 hours. Assume that the population standard deviation for freshmen is known to be 4.7 hours per week. A group of 39 seniors reported a mean study time per week of 21.0 hours. Assume that the population standard deviation for seniors is known to be 5.2 hours per week. Test the researcher's claim with a 0.05 level of significance.

Solution

Step 1: State the null and alternative hypotheses.

Let freshmen be Population 1 and seniors be Population 2. The researcher's claim is that freshmen study less than seniors. That is, the mean study time per week is lower for freshmen than for seniors. Thus, we have the following hypotheses.

$$H_0: \mu_1 - \mu_2 \ge 0$$
$$H_a: \mu_1 - \mu_2 < 0$$

Step 2: Determine which distribution to use for the test statistic, and state the level of significance.

We are looking at the difference between two population means when both population standard deviations are known. We also know that independent random samples were drawn, and both sample sizes are at least 30. Therefore the z-test statistic, which has a standard normal distribution, is appropriate.

We will draw a conclusion by computing the p-value for the calculated test statistic and comparing that value to the level of significance, α. For this hypothesis test, $\alpha = 0.05$.

Step 3: Gather data and calculate the necessary sample statistics.

The following values were given in the problem.

$$\bar{x}_1 = 19.1 \qquad \bar{x}_2 = 21.0$$
$$\sigma_1 = 4.7 \qquad \sigma_2 = 5.2$$
$$n_1 = 42 \qquad n_2 = 39$$

To use a TI-83/84 Plus to perform this test, press **STAT**, scroll over to `TESTS`, and choose option `3:2-SampZTest`. Choose `Stats`, since we know the sample statistics. Fill in the required values and choose the symbol for a left-tailed test, as shown in the following screenshots.

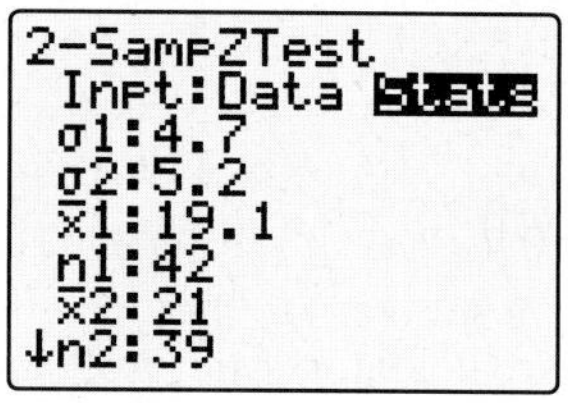

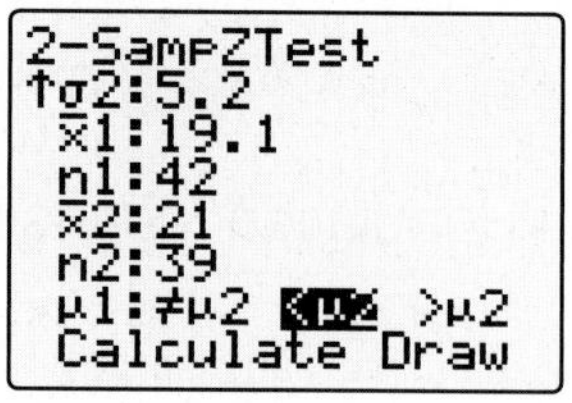

Choosing `Calculate` produces the following results.

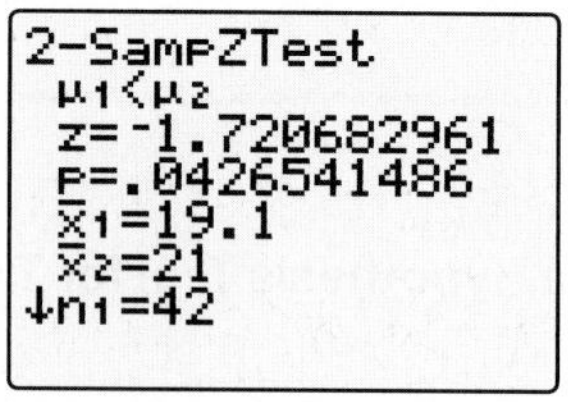

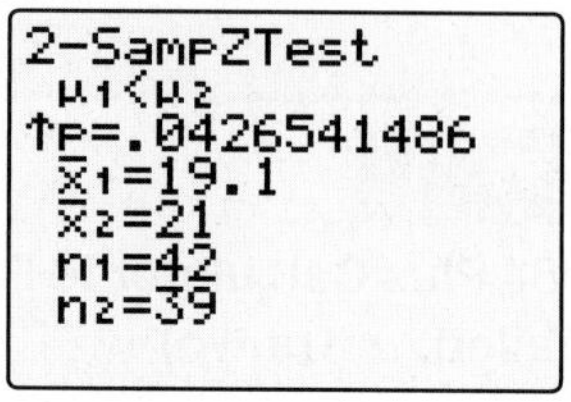

First, the alternative hypothesis is stated. Note that the form of the alternative hypothesis is different than, but equivalent to, the one we used in this chapter. The z-score is approximately -1.72 and the p-value is approximately 0.0427. The sample statistics are then repeated.

Step 4: **Draw a conclusion and interpret the decision.**

Since the p-value, 0.0427, is less than $\alpha = 0.05$, we have $p\text{-value} \le \alpha$, so we reject the null hypothesis. Therefore, there is sufficient evidence at the 0.05 level of significance to say that the mean study time per week is lower for freshmen than for seniors. Thus, the evidence supports the researcher's claim that freshmen study less than seniors.

Comparing Two Population Means (σ Unknown)

Let's look at how to use a TI-83/84 Plus to perform a hypothesis test that compares two population means using independent samples drawn from two populations for which neither standard deviation is known. The process is the same for pooled sample variances (population variances are assumed to be equal) as for sample variances that are not pooled (population variances are assumed to be unequal). Whether or not the sample variances are pooled is just one of the options in the menu.

Example T.2

Using a TI-83/84 Plus Calculator to Perform a Hypothesis Test for Two Population Means (Left-Tailed, σ Unknown, Equal Variances)

Obstetricians are concerned that a certain pain reliever may be causing lower birth weights. A medical researcher collects data from a random sample of 12 mothers who took the pain reliever while pregnant and calculates that the mean birth weight of their babies was 5.6 pounds with a standard deviation of 1.8 pounds. She also collects data from a random sample of 20 mothers who did not take the pain reliever while pregnant and calculates that the mean birth weight of their babies was 6.3 pounds with a standard deviation of 2.1 pounds. Assume that the distributions of birth weights are approximately normal for both populations. Assuming that the population variances are equal, test the hypothesis that using the pain reliever during pregnancy results in a lower mean birth weight. Use a 0.01 level of significance.

Solution

Step 1: State the null and alternative hypotheses.

Let babies whose mothers who took the pain reliever during pregnancy be Population 1. The claim being tested is that the mean birth weight is lower for babies whose mothers took the pain reliever during pregnancy, that is, $\mu_1 < \mu_2$. Thus, we have the following hypotheses.

$$H_0: \mu_1 - \mu_2 \ge 0$$
$$H_a: \mu_1 - \mu_2 < 0$$

Step 2: Determine which distribution to use for the test statistic, and state the level of significance.

We are looking at the difference between two population means when both population variances are unknown but assumed to be equal. We also know that independent random samples were drawn, and both population distributions are approximately normal. Therefore, the *t*-test statistic for equal variances is appropriate.

We will draw a conclusion by computing the *p*-value for the calculated test statistic and comparing that value to the level of significance, α. For this hypothesis test, $\alpha = 0.01$.

Step 3: Gather data and calculate the necessary sample statistics.

The following statistics were given in the problem.

$$\bar{x}_1 = 5.6 \qquad \bar{x}_2 = 6.3$$
$$s_1 = 1.8 \qquad s_2 = 2.1$$
$$n_1 = 12 \qquad n_2 = 20$$

We are also told that the population variances are assumed to be equal, so we will use the test statistic for equal variances. Now we are ready to enter the statistics in the calculator. Press STAT, scroll over to TESTS, and choose option `4:2-SampTTest`. Choose `Stats`, and then enter the requested statistics. This is a left-tailed test, so choose `<μ2` for the alternative hypothesis. Since the variances are assumed to be equal, choose `Yes` next to `Pooled`.

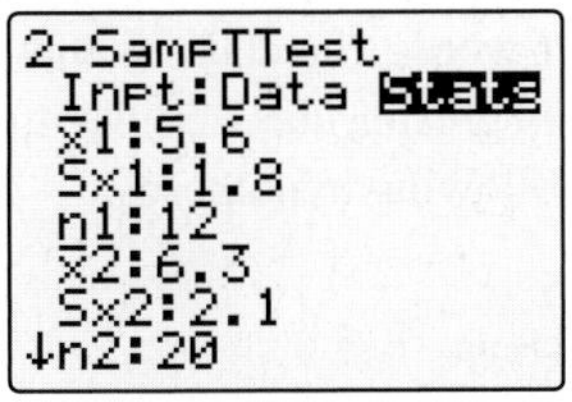

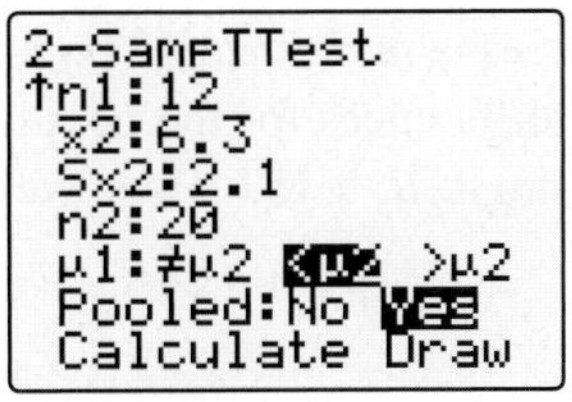

Choosing `Calculate` produces the following results.

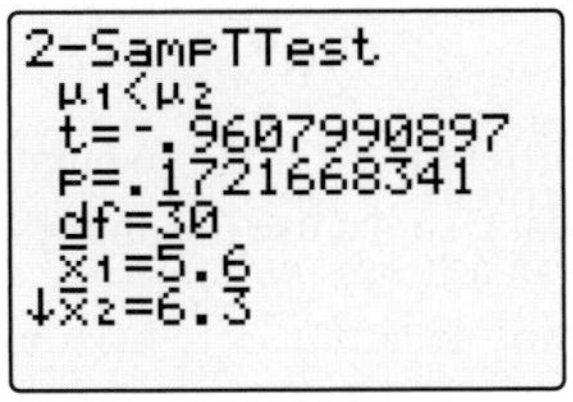

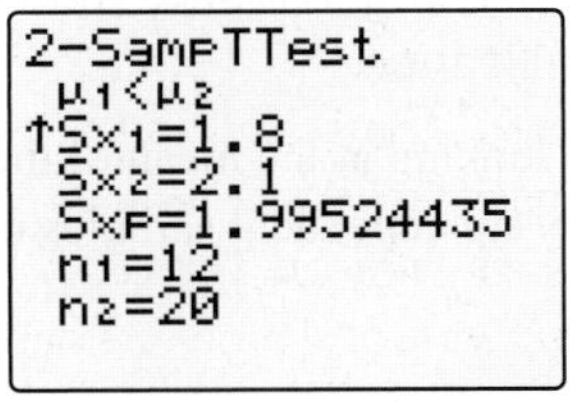

Step 4: **Draw a conclusion and interpret the decision.**

The p-value, which is approximately 0.1722, is greater than the significance level, $\alpha = 0.01$, so we fail to reject the null hypothesis. Thus, there is not sufficient evidence at the 0.01 level to say that the mean birth weight is lower for babies whose mothers took the pain reliever during pregnancy. This means that the evidence does not support the claim that the pain reliever causes lower birth weights.

Comparing Two Population Means (σ Unknown, Dependent Samples)

The TI-83/84 Plus calculator does not directly perform a hypothesis test that compares two population means using dependent samples, but we can perform a one-sample t-test using the paired differences as seen in the following example.

Example T.3

Using a TI-83/84 Plus Calculator to Perform a Hypothesis Test for the Mean of the Paired Differences for Two Populations (Left-Tailed, σ Unknown, Dependent Samples)

In a CPR class, students are given a pretest at the beginning of the class to determine their initial knowledge of CPR and then a posttest at the end of the class to determine their new knowledge of CPR. Educational researchers are interested in whether the class increases a student's knowledge of CPR. Test the claim that the mean increase in the students' CPR test scores is more than 15 points using a 0.05 level of significance. Data from 10 students who took the CPR class are listed in the following table.

Students' CPR Test Scores										
Pretest Score	60	63	68	70	71	68	72	80	83	79
Posttest Score	80	79	83	90	83	89	94	95	96	93

Solution

Step 1: **State the null and alternative hypotheses.**

Let the pretest scores be Population 1 and posttest scores be Population 2. We want to subtract the pretest score from the posttest score to calculate each paired difference. The claim being tested is that the mean increase in the students' CPR test scores is more than 15 points. In other words, the claim is that the mean of the paired differences for the population data is greater than 15. Written symbolically, this claim is $\mu_d > 15$. Thus, we have the following hypotheses.

$$H_0: \mu_d \le 15$$
$$H_a: \mu_d > 15$$

Step 2: **Determine which distribution to use for the test statistic, and state the level of significance.**

We are looking at the mean of the paired differences for the population data, and the samples are dependent samples of paired data, so we use t as the test statistic.

We will draw a conclusion by computing the p-value for the calculated test statistic and comparing that value to the level of significance, α. For this hypothesis test, $\alpha = 0.05$.

Step 3: **Gather data and calculate the necessary sample statistics.**

First, let's enter the pretest scores in L1 and the posttest scores in L2. Next, we want to calculate the paired differences in L3. To do so, highlight L3 and enter the formula to subtract the pretest scores from the posttest scores by pressing 2ND 2 − 2ND 1 and then ENTER. This will produce the formula L2−L1 and calculate the paired difference for each pair of data.

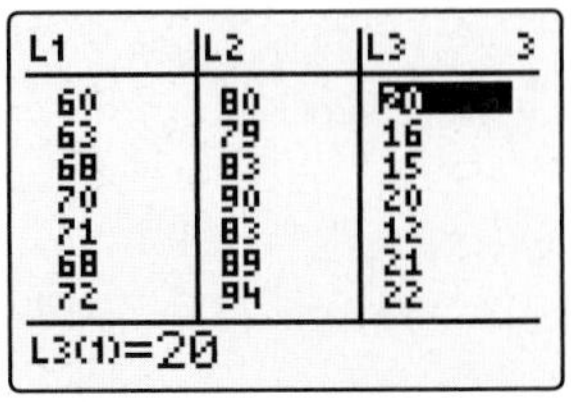

We now have the paired differences in L3, and can perform a one-sample t-test using the paired differences as our raw data. Press STAT, scroll to TESTS, and choose option 2:T-Test. We want to perform the hypothesis test using raw data, so choose the Data option. The presumed value of the mean of the paired differences from the null hypothesis, μ_0, is 15. The data are in List 3, so enter L3 by pressing 2ND and then 3. The frequency of the data (Freq) is the default value, which is 1. We have a right-tailed test, so choose $>\mu_0$ for the alternative hypothesis. The menu should then appear as it does in the following screenshot on the left. Choosing Calculate produces the results shown in the screenshot on the right.

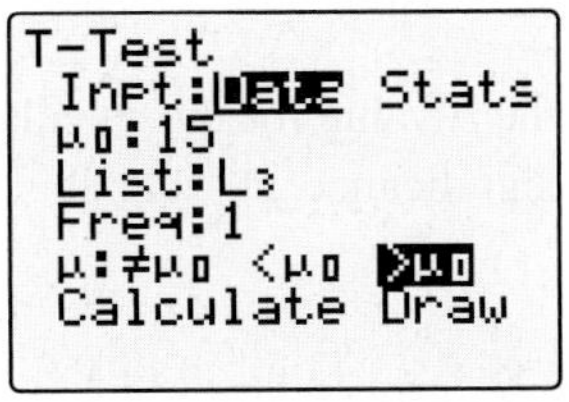

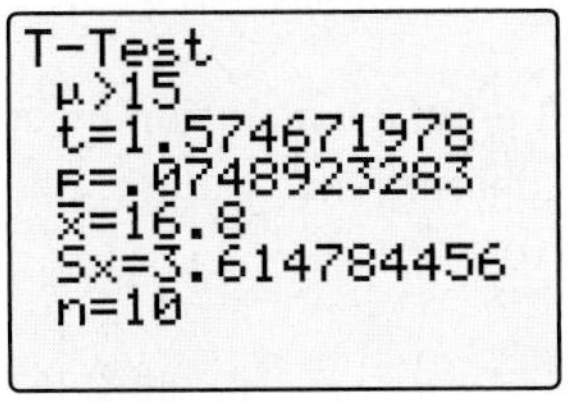

Step 4: **Draw a conclusion and interpret the decision.**

These results show that the p-value ≈ 0.0749. Since $0.0749 > 0.05$, we have p-value $> \alpha$, so we fail to reject the null hypothesis. Thus, there is not enough evidence at the 0.05 level of significance to support the claim that the mean increase in the students' CPR test scores is more than 15 points.

Comparing Two Population Proportions

A TI-83/84 Plus can perform a hypothesis test for two population proportions very simply. Let's look at an example.

Example T.4

Using a TI-83/84 Plus Calculator to Perform a Hypothesis Test for Two Population Proportions (Left-Tailed)

School administrators believe that the percentage of students at a school in a lower economic district (School A) who carry cell phones is smaller than the percentage of students who carry cell phones at a school in an upper economic district (School B). A survey of students is conducted at each school and the following results are tabulated.

Responses to Survey		
	School A	**School B**
Carry a Cell Phone	45	53
Do Not Carry a Cell Phone	31	25

Test the administrators' claim using p-values and a 0.10 level of significance.

Solution

Step 1: State the null and alternative hypotheses.

Let students at School A be Population 1 and students at School B be Population 2. The administrators' claim is that $p_1 < p_2$. Thus, we have the following hypotheses.

$$H_0: p_1 - p_2 \geq 0$$
$$H_a: p_1 - p_2 < 0$$

Step 2: Determine which distribution to use for the test statistic, and state the level of significance.

We are looking at the difference between two population proportions using large independent samples, so we use z as the test statistic.

We will draw a conclusion by computing the p-value for the calculated test statistic and comparing that value to the level of significance, α. For this hypothesis test, $\alpha = 0.10$.

Step 3: Gather data and calculate the necessary sample statistics.

We need two pieces of information from each sample to use the calculator to run a hypothesis test for two population proportions: x, the number in the sample with the characteristic we are evaluating, and n, the sample size. The value for x in each of the two samples is the number of students who carry cell phones. To find the sample size for each school, we simply add the numbers of students who do and do not carry cell phones. So, we have $x_1 = 45$, $n_1 = 45 + 31 = 76$ and $x_2 = 53$, $n_2 = 53 + 25 = 78$.

Enter the values into the calculator by pressing STAT, scrolling over to TESTS, and then choosing option 6:2-PropZTest. First, enter the values of x and n for the two samples. Since this is a left-tailed test, choose <p2 for the alternative hypothesis. The menu should then appear as it does in the following screenshot on the left. Choosing Calculate produces the results shown in the screenshot on the right.

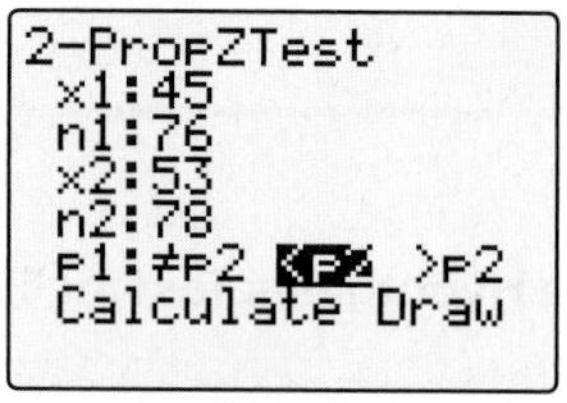

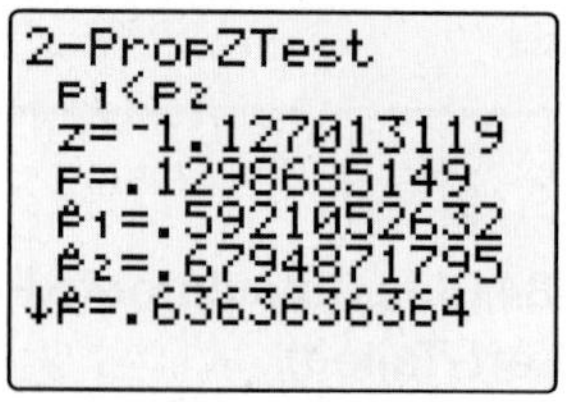

Step 4: Draw a conclusion and interpret the decision.

Since the p-value is approximately 0.1299, which is greater than the level of significance, we fail to reject the null hypothesis. Therefore, there is not sufficient evidence to support the school administrators' belief that a smaller percentage of students carry a cell phone at School A in the lower economic district at $\alpha = 0.10$.

Comparing Two Population Variances

The TI-83/84 Plus cannot construct a confidence interval for comparing two population variances, but it can perform a hypothesis test using the F-test statistic. Let's look at how to use the TI-83/84 Plus to solve the following problem.

Example T.5

Using a TI-83/84 Plus Calculator to Perform a Hypothesis Test for Two Population Variances (Left-Tailed)

A graduate student walking around a large college campus notices that the values of professors' cars seem to have a much smaller variance than the values of the cars owned by students. To test his hypothesis, he collects data from random samples of 15 professors and 14 students on the values of their cars and calculates the variance for each sample. The sample variance of the values of the professors' cars is 34,057 and the sample variance of the values of the students' cars is 45,923. Assuming that both populations are normally distributed, conduct a hypothesis test using a 0.10 level of significance to test the graduate student's claim. Does the evidence support the student's claim?

Solution

Step 1: **State the null and alternative hypotheses.**

Let the professors be Population 1 and the students be Population 2. The claim is that $\sigma_1^2 < \sigma_2^2$. Thus, we have the following hypotheses.

$$H_0: \sigma_1^2 \geq \sigma_2^2$$
$$H_a: \sigma_1^2 < \sigma_2^2$$

However, a TI-83/84 Plus calculator will only perform an F-test to compare two population standard deviations. Since variance is necessarily a positive number, we can convert these hypotheses regarding variance to hypotheses regarding standard deviation as follows.

$$H_0: \sigma_1 \geq \sigma_2$$
$$H_a: \sigma_1 < \sigma_2$$

Step 2: **Determine which distribution to use for the test statistic, and state the level of significance.**

Since we are comparing the standard deviations of two normally distributed populations using independent, simple random samples, we will use the ratio of the sample variances, F, as the test statistic.

We will draw a conclusion by computing the p-value for the calculated test statistic and comparing that value to the level of significance, α. For this hypothesis test, $\alpha = 0.10$.

Step 3: **Gather data and calculate the necessary sample statistics.**

The following values were given in the problem.

$$n_1 = 15 \qquad n_2 = 14$$
$$s_1^2 = 34{,}057 \qquad s_2^2 = 45{,}923$$

In addition to these sample statistics, we need the sample standard deviations. Taking the square root of each sample variance gives us the following.

$$s_1 \approx 184.545387 \qquad s_2 \approx 214.296524$$

Side Note

On TI-83 Plus calculators and older TI-84 Plus calculators, the function, `2-SampFTest`, is option `D` on the `TESTS` menu.

Press STAT. Scroll over to `TESTS` and choose option `E:2-SampFTest`. We want to enter the statistics, so choose `Stats` and enter the values as shown in the following screenshot. This is a left-tailed test, so choose the `<σ2` option for the alternative hypothesis. The menu should appear as it does in the following screenshot on the left. Then choose `Calculate` and press ENTER. The results are shown in the screenshot on the right.

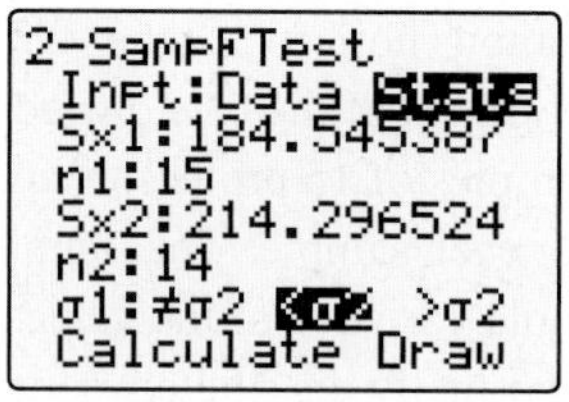

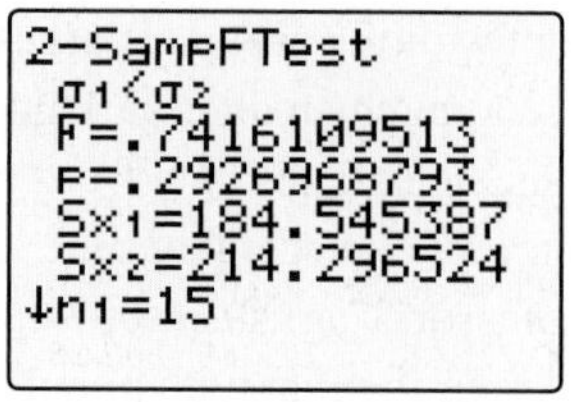

The first line is the alternative hypothesis. The second line is the F-test statistic. The third line is the p-value. The following lines repeat the statistics we entered.

Step 4: Draw a conclusion and interpret the decision.

Since the p-value is approximately 0.2927, which is greater than the level of significance, we fail to reject the null hypothesis. Therefore, there is not sufficient evidence at the 0.10 level of significance to support the graduate student's claim that the standard deviation (or variance) of the values of professors' cars is less than the standard deviation (or variance) of the values of students' cars.

ANOVA (Analysis of Variance)

In order to use either a TI-83/84 Plus or Microsoft Excel to calculate an ANOVA test, you must have the raw data.

Example T.6

Using a TI-83/84 Plus Calculator to Perform an ANOVA Test

Use a TI-83/84 Plus calculator to perform an ANOVA test to compare the means of three populations. Assume that the three population distributions are all approximately normal with equal population variances. The sample data from each of these populations are given in the following table. Use a 0.05 level of significance.

Sample Data							
Sample 1	2	3	4	5	5	3	6
Sample 2	2	2	3	4	5	6	2
Sample 3	3	5	5	3	6	4	2

Solution

To use a TI-83/84 Plus calculator to perform an ANOVA test, begin by entering the sample data in the list environment. Enter the data from Sample 1 in `L1`, the data from Sample 2 in `L2`, and the data from Sample 3 in `L3`. Next, press STAT, scroll to `TESTS`, and choose option `H:ANOVA(`. It will show `ANOVA(` on the screen. You need to enter the lists that contain the sample data for your ANOVA test. In our case, we enter `ANOVA(L1,L2,L3)`. Press ENTER. The following results will be displayed.

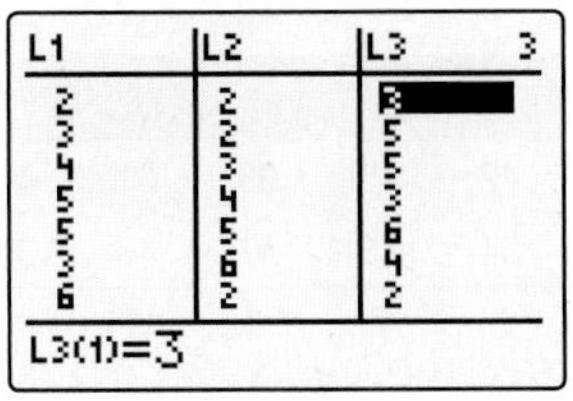

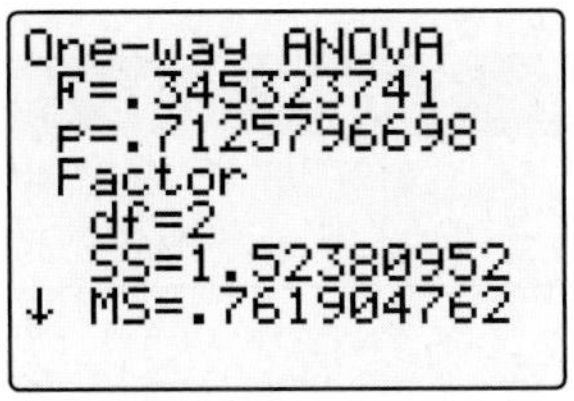

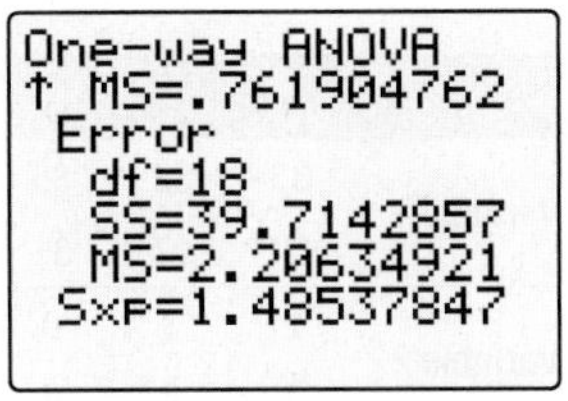

A traditional ANOVA table is not displayed; however, all of the values that would be in an ANOVA table are present. The first is the value of F, then the p-value. The values under `Factor` are the Treatment values and the values under `Error` are the Error values. Since the p-value, approximately 0.7126, is much larger than the level of significance, we fail to reject the null hypothesis. Thus, there is not enough evidence at the 0.05 level of significance to support the claim that at least one of the three population means is different from the others.

Side Note

On TI-83 Plus calculators and older TI-84 Plus calculators, the function, **ANOVA**, is option **F** on the **TESTS** menu.

Microsoft Excel

Hypothesis Tests for Two or More Populations

If you have the raw data available, Microsoft Excel has several data analysis tools that can simplify hypothesis testing for two or more populations, as demonstrated in Examples T.7 and T.9. The following table summarizes the options that are available in the Data Analysis menu under the Data tab.

Table T.1: Excel Data Analysis Tools

Hypothesis Test	Excel Data Analysis Tool
Two Population Means (σ Known)	z-Test: Two Sample for Means
Two Population Means (σ Unknown, Unequal Variances)	t-Test: Two-Sample Assuming Unequal Variances
Two Population Means (σ Unknown, Equal Variances)	t-Test: Two-Sample Assuming Equal Variances
Mean of the Paired Differences for Two Populations (σ Unknown, Dependent Samples)	t-Test: Paired Two Sample for Means
Two Population Variances	F-Test Two-Sample for Variances
ANOVA	ANOVA: Single Factor

The data analysis tools listed in Table T.1 can only be used with raw data. If only the sample statistics are known, or a data analysis tool is not available in Excel, you can use Excel formulas to calculate the test statistic and p-value or critical value to perform a hypothesis test, as shown in Examples T.8, T.10, and T.11.

	A	B
1	Variable 1	Variable 2
2	35	2
3	32	3
4	36	6
5	39	5
6	34	4
7	31	9
8	37	8
9	39	2
10	40	3
11	41	6
12	42	5
13	45	7
14	38	8
15	31	9
16	46	4
17	39	1
18	47	2
19	31	6
20	30	1
21	32	3
22	46	5
23	47	6
24	46	9
25	49	8
26	48	4
27	34	6
28	35	5
29	47	3
30	46	4
31	35	6

Side Note

If Data Analysis does not appear in your Excel menu under the Data tab, you can easily add this function. See the directions in Appendix B: Getting Started with Microsoft Excel.

Example T.7

Using Microsoft Excel to Perform a Hypothesis Test for Two Population Means (Right-Tailed, σ Known)

The following data were collected on two different independent variables. Test the claim that the difference between the two population means is greater than 34, with a level of significance of 0.05. Assume that the variance of the first population is known to be 37.93 and the variance of the second population is known to be 5.65.

Sample Data															
Variable 1	35	32	36	39	34	31	37	39	40	41	42	45	38	31	46
	39	47	31	30	32	46	47	46	49	48	34	35	47	46	35
Variable 2	2	3	6	5	4	9	8	2	3	6	5	7	8	9	4
	1	2	6	1	3	5	6	9	8	4	6	5	3	4	6

Solution

To begin, we need to enter our data in the worksheet. Enter the data for Variable 1 in column A and the data for Variable 2 in column B as shown in the screenshot in the margin. Let Variable 1 represent Population 1 and Variable 2 represent Population 2. The hypotheses for this test are as follows.

$$H_0: \mu_1 - \mu_2 \le 34$$
$$H_a: \mu_1 - \mu_2 > 34$$

We are looking at the difference between two population means when both population standard deviations are known. We also know that independent random samples were drawn, and both sample sizes are at least 30. Therefore the z-test statistic, which has a standard normal distribution, is appropriate.

Next, choose the **Data Analysis** option under the **Data** tab. Then, choose **z-Test: Two Sample for Means** and click **OK**. You are prompted to enter two ranges. The first (*Variable 1 Range*) corresponds to the data for Variable 1 and the second (*Variable 2 Range*) corresponds to the data for Variable 2. Enter **34** as the *Hypothesized Mean Difference*, **37.93** for *Variable 1 Variance (known)*, **5.65** for *Variable 2 Variance (known)*, **0.05** for *Alpha*, and **D2** for the *Output Range*. This dialog box is shown in the following screenshot.

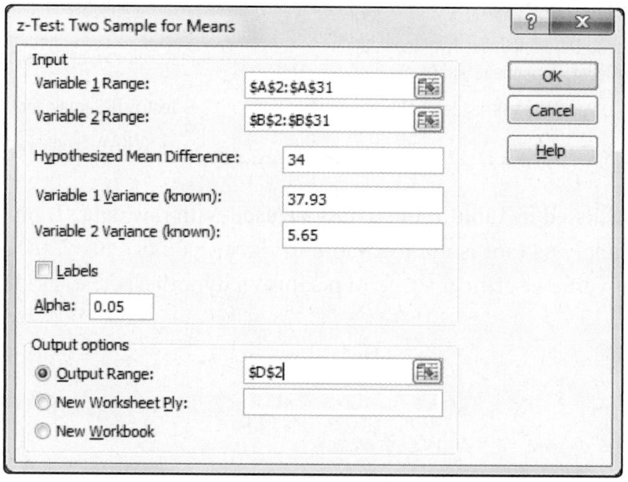

After clicking **OK**, the following results are displayed.

z-Test: Two Sample for Means		
	Variable 1	*Variable 2*
Mean	39.26666667	5
Known Variance	37.93	5.65
Observations	30	30
Hypothesized Mean Difference	34	
z	0.221251257	
P(Z<=z) one-tail	0.4124484	
z Critical one-tail	1.644853627	
P(Z<=z) two-tail	0.824896801	
z Critical two-tail	1.959963985	

This is a one-tailed test, so we will use the p-value for the one-tailed test, labeled "P(Z<=z) one-tail." Since the p-value ≈ 0.4124, which is greater than $\alpha = 0.05$, we fail to reject the null hypothesis. Thus, there is not sufficient evidence, at the 0.05 level of significance, to support the claim that the difference between the two population means is greater than 34.

Example T.8

Using Microsoft Excel to Perform a Hypothesis Test for Two Population Means (Two-Tailed, σ Unknown, Equal Variances)

A fleet supervisor tracks the mileage and fuel consumption of two types of vehicles. A sample of 24 vehicles of the first type from the fleet had a sample mean of 22.1 miles per gallon (mpg) with a standard deviation of 1.5 mpg. A sample of 36 vehicles of the second type had a mean of 19.5 mpg with a standard deviation of 2.3 mpg. Can the fleet supervisor conclude with 99% confidence that there is a difference in fuel efficiency between the two vehicle types? Assume that both population variances are the same and the distributions of gas mileage rates are approximately normal for both types of vehicles.

Solution

We are looking at the difference between two population means when both population variances are unknown but assumed to be equal. We also know that independent random samples were drawn, and both population distributions are approximately normal. Thus, we will use the t-test statistic for equal variances. We state the hypotheses.

$$H_0: \mu_1 - \mu_2 = 0$$
$$H_a: \mu_1 - \mu_2 \neq 0$$

This is a two-tailed test with the following statistics.

$$\bar{x}_1 = 22.1 \qquad \bar{x}_2 = 19.5$$
$$s_1 = 1.5 \qquad s_2 = 2.3$$
$$n_1 = 24 \qquad n_2 = 36$$

First, enter these statistics in cells B2 through C4 as shown in the screenshot. Recall that the test statistic is given by the following formula.

$$t = \frac{(\bar{x}_1 - \bar{x}_2) - (\mu_1 - \mu_2)}{\sqrt{\frac{(n_1 - 1)s_1^2 + (n_2 - 1)s_2^2}{n_1 + n_2 - 2}}\sqrt{\frac{1}{n_1} + \frac{1}{n_2}}}$$

Next, compute the test statistic by entering the following formula in cell B6.

=(B2-C2-0)/(SQRT(((B4-1)*B3^2+(C4-1)*C3^2)/(B4+C4-2))*SQRT(1/B4+1/C4))

Since this is a two-tailed test using pooled sample variances, the probability used in the T.INV.2T function is $\alpha = 0.01$ and the degrees of freedom are $n_1 + n_2 - 2$. Enter **=T.INV.2T(0.01, B4+C4-2)** to calculate the critical value $t_{\alpha/2} = t_{0.005}$.

	A	B	C
1		Vehicle Type 1	Vehicle Type 2
2	sample mean	22.1	19.5
3	sample standard deviation	1.5	2.3
4	*n*	24	36
5			
6	*t*	4.88186831	
7	critical *t*	2.66328695	

We see that the test statistic, $t \approx 4.882$, is greater than the critical value, $t_{0.005} \approx 2.663$. Therefore, the null hypothesis is rejected, and the fleet supervisor can conclude with 99% confidence that the mean gas mileage rates for the two populations of vehicles are not equal.

Example T.9

Using Microsoft Excel to Perform a Hypothesis Test for the Mean of the Paired Differences for Two Populations (Right-Tailed, σ Unknown, Dependent Samples)

Juan believes that physical exercise leads to a mean increase in pulse rate of more than 15 beats per minute. He measures the pulse rates of 15 volunteers before they run a mile and after they run a mile. The results are as follows. Is there evidence at the 0.10 level to support Juan's claim?

Pulse Rates (in Beats per Minute)															
Before	65	75	73	78	62	58	76	79	64	73	78	55	63	69	72
After	78	105	92	96	79	65	91	106	82	89	99	70	79	92	93

Solution

Since the data are in pairs, we will perform a paired two-sample test for means. We state the hypotheses.

$$H_0: \mu_d \le 15$$
$$H_a: \mu_d > 15$$

To begin, we need to enter our data in the worksheet. Enter the Before data in column A and the After data in column B as shown in the screenshot at the end of this example.

Under the **Data** tab, select **Data Analysis**. Next, choose **t-Test: Paired Two Sample for Means** and click **OK**. You are prompted to enter two ranges. Note that Excel calculates the paired differences by subtracting the values for Variable 2 from the values for Variable 1, which is the opposite of what we do when we calculate them by hand or using a TI-83/84 Plus calculator. Thus, for this example, *Variable 1 Range* corresponds to the pulse rates after exercise and *Variable 2 Range* corresponds to pulse rates before exercise. Select **Labels** if the first cells in your variable ranges are data labels. Enter **15** as the *Hypothesized Mean*

Difference, **0.1** for *Alpha*, and **D1** for the *Output Range*. This dialog box is shown in the following screenshot.

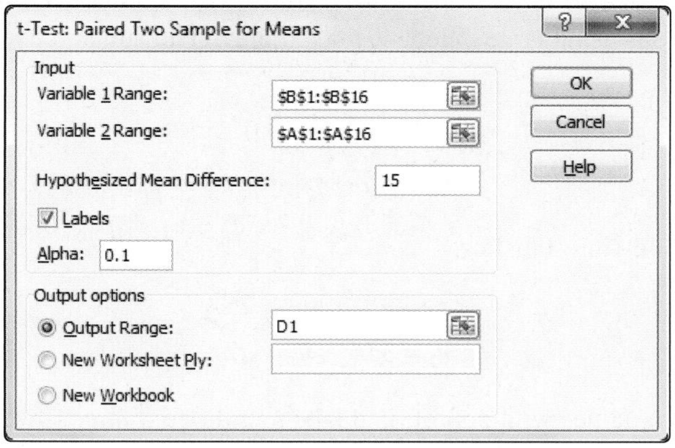

Clicking **OK** produces the completed test.

	A	B	C	D	E	F
1	Before	After		t-Test: Paired Two Sample for Means		
2	65	78				
3	75	105			*After*	*Before*
4	73	92		Mean	87.73333333	69.33333333
5	78	96		Variance	143.9238095	59.23809524
6	62	79		Observations	15	15
7	58	65		Pearson Correlation	0.930100732	
8	76	91		Hypothesized Mean Difference	15	
9	79	106		df	14	
10	64	82		t Stat	2.349955956	
11	73	89		P(T<=t) one-tail	0.016983142	
12	78	99		t Critical one-tail	1.345030374	
13	55	70		P(T<=t) two-tail	0.033966283	
14	63	79		t Critical two-tail	1.761310136	
15	69	92				
16	72	93				

Since the p-value for the one-tailed test, 0.0170, is less than 0.10, we have $p\text{-value} \le \alpha$, so we reject the null hypothesis. Therefore, there is sufficient evidence, at the 0.10 level of significance, to support the claim that exercise leads to a mean increase in pulse rates of more than 15 beats per minute for the population from which the participants were sampled.

Example T.10

Using Microsoft Excel to Perform a Hypothesis Test for Two Population Proportions (Two-Tailed)

Renaldo thinks that support for a referendum is stronger in one district than in a neighboring district. He takes a random sample of registered voters from each district and records the results. The sample size from the first district was 85, with 36 voters in favor of the referendum. Out of 72 voters surveyed in district two, only 25 supported the referendum. Using $\alpha = 0.05$, determine if there is sufficient evidence to reject the claim that support for the referendum is equal in the two districts.

Solution

Immediately, we recognize that we need to perform a z-test since we are comparing two population proportions using large independent samples. The null and alternative hypotheses are as follows.

$$H_0: p_1 - p_2 = 0$$
$$H_a: p_1 - p_2 \neq 0$$

We are given the following statistics.

$$x_1 = 36 \qquad x_2 = 25$$
$$n_1 = 85 \qquad n_2 = 72$$

First, enter the given values with appropriate labels in the first three rows of the spreadsheet as shown in the screenshot at the end of this example. Then use Excel to calculate the sample proportions (cells B4 and C4), $\bar{p}$ (cell B6), and $1-\bar{p}$ (cell B7). Enter **=B2/B3** and **=C2/C3** to produce the values of $\hat{p}_1$ and $\hat{p}_2$, respectively. Next, enter **=(B2+C2)/(B3+C3)** to calculate $\bar{p}$ and then **=1-B6** to calculate $1-\bar{p}$.

Recall that the test statistic is given by the following formula.

$$z = \frac{\left(\hat{p}_1 - \hat{p}_2\right) - \left(p_1 - p_2\right)}{\sqrt{\bar{p}\left(1-\bar{p}\right)\left(\dfrac{1}{n_1} + \dfrac{1}{n_2}\right)}}$$

Calculate the test statistic by entering the following formula in cell B8.

=(B4-C4-0)/SQRT(B6*B7*(1/B3+1/C3))

The result of this calculation is the test statistic, $z \approx 0.98$. Finally, calculate the p-value. Since this is a two-tailed test, the p-value is the sum of the areas under the standard normal curve to the left of $z_1 \approx -0.98$ and to the right of $z_2 \approx 0.98$. Recall from Chapter 6 that the NORM.S.DIST function calculates the area to the left of the z-value entered. Thus, enter **=2*(1-NORM.S.DIST(B8, TRUE))** in cell B9 to produce the p-value.

	A	B	C
1		District 1	District 2
2	*x*	36	25
3	*n*	85	72
4	*p-hat*	0.423529	0.347222
5			
6	*p-bar*	0.388535	
7	1 - *p-bar*	0.611465	
8	*z*	0.977441	
9	*p*-value	0.328351	

Since the p-value, 0.3284, is greater than the level of significance, $\alpha = 0.05$, we fail to reject the null hypothesis. Therefore, there is not sufficient evidence to reject the claim that equal proportions of registered voters in the two districts support the referendum.

Example T.11

Using Microsoft Excel to Perform a Hypothesis Test for Two Population Variances (Left-Tailed)

A commuter believes that the variance of her commute times is higher in the afternoon. She decides to test this claim. The sample variance of the morning commute times is 32.4 while the sample variance of the afternoon commute times is 38.7. The sample size is 15 for both morning and afternoon travel times. Assuming that both populations are normally distributed, test the commuter's claim using a 0.05 level of significance. Does the evidence support her claim?

Solution

We wish to compare two population variances. Let σ_1^2 be the morning variance. Thus, we have the following hypotheses.

$$H_0: \sigma_1^2 \geq \sigma_2^2$$
$$H_a: \sigma_1^2 < \sigma_2^2$$

The test statistic for comparing variances is $F = \frac{s_1^2}{s_2^2}$. Enter the sample variances and sample sizes into your spreadsheet as shown in the screenshot, and use the calculations **=B2/C2** and **=F.INV(0.05, B3-1, C3-1)** to compute the test statistic and critical value, respectively. The first parameter in the F.INV function is the area in the left tail of the *F*-distribution, and the latter two are the degrees of freedom for the numerator and denominator. The complete spreadsheet is shown in the following screenshot.

	A	B	C
1		Morning	Afternoon
2	sample variance	32.4	38.7
3	*n*	15	15
4			
5	*F*	0.837209302	
6	critical *F*	0.402620943	

Since the test statistic, $F \approx 0.8372$, is greater than the critical value, $F_{0.950} \approx 0.4026$, we do not reject the null hypothesis. Therefore, there is not sufficient evidence at the 0.05 level of significance to support the commuter's claim that her afternoon commute times have a larger variance than her morning commute times.

MINITAB

Example T.12

Using MINITAB to Perform a Hypothesis Test for the Mean of the Paired Differences for Two Populations (Two-Tailed, σ Unknown, Dependent Samples)

Cindy and her roommate both believe they have the longer commute to work. To settle this matter, both she and her roommate time their trips to and from work for 15 days. The total minutes that each spent commuting on each day is shown in the following table. Use a 95% level of confidence to determine whether there is a significant mean difference between their commute times.

Daily Commute Times (in Minutes)															
Cindy	24	25	28	32	19	22	27	36	32	28	22	24	21	29	31
Roommate	22	26	29	36	22	18	29	38	32	30	24	23	21	30	30

Solution

First, we identify the necessary procedure. We will perform a hypothesis test for the mean of the paired differences for the population data. The hypotheses are as follows.

$$H_0: \mu_d = 0$$
$$H_a: \mu_d \neq 0$$

This is a paired t-test. Next, enter the data into the first two columns in the worksheet, C1 and C2. Now select **Stat ▶ Basic Statistics ▶ Paired t**. Then select **Samples in columns**; enter **C1** for the first sample and **C2** for the second sample. Select **Options** in order to specify a confidence level of **95.0** and a test mean of **0**. Also select **not equal** for the alternative hypothesis. Note that MINITAB calculates the paired differences by subtracting the values for the second sample from the values for the first sample, which is the opposite of what we do when we calculate them by hand or using a TI-83/84 Plus calculator. The dialog boxes are shown in the following screenshots.

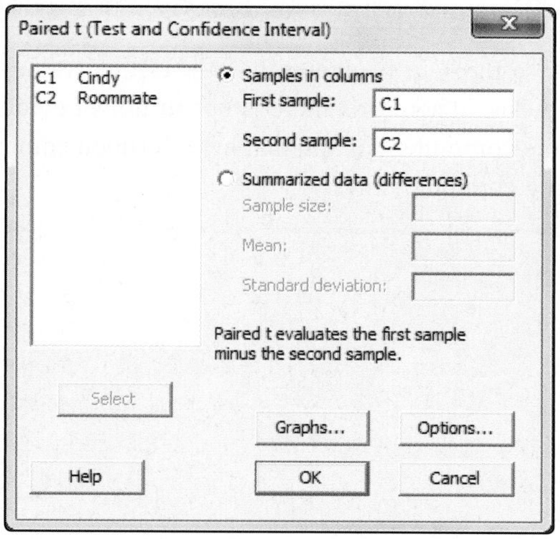

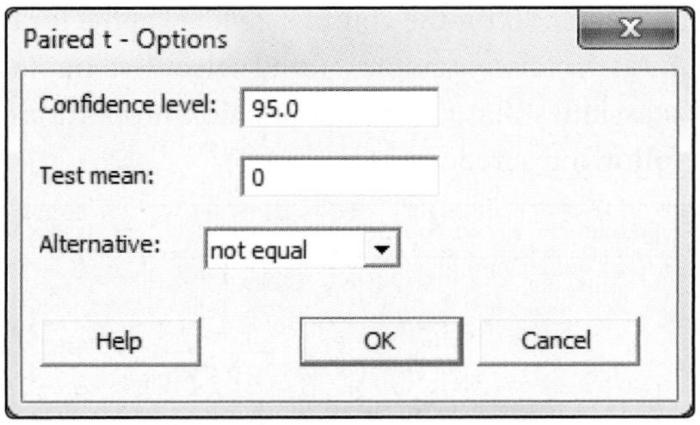

Click **OK** in both the options and main dialog windows and the p-value for the hypothesis test is produced in the Session window. The results in the Session window should appear as follows.

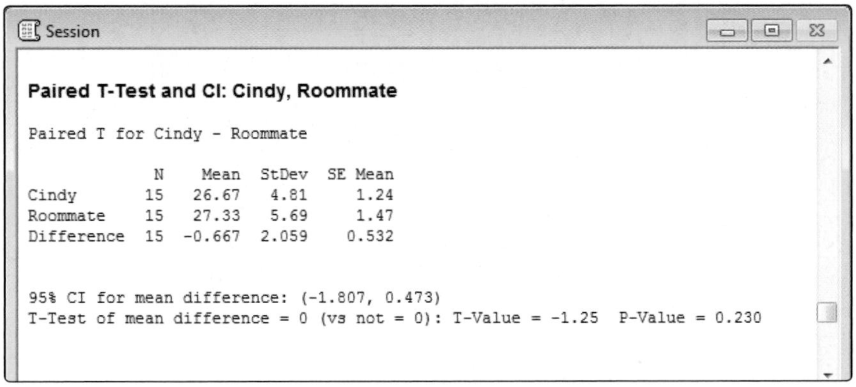

Session
Paired T-Test and CI: Cindy, Roommate
Paired T for Cindy - Roommate
N Mean StDev SE Mean
Cindy 15 26.67 4.81 1.24
Roommate 15 27.33 5.69 1.47
Difference 15 -0.667 2.059 0.532
95% CI for mean difference: (-1.807, 0.473)
T-Test of mean difference = 0 (vs not = 0): T-Value = -1.25 P-Value = 0.230

Since 0.230 > 0.05, we have p-value > α, so we cannot reject the null hypothesis. At the 95% level of confidence, there is not sufficient evidence to say that there is a significant mean difference between the daily commute times of Cindy and her roommate.

Example T.13

Using Minitab to Perform a Hypothesis Test for Two Population Proportions (Right-Tailed)

In random samples of 72 holes of golf recently played by Willis and Annie, Annie collects 10 birdies (which means scoring one point below par for the hole) and Willis shoots 8. Keeping in mind that in golf, the lowest score is the best, so more birdies would be better, can Annie claim that she is the better golfer? Use a 90% level of confidence.

Solution

We want to compare the proportions of birdies out of all holes of golf played by Annie and Willis. Let Population 1 be holes of golf played by Annie and Population 2 be those played by Willis. Since we are comparing two population proportions using large independent samples, we will perform a two-sample z-test for proportions. The following null and alternative hypotheses indicate that a right-tailed test will be utilized.

$$H_0: p_1 - p_2 \leq 0$$
$$H_a: p_1 - p_2 > 0$$

Select **Stat ► Basic Statistics ► 2 Proportions**. Choose **Summarized data** and enter **72** trials for both samples, **10** events for the first sample, and **8** events for the second sample.

Next, choose **Options**; then enter **90** for the confidence level and **0** for the test difference. Also select **greater than** for the alternative hypothesis, and select **Use pooled estimate of p for test** since the null hypothesis assumes that the two population proportions are equal. The dialog boxes are shown in the following screenshots.

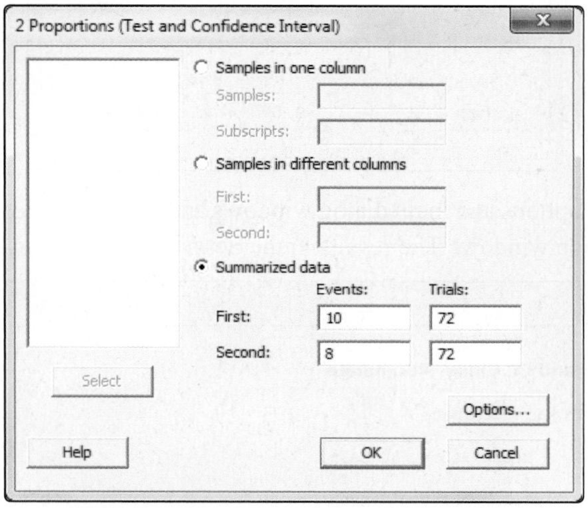

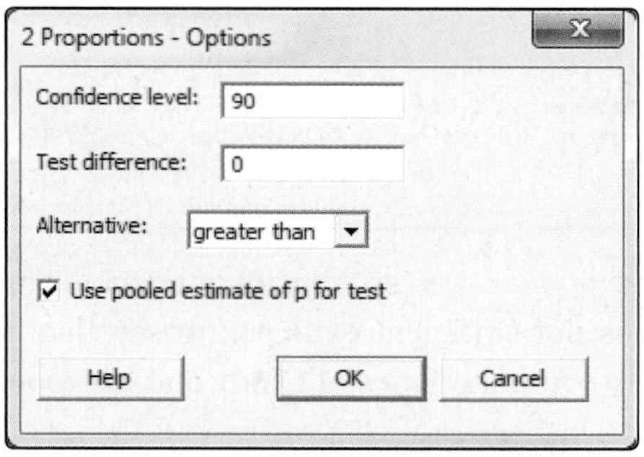

Click **OK** in both the options and main dialog windows and the results are displayed in the Session window, as shown in the following screenshot.

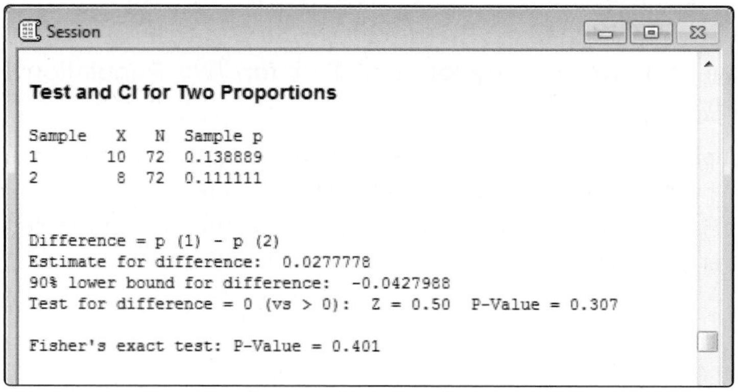

Since the p-value, 0.307, is greater than the level of significance, 0.10, we fail to reject the null hypothesis. At the 90% level of confidence, there is not sufficient evidence to support the claim that Annie's population proportion of birdies is higher than Willis's. Therefore, based on this test, Annie cannot claim that she is the better golfer.

Example T.14

Using Minitab to Perform a Hypothesis Test for Two Population Variances (Right-Tailed)

Ralph believes that the variance in batting averages of catchers is greater than that of outfielders. He decides to test this claim and takes a sample of 20 catchers and finds that their batting averages have a sample variance of 0.0062. A random sample of batting averages for 23 outfielders yields a sample variance of 0.0043. Assuming that both populations of batting averages are normally distributed, test Ralph's claim at a 0.10 level of significance. Does the evidence support his claim?

Solution

We wish to compare two population variances. Let σ_1^2 be the population variance of catchers' batting averages. Thus, we have the following hypotheses.

$$H_0\colon \sigma_1^2 \le \sigma_2^2$$
$$H_a\colon \sigma_1^2 > \sigma_2^2$$

Select **Stat ► Basic Statistics ► 2 Variances**. Choose **Summarized data** and enter the sample sizes and sample variances. For the first sample, enter **20** for the sample size and **0.0062** for the variance. For the second sample, enter **23** for the sample size and **0.0043** for the variance. Next, choose **Options**; then enter **90** for the confidence level. The dialog boxes are shown in the following screenshots.

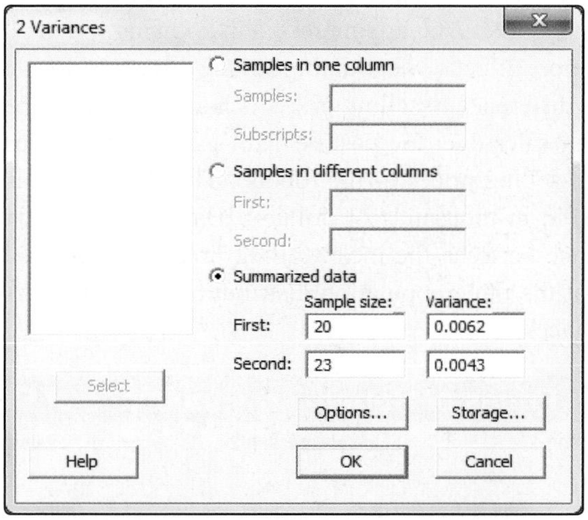

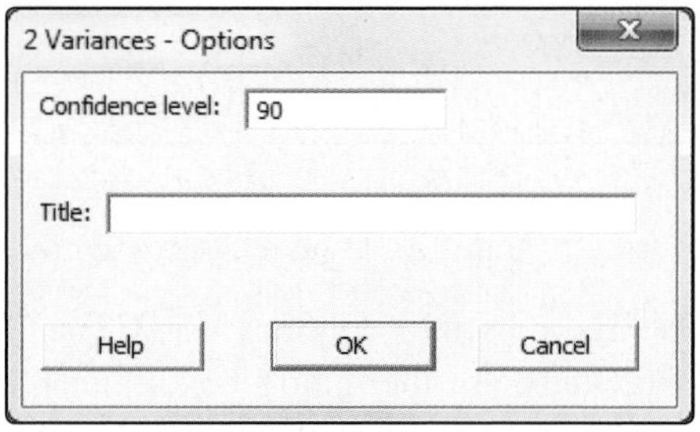

Click **OK** in both the options and main dialog windows and the results are displayed in the Session window, as shown in the following screenshot.

```
Session

Test for Equal Variances

90% Bonferroni confidence intervals for standard deviations

Sample   N      Lower      StDev     Upper
     1  20  0.0598811  0.0787401  0.115006
     2  23  0.0507149  0.0655744  0.092811


F-Test (Normal Distribution)
Test statistic = 1.44, p-value = 0.407
```

Since the p-value, 0.407, is greater than the level of significance, 0.10, the null hypothesis is not rejected. Therefore, at the 0.10 level of significance, the evidence does not support Ralph's claim that the variance in batting averages of catchers is greater than that of outfielders.

Example T.15

Using MINITAB to Perform an ANOVA Test

You have just been promoted to sales manager of a company manufacturing robots used to assemble automobiles. Although your sales force is given a suggested price at which to sell the robots, they have considerable leeway in negotiating the final price. Past sales records indicate that sometimes there is a large difference between the selling prices that different sales reps are able to negotiate. You are interested in knowing if this difference is significant, possibly because of a more effective negotiating strategy or exceptional interpersonal skills, or whether this observed difference in selling prices is just due to random variation. You decide to randomly select four sales over the past year for each of your three sales representatives and observe the actual selling prices of the robots. The following table shows the amounts at which the robots sold in thousands of dollars. Based on these data, can you conclude that there is a difference between the mean selling prices for these three salespeople? Use $\alpha = 0.05$. Assume that the three population distributions are all approximately normal with equal population variances.

Selling Prices of Robots (in Thousands of Dollars)		
Salesperson 1	**Salesperson 2**	**Salesperson 3**
10	11	11
14	16	13
13	14	12
12	15	15

Solution

Enter the sales data for the three reps into columns C1, C2, and C3. Then go to **Stat ▶ ANOVA ▶ One-way (Unstacked)**. In the dialog box, enter **C1 C2 C3** as the responses. Then enter **95** for the confidence level and click **OK**. Observe the output in the Session window for the ANOVA table.

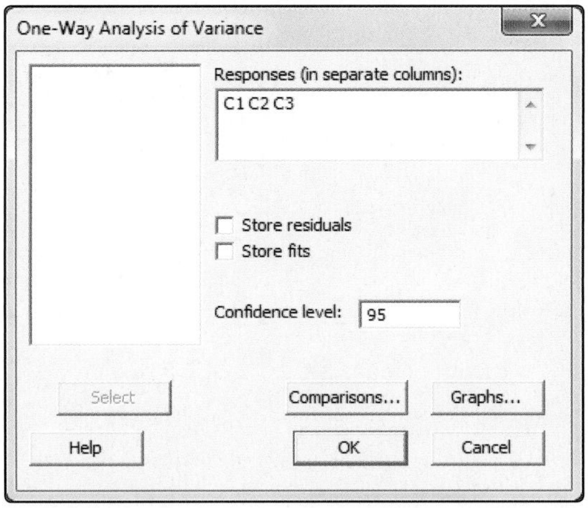

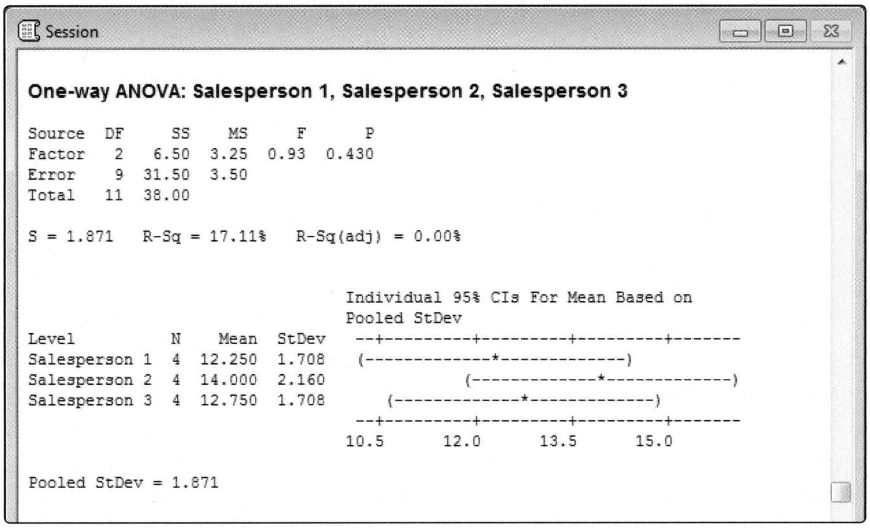

One-way ANOVA: Salesperson 1, Salesperson 2, Salesperson 3

```
Source  DF     SS    MS     F      P
Factor   2   6.50  3.25  0.93  0.430
Error    9  31.50  3.50
Total   11  38.00

S = 1.871   R-Sq = 17.11%   R-Sq(adj) = 0.00%


                                 Individual 95% CIs For Mean Based on
                                 Pooled StDev
Level          N    Mean  StDev  --+---------+---------+---------+-------
Salesperson 1  4  12.250  1.708   (-------------*-------------)
Salesperson 2  4  14.000  2.160              (-------------*-------------)
Salesperson 3  4  12.750  1.708      (-------------*-------------)
                                 --+---------+---------+---------+-------
                                 10.5      12.0      13.5      15.0

Pooled StDev = 1.871
```

Since the p-value, 0.430, is greater than the level of significance, 0.05, we fail to reject the null hypothesis. Thus, there is not sufficient evidence, at the 0.05 level of significance, to support the claim that there is a difference between the mean selling prices for these three salespeople.

Chapter Twelve

Regression, Inference, and Model Building

Sections

Objectives

1. Construct and interpret scatter plots.
2. Calculate and interpret the correlation between two variables.
3. Model a given set of data using a regression line.
4. Construct a prediction interval for an individual value of *y*.
5. Construct confidence intervals for the slope and the *y*-intercept of a regression line.
6. Determine and analyze the equation of a multiple regression model for a given set of data.

12

Introduction

In 2011, the online dating website Match.com claimed that one in five relationships began online. The website seeks to pair prospective partners according to such attributes as personal values, character traits, and physical attractiveness. This website, like many others, was created in order to give yet another avenue for people to connect with others. The fact is that relationships can be complicated, and it is often difficult not only to meet that special person, but also to know if there is a meaningful connection between the two of you.

Like dating, linear regression is all about relationships. Specifically, linear regression looks at the relationship between different variables. By using statistics, we can mathematically determine if two (or more) variables are related and, if so, how strong that relationship is. By knowing how the variables are related, we can make a model that allows us to make predictions regarding the variables. Too bad we don't have a social predictor that is this accurate!

12.1 Scatter Plots and Correlation

When analyzing the relationship between two characteristics of a population, it is helpful to look at the data visually first. The most common graph used to show the relationship between two quantitative variables is a *scatter plot*.

Suppose that we want to determine if a relationship exists between the numbers of hours a student spends studying for an exam and the student's score on the exam, both of which are quantitative variables. A **scatter plot** is a graph on the coordinate plane that contains one point for each pair of data. The horizontal axis (x-axis) represents one variable and the vertical axis (y-axis) represents the other. Thus, for our example, a point on the graph would represent the number of hours studied for a particular exam on one axis and the student's score on the exam on the other axis. It is important to note that, unlike a line graph, the points on a scatter plot are not connected. The reason the points are not connected is because each point represents a separate data pair; unlike the data depicted in a line graph, the data in a scatter plot do not necessarily represent a change in the value of a variable over time.

The choice of which variable to put on which axis depends on whether or not we believe that one variable *might* influence the other. If this type of relationship exists, we say that a change in the value of one variable, called the **explanatory variable**, influences a change in the value of the other variable, called the **response variable**. Other names for the explanatory variable are the *predictor variable* or the *independent variable*. Another name for the response variable is the *dependent variable*. For example, we suspect that the number of hours a student spends studying for an exam might influence the student's score on the exam. (We certainly would hope so!) In such a case, we would put the explanatory variable (the number of hours spent studying) on the x-axis and the response variable (grade on the exam) on the y-axis. If no such relationship is assumed to exist, such as with GPA and a student's weight, it does not matter which variable is placed on which axis.

Let's consider an example where creating a scatter plot would be useful in determining how two variables are related. Suppose a child development class wants to learn about the growth rate of children. The students visit a local preschool and measure the heights of the children. The mean height of all the children in the preschool may be interesting, but the students aim to go one step deeper and analyze the way in which children grow as they get older. The following table shows the heights of children of various ages.

Table 12.1: Ages and Heights of Children														
Age (in Years)	1.0	1.5	1.8	2.0	2.2	2.5	3.2	3.3	3.5	3.9	4.2	4.5	4.8	5.1
Height (in Inches)	29.4	28.2	30.1	33.7	33.5	34.3	37.5	36.0	37.5	39.0	41.4	43.9	44.5	45.8

In order to produce a scatter plot displaying the data from Table 12.1, the first step would be to determine if one variable might influence the other variable. Certainly, we would expect that as a child gets older, he or she would grow taller; thus the explanatory variable would be the age (in years) and should be placed on the x-axis. The response variable is then the height (in inches) and should be placed on the y-axis. The next step is to think of each age-height pair as a point on the graph. For instance, the height of 29.4 inches for the 1-year-old gives the point (1.0, 29.4), the height of 33.7 inches for the 2-year-old gives the point (2.0, 33.7), and so on. When we plot these points, we obtain the following scatter plot.

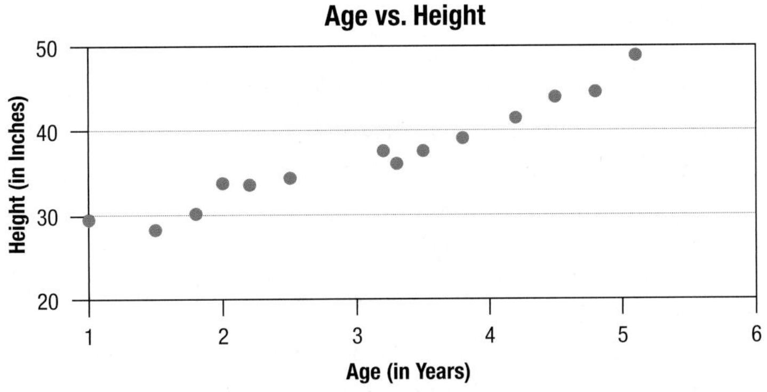

Figure 12.1: Scatter Plot

The importance of a scatter plot is that it enables us to identify trends in the data. For example, do the points fall in a linear pattern, a curved pattern, or no pattern at all? If the points have a linear shape, how close to a straight line are they? Does the linear pattern have a *positive slope* (the points rise from left to right) or a *negative slope* (the points fall from left to right)? Notice that the points in our scatter plot of age versus height in Figure 12.1 have a linear pattern that is very close to a straight line that has a positive slope. We will talk about the significance of the trend a little later in this section.

Let's now consider some data collected from a sample of 10 NFL quarterbacks for the 2011–2012 season. These data are given in the following table.

Table 12.2: Sample of NFL Quarterbacks (2011–2012 Season)

	Number of Passing Touchdowns	2012 Base Salary (in Millions of Dollars)	Quarterback Rating
Drew Brees	46	3.0	110.6
Michael Vick	18	12.5	84.9
Philip Rivers	27	10.2	88.7
Tony Romo	31	0.825	102.5
Aaron Rodgers	45	8.0	122.5
Jay Cutler	13	7.7	85.7
Alex Smith	17	5.0	90.7
Eli Manning	29	1.75	92.9
Tim Tebow	12	2.1	72.9
Tom Brady	39	0.95	105.6

Source: Yahoo! Sports. "NFL - Statistics by Position." http://sports.yahoo.com/nfl/stats/byposition?pos=QB&conference=NFL&year=season_2011&sort=49&timeframe=All (20 May 2012).
Source: Spotrac.com. "NFL Player Contracts, Salaries, and Transactions." http://www.spotrac.com/nfl/ (2 Oct. 2012).

Example 12.1

Creating a Scatter Plot to Identify Trends in Data

Use the data from Table 12.2 to produce a scatter plot that shows the relationship between the base salary of an NFL quarterback and the number of touchdowns the quarterback has thrown in one season.

Solution

We might expect for the number of touchdowns a quarterback throws in one season to influence his salary. Taking this into consideration, we will place the number of touchdowns on the x-axis and the base salary on the y-axis.

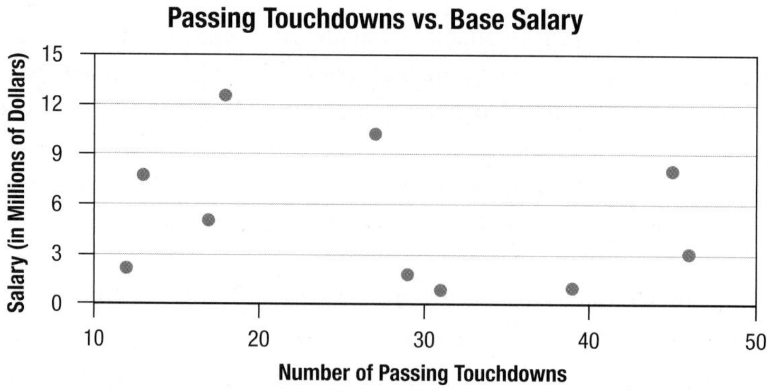

Looking at this scatter plot, we do not see a linear pattern. Actually, no pattern is evident. This probably indicates that these two variables do not have a relationship after all.

Example 12.2

Creating a Scatter Plot to Identify Trends in Data

Use the data in Table 12.2 to produce a scatter plot that shows the relationship between the number of touchdowns thrown in one season and the corresponding quarterback rating for the given sample of NFL quarterbacks.

Solution

In this case, we would expect that the number of touchdowns thrown by a quarterback does influence that quarterback's rating, since number of touchdowns is one of many factors used to determine the quarterback rating. Hence, the logical way to label the axes is to place the number of passing touchdowns on the *x*-axis and the quarterback rating on the *y*-axis.

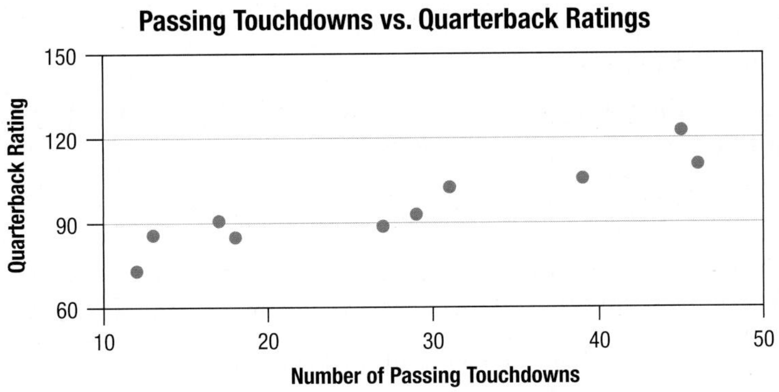

Notice that the points tend to go up from left to right, and fall close to a straight line. This pattern can be described as a linear pattern with a positive slope.

Let's compare the scatter plots that we have seen thus far. The scatter plot of age versus height shows a predictable pattern. That is, it shows that height seems to increase as age increases (not a very deep observation about child development). The two variables appear to have a **linear relationship**, meaning that the points in the scatter plot roughly follow a straight-line pattern. The third scatter plot, passing touchdowns versus quarterback ratings, also has a linear pattern with the points moving in an upward direction from left to right. This indicates, as you would expect, that as the number of touchdowns increases, the quarterback's rating increases as well.

When the points in a scatter plot do roughly follow a straight line, the direction of the pattern tells how the variables respond to each other. A **positive slope** indicates that as the values of one variable increase, so do the values of the other variable. This type of relationship between two variables is called a *positive linear relationship*. A **negative slope** indicates that as the values of one variable increase, the values of the other variable decrease. This type of relationship between two variables is called a *negative linear relationship*. Figure 12.2 demonstrates the difference between a positive and a negative linear relationship.

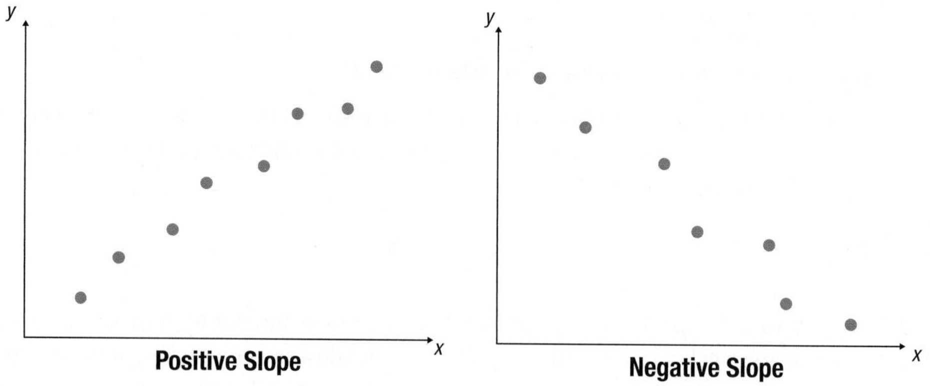

Figure 12.2: Direction of a Linear Pattern

Example 12.3

Determining Whether a Scatter Plot Would Have a Positive Slope, Negative Slope, or Not Follow a Straight-Line Pattern

Determine whether the points in a scatter plot for the two variables are likely to have a positive slope, negative slope, or not follow a straight-line pattern.

a. The number of hours you study for an exam and the score you make on that exam

b. The price of a used car and the number of miles on the odometer

c. The pressure on a gas pedal and the speed of the car

d. Shoe size and IQ for adults

Solution

a. As the number of hours you study for an exam increases, the score you receive on that exam is usually higher. Thus, the scatter plot would have a positive slope.

b. As the number of miles on the odometer of a used car increases, the price usually decreases. Thus, the scatter plot would have a negative slope.

c. The more you push on the gas pedal, the faster the car will go. Thus, the scatter plot would have a positive slope.

d. Common sense suggests that there is not a relationship, linear or otherwise, between a person's IQ and his or her shoe size.

Let's consider a new parameter that measures the strength of the linear relationship between two variables, that is, how strongly one of the variables is linearly dependent on the other. The strength of the linear relationship is determined by how closely the points in the scatter plot resemble a straight line. Thus, the stronger the relationship, the more the diagram looks like a straight line. The weaker the relationship, the more scattered the points are in the diagram. The following figure demonstrates the difference between a strong linear relationship and a weak linear relationship.

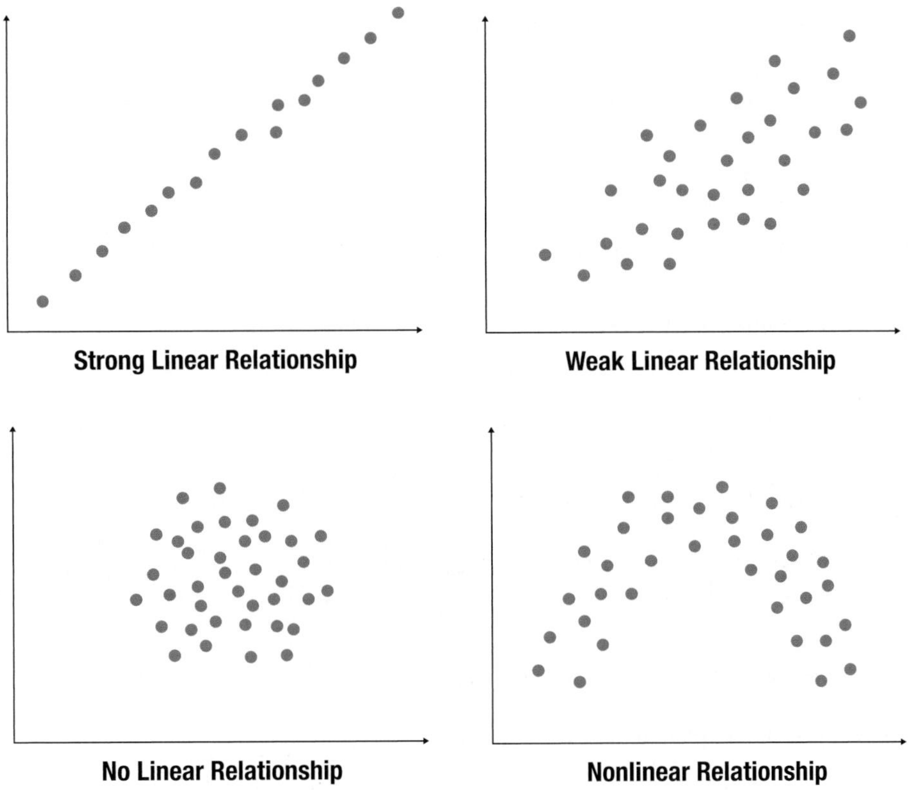

Figure 12.3: Strength of a Linear Relationship

A simple visual examination of a scatter plot is not a sufficient way to analyze the data. We need a parameter that measures the strength of the linear relationship between the two quantitative variables. This parameter is called the **Pearson correlation coefficient**, symbolically denoted as ρ, which is pronounced "rho." The correlation coefficient for a sample is denoted by r and is always a value between -1 and 1, inclusive.

Math Symbols

ρ: Pearson correlation coefficient for a population; Greek letter, rho

r: correlation coefficient for a sample

Definition

The **Pearson correlation coefficient**, ρ, is the parameter that measures the strength of a linear relationship between two quantitative variables in a population. The correlation coefficient for a sample is denoted by r. It always takes a value between -1 and 1, inclusive.

$$-1 \le r \le 1$$

If r is positive, then the scatter plot has a positive slope and the variables are said to have a positive linear relationship. Similarly, a negative value of r indicates that the scatter plot has a negative slope and there is a negative linear relationship between the variables. If $r = 0$, then no linear relationship exists between the two variables. If $r = 1$, then the data fall in a perfectly straight line with a positive slope. Similarly, if $r = -1$, the data fall in a perfectly straight line with a negative slope. If two variables have a strong positive or negative linear relationship, we say that the two variables are *correlated*. The strength of the correlation is expressed by $|r|$. The larger $|r|$ is, the stronger the correlation.

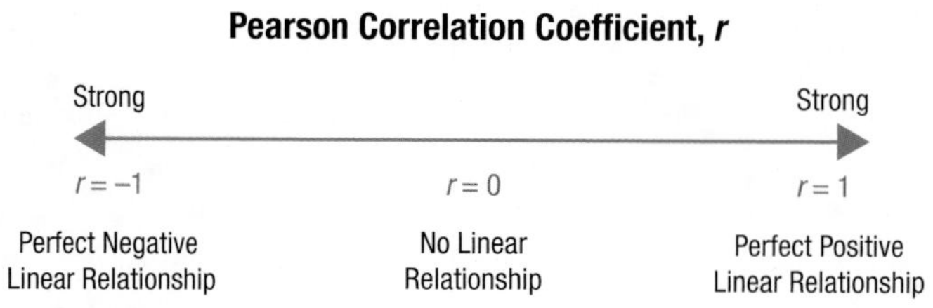

Figure 12.4: The Correlation Coefficient and Strength of a Linear Relationship

To calculate the correlation coefficient, r, the following formula is used.

Rounding Rule

Round the correlation coefficient, r, to three decimal places.

Formula

Pearson Correlation Coefficient

The **Pearson correlation coefficient** for paired data from a sample is given by

$$r = \frac{n\sum x_i y_i - \left(\sum x_i\right)\left(\sum y_i\right)}{\sqrt{n\sum x_i^2 - \left(\sum x_i\right)^2}\sqrt{n\sum y_i^2 - \left(\sum y_i\right)^2}}$$

where n is the number of data pairs in the sample,

x_i is the ith value of the explanatory variable, and

y_i is the ith value of the response variable.

Technology such as calculators, Excel, Minitab, and other software programs can also be used to calculate the correlation coefficient. We will use a TI-83/84 Plus calculator to find the correlation coefficient in the following examples.

A TI-83/84 Plus calculates the correlation coefficient simultaneously with a quantity to be defined later called the *coefficient of determination*, as well as the slope and y-intercept of the *regression line*, which we will also discuss in this chapter.

- Press STAT.
- Select option `1:Edit`.
- Enter the values for the x-variable in List 1 (`L1`) and the values for the y-variable in List 2 (`L2`).
- Press STAT.
- Select `CALC`.
- Choose option `4:LinReg(ax+b)`.
- We do not need to enter the lists where the data are located, as the data are in the default lists, `L1` and `L2`, so press ENTER twice.

Side Note

To ensure that the output for the function `LinReg(ax+b)` on a TI-83/84 Plus calculator will include the correlation coefficient and the coefficient of determination, begin by pressing 2ND and then 0, which will bring up the `CATALOG`. Scroll down to `DiagnosticOn`. Press ENTER twice. You only need to perform this step the first time that you use the function `LinReg(ax+b)`.

The output will include the correlation coefficient, r. (See the note in the margin for an additional step that must be performed on the calculator if the correlation coefficient is not displayed.)

Let's use a TI-83/84 Plus calculator to calculate the correlation coefficient for the data from Table 12.1 relating children's ages and heights.

- Press STAT.
- Select option `1:Edit`.

- Enter the values for age (x) in L1 and the values for height (y) in L2.
- Press STAT.
- Choose CALC.
- Choose option 4:LinReg(ax+b).
- Press ENTER twice.

The output should appear as shown in the screenshot in the margin.

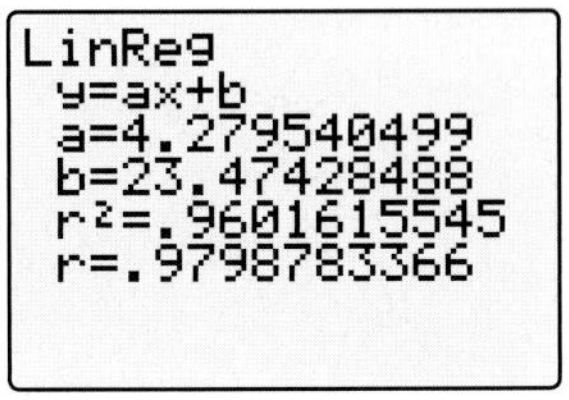

The output screen provides many values, one of which is the correlation coefficient. We can see that the correlation coefficient is $r \approx 0.980$, rounded to three decimal places. Notice that this value is positive, which matches with the positive slope of the scatter plot. Also, note that r is very close to 1, which corresponds to the points in the scatter plot closely following a straight line.

Example 12.4

Calculating the Correlation Coefficient Using a TI-83/84 Plus Calculator

Calculate the correlation coefficient, r, for the data from Table 12.2 relating touchdowns thrown and base salaries.

Solution

The data we need from Table 12.2 are reproduced in the following table.

NFL Quarterbacks	
Number of Passing Touchdowns	Base Salary (in Millions of Dollars)
46	3.0
18	12.5
27	10.2
31	0.825
45	8.0
13	7.7
17	5.0
29	1.75
12	2.1
39	0.95

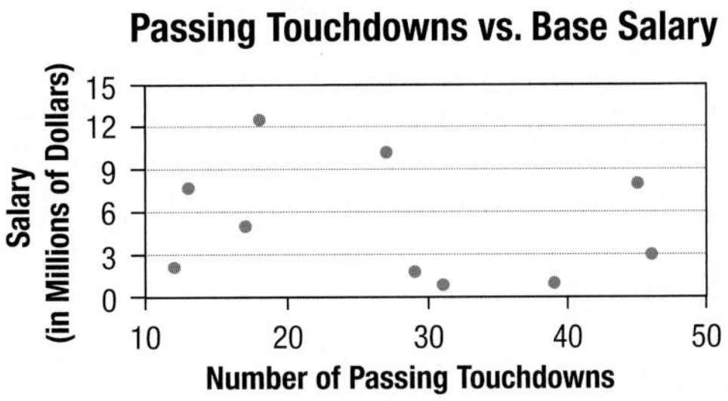

Let's enter these data into our calculator.

- Press STAT.
- Select option 1:Edit.
- Enter the values for number of touchdowns (x) in L1 and the values for base salary (y) in L2.
- Press STAT.
- Choose CALC.
- Choose option 4:LinReg(ax+b).
- Press ENTER twice.

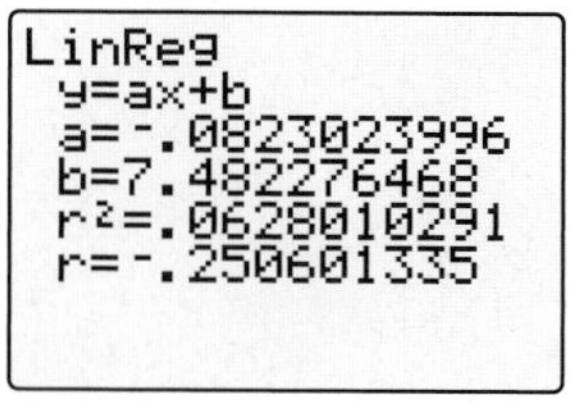

From the scatter plot, we would expect $|r|$ to be close to 0. The calculator confirms that the correlation coefficient for these two variables is $r \approx -0.251$, indicating a weak negative relationship, if any relationship exists at all.

Example 12.5

Calculating the Correlation Coefficient Using a TI-83/84 Plus Calculator

Calculate the correlation coefficient, r, for the data from Table 12.2 relating touchdowns thrown and quarterback ratings.

Solution

The data we need from Table 12.2 are reproduced in the following table.

NFL Quarterbacks	
Number of Passing Touchdowns	Quarterback Rating
46	110.6
18	84.9
27	88.7
31	102.5
45	122.5
13	85.7
17	90.7
29	92.9
12	72.9
39	105.6

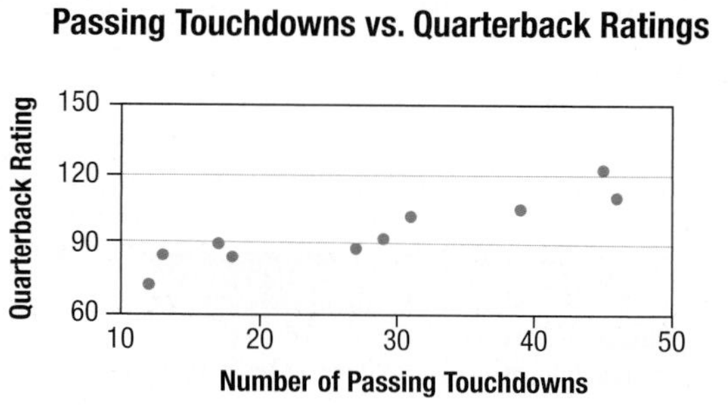

Let's enter these data into our calculator.

- Press STAT.
- Select option `1:Edit`.
- Enter the values for number of touchdowns (x) in L1 and the values for quarterback rating (y) in L2.
- Press STAT.
- Choose CALC.
- Choose option `4:LinReg(ax+b)`.
- Press ENTER twice.

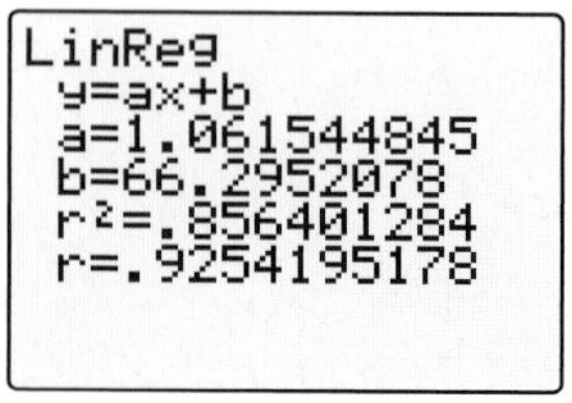

From the scatter plot, we would expect $|r|$ to be close to 1. The calculator confirms that the correlation coefficient for these two variables is $r \approx 0.925$. Since the value is close to 1, this indicates a very strong positive correlation between the variables.

Testing the Correlation Coefficient for Significance

For each of our examples, we have discussed the linear relationship between the two variables. However, we did not determine if the relationship occurred by chance, or if the relationship is statistically significant. The term *statistically significant* for the Pearson correlation coefficient indicates that a significant linear relationship exists between the two variables.

We know that the closer the correlation coefficient is to 1 or −1, the stronger the linear relationship between the two variables, but how close is close enough to imply statistical significance? Table I in Appendix A lists critical values for the Pearson correlation coefficient at particular levels of significance. To use this table, you need to know the level of significance, α, and the sample size, n, to look up the critical value for the correlation coefficient, r_α. The correlation coefficient, r, is statistically significant if the absolute value of the correlation coefficient is greater than or equal to the critical value, r_α, from the table.

Procedure

Using Critical Values of the Pearson Correlation Coefficient to Determine the Significance of a Linear Relationship

A sample correlation coefficient, r, is statistically significant if $|r| \geq r_\alpha$.

Example 12.6

Using a Table of Critical Values to Determine Significance of a Linear Relationship

Use the critical values in Table I to determine if the correlation between the number of passing touchdowns and base salary from Example 12.4 is statistically significant. Use a 0.05 level of significance.

Solution

Begin by finding the critical value for $\alpha = 0.05$ with $n = 10$ in Table I. Find the value in the table where the row for $n = 10$ intersects the column for $\alpha = 0.05$.

n	$\alpha = 0.05$	$\alpha = 0.01$
6	0.811	0.917
7	0.754	0.875
8	0.707	0.834
9	0.666	0.798
10	0.632	0.765
11	0.602	0.735
12	0.576	0.708

Thus, $r_\alpha = 0.632$. Comparing this critical value to the absolute value of the correlation coefficient we found for the data in Example 12.4, we have $0.251 < 0.632$, and thus $|r| < r_\alpha$. Therefore, the linear relationship between the variables is not statistically significant at the 0.05 level of significance. Thus, we do not have sufficient evidence, at the 0.05 level of significance, to conclude that a linear relationship exists between the number of passing touchdowns during the 2011–2012 season and the 2012 base salary of an NFL quarterback.

Testing the Correlation Coefficient for Significance Using Hypothesis Testing

If the linear relationship between the variables in the population is statistically significant, then the population's correlation coefficient, ρ, would not equal zero. That is, $\rho \neq 0$. We can use the principles of hypothesis testing to determine if the linear relationship is significant.

First, the null and alternative hypotheses for testing the statistical significance of a linear relationship are as follows.

H_0: $\rho = 0$ (Implies that there is no significant linear relationship)
H_a: $\rho \neq 0$ (Implies that there is a significant linear relationship)

If you wish to test if the linear relationship not only exists, but specifically if it is positive or negative, the null and alternative hypotheses can be adjusted as follows.

Procedure

Testing Linear Relationships for Significance

Significant Linear Relationship (Two-Tailed Test)

H_0: $\rho = 0$ (Implies that there is no significant linear relationship)

H_a: $\rho \neq 0$ (Implies that there is a significant linear relationship)

Significant Negative Linear Relationship (Left-Tailed Test)

H_0: $\rho \geq 0$ (Implies that there is no significant negative linear relationship)

H_a: $\rho < 0$ (Implies that there is a significant negative linear relationship)

Significant Positive Linear Relationship (Right-Tailed Test)

H_0: $\rho \leq 0$ (Implies that there is no significant positive linear relationship)

H_a: $\rho > 0$ (Implies that there is a significant positive linear relationship)

Second, the test statistic for this hypothesis test is the following t-test statistic with $n - 2$ degrees of freedom.

Formula

Test Statistic for a Hypothesis Test for a Correlation Coefficient

The test statistic for testing the significance of the correlation coefficient is given by

$$t = \frac{r}{\sqrt{\frac{1-r^2}{n-2}}}$$

where r is the sample correlation coefficient and

n is the number of data pairs in the sample.

The number of degrees of freedom for the t-distribution of the test statistic is given by $n - 2$.

When testing for significance of the correlation coefficient, we can use rejection regions to draw conclusions just as we did when conducting t-tests in previous chapters. Alternately, we can use technology, such as a TI-83/84 Plus calculator, to find the p-value, which can be compared to the level of significance to draw a conclusion.

Procedure

Rejection Regions for Testing Linear Relationships

Significant Linear Relationship (Two-Tailed Test)

Reject the null hypothesis, H_0, if $|t| \geq t_{\alpha/2}$.

Significant Negative Linear Relationship (Left-Tailed Test)

Reject the null hypothesis, H_0, if $t \leq -t_{\alpha}$.

Significant Positive Linear Relationship (Right-Tailed Test)

Reject the null hypothesis, H_0, if $t \geq t_{\alpha}$.

The only sample statistics we need in order to calculate the test statistic are the correlation coefficient, r, and the sample size, n. After we have calculated the test statistic, we can use either rejection regions or p-values to draw a conclusion, whichever we prefer.

Though the one-sided forms of the hypotheses were presented, we will only focus on testing whether a significant linear relationship exists (two-tailed test), as this is the most useful test. Let's look at an example of using a hypothesis test to determine if the linear relationship between two variables is significant.

Example 12.7

Performing a Hypothesis Test to Determine if the Linear Relationship between Two Variables Is Significant

Use a hypothesis test to determine if the linear relationship between the number of parking tickets a student receives during a semester and his or her GPA during the same semester is statistically significant at the 0.05 level of significance. Refer to the data presented in the following table.

GPA and Number of Parking Tickets

Number of Tickets	0	0	0	0	1	1	1	2	2	2	3	3	5	7	8
GPA	3.6	3.9	2.4	3.1	3.5	4.0	3.6	2.8	3.0	2.2	3.9	3.1	2.1	2.8	1.7

Solution

Step 1: State the null and alternative hypotheses.

We wish to test the claim that a significant linear relationship exists between the number of parking tickets a student receives during a semester and his or her GPA during the same semester. Thus, the hypotheses are stated as follows.

$$H_0: \rho = 0$$
$$H_a: \rho \neq 0$$

Step 2: **Determine which distribution to use for the test statistic, and state the level of significance.**

We will use the t-test statistic presented previously in this section along with a significance level of $\alpha = 0.05$ to perform this hypothesis test.

Step 3: **Gather data and calculate the necessary sample statistics.**

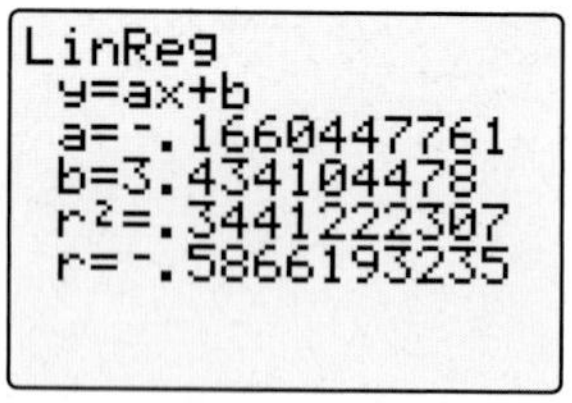

We need to begin by calculating the correlation coefficient, r. Since it is possible to argue for either of these two variables affecting the other, let's assign the number of tickets to be our explanatory variable (x), and thus the GPA as the response variable (y). Using a TI-83/84 Plus calculator, enter the values for the numbers of tickets (x) in L1 and the values for the GPAs (y) in L2. Then press STAT and choose CALC and option 4:LinReg(ax+b). Press ENTER twice. We get $r \approx -0.586619$ from the calculator and we know that $n = 15$. Note that we rounded r to six decimal places, rather than three decimal places, to avoid additional rounding error in the following calculation of the test statistic. Substituting these values into the formula for the t-test statistic yields the following.

$$\begin{aligned} t &= \frac{r}{\sqrt{\dfrac{1-r^2}{n-2}}} \\ &= \frac{-0.586619}{\sqrt{\dfrac{1-(-0.586619)^2}{15-2}}} \\ &\approx -2.612 \end{aligned}$$

Step 4: **Draw a conclusion and interpret the decision.**

We will use rejection regions in this example to draw the conclusion. Since the sample size for this example is 15, the number of degrees of freedom is $n - 2 = 15 - 2 = 13$. Using the t-distribution table or appropriate technology, we find the critical value for this test, $t_{\alpha/2} = t_{0.05/2} = t_{0.025} = 2.160$. So we will reject the null hypothesis, H_0, if $|t| \geq 2.160$.

Since $|t| \approx 2.612$ and $2.612 \geq 2.160$, the test statistic falls in the rejection region. Thus, we reject the null hypothesis. Therefore, there is sufficient evidence at the 0.05 level of significance to support the claim that there is a significant linear relationship between the number of parking tickets a student receives during a semester and his or her GPA during the same semester.

Something to Ponder...

Do you think it is always true that there is a linear relationship between a student's GPA in a semester and the number of parking tickets he or she receives in that semester? Would you ever say that a cause-and-effect relationship exists between the variables? Is it possible that a third factor is influencing both GPA and the number of parking tickets received? Is this sample necessarily representative of the entire population? It is always important to ask questions such as these when considering whether a linear relationship between two variables is statistically significant.

Example 12.8

Performing a Hypothesis Test to Determine if the Linear Relationship between Two Variables Is Significant

An online retailer wants to research the effectiveness of its mail-out catalogs. The company collects data from its eight largest markets with respect to the number of catalogs (in thousands) that were mailed out one fiscal year versus sales (in thousands of dollars) for that year. The results are as follows.

Number of Catalogs Mailed and Sales

Number of Catalogs (in Thousands)	2	3	3	3	4	4	5	6
Sales (in Thousands)	\$126	\$98	\$255	\$394	\$107	\$122	\$334	\$403

Use a hypothesis test to determine if the linear relationship between the number of catalogs mailed out and sales is statistically significant at the 0.01 level of significance.

Solution

Step 1: **State the null and alternative hypotheses.**

We wish to test the claim that a significant linear relationship exists between the number of catalogs mailed out and the corresponding sales for that area.

$$H_0: \rho = 0$$
$$H_a: \rho \neq 0$$

Step 2: **Determine which distribution to use for the test statistic and state the level of significance.**

We will use the t-test statistic with the given level of significance, $\alpha = 0.01$.

Step 3: **Gather data and calculate the necessary sample statistics.**

We first need to calculate the correlation coefficient, r. It is possible to infer that mailing a larger number of catalogs to a region will influence the number of sales in that region. Thus, the explanatory variable (x) will be the number of catalogs and the response variable (y) will be the sales. Using a TI-83/84 Plus calculator, enter the values for the numbers of catalogs mailed (x) in L1 and the sales values (y) in L2. Then press STAT and choose CALC and option 4:LinReg(ax+b). Press ENTER twice. From the calculator we see that $r \approx 0.504505$, and we know that $n = 8$. Note that we rounded r to six decimal places, rather than three decimal places, to avoid additional rounding error in the following calculation of the test statistic. Substituting these values into the equation for the test statistic, we have the following.

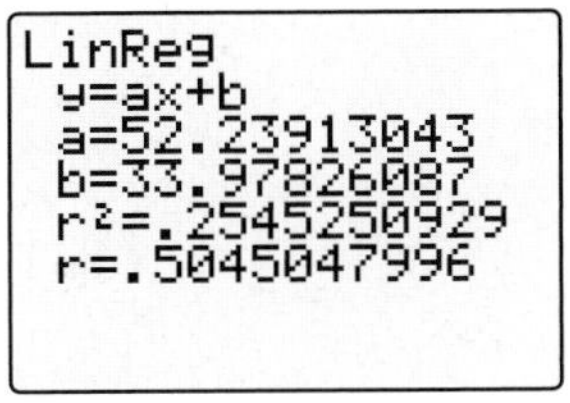

$$t = \frac{r}{\sqrt{\frac{1-r^2}{n-2}}}$$
$$= \frac{0.504505}{\sqrt{\frac{1-(0.504505)^2}{8-2}}}$$
$$\approx 1.431$$

Step 4: **Draw a conclusion and interpret the decision.**

We will use rejection regions to draw the conclusion. Since the sample size for this example is 8, the number of degrees of freedom is $n-2=8-2=6$. Using the t-distribution table or appropriate technology, we find the critical value for this test, $t_{\alpha/2}=t_{0.01/2}=t_{0.005}=3.707$. So, we will reject the null hypothesis, H_0, if $|t|\geq 3.707$.

Since the value of the test statistic, $t\approx 1.431$, is less than the critical value, $t_{\alpha/2}=3.707$, we fail to reject the null hypothesis. Hence, there is not enough evidence at the 0.01 level of significance to support the claim that there is a significant linear relationship between the number of catalogs distributed in a particular area and the corresponding sales in that area.

Interpreting Statistical Significance

Suppose that r is not statistically significant for a given set of data. Does this mean that there definitely is not a correlation between the variables? No—there is just not enough evidence based on the sample taken. Another sample may yield different results. On the other hand, it is possible that, regardless of the number of samples taken, there still may not be a significant correlation between the variables. As the number of trials completed increases, the more likely it is that the prediction regarding the correlation between the variables is accurate.

Before leaving the significance of the correlation coefficient, let's take a moment to consider what statistical significance does NOT tell us. Many times we are tempted to say that, because the linear relationship between two variables is statistically significant, a change in one variable must cause a change in the other variable. However, the world is much more complicated than that. When we do have a correlation between two variables that is statistically significant, one of at least four different things may actually be going on, as illustrated in the following figure.

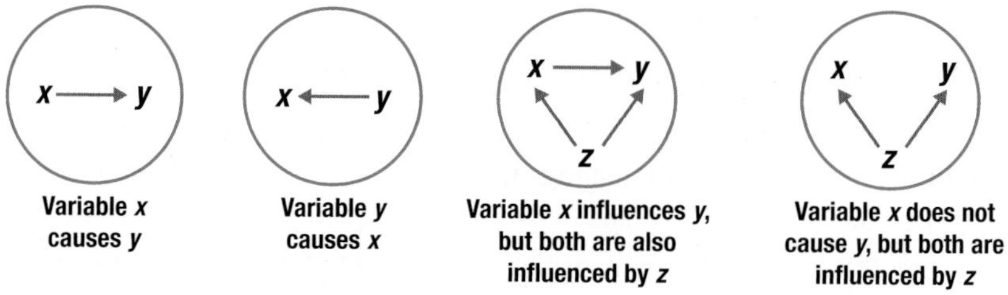

Figure 12.5: Types of Relationships between Two Correlated Variables

Suppose there is a correlation between two variables, x and y. The natural assumption is that x causes y. However, if we reversed the variables, in reality, y may be causing x to change. On the other hand, we might have a combination of factors influencing the response and/or explanatory variable. The x-variable may indeed be influencing the y-variable, but a third variable, z, might also be influencing them both. Take, for instance, the types of anesthesia used and death rates associated with various types of surgeries. It is possible that the type of anesthesia might influence the number of deaths. However, the type of surgery being performed affects both the type of anesthesia used and the death rate, so the apparent correlation between anesthesia and death rate could be misleading.

In addition, it could be the case that, even though a correlation exists between two variables, one is not influencing the other at all. Take, for instance, the correlation between monthly rates of watermelon consumption and drowning. There is a positive correlation between the two, but an outside variable, temperature, influences both. As the temperature goes up, so does watermelon consumption and the number of people swimming who might drown. Eating watermelon certainly does not increase your chances of drowning! Think for yourself about ice cream sales and flu cases, which have a negative correlation. Should we all eat loads of ice cream during flu season as a preventative measure, or could there be another variable causing this apparent negative correlation?

Coefficient of Determination

Beyond merely determining the significance of the linear relationship between two variables, we can actually determine the extent to which the explanatory variable influences the response variable. If we would like to say how much of the variation in children's heights can be associated with the variation in their ages, we can calculate the **coefficient of determination**. The coefficient of determination is a measure of the proportion of the variability in the response variable that can be associated with the variation in the explanatory variable. Furthermore, the coefficient of determination is the square of the correlation coefficient.

Rounding Rule

Round the coefficient of determination, r^2, to three decimal places, just as we do for the correlation coefficient.

Definition

The **coefficient of determination**, r^2, is a measure of the proportion of the variation in the response variable (y) that can be associated with the variation in the explanatory variable (x).

In our height versus age example, $r \approx 0.979878$, so $r^2 \approx 0.960$. Thus, 96.0% of the variation in the children's heights, y, can be associated with the variation in their ages. The other 4.0% of the variation is associated with other factors or even sampling error.

Note that the coefficient of determination can be found using a TI-83/84 Plus calculator using the same set of commands that we used to find the correlation coefficient. After entering the x and y data in lists L1 and L2, respectively, and then pressing STAT and choosing CALC and option 4:LinReg(ax+b), the coefficient of determination is given directly above the correlation coefficient in the output. The output from the TI-83/84 Plus for the age and height data from Table 12.1 is shown in the screenshot in the margin.

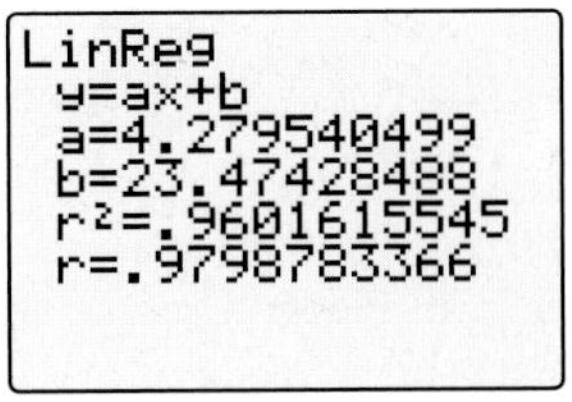

Example 12.9

Calculating and Interpreting the Coefficient of Determination

If the correlation coefficient for the relationship between the numbers of rooms in houses and their prices is $r = 0.65$, how much of the variation in house prices can be associated with the variation in the numbers of rooms in the houses?

Solution

Recall that the coefficient of determination tells us the amount of variation in the response variable (house price) that is associated with the variation in the explanatory variable (number of rooms). Thus, the coefficient of determination for the relationship between the numbers of rooms in houses and their prices will tell us the proportion or percentage of the variation in house prices that can be associated with the variation in the numbers of rooms in the houses. Also, recall that the coefficient of determination is equal to the square of the correlation coefficient. Since we know that the correlation coefficient for these data is $r = 0.65$, we can calculate the coefficient of determination as $r^2 = (0.65)^2 = 0.4225$. Thus, approximately 42.3% of the variation in house prices can be associated with the variation in the numbers of rooms in the houses.

12.1 Section Exercises

Type and Strength of a Linear Relationship

Directions: Predict the type and strength of the linear relationship between each pair of variables: weak negative, strong negative, weak positive, strong positive, or no linear relationship at all.

1. Height and shoe size

2. Cholesterol level and IQ

3. Hours of physical activity per week and body fat percentage

4. Number of fruits and vegetables eaten per day and risk of heart disease

5. Number of pets owned and household income

Statistical Significance of Correlation Coefficients

Directions: Determine whether the correlation coefficient is statistically significant at the specified level of significance for the given sample size. Assume that a scatter plot of the data shows a linear pattern.

6. $r = 0.731, \quad \alpha = 0.01, \quad n = 11$

7. $r = 0.638, \quad \alpha = 0.05, \quad n = 11$

8. $r = -0.499, \quad \alpha = 0.01, \quad n = 26$

9. $r = -0.443, \quad \alpha = 0.01, \quad n = 30$

10. $r = -0.462, \quad \alpha = 0.05, \quad n = 18$

11. $r = 0.305, \quad \alpha = 0.05, \quad n = 45$

Scatter Plots and Correlation

Directions: Complete the following objectives for each data set.

a. Draw a scatter plot.

b. Estimate the correlation in words (that is, positive, negative, or no correlation).

c. Calculate the correlation coefficient, *r*.

d. Determine whether *r* is statistically significant at the 0.01 level of significance.

12. Consider the relationship between the number of bids an item on eBay receives and the item's selling price. The following is a sample of 10 items sold through auctions on eBay.

Numbers of Bids and Selling Prices

Number of Bids	7	18	32	11	20	1	5	4	2	1
Price (in Dollars)	106.00	100.00	129.99	176.00	200.00	25.00	20.50	36.00	36.50	49.99

13. The following data represent the number of hours 10 students spent studying and their corresponding grades on their midterm exams.

Hours Spent Studying and Midterm Grades										
Hours Studying	0	0	0.5	1.5	2	3	3.5	3.5	4.5	6
Midterm Grade	64	83	72	74	85	89	79	93	98	95

14. The following table gives the average number of hours 12 junior high students were left unsupervised each day and their corresponding overall grade averages.

Hours Unsupervised and Overall Grade Averages												
Hours Unsupervised	0	0	0.5	1.0	1.0	1.5	2.0	3.0	3.0	4.0	4.5	5.5
Overall Grade Average	96	91	88	92	94	91	87	85	81	80	77	72

15. The following data represent the completion percentages and interception percentages of eight NFL quarterbacks.

Completion Percentages and Interception Percentages								
Completion %	62.5	67.0	60.2	65.4	57.8	57.5	59.1	59.2
Interception %	1.8	1.8	2.3	4.5	4.2	2.3	1.5	3.6

16. The following table gives the weeks of gestation and corresponding birth weights (in pounds) for a sample of 14 babies.

Weeks of Gestation and Birth Weights														
Weeks of Gestation	34	35	37	37	38	38	39	39	39	40	40	40	41	41
Weight (in Pounds)	4.8	5.9	5.3	6.7	5.6	7.3	7.7	8.4	8.2	9.5	7.6	8.6	8.9	9.4

17. A study on bone density at a women's hospital produced the following results.

Ages and Bone Density Levels					
Age (in Years)	34	43	49	51	65
Bone Density (in mg/cm^2)	946	875	804	723	691

Correlation Coefficients and Coefficients of Determination

Directions: Complete the following objectives for each data set.

a. Calculate the correlation coefficient, r.

b. Determine if r is statistically significant at the 0.05 level of significance.

c. Calculate the coefficient of determination, r^2.

d. Interpret the meaning of r^2 for the given set of data.

18. The following table contains data from 16 former elementary statistics students, where x represents the number of absences a student had for the semester and y represents the student's final class average.

Number of Absences and Class Average																
Absences	2	2	3	10	3	7	9	1	12	9	1	1	13	1	10	3
Class Average	86	83	81	53	92	71	68	79	53	78	77	85	62	97	54	79

19. The following table gives data collected on the birth weights of mothers and the birth weights of their babies. (**Hint:** To do calculations on these data, first convert the data to a single unit of measurement.)

Mothers' Birth Weights and Babies' Birth Weights

Mother	lb	6	6	7	8	7	7	7	6	9	6	9	8	6	5	6	5
	oz	6	2	9	10	6	15	7	5	3	10	3	9	1	0	1	0
Baby	lb	5	6	7	7	10	5	9	5	8	7	9	8	5	7	7	8
	oz	11	14	10	2	0	13	0	9	13	11	10	6	8	15	2	5

20. Many universities use a student's ACT score to help with admission decisions as well as placement decisions. From a sample of 20 students, the following data are obtained.

ACT Scores and College GPAs

ACT Score	College GPA
16	1.85
18	2.20
24	2.80
25	3.50
34	4.00
27	3.18
29	3.90
25	2.90
30	4.00
21	2.60
17	2.50
21	3.65
28	3.10
31	3.72
35	3.24
18	2.30
17	1.70
26	3.10
28	3.50
23	2.76

12.2 Linear Regression

If two variables, x and y, have a significant linear correlation, then we can predict what value y might take, given a specific value of x. Take, for instance, the relationship between class size and the average achievement test score for each class in one school district. Data from a sample of eight classes from the school district are given in the following table.

Table 12.3: Class Sizes and Average Test Scores

Class Size	15	17	18	20	21	24	26	29
Average Test Score	85.3	86.2	85.0	82.7	81.9	78.8	75.3	72.1

Let's go ahead and assume that there is, indeed, a significant correlation between these variables. It would be helpful to be able to make a *prediction* for the value of one variable based on the value of the other variable. For example, we may want to make a prediction as to what the average test score would be for a class of 25 students. To do this, we will need to use a *regression line* for the data. The regression line is the one line that fits the data set best. Look now at the following scatter plot depicting the data from Table 12.3. There are several lines shown that fall among the points on the scatter plot, but which one has the best fit?

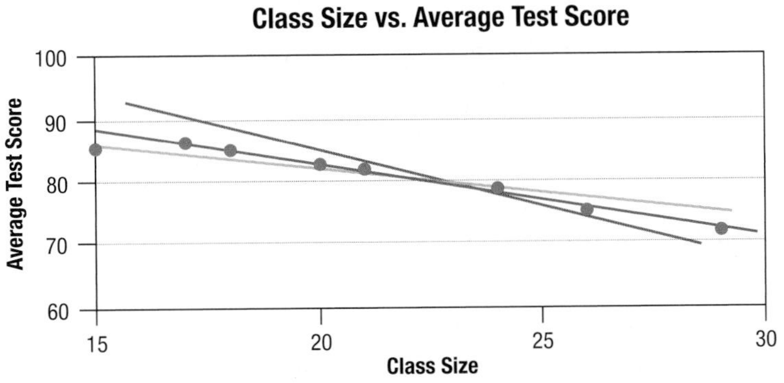

Figure 12.6: Scatter Plot of Data from Table 12.3

There is a particular line called the **least-squares regression line**, or the line of best fit, and it is the line for which the average variation from the data is the smallest. We can find the equation of this line by using some algebra. The regression line, $\hat{y} = b_0 + b_1x$, is determined by its slope, b_1, and its y-intercept, b_0. An alternate form for the regression line using standard algebraic notation is $\hat{y} = mx + b$. Also note that calculators and statistical software often omit the "hat" on y in the equation of a regression line, as in the following forms: $y = b + mx$, $y = ax + b$, or $y = a + bx$. The formulas used to find the slope and y-intercept are given in the boxes on the next page.

Math Symbols

Commonly used forms for the regression line:

$\hat{y} = b_0 + b_1x$

$\hat{y} = mx + b$

$y = b + mx$

$y = ax + b$

$y = a + bx$

Definition

The **least-squares regression line** is the line for which the average variation from the data is the smallest, also called the line of best fit, given by $\hat{y} = b_0 + b_1x$.

Formula

Slope of the Least-Squares Regression Line

The slope of the least-squares regression line for paired data from a sample is given by

$$b_1 = \frac{n\sum x_i y_i - \left(\sum x_i\right)\left(\sum y_i\right)}{n\sum x_i^2 - \left(\sum x_i\right)^2}$$

where n is the number of data pairs in the sample,

x_i is the ith value of the explanatory variable, and

y_i is the ith value of the response variable.

Rounding Rule

Round the slope and y-intercept of a regression line to three decimal places.

Formula

y-Intercept of the Least-Squares Regression Line

The y-intercept of the least-squares regression line for paired data from a sample is given by

$$b_0 = \frac{\sum y_i}{n} - b_1\frac{\sum x_i}{n}$$

where n is the number of data pairs in the sample,

x_i is the ith value of the explanatory variable,

y_i is the ith value of the response variable, and

b_1 is the slope of the least-squares regression line.

When calculating the correlation coefficient using a TI-83/84 Plus calculator, it produces the slope and y-intercept of the regression line. Thus, we do not need to do any additional work to find the regression line once we have calculated the correlation coefficient on a TI-83/84 Plus.

We have discussed the applications of r and r^2 for data whose correlation is statistically significant in the previous section. Determining if r is statistically significant is just the first hurdle in deciding whether or not a linear regression should be used to predict the value of the response variable. If r is statistically significant, we must still check that the scatter plot depicts a linear relationship between the two variables. In other words, we need to check if the data points fall somewhat in a linear fashion. We will leave more advanced techniques for determining linearity to higher-level textbooks.

What happens if the correlation coefficient between two variables is not statistically significant? If r fails to be statistically significant, then the results are not meaningful enough to use. In other words, there is not a strong enough linear relationship shown by the data to determine that there is a correlation between the two variables. *A linear regression should not be used to model the relationship between two variables if r is not statistically significant.*

Example 12.10

Finding a Least-Squares Regression Line Using a TI-83/84 Plus Calculator

The local school board wants to evaluate the relationship between class size and performance on the state achievement test. It decides to collect data from various schools in the district, and the data from a sample of eight classes are shown in the following table. Each pair of data values represents the class size and corresponding average score on the achievement test for one class.

Class Sizes and Average Test Scores								
Class Size	15	17	18	20	21	24	26	29
Average Test Score	85.3	86.2	85.0	82.7	81.9	78.8	75.3	72.1

Determine if there is a significant linear relationship between class size and average test score at the 0.05 level of significance. If the relationship is significant, find the least-squares regression line for these data.

Solution

First, we must decide which variable should be the x-variable and which variable should be the y-variable. Consider whether there is a *possibility* that one of these variables influences the values of the other. In this case, we are interested in the *possibility* that class size influences the average test score. Thus, class size is the explanatory variable, x, and average test score is the response variable, y.

- Press **STAT**.
- Select option `1:Edit`.
- Enter the values for class size (x) in List 1 (`L1`) and the values for average test score (y) in List 2 (`L2`).
- Press **STAT**.
- Select `CALC`.
- Choose option `4:LinReg(ax+b)`.
- Press **ENTER** twice.

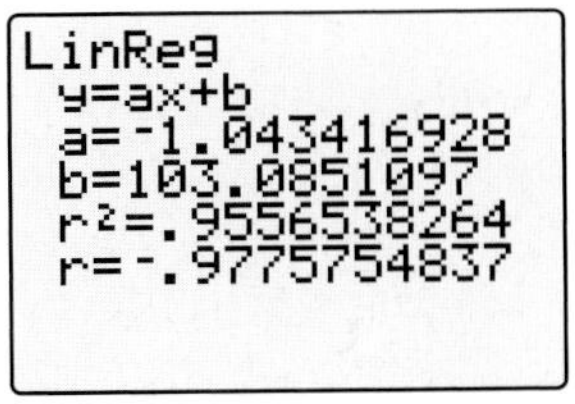

The output will include the correlation coefficient, r, and the coefficient of determination, r^2, that we discussed in the previous section. In addition, we see the slope, a, and the y-intercept, b, of the least-squares regression line.

Before we write out the equation for the regression line, we must be sure that there is sufficient evidence to claim that a linear relationship exists between these two variables. Based on the critical value for a 0.05 level of significance, the absolute value of the correlation coefficient must be greater than or equal to 0.707. The absolute value of the correlation coefficient, $|r| \approx 0.978$, is certainly greater than 0.707, so there is enough evidence at the 0.05 level of significance to conclude that the correlation between the two variables is statistically significant.

Next we need to consider the shape of the data in the scatter plot. Looking at the following graph, we can verify that the data points fall in a somewhat linear fashion. It is now appropriate to consider the linear regression model. The slope of the regression line is $a = b_1 \approx -1.043$ and the y-intercept of the regression line is $b = b_0 \approx 103.085$. Thus, the equation of the regression line, in the form $\hat{y} = b_0 + b_1 x$, is as follows.

$$\hat{y} = 103.085 - 1.043x$$

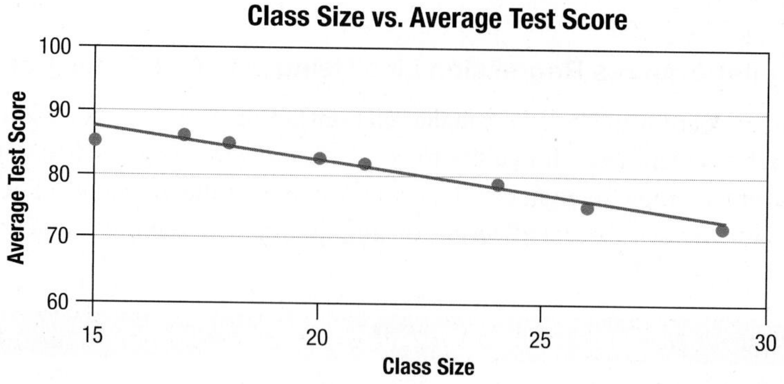

A few words of caution when using a regression model to make predictions. First, as stated previously, we must make sure that a scatter plot of the data reveals a possible linear relationship. Secondly, it is not appropriate to make a prediction if r is not statistically significant. Also, you should only make predictions for x-values that are *within the range of the sample data.* The regression model should not be applied outside of the known range. Finally, predictions should only be made *for the population from which data were drawn.* For example, if a regression model was created using data from Caucasian women ages 25 to 40, it would not be appropriate to use this model to make a prediction about an Asian woman or an 80-year-old Caucasian woman. This particular regression model should only be used to make predictions about Caucasian women between the ages of 25 and 40.

Properties

A prediction should <u>not</u> be made with a regression model if…

1. The data do not fall in a linear pattern when graphed on a scatter plot.
2. The correlation coefficient is not statistically significant.
3. You wish to make a prediction about a value outside the range of the sample data.
4. The population is different than that from which the sample data were drawn.

Example 12.11

Making Predictions Using a Least-Squares Regression Line

Use the equation of the regression line from the previous example,

$$\hat{y} = 103.085 - 1.043x,$$

to predict what the average achievement test score will be for the following class sizes.

a. 16

b. 19

c. 25

d. 45

Solution

a. $\hat{y} = 103.085 - 1.043(16) = 86.397$

b. $\hat{y} = 103.085 - 1.043(19) = 83.268$

c. $\hat{y} = 103.085 - 1.043(25) = 77.010$

d. It is not meaningful to predict the value of y for this class size because the value $x = 45$ is outside the range of the original data. The original data only considered class sizes between 15 and 29, so we should only predict the average achievement test scores for class sizes within this range.

Example 12.12

Finding the Least-Squares Regression Line for a Given Data Set

The following table lists data collected on the selling prices of used Land Rover Freelanders and their ages in years. Find a linear regression model for predicting the price of a used Land Rover Freelander based on its age in years, if appropriate at the 0.05 level of significance.

Selling Prices and Ages of Used Land Rover Freelanders										
Age (in Years)	3	4	1	2	2	3	4	3	4	1
Price (in Dollars)	15,500	14,995	30,795	28,995	23,995	20,900	20,500	19,995	19,888	29,995

Solution

To begin, we must determine which variable is the explanatory variable (x) and which variable is the response variable (y). We want to use the age of a used Freelander to predict its selling price, thus age (x) is the explanatory variable and price (y) is the response variable.

- Press STAT.
- Select option 1:Edit.
- Enter the values for age (x) in L1 and the values for price (y) in L2.
- Press STAT.
- Select CALC.
- Choose option 4:LinReg(ax+b).
- Press ENTER twice.

The output will include the correlation coefficient, r, the coefficient of determination, r^2, the slope, a, and the y-intercept, b, of the regression line.

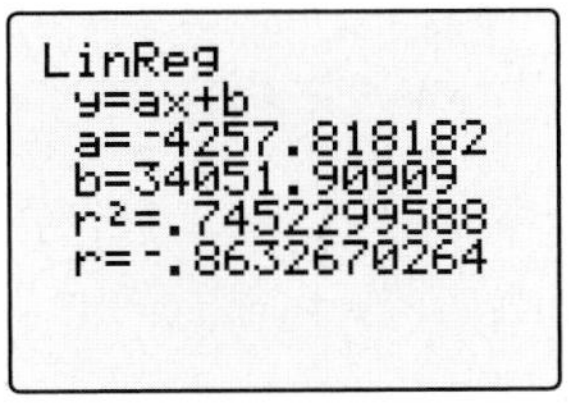

Before we write out the equation for the regression line, we must be sure that there is sufficient evidence to claim that a linear relationship exists between these two variables. First, notice that the scatter plot depicts a possible linear relationship.

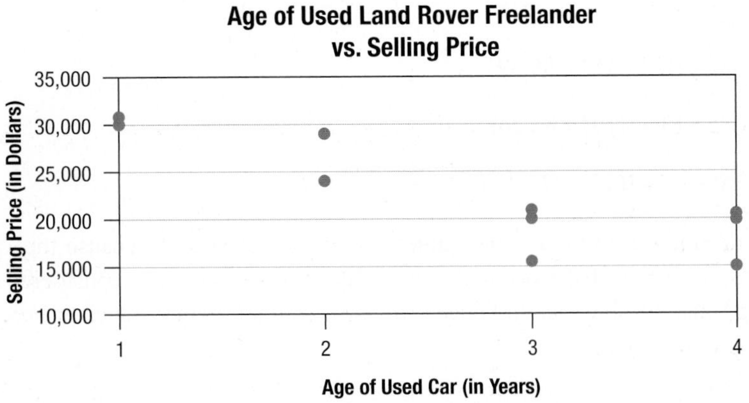

Next, based on the critical value for a 0.05 level of significance, the absolute value of the correlation coefficient must be greater than or equal to 0.632. The absolute value of the correlation coefficient, $|r| \approx 0.863$, is greater than 0.632, so there is enough evidence at the 0.05 level of significance to conclude that a possible linear relationship exists between these two variables.

Thus, it is appropriate to consider the linear regression model. The slope of the regression line is $a = b_1 \approx -4257.818$ and the y-intercept of the regression line is $b = b_0 \approx 34{,}051.909$. Therefore the regression line, in the form $\hat{y} = b_0 + b_1x$, is as follows.

$$\hat{y} = 34{,}051.909 - 4257.818x$$

Example 12.13

Making Predictions Using a Linear Regression Model

Use the linear regression model from the previous example,

$$\hat{y} = 34{,}051.909 - 4257.818x,$$

to predict the following.

a. The selling price of a Land Rover Freelander that is 2.5 years old

b. The selling price of a Land Rover Freelander that is 10 years old

c. The selling price of a Land Rover Range Rover that is 3 years old

Solution

a. Substitute the value $x = 2.5$ into the regression line and solve for $\hat{y}$.

$$\hat{y} = 34{,}051.909 - 4257.818x$$
$$\hat{y} = 34{,}051.909 - 4257.818(2.5)$$
$$\hat{y} = 23{,}407.364$$

Thus, we would predict that a 2.5-year-old Freelander would sell for approximately \$23,407.36.

b. The original sample only contains Land Rover Freelanders that are 1 to 4 years old; therefore, it is inappropriate to use this model to predict the price of a Freelander that is 10 years old.

c. The population is Freelanders, not Range Rovers. Thus, it is inappropriate to use this model to predict the price of a Range Rover, no matter how old it is.

Now, let's put all of the information we have learned about linear regression together for the last example.

Example 12.14

A Linear Regression Model, Start to Finish

The following table gives the average monthly temperatures and corresponding monthly precipitation totals for one year in Key West, Florida.

Average Temperatures and Precipitation Totals in Key West, Florida						
Average Temperature (in °F)	75	76	79	82	85	88
Precipitation (in Inches)	2.22	1.51	1.86	2.06	3.48	4.57
Average Temperature (in °F)	89	90	88	85	81	77
Precipitation (in Inches)	3.27	5.40	5.45	4.34	2.64	2.14

a. Create a scatter plot for the data. Does there appear to be a linear relationship between x and y?

b. Calculate the correlation coefficient, r.

c. Verify that the correlation coefficient is statistically significant at the 0.05 level of significance.

d. Determine the equation of the line of best fit.

e. Calculate and interpret the coefficient of determination, r^2.

f. If appropriate, predict the monthly precipitation total in Key West for a month in which the average temperature is 80 degrees.

g. If appropriate, predict the monthly precipitation in Destin, Florida, for a month in which the average temperature is 83 degrees.

Solution

a. In order to create a scatter plot, we must first decide which variable will be x and which will be y. Since it is not clear whether one variable influences another, we can choose either variable for the x-axis. Let's choose the average temperatures for the x-values and the inches of precipitation for the y-values. A scatter plot of the data is shown in the following figure. Notice that there does appear to be a possible positive linear relationship between precipitation and temperature. If r is statistically significant for the variables, a linear regression model would be appropriate.

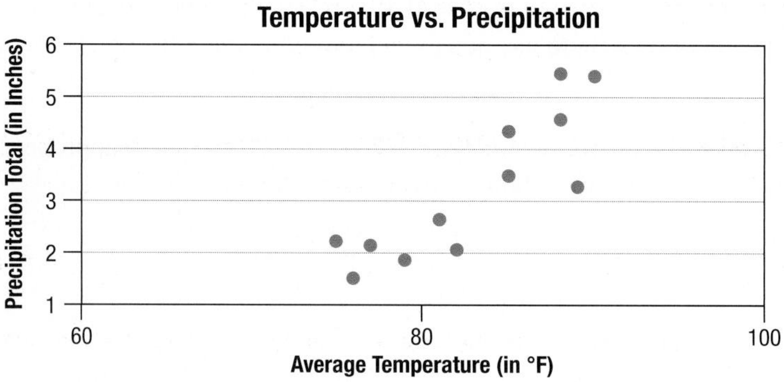

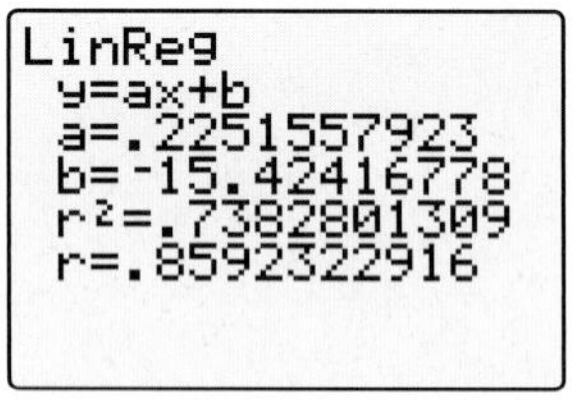

b. Using a TI-83/84 Plus calculator as we have in the previous examples, we calculate $r \approx 0.859$.

c. Using Table I in Appendix A, we find that the critical value at the 0.05 level of significance is 0.576. Since $|0.859| > 0.576$, we can conclude that r is indeed statistically significant.

d. From the calculator, we see that the equation of the line of best fit is as follows.

$$\hat{y} = -15.424 + 0.225x$$

e. The coefficient of determination is approximately 0.738. This tells us that approximately 73.8% of the variation in precipitation can be attributed to the linear relationship between temperature and precipitation. The remaining 26.2% of the variation is from unknown sources.

f. Because r is statistically significant, we can use the regression line to make predictions regarding the variables. In addition, 80 is within the range of the x-values from the sample data, so it is appropriate to predict the monthly precipitation total when the average temperature is 80 degrees. Because we designated the average temperatures as the x-values, substitute $x = 80$ into the regression equation to obtain an estimate for the precipitation total for a month when the average temperature is 80 degrees.

$$\hat{y} = -15.424 + 0.225x$$
$$\hat{y} = -15.424 + 0.225(80)$$
$$\hat{y} \approx 2.58$$

Thus, a reasonable estimate for the precipitation for a month in which the average temperature is 80 degrees is approximately 2.58 inches.

g. The data were collected in Key West—not Destin, Florida. Therefore, it is not appropriate to use the linear regression equation to make predictions regarding the precipitation in Destin.

12.2 Section Exercises

Finding Least-Squares Regression Lines and Making Predictions

Directions: Complete the following objectives for each data set.

a. Find the equation of the least-squares regression line.

b. Determine if the regression equation is appropriate to use for making predictions at the 0.05 level of significance. If so, make the indicated prediction.

1. **a.** The following table shows students' test scores on the first two tests in an introductory biology class.

Biology Test Scores												
First Test, *x*	55	40	71	82	90	50	83	75	65	52	77	93
Second Test, *y*	58	43	68	86	87	51	87	70	67	55	77	90

b. If a student scored a 70 on his first test, make a prediction for his score on the second test.

2. **a.** The following table lists the birth weights (in pounds), x, and the lengths (in inches), y, for a set of newborn babies at a local hospital.

Birth Weights and Lengths										
Birth Weight (in Pounds), *x*	5	3	8	6	9	7	7	6	6	10
Length (in Inches), *y*	16	15	18	19	19	18	20	17	15	21

b. Predict the length of an 8-pound baby.

3. **a.** The following table contains the weights of a group of 12 women, where x represents the woman's weight at age 17, and y represents the woman's weight after menopause.

Weights (in Pounds)												
Weight at Age 17, *x*	99	123	119	87	90	101	98	121	131	134	135	97
Weight after Menopause, *y*	121	156	145	109	111	137	128	142	149	152	155	128

b. For a woman who weighed 105 pounds at 17 years old, what would be her predicted weight after menopause?

4. **a.** Students studying for a state achievement test were asked to keep track of the total number of hours that they spent playing video games during the two weeks before the test. In the following table, x is the number of hours spent playing video games (on any type of device) and y is the student's score on the achievement test (out of 130).

Hours Playing Video Games and Achievement Test Scores								
Hours Playing Video Games, *x*	12	15	14	10	11	13	12	14
Test Score, *y*	99	63	79	115	108	82	98	73

b. If a student spent 11.5 hours playing video games, predict his achievement test score.

5. a. The following table shows the numbers of international tourists (in millions) as well as the income from those tourists (in billions of US dollars) for the eight countries that had the most international visitors in 2009.

International Tourism in 2009

	France	US	Spain	China	Italy	UK	Turkey	Germany
International Tourist Arrivals (in Millions), *x*	74.2	54.9	52.2	50.9	43.2	28.0	25.5	24.2
International Tourist Revenue (in Billions of US Dollars), *y*	48.7	94.2	53.2	39.7	40.2	30.1	21.3	34.7

Source: National Museum Directors' Conference. "Museums and Tourism." July 2010. http://www.nationalmuseums.org.uk/media/documents/what_we_do_documents/museums_tourism_briefing_jul10.pdf (20 Mar. 2012).

b. Predict the amount of revenue from international tourists for Malaysia in 2009 if the country had 23.6 million international tourists.

6. a. The following table compares print advertisement revenue to online advertisement revenue for US newspapers from 2003 to 2009.

Newspaper Ad Revenue (in Millions of Dollars)

	2003	2004	2005	2006	2007	2008	2009
Print, *x*	44,939	46,703	47,408	46,611	42,209	34,740	24,821
Online, *y*	1216	1541	2027	2664	3166	3109	2743

Source: Edmonds, R., E. Guskin, and T. Rosenstiel. "Newspapers: By the Numbers." The State of the News Media 2011: An Annual Report on American Journalism. Pew Research Center's Project for Excellence in Journalism. 2011. http://stateofthemedia.org/2011/newspapers-essay/data-page-6/ (21 Mar. 2012).

b. Predict the amount of online advertisement revenue for US newspapers in 2010 if the print advertisement revenue in 2010 was 22,795 million dollars.

7. a. The following table shows the average mathematics and science achievement test scores of eighth-grade students from 12 countries around the world.

Mathematics and Science Achievement Test Scores

Math, *x*	525	558	511	531	392	585	476	496	520	582	532	502
Science, *y*	540	535	518	533	420	569	450	538	535	530	552	515

b. Predict the average science score for a country with an average eighth-grade math score of 570.

8. a. The following table shows the average amount of time wasted at work per day by year of birth.

Birth Years and Time Wasted per Day

Birth Year, *x*	1940	1955	1965	1975	1985
Time Wasted (in Hours), *y*	0.5	0.68	1.19	1.61	1.95

b. Predict the amount of time that would be wasted by someone born in 1980.

12.3 Regression Analysis

In Section 12.1, we looked at how to evaluate whether or not a linear model is appropriate for a given set of data by using statistical significance. We also learned how to quantify the amount of variation in one variable that is associated with the amount of variation in the other variable by finding the coefficient of determination. In this section we will analyze the linear regression model in more detail.

One important point that we have yet to discuss is that the regression line, or line of best fit, that we calculated in the previous section is a sample statistic, because it results from calculations performed on a sample of the population's data. The true regression line would only come from a census of the entire population, and thus the variables, slope, and y-intercept contained in it are population parameters. We represent the true regression line using Greek letters, as we have with most population parameters.

$$y = \beta_0 + \beta_1 x \text{ (Population parameters)}$$

$$\hat{y} = b_0 + b_1 x \text{ (Sample statistics)}$$

Thus, y represents the actual value occurring in the population and $\hat{y}$ represents the predicted value based on sample data. As you might expect, these two values are often not the same. The difference between the actual value and the predicted value is called a **residual**.

Formula

Residual

A **residual** is the difference between the actual value of y from the original data and the predicted value of $\hat{y}$ found using the regression line, given by

$$\text{Residual} = y - \hat{y}$$

where y is the observed value of the response variable and

$\hat{y}$ is the predicted value of y using the least-squares regression model.

Example 12.15

Calculating Residuals Using an Estimated Regression Equation

The following table gives data from a local school district on children's ages (x) and reading levels (y). For these data, a reading level of 4.3 would indicate that the child's reading level is $\frac{3}{10}$ of the year through the fourth grade. The children's ages are given in years.

Ages and Reading Levels

Age (in Years), x	6	7	8	9	10	11	12	13	14	15
Reading Level, y	1.3	2.2	3.7	4.1	4.9	5.2	6.0	7.1	8.5	9.7

Using a TI-83/84 Plus calculator to determine the linear regression model, we calculate the regression line to be $\hat{y} = -3.811 + 0.865x$. Note that $r \approx 0.989$, which is greater than the critical value at the 0.05 level of significance, $r_{0.05} = 0.632$.

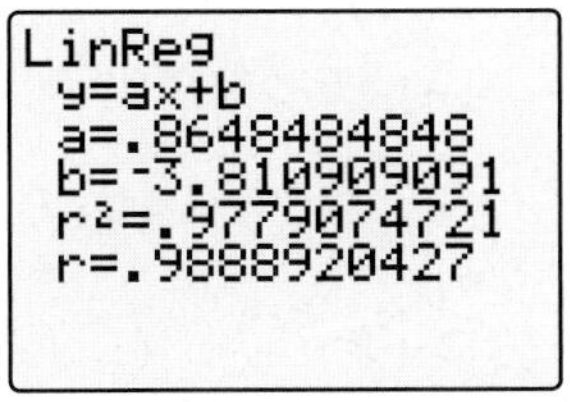

Furthermore, the following scatter plot depicts the linear pattern of the data values. Therefore, it is appropriate to use this linear regression model to make predictions.

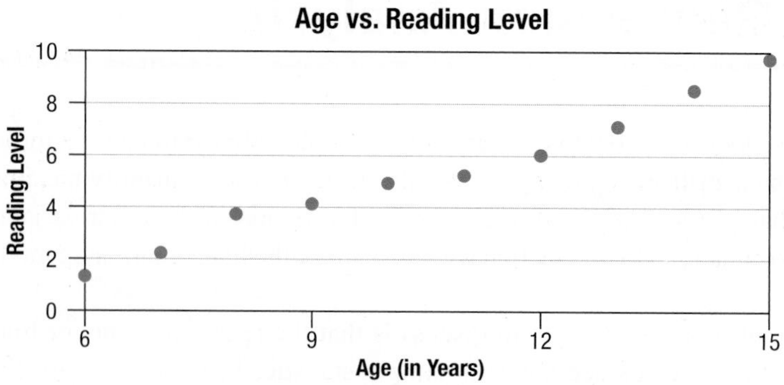

Use the regression equation to calculate an estimate, $\hat{y}$, for each value of x, and then use the estimate to calculate the residual for each value of y.

Solution

We can use a TI-83/84 Plus calculator to perform all of the necessary calculations at once. Age is the explanatory variable, x, and reading level is the response variable, y.

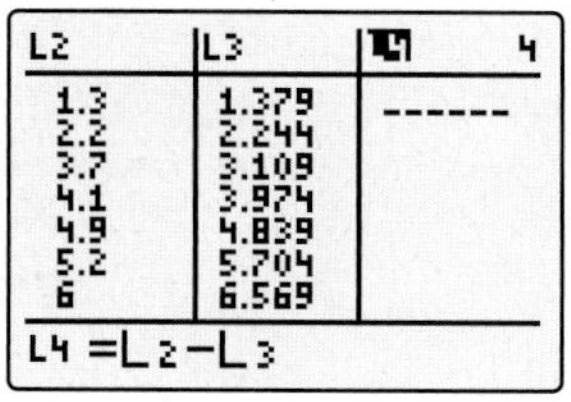

- Press STAT.
- Select option `1:Edit`.
- Enter the ages in `L1` and the reading levels in `L2`.
- Use the arrow keys to highlight `L3` and enter the formula `-3.811+Ø.865*L1`. This will calculate the predicted y-value for each x-value.
- Highlight `L4` and enter the formula `L2-L3`. This formula will calculate each of the residuals.

The results will be as follows.

Predicted Values and Residuals			
Age (in Years), x	Reading Level, y	Predicted Value, $\hat{y}$	Residual, $y-\hat{y}$
6	1.3	1.379	−0.079
7	2.2	2.244	−0.044
8	3.7	3.109	0.591
9	4.1	3.974	0.126
10	4.9	4.839	0.061
11	5.2	5.704	−0.504
12	6.0	6.569	−0.569
13	7.1	7.434	−0.334
14	8.5	8.299	0.201
15	9.7	9.164	0.536

The residual of each value reflects how far the original data point is from the point on the regression line. Graphically, the residual is the vertical distance from the original data point to the point on the regression line. If we graph the regression line on a scatter plot of the data, we can draw vertical bars from the regression line to the original points to indicate the residual for each point. Figure 12.7 depicts the residuals that we calculated in Example 12.15. Notice that the points in the scatter plot that lie below the regression line have negative residuals and the points that lie above the regression line have positive residuals.

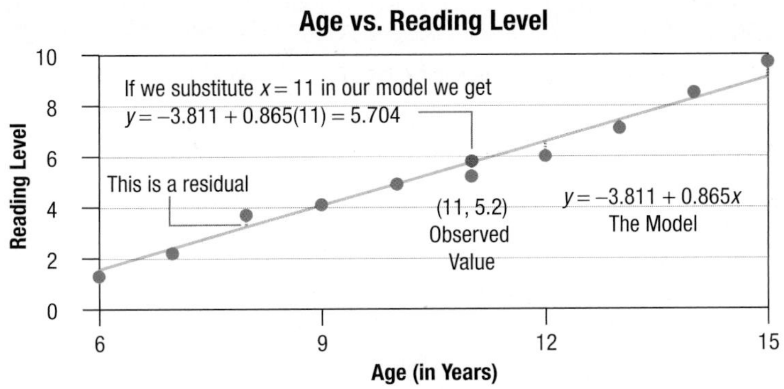

Figure 12.7: Residuals Shown Graphically

Another name for a residual is an *error* in the predicted value. It seems like we should be able to find the "total error" for the regression line by summing all of the errors; however, as was the case with standard deviation in Section 3.2, simply summing the errors would result in negative errors canceling out positive errors with a net result of 0. To eliminate this situation, we square the errors before adding them. The value calculated by summing the squares of the errors is called the **sum of squared errors**, which we abbreviate **SSE**.

Formula

Sum of Squared Errors (SSE)

The **sum of squared errors (SSE)** for a regression line is the sum of the squares of the residuals, given by

$$\text{SSE} = \sum \left(y_i - \hat{y}_i\right)^2$$

where y_i is the ith observed value of the response variable and

$\hat{y}_i$ is the predicted value of y_i using the least-squares regression model.

To interpret the sum of squared errors, consider the size of the value for SSE. If the data points are very far from the regression line, then the sum of squared errors will be large. If the data points are close to the regression line, then the sum of squared errors will be small. Thus, the larger the value of SSE, the worse the linear model will be at predicting the value of y. The smaller the value of SSE, the better the linear model will be at predicting the value of y. Therefore, the line that fits the data "best" would be the one with the smallest value of SSE. The following example simply focuses on how to calculate the SSE.

Example 12.16

Calculating the Sum of Squared Errors

Calculate the sum of squared errors, SSE, for the data on children's ages and reading levels from the previous example.

Solution

Using the values we calculated in the previous example, we begin by squaring each error as shown in the following table.

Squared Errors				
Age (in Years), x	Reading Level, y	Predicted Value, $\hat{y}$	Residual, $y-\hat{y}$	Squared Error, $\left(y-\hat{y}\right)^2$
6	1.3	1.379	–0.079	0.006241
7	2.2	2.244	–0.044	0.001936
8	3.7	3.109	0.591	0.349281
9	4.1	3.974	0.126	0.015876
10	4.9	4.839	0.061	0.003721
11	5.2	5.704	–0.504	0.254016
12	6.0	6.569	–0.569	0.323761
13	7.1	7.434	–0.334	0.111556
14	8.5	8.299	0.201	0.040401
15	9.7	9.164	0.536	0.287296

The last column lists the squares of the residual values. The sum of the squared errors is the sum of the values in this last column. Thus, SSE ≈ 1.394.

We can compare the calculation of the sum of squared errors to the calculation of the sum of squared deviations in the formula for the standard deviation of a set of data. This comparison is not a coincidence. We will use this value to calculate the **standard error of estimate**, which is a measure of how much the data points deviate from the regression line. This is analogous to how the standard deviation measures how much the data deviate from the sample mean. Thus, the smaller the value of the standard error of estimate is, the closer the data points are to the regression line. The formula is also similar to the standard deviation formula. The formula for the standard error of estimate, denoted by S_e, is as follows.

Formula

Standard Error of Estimate

The **standard error of estimate**, which is a measure of how much the sample data points deviate from the regression line, is given by

$$S_e = \sqrt{\frac{\sum\left(y_i - \hat{y}_i\right)^2}{n-2}} = \sqrt{\frac{\text{SSE}}{n-2}}$$

where y_i is the ith observed value of the response variable,

$\hat{y}_i$ is the predicted value of y_i using the least-squares regression model,

n is the number of data pairs in the sample, and

SSE is the sum of squared errors.

Note that we are dividing the sum of the squared errors by the sample size *minus 2*. Just as we divided by the number of degrees of freedom when calculating a sample standard deviation, we are dividing by the number of degrees of freedom in the formula for the standard error of estimate as well. However, the number of degrees of freedom for this calculation is given by $n-2$. The standard error of estimate can be calculated using a TI-83/84 Plus calculator.

Example 12.17

Calculating the Standard Error of Estimate Using a TI-83/84 Plus Calculator

Calculate the standard error of estimate for the data on children's ages and reading levels from Example 12.15 (repeated in the following table).

Ages and Reading Levels										
Age (in Years), x	6	7	8	9	10	11	12	13	14	15
Reading Level, y	1.3	2.2	3.7	4.1	4.9	5.2	6.0	7.1	8.5	9.7

Solution

Begin as follows.

- Press STAT.
- Choose 1:Edit.
- Enter the age data into L1 and the reading-level data in L2.
- Press STAT.
- Choose TESTS.
- Choose option F:LinRegTTest.
- Enter L1 for the Xlist and L2 for the Ylist. The value entered for the option Freq should be 1.
- Choose ≠Ø for the alternative hypothesis to test the significance of the linear relationship.
- Enter the regression equation into RegEQ if you have already calculated it. If not, you may leave this blank.
- Choose Calculate.
- Press ENTER.

Side Note

On TI-83 Plus calculators and older TI-84 Plus calculators, the function, LinRegTTest, is option E on the TESTS menu.

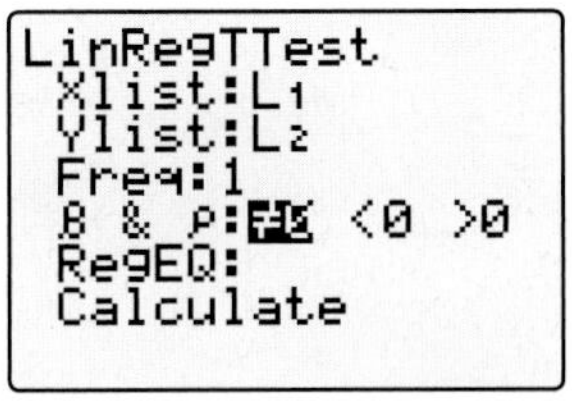

The results, shown in the following screenshots, include the t-test statistic for testing the significance of the linear relationship. The calculator also gives us the p-value for that hypothesis test and the number of degrees of freedom. The slope and y-intercept of the regression line are also given. Note that the regression line is given in the form y=a+bx, so a is the y-intercept and b is the slope, which is the opposite of the results that we get when we use the LinReg(ax+b) function. The last two values given are the coefficient of determination and the correlation coefficient. The standard error of estimate is s, the third-to-last value given.

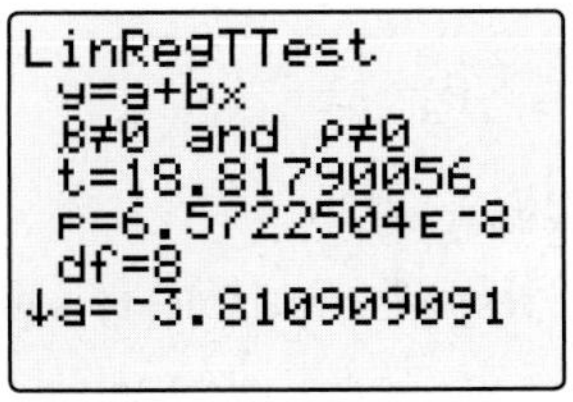

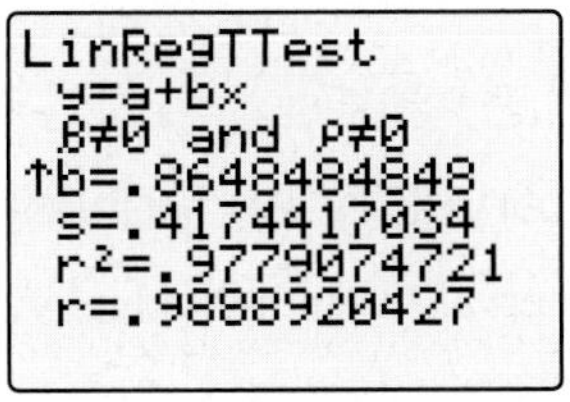

Thus, the standard error of estimate for the data on ages and reading levels is $S_e \approx 0.417$. Since this value is close to 0, we can conclude that the data points do not deviate very much from the regression line.

Prediction Interval for an Individual *y*-Value

In Chapter 8, we learned that it is preferable to use an interval estimate instead of a point estimate for a population parameter. So far in this chapter, we have only presented point estimates for the various parameters discussed. For example, the predicted value, $\hat{y}$, is a point estimate for the population parameter, y. We can use the standard error of estimate to create a confidence interval for y at a given fixed value of the explanatory variable, x. We call this confidence interval for the value of the response variable a **prediction interval**.

Definition

A **prediction interval** is a confidence interval for an individual value of the response variable, y, at a given fixed value of the explanatory variable, x.

Thus for a given fixed value, x_0, the prediction interval for an individual y can be found by calculating the margin of error, E, and then subtracting the margin of error from and adding the margin of error to the point estimate, $\hat{y}$. The formula for the margin of error is as follows.

Formula

Margin of Error of a Prediction Interval for an Individual *y*-Value

The margin of error of a prediction interval for an individual value of the response variable, y, is given by

$$E = t_{\alpha/2} S_e \sqrt{1 + \frac{1}{n} + \frac{n(x_0 - \bar{x})^2}{n\left(\sum x_i^2\right) - \left(\sum x_i\right)^2}}$$

where $t_{\alpha/2}$ is the critical value for the level of confidence, $c = 1 - \alpha$, such that the area under the t-distribution with $n - 2$ degrees of freedom to the right of $t_{\alpha/2}$ is equal to $\frac{\alpha}{2}$,

S_e is the standard error of estimate,

n is the number of data pairs in the sample,

x_0 is the fixed value of the explanatory variable, x,

$\bar{x}$ is the mean of the x-values for the data points in the sample, and

x_i is the ith value of the explanatory variable.

Formula

Prediction Interval for an Individual *y*-Value

The prediction interval for an individual value of the response variable, y, is given by

$$\hat{y} - E < y < \hat{y} + E$$

or

$$\left(\hat{y} - E,\ \hat{y} + E\right)$$

where $\hat{y}$ is the predicted value of the response variable, y, when $x = x_0$ and

E is the margin of error.

Example 12.18

Constructing a Prediction Interval for an Individual *y*-Value

Construct a 95% prediction interval for the reading level of a child who is 8 years old. Use the data from Example 12.15 on children's ages and reading levels as the sample data (repeated in the following table).

Ages and Reading Levels										
Age (in Years), *x*	6	7	8	9	10	11	12	13	14	15
Reading Level, *y*	1.3	2.2	3.7	4.1	4.9	5.2	6.0	7.1	8.5	9.7

Solution

Neither a TI-83/84 Plus calculator nor Microsoft Excel will directly calculate a prediction interval[1], so we must calculate the margin of error by hand and use this value to construct the prediction interval.

Step 1: Find the regression equation for the sample data.

We know from previous examples that the regression equation is as follows.

$$\hat{y} = -3.811 + 0.865x$$

Step 2: Use the regression equation to calculate the point estimate, $\hat{y}$, for the given value of x.

In this example, $x = 8$. Thus, we have the following.

$$\hat{y} = -3.811 + 0.865x$$
$$\hat{y} = -3.811 + 0.865(8)$$
$$\hat{y} = 3.109$$

Step 3: Calculate the sample statistics necessary to calculate the margin of error.

Using a TI-83/84 Plus calculator, we can enter the values for age in L1 and the values for reading level in L2. Next, press STAT, select CALC, and then choose option 2:2-Var Stats. This will give us many of the statistics we need.

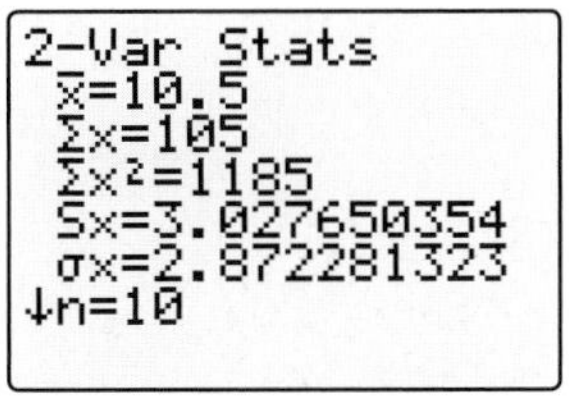

$$\bar{x} = 10.5, \quad \sum x_i = 105, \quad \sum x_i^2 = 1185, \quad n = 10$$

Next, recall that we found that $S_e \approx 0.417442$ in the previous example. This value was also found using a TI-83/84 Plus calculator. Lastly, using the t-distribution table or appropriate technology, we find the critical value for this test, $t_{\alpha/2} = t_{0.05/2} = t_{0.025} = 2.306$ for the t-distribution with $n - 2 = 10 - 2 = 8$ degrees of freedom.

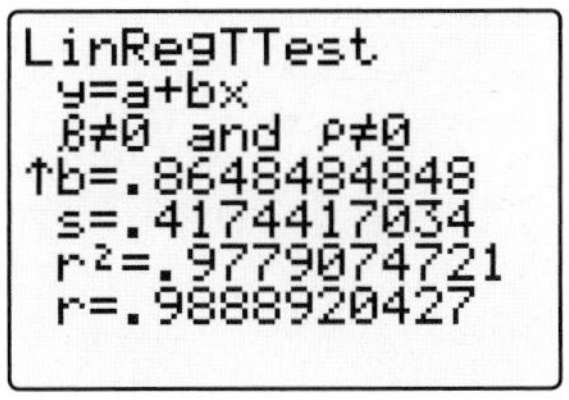

Step 4: Find the margin of error.

Substituting the necessary statistics into the formula for the margin of error, we obtain the following.

[1] *However, many statistical software packages, such as Minitab, will directly calculate a prediction interval.*

$$E = t_{\alpha/2} S_e \sqrt{1 + \frac{1}{n} + \frac{n(x_0 - \bar{x})^2}{n(\sum x_i^2) - (\sum x_i)^2}}$$

$$= (2.306)(0.417442)\sqrt{1 + \frac{1}{10} + \frac{10(8 - 10.5)^2}{10(1185) - (105)^2}}$$

$$\approx 1.043793$$

Step 5: **Subtract the margin of error from and add the margin of error to the point estimate.**

Subtracting the margin of error from the point estimate of $\hat{y} = 3.109$ gives us the lower endpoint for the prediction interval.

$$\text{Lower endpoint: } \hat{y} - E = 3.109 - 1.043793$$
$$\approx 2.065$$

By adding the margin of error to the point estimate, we obtain the upper endpoint for the prediction interval as follows.

$$\text{Upper endpoint: } \hat{y} + E = 3.109 + 1.043793$$
$$\approx 4.153$$

Thus the 95% confidence interval for the individual y-value ranges from 2.065 to 4.153. The confidence interval can be written mathematically using either inequality symbols or interval notation, as shown below.

$$2.065 < y < 4.153$$

or

$$(2.065, 4.153)$$

Thus, for an 8-year-old child, we can be 95% confident that he or she would have a reading level between 2.065 and 4.153, or be reading between the second and fourth grade levels.

Confidence Intervals for the Slope and *y*-Intercept of a Regression Equation

We can now construct a prediction interval for an individual value of y given a specific value of x, but $\hat{y}$ is not the only sample statistic in the regression equation. The values for the slope, b_1, and the y-intercept, b_0, are also sample statistics and thus, point estimates for the population parameters β_1 and β_0, respectively. We can construct confidence intervals for these parameters as well. We could construct the confidence intervals in the same manner as we have for all confidence intervals, by subtracting the margin of error from and adding the margin of error to the point estimate for that parameter. However, Microsoft Excel easily calculates these confidence intervals from the raw sample data, as we will see in Example 12.19.

Example 12.19

Constructing Confidence Intervals for β_1 and β_0 Using Microsoft Excel

Construct 95% confidence intervals for the slope, β_1, and the y-intercept, β_0, of the regression equation for age and reading level. Use the sample data from Example 12.15 (repeated in the following table).

Ages and Reading Levels										
Age (in Years), *x*	6	7	8	9	10	11	12	13	14	15
Reading Level, *y*	1.3	2.2	3.7	4.1	4.9	5.2	6.0	7.1	8.5	9.7

Solution

Begin by entering the sample data into Microsoft Excel as shown in the following screenshot.

	A	B
1	**Age**	**Reading Level**
2	6	1.3
3	7	2.2
4	8	3.7
5	9	4.1
6	10	4.9
7	11	5.2
8	12	6
9	13	7.1
10	14	8.5
11	15	9.7

Under the **Data** tab, choose **Data Analysis**. Select **Regression** from the options listed. Enter the necessary information into the Regression menu as shown in the following screenshot. Click **OK**.

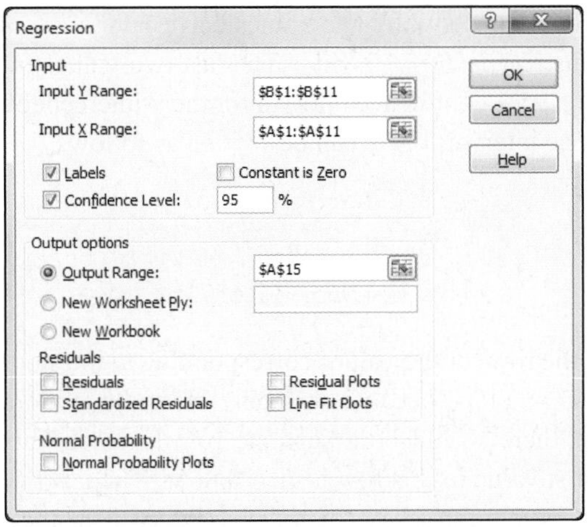

The results, shown in the following screenshot, provide an abundance of information, much of which we have discussed throughout this chapter.

Side Note

If Data Analysis does not appear in your Excel menu under the Data tab, you can easily add this function. See the directions in Appendix B: Getting Started with Microsoft Excel.

SUMMARY OUTPUT

Regression Statistics	
Multiple R	0.988892043
R Square	0.977907472
Adjusted R Square	0.975145906
Standard Error	0.417441703
Observations	10

ANOVA

	df	SS	MS	F	Significance F
Regression	1	61.70693939	61.70693939	354.1133814	6.57225E-08
Residual	8	1.394060606	0.174257576		
Total	9	63.101			

	Coefficients	Standard Error	t Stat	P-value	Lower 95%	Upper 95%	Lower 95.0%	Upper 95.0%
Intercept	-3.810909091	0.500297157	-7.617291121	6.20371E-05	-4.964596403	-2.657221778	-4.964596403	-2.657221778
Age	0.864848485	0.045958819	18.81790056	6.57225E-08	0.758867258	0.970829711	0.758867258	0.970829711

1 Multiple R is the absolute value of the correlation coefficient, $|r|$.

2 R Square is the coefficient of determination, r^2.

3 Standard Error is the standard error of estimate, S_e.

The ANOVA table will be discussed in the next section, since it is more meaningful when discussing more than one explanatory variable. However, it does contain a few of the important values we discussed so far in this section.

4 The intersection of the Residual row and the *SS* column is the sum of squared errors, SSE.

5 The *Lower 95.0%* and *Upper 95.0%* columns give the lower and upper endpoints of the 95% confidence intervals for the y-intercept and slope.

6 7 The *Coefficients* column gives the values for the coefficients, that is, the y-intercept and slope, of the regression line.

The lower and upper endpoints of the 95% confidence intervals for the y-intercept and slope are the values we are interested in for this example.

The row labeled Intercept is the row for the values corresponding to the y-intercept. Notice that the first value in this row is $b_0 \approx -3.811$. The last two values in this row are the lower and upper endpoints for a 95% confidence interval for the y-intercept of the regression line, β_0. Thus, the 95% confidence interval for β_0 can be written as follows.

$$-4.965 < \beta_0 < -2.657$$

or

$$(-4.965,\ -2.657)$$

The row labeled Age is the row for the values corresponding to the slope of the regression line. It is labeled Age instead of Slope because it is possible to have more than one explanatory variable, in which case there would be a separate row for each variable, labeled with the variable's name. The first value in this row is $b_1 \approx 0.865$. The last two values in this row are the lower and upper endpoints for a 95% confidence interval for the slope of the regression line, β_1. Thus, the 95% confidence interval for β_1 can be written as follows.

$$0.759 < \beta_1 < 0.971$$

or

$$(0.759,\ 0.971)$$

12.3 Section Exercises

Regression Analysis

Directions: Complete the following objectives for each data set.

a. Calculate the sum of squared errors, SSE.

b. Calculate the standard error of estimate, S_e.

c. Construct a 95% prediction interval for the given value of the explanatory variable.

Note: Microsoft Excel can be used to calculate the answers for parts a. and b.

1. The following table contains data from a sample of ten women regarding their weights and the times it took them to run/walk one mile. The times are in minutes and the weights are in pounds.

   Construct a 95% prediction interval for time given $x = 175$.

Weights and Times to Run/Walk One Mile										
Weight (in Pounds), x	178	182	180	165	159	170	189	193	195	203
Time (in Minutes), y	13.24	15.32	15.21	12.04	12.21	13.10	16.75	16.98	17.02	17.19

2. The following table contains data from a sample of ten students regarding their final exam scores in the first and second semesters of English Composition.

   Construct a 95% prediction interval for the final exam score in the second semester given $x = 70$.

Exam Scores										
Final Exam 1, x	87	65	99	78	81	86	61	63	90	100
Final Exam 2, y	89	72	93	81	75	87	72	69	88	89

3. The following table contains data from a sample of 15 couples regarding the weight gained by each partner during the first year of marriage (in pounds).

   Construct a 95% prediction interval for the husband's weight gain given $x = 20$.

Weight Gain (in Pounds)															
Weight Gain (Wife), x	15	5	32	24	33	20	19	8	12	40	21	15	24	3	10
Weight Gain (Husband), y	20	9	12	25	38	12	23	10	9	33	22	17	18	5	11

4. The following table contains data regarding golf scores and the average lengths of drives (in yards) for a sample of five golfers.

Construct a 95% prediction interval for the length of the golfer's drive given $x = 72$.

Golf Scores and Average Drive Lengths					
Golf Score, *x*	68	72	69	70	75
Length of Drive (in Yards), *y*	275	236	289	245	225

5. The following table contains data from a sample of eight people regarding the average number of fat grams consumed per day and the amount of weight lost (in pounds) over a period of six months.

Construct a 95% prediction interval for the amount of weight lost given $x = 55$.

Fat Consumption and Weight Lost								
Fat Consumed (in Grams), *x*	40	20	25	50	65	75	78	80
Weight Loss (in Pounds), *y*	23	30	38	15	10	8	7	5

6. The following table contains data from a sample of eight people regarding a person's physical fitness (measured on a scale of 1 to 10, with 10 being perfectly fit) and the number of days it took for the person to recover from gall bladder surgery.

Construct a 95% prediction interval for the length of the recovery time given $x = 5$.

Physical Fitness Levels and Recovery Times								
Physical Fitness, *x*	8	7	6	5	4	3	2	1
Recovery Time (in Days), *y*	5	5	6	7	7	9	10	12

7. The following table contains data from a sample of eight people regarding the number of bowls of chicken soup consumed while having cold symptoms and the number of days the symptoms persisted.

Construct a 95% prediction interval for the duration of the cold given $x = 3$.

Bowls of Soup and Duration of Cold Symptoms								
Bowls of Soup, *x*	5	5	4	4	3	2	1	0
Duration of Cold Symptoms, *y*	3	4	3	5	4	5	6	7

8. The following table contains data from a sample of five married couples regarding their weights (in pounds).

Construct a 95% prediction interval for the weight of the wife given $x = 225$.

Weights of Husbands and Wives					
Weight of Husband (in Pounds), *x*	220	230	267	200	195
Weight of Wife (in Pounds), *y*	154	167	186	132	125

Constructing Confidence Intervals for the *y*-intercept and the Slope of the Regression Line Using Microsoft Excel

Directions: Complete the following objectives for each data set.

a. Calculate the sum of squared errors, SSE.

b. Calculate the standard error of estimate, S_e.

c. Construct a 95% prediction interval for the given value of the explanatory variable.

d. Construct a 95% confidence interval for the *y*-intercept of the regression line.

e. Construct a 95% confidence interval for the slope of the regression line.

Note: Microsoft Excel can be used to calculate the answers for parts a., b., d., and e. simultaneously.

9. The following table contains data from a sample of ten students regarding the numbers of parking tickets they received during one semester and their monthly incomes (including allowances from parents as well as paychecks from employment as income).

Construct a 95% prediction interval for the monthly income given $x = 6$.

Parking Tickets and Monthly Income										
Number of Tickets, *x*	10	8	3	2	0	5	4	2	1	0
Monthly Income (in Dollars), *y*	4000	3800	1500	2000	870	2500	1800	1000	1200	1400

10. The following data were collected from a sample of fathers and sons. The heights are given in inches.

Construct a 95% prediction interval for the height of the son given $x = 70$.

Heights of Fathers and Sons (in Inches)									
Height of Father, *x*	72	70	73	68	74	73	69	70	75
Height of Son, *y*	73	69	73	70	73	70	68	71	76

11. The following data were collected from a sample of students on the numbers of times they were tardy for their world history class and their final exam grades in the course.

Construct a 95% prediction interval for the final grade given $x = 6$.

Numbers of Tardies and Final Exam Grades							
Number of Tardies, *x*	9	8	7	6	5	2	0
Final Exam Grade, *y*	57	64	69	70	72	84	91

12.4 Multiple Regression Equations

Thus far, we have only looked at linear regression models that have one explanatory variable. One example that we considered in detail was using a child's age to predict his or her reading level. However, as one might expect, there are other factors besides age that could influence a child's reading level. Two other possible factors that could affect the reading level of a child are the experience level of the child's current teacher and the education level of the child's parents. Could we create a model that uses all three factors (age, teacher's experience, and parents' education) to predict a child's reading level? The answer is yes! Actually, the techniques we have discussed so far can easily be expanded to include any number of explanatory variables.

Recall from the previous section that regression lines come in the following forms.

$$\text{Population parameters: } y = \beta_0 + \beta_1 x$$

$$\text{Sample statistics: } \hat{y} = b_0 + b_1 x$$

A linear regression model using two or more explanatory variables to predict a response variable is called a **multiple regression model**.

Definition

A **multiple regression model** is a linear regression model using two or more explanatory variables to predict a response variable, given by

$$\hat{y} = b_0 + b_1 x_1 + b_2 x_2 + \cdots + b_k x_k$$

where $x_1, x_2, \ldots, x_k$ are the explanatory variables in the model and

$b_1, b_2, \ldots, b_k$ are the coefficients of the explanatory variables.

The coefficients, $b_1, b_2, \ldots, b_k$, of the explanatory variables are the sample estimates of the corresponding population parameters, $\beta_1, \beta_2, \ldots, \beta_k$. As before, the y-intercept of the multiple regression equation is b_0, which is the sample estimate of the population parameter, β_0.

Constructing a multiple regression model from sample data, and analyzing the validity of the model, can be done with the help of statistical tools, such as those in Microsoft Excel. In the previous section, when we constructed the confidence intervals for the slope and y-intercept, we briefly mentioned that the ANOVA table produced by Microsoft Excel was more useful when we had multiple explanatory variables. This is the type of ANOVA table that we will use to construct and analyze multiple regression equations.

Let's consider the following data collected from ten children: the child's age, the teacher's experience, the parents' education, and the child's reading level. The child's age is measured to the nearest year. The teacher's experience is the number of years the child's current reading teacher has taught reading to children of this age. The parents' education is the average number of years of schooling obtained by parents with whom the child lives. Thus, 12 years indicates that the average education of the parents is a high school diploma. Lastly, the child's reading level is the grade level at which the child is reading. As before, a reading level of 4.3 indicates that the child is reading at a level equivalent to $\frac{3}{10}$ of the way through fourth grade. Table 12.4 displays the data collected for all ten children.

Table 12.4: Children's Reading Levels

Child's Age (in Years)	Teacher's Experience (in Years)	Parents' Education (in Years)	Child's Reading Level
6	5	13.1	1.3
7	10	14.5	2.2
8	7	16.1	3.7
9	3	12.5	4.1
10	2	11.8	4.9
11	4	11.1	5.2
12	1	10.2	6.0
13	3	12.6	7.1
14	6	12.1	8.5
15	5	14.3	9.7

Using these data, the ANOVA table shown in Figure 12.8 can be produced using the Regression tool in Microsoft Excel. The directions for creating this table will be presented in the next example. For now, we want to focus on interpreting the output.

SUMMARY OUTPUT

Regression Statistics	
Multiple R	0.996706497
R Square	0.993423841
Adjusted R Square	0.990135761
Standard Error	0.262983593
Observations	10

ANOVA

	df	*SS*	*MS*	*F*	*Significance F*
Regression	3	62.68603778	20.89534593	302.1288914	6.20571E-07
Residual	6	0.41496222	0.06916037		
Total	9	63.101			

	Coefficients	*Standard Error*	*t Stat*	*P-value*	*Lower 95%*	*Upper 95%*	*Lower 95.0%*	*Upper 95.0%*
Intercept	-6.983348818	0.932463961	-7.489135359	0.000292888	-9.265005934	-4.7016917	-9.265005934	-4.701691702
Age	0.89810517	0.031202521	28.78309636	1.1648E-07	0.821755351	0.974454989	0.821755351	0.974454989
Teacher's Experience	-0.033200086	0.050469878	-0.657819823	0.535061828	-0.156695429	0.090295257	-0.156695429	0.090295257
Parents' Education	0.231953619	0.075147789	3.086632629	0.021480082	0.048073602	0.415833635	0.048073602	0.415833635

Figure 12.8: Microsoft Excel Regression Summary Output

To begin, let's look at the regression statistics that are displayed above the ANOVA table, as shown in Figure 12.9.

	Regression Statistics	
1 →	Multiple R	0.996706497
2 →	R Square	0.993423841
3 →	Adjusted R Square	0.990135761
	Standard Error	0.262983593
	Observations	10

Figure 12.9: Regression Statistics

Notice that the first value given is Multiple R ①, which is the absolute value of the correlation coefficient of the multiple regression model. To denote the correlation coefficient, we use a capital R for a multiple regression model rather than the lowercase r that is used for a regression model with a single explanatory variable. So, in our example $|R| \approx 0.997$.

The next value is R Square ②. This is the **multiple coefficient of determination**, R^2. This value is analogous to the coefficient of determination for a linear regression with a single explanatory variable, r^2. Just as with r^2, the closer the value of R^2 is to 1, the better the multiple regression model fits the sample data. Similarly, the value of R^2 represents the proportion of variation in the response variable that is explained by the multiple regression model. For this example, $R^2 \approx 0.993$, which indicates that the model explains approximately 99.3% of the variation in the response variable.

Unfortunately, as more explanatory variables are added to the multiple regression model, the value of R^2 increases as a general rule. So should we continue to add explanatory variables until we obtain a very large value of R^2? Of course not! Not all of the possible explanatory variables affect the response variable equally, and not all of the possible explanatory variables are useful in predicting the value of the response variable. So the **adjusted coefficient of determination** is the value of the multiple coefficient of determination adjusted for the number of explanatory variables in the model and the sample size. This value appears as Adjusted R Square 3. For this example, adjusted $R^2 \approx 0.990$. Though this value is less than $R^2 \approx 0.993$, it still indicates that the multiple regression model fits the data well.

The next block of information is the ANOVA table, as shown in Figure 12.10.

ANOVA					
	df	*SS*	*MS*	*F*	*Significance F* (4)
Regression	3	62.68603778	20.89534593	302.1288914	6.20571E-07
Residual	6	0.41496222	0.06916037		
Total	9	63.101			

Figure 12.10: ANOVA Table

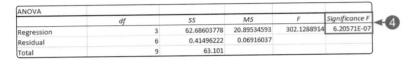

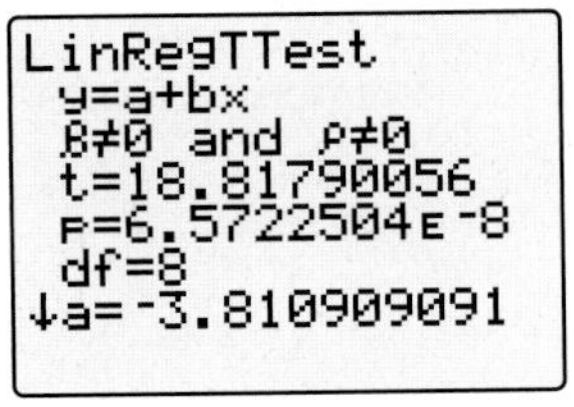

This ANOVA table further analyzes how well the regression model fits the sample data. Recall that for a single regression model, we ran a t-test to determine if the linear relationship between the two variables was statistically significant. We were testing the claim that the population's correlation coefficient did not equal 0, that is, $\rho \neq 0$. Equivalently, we could have used the same t-test to test the claim that the coefficient of the explanatory variable was not equal to 0, that is, $\beta_1 \neq 0$. You may have even noticed that the output for the t-test on the TI-83/84 Plus calculator listed *both* of these alternative hypotheses. When using the ANOVA table to analyze the statistical significance of the linear relationship between the variables in a multiple regression model, we test the claim that at least one of the explanatory variables' coefficients is not equal to 0. Thus, the null and alternative hypotheses for this ANOVA test are as follows.

Procedure

Null and Alternative Hypotheses for an ANOVA Test

The null and alternative hypotheses for an ANOVA test to analyze the statistical significance of the linear relationship between the variables in a multiple regression model are as follows.

$$H_0: \beta_1 = \beta_2 = \cdots = \beta_k = 0$$

H_a: At least one coefficient does not equal 0.

$\beta_1, \beta_2, \ldots, \beta_k$ are the coefficients of the explanatory variables, and

k is the number of explanatory variables in the model.

If all of the coefficients of the explanatory variables equal 0, then the sample data imply that there is not a significant linear relationship between the variables, and therefore, the regression model should not be used for predictions. However, if the null hypothesis is rejected, then we can conclude that there is enough evidence to support the claim that a significant linear relationship exists between at least one of the explanatory variables and the response variable. Therefore, the multiple regression equation can be used to predict values of the response variable. Note, however, that the hypothesis test does not identify which explanatory variable has a linear relationship with the response variable.

To test the null hypothesis, consider the p-value given under *Significance F* ❹. For our example, the p-value is so small that it is given in scientific notation. In standard notation, p-value ≈ 0.0000006, which is extremely small. Thus, we reject the null hypothesis and conclude that there is sufficient evidence at the 0.05 level of significance to support the claim that this multiple regression model fits the data well and can be used for predictions.

> **Memory Booster**
>
> **Conclusions Using p-Values**
>
> - If p-value $\le \alpha$, then reject the null hypothesis.
> - If p-value $> \alpha$, then fail to reject the null hypothesis.

The last block of information, shown in Figure 12.11, gives the coefficients of the multiple regression equation and the 95% confidence intervals for the y-intercept and the coefficients of the explanatory variables.

	Coefficients	*Standard Error*	*t Stat*	*P-value*	*Lower 95%*	*Upper 95%*	*Lower 95.0%*	*Upper 95.0%*
Intercept	-6.983348818	0.932463961	-7.489135359	0.000292888	-9.265005934	-4.7016917	-9.265005934	-4.701691702
Child's Age	0.89810517	0.031202521	28.78309636	1.1648E-07	0.821755351	0.974454989	[illegible]1755351	0.974454989
Teacher's Experience	-0.033200086	0.050469878	-0.657819823	0.535[illegible]	-0.156695429	0.090295257	-0.156695429	0.090295257
Parents' Education	0.231953619	0.075147789	3.086632629	0.021480082	0.048073602	0.415833635	[illegible]073602	0.415833635

Figure 12.11: Multiple Regression Model

The first row gives the predicted value ❺ and confidence interval ❻ for the y-intercept, β_0. The second row gives the predicted value ❼ and confidence interval ❽ for the coefficient of the first explanatory variable, child's age, β_1. The third row gives the predicted value ❾ and confidence interval ❿ for the coefficient of the second explanatory variable, teacher's experience, β_2. The fourth row gives the predicted value ⓫ and confidence interval ⓬ for the coefficient of the third explanatory variable, parents' education, β_3.

Substituting the estimated coefficients into the multiple regression model, we have the following equation for predicting a child's reading level based on the child's age, x_1, the teacher's experience, x_2, and the parents' education, x_3.

$$\hat{y} = -6.983 + 0.898x_1 - 0.033x_2 + 0.232x_3$$

Let's use this regression model to predict the reading level of a child who is 10 years old with a teacher who has 8 years of experience and parents with an average of 17.2 years of education. Thus, $x_1 = 10$, $x_2 = 8$, and $x_3 = 17.2$. Substituting these values into the multiple regression equation yields the following.

$$\hat{y} = -6.983 + 0.898x_1 - 0.033x_2 + 0.232x_3$$
$$\hat{y} = -6.983 + 0.898(10) - 0.033(8) + 0.232(17.2)$$
$$\hat{y} \approx 5.723$$

Based on this regression equation, we would predict that a child with these characteristics would be reading at an upper fifth-grade level.

Now let's go back and consider the individual explanatory variables more closely. Look at the p-values (13) for the explanatory variables shown in Figure 12.12.

	Coefficients	Standard Error	t Stat	P-value	Lower 95%	Upper 95%	Lower 95.0%	Upper 95.0%
Intercept	-6.983348818	0.932463961	-7.489135359	0.000292888	-9.265005934	-4.7016917	-9.265005934	-4.701691702
Child's Age	0.89810517	0.031202521	28.78309636	1.1648E-07	0.821755351	0.974454989	0.821755351	0.974454989
Teacher's Experience	-0.033200086	0.050469878	-0.657819823	0.535061828	-0.156695429	0.090295257	-0.156695429	0.090295257
Parents' Education	0.231953619	0.075147789	3.086632629	0.021480082	0.048073602	0.415833635	0.048073602	0.415833635

13

Figure 12.12: p-Values for Individual Coefficients

These p-values test the null hypothesis that the coefficient of a particular explanatory variable equals 0. Mathematically, the null and alternative hypotheses for the first explanatory variable would be as follows.

$$H_0: \beta_1 = 0$$
$$H_a: \beta_1 \neq 0$$

The null and alternative hypotheses for the other explanatory variables are similar. A small p-value (such as one less than 0.05) indicates that there is sufficient evidence to support the claim that the coefficient is not 0, and therefore, the linear relationship between this particular variable and the response variable is statistically significant. If the p-value is not small (for instance, those greater than 0.05) then this particular variable may not be useful in predicting the value of the response variable.

	Coefficients	Standard Error	t Stat	P-value	Lower 95%	Upper 95%	Lower 95.0%	Upper 95.0%
Intercept	-6.983348818	0.932463961	-7.489135359	0.000292888	-9.265005934	-4.7016917	-9.265005934	-4.701691702
Child's Age	0.89810517	0.031202521	28.78309636	1.1648E-07	0.821755351	0.974454989	0.821755351	0.974454989
Teacher's Experience	-0.033200086	0.050469878	-0.657819823	0.535061828	-0.156695429	0.090295257	-0.156695429	0.090295257
Parents' Education	0.231953619	0.075147789	3.086632629	0.021480082	0.048073602	0.415833635	0.048073602	0.415833635

Figure 12.13: Interpreting p-Values for Coefficients of Explanatory Variables

In our example of predicting a child's reading level, notice that the p-values for the explanatory variables of the child's age and parents' education are approximately 1.16E-07 (0.000000116) and 0.0215, respectively. This indicates that these explanatory variables are very likely to be effective in predicting the response variable. However, the p-value for teacher's experience is approximately 0.5351, which is much greater than 0.05. Thus, including teacher's experience in the multiple regression model may not be useful. We could recalculate the multiple regression model without this variable. You should be aware, however, that the p-values calculated in the ANOVA table measure the influence of each explanatory variable with all other variables taken into account. For example, it could be the choice of the other variables that were included, namely child's age and parents' education, that caused the p-value for the teacher's experience to be so high.

Example 12.20

Constructing and Analyzing a Multiple Regression Model

Construct and analyze a multiple regression equation for predicting a child's reading level based on the following sample data, which omits the variable of teacher's experience from the multiple regression model that we have been discussing. Use this new model to predict the reading level for a child who is 10 years old and has parents with an average of 17.2 years of education. Which of the two multiple regression models is better at predicting a child's reading level?

	A	B	C
1	Child's Age	Parents' Education	Reading Level
2	6	13.1	1.3
3	7	14.5	2.2
4	8	16.1	3.7
5	9	12.5	4.1
6	10	11.8	4.9
7	11	11.1	5.2
8	12	10.2	6
9	13	12.6	7.1
10	14	12.1	8.5
11	15	14.3	9.7

Solution

We will use Microsoft Excel to construct and analyze a multiple regression model for these data. Begin by entering the data as they appear in columns A, B and C. Next, under the **Data** tab, choose **Data Analysis**. Select **Regression** from the options listed. Enter the necessary information into the Regression menu as shown in the following screenshot. Then click **OK**.

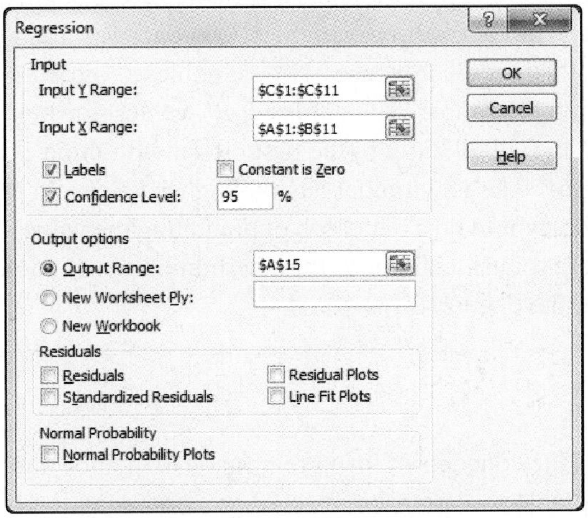

The output is shown in the following figure.

SUMMARY OUTPUT								
Regression Statistics								
Multiple R	0.996468545							
R Square	0.992949561							
Adjusted R Square	0.990935149							
Standard Error	0.252102523							
Observations	10							
ANOVA								
	df	*SS*	*MS*	*F*	*Significance F*			
Regression	2	62.65611023	31.32805511	492.9229633	2.94277E-08			
Residual	7	0.444889774	0.063555682					
Total	9	63.101						
	Coefficients	*Standard Error*	*t Stat*	*P-value*	*Lower 95%*	*Upper 95%*	*Lower 95.0%*	*Upper 95.0%*
Intercept	-6.728743363	0.813242326	-8.27397093	7.34694E-05	-8.651755889	-4.805730837	-8.651755889	-4.805730837
Child's Age	0.901985643	0.029372093	30.70893285	1.00214E-08	0.832531679	0.971439607	0.832531679	0.971439607
Parents' Education	0.197029939	0.050984389	3.864515026	0.006175322	0.076471015	0.317588862	0.076471015	0.317588862

To begin, notice that the p-value for the regression equation is $p\text{-value} \approx 2.94277\text{E-}08$, or approximately 0.00000003, which is extremely small. Thus, this multiple regression equation fits the sample data extremely well. This can also be seen in the adjusted value of $R^2 \approx 0.991$, which is close to 1. Notice that the p-values for the coefficients of the individual explanatory variables have changed. These p-values are both small, which indicates that both of these variables are useful in predicting the value of the response variable.

Using the coefficients listed in the table, the multiple regression equation for predicting a child's reading level based on the child's age, x_1, and the parents' education, x_2, is as follows.

$$\hat{y} = -6.729 + 0.902x_1 + 0.197x_2$$

We can now use this new equation to predict the reading level of a 10-year-old child with parents who have an average of 17.2 years of education. Note that $x_1 = 10$ and $x_2 = 17.2$. Substituting these values into the regression equation yields the following.

$$\hat{y} = -6.729 + 0.902x_1 + 0.197x_2$$
$$\hat{y} = -6.729 + 0.902(10) + 0.197(17.2)$$
$$\hat{y} \approx 5.679$$

Thus, we would again predict that this child would be reading at a fifth-grade reading level, but not quite as far along as the first prediction of 5.723.

This regression equation, and its predicted value of the response variable, is very similar to the one we calculated with three explanatory variables. So which regression equation does a better job of predicting the value of the response variable? Let's begin by looking at the overall picture. The first model includes three variables, but one was found to be not statistically significant. The second model only uses two variables, both of which were significant. Another consideration is to compare the adjusted R^2-values for both models. For this new model, the adjusted $R^2 \approx 0.990935$. For the first model with three explanatory variables, the adjusted $R^2 \approx 0.990136$. The new model has the higher value. In either case, it appears as though the second model would do a better job of predicting the value of the response variable. There are certainly more robust techniques for determining which model is better, but we will leave those for a higher-level course.

In this section we extended the concepts of linear regression to include multiple explanatory variables. We revisited the ANOVA table to determine if the regression equation was statistically significant. The ANOVA table also provided information on how closely each explanatory variable was related to the response variable. Lastly, we discussed possible ways to change the regression equation to eliminate unnecessary variables, and how to compare the new regression equation to the old equation to determine which equation is the best model for the given data. Most importantly, however, we should remember that a multiple regression model does not imply causation any more than a regression model with one explanatory variable does. Just as we know that an increase in a child's age does not *cause* his or her reading level to increase, we should not presume that adding more variables to a regression equation accounts for all the possible influences on the response variable, or that the variables included are even *causing* the changes in the response variable at all.

12.4 Section Exercises

Interpreting ANOVA Tables

Directions: The following ANOVA tables were generated using Microsoft Excel. Based on the ANOVA table given, is there sufficient evidence at the 0.05 level of significance to conclude that the linear relationship between the explanatory variables and the response variable is statistically significant?

1.	*df*	*SS*	*MS*	*F*	*Significance F*
Regression	2	2677.934566	1338.967283	167.6684672	5.21957E-10
Residual	13	103.8154339	7.985802609		
Total	15	2781.75			

2.	*df*	*SS*	*MS*	*F*	*Significance F*
Regression	2	4.272742323	2.136371162	8.88737913	0.004285466
Residual	12	2.88459101	0.240382584		
Total	14	7.157333333			

3.	*df*	*SS*	*MS*	*F*	*Significance F*
Regression	3	1691672.173	563890.7244	11.50954064	0.080992754
Residual	2	97986.66027	48993.33013		
Total	5	1789658.833			

4.	*df*	*SS*	*MS*	*F*	*Significance F*
Regression	2	8.553919903	4.276959952	3.555389644	0.061293877
Residual	12	14.43541343	1.202951119		
Total	14	22.98933333			

Directions: For each of the following computer outputs, write the multiple regression equation. Define each of the variables used in the multiple regression equation. Could one of the explanatory variables be eliminated from the model? If so, which one?

5. The following table was generated from a sample of 16 statistics students regarding each student's number of absences in the class, average number of hours spent studying for the class per week, and final grade in the class.

	Coefficients	*Standard Error*	*t Stat*	*P-value*
Intercept	63.88047809	4.914092338	12.99944602	7.95988E-09
Absences	−1.416204746	0.422223932	−3.354155552	0.005179159
Hours Studied per Week	4.168586523	0.73626183	5.661826206	7.77404E-05

6. The following table was generated from a sample of 16 babies regarding the weight of the mother at birth, the weight of the father at birth, and the weight of the baby at birth. All weights are measured in pounds.

	Coefficients	*Standard Error*	*t Stat*	*P-value*
Intercept	−2.882473037	0.9737576	−2.960154597	0.01105403
Mother's Weight	0.356055126	0.090343376	3.941131517	0.001689591
Father's Weight	1.02230869	0.099308161	10.29430688	1.28376E-07

7. The following table was generated from a sample of 15 babies regarding the number of weeks of gestation, the number of prenatal doctor visits the mother had, and the birth weight of the baby in pounds.

	Coefficients	*Standard Error*	*t Stat*	*P-value*
Intercept	−19.56998222	5.716106062	−3.423656245	0.005044399
Weeks of Gestation	0.684805526	0.12588231	5.44004576	0.000150053
Number of Prenatal Visits	0.063077042	0.126763644	0.497595684	0.627761713

8. The following table was generated from a sample of 20 college freshmen regarding their high school GPAs, their ACT scores, and their GPAs after their first year at college.

	Coefficients	*Standard Error*	*t Stat*	*P-value*
Intercept	−0.616112354	0.345507101	−1.783211841	0.092415337
High School GPA	0.975068267	0.186279465	5.234437779	6.7297E-05
ACT Score	0.015000337	0.019048101	0.787497806	0.441832554

Constructing and Analyzing a Multiple Regression Model

Directions: For each data set, use available technology, such as Microsoft Excel, to determine if a statistically significant linear relationship exists between the explanatory variables and the response variable at the 0.05 level of significance. If the linear relationship is significant, identify the multiple regression equation that best fits the data.

9. The following data were collected to explore how tuition costs, student-to-faculty ratios, and female-to-male ratios affect freshman enrollment. The student-to-faculty ratio is expressed as the number of students per faculty member. The female-to-male ratio is expressed as the number of female students per male student.

Effects on Freshman Enrollment

Tuition Cost (in Dollars)	Student-to-Faculty Ratio	Female-to-Male Ratio	Freshman Enrollment
5495	18	1.5	3219
7912	18	2	2781
5812	6	1.7	3556
7180	4	2.1	2987
8883	15	3	1794
6495	19	1.8	3014

10. The following data were collected to explore how the average number of hours a junior high student is unsupervised per night and the average number of hours per night the student watches television affect his class average.

Effects on Junior High Students' Class Averages		
Hours Unsupervised	Hours Watching TV	Class Average
0	2	96
0	3	91
0.5	3	88
1	2	92
1	1	94
1.5	2	91
2	4	87
3	3	85
3	4	81
4	5	80
4.5	4	77
5.5	5	72

11. The following data were collected to explore how women's ages and their average daily calcium intakes (measured in mg per day) affect their bone mineral density.

Effects on Bone Density		
Age (in Years)	Daily Calcium Intake (in mg per Day)	Bone Density (in mg/cm^2)
34	791	950
43	832	875
47	865	804
52	901	723
58	923	691

12. Common predictors for freshman grade point average are ACT scores and high school GPA. Let's explore how well two unusual factors, a student's age when entering college and the number of parking tickets he or she receives during his or her first semester at school, predict a student's GPA after one semester in college.

Effects on GPA

Student's Age	Number of Parking Tickets	GPA
20	0	3.6
21	0	3.9
17	0	2.4
18	0	3.1
19	1	3.5
35	1	4.0
23	1	3.6
19	2	2.8
18	2	3.0
18	2	2.2
24	3	3.9
19	3	3.1
18	5	2.1
17	7	2.8
18	8	1.7

Challenge Questions

Directions: Answer each of the following questions.

13. Which value in the ANOVA table tells you the number of explanatory variables in the multiple regression model?

14. Should you always use as many explanatory variables as possible?

15. What is the disadvantage of the multiple coefficient of determination?

R Chapter 12 Review

Section 12.1: Scatter Plots and Correlation

Definitions

Scatter plot
A graph on the coordinate plane that contains one point for each pair of data

Explanatory variable
The variable for which a change in its value influences a change in the value of the other variable, also called the predictor variable or the independent variable

Response variable
The variable that has its value change as a result of a change in the value of the explanatory variable, also called the dependent variable

Linear relationship
Describes the association between two variables when the points in a scatter plot for the two variables roughly follow a straight-line pattern

Positive slope
Describes a linear pattern of points in a scatter plot where as the values of one variable increase, so do the values of the other variable

Negative slope
Describes a linear pattern of points in a scatter plot where as the values of one variable increase, the values of the other variable decrease

Pearson correlation coefficient, *r*
Measures the strength of a linear relationship between two quantitative variables

Coefficient of determination, r^2
Measures the proportion of the variation in the response variable (y) that can be associated with the variation in the explanatory variable (x), the square of the correlation coefficient

Pearson Correlation Coefficient

Used for paired data from a sample, given by

$$r = \frac{n\sum x_i y_i - \left(\sum x_i\right)\left(\sum y_i\right)}{\sqrt{n\sum x_i^2 - \left(\sum x_i\right)^2}\sqrt{n\sum y_i^2 - \left(\sum y_i\right)^2}}$$

such that $-1 \le r \le 1$

Using Critical Values of the Pearson Correlation Coefficient to Determine the Significance of a Linear Relationship

A sample correlation coefficient, r, is statistically significant if $|r| \ge r_\alpha$.

Section 12.1: Scatter Plots and Correlation (cont.)

Testing Linear Relationships for Significance

Significant Linear Relationship (Two-Tailed Test)

H_0: $\rho = 0$ (Implies that there is no significant linear relationship)

H_a: $\rho \neq 0$ (Implies that there is a significant linear relationship)

Significant Negative Linear Relationship (Left-Tailed Test)

H_0: $\rho \geq 0$ (Implies that there is no significant negative linear relationship)

H_a: $\rho < 0$ (Implies that there is a significant negative linear relationship)

Significant Positive Linear Relationship (Right-Tailed Test)

H_0: $\rho \leq 0$ (Implies that there is no significant positive linear relationship)

H_a: $\rho > 0$ (Implies that there is a significant positive linear relationship)

Test Statistic for a Hypothesis Test for a Correlation Coefficient

Used for testing the significance of the correlation coefficient, given by

$$t = \frac{r}{\sqrt{\frac{1-r^2}{n-2}}} \text{ with } df = n-2$$

Rejection Regions for Testing Linear Relationships

Significant Linear Relationship (Two-Tailed Test)

Reject the null hypothesis, H_0, if $|t| \geq t_{\alpha/2}$.

Significant Negative Linear Relationship (Left-Tailed Test)

Reject the null hypothesis, H_0, if $t \leq -t_{\alpha}$.

Significant Positive Linear Relationship (Right-Tailed Test)

Reject the null hypothesis, H_0, if $t \geq t_{\alpha}$.

Section 12.2: Linear Regression

Definition

Least-squares regression line

The line for which the average variation from the data is the smallest, also called the line of best fit, given by $\hat{y} = b_0 + b_1 x$

Slope of the Least-Squares Regression Line

Used for paired data from a sample, given by

$$b_1 = \frac{n\sum x_i y_i - \left(\sum x_i\right)\left(\sum y_i\right)}{n\sum x_i^2 - \left(\sum x_i\right)^2}$$

Section 12.2: Linear Regression (cont.)

y-Intercept of the Least-Squares Regression Line

Used for paired data from a sample, given by

$$b_0 = \frac{\sum y_i}{n} - b_1 \frac{\sum x_i}{n}$$

A prediction should *not* be made with a regression model if...

1. The data do not fall in a linear pattern when graphed on a scatter plot.
2. The correlation coefficient is not statistically significant.
3. You wish to make a prediction about a value outside the range of the sample data.
4. The population is different than that from which the sample data were drawn.

Section 12.3: Regression Analysis

Definition

Prediction interval
A confidence interval for an individual value of the response variable, y, at a given fixed value of the explanatory variable, x

Regression Line (Line of Best Fit)

$$y = \beta_0 + \beta_1 x \text{ (Population parameters)}$$

$$\hat{y} = b_0 + b_1 x \text{ (Sample statistics)}$$

Residual

The difference between the actual value of y from the original data and the predicted value of $\hat{y}$ found using the regression line, also called an error, given by

$$\text{Residual} = y - \hat{y}$$

Sum of Squared Errors (SSE)

The sum of the squares of the residuals for a regression line, given by

$$\text{SSE} = \sum \left(y_i - \hat{y}_i\right)^2$$

Standard Error of Estimate

A measure of how much the sample data points deviate from the regression line, given by

$$S_e = \sqrt{\frac{\sum \left(y_i - \hat{y}_i\right)^2}{n-2}}$$

$$= \sqrt{\frac{\text{SSE}}{n-2}}$$

Section 12.3: Regression Analysis (cont.)

Margin of Error of a Prediction Interval for an Individual y-Value

$$E = t_{\alpha/2} S_e \sqrt{1 + \frac{1}{n} + \frac{n(x_0 - \bar{x})^2}{n\left(\sum x_i^2\right) - \left(\sum x_i\right)^2}} \text{ with } df = n - 2$$

Prediction Interval for an Individual y-Value

Can be written mathematically using either inequality symbols or interval notation, given by

$$\hat{y} - E < y < \hat{y} + E$$

or

$$\left(\hat{y} - E,\ \hat{y} + E\right)$$

Section 12.4: Multiple Regression Equations

Definitions

Multiple regression model

A linear regression model using two or more explanatory variables to predict a response variable, given by $\hat{y} = b_0 + b_1x_1 + b_2x_2 + \cdots + b_kx_k$

Multiple coefficient of determination, R^2

Analogous to the coefficient of determination for a linear regression with a single explanatory variable (r^2), measures the proportion of the variation in the response variable that is explained by the multiple regression model

Adjusted coefficient of determination

The value of the multiple coefficient of determination adjusted for the number of explanatory variables in the model and the sample size

Null and Alternative Hypotheses for an ANOVA Test

When using the ANOVA table to analyze the statistical significance of the linear relationship between the variables in a multiple regression model, the null and alternative hypotheses are as follows.

H_0: $\beta_1 = \beta_2 = \cdots = \beta_k = 0$

H_a: At least one coefficient does not equal 0.

E Chapter 12 Exercises

Directions: Answer each question.

1. The following table gives the heights and corresponding batting averages for a sample of 12 professional baseball players.

Heights and Batting Averages												
Height (in Inches)	70	73	73	71	68	76	71	75	73	73	75	73
Batting Average	.248	.273	.245	.202	.304	.349	.226	.250	.233	.273	.237	.229

a. Calculate the correlation coefficient.

b. Is the correlation coefficient statistically significant at the 0.05 level? At the 0.01 level?

c. Should a regression line be used to model the data and make predictions? Why or why not?

2. The following table gives the college GPAs and corresponding annual incomes after graduation for a sample of nine young adults.

College GPAs and Annual Incomes after Graduation									
GPA	3.1	3.8	2.9	1.8	2.7	3.5	3.2	2.2	4.0
Annual Income	\$38,000	\$36,500	\$39,500	\$29,000	\$33,000	\$40,500	\$26,000	\$30,000	\$61,000

a. Before calculating the correlation coefficient, what kind of relationship would you expect to see between the two variables?

b. Calculate the correlation coefficient and coefficient of determination.

c. Is r statistically significant at the 0.05 level? At the 0.01 level?

d. Suppose that your college GPA is 3.3. Using the regression model for these data, make a prediction for your future annual income.

3. The average monthly interest rate for a 30-year mortgage and corresponding number of new loans taken out at one bank are given for 12 consecutive months.

Interest Rates and Numbers of New Loans Taken Out												
Interest Rate (%)	7.12	7.23	7.25	7.04	6.99	6.87	6.68	6.91	7.03	7.09	7.15	7.11
New Loans	5	6	4	5	7	9	9	10	8	7	4	4

a. Before calculating the correlation coefficient, what kind of relationship would you expect to see between the two variables?

b. Calculate the correlation coefficient and coefficient of determination.

c. Is r statistically significant at the 0.05 level? At the 0.01 level?

d. Suppose that the average interest rate for a given month is 6.75%. Using the regression model for these data, make a prediction for the number of new loans taken out.

e. Suppose that the first column of data in the table represents the month of January, the second February, and so on. What factor, other than the change in interest rates, may have influenced the rise in the number of home purchases?

4. The following table gives the average price of gas and the corresponding SUV sales at a local car dealership for ten randomly selected months in 2003.

Gas Prices and SUV Sales										
Price of Gas (per Gallon)	\$1.26	\$1.24	\$1.51	\$1.37	\$1.53	\$1.43	\$1.35	\$1.49	\$1.42	\$1.58
SUVs Sold	45	44	36	45	40	43	47	39	47	37

a. Before calculating the correlation coefficient, what kind of relationship would you expect to see between the two variables?

b. Calculate the correlation coefficient and coefficient of determination.

c. Is r statistically significant at the 0.05 level? At the 0.01 level?

d. Suppose that the average gas price for one month is \$1.40. How many SUVs should the car dealership expect to sell?

e. Obviously, these data from 2003 should not be used for predictions this year. However, can you think of ways that the data might be useful for today's car dealerships?

5. A lifeguard collected the following data while she was observing the beach on the first day of each month, over the course of a year.

Numbers of Swimsuits and Seagulls												
Swimsuits	0	8	11	28	77	120	150	250	100	44	12	5
Seagulls	2	78	92	106	102	200	300	376	298	55	15	3

a. From analyzing the data above, can you conclude that there is a significant linear relationship between the two variables at the 0.05 level? At the 0.01 level?

b. Could you predict how many seagulls would be visible if you can see 200 swimsuits? What about 500 swimsuits?

c. What reasonable conclusions can you make regarding your findings?

6. Consider the following data from a recent study that included 14 participants who have used a cell phone consistently over the course of several years. For each participant, the number of years they had been using a cell phone was recorded, along with an estimate of the probability that they would get cancer in their lifetime, based on a complex system of multiple factors.

Cancer Risk and Number of Years Using a Cell Phone														
Cancer Risk	0.35	0.11	0.45	0.28	0.55	0.78	0.55	0.38	0.74	0.81	0.15	0.09	0.66	0.2
Years Using a Cell Phone	2	3.1	1.5	0.9	4	1.5	3.9	2	5	4.4	2.1	4.9	8	6.2

a. From analyzing the data, can you conclude that there is a significant linear relationship between the two variables at the 0.05 level? At the 0.01 level?

b. Is it reasonable to make predictions of cancer risk based on cell phone usage?

c. If you were a cell phone company, what would you do with the results of this study?

7. The following data were collected from a used car dealership regarding the selling prices of used cars and the numbers of miles on the cars when they were sold.

Mileages and Selling Prices of Used Cars	
Mileage	**Selling Price**
23,547	$21,803
48,710	$17,995
33,952	$18,995
14,876	$20,863
133,691	$4970
30,417	$28,500
29,104	$22,597
54,000	$17,600
63,079	$14,999
133,172	$10,995
110,000	$8000
63,263	$15,995

a. Can you conclude that there is a significant linear relationship between the two variables the 0.05 level? At the 0.01 level?

b. Predict the selling price of a vehicle that has 42,000 miles on it.

c. How much of the variation in the selling prices is associated with the variation in the mileages on the cars when they were sold?

8. Consider the following data collected on airline tickets. The following table lists the distance from Dallas, Texas to the destination and the corresponding price of the airline ticket.

Distances and Ticket Prices to Destinations from Dallas, TX		
Distance (in Miles)	**Destination**	**Ticket Price**
1573.61	New York	$212
1329.96	Washington, DC	$234
1316.38	Miami	$275
782.50	Atlanta	$231
2203.26	Seattle	$312
967.73	Chicago	$185
1368.18	Baltimore	$267
453.18	Memphis	$229
521.11	New Orleans	$345

a. Is there a significant linear relationship at the 0.05 level between distance traveled and price of a ticket from Dallas, Texas?

b. Predict the price of a ticket to San Francisco from Dallas assuming that there is a distance of 1465 miles between the two cities.

9. The following data represent the thread counts of various bed sheets and their corresponding prices.

Thread Counts and Prices of Bed Sheets								
Thread Count	150	200	225	250	275	300	350	400
Price (in Dollars)	18	21	25	28	30	31	35	45

a. Calculate the sum of squared errors, SSE.

b. Calculate the standard error of estimate, S_e.

c. Construct a 95% prediction interval for the price of 350-thread-count sheets.

d. Construct a 95% confidence interval for the *y*-intercept of the regression line.

e. Construct a 95% confidence interval for the slope of the regression line.

10. The following data were collected on the numbers of years of post-high-school education and the annual incomes of eight people ten years after graduation from high school.

Post-High-School Education and Annual Income								
Education after High School (in Years)	2	3	3	4	5	6	6	8
Annual Income (in Thousands of Dollars)	35	41	38	45	65	89	95	115

a. Calculate the sum of squared errors, SSE.

b. Calculate the standard error of estimate, S_e.

c. Construct a 95% prediction interval for the annual income of someone ten years after graduation from high school who had seven years of post-high-school education.

d. Construct a 95% confidence interval for the *y*-intercept of the regression line.

e. Construct a 95% confidence interval for the slope of the regression line.

11. In addition to the thread count of bed sheets, let's explore if the number of colors used is linearly related to the price of the sheets. Determine if a statistically significant linear relationship exists between these explanatory variables and the response variable at the 0.05 level. If the linear relationship is significant, write the multiple regression equation that best fits the data.

Thread Counts, Colors, and Prices of Bed Sheets

Thread Count	Number of Colors	Price (in Dollars)
150	2	18
200	1	21
225	2	25
250	4	28
275	3	30
300	2	31
350	1	35
400	4	45

12. In addition to the number of years of post-high-school education, the number of years of work experience also influences a person's salary. Determine if a statistically significant linear relationship exists between these explanatory variables and the response variable at the 0.05 level. If the linear relationship is significant, write the multiple regression equation that best fits the data.

Education, Work Experience, and Annual Income

Education after High School (in Years)	Work Experience (in Years)	Annual Income (in Thousands of Dollars)
2	8	35
3	7	41
3	5	38
4	6	45
5	6	65
6	5	89
6	4	95
8	4	115

P Chapter 12 Project

Project A: Arm Span vs. Height

Is it true that a person's arm span is equal to his or her height? If so, then there should be a perfect linear relationship between arm span and height. Let's find out if this is true by collecting data and creating a linear regression model.

Step 1: Collect data from 10 people. Measure each person's height (in inches) and then his or her arm span (in inches), which is the distance from fingertip to fingertip as he or she holds his or her arms outstretched. Record your data in a table similar to the one below. Your results will be more generalizable if you collect data from people of many different heights, ranging from small children to tall men and women.

Height										
Arm Span										

Step 2: Create a scatter plot of the data. Use height as the x-variable and arm span as the y-variable.

a. Does there appear to be a linear relationship between x and y?

b. Is the relationship positive or negative?

c. Is the relationship strong or weak?

d. Does the graph seem to support the claim that a person's height is equal to his or her arm span?

Step 3: Calculate the correlation coefficient, r.

a. Is the correlation coefficient statistically significant at the 0.05 level of significance?

b. What about the 0.01 level of significance?

c. Interpret what it means for the correlation coefficient to be statistically significant for this scenario.

Step 4: Determine the equation of the regression line, $\hat{y} = b_0 + b_1x$.

a. Draw the regression line on your scatter plot.

b. If it is true that a person's arm span is the same as his or her height, what would you expect the slope of the regression line to be?

c. How close is the actual slope of the regression line to the expected value?

d. Calculate a 95% confidence interval for the slope of the regression line.

e. Based on your confidence interval, could you conclude that it is likely that a person's arm span is equal to his or her height?

Step 5: Calculate the coefficient of determination, r^2.

a. Interpret what r^2 means for this scenario.

b. Does the value of r^2 support the claim that arm span is equal to height?

Step 6: Based on your answers for Steps 1 through 5, formulate a conclusion as to whether a person's arm span is equal to his or her height.

Project B: Attractiveness vs. Parental Attentiveness

Have you ever commented that someone is a "perfect 10," meaning that he or she is extremely attractive? A Canadian researcher believes that a person's attractiveness can predict how people respond to him or her. Specifically, he hypothesizes that a child's attractiveness is a predictor of parental attentiveness. He thinks that the cuter the child, the more attention the parents pay to the child, and the child thus receives better care. In a study performed in Canada, researchers rated a child's attractiveness on a scale of 1 to 10 and then counted the number of times the child was allowed to wander away from the parent in a ten-minute interval. In order to test the researcher's hypothesis, we will perform a modified version of this research study.

Source: Parker, Randall. "More Attractive Children Protected Better by Parents." FuturePundit. 12 Apr. 2005. http://www.futurepundit.com/archives/002711.html (12 Mar. 2012).

Step 1: Collect data on six children in a busy public setting. Begin by rating the child's overall attractiveness (how "cute" they are) on a scale of 1 to 10, where 1 is not attractive and 10 is extremely cute. Second, count the number of times in a ten-minute time period that the parent allows the child to wander more than 10 feet away. Record your results in a table similar to the one below.

Attractiveness						
Attention						

Step 2: Create a scatter plot of the data. Use attractiveness as the x-variable and attention as the y-variable.

a. Does there appear to be a linear relationship between x and y?

b. Is the relationship positive or negative?

c. Is the relationship strong or weak?

d. Does the graph seem to support the claim that a child's attractiveness is a predictor of his or her parents' attentiveness?

Step 3: Calculate the correlation coefficient, r.

a. Is the correlation coefficient statistically significant at the 0.05 level of significance?

b. What about the 0.01 level of significance?

c. Interpret what it means for the correlation coefficient to be statistically significant for this scenario.

Step 4: Determine the equation of the regression line, $\hat{y} = b_0 + b_1x$. Draw the regression line on your scatter plot.

Step 5: Calculate the coefficient of determination, r^2.

a. Interpret what r^2 means for this scenario.

b. Does the value of r^2 support the claim that a child's attractiveness is a predictor of the parents' attentiveness?

Step 6: How did any of the following factors affect your results, if at all?

a. The way in which you chose which children to watch

b. The different ages of the children

c. The location where you collected your data

d. The way you rated the children's attractiveness or the parental attentiveness

Step 7: Based on your answers for Steps 1 through 6, formulate a conclusion as to whether a child's attractiveness is a predictor of the parents' attentiveness.

T Chapter 12 Technology

Scatter Plots and Linear Regression

TI-83/84 Plus

The first step in creating any type of graph using technology is to enter the data. On a TI-83/84 Plus calculator, enter the values of the x-variable in L1 and the values of the y-variable in L2. Recall that to enter data in the List environment, press STAT and then choose EDIT and option 1:Edit.

After the data have been entered in L1 and L2, to create a scatter plot we must set up the STAT PLOTS menu. Press 2ND and then Y= to open the STAT PLOTS menu. Highlight Plot1 and then press ENTER. Turn Plot1 on by selecting On and then pressing ENTER. For the type of graph, choose the first one, which is a small picture of a scatter plot. The default entries for the rest of the menu should be correct if the data are in lists.

To create the scatter plot, press GRAPH. If a scatter plot does not immediately appear on the screen, which is likely, we may need to zoom in. The TI-83/84 Plus has a built-in zoom feature that will automatically zoom correctly for a statistical graph. Press ZOOM; then choose option 9:ZoomStat.

Side Note

The most recent TI-84 Plus calculators (Jan. 2011 and later) contain a new feature called STAT WIZARDS. This feature is not used in the directions in this text. By default this feature is turned ON. STAT WIZARDS can be turned OFF under the second page of MODE options.

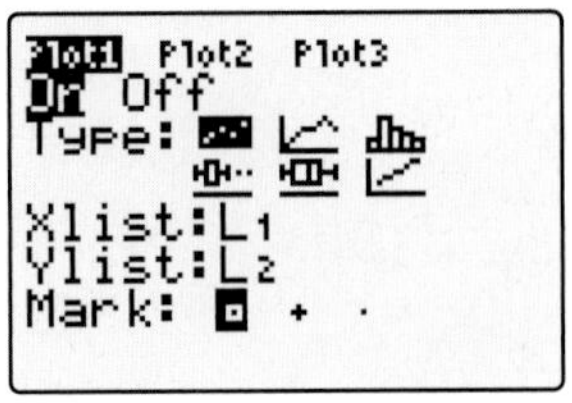

Example T.1

Using a TI-83/84 Plus Calculator to Create a Scatter Plot

Use a TI-83/84 Plus calculator to create a scatter plot for the following data relating children's ages and mean numbers of diaper changes per child per day.

Age Groups and Diaper Changes					
Age (in Years), x	1	2	3	4	5
Diaper Changes, y	3.5	2.3	1.8	0.65	0

Solution

We need to begin by entering the data into L1 and L2 as shown in the screenshot in the margin.

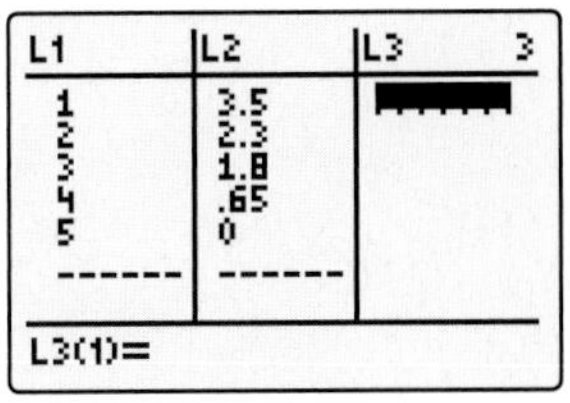

Once the data are entered, we can set up the STAT PLOTS menu as described previously. Next, press GRAPH. The screenshot in the margin shows what might appear on the screen, if anything.

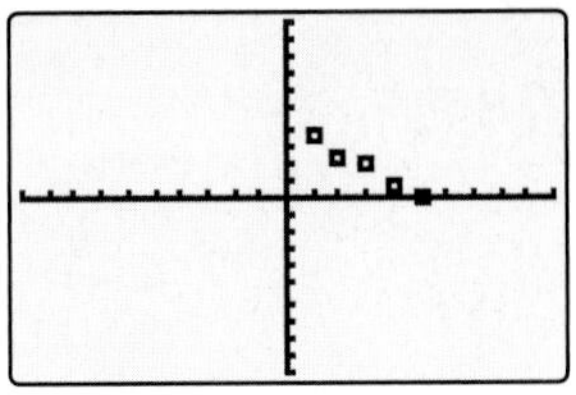

In order to adjust the screen to better view the scatter plot, we can press ZOOM and then choose option 9:ZoomStat. This will create the scatter plot shown in the following screenshot.

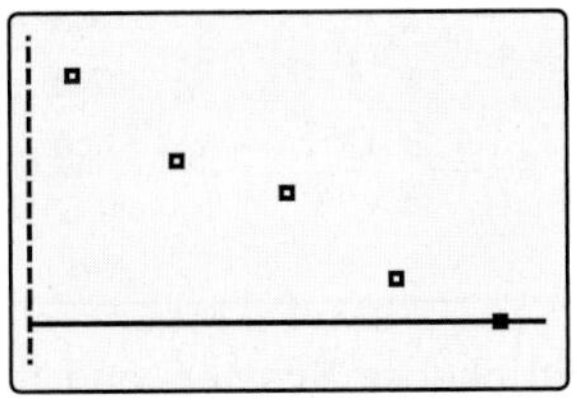

Microsoft Excel

To use Microsoft Excel to create a scatter plot, enter your data in columns A and B in the worksheet. Highlight the data in columns A and B and choose **Scatter** from the chart options under the **Insert** tab. You will choose the first chart option, **Scatter with only Markers**.

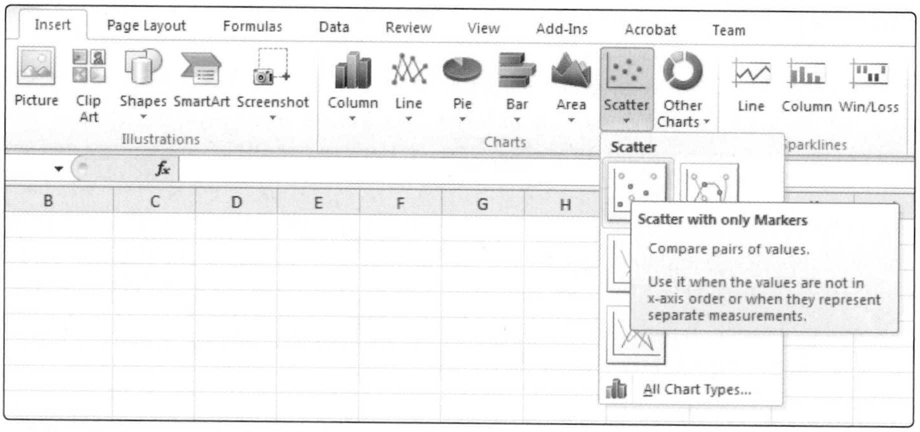

A scatter plot will then be created using the data in column A as the x-values and the data in column B as the y-values. You can then customize your scatter plot using the *Chart Tools* tabs: **Design**, **Layout**, and **Format**. Using these tabs you can edit the title of the chart, whether or not a legend appears, add labels for the x- and y-axes, among many other features. We will create a scatter plot using Microsoft Excel in the following example.

Example T.2

Using Microsoft Excel to Create a Scatter Plot

Use Microsoft Excel to create a scatter plot of the data given in Example T.1 relating children's ages and mean numbers of diaper changes per child per day. The data are reproduced in the following table.

Age Groups and Diaper Changes					
Age (in Years), *x*	1	2	3	4	5
Diaper Changes, *y*	3.5	2.3	1.8	0.65	0

Solution

Enter the age data in column A and the data for the number of diaper changes in column B as shown in the following screenshot.

	A	B
1	Age (in Years)	Diaper Changes
2	1	3.5
3	2	2.3
4	3	1.8
5	4	0.65
6	5	0

Highlight cells A1 through B6 and choose **Scatter**, and the first option, **Scatter with only Markers**, from the *Charts* section under the **Insert** tab. A scatter plot of the data is created. We will now use the chart tools to add a title, add axis labels, and remove the legend from the plot. Note that there are numerous ways to customize the scatter plot using the chart tools. When the chart is created, Microsoft Excel titles it "Diaper Changes" because this is the title of column B. Change this to read "Age vs. Diaper Changes" by clicking directly on the title in the chart area. Under the **Layout** tab, choose **Axis Titles**, **Primary Horizontal Axis Title**, and **Title Below Axis**. Enter **Age (in Years)** as the title for the x-axis. Again, choose **Axis Titles**, then **Primary Vertical Axis Title**, and **Rotated Title**. Enter **Number of Diaper Changes** as the title for the y-axis. Finally, under the **Layout** tab, choose **Legend**, and **None**. You should then have a scatter plot that looks similar to the one in the following screenshot.

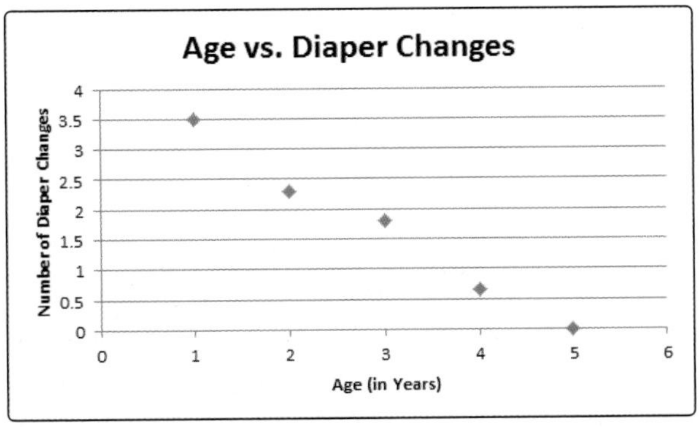

MINITAB

MINITAB can create a scatter plot and determine the linear regression model in a single operation.

Example T.3

Using MINITAB to Create a Scatter Plot and Analyze a Linear Regression Model

The following data describe the population of a hypothetical small city. Create a scatter plot and linear regression model for the given data.

Population of a Small City						
Year, x	2007	2008	2009	2010	2011	2012
Population, y	52,324	54,350	55,990	57,123	60,005	62,357

Solution

First, enter the years in column C1 and the corresponding population values in column C2. The title for column C1 should be **Year** and the title for column C2 should be **Population**.

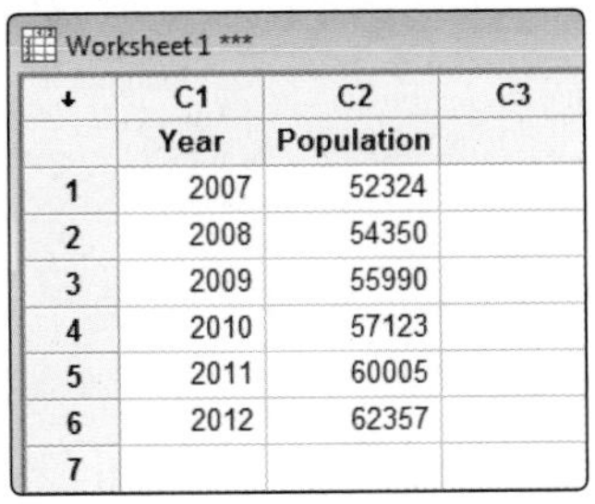

↓	C1	C2	C3
	Year	Population	
1	2007	52324	
2	2008	54350	
3	2009	55990	
4	2010	57123	
5	2011	60005	
6	2012	62357	
7			

Go to **Stat** ► **Regression** ► **Fitted Line Plot**. In the dialog box that appears, enter **C2** for Response (Y) and **C1** for Predictor (X). Make sure **Linear** is selected as the type of regression model and click **OK**.

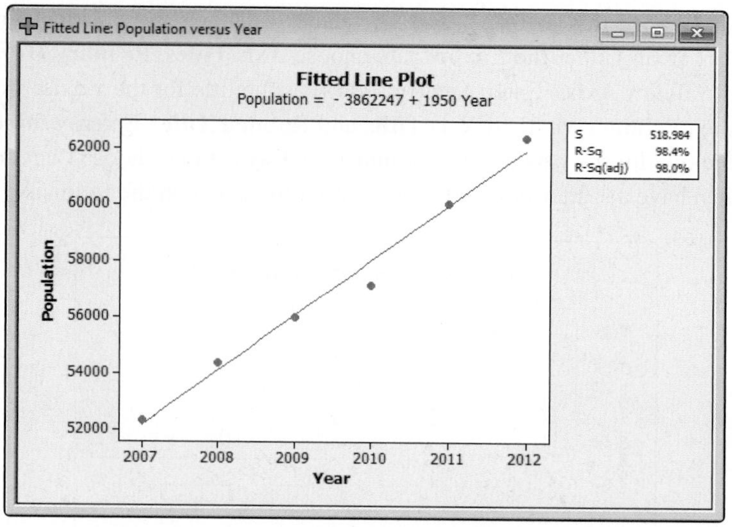

A fitted line plot is created, which includes a scatter plot of the data along with the least-squares regression line. The regression equation is displayed directly below the plot title. To the right of the plot, the R^2 and adjusted R^2 values are given along with the standard error of estimate. In the Session window, this information is also given, along with the ANOVA table. Notice that the p-value is also produced, which describes how well the model fits the data.

Session

Regression Analysis: Population versus Year

```
The regression equation is
Population = - 3862247 + 1950 Year


S = 518.984   R-Sq = 98.4%   R-Sq(adj) = 98.0%


Analysis of Variance

Source      DF        SS        MS       F      P
Regression   1  66569102  66569102  247.15  0.000
Error        4   1077376    269344
Total        5  67646479
```

Fitted Line: Population versus Year

Appendix A
Statistical Tables

A

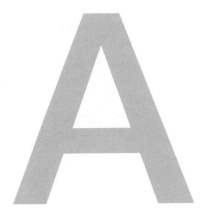

A Standard Normal Distribution

Numerical entries represent the probability that a standard normal random variable is between $-\infty$ and z where $z = \frac{x - \mu}{\sigma}$.

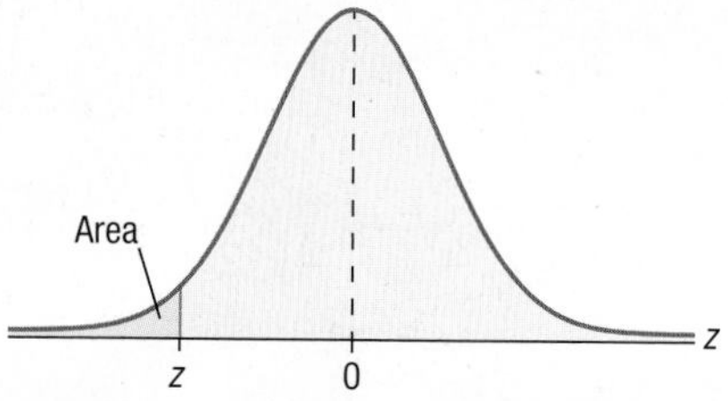

z	0.09	0.08	0.07	0.06	0.05	0.04	0.03	0.02	0.01	0.00
–3.4	0.0002	0.0003	0.0003	0.0003	0.0003	0.0003	0.0003	0.0003	0.0003	0.0003
–3.3	0.0003	0.0004	0.0004	0.0004	0.0004	0.0004	0.0004	0.0005	0.0005	0.0005
–3.2	0.0005	0.0005	0.0005	0.0006	0.0006	0.0006	0.0006	0.0006	0.0007	0.0007
–3.1	0.0007	0.0007	0.0008	0.0008	0.0008	0.0008	0.0009	0.0009	0.0009	0.0010
–3.0	0.0010	0.0010	0.0011	0.0011	0.0011	0.0012	0.0012	0.0013	0.0013	0.0013
–2.9	0.0014	0.0014	0.0015	0.0015	0.0016	0.0016	0.0017	0.0018	0.0018	0.0019
–2.8	0.0019	0.0020	0.0021	0.0021	0.0022	0.0023	0.0023	0.0024	0.0025	0.0026
–2.7	0.0026	0.0027	0.0028	0.0029	0.0030	0.0031	0.0032	0.0033	0.0034	0.0035
–2.6	0.0036	0.0037	0.0038	0.0039	0.0040	0.0041	0.0043	0.0044	0.0045	0.0047
–2.5	0.0048	0.0049	0.0051	0.0052	0.0054	0.0055	0.0057	0.0059	0.0060	0.0062
–2.4	0.0064	0.0066	0.0068	0.0069	0.0071	0.0073	0.0075	0.0078	0.0080	0.0082
–2.3	0.0084	0.0087	0.0089	0.0091	0.0094	0.0096	0.0099	0.0102	0.0104	0.0107
–2.2	0.0110	0.0113	0.0116	0.0119	0.0122	0.0125	0.0129	0.0132	0.0136	0.0139
–2.1	0.0143	0.0146	0.0150	0.0154	0.0158	0.0162	0.0166	0.0170	0.0174	0.0179
–2.0	0.0183	0.0188	0.0192	0.0197	0.0202	0.0207	0.0212	0.0217	0.0222	0.0228
–1.9	0.0233	0.0239	0.0244	0.0250	0.0256	0.0262	0.0268	0.0274	0.0281	0.0287
–1.8	0.0294	0.0301	0.0307	0.0314	0.0322	0.0329	0.0336	0.0344	0.0351	0.0359
–1.7	0.0367	0.0375	0.0384	0.0392	0.0401	0.0409	0.0418	0.0427	0.0436	0.0446
–1.6	0.0455	0.0465	0.0475	0.0485	0.0495	0.0505	0.0516	0.0526	0.0537	0.0548
–1.5	0.0559	0.0571	0.0582	0.0594	0.0606	0.0618	0.0630	0.0643	0.0655	0.0668
–1.4	0.0681	0.0694	0.0708	0.0721	0.0735	0.0749	0.0764	0.0778	0.0793	0.0808
–1.3	0.0823	0.0838	0.0853	0.0869	0.0885	0.0901	0.0918	0.0934	0.0951	0.0968
–1.2	0.0985	0.1003	0.1020	0.1038	0.1056	0.1075	0.1093	0.1112	0.1131	0.1151
–1.1	0.1170	0.1190	0.1210	0.1230	0.1251	0.1271	0.1292	0.1314	0.1335	0.1357
–1.0	0.1379	0.1401	0.1423	0.1446	0.1469	0.1492	0.1515	0.1539	0.1562	0.1587
–0.9	0.1611	0.1635	0.1660	0.1685	0.1711	0.1736	0.1762	0.1788	0.1814	0.1841
–0.8	0.1867	0.1894	0.1922	0.1949	0.1977	0.2005	0.2033	0.2061	0.2090	0.2119
–0.7	0.2148	0.2177	0.2206	0.2236	0.2266	0.2296	0.2327	0.2358	0.2389	0.2420
–0.6	0.2451	0.2483	0.2514	0.2546	0.2578	0.2611	0.2643	0.2676	0.2709	0.2743
–0.5	0.2776	0.2810	0.2843	0.2877	0.2912	0.2946	0.2981	0.3015	0.3050	0.3085
–0.4	0.3121	0.3156	0.3192	0.3228	0.3264	0.3300	0.3336	0.3372	0.3409	0.3446
–0.3	0.3483	0.3520	0.3557	0.3594	0.3632	0.3669	0.3707	0.3745	0.3783	0.3821
–0.2	0.3859	0.3897	0.3936	0.3974	0.4013	0.4052	0.4090	0.4129	0.4168	0.4207
–0.1	0.4247	0.4286	0.4325	0.4364	0.4404	0.4443	0.4483	0.4522	0.4562	0.4602
–0.0	0.4641	0.4681	0.4721	0.4761	0.4801	0.4840	0.4880	0.4920	0.4960	0.5000

B Standard Normal Distribution

Numerical entries represent the probability that a standard normal random variable is between $-\infty$ and z where $z = \dfrac{x - \mu}{\sigma}$.

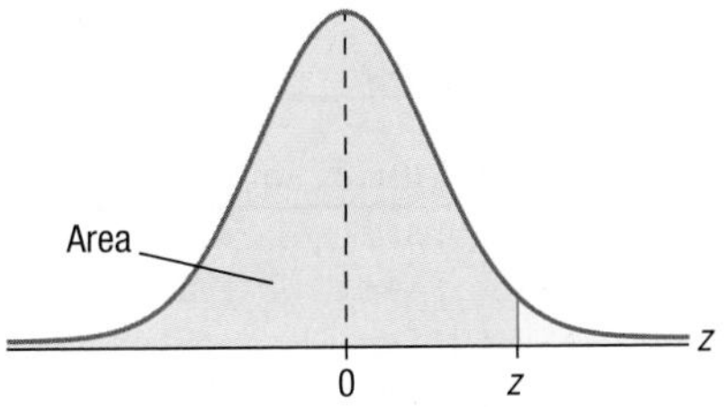

z	0.00	0.01	0.02	0.03	0.04	0.05	0.06	0.07	0.08	0.09
0.0	0.5000	0.5040	0.5080	0.5120	0.5160	0.5199	0.5239	0.5279	0.5319	0.5359
0.1	0.5398	0.5438	0.5478	0.5517	0.5557	0.5596	0.5636	0.5675	0.5714	0.5753
0.2	0.5793	0.5832	0.5871	0.5910	0.5948	0.5987	0.6026	0.6064	0.6103	0.6141
0.3	0.6179	0.6217	0.6255	0.6293	0.6331	0.6368	0.6406	0.6443	0.6480	0.6517
0.4	0.6554	0.6591	0.6628	0.6664	0.6700	0.6736	0.6772	0.6808	0.6844	0.6879
0.5	0.6915	0.6950	0.6985	0.7019	0.7054	0.7088	0.7123	0.7157	0.7190	0.7224
0.6	0.7257	0.7291	0.7324	0.7357	0.7389	0.7422	0.7454	0.7486	0.7517	0.7549
0.7	0.7580	0.7611	0.7642	0.7673	0.7704	0.7734	0.7764	0.7794	0.7823	0.7852
0.8	0.7881	0.7910	0.7939	0.7967	0.7995	0.8023	0.8051	0.8078	0.8106	0.8133
0.9	0.8159	0.8186	0.8212	0.8238	0.8264	0.8289	0.8315	0.8340	0.8365	0.8389
1.0	0.8413	0.8438	0.8461	0.8485	0.8508	0.8531	0.8554	0.8577	0.8599	0.8621
1.1	0.8643	0.8665	0.8686	0.8708	0.8729	0.8749	0.8770	0.8790	0.8810	0.8830
1.2	0.8849	0.8869	0.8888	0.8907	0.8925	0.8944	0.8962	0.8980	0.8997	0.9015
1.3	0.9032	0.9049	0.9066	0.9082	0.9099	0.9115	0.9131	0.9147	0.9162	0.9177
1.4	0.9192	0.9207	0.9222	0.9236	0.9251	0.9265	0.9279	0.9292	0.9306	0.9319
1.5	0.9332	0.9345	0.9357	0.9370	0.9382	0.9394	0.9406	0.9418	0.9429	0.9441
1.6	0.9452	0.9463	0.9474	0.9484	0.9495	0.9505	0.9515	0.9525	0.9535	0.9545
1.7	0.9554	0.9564	0.9573	0.9582	0.9591	0.9599	0.9608	0.9616	0.9625	0.9633
1.8	0.9641	0.9649	0.9656	0.9664	0.9671	0.9678	0.9686	0.9693	0.9699	0.9706
1.9	0.9713	0.9719	0.9726	0.9732	0.9738	0.9744	0.9750	0.9756	0.9761	0.9767
2.0	0.9772	0.9778	0.9783	0.9788	0.9793	0.9798	0.9803	0.9808	0.9812	0.9817
2.1	0.9821	0.9826	0.9830	0.9834	0.9838	0.9842	0.9846	0.9850	0.9854	0.9857
2.2	0.9861	0.9864	0.9868	0.9871	0.9875	0.9878	0.9881	0.9884	0.9887	0.9890
2.3	0.9893	0.9896	0.9898	0.9901	0.9904	0.9906	0.9909	0.9911	0.9913	0.9916
2.4	0.9918	0.9920	0.9922	0.9925	0.9927	0.9929	0.9931	0.9932	0.9934	0.9936
2.5	0.9938	0.9940	0.9941	0.9943	0.9945	0.9946	0.9948	0.9949	0.9951	0.9952
2.6	0.9953	0.9955	0.9956	0.9957	0.9959	0.9960	0.9961	0.9962	0.9963	0.9964
2.7	0.9965	0.9966	0.9967	0.9968	0.9969	0.9970	0.9971	0.9972	0.9973	0.9974
2.8	0.9974	0.9975	0.9976	0.9977	0.9977	0.9978	0.9979	0.9979	0.9980	0.9981
2.9	0.9981	0.9982	0.9982	0.9983	0.9984	0.9984	0.9985	0.9985	0.9986	0.9986
3.0	0.9987	0.9987	0.9987	0.9988	0.9988	0.9989	0.9989	0.9989	0.9990	0.9990
3.1	0.9990	0.9991	0.9991	0.9991	0.9992	0.9992	0.9992	0.9992	0.9993	0.9993
3.2	0.9993	0.9993	0.9994	0.9994	0.9994	0.9994	0.9994	0.9995	0.9995	0.9995
3.3	0.9995	0.9995	0.9995	0.9996	0.9996	0.9996	0.9996	0.9996	0.9996	0.9997
3.4	0.9997	0.9997	0.9997	0.9997	0.9997	0.9997	0.9997	0.9997	0.9997	0.9998

Critical Values of z

Level of Confidence, c	$\alpha = 1 - c$	$z_{\alpha/2}$
0.80	0.20	1.28
0.85	0.15	1.44
0.90	0.10	1.645
0.95	0.05	1.96
0.98	0.02	2.33
0.99	0.01	2.575

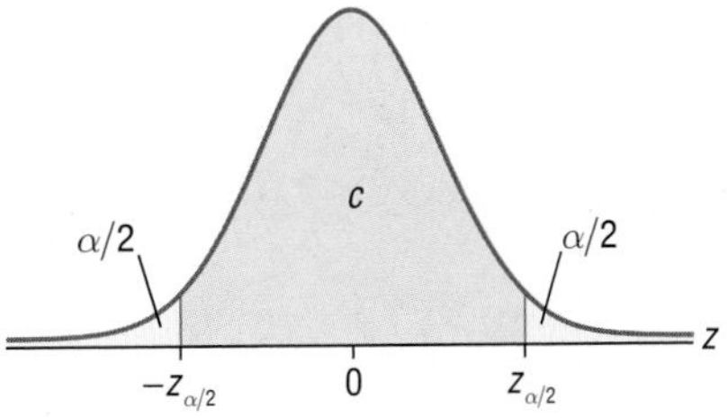

C Critical Values of *t*

	Area in One Tail				
	0.100	0.050	0.025	0.010	0.005
	Area in Two Tails				
df	0.200	0.100	0.050	0.020	0.010
1	3.078	6.314	12.706	31.821	63.657
2	1.886	2.920	4.303	6.965	9.925
3	1.638	2.353	3.182	4.541	5.841
4	1.533	2.132	2.776	3.747	4.604
5	1.476	2.015	2.571	3.365	4.032
6	1.440	1.943	2.447	3.143	3.707
7	1.415	1.895	2.365	2.998	3.499
8	1.397	1.860	2.306	2.896	3.355
9	1.383	1.833	2.262	2.821	3.250
10	1.372	1.812	2.228	2.764	3.169
11	1.363	1.796	2.201	2.718	3.106
12	1.356	1.782	2.179	2.681	3.055
13	1.350	1.771	2.160	2.650	3.012
14	1.345	1.761	2.145	2.624	2.977
15	1.341	1.753	2.131	2.602	2.947
16	1.337	1.746	2.120	2.583	2.921
17	1.333	1.740	2.110	2.567	2.898
18	1.330	1.734	2.101	2.552	2.878
19	1.328	1.729	2.093	2.539	2.861
20	1.325	1.725	2.086	2.528	2.845
21	1.323	1.721	2.080	2.518	2.831
22	1.321	1.717	2.074	2.508	2.819
23	1.319	1.714	2.069	2.500	2.807
24	1.318	1.711	2.064	2.492	2.797
25	1.316	1.708	2.060	2.485	2.787
26	1.315	1.706	2.056	2.479	2.779
27	1.314	1.703	2.052	2.473	2.771
28	1.313	1.701	2.048	2.467	2.763
29	1.311	1.699	2.045	2.462	2.756
30	1.310	1.697	2.042	2.457	2.750
31	1.309	1.696	2.040	2.453	2.744
32	1.309	1.694	2.037	2.449	2.738
34	1.307	1.691	2.032	2.441	2.728
36	1.306	1.688	2.028	2.434	2.719
38	1.304	1.686	2.024	2.429	2.712
40	1.303	1.684	2.021	2.423	2.704
45	1.301	1.679	2.014	2.412	2.690
50	1.299	1.676	2.009	2.403	2.678
55	1.297	1.673	2.004	2.396	2.668
60	1.296	1.671	2.000	2.390	2.660
70	1.294	1.667	1.994	2.381	2.648
80	1.292	1.664	1.990	2.374	2.639
90	1.291	1.662	1.987	2.368	2.632
100	1.290	1.660	1.984	2.364	2.626
120	1.289	1.658	1.980	2.358	2.617
200	1.286	1.653	1.972	2.345	2.601
300	1.284	1.650	1.968	2.339	2.592
400	1.284	1.649	1.966	2.336	2.588
500	1.283	1.648	1.965	2.334	2.586
750	1.283	1.647	1.963	2.331	2.582
1000	1.282	1.646	1.962	2.330	2.581
∞	1.282	1.645	1.960	2.326	2.576

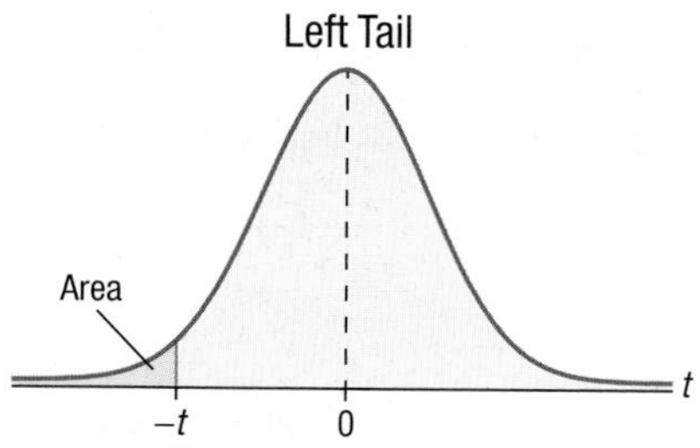

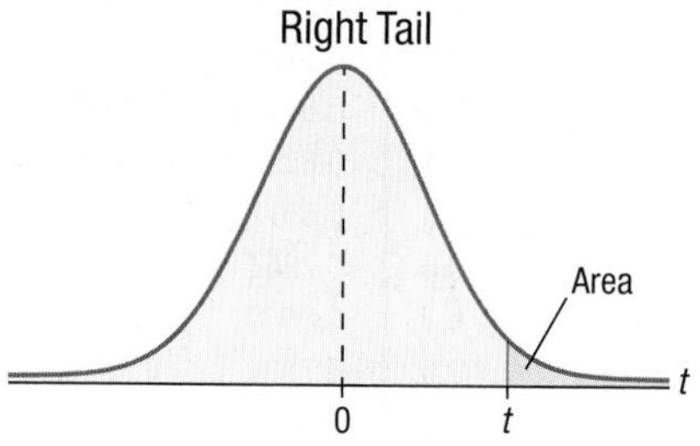

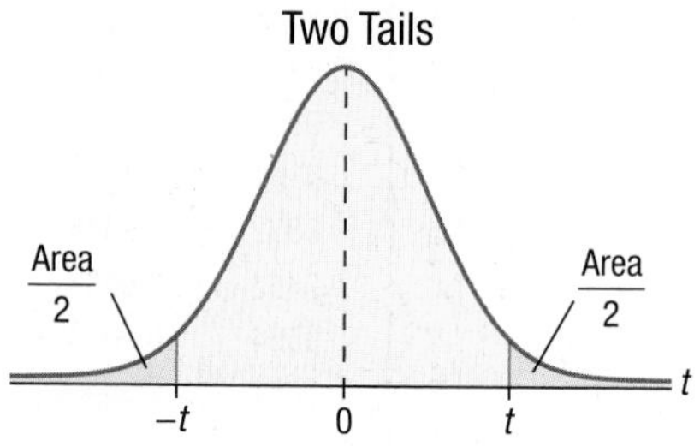

D Binomial Probabilities

Numerical entries represent $P(X = x)$.

						p				
n	x	0.1	0.2	0.3	0.4	0.5	0.6	0.7	0.8	0.9
1	0	0.9000	0.8000	0.7000	0.6000	0.5000	0.4000	0.3000	0.2000	0.1000
	1	0.1000	0.2000	0.3000	0.4000	0.5000	0.6000	0.7000	0.8000	0.9000
2	0	0.8100	0.6400	0.4900	0.3600	0.2500	0.1600	0.0900	0.0400	0.0100
	1	0.1800	0.3200	0.4200	0.4800	0.5000	0.4800	0.4200	0.3200	0.1800
	2	0.0100	0.0400	0.0900	0.1600	0.2500	0.3600	0.4900	0.6400	0.8100
3	0	0.7290	0.5120	0.3430	0.2160	0.1250	0.0640	0.0270	0.0080	0.0010
	1	0.2430	0.3840	0.4410	0.4320	0.3750	0.2880	0.1890	0.0960	0.0270
	2	0.0270	0.0960	0.1890	0.2880	0.3750	0.4320	0.4410	0.3840	0.2430
	3	0.0010	0.0080	0.0270	0.0640	0.1250	0.2160	0.3430	0.5120	0.7290
4	0	0.6561	0.4096	0.2401	0.1296	0.0625	0.0256	0.0081	0.0016	0.0001
	1	0.2916	0.4096	0.4116	0.3456	0.2500	0.1536	0.0756	0.0256	0.0036
	2	0.0486	0.1536	0.2646	0.3456	0.3750	0.3456	0.2646	0.1536	0.0486
	3	0.0036	0.0256	0.0756	0.1536	0.2500	0.3456	0.4116	0.4096	0.2916
	4	0.0001	0.0016	0.0081	0.0256	0.0625	0.1296	0.2401	0.4096	0.6561
5	0	0.5905	0.3277	0.1681	0.0778	0.0313	0.0102	0.0024	0.0003	0.0000
	1	0.3281	0.4096	0.3602	0.2592	0.1563	0.0768	0.0284	0.0064	0.0005
	2	0.0729	0.2048	0.3087	0.3456	0.3125	0.2304	0.1323	0.0512	0.0081
	3	0.0081	0.0512	0.1323	0.2304	0.3125	0.3456	0.3087	0.2048	0.0729
	4	0.0005	0.0064	0.0284	0.0768	0.1563	0.2592	0.3602	0.4096	0.3281
	5	0.0000	0.0003	0.0024	0.0102	0.0313	0.0778	0.1681	0.3277	0.5905
6	0	0.5314	0.2621	0.1176	0.0467	0.0156	0.0041	0.0007	0.0001	0.0000
	1	0.3543	0.3932	0.3025	0.1866	0.0938	0.0369	0.0102	0.0015	0.0001
	2	0.0984	0.2458	0.3241	0.3110	0.2344	0.1382	0.0595	0.0154	0.0012
	3	0.0146	0.0819	0.1852	0.2765	0.3125	0.2765	0.1852	0.0819	0.0146
	4	0.0012	0.0154	0.0595	0.1382	0.2344	0.3110	0.3241	0.2458	0.0984
	5	0.0001	0.0015	0.0102	0.0369	0.0938	0.1866	0.3025	0.3932	0.3543
	6	0.0000	0.0001	0.0007	0.0041	0.0156	0.0467	0.1176	0.2621	0.5314
7	0	0.4783	0.2097	0.0824	0.0280	0.0078	0.0016	0.0002	0.0000	0.0000
	1	0.3720	0.3670	0.2471	0.1306	0.0547	0.0172	0.0036	0.0004	0.0000
	2	0.1240	0.2753	0.3177	0.2613	0.1641	0.0774	0.0250	0.0043	0.0002
	3	0.0230	0.1147	0.2269	0.2903	0.2734	0.1935	0.0972	0.0287	0.0026
	4	0.0026	0.0287	0.0972	0.1935	0.2734	0.2903	0.2269	0.1147	0.0230
	5	0.0002	0.0043	0.0250	0.0774	0.1641	0.2613	0.3177	0.2753	0.1240
	6	0.0000	0.0004	0.0036	0.0172	0.0547	0.1306	0.2471	0.3670	0.3720
	7	0.0000	0.0000	0.0002	0.0016	0.0078	0.0280	0.0824	0.2097	0.4783

D Binomial Probabilities (cont.)

		p								
n	x	0.1	0.2	0.3	0.4	0.5	0.6	0.7	0.8	0.9
8	0	0.4305	0.1678	0.0576	0.0168	0.0039	0.0007	0.0001	0.0000	0.0000
	1	0.3826	0.3355	0.1977	0.0896	0.0313	0.0079	0.0012	0.0001	0.0000
	2	0.1488	0.2936	0.2965	0.2090	0.1094	0.0413	0.0100	0.0011	0.0000
	3	0.0331	0.1468	0.2541	0.2787	0.2188	0.1239	0.0467	0.0092	0.0004
	4	0.0046	0.0459	0.1361	0.2322	0.2734	0.2322	0.1361	0.0459	0.0046
	5	0.0004	0.0092	0.0467	0.1239	0.2188	0.2787	0.2541	0.1468	0.0331
	6	0.0000	0.0011	0.0100	0.0413	0.1094	0.2090	0.2965	0.2936	0.1488
	7	0.0000	0.0001	0.0012	0.0079	0.0313	0.0896	0.1977	0.3355	0.3826
	8	0.0000	0.0000	0.0001	0.0007	0.0039	0.0168	0.0576	0.1678	0.4305
9	0	0.3874	0.1342	0.0404	0.0101	0.0020	0.0003	0.0000	0.0000	0.0000
	1	0.3874	0.3020	0.1556	0.0605	0.0176	0.0035	0.0004	0.0000	0.0000
	2	0.1722	0.3020	0.2668	0.1612	0.0703	0.0212	0.0039	0.0003	0.0000
	3	0.0446	0.1762	0.2668	0.2508	0.1641	0.0743	0.0210	0.0028	0.0001
	4	0.0074	0.0661	0.1715	0.2508	0.2461	0.1672	0.0735	0.0165	0.0008
	5	0.0008	0.0165	0.0735	0.1672	0.2461	0.2508	0.1715	0.0661	0.0074
	6	0.0001	0.0028	0.0210	0.0743	0.1641	0.2508	0.2668	0.1762	0.0446
	7	0.0000	0.0003	0.0039	0.0212	0.0703	0.1612	0.2668	0.3020	0.1722
	8	0.0000	0.0000	0.0004	0.0035	0.0176	0.0605	0.1556	0.3020	0.3874
	9	0.0000	0.0000	0.0000	0.0003	0.0020	0.0101	0.0404	0.1342	0.3874
10	0	0.3487	0.1074	0.0282	0.0060	0.0010	0.0001	0.0000	0.0000	0.0000
	1	0.3874	0.2684	0.1211	0.0403	0.0098	0.0016	0.0001	0.0000	0.0000
	2	0.1937	0.3020	0.2335	0.1209	0.0439	0.0106	0.0014	0.0001	0.0000
	3	0.0574	0.2013	0.2668	0.2150	0.1172	0.0425	0.0090	0.0008	0.0000
	4	0.0112	0.0881	0.2001	0.2508	0.2051	0.1115	0.0368	0.0055	0.0001
	5	0.0015	0.0264	0.1029	0.2007	0.2461	0.2007	0.1029	0.0264	0.0015
	6	0.0001	0.0055	0.0368	0.1115	0.2051	0.2508	0.2001	0.0881	0.0112
	7	0.0000	0.0008	0.0090	0.0425	0.1172	0.2150	0.2668	0.2013	0.0574
	8	0.0000	0.0001	0.0014	0.0106	0.0439	0.1209	0.2335	0.3020	0.1937
	9	0.0000	0.0000	0.0001	0.0016	0.0098	0.0403	0.1211	0.2684	0.3874
	10	0.0000	0.0000	0.0000	0.0001	0.0010	0.0060	0.0282	0.1074	0.3487
11	0	0.3138	0.0859	0.0198	0.0036	0.0005	0.0000	0.0000	0.0000	0.0000
	1	0.3835	0.2362	0.0932	0.0266	0.0054	0.0007	0.0000	0.0000	0.0000
	2	0.2131	0.2953	0.1998	0.0887	0.0269	0.0052	0.0005	0.0000	0.0000
	3	0.0710	0.2215	0.2568	0.1774	0.0806	0.0234	0.0037	0.0002	0.0000
	4	0.0158	0.1107	0.2201	0.2365	0.1611	0.0701	0.0173	0.0017	0.0000
	5	0.0025	0.0388	0.1321	0.2207	0.2256	0.1471	0.0566	0.0097	0.0003
	6	0.0003	0.0097	0.0566	0.1471	0.2256	0.2207	0.1321	0.0388	0.0025
	7	0.0000	0.0017	0.0173	0.0701	0.1611	0.2365	0.2201	0.1107	0.0158
	8	0.0000	0.0002	0.0037	0.0234	0.0806	0.1774	0.2568	0.2215	0.0710
	9	0.0000	0.0000	0.0005	0.0052	0.0269	0.0887	0.1998	0.2953	0.2131
	10	0.0000	0.0000	0.0000	0.0007	0.0054	0.0266	0.0932	0.2362	0.3835
	11	0.0000	0.0000	0.0000	0.0000	0.0005	0.0036	0.0198	0.0859	0.3138

D Binomial Probabilities (cont.)

		p								
n	*x*	0.1	0.2	0.3	0.4	0.5	0.6	0.7	0.8	0.9
12	0	0.2824	0.0687	0.0138	0.0022	0.0002	0.0000	0.0000	0.0000	0.0000
	1	0.3766	0.2062	0.0712	0.0174	0.0029	0.0003	0.0000	0.0000	0.0000
	2	0.2301	0.2835	0.1678	0.0639	0.0161	0.0025	0.0002	0.0000	0.0000
	3	0.0852	0.2362	0.2397	0.1419	0.0537	0.0125	0.0015	0.0001	0.0000
	4	0.0213	0.1329	0.2311	0.2128	0.1208	0.0420	0.0078	0.0005	0.0000
	5	0.0038	0.0532	0.1585	0.2270	0.1934	0.1009	0.0291	0.0033	0.0000
	6	0.0005	0.0155	0.0792	0.1766	0.2256	0.1766	0.0792	0.0155	0.0005
	7	0.0000	0.0033	0.0291	0.1009	0.1934	0.2270	0.1585	0.0532	0.0038
	8	0.0000	0.0005	0.0078	0.0420	0.1208	0.2128	0.2311	0.1329	0.0213
	9	0.0000	0.0001	0.0015	0.0125	0.0537	0.1419	0.2397	0.2362	0.0852
	10	0.0000	0.0000	0.0002	0.0025	0.0161	0.0639	0.1678	0.2835	0.2301
	11	0.0000	0.0000	0.0000	0.0003	0.0029	0.0174	0.0712	0.2062	0.3766
	12	0.0000	0.0000	0.0000	0.0000	0.0002	0.0022	0.0138	0.0687	0.2824
13	0	0.2542	0.0550	0.0097	0.0013	0.0001	0.0000	0.0000	0.0000	0.0000
	1	0.3672	0.1787	0.0540	0.0113	0.0016	0.0001	0.0000	0.0000	0.0000
	2	0.2448	0.2680	0.1388	0.0453	0.0095	0.0012	0.0001	0.0000	0.0000
	3	0.0997	0.2457	0.2181	0.1107	0.0349	0.0065	0.0006	0.0000	0.0000
	4	0.0277	0.1535	0.2337	0.1845	0.0873	0.0243	0.0034	0.0001	0.0000
	5	0.0055	0.0691	0.1803	0.2214	0.1571	0.0656	0.0142	0.0011	0.0000
	6	0.0008	0.0230	0.1030	0.1968	0.2095	0.1312	0.0442	0.0058	0.0001
	7	0.0001	0.0058	0.0442	0.1312	0.2095	0.1968	0.1030	0.0230	0.0008
	8	0.0000	0.0011	0.0142	0.0656	0.1571	0.2214	0.1803	0.0691	0.0055
	9	0.0000	0.0001	0.0034	0.0243	0.0873	0.1845	0.2337	0.1535	0.0277
	10	0.0000	0.0000	0.0006	0.0065	0.0349	0.1107	0.2181	0.2457	0.0997
	11	0.0000	0.0000	0.0001	0.0012	0.0095	0.0453	0.1388	0.2680	0.2448
	12	0.0000	0.0000	0.0000	0.0001	0.0016	0.0113	0.0540	0.1787	0.3672
	13	0.0000	0.0000	0.0000	0.0000	0.0001	0.0013	0.0097	0.0550	0.2542
14	0	0.2288	0.0440	0.0068	0.0008	0.0001	0.0000	0.0000	0.0000	0.0000
	1	0.3559	0.1539	0.0407	0.0073	0.0009	0.0001	0.0000	0.0000	0.0000
	2	0.2570	0.2501	0.1134	0.0317	0.0056	0.0005	0.0000	0.0000	0.0000
	3	0.1142	0.2501	0.1943	0.0845	0.0222	0.0033	0.0002	0.0000	0.0000
	4	0.0349	0.1720	0.2290	0.1549	0.0611	0.0136	0.0014	0.0000	0.0000
	5	0.0078	0.0860	0.1963	0.2066	0.1222	0.0408	0.0066	0.0003	0.0000
	6	0.0013	0.0322	0.1262	0.2066	0.1833	0.0918	0.0232	0.0020	0.0000
	7	0.0002	0.0092	0.0618	0.1574	0.2095	0.1574	0.0618	0.0092	0.0002
	8	0.0000	0.0020	0.0232	0.0918	0.1833	0.2066	0.1262	0.0322	0.0013
	9	0.0000	0.0003	0.0066	0.0408	0.1222	0.2066	0.1963	0.0860	0.0078
	10	0.0000	0.0000	0.0014	0.0136	0.0611	0.1549	0.2290	0.1720	0.0349
	11	0.0000	0.0000	0.0002	0.0033	0.0222	0.0845	0.1943	0.2501	0.1142
	12	0.0000	0.0000	0.0000	0.0005	0.0056	0.0317	0.1134	0.2501	0.2570
	13	0.0000	0.0000	0.0000	0.0001	0.0009	0.0073	0.0407	0.1539	0.3559
	14	0.0000	0.0000	0.0000	0.0000	0.0001	0.0008	0.0068	0.0440	0.2288

D Binomial Probabilities (cont.)

						p				
n	*x*	0.1	0.2	0.3	0.4	0.5	0.6	0.7	0.8	0.9
15	0	0.2059	0.0352	0.0047	0.0005	0.0000	0.0000	0.0000	0.0000	0.0000
	1	0.3432	0.1319	0.0305	0.0047	0.0005	0.0000	0.0000	0.0000	0.0000
	2	0.2669	0.2309	0.0916	0.0219	0.0032	0.0003	0.0000	0.0000	0.0000
	3	0.1285	0.2501	0.1700	0.0634	0.0139	0.0016	0.0001	0.0000	0.0000
	4	0.0428	0.1876	0.2186	0.1268	0.0417	0.0074	0.0006	0.0000	0.0000
	5	0.0105	0.1032	0.2061	0.1859	0.0916	0.0245	0.0030	0.0001	0.0000
	6	0.0019	0.0430	0.1472	0.2066	0.1527	0.0612	0.0116	0.0007	0.0000
	7	0.0003	0.0138	0.0811	0.1771	0.1964	0.1181	0.0348	0.0035	0.0000
	8	0.0000	0.0035	0.0348	0.1181	0.1964	0.1771	0.0811	0.0138	0.0003
	9	0.0000	0.0007	0.0116	0.0612	0.1527	0.2066	0.1472	0.0430	0.0019
	10	0.0000	0.0001	0.0030	0.0245	0.0916	0.1859	0.2061	0.1032	0.0105
	11	0.0000	0.0000	0.0006	0.0074	0.0417	0.1268	0.2186	0.1876	0.0428
	12	0.0000	0.0000	0.0001	0.0016	0.0139	0.0634	0.1700	0.2501	0.1285
	13	0.0000	0.0000	0.0000	0.0003	0.0032	0.0219	0.0916	0.2309	0.2669
	14	0.0000	0.0000	0.0000	0.0000	0.0005	0.0047	0.0305	0.1319	0.3432
	15	0.0000	0.0000	0.0000	0.0000	0.0000	0.0005	0.0047	0.0352	0.2059
16	0	0.1853	0.0281	0.0033	0.0003	0.0000	0.0000	0.0000	0.0000	0.0000
	1	0.3294	0.1126	0.0228	0.0030	0.0002	0.0000	0.0000	0.0000	0.0000
	2	0.2745	0.2111	0.0732	0.0150	0.0018	0.0001	0.0000	0.0000	0.0000
	3	0.1423	0.2463	0.1465	0.0468	0.0085	0.0008	0.0000	0.0000	0.0000
	4	0.0514	0.2001	0.2040	0.1014	0.0278	0.0040	0.0002	0.0000	0.0000
	5	0.0137	0.1201	0.2099	0.1623	0.0667	0.0142	0.0013	0.0000	0.0000
	6	0.0028	0.0550	0.1649	0.1983	0.1222	0.0392	0.0056	0.0002	0.0000
	7	0.0004	0.0197	0.1010	0.1889	0.1746	0.0840	0.0185	0.0012	0.0000
	8	0.0001	0.0055	0.0487	0.1417	0.1964	0.1417	0.0487	0.0055	0.0001
	9	0.0000	0.0012	0.0185	0.0840	0.1746	0.1889	0.1010	0.0197	0.0004
	10	0.0000	0.0002	0.0056	0.0392	0.1222	0.1983	0.1649	0.0550	0.0028
	11	0.0000	0.0000	0.0013	0.0142	0.0667	0.1623	0.2099	0.1201	0.0137
	12	0.0000	0.0000	0.0002	0.0040	0.0278	0.1014	0.2040	0.2001	0.0514
	13	0.0000	0.0000	0.0000	0.0008	0.0085	0.0468	0.1465	0.2463	0.1423
	14	0.0000	0.0000	0.0000	0.0001	0.0018	0.0150	0.0732	0.2111	0.2745
	15	0.0000	0.0000	0.0000	0.0000	0.0002	0.0030	0.0228	0.1126	0.3294
	16	0.0000	0.0000	0.0000	0.0000	0.0000	0.0003	0.0033	0.0281	0.1853

D Binomial Probabilities (cont.)

		p								
n	x	0.1	0.2	0.3	0.4	0.5	0.6	0.7	0.8	0.9
17	0	0.1668	0.0225	0.0023	0.0002	0.0000	0.0000	0.0000	0.0000	0.0000
	1	0.3150	0.0957	0.0169	0.0019	0.0001	0.0000	0.0000	0.0000	0.0000
	2	0.2800	0.1914	0.0581	0.0102	0.0010	0.0001	0.0000	0.0000	0.0000
	3	0.1556	0.2393	0.1245	0.0341	0.0052	0.0004	0.0000	0.0000	0.0000
	4	0.0605	0.2093	0.1868	0.0796	0.0182	0.0021	0.0001	0.0000	0.0000
	5	0.0175	0.1361	0.2081	0.1379	0.0472	0.0081	0.0006	0.0000	0.0000
	6	0.0039	0.0680	0.1784	0.1839	0.0944	0.0242	0.0026	0.0001	0.0000
	7	0.0007	0.0267	0.1201	0.1927	0.1484	0.0571	0.0095	0.0004	0.0000
	8	0.0001	0.0084	0.0644	0.1606	0.1855	0.1070	0.0276	0.0021	0.0000
	9	0.0000	0.0021	0.0276	0.1070	0.1855	0.1606	0.0644	0.0084	0.0001
	10	0.0000	0.0004	0.0095	0.0571	0.1484	0.1927	0.1201	0.0267	0.0007
	11	0.0000	0.0001	0.0026	0.0242	0.0944	0.1839	0.1784	0.0680	0.0039
	12	0.0000	0.0000	0.0006	0.0081	0.0472	0.1379	0.2081	0.1361	0.0175
	13	0.0000	0.0000	0.0001	0.0021	0.0182	0.0796	0.1868	0.2093	0.0605
	14	0.0000	0.0000	0.0000	0.0004	0.0052	0.0341	0.1245	0.2393	0.1556
	15	0.0000	0.0000	0.0000	0.0001	0.0010	0.0102	0.0581	0.1914	0.2800
	16	0.0000	0.0000	0.0000	0.0000	0.0001	0.0019	0.0169	0.0957	0.3150
	17	0.0000	0.0000	0.0000	0.0000	0.0000	0.0002	0.0023	0.0225	0.1668
18	0	0.1501	0.0180	0.0016	0.0001	0.0000	0.0000	0.0000	0.0000	0.0000
	1	0.3002	0.0811	0.0126	0.0012	0.0001	0.0000	0.0000	0.0000	0.0000
	2	0.2835	0.1723	0.0458	0.0069	0.0006	0.0000	0.0000	0.0000	0.0000
	3	0.1680	0.2297	0.1046	0.0246	0.0031	0.0002	0.0000	0.0000	0.0000
	4	0.0700	0.2153	0.1681	0.0614	0.0117	0.0011	0.0000	0.0000	0.0000
	5	0.0218	0.1507	0.2017	0.1146	0.0327	0.0045	0.0002	0.0000	0.0000
	6	0.0052	0.0816	0.1873	0.1655	0.0708	0.0145	0.0012	0.0000	0.0000
	7	0.0010	0.0350	0.1376	0.1892	0.1214	0.0374	0.0046	0.0001	0.0000
	8	0.0002	0.0120	0.0811	0.1734	0.1669	0.0771	0.0149	0.0008	0.0000
	9	0.0000	0.0033	0.0386	0.1284	0.1855	0.1284	0.0386	0.0033	0.0000
	10	0.0000	0.0008	0.0149	0.0771	0.1669	0.1734	0.0811	0.0120	0.0002
	11	0.0000	0.0001	0.0046	0.0374	0.1214	0.1892	0.1376	0.0350	0.0010
	12	0.0000	0.0000	0.0012	0.0145	0.0708	0.1655	0.1873	0.0816	0.0052
	13	0.0000	0.0000	0.0002	0.0045	0.0327	0.1146	0.2017	0.1507	0.0218
	14	0.0000	0.0000	0.0000	0.0011	0.0117	0.0614	0.1681	0.2153	0.0700
	15	0.0000	0.0000	0.0000	0.0002	0.0031	0.0246	0.1046	0.2297	0.1680
	16	0.0000	0.0000	0.0000	0.0000	0.0006	0.0069	0.0458	0.1723	0.2835
	17	0.0000	0.0000	0.0000	0.0000	0.0001	0.0012	0.0126	0.0811	0.3002
	18	0.0000	0.0000	0.0000	0.0000	0.0000	0.0001	0.0016	0.0180	0.1501

D Binomial Probabilities (cont.)

n	x	p 0.1	0.2	0.3	0.4	0.5	0.6	0.7	0.8	0.9
19	0	0.1351	0.0144	0.0011	0.0001	0.0000	0.0000	0.0000	0.0000	0.0000
	1	0.2852	0.0685	0.0093	0.0008	0.0000	0.0000	0.0000	0.0000	0.0000
	2	0.2852	0.1540	0.0358	0.0046	0.0003	0.0000	0.0000	0.0000	0.0000
	3	0.1796	0.2182	0.0869	0.0175	0.0018	0.0001	0.0000	0.0000	0.0000
	4	0.0798	0.2182	0.1491	0.0467	0.0074	0.0005	0.0000	0.0000	0.0000
	5	0.0266	0.1636	0.1916	0.0933	0.0222	0.0024	0.0001	0.0000	0.0000
	6	0.0069	0.0955	0.1916	0.1451	0.0518	0.0085	0.0005	0.0000	0.0000
	7	0.0014	0.0443	0.1525	0.1797	0.0961	0.0237	0.0022	0.0000	0.0000
	8	0.0002	0.0166	0.0981	0.1797	0.1442	0.0532	0.0077	0.0003	0.0000
	9	0.0000	0.0051	0.0514	0.1464	0.1762	0.0976	0.0220	0.0013	0.0000
	10	0.0000	0.0013	0.0220	0.0976	0.1762	0.1464	0.0514	0.0051	0.0000
	11	0.0000	0.0003	0.0077	0.0532	0.1442	0.1797	0.0981	0.0166	0.0002
	12	0.0000	0.0000	0.0022	0.0237	0.0961	0.1797	0.1525	0.0443	0.0014
	13	0.0000	0.0000	0.0005	0.0085	0.0518	0.1451	0.1916	0.0955	0.0069
	14	0.0000	0.0000	0.0001	0.0024	0.0222	0.0933	0.1916	0.1636	0.0266
	15	0.0000	0.0000	0.0000	0.0005	0.0074	0.0467	0.1491	0.2182	0.0798
	16	0.0000	0.0000	0.0000	0.0001	0.0018	0.0175	0.0869	0.2182	0.1796
	17	0.0000	0.0000	0.0000	0.0000	0.0003	0.0046	0.0358	0.1540	0.2852
	18	0.0000	0.0000	0.0000	0.0000	0.0000	0.0008	0.0093	0.0685	0.2852
	19	0.0000	0.0000	0.0000	0.0000	0.0000	0.0001	0.0011	0.0144	0.1351
20	0	0.1216	0.0115	0.0008	0.0000	0.0000	0.0000	0.0000	0.0000	0.0000
	1	0.2702	0.0576	0.0068	0.0005	0.0000	0.0000	0.0000	0.0000	0.0000
	2	0.2852	0.1369	0.0278	0.0031	0.0002	0.0000	0.0000	0.0000	0.0000
	3	0.1901	0.2054	0.0716	0.0123	0.0011	0.0000	0.0000	0.0000	0.0000
	4	0.0898	0.2182	0.1304	0.0350	0.0046	0.0003	0.0000	0.0000	0.0000
	5	0.0319	0.1746	0.1789	0.0746	0.0148	0.0013	0.0000	0.0000	0.0000
	6	0.0089	0.1091	0.1916	0.1244	0.0370	0.0049	0.0002	0.0000	0.0000
	7	0.0020	0.0545	0.1643	0.1659	0.0739	0.0146	0.0010	0.0000	0.0000
	8	0.0004	0.0222	0.1144	0.1797	0.1201	0.0355	0.0039	0.0001	0.0000
	9	0.0001	0.0074	0.0654	0.1597	0.1602	0.0710	0.0120	0.0005	0.0000
	10	0.0000	0.0020	0.0308	0.1171	0.1762	0.1171	0.0308	0.0020	0.0000
	11	0.0000	0.0005	0.0120	0.0710	0.1602	0.1597	0.0654	0.0074	0.0001
	12	0.0000	0.0001	0.0039	0.0355	0.1201	0.1797	0.1144	0.0222	0.0004
	13	0.0000	0.0000	0.0010	0.0146	0.0739	0.1659	0.1643	0.0545	0.0020
	14	0.0000	0.0000	0.0002	0.0049	0.0370	0.1244	0.1916	0.1091	0.0089
	15	0.0000	0.0000	0.0000	0.0013	0.0148	0.0746	0.1789	0.1746	0.0319
	16	0.0000	0.0000	0.0000	0.0003	0.0046	0.0350	0.1304	0.2182	0.0898
	17	0.0000	0.0000	0.0000	0.0000	0.0011	0.0123	0.0716	0.2054	0.1901
	18	0.0000	0.0000	0.0000	0.0000	0.0002	0.0031	0.0278	0.1369	0.2852
	19	0.0000	0.0000	0.0000	0.0000	0.0000	0.0005	0.0068	0.0576	0.2702
	20	0.0000	0.0000	0.0000	0.0000	0.0000	0.0000	0.0008	0.0115	0.1216

E Cumulative Binomial Probabilities

Numerical entries represent $P(X \leq x)$.

		p								
n	*x*	0.1	0.2	0.3	0.4	0.5	0.6	0.7	0.8	0.9
1	0	0.9000	0.8000	0.7000	0.6000	0.5000	0.4000	0.3000	0.2000	0.1000
	1	1.0000	1.0000	1.0000	1.0000	1.0000	1.0000	1.0000	1.0000	1.0000
2	0	0.8100	0.6400	0.4900	0.3600	0.2500	0.1600	0.0900	0.0400	0.0100
	1	0.9900	0.9600	0.9100	0.8400	0.7500	0.6400	0.5100	0.3600	0.1900
	2	1.0000	1.0000	1.0000	1.0000	1.0000	1.0000	1.0000	1.0000	1.0000
3	0	0.7290	0.5120	0.3430	0.2160	0.1250	0.0640	0.0270	0.0080	0.0010
	1	0.9720	0.8960	0.7840	0.6480	0.5000	0.3520	0.2160	0.1040	0.0280
	2	0.9990	0.9920	0.9730	0.9360	0.8750	0.7840	0.6570	0.4880	0.2710
	3	1.0000	1.0000	1.0000	1.0000	1.0000	1.0000	1.0000	1.0000	1.0000
4	0	0.6561	0.4096	0.2401	0.1296	0.0625	0.0256	0.0081	0.0016	0.0001
	1	0.9477	0.8192	0.6517	0.4752	0.3125	0.1792	0.0837	0.0272	0.0037
	2	0.9963	0.9728	0.9163	0.8208	0.6875	0.5248	0.3483	0.1808	0.0523
	3	0.9999	0.9984	0.9919	0.9744	0.9375	0.8704	0.7599	0.5904	0.3439
	4	1.0000	1.0000	1.0000	1.0000	1.0000	1.0000	1.0000	1.0000	1.0000
5	0	0.5905	0.3277	0.1681	0.0778	0.0313	0.0102	0.0024	0.0003	0.0000
	1	0.9185	0.7373	0.5282	0.3370	0.1875	0.0870	0.0308	0.0067	0.0005
	2	0.9914	0.9421	0.8369	0.6826	0.5000	0.3174	0.1631	0.0579	0.0086
	3	0.9995	0.9933	0.9692	0.9130	0.8125	0.6630	0.4718	0.2627	0.0815
	4	1.0000	0.9997	0.9976	0.9898	0.9688	0.9222	0.8319	0.6723	0.4095
	5	1.0000	1.0000	1.0000	1.0000	1.0000	1.0000	1.0000	1.0000	1.0000
6	0	0.5314	0.2621	0.1176	0.0467	0.0156	0.0041	0.0007	0.0001	0.0000
	1	0.8857	0.6554	0.4202	0.2333	0.1094	0.0410	0.0109	0.0016	0.0001
	2	0.9842	0.9011	0.7443	0.5443	0.3438	0.1792	0.0705	0.0170	0.0013
	3	0.9987	0.9830	0.9295	0.8208	0.6563	0.4557	0.2557	0.0989	0.0159
	4	0.9999	0.9984	0.9891	0.9590	0.8906	0.7667	0.5798	0.3446	0.1143
	5	1.0000	0.9999	0.9993	0.9959	0.9844	0.9533	0.8824	0.7379	0.4686
	6	1.0000	1.0000	1.0000	1.0000	1.0000	1.0000	1.0000	1.0000	1.0000
7	0	0.4783	0.2097	0.0824	0.0280	0.0078	0.0016	0.0002	0.0000	0.0000
	1	0.8503	0.5767	0.3294	0.1586	0.0625	0.0188	0.0038	0.0004	0.0000
	2	0.9743	0.8520	0.6471	0.4199	0.2266	0.0963	0.0288	0.0047	0.0002
	3	0.9973	0.9667	0.8740	0.7102	0.5000	0.2898	0.1260	0.0333	0.0027
	4	0.9998	0.9953	0.9712	0.9037	0.7734	0.5801	0.3529	0.1480	0.0257
	5	1.0000	0.9996	0.9962	0.9812	0.9375	0.8414	0.6706	0.4233	0.1497
	6	1.0000	1.0000	0.9998	0.9984	0.9922	0.9720	0.9176	0.7903	0.5217
	7	1.0000	1.0000	1.0000	1.0000	1.0000	1.0000	1.0000	1.0000	1.0000

E Cumulative Binomial Probabilities (cont.)

		p								
n	x	0.1	0.2	0.3	0.4	0.5	0.6	0.7	0.8	0.9
8	0	0.4305	0.1678	0.0576	0.0168	0.0039	0.0007	0.0001	0.0000	0.0000
	1	0.8131	0.5033	0.2553	0.1064	0.0352	0.0085	0.0013	0.0001	0.0000
	2	0.9619	0.7969	0.5518	0.3154	0.1445	0.0498	0.0113	0.0012	0.0000
	3	0.9950	0.9437	0.8059	0.5941	0.3633	0.1737	0.0580	0.0104	0.0004
	4	0.9996	0.9896	0.9420	0.8263	0.6367	0.4059	0.1941	0.0563	0.0050
	5	1.0000	0.9988	0.9887	0.9502	0.8555	0.6846	0.4482	0.2031	0.0381
	6	1.0000	0.9999	0.9987	0.9915	0.9648	0.8936	0.7447	0.4967	0.1869
	7	1.0000	1.0000	0.9999	0.9993	0.9961	0.9832	0.9424	0.8322	0.5695
	8	1.0000	1.0000	1.0000	1.0000	1.0000	1.0000	1.0000	1.0000	1.0000
9	0	0.3874	0.1342	0.0404	0.0101	0.0020	0.0003	0.0000	0.0000	0.0000
	1	0.7748	0.4362	0.1960	0.0705	0.0195	0.0038	0.0004	0.0000	0.0000
	2	0.9470	0.7382	0.4628	0.2318	0.0898	0.0250	0.0043	0.0003	0.0000
	3	0.9917	0.9144	0.7297	0.4826	0.2539	0.0994	0.0253	0.0031	0.0001
	4	0.9991	0.9804	0.9012	0.7334	0.5000	0.2666	0.0988	0.0196	0.0009
	5	0.9999	0.9969	0.9747	0.9006	0.7461	0.5174	0.2703	0.0856	0.0083
	6	1.0000	0.9997	0.9957	0.9750	0.9102	0.7682	0.5372	0.2618	0.0530
	7	1.0000	1.0000	0.9996	0.9962	0.9805	0.9295	0.8040	0.5638	0.2252
	8	1.0000	1.0000	1.0000	0.9997	0.9980	0.9899	0.9596	0.8658	0.6126
	9	1.0000	1.0000	1.0000	1.0000	1.0000	1.0000	1.0000	1.0000	1.0000
10	0	0.3487	0.1074	0.0282	0.0060	0.0010	0.0001	0.0000	0.0000	0.0000
	1	0.7361	0.3758	0.1493	0.0464	0.0107	0.0017	0.0001	0.0000	0.0000
	2	0.9298	0.6778	0.3828	0.1673	0.0547	0.0123	0.0016	0.0001	0.0000
	3	0.9872	0.8791	0.6496	0.3823	0.1719	0.0548	0.0106	0.0009	0.0000
	4	0.9984	0.9672	0.8497	0.6331	0.3770	0.1662	0.0473	0.0064	0.0001
	5	0.9999	0.9936	0.9527	0.8338	0.6230	0.3669	0.1503	0.0328	0.0016
	6	1.0000	0.9991	0.9894	0.9452	0.8281	0.6177	0.3504	0.1209	0.0128
	7	1.0000	0.9999	0.9984	0.9877	0.9453	0.8327	0.6172	0.3222	0.0702
	8	1.0000	1.0000	0.9999	0.9983	0.9893	0.9536	0.8507	0.6242	0.2639
	9	1.0000	1.0000	1.0000	0.9999	0.9990	0.9940	0.9718	0.8926	0.6513
	10	1.0000	1.0000	1.0000	1.0000	1.0000	1.0000	1.0000	1.0000	1.0000
11	0	0.3138	0.0859	0.0198	0.0036	0.0005	0.0000	0.0000	0.0000	0.0000
	1	0.6974	0.3221	0.1130	0.0302	0.0059	0.0007	0.0000	0.0000	0.0000
	2	0.9104	0.6174	0.3127	0.1189	0.0327	0.0059	0.0006	0.0000	0.0000
	3	0.9815	0.8389	0.5696	0.2963	0.1133	0.0293	0.0043	0.0002	0.0000
	4	0.9972	0.9496	0.7897	0.5328	0.2744	0.0994	0.0216	0.0020	0.0000
	5	0.9997	0.9883	0.9218	0.7535	0.5000	0.2465	0.0782	0.0117	0.0003
	6	1.0000	0.9980	0.9784	0.9006	0.7256	0.4672	0.2103	0.0504	0.0028
	7	1.0000	0.9998	0.9957	0.9707	0.8867	0.7037	0.4304	0.1611	0.0185
	8	1.0000	1.0000	0.9994	0.9941	0.9673	0.8811	0.6873	0.3826	0.0896
	9	1.0000	1.0000	1.0000	0.9993	0.9941	0.9698	0.8870	0.6779	0.3026
	10	1.0000	1.0000	1.0000	1.0000	0.9995	0.9964	0.9802	0.9141	0.6862
	11	1.0000	1.0000	1.0000	1.0000	1.0000	1.0000	1.0000	1.0000	1.0000

E Cumulative Binomial Probabilities (cont.)

n	x	p								
		0.1	0.2	0.3	0.4	0.5	0.6	0.7	0.8	0.9
12	0	0.2824	0.0687	0.0138	0.0022	0.0002	0.0000	0.0000	0.0000	0.0000
	1	0.6590	0.2749	0.0850	0.0196	0.0032	0.0003	0.0000	0.0000	0.0000
	2	0.8891	0.5583	0.2528	0.0834	0.0193	0.0028	0.0002	0.0000	0.0000
	3	0.9744	0.7946	0.4925	0.2253	0.0730	0.0153	0.0017	0.0001	0.0000
	4	0.9957	0.9274	0.7237	0.4382	0.1938	0.0573	0.0095	0.0006	0.0000
	5	0.9995	0.9806	0.8822	0.6652	0.3872	0.1582	0.0386	0.0039	0.0001
	6	0.9999	0.9961	0.9614	0.8418	0.6128	0.3348	0.1178	0.0194	0.0005
	7	1.0000	0.9994	0.9905	0.9427	0.8062	0.5618	0.2763	0.0726	0.0043
	8	1.0000	0.9999	0.9983	0.9847	0.9270	0.7747	0.5075	0.2054	0.0256
	9	1.0000	1.0000	0.9998	0.9972	0.9807	0.9166	0.7472	0.4417	0.1109
	10	1.0000	1.0000	1.0000	0.9997	0.9968	0.9804	0.9150	0.7251	0.3410
	11	1.0000	1.0000	1.0000	1.0000	0.9998	0.9978	0.9862	0.9313	0.7176
	12	1.0000	1.0000	1.0000	1.0000	1.0000	1.0000	1.0000	1.0000	1.0000
13	0	0.2542	0.0550	0.0097	0.0013	0.0001	0.0000	0.0000	0.0000	0.0000
	1	0.6213	0.2336	0.0637	0.0126	0.0017	0.0001	0.0000	0.0000	0.0000
	2	0.8661	0.5017	0.2025	0.0579	0.0112	0.0013	0.0001	0.0000	0.0000
	3	0.9658	0.7473	0.4206	0.1686	0.0461	0.0078	0.0007	0.0000	0.0000
	4	0.9935	0.9009	0.6543	0.3530	0.1334	0.0321	0.0040	0.0002	0.0000
	5	0.9991	0.9700	0.8346	0.5744	0.2905	0.0977	0.0182	0.0012	0.0000
	6	0.9999	0.9930	0.9376	0.7712	0.5000	0.2288	0.0624	0.0070	0.0001
	7	1.0000	0.9988	0.9818	0.9023	0.7095	0.4256	0.1654	0.0300	0.0009
	8	1.0000	0.9998	0.9960	0.9679	0.8666	0.6470	0.3457	0.0991	0.0065
	9	1.0000	1.0000	0.9993	0.9922	0.9539	0.8314	0.5794	0.2527	0.0342
	10	1.0000	1.0000	0.9999	0.9987	0.9888	0.9421	0.7975	0.4983	0.1339
	11	1.0000	1.0000	1.0000	0.9999	0.9983	0.9874	0.9363	0.7664	0.3787
	12	1.0000	1.0000	1.0000	1.0000	0.9999	0.9987	0.9903	0.9450	0.7458
	13	1.0000	1.0000	1.0000	1.0000	1.0000	1.0000	1.0000	1.0000	1.0000
14	0	0.2288	0.0440	0.0068	0.0008	0.0001	0.0000	0.0000	0.0000	0.0000
	1	0.5846	0.1979	0.0475	0.0081	0.0009	0.0001	0.0000	0.0000	0.0000
	2	0.8416	0.4481	0.1608	0.0398	0.0065	0.0006	0.0000	0.0000	0.0000
	3	0.9559	0.6982	0.3552	0.1243	0.0287	0.0039	0.0002	0.0000	0.0000
	4	0.9908	0.8702	0.5842	0.2793	0.0898	0.0175	0.0017	0.0000	0.0000
	5	0.9985	0.9561	0.7805	0.4859	0.2120	0.0583	0.0083	0.0004	0.0000
	6	0.9998	0.9884	0.9067	0.6925	0.3953	0.1501	0.0315	0.0024	0.0000
	7	1.0000	0.9976	0.9685	0.8499	0.6047	0.3075	0.0933	0.0116	0.0002
	8	1.0000	0.9996	0.9917	0.9417	0.7880	0.5141	0.2195	0.0439	0.0015
	9	1.0000	1.0000	0.9983	0.9825	0.9102	0.7207	0.4158	0.1298	0.0092
	10	1.0000	1.0000	0.9998	0.9961	0.9713	0.8757	0.6448	0.3018	0.0441
	11	1.0000	1.0000	1.0000	0.9994	0.9935	0.9602	0.8392	0.5519	0.1584
	12	1.0000	1.0000	1.0000	0.9999	0.9991	0.9919	0.9525	0.8021	0.4154
	13	1.0000	1.0000	1.0000	1.0000	0.9999	0.9992	0.9932	0.9560	0.7712
	14	1.0000	1.0000	1.0000	1.0000	1.0000	1.0000	1.0000	1.0000	1.0000

E Cumulative Binomial Probabilities (cont.)

						p				
n	*x*	0.1	0.2	0.3	0.4	0.5	0.6	0.7	0.8	0.9
15	0	0.2059	0.0352	0.0047	0.0005	0.0000	0.0000	0.0000	0.0000	0.0000
	1	0.5490	0.1671	0.0353	0.0052	0.0005	0.0000	0.0000	0.0000	0.0000
	2	0.8159	0.3980	0.1268	0.0271	0.0037	0.0003	0.0000	0.0000	0.0000
	3	0.9444	0.6482	0.2969	0.0905	0.0176	0.0019	0.0001	0.0000	0.0000
	4	0.9873	0.8358	0.5155	0.2173	0.0592	0.0093	0.0007	0.0000	0.0000
	5	0.9978	0.9389	0.7216	0.4032	0.1509	0.0338	0.0037	0.0001	0.0000
	6	0.9997	0.9819	0.8689	0.6098	0.3036	0.0950	0.0152	0.0008	0.0000
	7	1.0000	0.9958	0.9500	0.7869	0.5000	0.2131	0.0500	0.0042	0.0000
	8	1.0000	0.9992	0.9848	0.9050	0.6964	0.3902	0.1311	0.0181	0.0003
	9	1.0000	0.9999	0.9963	0.9662	0.8491	0.5968	0.2784	0.0611	0.0022
	10	1.0000	1.0000	0.9993	0.9907	0.9408	0.7827	0.4845	0.1642	0.0127
	11	1.0000	1.0000	0.9999	0.9981	0.9824	0.9095	0.7031	0.3518	0.0556
	12	1.0000	1.0000	1.0000	0.9997	0.9963	0.9729	0.8732	0.6020	0.1841
	13	1.0000	1.0000	1.0000	1.0000	0.9995	0.9948	0.9647	0.8329	0.4510
	14	1.0000	1.0000	1.0000	1.0000	1.0000	0.9995	0.9953	0.9648	0.7941
	15	1.0000	1.0000	1.0000	1.0000	1.0000	1.0000	1.0000	1.0000	1.0000
16	0	0.1853	0.0281	0.0033	0.0003	0.0000	0.0000	0.0000	0.0000	0.0000
	1	0.5147	0.1407	0.0261	0.0033	0.0003	0.0000	0.0000	0.0000	0.0000
	2	0.7892	0.3518	0.0994	0.0183	0.0021	0.0001	0.0000	0.0000	0.0000
	3	0.9316	0.5981	0.2459	0.0651	0.0106	0.0009	0.0000	0.0000	0.0000
	4	0.9830	0.7982	0.4499	0.1666	0.0384	0.0049	0.0003	0.0000	0.0000
	5	0.9967	0.9183	0.6598	0.3288	0.1051	0.0191	0.0016	0.0000	0.0000
	6	0.9995	0.9733	0.8247	0.5272	0.2272	0.0583	0.0071	0.0002	0.0000
	7	0.9999	0.9930	0.9256	0.7161	0.4018	0.1423	0.0257	0.0015	0.0000
	8	1.0000	0.9985	0.9743	0.8577	0.5982	0.2839	0.0744	0.0070	0.0001
	9	1.0000	0.9998	0.9929	0.9417	0.7728	0.4728	0.1753	0.0267	0.0005
	10	1.0000	1.0000	0.9984	0.9809	0.8949	0.6712	0.3402	0.0817	0.0033
	11	1.0000	1.0000	0.9997	0.9951	0.9616	0.8334	0.5501	0.2018	0.0170
	12	1.0000	1.0000	1.0000	0.9991	0.9894	0.9349	0.7541	0.4019	0.0684
	13	1.0000	1.0000	1.0000	0.9999	0.9979	0.9817	0.9006	0.6482	0.2108
	14	1.0000	1.0000	1.0000	1.0000	0.9997	0.9967	0.9739	0.8593	0.4853
	15	1.0000	1.0000	1.0000	1.0000	1.0000	0.9997	0.9967	0.9719	0.8147
	16	1.0000	1.0000	1.0000	1.0000	1.0000	1.0000	1.0000	1.0000	1.0000

E Cumulative Binomial Probabilities (cont.)

		p								
n	*x*	0.1	0.2	0.3	0.4	0.5	0.6	0.7	0.8	0.9
17	0	0.1668	0.0225	0.0023	0.0002	0.0000	0.0000	0.0000	0.0000	0.0000
	1	0.4818	0.1182	0.0193	0.0021	0.0001	0.0000	0.0000	0.0000	0.0000
	2	0.7618	0.3096	0.0774	0.0123	0.0012	0.0001	0.0000	0.0000	0.0000
	3	0.9174	0.5489	0.2019	0.0464	0.0064	0.0005	0.0000	0.0000	0.0000
	4	0.9779	0.7582	0.3887	0.1260	0.0245	0.0025	0.0001	0.0000	0.0000
	5	0.9953	0.8943	0.5968	0.2639	0.0717	0.0106	0.0007	0.0000	0.0000
	6	0.9992	0.9623	0.7752	0.4478	0.1662	0.0348	0.0032	0.0001	0.0000
	7	0.9999	0.9891	0.8954	0.6405	0.3145	0.0919	0.0127	0.0005	0.0000
	8	1.0000	0.9974	0.9597	0.8011	0.5000	0.1989	0.0403	0.0026	0.0000
	9	1.0000	0.9995	0.9873	0.9081	0.6855	0.3595	0.1046	0.0109	0.0001
	10	1.0000	0.9999	0.9968	0.9652	0.8338	0.5522	0.2248	0.0377	0.0008
	11	1.0000	1.0000	0.9993	0.9894	0.9283	0.7361	0.4032	0.1057	0.0047
	12	1.0000	1.0000	0.9999	0.9975	0.9755	0.8740	0.6113	0.2418	0.0221
	13	1.0000	1.0000	1.0000	0.9995	0.9936	0.9536	0.7981	0.4511	0.0826
	14	1.0000	1.0000	1.0000	0.9999	0.9988	0.9877	0.9226	0.6904	0.2382
	15	1.0000	1.0000	1.0000	1.0000	0.9999	0.9979	0.9807	0.8818	0.5182
	16	1.0000	1.0000	1.0000	1.0000	1.0000	0.9998	0.9977	0.9775	0.8332
	17	1.0000	1.0000	1.0000	1.0000	1.0000	1.0000	1.0000	1.0000	1.0000
18	0	0.1501	0.0180	0.0016	0.0001	0.0000	0.0000	0.0000	0.0000	0.0000
	1	0.4503	0.0991	0.0142	0.0013	0.0001	0.0000	0.0000	0.0000	0.0000
	2	0.7338	0.2713	0.0600	0.0082	0.0007	0.0000	0.0000	0.0000	0.0000
	3	0.9018	0.5010	0.1646	0.0328	0.0038	0.0002	0.0000	0.0000	0.0000
	4	0.9718	0.7164	0.3327	0.0942	0.0154	0.0013	0.0000	0.0000	0.0000
	5	0.9936	0.8671	0.5344	0.2088	0.0481	0.0058	0.0003	0.0000	0.0000
	6	0.9988	0.9487	0.7217	0.3743	0.1189	0.0203	0.0014	0.0000	0.0000
	7	0.9998	0.9837	0.8593	0.5634	0.2403	0.0576	0.0061	0.0002	0.0000
	8	1.0000	0.9957	0.9404	0.7368	0.4073	0.1347	0.0210	0.0009	0.0000
	9	1.0000	0.9991	0.9790	0.8653	0.5927	0.2632	0.0596	0.0043	0.0000
	10	1.0000	0.9998	0.9939	0.9424	0.7597	0.4366	0.1407	0.0163	0.0002
	11	1.0000	1.0000	0.9986	0.9797	0.8811	0.6257	0.2783	0.0513	0.0012
	12	1.0000	1.0000	0.9997	0.9942	0.9519	0.7912	0.4656	0.1329	0.0064
	13	1.0000	1.0000	1.0000	0.9987	0.9846	0.9058	0.6673	0.2836	0.0282
	14	1.0000	1.0000	1.0000	0.9998	0.9962	0.9672	0.8354	0.4990	0.0982
	15	1.0000	1.0000	1.0000	1.0000	0.9993	0.9918	0.9400	0.7287	0.2662
	16	1.0000	1.0000	1.0000	1.0000	0.9999	0.9987	0.9858	0.9009	0.5497
	17	1.0000	1.0000	1.0000	1.0000	1.0000	0.9999	0.9984	0.9820	0.8499
	18	1.0000	1.0000	1.0000	1.0000	1.0000	1.0000	1.0000	1.0000	1.0000

E Cumulative Binomial Probabilities (cont.)

n	x	0.1	0.2	0.3	0.4	p 0.5	0.6	0.7	0.8	0.9
19	0	0.1351	0.0144	0.0011	0.0001	0.0000	0.0000	0.0000	0.0000	0.0000
	1	0.4203	0.0829	0.0104	0.0008	0.0000	0.0000	0.0000	0.0000	0.0000
	2	0.7054	0.2369	0.0462	0.0055	0.0004	0.0000	0.0000	0.0000	0.0000
	3	0.8850	0.4551	0.1332	0.0230	0.0022	0.0001	0.0000	0.0000	0.0000
	4	0.9648	0.6733	0.2822	0.0696	0.0096	0.0006	0.0000	0.0000	0.0000
	5	0.9914	0.8369	0.4739	0.1629	0.0318	0.0031	0.0001	0.0000	0.0000
	6	0.9983	0.9324	0.6655	0.3081	0.0835	0.0116	0.0006	0.0000	0.0000
	7	0.9997	0.9767	0.8180	0.4878	0.1796	0.0352	0.0028	0.0000	0.0000
	8	1.0000	0.9933	0.9161	0.6675	0.3238	0.0885	0.0105	0.0003	0.0000
	9	1.0000	0.9984	0.9674	0.8139	0.5000	0.1861	0.0326	0.0016	0.0000
	10	1.0000	0.9997	0.9895	0.9115	0.6762	0.3325	0.0839	0.0067	0.0000
	11	1.0000	1.0000	0.9972	0.9648	0.8204	0.5122	0.1820	0.0233	0.0003
	12	1.0000	1.0000	0.9994	0.9884	0.9165	0.6919	0.3345	0.0676	0.0017
	13	1.0000	1.0000	0.9999	0.9969	0.9682	0.8371	0.5261	0.1631	0.0086
	14	1.0000	1.0000	1.0000	0.9994	0.9904	0.9304	0.7178	0.3267	0.0352
	15	1.0000	1.0000	1.0000	0.9999	0.9978	0.9770	0.8668	0.5449	0.1150
	16	1.0000	1.0000	1.0000	1.0000	0.9996	0.9945	0.9538	0.7631	0.2946
	17	1.0000	1.0000	1.0000	1.0000	1.0000	0.9992	0.9896	0.9171	0.5797
	18	1.0000	1.0000	1.0000	1.0000	1.0000	0.9999	0.9989	0.9856	0.8649
	19	1.0000	1.0000	1.0000	1.0000	1.0000	1.0000	1.0000	1.0000	1.0000
20	0	0.1216	0.0115	0.0008	0.0000	0.0000	0.0000	0.0000	0.0000	0.0000
	1	0.3917	0.0692	0.0076	0.0005	0.0000	0.0000	0.0000	0.0000	0.0000
	2	0.6769	0.2061	0.0355	0.0036	0.0002	0.0000	0.0000	0.0000	0.0000
	3	0.8670	0.4114	0.1071	0.0160	0.0013	0.0000	0.0000	0.0000	0.0000
	4	0.9568	0.6296	0.2375	0.0510	0.0059	0.0003	0.0000	0.0000	0.0000
	5	0.9887	0.8042	0.4164	0.1256	0.0207	0.0016	0.0000	0.0000	0.0000
	6	0.9976	0.9133	0.6080	0.2500	0.0577	0.0065	0.0003	0.0000	0.0000
	7	0.9996	0.9679	0.7723	0.4159	0.1316	0.0210	0.0013	0.0000	0.0000
	8	0.9999	0.9900	0.8867	0.5956	0.2517	0.0565	0.0051	0.0001	0.0000
	9	1.0000	0.9974	0.9520	0.7553	0.4119	0.1275	0.0171	0.0006	0.0000
	10	1.0000	0.9994	0.9829	0.8725	0.5881	0.2447	0.0480	0.0026	0.0000
	11	1.0000	0.9999	0.9949	0.9435	0.7483	0.4044	0.1133	0.0100	0.0001
	12	1.0000	1.0000	0.9987	0.9790	0.8684	0.5841	0.2277	0.0321	0.0004
	13	1.0000	1.0000	0.9997	0.9935	0.9423	0.7500	0.3920	0.0867	0.0024
	14	1.0000	1.0000	1.0000	0.9984	0.9793	0.8744	0.5836	0.1958	0.0113
	15	1.0000	1.0000	1.0000	0.9997	0.9941	0.9490	0.7625	0.3704	0.0432
	16	1.0000	1.0000	1.0000	1.0000	0.9987	0.9840	0.8929	0.5886	0.1330
	17	1.0000	1.0000	1.0000	1.0000	0.9998	0.9964	0.9645	0.7939	0.3231
	18	1.0000	1.0000	1.0000	1.0000	1.0000	0.9995	0.9924	0.9308	0.6083
	19	1.0000	1.0000	1.0000	1.0000	1.0000	1.0000	0.9992	0.9885	0.8784
	20	1.0000	1.0000	1.0000	1.0000	1.0000	1.0000	1.0000	1.0000	1.0000

F Poisson Probabilities

Numerical entries represent $P(X = x)$.

	λ										
x	**0.02**	**0.03**	**0.04**	**0.05**	**0.06**	**0.07**	**0.08**	**0.09**	**0.10**	**0.20**	**0.30**
0	0.9802	0.9704	0.9608	0.9512	0.9418	0.9324	0.9231	0.9139	0.9048	0.8187	0.7408
1	0.0196	0.0291	0.0384	0.0476	0.0565	0.0653	0.0738	0.0823	0.0905	0.1637	0.2222
2	0.0002	0.0004	0.0008	0.0012	0.0017	0.0023	0.0030	0.0037	0.0045	0.0164	0.0333
3	0.0000	0.0000	0.0000	0.0000	0.0000	0.0001	0.0001	0.0001	0.0002	0.0011	0.0033
4	0.0000	0.0000	0.0000	0.0000	0.0000	0.0000	0.0000	0.0000	0.0000	0.0001	0.0003
x	**0.40**	**0.50**	**0.60**	**0.70**	**0.80**	**0.90**	**1.00**	**1.10**	**1.20**	**1.30**	**1.40**
0	0.6703	0.6065	0.5488	0.4966	0.4493	0.4066	0.3679	0.3329	0.3012	0.2725	0.2466
1	0.2681	0.3033	0.3293	0.3476	0.3595	0.3659	0.3679	0.3662	0.3614	0.3543	0.3452
2	0.0536	0.0758	0.0988	0.1217	0.1438	0.1647	0.1839	0.2014	0.2169	0.2303	0.2417
3	0.0072	0.0126	0.0198	0.0284	0.0383	0.0494	0.0613	0.0738	0.0867	0.0998	0.1128
4	0.0007	0.0016	0.0030	0.0050	0.0077	0.0111	0.0153	0.0203	0.0260	0.0324	0.0395
5	0.0001	0.0002	0.0004	0.0007	0.0012	0.0020	0.0031	0.0045	0.0062	0.0084	0.0111
6	0.0000	0.0000	0.0000	0.0001	0.0002	0.0003	0.0005	0.0008	0.0012	0.0018	0.0026
7	0.0000	0.0000	0.0000	0.0000	0.0000	0.0000	0.0001	0.0001	0.0002	0.0003	0.0005
8	0.0000	0.0000	0.0000	0.0000	0.0000	0.0000	0.0000	0.0000	0.0000	0.0001	0.0001
x	**1.50**	**1.60**	**1.70**	**1.80**	**1.90**	**2.00**	**2.10**	**2.20**	**2.30**	**2.40**	**2.50**
0	0.2231	0.2019	0.1827	0.1653	0.1496	0.1353	0.1225	0.1108	0.1003	0.0907	0.0821
1	0.3347	0.3230	0.3106	0.2975	0.2842	0.2707	0.2572	0.2438	0.2306	0.2177	0.2052
2	0.2510	0.2584	0.2640	0.2678	0.2700	0.2707	0.2700	0.2681	0.2652	0.2613	0.2565
3	0.1255	0.1378	0.1496	0.1607	0.1710	0.1804	0.1890	0.1966	0.2033	0.2090	0.2138
4	0.0471	0.0551	0.0636	0.0723	0.0812	0.0902	0.0992	0.1082	0.1169	0.1254	0.1336
5	0.0141	0.0176	0.0216	0.0260	0.0309	0.0361	0.0417	0.0476	0.0538	0.0602	0.0668
6	0.0035	0.0047	0.0061	0.0078	0.0098	0.0120	0.0146	0.0174	0.0206	0.0241	0.0278
7	0.0008	0.0011	0.0015	0.0020	0.0027	0.0034	0.0044	0.0055	0.0068	0.0083	0.0099
8	0.0001	0.0002	0.0003	0.0005	0.0006	0.0009	0.0011	0.0015	0.0019	0.0025	0.0031
9	0.0000	0.0000	0.0001	0.0001	0.0001	0.0002	0.0003	0.0004	0.0005	0.0007	0.0009
10	0.0000	0.0000	0.0000	0.0000	0.0000	0.0000	0.0001	0.0001	0.0001	0.0002	0.0002
x	**2.60**	**2.70**	**2.80**	**2.90**	**3.00**	**3.10**	**3.20**	**3.30**	**3.40**	**3.50**	**3.60**
0	0.0743	0.0672	0.0608	0.0550	0.0498	0.0450	0.0408	0.0369	0.0334	0.0302	0.0273
1	0.1931	0.1815	0.1703	0.1596	0.1494	0.1397	0.1304	0.1217	0.1135	0.1057	0.0984
2	0.2510	0.2450	0.2384	0.2314	0.2240	0.2165	0.2087	0.2008	0.1929	0.1850	0.1771
3	0.2176	0.2205	0.2225	0.2237	0.2240	0.2237	0.2226	0.2209	0.2186	0.2158	0.2125
4	0.1414	0.1488	0.1557	0.1622	0.1680	0.1733	0.1781	0.1823	0.1858	0.1888	0.1912
5	0.0735	0.0804	0.0872	0.0940	0.1008	0.1075	0.1140	0.1203	0.1264	0.1322	0.1377
6	0.0319	0.0362	0.0407	0.0455	0.0504	0.0555	0.0608	0.0662	0.0716	0.0771	0.0826
7	0.0118	0.0139	0.0163	0.0188	0.0216	0.0246	0.0278	0.0312	0.0348	0.0385	0.0425
8	0.0038	0.0047	0.0057	0.0068	0.0081	0.0095	0.0111	0.0129	0.0148	0.0169	0.0191
9	0.0011	0.0014	0.0018	0.0022	0.0027	0.0033	0.0040	0.0047	0.0056	0.0066	0.0076
10	0.0003	0.0004	0.0005	0.0006	0.0008	0.0010	0.0013	0.0016	0.0019	0.0023	0.0028
11	0.0001	0.0001	0.0001	0.0002	0.0002	0.0003	0.0004	0.0005	0.0006	0.0007	0.0009
12	0.0000	0.0000	0.0000	0.0000	0.0001	0.0001	0.0001	0.0001	0.0002	0.0002	0.0003
13	0.0000	0.0000	0.0000	0.0000	0.0000	0.0000	0.0000	0.0000	0.0000	0.0001	0.0001

F Poisson Probabilities (cont.)

x	λ 3.70	3.80	3.90	4.00	4.10	4.20	4.30	4.40	4.50	4.60	4.70
0	0.0247	0.0224	0.0202	0.0183	0.0166	0.0150	0.0136	0.0123	0.0111	0.0101	0.0091
1	0.0915	0.0850	0.0789	0.0733	0.0679	0.0630	0.0583	0.0540	0.0500	0.0462	0.0427
2	0.1692	0.1615	0.1539	0.1465	0.1393	0.1323	0.1254	0.1188	0.1125	0.1063	0.1005
3	0.2087	0.2046	0.2001	0.1954	0.1904	0.1852	0.1798	0.1743	0.1687	0.1631	0.1574
4	0.1931	0.1944	0.1951	0.1954	0.1951	0.1944	0.1933	0.1917	0.1898	0.1875	0.1849
5	0.1429	0.1477	0.1522	0.1563	0.1600	0.1633	0.1662	0.1687	0.1708	0.1725	0.1738
6	0.0881	0.0936	0.0989	0.1042	0.1093	0.1143	0.1191	0.1237	0.1281	0.1323	0.1362
7	0.0466	0.0508	0.0551	0.0595	0.0640	0.0686	0.0732	0.0778	0.0824	0.0869	0.0914
8	0.0215	0.0241	0.0269	0.0298	0.0328	0.0360	0.0393	0.0428	0.0463	0.0500	0.0537
9	0.0089	0.0102	0.0116	0.0132	0.0150	0.0168	0.0188	0.0209	0.0232	0.0255	0.0281
10	0.0033	0.0039	0.0045	0.0053	0.0061	0.0071	0.0081	0.0092	0.0104	0.0118	0.0132
11	0.0011	0.0013	0.0016	0.0019	0.0023	0.0027	0.0032	0.0037	0.0043	0.0049	0.0056
12	0.0003	0.0004	0.0005	0.0006	0.0008	0.0009	0.0011	0.0013	0.0016	0.0019	0.0022
13	0.0001	0.0001	0.0002	0.0002	0.0002	0.0003	0.0004	0.0005	0.0006	0.0007	0.0008
14	0.0000	0.0000	0.0000	0.0001	0.0001	0.0001	0.0001	0.0001	0.0002	0.0002	0.0003
15	0.0000	0.0000	0.0000	0.0000	0.0000	0.0000	0.0000	0.0000	0.0001	0.0001	0.0001

x	4.80	4.90	5.00	5.10	5.20	5.30	5.40	5.50	5.60	5.70	5.80
0	0.0082	0.0074	0.0067	0.0061	0.0055	0.0050	0.0045	0.0041	0.0037	0.0033	0.0030
1	0.0395	0.0365	0.0337	0.0311	0.0287	0.0265	0.0244	0.0225	0.0207	0.0191	0.0176
2	0.0948	0.0894	0.0842	0.0793	0.0746	0.0701	0.0659	0.0618	0.0580	0.0544	0.0509
3	0.1517	0.1460	0.1404	0.1348	0.1293	0.1239	0.1185	0.1133	0.1082	0.1033	0.0985
4	0.1820	0.1789	0.1755	0.1719	0.1681	0.1641	0.1600	0.1558	0.1515	0.1472	0.1428
5	0.1747	0.1753	0.1755	0.1753	0.1748	0.1740	0.1728	0.1714	0.1697	0.1678	0.1656
6	0.1398	0.1432	0.1462	0.1490	0.1515	0.1537	0.1555	0.1571	0.1584	0.1594	0.1601
7	0.0959	0.1002	0.1044	0.1086	0.1125	0.1163	0.1200	0.1234	0.1267	0.1298	0.1326
8	0.0575	0.0614	0.0653	0.0692	0.0731	0.0771	0.0810	0.0849	0.0887	0.0925	0.0962
9	0.0307	0.0334	0.0363	0.0392	0.0423	0.0454	0.0486	0.0519	0.0552	0.0586	0.0620
10	0.0147	0.0164	0.0181	0.0200	0.0220	0.0241	0.0262	0.0285	0.0309	0.0334	0.0359
11	0.0064	0.0073	0.0082	0.0093	0.0104	0.0116	0.0129	0.0143	0.0157	0.0173	0.0190
12	0.0026	0.0030	0.0034	0.0039	0.0045	0.0051	0.0058	0.0065	0.0073	0.0082	0.0092
13	0.0009	0.0011	0.0013	0.0015	0.0018	0.0021	0.0024	0.0028	0.0032	0.0036	0.0041
14	0.0003	0.0004	0.0005	0.0006	0.0007	0.0008	0.0009	0.0011	0.0013	0.0015	0.0017
15	0.0001	0.0001	0.0002	0.0002	0.0002	0.0003	0.0003	0.0004	0.0005	0.0006	0.0007
16	0.0000	0.0000	0.0000	0.0001	0.0001	0.0001	0.0001	0.0001	0.0002	0.0002	0.0002
17	0.0000	0.0000	0.0000	0.0000	0.0000	0.0000	0.0000	0.0000	0.0001	0.0001	0.0001

x	5.90	6.00	6.10	6.20	6.30	6.40	6.50	6.60	6.70	6.80	6.90
0	0.0027	0.0025	0.0022	0.0020	0.0018	0.0017	0.0015	0.0014	0.0012	0.0011	0.0010
1	0.0162	0.0149	0.0137	0.0126	0.0116	0.0106	0.0098	0.0090	0.0082	0.0076	0.0070
2	0.0477	0.0446	0.0417	0.0390	0.0364	0.0340	0.0318	0.0296	0.0276	0.0258	0.0240
3	0.0938	0.0892	0.0848	0.0806	0.0765	0.0726	0.0688	0.0652	0.0617	0.0584	0.0552
4	0.1383	0.1339	0.1294	0.1249	0.1205	0.1162	0.1118	0.1076	0.1034	0.0992	0.0952
5	0.1632	0.1606	0.1579	0.1549	0.1519	0.1487	0.1454	0.1420	0.1385	0.1349	0.1314

F Poisson Probabilities (cont.)

	λ										
x	**5.90**	**6.00**	**6.10**	**6.20**	**6.30**	**6.40**	**6.50**	**6.60**	**6.70**	**6.80**	**6.90**
6	0.1605	0.1606	0.1605	0.1601	0.1595	0.1586	0.1575	0.1562	0.1546	0.1529	0.1511
7	0.1353	0.1377	0.1399	0.1418	0.1435	0.1450	0.1462	0.1472	0.1480	0.1486	0.1489
8	0.0998	0.1033	0.1066	0.1099	0.1130	0.1160	0.1188	0.1215	0.1240	0.1263	0.1284
9	0.0654	0.0688	0.0723	0.0757	0.0791	0.0825	0.0858	0.0891	0.0923	0.0954	0.0985
10	0.0386	0.0413	0.0441	0.0469	0.0498	0.0528	0.0558	0.0588	0.0618	0.0649	0.0679
11	0.0207	0.0225	0.0244	0.0265	0.0285	0.0307	0.0330	0.0353	0.0377	0.0401	0.0426
12	0.0102	0.0113	0.0124	0.0137	0.0150	0.0164	0.0179	0.0194	0.0210	0.0227	0.0245
13	0.0046	0.0052	0.0058	0.0065	0.0073	0.0081	0.0089	0.0099	0.0108	0.0119	0.0130
14	0.0019	0.0022	0.0025	0.0029	0.0033	0.0037	0.0041	0.0046	0.0052	0.0058	0.0064
15	0.0008	0.0009	0.0010	0.0012	0.0014	0.0016	0.0018	0.0020	0.0023	0.0026	0.0029
16	0.0003	0.0003	0.0004	0.0005	0.0005	0.0006	0.0007	0.0008	0.0010	0.0011	0.0013
17	0.0001	0.0001	0.0001	0.0002	0.0002	0.0002	0.0003	0.0003	0.0004	0.0004	0.0005
18	0.0000	0.0000	0.0000	0.0001	0.0001	0.0001	0.0001	0.0001	0.0001	0.0002	0.0002
19	0.0000	0.0000	0.0000	0.0000	0.0000	0.0000	0.0000	0.0000	0.0001	0.0001	0.0001
x	**7.00**	**7.10**	**7.20**	**7.30**	**7.40**	**7.50**	**7.60**	**7.70**	**7.80**	**7.90**	**8.00**
0	0.0009	0.0008	0.0007	0.0007	0.0006	0.0006	0.0005	0.0005	0.0004	0.0004	0.0003
1	0.0064	0.0059	0.0054	0.0049	0.0045	0.0041	0.0038	0.0035	0.0032	0.0029	0.0027
2	0.0223	0.0208	0.0194	0.0180	0.0167	0.0156	0.0145	0.0134	0.0125	0.0116	0.0107
3	0.0521	0.0492	0.0464	0.0438	0.0413	0.0389	0.0366	0.0345	0.0324	0.0305	0.0286
4	0.0912	0.0874	0.0836	0.0799	0.0764	0.0729	0.0696	0.0663	0.0632	0.0602	0.0573
5	0.1277	0.1241	0.1204	0.1167	0.1130	0.1094	0.1057	0.1021	0.0986	0.0951	0.0916
6	0.1490	0.1468	0.1445	0.1420	0.1394	0.1367	0.1339	0.1311	0.1282	0.1252	0.1221
7	0.1490	0.1489	0.1486	0.1481	0.1474	0.1465	0.1454	0.1442	0.1428	0.1413	0.1396
8	0.1304	0.1321	0.1337	0.1351	0.1363	0.1373	0.1381	0.1388	0.1392	0.1395	0.1396
9	0.1014	0.1042	0.1070	0.1096	0.1121	0.1144	0.1167	0.1187	0.1207	0.1224	0.1241
10	0.0710	0.0740	0.0770	0.0800	0.0829	0.0858	0.0887	0.0914	0.0941	0.0967	0.0993
11	0.0452	0.0478	0.0504	0.0531	0.0558	0.0585	0.0613	0.0640	0.0667	0.0695	0.0722
12	0.0263	0.0283	0.0303	0.0323	0.0344	0.0366	0.0388	0.0411	0.0434	0.0457	0.0481
13	0.0142	0.0154	0.0168	0.0181	0.0196	0.0211	0.0227	0.0243	0.0260	0.0278	0.0296
14	0.0071	0.0078	0.0086	0.0095	0.0104	0.0113	0.0123	0.0134	0.0145	0.0157	0.0169
15	0.0033	0.0037	0.0041	0.0046	0.0051	0.0057	0.0062	0.0069	0.0075	0.0083	0.0090
16	0.0014	0.0016	0.0019	0.0021	0.0024	0.0026	0.0030	0.0033	0.0037	0.0041	0.0045
17	0.0006	0.0007	0.0008	0.0009	0.0010	0.0012	0.0013	0.0015	0.0017	0.0019	0.0021
18	0.0002	0.0003	0.0003	0.0004	0.0004	0.0005	0.0006	0.0006	0.0007	0.0008	0.0009
19	0.0001	0.0001	0.0001	0.0001	0.0002	0.0002	0.0002	0.0003	0.0003	0.0003	0.0004
20	0.0000	0.0000	0.0000	0.0001	0.0001	0.0001	0.0001	0.0001	0.0001	0.0001	0.0002
21	0.0000	0.0000	0.0000	0.0000	0.0000	0.0000	0.0000	0.0000	0.0000	0.0001	0.0001
x	**8.10**	**8.20**	**8.30**	**8.40**	**8.50**	**8.60**	**8.70**	**8.80**	**8.90**	**9.00**	**9.10**
0	0.0003	0.0003	0.0002	0.0002	0.0002	0.0002	0.0002	0.0002	0.0001	0.0001	0.0001
1	0.0025	0.0023	0.0021	0.0019	0.0017	0.0016	0.0014	0.0013	0.0012	0.0011	0.0010
2	0.0100	0.0092	0.0086	0.0079	0.0074	0.0068	0.0063	0.0058	0.0054	0.0050	0.0046
3	0.0269	0.0252	0.0237	0.0222	0.0208	0.0195	0.0183	0.0171	0.0160	0.0150	0.0140
4	0.0544	0.0517	0.0491	0.0466	0.0443	0.0420	0.0398	0.0377	0.0357	0.0337	0.0319

F Poisson Probabilities (cont.)

x	λ 8.10	8.20	8.30	8.40	8.50	8.60	8.70	8.80	8.90	9.00	9.10
5	0.0882	0.0849	0.0816	0.0784	0.0752	0.0722	0.0692	0.0663	0.0635	0.0607	0.0581
6	0.1191	0.1160	0.1128	0.1097	0.1066	0.1034	0.1003	0.0972	0.0941	0.0911	0.0881
7	0.1378	0.1358	0.1338	0.1317	0.1294	0.1271	0.1247	0.1222	0.1197	0.1171	0.1145
8	0.1395	0.1392	0.1388	0.1382	0.1375	0.1366	0.1356	0.1344	0.1332	0.1318	0.1302
9	0.1256	0.1269	0.1280	0.1290	0.1299	0.1306	0.1311	0.1315	0.1317	0.1318	0.1317
10	0.1017	0.1040	0.1063	0.1084	0.1104	0.1123	0.1140	0.1157	0.1172	0.1186	0.1198
11	0.0749	0.0776	0.0802	0.0828	0.0853	0.0878	0.0902	0.0925	0.0948	0.0970	0.0991
12	0.0505	0.0530	0.0555	0.0579	0.0604	0.0629	0.0654	0.0679	0.0703	0.0728	0.0752
13	0.0315	0.0334	0.0354	0.0374	0.0395	0.0416	0.0438	0.0459	0.0481	0.0504	0.0526
14	0.0182	0.0196	0.0210	0.0225	0.0240	0.0256	0.0272	0.0289	0.0306	0.0324	0.0342
15	0.0098	0.0107	0.0116	0.0126	0.0136	0.0147	0.0158	0.0169	0.0182	0.0194	0.0208
16	0.0050	0.0055	0.0060	0.0066	0.0072	0.0079	0.0086	0.0093	0.0101	0.0109	0.0118
17	0.0024	0.0026	0.0029	0.0033	0.0036	0.0040	0.0044	0.0048	0.0053	0.0058	0.0063
18	0.0011	0.0012	0.0014	0.0015	0.0017	0.0019	0.0021	0.0024	0.0026	0.0029	0.0032
19	0.0005	0.0005	0.0006	0.0007	0.0008	0.0009	0.0010	0.0011	0.0012	0.0014	0.0015
20	0.0002	0.0002	0.0002	0.0003	0.0003	0.0004	0.0004	0.0005	0.0005	0.0006	0.0007
21	0.0001	0.0001	0.0001	0.0001	0.0001	0.0002	0.0002	0.0002	0.0002	0.0003	0.0003
22	0.0000	0.0000	0.0000	0.0000	0.0001	0.0001	0.0001	0.0001	0.0001	0.0001	0.0001

x	9.20	9.30	9.40	9.50	9.60	9.70	9.80	9.90	10.00	11.00	12.00
0	0.0001	0.0001	0.0001	0.0001	0.0001	0.0001	0.0001	0.0001	0.0000	0.0000	0.0000
1	0.0009	0.0009	0.0008	0.0007	0.0007	0.0006	0.0005	0.0005	0.0005	0.0002	0.0001
2	0.0043	0.0040	0.0037	0.0034	0.0031	0.0029	0.0027	0.0025	0.0023	0.0010	0.0004
3	0.0131	0.0123	0.0115	0.0107	0.0100	0.0093	0.0087	0.0081	0.0076	0.0037	0.0018
4	0.0302	0.0285	0.0269	0.0254	0.0240	0.0226	0.0213	0.0201	0.0189	0.0102	0.0053
5	0.0555	0.0530	0.0506	0.0483	0.0460	0.0439	0.0418	0.0398	0.0378	0.0224	0.0127
6	0.0851	0.0822	0.0793	0.0764	0.0736	0.0709	0.0682	0.0656	0.0631	0.0411	0.0255
7	0.1118	0.1091	0.1064	0.1037	0.1010	0.0982	0.0955	0.0928	0.0901	0.0646	0.0437
8	0.1286	0.1269	0.1251	0.1232	0.1212	0.1191	0.1170	0.1148	0.1126	0.0888	0.0655
9	0.1315	0.1311	0.1306	0.1300	0.1293	0.1284	0.1274	0.1263	0.1251	0.1085	0.0874
10	0.1210	0.1219	0.1228	0.1235	0.1241	0.1245	0.1249	0.1250	0.1251	0.1194	0.1048
11	0.1012	0.1031	0.1049	0.1067	0.1083	0.1098	0.1112	0.1125	0.1137	0.1194	0.1144
12	0.0776	0.0799	0.0822	0.0844	0.0866	0.0888	0.0908	0.0928	0.0948	0.1094	0.1144
13	0.0549	0.0572	0.0594	0.0617	0.0640	0.0662	0.0685	0.0707	0.0729	0.0926	0.1056
14	0.0361	0.0380	0.0399	0.0419	0.0439	0.0459	0.0479	0.0500	0.0521	0.0728	0.0905
15	0.0221	0.0235	0.0250	0.0265	0.0281	0.0297	0.0313	0.0330	0.0347	0.0534	0.0724
16	0.0127	0.0137	0.0147	0.0157	0.0168	0.0180	0.0192	0.0204	0.0217	0.0367	0.0543
17	0.0069	0.0075	0.0081	0.0088	0.0095	0.0103	0.0111	0.0119	0.0128	0.0237	0.0383
18	0.0035	0.0039	0.0042	0.0046	0.0051	0.0055	0.0060	0.0065	0.0071	0.0145	0.0255
19	0.0017	0.0019	0.0021	0.0023	0.0026	0.0028	0.0031	0.0034	0.0037	0.0084	0.0161
20	0.0008	0.0009	0.0010	0.0011	0.0012	0.0014	0.0015	0.0017	0.0019	0.0046	0.0097
21	0.0003	0.0004	0.0004	0.0005	0.0006	0.0006	0.0007	0.0008	0.0009	0.0024	0.0055
22	0.0001	0.0002	0.0002	0.0002	0.0002	0.0003	0.0003	0.0004	0.0004	0.0012	0.0030
23	0.0001	0.0001	0.0001	0.0001	0.0001	0.0001	0.0001	0.0002	0.0002	0.0006	0.0016
24	0.0000	0.0000	0.0000	0.0000	0.0000	0.0000	0.0001	0.0001	0.0001	0.0003	0.0008

F Poisson Probabilities (cont.)

	λ										
x	13.00	14.00	15.00	16.00	17.00	18.00	19.00	20.00	21.00	22.00	23.00
0	0.0000	0.0000	0.0000	0.0000	0.0000	0.0000	0.0000	0.0000	0.0000	0.0000	0.0000
1	0.0000	0.0000	0.0000	0.0000	0.0000	0.0000	0.0000	0.0000	0.0000	0.0000	0.0000
2	0.0002	0.0001	0.0000	0.0000	0.0000	0.0000	0.0000	0.0000	0.0000	0.0000	0.0000
3	0.0008	0.0004	0.0002	0.0001	0.0000	0.0000	0.0000	0.0000	0.0000	0.0000	0.0000
4	0.0027	0.0013	0.0006	0.0003	0.0001	0.0001	0.0000	0.0000	0.0000	0.0000	0.0000
5	0.0070	0.0037	0.0019	0.0010	0.0005	0.0002	0.0001	0.0001	0.0000	0.0000	0.0000
6	0.0152	0.0087	0.0048	0.0026	0.0014	0.0007	0.0004	0.0002	0.0001	0.0000	0.0000
7	0.0281	0.0174	0.0104	0.0060	0.0034	0.0019	0.0010	0.0005	0.0003	0.0001	0.0001
8	0.0457	0.0304	0.0194	0.0120	0.0072	0.0042	0.0024	0.0013	0.0007	0.0004	0.0002
9	0.0661	0.0473	0.0324	0.0213	0.0135	0.0083	0.0050	0.0029	0.0017	0.0009	0.0005
10	0.0859	0.0663	0.0486	0.0341	0.0230	0.0150	0.0095	0.0058	0.0035	0.0020	0.0012
11	0.1015	0.0844	0.0663	0.0496	0.0355	0.0245	0.0164	0.0106	0.0067	0.0041	0.0024
12	0.1099	0.0984	0.0829	0.0661	0.0504	0.0368	0.0259	0.0176	0.0116	0.0075	0.0047
13	0.1099	0.1060	0.0956	0.0814	0.0658	0.0509	0.0378	0.0271	0.0188	0.0127	0.0083
14	0.1021	0.1060	0.1024	0.0930	0.0800	0.0655	0.0514	0.0387	0.0282	0.0199	0.0136
15	0.0885	0.0989	0.1024	0.0992	0.0906	0.0786	0.0650	0.0516	0.0395	0.0292	0.0209
16	0.0719	0.0866	0.0960	0.0992	0.0963	0.0884	0.0772	0.0646	0.0518	0.0401	0.0301
17	0.0550	0.0713	0.0847	0.0934	0.0963	0.0936	0.0863	0.0760	0.0640	0.0520	0.0407
18	0.0397	0.0554	0.0706	0.0830	0.0909	0.0936	0.0911	0.0844	0.0747	0.0635	0.0520
19	0.0272	0.0409	0.0557	0.0699	0.0814	0.0887	0.0911	0.0888	0.0826	0.0735	0.0629
20	0.0177	0.0286	0.0418	0.0559	0.0692	0.0798	0.0866	0.0888	0.0867	0.0809	0.0724
21	0.0109	0.0191	0.0299	0.0426	0.0560	0.0684	0.0783	0.0846	0.0867	0.0847	0.0793
22	0.0065	0.0121	0.0204	0.0310	0.0433	0.0560	0.0676	0.0769	0.0828	0.0847	0.0829
23	0.0037	0.0074	0.0133	0.0216	0.0320	0.0438	0.0559	0.0669	0.0756	0.0810	0.0829
24	0.0020	0.0043	0.0083	0.0144	0.0226	0.0328	0.0442	0.0557	0.0661	0.0743	0.0794
25	0.0010	0.0024	0.0050	0.0092	0.0154	0.0237	0.0336	0.0446	0.0555	0.0654	0.0731
26	0.0005	0.0013	0.0029	0.0057	0.0101	0.0164	0.0246	0.0343	0.0449	0.0553	0.0646
27	0.0002	0.0007	0.0016	0.0034	0.0063	0.0109	0.0173	0.0254	0.0349	0.0451	0.0551
28	0.0001	0.0003	0.0009	0.0019	0.0038	0.0070	0.0117	0.0181	0.0262	0.0354	0.0452
29	0.0001	0.0002	0.0004	0.0011	0.0023	0.0044	0.0077	0.0125	0.0190	0.0269	0.0359
30	0.0000	0.0001	0.0002	0.0006	0.0013	0.0026	0.0049	0.0083	0.0133	0.0197	0.0275
31	0.0000	0.0000	0.0001	0.0003	0.0007	0.0015	0.0030	0.0054	0.0090	0.0140	0.0204
32	0.0000	0.0000	0.0001	0.0001	0.0004	0.0009	0.0018	0.0034	0.0059	0.0096	0.0147
33	0.0000	0.0000	0.0000	0.0001	0.0002	0.0005	0.0010	0.0020	0.0038	0.0064	0.0102
34	0.0000	0.0000	0.0000	0.0000	0.0001	0.0002	0.0006	0.0012	0.0023	0.0041	0.0069
35	0.0000	0.0000	0.0000	0.0000	0.0000	0.0001	0.0003	0.0007	0.0014	0.0026	0.0045
36	0.0000	0.0000	0.0000	0.0000	0.0000	0.0001	0.0002	0.0004	0.0008	0.0016	0.0029
37	0.0000	0.0000	0.0000	0.0000	0.0000	0.0000	0.0001	0.0002	0.0005	0.0009	0.0018
38	0.0000	0.0000	0.0000	0.0000	0.0000	0.0000	0.0000	0.0001	0.0003	0.0005	0.0011
39	0.0000	0.0000	0.0000	0.0000	0.0000	0.0000	0.0000	0.0001	0.0001	0.0003	0.0006
40	0.0000	0.0000	0.0000	0.0000	0.0000	0.0000	0.0000	0.0000	0.0001	0.0002	0.0004
41	0.0000	0.0000	0.0000	0.0000	0.0000	0.0000	0.0000	0.0000	0.0000	0.0001	0.0002
42	0.0000	0.0000	0.0000	0.0000	0.0000	0.0000	0.0000	0.0000	0.0000	0.0000	0.0001
43	0.0000	0.0000	0.0000	0.0000	0.0000	0.0000	0.0000	0.0000	0.0000	0.0000	0.0001

G Critical Values of χ^2

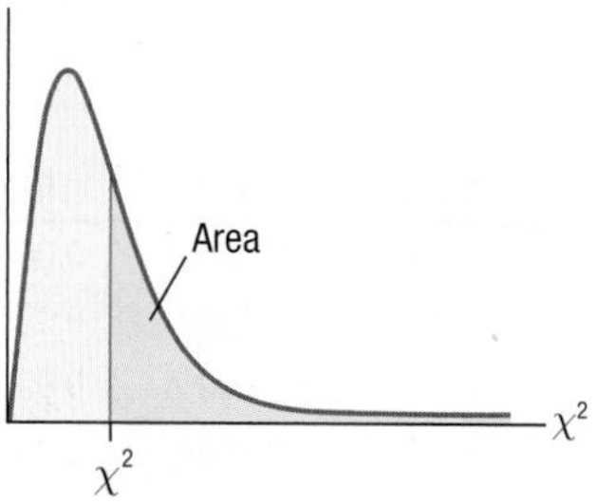

Area to the Right of the Critical Value of χ^2

df	0.995	0.990	0.975	0.950	0.900	0.100	0.050	0.025	0.010	0.005
1	0.000	0.000	0.001	0.004	0.016	2.706	3.841	5.024	6.635	7.879
2	0.010	0.020	0.051	0.103	0.211	4.605	5.991	7.378	9.210	10.597
3	0.072	0.115	0.216	0.352	0.584	6.251	7.815	9.348	11.345	12.838
4	0.207	0.297	0.484	0.711	1.064	7.779	9.488	11.143	13.277	14.860
5	0.412	0.554	0.831	1.145	1.610	9.236	11.070	12.833	15.086	16.750
6	0.676	0.872	1.237	1.635	2.204	10.645	12.592	14.449	16.812	18.548
7	0.989	1.239	1.690	2.167	2.833	12.017	14.067	16.013	18.475	20.278
8	1.344	1.646	2.180	2.733	3.490	13.362	15.507	17.535	20.090	21.955
9	1.735	2.088	2.700	3.325	4.168	14.684	16.919	19.023	21.666	23.589
10	2.156	2.558	3.247	3.940	4.865	15.987	18.307	20.483	23.209	25.188
11	2.603	3.053	3.816	4.575	5.578	17.275	19.675	21.920	24.725	26.757
12	3.074	3.571	4.404	5.226	6.304	18.549	21.026	23.337	26.217	28.300
13	3.565	4.107	5.009	5.892	7.042	19.812	22.362	24.736	27.688	29.819
14	4.075	4.660	5.629	6.571	7.790	21.064	23.685	26.119	29.141	31.319
15	4.601	5.229	6.262	7.261	8.547	22.307	24.996	27.488	30.578	32.801
16	5.142	5.812	6.908	7.962	9.312	23.542	26.296	28.845	32.000	34.267
17	5.697	6.408	7.564	8.672	10.085	24.769	27.587	30.191	33.409	35.718
18	6.265	7.015	8.231	9.390	10.865	25.989	28.869	31.526	34.805	37.156
19	6.844	7.633	8.907	10.117	11.651	27.204	30.144	32.852	36.191	38.582
20	7.434	8.260	9.591	10.851	12.443	28.412	31.410	34.170	37.566	39.997
21	8.034	8.897	10.283	11.591	13.240	29.615	32.671	35.479	38.932	41.401
22	8.643	9.542	10.982	12.338	14.041	30.813	33.924	36.781	40.289	42.796
23	9.260	10.196	11.689	13.091	14.848	32.007	35.172	38.076	41.638	44.181
24	9.886	10.856	12.401	13.848	15.659	33.196	36.415	39.364	42.980	45.559
25	10.520	11.524	13.120	14.611	16.473	34.382	37.652	40.646	44.314	46.928
26	11.160	12.198	13.844	15.379	17.292	35.563	38.885	41.923	45.642	48.290
27	11.808	12.879	14.573	16.151	18.114	36.741	40.113	43.195	46.963	49.645
28	12.461	13.565	15.308	16.928	18.939	37.916	41.337	44.461	48.278	50.993
29	13.121	14.256	16.047	17.708	19.768	39.087	42.557	45.722	49.588	52.336
30	13.787	14.953	16.791	18.493	20.599	40.256	43.773	46.979	50.892	53.672
40	20.707	22.164	24.433	26.509	29.051	51.805	55.758	59.342	63.691	66.766
50	27.991	29.707	32.357	34.764	37.689	63.167	67.505	71.420	76.154	79.490
60	35.534	37.485	40.482	43.188	46.459	74.397	79.082	83.298	88.379	91.952
70	43.275	45.442	48.758	51.739	55.329	85.527	90.531	95.023	100.425	104.215
80	51.172	53.540	57.153	60.391	64.278	96.578	101.879	106.629	112.329	116.321
90	59.196	61.754	65.647	69.126	73.291	107.565	113.145	118.136	124.116	128.299
100	67.328	70.065	74.222	77.929	82.358	118.498	124.342	129.561	135.807	140.169

H Critical Values of *F* (Area = 0.995)

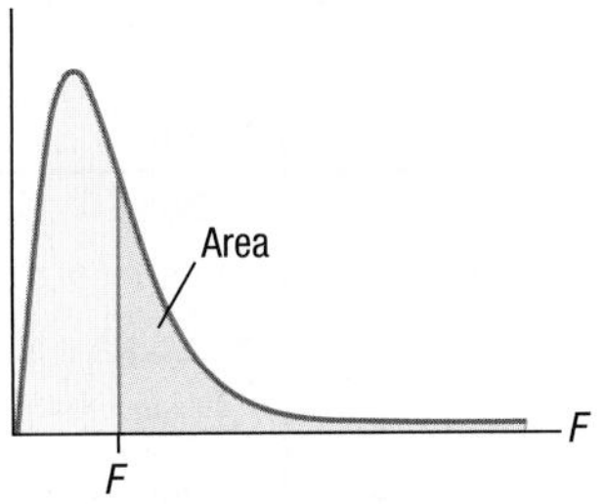

Numerator Degrees of Freedom

Denominator Degrees of Freedom

	1	2	3	4	5	6	7	8	9
1	0.0001	0.0050	0.0180	0.0319	0.0439	0.0537	0.0616	0.0681	0.0735
2	0.0001	0.0050	0.0201	0.0380	0.0546	0.0688	0.0806	0.0906	0.0989
3	0.0000	0.0050	0.0211	0.0412	0.0605	0.0774	0.0919	0.1042	0.1147
4	0.0000	0.0050	0.0216	0.0432	0.0643	0.0831	0.0995	0.1136	0.1257
5	0.0000	0.0050	0.0220	0.0445	0.0669	0.0872	0.1050	0.1205	0.1338
6	0.0000	0.0050	0.0223	0.0455	0.0689	0.0903	0.1092	0.1258	0.1402
7	0.0000	0.0050	0.0225	0.0462	0.0704	0.0927	0.1125	0.1300	0.1452
8	0.0000	0.0050	0.0227	0.0468	0.0716	0.0946	0.1152	0.1334	0.1494
9	0.0000	0.0050	0.0228	0.0473	0.0726	0.0962	0.1175	0.1363	0.1529
10	0.0000	0.0050	0.0229	0.0477	0.0734	0.0976	0.1193	0.1387	0.1558
11	0.0000	0.0050	0.0230	0.0480	0.0741	0.0987	0.1209	0.1408	0.1584
12	0.0000	0.0050	0.0230	0.0483	0.0747	0.0997	0.1223	0.1426	0.1606
13	0.0000	0.0050	0.0231	0.0485	0.0752	0.1005	0.1235	0.1441	0.1625
14	0.0000	0.0050	0.0232	0.0487	0.0757	0.1012	0.1246	0.1455	0.1642
15	0.0000	0.0050	0.0232	0.0489	0.0761	0.1019	0.1255	0.1468	0.1658
16	0.0000	0.0050	0.0233	0.0491	0.0764	0.1025	0.1263	0.1479	0.1671
17	0.0000	0.0050	0.0233	0.0492	0.0767	0.1030	0.1271	0.1489	0.1684
18	0.0000	0.0050	0.0233	0.0494	0.0770	0.1035	0.1278	0.1498	0.1695
19	0.0000	0.0050	0.0234	0.0495	0.0773	0.1039	0.1284	0.1506	0.1705
20	0.0000	0.0050	0.0234	0.0496	0.0775	0.1043	0.1290	0.1513	0.1715
21	0.0000	0.0050	0.0234	0.0497	0.0777	0.1046	0.1295	0.1520	0.1723
22	0.0000	0.0050	0.0234	0.0498	0.0779	0.1050	0.1300	0.1526	0.1731
23	0.0000	0.0050	0.0234	0.0499	0.0781	0.1053	0.1304	0.1532	0.1739
24	0.0000	0.0050	0.0235	0.0499	0.0782	0.1055	0.1308	0.1538	0.1745
25	0.0000	0.0050	0.0235	0.0500	0.0784	0.1058	0.1312	0.1543	0.1752
26	0.0000	0.0050	0.0235	0.0501	0.0785	0.1060	0.1315	0.1547	0.1758
27	0.0000	0.0050	0.0235	0.0501	0.0787	0.1063	0.1319	0.1552	0.1763
28	0.0000	0.0050	0.0235	0.0502	0.0788	0.1065	0.1322	0.1556	0.1768
29	0.0000	0.0050	0.0235	0.0502	0.0789	0.1067	0.1325	0.1560	0.1773
30	0.0000	0.0050	0.0235	0.0503	0.0790	0.1069	0.1327	0.1563	0.1778
40	0.0000	0.0050	0.0236	0.0506	0.0798	0.1082	0.1347	0.1590	0.1812
60	0.0000	0.0050	0.0237	0.0510	0.0806	0.1096	0.1368	0.1619	0.1848
120	0.0000	0.0050	0.0238	0.0514	0.0815	0.1111	0.1390	0.1649	0.1887
∞	0.0000	0.0050	0.0239	0.0517	0.0823	0.1126	0.1413	0.1681	0.1928

H Critical Values of *F* (Area = 0.995) (cont.)

Numerator Degrees of Freedom

Denominator Degrees of Freedom

	10	11	12	13	14	15	16	17	18
1	0.0780	0.0818	0.0851	0.0879	0.0904	0.0926	0.0946	0.0963	0.0979
2	0.1061	0.1122	0.1175	0.1222	0.1262	0.1299	0.1331	0.1360	0.1386
3	0.1238	0.1316	0.1384	0.1444	0.1497	0.1544	0.1586	0.1625	0.1659
4	0.1362	0.1453	0.1533	0.1604	0.1667	0.1723	0.1774	0.1819	0.1861
5	0.1455	0.1557	0.1647	0.1727	0.1798	0.1861	0.1919	0.1971	0.2018
6	0.1528	0.1639	0.1737	0.1824	0.1902	0.1972	0.2035	0.2093	0.2145
7	0.1587	0.1705	0.1810	0.1904	0.1988	0.2063	0.2131	0.2193	0.2250
8	0.1635	0.1760	0.1871	0.1970	0.2059	0.2139	0.2212	0.2278	0.2339
9	0.1676	0.1806	0.1922	0.2026	0.2120	0.2204	0.2281	0.2351	0.2415
10	0.1710	0.1846	0.1966	0.2075	0.2172	0.2261	0.2341	0.2414	0.2481
11	0.1740	0.1880	0.2005	0.2117	0.2218	0.2310	0.2393	0.2469	0.2539
12	0.1766	0.1910	0.2038	0.2154	0.2258	0.2353	0.2439	0.2518	0.2591
13	0.1789	0.1936	0.2068	0.2187	0.2294	0.2392	0.2481	0.2562	0.2637
14	0.1810	0.1960	0.2094	0.2216	0.2326	0.2426	0.2517	0.2601	0.2678
15	0.1828	0.1981	0.2118	0.2242	0.2355	0.2457	0.2551	0.2636	0.2715
16	0.1844	0.2000	0.2139	0.2266	0.2381	0.2485	0.2581	0.2669	0.2749
17	0.1859	0.2017	0.2159	0.2287	0.2404	0.2511	0.2608	0.2698	0.2780
18	0.1873	0.2032	0.2177	0.2307	0.2426	0.2534	0.2634	0.2725	0.2809
19	0.1885	0.2047	0.2193	0.2325	0.2446	0.2556	0.2657	0.2749	0.2835
20	0.1896	0.2060	0.2208	0.2342	0.2464	0.2576	0.2678	0.2772	0.2859
21	0.1906	0.2072	0.2221	0.2357	0.2481	0.2594	0.2698	0.2793	0.2881
22	0.1916	0.2083	0.2234	0.2371	0.2496	0.2611	0.2716	0.2813	0.2902
23	0.1925	0.2093	0.2246	0.2384	0.2511	0.2627	0.2733	0.2831	0.2922
24	0.1933	0.2103	0.2257	0.2397	0.2524	0.2641	0.2749	0.2848	0.2940
25	0.1941	0.2112	0.2267	0.2408	0.2537	0.2655	0.2764	0.2864	0.2957
26	0.1948	0.2120	0.2276	0.2419	0.2549	0.2668	0.2778	0.2879	0.2973
27	0.1954	0.2128	0.2285	0.2429	0.2560	0.2680	0.2791	0.2893	0.2988
28	0.1961	0.2135	0.2294	0.2438	0.2570	0.2692	0.2803	0.2906	0.3002
29	0.1967	0.2142	0.2302	0.2447	0.2580	0.2702	0.2815	0.2919	0.3015
30	0.1972	0.2149	0.2309	0.2455	0.2589	0.2712	0.2826	0.2930	0.3028
40	0.2014	0.2197	0.2365	0.2519	0.2660	0.2789	0.2909	0.3020	0.3124
60	0.2058	0.2250	0.2425	0.2587	0.2736	0.2873	0.3001	0.3119	0.3230
120	0.2105	0.2306	0.2491	0.2661	0.2819	0.2965	0.3102	0.3229	0.3348
∞	0.2156	0.2367	0.2562	0.2742	0.2910	0.3067	0.3214	0.3351	0.3480

H Critical Values of *F* (Area = 0.995) (cont.)

Numerator Degrees of Freedom

Denominator Degrees of Freedom

	19	20	24	30	40	60	120
1	0.0993	0.1006	0.1047	0.1089	0.1133	0.1177	0.1223
2	0.1410	0.1431	0.1501	0.1574	0.1648	0.1726	0.1805
3	0.1690	0.1719	0.1812	0.1909	0.2010	0.2115	0.2224
4	0.1898	0.1933	0.2045	0.2163	0.2286	0.2416	0.2551
5	0.2061	0.2100	0.2229	0.2365	0.2509	0.2660	0.2818
6	0.2192	0.2236	0.2380	0.2532	0.2693	0.2864	0.3044
7	0.2302	0.2349	0.2506	0.2673	0.2850	0.3038	0.3239
8	0.2394	0.2445	0.2613	0.2793	0.2985	0.3190	0.3410
9	0.2474	0.2528	0.2706	0.2898	0.3104	0.3324	0.3561
10	0.2543	0.2599	0.2788	0.2990	0.3208	0.3443	0.3697
11	0.2603	0.2663	0.2860	0.3072	0.3302	0.3550	0.3819
12	0.2657	0.2719	0.2924	0.3146	0.3386	0.3647	0.3931
13	0.2706	0.2769	0.2982	0.3212	0.3463	0.3735	0.4033
14	0.2749	0.2815	0.3034	0.3272	0.3532	0.3816	0.4127
15	0.2788	0.2856	0.3081	0.3327	0.3596	0.3890	0.4215
16	0.2824	0.2893	0.3124	0.3377	0.3654	0.3959	0.4296
17	0.2856	0.2927	0.3164	0.3423	0.3708	0.4023	0.4371
18	0.2886	0.2958	0.3200	0.3466	0.3758	0.4082	0.4442
19	0.2914	0.2987	0.3234	0.3506	0.3805	0.4137	0.4508
20	0.2939	0.3014	0.3265	0.3542	0.3848	0.4189	0.4570
21	0.2963	0.3039	0.3295	0.3577	0.3889	0.4238	0.4629
22	0.2985	0.3062	0.3322	0.3609	0.3927	0.4283	0.4684
23	0.3006	0.3083	0.3347	0.3639	0.3963	0.4326	0.4737
24	0.3025	0.3104	0.3371	0.3667	0.3997	0.4367	0.4787
25	0.3043	0.3123	0.3393	0.3693	0.4029	0.4406	0.4834
26	0.3059	0.3140	0.3414	0.3718	0.4059	0.4442	0.4879
27	0.3075	0.3157	0.3434	0.3742	0.4087	0.4477	0.4922
28	0.3090	0.3173	0.3452	0.3764	0.4114	0.4510	0.4964
29	0.3104	0.3188	0.3470	0.3785	0.4140	0.4542	0.5003
30	0.3118	0.3202	0.3487	0.3805	0.4164	0.4572	0.5040
40	0.3220	0.3310	0.3616	0.3962	0.4356	0.4810	0.5345
60	0.3333	0.3429	0.3762	0.4141	0.4579	0.5096	0.5725
120	0.3459	0.3564	0.3927	0.4348	0.4846	0.5452	0.6229
∞	0.3602	0.3717	0.4119	0.4596	0.5177	0.5922	0.6988

H Critical Values of *F* (Area = 0.990)

Numerator Degrees of Freedom

Denominator Degrees of Freedom	1	2	3	4	5	6	7	8	9
1	0.0002	0.0102	0.0293	0.0472	0.0615	0.0728	0.0817	0.0888	0.0947
2	0.0002	0.0101	0.0325	0.0556	0.0753	0.0915	0.1047	0.1156	0.1247
3	0.0002	0.0101	0.0339	0.0599	0.0829	0.1023	0.1183	0.1317	0.1430
4	0.0002	0.0101	0.0348	0.0626	0.0878	0.1093	0.1274	0.1427	0.1557
5	0.0002	0.0101	0.0354	0.0644	0.0912	0.1143	0.1340	0.1508	0.1651
6	0.0002	0.0101	0.0358	0.0658	0.0937	0.1181	0.1391	0.1570	0.1724
7	0.0002	0.0101	0.0361	0.0668	0.0956	0.1211	0.1430	0.1619	0.1782
8	0.0002	0.0101	0.0364	0.0676	0.0972	0.1234	0.1462	0.1659	0.1829
9	0.0002	0.0101	0.0366	0.0682	0.0984	0.1254	0.1488	0.1692	0.1869
10	0.0002	0.0101	0.0367	0.0687	0.0995	0.1270	0.1511	0.1720	0.1902
11	0.0002	0.0101	0.0369	0.0692	0.1004	0.1284	0.1529	0.1744	0.1931
12	0.0002	0.0101	0.0370	0.0696	0.1011	0.1296	0.1546	0.1765	0.1956
13	0.0002	0.0101	0.0371	0.0699	0.1018	0.1306	0.1560	0.1783	0.1978
14	0.0002	0.0101	0.0371	0.0702	0.1024	0.1315	0.1573	0.1799	0.1998
15	0.0002	0.0101	0.0372	0.0704	0.1029	0.1323	0.1584	0.1813	0.2015
16	0.0002	0.0101	0.0373	0.0707	0.1033	0.1330	0.1594	0.1826	0.2031
17	0.0002	0.0101	0.0373	0.0708	0.1037	0.1336	0.1603	0.1837	0.2045
18	0.0002	0.0101	0.0374	0.0710	0.1041	0.1342	0.1611	0.1848	0.2058
19	0.0002	0.0101	0.0374	0.0712	0.1044	0.1347	0.1618	0.1857	0.2069
20	0.0002	0.0101	0.0375	0.0713	0.1047	0.1352	0.1625	0.1866	0.2080
21	0.0002	0.0101	0.0375	0.0715	0.1050	0.1356	0.1631	0.1874	0.2090
22	0.0002	0.0101	0.0375	0.0716	0.1052	0.1360	0.1636	0.1881	0.2099
23	0.0002	0.0101	0.0376	0.0717	0.1054	0.1364	0.1641	0.1888	0.2107
24	0.0002	0.0101	0.0376	0.0718	0.1056	0.1367	0.1646	0.1894	0.2115
25	0.0002	0.0101	0.0376	0.0719	0.1058	0.1371	0.1651	0.1900	0.2122
26	0.0002	0.0101	0.0376	0.0720	0.1060	0.1374	0.1655	0.1905	0.2128
27	0.0002	0.0101	0.0377	0.0721	0.1062	0.1376	0.1659	0.1910	0.2135
28	0.0002	0.0101	0.0377	0.0721	0.1063	0.1379	0.1662	0.1915	0.2141
29	0.0002	0.0101	0.0377	0.0722	0.1065	0.1381	0.1666	0.1920	0.2146
30	0.0002	0.0101	0.0377	0.0723	0.1066	0.1383	0.1669	0.1924	0.2151
40	0.0002	0.0101	0.0379	0.0728	0.1076	0.1400	0.1692	0.1955	0.2190
60	0.0002	0.0101	0.0380	0.0732	0.1087	0.1417	0.1717	0.1987	0.2231
120	0.0002	0.0101	0.0381	0.0738	0.1097	0.1435	0.1743	0.2022	0.2274
∞	0.0002	0.0101	0.0383	0.0743	0.1109	0.1453	0.1770	0.2058	0.2320

H Critical Values of *F* (Area = 0.990) (cont.)

Numerator Degrees of Freedom

Denominator Degrees of Freedom	10	11	12	13	14	15	16	17	18
1	0.0996	0.1037	0.1072	0.1102	0.1128	0.1152	0.1172	0.1191	0.1207
2	0.1323	0.1388	0.1444	0.1492	0.1535	0.1573	0.1606	0.1636	0.1663
3	0.1526	0.1609	0.1680	0.1742	0.1797	0.1846	0.1890	0.1929	0.1964
4	0.1668	0.1764	0.1848	0.1921	0.1986	0.2044	0.2095	0.2142	0.2184
5	0.1774	0.1881	0.1975	0.2057	0.2130	0.2195	0.2254	0.2306	0.2354
6	0.1857	0.1973	0.2074	0.2164	0.2244	0.2316	0.2380	0.2438	0.2491
7	0.1923	0.2047	0.2155	0.2252	0.2338	0.2415	0.2484	0.2547	0.2604
8	0.1978	0.2108	0.2223	0.2324	0.2415	0.2497	0.2571	0.2638	0.2699
9	0.2023	0.2159	0.2279	0.2386	0.2482	0.2568	0.2645	0.2716	0.2780
10	0.2062	0.2203	0.2328	0.2439	0.2538	0.2628	0.2709	0.2783	0.2850
11	0.2096	0.2241	0.2370	0.2485	0.2588	0.2681	0.2765	0.2842	0.2912
12	0.2125	0.2274	0.2407	0.2525	0.2631	0.2728	0.2815	0.2894	0.2967
13	0.2151	0.2303	0.2439	0.2561	0.2670	0.2769	0.2859	0.2941	0.3015
14	0.2174	0.2329	0.2468	0.2592	0.2704	0.2806	0.2898	0.2982	0.3059
15	0.2194	0.2352	0.2494	0.2621	0.2735	0.2839	0.2933	0.3020	0.3099
16	0.2212	0.2373	0.2517	0.2647	0.2763	0.2869	0.2966	0.3054	0.3134
17	0.2229	0.2392	0.2539	0.2670	0.2789	0.2897	0.2995	0.3085	0.3167
18	0.2244	0.2409	0.2558	0.2691	0.2812	0.2922	0.3022	0.3113	0.3197
19	0.2257	0.2425	0.2576	0.2711	0.2833	0.2945	0.3046	0.3139	0.3224
20	0.2270	0.2440	0.2592	0.2729	0.2853	0.2966	0.3069	0.3163	0.3250
21	0.2281	0.2453	0.2607	0.2745	0.2871	0.2985	0.3090	0.3185	0.3273
22	0.2292	0.2465	0.2620	0.2761	0.2888	0.3003	0.3109	0.3206	0.3295
23	0.2302	0.2476	0.2633	0.2775	0.2903	0.3020	0.3127	0.3225	0.3315
24	0.2311	0.2487	0.2645	0.2788	0.2918	0.3036	0.3144	0.3243	0.3334
25	0.2320	0.2497	0.2656	0.2800	0.2931	0.3050	0.3160	0.3260	0.3352
26	0.2328	0.2506	0.2667	0.2812	0.2944	0.3064	0.3174	0.3276	0.3369
27	0.2335	0.2515	0.2676	0.2823	0.2956	0.3077	0.3188	0.3290	0.3385
28	0.2342	0.2523	0.2685	0.2833	0.2967	0.3089	0.3201	0.3304	0.3399
29	0.2348	0.2530	0.2694	0.2842	0.2977	0.3101	0.3213	0.3317	0.3413
30	0.2355	0.2537	0.2702	0.2851	0.2987	0.3111	0.3225	0.3330	0.3426
40	0.2401	0.2591	0.2763	0.2919	0.3062	0.3193	0.3313	0.3424	0.3527
60	0.2450	0.2648	0.2828	0.2993	0.3143	0.3282	0.3409	0.3528	0.3637
120	0.2502	0.2710	0.2899	0.3072	0.3232	0.3379	0.3515	0.3642	0.3760
∞	0.2558	0.2776	0.2975	0.3159	0.3329	0.3486	0.3633	0.3769	0.3897

H Critical Values of *F* (Area = 0.990) (cont.)

Numerator Degrees of Freedom

Denominator Degrees of Freedom

	19	20	24	30	40	60	120
1	0.1222	0.1235	0.1278	0.1322	0.1367	0.1413	0.1460
2	0.1688	0.1710	0.1781	0.1855	0.1931	0.2009	0.2089
3	0.1996	0.2025	0.2120	0.2217	0.2319	0.2424	0.2532
4	0.2222	0.2257	0.2371	0.2489	0.2612	0.2740	0.2874
5	0.2398	0.2437	0.2567	0.2703	0.2846	0.2995	0.3151
6	0.2539	0.2583	0.2727	0.2879	0.3039	0.3206	0.3383
7	0.2656	0.2704	0.2860	0.3026	0.3201	0.3386	0.3582
8	0.2754	0.2806	0.2974	0.3152	0.3341	0.3542	0.3755
9	0.2839	0.2893	0.3071	0.3261	0.3463	0.3679	0.3908
10	0.2912	0.2969	0.3156	0.3357	0.3571	0.3800	0.4045
11	0.2977	0.3036	0.3232	0.3442	0.3667	0.3908	0.4168
12	0.3033	0.3095	0.3299	0.3517	0.3753	0.4006	0.4280
13	0.3084	0.3148	0.3359	0.3585	0.3830	0.4095	0.4382
14	0.3130	0.3195	0.3413	0.3647	0.3901	0.4177	0.4476
15	0.3171	0.3238	0.3462	0.3703	0.3966	0.4251	0.4563
16	0.3209	0.3277	0.3506	0.3755	0.4025	0.4320	0.4643
17	0.3243	0.3313	0.3548	0.3802	0.4080	0.4384	0.4718
18	0.3274	0.3346	0.3585	0.3846	0.4131	0.4443	0.4788
19	0.3303	0.3376	0.3620	0.3886	0.4178	0.4498	0.4854
20	0.3330	0.3404	0.3652	0.3924	0.4221	0.4550	0.4915
21	0.3355	0.3430	0.3682	0.3959	0.4262	0.4598	0.4973
22	0.3378	0.3454	0.3710	0.3991	0.4301	0.4644	0.5028
23	0.3399	0.3476	0.3736	0.4022	0.4337	0.4687	0.5079
24	0.3419	0.3497	0.3761	0.4050	0.4371	0.4727	0.5128
25	0.3438	0.3517	0.3784	0.4077	0.4403	0.4766	0.5175
26	0.3455	0.3535	0.3805	0.4103	0.4433	0.4802	0.5219
27	0.3472	0.3553	0.3826	0.4127	0.4461	0.4836	0.5261
28	0.3487	0.3569	0.3845	0.4149	0.4488	0.4869	0.5301
29	0.3502	0.3584	0.3863	0.4171	0.4514	0.4900	0.5340
30	0.3516	0.3599	0.3880	0.4191	0.4538	0.4930	0.5376
40	0.3622	0.3711	0.4012	0.4349	0.4730	0.5165	0.5673
60	0.3739	0.3835	0.4161	0.4529	0.4952	0.5446	0.6040
120	0.3870	0.3973	0.4329	0.4738	0.5216	0.5793	0.6523
∞	0.4017	0.4130	0.4523	0.4984	0.5541	0.6247	0.7243

H Critical Values of *F* (Area = 0.975)

Numerator Degrees of Freedom (columns); Denominator Degrees of Freedom (rows)

	1	2	3	4	5	6	7	8	9
1	0.0015	0.0260	0.0573	0.0818	0.0999	0.1135	0.1239	0.1321	0.1387
2	0.0013	0.0256	0.0623	0.0939	0.1186	0.1377	0.1529	0.1650	0.1750
3	0.0012	0.0255	0.0648	0.1002	0.1288	0.1515	0.1698	0.1846	0.1969
4	0.0011	0.0255	0.0662	0.1041	0.1354	0.1606	0.1811	0.1979	0.2120
5	0.0011	0.0254	0.0672	0.1068	0.1399	0.1670	0.1892	0.2076	0.2230
6	0.0011	0.0254	0.0679	0.1087	0.1433	0.1718	0.1954	0.2150	0.2315
7	0.0011	0.0254	0.0684	0.1102	0.1459	0.1756	0.2002	0.2208	0.2383
8	0.0010	0.0254	0.0688	0.1114	0.1480	0.1786	0.2041	0.2256	0.2438
9	0.0010	0.0254	0.0691	0.1123	0.1497	0.1810	0.2073	0.2295	0.2484
10	0.0010	0.0254	0.0694	0.1131	0.1511	0.1831	0.2100	0.2328	0.2523
11	0.0010	0.0254	0.0696	0.1137	0.1523	0.1849	0.2123	0.2357	0.2556
12	0.0010	0.0254	0.0698	0.1143	0.1533	0.1864	0.2143	0.2381	0.2585
13	0.0010	0.0254	0.0699	0.1147	0.1541	0.1877	0.2161	0.2403	0.2611
14	0.0010	0.0254	0.0700	0.1152	0.1549	0.1888	0.2176	0.2422	0.2633
15	0.0010	0.0254	0.0702	0.1155	0.1556	0.1898	0.2189	0.2438	0.2653
16	0.0010	0.0254	0.0703	0.1158	0.1562	0.1907	0.2201	0.2453	0.2671
17	0.0010	0.0254	0.0704	0.1161	0.1567	0.1915	0.2212	0.2467	0.2687
18	0.0010	0.0254	0.0704	0.1164	0.1572	0.1922	0.2222	0.2479	0.2702
19	0.0010	0.0254	0.0705	0.1166	0.1576	0.1929	0.2231	0.2490	0.2715
20	0.0010	0.0253	0.0706	0.1168	0.1580	0.1935	0.2239	0.2500	0.2727
21	0.0010	0.0253	0.0706	0.1170	0.1584	0.1940	0.2246	0.2510	0.2738
22	0.0010	0.0253	0.0707	0.1172	0.1587	0.1945	0.2253	0.2518	0.2749
23	0.0010	0.0253	0.0708	0.1173	0.1590	0.1950	0.2259	0.2526	0.2758
24	0.0010	0.0253	0.0708	0.1175	0.1593	0.1954	0.2265	0.2533	0.2767
25	0.0010	0.0253	0.0708	0.1176	0.1595	0.1958	0.2270	0.2540	0.2775
26	0.0010	0.0253	0.0709	0.1178	0.1598	0.1962	0.2275	0.2547	0.2783
27	0.0010	0.0253	0.0709	0.1179	0.1600	0.1965	0.2280	0.2552	0.2790
28	0.0010	0.0253	0.0710	0.1180	0.1602	0.1968	0.2284	0.2558	0.2797
29	0.0010	0.0253	0.0710	0.1181	0.1604	0.1971	0.2288	0.2563	0.2803
30	0.0010	0.0253	0.0710	0.1182	0.1606	0.1974	0.2292	0.2568	0.2809
40	0.0010	0.0253	0.0712	0.1189	0.1619	0.1995	0.2321	0.2604	0.2853
60	0.0010	0.0253	0.0715	0.1196	0.1633	0.2017	0.2351	0.2642	0.2899
120	0.0010	0.0253	0.0717	0.1203	0.1648	0.2039	0.2382	0.2682	0.2948
∞	0.0010	0.0253	0.0719	0.1211	0.1662	0.2062	0.2414	0.2725	0.3000

H Critical Values of *F* (Area = 0.975) (cont.)

Numerator Degrees of Freedom

Denominator Degrees of Freedom

	10	11	12	13	14	15	16	17	18
1	0.1442	0.1487	0.1526	0.1559	0.1588	0.1613	0.1635	0.1655	0.1673
2	0.1833	0.1903	0.1962	0.2014	0.2059	0.2099	0.2134	0.2165	0.2193
3	0.2072	0.2160	0.2235	0.2300	0.2358	0.2408	0.2453	0.2493	0.2529
4	0.2238	0.2339	0.2426	0.2503	0.2569	0.2629	0.2681	0.2729	0.2771
5	0.2361	0.2473	0.2570	0.2655	0.2730	0.2796	0.2855	0.2909	0.2957
6	0.2456	0.2577	0.2682	0.2774	0.2856	0.2929	0.2993	0.3052	0.3105
7	0.2532	0.2661	0.2773	0.2871	0.2959	0.3036	0.3106	0.3169	0.3226
8	0.2594	0.2729	0.2848	0.2952	0.3044	0.3126	0.3200	0.3267	0.3327
9	0.2646	0.2787	0.2910	0.3019	0.3116	0.3202	0.3280	0.3350	0.3414
10	0.2690	0.2836	0.2964	0.3077	0.3178	0.3268	0.3349	0.3422	0.3489
11	0.2729	0.2879	0.3011	0.3127	0.3231	0.3325	0.3409	0.3485	0.3554
12	0.2762	0.2916	0.3051	0.3171	0.3279	0.3375	0.3461	0.3540	0.3612
13	0.2791	0.2948	0.3087	0.3210	0.3320	0.3419	0.3508	0.3589	0.3663
14	0.2817	0.2977	0.3119	0.3245	0.3357	0.3458	0.3550	0.3633	0.3709
15	0.2840	0.3003	0.3147	0.3276	0.3391	0.3494	0.3587	0.3672	0.3750
16	0.2860	0.3026	0.3173	0.3304	0.3421	0.3526	0.3621	0.3708	0.3787
17	0.2879	0.3047	0.3196	0.3329	0.3448	0.3555	0.3652	0.3741	0.3821
18	0.2896	0.3066	0.3217	0.3352	0.3473	0.3582	0.3681	0.3770	0.3853
19	0.2911	0.3084	0.3237	0.3373	0.3496	0.3606	0.3706	0.3798	0.3881
20	0.2925	0.3100	0.3254	0.3393	0.3517	0.3629	0.3730	0.3823	0.3908
21	0.2938	0.3114	0.3271	0.3410	0.3536	0.3649	0.3752	0.3846	0.3932
22	0.2950	0.3128	0.3286	0.3427	0.3554	0.3668	0.3773	0.3868	0.3955
23	0.2961	0.3140	0.3300	0.3442	0.3570	0.3686	0.3792	0.3888	0.3976
24	0.2971	0.3152	0.3313	0.3456	0.3586	0.3703	0.3809	0.3907	0.3996
25	0.2981	0.3163	0.3325	0.3470	0.3600	0.3718	0.3826	0.3924	0.4014
26	0.2990	0.3173	0.3336	0.3482	0.3614	0.3733	0.3841	0.3940	0.4031
27	0.2998	0.3183	0.3347	0.3494	0.3626	0.3746	0.3856	0.3956	0.4048
28	0.3006	0.3191	0.3357	0.3505	0.3638	0.3759	0.3869	0.3970	0.4063
29	0.3013	0.3200	0.3366	0.3515	0.3649	0.3771	0.3882	0.3984	0.4077
30	0.3020	0.3208	0.3375	0.3525	0.3660	0.3783	0.3894	0.3997	0.4091
40	0.3072	0.3267	0.3441	0.3598	0.3739	0.3868	0.3986	0.4095	0.4194
60	0.3127	0.3329	0.3512	0.3676	0.3825	0.3962	0.4087	0.4201	0.4308
120	0.3185	0.3397	0.3588	0.3761	0.3919	0.4063	0.4196	0.4319	0.4433
∞	0.3247	0.3469	0.3670	0.3853	0.4021	0.4175	0.4317	0.4450	0.4573

H Critical Values of F (Area = 0.975) (cont.)

Numerator Degrees of Freedom

Denominator Degrees of Freedom	19	20	24	30	40	60	120
1	0.1689	0.1703	0.1749	0.1796	0.1844	0.1892	0.1941
2	0.2219	0.2242	0.2315	0.2391	0.2469	0.2548	0.2628
3	0.2562	0.2592	0.2687	0.2786	0.2887	0.2992	0.3099
4	0.2810	0.2845	0.2959	0.3077	0.3199	0.3325	0.3455
5	0.3001	0.3040	0.3170	0.3304	0.3444	0.3589	0.3740
6	0.3153	0.3197	0.3339	0.3488	0.3644	0.3806	0.3976
7	0.3278	0.3325	0.3480	0.3642	0.3811	0.3989	0.4176
8	0.3383	0.3433	0.3598	0.3772	0.3954	0.4147	0.4349
9	0.3472	0.3525	0.3700	0.3884	0.4078	0.4284	0.4501
10	0.3550	0.3605	0.3788	0.3982	0.4187	0.4405	0.4636
11	0.3617	0.3675	0.3866	0.4069	0.4284	0.4513	0.4757
12	0.3677	0.3737	0.3935	0.4146	0.4370	0.4610	0.4867
13	0.3730	0.3792	0.3997	0.4215	0.4448	0.4698	0.4966
14	0.3778	0.3842	0.4052	0.4278	0.4519	0.4778	0.5057
15	0.3821	0.3886	0.4103	0.4334	0.4583	0.4851	0.5141
16	0.3860	0.3927	0.4148	0.4386	0.4642	0.4919	0.5219
17	0.3896	0.3964	0.4190	0.4434	0.4696	0.4981	0.5291
18	0.3928	0.3998	0.4229	0.4477	0.4747	0.5039	0.5358
19	0.3958	0.4029	0.4264	0.4518	0.4793	0.5093	0.5421
20	0.3986	0.4058	0.4297	0.4555	0.4836	0.5143	0.5480
21	0.4011	0.4084	0.4327	0.4590	0.4877	0.5190	0.5535
22	0.4035	0.4109	0.4356	0.4623	0.4914	0.5234	0.5587
23	0.4057	0.4132	0.4382	0.4653	0.4950	0.5275	0.5636
24	0.4078	0.4154	0.4407	0.4682	0.4983	0.5314	0.5683
25	0.4097	0.4174	0.4430	0.4709	0.5014	0.5351	0.5727
26	0.4115	0.4193	0.4452	0.4734	0.5044	0.5386	0.5769
27	0.4132	0.4210	0.4472	0.4758	0.5072	0.5419	0.5809
28	0.4148	0.4227	0.4491	0.4780	0.5098	0.5451	0.5847
29	0.4163	0.4243	0.4510	0.4802	0.5123	0.5481	0.5883
30	0.4178	0.4258	0.4527	0.4822	0.5147	0.5509	0.5917
40	0.4286	0.4372	0.4660	0.4978	0.5333	0.5734	0.6195
60	0.4406	0.4498	0.4808	0.5155	0.5547	0.6000	0.6536
120	0.4539	0.4638	0.4975	0.5358	0.5800	0.6325	0.6980
∞	0.4688	0.4795	0.5167	0.5597	0.6108	0.6747	0.7631

H Critical Values of F (Area = 0.950)

Numerator Degrees of Freedom (columns); Denominator Degrees of Freedom (rows)

	1	2	3	4	5	6	7	8	9
1	0.0062	0.0540	0.0987	0.1297	0.1513	0.1670	0.1788	0.1881	0.1954
2	0.0050	0.0526	0.1047	0.1440	0.1728	0.1944	0.2111	0.2243	0.2349
3	0.0046	0.0522	0.1078	0.1517	0.1849	0.2102	0.2301	0.2459	0.2589
4	0.0045	0.0520	0.1097	0.1565	0.1926	0.2206	0.2427	0.2606	0.2752
5	0.0043	0.0518	0.1109	0.1598	0.1980	0.2279	0.2518	0.2712	0.2872
6	0.0043	0.0517	0.1118	0.1623	0.2020	0.2334	0.2587	0.2793	0.2964
7	0.0042	0.0517	0.1125	0.1641	0.2051	0.2377	0.2641	0.2857	0.3037
8	0.0042	0.0516	0.1131	0.1655	0.2075	0.2411	0.2684	0.2909	0.3096
9	0.0042	0.0516	0.1135	0.1667	0.2095	0.2440	0.2720	0.2951	0.3146
10	0.0041	0.0516	0.1138	0.1677	0.2112	0.2463	0.2750	0.2988	0.3187
11	0.0041	0.0515	0.1141	0.1685	0.2126	0.2483	0.2775	0.3018	0.3223
12	0.0041	0.0515	0.1144	0.1692	0.2138	0.2500	0.2797	0.3045	0.3254
13	0.0041	0.0515	0.1146	0.1697	0.2148	0.2515	0.2817	0.3068	0.3281
14	0.0041	0.0515	0.1147	0.1703	0.2157	0.2528	0.2833	0.3089	0.3305
15	0.0041	0.0515	0.1149	0.1707	0.2165	0.2539	0.2848	0.3107	0.3327
16	0.0041	0.0515	0.1150	0.1711	0.2172	0.2550	0.2862	0.3123	0.3346
17	0.0040	0.0514	0.1152	0.1715	0.2178	0.2559	0.2874	0.3138	0.3363
18	0.0040	0.0514	0.1153	0.1718	0.2184	0.2567	0.2884	0.3151	0.3378
19	0.0040	0.0514	0.1154	0.1721	0.2189	0.2574	0.2894	0.3163	0.3393
20	0.0040	0.0514	0.1155	0.1723	0.2194	0.2581	0.2903	0.3174	0.3405
21	0.0040	0.0514	0.1156	0.1726	0.2198	0.2587	0.2911	0.3184	0.3417
22	0.0040	0.0514	0.1156	0.1728	0.2202	0.2593	0.2919	0.3194	0.3428
23	0.0040	0.0514	0.1157	0.1730	0.2206	0.2598	0.2926	0.3202	0.3438
24	0.0040	0.0514	0.1158	0.1732	0.2209	0.2603	0.2932	0.3210	0.3448
25	0.0040	0.0514	0.1158	0.1733	0.2212	0.2608	0.2938	0.3217	0.3456
26	0.0040	0.0514	0.1159	0.1735	0.2215	0.2612	0.2944	0.3224	0.3465
27	0.0040	0.0514	0.1159	0.1737	0.2217	0.2616	0.2949	0.3231	0.3472
28	0.0040	0.0514	0.1160	0.1738	0.2220	0.2619	0.2954	0.3237	0.3479
29	0.0040	0.0514	0.1160	0.1739	0.2222	0.2623	0.2958	0.3242	0.3486
30	0.0040	0.0514	0.1161	0.1740	0.2224	0.2626	0.2962	0.3247	0.3492
40	0.0040	0.0514	0.1164	0.1749	0.2240	0.2650	0.2994	0.3286	0.3539
60	0.0040	0.0513	0.1167	0.1758	0.2257	0.2674	0.3026	0.3327	0.3588
120	0.0039	0.0513	0.1170	0.1767	0.2274	0.2699	0.3060	0.3370	0.3640
∞	0.0039	0.0513	0.1173	0.1777	0.2291	0.2726	0.3096	0.3416	0.3695

H Critical Values of *F* (Area = 0.950) (cont.)

Numerator Degrees of Freedom

Denominator Degrees of Freedom	10	11	12	13	14	15	16	17	18
1	0.2014	0.2064	0.2106	0.2143	0.2174	0.2201	0.2225	0.2247	0.2266
2	0.2437	0.2511	0.2574	0.2628	0.2675	0.2716	0.2752	0.2784	0.2813
3	0.2697	0.2788	0.2865	0.2932	0.2991	0.3042	0.3087	0.3128	0.3165
4	0.2875	0.2979	0.3068	0.3146	0.3213	0.3273	0.3326	0.3373	0.3416
5	0.3007	0.3121	0.3220	0.3305	0.3380	0.3447	0.3506	0.3559	0.3606
6	0.3108	0.3231	0.3338	0.3430	0.3512	0.3584	0.3648	0.3706	0.3758
7	0.3189	0.3320	0.3432	0.3531	0.3618	0.3695	0.3763	0.3825	0.3881
8	0.3256	0.3392	0.3511	0.3614	0.3706	0.3787	0.3859	0.3925	0.3984
9	0.3311	0.3453	0.3576	0.3684	0.3780	0.3865	0.3941	0.4009	0.4071
10	0.3358	0.3504	0.3632	0.3744	0.3843	0.3931	0.4010	0.4082	0.4146
11	0.3398	0.3549	0.3680	0.3796	0.3898	0.3989	0.4071	0.4145	0.4212
12	0.3433	0.3587	0.3722	0.3841	0.3946	0.4040	0.4124	0.4201	0.4270
13	0.3464	0.3621	0.3759	0.3881	0.3988	0.4085	0.4171	0.4250	0.4321
14	0.3491	0.3651	0.3792	0.3916	0.4026	0.4125	0.4214	0.4294	0.4367
15	0.3515	0.3678	0.3821	0.3948	0.4060	0.4161	0.4251	0.4333	0.4408
16	0.3537	0.3702	0.3848	0.3976	0.4091	0.4193	0.4285	0.4369	0.4445
17	0.3556	0.3724	0.3872	0.4002	0.4118	0.4222	0.4316	0.4402	0.4479
18	0.3574	0.3744	0.3893	0.4026	0.4144	0.4249	0.4345	0.4431	0.4510
19	0.3590	0.3762	0.3913	0.4047	0.4167	0.4274	0.4371	0.4459	0.4539
20	0.3605	0.3779	0.3931	0.4067	0.4188	0.4296	0.4395	0.4484	0.4565
21	0.3618	0.3794	0.3948	0.4085	0.4207	0.4317	0.4417	0.4507	0.4589
22	0.3631	0.3808	0.3964	0.4102	0.4225	0.4336	0.4437	0.4528	0.4612
23	0.3643	0.3821	0.3978	0.4117	0.4242	0.4354	0.4456	0.4548	0.4632
24	0.3653	0.3833	0.3991	0.4132	0.4258	0.4371	0.4473	0.4567	0.4652
25	0.3663	0.3844	0.4004	0.4145	0.4272	0.4386	0.4490	0.4584	0.4670
26	0.3673	0.3855	0.4015	0.4158	0.4286	0.4401	0.4505	0.4600	0.4687
27	0.3681	0.3864	0.4026	0.4170	0.4299	0.4415	0.4520	0.4616	0.4703
28	0.3689	0.3874	0.4036	0.4181	0.4311	0.4427	0.4533	0.4630	0.4718
29	0.3697	0.3882	0.4046	0.4191	0.4322	0.4439	0.4546	0.4643	0.4732
30	0.3704	0.3890	0.4055	0.4201	0.4332	0.4451	0.4558	0.4656	0.4746
40	0.3758	0.3951	0.4122	0.4275	0.4412	0.4537	0.4650	0.4753	0.4848
60	0.3815	0.4016	0.4194	0.4354	0.4499	0.4629	0.4749	0.4858	0.4959
120	0.3876	0.4085	0.4272	0.4440	0.4592	0.4730	0.4857	0.4973	0.5081
∞	0.3940	0.4159	0.4355	0.4532	0.4693	0.4841	0.4976	0.5101	0.5217

H Critical Values of *F* (Area = 0.950) (cont.)

Numerator Degrees of Freedom

Denominator Degrees of Freedom	19	20	24	30	40	60	120
1	0.2283	0.2298	0.2348	0.2398	0.2448	0.2499	0.2551
2	0.2839	0.2863	0.2939	0.3016	0.3094	0.3174	0.3255
3	0.3198	0.3227	0.3324	0.3422	0.3523	0.3626	0.3731
4	0.3454	0.3489	0.3602	0.3718	0.3837	0.3960	0.4086
5	0.3650	0.3689	0.3816	0.3947	0.4083	0.4222	0.4367
6	0.3805	0.3848	0.3987	0.4131	0.4281	0.4436	0.4598
7	0.3932	0.3978	0.4128	0.4284	0.4446	0.4616	0.4792
8	0.4038	0.4087	0.4246	0.4413	0.4587	0.4769	0.4959
9	0.4128	0.4179	0.4347	0.4523	0.4708	0.4902	0.5105
10	0.4205	0.4259	0.4435	0.4620	0.4814	0.5019	0.5234
11	0.4273	0.4329	0.4512	0.4705	0.4908	0.5122	0.5350
12	0.4333	0.4391	0.4580	0.4780	0.4991	0.5215	0.5453
13	0.4386	0.4445	0.4641	0.4847	0.5066	0.5299	0.5548
14	0.4433	0.4494	0.4695	0.4908	0.5134	0.5376	0.5634
15	0.4476	0.4539	0.4745	0.4963	0.5196	0.5445	0.5713
16	0.4515	0.4579	0.4789	0.5013	0.5253	0.5509	0.5785
17	0.4550	0.4615	0.4830	0.5059	0.5305	0.5568	0.5853
18	0.4582	0.4649	0.4868	0.5102	0.5353	0.5623	0.5916
19	0.4612	0.4679	0.4902	0.5141	0.5397	0.5674	0.5974
20	0.4639	0.4708	0.4934	0.5177	0.5438	0.5721	0.6029
21	0.4665	0.4734	0.4964	0.5210	0.5477	0.5765	0.6080
22	0.4688	0.4758	0.4991	0.5242	0.5512	0.5806	0.6129
23	0.4710	0.4781	0.5017	0.5271	0.5546	0.5845	0.6174
24	0.4730	0.4802	0.5041	0.5298	0.5577	0.5882	0.6217
25	0.4749	0.4822	0.5063	0.5324	0.5607	0.5916	0.6258
26	0.4767	0.4840	0.5084	0.5348	0.5635	0.5949	0.6297
27	0.4784	0.4858	0.5104	0.5371	0.5661	0.5980	0.6333
28	0.4799	0.4874	0.5123	0.5393	0.5686	0.6009	0.6368
29	0.4814	0.4890	0.5141	0.5413	0.5710	0.6037	0.6402
30	0.4828	0.4904	0.5157	0.5432	0.5733	0.6064	0.6434
40	0.4935	0.5016	0.5286	0.5581	0.5907	0.6272	0.6688
60	0.5052	0.5138	0.5428	0.5749	0.6108	0.6518	0.6998
120	0.5181	0.5273	0.5588	0.5940	0.6343	0.6815	0.7397
∞	0.5325	0.5425	0.5770	0.6164	0.6627	0.7198	0.7975

H Critical Values of *F* (Area = 0.900)

Numerator Degrees of Freedom (columns); Denominator Degrees of Freedom (rows)

	1	2	3	4	5	6	7	8	9
1	0.0251	0.1173	0.1806	0.2200	0.2463	0.2648	0.2786	0.2892	0.2976
2	0.0202	0.1111	0.1831	0.2312	0.2646	0.2887	0.3070	0.3212	0.3326
3	0.0187	0.1091	0.1855	0.2386	0.2763	0.3041	0.3253	0.3420	0.3555
4	0.0179	0.1082	0.1872	0.2435	0.2841	0.3144	0.3378	0.3563	0.3714
5	0.0175	0.1076	0.1884	0.2469	0.2896	0.3218	0.3468	0.3668	0.3831
6	0.0172	0.1072	0.1892	0.2494	0.2937	0.3274	0.3537	0.3748	0.3920
7	0.0170	0.1070	0.1899	0.2513	0.2969	0.3317	0.3591	0.3811	0.3992
8	0.0168	0.1068	0.1904	0.2528	0.2995	0.3352	0.3634	0.3862	0.4050
9	0.0167	0.1066	0.1908	0.2541	0.3015	0.3381	0.3670	0.3904	0.4098
10	0.0166	0.1065	0.1912	0.2551	0.3033	0.3405	0.3700	0.3940	0.4139
11	0.0165	0.1064	0.1915	0.2560	0.3047	0.3425	0.3726	0.3971	0.4173
12	0.0165	0.1063	0.1917	0.2567	0.3060	0.3443	0.3748	0.3997	0.4204
13	0.0164	0.1062	0.1919	0.2573	0.3071	0.3458	0.3767	0.4020	0.4230
14	0.0164	0.1062	0.1921	0.2579	0.3080	0.3471	0.3784	0.4040	0.4253
15	0.0163	0.1061	0.1923	0.2584	0.3088	0.3483	0.3799	0.4058	0.4274
16	0.0163	0.1061	0.1924	0.2588	0.3096	0.3493	0.3812	0.4074	0.4293
17	0.0163	0.1060	0.1926	0.2592	0.3102	0.3503	0.3824	0.4089	0.4309
18	0.0162	0.1060	0.1927	0.2595	0.3108	0.3511	0.3835	0.4102	0.4325
19	0.0162	0.1059	0.1928	0.2598	0.3114	0.3519	0.3845	0.4114	0.4338
20	0.0162	0.1059	0.1929	0.2601	0.3119	0.3526	0.3854	0.4124	0.4351
21	0.0162	0.1059	0.1930	0.2604	0.3123	0.3532	0.3862	0.4134	0.4363
22	0.0162	0.1059	0.1930	0.2606	0.3127	0.3538	0.3870	0.4143	0.4373
23	0.0161	0.1058	0.1931	0.2608	0.3131	0.3543	0.3877	0.4152	0.4383
24	0.0161	0.1058	0.1932	0.2610	0.3134	0.3548	0.3883	0.4160	0.4392
25	0.0161	0.1058	0.1932	0.2612	0.3137	0.3553	0.3889	0.4167	0.4401
26	0.0161	0.1058	0.1933	0.2614	0.3140	0.3557	0.3894	0.4174	0.4408
27	0.0161	0.1058	0.1934	0.2615	0.3143	0.3561	0.3900	0.4180	0.4416
28	0.0161	0.1058	0.1934	0.2617	0.3146	0.3565	0.3904	0.4186	0.4423
29	0.0161	0.1057	0.1935	0.2618	0.3148	0.3568	0.3909	0.4191	0.4429
30	0.0161	0.1057	0.1935	0.2620	0.3151	0.3571	0.3913	0.4196	0.4435
40	0.0160	0.1056	0.1938	0.2629	0.3167	0.3596	0.3945	0.4235	0.4480
60	0.0159	0.1055	0.1941	0.2639	0.3184	0.3621	0.3977	0.4275	0.4528
120	0.0159	0.1055	0.1945	0.2649	0.3202	0.3647	0.4012	0.4317	0.4578
∞	0.0158	0.1054	0.1948	0.2659	0.3221	0.3674	0.4047	0.4362	0.4631

H Critical Values of *F* (Area = 0.900) (cont.)

Numerator Degrees of Freedom

Denominator Degrees of Freedom	10	11	12	13	14	15	16	17	18
1	0.3044	0.3101	0.3148	0.3189	0.3224	0.3254	0.3281	0.3304	0.3326
2	0.3419	0.3497	0.3563	0.3619	0.3668	0.3710	0.3748	0.3781	0.3811
3	0.3666	0.3759	0.3838	0.3906	0.3965	0.4016	0.4062	0.4103	0.4139
4	0.3838	0.3943	0.4032	0.4109	0.4176	0.4235	0.4287	0.4333	0.4375
5	0.3966	0.4080	0.4177	0.4261	0.4335	0.4399	0.4457	0.4508	0.4554
6	0.4064	0.4186	0.4290	0.4380	0.4459	0.4529	0.4591	0.4646	0.4696
7	0.4143	0.4271	0.4381	0.4476	0.4560	0.4634	0.4699	0.4758	0.4811
8	0.4207	0.4340	0.4455	0.4555	0.4643	0.4720	0.4789	0.4851	0.4907
9	0.4260	0.4399	0.4518	0.4621	0.4713	0.4793	0.4865	0.4930	0.4988
10	0.4306	0.4448	0.4571	0.4678	0.4772	0.4856	0.4931	0.4998	0.5058
11	0.4344	0.4490	0.4617	0.4727	0.4824	0.4910	0.4987	0.5056	0.5119
12	0.4378	0.4527	0.4657	0.4770	0.4869	0.4958	0.5037	0.5108	0.5172
13	0.4408	0.4560	0.4692	0.4807	0.4909	0.5000	0.5081	0.5154	0.5220
14	0.4434	0.4589	0.4723	0.4841	0.4945	0.5037	0.5120	0.5194	0.5262
15	0.4457	0.4614	0.4751	0.4871	0.4976	0.5070	0.5155	0.5231	0.5300
16	0.4478	0.4638	0.4776	0.4897	0.5005	0.5101	0.5187	0.5264	0.5334
17	0.4497	0.4658	0.4799	0.4922	0.5031	0.5128	0.5215	0.5294	0.5365
18	0.4514	0.4677	0.4819	0.4944	0.5054	0.5153	0.5241	0.5321	0.5394
19	0.4530	0.4694	0.4838	0.4964	0.5076	0.5176	0.5265	0.5346	0.5420
20	0.4544	0.4710	0.4855	0.4983	0.5096	0.5197	0.5287	0.5370	0.5444
21	0.4557	0.4725	0.4871	0.5000	0.5114	0.5216	0.5308	0.5391	0.5466
22	0.4569	0.4738	0.4886	0.5015	0.5131	0.5234	0.5327	0.5411	0.5487
23	0.4580	0.4750	0.4899	0.5030	0.5146	0.5250	0.5344	0.5429	0.5506
24	0.4590	0.4762	0.4912	0.5044	0.5161	0.5266	0.5360	0.5446	0.5524
25	0.4600	0.4773	0.4923	0.5056	0.5174	0.5280	0.5375	0.5462	0.5540
26	0.4609	0.4783	0.4934	0.5068	0.5187	0.5294	0.5390	0.5477	0.5556
27	0.4617	0.4792	0.4944	0.5079	0.5199	0.5306	0.5403	0.5491	0.5571
28	0.4625	0.4801	0.4954	0.5089	0.5210	0.5318	0.5415	0.5504	0.5584
29	0.4633	0.4809	0.4963	0.5099	0.5220	0.5329	0.5427	0.5516	0.5597
30	0.4639	0.4816	0.4971	0.5108	0.5230	0.5340	0.5438	0.5528	0.5610
40	0.4691	0.4874	0.5035	0.5177	0.5305	0.5419	0.5522	0.5616	0.5702
60	0.4746	0.4936	0.5103	0.5251	0.5384	0.5504	0.5613	0.5712	0.5803
120	0.4804	0.5001	0.5175	0.5331	0.5470	0.5597	0.5712	0.5817	0.5914
∞	0.4865	0.5071	0.5253	0.5417	0.5564	0.5698	0.5820	0.5932	0.6036

H Critical Values of *F* (Area = 0.900) (cont.)

Numerator Degrees of Freedom

Denominator Degrees of Freedom

	19	20	24	30	40	60	120
1	0.3345	0.3362	0.3416	0.3471	0.3527	0.3583	0.3639
2	0.3838	0.3862	0.3940	0.4018	0.4098	0.4178	0.4260
3	0.4172	0.4202	0.4297	0.4394	0.4492	0.4593	0.4695
4	0.4412	0.4447	0.4556	0.4668	0.4783	0.4900	0.5019
5	0.4596	0.4633	0.4755	0.4880	0.5008	0.5140	0.5275
6	0.4741	0.4782	0.4914	0.5050	0.5190	0.5334	0.5483
7	0.4859	0.4903	0.5044	0.5190	0.5340	0.5496	0.5658
8	0.4958	0.5004	0.5153	0.5308	0.5468	0.5634	0.5807
9	0.5041	0.5089	0.5246	0.5408	0.5578	0.5754	0.5937
10	0.5113	0.5163	0.5326	0.5496	0.5673	0.5858	0.6052
11	0.5176	0.5228	0.5397	0.5573	0.5757	0.5951	0.6154
12	0.5231	0.5284	0.5459	0.5641	0.5832	0.6033	0.6245
13	0.5280	0.5335	0.5514	0.5702	0.5900	0.6108	0.6328
14	0.5323	0.5380	0.5564	0.5757	0.5960	0.6175	0.6403
15	0.5363	0.5420	0.5608	0.5806	0.6015	0.6237	0.6472
16	0.5398	0.5457	0.5649	0.5851	0.6066	0.6293	0.6536
17	0.5431	0.5490	0.5686	0.5893	0.6112	0.6345	0.6595
18	0.5460	0.5521	0.5720	0.5930	0.6154	0.6393	0.6649
19	0.5487	0.5549	0.5751	0.5965	0.6194	0.6437	0.6700
20	0.5512	0.5575	0.5780	0.5998	0.6230	0.6479	0.6747
21	0.5535	0.5598	0.5807	0.6028	0.6264	0.6517	0.6792
22	0.5557	0.5621	0.5831	0.6056	0.6295	0.6554	0.6833
23	0.5576	0.5641	0.5854	0.6082	0.6325	0.6587	0.6873
24	0.5595	0.5660	0.5876	0.6106	0.6353	0.6619	0.6910
25	0.5612	0.5678	0.5896	0.6129	0.6379	0.6649	0.6945
26	0.5629	0.5695	0.5915	0.6150	0.6403	0.6678	0.6978
27	0.5644	0.5711	0.5933	0.6171	0.6427	0.6705	0.7010
28	0.5658	0.5726	0.5950	0.6190	0.6449	0.6730	0.7040
29	0.5672	0.5740	0.5966	0.6208	0.6469	0.6754	0.7068
30	0.5684	0.5753	0.5980	0.6225	0.6489	0.6777	0.7095
40	0.5781	0.5854	0.6095	0.6356	0.6642	0.6957	0.7312
60	0.5887	0.5964	0.6222	0.6504	0.6816	0.7167	0.7574
120	0.6003	0.6085	0.6364	0.6672	0.7019	0.7421	0.7908
∞	0.6132	0.6221	0.6524	0.6866	0.7263	0.7743	0.8385

H Critical Values of F (Area = 0.100)

Numerator Degrees of Freedom (columns); Denominator Degrees of Freedom (rows)

	1	2	3	4	5	6	7	8	9
1	39.8635	49.5000	53.5932	55.8330	57.2401	58.2044	58.9060	59.4390	59.8576
2	8.5263	9.0000	9.1618	9.2434	9.2926	9.3255	9.3491	9.3668	9.3805
3	5.5383	5.4624	5.3908	5.3426	5.3092	5.2847	5.2662	5.2517	5.2400
4	4.5448	4.3246	4.1909	4.1072	4.0506	4.0097	3.9790	3.9549	3.9357
5	4.0604	3.7797	3.6195	3.5202	3.4530	3.4045	3.3679	3.3393	3.3163
6	3.7759	3.4633	3.2888	3.1808	3.1075	3.0546	3.0145	2.9830	2.9577
7	3.5894	3.2574	3.0741	2.9605	2.8833	2.8274	2.7849	2.7516	2.7247
8	3.4579	3.1131	2.9238	2.8064	2.7264	2.6683	2.6241	2.5893	2.5612
9	3.3603	3.0065	2.8129	2.6927	2.6106	2.5509	2.5053	2.4694	2.4403
10	3.2850	2.9245	2.7277	2.6053	2.5216	2.4606	2.4140	2.3772	2.3473
11	3.2252	2.8595	2.6602	2.5362	2.4512	2.3891	2.3416	2.3040	2.2735
12	3.1765	2.8068	2.6055	2.4801	2.3940	2.3310	2.2828	2.2446	2.2135
13	3.1362	2.7632	2.5603	2.4337	2.3467	2.2830	2.2341	2.1953	2.1638
14	3.1022	2.7265	2.5222	2.3947	2.3069	2.2426	2.1931	2.1539	2.1220
15	3.0732	2.6952	2.4898	2.3614	2.2730	2.2081	2.1582	2.1185	2.0862
16	3.0481	2.6682	2.4618	2.3327	2.2438	2.1783	2.1280	2.0880	2.0553
17	3.0262	2.6446	2.4374	2.3077	2.2183	2.1524	2.1017	2.0613	2.0284
18	3.0070	2.6239	2.4160	2.2858	2.1958	2.1296	2.0785	2.0379	2.0047
19	2.9899	2.6056	2.3970	2.2663	2.1760	2.1094	2.0580	2.0171	1.9836
20	2.9747	2.5893	2.3801	2.2489	2.1582	2.0913	2.0397	1.9985	1.9649
21	2.9610	2.5746	2.3649	2.2333	2.1423	2.0751	2.0233	1.9819	1.9480
22	2.9486	2.5613	2.3512	2.2193	2.1279	2.0605	2.0084	1.9668	1.9327
23	2.9374	2.5493	2.3387	2.2065	2.1149	2.0472	1.9949	1.9531	1.9189
24	2.9271	2.5383	2.3274	2.1949	2.1030	2.0351	1.9826	1.9407	1.9063
25	2.9177	2.5283	2.3170	2.1842	2.0922	2.0241	1.9714	1.9292	1.8947
26	2.9091	2.5191	2.3075	2.1745	2.0822	2.0139	1.9610	1.9188	1.8841
27	2.9012	2.5106	2.2987	2.1655	2.0730	2.0045	1.9515	1.9091	1.8743
28	2.8938	2.5028	2.2906	2.1571	2.0645	1.9959	1.9427	1.9001	1.8652
29	2.8870	2.4955	2.2831	2.1494	2.0566	1.9878	1.9345	1.8918	1.8568
30	2.8807	2.4887	2.2761	2.1422	2.0492	1.9803	1.9269	1.8841	1.8490
40	2.8354	2.4404	2.2261	2.0909	1.9968	1.9269	1.8725	1.8289	1.7929
60	2.7911	2.3933	2.1774	2.0410	1.9457	1.8747	1.8194	1.7748	1.7380
120	2.7478	2.3473	2.1300	1.9923	1.8959	1.8238	1.7675	1.7220	1.6842
∞	2.7055	2.3026	2.0838	1.9449	1.8473	1.7741	1.7167	1.6702	1.6315

H Critical Values of *F* (Area = 0.100) (cont.)

Numerator Degrees of Freedom

Denominator Degrees of Freedom

	10	11	12	13	14	15	16	17	18
1	60.1950	60.4727	60.7052	60.9028	61.0727	61.2203	61.3499	61.4644	61.5664
2	9.3916	9.4006	9.4081	9.4145	9.4200	9.4247	9.4289	9.4325	9.4358
3	5.2304	5.2224	5.2156	5.2098	5.2047	5.2003	5.1964	5.1929	5.1898
4	3.9199	3.9067	3.8955	3.8859	3.8776	3.8704	3.8639	3.8582	3.8531
5	3.2974	3.2816	3.2682	3.2567	3.2468	3.2380	3.2303	3.2234	3.2172
6	2.9369	2.9195	2.9047	2.8920	2.8809	2.8712	2.8626	2.8550	2.8481
7	2.7025	2.6839	2.6681	2.6545	2.6426	2.6322	2.6230	2.6148	2.6074
8	2.5380	2.5186	2.5020	2.4876	2.4752	2.4642	2.4545	2.4458	2.4380
9	2.4163	2.3961	2.3789	2.3640	2.3510	2.3396	2.3295	2.3205	2.3123
10	2.3226	2.3018	2.2841	2.2687	2.2553	2.2435	2.2330	2.2237	2.2153
11	2.2482	2.2269	2.2087	2.1930	2.1792	2.1671	2.1563	2.1467	2.1380
12	2.1878	2.1660	2.1474	2.1313	2.1173	2.1049	2.0938	2.0839	2.0750
13	2.1376	2.1155	2.0966	2.0802	2.0658	2.0532	2.0419	2.0318	2.0227
14	2.0954	2.0729	2.0537	2.0370	2.0224	2.0095	1.9981	1.9878	1.9785
15	2.0593	2.0366	2.0171	2.0001	1.9853	1.9722	1.9605	1.9501	1.9407
16	2.0281	2.0051	1.9854	1.9682	1.9532	1.9399	1.9281	1.9175	1.9079
17	2.0009	1.9777	1.9577	1.9404	1.9252	1.9117	1.8997	1.8889	1.8792
18	1.9770	1.9535	1.9333	1.9158	1.9004	1.8868	1.8747	1.8638	1.8539
19	1.9557	1.9321	1.9117	1.8940	1.8785	1.8647	1.8524	1.8414	1.8314
20	1.9367	1.9129	1.8924	1.8745	1.8588	1.8449	1.8325	1.8214	1.8113
21	1.9197	1.8956	1.8750	1.8570	1.8412	1.8271	1.8146	1.8034	1.7932
22	1.9043	1.8801	1.8593	1.8411	1.8252	1.8111	1.7984	1.7871	1.7768
23	1.8903	1.8659	1.8450	1.8267	1.8107	1.7964	1.7837	1.7723	1.7619
24	1.8775	1.8530	1.8319	1.8136	1.7974	1.7831	1.7703	1.7587	1.7483
25	1.8658	1.8412	1.8200	1.8015	1.7853	1.7708	1.7579	1.7463	1.7358
26	1.8550	1.8303	1.8090	1.7904	1.7741	1.7596	1.7466	1.7349	1.7243
27	1.8451	1.8203	1.7989	1.7802	1.7638	1.7492	1.7361	1.7243	1.7137
28	1.8359	1.8110	1.7895	1.7708	1.7542	1.7395	1.7264	1.7146	1.7039
29	1.8274	1.8024	1.7808	1.7620	1.7454	1.7306	1.7174	1.7055	1.6947
30	1.8195	1.7944	1.7727	1.7538	1.7371	1.7223	1.7090	1.6970	1.6862
40	1.7627	1.7369	1.7146	1.6950	1.6778	1.6624	1.6486	1.6362	1.6249
60	1.7070	1.6805	1.6574	1.6372	1.6193	1.6034	1.5890	1.5760	1.5642
120	1.6524	1.6250	1.6012	1.5803	1.5617	1.5450	1.5300	1.5164	1.5039
∞	1.5987	1.5705	1.5458	1.5240	1.5046	1.4871	1.4714	1.4570	1.4439

H Critical Values of *F* (Area = 0.100) (cont.)

Numerator Degrees of Freedom

Denominator Degrees of Freedom	19	20	24	30	40	60	120
1	61.6579	61.7403	62.0020	62.2650	62.5291	62.7943	63.0606
2	9.4387	9.4413	9.4496	9.4579	9.4662	9.4746	9.4829
3	5.1870	5.1845	5.1764	5.1681	5.1597	5.1512	5.1425
4	3.8485	3.8443	3.8310	3.8174	3.8036	3.7896	3.7753
5	3.2117	3.2067	3.1905	3.1741	3.1573	3.1402	3.1228
6	2.8419	2.8363	2.8183	2.8000	2.7812	2.7620	2.7423
7	2.6008	2.5947	2.5753	2.5555	2.5351	2.5142	2.4928
8	2.4310	2.4246	2.4041	2.3830	2.3614	2.3391	2.3162
9	2.3050	2.2983	2.2768	2.2547	2.2320	2.2085	2.1843
10	2.2077	2.2007	2.1784	2.1554	2.1317	2.1072	2.0818
11	2.1302	2.1230	2.1000	2.0762	2.0516	2.0261	1.9997
12	2.0670	2.0597	2.0360	2.0115	1.9861	1.9597	1.9323
13	2.0145	2.0070	1.9827	1.9576	1.9315	1.9043	1.8759
14	1.9701	1.9625	1.9377	1.9119	1.8852	1.8572	1.8280
15	1.9321	1.9243	1.8990	1.8728	1.8454	1.8168	1.7867
16	1.8992	1.8913	1.8656	1.8388	1.8108	1.7816	1.7507
17	1.8704	1.8624	1.8362	1.8090	1.7805	1.7506	1.7191
18	1.8450	1.8368	1.8103	1.7827	1.7537	1.7232	1.6910
19	1.8224	1.8142	1.7873	1.7592	1.7298	1.6988	1.6659
20	1.8022	1.7938	1.7667	1.7382	1.7083	1.6768	1.6433
21	1.7840	1.7756	1.7481	1.7193	1.6890	1.6569	1.6228
22	1.7675	1.7590	1.7312	1.7021	1.6714	1.6389	1.6041
23	1.7525	1.7439	1.7159	1.6864	1.6554	1.6224	1.5871
24	1.7388	1.7302	1.7019	1.6721	1.6407	1.6073	1.5715
25	1.7263	1.7175	1.6890	1.6589	1.6272	1.5934	1.5570
26	1.7147	1.7059	1.6771	1.6468	1.6147	1.5805	1.5437
27	1.7040	1.6951	1.6662	1.6356	1.6032	1.5686	1.5313
28	1.6941	1.6852	1.6560	1.6252	1.5925	1.5575	1.5198
29	1.6849	1.6759	1.6465	1.6155	1.5825	1.5472	1.5090
30	1.6763	1.6673	1.6377	1.6065	1.5732	1.5376	1.4989
40	1.6146	1.6052	1.5741	1.5411	1.5056	1.4672	1.4248
60	1.5534	1.5435	1.5107	1.4755	1.4373	1.3952	1.3476
120	1.4926	1.4821	1.4472	1.4094	1.3676	1.3203	1.2646
∞	1.4318	1.4206	1.3832	1.3419	1.2951	1.2400	1.1686

H Critical Values of *F* (Area = 0.050)

Numerator Degrees of Freedom

Denominator Degrees of Freedom	1	2	3	4	5	6	7	8	9
1	161.4476	199.5000	215.7073	224.5832	230.1619	233.9860	236.7684	238.8827	240.5433
2	18.5128	19.0000	19.1643	19.2468	19.2964	19.3295	19.3532	19.3710	19.3848
3	10.1280	9.5521	9.2766	9.1172	9.0135	8.9406	8.8867	8.8452	8.8123
4	7.7086	6.9443	6.5914	6.3882	6.2561	6.1631	6.0942	6.0410	5.9988
5	6.6079	5.7861	5.4095	5.1922	5.0503	4.9503	4.8759	4.8183	4.7725
6	5.9874	5.1433	4.7571	4.5337	4.3874	4.2839	4.2067	4.1468	4.0990
7	5.5914	4.7374	4.3468	4.1203	3.9715	3.8660	3.7870	3.7257	3.6767
8	5.3177	4.4590	4.0662	3.8379	3.6875	3.5806	3.5005	3.4381	3.3881
9	5.1174	4.2565	3.8625	3.6331	3.4817	3.3738	3.2927	3.2296	3.1789
10	4.9646	4.1028	3.7083	3.4780	3.3258	3.2172	3.1355	3.0717	3.0204
11	4.8443	3.9823	3.5874	3.3567	3.2039	3.0946	3.0123	2.9480	2.8962
12	4.7472	3.8853	3.4903	3.2592	3.1059	2.9961	2.9134	2.8486	2.7964
13	4.6672	3.8056	3.4105	3.1791	3.0254	2.9153	2.8321	2.7669	2.7144
14	4.6001	3.7389	3.3439	3.1122	2.9582	2.8477	2.7642	2.6987	2.6458
15	4.5431	3.6823	3.2874	3.0556	2.9013	2.7905	2.7066	2.6408	2.5876
16	4.4940	3.6337	3.2389	3.0069	2.8524	2.7413	2.6572	2.5911	2.5377
17	4.4513	3.5915	3.1968	2.9647	2.8100	2.6987	2.6143	2.5480	2.4943
18	4.4139	3.5546	3.1599	2.9277	2.7729	2.6613	2.5767	2.5102	2.4563
19	4.3807	3.5219	3.1274	2.8951	2.7401	2.6283	2.5435	2.4768	2.4227
20	4.3512	3.4928	3.0984	2.8661	2.7109	2.5990	2.5140	2.4471	2.3928
21	4.3248	3.4668	3.0725	2.8401	2.6848	2.5727	2.4876	2.4205	2.3660
22	4.3009	3.4434	3.0491	2.8167	2.6613	2.5491	2.4638	2.3965	2.3419
23	4.2793	3.4221	3.0280	2.7955	2.6400	2.5277	2.4422	2.3748	2.3201
24	4.2597	3.4028	3.0088	2.7763	2.6207	2.5082	2.4226	2.3551	2.3002
25	4.2417	3.3852	2.9912	2.7587	2.6030	2.4904	2.4047	2.3371	2.2821
26	4.2252	3.3690	2.9752	2.7426	2.5868	2.4741	2.3883	2.3205	2.2655
27	4.2100	3.3541	2.9604	2.7278	2.5719	2.4591	2.3732	2.3053	2.2501
28	4.1960	3.3404	2.9467	2.7141	2.5581	2.4453	2.3593	2.2913	2.2360
29	4.1830	3.3277	2.9340	2.7014	2.5454	2.4324	2.3463	2.2783	2.2229
30	4.1709	3.3158	2.9223	2.6896	2.5336	2.4205	2.3343	2.2662	2.2107
40	4.0847	3.2317	2.8387	2.6060	2.4495	2.3359	2.2490	2.1802	2.1240
60	4.0012	3.1504	2.7581	2.5252	2.3683	2.2541	2.1665	2.0970	2.0401
120	3.9201	3.0718	2.6802	2.4472	2.2899	2.1750	2.0868	2.0164	1.9588
∞	3.8415	2.9957	2.6049	2.3719	2.2141	2.0986	2.0096	1.9384	1.8799

H Critical Values of *F* (Area = 0.050) (cont.)

Numerator Degrees of Freedom (columns); Denominator Degrees of Freedom (rows)

	10	11	12	13	14	15	16	17	18
1	241.8817	242.9835	243.9060	244.6898	245.3640	245.9499	246.4639	246.9184	247.3232
2	19.3959	19.4050	19.4125	19.4189	19.4244	19.4291	19.4333	19.4370	19.4402
3	8.7855	8.7633	8.7446	8.7287	8.7149	8.7029	8.6923	8.6829	8.6745
4	5.9644	5.9358	5.9117	5.8911	5.8733	5.8578	5.8441	5.8320	5.8211
5	4.7351	4.7040	4.6777	4.6552	4.6358	4.6188	4.6038	4.5904	4.5785
6	4.0600	4.0274	3.9999	3.9764	3.9559	3.9381	3.9223	3.9083	3.8957
7	3.6365	3.6030	3.5747	3.5503	3.5292	3.5107	3.4944	3.4799	3.4669
8	3.3472	3.3130	3.2839	3.2590	3.2374	3.2184	3.2016	3.1867	3.1733
9	3.1373	3.1025	3.0729	3.0475	3.0255	3.0061	2.9890	2.9737	2.9600
10	2.9782	2.9430	2.9130	2.8872	2.8647	2.8450	2.8276	2.8120	2.7980
11	2.8536	2.8179	2.7876	2.7614	2.7386	2.7186	2.7009	2.6851	2.6709
12	2.7534	2.7173	2.6866	2.6602	2.6371	2.6169	2.5989	2.5828	2.5684
13	2.6710	2.6347	2.6037	2.5769	2.5536	2.5331	2.5149	2.4987	2.4841
14	2.6022	2.5655	2.5342	2.5073	2.4837	2.4630	2.4446	2.4282	2.4134
15	2.5437	2.5068	2.4753	2.4481	2.4244	2.4034	2.3849	2.3683	2.3533
16	2.4935	2.4564	2.4247	2.3973	2.3733	2.3522	2.3335	2.3167	2.3016
17	2.4499	2.4126	2.3807	2.3531	2.3290	2.3077	2.2888	2.2719	2.2567
18	2.4117	2.3742	2.3421	2.3143	2.2900	2.2686	2.2496	2.2325	2.2172
19	2.3779	2.3402	2.3080	2.2800	2.2556	2.2341	2.2149	2.1977	2.1823
20	2.3479	2.3100	2.2776	2.2495	2.2250	2.2033	2.1840	2.1667	2.1511
21	2.3210	2.2829	2.2504	2.2222	2.1975	2.1757	2.1563	2.1389	2.1232
22	2.2967	2.2585	2.2258	2.1975	2.1727	2.1508	2.1313	2.1138	2.0980
23	2.2747	2.2364	2.2036	2.1752	2.1502	2.1282	2.1086	2.0910	2.0751
24	2.2547	2.2163	2.1834	2.1548	2.1298	2.1077	2.0880	2.0703	2.0543
25	2.2365	2.1979	2.1649	2.1362	2.1111	2.0889	2.0691	2.0513	2.0353
26	2.2197	2.1811	2.1479	2.1192	2.0939	2.0716	2.0518	2.0339	2.0178
27	2.2043	2.1655	2.1323	2.1035	2.0781	2.0558	2.0358	2.0179	2.0017
28	2.1900	2.1512	2.1179	2.0889	2.0635	2.0411	2.0210	2.0030	1.9868
29	2.1768	2.1379	2.1045	2.0755	2.0500	2.0275	2.0073	1.9893	1.9730
30	2.1646	2.1256	2.0921	2.0630	2.0374	2.0148	1.9946	1.9765	1.9601
40	2.0772	2.0376	2.0035	1.9738	1.9476	1.9245	1.9037	1.8851	1.8682
60	1.9926	1.9522	1.9174	1.8870	1.8602	1.8364	1.8151	1.7959	1.7784
120	1.9105	1.8693	1.8337	1.8026	1.7750	1.7505	1.7285	1.7085	1.6904
∞	1.8307	1.7887	1.7522	1.7202	1.6918	1.6664	1.6435	1.6228	1.6039

H Critical Values of *F* (Area = 0.050) (cont.)

Numerator Degrees of Freedom

Denominator Degrees of Freedom	19	20	24	30	40	60	120
1	247.6861	248.0131	249.0518	250.0951	251.1432	252.1957	253.2529
2	19.4431	19.4458	19.4541	19.4624	19.4707	19.4791	19.4874
3	8.6670	8.6602	8.6385	8.6166	8.5944	8.5720	8.5494
4	5.8114	5.8025	5.7744	5.7459	5.7170	5.6877	5.6581
5	4.5678	4.5581	4.5272	4.4957	4.4638	4.4314	4.3985
6	3.8844	3.8742	3.8415	3.8082	3.7743	3.7398	3.7047
7	3.4551	3.4445	3.4105	3.3758	3.3404	3.3043	3.2674
8	3.1613	3.1503	3.1152	3.0794	3.0428	3.0053	2.9669
9	2.9477	2.9365	2.9005	2.8637	2.8259	2.7872	2.7475
10	2.7854	2.7740	2.7372	2.6996	2.6609	2.6211	2.5801
11	2.6581	2.6464	2.6090	2.5705	2.5309	2.4901	2.4480
12	2.5554	2.5436	2.5055	2.4663	2.4259	2.3842	2.3410
13	2.4709	2.4589	2.4202	2.3803	2.3392	2.2966	2.2524
14	2.4000	2.3879	2.3487	2.3082	2.2664	2.2229	2.1778
15	2.3398	2.3275	2.2878	2.2468	2.2043	2.1601	2.1141
16	2.2880	2.2756	2.2354	2.1938	2.1507	2.1058	2.0589
17	2.2429	2.2304	2.1898	2.1477	2.1040	2.0584	2.0107
18	2.2033	2.1906	2.1497	2.1071	2.0629	2.0166	1.9681
19	2.1683	2.1555	2.1141	2.0712	2.0264	1.9795	1.9302
20	2.1370	2.1242	2.0825	2.0391	1.9938	1.9464	1.8963
21	2.1090	2.0960	2.0540	2.0102	1.9645	1.9165	1.8657
22	2.0837	2.0707	2.0283	1.9842	1.9380	1.8894	1.8380
23	2.0608	2.0476	2.0050	1.9605	1.9139	1.8648	1.8128
24	2.0399	2.0267	1.9838	1.9390	1.8920	1.8424	1.7896
25	2.0207	2.0075	1.9643	1.9192	1.8718	1.8217	1.7684
26	2.0032	1.9898	1.9464	1.9010	1.8533	1.8027	1.7488
27	1.9870	1.9736	1.9299	1.8842	1.8361	1.7851	1.7306
28	1.9720	1.9586	1.9147	1.8687	1.8203	1.7689	1.7138
29	1.9581	1.9446	1.9005	1.8543	1.8055	1.7537	1.6981
30	1.9452	1.9317	1.8874	1.8409	1.7918	1.7396	1.6835
40	1.8529	1.8389	1.7929	1.7444	1.6928	1.6373	1.5766
60	1.7625	1.7480	1.7001	1.6491	1.5943	1.5343	1.4673
120	1.6739	1.6587	1.6084	1.5543	1.4952	1.4290	1.3519
∞	1.5865	1.5705	1.5173	1.4591	1.3940	1.3180	1.2214

H Critical Values of *F* (Area = 0.025)

Numerator Degrees of Freedom

Denominator Degrees of Freedom	1	2	3	4	5	6	7	8	9
1	647.7890	799.5000	864.1630	899.5833	921.8479	937.1111	948.2169	956.6562	963.2846
2	38.5063	39.0000	39.1655	39.2484	39.2982	39.3315	39.3552	39.3730	39.3869
3	17.4434	16.0441	15.4392	15.1010	14.8848	14.7347	14.6244	14.5399	14.4731
4	12.2179	10.6491	9.9792	9.6045	9.3645	9.1973	9.0741	8.9796	8.9047
5	10.0070	8.4336	7.7636	7.3879	7.1464	6.9777	6.8531	6.7572	6.6811
6	8.8131	7.2599	6.5988	6.2272	5.9876	5.8198	5.6955	5.5996	5.5234
7	8.0727	6.5415	5.8898	5.5226	5.2852	5.1186	4.9949	4.8993	4.8232
8	7.5709	6.0595	5.4160	5.0526	4.8173	4.6517	4.5286	4.4333	4.3572
9	7.2093	5.7147	5.0781	4.7181	4.4844	4.3197	4.1970	4.1020	4.0260
10	6.9367	5.4564	4.8256	4.4683	4.2361	4.0721	3.9498	3.8549	3.7790
11	6.7241	5.2559	4.6300	4.2751	4.0440	3.8807	3.7586	3.6638	3.5879
12	6.5538	5.0959	4.4742	4.1212	3.8911	3.7283	3.6065	3.5118	3.4358
13	6.4143	4.9653	4.3472	3.9959	3.7667	3.6043	3.4827	3.3880	3.3120
14	6.2979	4.8567	4.2417	3.8919	3.6634	3.5014	3.3799	3.2853	3.2093
15	6.1995	4.7650	4.1528	3.8043	3.5764	3.4147	3.2934	3.1987	3.1227
16	6.1151	4.6867	4.0768	3.7294	3.5021	3.3406	3.2194	3.1248	3.0488
17	6.0420	4.6189	4.0112	3.6648	3.4379	3.2767	3.1556	3.0610	2.9849
18	5.9781	4.5597	3.9539	3.6083	3.3820	3.2209	3.0999	3.0053	2.9291
19	5.9216	4.5075	3.9034	3.5587	3.3327	3.1718	3.0509	2.9563	2.8801
20	5.8715	4.4613	3.8587	3.5147	3.2891	3.1283	3.0074	2.9128	2.8365
21	5.8266	4.4199	3.8188	3.4754	3.2501	3.0895	2.9686	2.8740	2.7977
22	5.7863	4.3828	3.7829	3.4401	3.2151	3.0546	2.9338	2.8392	2.7628
23	5.7498	4.3492	3.7505	3.4083	3.1835	3.0232	2.9023	2.8077	2.7313
24	5.7166	4.3187	3.7211	3.3794	3.1548	2.9946	2.8738	2.7791	2.7027
25	5.6864	4.2909	3.6943	3.3530	3.1287	2.9685	2.8478	2.7531	2.6766
26	5.6586	4.2655	3.6697	3.3289	3.1048	2.9447	2.8240	2.7293	2.6528
27	5.6331	4.2421	3.6472	3.3067	3.0828	2.9228	2.8021	2.7074	2.6309
28	5.6096	4.2205	3.6264	3.2863	3.0626	2.9027	2.7820	2.6872	2.6106
29	5.5878	4.2006	3.6072	3.2674	3.0438	2.8840	2.7633	2.6686	2.5919
30	5.5675	4.1821	3.5894	3.2499	3.0265	2.8667	2.7460	2.6513	2.5746
40	5.4239	4.0510	3.4633	3.1261	2.9037	2.7444	2.6238	2.5289	2.4519
60	5.2856	3.9253	3.3425	3.0077	2.7863	2.6274	2.5068	2.4117	2.3344
120	5.1523	3.8046	3.2269	2.8943	2.6740	2.5154	2.3948	2.2994	2.2217
∞	5.0239	3.6889	3.1161	2.7858	2.5665	2.4082	2.2876	2.1918	2.1137

H Critical Values of *F* (Area = 0.025) (cont.)

Numerator Degrees of Freedom

Denominator Degrees of Freedom	10	11	12	13	14	15	16	17	18
1	968.6274	973.0252	976.7079	979.8368	982.5278	984.8668	986.9187	988.7331	990.3490
2	39.3980	39.4071	39.4146	39.4210	39.4265	39.4313	39.4354	39.4391	39.4424
3	14.4189	14.3742	14.3366	14.3045	14.2768	14.2527	14.2315	14.2127	14.1960
4	8.8439	8.7935	8.7512	8.7150	8.6838	8.6565	8.6326	8.6113	8.5924
5	6.6192	6.5678	6.5245	6.4876	6.4556	6.4277	6.4032	6.3814	6.3619
6	5.4613	5.4098	5.3662	5.3290	5.2968	5.2687	5.2439	5.2218	5.2021
7	4.7611	4.7095	4.6658	4.6285	4.5961	4.5678	4.5428	4.5206	4.5008
8	4.2951	4.2434	4.1997	4.1622	4.1297	4.1012	4.0761	4.0538	4.0338
9	3.9639	3.9121	3.8682	3.8306	3.7980	3.7694	3.7441	3.7216	3.7015
10	3.7168	3.6649	3.6209	3.5832	3.5504	3.5217	3.4963	3.4737	3.4534
11	3.5257	3.4737	3.4296	3.3917	3.3588	3.3299	3.3044	3.2816	3.2612
12	3.3736	3.3215	3.2773	3.2393	3.2062	3.1772	3.1515	3.1286	3.1081
13	3.2497	3.1975	3.1532	3.1150	3.0819	3.0527	3.0269	3.0039	2.9832
14	3.1469	3.0946	3.0502	3.0119	2.9786	2.9493	2.9234	2.9003	2.8795
15	3.0602	3.0078	2.9633	2.9249	2.8915	2.8621	2.8360	2.8128	2.7919
16	2.9862	2.9337	2.8890	2.8506	2.8170	2.7875	2.7614	2.7380	2.7170
17	2.9222	2.8696	2.8249	2.7863	2.7526	2.7230	2.6968	2.6733	2.6522
18	2.8664	2.8137	2.7689	2.7302	2.6964	2.6667	2.6404	2.6168	2.5956
19	2.8172	2.7645	2.7196	2.6808	2.6469	2.6171	2.5907	2.5670	2.5457
20	2.7737	2.7209	2.6758	2.6369	2.6030	2.5731	2.5465	2.5228	2.5014
21	2.7348	2.6819	2.6368	2.5978	2.5638	2.5338	2.5071	2.4833	2.4618
22	2.6998	2.6469	2.6017	2.5626	2.5285	2.4984	2.4717	2.4478	2.4262
23	2.6682	2.6152	2.5699	2.5308	2.4966	2.4665	2.4396	2.4157	2.3940
24	2.6396	2.5865	2.5411	2.5019	2.4677	2.4374	2.4105	2.3865	2.3648
25	2.6135	2.5603	2.5149	2.4756	2.4413	2.4110	2.3840	2.3599	2.3381
26	2.5896	2.5363	2.4908	2.4515	2.4171	2.3867	2.3597	2.3355	2.3137
27	2.5676	2.5143	2.4688	2.4293	2.3949	2.3644	2.3373	2.3131	2.2912
28	2.5473	2.4940	2.4484	2.4089	2.3743	2.3438	2.3167	2.2924	2.2704
29	2.5286	2.4752	2.4295	2.3900	2.3554	2.3248	2.2976	2.2732	2.2512
30	2.5112	2.4577	2.4120	2.3724	2.3378	2.3072	2.2799	2.2554	2.2334
40	2.3882	2.3343	2.2882	2.2481	2.2130	2.1819	2.1542	2.1293	2.1068
60	2.2702	2.2159	2.1692	2.1286	2.0929	2.0613	2.0330	2.0076	1.9846
120	2.1570	2.1021	2.0548	2.0136	1.9773	1.9450	1.9161	1.8900	1.8663
∞	2.0483	1.9927	1.9447	1.9028	1.8657	1.8326	1.8028	1.7760	1.7515

H Critical Values of *F* (Area = 0.025) (cont.)

Numerator Degrees of Freedom

Denominator Degrees of Freedom

	19	20	24	30	40	60	120
1	991.7973	993.1028	997.2492	1001.4144	1005.5981	1009.8001	1014.0202
2	39.4453	39.4479	39.4562	39.4646	39.4729	39.4812	39.4896
3	14.1810	14.1674	14.1241	14.0805	14.0365	13.9921	13.9473
4	8.5753	8.5599	8.5109	8.4613	8.4111	8.3604	8.3092
5	6.3444	6.3286	6.2780	6.2269	6.1750	6.1225	6.0693
6	5.1844	5.1684	5.1172	5.0652	5.0125	4.9589	4.9044
7	4.4829	4.4667	4.4150	4.3624	4.3089	4.2544	4.1989
8	4.0158	3.9995	3.9472	3.8940	3.8398	3.7844	3.7279
9	3.6833	3.6669	3.6142	3.5604	3.5055	3.4493	3.3918
10	3.4351	3.4185	3.3654	3.3110	3.2554	3.1984	3.1399
11	3.2428	3.2261	3.1725	3.1176	3.0613	3.0035	2.9441
12	3.0896	3.0728	3.0187	2.9633	2.9063	2.8478	2.7874
13	2.9646	2.9477	2.8932	2.8372	2.7797	2.7204	2.6590
14	2.8607	2.8437	2.7888	2.7324	2.6742	2.6142	2.5519
15	2.7730	2.7559	2.7006	2.6437	2.5850	2.5242	2.4611
16	2.6980	2.6808	2.6252	2.5678	2.5085	2.4471	2.3831
17	2.6331	2.6158	2.5598	2.5020	2.4422	2.3801	2.3153
18	2.5764	2.5590	2.5027	2.4445	2.3842	2.3214	2.2558
19	2.5265	2.5089	2.4523	2.3937	2.3329	2.2696	2.2032
20	2.4821	2.4645	2.4076	2.3486	2.2873	2.2234	2.1562
21	2.4424	2.4247	2.3675	2.3082	2.2465	2.1819	2.1141
22	2.4067	2.3890	2.3315	2.2718	2.2097	2.1446	2.0760
23	2.3745	2.3567	2.2989	2.2389	2.1763	2.1107	2.0415
24	2.3452	2.3273	2.2693	2.2090	2.1460	2.0799	2.0099
25	2.3184	2.3005	2.2422	2.1816	2.1183	2.0516	1.9811
26	2.2939	2.2759	2.2174	2.1565	2.0928	2.0257	1.9545
27	2.2713	2.2533	2.1946	2.1334	2.0693	2.0018	1.9299
28	2.2505	2.2324	2.1735	2.1121	2.0477	1.9797	1.9072
29	2.2313	2.2131	2.1540	2.0923	2.0276	1.9591	1.8861
30	2.2134	2.1952	2.1359	2.0739	2.0089	1.9400	1.8664
40	2.0864	2.0677	2.0069	1.9429	1.8752	1.8028	1.7242
60	1.9636	1.9445	1.8817	1.8152	1.7440	1.6668	1.5810
120	1.8447	1.8249	1.7597	1.6899	1.6141	1.5299	1.4327
∞	1.7291	1.7085	1.6402	1.5660	1.4836	1.3883	1.2685

H Critical Values of F (Area = 0.010)

Numerator Degrees of Freedom (columns); Denominator Degrees of Freedom (rows)

	1	2	3	4	5	6	7	8	9
1	4052.1807	4999.5000	5403.3520	5624.5833	5763.6496	5858.9861	5928.3557	5981.0703	6022.4732
2	98.5025	99.0000	99.1662	99.2494	99.2993	99.3326	99.3564	99.3742	99.3881
3	34.1162	30.8165	29.4567	28.7099	28.2371	27.9107	27.6717	27.4892	27.3452
4	21.1977	18.0000	16.6944	15.9770	15.5219	15.2069	14.9758	14.7989	14.6591
5	16.2582	13.2739	12.0600	11.3919	10.9670	10.6723	10.4555	10.2893	10.1578
6	13.7450	10.9248	9.7795	9.1483	8.7459	8.4661	8.2600	8.1017	7.9761
7	12.2464	9.5466	8.4513	7.8466	7.4604	7.1914	6.9928	6.8400	6.7188
8	11.2586	8.6491	7.5910	7.0061	6.6318	6.3707	6.1776	6.0289	5.9106
9	10.5614	8.0215	6.9919	6.4221	6.0569	5.8018	5.6129	5.4671	5.3511
10	10.0443	7.5594	6.5523	5.9943	5.6363	5.3858	5.2001	5.0567	4.9424
11	9.6460	7.2057	6.2167	5.6683	5.3160	5.0692	4.8861	4.7445	4.6315
12	9.3302	6.9266	5.9525	5.4120	5.0643	4.8206	4.6395	4.4994	4.3875
13	9.0738	6.7010	5.7394	5.2053	4.8616	4.6204	4.4410	4.3021	4.1911
14	8.8616	6.5149	5.5639	5.0354	4.6950	4.4558	4.2779	4.1399	4.0297
15	8.6831	6.3589	5.4170	4.8932	4.5556	4.3183	4.1415	4.0045	3.8948
16	8.5310	6.2262	5.2922	4.7726	4.4374	4.2016	4.0259	3.8896	3.7804
17	8.3997	6.1121	5.1850	4.6690	4.3359	4.1015	3.9267	3.7910	3.6822
18	8.2854	6.0129	5.0919	4.5790	4.2479	4.0146	3.8406	3.7054	3.5971
19	8.1849	5.9259	5.0103	4.5003	4.1708	3.9386	3.7653	3.6305	3.5225
20	8.0960	5.8489	4.9382	4.4307	4.1027	3.8714	3.6987	3.5644	3.4567
21	8.0166	5.7804	4.8740	4.3688	4.0421	3.8117	3.6396	3.5056	3.3981
22	7.9454	5.7190	4.8166	4.3134	3.9880	3.7583	3.5867	3.4530	3.3458
23	7.8811	5.6637	4.7649	4.2636	3.9392	3.7102	3.5390	3.4057	3.2986
24	7.8229	5.6136	4.7181	4.2184	3.8951	3.6667	3.4959	3.3629	3.2560
25	7.7698	5.5680	4.6755	4.1774	3.8550	3.6272	3.4568	3.3239	3.2172
26	7.7213	5.5263	4.6366	4.1400	3.8183	3.5911	3.4210	3.2884	3.1818
27	7.6767	5.4881	4.6009	4.1056	3.7848	3.5580	3.3882	3.2558	3.1494
28	7.6356	5.4529	4.5681	4.0740	3.7539	3.5276	3.3581	3.2259	3.1195
29	7.5977	5.4204	4.5378	4.0449	3.7254	3.4995	3.3303	3.1982	3.0920
30	7.5625	5.3903	4.5097	4.0179	3.6990	3.4735	3.3045	3.1726	3.0665
40	7.3141	5.1785	4.3126	3.8283	3.5138	3.2910	3.1238	2.9930	2.8876
60	7.0771	4.9774	4.1259	3.6490	3.3389	3.1187	2.9530	2.8233	2.7185
120	6.8509	4.7865	3.9491	3.4795	3.1735	2.9559	2.7918	2.6629	2.5586
∞	6.6349	4.6052	3.7816	3.3192	3.0173	2.8020	2.6393	2.5113	2.4074

H Critical Values of *F* (Area = 0.010) (cont.)

Numerator Degrees of Freedom (columns); Denominator Degrees of Freedom (rows)

	10	11	12	13	14	15	16	17	18
1	6055.8467	6083.3168	6106.3207	6125.8647	6142.6740	6157.2846	6170.1012	6181.4348	6191.5287
2	99.3992	99.4083	99.4159	99.4223	99.4278	99.4325	99.4367	99.4404	99.4436
3	27.2287	27.1326	27.0518	26.9831	26.9238	26.8722	26.8269	26.7867	26.7509
4	14.5459	14.4523	14.3736	14.3065	14.2486	14.1982	14.1539	14.1146	14.0795
5	10.0510	9.9626	9.8883	9.8248	9.7700	9.7222	9.6802	9.6429	9.6096
6	7.8741	7.7896	7.7183	7.6575	7.6049	7.5590	7.5186	7.4827	7.4507
7	6.6201	6.5382	6.4691	6.4100	6.3590	6.3143	6.2750	6.2401	6.2089
8	5.8143	5.7343	5.6667	5.6089	5.5589	5.5151	5.4766	5.4423	5.4116
9	5.2565	5.1779	5.1114	5.0545	5.0052	4.9621	4.9240	4.8902	4.8599
10	4.8491	4.7715	4.7059	4.6496	4.6008	4.5581	4.5204	4.4869	4.4569
11	4.5393	4.4624	4.3974	4.3416	4.2932	4.2509	4.2134	4.1801	4.1503
12	4.2961	4.2198	4.1553	4.0999	4.0518	4.0096	3.9724	3.9392	3.9095
13	4.1003	4.0245	3.9603	3.9052	3.8573	3.8154	3.7783	3.7452	3.7156
14	3.9394	3.8640	3.8001	3.7452	3.6975	3.6557	3.6187	3.5857	3.5561
15	3.8049	3.7299	3.6662	3.6115	3.5639	3.5222	3.4852	3.4523	3.4228
16	3.6909	3.6162	3.5527	3.4981	3.4506	3.4089	3.3720	3.3391	3.3096
17	3.5931	3.5185	3.4552	3.4007	3.3533	3.3117	3.2748	3.2419	3.2124
18	3.5082	3.4338	3.3706	3.3162	3.2689	3.2273	3.1904	3.1575	3.1280
19	3.4338	3.3596	3.2965	3.2422	3.1949	3.1533	3.1165	3.0836	3.0541
20	3.3682	3.2941	3.2311	3.1769	3.1296	3.0880	3.0512	3.0183	2.9887
21	3.3098	3.2359	3.1730	3.1187	3.0715	3.0300	2.9931	2.9602	2.9306
22	3.2576	3.1837	3.1209	3.0667	3.0195	2.9779	2.9411	2.9082	2.8786
23	3.2106	3.1368	3.0740	3.0199	2.9727	2.9311	2.8943	2.8613	2.8317
24	3.1681	3.0944	3.0316	2.9775	2.9303	2.8887	2.8519	2.8189	2.7892
25	3.1294	3.0558	2.9931	2.9389	2.8917	2.8502	2.8133	2.7803	2.7506
26	3.0941	3.0205	2.9578	2.9038	2.8566	2.8150	2.7781	2.7451	2.7153
27	3.0618	2.9882	2.9256	2.8715	2.8243	2.7827	2.7458	2.7127	2.6830
28	3.0320	2.9585	2.8959	2.8418	2.7946	2.7530	2.7160	2.6830	2.6532
29	3.0045	2.9311	2.8685	2.8144	2.7672	2.7256	2.6886	2.6555	2.6257
30	2.9791	2.9057	2.8431	2.7890	2.7418	2.7002	2.6632	2.6301	2.6003
40	2.8005	2.7274	2.6648	2.6107	2.5634	2.5216	2.4844	2.4511	2.4210
60	2.6318	2.5587	2.4961	2.4419	2.3943	2.3523	2.3148	2.2811	2.2507
120	2.4721	2.3990	2.3363	2.2818	2.2339	2.1915	2.1536	2.1194	2.0885
∞	2.3209	2.2477	2.1848	2.1299	2.0815	2.0385	2.0000	1.9652	1.9336

H Critical Values of *F* (Area = 0.010) (cont.)

Numerator Degrees of Freedom

Denominator Degrees of Freedom

	19	20	24	30	40	60	120
1	6200.5756	6208.7302	6234.6309	6260.6486	6286.7821	6313.0301	6339.3913
2	99.4465	99.4492	99.4575	99.4658	99.4742	99.4825	99.4908
3	26.7188	26.6898	26.5975	26.5045	26.4108	26.3164	26.2211
4	14.0480	14.0196	13.9291	13.8377	13.7454	13.6522	13.5581
5	9.5797	9.5526	9.4665	9.3793	9.2912	9.2020	9.1118
6	7.4219	7.3958	7.3127	7.2285	7.1432	7.0567	6.9690
7	6.1808	6.1554	6.0743	5.9920	5.9084	5.8236	5.7373
8	5.3840	5.3591	5.2793	5.1981	5.1156	5.0316	4.9461
9	4.8327	4.8080	4.7290	4.6486	4.5666	4.4831	4.3978
10	4.4299	4.4054	4.3269	4.2469	4.1653	4.0819	3.9965
11	4.1234	4.0990	4.0209	3.9411	3.8596	3.7761	3.6904
12	3.8827	3.8584	3.7805	3.7008	3.6192	3.5355	3.4494
13	3.6888	3.6646	3.5868	3.5070	3.4253	3.3413	3.2548
14	3.5294	3.5052	3.4274	3.3476	3.2656	3.1813	3.0942
15	3.3961	3.3719	3.2940	3.2141	3.1319	3.0471	2.9595
16	3.2829	3.2587	3.1808	3.1007	3.0182	2.9330	2.8447
17	3.1857	3.1615	3.0835	3.0032	2.9205	2.8348	2.7459
18	3.1013	3.0771	2.9990	2.9185	2.8354	2.7493	2.6597
19	3.0274	3.0031	2.9249	2.8442	2.7608	2.6742	2.5839
20	2.9620	2.9377	2.8594	2.7785	2.6947	2.6077	2.5168
21	2.9039	2.8796	2.8010	2.7200	2.6359	2.5484	2.4568
22	2.8518	2.8274	2.7488	2.6675	2.5831	2.4951	2.4029
23	2.8049	2.7805	2.7017	2.6202	2.5355	2.4471	2.3542
24	2.7624	2.7380	2.6591	2.5773	2.4923	2.4035	2.3100
25	2.7238	2.6993	2.6203	2.5383	2.4530	2.3637	2.2696
26	2.6885	2.6640	2.5848	2.5026	2.4170	2.3273	2.2325
27	2.6561	2.6316	2.5522	2.4699	2.3840	2.2938	2.1985
28	2.6263	2.6017	2.5223	2.4397	2.3535	2.2629	2.1670
29	2.5987	2.5742	2.4946	2.4118	2.3253	2.2344	2.1379
30	2.5732	2.5487	2.4689	2.3860	2.2992	2.2079	2.1108
40	2.3937	2.3689	2.2880	2.2034	2.1142	2.0194	1.9172
60	2.2230	2.1978	2.1154	2.0285	1.9360	1.8363	1.7263
120	2.0604	2.0346	1.9500	1.8600	1.7628	1.6557	1.5330
∞	1.9048	1.8783	1.7908	1.6964	1.5923	1.4730	1.3246

H Critical Values of F (Area = 0.005)

Numerator Degrees of Freedom

Denominator Degrees of Freedom	1	2	3	4	5	6	7	8	9
1	16210.7227	19999.5000	21614.7414	22499.5833	23055.7982	23437.1111	23714.5658	23925.4062	24091.0041
2	198.5013	199.0000	199.1664	199.2497	199.2996	199.3330	199.3568	199.3746	199.3885
3	55.5520	49.7993	47.4672	46.1946	45.3916	44.8385	44.4341	44.1256	43.8824
4	31.3328	26.2843	24.2591	23.1545	22.4564	21.9746	21.6217	21.3520	21.1391
5	22.7848	18.3138	16.5298	15.5561	14.9396	14.5133	14.2004	13.9610	13.7716
6	18.6350	14.5441	12.9166	12.0275	11.4637	11.0730	10.7859	10.5658	10.3915
7	16.2356	12.4040	10.8824	10.0505	9.5221	9.1553	8.8854	8.6781	8.5138
8	14.6882	11.0424	9.5965	8.8051	8.3018	7.9520	7.6941	7.4959	7.3386
9	13.6136	10.1067	8.7171	7.9559	7.4712	7.1339	6.8849	6.6933	6.5411
10	12.8265	9.4270	8.0807	7.3428	6.8724	6.5446	6.3025	6.1159	5.9676
11	12.2263	8.9122	7.6004	6.8809	6.4217	6.1016	5.8648	5.6821	5.5368
12	11.7542	8.5096	7.2258	6.5211	6.0711	5.7570	5.5245	5.3451	5.2021
13	11.3735	8.1865	6.9258	6.2335	5.7910	5.4819	5.2529	5.0761	4.9351
14	11.0603	7.9216	6.6804	5.9984	5.5623	5.2574	5.0313	4.8566	4.7173
15	10.7980	7.7008	6.4760	5.8029	5.3721	5.0708	4.8473	4.6744	4.5364
16	10.5755	7.5138	6.3034	5.6378	5.2117	4.9134	4.6920	4.5207	4.3838
17	10.3842	7.3536	6.1556	5.4967	5.0746	4.7789	4.5594	4.3894	4.2535
18	10.2181	7.2148	6.0278	5.3746	4.9560	4.6627	4.4448	4.2759	4.1410
19	10.0725	7.0935	5.9161	5.2681	4.8526	4.5614	4.3448	4.1770	4.0428
20	9.9439	6.9865	5.8177	5.1743	4.7616	4.4721	4.2569	4.0900	3.9564
21	9.8295	6.8914	5.7304	5.0911	4.6809	4.3931	4.1789	4.0128	3.8799
22	9.7271	6.8064	5.6524	5.0168	4.6088	4.3225	4.1094	3.9440	3.8116
23	9.6348	6.7300	5.5823	4.9500	4.5441	4.2591	4.0469	3.8822	3.7502
24	9.5513	6.6609	5.5190	4.8898	4.4857	4.2019	3.9905	3.8264	3.6949
25	9.4753	6.5982	5.4615	4.8351	4.4327	4.1500	3.9394	3.7758	3.6447
26	9.4059	6.5409	5.4091	4.7852	4.3844	4.1027	3.8928	3.7297	3.5989
27	9.3423	6.4885	5.3611	4.7396	4.3402	4.0594	3.8501	3.6875	3.5571
28	9.2838	6.4403	5.3170	4.6977	4.2996	4.0197	3.8110	3.6487	3.5186
29	9.2297	6.3958	5.2764	4.6591	4.2622	3.9831	3.7749	3.6131	3.4832
30	9.1797	6.3547	5.2388	4.6234	4.2276	3.9492	3.7416	3.5801	3.4505
40	8.8279	6.0664	4.9758	4.3738	3.9860	3.7129	3.5088	3.3498	3.2220
60	8.4946	5.7950	4.7290	4.1399	3.7599	3.4918	3.2911	3.1344	3.0083
120	8.1788	5.5393	4.4972	3.9207	3.5482	3.2849	3.0874	2.9330	2.8083
∞	7.8795	5.2983	4.2794	3.7151	3.3499	3.0913	2.8968	2.7444	2.6211

H Critical Values of F (Area = 0.005) (cont.)

Numerator Degrees of Freedom

Denominator Degrees of Freedom

	10	11	12	13	14	15	16	17	18
1	24224.4868	24334.3581	24426.3662	24504.5356	24571.7673	24630.2051	24681.4673	24726.7982	24767.1704
2	199.3996	199.4087	199.4163	199.4227	199.4282	199.4329	199.4371	199.4408	199.4440
3	43.6858	43.5236	43.3874	43.2715	43.1716	43.0847	43.0083	42.9407	42.8804
4	20.9667	20.8243	20.7047	20.6027	20.5148	20.4383	20.3710	20.3113	20.2581
5	13.6182	13.4912	13.3845	13.2934	13.2148	13.1463	13.0861	13.0327	12.9850
6	10.2500	10.1329	10.0343	9.9501	9.8774	9.8140	9.7582	9.7086	9.6644
7	8.3803	8.2697	8.1764	8.0967	8.0279	7.9678	7.9148	7.8678	7.8258
8	7.2106	7.1045	7.0149	6.9384	6.8721	6.8143	6.7633	6.7180	6.6775
9	6.4172	6.3142	6.2274	6.1530	6.0887	6.0325	5.9829	5.9388	5.8994
10	5.8467	5.7462	5.6613	5.5887	5.5257	5.4707	5.4221	5.3789	5.3403
11	5.4183	5.3197	5.2363	5.1649	5.1031	5.0489	5.0011	4.9586	4.9205
12	5.0855	4.9884	4.9062	4.8358	4.7748	4.7213	4.6741	4.6321	4.5945
13	4.8199	4.7240	4.6429	4.5733	4.5129	4.4600	4.4132	4.3716	4.3344
14	4.6034	4.5085	4.4281	4.3591	4.2993	4.2468	4.2005	4.1592	4.1221
15	4.4235	4.3295	4.2497	4.1813	4.1219	4.0698	4.0237	3.9827	3.9459
16	4.2719	4.1785	4.0994	4.0314	3.9723	3.9205	3.8747	3.8338	3.7972
17	4.1424	4.0496	3.9709	3.9033	3.8445	3.7929	3.7473	3.7066	3.6701
18	4.0305	3.9382	3.8599	3.7926	3.7341	3.6827	3.6373	3.5967	3.5603
19	3.9329	3.8410	3.7631	3.6961	3.6378	3.5866	3.5412	3.5008	3.4645
20	3.8470	3.7555	3.6779	3.6111	3.5530	3.5020	3.4568	3.4164	3.3802
21	3.7709	3.6798	3.6024	3.5358	3.4779	3.4270	3.3818	3.3416	3.3054
22	3.7030	3.6122	3.5350	3.4686	3.4108	3.3600	3.3150	3.2748	3.2387
23	3.6420	3.5515	3.4745	3.4083	3.3506	3.2999	3.2549	3.2148	3.1787
24	3.5870	3.4967	3.4199	3.3538	3.2962	3.2456	3.2007	3.1606	3.1246
25	3.5370	3.4470	3.3704	3.3044	3.2469	3.1963	3.1515	3.1114	3.0754
26	3.4916	3.4017	3.3252	3.2594	3.2020	3.1515	3.1067	3.0666	3.0306
27	3.4499	3.3602	3.2839	3.2182	3.1608	3.1104	3.0656	3.0256	2.9896
28	3.4117	3.3222	3.2460	3.1803	3.1231	3.0727	3.0279	2.9879	2.9520
29	3.3765	3.2871	3.2110	3.1454	3.0882	3.0379	2.9932	2.9532	2.9173
30	3.3440	3.2547	3.1787	3.1132	3.0560	3.0057	2.9611	2.9211	2.8852
40	3.1167	3.0284	2.9531	2.8880	2.8312	2.7811	2.7365	2.6966	2.6607
60	2.9042	2.8166	2.7419	2.6771	2.6205	2.5705	2.5259	2.4859	2.4498
120	2.7052	2.6183	2.5439	2.4794	2.4228	2.3727	2.3280	2.2878	2.2514
∞	2.5188	2.4325	2.3583	2.2938	2.2371	2.1868	2.1417	2.1011	2.0643

H Critical Values of *F* (Area = 0.005) (cont.)

Numerator Degrees of Freedom

Denominator Degrees of Freedom

	19	20	24	30	40	60	120
1	24803.3549	24835.9709	24939.5653	25043.6277	25148.1532	25253.1369	25358.5735
2	199.4470	199.4496	199.4579	199.4663	199.4746	199.4829	199.4912
3	42.8263	42.7775	42.6222	42.4658	42.3082	42.1494	41.9895
4	20.2104	20.1673	20.0300	19.8915	19.7518	19.6107	19.4684
5	12.9422	12.9035	12.7802	12.6556	12.5297	12.4024	12.2737
6	9.6247	9.5888	9.4742	9.3582	9.2408	9.1219	9.0015
7	7.7881	7.7540	7.6450	7.5345	7.4224	7.3088	7.1933
8	6.6411	6.6082	6.5029	6.3961	6.2875	6.1772	6.0649
9	5.8639	5.8318	5.7292	5.6248	5.5186	5.4104	5.3001
10	5.3055	5.2740	5.1732	5.0706	4.9659	4.8592	4.7501
11	4.8863	4.8552	4.7557	4.6543	4.5508	4.4450	4.3367
12	4.5606	4.5299	4.4314	4.3309	4.2282	4.1229	4.0149
13	4.3008	4.2703	4.1726	4.0727	3.9704	3.8655	3.7577
14	4.0888	4.0585	3.9614	3.8619	3.7600	3.6552	3.5473
15	3.9127	3.8826	3.7859	3.6867	3.5850	3.4803	3.3722
16	3.7641	3.7342	3.6378	3.5389	3.4372	3.3324	3.2240
17	3.6372	3.6073	3.5112	3.4124	3.3108	3.2058	3.0971
18	3.5275	3.4977	3.4017	3.3030	3.2014	3.0962	2.9871
19	3.4318	3.4020	3.3062	3.2075	3.1058	3.0004	2.8908
20	3.3475	3.3178	3.2220	3.1234	3.0215	2.9159	2.8058
21	3.2728	3.2431	3.1474	3.0488	2.9467	2.8408	2.7302
22	3.2060	3.1764	3.0807	2.9821	2.8799	2.7736	2.6625
23	3.1461	3.1165	3.0208	2.9221	2.8197	2.7132	2.6015
24	3.0920	3.0624	2.9667	2.8679	2.7654	2.6585	2.5463
25	3.0429	3.0133	2.9176	2.8187	2.7160	2.6088	2.4961
26	2.9981	2.9685	2.8728	2.7738	2.6709	2.5633	2.4501
27	2.9571	2.9275	2.8318	2.7327	2.6296	2.5217	2.4079
28	2.9194	2.8899	2.7941	2.6949	2.5916	2.4834	2.3690
29	2.8847	2.8551	2.7594	2.6600	2.5565	2.4479	2.3331
30	2.8526	2.8230	2.7272	2.6278	2.5241	2.4151	2.2998
40	2.6281	2.5984	2.5020	2.4015	2.2958	2.1838	2.0636
60	2.4171	2.3872	2.2898	2.1874	2.0789	1.9622	1.8341
120	2.2183	2.1881	2.0890	1.9840	1.8709	1.7469	1.6055
∞	2.0307	1.9999	1.8983	1.7891	1.6692	1.5326	1.3638

I Critical Values of the Pearson Correlation Coefficient

n	$\alpha = 0.05$	$\alpha = 0.01$
4	0.950	0.990
5	0.878	0.959
6	0.811	0.917
7	0.754	0.875
8	0.707	0.834
9	0.666	0.798
10	0.632	0.765
11	0.602	0.735
12	0.576	0.708
13	0.553	0.684
14	0.532	0.661
15	0.514	0.641
16	0.497	0.623
17	0.482	0.606
18	0.468	0.590
19	0.456	0.575
20	0.444	0.561
21	0.433	0.549
22	0.423	0.537
23	0.413	0.526
24	0.404	0.515
25	0.396	0.505
26	0.388	0.496
27	0.381	0.487
28	0.374	0.479
29	0.367	0.471
30	0.361	0.463
35	0.334	0.430
40	0.312	0.403
45	0.294	0.380
50	0.279	0.361
55	0.266	0.345
60	0.254	0.330
65	0.244	0.317
70	0.235	0.306
75	0.227	0.296
80	0.220	0.286
85	0.213	0.278
90	0.207	0.270
95	0.202	0.263
100	0.197	0.256

Note: r is statistically significant if $|r|$ is greater than or equal to the value given in the table.

Appendix B

Getting Started with Microsoft Excel

The Basics of Excel 2010

Microsoft Excel is a spreadsheet program that allows users to track and analyze data. Spreadsheets such as those created with Microsoft Excel are widely used in the business world to perform various tasks such as accounting, budgeting, billing, reporting, planning, and tracking. When you open Excel, there are three tabs at the bottom of the worksheet labeled with Sheet1, Sheet2, and Sheet3, each opening a blank spreadsheet. These tabs can be renamed so that the data you require can be easily found. The smaller tab appearing to the right of the Sheet3 tab creates a new tab with a blank spreadsheet. Along the top, you will see various tabs such as File, Home, Insert, Page Layout, Formulas, Data, etc. The numerous commands and options available when working within Excel can be found under these tabs. Figure B.1 shows what a new workbook looks like when Excel is first opened.

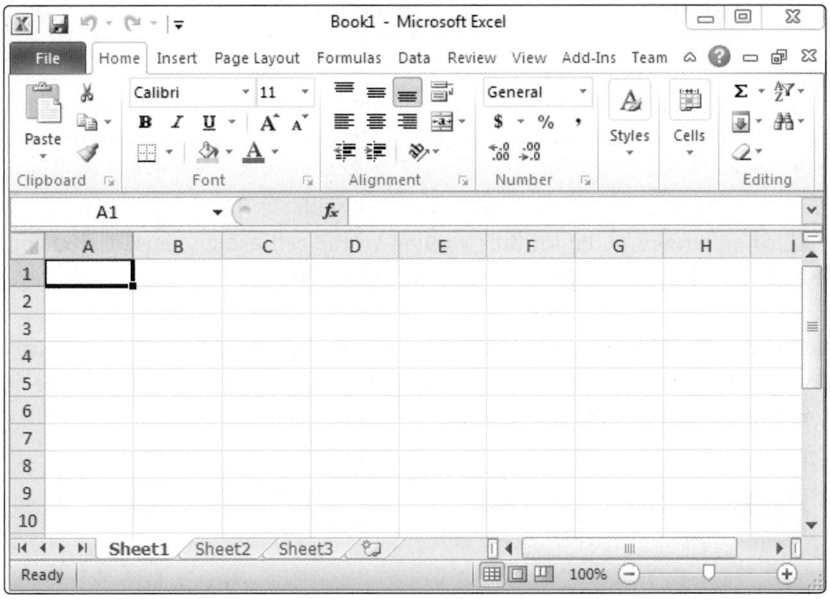

Figure B.1

Cells

Cells in Microsoft Excel may contain data and/or text. The cells' locations are described by their positions in terms of rows and columns. Rows are listed from top to bottom along the left side of the worksheet and are labeled with numbers. Columns are listed from left to right across the top of the worksheet and are labeled with letters. The colum width can be altered by placing your cursor between two column labels; when it turns into a black vertical line with arrows pointing to the left and right, you can then click and drag to the left or right to make the column width larger or smaller. The same can be done with row heights. Alternatively, double-clicking between two column or row labels will autofit the row height or column width to the data it contains.

A thick black border will outline the active cell. To help you identify the cell address, the column and row headers are also highlighted. Another way to identify the active cell's address is to look at the window just above the header for Column A. A column letter followed by a row number describes the active cell position. For example, B2 would be referring to the cell found in the second row of Column B (see Figure B.2).

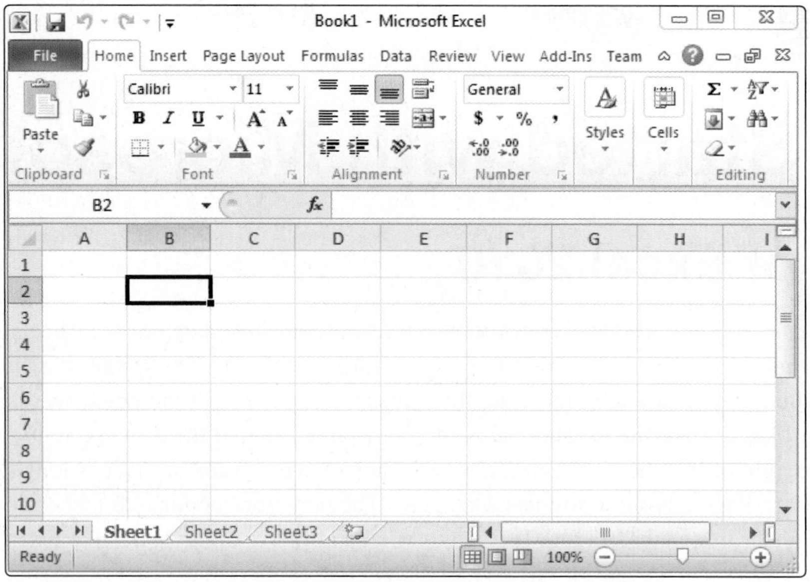

Figure B.2

To change the active cell, move the mouse to the desired cell and click. The border will now be surrounding the new cell and the address will have changed in the active cell address window. The arrow keys can also control navigation of the active cell.

To the right of the cell address window is the formula window (labeled with f_x). This displays the contents of the active cell. The contents of the cell can be edited either within the formula window or the cell itself. Thus, as you enter data in the active cell, it is displayed within the cell and in the formula window at the top of the worksheet.

Suppose you wanted to add the numbers 15, 10, and 3. To do this, enter the formula **=15+10+3** into cell A1. The = sign is the command to start a formula in Excel. Notice that the formula is seen within in the cell as well as in the formula window (see Figure B.3). When you press **enter**, you will see that the active cell moves to cell B1, and the solution of 28 is now displayed in cell A1.

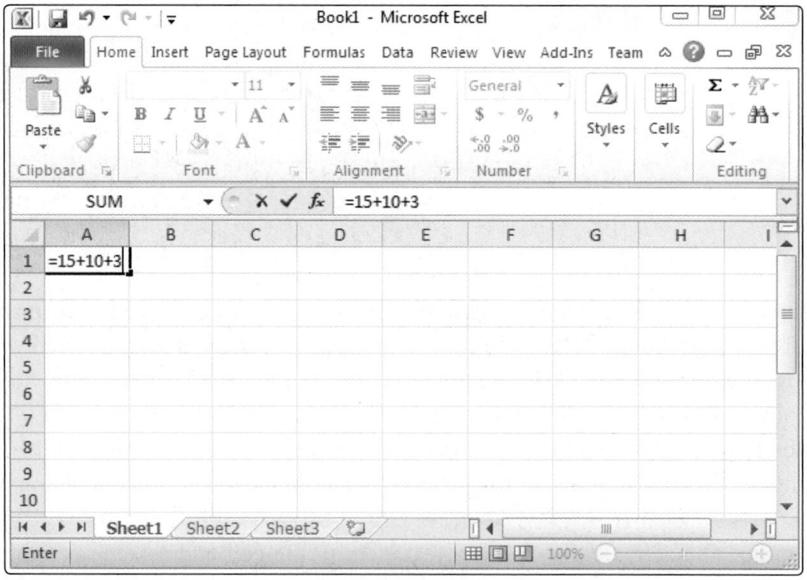

Figure B.3

When you finish entering data in a cell, press **Enter**. The active cell will now move one row down. The answer to the formula, rather than the formula itself, will be displayed in the cell. Cells or data can be added, subtracted, multiplied, divided, and manipulated in any combination through formulas.

Using Excel with Words or Phrases

Excel will try to anticipate a reoccurring word or phrase within a column. When you type a word in a cell, you can then move to a lower cell in that column. If you type the first letter of the word again, Excel will fill in the rest of the word or phrase. To accept the automatic completion of the word, just hit the Enter key or click in a new cell. If you do not want to use the automatic completion of the word, simply keep typing in the cell and the new word or phrase will appear.

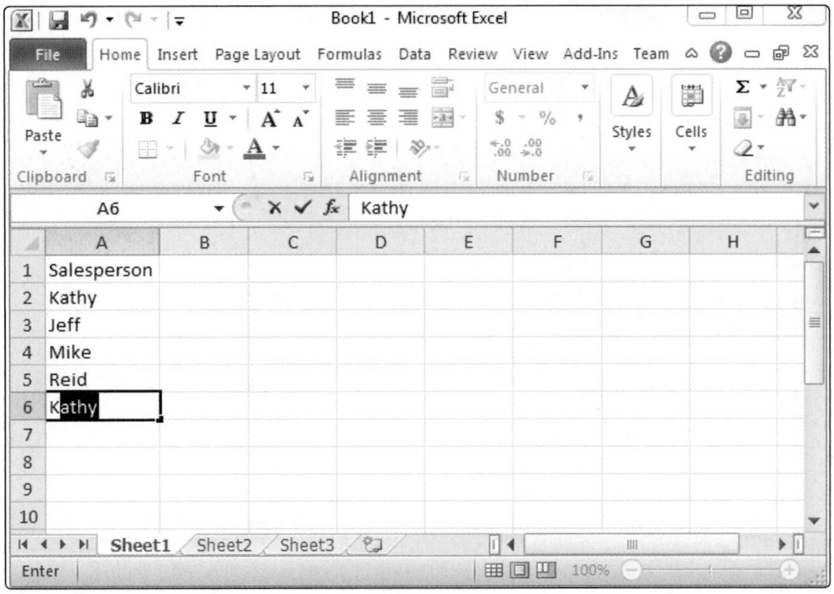

Figure B.4

Excel also has the ability to spell check your spreadsheet, just as you would in a word processor. You also have the ability to use Microsoft Excel to sort a long list of items alphabetically or numerically, or find a particular item you are looking for. The sort and find features can be found under the Home tab. Also under the Home tab, you can change the appearance of text by altering the font, size, color, or alignment of the text within the cells.

Filling Cells

Excel has a helpful tool when entering a series of data. Try entering **1** into cell A1 and **2** into cell A2. Then highlight both of the cells. You will see a small black square at the bottom right of the highlighted section.

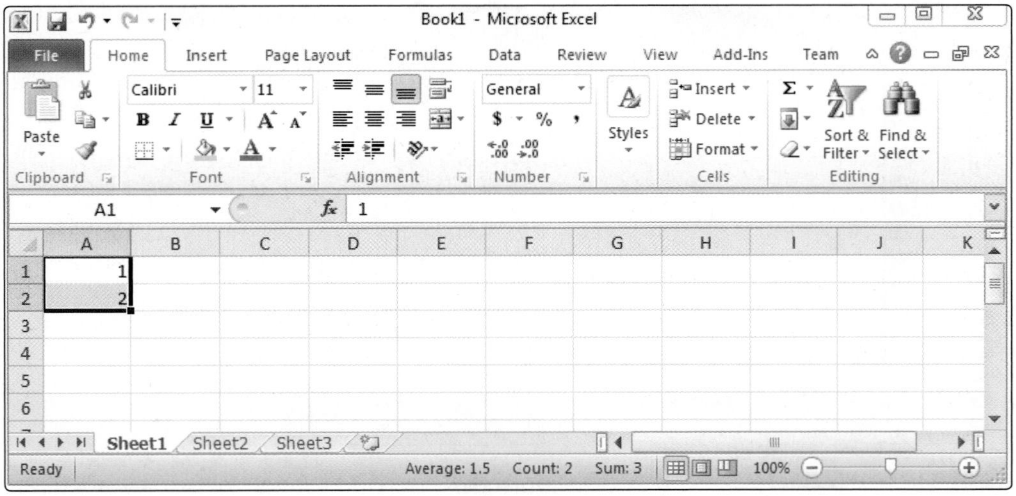

Figure B.5

If you move your cursor over this black square, it should change into a narrow plus sign. Now, click the left mouse button and drag the mouse down to include cells A3 through A10. When you release the mouse button, the cells should now be filled with values from 1 to 10. We will refer to this as *filling cells*. Also notice that at the bottom of the screen, Excel gives some summary statistics about the highlighted cells. The average, count of the number of values, and sum are displayed for any highlighted cells (see Figure B.6).

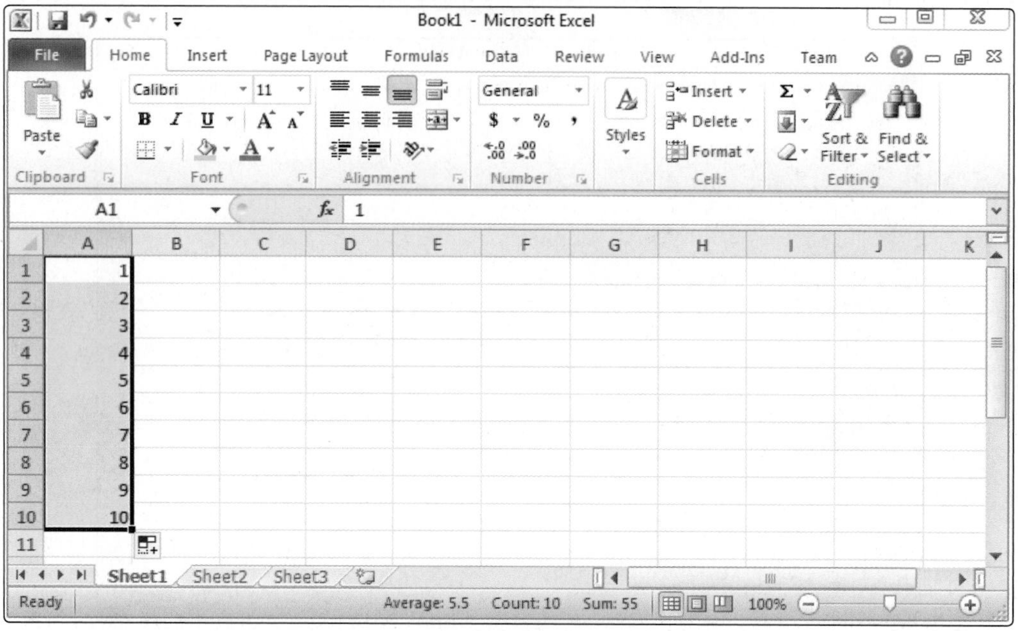

Figure B.6

The fill tool will work for other (nonsequential) values as well. Excel finds the relationship between two data points and replicates this when the tool is used. As an example, try this with multiples of 5. Enter **5** in cell A1 and **10** in cell A2. Highlight cells A1 and A2 again and put your cursor over the square in the bottom right of the highlighted section. Click the left mouse button and drag down to cell A10. Release the button. Now the values should have changed to 5 through 50, in increments of 5 (see Figure B.7).

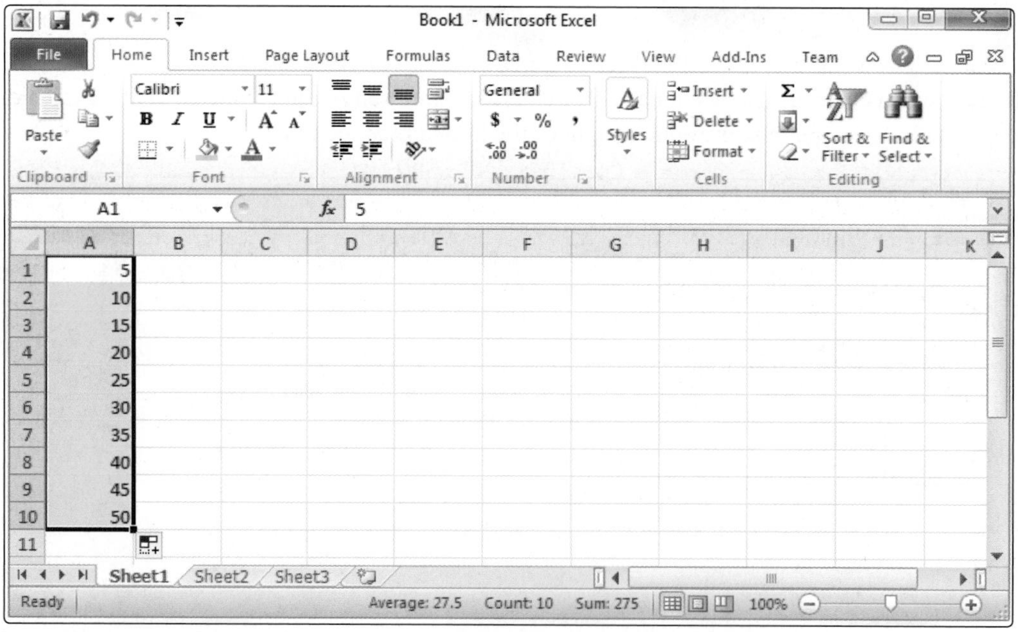

Figure B.7

You can also use this feature with sequential labels such as days of the week or months of the year. For example, if you type **Monday** into cell A1 and use the fill tool to drag to cell G1, the intermediate cells will populate with all of the days of the week.

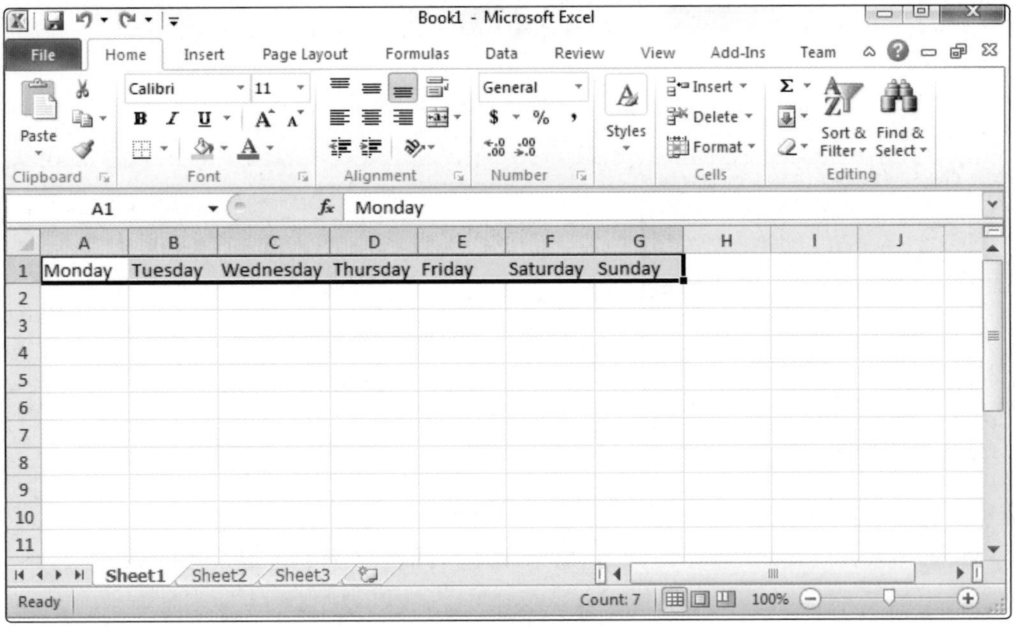

Figure B.8

Now that you have the basic skills to work with Excel, start a new worksheet and we will try some examples.

Formulas and Addressing Using Excel

Enter the labels **Checking Balance** and **Savings Balance** in cells A1 and A2, respectively. You will notice that the label you typed in cell A1 is cut off. We need to resize the columns to allow for the entire label. Move your cursor to the line that separates the column labels for A and B. Your cursor will change to an arrow pointing off to the left and right. You can click and drag the cursor to the right to increase the column width, or simply double click to have the width autofit the label. Repeat this for Column B.

Next select the column labels A and B and you should see both columns highlighted. Under the Home tab there is an area labeled "Number." The buttons in this area format any numbers you enter in the selected cells. With columns A and B highlighted, click on the **$** button. Now all of the values put in these columns will be formatted as dollars.

Next we will use the fill tool to get some values to work with. Start by entering **100** in cell A2 and **200** in cell A3. Now move to cell B2, enter **1000**, and enter **2000** in cell B3. Now highlight cells A2 through B3 by clicking in A2 and dragging down and to the right to include B3. With those cells highlighted, click on the narrow plus sign at the bottom right of the highlighted section. Click and drag down to row 11. When you release the mouse button, the cells should fill with data and end with $10,000.00 in cell B11. You should have a screen that looks like the following.

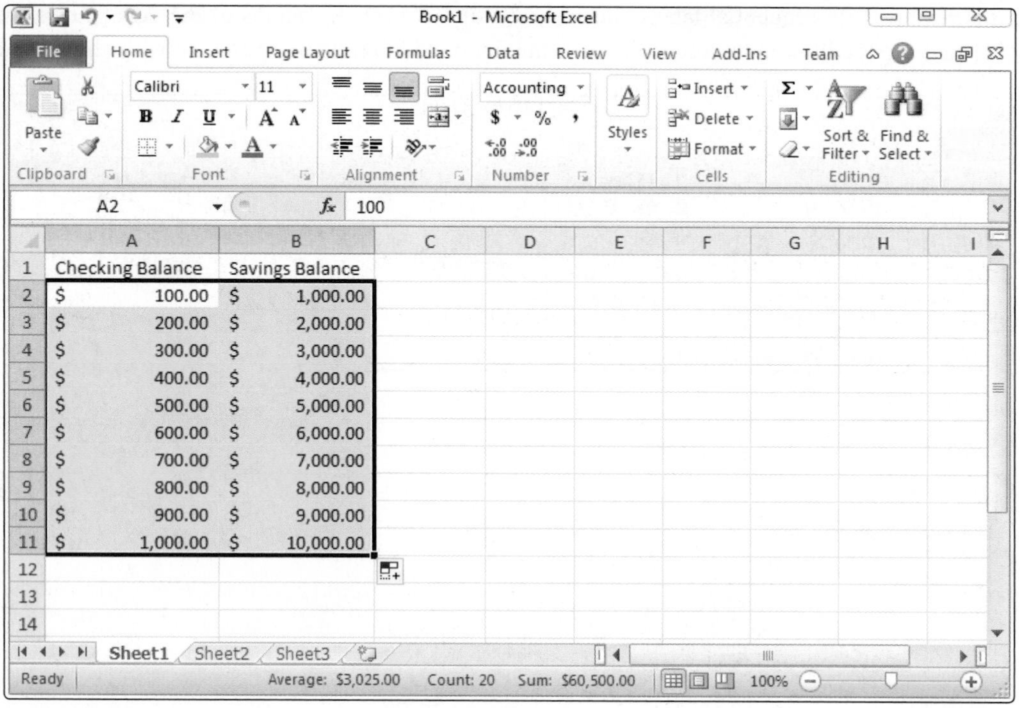

Figure B.9

Formulas can be applied to manipulate data between cells. We will see, however, that special attention needs to be paid to the copying of formulas. We will continue working with the above worksheet with the checking and savings balances while we try some formulas.

We want to see what your total bank balance would be if you had the amount listed in a particular row in your checking and savings accounts. In cell C2, type = to indicate you are entering a formula, and then click in cell A2, type +, and click in cell B2. Press **Enter**. You have just added cells A2 and B2 together, and the sum ($1,100.00) is displayed in cell C2. Under the Home tab, the "Clipboard" section contains buttons allowing you to cut, copy, and paste cells. With cell C2 selected, either press the **copy** button or use the keyboard shortcut **Ctrl+C** to copy the formula. Highlight cells C3 through C11. Under the Home tab, press the **Paste** button or use the keyboard shortcut **Ctrl+V** to paste the formula into these cells. Notice that the value $1,100.00 that was displayed in C2 was not copied, but the formula that made up that value. The formula changed for each of the rows and substituted the new row value in the formula. This is called *relative addressing*. The formula changes relative to its address. The last cell should show the value $11,000.00. If the value is not visible, the column needs to be resized.

Suppose we wanted to see what the annual interest would be on our savings account at the levels listed in Column B. Create a new label in cell A13 called **Interest Rate**. Now in cell B13, input the interest rate 0.045. We will need to format the cell for percentages rather than currency. To do this, highlight cell B13, and under the Home tab in the "Number" area, select the **%** button next to the $ button that we used to originally format the cells. The number will most likely be displayed as 5%. We can adjust the number of decimals displayed by clicking on the [.0→.00] button, also located in the "Number" section under the Home tab. This button increases the number of decimal places displayed by one decimal place each time it is clicked. (Notice that there is a similar button next to the one mentioned that decreases the number of places displayed.) The value should now display as 4.5%.

Now we want to use this value to compute interest for each level of savings. Label cell D1 **Savings Interest**. Resize the cell to fit the label. In cell D2, we will put our formula. Type = to start the formula, and then select cell B13 to get the percentage rate. Type * to indicate multiplication, and select cell B2. Upon pressing **Enter**, the resulting value will be displayed ($45.00). Thus, the annual interest gained on $1,000.00 in savings at 4.5% APR is $45.00.

Since we have the formula, we can copy it to the rest of the column and get the interest income for each of the savings levels. With cell D2 selected, click in the lower right hand corner of the cell and drag down to D3. Is this the result you expected? Probably not. Remember the rules of relative addressing. As you copy the formula down one row, the formula values change by one row. So in cell D3, the formula is pulling from cell B14 for the interest rate and cell B3 for the principal amount. You will notice that cell B14 is empty. Thus, Excel computes this formula as being equal to zero. However, want to use cell B13 for each of these formulas in Column D. So how do we lock in the address of cell B13? We use *absolute addressing*. The $ symbol is the key to locking the position in the formula. In cell D2, we need to change the formula to read **=B$13*B2**. This will "lock" cell B13 into the formula when it is copied throughout Column D.

Go back and copy the new formula into cells D3 through D11 using the fill tool. The results should now look like the following.

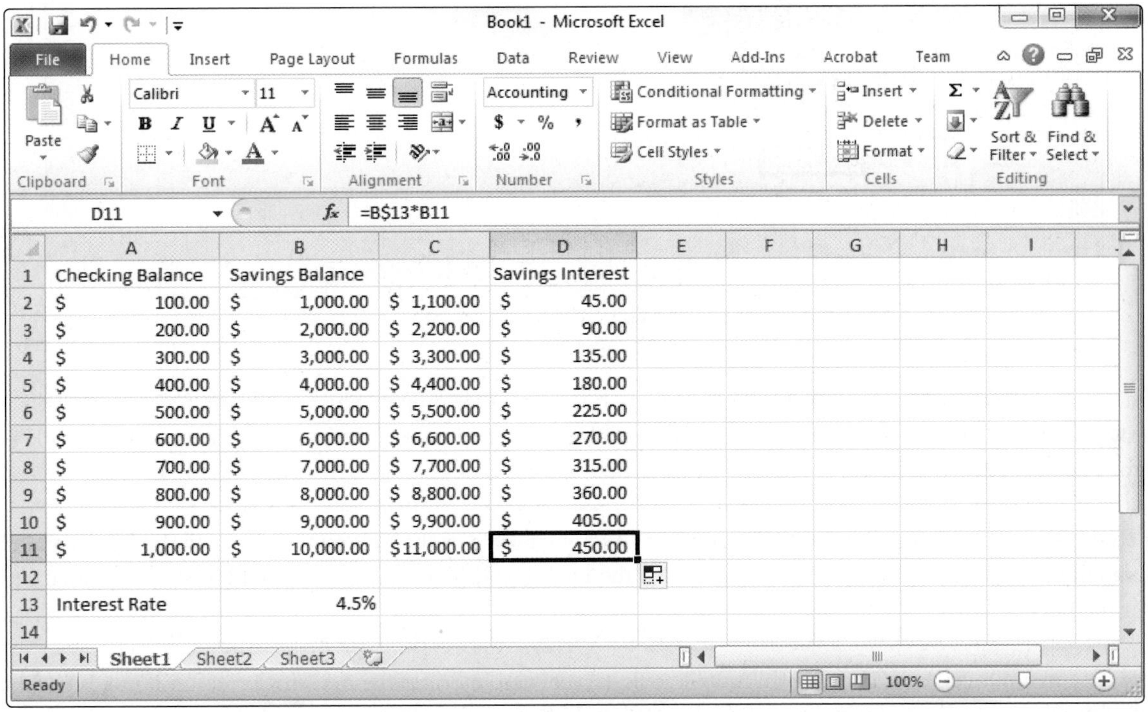

Figure B.10

Excel can compute a multitude of calculations on data values in the spreadsheet. If you press the f_x button next to the formula window, you can explore all of the functions and formulas that Excel has to offer. Next, we will highlight some of the chart tools that Excel provides.

Charts

Suppose you have the following data about ticket sales from your traveling circus. Adult tickets are $20 and child tickets are $12. The sales for the last week are listed in the following spreadsheet (see Figure B.11).

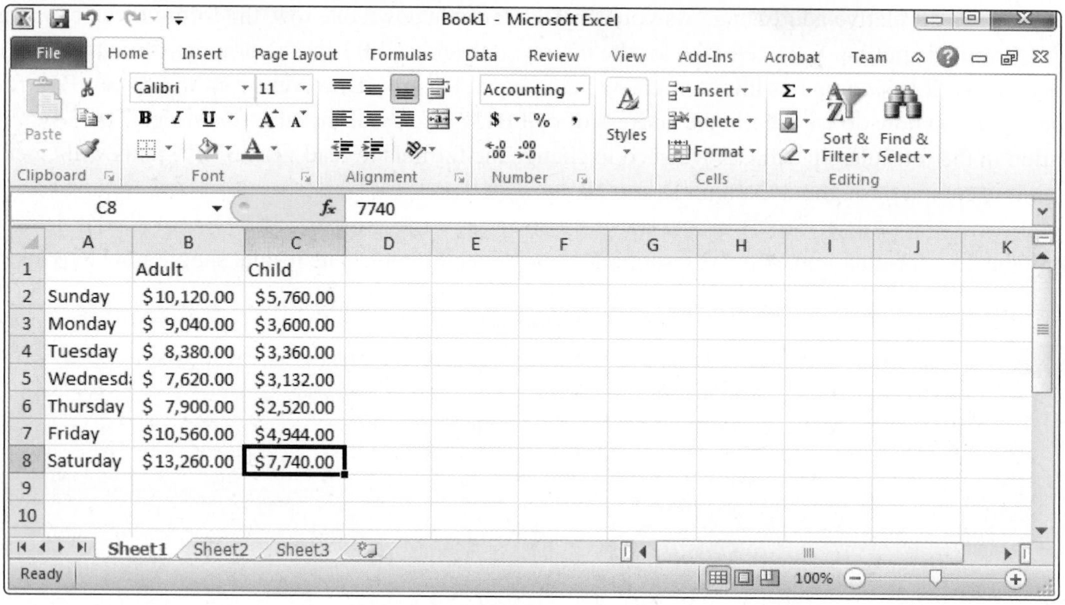

Figure B.11

From these data, you might see how the ticket sales change daily and wish to compare child and adult ticket sales. This information might be easier to understand if it were graphically displayed on a chart. Enter the data as seen above in a new worksheet and use a bar chart for this particular example.

To create a chart, we need to select the data to be graphed. We want to chart the ticket sales for both the adults and children. In a bar chart, the dollar amount will determine the height of the bar. The days of the week should also be included since they are the labels for the bars.

Highlight the data in columns A, B, and C from rows 1 to 8. (Excel can interpret the first row or column in a set of data as labels and not part of the data.) Under the Insert tab, there is a "Charts" group of icons.

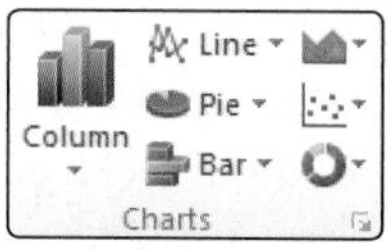

Figure B.12

Choose **Column**, and then the top left graph (the first one listed under the 2-D column heading). This is a clustered column graph. Excel creates a side-by-side bar graph based on the highlighted data. After the chart is created, a set of tabs labeled Chart Tools appears. These tabs can be used to edit the chart that has been made. For example, under the Layout tab, select **Chart Title**, and **Above Chart**. Give the graph a title of **Ticket Sales**. Under the Layout tab, you can edit the chart appearance by adding axis titles, data labels, or gridlines. Microsoft Excel makes it easy to create the chart you want.

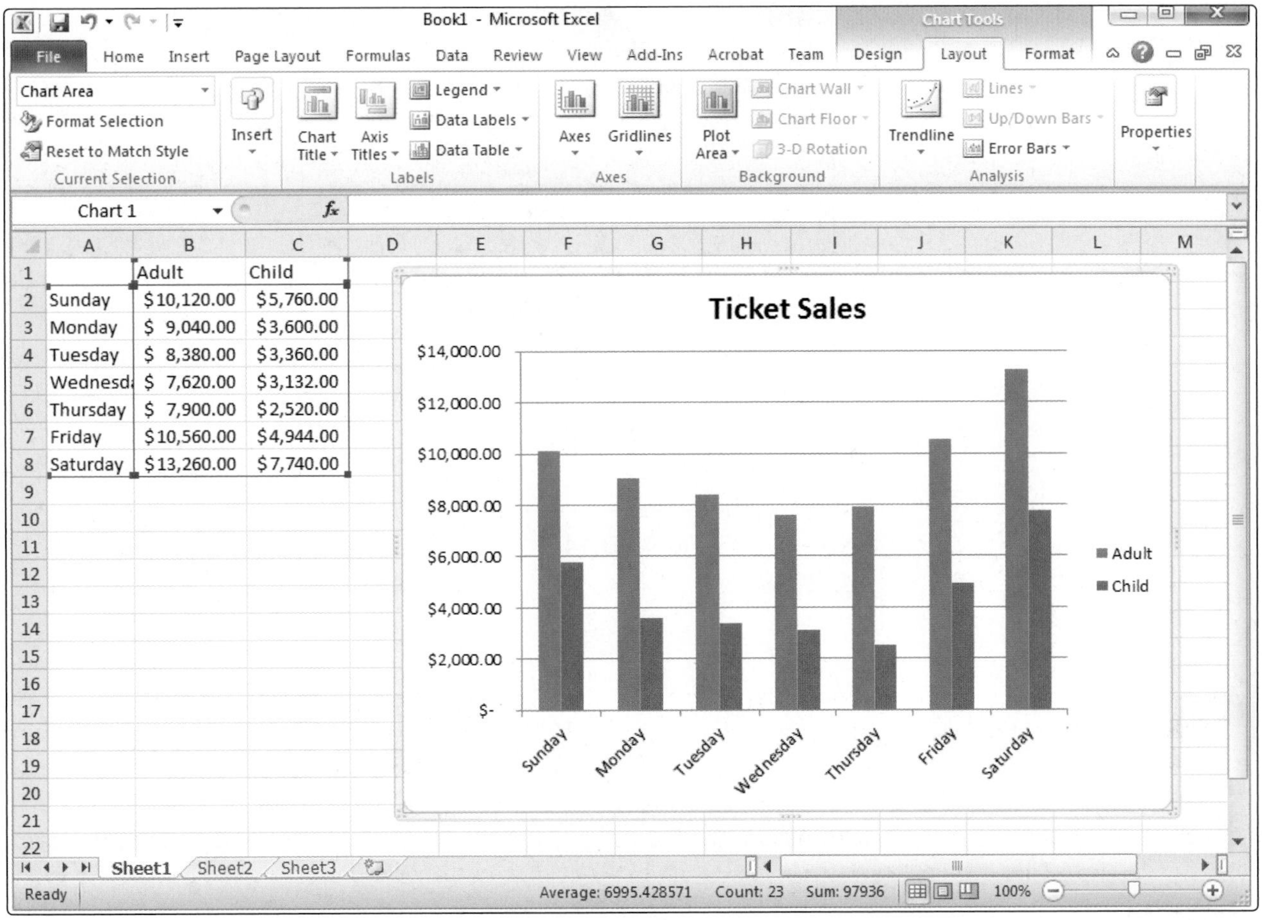

Figure B.13

Excel can create most types of charts, ranging from pie charts to line graphs to scatter plots. These types of charts are indicated by the icons seen when the Charts button is clicked on. Histograms can also be created using Excel, but require the **Data Analysis** tools, which will be discussed next.

Installing Data Analysis Tools

To get the most out of Excel as a statistical tool, you will want to use the Analysis ToolPak. When Microsoft Excel is first installed, this feature is not included. The ToolPak is an application add-in. To install the Analysis ToolPak, under the File tab, select **Options**, and then **Add-Ins** on the left-hand side. When **Add-Ins** is selected, at the bottom of the dialog box there is a Manage drop-down menu that has Excel Add-ins selected. Press the **Go** button next to this drop-down menu. A list of available add-ins will appear. Check the boxes next to **Analysis ToolPak** and **Analysis ToolPak – VBA** (see Figure B.14). Press **OK**. The data analysis tools will be installed.

With these tools installed, you can perform many useful analyses in statistics. The tool is found under the Data tab once it is installed. The button is located in the "Analysis" group of icons, and is labeled Data Analysis. When you click on the **Data Analysis** button under the Data tab, a dialog box will appear with the Excel analysis tools that are available (see Figure B.15). Using these tools, you can perform hypothesis tests, obtain basic descriptive statistics, create histograms, and run regression analysis. More detailed directions on using the Data Analysis tools to solve problems are found in the Technology sections throughout the text.

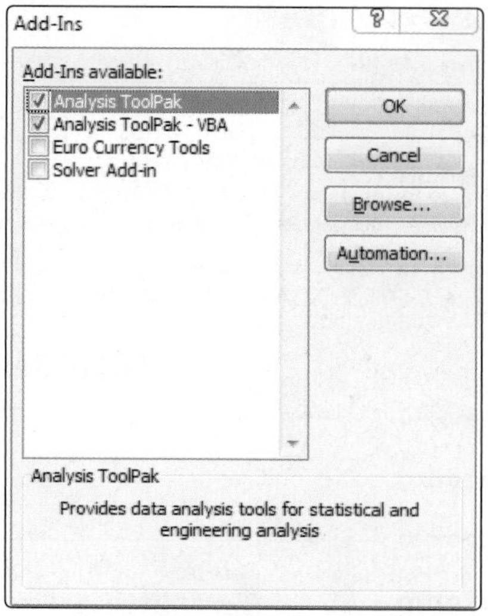

Figure B.14

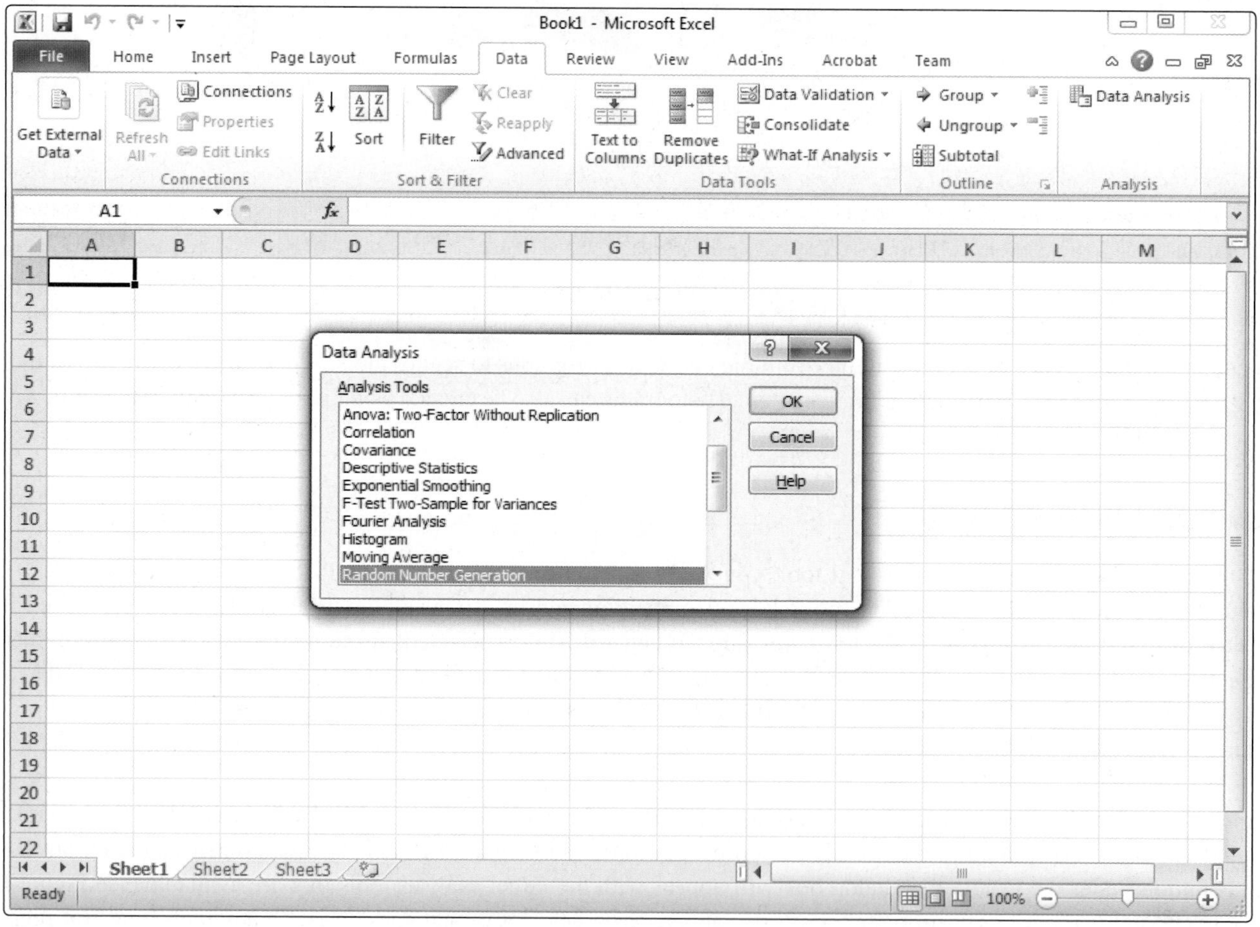

Figure B.15

Appendix C

Getting Started with Minitab

The Basics of Minitab

Minitab is a more advanced statistical software program that many businesses use for statistical analysis and quality control. In addition to these types of analytics, Minitab performs many of the basic functions of a spreadsheet program such as Microsoft Excel.

Opening a new Minitab file provides a blank worksheet. Just as in Microsoft Excel, the active cell is outlined with a thick black border. You can enter data into the active cell by typing into the active cell. To move to another cell, you can use the up/down/left/right arrow keys on the keyboard or simply click with your mouse. The first row of the worksheet (the grey cells) is reserved for column titles. Any numerical data values should be entered in the white cells of the worksheet. Cut, copy, and paste functions operate in Minitab in the same way as other spreadsheet programs. Data can be copied by highlighting the appropriate cell(s) and selecting **Copy Cells** from the **Edit** menu, or by using the keyboard shortcut **Ctrl+C**. The data can be pasted using **Paste Cells** in the **Edit** menu or by using the keyboard shortcut **Ctrl+V**. Cells can also be copied by clicking in the lower right corner of the selected cell and dragging to cover all cells for which you wish to have that value. This is similar to the fill feature in Microsoft Excel.

When you open Minitab you will notice a menu bar at the top of the screen. On that menu bar you will see menus such as Edit, Data, Calc, Stat, and Graph. The Stat menu contains many of the tools we utilize in this text. If you click on the **Stat** menu, you will see a list of concepts, such as Basic Statistics, Regression, and ANOVA. Clicking on any of these reveals other possibilities. For example, clicking on **Basic Statistics** leads to Display Descriptive Statistics, clicking on **Regression** allows you to make a Fitted Line Plot, and clicking on **ANOVA** allows you to perform a one- or two-way analysis of variance, to name a few.

If you choose **Calc** on the menu bar, you will see a list including options such as Random Data and Probability Distributions. Choosing **Random Data** will allow you to generate data according to some specified distribution. Clicking on **Probability Distributions** allows you to calculate probabilities for the binomial or Poisson distributions, among others.

Both Stat and Calc require inputs into the cells before clicking on that menu item. The Technology sections at the end of the chapters throughout the text explain how to provide these inputs for the particular application and give further explanation as to how to use the dialog boxes required along the way.

Above the worksheet is the Session window. The results of each session are displayed there. To print what is in the window, you can go to **File**, **Print Session Window**, and then press **OK**.

Cells

To create a new worksheet, you can select **New** from the **File** meanu, and then select **Minitab Worksheet**. This opens a new worksheet to use in the current project. Columns are listed from left to right and are labeled with the letter C and a number. Rows are listed from top to bottom and are labeled with numbers. To refer to a specific cell, we will use the format "Column name, Row name." For example, if we were referring to the third row in the second column, we would use "C2, 3" to reference the cell. Minitab provides a row that is not labeled for you to insert your column headings.

The cells in Minitab may contain data or text. If you are inputting text in the cells other than the cell for the column heading, the column label will add "-T" to the column name. For example, if we typed the days of the week in the first eight rows of Column C1, the column name would change to "C1-T" (see Figure C.1).

Worksheet 1 ***

↓	C1-T	C2	C3
	Days of the Week		
1	Monday		
2	Tuesday		
3	Wednesday		
4	Thursday		
5	Friday		
6	Saturday		
7	Sunday		
8	Monday		
9			
10			

Figure C.1

Filling Cells and Formulas

Now, let's input some data values into a new project to show some of the calculation features of Minitab. Choose **New** from the File menu, and select **Minitab Project**. Click OK.

Minitab allows you to fill cells when entering a series of data. Enter **100** into cell "C1, 1" and **200** into cell "C1, 2." Highlight both of the cells and locate the small black square in the bottom right corner of the highlighted section. When you move your cursor over the square, a plus sign should appear. Click the left mouse button and drag the mouse down to include "C1, 3" through "C1, 10." The cells should now be filled with values from 100 to 1000.

We can perform some basic functions on these values. From the menu bar, choose **Calc** and then **Column Statistics**. This allows you to find the sum, the mean, the standard deviation, and many other summary measures corresponding to a particular column. Choose the **mean** and click inside the box next to **Input variable**. While your cursor is in this box, all the columns that contain data will be listed in the box to the left. You can click on the column in this box and press **Select** to specify **C1** as the input variable. Click **OK** and the mean of C1 will appear in the session window (see Figure C.2).

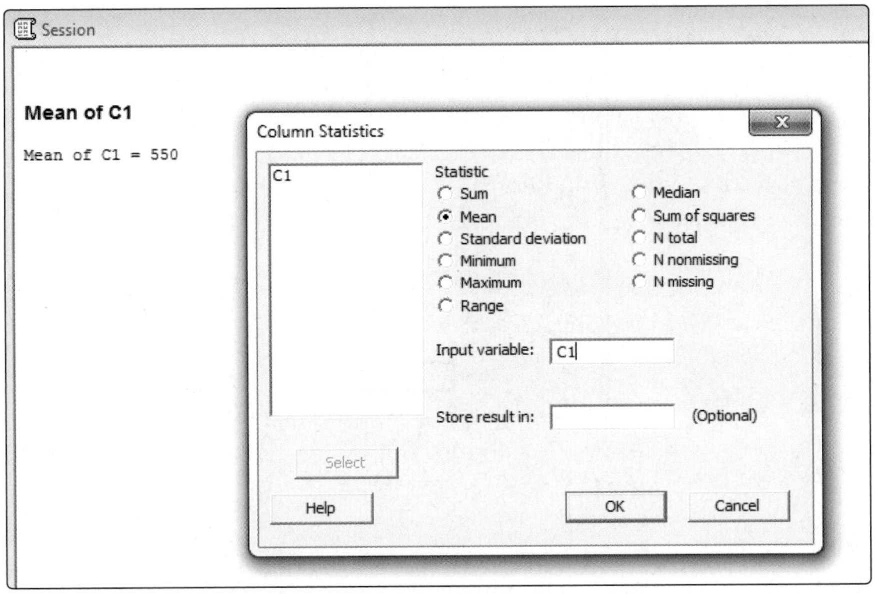

Figure C.2

To perform calculations on the data in the cells, select **Calc** and then **Calculator** from the menu bar. In the Store result in variable box, enter **C2**. You will see a drop-down menu of functions, as well as a list of functions on the right-hand side of the dialog box. In the list of functions, scroll down and select the function **Round**. Press **Select** and the text ROUND(number, decimals) will appear in the Expression box. For the parameter number, enter **C1*7/3**, and for the parameter decimals, enter **1**. This will multiply the data in Column C1 by $\frac{7}{3}$, round the result to one decimal place, and display the final value in Column C2.

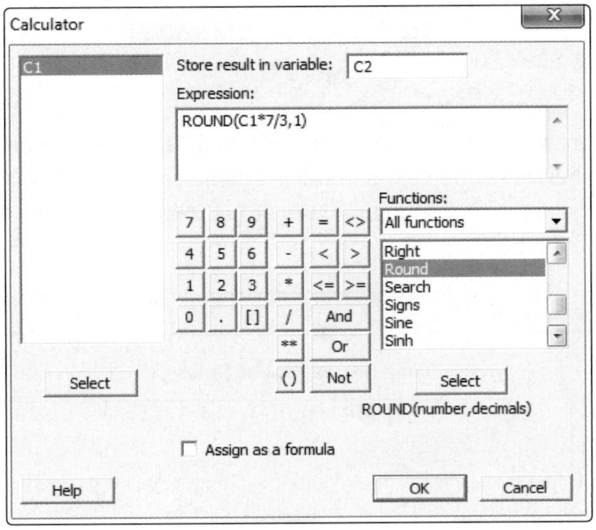

Figure C.3

Press **OK**, and you will see the values of this expression displayed in Column C2, rounded to one decimal place (see Figure C.4).

Worksheet 1 ***

↓	C1	C2	C3
1	100	233.3	
2	200	466.7	
3	300	700.0	
4	400	933.3	
5	500	1166.7	
6	600	1400.0	
7	700	1633.3	
8	800	1866.7	
9	900	2100.0	
10	1000	2333.3	
11			

Figure C.4

Graphs

Suppose you had the following data about a firm's sales per quarter for a particular year. Enter the data in a new worksheet as shown in Figure C.5.

Worksheet 1 ***

↓	C1-T	C2	C3
	Quarter	Sales	
1	First	356210	
2	Second	349800	
3	Third	370355	
4	Fourth	402775	
5			
6			

Figure C.5

To graphically display this information, select the **Graph** menu, and then **Time Series Plot**. Select the **Simple time series plot** and press **OK**. Put the cursor in the **Series** dialog box, and the available columns with data will appear in the box on the left. Since we only have one column with quantitative data in this case, only C2 appears. Select **C2** in the box and press **Select**. Press **OK**. A time series plot is created (See Figure C.6).

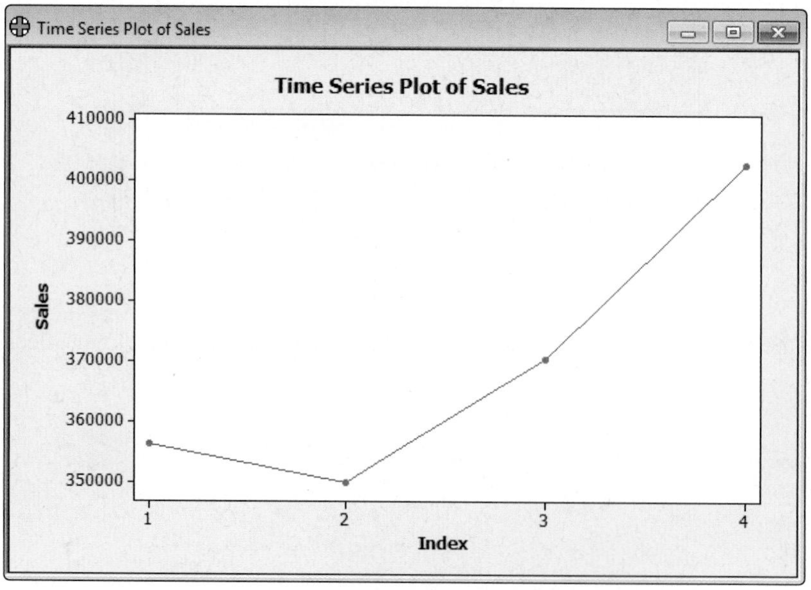

Figure C.6

The time series plot makes it easy to see that sales fell from the first quarter to the second quarter, but increased from the second quarter through the fourth quarter. Graphs in Minitab can be edited by simply double-clicking on the portion of the graph you would like to edit. This includes the graph title, axis titles, and graph area. For example, you can change the increments given on the horizontal or vertical axis by double clicking on one of the numbers on the axis.

Minitab can create a variety of graphs, some of which are not easily created with Microsoft Excel. For example, box plots and dot plots can be created using Minitab, which are not found under the graph options in Excel. Most of the available graphing options in Minitab are found under the Graph menu.

Data Analysis

One of the most important uses of Minitab is statistical analysis. Start a new Minitab project and enter the following data for the number of phone calls a company receives each hour of the work day.

Worksheet 1 ***

↓	C1	C2	C3
	Phone Calls		
1	10		
2	27		
3	21		
4	24		
5	30		
6	23		
7	36		
8	30		
9	15		
10			

Figure C.7

Select **Stat**, **Basic Statistics**, and **Display Descriptive Statistics**. In the Variables dialog box, enter **C1** and press **OK**. Basic descriptive statistics are displayed about the data in C1. This includes the number of data values, the mean, the standard deviation, the minimum value, the first quartile, the median, the third quartile, and the maximum value.

Descriptive Statistics: Phone Calls

Variable	N	N*	Mean	SE Mean	StDev	Minimum	Q1	Median	Q3
Phone Calls	9	0	24.00	2.67	8.00	10.00	18.00	24.00	30.00

Variable	Maximum
Phone Calls	36.00

Figure C.8

There are many other useful statistical features of Minitab that you will encounter throughout the text. Refer to the Technology sections at the end of the chapters throughout the text for detailed instructions on using Minitab for statistical analysis.

Answer Key

Chapter 1

Section 1.1

1. P: All Americans; S: Readers who mail in their ballots **3.** P: All shoppers at the large discount store; S: 100 shoppers chosen **5.** P: All Christmas trees sold in your city; S: 45 randomly selected Christmas trees **7.** P: All students in the 11:00 algebra class; S: All students in the 11:00 algebra class; parameter **9.** P: All professors at public universities in the United States; parameter **11.** P: All high school seniors in the Atlanta area; S: 230 seniors surveyed; statistic **13.** P: All adults in the viewing area; S: 1067 adults surveyed; statistic **15.** P: All condominiums sold in Okaloosa County; parameter **17.** P: All residents of the Northeast; S: 984 households that returned surveys; statistic **19.** Population: All coffee consumers; Sample: 6195 customers who complete the survey; Parameter: 77%; Statistics: 45%, 32%, and 23% **21.** Descriptive **23.** Descriptive **25.** Inferential **27.** The best choice is option c. because it best targets the group of women most likely to use the facility.

Section 1.2

1. a. Quantitative **b.** Discrete **c.** Ratio **3. a.** Quantitative **b.** Continuous **c.** Interval **5 a.** Quantitative **b.** Continuous **c.** Ratio **7. a.** Quantitative **b.** Continuous **c.** Ratio **9. a.** Quantitative **b.** Discrete **c.** Ratio **11. a.** Quantitative **b.** Continuous **c.** Ratio **13. a.** Quantitative **b.** Continuous **c.** Ratio **15. a.** Qualitative **b.** Neither **c.** Nominal **17. a.** Qualitative **b.** Neither **c.** Ordinal **19. a.** Qualitative **b.** Neither **c.** Ordinal **21. a.** Quantitative **b.** Discrete **c.** Ratio **23. a.** Quantitative **b.** Discrete **c.** Interval **25. a.** Quantitative **b.** Continuous **c.** Ratio **27.** One example is a person's age. Normally we would express this as a whole year, but it is really a measurement of continuous time. Answers will vary.

Section 1.3

1. True; before collecting data, a researcher must first state the question to be answered in a statistical study. **3.** False; if a researcher wishes to determine a cause-and-effect relationship, she should use an **experimental** study. **5.** True; an Institutional Review Board (IRB) reviews the design of a study to ensure that no unnecessary harm will come to the subjects involved, so an IRB will require researchers to get the informed consent of participants. **7.** False; participants in an experiment should be **assigned** to groups by researchers to ensure that similar characteristics are represented in both the treatment group and the control group. **9.** Subject or participant **11.** Treatment **13.** Single-blind **15.** Participant **17.** Observational study **19.** Observational study **21.** Experiment **23.** Systematic **25.** Simple random **27.** Cluster **29.** Stratified **31.** Convenience **33.** Cross-sectional **35.** Longitudinal **37.** Longitudinal **39.** Meta-analysis **41.** Case study **43.** The population would be restricted from all adults at risk for heart attacks to just African-American women over the age of 50 who are at risk for heart attacks. No, this would not apply to an uncle regardless of his ethnicity because he is not a woman and therefore not part of the population of the study. **45.** To get a sample with no errors, choose every 8th part and start with any part without an error. If you choose every 5th part, there will be parts with errors and without, but the sample may not be representative of the population. If you choose every 16th part, you will have the same situation as when choosing every 8th part since 16 is a multiple of 8.

Section 1.4

1. Processing error **3.** Bias **5.** Nonadherent **7.** Variables: amount of rain, air quality level; Possible ways to measure air quality: temperature, humidity, visibility, dust levels, oxygen levels, nitrogen levels, ozone levels, particle pollution, pollen levels; Terms that need more precise definition: quality of air, more rain **9.** Answers will vary. Group discussions could lead to students in varying fields of study taking the question to other forums for discussion. Nevertheless, instructors are encouraged to help students realize the difficult process of both defining and measuring such abstract and often controversial ideas.

11. Answers might include: lack of representative sample from store employees gathering data; participants might not want to tell a store employee that they don't like the store **13.** Answers might include: self-selected sample; if the call is not free, the cost could eliminate possible votes; not all Americans will be interested in watching the TV show; lines might be busy, preventing some votes from being counted **15.** Answers should include a method of sampling the state's voters (not just a convenient area), possible survey questions, and evidence of avoiding potential biases.

Chapter 1 Exercises

1. a. Statistic **b.** Inferential **3. a.** Qualitative; ordinal **b.** Quantitative; interval **c.** Quantitative; ratio **5.** No; yes **7. a.** All shoppers **b.** Convenience sampling **c.** If the population is only the customers of that store, then it is representative. **9.** Answers will vary. **11. a.** Population: All adults; Study: Experiment **b.** Answers will vary. **13. a.** Stratified **b.** Answers will vary. **15. a.** Adults in America **b.** 1067 adults surveyed **c.** 63% in favor **d.** Answers will vary.

Chapter 2

Section 2.1

1. a. 5

Class	b. Class Boundaries	c. Midpoint	d. Relative Frequency	e. Cumul Freq
15–19	14.5–19.5	17	$\frac{7}{30} \approx 23\%$	7
20–24	19.5–24.5	22	$\frac{4}{15} \approx 27\%$	15
25–29	24.5–29.5	27	$\frac{1}{3} \approx 33\%$	25
30–34	29.5–34.5	32	$\frac{1}{15} \approx 7\%$	27
35–39	34.5–39.5	37	$\frac{1}{10} = 10\%$	30

3. a. 4

Class	b. Class Boundaries	c. Midpoint	d. Relative Frequency	e. Cumul Freq
15–18	14.5–18.5	16.5	$\frac{1}{10} = 10\%$	2
19–22	18.5–22.5	20.5	$\frac{1}{4} = 25\%$	7
23–26	22.5–26.5	24.5	$\frac{1}{5} = 20\%$	11
27–30	26.5–30.5	28.5	$\frac{1}{4} = 25\%$	16
31–34	30.5–34.5	32.5	$\frac{1}{5} = 20\%$	20

5. a. 10

Class	b. Class Boundaries	c. Midpoint	d. Relative Frequency	e. Cumul Freq
15–24	14.5–24.5	19.5	$\frac{3}{11} \approx 27\%$	9
25–34	24.5–34.5	29.5	$\frac{8}{33} \approx 24\%$	17
35–44	34.5–44.5	39.5	$\frac{4}{11} \approx 36\%$	29
45–54	44.5–54.5	49.5	$\frac{1}{33} \approx 3\%$	30
55–64	54.5–64.5	59.5	$\frac{1}{11} \approx 9\%$	33

7. a. 7

Class	b. Class Boundaries	c. Midpoint	d. Relative Frequency	e. Cumul Freq
18–24	17.5–24.5	21	$\frac{2}{31} \approx 6\%$	2
25–31	24.5–31.5	28	$\frac{7}{31} \approx 23\%$	9
32–38	31.5–38.5	35	$\frac{4}{31} \approx 13\%$	13
39–45	38.5–45.5	42	$\frac{15}{31} \approx 48\%$	28
46–52	45.5–52.5	49	$\frac{3}{31} \approx 10\%$	31

9. a. 4

Class	b. Class Boundaries	c. Midpoint	d. Relative Frequency	e. Cumul Freq
16–19	15.5–19.5	17.5	$\frac{3}{14} \approx 21\%$	12
20–23	19.5–23.5	21.5	$\frac{1}{7} \approx 14\%$	20
24–27	23.5–27.5	25.5	$\frac{15}{56} \approx 27\%$	35
28–31	27.5–31.5	29.5	$\frac{3}{14} \approx 21\%$	47
32–35	31.5–35.5	33.5	$\frac{9}{56} \approx 16\%$	56

11.

Heights of Men (in Inches)	
Class	**Frequency**
69.0–69.9	3
70.0–70.9	4
71.0–71.9	5
72.0–72.9	9
73.0–73.9	3
74.0–74.9	3
75.0–75.9	3

13.

Number of Marbles in the Jar	
Class	**Frequency**
1100–1199	1
1200–1299	3
1300–1399	4
1400–1499	4
1500–1599	4
1600–1699	4

15.

Class	f	Class Boundaries	Mid	Relative Frequency	Cumul Freq
15-19	1	14.5–19.5	17	$\frac{1}{16} \approx 6\%$	1
20-24	1	19.5–24.5	22	$\frac{1}{16} \approx 6\%$	2
25-29	5	24.5–29.5	27	$\frac{5}{16} \approx 31\%$	7
30-34	5	29.5–34.5	32	$\frac{5}{16} \approx 31\%$	12
35-39	2	34.5–39.5	37	$\frac{1}{8} \approx 13\%$	14
40-44	2	39.5–44.5	42	$\frac{1}{8} \approx 13\%$	16

17.

Class	f	Class Boundaries	Mid	Relative Frequency	Cumul Freq
1800–2199	1	1799.5–2199.5	1999.5	$\frac{1}{15} \approx 7\%$	1
2200–2599	4	2199.5–2599.5	2399.5	$\frac{4}{15} \approx 27\%$	5
2600–2999	6	2599.5–2999.5	2799.5	$\frac{2}{5} = 40\%$	11
3000–3399	3	2999.5–3399.5	3199.5	$\frac{1}{5} = 20\%$	14
3400–3799	1	3399.5–3799.5	3599.5	$\frac{1}{15} \approx 7\%$	15

19. a. 27 **b.** 11–15 pounds **c.** 14.8% **d.** There is not enough information given to know this. **e.** 16 pounds **f.** 3 pounds **21. a.** 439 **b.** Approximately 34.9% **c.** 41 **d.** Approximately 22.3%

Section 2.2

1. Bar graph

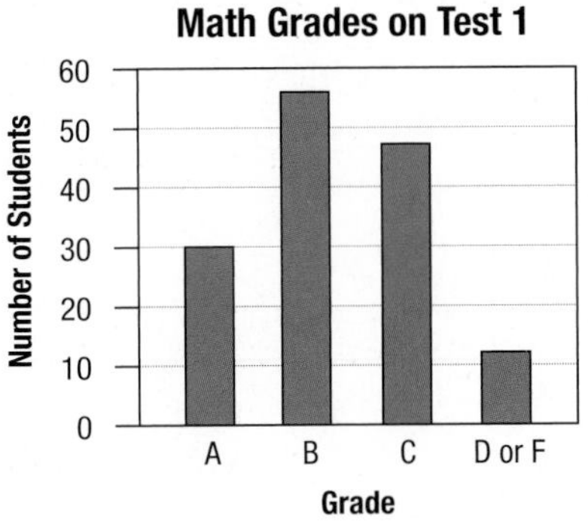

3. Pareto chart

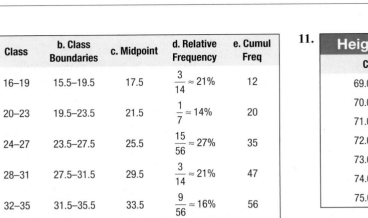

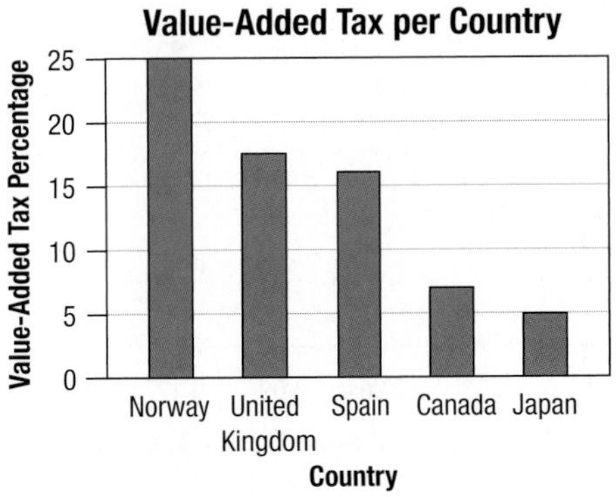

5.

Side-by-side bar graph

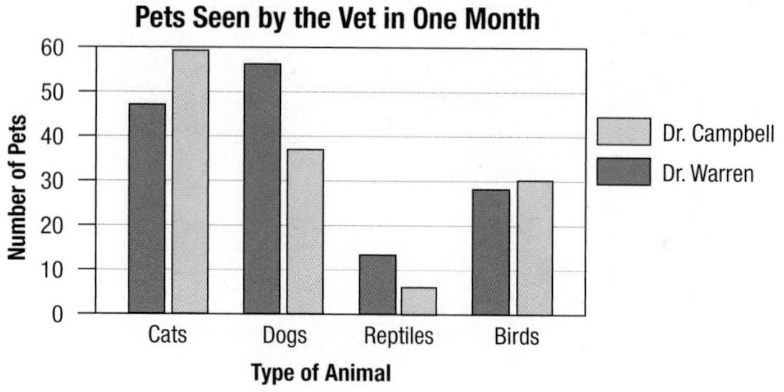

Stacked bar graph

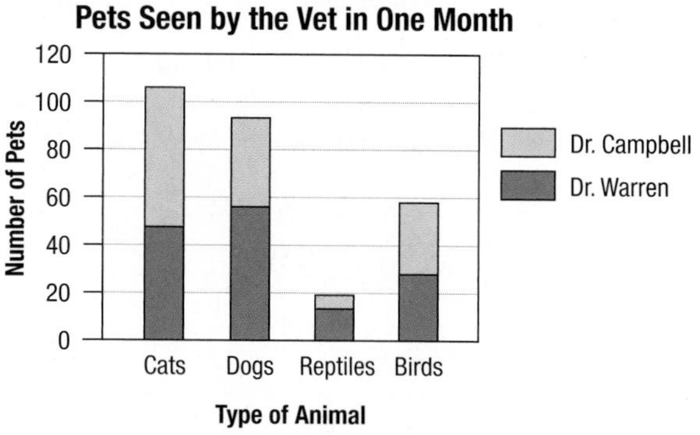

5. a. Dr. Warren, side-by-side bar graph **b.** Reptiles, stacked bar graph **c.** Dr. Campbell, side-by-side bar graph **d.** Reptiles, side-by-side bar graph **7.** English = 45°; Business = 149°; Education = 91°; Science = 75°

9. a.

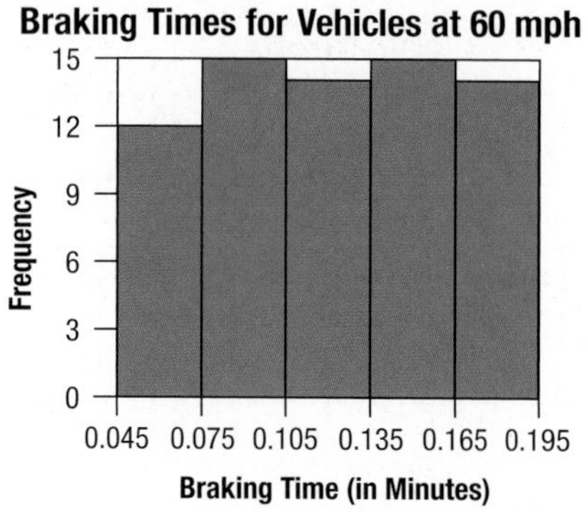

b.

Class	Relative Frequency
0.05–0.07	$\frac{6}{35} \approx 17\%$
0.08–0.10	$\frac{3}{14} \approx 21\%$
0.11–0.13	$\frac{1}{5} = 20\%$
0.14–0.16	$\frac{3}{14} \approx 21\%$
0.17–0.19	$\frac{1}{5} = 20\%$

c.

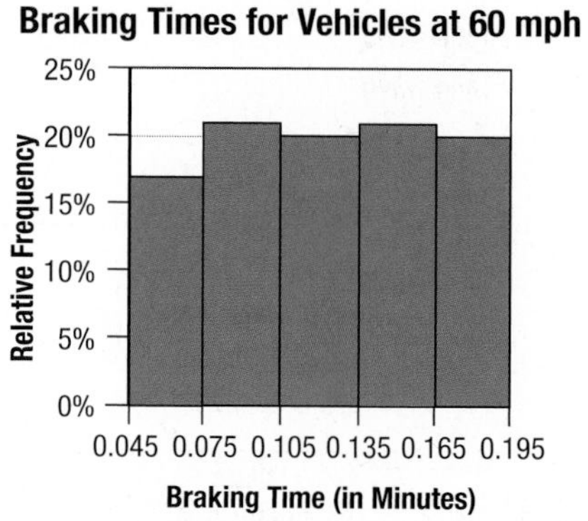

11. a.

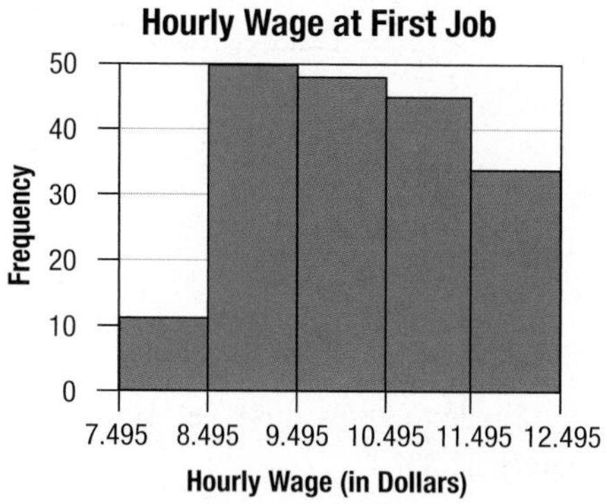

b.

Class	Relative Frequency
7.50–8.49	$\frac{4}{63} \approx 6\%$
8.50–9.49	$\frac{50}{189} \approx 26\%$
9.50–10.49	$\frac{16}{63} \approx 25\%$
10.50–11.49	$\frac{5}{21} \approx 24\%$
11.50–12.49	$\frac{34}{189} \approx 18\%$

c.

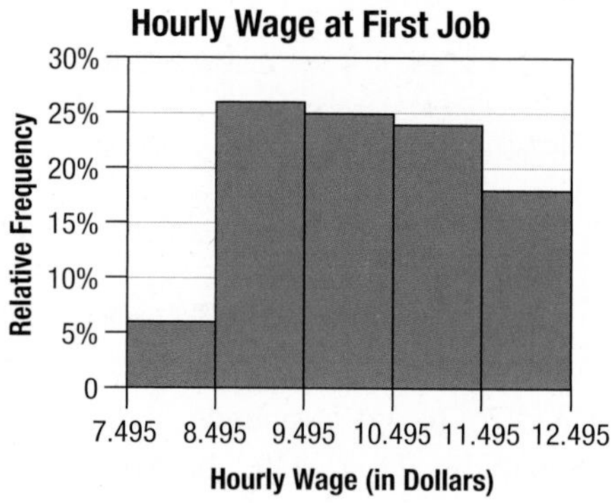

13.

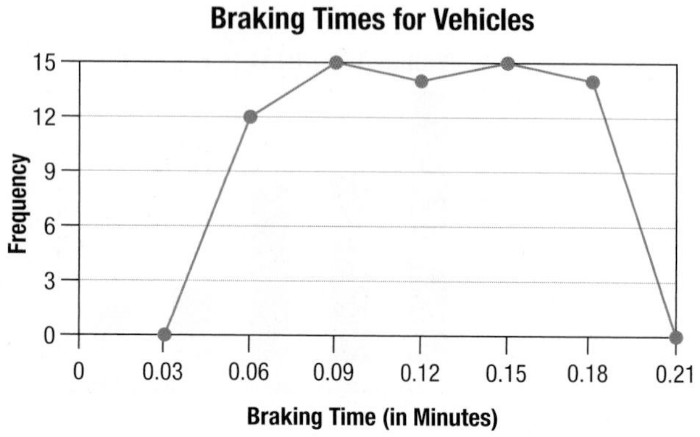

15.

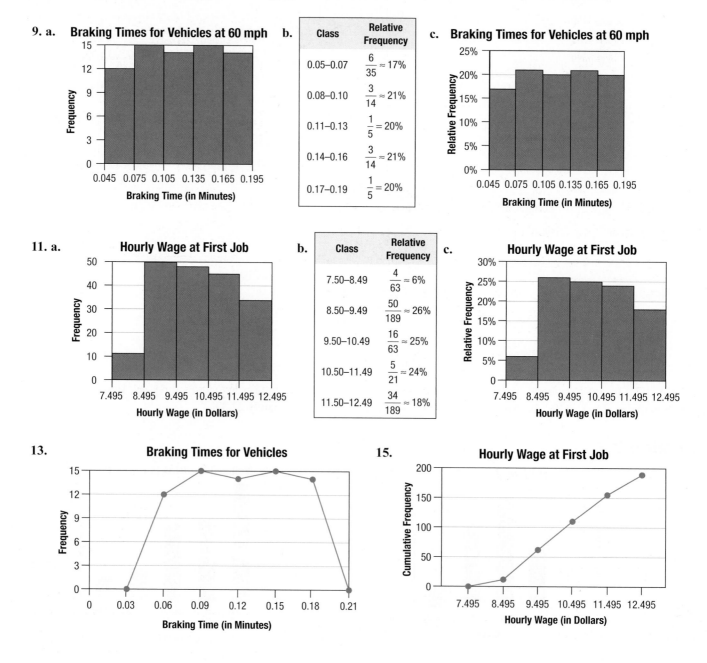

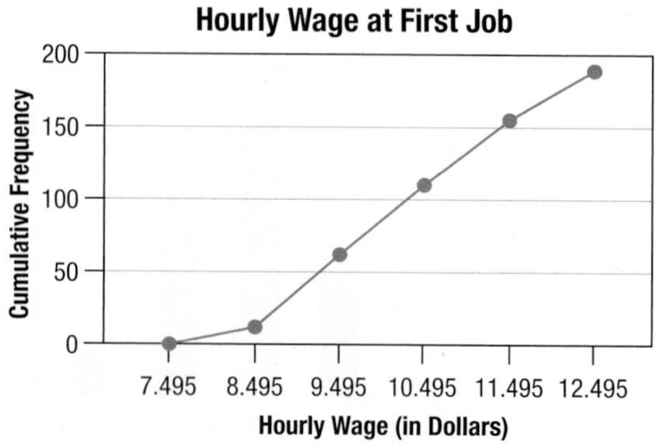

17. **Caloric Intake in One Day for Men Aged 20–39**

Stem	Leaves
1	8
2	7 2 5 8 6 6 2 8 5 9
3	0 1 5 0

Key: 1 | 8 = 1800

19. **September Precipitation in Inches for Towns in Alaska**

Stem	Leaves
25	0 4 5 6 9
26	1 2 4 4 7 8
27	0 3 4 7 8
28	1 1 6 8 9

Key: 25 | 0 = 2.50

21. **Plastic Bag Usage for a Single Shopping Trip**

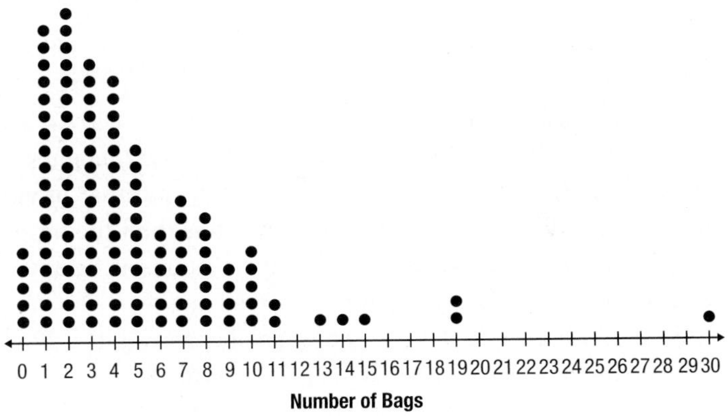

23. a. 74.000 cm **b.** 74.030 cm **c.** Answers will vary. Encourage answers that address the reader's difficulty in knowing the precise measurements due to the horizontal labels or lack of tick marks but acknowledge situations where this might not be a focus. Answers might also point to the fact that the overall range is still very small and the graph might be used to show how close the diameters are in reality. **25. a.** 2.38 dollars per gallon **b.** 2.56 dollars per gallon **c.** 23-Aug-2010 **d.** 27-Dec-2010

27. a. 12 **b.** 34 **c.** 18 **d.** 29 **29. a.** $N = 85$ **b.** Smallest: 1.82 mm; largest: 3.76 mm **c.** 2.99 mm and 3.08 mm **d.** Longer **31.** Answers will vary. Encourage thoughtful answers that point out the difficulty of displaying the wide range of data points as well as the outliers, although students have yet to be introduced formally to the concepts of "range" and "outliers."

Section 2.3

1. Cross-sectional **3.** Cross-sectional **5.** Time-series **7.** Answers will vary. A possible answer is that the scale is not a good choice since a dramatic change is shown even though the change is only a few cents. **9.** The class widths are not equal. **11.** Uniform **13.** Symmetric **15.** Skewed to the right **17.** Skewed to the left **19.** Uniform

Chapter 2 Exercises

1.

Class	*f*	Class Boundries	Midpoint	Rel Frequency	Cumul Freq
0–4	7	−0.5–4.5	2	$\frac{7}{24} \approx 29\%$	7
5–9	6	4.5–9.5	7	$\frac{1}{4} = 25\%$	13
10–14	5	9.5–14.5	12	$\frac{5}{24} \approx 21\%$	18
15–19	3	14.5–19.5	17	$\frac{1}{8} \approx 13\%$	21
20–24	2	19.5–24.5	22	$\frac{1}{12} \approx 8\%$	23
25–29	1	24.5–29.5	27	$\frac{1}{24} \approx 4\%$	24

3. **Saline Concentration (in Terms of Specific Gravity)**

Stem	Leaves
101	7 8 9 9
102	0 1 1 1 2 2 2 2 2 3 3 3 4 5

Key: 101 | 7 = 1.017

5. a. Line graph **b.** Bar graph **c.** Stacked or side-by-side bar graph **d.** Stem-and-leaf plot **7. a.** 7 **b.** 1 **c.** Wives who work less spend more time doing leisure activities. **d.** A pair of side-by-side bar graphs, one for husbands and one for wives **9. a.** 19% **b.** 53% **c.** 27% **d.** 47% **11. a.** Symmetric **b.** Skewed to the left **c.** Answers will vary. One possibility is that on the whole students knew less information prior to the material being taught. **d.** Answers will vary. One possibility is that the majority of students learned most of the material.

Chapter 3

Section 3.1

1. Mean = 5; Median = 4.5; Unimodal at 1 **3.** Mean ≈ 7.6; Median = 8; Unimodal at 10 **5.** Mean ≈ 4.883; Median = 4.9; No mode **7.** Mean = 0.75855; Median = 0.7353; No mode **9.** Mean ≈ −17.9; Median = −42; Unimodal at −42 **11.** Mean ≈ 7.49 pounds; Median = 7.5 pounds; Bimodal at 7.3 pounds and 7.5 pounds **13.** Mean ≈ 16.56 minutes; Median = 14.9 minutes; Unimodal at 14.9 minutes **15.** Mean ≈ 33.7 years; Median = 32 years; Multimodal at 29 years, 31 years, and 32 years **17.** 38.5 lb **19.** 2.78 **21.** $341.818 ≈ $341.82 (rounded for currency) **23.** Approximately 74.5% **25.** $10.201 ≈ $10.20 (rounded for currency) **27.** Median since the price of the Ferrari F430 is an outlier **29.** Mean **31.** Mode = *** **33.** Mean = B; Median = B; Mode = B

Section 3.2

1. Range = 5; $\sigma \approx 1.9$; $\sigma^2 \approx 3.7$ **3.** Range = 0; $\sigma \approx 0$; $\sigma^2 \approx 0$ **5.** Range = 8; $\sigma \approx 2.4$; $\sigma^2 \approx 5.7$ **7.** Range = 5; $s \approx 2.1$; $s^2 = 4.3$ **9.** Range = 0; $s = 0$; $s^2 = 0$ **11.** Range = 1.5; $s \approx 0.47$; $s^2 \approx 0.22$; **13.** 1.14 pounds **15.** 12.52 minutes **17.** 7.8 years **19.** False; the **population** variance is the sum of the squared deviations divided by the **number of values in the population**, while the **sample** variance is the sum of the squared deviations divided by **one less than the number of data values in the sample**, so these values are almost always different. **21.** True; the standard deviation can be any value greater than or equal to zero. The greater the standard deviation, the more the data values are spread out. **23.** A: CV ≈ 29.7%; B: CV ≈ 13.1%; Larger spread: A **25.** A: CV ≈ 16.8%; B: CV ≈ 9.6%; Smaller spread: B **27.** 95% **29.** 97.5% **31.** 97.5% **33.** 88.9% **35.** 8.4 points **37.** 0.004

Section 3.3

1. a. 7.4 pounds **b.** 73rd **3. a.** 6.55 Tweets per day **b.** 44th **5. a.** 9,611,568 **b.** 88th **7.** 32nd **9.** 6450, 11,201, 14,788, 15,692.5, 18,865 **11.** 1.2, 1.6, 1.75, 2.8, 4.1 **13.** 5.4, 7.15, 8.4, 9.2, 10.1 **15.** −12, −10, −5, 1, 4

17.

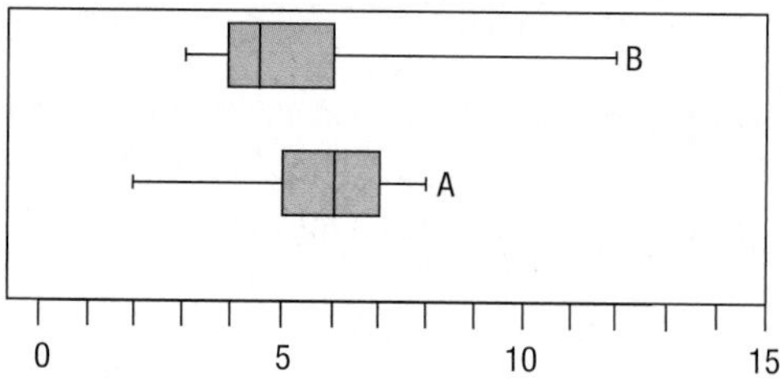

a. A **b.** A **c.** Neither; both data sets have IQR = 2

19.

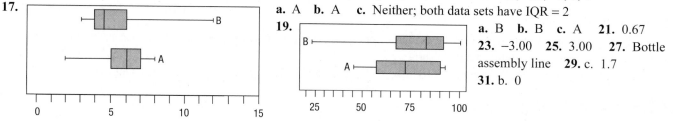

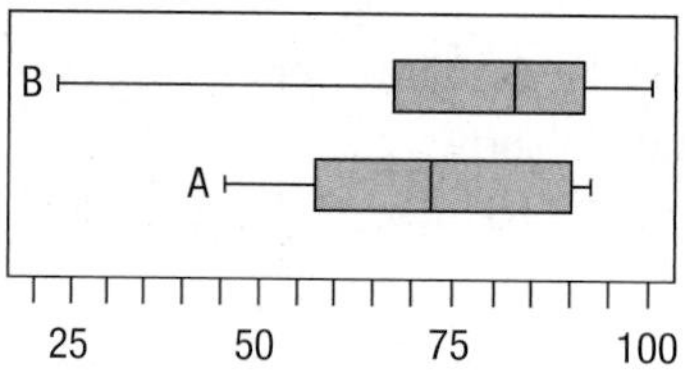

a. B **b.** B **c.** A **21.** 0.67 **23.** −3.00 **25.** 3.00 **27.** Bottle assembly line **29.** c. 1.7 **31.** b. 0

Chapter 3 Exercises

1. $10.92 **3. a.** False **b.** True **5.** 83 **7.** True **9.** At least 88.9% **11.** $s \approx \$17.710 \approx \$17.71$ (rounded for currency); $s^2 \approx 313.655$ **13.** 15.0 **15.** Yes, because if you put them back in numerical order from smallest to largest, you get the five-number summary, 8, 12.5, 14, 17.5, 21, so the first quartile is 12.5.

Chapter 4

Section 4.1

1. {HH, HT, TH, TT} **3.** {BBBB, BBBG, BBGB, BGBB, GBBB, BBGG, BGBG, GBBG, BGGB, GBGB, GGBB, BGGG, GBGG, GGBG, GGGB, GGGG} **5.** {RSunL, RSunC, RNoL, RNoC, BSunL, BSunC, BNoL, BNoC, SSunL, SSunC, SNoL, SNoC} **7.** {1A, 1N, 2A, 2N, 3A, 3N} **9.** Experimental **11.** Classical **13.** Subjective

15. a. $\frac{51}{175} \approx 0.2914$ **b.** $\frac{40}{175} = \frac{8}{35} \approx 0.2286$ **c.** $\frac{55+51}{175} = \frac{106}{175} \approx 0.6057$ **17.** $\frac{16}{66} = \frac{8}{33} \approx 0.2424$ **19.** $\frac{31}{365} \approx 0.0849$

21. $\frac{150}{350} = \frac{3}{7} \approx 0.4286$ **23.** $\frac{1}{45} \approx 0.0222$ **25.** $\frac{10}{1000} = \frac{1}{100} = 0.01$

Section 4.2

1. Complement: The 211 apple trees that are not ready for harvesting **3.** Complement: The 17 players who are not left-handed **5.** Complement: The 30% of viewers who are 30 years old or younger **7.** $\frac{10}{11} \approx 0.9091$ **9.** $\frac{31}{32} \approx 0.9688$

11. Not mutually exclusive **13.** Not mutually exclusive **15.** Not mutually exclusive **17.** Mutually exclusive

19. $\frac{12}{52} + \frac{13}{52} - \frac{3}{52} = \frac{22}{52} = \frac{11}{26} \approx 0.4231$ **21.** $\frac{12}{19} + \frac{9}{19} - \frac{5}{19} = \frac{16}{19} \approx 0.8421$

23. $\frac{15}{15+6+8+5+11} + \frac{5}{15+6+8+5+11} = \frac{20}{45} = \frac{4}{9} \approx 0.4444$ **25.** $\frac{1}{36} + \left(1 - \frac{11}{36}\right) = \frac{1}{36} + \frac{25}{36} = \frac{26}{36} = \frac{13}{18} \approx 0.7222$

27. $0.1181 + 0.0696 = 0.1877$ **29.** $\frac{18}{36} + \frac{10}{36} - \frac{6}{36} = \frac{22}{36} = \frac{11}{18} \approx 0.6111$

31. $\frac{23+40}{23+40+19+11} + \frac{23+19}{23+40+19+11} - \frac{23}{23+40+19+11} = \frac{82}{93} \approx 0.8817$

Section 4.3

1. Independent **3.** Dependent **5.** Independent **7.** $\frac{4}{52} \cdot \frac{13}{52} = \frac{1}{52} \approx 0.0192$ **9.** $\frac{4}{52} \cdot \frac{13}{52} + \frac{13}{52} \cdot \frac{4}{52} = \frac{1}{26} \approx 0.0385$

11. $(0.5)^5 \approx 0.0313$ **13.** $\frac{2}{26} = \frac{1}{13} \approx 0.0769$ **15.** $\frac{1}{3} \approx 0.3333$ **17. a.** $\frac{4}{4+5} = \frac{4}{9} \approx 0.4444$

b. $\frac{2}{13+5+2+3} = \frac{2}{23} \approx 0.0870$ **c.** $\frac{2}{4+2} = \frac{1}{3} \approx 0.3333$ **19.** $\frac{4}{52} \cdot \frac{11}{51} = \frac{11}{663} \approx 0.0166$

21. $\frac{16}{26+19+11+16} \cdot \frac{19}{26+19+11+16-1} = \frac{38}{639} \approx 0.0595$ **23.** $26 \cdot 25 \cdot 10 \cdot 9 \cdot 8 \cdot 7 = 3{,}276{,}000$ **25.** $10 \cdot 10 \cdot 10 \cdot 5 = 5000$

27. $8 \cdot 3 \cdot 4 = 96$ **29.** $\frac{2 \cdot 1 \cdot 3}{5 \cdot 4 \cdot 6} = \frac{1}{20} = 0.05$ **31.** $\frac{3}{11 \cdot 14} = \frac{3}{154} \approx 0.0195$

Section 4.4

1. 720 **3.** 30 **5.** 15 **7.** 15 **9.** 1 **11.** 10 **13.** 1 **15.** 12 **17.** 60 **19.** 8 **21.** 1 **23.** 56 **25.** $\frac{n!}{n!(n-n)!} = 1$

27. $\frac{n!}{(n-1)!} = n$ **29.** $\frac{n!}{[n-(n-1)]!} = n!$ **31.** ${}_{12}P_2 = 132$ **33.** ${}_{18}C_4 = 3060$ **35.** ${}_{8}P_3 = 336$ **37.** ${}_{15}C_{10} = 3003$

39. ${}_{12}C_4 = 495$ **41.** ${}_{15}P_4 = 32{,}760$ **43.** $\frac{10!}{3!\,3!\,1!\,2!\,1!} = 50{,}400$ **45.** $\frac{8!}{4!\,2!\,2!} = 420$ **47.** The only word without any repeated letters is the first word, BEAST. Therefore, it will have the most five-letter arrangements possible.

49. $\frac{1}{{}_{20}C_3} = \frac{1}{1140} \approx 0.0009$ **51.** $\frac{2}{\frac{4!}{2!2!}} = \frac{2}{6} = \frac{1}{3} \approx 0.3333$

Section 4.5

1. a. $4 \cdot 25 \cdot 24 \cdot 10 \cdot 9 \cdot 8 = 1{,}728{,}000$ **b.** $\frac{1 \cdot 25 \cdot 24 \cdot 10 \cdot 9 \cdot 8}{4 \cdot 25 \cdot 24 \cdot 10 \cdot 9 \cdot 8} = \frac{1}{4} = 0.25$ **3. a.** $5 \cdot 4 \cdot 8 \cdot 7 = 1120$

b. $\frac{1 \cdot 4 \cdot 8 \cdot 7}{5 \cdot 4 \cdot 8 \cdot 7} = \frac{1}{5} = 0.2$ **5. a.** ${}_{13}C_2 \cdot {}_{13}C_3 = 22{,}308$ **b.** $\frac{22{,}308}{{}_{52}C_5} = \frac{22{,}308}{2{,}598{,}960} = \frac{143}{16{,}660} \approx 0.0086$

7. $5 \cdot {}_{10}C_6 \cdot {}_6C_2 = 15{,}750$ **9. a.** ${}_{20}C_2 \cdot {}_8C_3 \cdot {}_5C_2 = 106{,}400$

b. $\frac{{}_1C_1 \cdot {}_{19}C_1 \cdot {}_8C_3 \cdot {}_5C_2}{{}_{20}C_2 \cdot {}_8C_3 \cdot {}_5C_2} + \frac{{}_1C_1 \cdot {}_{19}C_1 \cdot {}_8C_3 \cdot {}_5C_2}{{}_{20}C_2 \cdot {}_8C_3 \cdot {}_5C_2} - \frac{1 \cdot {}_8C_3 \cdot {}_5C_2}{{}_{20}C_2 \cdot {}_8C_3 \cdot {}_5C_2} = \frac{{}_1C_1 \cdot {}_{19}C_1 + {}_1C_1 \cdot {}_{19}C_1 - 1}{{}_{20}C_2} = \frac{37}{190} \approx 0.1947$

11. ${}_7C_7 \cdot {}_{(6+6+8)}C_4 = 4845$ **13. a.** ${}_{41}C_4 \cdot {}_{104}C_4 \cdot {}_8C_4 \cdot {}_{12}P_{12} = 15{,}613{,}371{,}090{,}319{,}242{,}240{,}000 \approx 1.56 \times 10^{22}$

b. $\frac{{}_{41}C_{12}}{{}_{(41+104+8)}C_{12}} = \frac{7{,}898{,}654{,}920}{220{,}667{,}975{,}965{,}944{,}780} = \frac{2542}{71{,}016{,}901{,}053} \approx 3.58 \cdot 10^{-8} \approx 0.00000004$

Chapter 4 Exercises

1. ${}_{12}C_5 = 792$ **3.** ${}_6C_2 \cdot {}_5C_2 \cdot {}_7C_2 = 3150$ **5. a.** $7 \cdot 10 \cdot 10 \cdot 5 = 3500$ **b.** $6 \cdot 10 \cdot 10 \cdot 5 - 1 = 2999$ **c.** $5 \cdot 10 \cdot 10 \cdot 5 = 2500$

7. ${}_{10}C_4 = 210$ **9.** $({}_{12}C_3 \cdot {}_4C_1) \cdot ({}_9C_3 \cdot {}_3C_1) \cdot ({}_6C_3 \cdot {}_2C_1) \cdot ({}_3C_3 \cdot {}_1C_1) = 8{,}870{,}400$ **11.** $\frac{{}_{13}C_5}{{}_{52}C_5} = \frac{1287}{2{,}598{,}960} = \frac{33}{66{,}640} \approx 0.0005$

13. $1 - \frac{6}{6 \cdot 6 \cdot 6 \cdot 6 \cdot 6} = \frac{7770}{7776} = \frac{1295}{1296} \approx 0.9992$ **15.** $1 - \left[\left(1 - \frac{32}{87}\right) + \left(1 - \frac{27}{87}\right) - \frac{38}{87}\right] = \frac{10}{87} \approx 0.1149$

17. $1 - (0.2 + 0.35 + 0.15) = 0.3$ **19.** $1 - \left[\left(1 - \frac{22}{62}\right) + \left(1 - \frac{39}{62}\right) - \frac{9}{62}\right] = \frac{8}{62} = \frac{4}{31} \approx 0.1290$ **21.** $\frac{3}{6 \cdot 6} = \frac{1}{12} \approx 0.0833$

23. $\frac{13}{52} \cdot \frac{13}{51} = \frac{13}{204} \approx 0.0637$ **25.** $\frac{({}_{28}C_2)({}_{41}C_2)({}_{35}C_0)({}_{18}C_0)}{{}_{122}C_4} = \frac{4428}{125{,}477} \approx 0.0353$

Chapter 5

Section 5.1

1. Valid **3.** Not valid; $P(X = x)$ cannot have a negative value.

5.

x	P(X = x)
0	$\frac{1}{16} = 0.0625$
1	$\frac{4}{16} = \frac{1}{4} = 0.25$
2	$\frac{6}{16} = \frac{3}{8} = 0.375$
3	$\frac{4}{16} = \frac{1}{4} = 0.25$
4	$\frac{1}{16} = 0.0625$

7.

x	P(X = x)
0	$\frac{6}{36} = \frac{1}{6}$
1	$\frac{10}{36} = \frac{5}{18}$
2	$\frac{8}{36} = \frac{2}{9}$
3	$\frac{6}{36} = \frac{1}{6}$
4	$\frac{4}{36} = \frac{1}{9}$
5	$\frac{2}{36} = \frac{1}{18}$

9. $\mu = E(X) = 17.8$; $\sigma \approx 3.4$ **11.** $\mu = E(X) = 12.2$; $\sigma \approx 9.80$ **13. a.** –\$3.24 **b.** –\$324 **15. a.** $-\frac{325}{72} \approx -\$4.51$ **b.** $\frac{325}{72} \approx \$4.51$ **c.** $-\frac{1625}{12} \approx -\$135.42$ **17. a.** \$73.20 **b.** \$732,000 **c.** California **19. a.** 1 ticket: –\$0.50, 3 tickets: –\$1.00, 5 tickets: –\$1.50 **b.** 1 **21. a.** $-\frac{160}{3} \approx -\$53.33$ **b.** \$4800 **23.** Answers will vary. Carnival Game 2 has a positive expected value (\$0.20) with a small standard deviation (\$1.12), but the standard deviation still puts the carnival at risk for a possible loss over time. Since Carnival Game 1 has a negative expected value (–\$0.98), few would choose this game. However, with such a large standard deviation (\$7.70) and the expected value being close to \$0.00, one might take the riskier game and end up with a bigger positive income.

Section 5.2

1. No; more than two possible outcomes for each trial, no fixed number of trials **3.** Yes; it can be modeled using a binomial distribution. **5.** 0.2304 **7.** 1.0000 (From table: 1.0000 or 0.9999) **9.** 0.3743 (From table: 0.3743 or 0.3742) **11.** 0.4096 **13.** 0.0004 (From table: 0.0004 or 0.0003) **15. a.** 0.4199 (From table: 0.4199 or 0.4200) **b.** 0.028 **17. a.** 0.1574 **b.** 0.3075 (From table: 0.3075 or 0.3076) **19. a.** 0.0256 **b.** 0.5904 **c.** 0.4096 **21. a.** 1.0000 **b.** 0.6733 **c.** 0.00000000000005 (From table: 0.0000) **23. a.** 0.0727 **b.** 0.7307 **25. a.** 0.9950 **b.** 0.4491 **c.** 0.00003

Section 5.3

1. 12 **3.** 187.5 **5.** $\frac{35}{3} \approx 11.67$ **7.** 0.0743 **9.** 0.1269 **11.** 0.2881 (From table: 0.2882) **13.** 0.9901 (From table: 0.9902) **15.** 0.2588 (From table: 0.2586 or 0.2587) **17. a.** 0.0067 **b.** 0.5438 **19. a.** 0.2381 **b.** 0.7851 (From table: 0.7852) **21. a.** 0.2510 **b.** 0.5578 **23. a.** 0.0233 **b.** 0.0000

Section 5.4

1. a. $\frac{9139}{33{,}320} \approx 0.2743$ **b.** $\frac{33}{66{,}640} \approx 0.0005$ **3. a.** $\frac{4312}{20{,}995} \approx 0.2054$ **b.** $\frac{7}{1938} \approx 0.0036$ **5. a.** $\frac{3}{13} \approx 0.2308$ **b.** $\frac{1}{2} \approx 0.5$ **7. a.** $\frac{33}{646} \approx 0.0511$ **b.** $\frac{231}{646} \approx 0.3576$ **9. a.** $\frac{77}{260} \approx 0.2962$ **b.** $\frac{153}{703} \approx 0.2176$ **11.** $\frac{1}{5} = 0.2$ **13.** Poisson; 0.0177 **15.** Binomial; 0.1172 **17.** Hypergeometric; $\frac{3}{10} = 0.3$ **19.** Binomial; 0.0834

Chapter 5 Exercises

1. It is possible for this to be part of a probability distribution. Although the probabilities on the poster only add up to 95%, the remaining 5% could be the probability of "losing your shirt." The casino just chose not to advertise that particular part of the distribution. However, we cannot say for certain that it is a probability distribution without more information. As it stands, it is not a probability distribution. **3. a.** $-\frac{55}{13} \approx -\$4.23$ **b.** $\frac{19{,}250}{13} \approx \$1480.77$ **5. a.** $\sigma \approx 1.118034$; Thus the average distance from the mean for the number of heads obtained in five coin tosses is approximately 1.1 heads. **b.** $\sigma = \sqrt{np(1-p)} = \sqrt{5 \cdot 0.5(1-0.5)} \approx 1.118034 \approx 1.1$ heads; This is the same value since the distribution is a binomial distribution. **7.** Hypergeometric; $\frac{3}{14} \approx 0.2143$ **9.** Binomial; 0.0870 **11.** Binomial; 0.9437 **13.** Binomial; **a.** 0.1752 **b.** 0.0520

Chapter 6

Section 6.1

1. False; There are an **unlimited** number of normal distributions with different means and standard deviations. **3.** False; The mean of the **standard** normal distribution is always 0. **5.** False; The standard deviation of the standard normal distribution is always **1**. **7.** True; A normal distribution is symmetric about its mean, so the line of symmetry is the vertical line $x = \mu$. **9.** True; The distance along the x-axis from one inflection point to the mean is equal to the value of the standard deviation of the particular normal distribution.

11.

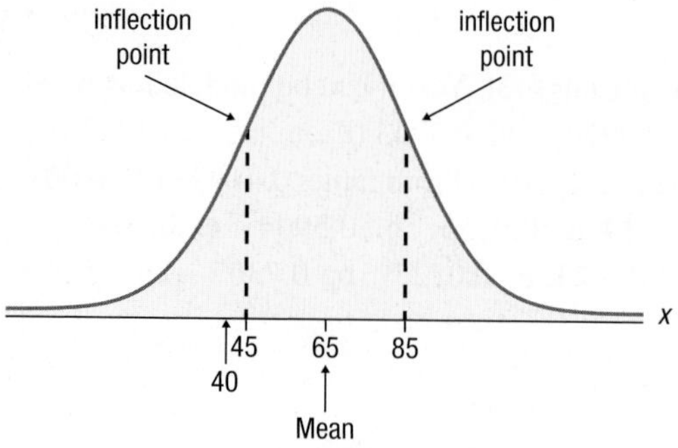

13.

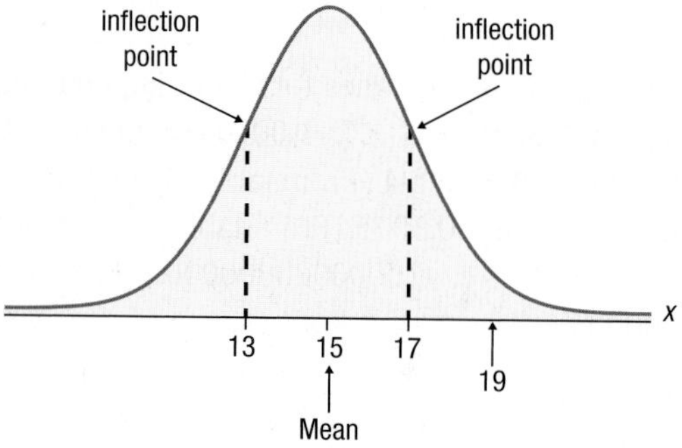

15.

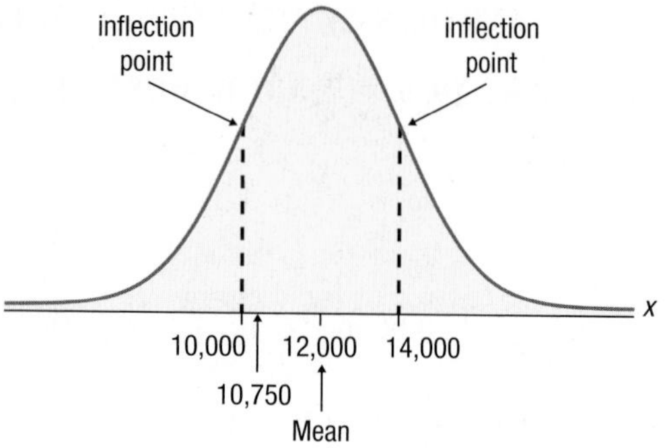

17.

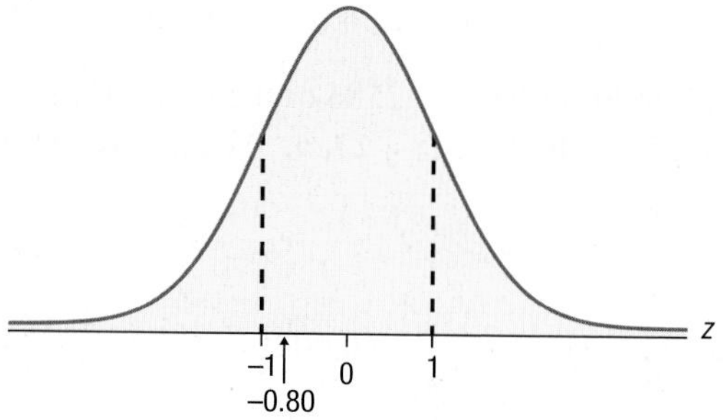

19.

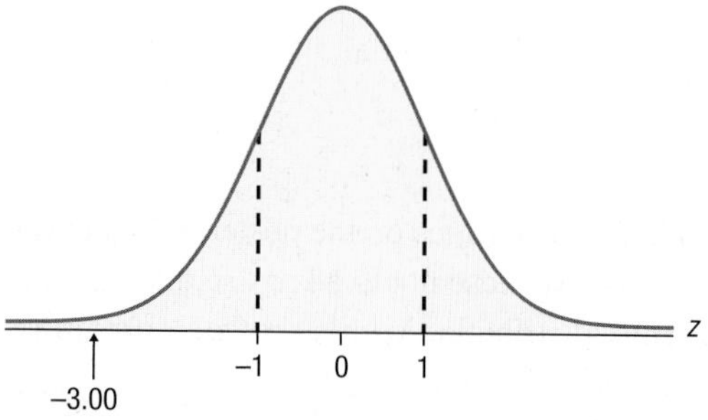

21. a. −1.25 **b.** 7.4 **c.** 2.78 **d.** 20 **23.** $x = z \cdot \sigma + \mu$
25. Answers will vary. For example:

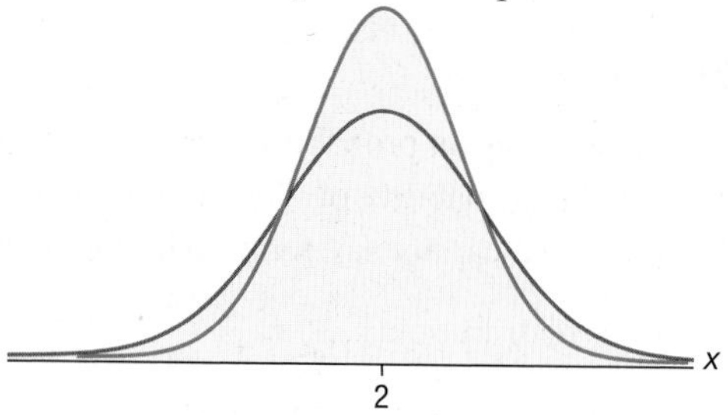

Section 6.2

1. 0.9906 **3.** 0.1056 **5.** 0.9772 **7.** 0.9998 (From table: ≈ 1.0000) **9.** 0.0885 **11.** 0.9940 **13.** 0.8413 **15.** 0.3310 **17.** 0.1335 (From table: 0.1336) **19.** 0.9677 (From table: 0.9678) **21.** 0.9545 − 0.95 = 0.0045 (From table: 0.9544 − 0.95 = 0.0044); 0.9973 − 0.997 = 0.0003 (From table: 0.9974 − 0.997 = 0.0004) **23.** 0.0349 (From table: 0.0348) **25.** 0.0579 **27.** 0.1814 (From table: 0.1815) **29.** 0.9317 **31.** 0.9630 (From table: 0.9629) **33.** 0.0008 **35.** 0.0033 **37.** 0.9677 (From table: 0.9678) **39.** 0.4973 **41.** 0.2077 (From table: 0.2076) **43.** 0.3173 (From table: 0.3174) **45.** 0.9545 (From table: 0.9544) **47.** 1 **49.** 0.00002 (From table: ≈ 0.0000)

Section 6.3

1. a. 0.7421 (From table: 0.7422) **b.** 0.4363 (From table: 0.4364) **c.** 0.6465 (From table: 0.6471) **d.** 0.0505 (From table: 0.0507) **3. a.** 59.35% (From table: 59.48%) **b.** 80.14% (From table: 80.23%) **c.** 40.15% (From table: 40.40%) **d.** 59.85% (From table: 59.60%) **5. a.** 0.9234 (From table: 0.9236) **b.** 0.0008 **c.** 0.3867 (From table: 0.3851) **7. a.** 4.44% (From table: 4.46%) **b.** 0.53% (From table: 0.54%) **c.** 28.74% (From table: 28.92%) **9. a.** 84.13% **b.** Answers will vary. For example, lowering the number of years in the warranty would decrease the company's expense on returned batteries. **11.** a. and b.

Section 6.4

1. −2.67 **3.** −2.03 **5.** −0.52 **7.** 1.38 **9.** −0.25 **11.** 1.28 **13.** 0.57 **15.** 2.88 **17.** −0.67 **19.** 58.3 **21.** 97.6 inches (From table: 97.5 inches) **23.** 137 pounds **25.** 79.5 °F **27.** 34 hours **29.** 29 months

Section 6.5

1. Area to the right of 39.5 **3.** Area to the right of 500.5 **5.** Area to the left of 30.5 **7.** Area to the left of 11.5 **9.** Area between 34.5 and 35.5 **11.** Conditions are met **13.** Conditions are not met; $np = 1.5 < 5$ **15.** 0.2222 (From table: 0.2236) **17.** ≈ 1.0000 **19.** 0.9937 (From table: 0.9938) **21.** 0.9976 **23.** 0.8788 (From table: 0.8790) **25.** 0.0732 (From table: 0.0735) **27.** 0.0243 (From table: 0.0244) **29.** 0.0449 (From table: 0.0453)

Chapter 6 Exercises

1. 1 **3.** Bell **5.** 1 **7.** 47.72% **9.** 0.1573 (From table: 0.1574) **11.** 25% **13.** 2.28% **15. a.** 0.8667 (From table: 0.8665) **b.** 116.11 **c.** 0.25 **d.** 972 **17.** Answers will vary. This question is intended to get students to think about the shape that the distribution might have. **19.** 98.81% of women, 99.87% of men **21.** 0.9882 (From table: 0.9881)

Chapter 7

Section 7.1

1. False; A sampling distribution refers to **groups** rather than **individuals**. **3.** True; According to the Central Limit Theorem, the standard deviation of a sampling distribution of sample means, $\sigma_{\bar{x}}$, equals the population standard deviation divided by the square root of the sample size. **5.** $\mu_{\bar{x}} = 35$, $\sigma_{\bar{x}} \approx 1.1$ **7.** $\mu_{\bar{x}} = 12.0$, $\sigma_{\bar{x}} \approx 0.38$ **9.** $\mu_{\bar{x}} = 9.5$, $\sigma_{\bar{x}} \approx 1.60$ **11.** $\mu_{\bar{x}} = \$2.98$ per gallon **13.** $\mu_{\bar{x}} = 7$ days **15.** $\sigma_{\bar{x}} \approx 0.02\%$ **17.** $\sigma_{\bar{x}} = 0.58$ pounds **19.** $\sigma_{\bar{x}} \approx \$191.00$ per year

Section 7.2

1. 0.44 **3.** −2.06 **5.** −1.33 **7. a.** 0.1587 **b.** 0.000000003 (From table: ≈ 0.0000) **c.** 0.0260 (From table: 0.0262) **d.** 0.2435 (From table: 0.2420) **9. a.** 0.3694 (From table: 0.3707) **b.** 0.0092 (From table: 0.0091) **c.** 0.9908 (From table: 0.9909) **d.** 0.0092 (From table: 0.0091) **e.** 0.0184 (From table: 0.0182) **11. a.** 0.2881 **b.** 0.0057 **c.** 0.0090 (From table: 0.0089) **d.** 0.0833 (From table: 0.0836) **13.** 0.9505 (From table: 0.9500) **15.** 0.00004 (From table: ≈ 0.0000) **17.** 0.7248 (From table: 0.7258) **19.** 0.0372 (From table: 0.0376) **21.** 0.0402 (From table: 0.0401) **23.** 0.9923 **25.** 0.0040 **27.** 0.9487 (From table: 0.9490) **29.** 0.9886

Section 7.3

1. $\hat{p} = 0.35, z \approx 0.21$ **3.** $\hat{p} = \frac{24}{29} \approx 0.827586, z \approx 0.37$ **5.** $\hat{p} = 0.09216, z \approx -18.11$ **7.** 0.1367 (From table: 0.1357) **9.** 0.6865 (From table: 0.6879) **11.** 0.0432 (From table: 0.0436) **13.** 0.6726 (From table: 0.6744) **15.** 0.2227 (From table: 0.2206)

Chapter 7 Exercises

1. 0.1483 years **3.** 0.7084 (From table: 0.7088) **5.** 39.0 grams **7.** 0.9493 (From table: 0.9488) **9.** 0.0741 (From table: 0.0734) **11.** 0.0014 **13.** 0.3050 (From table: 0.3030) **15.** The group of 15 employees would be more likely to have a mean closer to \$27,500 since the sampling distribution of sample means for samples of size 15 has a smaller standard deviation than the sampling distribution for samples of size 8.

Chapter 8

Section 8.1

1. \$18 **3.** (16.30, 19.70) **5.** 0.690181 (From table: 0.690242) **7.** 0.254048 (From table: 0.254053) **9.** (44.58, 45.42) **11.** (18.25, 18.65) **13.** 424 **15.** 67 **17.** (14.6, 15.4); The professor can be 90% confident that the mean amount of time that her students spend studying is between 14.6 and 15.4 hours per week. **19.** (18.6, 19.8); The writer can be 95% confident that the mean computer usage time for American households is between 18.6 and 19.8 hours per week.
21. (691, 749); We can be 98% confident that the mean amount of money that homeowners spend on lawn service each year is between \$691 and \$749. **23.** (21.1, 24.1); The physical therapist can be 99% confident that the mean recovery time for patients using the new therapy after ACL surgery is between 21.1 and 24.1 days. **25.** 58 students **27.** 97 students **29.** Yes; add the endpoints together and divide by two to find the midpoint of the interval, which is the original point estimate of 1.49. **31.** Answers will vary. Encourage students to consider that the interval contains values both above and below last quarter's sales. A more conservative view would not report an increase in sales because of the lower bound of the interval. A less conservative view would lean towards an estimate of gain because the majority of the interval is above last quarter's sales figure.

Section 8.2

1. 1.753 **3.** 2.779 **5.** 2.462 **7.** 1.833 **9.** 3.365 **11.** −1.812 **13.** −3.055 **15.** 3.012 **17.** 2.681 **19.** 2.101 **21.** 1.708 **23.** 2.878 **25.** These essentially ask for the same thing for two different *t*-distributions; the answers are different because the *t*-distributions have different numbers of degrees of freedom.

Section 8.3

1. (92.2, 97.8) **3.** (5.5, 8.5) **5.** (246.9, 443.3) or (247.0, 443.3) **7.** (32.9, 48.8) **9.** 1.7 miles **11.** (555, 645); Conservationists can be 98% confident that the mean weight of adult male grizzly bears in the United States is between 555 and 645 pounds. **13.** (78.70, 81.30); We can say with 95% confidence that the mean fastball pitching speed of all high school baseball pitchers in the county is between 78.70 and 81.30 mph. **15.** (3837.7, 4412.3) or (3837.6, 4412.4); With 95% confidence, the university can say that the mean number of fans at men's basketball games is between 3837.7 and 4412.3 people. **17.** (39,978, 55,922); The mean annual salary of high school counselors in the United States, with 90% confidence, is between \$39,978 and \$55,922. **19.** No, not enough information is given. **21.** Answers will vary. Encourage students to consider the point estimate, the sample size, and the sample standard deviation in their answers. **23.** The width will increase. **25.** 99% **27.** The width will decrease.

Section 8.4

1. $\frac{41}{50} = 0.82$ **3.** $\frac{48}{112} = \frac{3}{7} \approx 0.428571$ **5.** (0.731, 0.909) **7.** (0.050, 0.127); With 90% confidence, we can say that the proportion of all faculty members at the community college who know sign language is between 0.050 and 0.127.
9. (17.0%, 28.7%); We can say with 90% confidence that the percentage of all kindergartners who say pancakes are their favorite breakfast food is between 17.0% and 28.7%. **11.** (33.3%, 46.7%); With 90% confidence, the percentage of all students at that college who do not regularly check their campus e-mail accounts is between 33.3% and 46.7%.
13. (0.305, 0.442); We can say with 95% confidence that the proportion of all adults in the United States who exercise on a regular basis is between 0.305 and 0.442. **15.** (40.6%, 47.8%); The wireless phone company can be 95% confident that between 40.6% and 47.8% of its smartphone customers would switch to the iPhone if it is offered. **17.** 929 college students in the United States **19.** 7007 patients using the drug (From table: 7002 patients using the drug) **21.** The confidence interval for the stated index of 85.3 is actually (80.92, 89.68). Since the claim was that the new index was 5.4 points higher than the prior month, we can calculate the prior month's index to be 85.3 − 5.4 = 79.9. Since the entire confidence interval is above 79.9 (the prior month's index), we can say with 95% confidence that the index did indeed rise.

Section 8.5

1. 3.5 **3.** 0.12 **5.** $\chi^2_{0.025} = 39.364$, $\chi^2_{0.975} = 12.401$ **7.** $\chi^2_{0.05} = 32.671$, $\chi^2_{0.95} = 11.591$ **9.** $(9.6, 55.3)$ **11.** $(6.1, 30.3)$ **13.** $(2.0, 3.2)$ **15.** $(5.5, 8.7)$ **17.** $(2.25, 8.10)$; The grocer can be 90% confident that the variance in weights of all packages of strawberries is between 2.25 and 8.10. **19.** $(12.1, 46.6)$; With 98% confidence, the variance in speeds of all fastballs thrown by major league pitchers is between 12.1 and 46.6. **21.** $(6.0, 10.2)$; The standard deviation of completion times for all circuits driven for the transit system at the theme park is between 6.0 and 10.2 minutes, with a 98% level of confidence. **23.** $(0.02, 0.15)$; We are 99% confident that the population variance is between 0.02 and 0.15. **25.** $(1.37, 1.84)$; The factory can say with 95% confidence that the population standard deviation of the weights of all new truck engines is between 1.37 and 1.84 pounds. **27.** 212 **29.** 337

Chapter 8 Exercises

1. $(43.00, 47.00)$; The mean amount of money that accountants spend on lunch each week is between \$43.00 and \$47.00. **3.** 0.015842 **5.** 0.35 minutes **7.** $(38.1, 43.9)$; We can say with 98% confidence that the mean weight of all stray dogs in this area is between 38.1 and 43.9 pounds. **9.** $(0.001, 0.003)$; We are 95% confident that the population variance for the volumes of soda in all the soda cans that come off that particular assembly line is between 0.001 and 0.003. **11.** 198 soldiers **13.** 135 applicants

Chapter 9

Section 9.1

1. 3 **3.** 0.20 minutes **5.** 1.796537 °C **7.** 1.166273 **9.** 0.246268 **11.** (–5, 1); We are 90% confident that the mean delivery time for the local pizza store is between 5 minutes less than and 1 minute more than the mean delivery time for the national chain store. Since the confidence interval contains 0, the data do not provide evidence that the two population means are unequal at this level of confidence. **13.** (–6, 0); We are 95% confident that the mean exam score for third graders in the first school is between 0 and 6 points less than the mean exam score for the third graders in the second school. **15.** (–13,719, 7119) or (–13,720, 7120); We are 90% confident that the mean home price for houses in the first area is between \$13,719 less than and \$7119 more than the mean home price for houses in the second area. Since the confidence interval contains 0, the data do not provide evidence that the two population means are unequal at this level of confidence.

Section 9.2

1. $t_{0.025} = 2.045$ **3.** $t_{0.05} = 1.796$ **5.** $t_{0.025} = 3.182$ **7.** $t_{0.01} = 2.539$ **9.** 1.576539 **11.** 5.075175 **13.** $(-2.8, -1.2)$ or (–2.9, –1.1); We are 80% confident that the mean pain score for mothers who had a water birth is between 1.2 and 2.8 points lower than the mean pain score for mothers who did not have a water birth. **15.** (0.4, 2.0); We are 99% confident that the mean number of cavities for children whose diets are high in sugar is between 0.4 and 2.0 cavities greater than the mean number of cavities for children whose diets are low in sugar. **17.** (–0.95, 13.97) or (–1.05, 14.07); We are 80% confident that the mean amount of money spent per month on Internet purchases for people 18–24 years old is between \$0.95 less than and \$13.97 more than the mean amount of money spent per month on Internet purchases for people 25–30 years old. Since the confidence interval contains 0, the data do not provide evidence that the two population means are unequal at this level of confidence. **19.** (–2.5, –0.1); We are 95% confident that the mean rating for the literature professor in the first semester is between 0.1 and 2.5 points lower than the mean rating for the literature professor in the second semester. **21.** (0.7, 3.1); We are 95% confident that the mean amount of weight lost by dieters who supplement with calcium is between 0.7 and 3.1 pounds greater than the mean amount of weight lost by dieters who do not take the calcium supplement. **23.** (–30, 12); We are 99% confident that the mean amount spent by customers listening to slow instrumental music is between \$30 less than and \$12 more than the mean amount spent by customers listening to upbeat instrumental music. Since the confidence interval contains 0, the data do not provide evidence that the two population means are unequal at this level of confidence.

Section 9.3

1. $\bar{d} \approx 1.888889$; $s_d \approx 1.615893$ **3.** $\bar{d} = 1.25$; $s_d \approx 5.994045$ **5.** $(1.991, 2.469)$ **7.** $(-0.10, 2.38)$ **9.** $(0.1, 2.2)$; We are 99% confident that the mean difference between the durations of a cold for people who take the traditional cold medicine and those who take the new medicine is between 0.1 and 2.2 days for the population from which the participants in the study were sampled. **11.** (–12.2, –7.4); We are 99% confident that the mean amount of weight lost after being on the diet for 30 days is between 7.4 and 12.2 pounds for the population from which the participants in the study were sampled. **13.** (7.2, 10.3); We are 98% confident that, after learning the memory method, the mean increase in the number of words that people can memorize is between 7.2 and 10.3 for the population from which the participants in the study were sampled. **15.** (–9.6, 3.3); We are 90% confident that, after using the new device, the mean change in the amounts of time spent cooking dinner is between –9.6 and 3.3 minutes for the population from which the participants were sampled. Since the confidence interval contains 0, the data do not provide evidence that the population mean decreased at this level of confidence. Thus, we cannot conclude that the device actually reduces the mean amount of time that people spend preparing dinner. **17.** (–8.24, 0.44); We are 90% confident that the mean difference between the lengths of time required to sell a house for homeowners who follow the show's tips and those who do not follow the tips is between –8.24 and 0.44 weeks. Since the confidence interval contains 0, the data do not provide evidence that the population mean decreased at this level of confidence. **19.** (5.3, 16.2); We are 95% confident that the mean increase in the number of sit-ups completed in one minute after training to do sit-ups properly is between 5.3 and 16.2 for the population from which the students were sampled. **21.** (–3.0, 19.6); We are 99% confident that the mean change in the test scores after Ms. Comeaux started using positive reinforcement in her math class is between –3.0 and 19.6 points. Since the confidence interval contains 0, the data do not provide evidence that the population mean changed at this level of confidence.

Section 9.4

1. Conditions are met **3.** Conditions are not met; $n_2\left(1-\hat{p}_2\right) = 4.004 < 5$ **5.** –0.175887 **7.** 0.019445 **9.** 0.085037 **11.** 0.199600 **13.** 0.107037 **15.** (–0.185, 0.266); We are 99% confident that the proportion of multiple births for women taking fertility drugs is between 0.185 less than and 0.266 greater than the proportion of multiple births for women not taking fertility drugs. Since the confidence interval contains 0, the data do not provide evidence that the two population proportions are unequal at this level of confidence. **17.** (–57.6%, –27.5%); We are 90% confident that the proportion of Northerners that attend church on a weekly basis is between 27.5% and 57.6% less than the proportion of Southerners that attend church on a weekly basis. Since both endpoints of the interval are negative, there is sufficient evidence to conclude with 90% confidence that a larger percentage of Southerners than Northerners attend church on a weekly basis. **19.** (–0.295, 0.022); We are 90% confident that the proportion of men who will return a lost wallet is between 0.295 less than and 0.022 greater than the proportion of women who will return a lost wallet. Since the confidence interval contains 0, the data do not provide evidence that the two population proportions are unequal at this level of confidence. **21.** (–0.098, 0.405); We are 99% confident that the proportion of gerbera daisy plants that bloom after being treated with FertiGro fertilizer is between 0.098 less than and 0.405 greater than the proportion of gerbera daisies that bloom after being treated with regular fertilizer. Since the confidence interval contains 0, the data do not provide evidence that the two population proportions are unequal at this level of confidence. **23.** (–0.070, 0.178); We are 95% confident that the proportion of customers who order water to drink at Restaurant A is between 0.070 less than and 0.178 greater than the proportion of customers who order water to drink at Restaurant B. Since the confidence interval contains 0, the data do not provide evidence that the two population proportions are unequal at this level of confidence. **25.** (0.010, 0.355); We are 98% confident that the proportion of children who start reading by age 4 when read to at least three times per week is between 0.010 and 0.355 greater than the proportion of children who start reading by age 4 when read to less often. Since both endpoints of the interval are positive, there is sufficient evidence for the parents' group to conclude with 98% confidence that children who are read to at least three times per week are more likely to begin reading by age 4.

Section 9.5

1. 1.194282 **3.** 1.113022 **5.** $F_{0.025} = 2.4478$, $F_{0.975} = 0.3868$ **7.** $F_{0.05} = 2.1555$, $F_{0.95} = 0.4679$ **9.** $F_{0.005} = 4.9884$, $F_{0.995} = 0.1910$ **11.** (0.5021, 3.4589) **13.** (0.3324, 3.7861) **15.** (0.2165, 8.7358) **17.** (1.0094,8.6067) **19.** (0.4836, 2.1496); We are 90% confident that the ratio of the population variances of systolic blood pressure levels for patients who do not take the new medication for high blood pressure and patients who take the new medication is between 0.4836 and 2.1496 for the populations of patients from which the participants in the study were sampled. Since the confidence interval contains 1, the data do not provide evidence that the two population variances are unequal at this level of confidence. **21.** (0.1824, 6.9829); We are 99% confident that the ratio of the population variances of the weights of the bags of flour for Machines A and B is between 0.1824 and 6.9829. Since the value of 1 is in the interval, the data do not provide evidence that the two population variances are unequal at this level of confidence. Thus, it is not necessary to adjust the machines. **23.** (1.0357, 3.2750); We are 90% confident that the ratio of the population variances of driving distances for the old brand and the new brand of golf balls is between 1.0357 and 3.2750. Since both endpoints of the confidence interval are greater than 1, there is sufficient evidence to conclude that the population variance of driving distances is greater for the old brand. Thus, the new brand of golf ball significantly increases precision. **25.** (1.0025, 4.9195); We are 95% confident that the ratio of the population variances of prices of students' and professors' cars at this college is between 1.0025 and 4.9195. Since both endpoints of the confidence interval are greater than 1, the sample data provide evidence that the population variance of car prices is greater for professors' cars than for students' cars at this college. **27.** (0.2512, 4.6427); We are 99% confident that the ratio of the population variances of the thicknesses of the painted coatings inside these two tankers is between 0.2512 and 4.6427. Since the confidence interval contains 1, there is not sufficient evidence to conclude that there was a significant difference in how consistently the equipment applied the coating in one tanker versus the other.

Chapter 9 Exercises

1. (0.5, 5.9); We are 95% confident that the mean exam score for Class A is between 0.5 and 5.9 points higher than the mean exam score for Class B. **3.** (−3.65, −2.01); We are 99% confident that, 30 minutes after taking the new drug for a fever, the mean decrease in patients' temperatures is between 2.01 °F and 3.65 °F for the population from which the patients in the study were sampled. **5.** (−1.4, 0.2); We are 95% confident that, after children eat an apple every day for six months, the true mean change in the number of doctor's visits during a six-month period is between −1.4 and 0.2 for the population from which the children in the study were sampled. Since the confidence interval contains 0, the data do not provide evidence that the population mean decreased at this level of confidence. Thus, there is not sufficient evidence to conclude that an apple a day keeps the doctor away. **7.** (0.2, 0.6); We are 95% confident that the mean exercise rate for citizens of State B is between 0.2 and 0.6 times per week more than the mean exercise rate for citizens of State A for the populations of citizens in these two states from which the survey respondents were sampled. **9.** (7.3, 9.7); We are 90% confident that the population mean of students' GRE verbal scores in 2010 was between 7.3 and 9.7 points higher for University A than for University B. **11.** (0.7, 7.3); We are 95% confident that, after using the engine-cleaning product, the mean improvement in gas mileage is between 0.7 and 7.3 miles per gallon for the population of vehicles from which the cars in the study were sampled. **13.** (0.3701, 3.2841); We are 95% confident that the ratio of the population variances of the numbers of times per game that John and Mark get on base is between 0.3701 and 3.2841. Since the confidence interval contains 1, the data do not provide evidence that the two population variances are unequal at this level of confidence. **15.** (0.4407, 2.8219); We are 90% confident that the ratio of the population variances of the annual salaries of full professors and associate professors is between 0.4407 and 2.8219. Since the confidence interval contains 1, the data do not provide evidence for a difference in the population variances at this level of confidence.

Chapter 10

Section 10.1

1. False; We can **never** prove **either** hypothesis to be true. **3.** True; The null and alternative hypotheses are mathematically opposite statements about the numerical value of some population parameter. **5.** False; A Type II error is made when we fail to reject a **false** null hypothesis. **7.** True; If we decrease the probability of making a Type I error, which is equal to the level of significance, α, then the probability of making a Type II error, denoted by β, increases. **9.** H_0: $\mu = 7.50$; H_a: $\mu \neq 7.50$ **11.** H_0: $\mu \leq 3.0$; H_a: $\mu > 3.0$ **13.** H_0: $p \geq 0.45$; H_a: $p < 0.45$ **15.** H_0: $\mu \geq 5$; H_a: $\mu < 5$ **17.** H_0: $p \leq 0.5$; H_a: $p > 0.5$ **19.** The evidence does not support the researcher's claim that the percentage of people with significant side effects is more than 4% with $\alpha = 0.01$. **21.** The evidence does not support the college student's claim that the mean is higher than \$60 per month at the 0.01 level of significance. **23.** Yes; Type II error **25.** No; correct decision **27.** Yes; Type I error **29.** Yes; Type II error

Section 10.2

1.

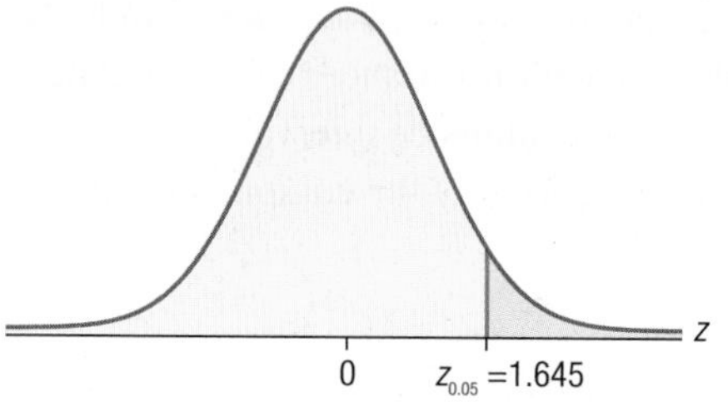

3.

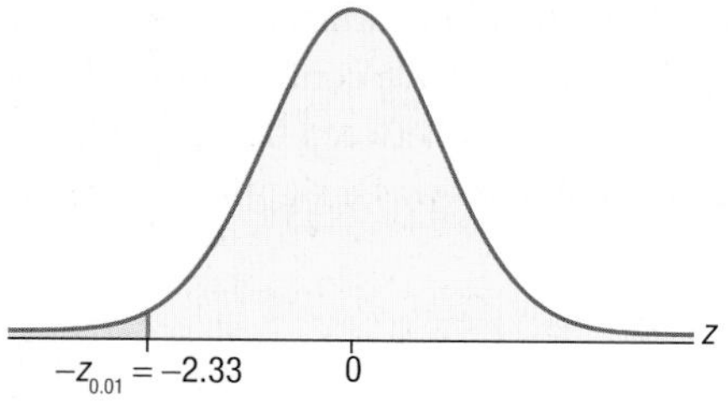

5.

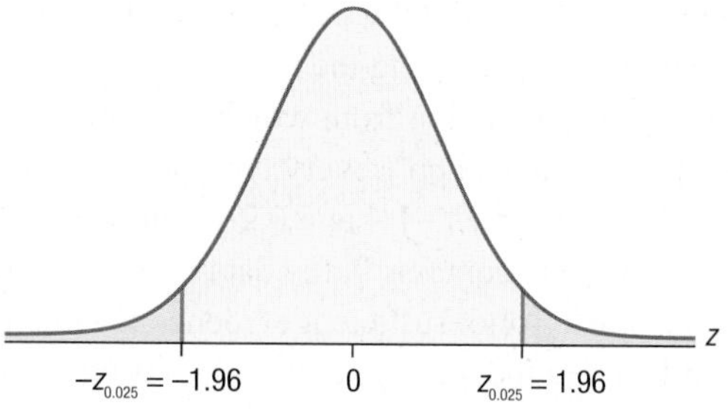

7. 0.0694 **9.** 0.0271 (From table: 0.0272) **11.** 0.0537 **13.** 0.0414 **15.** Fail to reject H_0 **17.** Reject H_0 **19.** Fail to reject H_0 **21.** Reject H_0 **23.** Fail to reject H_0 **25. a.** H_0: $\mu = 2.00$; H_a: $\mu \neq 2.00$ **b.** Normal distribution; $\alpha = 0.05$ **c.** $z = -2$ **d.** p-value ≈ 0.0455 (From table: p-value $= 0.0456$); therefore reject H_0. At the 0.05 level of significance, the evidence supports the conclusion that the bolts are not 2.00 cm long and the manufacturer needs to recalibrate the machines. **27. a.** H_0: $\mu \leq 839.00$; H_a: $\mu > 839.00$ **b.** Normal distribution; $\alpha = 0.01$ **c.** $z \approx 2.22$ **d.** p-value ≈ 0.0131 (From table: p-value $= 0.0132$); therefore fail to reject H_0. At the 0.01 level of significance, the evidence does not support the statement that the mean amount spent by theme park travelers is more than \$839 per trip. **29. a.** H_0: $\mu \geq 164.7$; H_a: $\mu < 164.7$ **b.** Normal distribution; $\alpha = 0.01$ **c.** $z \approx -3.23$ **d.** p-value ≈ 0.0006; therefore reject H_0. At the 0.05 level of significance, the evidence supports the group's claim that the mean weight of women in this community is lower than 164.7 pounds.

Section 10.3

1. Reject H_0 if $t \leq -1.761$. **3.** Reject H_0 if $|t| \geq 1.734$. **5.** Reject H_0 if $|t| \geq 1.833$.

7. ***t*-Distribution, *df* = 19**

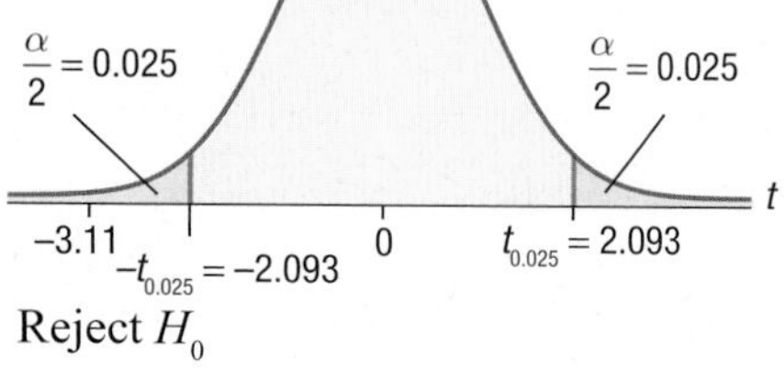

9. ***t*-Distribution, *df* = 28**

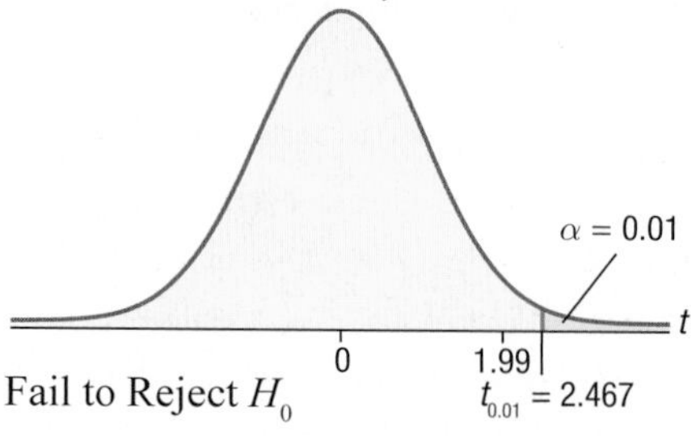

11. ***t*-Distribution, *df* = 9**

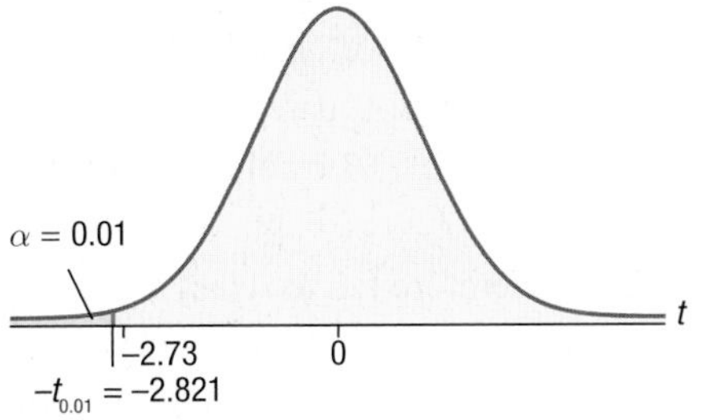

Fail to Reject H_0

13. a. H_0: $\mu \geq 15.0$; H_a: $\mu < 15.0$ **b.** Student's t-distribution; $\alpha = 0.10$
c. $t \approx -0.577$
d. ***t*-Distribution, *df* = 11**

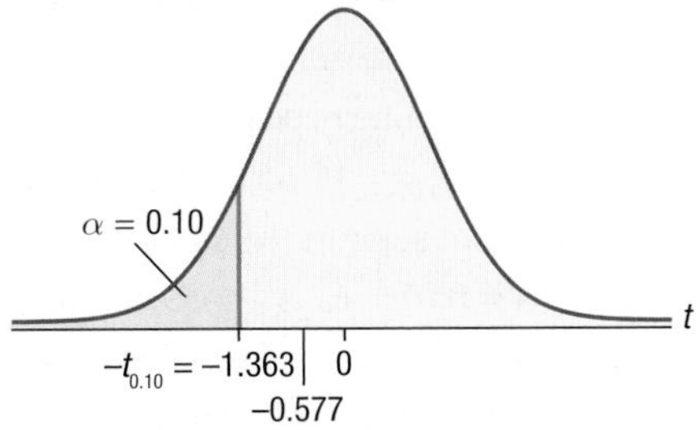

p-value ≈ 0.2877; fail to reject H_0. At the 0.10 level of significance, the evidence does not support the reporter's claim that the mean is lower than 15.0 calls per semester.

15. a. H_0: $\mu \geq 4$; H_a: $\mu < 4$ **b.** Student's t-distribution; $\alpha = 0.01$ **c.** $t \approx -3.333$
d. ***t*-Distribution, *df* = 24**

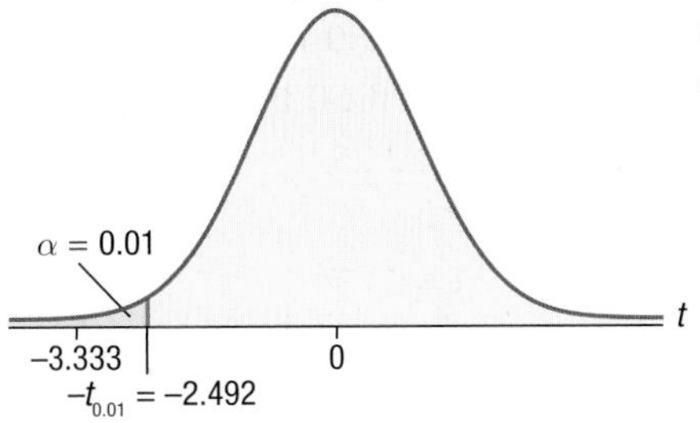

p-value ≈ 0.0014; reject H_0. At the 0.01 level of significance, the evidence supports Ella's claim that the mean for teenage girls in her area is less than 4 calls per night.

17. a. H_0: $\mu \leq 20$; H_a: $\mu > 20$ **b.** Student's t-distribution; $\alpha = 0.05$ **c.** $t \approx 1.661$
d. ***t*-Distribution, *df* = 6**

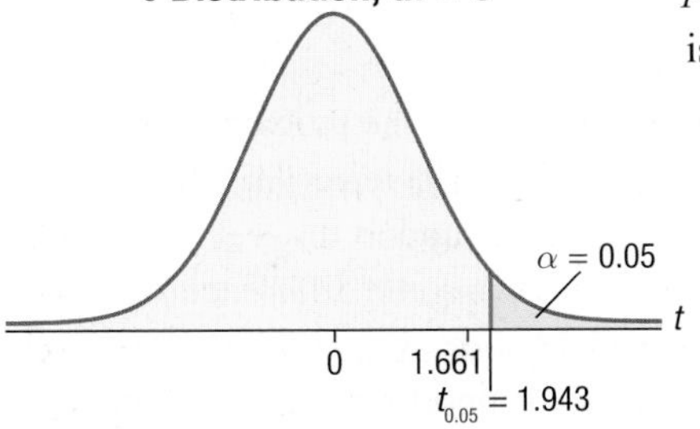

p-value ≈ 0.0739; fail to reject H_0. At the 0.05 level of significance, the evidence is not strong enough to say that the mean delivery time is more than 20 minutes.

19. a. H_0: $\mu = 15$; H_a: $\mu \neq 15$ **b.** Student's t-distribution; $\alpha = 0.05$ **c.** $t \approx -2.118$
d. ***t*-Distribution, *df* = 19**

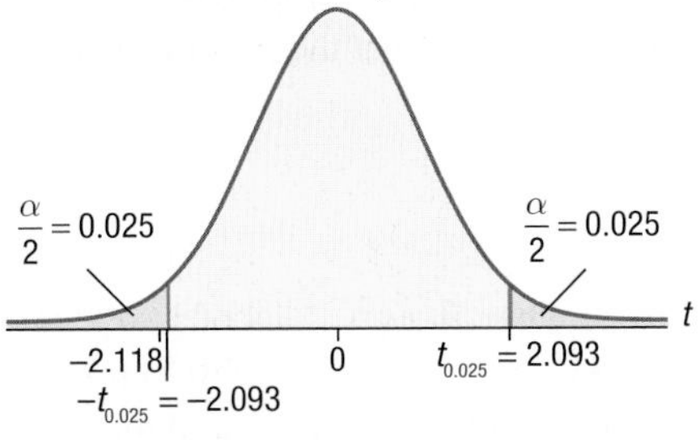

p-value ≈ 0.0475; reject H_0. At the 0.05 level of significance, the evidence collected supports the claim that the mean installation time is no longer 15 minutes. The company may need to change its advertising.

Section 10.4

1. No; $np < 5$ **3.** Yes **5. a.** H_0: $p \geq 0.74$; H_a: $p < 0.74$ **b.** Normal distribution; $\alpha = 0.10$ **c.** $z \approx -0.31$ **d.** p-value ≈ 0.3799 (From table: p-value = 0.3783); therefore fail to reject H_0. At the 0.10 level of significance, the evidence does not support Patrice's claim that less than 74% of American women have been married by the age of 30. **7. a.** H_0: $p = 0.058$; H_a: $p \neq 0.058$ **b.** Normal distribution; $\alpha = 0.01$ **c.** $z \approx 2.22$ **d.** p-value ≈ 0.0265 (From table: p-value = 0.0264); therefore fail to reject H_0. At the 0.01 level of significance, the evidence does not support the claim that the percentage of commercial truck drivers with sleep apnea is not 5.8%. **9. a.** H_0: $p \leq 0.15$; H_a: $p > 0.15$ **b.** Normal distribution; $\alpha = 0.05$ **c.** $z \approx 0.44$ **d.** p-value ≈ 0.3290 (From table: p-value = 0.3300); therefore fail to reject H_0. The evidence does not support the cost effectiveness of the direct mail campaign at the 0.05 level of significance.

Section 10.5

1. $\chi^2_{0.950} = 15.379$; reject H_0 if $\chi^2 \le 15.379$. **3.** $\chi^2_{0.050} = 27.587$; reject H_0 if $\chi^2 \ge 27.587$. **5.** $\chi^2_{0.975} = 12.401$, $\chi^2_{0.025} = 39.364$; reject H_0 if $\chi^2 \le 12.401$ or $\chi^2 \ge 39.364$. **7. a.** H_0: $\sigma^2 \le 4$; H_a: $\sigma^2 > 4$ **b.** Chi-square distribution; $\alpha = 0.01$ **c.** $\chi^2 = 37.5$ **d.** $\chi^2_{0.01} = 42.980$; reject H_0 if $\chi^2 \ge 42.980$. Therefore fail to reject H_0. At the 0.01 level of significance, the evidence does not support the claim that the machine needs servicing. **9. a.** H_0: $\sigma^2 \ge 0.33$; H_a: $\sigma^2 < 0.33$ **b.** Chi-square distribution; $\alpha = 0.10$ **c.** $\chi^2 \approx 12.091$ **d.** $\chi^2_{0.90} = 11.651$; reject H_0 if $\chi^2 \le 11.651$. Therefore fail to reject H_0. At the 0.10 level of significance, the evidence does not support the student's claim that the gun's sight is off. **11. a.** H_0: $\sigma^2 \ge 0.1$; H_a: $\sigma^2 < 0.1$ **b.** Chi-square distribution; $\alpha = 0.01$ **c.** $\chi^2 = 89.1$ **d.** $\chi^2_{0.99} = 70.065$; reject H_0 if $\chi^2 \le 70.065$. Therefore fail to reject H_0. At the 0.01 level of significance, the evidence does not support the claim that the variance of the amounts of active ingredient per dose is less than 0.1. **13. a.** H_0: $\sigma^2 = 0.50$; H_a: $\sigma^2 \ne 0.50$ **b.** Chi-square distribution; $\alpha = 0.05$ **c.** $\chi^2 = 14.52$ **d.** $\chi^2_{0.975} = 12.401$, $\chi^2_{0.025} = 39.364$; reject H_0 if $\chi^2 \le 12.401$ or $\chi^2 \ge 39.364$. Therefore fail to reject H_0. At the 0.05 level of significance, the evidence does not support the manager's claim that the variance of the temperatures in the chicken incubators is not 0.50. **15. a.** H_0: $\sigma^2 = 2.25$; H_a: $\sigma^2 \ne 2.25$ **b.** Chi-square distribution; $\alpha = 0.10$ **c.** $\chi^2 \approx 23.707$ **d.** $\chi^2_{0.95} = 6.571$, $\chi^2_{0.05} = 23.685$; reject H_0 if $\chi^2 \le 6.571$ or $\chi^2 \ge 23.685$. Therefore reject H_0. At the 0.10 level of significance, the evidence supports the manager's claim that the variance in the water temperature is not 2.25.

Section 10.6

1. H_0: The proportions of prescriptions filled at the pharmacy do not vary by the day of the week. H_a: The proportions of prescriptions filled at the pharmacy do vary by the day of the week. **3.** H_0: The proportions of customers at the salon do not vary by the day of the week. H_a: The proportions of customers at the salon do vary by the day of the week. **5.** H_0: The proportions of customers who prefer various brands of cola do not vary by the brand. H_a: The proportions of customers who prefer various brands of cola do vary by the brand. **7.** 4.907 **9.** 3.379 **11.** 1.061 **13.** $\chi^2_{0.050} = 32.671$ **15.** $\chi^2_{0.010} = 20.090$ **17.** $\chi^2_{0.005} = 31.319$; fail to reject H_0. **19.** $\chi^2_{0.050} = 41.337$; reject H_0.
21. a. H_0: $p_1 = p_2 = p_3 = p_4 = p_5 = p_6 = p_7 = p_8 = p_9 = p_{10}$; H_a: There is some difference amongst the probabilities. **b.** Chi-square distribution; $\alpha = 0.05$ **c.** 12.5 for each month; $\chi^2 = 15.56$ **d.** $\chi^2_{0.05} = 16.919$; reject H_0 if $\chi^2 \ge 16.919$. p-value ≈ 0.0767; fail to reject H_0. At the 0.05 level of significance, the evidence does not support the teacher's claim that the numbers of tardy students vary by the month. **23. a.** H_0: $p_1 = p_2 = p_3 = p_4$; H_a: There is some difference amongst the probabilities. **b.** Chi-square distribution; $\alpha = 0.10$ **c.** 79 for each phase; $\chi^2 \approx 8.228$ **d.** $\chi^2_{0.10} = 6.251$; reject H_0 if $\chi^2 \ge 6.251$. p-value ≈ 0.0415; reject H_0. At the 0.10 level of significance, the evidence supports the nurses' claim that the number of patients seen in the emergency room during the midnight shift varies by the phase of the moon.
25. a. H_0: $p_1 = p_2 = p_3 = p_4 = p_5$; H_a: There is some difference amongst the probabilities. **b.** Chi-square distribution; $\alpha = 0.01$ **c.** 48.6 for each weekday; $\chi^2 \approx 1.835$ **d.** $\chi^2_{0.01} = 13.277$; reject H_0 if $\chi^2 \ge 13.277$. p-value ≈ 0.7660; fail to reject H_0. At the 0.01 level of significance, the evidence does not support the manager's claim that the proportions of swimmers vary on weekdays. He does not need to hire extra lifeguards.

Section 10.7

1.

$\frac{1480}{73} \approx 20.273973$	$\frac{1200}{73} \approx 16.438356$	$\frac{760}{73} \approx 10.410959$	$\frac{2400}{73} \approx 32.876712$
$\frac{1221}{73} \approx 16.726027$	$\frac{990}{73} \approx 13.561644$	$\frac{627}{73} \approx 8.589041$	$\frac{1980}{73} \approx 27.123288$

3.

$\frac{957}{119} \approx 8.042017$	$\frac{1584}{119} \approx 13.310924$	$\frac{33}{7} \approx 4.714286$	$\frac{825}{119} \approx 6.932773$
$\frac{145}{17} \approx 8.529412$	$\frac{240}{17} \approx 14.117647$	5	$\frac{125}{17} \approx 7.352941$
$\frac{116}{17} \approx 6.823529$	$\frac{192}{17} \approx 11.294118$	4	$\frac{100}{17} \approx 5.882353$
$\frac{667}{119} \approx 5.605042$	$\frac{1104}{119} \approx 9.277311$	$\frac{23}{7} \approx 3.285714$	$\frac{575}{119} \approx 4.831933$

5. 0.005 **7.** 13.336 **9.** $\chi^2_{0.100} = 4.605$; reject H_0; not independent **11.** $\chi^2_{0.005} = 45.559$; fail to reject H_0; independent **13. a.** H_0: The grades and the combination of particular subject and gender are independent. H_a: The grades and the combination of particular subject and gender are not independent. **b.** Chi-square distribution; $\alpha = 0.005$

c.

10.75	33.75	38.75	20	6.75
10.75	33.75	38.75	20	6.75
10.75	33.75	38.75	20	6.75
10.75	33.75	38.75	20	6.75

$\chi^2 \approx 34.225$

d. $\chi^2_{0.005} = 28.299$; reject H_0 if $\chi^2 \geq 28.299$. p-value ≈ 0.00006; reject H_0. At the 0.005 level of significance, the evidence supports the claim that an association exists between the grades and the combination of subject and gender. **15. a.** H_0: Cola preference and age are independent. H_a: Cola preference and age are not independent. **b.** Chi-square distribution; $\alpha = 0.005$

c.

$\frac{78,561}{800} = 98.20125$	$\frac{5117}{50} = 102.34$	$\frac{80,367}{800} = 100.45875$
$\frac{36,801}{400} = 92.0025$	$\frac{2397}{25} = 95.88$	$\frac{37,647}{400} = 94.1175$
$\frac{56,637}{800} = 70.79625$	$\frac{3689}{50} = 73.78$	$\frac{57,939}{800} = 72.42375$

$\chi^2 \approx 1.927$ **d.** $\chi^2_{0.005} = 14.860$; reject H_0 if $\chi^2 \geq 14.860$. p-value ≈ 0.7492; fail to reject H_0. At the 0.005 level of significance, the evidence does not support the claim that an association exists between cola preference and age. **17. a.** H_0: Book preference and gender are independent. H_a: Book preference and gender are not independent. **b.** Chi-square distribution; $\alpha = 0.005$

c.

$\frac{3630}{133} \approx 27.293233$	$\frac{10,769}{266} \approx 40.484962$	$\frac{8349}{266} \approx 31.387218$	$\frac{2904}{133} \approx 21.834586$
$\frac{4350}{133} \approx 32.706767$	$\frac{12,905}{266} \approx 48.515038$	$\frac{10,005}{266} \approx 37.612782$	$\frac{3480}{133} \approx 26.165414$

$\chi^2 \approx 10.418$ **d.** $\chi^2_{0.005} = 12.838$; reject H_0 if $\chi^2 \geq 12.838$. p-value ≈ 0.0153; fail to reject H_0. At the 0.005 level of significance, the evidence does not support the claim that an association exists between book preference and gender.

19. a. H_0: The commercial chosen and the combination of age and gender are independent. H_a: The commercial chosen and the combination of age and gender are not independent. **b.** Chi-square distribution; $\alpha = 0.10$

c.

10.25	9.75
10.25	9.75
10.25	9.75
10.25	9.75

$\chi^2 \approx 7.355$ **d.** $\chi^2_{0.10} = 6.251$; reject H_0 if $\chi^2 \geq 6.251$. p-value ≈ 0.0614; reject H_0. At the 0.10 level of significance, there is enough evidence to conclude that an association exists between the commercial chosen and the combination of age and gender.

Chapter 10 Exercises

1. a. H_0: $\mu = 55$; H_a: $\mu \neq 55$ **b.** Student's t-distribution; $\alpha = 0.01$ **c.** $t \approx -1.949$ **d.** $t_{0.005} = 2.779$; reject H_0 if $|t| \geq 2.779$. p-value ≈ 0.0622; fail to reject H_0. Based on the sample chosen, the manufacturer can be 99% confident that the barrels meet the necessary standards. **3. a.** H_0: $\mu \geq 6.83$; H_a: $\mu < 6.83$ **b.** Student's t-distribution; $\alpha = 0.10$ **c.** $t \approx -1.846$ **d.** $-t_{0.10} = -1.415$; reject H_0 if $t \leq -1.415$. p-value ≈ 0.0537; reject H_0. Based on the sample chosen, there is enough evidence for the agency to be 90% confident that the new jumbo jet will decrease the mean transatlantic flight time from Newark to London Heathrow. **5. a.** H_0: $\mu \leq 264$; H_a: $\mu > 264$ **b.** Normal distribution; $\alpha = 0.01$ **c.** $z \approx 2.93$ **d.** p-value ≈ 0.0017; reject H_0. At the 0.01 level of significance, the evidence supports the claim that the mean usable lifetime of the batteries is increased with the new process. **7. a.** H_0: $p \leq 0.50$; H_a: $p > 0.50$ **b.** Normal distribution; $\alpha = 0.10$ **c.** $z \approx 0.26$ **d.** p-value ≈ 0.3958 (From table: p-value $= 0.3974$); fail to reject H_0. At the 0.10 level of significance, the evidence does not support Bren's claim that more than half of all sophomores live off campus. Bren failed to reject a true null hypothesis, so she made the correct decision. **9. a.** H_0: $p = 0.35$; H_a: $p \neq 0.35$ **b.** Normal distribution; $\alpha = 0.01$ **c.** $z \approx 1.26$ **d.** p-value ≈ 0.2078 (from table: p-value $= 0.2076$); fail to reject H_0. At the 0.01 level of significance, the evidence does not support the organization's claim that the percentage of minority students is different from 35%. The organization failed to reject a false null hypothesis, so it made a Type II error. **11. a.** H_0: $\sigma^2 \leq 15.5$; H_a: $\sigma^2 > 15.5$ **b.** Chi-square distribution; $\alpha = 0.05$ **c.** $\chi^2 \approx 26.787$ **d.** $\chi^2_{0.05} = 36.415$; reject H_0 if $\chi^2 \geq 36.415$. Fail to reject H_0. At the 0.10 level of significance, the evidence does not support the historian's claim that the variance in the hem lengths of skirts and dresses is more than 15.5. **13. a.** H_0: $p_1 = p_2 = p_3 = p_4$; H_a: There is some difference amongst the probabilities. **b.** Chi-square distribution; $\alpha = 0.05$ **c.** $\chi^2 \approx 6.414$ **d.** $\chi^2_{0.05} = 7.815$; reject H_0 if $\chi^2 \geq 7.815$. p-value ≈ 0.0931; fail to reject H_0. At the 0.05 level of significance, the evidence does not support the claim that there is some difference between the numbers of calls for the four beats. **15. a.** H_0: Education level and region in the United States are independent. H_a: Education level and region in the United States are not independent. **b.** Chi-square distribution; $\alpha = 0.05$ **c.** $\chi^2 \approx 4.997$ **d.** $\chi^2_{0.05} = 16.919$; reject H_0 if $\chi^2 \geq 16.919$. p-value ≈ 0.8346; fail to reject H_0. At the 0.05 level of significance, the evidence does not support the claim that education levels and regions in the United States are not independent.

Chapter 11

Section 11.1

1. $z \approx 1.16$ **3.** $z \approx -0.88$ **5.** H_0: $\mu_1 - \mu_2 \geq -30$; H_a: $\mu_1 - \mu_2 < -30$ **7.** H_0: $\mu_1 - \mu_2 \leq 4$; H_a: $\mu_1 - \mu_2 > 4$ **9. a.** H_0: $\mu_1 - \mu_2 \leq 0$; H_a: $\mu_1 - \mu_2 > 0$ **b.** Normal distribution; $\alpha = 0.05$ **c.** $z \approx 2.31$ **d.** p-value $= 0.0104$; therefore reject H_0. There is sufficient evidence at the 0.05 level to support the car company's claim that its new SUV has a better mean gas mileage than its competitor's SUV. **11. a.** H_0: $\mu_1 - \mu_2 = 0$; H_a: $\mu_1 - \mu_2 \neq 0$ **b.** Normal distribution; $\alpha = 0.05$ **c.** $z \approx 1.57$ **d.** p-value ≈ 0.1171 (From table: p-value $= 0.1164$); therefore fail to reject H_0. There is not sufficient evidence at the 0.05 level to say that there is a difference between the mean numbers of miles run each week by group runners and individual runners who are training for marathons. **13. a.** H_0: $\mu_1 - \mu_2 \leq 0$; H_a: $\mu_1 - \mu_2 > 0$ **b.** Normal distribution; $\alpha = 0.05$ **c.** $z \approx 1.49$ **d.** p-value ≈ 0.0685 (From table: p-value $= 0.0681$); therefore fail to reject H_0. There is not sufficient evidence at the 0.05 level to say that, on average, clients lose more weight with the company's help than without it. **15. a.** H_0: $\mu_1 - \mu_2 \geq -5$; H_a: $\mu_1 - \mu_2 < -5$ **b.** Normal distribution; $\alpha = 0.05$ **c.** $z \approx -1.67$ **d.** p-value ≈ 0.0470 (From table: p-value $= 0.0475$); therefore reject H_0. There is sufficient evidence at the 0.05 level of significance to say that the mean score for evening classes is more than 5 points lower than the mean score for morning classes on art history tests. **17. a.** H_0: $\mu_1 - \mu_2 \geq 0$; H_a: $\mu_1 - \mu_2 < 0$ **b.** Normal distribution; $\alpha = 0.10$ **c.** $z \approx -0.80$ **d.** p-value ≈ 0.2121 (From table: p-value $= 0.2119$); therefore fail to reject H_0. There is not sufficient evidence at the 10% level of significance to say that the mean ACT score of first-born children is lower than the mean ACT score of second-born children.
19. a. H_0: $\mu_1 - \mu_2 \leq 0$; H_a: $\mu_1 - \mu_2 > 0$ **b.** Normal distribution; $\alpha = 0.01$ **c.** $z \approx 0.66$ **d.** p-value ≈ 0.2559 (From table: p-value $= 0.2546$); therefore fail to reject H_0. There is not enough evidence at the 0.01 level to say that the mean monthly residential electric bill is higher for Lauren's state than for Keri's state. **21. a.** H_0: $\mu_1 - \mu_2 = 0$; H_a: $\mu_1 - \mu_2 \neq 0$ **b.** Normal distribution; $\alpha = 0.10$ **c.** $z \approx 1.33$ **d.** p-value ≈ 0.1852 (From table: p-value $= 0.1836$); therefore fail to reject H_0. There is not sufficient evidence at the 0.10 level to say that there is a difference between the mean batting averages of the SEC East and SEC West baseball players.

Section 11.2

1. $t \approx 1.192$; $df = 17$ **3.** $t \approx 2.452$; $df = 23$ **5. a.** H_0: $\mu_1 - \mu_2 = 0$; H_a: $\mu_1 - \mu_2 \neq 0$ **b.** Student's t-distribution; $\alpha = 0.05$ **c.** $t \approx -0.883$ **d.** $t_{0.025} = 2.060$; reject H_0 if $|t| \geq 2.060$. p-value ≈ 0.3857; fail to reject H_0. At the 0.05 level, there is not sufficient evidence to say that the mean exam scores for the two classes are different. **7. a.** H_0: $\mu_1 - \mu_2 \leq 0$; H_a: $\mu_1 - \mu_2 > 0$ **b.** Student's t-distribution; $\alpha = 0.01$ **c.** $t \approx 2.773$ **d.** $t_{0.010} = 3.365$; reject H_0 if $t \geq 3.365$. p-value ≈ 0.0128; fail to reject H_0. At the 0.01 level, there is not sufficient evidence to say that the mean thickness of briquettes from Brand A is larger than the mean thickness of those from Brand B. **9. a.** H_0: $\mu_1 - \mu_2 \leq 0$; H_a: $\mu_1 - \mu_2 > 0$ **b.** Student's t-distribution; $\alpha = 0.10$ **c.** $t \approx 0.797$ **d.** $t_{0.100} = 1.397$; reject H_0 if $t \geq 1.397$. p-value ≈ 0.2242; fail to reject H_0. At the 0.10 level, there is not sufficient evidence to say that there is a reduction in Gary's mean time to paint a room using the new tool. **11. a.** H_0: $\mu_1 - \mu_2 \geq 0$; H_a: $\mu_1 - \mu_2 < 0$ **b.** Student's t-distribution; $\alpha = 0.10$ **c.** $t \approx -0.455$ **d.** $-t_{0.100} = -1.345$; reject H_0 if $t \leq -1.345$. p-value ≈ 0.3262; fail to reject H_0. At the 0.10 level, there is not sufficient evidence to say that the mean weight of bags from Line A is lower than the mean weight of bags from Line B.
13. a. H_0: $\mu_1 - \mu_2 \leq 0$; H_a: $\mu_1 - \mu_2 > 0$ **b.** Student's t-distribution; $\alpha = 0.01$ **c.** $t \approx 3.238$
d. $t_{0.010} = 2.492$; reject H_0 if $t \geq 2.492$. p-value ≈ 0.0012; reject H_0. At the 0.01 level, there is sufficient evidence to say that the mean decrease in cholesterol level is greater for patients who take Praxor than for those who take a placebo. Thus, the evidence supports the company's claim that Praxor lowers cholesterol levels more effectively than a placebo.
15. a. H_0: $\mu_1 - \mu_2 \geq 0$; H_a: $\mu_1 - \mu_2 < 0$ **b.** Student's t-distribution; $\alpha = 0.01$ **c.** $t \approx -0.433$ **d.** $-t_{0.010} = -2.552$; reject H_0 if $t \leq -2.552$. p-value ≈ 0.3350; fail to reject H_0. At the 0.01 level, there is not sufficient evidence to say that the mean monthly cost for car insurance is lower for customers of Company A than for customers of Company B.
17. a. H_0: $\mu_1 - \mu_2 = 0$; H_a: $\mu_1 - \mu_2 \neq 0$ **b.** Student's t-distribution; $\alpha = 0.05$ **c.** $t \approx -1.585$ **d.** $t_{0.025} = 2.080$; reject H_0 if $|t| \geq 2.080$. p-value ≈ 0.1278; fail to reject H_0. At the 0.05 level, there is not sufficient evidence to say that the mean lengths of time spent to complete a sculpture are different for male and female artists.

Section 11.3

1. a. $H_0: \mu_d \geq 0$; $H_a: \mu_d < 0$ **b.** Student's t-distribution; $\alpha = 0.05$ **c.** $\bar{d} \approx -1.583333$; $s_d \approx 2.609714$; $t \approx -2.102$ **d.** $-t_{0.050} = -1.796$; reject H_0 if $t \leq -1.796$. p-value ≈ 0.0297; reject H_0. There is sufficient evidence, at the 0.05 level of significance, to support the claim that participants in the anger-management course will lose their temper less often during the two-week period after completing the course than during the two weeks prior to taking the course.
3. a. $H_0: \mu_d \leq 60$; $H_a: \mu_d > 60$ **b.** Student's t-distribution; $\alpha = 0.01$ **c.** $\bar{d} \approx 81.111111$; $s_d \approx 86.954650$; $t \approx 0.728$ **d.** $t_{0.010} = 2.896$; reject H_0 if $t \geq 2.896$. p-value ≈ 0.2436; fail to reject H_0. There is not sufficient evidence, at the 0.01 level of significance, to support the claim that students' SAT scores increase by a mean of more than 60 points after completing the SAT prep course. **5. a.** $H_0: \mu_d = 0$; $H_a: \mu_d \neq 0$ **b.** Student's t-distribution; $\alpha = 0.10$ **c.** $\bar{d} \approx 2.888889$; $s_d \approx 5.348936$; $t \approx 1.620$ **d.** $t_{0.050} = 1.860$; reject H_0 if $|t| \geq 1.860$. p-value ≈ 0.1438; fail to reject H_0. There is not sufficient evidence, at the 0.10 level of significance, to say that the mean difference between the weights of two boys with the same parents is not 0. That is, the evidence does not support the claim that boys with the same parents do not have the same weight. **7. a.** $H_0: \mu_d \leq 30$; $H_a: \mu_d > 30$ **b.** Student's t-distribution; $\alpha = 0.10$ **c.** $\bar{d} = 31.625$; $s_d \approx 1.505941$; $t \approx 3.052$ **d.** $t_{0.100} = 1.415$; reject H_0 if $t \geq 1.415$. p-value ≈ 0.0093; reject H_0. There is sufficient evidence, at the 0.10 level of significance, to support the teacher's claim that after a year of lessons, 5-year-old students will increase their stamina for standing in the correct position by a mean of more than 30 minutes. **9. a.** $H_0: \mu_d \leq 10$; $H_a: \mu_d > 10$ **b.** Student's t-distribution; $\alpha = 0.05$ **c.** $\bar{d} = 11.5$; $s_d \approx 0.755929$; $t \approx 5.612$ **d.** $t_{0.050} = 1.895$; reject H_0 if $t \geq 1.895$. p-value ≈ 0.0004; reject H_0. There is sufficient evidence, at the 0.05 level of significance, to support the publisher's claim that the new textbook increases second-grade students' standardized test scores by a mean of more than 10 points.

Section 11.4

1. $z \approx -0.80$ **3.** $z \approx -0.86$ **5. a.** $H_0: p_1 - p_2 \leq 0$; $H_a: p_1 - p_2 > 0$ **b.** Normal distribution; $\alpha = 0.01$ **c.** $z \approx 2.35$ **d.** p-value ≈ 0.0095 (From table: p-value $= 0.0094$); therefore reject H_0. There is sufficient evidence at the 0.01 level to support the newspaper's claim that single women purchase more houses than single men. **7. a.** $H_0: p_1 - p_2 \geq 0$; $H_a: p_1 - p_2 < 0$ **b.** Normal distribution; $\alpha = 0.05$ **c.** $z \approx -1.64$ **d.** p-value $= 0.0505$; therefore fail to reject H_0. There is not sufficient evidence at the 0.05 level to support the university's claim that the retention rate improved between 2010 and 2011. **9. a.** $H_0: p_1 - p_2 \leq 0$; $H_a: p_1 - p_2 > 0$ **b.** Normal distribution; $\alpha = 0.01$ **c.** $z \approx 0.40$ **d.** p-value ≈ 0.3441 (From table: p-value $= 0.3446$); therefore fail to reject H_0. There is not sufficient evidence at the 0.01 level to support the old wives' tale that women who eat chocolate during pregnancy have happier babies. **11. a.** $H_0: p_1 - p_2 = 0$; $H_a: p_1 - p_2 \neq 0$ **b.** Normal distribution; $\alpha = 0.05$ **c.** $z \approx 0.50$ **d.** p-value ≈ 0.6192 (From table: p-value $= 0.6170$); therefore fail to reject H_0. There is not sufficient evidence at the 0.05 level to say that women and men are unequally likely to get speeding tickets in the area from where the participants in the survey were sampled.

Section 11.5

1. $H_0: \sigma_1^2 \geq \sigma_2^2$; $H_a: \sigma_1^2 < \sigma_2^2$ **3.** $H_0: \sigma_1^2 \leq \sigma_2^2$; $H_a: \sigma_1^2 > \sigma_2^2$ **5.** $H_0: \sigma_1^2 = \sigma_2^2$; $H_a: \sigma_1^2 \neq \sigma_2^2$ **7.** $F \approx 1.0380$ **9.** $F \approx 0.8211$ **11.** $F_{0.950} = 0.4632$; reject H_0 if $F \leq 0.4632$; reject H_0. **13.** $F_{0.990} = 0.2062$; reject H_0 if $F \leq 0.2062$; fail to reject H_0. **15.** $F_{0.100} = 1.7675$; reject H_0 if $F \geq 1.7675$; reject H_0. **17.** $F_{0.975} = 0.3629$, $F_{0.025} = 2.5731$; reject H_0 if $F \leq 0.3629$ or $F \geq 2.5731$; fail to reject H_0. **19.** $F_{0.950} = 0.4550$, $F_{0.050} = 2.2429$; reject H_0 if $F \leq 0.4550$ or $F \geq 2.2429$; fail to reject H_0. **21. a.** $H_0: \sigma_1^2 \geq \sigma_2^2$; $H_a: \sigma_1^2 < \sigma_2^2$ **b.** F-distribution; $\alpha = 0.05$ **c.** $F \approx 0.5531$ **d.** $F_{0.950} = 0.3146$; reject H_0 if $F \leq 0.3146$. p-value ≈ 0.1955; fail to reject H_0. At the 0.05 level of significance, there is not sufficient evidence to support the golf pro's claim that his driving distances have a smaller variance when he uses Titleist golf balls than when he uses the store brand. **23. a.** $H_0: \sigma_1^2 \leq \sigma_2^2$; $H_a: \sigma_1^2 > \sigma_2^2$ **b.** F-distribution; $\alpha = 0.10$ **c.** $F \approx 3.1124$ **d.** $F_{0.100} = 3.0145$; reject H_0 if $F \geq 3.0145$. p-value ≈ 0.0939; reject H_0. At the 0.10 level of significance, there is sufficient evidence to support the researcher's belief that the variance of total cholesterol levels in men is greater than that for women.
25. a. $H_0: \sigma_1^2 = \sigma_2^2$; $H_a: \sigma_1^2 \neq \sigma_2^2$ **b.** F-distribution; $\alpha = 0.01$ **c.** $F \approx 5.2565$ **d.** $F_{0.995} = 0.0432$, $F_{0.005} = 23.1545$; reject H_0 if $F \leq 0.0432$ or $F \geq 23.1545$. p-value ≈ 0.1369; fail to reject H_0. At the 0.01 level of significance, there is not sufficient evidence to support the study's claim that the variance in resting heart rates of smokers is different than that of nonsmokers.

Section 11.6

1. $F_{0.05} = 2.7534$ **3.** $F_{0.10} = 2.6241$ **5.** $F_{0.10} = 4.0604$

7.	*SS*	*df*	*MS*	*F*
Treatments (T)	4	5	0.8	0.4364
Error (E)	11	6	1.833333	
Total	15	11		

9.	*SS*	*df*	*MS*	*F*
Treatments (T)	50	4	12.5	1.5625
Error (E)	24	3	8	
Total	74	7		

11.	*SS*	*df*	*MS*	*F*
Treatments (T)	313	6	52.166667	1.2421
Error (E)	336	8	42	
Total	649	14		

13. a.	*SS*	*df*	*MS*	*F*	*P-value*	*F crit*
Treatments (T)	5.733333	2	2.866667	4.7778	0.0298	2.8068
Error (E)	7.2	12	0.6			
Total	12.933333	14				

b. The *p*-value is less than 0.10, so reject the null hypothesis. Thus, there is enough evidence, at the 0.10 level of significance, to support the claim that there is a difference between the mean numbers of defective parts produced each day by these three workers.

15. a.	*SS*	*df*	*MS*	*F*	*P-value*	*F crit*
Treatments (T)	10.666667	2	5.333333	3.9070	0.0390	6.0129
Error (E)	24.571429	18	1.365079			
Total	35.238095	20				

b. The *p*-value is greater than 0.01; thus fail to reject the null hypothesis. Therefore, there is not enough evidence, at the 0.01 level of significance, to support the claim that the mean numbers of calls received per shift are different for the morning, afternoon, and night shifts.

Chapter 11 Exercises

1. H_0: $\mu_1 - \mu_2 \geq 0$; H_a: $\mu_1 - \mu_2 < 0$; $z = -3.05$; *p*-value ≈ 0.0012 (From table: *p*-value $= 0.0011$); reject H_0. Therefore, there is sufficient evidence at the 0.05 level of significance to support the psychologists' claim that the mean exam score of students with test anxiety is lower than the mean score of students without test anxiety. **3.** H_0: $\mu_d \leq 0$; H_a: $\mu_d > 0$; $t \approx 2.951$; $t_{0.05} = 1.833$; reject H_0 if $t \geq 1.833$. *p*-value ≈ 0.0081; reject H_0. At the 0.05 level of significance, there is sufficient evidence to support the claim that husbands gain more weight than wives during the first year of marriage. **5.** H_0: $\mu_1 - \mu_2 \geq 0$; H_a: $\mu_1 - \mu_2 < 0$; $t \approx -2.818$; $-t_{0.01} = -2.624$; reject H_0 if $t \leq -2.624$. *p*-value ≈ 0.0045; reject H_0. At the 0.01 level of significance, there is sufficient evidence to say that cuts treated with the new ointment have a faster mean healing time than those without medication. **7.** H_0: $\mu_d = 0$; H_a: $\mu_d \neq 0$; $t \approx 2.377$; $t_{0.005} = 3.250$; reject H_0 if $|t| \geq 3.250$. *p*-value ≈ 0.0414; fail to reject H_0. At the 0.01 level of significance, there is not sufficient evidence to say that the mean difference between the numbers of times per month that each student is disciplined in single-gender and mixed-gender classes is not 0. Thus, the evidence does not support the professor's claim that single-gender classrooms do not have the same number of discipline problems as mixed-gender classrooms. **9.** H_0: $\mu_1 - \mu_2 \geq 0$; H_a: $\mu_1 - \mu_2 < 0$; $z \approx -2.71$; *p*-value ≈ 0.0034; reject H_0. Therefore, there is sufficient evidence at the 0.05 level of significance to support the claim that the mean number of doctor's visits per year is lower for adults who attend church regularly than for those who do not.

Chapter 12

Section 12.1

1. Strong positive **3.** Strong negative **5.** None **7.** Yes **9.** No **11.** Yes

13. a.

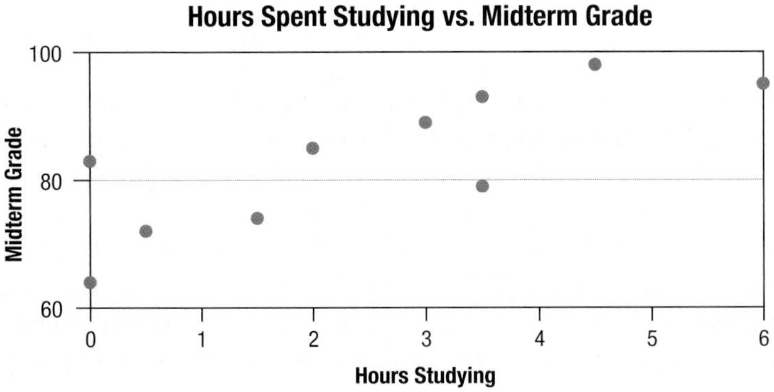

b. Positive **c.** 0.799 **d.** Yes

15. a.

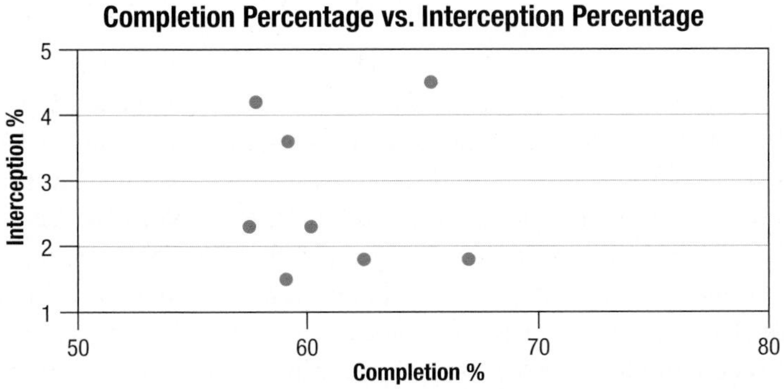

b. No correlation **c.** –0.044 **d.** No

17. a.

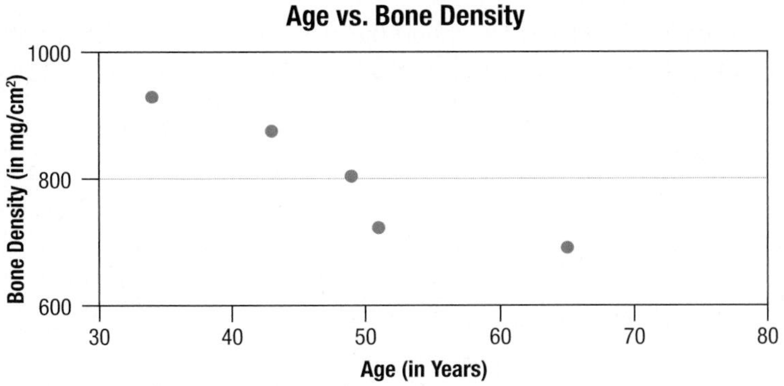

b. Negative **c.** –0.940 **d.** No

19. a. 0.375 **b.** No **c.** 0.140

d. 14.0% of the variation in the babies' birth weights can be associated with the variation in the birth weights of the mothers.

Section 12.2

1. a. $\hat{y} = 6.436 + 0.914x$ **b.** $r \approx 0.984 \geq 0.576$ and a scatter plot of the data shows a linear pattern, so it is appropriate to make predictions. 70.416 **3. a.** $\hat{y} = 38.536 + 0.877x$ **b.** $r \approx 0.937 \geq 0.576$ and a scatter plot of the data shows a linear pattern, so it is appropriate to make predictions. 130.621 pounds **5. a.** $\hat{y} = 12.972 + 0.732x$ **b.** $r \approx 0.576 < 0.707$, so it is not appropriate to make this prediction. **7. a.** $\hat{y} = 153.319 + 0.708x$ **b.** $r \approx 0.857 \geq 0.576$ and a scatter plot of the data shows a linear pattern, so it is appropriate to make predictions. 557

Section 12.3

1. a. 3.733 **b.** 0.683 **c.** (12.334, 15.674) or (12.328, 15.667) **3. a.** 479.905 **b.** 6.076 **c.** (4.901, 32.023) or (4.910, 32.031) **5. a.** 83.622 **b.** 3.733 **c.** (6.914, 26.294) or (6.887, 26.267) **7. a.** 3.208 **b.** 0.731 **c.** (2.726, 6.522) or (2.727, 6.523) **9. a.** 1,148,767.96 **b.** 378.941 **c.** (1849.911, 3733.989) **d.** (496.446, 1319.694) **e.** (226.814, 401.146) **11. a.** 15.660 **b.** 1.770 **c.** (65.028, 74.790) or (65.026, 74.789) **d.** (87.609, 94.558) **e.** (–4.100, –2.958)

Section 12.4

1. Yes, because $0.0000000005 \le 0.05$, that is, p-value $\le \alpha$. **3.** No, because $0.0810 > 0.05$, that is, p-value $> \alpha$. **5.** $\hat{y} = 63.880 - 1.416x_1 + 4.169x_2$; $x_1 =$ Absences; $x_2 =$ Hours studied per week; Neither variable should be eliminated from the regression model. **7.** $\hat{y} = -19.570 + 0.685x_1 + 0.063x_2$; $x_1 =$ Weeks of gestation; $x_2 =$ Number of prenatal visits; The variable "Number of prenatal visits" could be eliminated from the regression model. **9.** p-value $= 0.0259$; thus, there is enough evidence to support the claim that a statistically significant linear relationship exists between the explanatory variables and the response variable. $\hat{y} = 5476.596 - 0.012x_1 - 29.124x_2 - 1045.990x_3$ **11.** p-value ≈ 0.0002; thus, there is enough evidence to support the claim that a statistically significant linear relationship exists between the explanatory variables and the response variable. $\hat{y} = 3010.562 + 4.621x_1 - 2.804x_2$ **13.** *df* for regression **15.** It usually increases as the number of explanatory variables increases.

Chapter 12 Exercises

1. a. 0.182 **b.** No; no **c.** No; the correlation coefficient is not significant. **3. a.** Negative linear relationship **b.** $r \approx -0.784$; $r^2 \approx 0.615$ **c.** Yes; yes **d.** 10 **e.** Historically, housing sales are stronger in the spring and summer months than in the winter months. Thus, the season may be another factor influencing the rise in home purchases.
5. a. Because both the scatter plot is linear and value of r is significant at the 0.05 and 0.01 levels, yes. **b.** Yes, 339; no, 500 swimsuits is outside the range of the data. **c.** Seagulls flock to warmer climates and abundant food sources. Humans also "flock" to the beach during warmer months. Furthermore, humans bring food to the seagulls, even if it is in the form of trash they leave behind. However, seagulls also flock to the shrimp boats on the coast. Thus, though there is a significant linear relationship between these two variables, it is doubtful that it is a causal relationship. **7. a.** Yes; yes **b.** $19,680.11 **c.** 80.5% **9. a.** 16.724 **b.** 1.670 **c.** (32.690, 41.894) or (32.847, 42.051) **d.** (–8.593, 11.926) **e.** (0.069, 0.146) **11.** $p \approx 0.00001$; thus, there is enough evidence to support the claim that a statistically significant linear relationship exists between the explanatory variables and the response variable. $\hat{y} = 0.325 + 0.097x_1 - 1.156x_2$

Index

E

F

G

H

O

P

Q

R

S

T

C—Critical Values of *t*

	Area in One Tail				
	0.100	0.050	0.025	0.010	0.005
	Area in Two Tails				
df	**0.200**	**0.100**	**0.050**	**0.020**	**0.010**
1	3.078	6.314	12.706	31.821	63.657
2	1.886	2.920	4.303	6.965	9.925
3	1.638	2.353	3.182	4.541	5.841
4	1.533	2.132	2.776	3.747	4.604
5	1.476	2.015	2.571	3.365	4.032
6	1.440	1.943	2.447	3.143	3.707
7	1.415	1.895	2.365	2.998	3.499
8	1.397	1.860	2.306	2.896	3.355
9	1.383	1.833	2.262	2.821	3.250
10	1.372	1.812	2.228	2.764	3.169
11	1.363	1.796	2.201	2.718	3.106
12	1.356	1.782	2.179	2.681	3.055
13	1.350	1.771	2.160	2.650	3.012
14	1.345	1.761	2.145	2.624	2.977
15	1.341	1.753	2.131	2.602	2.947
16	1.337	1.746	2.120	2.583	2.921
17	1.333	1.740	2.110	2.567	2.898
18	1.330	1.734	2.101	2.552	2.878
19	1.328	1.729	2.093	2.539	2.861
20	1.325	1.725	2.086	2.528	2.845
21	1.323	1.721	2.080	2.518	2.831
22	1.321	1.717	2.074	2.508	2.819
23	1.319	1.714	2.069	2.500	2.807
24	1.318	1.711	2.064	2.492	2.797
25	1.316	1.708	2.060	2.485	2.787
26	1.315	1.706	2.056	2.479	2.779
27	1.314	1.703	2.052	2.473	2.771
28	1.313	1.701	2.048	2.467	2.763
29	1.311	1.699	2.045	2.462	2.756
30	1.310	1.697	2.042	2.457	2.750
31	1.309	1.696	2.040	2.453	2.744
32	1.309	1.694	2.037	2.449	2.738
34	1.307	1.691	2.032	2.441	2.728
36	1.306	1.688	2.028	2.434	2.719
38	1.304	1.686	2.024	2.429	2.712
40	1.303	1.684	2.021	2.423	2.704
45	1.301	1.679	2.014	2.412	2.690
50	1.299	1.676	2.009	2.403	2.678
55	1.297	1.673	2.004	2.396	2.668
60	1.296	1.671	2.000	2.390	2.660
70	1.294	1.667	1.994	2.381	2.648
80	1.292	1.664	1.990	2.374	2.639
90	1.291	1.662	1.987	2.368	2.632
100	1.290	1.660	1.984	2.364	2.626
120	1.289	1.658	1.980	2.358	2.617
200	1.286	1.653	1.972	2.345	2.601
300	1.284	1.650	1.968	2.339	2.592
400	1.284	1.649	1.966	2.336	2.588
500	1.283	1.648	1.965	2.334	2.586
750	1.283	1.647	1.963	2.331	2.582
1000	1.282	1.646	1.962	2.330	2.581
∞	1.282	1.645	1.960	2.326	2.576

Left Tail

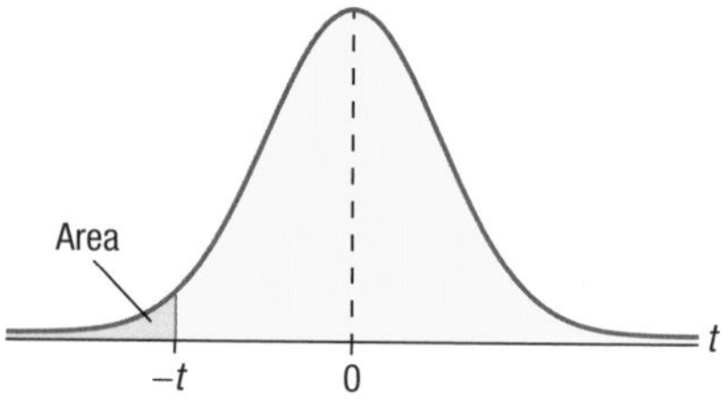

Right Tail

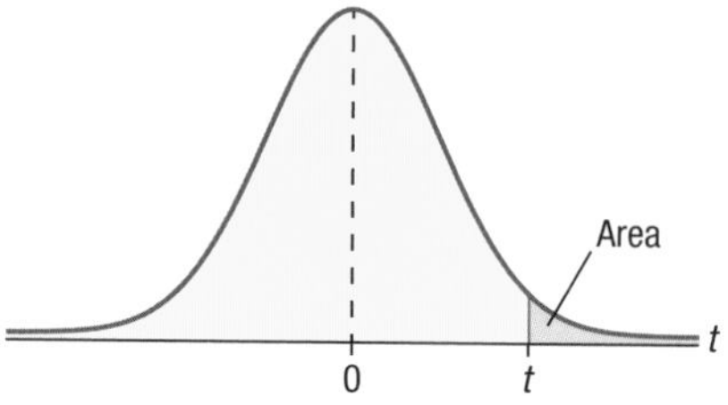

Two Tails

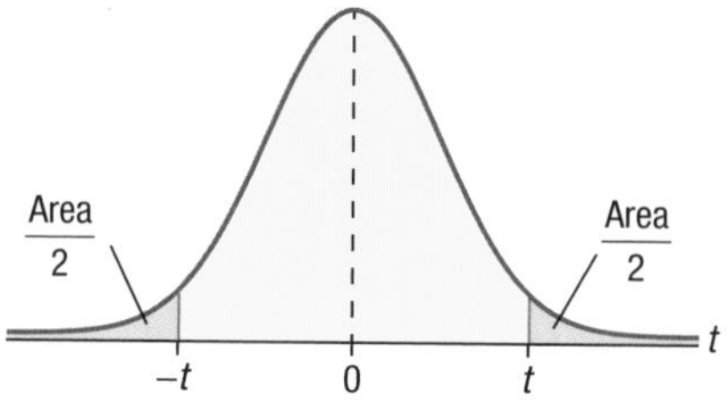